Microorganisms and Nitrogen Sources

Microorganisms and Nitrogen Sources

Transport and Utilization of Amino Acids, Peptides, Proteins, and Related Substrates

Edited by

J. W. PAYNE

Department of Botany, University of Durham, Durham, UK

A Wiley–Interscience Publication

JOHN WILEY & SONS

Chichester · New York · Brisbane · Toronto

British Library Cataloguing in Publication Data:

Microorganisms and nitrogen sources.
1. Amino acids 2. Microbial peptides
3. Microbial proteins
I. Payne, John Weston
576'.11'9245 QR92.A6 79-42900

ISBN 0 471 27697 9

Typeset by Preface Ltd, Salisbury, Wilts. and printed in Great Britain by Page Bros. (Norwich) Ltd.

Editor's Preface

The idea behind this book was simple and clear: to present an account of the 'transmembrane metabolism' of certain related nitrogen sources by various microorganisms. By integrating descriptions of the characteristics of their transport with accounts of the metabolic fate and utilization of the different substrates it was planned to emphasize and to explore the cellular coordination of these events. Reviews dealing with, for example, aspects of amino acid transport or utilization appear often in separate texts on these specific topics but rarely, if ever, are they placed together; the same may be said for an integrated coverage of these features in prokaryotic bacteria through to eukaryotic fungi and algae. In this book these various aspects are brought together. However, the book overall was also envisaged as one of general reference for the area and, to this end, other closely related topics such as peptide ionophores and antimicrobial agents are included. Each section was to be self-contained and occasionally, as a result some overlaps occurred during their preparation. In several instances these are preserved, partly to maintain the unity of the section, partly because the context of the discussions are different, and partly because they serve to emphasize the interrelatedness of the topics.

Certain aspects of nitrogen utilization have been intentionally excluded. Nitrogen fixation, for example, is not considered here, for it has been the subject of several other texts. So, although it is convenient for a book to have a short title such as 'Microorganisms and Nitrogen Sources', the full title gives a stricter indication of the material to be found here. Even amongst the topics selected the information currently available varies considerably, and several authors responded bravely to the request to combine fact and speculation by presenting an analysis of the questions that need to be asked rather than just the answers obtained. Some of these speculations may have borne fruit and others withered even before publication but they should remain a source for reference and for ideas for some time to come. In particularly active research areas the addenda go some way to bridge the gap that inevitably arises in the production of a large, multi-author volume such as this.

I am particularly grateful to the contributors for the enthusiastic way in which they greeted the idea for the book, for their willingness to give their time and expertise in writing the sections, and not least for their patience once these had been handed over to the editor. Mrs Ethne Ellis deserves especial thanks for so expertly typing the bulk of the material. Finally, I thank my wife, and Clare and Adrian, for their encouragement and tolerance; only on rare occasions did it seem that, whereas microorganisms and man might manage perfectly well on a diet of amino acids and peptides alone, women and children might need more varied fare.

Contributors

J. J. ANDERSON — *Department of Biological Chemistry, The University of Michigan Medical School, Ann Arbor, Michigan 48109, USA*

Y. ANRAKU — *Department of Biology, Faculty of Science, University of Hongo, Bunkyo-ku, Tokyo 113, Japan*

J. M. BECKER — *Department of Microbiology, University of Tennessee, Knoxville, Tennessee 37916, USA*

I. R. BOOTH — *Department of Microbiology, University of Aberdeen, Marischal College, Aberdeen AB9 1AS, UK*

C. M. BROWN — *Department of Biological Sciences, University of Dundee, Dundee DD1 4HU, UK (Present address: Department of Brewing and Biological Sciences, Heriot-Watt University, Chambers Street, Edinburgh EH1 1HX, UK)*

P. H. CLARKE — *Department of Biochemistry, University College, Gower Street, London WC1E 6BT, UK*

B. L. COHEN — *Department of Genetics, University of Glasgow, Glasgow G11 5JS, UK*

A. A. EDDY — *Department of Biochemistry, University of Manchester Institute of Science and Technology, Manchester M60 1QD, UK*

M. DE FELICE — *International Institute of Genetics and Biophysics, Via G. Marconi 10 80125 Naples, Italy*

C. E. FURLONG — *Department of Genetics, SK-50, University of Washington, Seattle, Washington 98195, USA*

J. GUARDIOLA — *International Institute of Genetics and Biophysics, Via G. Marconi 10 80125 Naples, Italy*

W. A. HAMILTON — *Department of Microbiology, University of Aberdeen, Marischal College, Aberdeen AB9 1AS, UK*

C. F. HIGGINS	*Department of Botany, University of Durham, South Road, Durham DH1 3LE, UK (Present address: Department of Biochemistry, University of California, Berkeley, California 94720, USA)*
C. L. HERMSDORF	*Department of Molecular Biophysics and Biochemistry, Yale University, New Haven, Connecticut 06510, USA*
H. HOLZER	*Biochemisches Institut der Universität Freiburg im Breisgau und Institut für Toxikologie und Biochemie der Gesellschaft für Strahlenund, Umweltforschung D-7800, Freiburg im Breisgau, West Germany*
M. IACCARINO	*The International Institute of Genetics and Biophysics, Via G. Marconi 10 80125 Naples, Italy*
B. A. LAW	*National Institute for Research in Dairying, Shinfield, Reading RG2 9AT, UK*
A. MEISTER	*Department of Biochemistry, Cornell University Medical College, 1300 York Avenue, New York, New York 10021, USA*
F. NAIDER	*Department of Chemistry, The College of Staten Island of The City University of New York, Staten Island, New York 10301, USA*
D. L. OXENDER	*Department of Biological Chemistry, The University of Michigan Medical School, Ann Arbor, Michigan 48109, USA*
J. W. PAYNE	*Department of Botany, University of Durham, South Road, Durham DH1 3LE, UK*
M. J. PINE	*Department of Experimental Therapeutics, Grace Cancer Drug Center, Roswell Park Memorial Institute, Buffalo, New York 14263, USA*
S. C. QUAY	*Department of Pathology, Harvard Medical School and Massachusetts General Hospital, Boston, Massachusetts 02114, USA*
C. RATLEDGE	*Department of Biochemistry, University of Hull, Hull HU6 7RX, UK*
P. S. RINGROSE	*Roche Products Ltd, P.O. Box 8, Welwyn Garden City, Hertfordshire AL7 3AY, UK (Present address: Chemotherapy Division, Sandoz Research Institute, Brunner Strasse 59, A-1235 Vienna, Austria)*

G. D. SCHELLENBERG — *Department of Medicine, Division of Medical Genetics, Center for Inherited Diseases, and Department of Genetics, SK50, Seattle, Washington 98195, USA*

S. SIMMONDS — *Department of Molecular Biophysics and Biochemistry, Yale University, New Haven, Connecticut 06510, USA*

E. E. SNELL — *Department of Microbiology, University of Texas at Austin, Austin, Texas 78712, USA*

M. C. STEPHENSON — *Department of Biochemistry, University of Hull, Hull HU6 7RX, UK*

W. D. P. STEWART — *Department of Biological Sciences, University of Dundee, Dundee DD1 4HN, UK*

J. M. TUFFNELL — *Department of Botany, University of Durham, South Road, Durham DH1 3LE, UK*

H. E. WADE — *Centre for Applied Microbiology and Research, Porton Down, Wiltshire SP4 0JG, UK*

D. H. WOLF — *Biochemisches Institut der Universität Freiburg im Breisgau und Institut für Toxicologie und Biochemie der Gesellschaft für Strahlenund, Umweltforschung D-7800, Freiburg im Breisgau, West Germany*

L. WOLFINBARGER, JR — *Department of Biological Sciences, Old Dominion University, Norfolk, Virginia 23508, USA*

Contents

Microorganisms and Nitrogen Sources
Edited by J. W. Payne

Introduction

E. E. SNELL

Department of Microbiology, University of Texas at Austin, Austin, Texas 78712, USA

The late D. D. Woods (Woods, 1953), paraphrasing an unpublished lecture by Marjory Stephenson, differentiated five developmental levels, characterized by a decreasing complexity of interactions, in the study of microbial metabolism. These levels included the study of: (1) mixed populations in 'natural media'; (2) pure cultures in complex, undefined media; (3) pure cultures in defined media; (4) viable, non-growing cell suspensions of a single cell type with defined substrates; and finally (5) non-viable, fractionated cell extracts—down to the stage of pure enzymes—with defined substrates for the elucidation of individual steps in metabolism.

The reader will recognize in these overlapping investigational levels an encapsulated history of microbial biochemistry. Progression from level (1) to level (2) could occur only as pure culture techniques were evolved, including the development of appropriate crude culture media. The change from level (2) to level (3) depended upon the painfully slow but cumulatively tremendous increase in knowledge of microbial and animal nutrition that occurred mostly from the 1930s through the 1950s and resulted in identification of the vitamins, amino acids, many of the inorganic ions, and other factors required for growth. (For reviews reflecting our increasing understanding of this subject, see Woods, 1953; Williams, 1941; Peterson and Peterson, 1945; Knight, 1945; Guirard and Snell, 1963.) Such identification in turn stimulated development of synthetic procedures for these and related substances, and of microbiological, chromatographic, and other methods for their determination. It also made possible recognition of the first biochemical (nutritional) mutants of microorganisms. With the parallel development of isotopic methods and of biochemical supply houses that made these unlabelled and labelled metabolites generally available, the stage was set for the sophisticated studies at levels (3), (4), and (5) characteristic of biochemistry today. These studies have resulted in a rather detailed understanding of intracellular pathways leading to and from most of the now-recognized components of living cells.

The same cannot yet be said for transport of amino acids, peptides, and proteins, which is the subject of this volume. Since, in the growing cell, transport must precede metabolism, a substantial if diminishing gap remains in our understanding of utilization of these substances for growth. Judging from recent reviews (Holden, 1962; Boos, 1974; Simoni and Postma, 1975; Wilson, 1978), the leading edge of experimentation here seems to lie principally at level (4) and the early stages of level (5). Particularly sketchy, in part because of the limited availability of pure substrates, has been our knowledge of peptide transport and utilization, and as an introduction to the many articles in this book that deal with various aspects of this topic, it may be useful to review briefly the experiments, mostly with growing cultures (level (3)) which led to the realization that peptides sometimes are far superior to their component amino acids as sources of amino acids for growth, that such peptides are nevertheless hydrolysed intracellularly prior to utilization, and that uptake systems exist for peptides that differ from those for amino acids.

Gale (1945) observed that partial hydrolysates of salmine were more effective than equimolar amounts of arginine in supplying the growth requirement of a Group D streptococcus that was auxotrophic for this amino acid. He attributed this superiority to the presence of arginine dihydrolase, which converts free arginine, but not its peptides, to the inactive ornithine. Since pure peptides were not available for testing, it could not be determined whether the active peptides of the partial hydrolysate were utilized directly, or were hydrolysed prior to use.

These points were later tested in a system more amenable to experimental analysis by Kihara *et al.* (1952). *Streptococcus faecalis* 8043 is auxotrophic for tyrosine; it also contains an active tyrosine decarboxylase when grown in the presence of vitamin B_6, but not when grown without this vitamin. In the absence of vitamin B_6, tyrosine peptides were equally (Leu-Tyr) or less (Gly-Tyr) active than free tyrosine in supporting growth and no tyramine was formed; when pyridoxamine was added to the growth medium, both peptides were more active than free tyrosine. Appropriate comparisons showed that the cell yield per unit of added tyrosine was greatly decreased by the presence of pyridoxamine, while that from tyrosine peptides decreased only slightly. Cell suspensions hydrolysed both peptides; the authors suggested that when peptides are being hydrolysed, tyrosine became available continously, at a relatively constant rate per cell, in concentrations sufficient for protein synthesis, but not sufficient to permit a large fraction of it to be destroyed by the decarboxylase before it could be utilized for growth (Kihara *et al.*, 1952).

In an analogous fashion, the presence of proline oxidase apparently accounts for the observed superiority of proline peptides to free proline in supporting growth of a mutant strain of *Escherichia coli* (Stone and Hoberman, 1953), a superiority, that, when first observed, had been ascribed to direct utilization of these peptides (Fruton and Simmons, 1948). Since numerous catabolic enzymes that destroy free amino acids are known, circumstances analogous to those described in the preceding examples could occur frequently and find their explanation in similar terms.

Discovery of a second set of circumstances under which peptides surpass free amino acids in growth-promoting activity was partly serendipitous. Although vitamin B_6 is normally required for growth of *Streptococcus faecalis*, we found that it was not required if sufficient alanine was added to a medium that contained each of the other amino acids (Snell and Guirard, 1943). To our surprise (since no D-amino acid had previously been implicated in growth processes), D-alanine proved more effective than L-alanine in supporting growth under these conditions (Snell, 1945). However, D-alanine did not 'replace' the growth requirement of *Lactobacillus casei* for vitamin B_6 under similar conditions (Snell, 1945). Reasoning from structural similarities, we speculated that D-alanine might serve as a biosynthetic precursor of vitamin B_6 by supplying the nitrogen at position 1, and the carbons at positions 2, 2′, and 3 (Snell and Guirard, 1943). Would *L. casei* also grow without vitamin B_6 if D-alanine and an unknown precursor for the other 'half' of the vitamin were supplied? We assayed for the missing 'precursor' and found indeed that *L. casei* grew well without vitamin B_6 if D-alanine and a partial hydrolysate of casein (vitamin B_6-free!) were supplied together in a medium that contained all of the L-amino acids (Snell, 1945).

In the meantime, however, our theory concerning D-alanine as a precursor of vitamin B_6 had collapsed—appropriate assays showed instead that vitamin B_6 was necessary for synthesis of D-alanine (Holden and Snell, 1949), which was present in all cells, however grown, and which we eventually identified as an essential component of the bacterial cell wall (Snell *et al*., 1955)—and we were left to ponder the role of the partial casein hydrolysate in promoting the response of *L. casei* to D-alanine. We fractionated the hydrolysate until we obtained active fractions that contained chiefly alanyl and tyrosyl peptides of valine, isoleucine, and leucine (Kihara *et al*., 1952). Tests then showed that synthetic dipeptides of L-alanine were fully active (Kihara and Snell, 1952). Each of the active peptides was hydrolysed by cell suspensions, and there was no evidence for destruction of L-alanine by catabolic processes.

To summarize the findings: (a) peptides of L-alanine are required for growth in the presence of D-alanine and the absence of vitamin B_6, (b) such peptides are hydrolysed by the cells, but (c) added D-alanine does not induce a requirement for L-alanine peptides in the presence of vitamin B_6 where both D- and L-alanine are synthesized internally. The simplest explanation of these findings, which we adopted (Kihara and Snell, 1952), was that D-alanine interfered with *absorption* of free L-alanine, but not of L-alanine peptides from the external medium; no interference, however, occurred with utilization of intracellular L-alanine. Experimental evidence for the postulated interference of D-alanine with L-alanine absorption was also presented. The existence of specific absorption sites for amino acids that differed from those for peptides, and that could be separately inhibited without inhibiting intracellular processes, was implicit in this explanation of the findings.

These conclusions are notable because they antedate the subsequent demonstration by currently acceptable procedures of specific transport systems for amino acids in bacterial cells, (Cohen and Rickenberg, 1956; Cohen and Monod, 1957) and because they focussed attention on competition between external nutrients for common absorption sites as a possible explanation for other cases (many of which were known; e.g. Sprince and Woolley, 1944; Malin *et al.*, 1951) in which partial protein digests stimulated growth of bacteria in media that contained a full complement of free amino acids. In general terms, if peptides were absorbed at sites separate from amino acids, inhibition of amino acid absorption by an antagonistic amino acid should not interfere with absorption of peptides, which, since they entered the cell by independent pathways, could then be hydrolysed and thus supply the amino acid for protein synthesis even when absorption of external free amino acid was blocked. This possibility seemed especially attractive to us, since many instances of one amino acid antagonizing utilization of structurally related amino acids had been observed both in the early literature (e.g. Gladstone, 1939; Doudoroff, 1943) and during studies leading to the development of quantitative microbiological assays for amino acids (Snell and Guirard, 1943; Malin *et al.*, 1951; Brickson *et al.*, 1948). We studied a sufficient number of such instances to convince us that this set of circumstances occurred frequently and that the explanation was generally valid (Prescott *et al.*, 1953; Kihara and Snell, 1955, 1960*b*). We later demonstrated (Leach and Snell, 1960) the validity of these postulates by direct tests of the effects of D-alanine on uptake of [^{14}C]-L-alanine and its peptides by resting cells of *L. casei*.

Especially instructive in this connection were experiments with *Leuconostoc mesenteroides*, in which antagonists to two different essential amino acids, A and B, were added simultaneously to the growth medium (Kihara and Snell, 1960*b*). Growth then occurred on addition either of comparatively large amounts of the free amino acids A + B, of small amounts of a mixture of dipeptides, AC + BD, or of equally small amounts of the dipeptide AB; all active peptides were hydrolysed by the cells. The activity of AB as a source of both A and B for growth under these conditions appears to preclude the necessity for mechanisms other than hydrolysis for the redistribution of these amino acid residues into other structures.

These observations supplied the background to our identification (Kihara and Snell 1960*a*) of 'strepogenin', an unidentified peptide growth factor required for rapid growth of *Lactobacillus casei* from small inocula in media based on acid hydrolysates of casein and vitamins. A considerable literature had accumulated on this subject; several different compounds, including peptides, showed some activity, but none of them was as effective as an enzymic protein digest in promoting growth. We adopted the working hypothesis that impediments to the utilization of several different amino acids, analogous to those discussed earlier, might be present in this medium. Correction of one such impediment by addition of a single limiting amino acid in excess or of smaller amounts of a peptide of that amino acid would not correct the overall deficiency: at best it could stimulate the growth rate to the point where availability of a second amino acid limited the growth rate. An enzymic protein digest would contain peptides that corrected all of the deficiencies; only in the unlikely event that one isolated a single peptide that contained each of the limiting amino acids in a structure that could be both rapidly absorbed by the cell and rapidly hydrolysed could one expect to duplicate the effects of an enzymic casein digest with a single peptide. Even then the peptide would account for only a small portion of the activity of the digest. Guided by this hypothesis, we showed that the availability of serine, cysteine, and glutamic acid limited growth in the strepogenin assay medium; simultaneous addition of large amounts of these three amino acids, or small amounts of a mixture of peptides that supplied them, greatly stimulated the growth rate; in their presence, addition of small amounts of other known growth factors (uracil, guanylic acid, spermine, fatty acids) gave rates of growth as great as those obtained with enzymic casein digests (Kihara and Snell, 1960*a*).

The foregoing discussion has given examples of three circumstances that result in increased activity of peptides as compared with their component amino acids for amino acid auxotrophs: (1) peptides may protect the amino acid from degradation, or (2) because peptides commonly enter the cell by routes other than those used by amino acids, they may bypass transport systems for free amino acids that are not fully operative because of (a) partial saturation of that system by a competing natural or synthetic amino acid antagonist or (b) loss of transport capability for other reasons (Peters *et al*., 1953).

In none of the work described thus far has there been any indication that peptides serve any role other than as a source of limiting amino acids. On the other hand, if one employs nutritional criteria, one should not be surprised if only nutritional answers are forthcoming. Nevertheless, considering the profusion of peptide hormones that operate in mammalian systems, one wonders if similar considerations may sometimes apply to their action. My colleagues indicate that far too little hormone is required to serve as a useful amino acid source in target cells. However, in these days of cascade amplification and differential rates of protein turnover, this objection is hardly compelling; the possibility at least needs study. It is clear in any case that transport phenomena, which control access to the cell of all types of compounds in its external environment, occupy a focal point in the utilization of nutrients. Volumes such as the present one play an important role in summarizing and systematizing our knowledge of such events and thus in directing future effort in the field.

REFERENCES

Boos, W. (1974). *Annu. Rev. Biochem.*, **43**, 123–146

Brickson, W. L., Henderson, L. M., Solhjell, I., and Elvehjem, C. A. (1948). *J. Biol. Chem.*, **176**, 517–528.

Cohen, G. N. and Monod, J. (1957). *Bacteriol. Rev.*, **21**, 169–194.

Cohen, G. N. and Rickenberg, H. V. (1956). *Ann. Inst. Pasteur, Paris*, **91**, 693–720.

Doudoroff, M. (1943). *Proc. Soc. Exp. Biol. Med.*, **53**, 73

Fruton, J. and Simmons, S. (1948). *J. Biol. Chem.*, **174**, 705–715 (cf. also *Science, N.Y.*, **109**, 561 (1949) and **111**, 329 (1950))

Gale, E. F. (1945). *Br. J. Exp. Path.*, **26**, 225–233

Gladstone, G. P. (1939). *Br. J. Exp. Path.*, **20**, 189

Guirard, B. M. and Snell, E. E. (1963). In *The Bacteria*, Vol. 4 (I.C. Gunsalus and R. Y. Stanier, eds), Academic Press, New York, pp. 33–93

Holden, J. T. (ed.) (1962). *Amino Acid Pools*, Elsevier, Amsterdam, London, New York, pp. 527–616

Holden, J. T. and Snell, E. E. (1949) *J. Biol. Chem.*, **178**, 799–809

Kihara, H. and Snell, E. E. (1952). *J. Biol. Chem.*, **197**, 791–800

Kihara, H. and Snell, E. E. (1955). *J. Biol. Chem.*, **212**, 83–94

Kihara, H. and Snell, E. E. (1960*a*). *J. Biol. Chem.*, **235**, 1409–1414.

Kihara, H. and Snell, E. E. (1960*b*). *J. Biol. Chem.*, **235**, 1415–1418.

Kihara, H., Klatt, O. A. and Snell, E. E. (1952). *J. Biol. Chem.*, **197**, 801–807.

Kihara, H., McCullough, W. G. and Snell, E. E. (1952). *J. Biol. Chem.*, **197**, 783–789

Knight, B. C. J. G. (1945). *Vitamins Hormones*, **2**, 105–228

Leach, F. R. and Snell, E. E. (1960). *J. Biol. Chem.*, **235**, 3523–3531

Malin, R. B., Camien, M. N. and Dunn, M. S. (1951). *Arch. Biochem. Biophys.*, **32**, 106–112

Peters, V. J., Prescott, J. M. and Snell, E. E. (1953). *J. Biol. Chem.*, **202**, 521–532

Peterson, W. H. and Peterson, M. F. (1945). *Bacteriol. Rev*, **9**, 49–109

Prescott, J. M., Peters, V. J., and Snell, E. E. (1953). *J. Biol. Chem.*, **202**, 533–540

Simoni, R. D. and Postma, P. W. (1975) *Annu. Rev. Biochem.*, **44**, 523–554

Snell, E. E. (1945). *J. Biol. Chem.*, **158**, 497–503

Snell, E. E. and Guirard, B. M. (1943). *Proc. Natl. Acad. Sci. U.S.A.*, **29**, 66–73

Snell, E. E., Radin, N. and Ikawa, M. (1955). *J. Biol. Chem.*, **217**, 803–818

Sprince, H. and Woolley, D. W. (1944). *J. Exp. Med.* **80**, 213–221

Stone, D. and Hoberman, H. D. (1953). *J. Biol. Chem.*, **202**, 203–212

Williams, R. J. (1941). *Biol. Rev.*, **16**, 49–80

Wilson, D. B. (1978). *Annu. Rev. Biochem.*, **47**, 933–965

Woods, D. D. (1953). *J. Gen. Microbiol.*, **9**, 151–173

Part 1

Amino Acids

Microorganisms and Nitrogen Sources
Edited by J. W. Payne

Section 1.1

Transport and Utilization of Amino Acids by Bacteria

YASUHIRO ANRAKU
Department of Biology, Faculty of Science,
University of Tokyo, Hongo Bunkyo-Ku, Tokyo 113, Japan

1 INTRODUCTION

Most bacteria can synthesize all the twenty amino acids necessary for protein biosynthesis utilizing inorganic ammonium salts as nitrogen source. In general, the biosyntheses of amino acids are strictly controlled by mechanisms such as end-product regulation (Umbarger, 1969) and the organisms do not normally overproduce them. These facts imply that the control mechanisms are physiologically effective and also that the bacterial envelopes form highly impermeable barriers for the free amino acids which are synthesized endogenously. On the other hand, bacteria can also exclusively utilize amino acids from the medium, by taking them up actively (Kepes and Cohen, 1962).

To date, no one doubts that the accumulation and transport of amino acids in cells are catalysed by functionally specific transport systems which are located in the cytoplasmic membrane. In a broader sense, transport systems serve dual roles in cell physiology: one as chemical sensing systems for detecting nutritional information in the medium and the other as functional barriers for maintaining pool amino acids in the cells (Anraku, 1975*a,b*).

Cohen and Monod (1957) founded modern biological studies on the transport and utilization of amino acids in bacteria. Independently, Britten and McClure (1962*a*) studied the properties of the amino acid pool in *Escherichia coli* in some detail. These two achievements provided the stimulus for the investigation of amino acid transport in its early phase. In the mid 1960s two new methods were developed which expedited transport studies. First, Neu and Heppel (1965) established a mild osmotic shock procedure which produced slight alterations in the cell envelopes of Gram-negative bacteria and turned out to be a useful technique in analysing the function of transport systems (Anraku, 1967, 1978; Anraku and Heppel, 1967; Heppel, 1971; Rosen and Heppel, 1973). Second, Kaback and Stadtman (1966) reported a procedure for preparing membrane vesicles after osmotic lysis of spheroplasts, and first described the properties of proline uptake by such vesicles. These vesicle preparations have been widely used (Kaback, 1971), particularly in studies of the mechanism of energy coupling for concentrative uptake (Harold, 1972, 1977; Kaback, 1970, 1972, 1974).

The general properties and information on the molecular structure of amino acid transport systems have been reviewed recently (Anraku, 1978; Rosen and Kashket, 1978). Readers are referred to other comprehensive reviews for further information on bacterial active transport of amino acids (Halpern, 1974; Lin, 1971; Oxender, 1972*a,b*).

This chapter summarizes the substrate specificities of the active transport systems for amino acids in bacteria. Molecular properties of the transport systems and regulation of the activities are also mentioned briefly. Readers are referred to other sections of this book for further information (Sections 1.2–1.7). Finally, intracellular homeostasis of amino acids is discussed in the light of cross-control by biosynthesis and transport.

2 DEFINITIONS

I will use the term *active transport system* as a representative molecular apparatus capable of translocating substrate amino acids across the cytoplasmic membrane stereospecifically and catalytically, and resulting in the accumulation of substrates against a concentration gradient in osmotically active forms in the cells. The transport system contains a *carrier* as an essential component and has regulatory ability for the carrier, especially in relation to the process of energy coupling. Thus, in this definition, the transport system seems to be like the *permease* defined by Winkler and Wilson (1966) but not similar to that defined by Cohen and Monod (1957).

The term carrier was defined thermodynamically by Wilbrandt and Rosenberg (1961) and was also considered bioenergetically by Winkler and Wilson (1966), and is used here with essentially the same meaning. Biochemical properties of the carriers in bacteria are not yet known, except for a proline carrier. The proline carrier of *E. coli* K12 is an intrinsic membrane protein located in the cytoplasmic membrane (Motojima *et al.*, 1978); it forms a ternary complex with a proton and a proline molecule (Amanuma *et al.*, 1977*b*), and translocates substrate across the membrane of reconstituted vesicles when provided with a supply of chemiosmotic energy (Amanuma *et al.*, 1977*a*).

The term *binding protein* refers to periplasmic binding protein as described by Heppel (1971). Binding proteins themselves have no ability to transfer substrates across the cytoplasmic membrane but they do stimulate stereospecifically the activities of certain transport systems (Anraku, 1968*c*; Anraku *et al.*, 1973*a*; Yamato and Anraku, 1977; Kustu and Ames, 1974). Therefore, they are not carriers in the sense defined as above but could be components of some transport systems. Possible roles of binding proteins in active transport (Anraku, 1975*a*, 1978), and their occurrence and binding specificities have been discussed elesewhere in greater detail (Oxender and Quay, 1976).

3 SUBSTRATE SPECIFICITIES OF TRANSPORT SYSTEMS

1 Transport systems of Escherichia coli

In *E. coli* K12, there are twelve kinetically-defined transport systems for groups of structurally related amino acids. These are glycine–alanine, threonine–serine, leucine–isoleucine–valine, phenylalanine–tyrosine–tryptophan, methionine, proline, lysine–arginine–ornithine, cystine, asparagine, glutamine, aspartate, and glutamate. Beside these main systems, the organism has transport systems specific for phenylalanine, tyrosine, tryptophan, lysine, and arginine. Transport systems for histidine and cysteine in intact *E. coli* K12 have not been studied, but these two amino acids were taken up separately by membrane vesicles prepared from *E. coli* ML308–225 (Lombardi and Kaback, 1972).

(i) Glycine–Alanine

A transport system for glycine and alanine was first described by Maas and Davis (1950). Schwartz *et al*. (1959) isolated mutants resistant to D-serine and found that they were defective in the uptake of glycine, alanine, and D-serine. Lubin and his associates studied the glycine–alanine transport system genetically and concluded that the system is also used for D-cycloserine uptake (Kessel and Lubin, 1965; Lubin *et al*., 1960). Essentially the same result was obtained by Cosloy (1973). The gene which effects the transport activity is located at 94 min on the *E. coli* genetic map (Cosloy, 1973; Curtiss *et al*., 1965; Russel, 1972; Wargel *et al*., 1971).

Robbins and Oxender (1973) studied the transport system kinetically. The K_t values for entry of glycine and alanine were determined as $2.8 \sim 4.8\ \mu M$, respectively (Table 1). Existence of a transport system specific for alanine but not for glycine was reported by several investigators (Kessel and Lubin, 1965; Piperno and Oxender, 1968; Wargel *et al*., 1970, 1971). This transport system has a K_t value of $2\ \mu M$ for alanine and is inhibited by leucine competitively, with a K_i value of $0.2\ \mu M$ (Robbins and Oxender, 1973).

(ii) Threonine–Serine

Templeton and Savageau (1974) reported a specific transport system for threonine, which shows a K_t value of $2\ \mu M$ and is inhibited competitively by serine (K_i value of $1.4\ \mu M$). Yamato *et al*. (1976) isolated transport mutants which lack the uptake activities for leucine, isoleucine, and valine but are normal in threonine uptake. Kinetic studies using these mutants indicated the existence of the threonine–serine transport system which is distinct from that for branched-chain amino acids (Fujimura, I. Yamato and Y. Anraku, unpublished result).

Smulson *et al*. (1967) studied threonine uptake in a wild-type strain and suggested that threonine could be transported into the cells via a branched-chain amino acid transport system. Robbins and Oxender (1973) determined a K_t value of $0.39\ \mu M$ for uptake of threonine via a leucine–isoleucine–valine system (see next section).

Table 1 Transport systems for amino acids in *Escherichia coli*

	Transport system	Substrate	K_t (μM)	Reference
1.	Glycine–alanine	Glycine	2.8–4.8	Robbins and Oxender (1973)
		Alanine	27	
1a.	Alanine	Alanine	2	
2.	Threonine–serine	Threonine	2	Templeton and Savageau (1974)
2a.	Threonine	Threonine	0.39	Robbins and Oxender (1973)
3.	Leucine–isoleucine–valine			
	LIV-1	Isoleucine	0.4	Yamato *et al*. (1975*b*)
	LIV-2	Isoleucine	6	
	LIV-3	Isoleucine	6	
4.	Phenylalanine–tyrosine–tryptophan	Phenylalanine	0.47	Brown (1970)
		Tyrosine	0.57	
		Tryptophan	0.40	
4a.	Phenylalanine	Phenylalanine	2.0	
4b.	Tyrosine	Tyrosine	2.2	
4c.	Tryptophan	Tryptophan	3.0	
5.	Methionine-1	Methionine	0.075	Kadner (1974)
	Methionine-2	Methionine	40	
6.	Proline	Proline	0.44 and 40	Morikawa *et al*. (1974)
7.	Lysine–arginine–ornithine	Lysine	0.5	Rosen (1971)
		Ornithine	1.4	
7a	Lysine	Lysine	10	
7b.	Arginine	Arginine	0.026	
8.	Cystine-1	Cystine	0.3	Berger and Heppel (1972)
	Cystine-2	Cystine	0.02	
9.	Asparagine-1	Asparagine	3.5	Willis and Woolfolk (1975)
	Asparagine-2	Asparagine	80	
10.	Glutamine-1	Glutamine	0.15	Weiner *et al*. (1971)
	Glutamine-2	Glutamine	2	
11.	Aspartate-1	Aspartate	3.7	Kay and Kornberg (1971)
	Aspartate-2	Aspartate	39	
12.	Glutamate	Glutamate	0.7	Willis and Furlong (1975*a*)
		Glutamate	10	

(iii) Leucine–Isoleucine–Valine

Cohen and Rickenberg (1956) first showed the presence of a high-affinity common transport system for these three branched-chain amino acids. At present there have been at least three independent transport systems characterized in *E. coli* K12, which show different specificities and different biochemical properties (see Section 1.6.4) (Rahmanian *et al*., 1973; Wood, 1975; Yamato *et al*., 1975*b*). All the transport systems seem to require periplasmic binding proteins for branched-chain amino acids (Amanuma and Anraku, 1974; Amanuma *et al*., 1976; Anraku, 1968*a,b,c*; Furlong and Weiner, 1970; Furlong *et al*., 1973; Penrose *et al*., 1968) as evidenced by sensitivity to osmotic shock, and the fact that rates of uptake are affected by the presence of the binding proteins (Anraku *et al*., 1973*a*; Yamato and Anraku, 1977).

Rahmanian *et al*. (1973) classified the three systems as LIV-I, LIV-II, and L-systems, based on their kinetic specificities in intact cells. The LIV-I system corresponds to the common transport system of Cohen and Rickenberg (1956) but it also transports threonine and alanine as additional substrates. The LIV-II system is minor and takes up the three branched-chain amino acids with a low affinity. On the other hand, the L-system is specific for leucine only and is not inhibited by isoleucine. Essentially the same conclusion was obtained from a kinetic analysis carried out by Wood (1975).

By studying the specificities of carriers for branched-chain amino acids directly, Anraku and his associates showed that three independent carriers were kinetically distinguishable in cytoplasmic membrane preparations and intact cells (Yamato and Anraku, 1977; Yamato *et al*., 1975*b*). All the carriers are equally specific for the three branched-chain amino acids but apparent transport activities are differently affected by binding proteins. Thus, they classified the systems as LIV-1, LIV-2, and LIV-3. Table 1 shows the K_t values for isoleucine uptake by these systems. LIV-1 and LIV-2 systems require ATP as a source of coupling energy, and they are stimulated, respectively, by binding protein obligatorily and additionally. The LIV-3 system is energized via a proton motive force and is detectable in cytoplasmic membrane vesicles; its activity is additionally stimulated by the binding proteins.

Evidence used for the classification of branched-chain amino acid transport systems and apparent differences in their specificities are discussed in greater detail elsewhere (Anraku, 1978) (see also Section 1.6.4).

(iv) Phenylalanine–Tyrosine–Tryptophan

A common transport system for these three amino acids is referred to as the general aromatic amino acid transport system, and has been studied extensively by Brown (1970, 1971). The K_t values for the substrates are shown in Table 1 (Brown, 1970). The transport activity is inhibited by high concentrations of histidine, leucine, and methionine. The gene *aroP*, which determines this activity, was mapped at 2 min on the *E. coli* genetic map (Brown, 1970, 1971; Guest, 1974).

Brown (1970) also showed that there are three further, related transport systems, each of which is specific for a particular aromatic amino acid and with affinities of the order of 10^{-6} M. A tryptophan-specific transport system was studied genetically by Kuhn and Somerville, (1971).

(v) Methionine

Kadner and his associates (Kadner, 1974; Kadner and Watson, 1974) reported two transport systems specific for methionine. The methionine-1 system has a high affinity (K_t value of 0.075 μM) and was reported to be inhibited by methionyl peptides (see Section 2.6). The methionine-2 system has a low affinity (K_t, 40 μM) and is inhibited by selenomethionine. Transport and utilization of D-methionine and other methionine derivatives has been reported recently (Kadner, 1977).

Genetical studies indicated the presence of the *metD* and *metP* genes for the methionine-1 and methionine-2 systems, respectively (Cooper, 1966; Kadner, 1974; Kadner and Watson, 1974).

(vi) Proline

Anraku and his associates showed that the transport system for proline is highly specific and the proline carrier activity is regulated by a mechanism of negative cooperativity (Morikawa *et al*., 1974; Awazu *et al*., 1976). The apparent K_t values were estimated as 0.44 and 40 μM for the high-affinity and low-affinity transport processes, respectively (Morikawa *et al*., 1974). The mutation in the gene *proT*, which seems to determine the proline carrier and is located at 82 min on the *E. coli* genetic map, results in alteration in both high- and low-affinity transport processes (Motojima *et al*., 1978).

Substrate specificity of the transport system has been studied in several laboratories (Kessel and Lubin, 1962; Rowland and Tristram, 1974; Tristram and Neale, 1968). Tristram and Neale (1968) reported a K_t value for high-affinity proline transport of 0.46 μM and a K_i value of 24 μM for L-azetidine-2-carboxylate.

(vii) Lysine–Arginine–Ornithine

Schwartz *et al*. (1959) found a basic amino acid transport system, which is common for lysine, arginine, and ornithine uptake, by studying mutants resistant to canavanine. Later, the properties of this common transport system were studied genetically and biochemically in two laboratories. Maas and his associates showed that the gene *argP*, which is located at 62 min on the *E. coli* genetic map, is responsible for the activity (Celis, 1977*a*; Celis *et al*., 1973; Maas, 1972). Rosen (1971, 1973*a,b*) showed that the K_t values for lysine and ornithine are 0.5 and 1.4 μM, respectively, and for canavanine 400 μM (Table 1).

Rosen (1971) further demonstrated the existence of a highly specific arginine transport system (K_t value of 0.026 μM) which is not affected by a 500-fold molar excess of lysine or ornithine. A lysine-specific transport system was also reported in *E. coli* K12, which shows an apparent K_t value of 10 μM and is insensitive to arginine and ornithine (Rosen, 1971).

(viii) Cystine

Cystine transport systems have been studied by Leive and Davis (1965*a,b*) and by Berger and Heppel (1972). The cystine-1 system is also used for the transport of diaminopimelic acid; the K_t value for cystine via this sytem is 0.3 μM and the K_i value of diaminopimelic acid is 14 μM. The cystine-2 system is specific for cystine (K_t value, 0.02 μM) and the activity is inhibited by selenocystine (Berger and Heppel, 1972).

(ix) Asparagine

Willis and Woolfolk (1975) described the properties of two transport systems specific for asparagine. The asparagine-1 system (K_t value of 3.5 μM) is inhibited competitively by the following structurally related compounds: 5-diazo-4-oxo-L-norvaline (K_i, 4.6 μM), β-hydroxyamyl-L-aspartic acid (K_i, 10 μM), and L-aspartic acid (K_i, 50 μM). The activity of this system is repressible by asparagine. The asparagine-2 system (K_t value of 80 μM) is necessary for utilization of asparagine as a nitrogen source.

(x) Glutamine

Heppel and his coworkers investigated the properties of two transport systems for glutamine (Weiner *et al*., 1971; Weiner and Heppel, 1971). The glutamine-1 system has a K_t value of 0.15 μM and is inhibited competitively by α-glutamylhydrazide and α-glutamylhydroxamate. Willis *et al*. (1975) found that the activity of this system was derepressed when the cells were grown in a nitrogen-limited medium and was repressed about 20-fold when they were grown in a glutamine-depleted nutrient broth. The glutamine-2 system, which has a K_t value larger than 2 μM, appeared only in these repressed cells.

(xi) Aspartate

Transport of aspartate is catalysed by two distinct systems (Kay, 1971; Kay and Kornberg, 1969, 1971). The aspartate-1 system (K_t value of 3.7 μM) is inhibited by β-hydroxyaspartate but affected only slightly by glutamate (K_i, 640 μM), glutamine (K_i, 840 μM), and D-aspartate (K_i, 200 μM). A mutant, which was resistant to inhibition of growth by D,L-*threo*-β-hydroxyaspartic acid, was found to be defective in this activity. The aspartate-2 system is used in common for the transport of dicarboxylic acids such as succinate, fumarate, D,L-malate, and oxaloacetate (Kay and Kornberg, 1971).

(xii) Glutamate

The glutamate transport systems of *E. coli* K12 and *E. coli* B are apparently unique since the uptake activities require sodium ions for regulation (Frank and Hopkins, 1969; Miner and Frank, 1974; Willis and Furlong, 1975*a*) and to provide the driving force for uphill transport (Tsuchiya *et al*., 1977).

Studies on the substrate specificity and regulation of the glutamate transport system was initiated by Halpern and his coworkers (Halpern and Umbarger, 1961; Halpern and Even-Shoshan, 1967; Halpern and Lupo, 1965; Marcus and Halpern, 1967, 1969). The transport system is highly specific for glutamate. γ-Methylglutamate, β-hydroxyglutamate, and α-methylglutamate were reported to be competitive inhibitors (Halpern and Even-Shoshan, 1967). Marcus and Halpern (1969) found two genes controlling the transport system: the gene *gltS* (81 min) codes for a structural component of the system, and the gene *gltR* (91 min) regulates the synthesis of the system.

Recently, the cotransport mechanism of glutamate uptake with sodium ion was studied biochemically. Willis and Furlong (1975*a*) demonstrated that sodium ion strikingly changed the K_t value for entry of glutamate in *E. coli* K12, strain W3092: it was 10 μM in the absence of sodium ion but only 0.7 μM in the presence of an optimal level of sodium ion (Table 1).

2 Transport Systems of Other Bacteria

The properties of amino acid transport systems in several other bacteria have been investigated. Table 2 shows a summary of the transport systems examined in the respective microorganisms. The substrate specificities of the systems are essentially the same as in *E. coli*. Transport systems for threonine, serine, cysteine, glutamine, and asparagine have not been described.

Among the systems studied, the histidine transport system of *Salmonella typhimurium* has been analysed most extensively, mainly by Ames and her associates (see Ames and Spudich, 1976). Genetical studies showed that a

Table 2 Transport systems for amino acids in bacteria

Transport system	Organism	References
Glycine–alanine	*Bacillus megaterium* KM	Marquis and Gerhardt (1964)
	Lactobacillus casei	Leach and Snell (1960)
	Streptococcus faecalis	Mora and Snell (1963), Ashghar *et al*. (1973)
Alanine	Thermobacterium PS-3	Hirata *et al*. (1976*a*)
Leucine–isoleucine–valine	*Halobacterium halobium*	MacDonald and Lanyi (1975)
	Marine pseudomonad B16	Pearce *et al*. (1977)
	Salmonella typhimurium	Kiritani (1974), Kiritani and Ohnishi (1977), Thorne and Corwin (1972)
Phenylalanine–tyrosine–tryptophan	*Bacillus subtilis*	D'Ambrosio *et al*. (1973), Glover *et al*. (1975)
	Pseudomonas aeruginosa	Kay and Gronlund (1971)
	Pseudomonas Sp. 11299a	Guroff *et al*. (1969), Guroff and Bromwell (1970)
	Pseudomonas acidovorans	Rosenfeld and Feigelson (1969)
Methionine	*Salmonella typhimurium*	Ayling and Bridgeland (1972)
Cystine	*Salmonella typhimurium*	Baptist and Kredich (1977)
Proline	*Mycobacterium phlei*	Israeli *et al*. (1977)
	Pseudomonas aeruginosa	Kay and Gronlund (1969*a*, *b*)
	Pseudomonas putida	Gryder and Adams (1970)
Histidine	*Salmonella typhimurium*	Ames (1964), Ames and Lever (1970, 1972), Ames and Roth (1968), Ames and Spudich (1976), Kustu and Ames (1973, 1974), Rosen and Vasington (1971)
Lysine–arginine–ornithine	*Arthrobacter pyridinolis*	Krulwich *et al*. (1977)
	Pseudomonas putida	Miller and Rodwell (1971), Fan *et al*. (1972)
Glutamate–aspartate	*Halobacterium salinarium*	Stevenson (1966)
	Lactobacillus plantarum	Holden *et al*. (1968)
	Marine pseudomonad	Sprott and Macleod (1972)
	Mycobacterium avium	Yabu (1967)
	Mycobacterium smegmatis	Yabu (1971)
	Streptococcus faecalis	Holden *et al*. (1968), Reid *et al*. (1970), Utech *et al*. (1970)
	Thiobacillus novellus	Haban and Lyric (1977)

periplasmic binding protein (*hisJ* protein) and a membrane component (*hisP* protein) are both essential for histidine transport across the cytoplasmic membrane. Molecular mechanisms of solute transport that involve functional association with periplasmic binding proteins have recently been reviewed in detail (Ames and Spudich, 1976; Anraku, 1978).

3 Transport Systems of Membrane Vesicles

Lombardi and Kaback (1972) examined the specificities of amino acid transport in membrane vesicles of *E. coli* ML308-225, prepared after osmotic lysis of spheroplasts (Kaback, 1971). It was found that this membrane preparation actively took up sixteen different amino acids through mechanisms that could be coupled to the respiratory substrate D-lactate and the artificial electron donor system, ascorbate-phenazine methosulphate, and that could be categorized into nine kinetically different transport systems (Table 3). These membrane vesicles take up histidine and cysteine but not methionine, cystine, glutamine, or asparagine. A basic amino acid transport system was not detectable in the vesicles.

Anraku and his associates have established a procedure for preparing the cytoplasmic membrane vesicles of *E. coli* K12 (Yamato and Anraku, 1977; Yamato *et al*., 1975*a*, 1978). This membrane preparation shows differences from that above in that it takes up glutamine and cystine, and shows high transport activities for proline, serine, branched-chain amino acids, and basic amino acids (Yamato and Anraku, 1977).

Recently, Lanyi and his coworkers prepared envelope membranes of *Halobacterium halobium* by sonication, and studied the energetics of amino acid transport in these membrane vesicles (Lanyi, 1977; Lanyi *et al*., 1976*a,b*; MacDonald and Lanyi, 1975; MacDonald *et al*., 1977). All the amino acids except cysteine are taken up actively in the vesicles in response to light-induced membrane potential and an artificially created sodium ion gradient (MacDonald *et al*., 1977): glutamate transport is driven only by a chemical gradient of sodium ions. Inhibition studies indicated that the vesicles possess nine kinetically different transport systems (Table 3).

The membrane vesicles of *Bacillus subtilis* take up eighteen amino acids via ten transport systems for glycine–alanine, threonine–serine, leucine–isoleucine–valine, phenylalanine–tyrosine, methionine, cysteine, proline, lysine–arginine, asparagine–glutamine, and aspartate–glutamate (Konings *et al*., 1971, 1972; Konings and Freese, 1972). Kaback and his associates showed that the membrane vesicles of *Staphylococcus aureus* have twelve transport systems covering the uptake of twenty amino acids; these are glycine–alanine, threonine–serine, leucine–isoleucine–valine, phenylalanine–tyrosine–tryptophan, methionine, cysteine, proline, lysine, arginine, histidine, asparagine–glutamine, and aspartate–glutamate (Short *et al*., 1972*a,b*).

Table 3 Transport systems for amino acids detectable in membrane vesicles prepared from *E. coli* ML308-225 and *H. halobium*

E. coli ML308-225[a]			*H. halobium*[b]		
System	Substrate	K_t (μM)	System	Substrate	K_t (μM)
Glycine–alanine	Glycine	1.6	Glycine–alanine–threonine–serine	Glycine	4.5
	Alanine	8.4		Alanine	4.2 and 9.6
				Threonine	5.0
Threonine–serine	Threonine	2.6		Serine	2.9 and 3.2
	Serine	5.4			
Leucine–isoleucine–valine	Leucine	1.1 and 18	Leucine–isoleucine valine–methionine	Leucine	1.4
	Isoleucine	1.7 and 21		Isoleucine	2.4
	Valine	2.0 and 29		Valine	1.1
				Methionine	3.7
Phenylalanine–tyrosine–tryptophan	Phenylalanine	0.42	Phenylalanine–tyrosine–tryptophan	Phenylalanine	6.4 and 8.3
	Tyrosine	0.68		Tyrosine	33
	Tryptophan	0.33		Tryptophan	2.3
Cysteine	Cysteine	38			
Proline	Proline	1.0	Proline	Proline	0.8 and 0.7
Lysine	Lysine	0.94	Arginine–lysine	Arginine	1.0
				Lysine	2.0
Histidine	Histidine	0.15 and 4.0	Histidine	Histidine	3.9 and 6.8
Glutamate–aspartate	Glutamate	1.6	Glutamate	Glutamate	0.13
	Aspartate	8.4			
			Aspartate	Aspartate	25
			Glutamine–asparagine	Glutamine	13 and 32
				Asparagine	10.3

[a]Lombardi and Kaback (1972).
[b]MacDonald *et al*. (1977).

A number of membrane vesicles capable of transporting amino acids have been prepared from bacteria such as thermophilic bacterium PS-3 (Hirata *et al.*, 1976*a*), *Veillonells alcalences* (Konings *et al.*, 1975), *Salmonella typhimurium, Micrococcus denitrificans, Arthrobacter pyridinolis, Mycobacterium phlei, Pseudomonas putida, Proteus mirabilis, Bacillus megaterium* (see Kaback, 1974).

4 GENERAL PROPERTIES OF TRANSPORT SYSTEMS

1 Molecular Structure

Although the substrate specificities of various transport systems in bacteria are kinetically characterized, biochemical knowledge on the whole structure of the systems mediating transport of amino acids is still very incomplete. Early studies, using the procedure of osmotic shock, showed that this treatment causes the transport activities for branched-chain amino acids, arginine, and glutamate to be decreased and gives concomitant release of the corresponding binding proteins into the medium (Anraku, 1968*c*; Anraku *et al.*, 1973*a*; Barash and Halpern, 1971; Wilson and Holden, 1969). In several cases the decreased transport activities could be restored by incubating the shocked cells with the appropriate purified binding protein and other protein components: the effect of the binding proteins is stereospecific and catalytic (see Anraku, 1978).

Berger and Heppel (1974) studied the effect of osmotic shock in detail and found that the transport systems for glutamine, diaminopimelic acid, arginine, and histidine are shock-sensitive while those for proline, phenylalanine, serine–glycine, and cysteine are shock-resistant. These, and other investigators, (Heppel, 1971; Rosen and Heppel, 1973; Oxender, 1972*a,b*) have suggested that the transport systems for amino acids are structurally different and that in certain of them the periplasmic binding proteins act as effectors for their transport activities. Genetic studies on the transport systems for histidine (Ames and Lever, 1970, 1972; Ames and Spudich, 1976; Kustu and Ames, 1974) and the branched-chain amino acids (Anderson *et al.*, 1976; Anderson and Oxender, 1977; Rahmanian and Oxender, 1972) have also shown that the genes determining the structure and biosynthesis of binding proteins are involved in regulating the transport activities (see Section 1.6.4).

Recently, evidence that a periplasmic binding protein has no ability to translocate substrate across the cytoplasmic membrane but can stimulate stereospecifically the rate of uptake with turnover ability has been obtained. Thus, Yamato and Anraku (1977) reported that the LIVT (leucine–isoleucine–valine–threonine)-binding protein does stimulate, but the Ls (leucine specific)-binding protein does not stimulate the initial rate of isoleucine uptake by cytoplasmic membrane vesicles; the stimulation occurs nout changing the K_t value for entry, and is observed only when the vesicles possess the carrier activity for branched-chain amino acids.

It is presently believed that the binding proteins for the branched-chain amino acids (Yamato and Anraku, 1977; Yamato *et al*., 1975*b*) and for glutamate (Kahane *et al*., 1976*a,b*; Willis and Furlong, 1975*b*) in *E. coli*, and for histidine in *S. typhimurium* (Ames and Lever, 1972; Ames and Spudich, 1976) are components of the respective transport systems. However, the precise mechanisms by which a binding protein stimulates carrier activity is not yet known. Possible mechanisms by which a carrier and a binding protein may mutually interact were discussed recently (Anraku, 1978).

2 Energy Coupling

Kaback and coworkers (Kaback, 1971) established experimental techniques for measuring the activity of respiration-driven amino acid uptake using membrane vesicles prepared after osmotic lysis of spheroplasts. Subsequently coupling of chemiosmotic energy (proton motive force) to the transport systems for proline and alanine was demonstrated, using membrane vesicles (Hirata *et al*., 1973, 1974, 1976*a*; Kasahara and Anraku, 1974) and reconstituted proteoliposomes (Hirata *et al*., 1976*b*, 1977; Amanuma *et al*., 1977*b*). Recently, Tsuchiya *et al.* (1977) studied the role of sodium ion in glutamate transport by *Escherichia coli* B. In this study, the cells were first depleted of endogenous energy by starvation, and then incubated with glutamate, uptake of which was driven by a supply of chemical energy provided in a form of sodium ion gradient; uphill transport of glutamate, showing an accumulation of more than 100-fold, took place under these conditions. It is interesting to note that this Na^+-driven glutamate transport was observed in *E. coli* B but not in *E. coli* K12.

On the other hand, Berger (1973) found that the energy present in the high energy phosphate bond (ATP) is absolutely required for glutamine uptake by intact cells of *E. coli*. Subsequent studies confirmed this observation, and further indicated that the intracellular concentration of ATP is directly related to the active transport of several amino acids (Berger and Heppel, 1974; Kobayashi *et al*., 1974; Wood, 1975). The mechanism of ATP-dependent active transport has not yet been elucidated. Comprehensive articles on energy coupling in active transport have been published by several investigators (Harold, 1972, 1977; Rosen and Kashket, 1978; Simoni and Postma, 1975) (see also Section 1.7).

3 Regulation

(i) *Asymmetrical affinity*

When energy is supplied, the proline transport system of *E. coli* K12 takes up proline actively and maintains it in the cells at concentrations more than

1000-fold against a concentration gradient (Morikawa *et al*., 1974). These authors also found that under these conditions, the $K_{t_{entry}}$ and $K_{t_{exit}}$ for proline differed by more than three orders of magnitude and that 2,4-dinitrophenol and potassium cyanide decreased the rate of uptake (V_{max} value) greatly but did not change either the $K_{t_{entry}}$ or $K_{t_{exit}}$ appreciably. Therefore, these findings suggested that the proline carrier uses energy mainly for translocating substrate but not for binding it. This possibility was later confirmed directly by assaying the binding of proline to its carrier using sonicated membranes (Amanuma *et al*., 1977*a*). No clear explanation is yet available for the mechanism of regulation of the asymmetrical affinity of a carrier on the different sides of the cytoplasmic membrane. However, the observation by Amanuma *et al*. (1977*a*) that protonation of the proline carrier results in the appearance of a high-affinity binding site seems likely to be relevant to the regulation of the binding affinity of carrier molecule. Also relevant to this may be a report on the effects of zinc ions on the exchange-exit reaction for proline and in increasing the maximal level of accumulation (Anraku *et al*., 1975).

(ii) Negative cooperativity

Anraku and his coworkers reported kinetic evidence indicating that the transport systems for the branched-chain amino acids and for proline seemed to be regulated by a negative cooperative mechanism (Anraku *et al*., 1973*b*; Awazu *et al*., 1976; Morikawa *et al*., 1974). The physiological and biochemical meanings of this regulation are not known but this type of regulation may be teleologically important for transport systems performing unidirectional uptake of amino acids from media with different substrate concentrations of many orders of magnitude (Awazu *et al*., 1976).

Jensen and associates (D'Ambrosio *et al*., 1973; Glover *et al*., 1975) examined the single transport system for tyrosine–phenylalanine of *B. subtilis* and obtained evidence supporting a mechanism for controlling the activity by a form of negative cooperativity. The biochemical basis of the mechanism has not been examined in any detail, but it seems attractive to study the mechanism further, because no one knows the attributes of a carrier molecule in a phospholipid bilayer.

(iii) Repression and derepression

Anderson *et al*. (1976) found two genes for controlling the biosynthesis of branched-chain amino acid-binding proteins and mapped them at 20 min on the *E. coli* genetic map. They showed that a lesion of one of the genes *lstR*, results in a striking increase of Ls-binding protein, while the other gene, *livR*, derepresses the LIVT-binding protein (see also Sections 1.6.3, 1.6.4).

Kanzaki and Anraku (1971) compared the regulation of the transport activities for branched-chain amino acids in a valine-sensitive strain (W3092) and a valine-resistant strain (WV-5) of *E. coli* K12. It was found that when valine was added to the culture of strain W3092 growing in a synthetic salt medium it derepressed greatly the transport activities, while addition of isoleucine and leucine showed opposite effects to those seen with valine and repressed transport activity. Under these experimental conditions and with a supplement of 1 mM valine, no further synthesis of acetohydroxyacid synthetase occurred but the specific activities of the branched-chain amino acid-binding proteins that could be released into an osmotic shock medium were reduced noticeably (Anraku *et al*., 1973*b*). In strain WV-5, both valine and leucine repressed the enzyme and the transport activities. It was noted that in strain W3092, which is also cysteine-sensitive, cysteine derepressed the transport activities for branched-chain amino acids (Kanzaki and Anraku, 1971). These observations suggest that transport and biosynthetic activities for these amino acids seem not to be regulated in a concerted manner. Similarly, Celis (1977*b*) demonstrated that arginine did not repress arginine transport activity but it did control its own biosynthetic activity in *E. coli.* However, the transport activity for proline was shown to be derepressible by proline (Morikawa *et al*., 1974) while that of the asparagine-1 system was repressible by asparagine (Willis and Woolfolk, 1975). On the other hand, the activities of the methionine-1 and methionine-2 systems were found to be under repressible control by an internal methionine pool (Kadner, 1974; Kadner and Watson, 1974).

Two laboratories have reported on the regulation of glutamate transport activity of *E. coli.* Kahane *et al*. (1976*a*), examined the effect of growth conditions on the transport activity: when strain CS101 was grown in a synthetic medium containing aspartate as the sole source of carbon and nitrogen, the activities in intact cells and in membrane vesicles increased several-fold, while the synthesis of glutamate-binding protein was not affected. When the strain was grown in a rich broth the level of the binding protein was reduced whereas the activities were not affected appreciably. Willis and Furlong (1975*b*) found that the synthesis of the glutamate–aspartate binding protein was repressed two- to threefold when strain D_2W was grown in the presence of glucose while the glutamate transport activity did not change essentially under this growth condition.

At present, studies on the regulation of transport activities are clearly inadequate to draw clear, general conclusions. However, the findings cited above suggest the presence of diverse mechanisms for regulating the biosynthesis and functioning of the components of the amino acid transport systems. Undoubtedly, both general and specific studies on these problems are required before we can hope for further understanding of the physiological roles of amino acid transport systems (see also Sections 1.6 and 2.6).

5 AMINO ACID POOL

Many bacteria maintain an internal amino acid pool that is involved in the biosyntheses of proteins and various nitrogen-containing organic metabolites. The pool amino acids seem to be osmotically active and are maintained against a large concentration gradient. The size of the pool depends on the energy supply and also on the growth environment—particularly the available nutrients and the osmolarity in the medium. The concept, formation, and stability of pool amino acids have been reviewed (Britten and McClure, 1962*a,b*; Holden, 1962; Mandelstam, 1958).

The amount and composition of pool amino acids are important effectors in the regulation of microbial growth. Tempest and Meers (1970) studied the free amino acid pools of various microorganisms grown in chemostat cultures under various growth limitations. They found that the contents and composition of the pools did not vary with time, but did change significantly with alterations in either growth rate or the nature of the growth limitation. It has been known for some time that depletion of one or several amino acids in the pool, resulting, for example, from failure to supplement the medium for a mutant with a biosynthetic defect, causes cessation of growth with a concomitant initiation of intracellular turnover of proteins (Mandelstam, 1960) (see also Section 3.4). In another instance, Kanzaki and Anraku (1971) demonstrated dramatic changes in the contents and composition of the pool amino acids in a valine-sensitive strain of *E. coli* upon addition of valine to an exponentially growing culture. These results indicate that the homeostatic regulation of pool amino acids, both in their contents and composition, is physiologically essential for growth of microorganisms. In fact, Alim and Ring (1976) demonstrated a close correlation between the transport capacity and growth rate of *Streptomyces hydrogenans*, and concluded that the growth-rate-dependent modulation of transport capacities for amino acids is partially due to the variation of the concentrations of active transport components and seems to be affected by endogenous effectors.

Obviously, the formation and homeostasis of pool amino acids should be regulated through cross-control of the activities of biosynthetic enzymes and of membrane transport of exogenous nutrients. As is well known, the biosyntheses of the twenty protein amino acids are carried out through six independent routes, using different initial precursors. Amino acids belonging to the alanine family (alanine, valine, and leucine) and serine family (serine and cysteine) are synthesized from pyruvate and 3-phosphoglycerate as precursors, respectively. The aspartate family (aspartate, asparagine, methionine, threonine, isoleucine, and lysine) derives from oxaloacetate whereas the glutamate family (glutamine, arginine, and proline) originates from α-ketoglutarate. Tryptophan, phenylalanine, and tyrosine comprise the aromatic amino acid family and are synthesized from phosphoenolpyruvate and erythrose-4-phosphate. Histidine, which occupies a group of its own, is synthesized from phosphoribosylpyrophosphate.

It should be pointed out that the structures of the different amino acids belonging to a particular biosynthetic family are not necessarily very similar to each other (e.g. glutamate, glutamine, arginine, and proline) and their transport is often catalysed by independent systems with particular specificities (see Table 1). However, the aromatic amino acids do provide an exception in that they share a common biosynthetic precursor and a common transport system. In general, therefore, the contents of most amino acids in the pool may be maintained through the mutual but independent regulation of the processes of biosynthesis and transport. Clearly, extensive studies on the mechanisms for regulating the homeostasis of the pool amino acids are necessary before we have a clear understanding of the teleological utilization of amino acids *in vivo*.

6 CONCLUDING REMARKS

This chapter summarizes the present knowledge on the characteristics of the active transport systems for amino acids in bacteria. It can be said that the transport systems generally have strict specificities and are of high physiological activity.

The author discussed briefly the common terminology used in the field of membrane transport, as an aid to understanding the biochemical information on the protein components that catalyse active transport that has been accumulated largely during the last decade. On the basis of the definitions given in the text, brief comments on the functions and regulation of the transport systems are made partly as introductory remarks for the following topics in this book.

The concept of homeostasis of pool amino acids and its importance for microbiological physiology are well accepted. However, insufficient basic research has been done on the problem to date, especially from the points of view discussed in the text, and it is hoped that further studies on membrane transport will help to correct these deficiencies in our knowledge of the utilization of amino acids in bacteria.

ACKNOWLEDGEMENTS

I am grateful to Drs. M. Kasahara and I. Yamato for helpful comments. The original work carried out in the laboratory of the author, discussed in this article, was supported by grants (C758034, C858069, 048095, and 148319) from the Ministry of Education, Science, and Culture of Japan and the Toray Science Foundation, Japan.

REFERENCES

Alim, S. and Ring, K. (1976). Regulation of amino acid transport in growing cells of *Streptomyces hydrogenans*. II. Correlation between transport capacity and growth rate in chemostat cultures. *Arch. Microbiol.*, **111**, 105–110.

Amanuma, H. and Anraku, Y. (1974). Transport of sugars and amino acids in bacteria. XII. Substrate specificities of the branched chain amino acid-binding proteins of *Escherichia coli. J. Biochem., Tokyo*, **76**, 1165–1173.

Amanuma, H., Itoh, J., and Anraku, Y. (1976). Transport of sugars and amino acids in bacteria. XVII. On the existence and nature of substrate amino acids bound to purified branched chain amino acid-binding proteins of *Escherichia coli. J. Biochem., Tokyo*, **79**, 1167–1182.

Amanuma, H., Itoh, J. and Anraku, Y. (1977*b*). Proton-dependent binding of proline to carrier in *Escherichia coli* membrane. *FEBS Lett.*, **78**, 173–176.

Amanuma, H., Motojima, K., Yamaguchi, A., and Anraku, Y. (1977*a*). Solubilization of a functionally active proline carrier from membrane of *Escherichia coli* with an organic solvent. *Biochem. Biophys. Res. Commun.*, **74**, 366–373.

Ames, G. F.-L. (1964). Uptake of amino acids by *Salmonella typhimurium. Arch. Biochem. Biophys.*, **104**, 1–18.

Ames, G. F.-L. and Lever, J. (1970). Components of histidine transport: Histidine-binding proteins and hisP protein. *Proc. Natl. Acad. Sci. U.S.A.*, **66**, 1096–1103.

Ames, G. F.-L. and Lever, J. (1972). The histidine-binding protein J is a component of histidine transport. *J. Biol. Chem.*, **247**, 4309–4316

Ames, G. F.-L. and Roth, J. H. (1968). Histidine and aromatic permeases of *Salmonella typhimurium. J. Bacteriol.*, **96**, 1742–1749

Ames, G. F.-L. and Spudich, E. N. (1976). Protein–protein interaction in transport: the periplasmic histidine-binding protein J interacts with the P protein. *Proc. Natl. Acad. Sci. U.S.A.*, **73**, 1877–1881

Anderson, J. J. and Oxender, D. L. (1977). *Escherichia coli* transport mutants lacking binding proteins and other components of the branched-chain amino acid transport systems. *J. Bacteriol.*, **130**, 384–392.

Anderson, J. J., Quay, S. C., and Oxender, D. L. (1976). Mapping of two loci affecting the regulation of branched chain amino acid transport in *Escherichia coli* K-12. *J. Bacteriol.*, **126**, 80–90

Anraku, Y. (1967). The reduction and restoration of galactose transport in osmotically shocked cells of *Escherichia coli. J. Biol. Chem.*, **242**, 793–800.

Anraku, Y. (1968*a*). Transport of sugars and amino acids in bacteria. I. Purification and specificity of the galactose- and leucine- binding proteins. *J. Biol. Chem.*, **243**, 3116–3122

Anraku, Y. (1968*b*). Transport of sugars and amino acids in bacteria. II. Properties of galactose- and leucine-binding proteins. *J. Biol. Chem.*, **243**, 3123–3127.

Anraku, Y. (1968*c*). Transport of sugars and amino acids in bacteria. III. Studies on the restoration of active transport. *J. Biol. Chem.*, **243**, 3128–3135

Anraku, Y. (1975*a*). On the roles of binding proteins in the membrane transport and cellular homeostasis of amino acids. In *Proceedings of the 1st Intersectional Congress of IAMS*, Vol. 1 (T. Hasegawa, ed.), Science Council of Japan, Tokyo, pp. 563–572

Anraku, Y. (1975*b*). Membrane transport and metabolic regulation. In *Modern Biological Sciences*, Vol. 5 (R. Sato and Y. Nishizaka, eds). Iwanami Shoten, Tokyo, pp. 17–53 (in Japanese)

Anraku, Y. (1978). Active transport of amino acids. In *Bacterial Transport* (B. P. Rosen, ed.), Marcel Dekker, New York, pp. 171–219

Anraku, Y. and Heppel, L. A. (1967). On the nature of the changes induced in *Escherichia coli* by osmotic shock. *J. Biol. Chem.*, **242**, 2561–2569

Anraku, Y., Kin, E., and Tanaka, Y. (1975). Transport of sugars and amino acids in bacteria. XV. Comparative studies on the effects of various energy poisons on the oxidative and phosphorylating activities, and energy coupling reactions for the active transport systems for amino acids in *Escherichia coli. J. Biochem., Tokyo*, **78**, 165–179.

Anraku, Y., Naraki, T., and Kanzaki, S. (1973*b*). Transport of sugars and amino acids in bacteria. VI. Changes induced by valine in the branched chain amino acid transport systems of *Escherichia coli. J. Biochem., Tokyo*, **73**, 1149–1161.

Anraku, Y., Kobayashi, H., Amanuma, H., and Yamaguchi, A. (1973*a*). Transport of sugars and amino acids in bacteria. VII. Characterization of the reaction of restoration of active transport mediated by binding protein. *J. Biochem., Tokyo*, **74**, 1249–1261.
Ashghar, S. S., Levin, E., and Harold, F. M. (1973). Accumulation of neutral amino acids by *Streptococcus faecalis*: energy coupling by a proton motive force. *J. Biol. Chem.*, **248**, 5225–5233
Awazu, S., Amanuma, H., Morikawa, A., and Anraku, Y. (1976). Transport of sugars and amino acids in bacteria. XVI. Theory and evaluation of a model for the membrane transport reaction mediated by a single carrier with three binding sites for substrate. *J. Biochem., Tokyo*, **78**, 1047–1056
Ayling, P. D. and Bridgeiand, E. S. (1972). Methionine transport in wild-type and transport-defective mutants of *Salmonella typhimurium. J. Gen. Microbiol.*, **73**, 127–141
Baptist, E. W. and Kredich, N. M. (1977). Regulation of L-cystine transport in *Salmonella typhimurium. J. Bacteriol.*, **131**, 111–118
Barash, H. and Halpern, Y. S. (1971). Glutamate-binding protein and its relation to glutamate transport in *Escherichia coli* K-12. *Biochem. Biophys. Res. Commun.*, **45**, 681–688
Berger, E. A. (1973). Different mechanisms of energy coupling for the active transport of proline and glutamine in *Escherichia coli. Proc. Natl. Acad. Sci. U.S.A.*, **70**, 1514–1518
Berger, E. A. and Heppel, L. A. (1972). A binding protein involved in the transport of cystine and diaminopimelic acid in *Escherichia coli. J. Biol. Chem.*, **247**, 7684–7694
Berger, E. A. and Heppel, L. A. (1974). Different mechanisms of energy coupling for the shock-sensitive and shock-resistant amino acid permeases of *Escherichia coli. J. Biol. Chem.*, **249**, 7747–7755
Britten, R. J. and McClure, F. T. (1962*a*). The amino acid pool in *Escherichia coli. Bacteriol. Rev.*, **26**, 292–335
Britten, R. J. and McClure, F. T. (1962*b*). The mechanism of amino acid pool formation in *Escherichia coli.* In *Amino Acid Pools* (J. T. Holden ed.), Elsevier, Amsterdam and New York, pp. 595–609
Brown, K. D. (1970). Formation of aromatic amino acid pools in *Escherichia coli* K12. *J. Bacteriol.*, **104**, 177–188
Brown, K. D. (1971). Maintenance and exchange of the aromatic amino acid pool in *Escherichia coli. J. Bacteriol.*, **106**, 70–81
Celis, T. F. R. (1977*a*). Properties of an *Escherichia coli* K12 mutant defective in the transport of arginine and ornithine. *J. Bacteriol*, **130**, 1234–1243
Celis, T. F. R. (1977*b*). Independent regulation of transport and biosynthesis of arginine in *Escherichia coli* K12. *J. Bacteriol.*, **130**, 1244–1252
Celis, T. F. R., Rosenfeld, H. J., and Maas, W. K. (1973). A mutant of *Escherichia coli* K12 deficient in the transport of basic amino acids. *J. Bacteriol.*, **116**, 619–626
Cohen, G. N. and Monod, J. (1957). Bacterial permeases. *Bacteriol. Rev.*, **21**, 169–194
Cohen, G. N. and Rickenberg, H. V. (1956). Concentration specifique reversible des aminoacides chez *Escherichia coli. Ann. Inst. Pasteur, Paris*, **91**, 693–720
Cooper, S. (1966). Utilization of D-methionine by *Escherichia coli. J. Bacteriol.*, **92**, 328–332
Cosloy, S. D. (1973). D-Serine transport system in *Escherichia coli* K-12. *J. Bacteriol.*, **114**, 679–684
Curtiss, R., Charamella, L. J., III, Berg, C. M., and Harris, P. E. (1965). Kinetic and genetic analyses of D-cycloserine inhibition and resistance in *Escherichia coli. J. Bacteriol.*, **90**, 1238–1250
D'Ambrosio, S. M., Glover, G. I., Nelson, S. O., and Jensen, R. A. (1973). Specificity of the tyrosine–phenylalanine transport system in *Bacillus subtilis. J. Bacteriol.*, **115**, 673–681
Fan, C. L., Miller, D. L., and Rodwell, V. W. (1972). Metabolism of basic amino acids in *Pseudomonas putida. J. Biol. Chem.*, **247**, 2283–2288
Frank, L. and Hopkins, I. (1969). Sodium-stimulated transport of glutamate in *Escherichia coli. J. Bacteriol.*, **100**, 329–336
Furlong, C. E. and Weiner, J. H. (1970). Purification of a leucine-specific binding protein from *Escherichia coli. Biochem. Biophys. Res. Commun.*, **38**, 1076–1083
Furlong, C. E., Cirakoglu, C., Willis, R. C., and Santy, P. A. (1973). A simple preparative polyacrylamide disc gel electrophoresis apparatus: purification of three branched chain amino acid binding proteins from *Escherichia coli. Anal. Biochem.*, **5**, 297–311

Glover, G. I., D'Ambrosio, S. M., and Jensen, R. A. (1975). Versatile properties of a nonsaturable homogeneous transport system in *Bacillus subtilis*: genetic, kinetic and affinity labeling studies. *Proc. Natl. Acad. Sci. U.S.A.*, **72**, 814–818

Gryder, R. M. and Adams, E. (1970). Properties of the inducible hydroxyproline transport system of *Pseudomonas putida*. *J. Bacteriol.*, **101**, 948–958

Guest, J. R. (1974) Gene–protein relationships of the α-keto acid dehydrogenase complexes of *Escherichia coli* K12: chromosomal location of the lipoamide dehydrogenase gene. *J. Gen. Microbiol.*, **80**, 523–532

Guroff, G. and Bromwell, K. (1970). Phenylalanine uptake and phenylalanine-binding material in *Comamonas* sp. (ATCC 11299a). *Arch. Biochem. Biophys.*, **137**, 379–387

Guroff, G., Bromwell, K., and Abramowitz, A. (1969). Mode of action of *p*-chlorophenylalanine on *Pseudomonas* species (11299a). *Arch. Biochem. Biophys.*, **131**, 543–550

Haban, D. J. and Lyric, R. M. (1977). Glutamate uptake in *Thiobacillus novellus*. *Can. J. Microbiol.*, **23**, 271–277

Halpern, Y. S. (1974). Genetics of amino acid transport in bacteria. *Annu. Rev. Genetics*, **8**, 103–133

Halpern, Y. S. and Even-Shoshan, A. (1967). Properties of the glutamate transport system in *Escherichia coli*. *J. Bacteriol.*, **93**, 1009–1016

Halpern, Y. S. and Lupo, M. (1965). Glutamate transport in wild-type and mutant strains of *Escherichia coli*. *J. Bacteriol.*, **90**, 1288–1295

Halpern, Y. S. and Umbarger, H. E. (1961). Utilization of L-glutamic and 2-oxoglutaric acid as sole sources of carbon by *Escherichia coli*. *J. Gen. Microbiol.*, **26**, 175–183

Harold, F. M. (1972). Conservation and transformation of energy by bacterial membranes. *Bacteriol. Rev.*, **36**, 172–230

Harold, F. M. (1977). Membranes and energy transduction in bacteria. *Curr. Topics Bioenerg.*, **6**, 83–149

Heppel, L. A. (1971). The concept of periplasmic enzymes. In *Structure and Function of Biological Membranes* (L. I. Rothfield, ed.), Academic Press, New York, pp. 223–247

Hirata, H., Altendorf, K., and Harold, F. M. (1973). Role of an electrical potential in the coupling of metabolic energy to active transport by membrane vesicles of *Escherichia coli*. *Proc. Natl. Acad. Sci. U.S.A.*, **70**, 1804–1808

Hirata, H., Altendorf, K., and Harold, F. M. (1974). Energy coupling in membrane vesicles of *Escherichia coli*. I. Accumulation of metabolites in response to an electrical potential. *J. Biol. Chem.*, **249**, 2939–2945

Hirata, H., Sone, N., Yoshida, M., and Kagawa, Y. (1976*a*). Active transport of alanine by thermostable membrane vesicles isolated from a thermophilic bacterium. *J. Biochem., Tokyo*, **79**, 1157–1166

Hirata, H., Sone, N., Yoshida, M., and Kagawa, Y. (1976*b*). Solubilization and partial purification of alanine carrier from membranes of a thermophilic bacterium and its reconstitution into functional vesicles. *Biochem. Biophys. Res. Commun.*, **69**, 665–671

Hirata, H., Sone, N., Yoshida, M., and Kagawa, Y. (1977). Isolation of the alanine carrier from the membranes of a thermophilic bacterium and its reconstitution into vesicles capable of transport. *J. Supramol. Structure*, **6**, 77–84

Holden, J. T. (1962). Transport and accumulation of amino acids by microorganisms. In *Amino Acid Pools* (J. T. Holden, ed.), Elsevier, New York, pp. 566–594

Holden, J. T., van Balgooy, J. N. A., and Kittredge, J. S. (1968). Transport of aminophosphonic acids in *Lactobacillus plantarum* and *Streptococcus faecalis*. *J. Bacteriol.*, **96**, 950–957

Israeli, E., Kalra, V. K. and Brodie, A. F. (1977). Differential binding sites for entry and exit of amino acids in whole cells of *Mycobacterium phlei*. *J. Bacteriol.*, **130**, 729–735

Kaback, H. R. (1970). Transport. *Annu. Rev. Biochem.*, **39**, 561–598

Kaback, H. R. (1971). Bacterial membranes. In *Methods in Enzymology*, Vol. 22 (W. B. Jakoby, ed.). Academic Press, New York, pp. 99–120

Kaback, H. R., (1972). Transport across isolated bacterial cytoplasmic membranes. *Biochim. Biophys. Acta*, **265**, 367–416

Kaback, H. R. (1974). Transport studies in bacterial membrane vesicles. *Science, N.Y.*, **186**, 882–892

Kaback, H. R. and Stadtman, E. R. (1966). Proline uptake by an isolated cytoplasmic membrane preparation of *Escherichia coli*. *Proc. Natl. Acad. Sci. U.S.A.*, **55**, 920–927
Kadner, R. J. (1974). Transport systems for L-methionine in *Escherichia coli*. *J. Bacteriol.*, **117**, 232–241
Kadner, R. J. (1977). Transport and utilization of D-methionine and other methionine sources in *Escherichia coli*. *J. Bacteriol.*, **129**, 207–216
Kadner, R. J. and Watson, W. J. (1974). Methionine transport in *Escherichia coli*: physiological and genetic evidence for two uptake systems. *J. Bacteriol.*, **119**, 401–409
Kahane, S., Marcus, M., Metzer, E., and Halpern, Y. S. (1976*a*). Effect of growth conditions on glutamate transport in the wild-type strain and glutamate-utilizing mutants of *Escherichia coli*. *J. Bacteriol.*, **125**, 762–769
Kahane, S., Marcus, M., Metzer, E., and Halpern, Y. S. (1976*b*). Glutamate transport in membrane vesicles of the wild-type strain and glutamate-utilizing mutants of *Escherichia coli*. *J. Bacteriol.*, **125**, 770–775
Kanzaki, S. and Anraku, Y. (1971). Transport of sugars and amino acids in bacteria. IV. Regulation of valine transport activity by valine and cysteine. *J. Biochem., Tokyo*, **70**, 215–224
Kasahara, M. and Anraku, Y. (1974). Transport of sugars and amino acids in bacteria. XI. Mechanism of energy coupling reaction for the concentrative uptake of proline by *Escherichia coli* membrane vesicles. *J. Biochem., Tokyo*, **76**, 977–983
Kay, W. W. (1971). Two aspartate transport systems in *Escherichia coli*. *J. Biol. Chem.*, **246**, 7373–7382
Kay, W. W. and Gronlund, A. F. (1969*a*). Proline transport by *Pseudomonas aeruginosa*. *Biochim. Biophys. Acta*, **193**, 444–445
Kay, W. W. and Gronlund, A. F. (1969*b*). Isolation of amino acid transport-negative mutants of *Pseudomonas aeruginosa* and cells with repressed transport activity. *J. Bacteriol.*, **98**, 116–123
Kay, W. W. and Gronlund, A. F. (1971). Transport of aromatic amino acid by *Pseudomonas aeruginosa*. *J. Bacteriol.*, **105**, 1039–1046
Kay, W. W. and Kornberg, H. L. (1969). Genetic control of the uptake of C_4-dicarboxylic acids. *FEBS Lett.*, **3**, 93–96
Kay, W. W. and Kornberg, H. L. (1971). The uptake of C_4-dicarboxylic acids by *Escherichia coli*. *Eur. J. Biochem.*, **18**, 274–281
Kepes, A. and Cohen, G. N. (1962). Permeation. In *The Bacteria*, Vol. 4 (I. C. Gunsalus and R. Y. Stanier, eds), Academic Press, New York, pp. 179–221
Kessel, D. and Lubin, M. (1962). Transport of proline in *Escherichia coli*. *Biochim. Biophys. Acta*, **57**, 32–43
Kessel, D. and Lubin, M. (1965). Stability of α-hydrogen of amino acids during active transport. *Biochemistry*, **4**, 561–565
Kiritani, K. (1974). Mutants of *Salmonella typhimurium* defective in transport of branched chain amino acids. *J. Bacteriol.*, **120**, 1093–1101
Kiritani, K. and Ohnishi, K. (1977). Repression and inhibition of transport systems for branched chain amino acids in *Salmonella typhimurium*. *J. Bacteriol.*, **129**, 589–598
Kobayashi, H., Kin, E., and Anraku, Y. (1974). Transport of sugars and amino acids in bacteria. X. Sources of energy and energy coupling reaction of the active transport systems for isoleucine and proline in *Escherichia coli*. *J. Biochem., Tokyo*, **76**, 251–261
Konings, W. N. and Freese, E. (1972). Amino acid transport in membrane vesicles of *Bacillus subtilis*. *J. Biol. Chem.*, **247**, 2408–2418
Konings, W. N., Barnes, E. M., Jr, and Kaback, H. R. (1971). Mechanism of active transport in isolated membrane vesicles. III. The coupling of reduced phenazine methosulfate to the concentrative uptake of β-galactosides and amino acids. *J. Biol. Chem.*, **246**, 5857–5861
Konings, W. N., Bisschop, A., and Daarselaar, M. C. C. A. (1972). Transport of L-glutamate and L-aspartate by membrane vesicles of *Bacillus subtilis* W23. *FEBS Lett.*, **24**, 260–264
Konings, W. N., Boonstra, J., and De Vries, W. (1975). Amino acid transport in membrane vesicles of obligately anaerobic *Veillonells alcalences*. *J. Bacteriol.*, **122**, 245–249
Krulwich, T. A., Blanco, R., and McBride, P. A. (1977). Amino acid transport in whole cells and membrane vesicles of *Arthrobacter pyridinolis*. *Arch. Biochem. Biophys.*, **178**, 108–117
Kuhn, J. and Somerville, R. L. (1971). Mutant strains of *Escherichia coli* K12 that use D-amino acids. *Proc. Natl. Acad. Sci. U.S.A.*, **68**, 2482–2484

Kustu, S. G. and Ames, G. F.-L. (1973). The *hisP* protein, a known histidine transport component in *Salmonella typhimurium*, is also an arginine transport component. *J. Bacteriol.*, **116**, 107–113

Kustu, S. G. and Ames, G. F.-L. (1974). The histidine-binding protein J, a histidine transport component, has two different functional sites. *J. Biol. Chem.*, **249**, 6976–6983

Lanyi, J. K. (1977). Transport in *Halobacterium halobium*: light-induced cation-gradients, amino acid transport kinetics, and properties of transport carriers. *J. Supramol. Structure*, **6**, 169–178

Lanyi, J. K., Renthal, R., and MacDonald, R. E. (1976*b*). Light-induced glutamate transport in *Halobacterium halobium* envelope vesicles. II. Evidence that the driving force is a light-dependent sodium gradient. *Biochemistry*, **15**, 1603–1610

Lanyi, J. K., Yearwood-Drayton, V., and MacDonald, R. E. (1976*a*). Light-induced glutamate transport in *Halobacterium halobium* envelope vesicles. I. Kinetics of the light-dependent and the sodium gradient-dependent uptake. *Biochemistry*, **15**, 1595–1603

Leach F. R. and Snell, E. E. (1960). The absorption of glycine and alanine and their peptides by *Lactobacillus casei*. *J. Biol. Chem.*, **235**, 3523–3531

Leive, L. and Davis, B. D. (1965*a*). The transport of diaminopimelate and cystine in *Escherichia coli*. *J. Biol. Chem.*, **240**, 4362–4369

Leive, L. and Davis, B. D. (1965*b*). Evidence for a gradient of exogenous and endogenous diaminopimelate in *Escherichia coli*. *J. Biol. Chem.*, **240**, 4370–4376

Lin, E. C. C. (1971). The molecular basis of membrane transport systems. In *Structure and Function of Biological Membranes* (L. I. Rothfield, ed.), Academic Press, New York, pp. 285–341

Lombardi, F. J. and Kaback, H. R. (1972). Mechanisms of active transport in isolated bacterial membrane vesicles. VIII. The transport of amino acids by membranes prepared from *Escherichia coli*. *J. Biol. Chem.*, **247**, 7844–7857

Lubin, M., Kessel, D., Budreau, A., and Gross, J. D. (1960). The isolation of bacterial mutants defective in amino acid transport. *Biochim. Biophys. Acta*, **42**, 535–538

Maas, W. K. (1972). Mapping of the genes involved in the synthesis of spermidine in *Escherichia coli*. *Mol. Gen. Genet.*, **119**, 1–9

Maas, W. K. and Davis, B. D. (1950). Interference by D-serine and L-aspartic acid with pantothenate synthesis in *Escherichia coli*. *J. Bacteriol.*, **60**, 733–745

MacDonald, R. E. and Lanyi, J. K. (1975). Light-induced leucine transport in *Halobacterium halobium* envelope vesicles: a chemiosmotic system. *Biochemistry*, **14**, 2882–2889

MacDonald, R. E., Greene, R. V., and Lanyi, J. K. (1977). Light-activated amino acid transport systems in *Halobacterium halobium* envelope vesicles: role of chemical and electrical gradients. *Biochemistry*, **16**, 3227–3235

Mandelstam, J. (1958). The free amino acids in growing and non-growing populations of *Escherichia coli*. *Biochem. J.*, **69**, 103–110

Mandelstam, J. (1960). The intracellular turnover of protein and nucleic acids and its role in biochemical differentiation. *Bacteriol Rev.*, **24**, 289–308

Marcus, M. and Halpern, Y. S. (1967). Genetic analysis of glutamate transport and glutamate decarboxylase in *Escherichia coil*. *J. Bacteroil.*, **93**, 1409–1415.

Marcus, M. and Halpern, Y. S. (1969). Genetic analysis of the glutamate permease in *Escherichia coli*. K12. *J. Bacteriol.*, **97**, 1118–1128

Marquis, R. E. and Gerhardt, P. (1964). Respiration-coupled and passive uptake of α-aminoisobutyric acid, a metabolically inert transport analogue, by *Bacillus megaterium*. *J. Biol. Chem.*, **239**, 3361–3371

Miller, D. L. and Rodwell, V. W. (1971). Metabolism of basic amino acids in *Pseudomonas putida*; properties of the inducible lysine transport system. *J. Biol. Chem.*, **246**, 1765–1771

Miner, K. M. and Frank, L. (1974). Sodium-stimulated glutamate transport in osmotically shocked cells and membrane vesicles of *Escherichia coli*. *J. Bacteriol.*, **117**, 1093–1098

Mora, J. and Snell, E. E. (1963). The uptake of amino acids by cells and protoplasts of *S. faecalis*. *Biochemistry*, **2**, 136–141

Morikawa, A., Suzuki, H., and Anraku, Y. (1974). Transport of sugars and amino acids in bacteria. VIII. Properties and regulation of the active transport reaction of proline in *Escherichia coli*. *J. Biochem., Tokyo*, **75**, 229–241

Motojima, K., Yamato, I., and Anraku, Y. (1978). Proline transport carrier defective mutants of *Escherichia coli*. K12: properties and mapping. *J. Bacteriol.*, **136**, 5–9

Neu, H. C. and Heppel, L. A. (1965). The release of enzymes from *Escherichia coli* by osmotic shock and during the formation of spheroplasts. *J. Biol. Chem.*, **240**, 3685–3692

Oxender, D. L. (1972*a*). Membrane transport. *Annu. Rev. Biochem.*, **41**, 777–814

Oxender, D. L. (1972*b*) Amino acid transport in microorganisms. In *Metabolic Pathways*, Vol. 6 (L. E. Hokin, ed.) Academic Press, New York, pp. 133–185

Oxender, D. L. and Quay, S. C. (1976). Isolation and characterization of membrane binding proteins. In *Methods in Membrane Biology*, Vol. 6 (E. D. Korn, ed.). Plenum Press, New York, pp. 183–242

Pearce, S. M., Hindebrandt, V. A., and Lee, T. (1977). Third system for neutral amino acid transport in marine pseudomonad. *J. Bacteriol.*, **130**, 37–47

Penrose, W. R., Nichoalds, G. E., Piperno, J. R., and Oxender, D. L. (1968). Purification and properties of a leucine-binding protein from *Escherichia coli. J. Biol. Chem.*, **243**, 5921–5928

Piperno, J. R. and Oxender, D. L. (1968). Amino acid transport systems in *Escherichia coli.* K12. *J. Biol. Chem.*, **243**, 5914–5920

Rahmanian, M. and Oxender, D. L. (1972). Derepressed leucine transport activity in *Escherichia coli. J. Supramol. Structure*, **1**, 55–59

Rahmanian, M., Claus, D. R., and Oxender, D. L. (1973). Multiplicity of leucine transport systems in *Escherichia coli* K12. *J. Bacteriol.*, **116**, 1258–1266

Reid, K. G., Utech, N. M., and Holden, J. T. (1970). Multiple transport components for dicarboxylic amino acids in *Streptococcus faecalis. J. Biol. Chem.*, **245**, 5261–5272

Robbins, J. C. and Oxender, D. L. (1973). Transport systems for alanine, serine, and glycine in *Escherichia coli.* K12. *J. Bacteriol.*, **116**, 12–18

Rosen, B. P. (1971). Basic amino acid transport in *Escherichia coli. J. Biol. Chem.*, **246**, 3653–3662

Rosen, B. P. (1973*a*). Basic amino acid transport in *Escherichia coli.* II. Purification and properties of an arginine specific binding protein. *J. Biol. Chem.*, **248**, 1211–1218

Rosen, B. P. (1973*b*). Basic amino acid transport in *Escherichia coli*: properties of canavanine-resistant mutants. *J. Bacteriol.*, **116**, 627–635

Rosen, B. P. and Heppel, L. A. (1973). Present status of binding proteins that are released from Gram-negative bacteria by osmotic shock. In *Bacterial Membranes and Walls* (L. Leive, ed.), Marcel Dekker, New York, pp. 209–239

Rosen, B. P. and Kashket, E. R. (1978). Energetics of active transport. In *Bacterial Transport* (B. P. Rosen, ed), Marcel Dekker, New York, pp. 559–620

Rosen, B. P. and Vasington, R. D. (1971). Purification and characterization of a histidine-binding protein from *Salmonella typhimurium* LT-2 and its relationship to the histidine permease system. *J. Biol. Chem.*, **246**, 5351–5360

Rosenfeld, H. and Feigelson, P. (1969). Product induction in *Pseudomonas acidovorans* of a permease system which transports L-tryptophan. *J. Bacteriol.*, **97**, 705–714

Rowland, I. and Tristram, H. (1974). Specificity of the *Escherichia coli* proline transport system. *J. Bacteriol.*, **123**, 871–877

Russel, R. P. B. (1972). Mapping of D-cycloserine resistance locus in *Escherichia coli* K12. *J. Bacteriol.*, **111**, 622–624

Schwartz, J. H., Maas, W. K., and Simon, E. J. (1959). An impaired concentrating mechanism for amino acids in mutants of *Escherichia coli* resistant to L-canavanine and D-serine. *Biochim. Biophys. Acta*, **32**, 582–583

Short, S. A., White, D. C., and Kaback, H. R. (1972*a*). Active transport in isolated bacterial membrane vesicles. V. The transport of amino acids by membrane vesicles prepared from *Staphylococcus aureus. J. Biol. Chem.*, **247**, 298–304

Short, S. A., White, D. C., and Kaback, H. R. (1972*b*). Mechanisms of active transport in isolated bacterial membrane vesicles. IX. The kinetics and specificity of amino acid transport in *Staphylococcus aureus* membrane vesicles. *J. Biol. Chem.*, **247**, 7452–7458

Simoni, R. D. and Postma, P. W. (1975). The energetics of bacterial active transport. *Annu. Rev. Biochem.*, **44**, 523–554

Smulson, M. E., Rabinowitz, M., and Breitman, T. R. (1967). *O*-Methylthreonine inhibition of growth and of threonine deaminase in *Escherichia coli. J. Bacteriol.*, **94**, 1890–1895

Sprott, G. and Macleod, R. A. (1972). Na^+-dependent amino acid transport in isolated membrane vesicles of a marine pseudomonad energized by electron donors. *Biochem. Biophys. Res. Commun.*, **47**, 838–845

Stevenson, J. (1966). The specific requirement for sodium chloride for the active uptake of L-glutamate by *Halobacterium salinarium. Biochem. J.*, **99**, 257–265

Tempest, D. W. and Meers, J. L. (1970). Influence of environment on the content and composition of microbial free amino acid pools. *J. Gen. Microbiol.*, **64**, 171–185
Templeton, B. A. and Savageau, M. A. (1974). Transport of biosynthetic intermediates: homoserine and threonine uptake in *Escherichia coli J. Bacteriol.*, **117**, 1002–1009
Thorne, G. M. and Corwin, L. M. (1972). Genetic locus of a gene affecting leucine transport in *Salmonella typhimurium. J. Bacteriol.*, **110**, 784–785
Tristram, H. and Neale, S. (1968). The activity and specificity of the proline permease in wild-type and analogue-resistant strains of *Escherichia coli. J. Gen. Microbiol.*, **50**, 121–137
Tsuchiya, T., Hasan, S. M., and Raven, J. (1977). Glutamate transport driven by an electrochemical gradient of sodium ions in *Escherichia coli. J. Bacteriol.*, **131**, 848–853
Umbarger, H. E. (1969). Regulation of amino acid metabolism. *Annu. Rev. Biochem.*, **38**, 323–370
Utech, N. M., Reid, K. G., and Holden, J. T. (1970). Properties of a dicarboxylic amino acid transport-deficient mutant of *Streptococcus faecalis. J. Biol. Chem.*, **245**, 5273–5280
Wargel, R. J., Shadur, C. A., and Neuhaus, F. C. (1970). Mechanism of D-cycloserine action: transport systems for D-alanine, D-cycloserine, alanine, and glycine. *J. Bacteriol.*, **103**, 778–788
Wargel, R. J., Shadur, C. A., and Neuhaus, F. C. (1971). Mechanism of D-cycloserine action: transport mutants for D-alanine, D-cycloserine, and glycine. *J. Bacteriol.*, **105**, 1028–1035
Weiner, J. H. and Heppel, L. A. (1971). A binding protein for glutamine and its relation to active transport in *Escherichia coli. J. Biol. Chem.*, **246**, 6933–6941
Weiner, J. H., Furlong, C. E., and Heppel, L. A. (1971). A binding protein for L-glutamine and its relation to active transport in *Escherichia coli. Arch. Biochem. Biophys.*, **142**, 715–717
Wilbrandt, W. and Rosenberg, T. (1961). The concept of carrier transport and its corollaries in pharmacology. *Pharmacol. Rev.*, **13**, 109–183
Willis, R. C. and Furlong, C. E. (1975*a*). Purification and properties of a periplasmic glutamate–aspartate binding protein from *Escherichia coli* K-12 strain W3092. *J. Biol. Chem.*, **250**, 2574–2580
Willis, R. C. and Furlong, C. E. (1975*b*) Interactions of a glutamate–aspartate binding protein with the glutamate transport system of *Escherichia coli. J. Biol. Chem.*, **250**, 2581–2586
Willis, R. C. and Woolfolk, C. A. (1975). L-Asparagine uptake in *Escherichia coli. J. Bacteriol.*, **123**, 937–945
Willis, R. C., Iwata, K. K., and Furlong, C. E. (1975). Regulation of glutamine transport in *Escherichia coli. J. Bacteriol.*, **122**, 1032–1037
Wilson, O. H. and Holden, J. T. (1969). Stimulation of arginine transport in osmotically shocked *Escherichia coli* W cells by purified arginine-binding protein fractions. *J. Biol. Chem.*, **244**, 2743–2749
Winkler, H. H. and Wilson, T. H. (1966). The role of energy coupling in the transport of β-galactosides by *Escherichia coli. J. Biol. Chem.*, **241**, 2200–2211
Wood, J. M. (1975). Leucine transport in *Escherichia coli*: the resolution of multiple transport systems and their coupling to metabolic energy. *J. Biol. Chem.*, **250**, 4477–4485
Yabu, K. (1967). The uptake of D-glutamic acid by *Mycobacterium avium. Biochim. Biophys. Acta*, **135**, 181–183
Yabu, K. (1971). Aspartic acid transport in *Mycobacterium smegmatis. Jap. J. Microbiol.*, **15**, 449–456
Yamato, I. and Anraku, Y. (1977). Transport of sugars and amino acids in bacteria. XVIII. Properties of an isoleucine transport carrier in the cytoplasmic membrane vesicles of *Escherichia coli. J. Biochem., Tokyo*, **81**, 1517–1523
Yamato, I., Amanuma, H., and Anraku, Y. (1975*b*). On the molecular mechanism of active transport of branched chain amino acids in *Escherichia coli. Abst. Japan Bioenergetics Group*, **1**, 19–21 (in Japanese)
Yamato, I., Anraku, Y., and Hirosawa, K. (1975*a*). Cytoplasmic membrane vesicles of *Escherichia coli.* I. A simple method for preparing the cytoplasmic and outer membranes. *J. Biochem., Tokyo*, **77**, 705–718
Yamato, I., Anraku, Y., and Ohki, M. (1976). Genetic analysis of the active transport systems for branched chain amino acids in *Escherichia coli. Seikagaku*, **48**, 768 (abstract in Japanese)
Yamato, I., Futai, M., Anraku, Y., and Nonomura, Y. (1978). Cytoplasmic membrane vesicles of *Escherichia coli*. II. Orientation of the vesicles studied by localization of enzymes. *J. Biochem., Tokyo*, **83**, 117–128

Microorganisms and Nitrogen Sources
Edited by J. W. Payne

Section 1.2

Some Aspects of Amino Acid Transport in Yeast

A. A. EDDY
Department of Biochemistry, University of Manchester Institute of Science and Technology, Manchester M60 1QD, UK

1 INTRODUCTION

The demonstration that common yeasts contain a large pool of endogenous amino acids and concentrate glutamate (Taylor, 1949), arginine and lysine (Halvorson *et al.*, 1955), and threonine (Cowie and McClure, 1959) focussed attention on two general problems: first, that of the mechanism of amino acid transport in these organisms; second, the nature of the processes integrating the biosynthesis of the amino acids with their utilization and transport. Taylor (1949) found that not only was the assimilation of glutamate slower when an ammonium salt was present in the growth medium, but the total amount of glutamate absorbed was also smaller.

Aspartate and arginine had a greater sparing effect than NH_4^+ itself. Moreover, the yeast cells contained a relatively large pool of amino acids, including glutamate, when NH_4^+ was the sole source of nitrogen for growth. Thus, the composition and size of the amino acid pool (0.2–2.0 μmol (mg yeast dry weight)$^{-1}$) is a complex function not only of the processes involved in the biosynthesis and degradation of the constituent amino acids, but also of the mechanisms that govern the influx and efflux of these amino acids.

2 AMINO ACIDS AS AN ALTERNATIVE NITROGEN SOURCE TO AMMONIUM IONS

Early work on the production of fusel-oil during yeast fermentation showed that yeast used certain hydrophobic amino acids as a source of nitrogen for growth, meanwhile excreting the degraded carbon skeleton as fusel alcohol. For instance, leucine gave rise to iso-amyl alcohol whereas tyrosine gave rise to tyrosol (Ehrlich, 1912; Thorne, 1937). Ehrlich had been unable to detect the formation of NH_4^+ as an intermediate product of these reactions and it later became clear that the absorbed amino acid underwent transamination with 2-oxoglutarate (Senthe Shanmuganathan and Elsden, 1958). The glutamate so formed would serve as the nitrogen donor in the synthesis of various, other amino acids (Sims and Folkes, 1964; Jones *et al*., 1969), whereas the α-ketoacid derived from the original amino acid would be decarboxylated and reduced to the fusel alcohol. The rapid catabolism of, for instance, valine or tyrosine in the presence of glucose complicates the study of their transport across the yeast cell membrane (Woodward and Cirillo, 1977).

Selected strains of *Saccharomyces cerevisiae* use a wide range of amino acids as solc nitrogen source (see Watson, 1976). However, most nitrogen nutrients other than glutamine and asparagine (Dubois *et al*., 1974) and, in other yeast strains, aspartate (Thomulka and Moat, 1972) support a slower rate of growth than occurs with NH_4^+. Thus, (a) in addition to the examples referred to above, arginine is readily absorbed and degraded by intracellular arginase to produce ornithine and urea, both of which can themselves serve as sole nitrogen source for the growth of appropriate strains of yeast. Ornithine is subsequently degraded to glutamate (Whitney and Magasanik, 1973; Bossinger and Cooper, 1977). The further catabolism of urea involves its carboxylation to allophanate which in turn yields CO_2 and NH_3 under the action of allophanate hydrolase (Waheed and Castric, 1977). (b) Urea also serves as a sole source of nitrogen for growth, being absorbed by two distinct mechanisms, one of which requires metabolic energy to function (Sumadra *et al*., 1976). (c) Allantoin acts as a sole source of nitrogen for yeast growth,, being degraded to NH_3, CO_2, and glyoxylate (Sumadra and Cooper, 1977). (d) When glutamate or aspartate are the sole source of nitrogen for growth, they are first absorbed as such. Aspartate assimilation probably involves transamination, mediated by glutamate–oxaloacetate transaminase, rather than the action of aspartase. (e) Assimilation of NH_4^+ (see also Section 4.2.4) involves initially a specific permease for which methylamine is also a substrate (Roon *et al.*, 1975*a*, 1977). Studies with $^{15}NH_4^+$ indicate that in *Candida utilis* the absorbed NH_4^+ is initially converted to two compounds, glutamate and glutamine (Sims and Folkes, 1964). The former arises through the action of the anabolic NADP-linked glutamate dehydrogenase and the latter through the action of glutamine synthetase. The NAD-linked glutamate dehydrogenase serves a catabolic function. The regulation of these important enzymes has been extensively studied (Drillien *et al.*, 1973; Sims and Ferguson, 1974). There is some evidence that certain *Schizosaccharomyces* spp. may assimilate NH_4^+ by another mechanism based on the combined actions of glutamine synthetase and glutamate synthase, as well as by the classical pathway dependent on the anabolic glutamate dehydrogenase (Johnson and Brown, 1974). However, mutant strains of *Saccharomyces cerevisiae* lacking a functional NADP-linked glutamate dehydrogenase grow slowly with NH_4^+ as sole nitrogen source, possibly by utilizing the NAD-linked glutamate dehydrogenase to assimilate it (Grenson *et al.*, 1974). (f) The assimilation of nitrate, a characteristic property of certain yeasts such as *Candida utilis*, involves an inducible nitrate reductase (Choudary and Rao, 1976). Little is known about the transport of nitrate in yeast. (g) Two modes of utilization of asparagine have been demonstrated, dependent respectively on the presence of (i) an intracellular asparaginase that hydrolyses intracellular L-asparagine to yield aspartate and NH_4^+ and (ii) an asparaginase that hydrolyses both L- and D-asparagine when one of these is present in the extracellular medium, releasing ammonia which the yeast assimilates (Dunlop and Roon, 1975; Dunlop *et al.*, 1976). The latter enzyme is formed under conditions of nitrogen deficiency. It appears to be located at or in the cell surface and may be a mannan–protein complex analogous to extracellular invertase. Whereas NH_4^+ is thus made available from D-asparagine for yeast growth, by the action of the superficial asparaginase, other D-amino acids are in general found to inhibit the growth of *Saccharomyces cerevisiae*. (h) Proline is absorbed as such and degraded by proline oxidase (Duteurtre *et al.*, 1971). (i) Selected strains of *Saccharomyces cerevisiae* utilize cytosine and oxypyrimidines, but not purines, as a sole source of nitrogen for growth (Grenson, 1969; Milstein and Bekker, 1976; Di Carlo *et al.*, 1951).

It is now clear that the amino acids supplied to a yeast culture are not in general incorporated into yeast protein intact. Rather they participate in a complex series of transamination reactions and other chemical processes. The net outcome is that simple sugars such as glucose contribute significantly to the carbon skeletons of the amino acid residues eventually incorporated into the yeast protein. For instance, Jones *et al.* (1969) found that during the first 12 hours of fermentation of a complex nutrient solution containing [^{14}C]-L-glutamate and several other amino acids, only about 43 per cent of the glutamate residues in the cellular protein were derived directly from that source. Parallel studies with [^{15}N]-L-glutamate showed that only about 17 per cent of the glutamate nitrogen in the protein was derived from glutamate in the growth medium. Clearly, the nitrogen atoms and the carbon atoms of the glutamate provided were metabolized by different pathways. Analogous observations, made with several other amino acids, labelled with either ^{15}N or ^{14}C, confirmed the central role of glutamate in the cellular transamination reactions. There is some evidence from studies of this kind, on the flow of isotopic carbon or nitrogen from one compound to another, that more than one kinetically distinct pool of amino acids exists in the yeast cell (Sims and Folkes, 1964; Cowie and McClure, 1959).

3 GENETIC ANALYSES OF THE PATHWAYS OF AMINO ACID ABSORPTION

Very extensive work by Grenson and her collaborators has clearly established that the wild-type yeast, designated strain Σ1278b of *Saccharomyces cerevisiae*, carries a so-called general permease, catalysing the uptake of a wide range of natural amino acids, as well as a series of more selective permeases (Table 1). Earlier studies had tended to emphasize the relatively broad specificity and unselective character of the amino acid permeases of yeast (Surdin *et al.*, 1965; Halvorson and Cohen, 1958). The preparations studied by Surdin *et al.* (1965) were starved of nitrogen, a circumstance that may have led to derepression of a general permease. The preparations studied by Halvorson and Cohen (1958) were grown with NH_4^+ and no special steps appear to have been taken to starve the yeast cells and thereby amplify the subsequent uptake of amino acids. Whether the yeast strains used in the above work carry the same range of permeases as the Belgian series is not known.

Discrimination between the various types of permease only became feasible when it was realized that the activity of the general permease was virtually abolished when the yeast strain Σ1278b was growing with NH_4^+ as sole source of nitrogen (Grenson and Hou, 1972). Replacement of NH_4^+ by either proline or glutamate derepressed the general permease, thereby increasing the rate of uptake of a wide range of amino acids by 10-fold or more (Grenson *et al.*, 1970). In general, amino acids absorbed both by the general permease and by one or more specific permeases would enter the cells mainly by the former system when both were functioning together.

The discovery of the arginine permease and the lysine permease illustrates the principles involved in the characterization of these systems. (a) During

Table 1 Genes associated with various permeases of *Saccharomyces cerevisiae*

	Gene	Representative substrates	References
General amino acid permease	*gap*	Various amino acids, not proline	Grenson *et al*. (1970), Rytka (1975)
Lysine permease	*lys-p*$_1$	L-Lysine	Grenson (1966)
Arginine permease[a]	*arg-p*$_1$	L-Arginine, L-Lysine	Grenson *et al*. (1966); see also Chan and Cossins (1976)
Histidine permease[a]	*his-p*$_1$	L-Histidine	Crabeel and Grenson (1970)
Methionine permease[a]	*met-p*$_1$	L-Methionine	Gits and Grenson (1967)
Dicarboxylate permease[a]	*dic-p*$_1$	L-Glutamate	Darte and Grenson (1975), Joiris and Grenson (1969)
Proline permease		L-Proline	Seaston *et al*. (1973), see also Magaña-Schwencke *et al*. (1973)
	apf	A gene affecting the activity of several of the above systems	Grenson and Hennaut (1971), Surdin *et al*. (1965)
Ammonia permease	*amt-1*	Ammonia, methylamine	Roon *et al*. (1975a), Dunlop *et al*. (1976)
Urea permease[a]	*dur-3*	Urea	Sumadra *et al*. (1976)
Allantoin permease		Allantoin	Sumadra and Cooper (1977)
S-Adenosylmethionine permease	*samP*	Adenosylmethionine	Petrotta-Simpson *et al*. (1975)
Cytosine–purine permease	*cyt-p*	Cytosine, adenine, hypoxanthine	Grenson (1969), Polak and Grenson (1973), Chevallier *et al*. (1975)
Uracil permease	*ura-p*	Uracil	Grenson (1969), Jund *et al*. (1977)
Uridine permease	*urid-p*	Uridine	Grenson (1969)

[a]In these instances more than one specific system is present, but the other systems have not been characterized genetically.

growth with NH_4^+, lysine uptake exhibited two kinetic components with K_m values of about 0.2 mM and 0.02 mM (Grenson, 1966). (b) Arginine competitively inhibited lysine uptake by the former system, the K_i value being 0.01 mM. (c) The uptake of arginine itself exhibited a K_m of 0.01 mM. (d) Mutant yeast strains carrying the *arg-p₁* gene selected on the basis of their resistance to the arginine antagonist L-canavanine, failed both to absorb arginine by the above route and lacked the lysine system that interacted with arginine. (e) Yeast strains carrying the mutant *lys-p₁* gene, selected on the basis of their resistance to L-thiosine, an antagonist of lysine, failed to absorb lysine by the route showing a K_m of 0.02 mM and scarcely absorbed lysine at all when the *arg-p₁* gene was also present. (f) Such double mutants absorbed lysine rapidly, however, when proline was the nitrogen source and the general amino acid permease was derepressed.

Somewhat similar arguments were used to demonstrate the existence of a specific methionine permease exhibiting a K_m of about 0.01 mM. A mutant yeast strain bearing the *met-p₁* gene, selected in the presence of the toxic analogue ethionine, lacked the specific permease. A second relatively specific methionine permease (K_m about 0.8 mM) was not genetically characterized (Gits and Grenson, 1967). Histidine uptake in the presence of NH_4^+ likewise exhibited two kinetic components, one of which was suppressed in the presence of the *his-p₁* gene (Crabeel and Grenson, 1970). The permease in question is highly specific for histidine. Histidine uptake through the second system was competitively inhibited by tyrosine.

The genetic basis of the general amino acid permease was studied by selecting a canavanine resistant mutant from an *arg-p₁* mutant growing in the presence of proline, circumstances in which the general permease would otherwise be expressed (Grenson *et al.*, 1970). The relevant gene *gap* also conferred resistance to ethionine and thiosine in the presence, respectively, of *met-p₁* and *lys-p₁*. The substrates of the general amino acid permease comprise a wide range of D- and L-amino acids but not proline. Thus, a different method of selecting *gap* mutants is to look for yeast strains that are resistant to the presence of D-amino acids, such as D-histidine, in a growth medium containing proline as nitrogen source (Rytka, 1975). L-Citrulline uptake appears to occur mainly through the general permease. Thus, strains carrying the *gap* mutation grow very slowly with citrulline as nitrogen source.

It is interesting that the apparent affinities of arginine, lysine, and methionine for their specific permeases is similar to their affinities for the general permease. Hence derepression of the general permease gives rise to a situation in which the uptake of each of these amino acids shows a small change in K_m, a large increase in V_{max}, and an apparent loss of specificity in that a wider range of the common amino acids become competitive inhibitors of the system (Grenson *et al.*, 1970).

The term permease has been used above in the general sense of a carrier system catalysing the absorption of a certain spectrum of substrates and whose activity can be modulated both by genetic mutation and by physiological circumstances. The molecular basis of the several permeases listed in Table 1 has not been established. However, the work outlined below on the role of proton gradients and K^+ ions in the functioning of the permeases, indicates that each of them is probably not a single physical entity but comprises a carrier component that is functionally coupled to the ionic pumps of the yeast plasmalemma.

4 THE ROLE OF PROTONS, POTASSIUM IONS, AND SODIUM IONS IN AMINO ACID TRANSPORT IN YEAST: THE ION PUMPS OF YEAST

The cells of typical strains of *Saccharomyces cerevisiae* have long been known to contain a large concentration of intracellular solutes representing an osmotic pressure of the order of 0.5 M and including 0.15–0.25 M potassium ions (Rothstein, 1960; Conway and Armstrong, 1961). At pH 4.5 the large cellular content of K^+ is maintained over long periods, in the presence of a fermentable substrate such as glucose, provided the extracellular concentration of K^+ is not smaller than about 50–100 μM. The magnitude of the gradient of K^+ activity across the plasmalemma thus corresponds to a factor of the order of $10^3 - 3 \times 10^3$, if it is correct to treat the cellular contents as a single compartment in which concentrations and activities are identical. Hence a membrane potential of the order of 200 mV would be needed to maintain a steady state, if the movements of K^+ across the plasmalemma occur simply in response to the difference in chemical potential of the K^+ ions on the two sides of the cell membrane.

During energy metabolism a marked pH gradient is also maintained across the plasmalemma, the magnitude of which has been studied in various ways. For instance, the distribution of the weak acid propionate between the cell water and the extracellular medium (Ryan and Ryan, 1972; Seaston *et al*., 1976), and other methods based on the apparent pH of disrupted cells (Conway and O'Malley, 1946), or on high-resolution ^{31}P nuclear magnetic resonance studies of intact yeast (Salhany *et al*., 1975) show that the intracellular acidity is near pH 6 when the extracellular pH is near 4.5. It follows that if the distribution of K^+ were a reliable indicator of the magnitude of the membrane potential across the plasmalemma, the gradient of chemical potential of protons across the cell membrane would correspond to a factor of the order of 3×10^4–10^5. It is the very large driving force corresponding to this factor that is now believed to energize the yeast amino acid permeases (Eddy, 1978). In other words amino acid accumulation appears to be driven by the proton gradient.

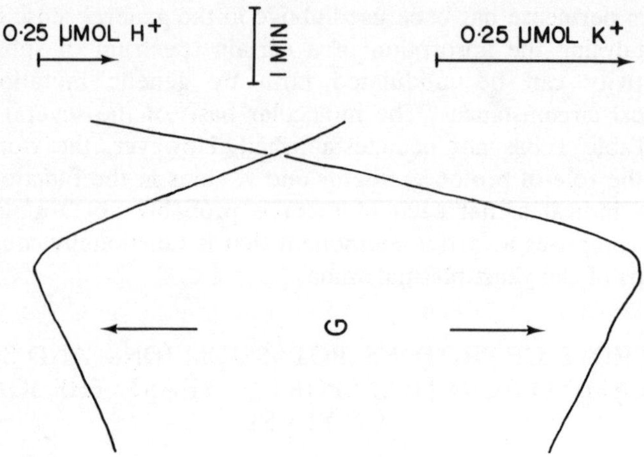

Figure 1 The ejection of protons and the uptake of potassium ions following the addition of glucose to a preparation of washed yeast cells at pH 4.8.

The yeast strain N.C.Y.C. 240 was grown with glucose as the principal carbon source in the Solution 5 described by Seaston *et al.* (1973). The figure shows, as a function of time, the recordings obtained from glass electrodes selective towards protons and K^+ respectively. Fermentation was initiated at G by the addition of 200 μmol of glucose to the yeast (50 mg dry weight) suspended in buffer solution (5 ml), containing 0.5 mM KC1 at 30 °C (Seaston *et al.*, 1973). Ionic pumping started about 0.7 min later

The proton gradient arises in the following way. When glucose is added to yeast that is partially depleted of K^+ ions, the latter are absorbed from solution in exchange for protons (Figure 1). It has long been known that in the presence of K^+, at low pH values, the amount of acid excreted is equivalent to the amount of K^+ absorbed, whereas in the absence of K^+ the flow of protons out of the yeast is neutralized by succinate and bicarbonate ions (Conway and Brady, 1950). Proton ejection under these conditions depends on energy metabolism and can be maintained either by respiration (Misra and Höfer, 1975) or by glycolysis. It is inhibited by proton conductors such as azide or dinitrophenol (see for example Peña, 1975), although glycolysis itself is not so inhibited. A plausible interpretation is that proton ejection at the plasmalemma is an electrogenic process. This might be actuated by an electrogenic ATPase system like that which is found in *Neurospora* (Bowman and Slayman, 1977). Studies on the absorption of lipophilic anions and cations by yeast have also been taken to imply that proton pumping is electrogenic (Riemersma and Alsbach, 1974; Hoeberichts and Borst-Pauwels, 1975; Eddy *et al.*, 1977).

In the scheme shown in Figure 2, yeast cells depleted of ATP would be expected to lose K^+ ions, at a rate that might be limited by the availability of counterions to neutralize the efflux of positive charge. Supporting evidence has come from studies in which the acceleration of the rate of uptake of

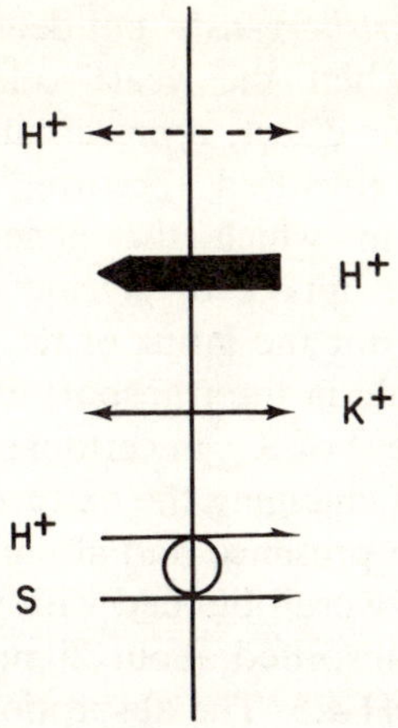

Figure 2 The general relations involved in the flow of protons and of K^+ across the yeast plasmalemma.

The broad arrow represents proton ejection through the proton pump. The K^+ ions diffuse across the membrane through a selective channel, in response to the prevailing electric field and the activities of K^+ on the two sides of the membrane. The proton pump is believed to be electrogenic and may be driven by ATP. The net absorption of K^+ illustrated in Figure 1 would thus reflect the hyperpolarization of the cell membrane caused by proton pumping. The diagram also shows the absorption of protons in a symport relationship with the substrate S. The interrupted line represents the conduction of protons catalysed by dinitrophenol or another proton conductor. All these flows are coupled by the requirement to maintain electroneutrality. For instance, the proton symport might be neutralized either by proton efflux through the proton pump or by the efflux of K^+

protons, caused by the addition of various proton conductors, was paralleled by a similar acceleration in the efflux of K^+ ions from the energy-depleted yeast (Eddy, 1978). It is significant that the maximum rate of leakage of K^+ under these conditions was similar to the maximum rate of absorption of K^+ ions during energy metabolism. All these arguments are thus consistent in a general way with the model shown in Figure 2.

5 IONIC DISPLACEMENTS DURING AMINO ACID ABSORPTION: STUDIES WITH ENERGY-DEPLETED YEAST

The absorption of lysine and arginine, on the one hand, or glutamate and aspartate on the other hand, might be expected to lead to the efflux or influx, respectively, of K^+ or Na^+, as indeed has been observed (Davies *et al*., 1953; Conway and Duggan, 1958; Eddy *et al*., 1970*a*). However, Conway and Duggan (1958) found that several, but not all, of the common neutral amino acids accelerated the efflux of Na^+ ions from the cells of a strain of *Saccharomyces cerevisiae* containing an excess of Na^+ over K^+. Eddy *et al.* (1970*a*) were unable to reproduce this effect using another yeast, a strain of *Saccharomyces carlsbergensis*, but detected an efflux of K^+ ions in the presence of glycine when the yeast cells contained their normal complement of K^+ ions. This effect, representing about 2–3 equivalents of K^+ per mole of glycine absorbed, occurred in the range of glycine concentration (<100 μM) in which the general amino acid permease functions. Interestingly, the uptake of glycine in the presence of glucose increased neither the efflux nor the influx of K^+. It was inferred that K^+ did not play an indispensable role in the transport of glycine.

Support for the involvement of K^+, in certain specific circumstances, came from studies of the effect of changing the extracellular pH from 4.5 to 7 on the uptake of glycine in the presence and absence of extracellular K^+ ions. The most significant results were obtained with yeast preparations depleted of ATP (Figure 3). These absorbed about 20 nmol glycine (mg dry weight yeast)$^{-1}$ in about 2 min at pH 4.5. The absorption was much slower at pH 7 and when extracellular K^+ was present. It was inferred that the accumulation of glycine under these conditions might be driven by the combined gradients of H^+, acting inwards, and K^+ acting outwards across the plasmalemma (Eddy *et al*., 1970*a,b*). The maximum concentration gradient of glycine formed under these conditions was about 4×10^4 (Seaston *et al*., 1976). The process appeared not to involve exchange of glycine with the cellular amino acids. The absorption of glycine by energy-depleted yeast appears to be limited by small changes in intracellular pH (Seaston *et al*., 1976). There is a dramatic fall, therefore, in the capacity of the yeast to take up amino acids once energy metabolism has stopped. If the gradients of H^+ and K^+ do indeed drive the amino acid pump in the absence of ATP, then the flow of

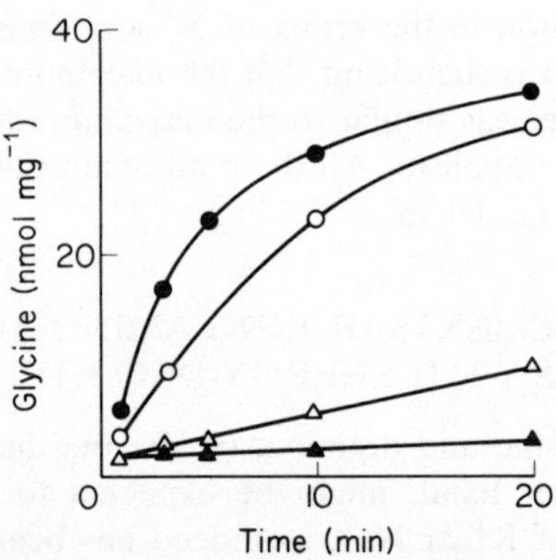

Figure 3 Effects of pH and [K^+] on the absorption of glycine by yeast cells depleted of ATP. The preparations of washed cells of strain N.C.Y.C. 74 were depleted of ATP in the presence of antimycin and 2-deoxyglucose. Glycine uptake was assayed at either pH 4.5 (●, ○) or pH 7 (△, ▲), in the presence of (○, ▲) or absence (●, △) of 100 mM KC1. The proton gradient driving the uptake of glycine is largest at pH 4.5 in the absence of added KC1. (Adapted from Eddy *et al*., 1970*b*)

both ions would be expected to be accelerated during the uptake of glycine. The existence of this phenomenon was subsequently verified (Eddy and Nowacki, 1971). The manner in which protons are absorbed is illustrated in Figure 4 for two contrasting conditions: (a) where the cells are depleted of ATP, owing to the presence of antimycin and 2-deoxyglucose, and (b) where the yeast respires its endogenous carbon reserves, in the absence of an external carbon substrate. Figure 4 shows that, in the former situation, glycine caused the solution to become more alkaline, as though protons were absorbed as cosubstrates with the amino acid. The response was different when the metabolic inhibitors were omitted so that endogenous energy metabolism could occur. Successive additions of glycine to the system then resulted in a dual response whereby protons were first absorbed and a fraction of them were then ejected (Figure 4). The ejection of protons was prevented by the presence of antimycin and deoxyglucose. It seemed likely, therefore, that the uptake of protons with the amino acid occurred through the amino acid carrier and that the ejection of protons took place through the proton pump. (Ramos *et al.* (1975) have shown that the overall process utilizes ATP formed either by oxidative phosphorylation or during glycolysis.) The system is illustrated in Figures 2 and 5. The choice of glycine in the above work was partly based on its relative metabolic stability. Thus, glycine is a poor nitrogen source for many yeast strains and it was apparently absorbed as such by the cell preparations in which the ionic fluxes were assayed.

6 THE ROLE OF K^+

The circumstance referred to earlier, namely that the influx and efflux of K^+ was not accelerated during the uptake of glycine in the presence of glucose, is consistent with the behaviour expected of the electrogenic mode of glycine absorption illustrated in Figure 5, whereby protons absorbed with the amino acid are ejected through the proton pump. However, when the proton pump is not functioning the flow of K^+ out of the yeast provides the natural means of neutralizing the influx of protons with the amino acid. The presence of extracellular K^+ would inhibit the efflux of K^+ and indirectly inhibit the uptake of glycine by lowering the membrane potential. The inhibition of glycine uptake by various proton conductors is also consistent with the electrogenic model of amino acid absorption (Seaston *et al.*, 1976). Further support, of an indirect kind, for the notion that the absorption of protons with various carbohydrates depolarizes the yeast cells, has come from studies with fluorescent carbocyanine dyes (Eddy *et al.*, 1977). A similar relatively small fluorescence change is caused by glycine (Seaston *et al.*, 1976). While these fluorescence effects are clear-cut and reproducible, their interpretation is complex (Eddy *et al.*, 1977). Further work thus seems to be

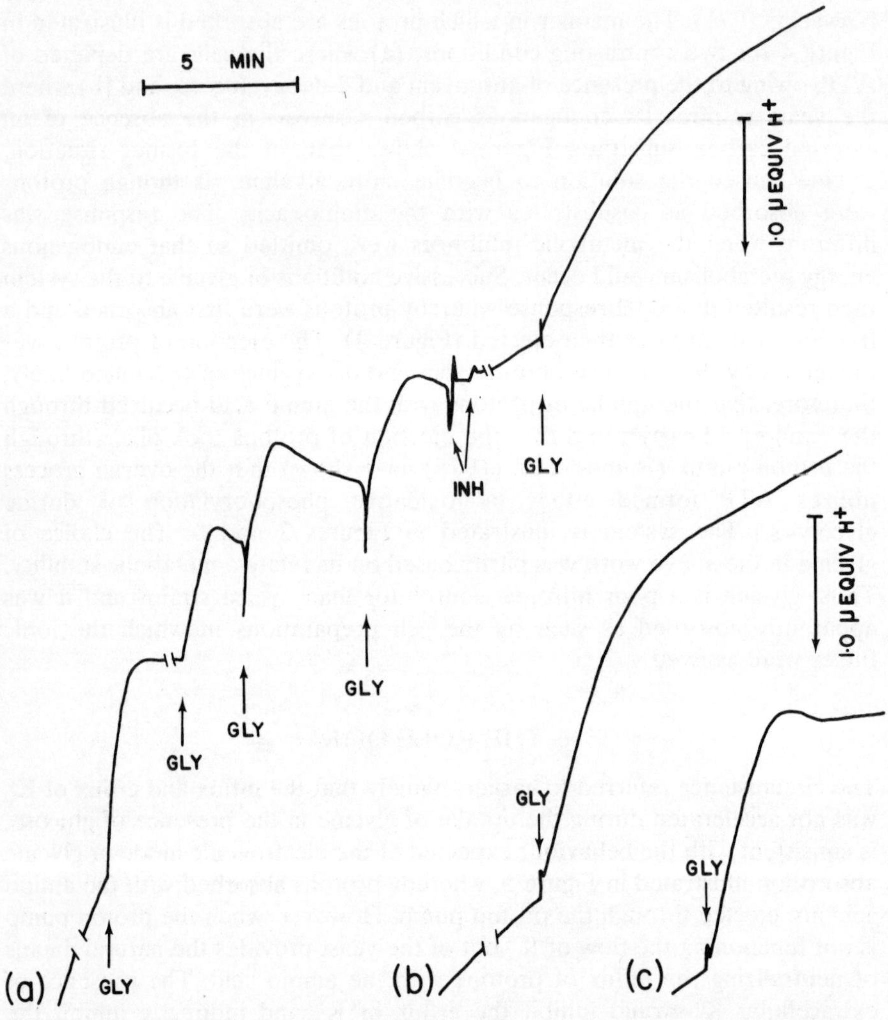

Figure 4 Proton uptake by yeast during the absorption of successive small amounts of glycine. In trace (a) the yeast could respire and it ejected some of the acid initially absorbed with each of the three portions (0.5 μmol) of glycine added after the first one. The metabolic inhibitors, antimycin and deoxyglucose, were then added to stop further proton ejection, which depends on a supply of ATP. The next addition of glycine thus caused a proton uptake which was not apparently recycled. In trace (b) a larger amount of glycine (1 μmol) was added in the presence of the metabolic inhibitors. In trace (c) the inhibitors were omitted and some proton recycling occurred following the addition of 1 μmol of glycine. (Adapted from Eddy and Nowacki (1971))

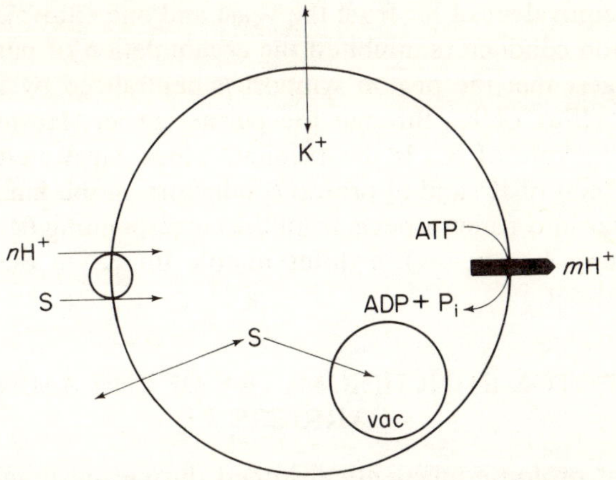

Figure 5 Some factors influencing the distribution of a substrate absorbed by symport with protons.

The number of protons (n) accompanying each equivalent of substrate S traversing the cell membrane through the symport is one factor governing the energy input into the substrate carrier. The symport mechanism is in principle reversible; additionally, the substrate S might leak out of the yeast by a route that was independent of the proton gradient. The uptake of S into the vacuole is another factor to be considered. The proton pump is shown as an ATPase, by analogy with other microorganisms. The value of m would be 2 or possibly 1. The flow of K^+ would be by diffusion through channels selecting K^+ in preference to Na^+ or protons

required to prove that the different symport mechanisms found in yeast are indeed electrogenic.

When cellular K^+ was largely but not completely replaced by Na^+ ions, rapid glycine uptake was found to depend on continuing energy metabolism, in the sense that neither proton uptake nor rapid amino acid uptake occurred after energy depletion (Eddy and Nowacki, 1971). This behaviour is consistent with the relative slowness of the electrogenic efflux of Na^+ from yeast (Eddy, 1978). During the uptake of each equivalent of glycine and various other amino acids by such 'sodium yeast', about 1 equivalent of protons were simultaneously absorbed. In contrast about 2 proton equivalents were absorbed per glycine equivalent when K^+ was the principal cellular cation. The explanation of this puzzling phenomenon is not known (Eddy, 1978).

Kotyk *et al*. (1971*c*) also studied the effects of Na^+ and K^+ on the uptake of various amino acids during energy metabolism. Their conclusion that the transport of amino acids in bakers' yeast is not essentially coupled with the transport of Na^+ or K^+ seems consistent with the scheme shown in Figure 5.

Purine transport in *Saccharomyces cerevisiae* shows some analogies with amino acid transport in that the uptake of purines, such as hypoxanthine, is activated by protons and inhibited by K^+ or Na^+ (Reichert *et al*., 1975; Reichert and Forêt, 1977). Furthermore, the absorption of hypoxanthine displaced one equivalent of K^+ from the yeast and one equivalent of protons into cells. Proton conductors inhibited the accumulation of purines. Forêt *et al.* (1978) suggest that the proton symport is neutralized by a K^+ antiport, that is by the efflux of K^+ through the purine carrier. However, it seems possible that the role of K^+ is the facultative one illustrated in Figure 5. Because the effects of K^+ and of proton conductors on the kinetics of purine absorption differ in certain respects from the corresponding behaviour of the amino acid pump (see below), a different role for K^+ in the two systems cannot be ruled out.

7 THE PROTON STOICHEIOMETRY OF THE AMINO ACID CARRIERS

The number of proton equivalents absorbed during the uptake of various amino acids by the general amino acid permease and by certain of the specific permeases has been studied in energy-depleted preparations of various genetically defined yeasts related to the strain Σ1278b. Analogous

Table 2 The number of equivalents of protons (n) absorbed with each equivalent of a given substrate by different routes

System	Number of equivalents
Saccharomyces	
General amino acid permease (Eddy and Nowacki, 1971; Seaston *et al.*, 1973, 1976)	
Glycine	2
Methionine	2
Lysine	2
Citrulline	2
Phenylalanine	2
Leucine	2
Specific permeases (Eddy *et al.*, 1977; Cockburn *et al.*, 1975; Reichert and Forêt. 1977; Serrano, 1977)	
Lysine	1
Methionine	1
Proline	1
Glutamate	2 or 3
Maltose	1
Phosphate	3
Hypoxanthine	1
Candida (Eddy *et al.*, 1977)	
Glycine	1
Lysine	1
Arginine	1
Glutamate	2

studies have been made with other strains of *Saccharomyces* and with a strain of *Candida utilis* (Table 2). Whereas the general permease of *Saccharomyces cerevisiae* exhibits a stoicheiometry of about 2, the values found for the specific permeases are nearer 1. The absorption of various carbohydrates, of phosphate and of certain purine bases is also proton-linked (Table 2).

The flow of protons and K^+ ions during the uptake of glutamate, glycine, and lysine by energy-depleted preparations of *Candida utilis* illustrates the influence of substrate charge on these processes (Figure 6). In each case, the substrate is effectively absorbed as a positively charged complex, so that K^+ ions are expelled from the yeast. The intervention of the proton pump in these processes leads to quite a different pattern of ionic flows. This has been studied by introducing air into the cell suspension to initiate energy metabolism. Lysine uptake was then accompanied by the rapid ejection of protons with a smaller number of K^+ ions (Figure 7). In contrast glutamate uptake initially caused a small loss of K^+, but these ions were then absorbed, so that within about 1 min no net uptake of protons was apparent (Figure 8). The nature of the factors governing the interaction between the proton pump and the proton-linked amino acid absorption requires further study. Recent work with *Neurospora crassa* suggests that cyclic AMP may be involved (Pall, 1977). The behaviour shown towards glutamate implies that the plasmalemma became hyperpolarized after about 0.2 min. On the other hand, depolarization appears to have occurred almost immediately in the presence of lysine.

8 ENERGETICS OF AMINO ACID ABSORPTION

The energy input into the general amino acid permease, which absorbs 2 equivalents of protons per mole of amino acid absorbed, is twice that flowing into the specific permeases exhibiting a proton stoicheiometry of 1. Thus, if the glycine distribution across the plasmalemma reached equilibrium with the proton gradient operating across the general amino acid permease, the glycine gradient would reach the very large magnitude of about 10^{10}. On the other hand, the specific methionine permease would be expected, in these terms, to concentrate its substrate about 10^5-fold. While there is a dearth of detailed information in the literature about the extent to which yeasts in fact concentrate various amino acids during energy metabolism, the maximum ratios observed do not exceed 10^5 and are usually nearer 10^3 (Halvorson *et al*., 1955; Surdin *et al*., 1965; Kotyk *et al*., 1971*a*; Saiyid and Kotyk, 1972). The largest ratios are observed in the lower range of amino acid concentration. For instance, Kotyk *et al.* (1971*a*) found that the ratio for glycine fell from 800 at 10 μM to about 50 at 10 mM. Indge *et al.* (1977) found that glycine was concentrated by a roughly constant factor,

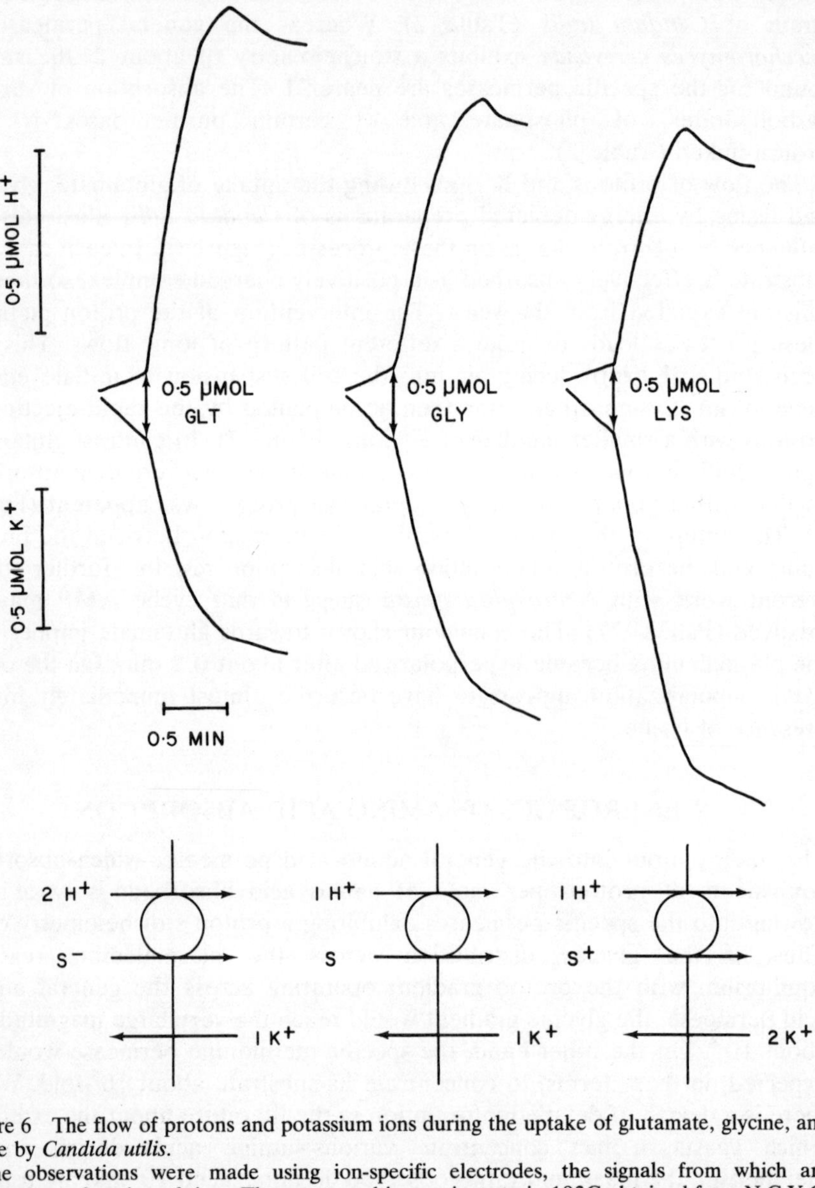

Figure 6 The flow of protons and potassium ions during the uptake of glutamate, glycine, and lysine by *Candida utilis*.

The observations were made using ion-specific electrodes, the signals from which are depicted as a function of time. The yeast used is a variant strain 193C obtained from N.C.Y.C. 193 by cultivation in a chemostat with glutamate as the limiting nitrogen source. Strain 193C was then grown to early log phase in a glucose mineral salts medium containing 2.5 mM glutamate. The cell preparations were depleted of ATP in the presence of antimycin and deoxyglucose and assayed at about pH 4.8 (Eddy *et al.*, 1977). The lower part of the diagram shows how electroneutrality appeared to have been preserved

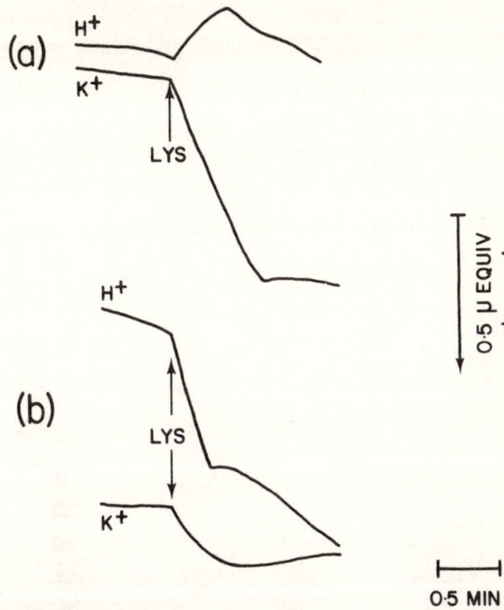

Figure 7 Effect of energy metabolism on the flows of protons and potassium ions during the uptake of lysine by preparations of *Candida utilis*.

The yeast strain 193C was grown for 40 h in glucose–mineral salts solution containing 2.5 mM glutamate. In trace (a) antimycin and deoxyglucose were present to stop energy metabolism, whereas in trace (b) these compounds were omitted and the yeast respired. In the former case neutralization was achieved as illustrated in Figure 6. In the latter case proton ejection mainly replaced K^+ efflux as the means of neutralizing the inflow of lysine cations. (Adapted from Eddy *et al.* (1977) and unpublished observations)

of about 4×10^4-fold, in the concentration range 2–10 μM when glucose was present to maintain energy metabolism. A similar ratio was transiently maintained, in the presence of 0.1 μM glycine, with the same strain of yeast cells, when these had been depleted of ATP. The depletion was carried out in circumstances where a large proton gradient was maintained across the plasmalemma. As noted above, the amount of glycine absorbed by such preparations is limited by the fall in intracellular pH which tends to occur when the proton pump is not functioning. However, it is clear that the maximum accumulation ratio observed during energy metabolism is similar to the maximum ratio that the depleted preparations exhibited. Amino acid accumulation in both systems appears to be driven by the proton gradient.

The limited evidence available thus suggests that the maximum accumulation ratios for glycine, observed at pH values near 4.5, are considerably smaller than the general amino acid permease would maintain at thermodynamic equilibrium. This inference depends of course on the likely, but as yet unproven, assumption that, when the concentration of extracellular K^+ is small, the membrane potential during energy metabolism

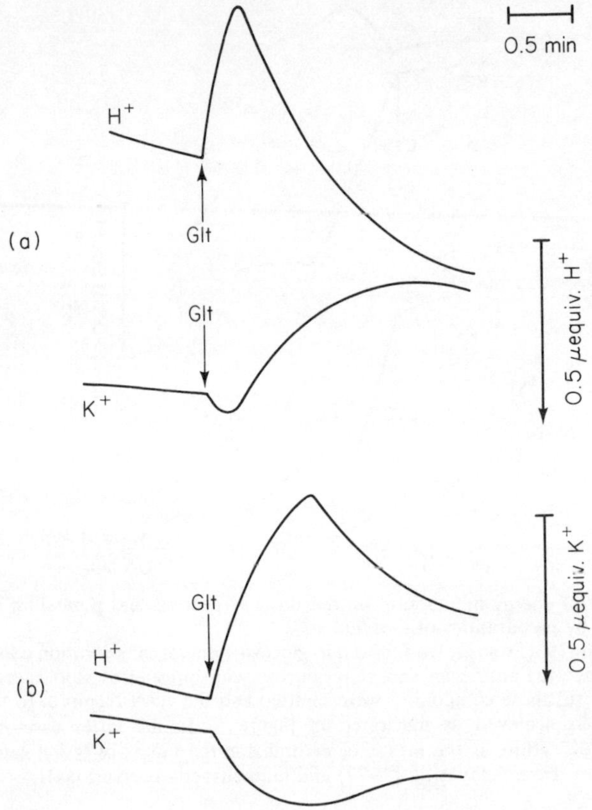

Figure 8 Effect of energy metabolism on the flows of protons and potassium ions during the uptake of glutamate by preparations of *Candida utilis*.

The conditions resembled those outlined in the legend to Figure 7. In trace (a) energy metabolism eventually resulted in the absorption of almost 1 equivalent of K^+ with the glutamate. In trace (b) the action of the proton pump was restricted, but not abolished, by the presence of the metabolic inhibitors. The type of response shown in (b) is typical of the cultures grown for 40 h. Those grown for 17.5 h show no net ejection of protons in the presence of the metabolic inhibitors

can be inferred from the distribution of K^+ between the yeast cells and their environment (Figure 5). The failure to reach thermodynamic equilibrium would be consistent with the physiological role of the general permease, namely that of providing a fast flow of assimilable amino groups into the yeast cells. In general, two types of mechanism might limit the amount of amino acid absorbed through the various permeases. The first concerns the possibility that the amino acid pumps are bypassed by leak pathways traversed without the participation of protons as cosubstrates (Figure 5). The second depends on the extensive evidence that the operation of the permeases is restricted by feedback inhibition dependent on the composition of the cellular pool of amino acids.

1 Amino acid 'leaks'

The ability of yeast to retain most of its normal content of amino acids, whether these were formed by biosynthesis or had been accumulated from the environment, was early recognized (Taylor, 1949; Halvorson *et al.*, 1955; Surdin *et al.*, 1965; Crabeel and Grenson, 1970). Indeed, even the addition of metabolic inhibitors such as dinitrophenol, or the exhaustive depletion of the cellular ATP content, results in only a very slow rate of efflux of amino acids such as glycine (Kotyk *et al.*, 1971*a,b*; Seaston *et al.*, 1976). This behaviour indicates that the hypothetical leak pathways, referred to above, are kinetically unimportant except at very small external amino acid concentrations. Seaston *et al.* (1976) estimated that the rate of leakage of ninhydrin-positive material from the yeast cells was about 0.4 nmol mg^{-1} min^{-1}, or about 0.1 per cent of the pool complement.

2 Feedback regulation

The main factor limiting the uptake of amino acids by washed yeast cell suspensions is expressed as a rapid fall in the rate of uptake of a given amino acid as the cellular content rises. Figure 9 shows how the rate of uptake of glycine, probably through the general amino acid permease, diminished as a result of first incubating the yeast with glycine. The proper interpretation of this behaviour takes into account the fact that uptake of glycine by the controls increased as these were starved. The increase in permease activity (derepression) during nitrogen starvation may represent the biosynthesis of new carrier molecules. Alternatively, starvation may have removed an endogenous inhibitor. A number of recent studies have provided further details.

(a) Incubation with either L- or D-amino acids inhibits the general permease (Rytka, 1975). Since the D-amino acids are only slowly metabolized, it therefore seems likely that they interact directly with the system, in one or other of two ways. The absorbed amino acids might either bind to a stable carrier and inhibit it reversibly (transinhibition) or they might activate a reaction which destroyed thc carrier (Hunter and Segel, 1973; Roon *et al.*, 1977). Indge *et al.* (1977) found that preliminary loading of the yeast with glycine failed to stop the further uptake of glycine (Figure 9) when the yeast had already been incubated with glucose for 1–2 h. The inhibition of amino acid uptake by cellular glycine thus appears to depend on factors ancillary to the normal working of the amino acid pump. (b) NH_4^+ ions inhibit the activity of the general amino acid permease (Roon *et al.*,

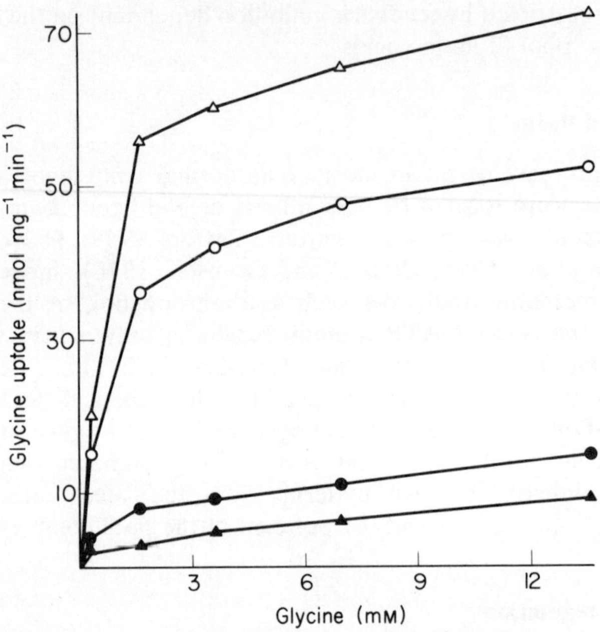

Figure 9 The inhibition of glycine uptake caused by loading the yeast with glycine. The washed cell preparations of yeast N.C.Y.C. 74 were incubated with glucose for 30 min (○, ●) or 60 min (△, ▲) with (●, ▲) or without (△, ○) 13mM glycine present. The yeast was washed in each case and the rate of uptake of glycine determined at several concentrations as shown. The presence of glycine in the preincubation markedly lowered its subsequent uptake. In contrast the glycine uptake rate in the controls increased after starvation for 60 min as opposed to 30 min (Redrawn from Indge *et al.*, 1977)

1975*b*). Because a mutant yeast lacking the anabolic NADP–glutamate dehydrogenase also lost its sensitivity to NH_4^+, Roon *et al.* (1975*b*) concluded that NH_4^+ inhibition results from the assimilation of the compound into the α-amino groups of glutamate and other amino acids. (c) The feedback inhibition of the specific amino acid permeases has also been demonstrated. Both histidine permeases, especially the one with a higher affinity for histidine, were inhibited by the preliminary loading of the yeast with histidine (Crabeel and Grenson, 1970). Loading with other amino acids inhibited their own further absorption but not that of histidine. The lysine-specific and the arginine–lysine transport systems were likewise inhibited by intracellular lysine or arginine, on the one hand, and, on the other hand, stimulated by the depletion of the cellular content of these amino acids (Morrison and Lichstein, 1976). The effects of preloading with amino acids that were not substrates of either permease could be accounted for in terms of their effects on the pool size of arginine or lysine. Feedback inhibition of leucine uptake was described by Bussey and Umbarger

(1970*a*). In one mutant yeast, leucine pool formation was accompanied by an overshoot leading to loss of leucine from the pool. A study by Dürr *et al.* (1976) of the properties of the arginine transport system of purified yeast vacuoles is interesting in the above context. These workers observed that arginine uptake became several times faster when the vacuoles were treated with certain proteolytic enzymes, as though arginine uptake was initially restrained by the presence of a modifying protein.

3 Modulation of the proton gradient

Reference was made above to the changes which occur in the rate of uptake of amino acids, such as glycine, when either the extracellular pH or the K^+ ion concentration is varied. Such effects are larger in energy-depleted yeast preparations (Eddy *et al*., 1970*a*) in which the proton pump is not functioning. It is interesting that the presence of cyclic AMP stimulates the ejection of protons from preparations of *Schizosaccharomyces pombe* under certain restricted conditions, and that the absorption of nucleosides and amino acids was also stimulated, but without the elevation of the cellular ATP content (Foury and Goffeau, 1975). These workers suggested that the cyclic nucleotide modulated the activity of the plasma membrane ATPase that is believed to drive the extrusion of protons. In *Neurospora crassa* membrane depolarization appears to increase the cellular content of cyclic AMP by stimulating the adenylate cyclase associated with the plasma membrane (Pall, 1977). Cyclic AMP may therefore be involved in the interactions between the symport of protons with an amino acid and their subsequent ejection through the proton pump. However, the subcellular localization of adenyl cyclase in yeast is controversial (Wheeler *et al*., 1974).

An important general question concerns the way in which changes in the magnitude of the proton gradient, or more specifically, changes in extracellular pH, intracellular pH or in the membrane potential, are expected to affect the rate of amino acid uptake. This complex problem was briefly discussed by Seaston *et al.* (1976). Early observations indicated that amino acid uptake was fastest at about pH 4.5 to pH 6 (see, for example, Surdin *et al*., 1965). In energy-depleted yeast the V_{max} for glycine uptake was lower at pH 7 with the K_m roughly constant. The presence of 0.1 M KCl, which would depolarize the yeast cells, lowered the V_{max} without significantly changing the K_m. Seaston *et al.* (1976) suggested that the very low natural rate of efflux of glycine might be understood in terms of these ionic and electrical factors. Briefly, the low rate of efflux is supposed to be the result of the smallness of the V_{max} of the carrier system, under the electrical and ionic conditions in which efflux occurs. The detailed hypothesis also accounts for the low rate of exchange of glycine through the general permease. The apparent failure of the amino acid carriers of yeast to permit anything but a relatively slow efflux of their substrates is well documented (Crabeel and Grenson, 1970; Kotyk *et al.*, 1971*a*; Magaña-Schwencke *et al.*, 1973; Morrison and Lichstein, 1976). The low rate of efflux does not appear to be the result simply of feedback inhibition incurred during the initial loading of the cells with the relevant amino acid. Thus Seaston *et al.* (1976) observed a low rate of efflux of glycine from yeast preparations that could still absorb glycine rapidly. Nor is the low rate of efflux due to competition with endogenous amino acids, because the efflux of these is also slow. Some workers have attributed the apparent lack of efflux of various amino acids to the possibility that the amino acid pool might be segregated in the yeast vacuole (see below), but it seems more likely that the low natural rate of efflux is an intrinsic property of the glycine permease which can be explained in terms of the complex response of the system to the electrical and ionic conditions acting at the yeast plasmalemma.

The characteristics of the cytosine and purine permease are relevant to the understanding of the behaviour of the amino acids (Chevallier *et al.*, 1975; Reichert and Forêt, 1977). (a) The system concentrated cytosine about 3×10^3-fold from a 1 μM solution. (b) The presence of dinitrophenol considerably accelerated the efflux of cytosine and slowed its influx. (c) The carrier functions with a single proton as cosubstrate (Reichert and Forêt, 1977). (d) Raising the concentrations of extracellular K^+ or Na^+ increased the K_m for cytosine competitively, whereas Eddy *et al.* (1970*a*) observed that it lowered the V_{max} at constant K_m in their study of the effect of K^+ on glycine uptake. (e) It seems possible therefore that purine efflux is relatively fast because, in this instance, V_{max} is relatively large even when the membrane potential is small.

9 CELLULAR COMPARTMENTS: THE ROLE OF THE VACUOLE

Work by Cowie and McClure (1959) early led to the idea that the cells of *Candida utilis* contain two functionally distinct amino acid pools. Studies on the biosynthesis of various amino acids have been interpreted in a similar way (Sims and Folkes, 1964), a mitochondrial compartment, for instance, being involved (Ryan and Kohlhaw, 1974). Recent work has revealed the complex nature of the interactions between the mitochondria, the vacuole, and the cytoplasmic pool that are involved in the regulation of arginine biosynthesis and catabolism (Wipf and Leisinger, 1977).

Cowie and McClure (1959) believed that exogenous amino acids were absorbed into a *concentrating pool* ('expandable pool'), from which they were released either when the yeast was subjected to an osmotic shock or in exchange for extracellular amino acids. A *conversion pool* ('internal pool') on the other hand, contained mainly amino acids that were the product of endogenous biosynthesis, together with a smaller fraction derived from the concentrating pool. Amino acids were incorporated into protein from the internal pool. The correct interpretation of this work in terms of current ideas about the functions of the vacuole, mitochondria, and cytosol is problematical. The apparent ease of removal from the yeast of the absorbed amino acids is also puzzling. It seems possible that the cell membrane of the yeast strain used was unstable in certain conditions and underwent transient changes in permeability leading to differential extraction of the cellular contents (Lewis and Phaff, 1965). An analogy is provided by the phenomenon of shock excretion, shown by certain yeast strains. In the presence of glucose, these release some of their amino acids and then reabsorb them. A superficial adenosine triphosphatase activity is meanwhile exposed (Lewis and Phaff, 1965; Lewis and Stephanopoulos, 1967). Bussey and Umbarger (1970*b*) describe the differential extraction of [^{14}C]-leucine from yeast kept at 40 °C in the presence of 10 mM sodium azide and 50 mM leucine. Such a treatment removes about 75 per cent of the ethanol-extractable pool in 90 min. These workers refer to this fraction as being analogous to the expandable pool of Cowie and McClure (1959). The need to use these relatively harsh conditions in order to extract the leucine may mean, in this instance too, that the cell membrane was perturbed.

A great deal of subsequent work has focussed attention on the role of the main yeast vacuole in solute transport. The marked ability of the vacuole to accumulate purines from the environment was detected by ultraviolet microscopy in a strain of *Candida utilis* (Svihla *et al*., 1963). Indge (1968) observed that yeast fed serine or lysine accumulated 30 and 70 per cent, respectively, of it in a crude vacuolar fraction that he isolated by controlled lysis of the protoplasts. The technique of isolating vacuoles was improved by Wiemken and Dürr (1974) and Dürr *et al*. (1975) who concluded that at least about 60 per cent of the total amino acid pool was located in the vacuoles of *Saccharomyces cerevisiae*. Furthermore, when the protoplasts were exposed to tracer glucose, the non-vacuolar, cytoplasmic fraction of the amino acids was preferentially labelled. The overall amino acid concentration of the vacuolar pool (0.97 M) was estimated to be about five times that in the cytoplasm (0.2 M). The vacuolar pool contained relatively more glutamine, citrulline, arginine, and lysine and much less glutamate than the cytoplasmic pool. Growth with arginine as nitrogen source resulted in the accumulation of a large amount of that amino acid in the vacuole, which evidently acts as a storage compartment.

Boller *et al*. (1975) viewed the above observations on the vacuolar fraction as showing that 'the bulk of the soluble amino acids has been found to be located and stored within this cell organelle, separated from the small metabolic pool of amino acids in the cytoplasm'. While this statement somewhat exaggerates the extent to which the vacuole segregates amino acids, it leads to the question: where are the amino acid pumps located, at the plasmalemma or in the vacuolar membrane? It will be recalled that glutamate, a major pool component and one which many yeasts concentrate, is largely excluded from the vacuole. Moreover, the ratio of the amino acid concentration found in the vacuolar compartment to the concentration found in the cytoplasmic compartment was apparently not greater than about 20-fold even in the case of arginine. Now the total cytoplasmic amino acid pool represented a 0.2 M solution (Wiemken and Dürr, 1974); and the general amino acid permease appears to concentrate glycine about 5×10^4-fold, from a 2 μM solution, to produce a 0.1 M solution mainly in the cytosol (Indge *et al*., 1977). Taking all these facts into account, it seems clear that the amino acid gradients formed across the plasmalemma are very much larger than those existing across the membrane of the vacuole. Indge *et al*. (1977) found that certain yeast preparations concentrated both glycine and lysine, but only the latter entered the vacuole in amounts sufficient to make it swell. However, glycine was also absorbed into the vacuole, causing it to swell, following a preliminary treatment of the yeast with glucose. Indeed the uncontrolled uptake of glycine into the vacuole eventually ruptured many of the yeast cells.

The existence of a specific transport system for arginine in isolated yeast vacuoles was elegantly shown by Boller *et al.* (1975). The kinetic characteristics and substrate specificity of this system differed from those exhibited by the spheroplasts from which the vacuoles were isolated. Notably, arginine uptake by the vacuoles proceeded strictly by exchange with intravacuolar arginine, and was not inhibited by proton conductors, whereas net absorption of arginine occurred at the plasmalemma and was retarded in the presence of the proton conductors. The mechanism whereby the vacuole concentrates arginine *in vivo* is unknown. Proteinases activate arginine transport into isolated vacuoles (Dürr *et al*., 1976).

Schwencke and De Robichon-Szulmajster (1976) found that isolated vacuoles concentrated the cation *S*-adenosylmethionine by a specific saturable mechanism that, unlike the corresponding permease located in the plasmalemma (Petrotta-Simpson *et al*., 1975), was insensitive to dinitrophenol or azide. The way in which the uptake of the adenosylmethionine cations was neutralized was not established. The presence of certain polyphosphates in the yeast vacuole was demonstrated by Indge (1968). It seems possible that these compounds may play a part in the retention of basic amino acids and other cations in the vacuole (see Ludwig *et al*., 1977).

10 STUDIES WITH MEMBRANE FRACTIONS

There have been attempts to study solute transport in sealed vesicles derived from the yeast plasmalemma. Christensen and Cirillo (1972) isolated such vesicles by grinding yeast with glass beads under carefully controlled conditions. The diameter of the product varied from 10 to 50 per cent of that of the parent cells. The vesicles contained mitochondria and respired. Galactose or glucose transport into the vesicles occurred by facilitated diffusion. However, amino acid or uracil uptake failed to occur in assays made under various conditions. Fuhrmann *et al.* (1976) succeeded in preparing vesicles free from mitochondria and lacking respiration. The presence of the glucose and galactose carrier systems was demonstrated. Evidence was also obtained that the plasma membrane was more permeable to K^+ than to Na^+.

There is a great deal of evidence showing that plasma membrane fragments from yeast contain an ATPase system with properties distinguishing it from the corresponding mitochondrial enzyme (Delhez *et al.*, 1977; Foury *et al.*, 1977). For instance, the former system is not inhibited by concentrations of oligomycin that strongly inhibit the mitochondrial ATPase. Marriott (1975) has implied that the plasmalemma enzyme, like its mammalian counterpart, is selectively activated by Na^+ and K^+ and inhibited by ouabain. Other workers do not appear to have encountered these effects. The notion that the oligomycin-insensitive ATPase of yeast is an electrogenic proton pump is based on an analogy with the behaviour of *Neurospora* where the evidence associating the enzyme with proton pumping is rather convincing (Bowman and Slayman, 1977). Also both organisms pump ions in a superficially similar fashion (Eddy, 1978).

Another approach to the molecular basis of amino acid transport has been through the search for membrane components that bind amino acids reversibly. (a) Voříšek (1972, 1973) extracted isolated yeast membrane preparations with Tween 80 in the presence of 0.1 M $MgCl_2$. Fractionation of the extract gave a material which reproducibly bound phenylalanine in an equilibrium dialysis assay and irregularly bound a series of other amino acids. (b) In a study of leucine binding by dialysed extracts of yeast, binding was found to increase in parallel with the rate of uptake of leucine by the parent yeast (Bussey and Umbarger, 1970*a,b*). Further, a mutant yeast with a lower affinity than the parent strain for leucine in a transport assay also exhibited a lower affinity in the binding assay. (c) An arginine-binding protein was isolated from the wild-type yeast strain Σ1278b (Opekarová *et al.*, 1975). The exponentially growing cells were found to release various superficially located proteins, with a concomitant loss of ability to transport arginine and lysine. One protein which was purified to homogeneity bound arginine, with a K_d of about 0.4 mM. Lysine was a competitor but three other amino acids were not. The molecular mass of the protein was about 5000 daltons. A mutant yeast carrying the *arg-p₁* gene also contained the binding protein.

A different approach was used to detect polypeptides associated with the cytosine–purine permease (Parlebas and Chevallier, 1976). Plasma membranes from permease-less mutants were found to lack a polypeptide, of molecular mass about 80 000 daltons, which was present in the corresponding wild-type yeast. The presumptive permease protein represented about 1 per cent of the protein content of the membrane.

REFERENCES

Boller, T., Dürr, M., and Wiemken, A. (1975). *Eur. J. Biochem.*, **54**, 81–91
Bossinger, J. and Cooper, T. G. (1977). *J. Bacteriol.*, **131**, 163–173
Bowman, B. J. and Slayman, C. W. (1977). *J. Biol. Chem.*, **252**, 3357–3363
Bussey, H. and Umbarger, H. E. (1970*a*). *J. Bacteriol.*, **103**, 277–285
Bussey, H. and Umbarger, H. E. (1970*b*). *J. Bacteriol.*, **103**, 286–294
Chan, P. Y. and Cossins, E. A. (1976). *J. Cell. Comp. Physiol.*, **17**, 311–353.
Chevallier, M. R., Jund, R., and Lacroute, F. (1975). *J. Bacteriol.*, **122**, 629–641
Choudary, V. P. and Rao, G. R. (1976). *Biochem. Biophys. Res. Commun.*, **72**, 598–602
Christensen, M. S. and Cirillo, V. P. (1972). *J. Bacteriol.*, **110**, 1190–1205
Cockburn, M., Earnshaw, P., and Eddy, A. A. (1975). *Biochem. J.*, **146**, 705–712
Conway, E. J. and Armstrong, W. McD. (1961). *Biochem. J.*, **81**, 631–639
Conway, E. J. and Brady, T. G. (1950). *Biochem. J.*, **47**, 360–369
Conway, E. J. and Duggan, F. (1958). *Biochem. J.*, **69**, 265–274
Conway, E. J. and O'Malley, E. (1946). *Biochem. J.*, **40**, 59–67
Cowie, D. B. and McClure, F. T. (1959). *Biochim. Biophys. Acta*, **31**, 236–245
Crabeel, M. and Grenson, M. (1970). *Eur. J. Biochem.*, **14**, 197–204
Darte, C. and Grenson, M. (1975). *Biochem. Biophys. Res. Commun.*, **67**, 1028–1033
Davies, R., Folkes, J. P., Gale, E. F., and Bigger, L. C. (1953). *Biochem. J.*, **54**, 430–437
Delhez, J., Dufour, J.-P., Thines, D., and Goffeau, A. (1977). *Eur. J. Biochem.*, **79**, 319–328
Di Carlo, F. I., Schultz, A. S., and McManus, D. K. (1951). *J. Biol. Chem.*, **189**, 151–157
Drillien, R., Aigle, M., and Lacroute, F. (1973). *Biochem. Biophys. Res. Commun.*, **53**, 367–372
Dubois, E., Grenson, M., and Wiame, J.-M. (1974). *Eur. J. Biochem.*, **48**, 603–616
Dunlop, P. C. and Roon, R. J. (1975). *J. Bacteriol.*, **122**, 1017–1024
Dunlop, P. C., Roon, R. J., and Even, H. L. (1976). *J. Bacteriol.*, **125**, 999–1004
Dürr, M., Boller, T., and Wiemken, A. (1975). *Arch. Microbiol.*, **105**, 319–327
Dürr, M., Boller, Th., and Wiemken, A. (1976). *Biochem. Biophys. Res. Commun.*, **73**, 193–199
Duteurtre, B., Bourgeois, C., and Chollot, B. (1971). *J. Inst. Brew.*, **77**, 28–35
Eddy, A. A. (1978). *Curr. Topics Membrane Transport*, **10**, 279–360
Eddy, A. A. and Nowacki, J. A. (1971). *Biochem. J.*, **122**, 701–711
Eddy, A. A., Indge, K. J., Backen, K., and Nowacki, J. A. (1970*a*). *Biochem. J.*, **120**, 845–852
Eddy, A. A., Backen, K., and Watson, G. (1970*b*). *Biochem. J.*, **120**, 853–858
Eddy, A. A., Philo, R., Earnshaw, P., and Brocklehurst, R. (1977). *Fed. Eur. Biochem. Soc. Symp.*, **42**, 250–260
Ehrlich, F. (1912). *Ber. Dt. Chem. Ges.*, **45**, 883–889
Forêt, M., Schmidt, R., and Reichert, U. (1978). *Eur. J. Biochem.*, **82**, 33–43
Foury, F. and Goffeau, A. (1975). *J. Biol. Chem.*, **250**, 2354–2362
Foury, F., Boutry, M., and Goffeau, A. (1977). *J. Biol. Chem.*, **252**, 4577–4583
Fuhrmann, G. F., Boehm, C., and Theuvenet, A. P. R. (1976). *Biochim. Biophys. Acta*, **433**, 583–596
Gits, J. J. and Grenson, M. (1967). *Biochim. Biophys. Acta*, **135**, 507–516
Grenson, M. (1966). *Biochim. Biophys. Acta*, **127**, 339–346
Grenson, M. (1969). *Eur. J. Biochem.*, **11**, 249–260
Grenson, M. and Hennaut, C. (1971). *J. Bacteriol.*, **105**, 477–482
Grenson, M. and Hou, C. (1972). *Biochem. Biophys. Res. Commun.*, **48**, 749–756

Grenson, M., Mousset, M., Wiame, J. M., and Bechet, J. (1966). *Biochim. Biophys. Acta*, **127**, 325–338
Grenson, M., Hou, C., and Crabeel, M. (1970). *J. Bacteriol.*, **103**, 770–777
Grenson, M., Dubois, E., Piotrowska, M., Drillien, R., and Aigle, M. (1974). *Mol. Gen. Genet.*, **128**, 73–85.
Halvorson, H. O. and Cohen, G. N. (1958). *Ann. Inst. Pasteur, Paris*, **95**, 73–87
Halvorson, H., Fry, W., and Schwemmin, D. (1955). *J. Gen. Physiol.*, **38**, 549–573
Hoeberichts, J. A. and Borst-Pauwels, G. W. F. H. (1975). *Biochim. Biophys. Acta*, **413**, 248–251
Hunter, D. R. and Segel, I. H. (1973). *Arch. Biochem. Biophys.*, **154**, 387–399
Indge, K. J. (1968). *J. Gen. Microbiol.*, **51**, 447–455
Indge, K., Seaston, A., and Eddy, A. A. (1977). *J. Gen. Microbiol.*, **99**, 243–255
Johnson, B. and Brown, C. M. (1974). *J. Gen. Microbiol.*, **85**, 169–172
Joiris, C. R. and Grenson, M. (1969). *Arch. Intern. Physiol.*, **77**, 154–156
Jones, M., Pragnell, M., and Pierce, J. S. (1969). *J. Inst. Brew.*, **75**, 520–536
Jund, R., Chevallier, M. R., and Lacroute, F. (1977). *J. Membrane Biol.*, **36**, 233–251
Kotyk, A., Ponec, M., and Říhová, L. (1971*a*). *Folia Microbiol., Praha*, **16**, 432–444
Kotyk, A., Říhová, L., and Ponec, M. (1971*b*). *Folia Microbiol., Praha*, **16**, 445–450
Kotyk, A., Říhová, L., and Ponec, M. (1971*c*). *Folia Microbiol., Praha*, **16**, 451–453
Lewis, M. J. and Phaff, H. J. (1965). *J. Bacteriol*, **89**, 960–966
Lewis, M. J. and Stephanopoulos, D. (1967). *J. Bacteriol*, **93**, 976–984
Ludwig, J. R., II, Oliver, S. G., and McLaughlin, C. S. (1977). *Biochem. Biophys. Res. Commun.*, **79**, 16–23
Magaña-Schwencke, N., Kuznar, J., and Schwencke, J. (1973). *Biochim. Biophys. Acta*, **318**, 281–288
Marriott, M. S. (1975). *J. Gen. Microbiol.*, **89**, 345–352
Milstein, O. A. and Bekker, M. L. (1976). *J. Bacteriol.*, **126**, 1–6
Misra, P. C. and Höfer, M. (1975). *FEBS Lett.*, **52**, 95–99
Morrison, C. E. and Lichstein, H. C. (1976). *J. Bacteriol.*, **125**, 864–871
Opekarová, M., Kotyk, A., Horák, J., and Kholodenko, V. P. (1975). *Eur. J. Biochem.*, **59**, 373–376
Pall, M. L. (1977). *J. Biol. Chem.*, **252**, 7146–7150
Parlebas, N. and Chevallier, M. R. (1976). *FEBS Lett.*, **65**, 327–334
Peña, A. (1975). *Arch. Biochem. Biophys.*, **167**, 397–409
Petrotta-Simpson, T. F., Talmadge, J. E., and Spence, K. D. (1975). *J. Bacteriol.*, **123**, 516–522
Polak, A. and Grenson, M. (1973). *Eur. J. Biochem.*, **32**, 276–282
Ramos, E. H., Bongioanni, L. C. de, Claisse, L. M., and Stoppani, A. O. M. (1975). *Biochim. Biophys. Acta*, **394**, 470–481
Reichert, U. and Forêt, M. (1977). *FEBS Lett.*, **83**, 325–328
Reichert, U., Schmidt, R., and Forêt, M. (1975). *FEBS Lett.*, **52**, 100–102
Riemersma, J. C. and Alsbach, E. J. J. (1974). *Biochim. Biophys. Acta*, **339**, 274–284
Roon, R. J., Larimore, F., and Levy, J. S. (1975*b*). *J. Bacteriol.*, **124**, 325–331
Roon, R. J., Levy, J. S., and Larimore, F. (1977). *J. Biol. Chem.*, **252**, 3599–3604
Roon, R. J., Even, H. L., Dunlop, P. C., and Larimore, F. (1975*a*). *J. Bacteriol.*, **122**, 502–509
Rothstein, A. (1960). In *Regulation of the Inorganic Ion Content of Cells* (G. E. W. Wolstenholme and M. O'Connor, eds), Churchill, London, pp. 53–68
Ryan, E. D. and Kohlhaw, G. B. (1974). *J. Bacteriol.*, **120**, 631–637
Ryan, J. P. and Ryan, H. (1972). *Biochem. J.*, **128**, 139–146
Rytka, J. (1975). *J. Bacteriol.*, **121**, 562–570
Saiyid, N. H. and Kotyk, A. (1972). *Folia Microbiol., Praha*, **17**, 261–268
Salhany, J. M., Yamane, T., Shulman, R. G., and Ogawa, S. (1975). *Proc. Natl. Acad. Sci. U.S.A.*, **72**, 4966–4970
Schwencke, J. and De Robichon-Szulmajster, H. (1976). *Eur. J. Biochem.*, **65**, 49–60
Seaston, A., Inkson, C., and Eddy, A. A. (1973). *Biochem. J.*, **134**, 1031–1043
Seaston, A., Carr, G., and Eddy, A. A. (1976). *Biochem. J.*, **154**, 669–676
Senthe Shanmuganathan, S. and Elsden, S. R. (1958). *Biochem. J.*, **69**, 210–218
Serrano, R. (1977). *Eur. J. Biochem.*, **80**, 97–102

Sims, A. P. and Ferguson, A. R. (1974). *J. Gen. Microbiol.*, **80**, 143–158
Sims, A. P. and Folkes, B. F. (1964). *Proc. R. Soc. B*, **159**, 479–502
Sumadra, R. and Cooper, T. G. (1977). *J. Bacteriol.*, **131**, 839–847
Sumadra, R., Gorski, M., and Cooper, T. G. (1976). *J. Bacteriol.*, **125**, 1048–1056
Surdin, Y., Sly, W., Sire, J., Bordes, A. M., and De Robichon-Szulmajster, H. (1965). *Biochim. Biophys. Acta,* **107**, 546–566
Svihla, G., Dainko, J. L., and Schlenk, F. (1963). *J. Bacteriol.,* **85**, 399–409
Taylor, E. S. (1949). *J. Gen. Microbiol.*, **3**, 211–223
Thomulka, K. W. and Moat, A. G. (1972). *J. Bacteriol.*, **109**, 25–33
Thorne, R. S. W. (1937). *J. Inst. Brew.*, **43**, 288–293
Voříšek, J. (1972). *Biochim. Biophys. Acta*, **290**, 256–266
Voříšek, J. (1973). *Folia Microbiol., Praha*, **18**, 17–21
Waheed, A. and Castric, P. A. (1977). *J. Biol. Chem.*, **252**, 1628–1632
Watson, T. G. (1976). *J. Gen. Microbiol.*, **96**, 263–268
Wheeler, G. .E., Schibeci, A., Epand, R. M., Rattray, J. B. M., and Kidby, D. K. (1974). *Biochim. Biophys. Acta*, **372**, 15–22
Whitney, P. A. and Magasanik, B. (1973). *J. Biol. Chem.*, **248**, 6197–6202
Wiemken, A. and Dürr, M. (1974). *Arch. Microbiol.*, **101**, 45–47
Wipf, B. and Leisinger, Th. (1977). *FEMS Lett.*, **2**, 239–242
Woodward, J. R. and Cirillo, V. P. (1977). *J. Bacteriol.*, **130**, 714–723

Microorganisms and Nitrogen Sources
Edited by J. W. Payne

Section 1.3

Transport and Utilization of Amino Acids by Fungi

L. WOLFINBARGER, JR
Department of Biological Sciences, Old Dominion University, Norfolk, Virginia 23508, USA

1 INTRODUCTION

In beginning to prepare a review article on amino acid transport in fungi one cannot help but be overwhelmed by the magnitude of research that has been performed in the last 10–15 years. What began in the late 1950s and early 1960s as growth assays for amino acids using amino acid auxotrophs has evolved into complex biophysical/biochemical studies on membrane potentials, binding sites of transport components, membrane lipid phase transitions, etc. Indeed, it is a widely held belief that the main amino acid transport systems in many fungi have been delineated, and attention is now being focussed on aspects of these individual transport systems and their relationships with the lipid matrix of the plasma membrane.

In this review, attention will be focussed primarily on amino acid transport by *Neurospora crassa*, *Penicillium*, *Aspergillus nidulans*, and *Achlya* with specific reference to *Claviceps*, *Arthrobotrys*, and *Botrytis*. Because the yeasts are being considered separately in this text (Section 1.2) they will not be considered here under this general heading of fungi. Unlike the yeasts, filamentous fungi characteristically exist in various morphologically distinct stages, viz. asexual spores (conidia), germinating conidia, and mycelia. As a general rule, comparable transport systems for amino acids may be found in each stage of development although these systems may exhibit considerable differences in activities, substrate specificities, and patterns of regulation.

The methods employed in transport assays have been extensively reviewed (Christensen, 1975; Matthews and Payne, 1975) and will not be considered here in other than a cursory fashion. Transport assays in fungi have at any rate been limited essentially to an incubation of cells in the presence of a radioactively labelled amino acid followed by filtration, washing, and counting of radioactivity remaining associated with the cells. In most instances extraction of the radiolabel from the cells and confirmation of its chemical identity has permitted a calculation of concentrative capabilities and assignment of an active transport mechanism to the accumulation of amino acids by specific transport systems. Confusion over transport of individual amino acids has usually resulted from the presence of multiple transport systems with overlapping substrate specificities. Unlike comparable studies in animal cell lines, however, it has usually been possible to resolve this confusion by genetic studies, isolating mutant strains deficient in one or more transport systems. This latter technique has permitted a kinetic characterization of individual transport systems unclouded by other transport activities. Unfortunately, as we shall see later, the isolation of strains deficient in one or more transport systems has contributed its own complications in the study of transport systems.

2 INDIVIDUAL AMINO ACID TRANSPORT SYSTEMS

Amino acid transport in fungi appears to represent a compromise between bacterial transport systems, with specificities for individual amino acids, and animal cell systems, with broad specificities for classes of structurally related amino acids. Thus, *Penicillium chrysogenum* reportedly possesses up to ten amino acid transport systems (Hunter and Segel, 1971) (Table 1), whereas *Achlya* has nine (Singh and Le John, 1975) (Table 2), *Neurospora* approximately five (Table 3), and *Aspergillus nidulans* four or five (Table 4).

1 Transport in *Penicillium chrysogenum*

Elucidation of the individual amino acid transport systems in *Penicillium* has been primarily undertaken in Dr Irwin H. Segel's laboratory at the

University of California, Davis, California. As indicated in Table 1, these systems include specific systems for L-methionine (Benko *et al.*, 1967) and L-cystine (Skye and Segel, 1970) that appear under sulphur starvation conditions, and a constitutive system for L-cysteine (Skye and Segel, 1970). A transport system for reduced glutathione also appears under sulphur starvation conditions. Starvation for either carbon or nitrogen appears to derepress a transport system for neutral and basic amino acids (a general–non-specific amino acid transport system) (Benko *et al.*, 1969), for acidic amino acids (Hunter and Segel, 1971), and for proline (Hunter and Segel, 1971). In addition to these regulated transport systems, kinetic analysis of that transport activity remaining in the presence of competing amino acids revealed the presence of constitutive transport systems for the basic amino acids, L-arginine, and L-lysine (Hunter and Segel, 1971).

All nine amino acid transport systems in *Penicillium* are active transport systems, which are subject to inhibition by metabolic inhibitors, and accumulate the substrate molecules against their respective chemical gradients. Transport of amino acids bearing a net charge is inhibited at ionic strengths greater than 0.1 M and transport assays carried out at various pH values revealed that the neutral and basic amino acid transport system transported the zwitterionic (no net charge) species of the acidic amino acids (Hunter and Segel, 1971). The acidic amino acid transport system appears

Table 1 Amino acid transport systems[a] in *Penicillium chrysogenum* (Hunter and Segel, 1971)

Transport system	Apparent K_m (M)	V_{max} (nmol mg^{-1} min^{-1})	Comments
Methionine	10^{-5}	1	Appears under sulphur deficiency
Cystine	2×10^{-5}	1.2	Appears under sulphur deficiency
Cysteine	1.4×10^{-4}	2.1	Constitutive
Glutathione (reduced)	1.7×10^{-5}	2.3	Appears under sulphur deficiency
Neutral and basic amino acids	2×10^{-5}	10	Appears under carbon or nitrogen deficiency, cysteine is a substrate of this system, cystine is not
Acidic amino acids	10^{-4}	10	Appears under carbon or nitrogen deficiency, does not transport D-stereoisomers
Basic amino acids	5×10^{-6}	1	Constitutive
Arginine	—	—	Constitutive
Lysine	—	—	Constitutive
Proline	—	—	Appears under carbon or nitrogen deficiency

[a]Kinetic constants given were obtained using 'only fine hairy microfilamentous mycelium for transport' (Hunter and Segel, 1971).

to transport both the zwitterionic and the mono-anionic species of dicarboxylic amino acids. No systems specific for neutral, and aromatic amino acids have been reported.

Most of the amino acid transport systems in *Penicillium* are subject to a feedback inhibition phenomenon, termed transinhibition by Ring *et al.* (1970). However, the effector molecules involved in regulation of individual systems are variable. For example, aspartate and glutamate are more potent feedback inhibitors of the non-specific neutral and basic transport system than they are of the acidic transport system. It was suggested by Hunter and Segel (1970) that this effect might arise from the rapid metabolism of aspartate and glutamate, providing for the synthesis of many other amino acids that might actually effect the repression (or 'transinhibition') of the non-specific system. The strong feedback inhibition of the acidic transport system by L-α-amino adipate, suggested that this system could be regulated by the unchanged, intracellular substrates of this system (Hunter and Segel, 1971).

2 Transport in *Achyla*

As with *Penicillium*, the bulk of the work on amino acid transport in the water mould *Achlya* has been performed in a single laboratory. Dr Herb B. LeJohn and coworkers have presented evidence that as many as nine kinetically separable amino acid transport systems are present in *Achlya* (Singh and LeJohn, 1975) (Table 2).

Three amino acids—lysine, methionine, and threonine—exhibit complex saturation curves suggesting the presence of more than one saturable component for each solute. By analogy with other fungal systems it might be assumed that methionine could be transported by the specific methionine transport system and by the neutral–aromatic amino acid transport system (system designated (ix) in Table 2); this possibility might explain the presence of two saturable components for methionine transport. Based on the specificities of the nine proposed transport systems an analogous proposal is not, however, applicable to lysine.

Studies on inhibition of amino acid transport by other amino acids have sometimes been difficult to evaluate. For example, methionine inhibits in a non-competitive manner transport of every other amino acid, yet none of these same amino acids inhibit the transport of methionine. Since methionine did not alter the 'affinities' (K_m values) of the various transport systems for their respective amino acids it must be assumed that the inhibition of amino acid transport by methionine is via an indirect mechanism (but see also Section 1.6.2.1).

Kinetic analyses of amino acid inhibition of transport revealed the following: (a) tyrosine is a competitive inhibitor of phenylalanine and valine

Table 2 Amino acid transport systems[a] in *Achlya* (Singh and LeJohn, 1975)

Amino acids transported	Apparent K_m (M)	Comments	System notation[b]
Methionine	5.33×10^{-6} and 2.0×10^{-4}	Two saturable components	(i)
Cysteine	7.5×10^{-5}		(ii)
Proline	1.5×10^{-4}		(iii)
Serine, Threonine	1.25×10^{-4} 2.5×10^{-5} and 2.0×10^{-4}	Two saturable components for threonine	(iv)
Aspartate, Glutamate	4.0×10^{-6} 2.5×10^{-5}		(v)
Glutamine, Asparagine	1.5×10^{-4} 7.5×10^{-5}		(vi)
Glycine, Alanine	1.5×10^{-4} 1.67×10^{-4}		(vii)
Histidine, Lysine, Arginine	1.67×10^{-4} 8.33×10^{-6} and 1.0×10^{-4} 3.33×10^{-5}	Two saturable components for lysine	(viii)
Phenylalanine, Tyrosine, Tryptophan, Leucine, Isoleucine, Valine	5.0×10^{-5} 3.33×10^{-5} 1.67×10^{-4} 1.11×10^{-4} 6.67×10^{-5} 8.3×10^{-5}	System (ix) transports these amino acids as two overlapping groups	(ix)

[a] Ca^{2+} bound to a membrane-localized glycopeptide permits concentrative transport to take place (see text for discussion).
[b] A definitive system notation has not been proposed. The symbols employed here correspond only to those used to delineate one system from another (Singh and LeJohn, 1975).

transport, but a non-competitive inhibitor of tryptophan and glutamate transport; (b) phenylalanine is a competitive inhibitor of tryptophan and valine transport; (c) valine, isoleucine, and leucine are competitive inhibitors of phenylalanine, tryptophan, and valine transport; and (d) lysine, arginine, and histidine are competitive inhibitors of each other's transport (Singh and LeJohn, 1975). The interrelationships between the individual amino acid transport systems and the amino acids transported have still not been clearly resolved. The possibility that such competition involves the use of a common energy source for transport would, at present, seem a reasonable hypothesis.

Transport assays in *Achlya* were undertaken in germinating zoospores. The optimal pH for transport of twenty protein amino acids varied between pH 6.5 and 7.5. Acidic amino acids were not well transported at the more acidic pH values, unlike acidic amino acid transport in the other fungi (Wolfinbarger *et al.*, 1971) and yeast (Kotyk and Janacek, 1975), and indeed, aspartic acid was optimally transported at pH 7.5 whereas glutamate was optimally transported at pH 6.5—one full pH unit difference. In *Neurospora*, as will be seen later, the pH for optimal acidic amino acid transport closely parallels the p*K* value of the side-chain carboxyl group with the result that the optimal transport of glutamate is at a higher pH than that for optimal aspartate transport (Wolfinbarger and DeBusk, 1972). No information is available for amino acid transport in *Achlya* under conditions of carbon and/or nitrogen starvation. However, the interesting involvement of calcium ions in regulation of amino acid transport has led to the finding of a calcium-binding glycopeptide associated with the cell wall and/or membrane (Cameron and LeJohn, 1972). The glycopeptide can be removed from germinating sporangiospores by cold osmotic shock treatment of the cells. The molecular weight of this molecule is estimated at 6000 ± 150 daltons and it consists of 65 per cent carbohydrate and 35 per cent protein (55 per cent of the amino acid residues are glutamic acid). No detectable phospholipid was associated with the glycopeptide (LeJohn *et al.*, 1974).

The presence of calcium ions (Ca^{2+}) appears to be essential for amino acid transport in *Achlya*. Calcium ion uptake is continuous throughout the growth phase of the organism and any termination of uptake results in 'immediate' cessation of growth (Singh and LeJohn, 1975). Addition of chelators, such as ethylenediaminetetraacetic acid (EDTA), ethylene glycol-*bis*-(β-aminoethyl ether)-*N*,*N*′-tetraacetic acid (EGTA), and di- and tricarboxylic acids (which are not transportable in *Achlya*) resulted in inhibition of both amino acid transport and protein synthesis (Singh and LeJohn, 1975). It has been suggested (Singh and LeJohn, 1975) that calcium may play a secondary role in amino acid transport; the membrane-localized glycopeptide acting as a 'Ca^{2+} sink' (binding up to 20 atoms of Ca^{2+} per mole of glycopeptide (Le John and Cameron, 1973)) and permitting concentrative transport of amino acids to occur. The manner in which this regulation of amino acid transport by Ca^{2+} and the membrane-localized glycopeptide may occur is at present unknown. It has, however, been reported that cytokinins regulate calcium binding to the glycopeptide (Singh and LeJohn, 1975; LeJohn and Stevenson, 1973) and thus this phenomenon may eventually be found to represent a major controlling element in fungal cell development. Maintaining the structural and functional integrity of the cell membrane, as proposed by Cameron and LeJohn (1972), may well prove to be the principal aspect of regulation by the controlling element.

3 Transport in *Neurospora crassa*

Unlike *Penicillium* and *Achlya*, the characterization of amino acid transport in *Neurospora* has been undertaken in a large number of laboratories and in various stages of development of this fungus. It is generally agreed that

Neurospora possesses only five genetically and biochemically distinct amino acid transport systems; one for neutral and aromatic amino acids, one for basic amino acids, a general system which transports neutral, basic, and acidic amino acids, an acidic amino acid transporting system, and one specific for methionine (Table 3).

Growth antagonisms were early noted in amino acid-requiring mutants of *Neurospora.* The growth of lysine auxotrophs on lysine is inhibited by excess arginine, but not by neutral and aromatic amino acids (Doermann, 1944), arginine auxotrophs are similarly inhibited by excess lysine (Srb, 1953). A mutant auxotrophic for isoleucine and valine fails to grow on these supplements in the presence of excess concentrations of phenylalanine, norleucine, or norvaline (Bonner *et al.*, 1943). Until it was finally discovered that histidine is transported in *Neurospora* by three separate transport systems (the neutral, general, and basic amino acid transport systems), it was not understood why a combination of a neutral amino acid, i.e. phenylalanine, leucine, etc., and a basic one, i.e. arginine, lysine, etc., was required to block growth of a histidine auxotroph on histidine (Haas *et al.*, 1952; Wolfinbarger and Marzluf, 1974). Similarly, mutants auxotrophic for either phenylalanine or tyrosine would not grow on these supplement amino acids in the presence of excess neutral and aromatic amino acids such as leucine and tryptophan (Brockman *et al.*, 1959). These early growth assays were sufficient to initially characterize three of the five known transport systems. The existence of a methionine-specific transport system was also suggested when it was found that growth of methionine auxotrophs on methionine could be reduced, but not completely blocked, by the inclusion of competing amino acids such as phenylalanine.

Early transport studies, using radioactively labelled amino acids (DeBusk and DeBusk, 1965; Stadler, 1966; Kinsey and Stadler, 1969; Lester, 1966) demonstrated a transport system specific for neutral and aromatic amino acids. Only the L-stereoisomers with free terminal α-amino and α-carboxyl groups were substrates for transport, and interactions between members of this structurally related family of amino acids were shown to be competitive for the transport process (Wiley and Matchett, 1966, 1968). Mutant strains deficient in the neutral amino acid transport system have been isolated by using as selection procedure, resistance to the growth inhibitory properties of 4-methyltryptophan (*mtr*) (Lester, 1966), resistance to *p*-fluorophenylalanine (*pm-n*) (Wolfinbarger and DeBusk, 1971*b*), and inability to grow on histidine in the presence of excess arginine (*neu*^{a}) (Magill *et al.*, 1972). In each of these studies, residual transport activity for neutral amino acids was reported, and this observation together with the ability of basic amino acids to inhibit this transport suggested the presence of a second general amino acid transport system.

The carrier system for basic amino acids was shown by Baurle and Garner

Table 3 Amino acid transport systems in *Neurospora crassa*

Amino acids transported	Apparent K_m (M)	V_{max} (nmol min^{-1} mg^{-1})	Comments	System notation
Phenylalanine	$2–4 \times 10^{-5}$	—	System also transports val, ala, gly, ser, met; asp and glu transported optimally at acidic pH values	Neutral amino acid transport system
Tyrosine	—	—		
Tryptophan	5×10^{-5}	—		
Leucine	1×10^{-4}	—		
Histidine	6.5×10^{-4}	12.4		
Aspartate	8×10^{-5}	1.5		
Glutamate	—	—		
Tryptophan	4×10^{-5}	—	Aspartate and glutamate optimally transported at acidic pH values	General amino acid transport system
Methionine	3.0×10^{6}	—		
Phenylalanine	4×10^{-5} (C);[a] 2×10^{-6} (M)[b]	0.85		
Tyrosine	—	—		
Leucine	4.9×10^{-6} (M)	—		
Asparagine	2×10^{-5}	—		
Aspartate	3.4×10^{-3} (C); 1×10^{-5} (M)	20 (C); 25 (M)		
Glutamate	4×10^{-5} (M)	15 (M)		
Citrulline	—	—		
Arginine	3.2×10^{-6}	—		
Lysine	—	—		
Glycine	8×10^{-6}	—		
Histidine	1.2×10^{-3}	3.6		
Arginine	2×10^{-6}	—		Basic amino acid transport system
Lysine	5×10^{-6}	—		
Histidine	3.5×10^{-3}, 1.6×10^{-6}	30.5		
Canavanine	7×10^{-6} (K_i)	—		
Cysteic acid	7×10^{-6}	—	Active under nitrogen-starvation conditions	Acidic amino acid transport system
Aspartate	1.3×10^{-5}	0.2		
Glutamate	1.6×10^{-5}	—		
Methionine	2.3×10^{-6}	—	Active under sulphur starvation conditions	Methionine transport system
Proline	?	?	?	Proline transport system

(1964) to be specific for transport of arginine, lysine, and the arginine analogue canavanine sulphate. Mutants with reduced transport for these basic amino acids were reported by Woodward *et al.* (1967), who also showed them to be reduced in transport ability for histidine; the mutant locus was located on the right arm of linkage group V. Roess and DeBusk (1968) examined basic amino acid transport in greater detail and reported another mutant, *Cr-10* which was deficient in transport of basic amino acids by the basic amino acid transport system. Wolfinbarger and DeBusk (1972) further evaluated the *Cr-10* mutant, renaming it *pm-b*. The *pm-b* locus has been tentatively mapped in linkage group V and shown to be deficient in the basic amino acid transport system. Thwaites and Pendyala (1969) isolated a further mutant, *bat*, as deficient in basic amino acid transport. The *bat* locus has not yet been mapped but has been shown to be missing the basic amino acid transport system (Thwaites and Pendyala, 1969; Pall, 1970*a*). Both *bat* and *pm-b* have been assumed to be allelic, although there is no genetic data to support this assumption. Each of these studies also reported the presence of residual transport activity for basic amino acids in the mutants deficient in the basic amino acid transport system. The observation that neutral amino acids compete with basic amino acid transport by this system again supported the presence of a general amino acid transport system.

Although the general amino acid transport system was initially characterized by examination of that portion of phenylalanine transport subject to competitive inhibition by arginine (Pall, 1969), a clear understanding of its developmental, kinetic, and specificity properties required a mutant strain deficient in both the neutral and basic amino acid transport systems. Recently, a mutant strain, *pm-g*, was reported as deficient in transport activity by the general system (Rao *et al.*, 1975). The addition of this new mutation to combinations of other mutations, i.e. *pm-n* and *pm-b*, has permitted the selection of a triple mutant deficient in the neutral, the basic, and the general amino acid transport systems, i.e. *pm-n*; *pm-b*; *pm-g* (Rao *et al.*, 1975).

The acidic amino acid transport system in *Neurospora* was reported by Pall (1970*b*) to be present in mycelia starved for nitrogen, carbon, or sulphur. It transports cysteic acid, L- and D-aspartate, and glutamate. Typical K_m values for transport of acidic amino acids by this system centre around 10^{-5} M, with maximal transport velocities of 0.22 nmol min^{-1} (mg $cells)^{-1}$. Until it was found that at low pH values acidic amino acids are also transported by the neutral and general amino acid transport systems, it was thought that the acidic transport system was the sole route for transport of dicarboxylic amino acids. Thus, by lowering the pH of the surrounding medium from pH 5.8 to pH 3.8 (p*K* of side-chain carboxyl groups about 3.5 and α-carboxyl groups about 2.1) it becomes possible to transport aspartate and glutamate by the neutral and general amino acid transport systems in both conidia and mycelia (Wolfinbarger *et al.*, 1971; Wolfinbarger and Kay, 1973). In conidia, it was suggested that these systems transported only the neutral (zwitterionic) species of aspartate and glutamate, while in germinated conidia these same systems appear to transport both the neutral and anionic species of both the D- and L-stereoisomers (Wolfinbarger and Kay, 1973). In addition, after germination the general transport system comprises roughly 85–90 per cent of the total acidic amino acid transport activity. In conidia it constitutes approximately 40 per cent of the total acidic amino acid transport (Wolfinbarger *et al.*, 1971). Aspartate transport in conidia via the neutral and general systems has K_m values of 8×10^{-5} M and 3.4×10^{-3} M respectively, and V_{max} values of 1.5 and 20 nmol min^{-1} (mg cells)$^{-1}$. In mycelia, the values for the general system change to a K_m of 1×10^{-5} M and a V_{max} of 25.0 nmol min^{-1} (mg cells)$^{-1}$. Because the acidic amino acid transport system would constitute less than 1 per cent of the total available transport activity its assay was not attempted (Wolfinbarger *et al.*, 1971; Wolfinbarger and Kay, 1973).

Deprivation of *Neurospora* for a readily usable source of sulphur results in the derepression of numerous enzymes involved in sulphur utilization and of a methionine-specific transport system (Pall, 1971). This system exhibits a K_m of 2.3×10^{-5} M for methionine and a lower 'affinity' for *S*-ethylcysteine, ethionine, *S*-methylcysteine, and norleucine. Synthesis of a protein necessary for the activity of the methionine transport system is regulated by the *cys-3* locus (Pall, 1971); synthesis is induced in wild-type cells by sulphur starvation. Methionine is also readily transported by those systems which also transport neutral and aromatic amino acids, as evidenced by the ability of excess phenylalanine to block growth of *cys-3*, but not methionine auxotrophs, on extracellularly supplied methionine (Wolfinbarger and Marzluf, 1974).

4 Transport in *Aspergillus nidulans*

The amino acid transport systems in *Aspergillus* would appear to resemble closely those systems reported in *Neurospora.* Table 4 lists a system for transporting neutral amino acids and methionine (and presumably acidic amino acids at lower pH values) (Piotrowska *et al.*, 1976), a system for basic amino acids, an acidic amino acid transporting system that is regulated by the available nitrogen source, and a methionine-specific system that is derepressible by starvation for sulphur. A general amino acid transport system has been suggested to be derepressed under nitrogen starvation conditions (Piotrowska *et al.*, 1976).

Acidic amino acids are transported optimally at pH 5.0 by the acidic transport system, and the anionic species is presumably the preferred substrate (Robinson *et al.*, 1973*a*). Characterization of this system was

Table 4 Amino acid transport systems[a] in *Aspergillus nidulans*

Amino acids transported	Apparent K_m (M)	V_{max} (nmol min^{-1} mg^{-1})	Comments	Reference
Aspartate Glutamate Cysteate	1.0×10^{-5} 1.8×10^{-5} 1.9×10^{-5}	1.72 4.35 0.03	Regulated by ammonium ion, missing in mutant *aau*	Robinson *et al.* (1973*a*, *b*), Pateman *et al.* (1974)
Lysine Arginine Ornithine	8×10^{-6} 1×10^{-5} 3×10^{-3} (K_i)	— 2.8 —	His is not an effective inhibitor of Arg transport	Piotrowska *et al.* (1976)
Phenylalanine Methionine Serine Leucine	— 1×10^{-4} — —	— 1 — —	Missing in *nap-3*	Sinha (1969), Piotrowska *et al.* (1976)
Methionine	—	—	Derepressed under sulphur starvation	Piotrowska *et al.* (1976)
General amino acids	—	—	Derepressed under nitrogen starvation conditions	Piotrowska *et al.* (1976)

[a]Transport systems characterized in germinated conidia and mycelia.

typically carried out using transport assays run at pH 6.5 to preclude transport of the zwitterionic species by the neutral and general amino acid transport systems (Robinson *et al.*, 1973*b*). Germination of conidia in a glutamate medium (25 mM) increased the specific activity of the acidic transport system over 100-fold, maximum activity being observed when germ tubes were just emerging from the conidia (8–10 h) and declining thereafter to 20–30 per cent of the maximum level attained (Robinson *et al.*, 1973*b*). The system is regulated by ammonium repression. A mutant, *aauA*, impaired in acidic amino acid transport by the acidic amino acid transport system, has been isolated (Pateman *et al.*, 1974).

The basic amino acid transport system appears to be constitutive, to transport arginine optimally at pH 5.0, and to have a relatively large K_m value for histidine, i.e. a 100-fold excess of histidine failed to inhibit arginine transport by this sytem (Piotrowska *et al.*, 1976). The frequent observation that arginine can block transport of other amino acids would seem to imply that basic amino acids may also be transported by a second system (a 'general permease') in *Aspergillus*, although no hard data have been presented to date. By analogy with other fungal systems such reasoning would not seem unreasonable.

A phenylalanine transport system, first reported by Sinha (1969), has not been well characterized with regard to substrate specificities and kinetics of transport. Inhibition studies, and the isolation of a mutant deficient in this system, *nap3* (Piotrowska *et al.*, 1976), have suggested a transport activity shared by phenylalanine, methionine, serine, and leucine; transport of arginine and proline was not affected by this mutation. Using *nap3* it became possible to characterize other transport systems more clearly, and it was discovered that, following nitrogen starvation, *nap3* regained the capability of transporting methionine and phenylalanine. This derepressed transport activity was, however, 90 per cent inhibited by a 10-fold excess concentration of arginine (Piotrowska *et al*., 1976). That a methionine-sensitive increase in arginine transport also occurred under nitrogen starvation suggests that this new transport activity might be a 'general permease'.

Starvation of *nap3* for sulphur elicited a capability for transport of methionine without an increase in phenylalanine transport (Piotrowska *et al*., 1976). This new transport activity is presumably the methionine-specific transport system that is not present under normal conditions of growth on sulphur-sufficient medium. The understanding that accrued from isolation of *nap3* serves to underscore the importance of genetic techniques in the characterization of biochemically distinct transport systems.

3 REGULATION OF TRANSPORT

1 Effects of Ammonium Ion

One underlying theme in regulation of amino acid transport in fungi appears to be the ability of ammonium ion to 'repress' the activity on formation of individual amino acid transport systems (see also Section 3.2.3). Chief among these NH_4^+-regulated systems are the general and acidic amino acid transport systems. On the other hand, the basic and neutral amino acid transport systems, as a generality, appear to be constitutive in those stages of fungal development where they are present. The models proposed for ammonium ion regulation of amino acid transport systems collectively involve NAD- and NADP-dependent glutamate dehydrogenases. A number of mutants altered in glutamate dehydrogenase (GDH) activity have been reported, each providing different responses to ammonium ion repression of transport. Certain mutants of *Aspergillus*, such as *meaA8*, *meaB6* (*mea* signifying methylammonium resistant) (Arst and Cove, 1969), DER3, *amrA1* (ammonium transport), and *xprD1* (derepressed for extracellular protease (Cohen, 1972)) are derepressed with respect to external ammonium, whereas *gdhA* mutants, lacking normal NADP-dependent GDH activity, are derepressed with respect to both external and internal ammonium (Pateman *et al*., 1973). Similar mutants have been reported in *Neurospora* (Chang and Sorger, 1976) and in yeast (Grenson and Hou, 1972). The *gdhA* mutants of *Aspergillus*, unlike the *mea*, *amr* and *xpr* strains, are deficient in NADP–GDH activity, are repressed for extracellular protease release, and are sensitive to methylamine (an analogue of ammonia) (Pateman *et al.*, 1973). They are unable to utilize inorganic nitrogen efficiently for synthesis of glutamic acid and require supplementation for normal growth. Pateman and his associates have proposed a model for the relationship between NADP–GDH and ammonium regulation, in which it is suggested that the NADP–GDH enzyme, residing in the plasma membrane or within the cytoplasm, may, on binding NH_4^+, act as a regulatory protein mediating control of other ammonia regulated enzymes (Pateman *et al.*, 1973) (Figure 1). Models for inhibition of amino acid transport by ammonium ions in yeast suggest that following uptake of ammonia, this metabolite interacts with NADP–GDH which transmits the signal to an NH_4^+-regulated gene by participation of a GDH–CR molecule (CR is catabolite repression) issued from the *gdhCR* gene (Dubois *et al.*, 1977). The recent observation that extracellularly supplied NH_4^+ transiently depolarizes the membrane potential of *Neurospora* (Slayman and Gradman, 1975) lends support to the notion that ammonia regulation of nitrogen metabolizing enzymes may be quite complex, being variable with individual systems, and deserving of considerable effort in future research.

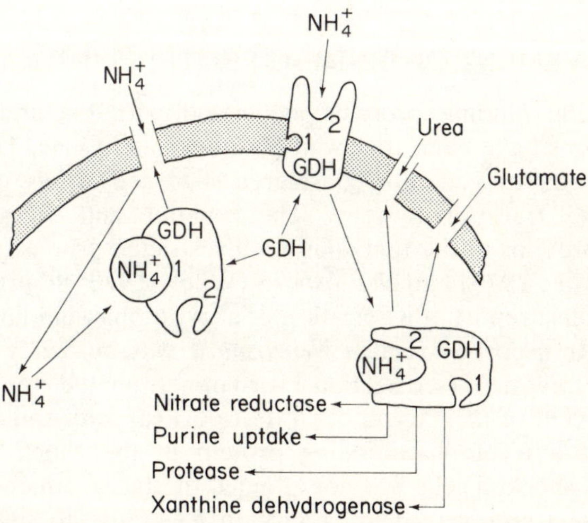

Figure 1 Hypothesis for ammonium regulation of various uptake and enzyme systems. One possible mechanism which would allow the formation of two different GDH regulatory complexes is the existence of two different ammonium binding sites on the molecule; this possibility is illustrated. Further explanation is given in the text. (1) Indicates interaction site with intracellular ammonium; (2) indicates interaction site with extracellular ammonium. (Reproduced by courtesy of the authors from Pateman *et al.* (1973) and with permission of the American Society for Microbiology)

2 Effects of Pool Amino Acids

A second, not too well understood, regulatory mechanism for amino acid transport involves the ability of intracellular pools of amino acids to transinhibit individual transport systems (see also Sections 1.1.4.3 and 1.6.2). In *Neurospora*, the regulation of the methionine transport system by intracellular methionine prompted Pall (1971) to propose that methionine exerted its regulatory effect by binding with a presumed membrane-localized carrier molecule, restricting it to the interior surface of the plasma membrane, i.e. preventing recycling of the membrane-localized carrier to the exterior surface of the plasma membrane. Other workers (Tisdale and DeBusk, 1970; Nell and DeBusk, 1974) reported that intracellular pools of amino acids prevented development of 'post-conidial' transport activity, and Wolfinbarger and DeBusk (1971*a*) described a regulation of general amino acid transport activity that was effected by transported phenylalanine and arginine. Transport of arginine via the general system caused an increase in transport rate with time whereas transport of phenylalanine via this same system reduced the transport rate with time (Wolfinbarger and DeBusk, 1971*a*). No explanation for this mode of regulation was offered, but it was presumed to function as a separate regulatory mechanism to that of ammonium ion for the general amino acid transport system.

4 INVOLVEMENT OF BINDING PROTEINS IN TRANSPORT

Substrate-specific binding proteins associated with bacterial amino acid transport systems have been reviewed (Oxender and Quay, 1975) (see also Section 1.4.2) and are generally considered to be cell surface components of certain complex transport systems. The isolation and characterization of such binding proteins in the fungi has been restricted primarily to the yeasts (Opekarova *et al.*, 1975) and *Neurospora* (Wiley, 1970; Stuart and DeBusk, 1971).The earliest report of association of a tryptophan-binding protein with a tryptophan transport system in *Neurospora* was by Wiley (1970). Cold osmotic shock treatment (see Section 1.4.1) of germinated conidia caused the loss of 90 per cent of their tryptophan transport capacity and also led to the appearance of a tryptophan-binding protein in the shock medium. The viability of the shocked cells was not affected by this treatment. The binding constant of the tryptophan-binding protein was close to the reported K_m (5×10^{-5} M) for tryptophan transport by the neutral amino acid transport system. Using Sephadex G-100 chromatography the molecular mass of the binding protein was estimated at 200 000 daltons, a value that is unusually high compared with binding proteins from bacterial sources (see Section 1.4.2). Further evidence that this binding protein is a component of the neutral amino acid transport system was the finding that in an *mtr* mutant (deficient in this transport system) the amount of binding protein released by osmotic shock was considerably decreased.

Employing a salt extraction procedure, Stuart and DeBusk (1971) were able to isolate three fractions with the capability of binding amino acids. Following extraction of conidia with saturated solutions of KCl, passage of the extract over an affinity column charged with arginine (as a ligand) resulted in one peak of material with little to no affinity for the ligand (fraction A), and two peaks of materials retained for different periods on the column before eluting (fractions B and C). Similar extraction of the three transport deficient mutants, *pm-b*, *pm-n*, and *pm-n*; *pm-b* (Wolfinbarger and DeBusk, 1971*b*) resulted in extracts that were missing (a) that material eluting in fractions designated B and C (*pm-b*), (b) that material eluting in fraction C, (*pm-n*), and (c) those fractions designated B and C (*pm-n*; *pm-b*). Transport of arginine by the extracted cells was reduced by approximately 30 per cent without loss in viability. One interesting byproduct of this study was the observation that KCl extraction of the double transport deficient mutant *pm-n*; *pm-b* resulted in severe disruption of the conidia whereas the wild-type and single transport deficient mutants were not disrupted (Travis *et al.*, 1972). The proteins released by KCl extraction comprised roughly 0.2–0.3 per cent of the total cell dry weight, and part of the extracted material was separated and purified by affinity chromatography to give a single glycoprotein band in disc gel electrophoresis (Stuart and DeBusk, 1971). The peak B material was proposed to be the product of a genetic locus controlling basic amino acid transport. Binding of arginine by glycoprotein B (peak B) *in vitro* was enhanced in the presence of phenylalanine while binding of phenylalanine was reduced in the presence of arginine. Since glycoprotein C (peak C material) did not appear in chromatograms of extracts obtained from a transport deficient mutant it was suggested that glycoprotein C was a complex formed by the products of the *pm-n* and *pm-b* loci (or a complex assembled by enzymes coded by these loci) (Stuart and DeBusk, 1971). It has been reported (Stuart and DeBusk, 1975*b*) that glycoprotein B contains 113 amino acid residues per minimum molecular weight, and eight types of neutral sugar residues, although no amino sugars were detected. Furthermore, tryptic 'fingerprints' of the glycoprotein extracts indicated that mutations that reduce amino acid transport result in amino acid substitutions in the extractable glycoproteins. Addition of the glycoprotein extracts to a lipid film, in an *in vitro* transport system, enhanced permeation of arginine at an initial rate 1000 times the rate of permeation through lipid alone; the initial rate of passage exceeded the rate of unhindered passage (no lipid film) through the same cross-sectional area by 10-fold (Stuart and DeBusk, 1973) (Figure 2).

An interesting extension of the studies on these amino acid binding

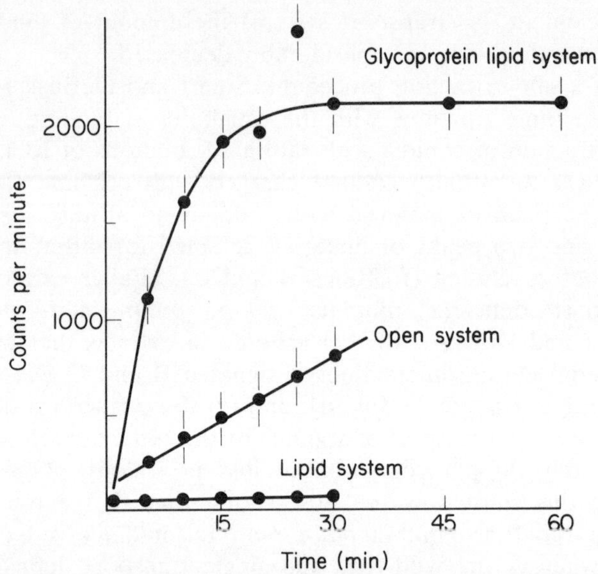

Figure 2 The appearance of arginine in the negative chamber with time. The top curve represents appearance of arginine after passage through the crude lipid-glycoprotein film, the middle curve represents unhindered passage of substrate while the bottom line represents restricted passage across a lipid barrier. The arginine concentration is measured experimentally as counts per minute per 0.05 ml of the contents of the negative chamber. (1000 counts $min^{-1} = 3.3 \times 10^{-6}$ M arginine or a total of 9.9×10^{-9} mol arginine in the negative chamber.) All experiments were performed at 25 °C. (Reproduced by courtesy of the authors from Stuart and DeBusk (1973) and with permission of Academic Press Inc)

glycoproteins was the finding that they could be radioactively labelled using either [^{14}C]-phenylalanine or [^{14}C]-uridine (Stuart and DeBusk, 1975*a*). Furthermore, the KCl-extractable glycoproteins became dialysable after digestion with ribonuclease, and similar enzymic treatment of intact cells reduced the intial rate of phenylalanine transport by conidia. The site of this enzymic reaction was shown to be confined to the cell surface and the reduction in transport of phenylalanine was reportedly due to decreased activity of the general amino acid transport system (Stuart and DeBusk, 1975*a*). Activity of the neutral amino acid transport system could also be reduced but this required a higher concentration of ribonuclease.

5 ROLE OF MEMBRANE LIPIDS IN TRANSPORT

The interrelationship(s) between the protein (glycoprotein) components of transport systems and the lipids of the membrane has become a fertile, if poorly understood, field for research. Although the bulk of research in this area has been undertaken in bacterial systems (see Cronan and Gelman, 1975) comparable studies in fungal systems are increasing.

One of the earliest studies on the role of the various lipid classes in metabolite transport was undertaken by Scarborough (1971) who showed that starvation of an inositol auxotroph of *Neurospora* (*inos*) for inositol caused a 'profound decline' in transport of glucose by the glucose transport system (Glu II). Inositol supplementation of germinated conidia from a wild-type strain resulted in maintenance of normal levels of Glu II activity. During the 90 min of inositol deprivation of the *inos* strain, its growth rate, ATP levels, and K^+levels were the same as found for inositol-supplemented cells. Addition of inositol to inositol-starved cells resulted in complete restoration of Glu-II transport activity following a short period of incubation.

In addition to inositol- (Williams, 1971), and choline- (Scarborough and Nye, 1967) requiring auxotrophs, mutants requiring unsaturated fatty acids (*ufa*) (Scott, 1977), as well as fatty acid chain elongation (*cel*) mutants (Henry and Keith, 1971) have been isolated and described. The *ufa* mutants recently reported by Scott (1977) could be assigned to two complementation classes, *ufa-1* and *ufa-2*, with the *ufa* locus mapping on linkage group V. The importance of such mutant strains to a study of amino acid transport lies in the potential for altering the fatty acid make-up of the plasma membrane and/or the relative percentage of charged versus non-charged glycerophosphatides. The influence of saturated/unsaturated fatty acids and of membrane surface charge on the gel–fluid transition of bilayer membranes has been extensively studied in bacterial transport systems (see Cronan and Gelman (1975) for a review of this topic). A main conclusion of these studies was that a membrane in its gel state may restrict to a greater extent than a membrane in its fluid state the flux of a metabolite across that membrane.

In a recent publication (Wolfinbarger, 1976), the effect of temperature on the transport of acidic amino acids by *Neurospora* was examined. An Arrhenius plot of the transport data revealed a break in the data points such that two energies of activation could be calculated (Figure 3). Between 5 °C and 20 °C the energy of activation for aspartate transport (at pH 3.8) by the wild type was 25.65 kcal mol^{-1} but it was only 8.82 kcal mol^{-1} within the temperature range of 25 °C to 40 °C. In bacterial systems a biphasic line in such an Arrhenius plot has been interpreted as due to a gel–liquid phase transition of the cytoplasmic membrane (Cronan and Gelman, 1975). However, in *Neurospora* there was a strong possibility that two energies of activation might merely be the result of a differential effect of temperature on the neutral and general amino acid transport systems—two systems known to be able to transport aspartate at this pH. Indeed, examination of aspartate transport as a function of temperature in strains deficient in one or the other system, i.e. *pm-n* and *pm-g*, revealed that the neutral amino acid transport system exhibited a single energy of activation (13.5 kcal mol^{-1})

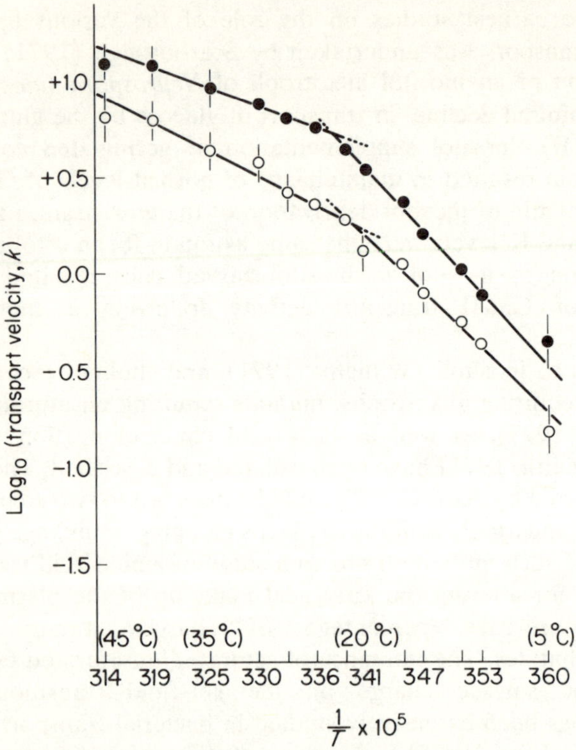

Figure 3 Arrehenius plot of aspartate transport, pH 3.8, by wild-type *Neurospora* (●–●) and *hgu-4* (○–○) as a function of temperature. Transport velocity was given in nmol min^{-1} (mg dry wt cells)$^{-1}$. The temperatures in degrees Celsius, given in parentheses, were included for reference points only. The error bars passing vertically through the experimental points describe the deviation from the mean of two, or more, independent determinations. If a point does not possess an error bar, the circle size simply exceeded its length. The slope of the best-fit straight line was calculated by a least squares analysis using a Texas Instrument SR-51A pocket calculator. The cells were grown for 6 h at 28 °C prior to the assays (see text for details). (Reproduced from Wolfinbarger (1976) with permission of Elsevier/North-Holland Biomedical Press)

between 5 °C and 40 °C whereas the general amino acid transport system (assayed in *pm-n*) exhibited two energies of activation, 16.8 kcal mol^{-1} between 20 °C and 40 °C and 49.46 kcal mol^{-1} between 5 °C and 15 °C. The break (transition) in straight-line plots of the data on an Arrhenius plot occurred at 17 °C when aspartate transport via the general amino acid transport system was measured (Wolfinbarger, 1976) (Figure 4).

One of the reasons for initially undertaking an analysis of transport versus temperature was to try and characterize a particular class of amino acid transport deficient mutants. Thus, several transport deficient mutants have been described in *Neurospora* as being simultaneously reduced in transport

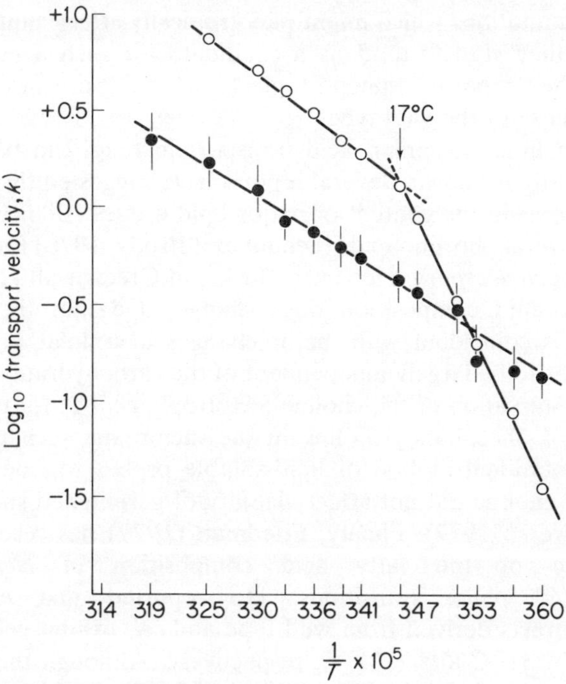

Figure 4 Arrhenius plot of L-aspartate transport, pH 3.8, by *pm-n* (○–○) and *pm-g* (●–●) as a function of temperature. Assay conditions as described in the legend to Figure 3. (Reproduced from Wolfinbarger (1976) with permission of Elsevier/North Holland Biomedical Press)

activity of multiple amino acid transport systems. The mutants *fpr-1* (Kinsey and Stadler, 1969), *un-3* (Kappy and Metzenberg, 1967), *hgu-4* (Wolfinbarger and Marzluf, 1974), and *nap* (Jacobson and Metzenberg, 1968) (not to be confused with nap^3 in *Aspergillus*) have all been reported to be resistant to a variety of structurally unrelated amino acid analogues and to be reduced (usually about 50 per cent) in transport of amino acids via the neutral, general, and basic amino acid transport systems. An obvious question that followed from these results was as to which component(s) might these transport systems have in common and how might a mutational (or physiological) alteration of this component simultaneously affect the individual amino acid transport systems? The lipid membrane matrix was one obvious choice as an answer. Might such a mutant strain be altered in its ability to synthesize, for example, some saturated or unsaturated fatty acid or other membrane associated lipid, and as a result suffer a general alteration in the transport kinetics of those proteins (or glycoproteins) associated with the membrane? The mutant loci *nap*, *fpr-1*, and *hgu-4* all mapped to linkage group V (their relation to the *ufa* loci also on linkage group V is not known). Kappy and Metzenberg (1967) early postulated the possibility of mutations which might pleiotropically affect multiple transport systems and they studied *un-3* as a candidate for such a mutant. In this connection, they reported that protoplasts of *un-3* were more osmotically fragile than those of the wild type.

The role of lipids in amino acid transport in fungi can be summarized fairly easily: little is known. Several reports have suggested that there are no major differences in the content of major lipid classes (Bianchi and Turian, 1967) or of several phospholipids (Schmit and Brody, 1976) found in conidia and mycelia of *Neurospora.* However, Beck and Greenawalt (1977) recently reported that lipid composition does change, and that the synthesis of phospholipid is coincident with major changes in cellular proteins. These changes appear to be largely independent of the carbohydrate supplement of the medium. Starvation of the choline auxotroph, *chol-1*, for choline results in depletion of phosphatidylcholine in the membrane such that increased freedom of rotational motion of lipid-soluble probes was detectable. This starvation for choline did not affect viability of germinated spores or whole mycelium (Juretic, 1977). Finally, Friedman (1977) has recently explored the influence of the fatty acid composition of *Neurospora* on membrane–lipid phase transitions. He reported that ergosterol-free membrane extracts derived from wild type and *cel*, exhibit gel–liquid phase transitions at −31 °C and −11 °C, respectively. Although these values are markedly different from the temperature (17 °C) at which the transition occurs in an Arrhenius plot of aspartate transport via the general amino acid transport system (Wolfinbarger, 1976) an observed endothermic peak at 30 °C was noted in intact cells of both wild type and *cel*. Much work remains to be done in assessing the interrelationships between metabolite transport and membrane composition in fungi.

6 ENERGETICS OF TRANSPORT

In most review articles a discussion of the energetics of transport is reserved for the later part perhaps because it is the most complex aspect to evaluate. The majority of amino acid transport systems described in fungi have been shown, or presumed, to be, 'active', in that the substrate molecule is accumulated inside the cell chemically unchanged at a greater concentration than it existed outside the cell. Mechanistically it is generally assumed that translocation of a substrate amino acid across the plasma membrane involves a variety of molecular components (usually proteins or glycoproteins) which may or may not be directly involved in binding with the transported molecule. For example, one might consider the possibility that binding proteins, rather than conceivably being involved in the actual translocation process, may serve to increase the effective concentration of substrate molecules within a restricted binding sphere of a mobile membrane carrier molecule which has a rather low 'affinity' for that substrate molecule. Once the substrate was bound the membrane carrier would shift orientation towards the membrane interior where due to its low 'affinity' for the substrate molecule the latter would simply diffuse into the cell interior. Such a mechanism could help explain the seemingly rather loose association of the binding proteins with the membrane, provide a convenient means of altering specificity of a given translocation mechanism simply by changing the associated binding protein, and permit the isolation of a mutant or mutants reduced in transport activity of multiple transport systems. One need only to consider the glycoprotein species described by Stuart and DeBusk (1971) in *Neurospora* in relation to their tendencies to aggregate and to bind with different specificities upon aggregation, i.e. glycoprotein B versus glycoprotein C, to realize that the proposed mechanism is possible.

Alternatively, in some instances it may just as easily be proposed that binding proteins are directly involved in translocation of substrate amino acids across the plasma membrane. Indeed, it may just be that a variety of somewhat different mechanisms may exist for each amino acid transport system. Given these possible circumstances it will be very difficult to effectively evaluate a potential mechanism for coupling metabolic energy to a given transport mechanism until that transport mechanism is better understood.

At present, there are a variety of sources of energy that are being considered for a possible role in amino acid transport. The one mechanism receiving the greatest attention involves a proton (or ion) gradient. The reason for this attention lies in the ability of a variety of cell types to generate an inside-negative environment (low H^+ concentration, hence an anionic character for most weak acids and bases) by pumping protons out of the cell during the enzymic hydrolysis of ATP by membrane-localized ATPases (see also Section 1.7).

The ability of an ATPase-generated membrane potential difference to energize transport implies that the 'uphill' accumulation of an amino acid coupled with the 'downhill' flow of protons involves cotransport of a proton (or protons) with the amino acid. No mention is usually made of whether the amino acid or the membrane carrier directly binds the proton(s) (but see Section 1.1.4.3). It is for this reason that transport of the acidic amino acids via the neutral and general amino acid transport systems in *Neurospora* has been of interest. The available evidence strongly suggests that in conidia of *Neurospora* aspartate and glutamate are transported via these two systems only so long as the side-chain carboxyl groups ($pK \approx 3.6$) of the amino acids are protonated, i.e. the zwitterions carry no net charge. Protonation of these molecules can be easily accomplished simply by lowering the pH of the medium. This act may have a dual function in that it not only protonates the acidic amino acid substrates but it may also increase the potential difference or hydrogen ion concentration gradient across the membrane. Furthermore, the negative charge of the membrane surface would be reduced, due to protonation of the cell surface anions, and (all other things being equal) a 'neutral' molecule might more easily approach a site on the membrane-localized carrier molecule and get carried into the hydrophobic environment of the membrane. It is well established that neutral zwitterionic amino acids are more readily partitioned into non-polar solvents from polar solvents than charged amino acids (Spratt *et al.*, 1977). Thus, although it is possible that the system under study is artificial in that the low pH imposes an unnatural flux of acidic amino acids over a transport system that is normally energized differently, there is hope for use of the system in studying proton/amino acid cotransport.

Slayman *et al.*, (1976) have provided extensive evidence that *Neurospora* maintains a membrane potential of approximately 200 mV (inside negative) via the hydrolysis of ATP, and Scarborough (1976) has reported that the plasma membrane ATPase is an electrogenic pump. Indeed, it is of interest that membrane ATPases in some instances exhibit biphasic kinetics when enzyme activity is plotted as a function of the reciprocal of the absolute temperature, i.e. on an Arrhenius plot (Charnock *et al.*, 1971). One wonders whether such behaviour may be related to the effects of temperature on acidic amino acid transport in *Neurospora* discussed earlier.

The studies of transport and membrane potential differences in Slayman's laboratory have been confined primarily to sugar transport. They have, however, provided a convenient working model for substrate transport across the plasma membrane of *Neurospora* and probably other fungi. This model (Figure 5) is supported primarily by the speed and extent of membrane depolarizations caused by certain ions, metabolic inhibitors, and antibiotics (Slayman and Gradman, 1975). Among the common ions, hydrogen ions have a strong depolarizing capacity such that over the pH range 5–3, a 30–40 mV drop in potential occurs for each unit decrease in pH_0 (Slayman, 1965). Cyanide depolarizes about 75 per cent with an exponential time constant of 5.7 s (Slayman *et al.*, 1973), EDTA depolarizes by 60 per cent with a time constant of 3–5 s, and as mentioned earlier 50–100 mM NH_4^+ (perhaps acting as a lipid-soluble ion) depolarizes 75 per cent with a time constant of *c.* 0.55 s. Glucose, at concentrations sufficient to saturate the active glucose transport system (Glu II) depolarizes the plasma membrane by 60 per cent with a time constant of 0.5 s (Slayman and Slayman, 1974). Comparable studies with amino acid transport, primarily tryptophan, in *Neurospora* have been slow to yield usable information. These problems may be attributable to the slower flux rates of amino acids across the plasma membrane (C. L. Slayman, personal communication).

That protons do not form the only ion gradient capable of energizing active amino acid transport systems is suggested by transport studies in

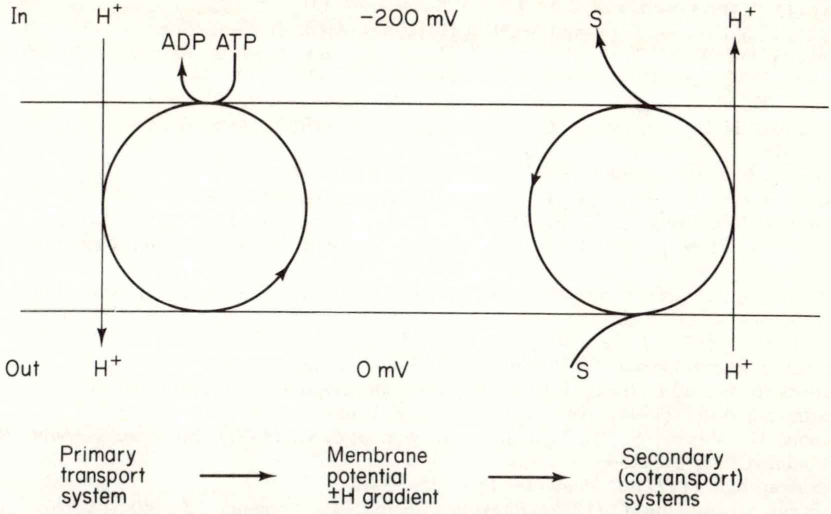

Figure 5 A chemiosmotic model for transport in the plasma membrane of *Neurospora*. (Reproduced by courtesy of the authors from Slayman and Gradman (1975) and with permission of the Biophysical Society, Inc.)

'higher' eukaryotes. For example, work in Christensen's lab, using Ehrlich ascites cells, has strongly implicated sodium and potassium ions in amino acid transport, and the reader is referred to the review (Christensen *et al.*, 1974) for a discussion on the use of transport substrates as probes for intramembrane ion gradients and their role in energizing transport. The reader is also referred to the excellent review article on 'Energetics and control of transport in *Neurospora*' (Slayman, 1977) for a more detailed discussion on energization of transport (see also Section 1.7).

ACKNOWLEDGEMENTS

During the writing of this manuscript the author was supported in part by a grant from the Old Dominion University Research Foundation. The author wishes to thank Drs. C. L. Slayman and A. G. DeBusk for candid discussions of their unpublished data. Apologies are extended to those individuals who have contributed to the knowledge in this field, but who, due to space limitations, were not cited in this reviews. Appreciation is also extended to Nadean Salalila for excellent technical assistance.

REFERENCES

Arst, H. N. and Cove, D. J. (1969). *J. Bacteriol*, **98**, 1284–1293
Baurle, R. H. and Garner H. R. (1964). *Biochim. Biophys. Acta*, **93**, 316–322

Beck, D. P. and Greenawalt, J. W. (1977). *J. Bacteriol*, **131**, 188–193
Benko, P. V., Wood, T. C., and Segel, I. H. (1967). *Arch. Biochem. Biophys.*, **122**, 783–784
Benko, P. V., Wood, T. C., and Segel, I. H. (1969). *Arch. Biochem. Biophys.*, **129**, 498–508
Bianchi, D. E. and Turian, G. (1967). *Nature, Lond.*, **214**, 1344–1345
Bonner, D. M., Tatum, E. L., and Beadle, G. W. (1943). *Arch. Biochem.*, **3**, 71
Brockman, H. E., DeBusk, A. G., and Wagner, R. P. (1959). *Arch. Biochem. Biophys.*, **84**, 455–461
Cameron, L. E. and LeJohn, H. B. (1972). *J. Biol. Chem.*, **247**, 4729–4739
Chang, H. C. P. and Sorger, G. J. (1976). *J. Bacteriol.*, **126**, 1002–1004
Charnock, J. S., Cook, D. A., and Opit, L. J. (1971). *Nature New Biol.*, **233**, 171–172
Christensen, H. N. (1975). *Biological Transport*, 2nd edn, W. A. Benjamin, New York and London
Christensen, H. N., De Cespedes, C., Handlogten, M. E., and Ronquist, G. (1974). *Ann. N.Y. Acad. Sci.*, **227**, 355–379
Cohen, B. L. (1972). *J. Gen. Microbiol.*, **71**, 293–299
Cronan, J. E. and Gelman, E. P. (1975). *Bacteriol. Rev.*, **39**, 232–256
DeBusk, B. G. and DeBusk, A. G. (1965). *Biochim. Biophys. Acta*, **104**, 139–159
Doermann, A. H. (1944). *Arch. Biochem.*, **5**, 373–384
Dubois, E., Vissers, S., Grenson, M., and Wiame, J. M. (1977). *Biochem. Biophys. Res. Commun.*, **75**, 233–239
Friedman, K. J. (1977). *J. Membrane Biol.*, **32**, 33–47
Grenson, M. and Hou, C. (1972). *Biochem. Biophys. Res. Commun.*, **48**, 749–756
Haas, F., Mitchell, M. B., Ames, B. N., and Mitchell, H. K. (1952). *Genetics*, **37**, 217
Henry, S. A. and Keith, A. D. (1971). *J. Bacteriol.*, **106**, 174–179
Hunter, D. R. and Segel, I. H. (1971). *Arch. Biochem. Biophys.*, **144**, 168–183
Jacobson, E. S. and Metzenberg, R. L. (1968). *Biochim. Biophys. Acta*, **156**, 140–147
Juretic, D. (1977). *Biochim. Biophys. Acta*, **469**, 137–150
Kappy, M. S. and Metzenberg, R. L. (1967). *J. Bacteriol.*, **94**, 1629–1637
Kinsey, J. A. and Stadler, D. R. (1969). *J. Bacteriol.*, **97**, 1114–1117
Kotyk, A. and Janacek, K. (1975). *Cell Membrane Transport*, 2nd edn, Plenum Press, New York, p. 353
LeJohn, H. B. and Cameron, L. E. (1973). *Biochem. Biophys. Res. Commun.*, **54**, 1053–1060
LeJohn, H. B. and Stevenson, R. M. (1973). *Biochem. Biophys. Res. Commun.*, **54**, 1061–1066
LeJohn, H. B., Cameron, L. E., Stevenson, R. M., and Menser, R. U. (1974). *J. Biol. Chem.*, **249**, 4016–4022
Lester, G. J. (1966). *J. Bacteriol.*, **91**, 677–684
Magill, C. W., Sweeney, H., and Woodward, V. W. (1972). *J. Bacteriol.*, **110**, 313–320
Matthews, D. M. and Payne, J. W. (eds) (1975). *Peptide Transport in Protein Nutrition*, North Holland, Amsterdam
Nell, J. H. and DeBusk, A. G. (1974). *Biochem. Genet.*, **11**, 183–203
Operkarova, M., Kotyk, A., Horak, J., and Kholodenko, V. P. (1975). *Eur. J. Biochem.*, **59**, 373–376
Oxender, D. and Quay, S. C. (1975). *Ann. N.Y. Acad. Sci.*, **264**, 358–372
Pall, M. L. (1969). *Biochim. Biophys. Acta*, **173**, 113–127
Pall, M. L. (1970*a*). *Biochim. Biophys. Acta*, **203**, 139–149
Pall, M. L. (1970*b*). *Biochim. Biophys. Acta*, **211**, 513–520
Pall, M. L. (1971). *Biochim. Biophys. Acta*, **233**, 201–214
Pateman, J. A., Kinghorn, J. R., and Dunn, E. (1974). *J. Bacteriol.*, **119**, 534–542
Pateman, J. A., Kinghorn, J. R., Dunn, E., and Forbes, E. (1973). *J. Bacteriol.*, **114**, 943–950
Piotrowska, M., Stepien, P. P., Bartnik, E., and Zahrewska, E. (1976). *J. Gen. Microbiol.*, **92**, 89–96
Rao, E. Y. T., Rao, T. K., and DeBusk, A. G. (1975). *Biochim. Biophys. Acta*, **413**, 45–51
Ring, K., Gross, W., and Heinz, E. (1970). *Arch. Biochem. Biophys.*, **137**, 243–252
Robinson, J. H., Anthony, C., and Drabble, W. T. (1973*a*). *J. Gen. Microbiol.*, **79**, 53–63
Robinson, J. H., Anthony, C., and Drabble, W. T. (1973*b*). *J. Gen. Microbiol.*, **79**, 65–80
Roess, W. B. and DeBusk, A. G. (1968). *J. Gen. Microbiol.*, **52**, 421–432
Scarborough, G. A. (1971). *Biochem. Biophys. Res. Commun.*, **43**, 968–975

Scarborough, G. A. (1976). *Proc. Natl. Acad. Sci. U.S.A.*, **73**, 1485–1488
Scarborough, G. A. and Nye, J. F. (1967). *J. Biol. Chem.*, **242**, 238–242
Schmit, J. C. and Brody, S. (1976). *Bacteriol. Rev.*, **40**, 1–41
Scott, W. A. (1977). *J. Bacteriol.*, **130**, 1144–1148
Singh, D. P. and LeJohn, H. B. (1975). *Can. J. Biochem.*, **53**, 975–988
Sinha, U. (1969). *Genetics*, **62**, 495–505
Skye, G. E. and Segel, I. H. (1970). *Arch. Biochem. Biophys.*, **138**, 306–318
Slayman, C. L. (1965). *J. Gen. Physiol.*, **49**, 69–76
Slayman, C. L. (1977). In *Water Relations in Membrane Transport in Plants and Animals* (A. M. Jangreis, T. K. Hodges, A. Kleinzeller, and S. G. Schultz, eds) Academic Press, New York, pp. 69–86
Slayman, C. L. and Gradman, D. (1975). *Biophys. J.*, **15**, 968–971
Slayman, C. L. and Slayman, C. W. (1974). *Proc. Natl. Acad. Sci. U.S.A.*, **71**, 1935–1938
Slayman, C. L., Long, W. S., and Gradman, D. (1976). *Biochim. Biophys. Acta*, **426**, 732–744
Slayman, C. L., Long, W. S., and Hu, C. Y. H. (1973). *J. Membrane Biol.*, **14**, 305
Spratt, G. D., Wood, J. M., Martin, W. G., and Schneider, H. (1977). *Biochem. Biophys. Res. Commun.*, **76**, 1099–1106
Srb, A. M. (1953). *Genetics*, **38**, 694–703
Stadler, D. R. (1966). *Genetics*, **54**, 677–685
Stuart, W. D. and DeBusk, A. G. (1971). *Arch. Biochem. Biophys.*, **144**, 512–518
Stuart, W. D. and DeBusk, A. G. (1973). *Biochem. Biophys. Res. Commun.*, **52**, 1046–1050
Stuart, W. D. and DeBusk, A. G. (1975*a*). *J. Bacteriol.*, **122**, 41–46
Stuart, W. D. and DeBusk, A. G. (1975*b*). *Arch. Biochem. Biophys.*, **166**, 213–222
Thwaites, W. M. and Pendyala, L. (1969). *Biochim. Biophys. Acta*, **192**, 455–461
Tisdale, J. H. and DeBusk, A. G. (1970). *J. Bacteriol.*, **104**, 689–697
Travis, G. L., Wolfinbarger, L., Jr, Stuart, W. D., and DeBusk, A. G. (1972). *J. Bacteriol.*, **109**, 933–935
Wiley, W. R. (1970). *J. Bacteriol.*, **103**, 656–662
Wiley, W. R. and Matchett, W. H. (1966). *J. Bacteriol.*, **92**, 1698–1705
Wiley, W. R. and Matchett, W. H. (1968). *J. Bacteriol.*, **95**, 959–966
Williams, S. (1971). *Aust. J. Biol. Sci.*, **24**, 1181–1188
Wolfinbarger, L., Jr (1976). *Biochim. Biophys. Acta*, **436**, 774–788
Wolfinbarger, L., Jr and DeBusk, A. G. (1971*a*). *Biochim. Biophys. Acta*, **241**, 677–681
Wolfinbarger, L., Jr and DeBusk, A. G. (1971*b*). *Arch. Biochem. Biophys.*, **144**, 503–511
Wolfinbarger, L., Jr and DeBusk, A. G. (1972). *Biochim. Biophys. Acta*, **290**, 355–367
Wolfinbarger, L., Jr and Kay, W. W. (1973). *Biochim. Biophys. Acta*, **330**, 335–343
Wolfinbarger, L., Jr and Marzluf, G. A. (1974). *J. Bacteriol.*, **119**, 371–378
Wolfinbarger, L, Jr, Jervis, H. H., and DeBusk, A. G. (1971). *Biochim. Biophys. Acta*, **249**, 63–68
Woodward, C. K., Read, C. P., and Woodward, V. W. (1967). *Genetics*, **56**, 598–606

Microorganisms and Nitrogen Sources
Edited by J. W. Payne

Section 1.4

Characterization of Membrane Proteins Involved in Transport

C. E. FURLONG AND G. D. SCHELLENBERG

Department of Medicine, Division of Medical Genetics, Center for Inherited Diseases, and Department of Genetics SK 50, Seattle, Washington 98195, USA

1 INTRODUCTION

Nutrient transport systems in microorganisms are most conveniently categorized on the basis of the mechanism employed by the system in question. For the transport systems which have been studied in detail, at least five different mechanisms of substrate accumulation are apparent: (a) phosphonenolpyruvate phosphotransferase-type systems which function in carbohydrate uptake and require both cytoplasmic and membrane-bound protein components (Postma and Roseman, 1976); (b) binding protein-dependent transport systems which require at least a periplasmic, substrate recognition component and have been shown to catalyse the uptake of amino acids (Heppel, 1969), sugars (e.g. Anraku, 1968*a*, *b*, *c*), inorganic ions (Pardee, 1966; Medveczky and Rosenberg, 1970, 1971), and vitamins (Nishimune and Hayashi, 1971; Taylor *et al.*, 1972); (c) membrane-bound, sodium-independent transport systems which require only membrane-bound components and have been shown to transport amino acids (Kaback, 1970), sugars (Kaback, 1974), sugar phosphates (Kaback and Barnes, 1971; Dietz, 1972), and inorganic ions (Rosenberg *et al.*, 1977); (d) membrane-bound, sodium-dependent transport systems which in addition to sodium require only membrane-bound components and have been shown to be involved in amino acid (Frank and Hopkins, 1969) and sugar (Tokuda and Kaback, 1977) uptake; and (e) cation-stimulated ATP-hydrolysing systems (Mg^{2+}, Ca^{2+} ATPases) which are postulated to be proton transport systems (Mitchell, 1970).

The scope of this section is limited to the discussion of proteins involved in amino acid transport. Since the only proteins so far characterized are those associated with systems which fall into categories (b) or (c), those which fall into the other categories will not be discussed in detail. For a more general treatment of microbial transport systems, the reader is referred to a number of excellent reviews which have appeared in recent years (Halpern, 1974; Boos, 1974; Oxender, 1972; Oxender and Quay, 1976; Slayman, 1973; Simoni and Postma, 1975; Heppel, 1969).

It should be emphasized at the start that many nutrients enter the cell through more than one transport system (e.g. Ames, 1964). Further, the multiple systems for a given substrate may operate either by a similar mechanism and share one or more components, or they may function by entirely different mechanisms and not share common protein components. Thus, for a given nutrient, one or more proteins associated with a given system may be well characterized, while protein components and even the general properties of other systems for the same substrate may be unknown or only partially understood. We will attempt to point out what is known about the multiplicity or redundancy of the amino acid transport systems which are pertinent to this section.

The amino acid transport proteins which are the best understood in terms of their physical properties, and which will be discussed most extensively below, are the osmotic shock-releasable binding proteins, which function as substrate recognition components for the binding protein-dependent class of transport systems. The binding proteins are thought to be located between the inner and outer cell membranes of Gram-negative bacteria (Nakane *et al*, 1968; Heppel, 1969) and are selectively released from such cells by an osmotic shock procedure developed by Neu and Heppel (1965). Although binding proteins have been extensively studied and are reasonably well understood biochemically, it is apparent that these proteins alone are not sufficient for substrate accumulation. Genetic evidence suggests that, in addition to the binding proteins, other specific proteins are required for the function of this class of systems (Ames and Roth, 1968; Kellerman and Szmelcman, 1974; Robbins, *et al.*, 1976; Gerdes *et al.*, 1977). Unfortunately, none of these additional proteins has been purified or even physically identified. Reports have only recently begun to appear of proteins associated with systems which are classified as membrane-bound (Kusaka *et al.*, 1976; Amanuma *et al.*, 1977; Hirata *et al.*, 1976, 1977; Lanyi, 1977). Also, both membrane-bound and detergent solubilized substrate-binding activities have been described (Gordon *et al.*, 1972; Aksamit *et al.*, 1975; Lanyi, 1977). Below, we will discuss the transport systems associated with given amino acids or groups of structurally related amino acids.

2 AMINO ACID TRANSPORT SYSTEMS

1 Glutamine

The least complicated case of a binding protein-dependent amino acid transport system is perhaps that of the glutamine transport system of *Escherichia coli*. There appears to be only one system (K_m = 0.05 μM) which transports glutamine in this organism, from which a glutamine-binding protein (K_D = 0.3 μM) has been purified (Weiner *et al.*, 1971; Weiner and Heppel, 1971). Results from genetic and kinetic studies have suggested that this protein is the substrate recognition component of the glutamine transport system. Mutants were isolated which had an elevated level of glutamine transport and a corresponding increase in the level of the glutamine-binding protein. Other mutants, isolated on the basis of their resistance to the glutamine structural analogue γ-glutamyl hydrazide, had only about 3 per cent of the wild-type glutamine transport together with a corresponding 95 per cent decrease in glutamine binding activity. Consistent with the binding protein being the substrate recognition component of the glutamine transport system was the finding that γ-glutamyl hydrazide and γ-glutamyl hydroxamate inhibit both glutamine transport in whole cells and glutamine binding to the isolated protein, with the later analogue having K_I values of 75 μM for both transport and binding.

The glutamine-binding protein is conveniently purified by several different procedures. The original procedure of Weiner *et al.* (1971), and the modified procedure of Weiner and Heppel (1971), utilize DEAE–cellulose chromatography of the osmotic shock fluid proteins, followed by isoelectric focusing. The large-scale, cell growth–osmotic shock procedure (Willis *et al.*, 1974) followed by CM–cellulose, DEAE–cellulose, and phosphocellulose chromatography (Schellenberg and Furlong, 1977) also provides homogeneous protein, and avoids the use of—and therefore the problem of removing—ampholytes introduced in the electrofocusing step.

The availability of a reasonable quantity of homogeneous protein has allowed for a fairly detailed physical characterization of the glutamine-binding protein (Weiner and Heppel, 1971). A molecular mass of 26 000 daltons was calculated from the amino acid composition, assuming one histidine residue per molecule, while estimates of 24 000 and 29 000 daltons were obtained by sedimentation equilibrium and gel filtration, respectively. This protein is quite stable and readily renatures after boiling at 100 °C for 15 min or following treatment with 7 M urea or 6 M guanidine hydrochloride. Optical rotatory dispersion measurements indicated an α-helix content of approximately 45 per cent for the native protein and a random coil conformation upon denaturation with 6 M guanidine hydrochloride. The tryptophan fluorescence emission maximum at 336 nm with the native protein suggested a somewhat hydrophobic environment for the tryptophan residues. A shift of this maximum to an even lower wavelength upon binding of glutamine suggested a ligand-induced conformational change which placed the tryptophan residues in an even more hydrophobic environment. In addition, quenching of tryptophan fluorescence was also noted when glutamine was present, and this property of the protein was used to determine the on and off rates of binding in stopped-flow kinetic experiments. The rate of the forward reaction was 9.8×10^{-7} M^{-1} s^{-1} and that of the back reaction was 16 M^{-1} s^{-1}, yielding an equilibrium dissociation constant of 1.6×10^{-7} M. A similar value of 3×10^{-7} M was obtained by equilibrium dialysis, with one mole of glutamine bound per mole of protein. Titration of tryptophan residues with *N*-bromosuccinimide showed that both residues present in the protein could be oxidized by a sixfold molar excess of this reagent. Glutamine protected the tryptophan residues against oxidation, which suggested that a ligand-induced conformational change resulted in the movement of the tryptophan residues to a more protected environment. The tryptophan residues did not appear to be at the ligand binding site since complete oxidation by *N*-bromosuccinimide did not result in the loss of glutamine-binding activity. Proton NMR studies (Kreischman *et al.*, 1973) have also indicated that ligand-induced conformational changes occur in this protein. Upon binding of glutamine, a large number of changes were observed over the entire spectrum, indicating a shift in the position and environment of a substantial number of residues. Of particular interest, in light of the fluorescence studies of Weiner and Heppel (1971), was the change observed in the position of signals from ring current shifted methyl groups. The observed change was interpreted as evidence that the positions of several methyl groups were altered with respect to adjacent aromatic residues.

Immunological experiments suggest that the glutamine-binding protein does not appear to be closely related to a number of other osmotic shock-releasable binding proteins. Antisera prepared against the leucine–

isoleucine–valine-binding protein, the leucine-specific binding protein, the lysine–arginine–ornithine-binding protein, or the cystine-binding protein did not cross-react with the glutamine-binding protein (Weiner and Heppel, 1971). More recent work (Willis and Furlong, 1975; Schellenberg, 1978) showed that antiserum prepared against the glutamate–aspartate-binding protein did not cross-react with the glutamine-binding protein nor did antiserum prepared against the glutamine-binding protein cross-react against the glutamate–aspartate-binding protein. This is interesting, in that both proteins have basic isoeletric points (pH 8.6) for the glutamine-binding protein, and 9.7 for the glutamate–aspartate-binding protein) and both bind structurally related substrates.

2 Leucine, Isoleucine, Valine, and Threonine

Historically, the first protein associated with an amino acid transport system to be characterized was the so-called leucine-binding protein, which is now referred to as the LIV-BP or the LIVT-BP (leucine–isoleucine–valine–threonine-binding protein) (see also Section 1.6.4). We will refer to this protein as the LIVT-BP in the following section. Piperno and Oxender (1966) reported that the transport of branched-chain amino acids was considerably reduced in *E. coli* K12 cells when they were subjected to the osmotic shock procedure of Neu and Heppel (1965). Further, they achieved a 10-fold purification of a branched-chain amino acid binding protein by ammonium sulphate precipitation followed by column chromatography. They also noted that the K_D values for ligand binding to the isolated protein were indistinguishable from the respective K_m values for the uptake of these amino acids by whole cells. Oxender *et al*. (1967) reported that the presence of leucine in the growth medium repressed the transport of leucine and isoleucine, and further, leucine-binding activity was not present in the osmotic shock fluid obtained from repressed cells. Penrose *et al.* (1968*a*) described the crystallization of the LIVT-BP and reported a molecular weight value of 36 000 daltons determined by ultracentrifugation, gel filtration, and amino acid analysis. They also presented immunological data which suggested that this protein was located at the cell surface.

These studies were independently confirmed by Anraku and Heppel (1967). Anraku (1968*a*) purified the LIVT-BP to homogeneity and crystallized it from ammonium sulphate. In agreement with Piperno and Oxender (1966) he found that the protein was specific for leucine, isoleucine, and valine, with K_D values of 2.2 and 1.6 μM for leucine and isoleucine, respectively. Incubation of the protein with leucine did not result in modification of the ligand. Studies on the physical properties of the protein showed that it was stable at 4 °C or 100 °C, only partially stable to freezing and thawing, resistant to pronase digestion, and more stable at acid than alkaline pH values. A broad pH optimum of from pH 4–10 was observed for leucine binding, for which the stoicheiometery was one molecule of leucine bound per molecule of binding protein.

Penrose *et al.* (1968*b*) also reported an extensive characterization of the LIVT-BP. The protein was crystallized from 2-methyl-2,4-pentanediol and molecular mass values of 24 000, 36 000, and 34 000 daltons were obtained by use of ultracentrifugation, gel filtration, and amino acid analysis, respectively. The isoelectric point was 5.15 and the amino acid analysis was essentially in agreement with that obtained by Anraku (1968*b*). The K_D values were 0.2–0.4 μM and 0.1–0.3 μM for leucine and isoleucine, respectively. These were in close agreement with K_m values for the whole cell uptake of these amino acids. Extensive studies on the inhibition of ligand binding by substrate analogues demonstrated an absolute requirement for a free amino group although some modification to the carboxyl group was thought to be possible. However, contaminating leucine in the analogue preparations was not ruled out as an explanation for this observation. L-Alloisoleucine was shown to be a potent inhibitor of leucine binding.

A number of alternate methods of purifying the LIVT-BP have been reported (Furlong and Heppel, 1970; Antonov *et al.*, 1974; Willis *et al.*, 1974; Amanuma and Anraku, 1974).

Two separate studies aimed at resolving the differences in reported K_D values for the LIVT-BP have been reported. Antonov et al. (1976) studied binding by equilibrium dialysis at different concentrations of protein and substrate. They also studied the behaviour of the LIVT-BP in gel filtration column chromatography, in the analytical ultracentrifuge, and by fluorescence polarization. From their data, they concluded that leucine binding was dependent on the degree of oligomerization of the protein, which in turn was determined by the concentration of the ligand and protein.

The isolated LIVT-BP as well as the L-BP (see below) have been found by use of a dansylation procedure to contain tightly bound substrate (Amanuma *et al.*, 1976). While the bound ligand was readily exchanged in the presence of exogenous substrate, its release was very slow in the absence of the exchange reaction when measured either by dialysis or gel filtration. The amount of substrate retained was about equimolar with the protein. True dissociation constants were obtained with protein from which the bound ligand had been removed by extensive dialysis in 6 M urea. The 'true' K_D values were 0.12 and 0.19 μM for isoleucine and leucine binding, respectively. These authors point out the importance of considering the ratio of total free substrate to that possibly bound to the isolated protein when determining K_D values. Thus, biphasic binding curves were obtained with microdialysis cells where the volume of the protein solution was equal to that of the labelled free ligand solution and the bound unlabelled substrate diluted the specific activity of the added ligand. However, when the dialysis bag procedure was used, the labelled free substrate molecules were greatly in excess of the unlabelled 'protein donated' substrate (see also Lever, 1972*b*).

The slow rate of dialysis of bound substrate has been noted in several previous investigations of binding protein–ligand interactions (Hogg and Englesberg, 1969; Richarme and Kepes, 1974). Willis and Furlong (1974) proposed that the retention of ligand in a confined space by binding proteins could account for the 'memory component' of chemotactic responses; a number of binding proteins have been shown to play a role in chemotactic responses in addition to their transport function. A general theoretical treatment of ligand retention was presented by Silhavy *et al.* (1975). These observations on ligand retention by proteins are certainly a general feature of high-affinity binding sites and should be borne in mind when characterizing the interaction of any protein with its substrate if the K_D value is in the micromolar region or below (see also Section 1.4.3.3).

Ovchinnikov *et al.* (1977) have recently determined the amino acid sequence of the LIVT-BP. They found that the seven arginine residues were all located in the central part of the polypeptide chain and the three tryptophan residues were near the *C*-terminus of the 344 residue molecule.

Anraku (1968*c*) described the reconstitution of leucine transport in osmotically shocked cells, which was achieved by adding pure LIVT-BP to transport-deficient, osmotically shocked cells. The effect seemed to be specific since galactose-binding protein did not restore leucine transport. However, the reconstitution of transport in osmotically shocked cells has been difficult to reproduce in other laboratories (Oxender, 1972; Boos, 1974). There are several important points that should be made regarding the reconstitution of nutrient transport in osmotically shocked cells. First, it should be noted that cold osmotic shock removes virtually the entire small molecule pool of a cell (Britten and McClure, 1959, 1962; Hayashi *et al.*, 1964), thus depriving it of intermediary metabolites and probably cofactors which are necessary for the energy production used to drive active transport (Britten and McClure, 1962). Figure 1 demonstrates the problems encountered in comparing transport in osmotically shocked cells with unshocked cells. Simply adding an energy source to unshocked cells stimulates leucine transport. Even more problematical is the observation that the addition of an energy source to shocked cells often brings the leucine transport activity to a higher level than that observed with unshocked cells minus an energy source (Figure 1). The best measure of the effect of the osmotic shock treatment on transport is achieved by comparing unshocked cells plus an energy source with shocked cells plus an energy

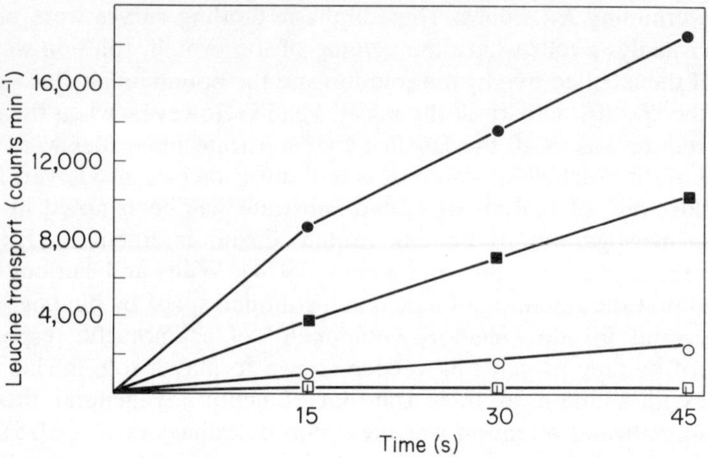

Figure 1 Leucine transport in osmotically shocked and unshocked cells, with and without an added exogenous energy source (30 mM glycerol). (○—○), unshocked cells −glycerol; (●—●), unshocked cells + glycerol; (□—□), shocked cells − glycerol; (■—■), shocked cells + glycerol

source. When this is done, usually a two- to fourfold difference in transport of leucine is observed (Figure 1). As will be discussed below, the transport remaining in the shocked cells is probably mediated by a membrane-bound, branched-chain amino acid transport system.

A third potential problem in reconstitution studies is that the nitrocellulose filters, which are commonly used in the transport measurements, also bind protein–ligand complexes (Furlong and Weiner, 1970), and indeed they are used to measure this association qualitatively (Lever, 1972*b*) as well as quantitatively (Willis and Furlong, 1975). The interference or background contributed to reconstitution experiments from this cause may possibly be eliminated or at least greatly reduced by pretreating the filters with a solution of another protein (such as bovine serum albumin) which will not bind ligand or interfere with the transport measurements. In any case, the interaction of the binding protein–ligand complex with the filter used to measure transport in the reconstitution assay should be checked directly.

A fourth problem stems from the multiplicity of leucine transport systems present in *E. coli* (discussed below). Recent studies have shown that the 'background' leucine transport observed in osmotically shocked cells is attributable to a membrane-bound, osmotic shock-resistant, leucine–isoleucine–valine transport system, termed the LIV-II system (Rahmanian *et al.*, 1973; Wood, 1975; Kaback and Milner, 1970) (see also Section 1.6.4). Ideally, it would be desirable to study the restoration of leucine transport in a mutant which is devoid of the shock-resistant, membrane-bound system(s), and under conditions where the energy supply was not limited and where the interaction of the protein–ligand complex with the filter was eliminated or minimized. A more recent study of reconstitution of leucine transport by Anraku *et al.* (1973) has taken into account the energy requirements for uptake. However, the problem of the membrane-bound, branched-chain transport system, and perhaps the binding-protein filter interaction still exist.

Gerdes *et al.* (1977) have recently described a successful reconstitution of binding-protein-dependent phosphate transport in *E. coli.* Their approach is well worth describing since it seems to avoid most of the objections raised to the reports of reconstitution of binding-protein-dependent transport in osmotically shocked cells (Oxender, 1972; Boos, 1974). They avoided the complications introduced by the presence of membrane-bound systems by utilizing a mutant which was devoid of the membrane-bound phosphate transport system and which had only the binding-protein-dependent phosphate transport system. They avoided depleting the cells of their small molecule pool by forming spheroplasts rather than shocking the cells. Also, in the spheroplasts, the site of interaction of the binding protein is probably much more accessible to the added binding protein than in shocked cells. They also included glucose as an energy source. The question of the binding-protein–substrate complex sticking to the filter, as well as other possible artefacts, were checked by attempting the restoration of transport in cells which were genetically deficient in another component of the binding-protein-dependent transport system. Using this approach, they achieved a reproducible 14-fold increase in binding-protein-dependent phosphate transport in their spheroplast preparations. It is important to note that they did see stimulation of transport in shocked cells. However, they attributed this stimulation to artefacts arising from spontaneous recovery of the cells or to non-specific stimulation of transport. (See also Section 1.4.2.5 on reconstitution of glutamate–aspartate transport systems.)

Of course, it is entirely possible that strain, media, and procedural differences could account for variation in the success of reconstituting binding-protein-dependent transport; however, the points mentioned above are worth keeping in mind when assessing these experiments.

A second osmotic shock-releasable binding-protein from *E. coli* which was specific for leucine (L-BP) has been purified and crystallized (Furlong and Weiner, 1970). Molecular mass values of 36 000, 37 000, and 39 000 daltons were determined by the use of gel electrophoresis, gel filtration, and ultracentrifugation, respectively (Furlong and Weiner, 1970; Furlong and Heppel, 1970). Leucine binding activity was not inhibited by valine or isoleucine; however, it was effectively inhibited by trifluoroleucine (TFL). This was interesting since TFL did not inhibit substrate binding to the LIVT-BP. Also, whole cells were shown to have a component of leucine transport which was resistant to inhibition by isoleucine but was sensitive to inhibition by TFL. Further, two K_m values were obtained for leucine transport in whole cells (0.2 and 2 μM). In the presence of 20 μM unlabelled isoleucine, K_m values for leucine uptake of 0.2 and 20 μM were observed (Furlong and Weiner, 1970). Thus, the high-affinity system in whole cells appears to be dependent on L-BP. Antiserum prepared against either of the two binding proteins cross-reacted against the other protein, while neither of the sera reacted with binding proteins for galactose, arginine (three proteins), cystine, or glutamine (Furlong and Heppel, 1970). The existence of the L-BP has been confirmed by Amanuma and Anraku (1974) and by Rahmanian *et al.* (1973).

A third branched-chain amino acid binding protein (referred to below as the ILVT-BP) was purified by Furlong *et al.* (1973*a*). This protein preferentially bound isoleucine and cross-reacted with antiserum prepared against the LIVT-BP (Furlong *et al.*, 1973*b*). The electrophoretic mobility of this isoleucine-preferring binding protein was intermediate to those of the L-BP and LIVT-BP. Rahmanian *et al.* (1973) also resolved this protein in DEAE–cellulose column eluates. Amanuma *et al.* (1976) recently purified a third osmotic shock-releasable branched-chain amino acid binding protein which they refer to as 'the second LIVT-BP'. However, judging from their DEAE–cellulose column elution profile and their acrylamide gel data, the second LIVT-BP appears to be identical to the protein termed the isoleucine-preferring binding protein by the other two laboratories (see also Sections 1.1.3.1 and 1.6.4). Amanuma *et al.* (1976) were able to purify a sufficient quantity of this new protein for a partial characterization of its properties. Use of sodium dodecyl sulphate polyacrylamide gel electrophoresis (SDS-PAGE) indicated that the second LIVT-BP had the same molecular weight as the LIVT-BP. Analogues which inhibited isoleucine binding, listed in the order of their effectiveness, were: alloisoleucine, leucine, valine, norvaline, α-aminobutyrate, threonine, and norleucine. They also noted that this protein cross-reacted with antibodies prepared against the LIVT-BP. In the light of the properties of this protein, and in the interest of avoiding confusion, it would seem appropriate to refer to this protein as simply the ILVT-BP (see also Section 1.4.3.1 on discussion of nomenclature).

It thus appears that there is a family of three closely related binding proteins (LIVT-BP, ILVT-BP, and L-BP) which are involved in the uptake of the branched-chain amino acids. The immunological cross-reactivity observed with the three branched-chain amino acid binding proteins, along with the finding that the LIVT-BP and the L-BP have approximately half of their tryptic peptides in common, led Furlong *et al.* (1973*b*, *c*) to propose that this family of transport proteins arose through gene duplication and genetic drift.

Although additional protein components involved in the binding-protein-dependent uptake of branched-chain amino acids have not yet been purified, a considerable body of evidence indicates that there are probably one or several more transport-related proteins which have yet to be identified. Anderson and Oxender (1977) isolated a mutant which retained functional leucine-binding proteins but was deficient in high-affinity leucine transport, presumably because it was missing an additional required protein component of the binding protein dependent-transport systems (see also Section 1.6.4) The branched-chain amino acids are also transported through a separately regulated membrane-bound transport system. Lombardi and Kaback (1972) have measured the uptake of the branched-chain amino acids in membrane vesicles which are devoid of binding-protein-dependent transport systems. Work by Rahmanian *et al.* (1973) on the multiplicity of leucine transport systems in *E. coli* has clearly shown the existence of an osmotic shock-resistant branched-chain amino acid transport system, referred to as the LIV-II system. This system, in contrast to the binding-protein-dependent system, is not repressed by growth on leucine. Further, threonine, alanine, and serine, which inhibit binding-protein-dependent transport, are not effective inhibitors of the membrane-bound system. Wood (1975) also confirmed the existence of the membrane-bound system in her studies on the multiplicity of leucine transport systems and their energy coupling mechanisms. From her kinetic results, she was able to estimate that the proportion of leucine transport mediated by the separate systems is as follows: L-BP, 15 per cent; LIVT-BP, 30 per cent; and LIV-II 40 per cent; with 15 per cent being so called non-specific transport.

From the available results it is possible to construct a working model for the proteins involved in the transport of the branched-chain amino acids in *Escherichia coli* (Figure 2; see Section 1.4.3.1 for nomenclature of the different systems). The binding-protein-dependent transport systems including auxillary protein component(s) comprise one route of entry for the branched-chain amino acids. The membrane-bound transport system(s) provides yet another mechanism for transporting this family of amino acids. Since the binding proteins are repressed when leucine is present in the medium (Oxender *et al.*, 1967), it could be argued that the systems with which they are associated function in retaining the amino acids synthesized by the cell. The work of Berger and Heppel (1974) suggests that the binding protein systems use quite a different mode of energy coupling than do the membrane-bound systems (Simoni and Postma, 1975). The availability of appropriate mutant strains should provide more insight into the reasons for the existence of the parallel systems which operate by such different mechanisms.

Genetic studies by Anderson and Oxender (1977) (see also Section 1.6.4) have suggested that the *livJ* locus is responsible for the synthesis of the

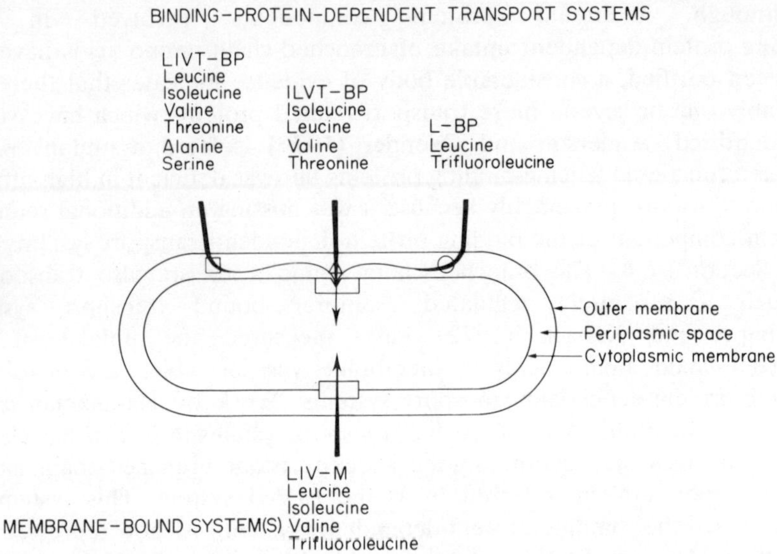

Figure 2 Working model for leucine transport in *E. coli*

LIVT-BP and the *livK* locus is the structural gene for the L-BP. An inactive gene product, which appears to be an altered L-BP, has been fractionated by DEAE–cellulose chromatography. An additional locus, *livH*, apparently specifies additional protein components required for the function of the branched-chain amino acid binding protein-dependent systems. These mutations are clustered, and map at minute 74 on the *E. coli* chromosome. Studies by Iaccarino and coworkers (Guardiola and Iaccarino, 1971, 1974; Guardiola *et al*., 1974) (see also Section 1.5.2.2) suggest that the low affinity (membrane-bound?) transport of the branched-chain amino acids may be mediated by more than one system.

From *Salmonella typhimurium* a partial fractionation of the branched-chain amino acid binding activities has also been reported (Willis *et al*., 1974).

From *Saccharomyces*, strain 60615, a leucine-binding activity has been solubilized by sonication (Bussey and Umbarger, 1970). A mutant with a four- or fivefold elevation in the levels of leucine, valine, and tryptophan transport had a threefold increase in this binding activity. The leucine binding activity, which the authors called a protein only with some reservations, was purified approximately eightfold by a detergent–DEAE column step. Another cell line with a 30-fold lower affinity for leucine also had a 30-fold decrease in the binding affinity for leucine, although the total binding units in this mutant were increased.

3 Histidine

The histidine transport systems of *Salmonella typhimurium* have been characterized by Ames and coworkers. Histidine transport in this organism appears to be mediated by a general aromatic amino acid permease as well as by several binding-protein-dependent transport systems (Ames, 1964; Ames and Lever, 1970). The work of Lombardi and Kaback (1972) suggests that at least one membrane-bound type of histidine transport system is present in *E. coli* membrane vesicles. Several proteins believed to be involved in the transport of histidine have been purified from osmotic shock fluids of *S. typhimurium* (Ames and Lever, 1970; Lever, 1972*a*; Rosen and Vasington, 1971). A partial purification of a histidine-binding protein from *E. coli* has been reported (Willis *et al.*, 1974), and more recently this protein has been purified to homogeneity (Beck and C. E. Furlong, unpublished results). The most thoroughly characterized protein component involved in histidine transport is the *S. typhimurium* histidine-binding J protein (Ames and Lever, 1970; Lever, 1972*a*). The major (J) and minor (K) histidine-binding proteins were resolved by use of DEAE–cellulose column chromatography. The major histidine-binding protein was resolved further into the following three histidine-binding activities by preparative electrofocusing: a major histidine-binding activity (J protein) with a p*I* of 5.5, and two minor histidine-binding activities with p*I* values of 5.05 and 5.8. Subsequent chromatography of the J protein on Sephadex G-100 removed the carrier ampholytes introduced during the electrofocusing step and resulted in pure J protein, as judged by a single protein band upon electrophoresis in buffer systems with pH values of 4.3 or 8.9. The minor histidine-binding activities have not been purified to homogeneity or further characterized. An alternate purification scheme for the J protein has been developed which is based on ω-aminodecyl agarose chromatography (Shaltiel *et al.*, 1973). Most of the osmotic shock-releasable proteins are retained by this column, whereas the J protein passes straight through and is thus purified seven- to 10-fold in a single step from osmotic shock fluid. Since the binding of proteins to this type of column depends not only on ionic interactions with the ω-amino group but also on hydrophobic interactions with the 10-carbon alkyl chain, the lack of interaction of the J protein with this type of column suggests that this protein may be somewhat more hydrophilic than most other shock-releasable proteins.

The physical properties of the J protein (Lever, 1972*a*) are typical of other binding proteins. It is a monomer and molecular weight values of 25 000 and 26 000 daltons have been determined by use of sedimentation equilibrium and gel filtration, respectively. The binding of histidine to this protein is fully reversible and a K_D value of 0.15 μM was obtained by both equilibrium dialysis and a filter-binding assay. The protein appears to have one histidine binding site per molecule. Specificity studies indicate that of the naturally occurring L-amino acids only arginine and lysine inhibit binding *in vitro*. Binding of histidine is also weakly inhibited by D-histidine, azaserine, L-citrulline, D-ornithine, and D-2-hydrazino-3-(4-imidazoyl)-propionic acid (Ames and Lever, 1972). The purified J protein is quite stable to heat inactivation, retaining 50 per cent of its activity after 20 min of boiling (Lever, 1972*a*).

Proton NMR, ^{19}F NMR, and fluorescence spectroscopy have been used recently (Robertson *et al.*, 1977) to study the binding properties of both the native J protein and of J protein which contained 5-fluoro-D,L-tryptophan in place of L-tryptophan (5F-Trp J protein). The 5F-Trp J protein was prepared from a strain of *S. typhimurium* which was auxotrophic for tryptophan and which had been cultured in the presence of 5-fluorotryptophan for several generations prior to harvest. As with the glutamine-binding protein (Kreishman *et al.*, 1973), the binding of histidine to the native J protein resulted in a dramatic change in the proton NMR spectrum, including changes in the aromatic, aliphatic, and ring current-shifted methyl proton resonances. The large number of changes observed led the authors to conclude that a general conformational change occurred upon the binding of histidine, and that these changes involved a large number of amino acid residues. Similar, but not identical, spectral changes were observed when L-arginine or D-histidine was bound to the protein, even though these two amino acids and L-histidine all bind to the same site. Studies of the 5F-Trp protein demonstrated that upon binding of histidine, at least one and probably two tryptophan residues undergo a conformational change, as indicated by a downfield shift of a fluorotryptophan-generated ^{19}F resonance. The observed shift was thought to be indicative of the movement of a tryptophan residue(s) to a more hydrophobic environment. No change in line-width (of the shifted resonance) was observed, and thus there was probably no change in the rotational freedom of the residue involved. The ^{19}F signal from a third fluorotryptophan residue was not shifted upon histidine binding. Complementary fluorescence studies of histidine binding to the native J protein and the 5F-Trp J protein showed a ligand-induced increase in the quantum yields of intrinsic tryptophan fluorescence emission for both proteins. This was interpreted as indicating a shift of at least some tryptophan residues to a more hydrophobic environment. The presence of substrate protected the tryptophan residues from oxidation by *N*-bromosuccinimide, although at least one residue was still susceptible to this reagent.

As mentioned in the introduction to this section, genetic evidence suggests that the binding protein-dependent transport systems require not only the osmotic shock releasable binding proteins, but also an additional protein component(s) in order to move nutrients across the cell membrane. In the case of histidine, the J protein has been shown to be a component of a high-affinity histidine transport system ($K_m = 10^{-8}$ M) in a series of genetic experiments by Ames and coworkers, which will not be reviewed here (Ames and Lever, 1972; Ames and Spudich, 1976; Ames *et al.*, 1977).

Work from the same laboratory has also shown that the product of a second gene, the *hisP* gene, is also required for the function of the high-affinity histidine transport system (Ames and Lever, 1970). Although the *hisP* gene product has yet to be identified, it is presumed to be the membrane component with which the J protein interacts during the active transport of histidine. Kustu and Ames (1974) have shown that the J protein contains a site which is required for transport activity but which is distinct from the histidine-binding site of the protein. These workers isolated a strain (TA 300) with a mutation in the structural gene for the J protein, as judged by the J protein's altered electrophoretic mobility in both non-denaturing and sodium dodecyl sulphate-containing gels. The mutant protein did not function properly in whole cell histidine transport, as evidenced by a sixfold increase in the K_m value of the high-affinity histidine transport system relative to the parent strain. (The K_m values of the parent and mutant strain were 0.6×10^{-8} M and 3.7×10^{-8} M, respectively.) However, the defect in the J protein responsible for the altered transport did not appear to involve the ligand binding site of the protein, since the K_m value for histidine binding to the mutant protein (0.8×10^{-7} M) was similar to the K_D for histidine binding to the parental J protein (0.6×10^{-7} M). The authors postulated that the portion of the protein which was altered by the TA 300 mutation was that which interacts with the *hisP* component of the high-affinity transport system.

In subsequent work, Ames and Spudich (1976) isolated a mutant in the *hisP* gene which showed a partial suppression of the histidine-transport defect caused by the altered J protein from the TA 300 mutant. Their interpretation of these results was that the restoration of correct J protein–P protein interaction was made possible by a change in the P protein which compensated for the altered J protein. A complete understanding of the protein–protein interactions between the binding proteins and other required protein components awaits the identification, purification, and characterization of the elusive additional protein components.

Rosen and Vasington (1971) have purified an osmotic shock-releasable protein from *S. typhimurium* which bound histidine with a K_D of 1.5 μM. This protein also bound arginine weakly, and histidine binding was slightly inhibited by a mixture of sulphur-containing amino acids. A molecular mass of 25 250 daltons was obtained by use of sedimentation equilibrium ultracentrifugation. This protein could be maintained at 95 °C for up to

60 min without loss of activity. Ligand binding was not affected by high ionic strength, nor was there any pH effect on binding between pH values of 5–9. It is not clear whether this histidine-binding protein is the same as the J protein described by Lever (1972*a*), or whether it is one of the other histidine-binding activities present in *S. typhimurium* shock fluid. While both proteins have identical molecular weights, bind histidine and arginine, and are quite stable to heat, they do differ in other properties, including the K_D of histidine binding, the pH optimum of binding, and possibly the substrate specificity of binding.

Osmotic shock of *E. coli* also releases histidine-binding activity, which has been fractionated by Willis *et al.* (1974) and subsequently purified by Beck and C. E. Furlong (unpublished results). Molecular mass values of 31 000 and 32 400 daltons were determined by use of SDS-PAGE and gel filtration, respectively. A K_D of 0.8 μM for histidine binding was determined from equilibrium dialysis, and 1 mol histidine was bound per mole of protein. A pH optimum of 5–7 was found, and the binding activity decreased substantially at either more acidic or basic pH values.

4 Arginine, Lysine, and Ornithine

Wilson and Holden (1969*a*,*b*) working with *E. coli*, were the first to report the fractionation of proteins believed to be involved in the transport of arginine and lysine. They described the partial resolution of a number of osmotic shock-releasable, arginine-binding activities by use of DEAE–cellulose chromatography. Limited data on the specificity of the various fractions was presented. Osmotic shock was found to reduce arginine uptake by 25–33 per cent, and two of the DEAE fractions were found to give a slight by reproducible increase in arginine transport in the shocked cells. Both of these fractions had approximately the same affinity for arginine (0.5–1.0 μM), which was close to the K_m value for arginine uptake in whole cells.

Rosen (1971, 1973*a*) subsequently purified to homogeneity two osmotic shock-releasable binding proteins from *E. coli*. One protein bound lysine, arginine, and ornithine (LAO-BP), while the second was specific for arginine (Arg-BP). He also noted a third, low-affinity, arginine-binding activity which co-migrated with the histidine-binding activity during DEAE–cellulose chromatography. (As noted above, the *S. typhimurium* histidine-binding protein(s) (Lever, 1972; Rosen and Vasington, 1971) also bound arginine with a low affinity.)

The pure LAO-BP had K_D values (determined by use of microequilibrium dialysis) of 1.5, 3.0, and 5.0 μM for arginine, lysine, and ornithine, respectively. The binding of these ligands was not inhibited by other naturally occurring amino acids, although a number of structurally related analogues were effective inhibitors. The most potent inhibitors, listed in order of effectiveness, were: dehydrolysine, α-*N*-acetylornithine, L-canavanine, and D,L-4-oxalysine. Competition experiments suggested a single ligand-binding site. Redetermination of the arginine K_D value (Rosen, 1973*a*) by the flow dialysis method of Colowick and Womak (1969) yielded a value of 0.2 μM, approximately sevenfold lower than that originally determined by use of the microequilibrium dialysis method. Rosen attributed the difference in K_D values to the different temperatures used for the two methods. However, it is possible that the isolated LAO-BP contained tightly bound ligand and that the presence of this ligand interferred with the microequilibrium dialysis method but not with the flow dialysis method (see the above discussion of this problem in Section 1.4.2.2). Molecular mass values for the LAO-BP of 26 200, 30 200, and 27 500 daltons were determined by use of ultracentrifugation, gel filtration, and disc-gel electrophoresis, respectively. The isoelectric point of the protein was 5.1. Results of transport experiments in whole cells suggested that the LAO-BP does not function in arginine transport but does function in a binding-protein-dependent transport system for lysine and ornithine (Rosen, 1971). In addition to the LAO-BP-dependent system, Rosen reported the existence of an osmotic shock-resistant, lysine-specific transport system which is probably a membrane-bound type of system. There does not appear, however, to be a significant amount of membrane-bound arginine transport activity (Lombardi and Kaback, 1972).

The arginine-specific binding protein (Arg-BP) was found to have a K_D value for arginine binding of 3×10^{-8} M, a p*I* of 5.1, and bound one molecule of arginine per molecule of protein (Rosen, 1973*a*). Molecular mass values of 33 500 and 27 000 daltons were determined by use of gel filtration and sedimentation equilibrium centrifugation, respectively. This second binding protein was extremely specific for arginine—binding was not inhibited by other amino acids or structural analogues. Antiserum prepared against either of the arginine-binding proteins did not cross-react with the other protein. The Arg-BP has two half-cystine residues, which are apparently in disulphide linkage. However, it can be reversibly denatured by urea treatment. Whole-cell experiments indicated that the Arg-BP probably functions in a binding protein-dependent, arginine-specific system which is not inhibitable by lysine or ornithine. Rosen was not able to reproduce the reconstitution of binding protein-dependent arginine transport described by Wilson and Holden (1969*b*).

Analysis of several canavanine-resistant mutants led Rosen (1973*b*) to postulate that an additional protein(s) served as a common second component for two binding-protein-dependent, basic amino acid transport systems. A mutation in the locus coding for canavanine resistance led to the loss of the activity of both the arginine-specific and the LAO-transport systems, in spite of the fact that both binding proteins were still produced.

Kustu and Ames (1973) have shown that the wild-type *hisP* gene of *S. typhimurium* (thought to code for the second component of the binding protein-dependent histidine transport systems) is required for the utilization of arginine as a nitrogen source by this organism. However, the *hisP* gene product is not required for the function of the high-affinity arginine transport system. This high-affinity system has a sufficient velocity to supply arginine for protein synthesis in an arginine-requiring cell, but apparently not an adequate velocity to allow for the utilization of arginine as a nitrogen source. Thus, there are probably at least two arginine transport systems in *S. typhimurium*. Mutations in the histidine J binding protein had no effect on the utilization of arginine as a nitrogen source, and this protein does not appear to function in arginine transport, despite the fact that it does bind arginine.

Basic amino acid recognition components have also been reported for transport systems of higher organisms (see also Section 1.3.4). Stuart and DeBusk (1971) reported that treatment of *Neurospora* cells with 4.82 M KCl resulted in a decrease in arginine transport as well as a release of arginine-binding activity with no loss in viability. The arginine-binding activity was resolved into three peaks by use of affinity chromatography on an arginine-coupled Sephadex column. The first peak (A) was observed in all strains studied; the second peak (B) was missing in a mutant which was devoid of the basic amino acid permease system; and the third peak (C) was absent in mutants missing either the basic amino acid permease or the neutral amino acid permease. The crude, KCl-extracted glycoprotein mixture was used to reconstitute arginine transport through a lipid film stretched over a 43 μm filter screen in a specially constructed apparatus (Stuart and DeBusk, 1973). The rate of arginine transport under their conditions was 1000-fold greater than that observed when no glycoprotein was included in the lipid film. In more recent work (Stuart and DeBusk, 1975) the amino acid and carbohydrate composition, as well as the affinity of fraction B were determined. The protein contained no cyst(e)ine but did contain eight neutral sugars. No amino sugars were detected, and the authors suggested that the carbohydrate linkages may involve the serine and threonine residues of the protein. Cells grown in labelled uridine were shown to incorporate label into fraction B, thus suggesting the presence of a ribonucleic acid component. This possibility was further explored by treatment of whole cells with RNase which resulted in a reduced initial rate of amino acid transport.

A protein has been purified from *Saccharomyces cerevisiae* which binds arginine with a K_D of 3.8×10^{-4} M and lysine with a K_D of 4.2×10^{-4} M

(Opekarova *et al.*, 1975). The protein had a molecular mass of less than 5000 daltons. It did not restore transport to protein-depleted cultures, and was still present in a mutant with impaired arginine transport. Binding was inhibited by inorganic phosphate and azide, and decreased at high concentrations of protein suggesting the possibility of self-association.

5 Glutamate and Aspartate

Glutamate-binding activity was first detected by Barash and Halpern (1971) in the class of proteins released from *E. coli* during spheroplast formation. Subsequently, Miner and Frank (1974) demonstrated that a glutamate-binding activity could be released from *E. coli* by the osmotic shock procedure of Neu and Heppel (1965). Following this work, glutamate–aspartate-binding proteins were purified to homogeneity from *E. coli* K12 by Barash and Halpern (1975), from *E. coli* W by Willis and Furlong (1975), and from *Salmonella typhimurium* by Aksamit *et al.* (1975). However, considerable difficulty was encountered in assigning the glutamate–aspartate binding activity to a specific transport system. As discussed below, this was due to the presence of several different systems which were simultaneously taking up these two amino acids.

The *E. coli* K12 glutamate–aspartate-binding protein characterized by Barash and Halpern (1975) bound glutamate with a K_D of 6.7 μM, and glutamate binding was inhibited by aspartate with a K_I of 78 μM. L-Glutamate-γ-ethylester, reduced glutathione, and L-glutamate-γ-benzyl esters were also inhibitors of glutamate binding. Other naturally occurring amino acids and glutamate analogues tested did not inhibit binding. Molecular mass values of 29 700, 27 800, 32 000, and 28 797 daltons were determined from use of sedimentation equilibrium, gel filtration, SDS-PAGE, and amino acid analysis, respectively. The isoelectric point was 9.1 and a broad pH optimum from 4–8 was observed for glutamate binding.

The *E. coli* W glutamate–aspartate-binding protein described by Willis and Furlong (1975) had an isoelectric point of 9.7 and bound one molecule of glutamate (K_D = 0.8 μM) or one molecule of aspartate (K_D = 1.2 μM) per molecule of protein. Competition data were consistent with a single binding site for both ligands. Weak inhibition of binding was observed with glutamine, asparagine, and alanine, although the inhibition was at sufficiently high levels of inhibitors that contaminating glutamate or aspartate could have accounted for these observations. L-Glutamyl-γ-esters and L-glutamyl-γ-hydroxamate were also reported to be weak inhibitors. However, more recent work (Schellenberg and Furlong, 1977) has indicated that the inhibition observed with the esters was probably due to contaminating glutamate. Molecular mass values of 29 800, 31 000, and 31 500 daltons were determined by use of gel electrophoresis at pH 3.5, pH

4.5, and in the presence of sodium dodecyl sulphate, respectively. Values of 29 500 and 30 400 daltons were determined by use of gel filtration and sedimentation equilibrium, respectively. A minimum molecular mass of 32 000 daltons was calculated from the amino acid analysis, assuming two histidine residues per molecule. The occurrence of two cysteic acid residues in each performic acid-oxidized molecule suggested the possibility of an intramolecular disulphide bridge. The finding that dithiothreitol could inactivate only the unfolded protein suggested that a buried disulphide group might be important in maintaining an active conformation of the protein. The protein lost about 50 per cent of its activity after 2 min at 100 °C, and thus seems to be more heat labile than other binding proteins. A substrate-induced conformational change of the glutamate–aspartate-binding protein was suggested by fluorescence changes observed during glutamate binding; a blue shift in native tryptophan fluorescence, as well as a 7 per cent decrease in amplitude was observed when glutamate was bound. From recent reports that retained ligand can occur in binding protein preparations (see above discussions, Section 1.4.2.2), it would not be surprising to see even larger changes with protein that had been exhaustively dialysed against urea to remove bound ligand before doing the fluorescence experiments. Antibody prepared against the *E. coli* glutamate–aspartate-binding protein did not cross-react with the glutamine-binding protein. The serum did, however, cross-react against crude osmotic shock fluid obtained from *Salmonella typhimurium*.

Aksamit *et al.* (1975) reported both soluble and membrane-bound aspartate-binding activities in *Salmonella typhimurium.* They compared the specificities of both binding activities with the specificity of aspartate chemotaxis, and concluded that the binding activities differed in substrate specificity from the aspartate chemotaxis attractant receptor. However, they did note that these activities may be similar in properties to the two aspartate transport systems described by Kay (1971). The pure, soluble binding protein was assumed to have a high isoelectric point, like the *E. coli* glutamate–aspartate-binding protein, based on its behaviour on chromatographic columns and gels. A molecular mass of 30 000 daltons was determined by use of SDS-PAGE. A K_D of 1 μM was determined for aspartate binding which was inhibited by glutamate, cysteate, and 2-amino-3-phosphonopropionate. Cysteate was also found to inhibit aspartate transport in whole cells. Aspartate binding exhibited a broad pH optimum and was unaffected by EDTA or dithiothreitol. The membrane-bound aspartate binding activity was not inhibited by glutamate, but was inhibited by aspartylphenylalanine, aspartyl-β-benzyl ester, and cysteate. From the nature of the inhibition of substrate binding to the soluble and membrane-bound activities, the authors suggested that there was more than one aspartate transport system present in this organism.

The role of the glutamate–aspartate-binding protein in the transport of these amino acids in *E. coli* was initially difficult to rationalize, since the following perplexing observations needed to be accounted for:

(a) Although sodium had been shown to stimulate glutamate transport (Frank and Hopkins, 1969), it had no effect on substrate binding to the isolated protein (Willis and Furlong, 1975).
(b) Aspartate was a potent competitive inhibitor of glutamate binding to the isolated protein (Barash and Halpern, 1975; Willis and Furlong, 1975), but only a weak, non-competitive inhibitor of glutamate transport in whole cells (Halpern and Even-Shoshan, 1967).
(c) Glutamate was a weak inhibitor of aspartate uptake in whole cells (Kay, 1971), but was an effective inhibitor of aspartate binding to the isolated protein (Barash and Halpern, 1975; Willis and Furlong, 1975).
(d) Some vesicle preparations had marked aspartate transport and little glutamate transport (Willis and Furlong, 1975), while other vesicle preparations had both glutamate and aspartate transport (Lombardi and Kaback, 1972).
(e) Binding protein levels in mutants with elevated glutamate transport were not consistent with the fold increase in glutamate uptake (Barash and Halpern, 1975).
(f) Culture conditions which altered the level of glutamate transport did not effect corresponding changes in the level of the glutamate–aspartate-binding protein (Kahane *et al*., 1976).
(g) α-Methylglutamate, an effective inhibitor of glutamate transport, did not inhibit substrate binding to the isolated binding protein (Schellenberg and Furlong, 1977).
(h) β-Hydroxyaspartate, an inhibitor of aspartate transport in whole cells (Kay, 1971), did not inhibit substrate binding to the isolated protein (Schellenberg and Furlong, 1977).

Recently, Schellenberg and Furlong (1977) presented the model for glutamate and aspartate transport shown below (Figure 3), which was based on genetic, kinetic, and biochemical observations. The model appears to be consistent with most of the existing observations on the transport of these two amino acids. Five separate systems appear to recognize glutamate and/or aspartate in *E. coli*. The aspartate-specific system (System 1) recognizes aspartate (K_m = 3.7 μM) (Kay, 1971) but not glutamate; this system appears to be of the membrane-bound type. System 2 (Figure 3) recognizes both glutamate (K_m = 5 μM) and aspartate (K_m = 5 μM), is inhibited by β-hydroxyaspartate, and also appears to be a membrane bound type of system. System 3 is specific for glutamate and has the unusual property for an *E. coli* transport system of being highly dependent on

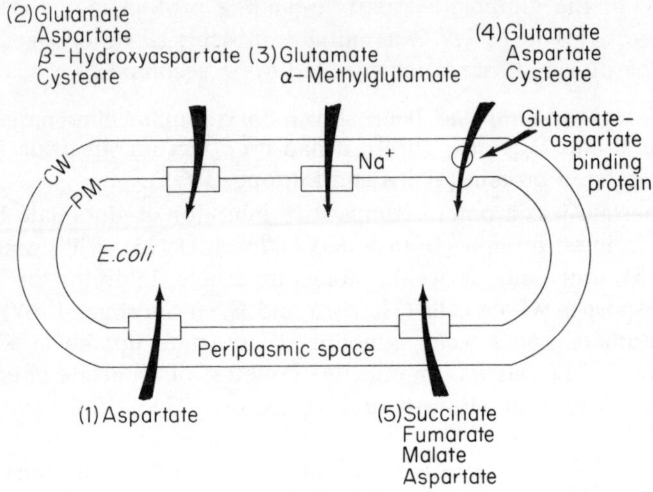

Figure 3 A model for the multiplicity of glutamate and aspartate transport in *E. coli* W. The model presents the specificity of each of the five transport systems which recognize glutamate and/or aspartate. (Reproduced from Schellenberg and Furlong (1977) with permission of the American Society of Biological Chemists, Inc.)

sodium ion (Frank and Hopkins, 1969; Tsuchiya *et al.*, 1977; MacDonald *et al.*, 1977). In the presence of sodium ion, this system has a K_m of 1.5 μM, while at lower concentrations of sodium ion, the K_m appears to be substantially higher (Kahane *et al.*, 1975). The glutamate–aspartate-binding protein (GA-BP) is associated with only one of the five separate transport systems (System 4) for these two amino acids. Transport through the binding-protein-dependent system was characterized through the use of a mutant which was missing Systems 1 and 2, and which had an eight-fold increase in the level of the GA-BP and its corresponding transport system. K_m values of 0.5 μM were observed for the uptake of both amino acids in this mutant, and these values were in good agreement with the respective K_D values reported for the homogeneous binding protein (Willis and Furlong, 1975). Under the conditions used for the K_m determination, the membrane-bound glutamate-specific system did not interfere, since it requires sodium ion. The dicarboxylic acid transport system (*dct* system) (System 5) is distinct from the above four systems and recognizes succinate, malate, fumarate, and aspartate. The *dct* system apparently has a low affinity for aspartate when compared with the other aspartate systems described above. Growth experiments, however, have shown that the *dct* system is important for growth on aspartate as a nitrogen and carbon source (Kay, 1971; Leonard, 1978). Gene products associated with Systems 1, 2, and 3 have not yet been identified, however, proteins have been described which appear to be associated with the *dct* system. Thus,

Lo and Sanwal (1975*a,b*) used affinity chromatography to isolate three proteins which appear to be associated with dicarboxylic acid transport in *E. coli*. A protein with a molecular mass of 15 000 daltons was isolated from osmotic shock fluid (Lo and Sanwal, 1975*a*). The reported K_D values were: 40 μM for succinate, 55 μM for fumarate, 34 μM for malate, and 1.3 mM for lactate. Other dicarboxylic acids such as aspartate, malonate, tartrate, and *N*-hydroxy succinimide had no significant effect on succinate binding. *N*-Ethylmaleimide, which was found to inhibit succinate transport in whole cells but not in membrane vesicles, did inhibit succinate binding to the isolated protein. Zinc, which is a potent inhibitor of succinate transport in whole cells and in membrane vesicles, had no effect on succinate binding to the isolated binding protein. A further interesting finding was that sulphydryl reducing agents decreased the affinity of the isolated protein for succinate. A succinate transport mutant, which mapped at the *cbt* locus, appeared to have a very low level of the binding protein.

Lo and Sanwal (1975*b*) also used affinity chromatography to purify two detergent-solubilized, membrane-bound, succinate-binding activities, SBP 1 and SBP 2, which bound succinate with K_D values of 23 and 2.3 μM, respectively. SBP 1 also bound malate and fumarate, with K_D values around 47 μM as did SBP 2, with K_D values around 7 μM. D-Lactate, citrate, acetate, tartrate, aspartate, and malonate did not inhibit succinate binding. A mutant which maps at the *dctA* locus had a low level of the SBP 2 protein.

The failure of aspartate to inhibit succinate binding to the three isolated succinate-binding proteins is curious, since aspartate has been shown to be transported by the *dct* system (Kay, 1971). Another curious point is that while the authors suggest that there is only a single *dct* transport system, binding-protein-dependent systems for other substrates which have been studied in detail are inactive in membrane vesicles, and yet succinate is transported by vesicle preparations (Rayman *et al.*, 1972). The *dct* system may of course be unique; however, the possibility of more than one system should perhaps be considered. Recent studies by Leonard (1978) suggest that malate also may be transported by the membrane-bound aspartate and glutamate–aspartate transport systems in *E. coli*.

A membrane-bound, glutamate-binding activity has been solubilized from *Halobacterium halobium* (Lanyi, 1977). The solubilized binding activity had a K_D for glutamate binding of 6×10^{-8} M and an apparent molecular mass of 50 000 daltons, as judged by agarose column chromatography in the presence of detergent. The solubilized glutamate-binding activity is capable of supporting glutamate transport in synthetic vesicles prepared from soybean phospholipids. The 'reconstituted' transport activity appeared to be facilitated rather than active. There was no stimulation of either binding or uptake by sodium ion, which appears to drive glutamate uptake in whole cells or membrane vesicles of this organism (see also Section 1.7.3.3).

6 Cystine

The understanding of cystine transport began with the observation that an *E. coli* diaminopimelic (DAP) auxotroph, actually required both DAP and lysine, although DAP alone should have fulfilled this requirement, since it is the immediate precursor of lysine (Davis, 1952). Subsequently, it was found that this observation could be explained by the low velocity of the DAP transport system which could not transport DAP at a rate sufficient to supply the cell's need for lysine (Leive and Davis, 1965). A mutant with a fivefold increase in the velocity of DAP transport lost the partial requirement for lysine. Interestingly, cystine transport also increased fivefold in the mutant. Kinetic studies showed that cystine was an effective inhibitor of DAP transport, and that some strains also possessed an additional cystine transport system which was not shared with DAP.

Berger and Heppel (1972) subsequently characterized an osmotic shock-sensitive, cystine–DAP transport system (cystine general system, $K_m = 1 \times 10^{-7}$ M), and an osmotic shock-resistant, cystine-specific system ($K_m = 2 \times 10^{-8}$ M). They also purified and characterized a cystine–DAP-binding protein (C-BP). DEAE–cellulose chromatography of osmotic shock–releasable proteins was used to resolve two cystine-binding peaks (C-BP I and II). C-BP I was further purified to homogeneity by use of gel filtration chromatography, followed by preparative electrofocusing. C-BP II was purified to homogeneity by use of preparative disc-gel electrophoresis and electrofocusing. Both proteins were found to be homogeneous by disc-gel electrophoresis at two pH values using gels made with several different percents acrylamide, as well as by electrophoresis in the presence of sodium dodecyl sulphate. However, it was impossible to distinguish the two homogeneous proteins from each other by a number of physical, chemical, and immunological criteria, and the authors were unable to explain why two peaks were resolved using DEAE–cellulose chromatography. (Oshima *et al.* (1974) found only one cystine-binding protein peak following DEAE–cellulose chromatography, using a somewhat different purification procedure—see below.) The pure C-BP had a broad pH optimum (pH 3–10) for cystine binding. The binding of substrate was insensitive to ionic strength, and readily renatured after boiling for 10 min, or after treatment with 8 M urea. Molecular mass values of 27 000, 28 000, and 27 000 daltons were determined by use of sedimentation equilibrium, SDS–PAGE and gel filtration, respectively. Cystine binding and transport were not inhibited by other naturally occurring amino acids, although both transport and binding were inhibited by cystine hydroxamate, and cystine dimethylester, as well as by several analogues with an α,α'-diaminodicarboxylate structure. Mutants selected from strain D2W (which had only the C-BP-dependent transport system) on the basis of their resistance to selenocystine, lost both cystine and DAP transport, as well as the C-BP. A cystine-specific binding activity, which was apparently distinct from the C-BP, was observed in cells grown on rich medium although it was not further characterized. Under these culture conditions, the cystine–DAP general system was repressed, and curiously, it was now sensitive to osmotic shock.

Oshima *et al*. (1974) reported the use of the C-BP in assaying amino acid pool sizes in cultured fibroblasts derived from normal subjects and individuals with nephropathic cystinosis. As noted above, cystine binding to the C-BP was not inhibited by other amino acids commonly found in proteins, and thus it has the specificity required for radiodilution assays. They also described the adaptation of the high yield, cell-growth, osmotic shock procedure to the purification of C-BP. Their purification resulted in 86 mg of high specific activity binding protein after only two column chromatographic steps, with a yield of 80 per cent.

7 Proline

The transport of proline in *E. coli* appears to be mediated by a membrane-bound type of transport system (Kaback, 1970). Proline-binding activity has been solubilized from *E. coli* membrane vesicles with the non-ionic detergent, Brij-36T (Gordon *et al.*, 1972). Fractionation of the solubilized material by Sephadex G-100 in the presence of the same detergent revealed three proline-binding fractions. Fraction I, which eluted in the void volume of the column, had D-lactate dehydrogenase activity coincident with the proline-binding activity. Fraction II also had both the binding and dehydrogenase activities, while Fraction III, which had the highest specific activity of binding, did not have the dehydrogenase activity. Proline binding to this third fraction was not inhibited by isoleucine, alanine, histidine, valine, lysine, serine, glycine, or leucine, although it was inhibited by unlabelled proline, and by heating at 100 °C for 5 min, or by treatment with 2.7 M urea. This fraction also bound lysine, serine, tyrosine, and glycine. The sulphydryl reagents, *N*-ethylmaleimide and *p*-chloromercuribenzoate inhibited binding, and the inhibition by the latter compound was reversed by dithiothreitol. This pattern of sulphydryl reactivity is characteristic of membrane-bound transport systems when transport is assayed in membrane vesicles. In contrast, other inhibitors of proline transport, such as sodium amytal, 2-heptyl-4-hydroxyquinoline-*N*-oxide, 2,4-dinitrophenol, carbonyl-cyanide-*m*-chlorophenylhydrazone and valinomycin, did not inhibit binding. The proline-binding activities were not further fractionated and the relationship of these proteins to proline transport is currently not clear.

A quite different approach was used in studying the proline transport system of *Bacillus subtilis* (Kusaka *et al.*, 1976). The transport of both alanine and proline was reconstituted in phospholipid vesicles by incorporating into the vesicles the heavy membrane subfraction obtained from haemin-grown *B. subtilis* cells. (Previous work (Kusaka, 1974) had shown that membranes from this organism could be separated into heavy, medium, and light subfractions by sucrose gradient centrifugation.) Either NADH, or ascorbate-reduced phenazine methosulphate, would serve as an energy source for transport. NADH was the best electron donor for alanine uptake, and reduced phenazine methosulphate was best for proline transport. Addition of valinomycin to K^+-loaded vesicles also energized amino acid uptake (see Section 1.7), and this transport was effectively blocked by carbonyl cyanide *m*-chlorophenylhydrazone.

A similar approach was used to extract a proline carrier from *E. coli* membranes (Amanuma *et al.,* 1977). Proline-binding activity could not be extracted from *E. coli* membranes with a number of detergents, but an acidic butanol extract was active in restoring proline uptake in vesicles prepared from purified *E. coli* phospholipids, provided that a membrane potential (negative inside) was induced by valinomycin-mediated K^+ diffusion from K^+-loaded vesicles. The protein fraction from one proline transport-deficient mutant was inactive in restoring transport, and that from another was partially active. The reconstituted transport activity was inhibitable by L-proline or D,L-3,4-dehydroproline, as well as with the proton uncoupler SF 6847(3,5,-di-tert-butyl-4-hydroxybenzylidene-malononitrile). The reconstituted vesicles also transported glutamate and aspartate.

The induction of amino acid binding proteins, which appear to be related to transport systems needed for synthesizing the antibiotic tyrocidine in *Bacillus brevis*, has been reported by Lipmann and Lee (1976). This antibiotic is composed of the following amino acids: D-phenylalanine, proline, phenylalanine, asparagine, glutamine, tyrosine, valine, ornithine, and leucine. The induction of osmotic shock-releasable binding proteins for proline, ornithine, and β-alanine appeared to be correlated with the onset of antibiotic synthesis. The β-alanine binding activity had a molecular mass of 10 000 daltons, as judged by results from SDS-PAGE.

8 Alanine

The alanine carrier of the thermophylic bacterium PS3 has been extensively purified and used to reconstitute alanine transport in membrane vesicles (Hirata *et al*., 1976, 1977). The purification procedure involved solubilizing the alanine carrier by sonicating a suspension of membranes in the presence of cholate, deoxycholate, sodium sulphate, and Tris buffer. The solubilized alanine carrier activity was further purified by using two detergent–DEAE–cellulose chromatographic steps (the second in the presence of urea), ammonium sulphate fractionation, and a detergent–urea CM–cellulose column step. Sodium dodecyl sulphate polyacrylamide gel electrophoresis of the purified carrier activity showed only two major bands with molecular mass values of 36 000 and 9400 daltons. Several minor bands were also seen (Hirata *et al.*, 1977).

Reconstitution was achieved by sonicating a mixture of PS3 bacterial phospholipids (other lipids were not effective) and purified protein in the presence of cholate, deoxycholate, dithiothreitol, and Tris buffer. The sonicated mixture was dialysed to remove the detergents. Alanine uptake in the reconstituted vesicles could be driven by a membrane potential generated by adding valinomycin to K^+-loaded vesicles. The reconstituted transport (both initial rate and extent) was linearly related to added protein and was inhibited by the uncoupler, carbonylcyanide *p*-trifluoromethoxyphenylhydrazone, but not by dicyclohexylcarbodiimide (Hirata *et al.*, 1976). To overcome the transient nature of the K^+-loaded vesicle system, beef heart cytochrome oxidase was also incorporated into the vesicles. The addition of cytochrome *c* and ascorbic acid to the assay mixture resulted in a proton motive force, interior negative, which was also able to drive alanine uptake. Incorporation of the TF_0 component of the ATPase complex of the same organism decreased the specific activity of alanine uptake, presumably by conducting protons. The resulting decrease in specific activity was partially reversed by dicyclohexylcarbodiimide, a known TF_0 inhibitor (Hirata *et al.*, 1977).

9 Tryptophan

Wiley (1970) reported the partial purification of an osmotic shock-releasable tryptophan binding activity from *Neurospora crassa*. The K_D for tryptophan binding was 8×10^{-5} M, which was close to the previously determined K_m value of 5×10^{-5} M for the cellular transport system (Wiley and Matchett, 1968). Leucine and phenylalanine, which were inhibitors of tryptophan transport, inhibited tryptophan binding to the partially resolved protein fraction. Unlike osmotic shock-releasable binding activities derived from bacterial sources, the *N. crassa* tryptophan-binding activity appeared to be quite large, with a molecular mass greater than 200 000 daltons.

10 Methionine

A methionine-binding activity has been extracted from *Aspergillus nidulans* with Triton X-100 by Stepien (1976). The extracted protein was partially purified on an affinity column of L-methionine–CH-Sepharose. The molecular mass, estimated by use of SDS-PAGE was 15 500 daltons. The protein was apparently absent in a methionine transport-deficient mutant, isolated on the basis of its resistance to D,L-selenomethionine and *p*-fluorophenylalanine. It was postulated that the binding protein is a component of the neutral amino acid permease of this organism.

11 Phenylalanine

Kuzuya *et al.* (1971) have described the purification of a phenylalanine-binding protein (P-BP) from *Comamonas sp.* (ATTC 11299a). The isolated P-BP had an isoelectric point of 9 and appeared to be homogeneous as judged by electrophoresis at pH 4.3 and by SDS-PAGE. Molecular mass values of 24 000, 27 000, and 25 000 daltons were determined by use of gel filtration (in the presence of ligand), sedimentation velocity, and SDS-PAGE, respectively. A broad pH optimum of pH 5–9 was observed, and binding was inhibited by tryptophan and tyrosine as well as by several structural analogues. Binding was stimulated by sodium chloride, and reversibly denatured by extreme pH values or by urea. Chemical studies indicated that particular lysine residues of the protein were important for activity. From their studies, these authors concluded that the properties of the isolated protein were different from those of the whole-cell transport system (Guroff and Bromwell, 1970). However, they seem not to have considered the possibility of multiple transport systems for this amino acid. In the light of the high multiplicity of aromatic amino acid transport systems observed in other organisms (Ames, 1964; Ames and Roth, 1968), it may well be that the binding protein is a component of one of several aromatic amino acid transport systems.

An osmotic shock-releasable phenylalanine-binding protein from *E. coli* has also been reported (Klein *et al.*, 1970).

3 GENERAL COMMENTS

1 Nomenclature

Some confusion exists with respect to the nomenclature of the proteins associated with specific transport systems, as well as with the nomenclature of the transport systems themselves. A simple way of naming the binding proteins would be based on the ligands that they bind and their relative affinities for these ligands. In some cases this has already been done. For example, the leucine–isoleucine–valine–threonine-binding protein (LIVT-BP) was first called the leucine-binding protein (Anraku, 1968*a*; Penrose *et al.*, 1968*b*); then, the leucine–isoleucine–valine-binding protein (LIV-BP) (Furlong and Weiner, 1970); and most recently, the current designation (LIVT-BP) (Amanuma and Anraku, 1974). The second protein member of this family of binding proteins is appropriately referred to as simply the leucine-binding protein (L-BP) (Furlong and Weiner, 1970). The third protein member of this family was first referred to as the isoleucine-preferring binding protein (Furlong *et al.*, 1973; Rahmanian *et al.*, 1973). Following further characterization of this protein, it was referred to as the second leucine–isoleucine–valine–threonine-binding protein (sLIVT-BP) (Amanuma *et al.*, 1976). However, this becomes somewhat confusing, since there is a separate, membrane-bound, branched-chain amino acid transport system which is called the LIV-II system (Rahmanian *et al.*, 1973). It would seem to be more appropriate to refer to the isoleucine-preferring protein as the ILVT-BP. Thus, the branched-chain amino acid binding proteins would be referred to simply as the LIVT-BP, L-BP, and ILVT-BP. The respective transport systems would simply be referred to as the X-BP-dependent transport system or the X-BP transport system.

A simple, uniform nomenclature for the membrane-bound transport systems (and their respective protein components, once they become identified) would denote the specificity of the system, followed by the capital letter 'M' to indicate that the system in question was a membrane-bound type of transport system. Thus, the LIV-II system would be the LIV-M system, readily distinguishing it from the binding protein-dependent transport system. In the case of the sodium-dependent transport systems, a superscript 'Na^+' could be used.

In higher organisms, where systems of broader specificity are often found, more general appelations of transport systems are probably appropriate. However, as noted above, the tryptophan-binding protein of *Neurospora crassa* also binds leucine and phenylalanine (Wiley, 1970), and the arginine-binding activites of the same organism also appear to bind other amino acids (Stuart and DeBusk, 1971) (see also Section 1.3.2.3).

In short, it seems reasonable to name a binding protein or transport system in such a way as to reflect the specificity of nutrient recognition, and if possible, the basic mechanism employed by the given transport system.

2 General Approach to Isolating Protein Components of Transport Systems

So far, the protein components of the membrane-bound type of transport systems have been difficult to identify and purify. However, as discussed above, some progress in this area has been made recently. We would propose that the following general approach may well prove to be a useful one for identifying the more elusive proteins.

In a wild-type cell, the protein(s) associated with a given transport system often represent only a small fraction of the total proteins present. Therefore, one of the first efforts should be directed at generating a mutant which overproduces the transport system in question. This is conveniently done by requiring the organism to grow on a given nutrient (or limiting concentration of the given nutrient) as a sole source of carbon, nitrogen, phosphorous, sulphur, *etc*., or requiring an auxotroph to grow on a low concentration of the required nutrient. Once a mutant with an increased level of transport is isolated, it should be characterized with respect to specificity and basic mechanism (i.e., binding protein-dependent or membrane-bound). Following this step, attempts should be made to isolate a frame-shift, mutagen-generated, transport-negative mutant (e.g., on the basis of analogue resistance, or through the use of a penicillin counter-selection procedure). The next step would be a comparison of the protein profiles of the parent and two mutants by two-dimensional gel analysis. The overproducer mutant should facilitate the locating of a protein spot, which should be present in the parent and missing in the transport negative mutant.

Figure 4 illustrates part of this general approach. Figure 4(a) is a two-dimensional gel (O'Farrell, 1975) of proteins from a mutant which overproduces the glutamate–aspartate binding protein (Schellenberg, 1978). Figure 4(b) is an identical gel of the parent strain. It is easy to spot the overproduction of the binding protein, which has a very basic isoelectric point (9.7). A transport-negative mutant has not yet been isolated for this system. However, it is clear from the gel profile that such a mutant would not be detectable in a single-dimension SDS gel since there are so many proteins with the same molecular weight. A transport-negative, binding-protein negative mutant might possibly be detectable on a one-dimensional isoelectric focusing gel. However, as a general approach, two-dimensional gel analysis of the parent, transport-super, and frame-shift, mutagen-generated, transport-negative mutants seems to be a reasonable approach for identifying the protein products associated with specific transport systems. The example used was for a soluble protein, however, it should work as well for a membrane-bound protein since these proteins are also readily resolved by two-dimensional gel analysis (Ames and Nikaido, 1976).

Another approach, that of using the reconstitution of transport function as an assay for the protein carrier, appears to be working well in some of the cases discussed above.

3 Dissociation Constants

Depending on the method of isolation, a number of substrate-recognition components of transport systems may be isolated with ligands attached (Amanuma *et al.*, 1976; Richarme and Kepes, 1974; Silhavy *et al.*, 1975; Silhavy and Boos, 1975). Thus, it may be important to re-evaluate K_D values which were determined under conditions where the binding protein-donated

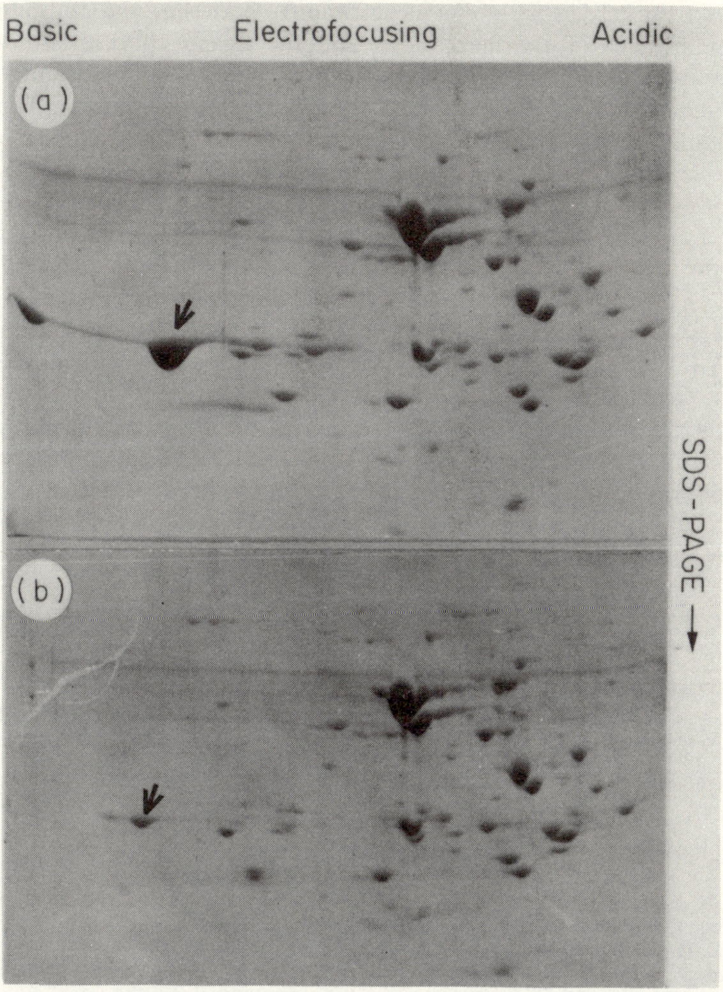

Figure 4 Two-dimensional gels of (a) an *E. coli* mutant which overproduces the glutamate–aspartate binding protein and (b) the parent strain, *E. coli* D2W

substrate may have significantly diluted the specific activity of the added labelled ligand. In future characterizations of high-affinity substrate binding, it would seem advisable to assume that the binding protein has an equimolar amount of attached ligand (unless, for example, it has been removed by dialysis against urea) and to carry out the binding assays under conditions where the added labelled ligand is 10–20-fold in excess of the possible binding protein-donated, unlabelled ligand. Since the protein concentration is more or less fixed at a value that provides a reasonable difference in counts (or total substrate concentration) between the protein and non-protein sides of the membrane, and the ligand concentration varies between approximately $0.1 \times K_D$ and $10 \times K_D$, the most important variable is the ratio of the volumes on the two sides of the membrane. Thus, the experimental design should include considerations for possible protein donated cold ligand. Assay volume ratios should be adjusted to provide a sufficient excess of added labelled ligand, so that any protein-donated ligand will not dilute the specific activity of the added ligand.

Any protein-bound substrate could also be important in the studies on physical changes in the proteins which occur during ligand binding. Some of the observed physical changes have been rather small (Weiner and Heppel, 1971; Willis and Furlong, 1975), and it would not be surprising to see more dramatic changes on ligand binding if the proteins were first stripped of ligand before the measurements were made. As far as the ligand protein interaction itself is concerned, it seems possible that the ligand interaction with 'apoprotein' might be quite different from simply a ligand exchange reaction, which possibly has been the reaction characterized in many of the cases investigated so far.

ACKNOWLEDGEMENTS

The experiments described in Figures 1, 3 and 4 were supported by Research Grant No. GM-18311 from the United States Public Health Service. C. E. Furlong is the recipient of United States Public Health Service Research Career Development Award No. 5 K04 GM 00452.

REFERENCES

Aksamit, R. R., Howlett, B. J., and Koshland, D. E. (1975). *J. Bacteriol.*, **123**, 1000–1005
Amanuma, H. and Anraku, Y. (1974). *J. Biochem., Tokyo*, **76**, 1165–1173
Amanuma, H., Itoh, J., and Anraku, Y. (1976). *J. Biochem., Tokyo*, **79**, 1167–1182
Amanuma, H., Motojima, K., Yamaguchi, A., and Anraku, Y. (1977). *Biochem. Biophys. Res. Commun.*, **74**, 366–373
Ames, G. F.-L. (1964). *Arch. Biochem. Biophys*., **104**, 1–18
Ames, G. F.-L. and Lever, J. (1970). *Proc. Natl. Acad. Sci. U.S.A*., **66**, 1096–1103
Ames, G. F.-L. and Lever, J. E. (1972). *J. Biol. Chem.*, **247**, 4309–4316
Ames, G. F.-L. and Nikaido, K. (1976). *Biochemistry*, **15**, 616–623
Ames, G. F. and Roth, J. R. (1968). *J. Bacteriol.,* **96**, 1742–1749
Ames, G. F.-L. and Spudich, E. N. (1976). *Proc. Natl. Acad. Sci. U.S.A.,* **73**, 1877–1881
Ames, G. F.-L., Noel, K. D., Taber, H., Spudich, E. N., Nikaido, K., Afong, J., and Ardeshir, F. (1977). *J. Bacteriol.*, **129**, 1289–1297.
Anderson, J. J. and Oxender, D. L. (1977). *J. Bacteriol*., **130**, 384–392
Anraku, Y. (1968*a*). *J. Biol. Chem.*, **243**, 3116–3122
Anraku, Y. (1968*b*). *J. Biol. Chem.*, **243**, 3123–3127
Anraku, Y. (1968*c*). *J. Biol. Chem.*, **243**, 3128–3135
Anraku, Y. and Heppel, L. A. (1967). *Fed. Proc. Fed. Am. Soc. Exp. Biol.*, **26**, 393
Anraku, Y., Kobayashi, H., Amanuma, H., and Yamaguchi, A. (1973). *J. Biochem., Tokyo*, **74**, 1249–1261

Antonov, V. K., Alexandrov, S. L., and Vorotyntsvea, T. I. (1976). *Adv. Enzyme. Regul.*, **14**, 269

Antonov, V. K., Arsen'eva, E. L., Gavrilova, N. A., Ginodman, L. M., and Krylova, Y. I. (1974). *Biokhimiya*, **38**, 1088–1091

Barash, H. and Halpern, Y. S. (1971). *Biochem. Biophys. Res. Commun.*, **45**, 681–688

Barash, H. and Halpern, Y. S. (1975). *Biochim. Biophys. Acta*, **386**, 168–180

Berger, E. A. and Heppel, L. A. (1972). *J. Biol. Chem.*, **247**, 7684–7694

Berger, E. A. and Heppel, L. A. (1974). *J. Biol. Chem.*, **249**, 7747–7755

Boos, W. (1974). *Ann Rev. Biochem.*, **43**, 123–146

Britten, R. J. and McClure, F. T. (1959). In *First National Biophysics Conference*, Yale University Press, New Haven, Conn., pp. 414–423

Britten, R. J. and McClure, F. T. (1962). *Bacteriol. Rev.*, **26**, 292–335

Bussey, H. and Umbarger, E. (1970). *J. Bacteriol.*, **103**, 277–285

Colowick, S. P. and Womack, F. C. (1969). *J. Biol. Chem.*, **244**, 774–777

Davis, B. D. (1952). *Nature, Lond.*, **169**, 534

Dietz, G. W. (1972). *J. Biol. Chem.*, **247**, 4561–4565

Frank, L. and Hopkins, L. (1969). *J. Bacteriol.*, **100**, 329–336

Furlong, C. E. and Heppel, L. A. (1970). *Methods Enzymol.*, **17B**, 639–643

Furlong, C. E. and Weiner, J. H. (1970). *Biochem. Biophys. Res. Commun.*, **38**, 1076–1083

Furlong, C. E., Willis, R. C., and Gerber, N. H. (1973*b*). *Abs. 9th Int. Cong. Biochem., Stockholm, Sweden*, p. 278

Furlong, C. E., Cirakoglu, C., Willis, R. C., and Santy, P. A. (1973*a*). *Anal. Biochem.*, **51**, 297–311

Furlong, C. E., Willis, R. C., Morris, R. G., Schellenberg, G. D., and Gerber, N. H. (1973*c*) *Fed. Proc. Fed. Am. Soc. Exp. Biol.*, **32**, 517 (abstract)

Gerdes, R. G., Strickland, K. P., and Rosenberg, H. (1977). *J. Bacteriol.*, **131**, 512–518

Gordon, A. S., Lombardi, F. J., and Kaback, H. R. (1972). *Proc. Natl. Acad. Sci. U.S.A.*, **69**, 358–362

Guardiola, J. and Iaccarino, M. (1971). *J. Bacteriol.*, **108**, 1034–1044

Guardiola, J. and Iaccarino, M. (1974). *J. Bacteriol.*, **117**, 393–405

Guardiola, J., DeFelice, M., Klopotowski, T., and Iaccarino, M. (1974). *J. Bacteriol.*, **117**, 382–392

Guroff, G. and Bromwell, K. E. (1970). *Arch. Biochem. Biophys.*, **137**, 379–387

Halpern, Y. S. (1974). *Annu. Rev. Genet.*, **8**, 103–133

Halpern, Y. S. and Even-Shoshan, A. (1967). *J. Bacteriol.*, **93**, 1009–1016

Hayashi, S., Kochi, J. P., and Lin, E. C. C. (1964). *J. Biol. Chem.,*, **239**, 3098–3105

Heppel, L. A. (1969). *J. Gen. Physiol.*, **54**, 95_s–109_s

Hirata, H., Sone, N., Yoshida, M., and Kagawa, Y. (1976). *Biochem. Biophys. Res. Commun.*, **69**, 665–671

Hirata, H., Sone, N., Yoshida, M., and Kagawa, Y. (1977). *J. Supramol. Struct.*, **6**, 77–84

Hogg, R. and Englesberg, E. (1969). *J. Bacteriol.*, **100**, 423–432

Kaback, H. R. (1970). *Annu. Rev. Biochem.*, **39**, 561–598

Kaback, H. R. (1974). *Science, N.Y.*, **186**, 882–892

Kaback, H. R. and Barnes, E. M., Jr (1971). *J. Biol. Chem.*, **246**, 5523–5531

Kaback, H. R. and Milner, L. S. (1970). *Proc. Natl. Acad. Sci. U.S.A.*, **66**, 1008–1015

Kahane, S., Marcus, M., Barash, H., and Halpern, Y. S. (1975). *FEBS Lett.*, **56**, 235–239

Kahane, S., Marcus, M., Metzer, G., and Halpern, Y. S. (1976). *J. Bacteriol.*, **125**, 762–769

Kay, W. W. (1971). *J. Biol. Chem.*, **246**, 7373–7382

Kellerman, O. and Szmelcman, S. (1974). *Eur. J. Biochem.*, **47**, 139–149

Klein, W. L., Dahms, A. S., and Boyer, P. O. (1970). *Fed. Proc. Fed. Am. Soc. Exp. Biol.*, **28**, 341

Kreishman, G. P., Robertson, D. E., and Ho, C. (1973). *Biochem. Biophys. Res. Commun.*, **53**, 18–23

Kusaka, I. (1974). *Biochim. Biophys. Acta*, **345**, 62–73

Kusaka, I., Hayakawa, K., Kanai, K., and Fukui, S. (1976). *Eur. J. Biochem.*, **71**, 451–458

Kustu, S. G. and Ames, G. F.-L. (1973). *J. Bacteriol.*, **116**, 107–113

Kustu, S. G. and Ames, G. F.-L. (1974). *J. Biol. Chem.*, **249**, 6976–6983

Kuzuya, H., Bromwell, K., and Guroff, G. (1971). *J. Biol. Chem.*, **246**, 6371–6380
Lanyi, J. K. (1977). *J. Supramol. Struct.*, **6**, 169–177
Leive, L. and Davis, B. D. (1965). *J. Biol. Chem.*, **240**, 4362–4369
Leonard, J. E. (1978). Ph.D. Dissertation, University of California, Riverside, Calif.
Lever, J. E. (1972*a*). *J. Biol. Chem.*, **247**, 4317–4326
Lever, J. E. (1972*b*). *Anal. Biochem.*, **50**, 73–83
Lipmann, F. and Lee, S. G. (1976). *J. Cell Physiol.* **89**, 523–528
Lo, T. C. Y. and Sanwal, B. D. (1975*a*). *J. Biol. Chem.*, **250**, 1600–1602
Lo, T. C. Y. and Sanwal, B. D. (1975*b*). *Biochem. Biophys. Res. Commun.*, **63**, 278–285
Lombardi, F. J. and Kaback, H. R. (1972). *J. Biol. Chem.*, **247**, 7844–7857
MacDonald, R. E., Lanyi, J. K., and Greene, R. V. (1977). *Proc. Natl. Acad. Sci. U.S.A.*, **74**, 3167–3170
Medveczky, N. and Rosenberg, H. (1970). *Biochim. Biophys. Acta*, **211**, 158–168
Medveczky, N. and Rosenberg, H. (1971). *Biochim. Biophys. Acta*, **241**, 494–506
Miner, K. M. and Frank, L. (1974). *J. Bacteriol.*, **117**, 1093–1098
Mitchell, P. (1970). *Symp. Soc. Gen. Microbiol.*, **20**, 121–166
Nakane, P. K., Nichoalds, G. E., and Oxender, D. L. (1968). *Science, N.Y.*, **161**, 182
Neu, H. C. and Heppel, L. A. (1965). *J. Biol. Chem.*, **240**, 3685–3692
Nishimune, T. and Hayashi, R. (1971). *Biochim. Biophys. Acta*, **244**, 573–583
O'Farrell, P. H. (1975). *J. Biol. Chem.*, **250**, 4007–4021
Opekarová, M., Kotyk, A., Horak, J., and Kholodenko, V. P. (1975). *J. Biochem., Tokyo*, **59**, 373–376
Oshima, R. G., Willis, R. C., Furlong, C. E., and Schneider, J. A. (1974). *J. Biol. Chem.*, **249**, 6033–6039
Ovchinnikov, Y. A., Aidanova, N. A., Grinkevich, V. A., Arzamazova, N. M., and Movoz, I. N. (1977). *FEBS Lett.*, **78**, 313–316
Oxender, D. L. (1972). *Annu. Rev. Biochem.*, **41**, 777–814
Oxender, D. L. and Quay, S. C. (1976). In *Methods of Membrane Biology*, Vol. 6 (E. D. Korn, ed.), Plenum Press, New York, pp. 183–242
Pardee, A. B. (1966). *J. Biol. Chem.*, **241**, 5886–5892
Penrose, W. R., Nichoalds, G. E., and Oxender, D. L. (1968*a*). *Fed. Proc. Fed. Am. Soc. Exp. Biol.*, **27**, 643
Penrose, W. R., Nichoalds, G. E., Piperno, J. R., and Oxender, D. L. (1968b). *J. Biol. Chem.*, **243**, 5921–5928
Piperno, J. R. and Oxender, D. L. (1966). *J. Biol. Chem.*, **241**, 5732–5743
Postma, P. W. and Roseman, S. (1976). *Biochim. Biophys. Acta*, **457**, 213–257
Rahmanian, M., Claus, D. R., and Oxender, D. L. (1973). *J. Bacteriol.*, **116**, 1258–1266
Rayman, K. M., Lo, T. C. Y., and Sanwal, B. D. (1972). *J. Biol. Chem.*, **247**, 6332–6339
Richarme, G. and Kepes, A. (1974). *Eur. J. Biochem.*, **45**, 127–133
Robbins, A. R., Guzman, R., and Rotman, B. (1976). *J. Biol. Chem.*, **251**, 3112–3116
Robertson, D. E., Kroon, P. A., and Ho, C. (1977). *Biochemistry*, **16**, 1443–1451
Rosen, B. P. (1971). *J. Biol. Chem.*, **246**, 3653–3662
Rosen, B. P. (1973*a*). *J. Biol. Chem.*, **248**, 1211–1218
Rosen, B. P. (1973*b*). *J. Bacteriol.*, **116**, 627–635
Rosen, B. P. and Vasington, F. D. (1971). *J. Biol. Chem.*, **246**, 5351–536
Rosenberg, H., Gerdes, R. G., and Chegwidden, K. (1977). *J. Bacteriol.*, **131**, 505–511
Schellenberg, G. D. (1978). Ph.D. Dissertation, University of California, Riverside, Calif.
Schellenberg, G. D. and Furlong, C. E. (1977). *J. Biol. Chem.*, **252**, 9055–9064
Shaltiel, S., Ames, G. F.-L., and Noel, K. D. (1973). *Arch. Biochem. Biophys.*, **159**, 174–179
Silhavy, T. J. and Boos, W. (1975). *Eur. J. Biochem.*, **54**, 163–167
Silhavy, T. J., Szmelcman, S., Boos, W., and Schwartz, M. (1975). *Proc. Natl. Acad. Sci. U.S.A.*, **72**, 2120–2124
Simoni, R. D. and Postma, P. W. (1975). *Annu. Rev. Biochem.*, **44**, 523–554
Slayman, C. W. (1973). *Curr. Topics Membranes Transport*, **4**, 1–174
Stepien, P. O. (1976). *Biochim. Biophys. Acta*, **439**, 154–159
Stuart, W. D. and De Busk, A. G. (1971). *Archs Biochem. Biophys.*, **144**, 512–518
Stuart, W. D. and De Busk, A. G. (1975). *Archs Biochem. Biophys.*, **166**, 213–222

Stuart, W. D. and Stuart, A. G. (1973). *Biochem. Biophys. Res. Commun.*, **52**, 1046–1050
Taylor, R. T., Norell, S. A., and Hanna, M. L. (1972). *Arch. Biochem. Biophys.*, **148**, 366–381
Tokuda, H. and Kaback, H. R. (1977). *Biochemistry*, **16**, 2130–2136
Tsuchiya, T., Hasan, S. M., and Raven, J. (1977). *J. Bacteriol.*, **131**, 848–853
Weiner, J. H. and Heppel, L. A. (1971). *J. Biol. Chem.*, **246**, 6933–6941
Weiner, J. H., Furlong, C. E., and Heppel, L. A. (1971). *Arch. Biochem. Biophys.*, **142**, 715–717
Wiley, W. R. (1970). *J. Bacteriol.*, **103**, 656–662
Wiley, W. R. and Matchett, W. H. (1968). *J. Bacteriol.*, **95**, 959–966
Willis, R. C. and Furlong, C. E. (1974). *J. Biol. Chem.*, **249**, 6926–6929
Willis, R. C. and Furlong, C. E. (1975). *J. Biol. Chem.*, **250**, 2574–2580
Willis, R. C., Morris, R. G., Cirakoglu, C., Schellenberg, G. D., Gerber, N. H., and Furlong, C. E. (1974). *Arch. Biochem. Biophys.*, **161**, 64–75
Wilson, O. H., and Holden, J. T. (1969*a*). *J. Biol. Chem.*, **244**, 2737–2742
Wilson, O. H. and Holden, J. T. (1969*b*). *J. Biol. Chem.*, **244**, 2743–2749
Wood, J. M. (1975). *J. Biol. Chem.*, **250**, 4477–4485

Microorganisms and Nitrogen Sources
Edited by J. W. Payne

Section 1.5

Genetics of Amino Acid Transport

MAURIZIO IACCARINO, JOHN GUARDIOLA, AND MAURILIO DE FELICE
The International Institute of Genetics and Biophysics, Via G. Marconi 10, 80125 Napoli, Italy

1 INTRODUCTION

In this chapter we review the ways in which genetic approaches have been used for the analysis of amino acid transport in microorganisms. During the initial phases of the studies on transport, mutant analysis contributed to the idea of the presence of specific permeases (Cohen and Monod, 1957) as opposed to the view of a semipermeable membrane allowing 'diffusion' of small molecular weight substances. We will use here the term transport system (and not permease) to define an object catalysing the passage of a specific set of amino acids through the membrane (see also Section 1.1.2). Mutant strains have been very useful (and sometimes essential) to define a specific transport system. Physiological studies on these mutant strains have increased our understanding of the relevance of a transport system in the cell economy.

Most of the genetic studies on amino acid transport concern the Gram-negative bacteria *Escherichia coli* K-12 and *Salmonella typhimurium*. Since these two species are very similar to each other (Sanderson, 1976) we will discuss them in the same section. A major distinction among bacterial species, namely their classification as Gram-negative or Gram-positive, is related to differences in their envelope structure (Mandelstam and McQuillen, 1973). The outermost layer of the Gram-negative species, namely the outer membrane, is very different from the outermost layers of the Gram-positive species which uniquely contain teichoic or teichuronic acid (Archibald, 1974). These differences probably affect the permeability properties and therefore we discuss Gram-negative bacteria separately from Gram-positive bacteria.

Recent reviews on the genetics of amino acid transport in microorganisms are those of Lin (1970), Slayman (1973), and Halpern (1974).

2 GRAM-NEGATIVE BACTERIA

Table 1, derived from data of Bachmann *et al*. (1976), shows the genetic markers involved in amino acid transport in *E. coli* K-12. Table 2, derived from data of Sanderson and Hartman (1978), shows the genetic markers involved in amino acid transport in *S. typhimurium*.

Early information relating to genetic aspects of transport systems for amino acids in Gram-negative bacteria has been recently reviewed (Lin, 1970; Slayman, 1973; Halpern, 1974) and therefore, in this section, we will concentrate on the papers that have appeared after the publication of these reviews.

1 Histidine and Aromatic Amino Acids

A summary of the transport systems involved in the uptake of these amino acids in *S. typhimurium* is given in Figure 1, which is adapted from a diagram of Ames (1972) (see also Section 1.4.2.3). The transport system for histidine having the highest affinity requires the presence of two genes, *hisJ* and *hisP*. Most of the studies on this system have been carried out in a strain containing the *dhuA*, operator site-mutation (Krajeska-Grinkiewicz *et al*., 1971) which causes derepression of the transport rate (Ames and Lever, 1970). A fine structure map of the *hisJ* and *hisP* genes was recently reported (Ames *et al*., 1977). These authors described the isolation of 25 deletion mutants, dividing the *hisJ* and the *hisP* genes into eight and 13 regions respectively; a total of 308 mutations were ordered in these regions

Table 1 Genetic markers for amino acid transport in *Escherichia coli* K-12[a]

Genetic symbol	Map position (min)	Phenotype of mutants and functions involved	References
argP	62	Transport of arginine, lysine, and ornithine	Celis *et al*., 1973; Maas, 1965, 1972
aroP	2	Aromatic amino acid transport	Brown, 1970, 1971; Guest, 1974
aroR	27	Aromatic amino acids, alanine, and glycine transport	Thorne and Corwin, 1975
aroS	27	Aromatic amino acids, alanine and glycine, transport	Thorne and Corwin, 1975
brnQ	8	High-affinity-1 transport system for branched-chain amino acids	Guardiola *et al*., 1974*b*; Guardiola and Iaccarino, 1971
brnR	8	Very high-affinity, high-affinity-1 and high-affinity-2 transport systems for branched-chain amino acids	Guardiola *et al*., 1974*b*
brnS	1	High-affinity-2 transport system for branched-chain amino acids	Guardiola *et al*., 1974*b*
brnT	61	Very high-affinity transport system for branched-chain amino acids, and isoleucine-specific low-affinity transport system	Guardiola *et al*., 1974*b*; Iaccarino *et al*., 1978
cbt	16	Uptake of dicarboxylic acids	Lo *et al*., 1972; Lo and Sanwal, 1975*a*
cycA	94	Resistance to D-cycloserine and D-serine; transport of D-alanine, glycine, and D-serine	Curtiss *et al*., 1965; Russel, 1972; Wargel *et al*., 1971
dagA	94	Resistance to D-cycloserine and D-serine; transport of D-alanine, glycine, and D-serine	Cosloy, 1973; Robbins and Oxender, 1973
dctA	79	Uptake of C_4-dicarboxylic acids	Kay and Kornberg, 1969; Lo *et al*., 1972
dctB	16	Uptake of C_4-dicarboxylic acids	Kay and Kornberg, 1969; Lo *et al*., 1972; Lo and Sanwal, 1975*a*
gltC	81	Operator locus for glutamate transport	Marcus and Halpern, 1967, 1969
gltR	91	Regulatory gene for glutamate transport	Marcus and Halpern, 1967
gltS	81	Glutamate transport system	Marcus and Halpern, 1967
livH	74	Very high-affinity transport system for branched-chain amino acids	Anderson and Oxender, 1977
livJ	74	Very high-affinity transport system for branched-chain amino acids. Coding for the isoleucine, leucine, and valine binding protein	Anderson and Oxender, 1977

Table 1 (*Continued*)

Genetic symbol	Map position (min)	Phenotype of mutants and functions involved	References
livK	74	Leucine-specific high-affinity transport system. Coding for the leucine-binding protein	Anderson and Oxender, 1977
livR	20	Very high-affinity transport system for branched-chain amino acids: regulatory gene	Anderson *et al.*, 1976
lstR	20	Leucine-specific high-affinity transport system: regulatory gene	Anderson *et al.*, 1976
metD	5	Uptake of D- and L-methionine	Cooper, 1966; Kadner, 1974; Kadner and Watson, 1974
trpP	27	Uptake of tryptophan	Kuhn and Somerville, 1971

[a]Derived from Bachmann *et al.* (1976), with more recent additions.

by deletion mapping. The orientation of the operon is *purF*, *dhuA* (operator) *hisJ*, *hisP*. This map, together with that of the lactose transport system of *E. coli* (Hobson *et al.*, 1977), is presently the most detailed for any transport system.

The product of the *hisJ* gene is the histidine-binding protein (Ames and Lever, 1970) which can be released from *S. typhimurium* cells by the cold, osmotic-shock procedure of Heppel (1967) (see Section 1.4.1). This protein has been purified to homogeneity from both wild-type *S. typhimurium* (Rosen and Vasington, 1971) and from a mutant strain showing a fivefold derepressed transport rate because of the presence of the *dhuA* mutation (Lever, 1972). The histidine-binding protein is water-soluble, has a molecular mass of 25 000 daltons and binds one molecule of histidine with a K_D of 1.5 μM. Ames and Lever (1972) isolated a *hisJ* mutant with an altered histidine-binding protein. The protein isolated from this mutant strain showed increased temperature sensitivity and altered chromatographic and electrophoretic mobilities. Since the *hisJ*

Table 2 Genetic markers for amino acid transport in *Salmonella typhimurium*[a]

Genetic symbol	Map position (min)	Phenotype of mutants and functions involved	References
argP	Not mapped	Arginine transport	Quay and Christensen, 1974
aroP	3	Aromatic amino acid transport	Bahramian and Middleton, 1973

Table 2 (*Continued*)

Genetic symbol	Map position (min)	Phenotype of mutants and functions involved	References
aroT	34	Ability to transport phenylalanine, tryptophan, tyrosine, alanine, and glycine	Thorne and Corwin, 1975
brnQ	8	Branched-chain amino acid transport	Kiritani, 1974; Kiritani and Ohnishi, 1977
cysB	34	Positive control of cystine transport	Borum and Monty, 1976, and unpublished data quoted by Sanderson and Hartman (1978); Cheney and Kredich, 1975; Qureshi *et al*., 1975
dctA	78	Transport of dicarboxylic acids	Parada *et al*., 1973; Straus and Hoffman, 1975
dhuA	49	Utilization of D-histidine; increased activity of histidine-binding protein J	Ames, 1972, 1973, 1976; Ames *et al*., 1977; Kustu and Ames, 1973
glnP	17	High-affinity transport system for glutamine	Betteridge and Ayling, 1975, 1976
hisJ	49	Histidine-binding protein J for histidine transport	Ames, 1972, 1973, 1976; Ames *et al*., 1977; Ames and Spudich, 1976; Shaltiel *et al*., 1973; Willis *et al*., 1974
hisP	49	Histidine transport	Ames, 1972, 1973, 1976; Ames *et al*., 1977; Ames and Spudich, 1976; Bahramian and Middleton, 1973; Kustu and Ames, 1974
leuT	35	Leucine transport	Thorne and Corwin, 1975; L. M. Corwin, personal communication, quoted in Sanderson and Hartman (1978)
metP	6	α-Methylmethionine resistance; methionine transport	Ayling and Bridgeland, 1972; Betteridge and Ayling, 1975
nol	60	Norleucine resistance; possible defect in valine uptake or regulation	Hobson, 1974
proP	Not mapped	Minor proline transport process	R. Menzel, in Kleckner *et al*., 1977
putP	21	Utilization of proline; major proline transport process	R. Menzel, in Kleckner *et al*., 1977

[a]Derived from Sanderson and Hartman (1978).

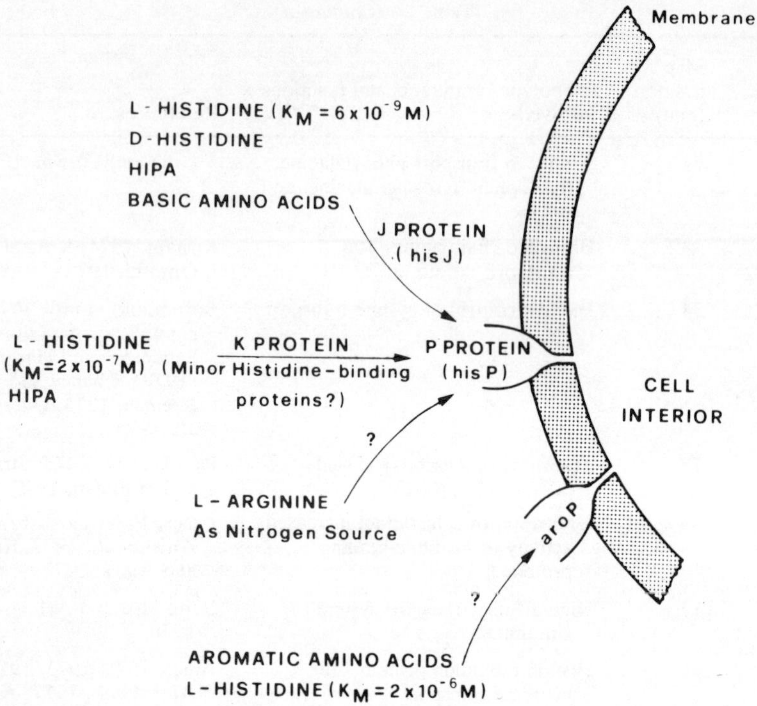

Figure 1 Histidine transport systems of *Salmonella typhimurium*. Adapted with permission from Ames (1972)

mutant also showed a temperature-sensitive defect in histidine transport, this elegant paper demonstrates that the histidine-binding protein is an obligatory component of histidine transport. The histidine-binding protein purified from another *hisJ* mutant showed (Kustu and Ames, 1974) normal affinity of histidine-binding but very low transport rate, leading the authors to suggest that the histidine-binding protein has, in addition to the histidine-binding site, a second site essential for transport.

The *hisP* gene also codes for a protein, as indicated by the isolation of amber mutants in this gene (Ames and Roth, 1968). Strains having mutations in the *hisP* gene, though defective in transport, have normal levels of the histidine-binding protein. The *hisP* gene product interacts with the *hisJ* gene product (Ames and Spudich, 1976). In fact, the low rate of histidine uptake of a specific *hisJ* mutant can be suppressed by a mutation in the *hisP* gene, showing that the interaction between the *hisJ* and the *hisP* gene products is an obligatory step in transport.

The *hisP* gene product is also an essential component of an arginine transport system (Kustu and Ames, 1973), that is involved in the transport of arginine as a nitrogen source, but is not involved in the high-affinity transport of arginine.

When the histidine transport system with highest affinity is missing as a result of a *hisJ* mutation, a different transport system, the K-system, becomes evident (Figure 1). No mutations affecting specifically this transport system have been reported, but *hisP* mutations were shown (Ames and Lever, 1970) to abolish both this and the high-affinity transport system. Therefore, these two systems work apparently in parallel and the transport through either of them requires the presence of the *hisP* gene product, as shown in Figure 1. An analogous situation is found in the case of the branched-chain amino acids and is described below.

Mutations in the *aroP* gene abolish a transport system for the aromatic amino acids (tryptophan, phenylalanine, and tyrosine) and also for histidine. Kinetic evidence (Ames, 1964; Ames and Roth, 1968) has indicated the presence of three more transport systems specific for tryptophan, phenylalanine, and tyrosine. The *aroP* gene has also been described in *E. coli* (Brown, 1970; 1971). It has a similar chromosomal location and similar effects are shown by mutations in this gene. The *aroP* mutation described by Brown (1970) leaves unaltered an inducible transport system for tryptophan, previously described by Boezi and De Moss (1961). However, it shows a defect in the exchange of the aromatic amino acids and in their efflux (Brown, 1971), thus indicating that the presence of a transport system is required not only for uptake, but also for exchange and efflux.

Other loci involved in aromatic amino acid transport are *aroT* in *S. typhimurium*, and *aroR* and *aroS* in *E. coli*. Their location is near *trp* in both species and the phenotype of strains carrying lesions in these genes is similar to that shown by *aroP* mutants (Thorne and Corwin, 1975).

Some *S. typhimurium* strains altered in the lipopolysaccharide component of the outer membrane are altered in tryptophan uptake (quoted by Thorne and Corwin, 1975). Although the authors only assayed tryptophan uptake, these strains might well have more generalized permeability defects and conceivably might be of help in trying to understand the reason for the multiplicity of transport systems observed with certain groups of amino acids. In fact, a similar multiplicity of transport systems for the aromatic amino acids has been described for *Pseudomonas aeruginosa* (Kay and Gronlund, 1971) and of *Yersinia pestis* (previously called *Pasteurella pestis*; Blaise Smith and Montie, 1975), but only in the first case has this analysis been helped by the study of specific mutants.

2 Branched-chain Amino Acids and Threonine

A combination of kinetic and genetic analyses has been used in the characterization of several transport systems for the branched-chain amino acids, mainly in *E. coli* K-12, and for the identification of many loci coding for structural components of transport systems, or involved in the regulation of their expression. Since different laboratories have used different nomenclature to describe these systems, we will give some space to their description so as to help the reader in relating the various results to the particular transport systems. (See also Sections 1.1.3.1.iii, 1.4.2.2, 1.4.3.1, and 1.6.4.1.)

The system with highest affinity is called 'LIV-I' (Rahmanian *et al.*, 1973; Robbins and Oxender, 1973; Wood, 1975) or 'very high affinity' (Guardiola *et al.*, 1974*a*,*b*). Isoleucine, leucine, and valine show apparent K_m values in the 10^{-7}–10^{-8} M range. Methionine, threonine, and alanine inhibit the uptake of these radioactive substrates, being probably themselves substrates. This transport system is repressed by the substrates (see Section 1.6.4). The isolation of mutants specifically altered in this transport system was described by Rahmanian *et al.* (1973), and by Anderson and Oxender (1977) who defined the *livH* locus mapping at 74 min on the *E. coli* map. Another locus, *livR*, located at 20 min on the *E. coli* map is involved in regulation of expression of this transport system (Anderson and Oxender, 1977). A protein that binds isoleucine, leucine, and valine was isolated (Anraku, 1968; Penrose *et al.*, 1968) and shown to be missing in a strain mutated in the *livJ* locus (Anderson and Oxender, 1977), located at 74 min on the *E. coli* map.

The 'leucine-specific' high-affinity transport system has a K_m of 2×10^{-7} M. Kinetic evidence for the presence of this system has been reported by several authors (Furlong and Weiner, 1970; Rahmanian *et al.*, 1973; Guardiola *et al.*, 1974*a*; Wood, 1975). The isolation of mutants specifically altered in this transport system was described by Rahmanian *et al.* (1973) and by Anderson and Oxender (1977) who defined the *livK* locus, mapping at 74 min on the *E. coli* map, and devoid of the leucine-specific binding protein isolated by Furlong and Weiner (1970). A mutant altered in the *lstR* locus, located at 20 min on the *E. coli* map, is altered in the regulation of expression of this transport system (Anderson *et al.*, 1976).

When the 'very high affinity' (or LIV-I) and the leucine-specific transport systems are absent, because of inhibition or repression, residual uptake of isoleucine, leucine, and valine can be measured. This uptake arises from the sum of two different transport systems: 'high affinity-1', and 'high affinity-2' (Guardiola *et al.*, 1974*b*), also jointly called the 'LIV-II' system (Rahmanian *et al.*, 1973; Wood, 1975) (see also Section 1.6.4.1). The high affinity-1 transport system needs the activity of a protein coded by the *brnQ* gene: it transports isoleucine, leucine, and valine, and is unaffected by threonine. The high affinity-2 transport system needs the activity of a protein coded by the *brnS* gene: it transports isoleucine, leucine, and valine, and it is inhibited by threonine (which is probably a substrate). The characterization of these two systems illustrates the

Table 3 Inhibition of the 'high-affinity-1' and 'high-affinity-2' transport systems by threonine

Strain	Inhibitor	Substrates		
		Isoleucine	Leucine	Valine
MI148a (wild type)	None	0.70	0.64	0.71
	2×10^{-4} M threonine	0.53	0.47	0.53
MI174b (*brnQ*$^-$)	None	0.43	0.55	0.45
	2×10^{-4} M threonine	0.01	0.01	0.03

The concentration of [^{14}C]-isoleucine, leucine, and valine was 2 μM. Values are given in micromoles of amino acid transported per 0.5 min per gram of cells. The 'very high affinity' transport system was repressed because of previous growth in the presence of methionine, while the leucine-specific transport system (Guardiola *et al*., 1974*a*) is absent in these two strains. (Reproduced from Guardiola *et al*. (1974*b*) with permission of the American Society for Microbiology.)

advantages of mutant isolation over kinetic analysis in defining a transport system, for, kinetic evidence suggests the presence of a single transport system and only mutant isolation has permitted the resolution of the two. As shown in Table 3, threonine inhibits only a fraction of the total isoleucine, leucine, and valine uptake in the wild type while it causes complete inhibition in a *brnQ* mutant (which leaves intact the high affinity-2 system). A small decrease in 'high affinity' uptake is observed in a strain containing either a *brnQ* or a *brnS* mutation. When the two mutations are both present there is no 'high affinity' uptake (Table 4). Mutations in another gene, *brnR*, abolish both the high affinity-1 and the high affinity-2 transport systems. Therefore, these two transport systems appear to work in parallel, *brnR* being a branching point, similar to the situation in the high-affinity transport systems for histidine, shown in Figure 1. Because several amber mutations have been isolated in each locus, the

Table 4 'High-affinity' transport of isoleucine and valine in strains carrying a *brnQ*$^-$, a *brnS*$^-$, or both mutations

Strain	Substrates		
	Isoleucine	Leucine	Valine
Ca85, wild type	0.46	0.46	0.48
MI238, *brnQ*$^-$	0.36	0.43	0.29
MI248, *brnS*$^-$	0.45	0.45	0.32
MI249, *brnQ*$^-$, *brnS*$^-$	0.04	0.06	0.06

The concentration of [^{14}C]-isoleucine, leucine, or valine was 2 μM. Values are given in micromoles of amino acid transported per 0.5 min per gram of cells. The 'very high affinity' transport system was repressed because of previous growth in the presence of methionine which, in this experiment, was present also in the assay tubes. (Reproduced from Guardiola *et al*. (1974*b*) with permission from the American Society for Microbiology.)

phenotype is assumed to arise from lack of the gene product. In the case of the *brnQ* gene this was confirmed by the isolation of deletion mutants (Brickman and Beckwith, 1975).

No discrepancy exists between the data of Guardiola *et al.* (1974*a,b*) and those of Wood (1975) (although a claim to the contrary is made by the latter author), provided that the assumption is made that the LIV-II transport system corresponds to the sum of the high affinity-1 and -2 transport systems.

When the systems with higher affinity are missing, residual uptake of isoleucine, leucine, and valine, with a K_m of about 10^{-4} M, can be measured. This uptake is due to the 'low-affinity' transport systems, specific for isoleucine or leucine or valine. A mutation, *brnT8*, located at 61 min on the map of *E. coli* (Guardiola *et al.*, 1974*b*; Iaccarino *et al.*, 1978), abolishes the low-affinity transport system for isoleucine.

The very high-affinity, or LIV-1, transport system is present also in *S. typhimurium*. Thorne and Corwin (1972) showed that a locus coding for a component involved in transport of leucine is located near the *trp* genes. Results of a kinetic analysis (Kiritani and Ohnishi, 1977) showed the presence of a transport system with a specificity similar to that of the LIV-1. Mutations in the *ilvT* locus (Kiritani, 1974; Kiritani and Ohnishi, 1977) affect a different transport system, analogous to the high affinity-1 and -2 (LIV-II) transport systems. The location of the *ilvT* gene, near *proC* and *purE*, is similar to that of the *brnQ* and *brnR* genes of *E. coli* K-12. Transport systems similar to the LIV-I and LIV-II, but with broader specificity, are described in *E. coli* B by Templeton and Savageau (1974*a,b*).

3 Methionine

Genetic and kinetic characterization of methionine uptake in *E. coli* revealed the presence of at least two transport systems (Kadner, 1974, 1975, 1977). A high-affinity system shows a K_m of 1.3×10^{-7} M and is missing in strains mutated in the *metD* gene, located at 5 min on the genetic map of *E. coli.* A low-affinity system shows a K_m of 2×10^{-5} M and is altered in strains mutated at a locus distinct from *metD*, but not yet localized. In *S. typhimurium* a highly specific methionine transport system has been described (Ayling and Bridgeland, 1972). This system is altered in strains containing the *metP760* mutation located at 7 min, or the *metP761* mutation located near the *gal* locus.

4 Glycine, Serine, and Alanine

Studies on the transport systems for glycine, L-alanine, D-alanine, and serine, and on the genetic markers involved with them have been reviewed by Halpern (1974). A recent paper (Thorne and Corwin, 1975) shows that lesions at the *aroT* locus of *S. typhimurium*, and at the *aroR* and *aroS* loci of *E. coli* cause a decrease in the uptake of alanine and glycine as well as of aromatic amino acids (see Section 1.5.2.1 above).

5 Arginine, Lysine, and Ornithine

Three transport systems for these amino acids are present in *E. coli* K-12: a high-affinity system specific for arginine, a system specific for lysine, and a general system for lysine, ornithine, and arginine (Rosen, 1971, 1973*a*). Mutants in the *argP* locus show a reduction in uptake through both the arginine-specific transport system and the general transport system. More details will be found in the reviews by Slayman (1973) and Halpern (1974) (see also Section 1.4.2.4).

In *S. typhimurium* there are at least two transport systems for arginine: a high-affinity one, analogous to that of *E. coli*, and a transport system allowing *S. typhimurium* to use arginine as a sole nitrogen source. The latter system is missing in strains mutated in the *hisP* gene (Kustu and Ames, 1973). A comparison of the results for *E. coli* and *S. typhimurium* does not exclude the possibility that both species actually possess all four transport systems.

6 Proline

Proline uptake in *E. coli* and *S. typhimurium* is unusual, compared to the uptake of other amino acids, because it apparently takes place through a single transport system. This uniqueness is found also in other species. In *Pseudomonas aeruginosa* a transport system has been described which takes in proline only when it is being utilized in a catabolic way (Kay and Gronlund, 1969). This transport system is repressed by glucose, and if an analogous transport system were to be present in *E. coli* and *S. typhimurium* it has apparently been functionally absent in the experiments reported for these species.

The presence of a single transport system for proline permits the isolation of transport mutants using a selection procedure that is usually not possible when a multiplicity of transport systems exists. Thus, such mutants were isolated from proline auxotrophs of *E. coli* W (Lubin *et al.*, 1960; Kessel and Lubin, 1962) as strains requiring high concentrations of proline for growth. Similar mutants of *E. coli* K-12 have also been described (Tristram and Neale, 1968). A different mutant, isolated by Eisenberg *et al.* (1975), shows pleiotropic effects, and might have a generalized membrane defect.

The substrate specificity of the proline transport system of *E. coli* has been carefully analysed (Rowland and Tristram, 1975). The osmotic shock procedure of Heppel (1967) does not release a proline-binding protein, as occurs for many other amino acids (see Section 1.4.1). However, a proline-binding protein was solubilized by Gordon *et al*. (1972) from purified membranes of *E. coli* ML, but solubilization required use of a non-ionic detergent and therefore this proline-binding protein is different from the others described in this chapter (see also Sections 1.1.3.1 and 1.4.2.7).

7 Aspartate and Glutamate

Two transport systems for aspartate exist in *E. coli* K-12 (Kay and Kornberg, 1971; Kay, 1971; Lo *et al*., 1972) (see also Sections 1.1.3.1 and 1.4.2.5). Some mutants, named *ast*, but different from the *ast* mutants quoted by Bachmann *et al*. (1976), lack a transport system for aspartate, while mutants altered at other loci (see below) lack a transport system for dicarboxylic acids, aspartate, fumarate, malate, and succinate, and for glutamate (Kahane *et al*., 1976). The latter mutants are unable to use malate, succinate, or fumarate as sole carbon source, but will reacquire this ability as a consequence of mutations at an unlinked locus, *prp*, causing the simultaneous increase in activity of a number of tricarboxylic acid cycle enzymes concerned with dicarboxylic acid metabolism (Kay, 1972). Mutants unable to transport dicarboxylic acids have their genetic defects localized at three different loci: *dctA*, linked to *xyl* at 76 min on the *E. coli* map, and *dctB* and *cbt*, both linked to *gal* at 16 min (Lo and Sanwal, 1975*a*). A soluble protein was isolated by use of the osmotic shock procedure and purified by affinity chromatography on Sepharose 4B. This protein has a molecular weight of about 15 000 daltons, binds aspartate, succinate, malate, and fumarate and is missing in mutants altered at the *cbt* locus.

Glutamate is taken up by *E. coli* cells through a transport system inhibited by D-glutamate, L-glutamine, and several glutamate derivatives (Halpern and Even-Shoshan, 1967) (see also Sections 1.1.3.1 and 1.4.2.5). Mutants altered at the *gltC* locus (Marcus and Halpern, 1967) show increased glutamate transport, while, in contrast, mutants altered at the closely linked *gltS* locus are unable to transport glutamate (Marcus and Halpern, 1969). From these results therefore, it appears that *gltC* is an operator locus for *gltS*, the structural gene for glutamate transport. Another locus, *gltR*, codes for a protein behaving as the repressor for the glutamate transport operon. A more detailed review of these studies on glutamate uptake and exit has been given by Halpern (1974; see also Halpern and Ennis, 1975), who also reviewed the related papers describing a transport system specific for glutamine (Weiner *et al*., 1971; Weiner and Heppel, 1971).

3 GRAM-POSITIVE BACTERIA

Although much of the early work on amino acid transport in bacteria was with Gram-positive organisms, such as *Staphylococcus aureus* and *Streptococcus faecalis* (see review by Gale, 1971), in recent years these bacteria have received relatively little attention and only in a few of these studies have genetic techniques been used.

In *Bacillus subtilis* a transport system highly specific for tyrosine and phenylalanine has been described (D'Ambrosio *et al.*, 1973; Glover *et al.*, 1975). This system shows a biphasic dependence of transport velocity upon tyrosine and phenylalanine concentration. The usual interpretation of such a finding is (see above) that multiple transport systems are present that have different affinities for the substrates. In this case, however, the authors presented evidence suggesting the presence of a single transport system showing a negative cooperative effect. This finding suggests that the evidence on at least some of the apparent multiple systems described in other species might possibly be open to reinterpretation to see if it fits a single transport system.

Two transport systems for D-alanine and L-alanine have been described in *B. subtilis* (Clark and Young, 1975). A high-affinity one is specific for these two amino acids, while a low-affinity one also transports D-serine and glycine. Hayakawa *et al.* (1976) have presented results on the reconstitution *in vitro* of one of these transport systems. These authors described the isolation of 'hydrophobic proteins' from membranes, and the preparation of membrane vesicles from these proteins and added phospholipids; these vesicles were reported to concentrate L-alanine in the presence of added NADH or ascorbate-phenazine methosulphate as an energy source. This work might open a new way of analysing transport systems from wild-type and mutant strains.

In *Bacillus megaterium* a tryptophan transport system has been described that is repressed by arginine and induced by tryptophan and kynurenine (Bouknight and Sadoff, 1975).

The uptake of α-amino-isobutyrate was studied in a thermophilic microorganism by Reizer and Grossowicz (1974), who reported the presence of a transport system for α-amino-isobutyrate, alanine, serine, and glycine which was altered in a specific mutant.

In *S. faecalis* two transport systems for glutamate and aspartate have been described (Holden *et al.*, 1968; 1975; Utech *et al.*, 1970) that differ in their affinity for the substrates. The high-affinity system, specifically missing in a mutant strain, also transports α-methylglutamate.

4 YEASTS AND OTHER FUNGI

When comparing amino acid transport in bacteria with amino acid transport in yeasts and other fungi it is important to keep in mind that the prokaryotic cells of the former are enormously different from the enkaryotic cells of the latter (Stainer *et al.*, 1970). Many compartments are observed with the electron microscope in thin sections of eukaryotic cells and this compartmentalization might explain why the amino acids taken up by these cells often do not exchange with external ones, while in the case of prokaryotic cells they usually do. One approach to the study of this problem is found in an interesting paper by Boller *et al.* (1975) who reported on the transport of arginine in vacuoles isolated from *Saccharomyces cerevisiae*. These vacuoles can accumulate arginine and exchange it with external substrate, while such an exchange is not observed in intact spheroplasts or cells.

In this section we will describe genetic studies related first to amino acid transport in yeasts (mostly *S. cerevisiae*) and then in other fungi (see also Sections 1.2 and 1.3).

1 *Saccharomyces Cerevisiae*

A summary of the genetic markers for amino acid transport in *S. cerevisiae*, derived from Plischke *et al.* (1976), is shown in Table 5.

One distinctive feature of *S. cerevisiae* is the presence of a transport system for all the amino acids except proline that is inhibited by ammonium ions (Grenson *et al.*, 1970). This general amino acid transport system is constitutive and is altered in *gap* mutants. Mutations at the *ura* locus (Drillien and Lacroute, 1972) may cause alterations in either the structure or the regulation of the *gap* transport system. D-Amino acids are

Table 5 Genetic markers for amino acid transport in *Saccharomyces cerevisiae*[a]

Genetic symbol	Phenotype of mutants and functions involved	References
aap1	Reduced uptake of amino acids; possibly a regulatory gene	Surdin *et al.*, 1965; Grenson and Hennaut, 1971
can1	Resistance to canavanine; arginine uptake	Grenson *et al.*, 1966
dip1	Reduced uptake of dicarboxylic amino acids	Crabeel and Grenson, 1970
FLE2	Resistance to trifluoroleucine; decreased uptake of leucine and trifluoroleucine	Bussey and Umbarger, 1970
gap1	Reduced uptake of basic and neutral amino acids	Grenson *et al.*, 1970
hip1	Reduced uptake of histidine	Crabeel and Grenson. 1970
lyp1	Reduced uptake of lysine	Grenson, 1966
mep1	Reduced uptake of methionine	Gits and Grenson, 1967

[a]Derived from Plischke *et al.* (1976).

accumulated inside the cell via the *gap* transport system and, since they inhibit growth, their use provides a method (Rytka, 1975) for positive selection of *gap* mutants, and also of *amc* mutants which are insensitive to the presence of ammonium ions in culture media. The use of *gap* mutants permitted Seaston *et al*. (1973) to show that two protons are taken up with one molecule of amino acid through the *gap* transport system, while one proton is taken up per molecule of amino acid through other (see below) transport systems (see also Section 1.2).

When the *gap* transport system is inhibited by ammonium ions or is absent because of mutation, then other specific systems can conveniently be measured. Histidine can be accumulated by two specific systems (Crabeel and Grenson, 1970), one of which is missing in *hip1* mutants. The accumulated histidine does not exchange with the external substrate and it also inhibits further uptake. A leucine-specific transport system was studied in yeast by Bussey and Umbarger (1970), who also described a mutant possessing 30-fold lower affinity for leucine uptake. Leucine-binding proteins were isolated from both the parent and the mutant strains and were shown to differ in affinity of leucine binding. Methionine is taken up into *S. cerevisiae* through two specific transport systems (Gits and Grenson, 1967), a high-affinity one which is altered in *mep1* mutants, and a low-affinity one. Two transport systems for lysine and arginine have been described in *S. cerevisiae*. One is specific for lysine and is altered in *lyp1* mutants, and the other is specific for arginine and is altered in *can1* mutants (Grenson, 1966; Grenson *et al*., 1966). Several transport systems for glutamate were described (Darte and Grenson, 1975), which show different substrate specificities and regulatory properties and are used for different functions under different physiological conditions.

2 *Neurospora crassa* and *Aspergillus nidulans*

Most of the work on the genetics of amino acid transport in fungi concerns these two species. Slayman (1973) has reviewed the work on *N. crassa* (see also Section 1.3). In Table 6 is given a list, derived from Radford (1976), of genetic markers for amino acid transport in this organism. Although the nomenclature for the various transport systems is quite different from that used for *S. cerevisiae*, the characteristics of their respective transport are probably not so different. Most of the results may be assigned to four transport systems, that were at first characterized mainly by kinetic experiments and later on by the use of specific mutants. These transport systems are; (i) a general amino acid transport system that becomes apparent under ammonium starvation—mutants lacking this system have been described by Rao *et al*. (1975); (ii) a transport system for neutral

Table 6 Genetic markers for amino acid transport in *Neurospora crassa*[a]

Genetic symbol	Linkage group	Phenotype of mutants and functions involved	References
arg-R	IV	Resistance to arginine; confers arginine-resistant lysine uptake	Hoffman *et al*., 1973
fpr	V	Fluorophenylalanine resistance; alteration in the uptake of some amino acids	Kinsey, 1967
hlp-1	VII	Histidinol permeable; uptake of basic amino acids	Ho, 1969
hlp-2	VII	Histidinol permeable; uptake of basic amino acids	Ho, 1969
lys^R	I	Resistance to lysine; confers lysine-resistant arginine uptake	Perkins, unpublished data, reported in Radford (1976)
mod-5	VI	Increased permeability to certain compounds, including some amino acids	St. Lawrence *et al*., 1964
mtr	IV	Resistance to methyl-tryptophan; retarded aromatic amino acid uptake	Stadler, 1967
nap	V	Resistance to 4-methyl tryptophan, aminopterin and glycine; uptake of neutral and acidic amino acids	Jacobson and Metzenberg, 1968
su-mtr	I	Suppressor of neutral amino acid uptake	Stadler, 1967

[a]Derived from Radford (1976).

amino acids, that is altered in certain mutants described by Wolfinbarger and De Busk (1971), by Magill *et al*. (1972), and by Sanchez *et al*. (1972); (iii) a transport system for basic amino acids, that is altered in certain mutants described by Pall (1970), by Magill *et al*. (1972), and by Sanchez *et al*. (1972); (iv) a transport system for acidic amino acids, that is altered in certain mutants described by Wolfinbarger *et al*. (1971), by Wolfinbarger and De Busk (1972), and by Wolfinbarger and Kay (1973).

Basic amino acid-binding macromolecules extracted with KCl from *N. crassa* have been described (Stuart and De Busk, 1975; Stuart and Woodward, 1975), these molecules were composed of amino acids, sugars, and probably also ribonucleotides. Their involvement in transport is strongly suggested by the fact that their structure is altered in specific transport mutants (see also Sections 1.3.4 and 1.4.2.4).

In *Aspergillus nidulans*, the presence of a general amino acid transport system like those described in *S. cerevisiae* and *N. crassa* has not been well documented, although Sinha (1969) described mutants resistant to *p*-fluorophenylalanine and altered in the *fpaD* locus of the linkage group VIII that showed an impaired capacity for the uptake of most amino acids. A transport system for neutral amino acids described by Piotrowska *et al.* (1976), was altered in the recessive *nap3* mutant. Other mutants, described by Kinghorn and Pateman (1975), were altered in the uptake of neutral amino acids, but their defects were not specific, one also being impaired in uptake of the acidic amino acids, and another also being defective in uptake of the acidic and aromatic amino acids. Further studies on these mutants could aid our understanding of amino acid transport in *A. nidulans*. A methionine-binding protein that was partially purified by use of affinity chromatography has been described by Stepién (1976); the protein was absent from a methionine transport mutant. A single transport system has been described for the basic amino acids arginine, lysine, and ornithine (Stepién, 1976). Glutamate, aspartate, and cysteate are taken up through a single transport system, that is missing in mutants altered in the *aanA* locus (Pateman *et al.*, 1974; Kinghorn and Pateman, 1975). The apparent K_m values are 180 μM, 100 μM, and 190 μM respectively (Robinson *et al.*, 1973*a*). This transport system is regulated from within the cell by ammonium inhibition and/or repression (Robinson *et al.*, 1973*b*). Mutants that are pleiotropically insensitive to ammonium control show lack of repression by ammonium of the glutamate transport system (Pateman *et al.*, 1974) (see also Section 4.2.4). The glutamate transport system is also sensitive to glucose repression: a mutant, altered in the *amdT* locus, is hypersensitive to glucose repression, while a different allele in the same locus is responsible for insensitivity to glucose repression (Hynes, 1973).

5 PLEIOTROPIC MUTATIONS

Most of the mutants described above are altered in one or more specific transport systems and any pleiotropic effects shown by them are interpreted as a transport system having the capacity to transport multiple substrates. In most of these cases, in fact, it has been shown that the transport of other, non-related metabolites is not impaired. However, in other cases, the pleiotropic effect of the mutation is thought to arise from an effect on more than one transport system. One of these cases might be that, already quoted above, of the *S. typhimurium* strains altered in the lipopolysaccharide component of the outer membrane (Thorne and Corwin, 1975). Most of the other mutants, except the one described by Eisenberg *et al.* (1975), for which a generalized permeability defect was suggested, are altered in the mechanism(s) by which the transport systems use cellular energy to concentrate the substrate.

In regard to their mechanisms of energy coupling (see also Section 1.7), transport systems for amino acids in Gram-negative bacteria may broadly be divided into two categories: the shock-sensitive systems which are associated with periplasmic binding proteins, and the shock-insensitive systems (Berger and Heppel, 1974). The former transport systems are absent from isolated membrane vesicles while the latter are present. The shock-sensitive transport systems obtain their energy from phosphate bonds and are inactive in mutants defective in the Ca^{2+}, Mg^{2+}-ATPase (Simoni and Schallenberger, 1972; Berger and Heppel, 1974; Rosen, 1973*b*; Yamamoto *et al*., 1973; Rosenberg *et al*., 1975). The activity of the shock-insensitive transport systems, when assayed in membrane vesicles, is stimulated by D-lactate (Kaback, 1972), and is defective in mutants altered in D-lactate dehydrogenase (Simoni and Schallenberger, 1972; Reeves *et al*., 1973) or in energy coupling (Lieberman and Hong, 1974, 1976). A recent study with a cytochrome-deficient mutant (Singh and Bragg, 1976) presented evidence that the latter transport systems, when under anaerobic conditions, may also use energy derived from ATP.

Few mutants deficient in energy coupling have been reported in other species. Results obtained with a mutant of *B. subtilis* (Bisschop *et al*., 1975) show an involvement of menaquinone in the transfer of energy to amino acid transport systems in this organism, while a mutant of *N. crassa* appears to be altered in some common component of all transport systems that may be at the level of energy coupling (Rao and De Busk, 1973).

6 MULTIPLICITY OF TRANSPORT SYSTEMS

One striking characteristic of amino acid uptake in microorganisms is the multiplicity of transport systems (Ames, 1972, 1974). In kinetic experiments, this may become apparent as a biphasic double reciprocal plot, in which each part of the curve represents one component. However, alternatively, this behaviour may sometimes be interpreted as evidence for a single transport system with complex kinetics (Glover *et al*., 1975; Denizeau *et al*., 1976) (see Section 1.6.3 below). However, to resolve any ambiguities in the interpretation of kinetic data requires the use of inhibitors that act specifically on one of the transport components and, more importantly, the characterization of mutations that abolish one component specifically, such evidence being a good indication of the presence of multiple transport systems. Thus, studies on mutant strains have in some cases shown that a single kinetic component is actually the average of two very similar processes (see Tables 3, 4).

Although the multiplicity of transport systems is particularly evident for amino acid transport, multiple systems also occur for other classes of substances, as reviewed by Slayman (1973). With carbohydrates it is less immediately evident because some of the transport systems are inducible. For example, Messer (1974) reported that although lactose is a low-affinity substrate of the arabinose transport system, the system is not induced by lactose and it can only be used for lactose accumulation after it has been induced by a different substrate.

The mode of amino acid uptake with lowest affinity is often interpreted as being due to 'diffusion' (see, for example, Denizeau *et al*., 1976). That this may not be so is suggested by studies with mutants having defects in a transport system that give rise to impermeability to a specific substance. This phenomenon is found more frequently with carbohydrates, probably because (a) less mediated transport multiplicity is present and, (b) a measure of impermeability is the lack of growth on the carbohydrate as a source of carbon which usually requires a higher transfer rate of carbohydrate molecules through the membrane as compared to amino acid uptake. Some relevant examples concerning amino acids are the following: (a) loss of ability to grow on arginine as a nitrogen source in a *S. typhimurium hisP* mutant (Kustu and Ames, 1973); (b) poor growth on proline of an *E. coli* proline auxotroph altered in the proline transport system (Lubin *et al*., 1960; Kessel and Lubin, 1962); (c) complete impermeability to isoleucine of a strain of *E. coli* containing multiple mutations in the transport systems for this amino acid (Guardiola *et al*., 1974*a,b*; Iaccarino *et al*., 1978).

The multiplicity of amino acid transport systems suggests that they serve varied functions. Furthermore, the phenotypes of the mutants described in this chapter suggests that at least two classes of transport systems for amino acids may be distinguished: (a) a set of transport systems that concentrate amino acids which then function mainly in feedback inhibition of the amino acid biosynthetic enzymes, and also for protein synthesis; (b) another class of transport systems, often inducible, that is used to concentrate amino acids that function primarily as a source of carbon and/or nitrogen. In the latter case, more amino acid is required to pass the membrane per unit time and this might be one evolutionary reason for the necessity of specialization of function. Certainly, many of the examples quoted above are in agreement with these ideas for bacteria, and similar conclusions are reached on a paper dealing with the three transport systems for glutamate in yeast (Darte and Grenson, 1975).

7 ISOLATION OF MUTANTS

Slayman (1973) has reviewed this subject extensively and therefore we will discuss it only briefly.

The multiplicity of transport systems for many amino acids may be the main cause of failure in attempts to isolate mutants altered in amino acid uptake: thus, if the phenotype selected for can only be found in a double mutant the event will be too rare to be found under normal selection conditions. A way to combat this problem is to perform an initial analysis of the kinetics of uptake in order to define a tentative number of transport systems and to determine a range of specific inhibitors. Then if all but one can be inhibited, the frequency of isolating permeability mutants in this system is likely to be adequately high (Guardiola *et al*., 1974*b*). One method that is often used is to search for mutants that are resistant to an analogue that is a growth inhibitor; usually, these mutants are either altered in the intracellular target of the inhibitor or are permeability mutants. Therefore, the next step is a rapid screening to distinguish the permeability mutants. Such screening may be: (a) to use competitive inhibitors of other transport systems such that their removal will make the permeability mutant sensitive to the growth inhibitor, as in the case of the transport mutants of *E. coli* K-12 that are resistant to the growth inhibitor valine only when inhibitors of other valine transport systems are also used, such as methionine, threonine, or leucine (Guardiola *et al*., 1974*b*); (b) to make use of the fact that the permeability mutant will be sensitive if the growth inhibitor is supplied in a different chemical form, e.g. a peptide that enters the cell through a different transport system and is then converted intracellularly into the growth inhibitor (Ames *et al*., 1973; De Felice *et al*., 1973; Fickel and Gilvarg, 1973); (c) to carry out direct screening for defective uptake. In this case, a possible difficulty of interpretation is caused by the amino acid excretion of some mutants; if the transport assay involves a radioactively-labelled amino acid, dilution of the specific radioactivity will mimic a lowered activity (Ames, 1964).

A different method of isolating transport-defective mutants is the selection of slow growing derivatives of an amino acid auxotroph (or non-growers). As already discussed above, this approach was used to isolate a mutant in the proline transport system (Lubin *et al*., 1960; Kessel and Lubin, 1962). Similarly, one can isolate mutants that do *not* use an amino acid as a carbon or nitrogen source (Marcus and Halpern, 1969).

Mutants with higher transport rates, e.g. derepressed mutants or mutants possessing a higher turnover, may be isolated using as selection procedure growth on a poor substrate of the transport system. For example, mutants derepressed in histidine transport become able to use D-histidine as a source of histidine for protein synthesis (Krajewska-Grynkiewicz *et al*., 1971).

ACKNOWLEDGEMENTS

We thank T. Klopotowski, L. Alberghina, M. Bandiera, M. Bazzicalupo, and M. Polsinelli for useful suggestions, A. Lamberti for help in the editing, and K. E. Sanderson for supplying unpublished information.

REFERENCES

Ames, B. N., Ames, G. F. L., Young, J. D., Tsuchiya, D., and Lecocq, J. (1973). Illicit transport: the oligopeptide permease. *Proc. Natl. Acad. Sci. U.S.A.*, **70**, 456–458

Ames, G. F. L. (1964). Uptake of amino acids by *Salmonella typhimurium. Arch. Biochem. Biophys.*, **104**, 1–18

Ames, G. F. L. (1972). Components of histidine transport In *Membrane Research*, First ICN-UCLA Symposium on Molecular Biology (C. F. Fox, ed.), Academic Press, New York and London, pp. 409–426

Ames, G. F. L. (1973). Resolution of bacterial proteins by polyacrylamide gel electrophoresis on slabs. Membrane-soluble and periplasmic fractions. *J. Biol. Chem.*, **249**, 634–644

Ames, G. F. L. (1976). Protein–protein interaction in transport: periplasmic histidine-binding protein J interacts with P protein. *Proc. Natl. Acad. Sci. U.S.A.*, **73**, 1871–1876

Ames, G. F. L. and Lever, J. E. (1970). Components of histidine transport: histidine-binding protein and *hisP* protein. *Proc. Natl. Acad. Sci. U.S.A.*, **66**, 1096–1103

Ames, G. F. L. and Lever, J. E. (1972). The histidine-binding protein J is a component of histidine transport. *J. Biol. Chem.*, **247**, 4309–4316

Ames, G. F. L. and Roth, J. R. (1968). Histidine and aromatic permeases of *Salmonella typhimurium. J. Bacteriol.*, **96**, 1742–1749

Ames, G. F. L. and Spudich, E. N. (1976). Protein–protein interaction in transport: periplasmic histidine-binding protein J interacts with P protein. *Proc. Natl. Acad. Sci. U.S.A.*, **73**, 1877–1881

Ames, G. F. L., Noel, K. D., Taber, H., Spudich, E. N., Nikaido, K., Afong, J., and Ardeshir, F. (1977). Fine-structure map of the histidine transport genes in *Salmonella typhimurium. J. Bacteriol.*, **129**, 1289–1297

Anderson, J. J. and Oxender, D. L. (1977). *Escherichia coli* transport mutants lacking binding protein and other components of the branched-chain amino acid transport systems. *J. Bacteriol.*, **130**, 384–392

Anderson, J. J., Quay, S. C., and Oxender, D. L. (1976). Mapping of two loci affecting the regulation of branched-chain amino acid transport in *Escherichia coli* K-12. *J. Bacteriol.*, **126**, 80–90

Anraku, Y. (1968). Transport of sugars and amino acids in bacteria. II. Properties of galactose and leucine-binding proteins. *J. Biol. Chem.*, **243**, 3123–3127

Archibald, A. R. (1974). The structure, biosynthesis and function of teichoic acid. *Adv. Microbial Physiol.*, **11**, 53–96

Ayling, P. D. and Bridgeland, E. S. (1972). Methionine transport in wild-type and transport-defective mutants of *Salmonella typhimurium. J. Gen. Microbiol.*, **73**, 127–141

Bachmann, B. J., Low, B. K., and Taylor, A. L. (1976). Recalibrated linkage map of *Escherichia coli* K-12. *Bacteriol. Rev.*, **40**, 116–167

Bahramian, M. B. and Middleton, R. B. (1973). Reversal by aromatic amino acids of 2-thiazole-D,L-alanine inhibition of *Salmonella typhimurium. J. Bacteriol.*, **113**, 504–507

Berger, E. A. and Heppel, L. A. (1974). Different mechanisms of energy coupling for the shock-sensitive and shock-resistant amino acid permeases of *Escherichia coli. J. Biol. Chem.*, **249**, 7747–7755

Betteridge, P. R. and Ayling, P. D. (1975). The role of methionine transport-defective mutations in relation to methionine sulfoxime in *Salmonella typhimurium. Mol. Gen. Genet.*, **138**, 41–52

Betteridge, P. R. and Ayling, P. D. (1976). The regulation of glutamine transport and glutamine synthetase in *Salmonella typhimurium. J. Gen. Microbiol.*, **95**, 324–334

Bisschop, A., De Jong, L., and Lima Costa, M. E. (1975). Relation between reduced nicotinamide adenine dinucleotide oxidation and amino acid transport in membrane vesicles from *Bacillus subtilis. J. Bacteriol.*, **121**, 807–813

Blaise Smith, P. and Montie, T. C. (1975). Aromatic amino acid transport in *Yersinia pestis. J. Bacteriol.*, **122**, 1045–1052

Boezi, J. A. and De Moss, R. D. (1961). Properties of a tryptophan transport system in *Escherichia coli. Biochim. Biophys. Acta*, **49**, 471–484

Boller, T., Durr, M. and Wiemken, A. (1975). Characterization of a specific transport system for arginine in isolated yeast vacuoles. *Eur. J. Biochem.*, **54**, 81–92

Borum, P. R. and Monty, K. J. (1976). Regulatory mutants and control of cysteine biosynthetic enzymes in *Salmonella typhimurium. J. Bacteriol.*, **125**, 94–101

Bouknight, R. R. and Sadoff, H. L. (1975). Transport of D- and L-tryptophan in *Bacillus megaterium* by an inducible permease. *J. Bacteriol.*, **121**, 65–69

Brickman, E. and Beckwith, J. (1975). Analysis of the regulation of *Escherichia coli* alkaline phosphatase synthesis using deletions and $\phi 80$ transducing phages. *J. Mol. Biol.*, **96**, 307–316

Brown, K. D. (1970). Formation of aromatic amino acid pools in *Escherichia coli* K-12. *J. Bacteriol.*, **104**, 177–188

Brown, K. D. (1971). Maintenance and exchange of the aromatic amino acid pool in *Escherichia coli. J. Bacteriol.*, **106**, 70–81

Bussey, H. and Umbarger, H. E. (1970). Biosynthesis of the branched-chain amino acids in yeast: a leucine-binding component and regulation of leucine uptake. *J. Bacteriol.*, **103**, 277–285

Celis, T. F. R., Rosenfeld, H. J., and Maas, W. K. (1973). Mutant of *Escherichia coli* K-12 defective in the transport of basic amino acids. *J. Bacteriol.*, **116**, 619–626

Cheney, R. W., Jr and Kredich, N. M. (1975). Fine-structure genetic map of the *cysB* locus in *Salmonella typhimurium. J. Bacteriol.*, **124**, 1273–1281

Clark, V. and Young, F. E. (1975). Active transport of D-alanine and related amino acids by whole cells of *Bacillus subtilis. J. Bacteriol.*, **120**, 1085–1092

Cohen, G. N. and Monod, J. (1957). Bacterial permeases. *Bacteriol. Rev.*, **21**, 169–194

Cooper, S. (1966). Utilization of D-methionine by *Escherichia coli. J. Bacteriol.*, **92**, 328–332

Cosloy, S. D. (1973). D-Serine transport system in *Escherichia coli* K-12. *J. Bacteriol.*, **114**, 679–684

Crabeel, M. and Grenson, M. (1970). Regulation of histidine uptake by specific feedback inhibition of two histidine permeases in *Saccharomyces cerevisiae. Eur. J. Biochem.*, **14**, 197–204

Curtiss, R., Charamella, L. J., Berg, C. M., and Harris, P. E. (1965). Kinetic and genetic analysis of D-cycloserine inhibition and resistance in *Escherichia coli. J. Bacteriol.*, **90**, 1238–1250

D'Ambrosio, M. S., Glover, G. I., Nelson, S. O., and Jensen, R. A. (1973). Specificity of the tyrosine–phenylalanine transport system in *Bacillus subtilis. J. Bacteriol.*, **115**, 673–681

Darte, C. and Grenson, M. (1975). Evidence for three glutamic acid transporting systems with specialized physiological functions in *Saccharomyces cerevisiae. Biochem. Biophys. Res. Commun.*, **67**, 1028–1033

De Felice, M., Guardiola, J., Lamberti, A., and Iaccarino, M. (1973). *Escherichia coli* K-12 mutants altered in the transport systems for oligo- and dipeptides. *J. Bacteriol.*, **116**, 751–756

Denizeau, F., Wyse, J., and Sourkes, T. L. (1976). Kinetics of multiple transport systems for amino acids in the brain. *J. Theor. Biol.*, **63**, 99–110

Drillien, R. and Lacroute, F. (1972). Ureidosuccinic acid uptake in yeast and some aspects of its regulation. *J. Bacteriol.*, **109**, 203–208

Eisenberg, M. A., Mee, B., Prakash, O., and Eisenberg, M. R. (1975). Properties of α-dehydrobiotin-resistant mutants of *Escherichia coli* K-12. *J. Bacteriol.*, **122**, 66–72

Fickel, T. E. and Gilvarg, C. (1973). Transport of impermeant substances in *Escherichia coli* by way of oligopeptide permease. *Nature New Biol.*, **241**, 161–163

Furlong, C. E. and Weiner, J. H. (1970). Purification of a leucine-specific binding protein from *Escherichia coli. Biochem. Biophys. Res. Commun.*, **38**, 1076–1083

Gale, E. F. (1971). 'Don't talk to me about permeability'. *J. Gen. Microbiol.*, **68**, 1–14

Gits, J. J. and Grenson, M. (1967). Multiplicity of the amino acid permeases in *Saccharomyces cerevisiae*. III. Evidence for a specific methionine-transporting system. *Biochim. Biophys. Acta*, **135**, 507–516

Glover, G. I., D'Ambrosio, S. M., and Jensen, R. A. (1975). Versatile properties of a non-saturable, homogeneous transport system in *Bacillus subtilis*: genetic, kinetic, and affinity labeling studies. *Proc. Natl. Acad. Sci. U.S.A.*, **72**, 814–818

Gordon, A. S., Lombardi, F. J., and Kaback, H. R. (1972). Solubilization and partial purification of amino acid-specific components of the D-lactate dehydrogenase-coupled amino acid-transport systems. *Proc. Natl. Acad. Sci. U.S.A.*, **64**, 358–362

Grenson, M. (1966). Multiplicity of the amino acid permeases in *Saccharomyces cerevisiae*. II. Evidence for a specific lysine-transporting system. *Biochim. Biophys. Acta*, **127**, 339–346

Grenson, M. and Hennaut, C. (1971). Mutation affecting activity of several distinct amino acid transport systems in *Saccharomyces cerevisiae. J. Bacteriol.*, **105**, 477–482

Grenson, M., Hou, C. and Crabeel, M. (1970). Multiplicity of the amino acid permeases in *Saccharomyces cerevisiae*. IV. Evidence for a general amino acid permease. *J. Bacteriol.*, **103**, 770–777

Grenson, M., Mousset, M., Wiame, J. M., and Bechet, J. (1966). Multiplicity of the amino acid permeases in *Saccharomyces cerevisiae*. I. Evidence for a specific arginine-transporting system. *Biochim. Biophys. Acta*, **127**, 325–338

Guardiola, J. and Iaccarino, M. (1971). *Escherichia coli* K-12 mutants altered in the transport of branched-chain amino acids. *J. Bacteriol.*, **108**, 1034–1044

Guardiola, J., De Felice, M., Klopotowski, T., and Iaccarino, M. (1974*a*). Multiplicity of isoleucine, leucine, and valine transport systems in *Escherichia coli.* K-12. *J. Bacteriol.*, **117**, 382–392

Guardiola, J., De Felice, M., Klopotowski, T., and Iaccarino, M. (1974*b*). Mutation affecting the different transport systems for isoleucine, leucine, and valine in *Escherichia coli* K-12. *J. Bacteriol.*, **117**, 393–405

Guest, J. R. (1974). Gene–protein relationships of the α-keto acid dehydrogenase complexes of *Escherichia coli* K-12: chromosomal location of the lipoamide dehydrogenase gene. *J. Gen. Microbiol.*, **80**, 523–532

Halpern, S. Y. (1974). Genetics of amino acid transport in bacteria. *Annu. Rev. Genet.*, **8**, 103–133

Halpern, Y. S. and Ennis, H. L. (1975). Analysis of glutamate exit in *Escherichia coli. J. Bacteriol.*, **122**, 332–334

Halpern, Y. S. and Even-Shoshan, A. (1967). Properties of the glutamate transport system in *Escherichia coli. J. Bacteriol.*, **93**, 1009–1016

Hayakawa, K., Ueda, T., Kusaka, I., and Fukui, S. (1976). Preparation of artificial vesicles having an L-alanine uptake activity which requires NADH as energy source. *Biochem. Biophys. Res. Commun.*, **72**, 1548–1553

Heppel, L. A. (1967). Selective release of enzymes from bacteria. *Science, N.Y.*, **156**, 1451–1455

Ho, C. C. (1969). Mutants of *Neurospora crassa* permeable to histidinol. *Genetics*, **62**, 725–733

Hobson, A. C. (1974). A norleucine-resistant mutant of *Salmonella typhimurium* with a possible defect in valine uptake or regulation. *J. Gen. Microbiol.*, **82**, 425–429

Hobson, A. C., Gho, D., and Muller-Hill, B. (1977). Isolation, genetic analysis, and characterization of *Escherichia coli* mutants with defects in the *lacY* gene. *J. Bacteriol.*, **131**, 830–838

Hoffmann, G. R., Malling, H. V., and Mitchell, T. J. (1973). Genetics of 5-fluorodeoxyuridine-resistant mutants of *Neurospora crassa. Can. J. Genet. Cytol.* **15**, 831–844

Holden, J. T., Utech, N. M., and Reid, K. G. (1975). α-Methyl-L-glutamic acid uptake by high affinity dicarboxylic amino acid transport system in *Streptococcus faecalis. Biochim. Biophys. Acta*, **394**, 55–64

Holden, J. T., van Balgooy, J. N. A., and Kittredge, J. S. (1968). Transport of aminophosphonic acid in *Lactobacillus plantarum* and *Streptococcus faecalis. J. Bacteriol.*, **96**, 950–957

Hynes, M. J. (1973). Alterations in the control of glutamate uptake in mutants of *Aspergillus nidulans. Biochem. Biophys. Res. Commun.*, **54**, 685–689

Iaccarino, M., Guardiola, J., and De Felice, M. (1978). On the permeability of biological membranes. *J. Membrane Sci.*, **3**, 287–302

Jacobson, E. S. and Metzenberg, R. L. (1968). A new gene which affects uptake of neutral and acidic amino acids in *Neurospora crassa. Biochim. Biophys. Acta*, **156**, 140–147

Kaback, H. R. (1972). Transport across isolated bacterial cytoplasmic membranes. *Biochim. Biophys. Acta*, **265**, 367–416

Kadner, R. J. (1974). Transport system for L-methionine in *Escherichia coli. J. Bacteriol.*, **117**, 232–241

Kadner, R. J. (1975). Regulation of methionine transport activity in *Escherichia coli. J. Bacteriol.*, **122**, 110–119

Kadner, R. J. (1977). Transport and utilization of D-methionine and other methionine sources in *Escherichia coli. J. Bacteriol.*, **129**, 207–216

Kadner, R. J. and Watson, W. J. (1974). Methionine transport in *Escherichia coli*: physiological and genetic evidence for two uptake systems. *J. Bacteriol.*, **119**, 401–409

Kahane, S., Metzer, E., and Halpern, Y. S. (1976).. Spatial characteristics of the aspartate membrane carrier and its relationship to the glutamate transport system in membrane vesicles of *Escherichia coli* K-12 Cavalli. *Eur. J. Biochem.*, **66**, 583–589

Kay, W. W. (1971). Two aspartate transport systems in *Escherichia coli. J. Biol. Chem.*, **246**, 7373–7382

Kay, W. W. (1972). Suppression of a dicarboxylic acid transport mutant phenotype in *Escherichia coli* K-12. *Biochim. Biophys. Acta*, **264**, 522–529

Kay, W. W. and Gronlund, A. F. (1969). Proline transport by *Pseudomonas aeruginosa. Biochim. Biophys. Acta*, **193**, 444–455

Kay, W. W. and Gronlund, A. F. (1971). Transport of aromatic amino acids by *Pseudomonas aeruginosa. J. Bacteriol.*, **105**, 1039–1046

Kay, W. W. and Kornberg, H. L. (1969). Genetic control of the uptake of C_4-dicarboxylic acids by *Escherichia coli. FEBS Lett.*, **3**, 93–96

Kay, W. W. and Kornberg, H. L. (1971). The uptake of C_4-dicarboxylic acids by *Escherichia coli. Eur. J. Biochem.*, **18**, 274–281

Kessel, D. and Lubin, M. (1962). Transport of proline in *Escherichia coli. Biochim. Biophys. Acta,* **57**, 32–43

Kinghorn, J. R. and Pateman, J. A. (1975). Mutations which affect amino acid transport in *Aspergillus nidulans, J. Gen. Microbiol.,* **86**, 174–184

Kinsey, J. A. (1967). Phenotypic suppression of a resistance mutant in *Neurospora. Genetics,* **56**, 570–571

Kiritani, K. (1974). Mutants of *Salmonella typhimurium* defective in transport of branched-chain amino acids. *J. Bacterol.,* **120**, 1093–1101

Kiritani, K. and Ohnishi, K. (1977). Repression and inhibition of transport systems for branched-chain amino acids in *Salmonella typhimurium. J. Bacteriol.,* **129**, 589–598

Kleckner, N., Roth, J., and Botstein, D. (1977). Genetic engineering *in vivo* using translocatable drug-resistance elements. New methods in bacterial genetics. *J. Mol. Biol.*, **116**, 125–159

Krajewska-Grynkiewicz, K., Walezak, W., and Klopotowski, T. (1971). Mutants of *Salmonella typhimurium* able to utilize D-histidine as a source of L-histidine. *J. Bacteriol.,* **105**, 28–37

Kuhn, J. and Somerville, R. L. (1971). Mutant strains of *Escherichia coli* K-12 that use D-amino acids. *Proc. Natl. Acad. Sci. U.S.A.*, **68**, 2482–2484

Kustu, S. G. and Ames, G. F. L. (1973). The *hisP* protein, a known histidine transport component in *Salmonella typhimurium*, is also an arginine transport component. *J. Bacteriol.*, **116**, 107–113

Kustu, S. G. and Ames, G. F. L. (1974). The histidine-binding protein J, a histidine transport component, has two different functional sites. *J. Biol. Chem.*, **249**, 6976–6983

Lever, J. E. (1972). Purification and properties of a component of histidine transport in *Salmonella typhimurium*: the histidine-binding protein J. *J. Biol. Chem.*, **247**, 4317–4326

Lieberman, M. A. and Hong, J. (1974). A mutant of *Escherichia coli* defective in the coupling of metabolic energy to active transport. *Proc. Natl. Acad. Sci. U.S.A.*, **71**, 4395–4399

Lieberman, M. A. and Hong, J. (1976). Energization of osmotic shock-sensitive transport systems in *Escherichia coli* requires more than ATP. *Arch. Biochem. Biophys.*, **172**, 312–315

Lin, E. C. C. (1970). The genetics of bacterial transport systems. *Ann. Rev. Genet.*, **4**, 225–262

Lo, T. C. Y. and Sanwal, B. D. (1975*a*). Genetic analysis of mutants of *Escherichia coli* defective in dicarboxylate transport. *Mol. Gen. Genet.*, **140**, 303–307

Lo, T. C. Y. and Sanwal, B. D. (1975*b*). Isolation of the soluble substrate recognition component of the dicarboxylate transport system of *Escherichia coli*. *J. Biol. Chem.*, **250**, 1600–1602

Lo, T. C. Y., Rayman, M. K., and Sanwal, B. D. (1972). Transport of succinate in *Escherichia coli*. I. Biochemical and genetic studies of transport in whole cells. *J. Biol. Chem.*, **247**, 6323–6331

Lubin, M., Kessel, D. H., Budreau, A., and Gross, J. D. (1960). The isolation of bacterial mutants defective in amino acid transport. *Biochim. Biophys. Acta*, **42**, 535–538

Maas, W. K. (1965). Genetic defects affecting an arginine permease and repression of arginine synthesis in *Escherichia coli*. *Fed. Proc. Fed. Am. Soc. Exp. Biol.*, **24**, 1239–1242

Maas, W. K. (1972). Mapping of genes involved in the synthesis of spermidine in *Escherichia coli*. *Mol. Gen. Genet.*, **119**, 1–9

Magill, C. W., Sweeney, H., and Woodward, V. W. (1972). Histidine uptake in strains of *Neurospora crassa* with normal and mutant transport systems. *J. Bacteriol.*, **110**, 313–320

Mandelstam, J. and McQuillen, K. (eds) (1973). *Biochemistry of Bacterial Growth*, Blackwell Scientific Publications, Oxford

Marcus, M. and Halpern, Y. S. (1967). Genetic analysis of glutamate transport and glutamate decarboxylase in *Escherichia coli*. *J. Bacteriol.*, **93**, 1409–1415

Marcus, M. and Halpern, Y. S. (1969). Genetic analysis of the glutamate permease in *Escherichia coli* K-12. *J. Bacteriol.*, **97**, 1118–1128

Messer, A. (1974). Lactose permeation via the arabinose transport system in *Escherichia coli* K-12. *J. Bacteriol.*, **120**, 262–272

Pall, M. L. (1970). Amino acid transport in *Neurospora crassa*. II. Properties of a basic amino acid transport system. *Biochim. Biophys. Acta*, **203**, 139–149

Parada, J. L., Ortega, M. V., and Carrillo-Castaneda, G. (1973). Biochemical and genetic characteristics of the C_4-dicarboxylic acid transport system of *Salmonella typhimurium*. *Arch. Mikrobiol.*, **94**, 65–76

Pateman, J. A., Kinghorn, J. R., and Dunn, E. (1974). Regulatory aspects of L-glutamate transport in *Aspergillus nidulans*. *J. Bacteriol.*, **119**, 534–542

Penrose, W. R., Nichoalds, G. E., Piperno, J. R., and Oxender, D. L. (1968). Purification and properties of a leucine-binding protein from *Escherichia coli*. *J. Biol. Chem.*, **243**, 5921–5928

Piotrowska, M., Stepien, P. P., Bartnik, E., and Zakrzewska, E. (1976). Basic and neutral amino acid transport in *Aspergillus nidulans*. *J. Gen. Microbiol.*, **92**, 89–96

Plischke, M. E., von Borstel, R. C., Mortimer, R. K., and Cohn, W. E. (1976). In *Handbook of Biochemistry and Molecular Biology*, Vol. 2 (G. D. Fasman, ed.), The Chemical Rubber Co., Cleveland, Ohio, pp. 767–832

Quay, S. C. and Christensen, H. N. (1974). Basis of transport discrimination of arginine from basic amino acids in *Salmonella typhimurium*. *J. Biol. Chem.*, **249**, 7011–7017

Qureshi, M. A., Smith, D. A., and Kingsman, A. J. (1975). Mutants of *Salmonella typhimurium* responding to cysteine or methionine: their nature and possible role in the regulation of cysteine biosynthesis. *J. Gen. Microbiol.*, **89**, 353–370

Radford, A. (1976). Linkage maps of *Neurospora crassa*. In *Handbook of Biochemistry and Molecular Biology*, Vol. 2 (G. D. Fasman, ed.), The Chemical Rubber Co., Cleveland, Ohio, pp. 739–761

Rahmanian, M., Claus, D. R. and Oxender, D. L. (1973). Multiplicity of leucine transport system in *Escherichia coli* K-12. *J. Bacteriol.*, **116**, 1258–1266

Rao, E. Y. T., Rao, T. K., and De Busk, A. G. (1975). Isolation and characterization of a mutant of *Neurospora crassa* deficient in general amino acid permease activity. *Biochim. Biophys. Acta*, **413**, 45–51

Rao, T. K. and De Busk, A. G. (1973). Characteristics of a transport-deficient mutant (*nap*) of *Neurospora crassa*. *Biochim. Biophys. Acta*, **323**, 619–626

Reeves, J. P., Hong, J., and Kaback, H. R. (1973). Reconstitution of D-lactate-dependent transport in membrane vesicles from a D-lactate dehydrogenase mutant of *Escherichia coli*. *Proc. Natl. Acad. Sci. U.S.A.*, **70**, 1917–1921

Reizer, J. and Grossowicz, N. (1974). Properties of α-amino-isobutyric acid transport in a thermophilic microorganism. *J. Bacteriol.*, **118**, 414–424

Robbins, J. C. and Oxender, D. L. (1973). Transport systems for alanine, serine and glycine in *Escherichia coli* K-12. *J. Bacteriol.*, **116**, 12–18

Robinson, J. H., Anthony, C., and Drabble, W. T. (1973*a*). The acidic amino acid permease of *Aspergillus nidulans*. *J. Gen. Microbiol.*, **79**, 53–63

Robinson, J. H., Anthony, C., and Drabble, W. T. (1973*b*). Regulation of the acidic amino acid permease of *Aspergillus nidulans*. *J. Gen. Microbiol.*, **79**, 65–80

Rosen, B. P. (1971). Basic amino acid transport in *Escherichia coli*. *J. Biol. Chem.*, **246**, 3653–3662

Rosen, B. P. (1973*a*). Basic amino acid transport in *Escherichia coli*: properties of canavanine-resistant mutants. *J. Bacteriol.*, **116**, 627–635

Rosen, B. P. (1973*b*). Restoration of active transport in an Mg^{2+}-adenosine triphosphate-deficient mutant of *Escherichia coli*. *J. Bacteriol.*, **116**, 1124–1129

Rosen, B. P. and Vasington, F. D. (1971). Purification and characterization of a histidine-binding protein from *Salmonella typhimurium* LT-2 and its relationship to the histidine permease system. *J. Biol. Chem.*, **246**, 5351–5360

Rosenberg, H., Cox, G. B., Butlin, J. D., and Gutowski, S. J. (1975). Metabolite transport in mutants of *Escherichia coli* K-12 defective in electron transport and coupled phosphorylation. *Biochem. J.*, **146**, 417–423

Rowland, I. and Tristram, H. (1975). Specificity of the *Escherichia coli* proline transport system. *J. Bacteriol.*, **123**, 871–877

Russel, R. R. B. (1972). Mapping of a D-cycloserine-resistance locus in *Escherichia coli* K-12. *J. Bacteriol.*, **111**, 622–624

Rytka, J. (1975). Positive selection of general amino acid permease mutants in *Saccharomyces cerevisiae*. *J. Bacteriol.*, **121**, 562–570

St. Lawrence, P., Maling, B. D., Altwerger, L., and Rachmeler, M. (1964). Mutational alteration of permeability in *Neurospora*: effects on growth and the uptake of certain amino acids and related compounds. *Genetics*, **50**, 1383–1402

Sanchez, S., Martinez, L., and Mora, J. (1972). Interactions between amino acid transport systems in *Neurospora crassa*. *J. Bacteriol.*, **112**, 276–284

Sanderson, K. E. (1976). Genetic relatedness in the family enterobacteriaceae. *Annu. Rev. Microbiol.*, **30**, 327–349

Sanderson, K. E. and Hartman, P. E. (1978). Linkage map of *Salmonella typhimurium*, Edition V. *Bacteriol. Rev.*, **42**, 471–519

Seaston, A., Inkson, C., and Eddy, A. A. (1973). The absorption of protons with specific amino acids and carbohydrates by yeast. *Biochem. J.*, **134**, 1031–1043

Shaltiel, S., Ames, G. F. L., and Noel, K. D. (1973). Hydrophobic chromatography in the purification of the histidine-binding protein J from *Salmonella typhimurium*. *Arch. Biochem. Biophys.*, **159**, 174–179

Simoni, R. D. and Shallenberger, M. K. (1972). Coupling of energy to active transport of amino acids in *Escherichia coli*. *Proc. Natl. Acad. Sci. U.S.A.*, **69**, 2663–2667

Singh, A. P. and Bragg, P. D. (1976). Anaerobic transport of amino acids coupled to the glycerol-3-phosphate–fumarate oxidoreductase system in a cytochrome-deficient mutant of *Escherichia coli*. *Biochim. Biophys. Acta*, **423**, 450–461

Sinha, U. (1969). Genetic control of the uptake of amino acids in *Aspergillus nidulans*. *Genetics*, **62**, 495–505

Slayman, C. W. (1973). The genetic control of membrane transport. *Curr. Top. Mem. Trans.*, **4**, 1–174

Stadler, D. R. (1967). Suppressors of amino acid uptake mutants of *Neurospora*. *Genetics*, **57**, 935–942

Stanier, R. Y., Doudoroff, M., and Adelberg, E. A. (1970). *The Microbial World*, Prentice Hall, New York

Stepien, P. P. (1976). Partial purification of methionine-binding protein from *Aspergillus nidulans* by affinity chromatography. *Biochim. Biophys. Acta*, **439**, 154–159

Straus, D. S. and Hoffmann, G. R. (1975). Selection for a large genetic duplication in *Salmonella typhimurium*. *Genetics*, **80**, 227–237

Stuart, W. D. and De Busk, A. G. (1975). Genetic alterations of ribonuclease-sensitive glycoprotein subunits of amino acid transport system in *Neurospora crassa* conidia. *Arch. Biochem. Biophys.*, **166**, 213–222

Stuart, W. D. and Woodward, D. O. (1975). Effects of ribonuclease A on amino acid transport in *Neurospora crassa. J. Bacteriol.*, **122**, 41–46

Surdin, Y., Sly, J., Sire, J., Bordes, A. M., and De Robichon-Szulmajster, H. (1965). Proprietes et contrôle génétique du système d'accumulation des acides aminés chez *Saccharomyces cerevisiae. Biochim. Biophys. Acta*, **107**, 546–566

Templeton, B. A. and Savageau, M. A. (1974*a*). Transport of biosynthetic intermediates: homoserine and threonine uptake in *Escherichia coli. J. Bacteriol.*, **117**, 1002–1009

Templeton, B. A. and Savageau, M. A. (1974*b*). Transport of biosynthetic intermediates: regulation of homoserine and threonine uptake in *Escherichia coli. J. Bacteriol.*, **120**, 114–120

Thorne, G. M. and Corwin, L. M. (1972). Genetic locus of a gene affecting leucine transport in *Salmonella typhimurium. J. Bacteriol.*, **110**, 784–785

Thorne, G. M. and Corwin, L. M. (1975). Mutations affecting aromatic amino acid transport in *Escherichia coli* and *Salmonella typhimurium. J. Gen. Microbiol.*, **90**, 203–216

Tristram, H. and Neale, S. (1968). The activity and specificity of the proline permease in wild-type and analogue-resistant strains of *Escherichia coli. J. Gen. Microbiol.*, **50**, 121–137

Utech, N. M., Reid, K. G., and Holden, J. T. (1970). Properties of a dicarboxylic amino acid transport-deficient mutant of *Streptococcus faecalis. J. Biol. Chem.*, **25**, 5273–5280

Wargel, R. J., Shadur, C. A., and Neuhaus, F. C. (1971). Mechanism of D-cycloserine action: transport mutants for D-alanine, D-cycloserine, and glycine. *J. Bacteriol.*, **105**, 1028–1035

Weiner, J. H. and Heppel, L. A. (1971). A binding protein for glutamine and its relation to active transport in *Escherichia coli. J. Biol. Chem.*, **246**, 6933–6941

Weiner, J. H., Furlong, C. E., and Heppel, L. A. (1971). A binding protein for L-glutamine and its relation to active transport in *Escherichia coli. Arch. Biochem. Biophys.*, **142**, 715–717

Willis, R. C., Morris, R. C., Cirakoglu, C., Schellenberg, G. D., Gerber, N. H., and Furlong, C. E. (1974). Preparation of the periplasmic binding proteins from *Salmonella typhimurium* and *Escherichia coli. Arch. Biochem. Biophys.*, **161**, 64–75

Wolfinbarger, L., Jr and De Busk, A. G. (1971). Molecular Transport. I. *In vivo* studies of transport mutants of *Neurospora crassa* with altered amino acid competition patterns. *Arch. Biochem. Biophys.*, **144**, 503–511

Wolfinbarger, L., Jr and De Busk, A. G. (1972). The kinetics of L-aspartate transport in *Neurospora crassa* conidia. *Biochim. Biophys. Acta*, **290**, 355–367

Wolfinbarger, L., Jr and Kay, W. W. (1973). Acidic amino acid transport in *Neurospora crassa* mycelia. *Biochim. Biophys. Acta*, **330**, 335–343

Wolfinbarger, L., Jr, Jervis, H. H., and De Busk, A. G. (1971). Active transport of L-aspartic acid in *Neurospora crassa. Biochim. Biophys. Acta*, **249**, 63–68

Wood, J. M. (1975). Leucine transport in *Escherichia coli. J. Biol. Chem.*, **250**, 4477–4485

Yamamoto, T. H., Mevel-Ninio, M., and Valentine, R. C. (1973). Essential role of membrane ATPase or coupling factor for anaerobic growth and anaerobic active transport in *Escherichia coli. Biochim. Biophys. Acta,* **314**, 267–275

Microorganisms and Nitrogen Sources
Edited by J. W. Payne

Section 1.6

Regulation of Amino Acid Transport

DALE L. OXENDER
Department of Biological Chemistry,
The University of Michigan Medical School,
Ann Arbor, Michigan 48109, USA

STEVEN C. QUAY
Department of Pathology,
Harvard Medical School and Massachusetts General Hospital,
Boston, Massachusetts 02114, USA

AND

JAMES J. ANDERSON
Department of Biological Chemistry
The University of Michigan Medical School
Ann Arbor, Michigan 48109, USA

1 INTRODUCTION

The intracellular levels of most amino acids are carefully controlled to meet the varying demands by the normal growth cycle of the cell. To control the intracellular levels the cell must balance the processes that lead to increases in intracellular levels such as biosynthesis with those that decrease the level such as metabolism or protein synthesis. Since transport activities alter the cellular level of amino acids it is important to the cell to have a way of regulating them. The regulation of the biosynthesis of various amino acids has been studied extensively in the past but the regulation of transport systems for amino acids has been largely ignored until recently.

This chapter is not intended to be a comprehensive review of the regulation of various amino acid transport systems. Instead it is our intent to identify a few systems where recent progress has been made in understanding the molecular basis of the regulation of amino acid transport. We have drawn heavily on studies in our laboratory on the regulation of branched-chain amino acid transport in the bacterium *Escherichia coli*, where we have found a complex regulatory scheme similar to that found for the leucine biosynthetic pathway. The regulation of leucine transport appears to be important to the physiology of the *E. coli* cell. It is hoped that these studies on the regulation of transport in prokaryotic organisms will provide some insight into the regulatory processes of eukaryotic organisms. Recent studies in our laboratory have suggested there are similarities in the regulation of amino acid transport in bacteria and various tissue culture cells (Moore *et al*., 1977).

2 REGULATION OF MEMBRANE TRANSPORT ACTIVITY: *CIS—TRANS* EFFECTS

Perhaps, one of the simplest examples of alterations of uptake of a particular amino acid involves the inhibition of uptake by a structurally related amino acid present in the medium surrounding the cell. A quantitative treatment of the inhibition by a structural analogue of a substrate for a transport system often reveals a competitive mode of inhibition (Quay and Christensen, 1974). Two types of *cis*-inhibitory interactions can be imagined: (a) the inhibitory compound both blocks the uptake of the substrate and is itself taken up; (b) the inhibitor binds at the substrate site but is not transported. Both kinds of inhibitors have been found among the amino acid transport systems. The branched-chain amino acid transport system designated LIV-I serves primarily for the uptake of leucine, isoleucine, and valine although threonine, serine, and alanine are competitive inhibitors for the LIV-I system (Rahmanian *et al*., 1973). For this system all compounds which serve as inhibitors have also been shown to be substrates. In the case of the arginine-specific transport system, although homoserine and *trans*-hydroxyproline were able to inhibit arginine transport completely, arginine does not inhibit homoserine uptake. The conclusion was drawn that homoserine and *trans*-hydroxyproline inhibit the arginine transport system, but are not accumulated by this system (Quay and Christensen, 1974). *Trans* effects refer to alterations in the transport activity of a test solute produced by a related solute that has accumulated on the opposite side of the membrane. Examples of both *trans*-stimulation and *trans*-inhibition have been reported for various organisms (Oxender, 1972). *Trans*-inhibition of amino acid uptake has been observed in *Penicillium chrysogenum* by Benko *et al.* (1967), in yeast by Crabeel and Grenson (1970), in *Streptomyces hydrogenans* by Ring *et al.* (1970), in *Neurospora crassa* by Wiley and Matchett (1966) and Pall (1971), and in *Escherichia coli* by Kadner (1975). *Trans*-inhibition effects can serve as a feedback mechanism to regulate transport activity and prevent excessive accumulation of amino acids. *Trans*-stimulation effects can serve to produce rapid exchange of one amino acid for another, thus maintaining adequate levels of a variety of amino acids required by the cell. An example of this phenomenon is the rapid exchange of internal L-isoleucine in *E. coli* for external L-leucine (Quay *et al.*, 1977). In special circumstances this process can lead to a transitory L-isoleucine deprivation, thus accounting for leucine toxicity in *E. coli*.

3 REGULATION OF MEMBRANE TRANSPORT PROTEIN SYNTHESIS

1 Induction of Transport Systems

Much of the current work on regulation of transport of nutrients in bacteria arose from concepts formulated before the turn of the century to explain the preferential utilization of one carbon source when two or more carbon sources are present. The biphasic pattern of growth under these conditions has been termed diauxie. This phenomena was initially investigated by Monod (1947) and Gale (1943).

Many of the sugar transport systems are inducible and, therefore, subject to catabolite repression in the presence of glucose. Among the amino acid transport systems in *E. coli* induction of the transport system seems to be the exception rather than the rule and generally represents certain cases where the amino acids are used mainly as either nitrogen or carbon sources. A specific tryptophan transport system can be induced in *E. coli* in response to tryptophan in the medium (Boezi and De Moss, 1961; Burrous and De Moss, 1963). This system is catabolite-repressible and maximum activity is achieved when *E. coli* is grown on Casamino acid medium. A mutant which is defective in the inducible tryptophan transport system has been obtained (Burrous and DeMoss, 1963) and it maps at min 82 on the *E. coli* chromosome near the locus for the inducible tryptophanase allele (D. L. Oxender, unpublished observation). Recently, Wood *et al.* (1978) have reported the induction of a proline uptake system in *E. coli* when proline was used as a nitrogen source in a defined medium. The transport of L-aspartate by the general four-carbon carboxylic acid system (*dct*) is induced when *E. coli* is grown on acids such as succinate, malate or fumarate as a carbon source (Kay and Kornberg, 1971). Again these inducible transport systems serve mainly for solutes that can be used as alternate carbon or nitrogen sources.

2 Regulation of Membrane Transport by Derepression

Many of the amino acid transport systems are repressed by their cognate amino acids (or some metabolically related derivative); for example, the transport of leucine, isoleucine, and valine (Quay and Oxender, 1976); arginine, lysine, and ornithine (Celis *et al.*, 1973; Rosen, 1973; Quay and Christensen, 1974); aromatic amino acids (Whipp and Pittard, 1977); threonine and serine (Templeton and Savageau, 1974) are all repressible. The repression of transport of the aromatic amino acids is particularly complex. Harrison *et al.* (1975) showed that each of the aromatic amino acids repressed the common aromatic transport system (*aroP*) and Whipp and Pittard (1977) showed that this repression required the product of the gene *tyrR* which is involved in the regulation of aromatic biosynthetic enzyme synthesis. Each of the aromatic amino acids has a specific transport system in addition to the common transport system. Tyrosine and tryptophan cause repression of their respective specific systems, but the phenylalanine-specific system is apparently constitutively synthesized and not repressed by the cognate amino acid.

Glutamine transport activity in *E. coli* (Willis *et al.*, 1975) and in *S. typhimurium* (Betteridge and Ayling, 1976) is very low in rich medium and only maximally derepressed when the medium contains a poor source of nitrogen such as glutamate. Glutamine itself does not cause repression, but free NH_4^+ is an excellent repressor. The close correlation between the levels of glutamine synthetase and the high-affinity glutamine uptake system suggest a common regulation in response to the nitrogen balance of the cell (J. E. Brenchley, personal communication).

In the case of branched-chain amino acid uptake systems of *E. coli* early studies indicated a decrease in transport capacity upon growth on complex media (Inui and Akedo, 1965). The early observations were the stimulus for our extensive studies on the regulation of these systems.

4 REGULATION OF BRANCHED-CHAIN AMINO ACID TRANSPORT ACTIVITY

1 Characterization of the Transport Systems and their Components

The existence of multiple systems for entry of the branched-chain amino acids was inferred from our kinetic data (Rahmanian *et al.*, 1973) and later confirmed by others (Wood, 1975; Guardiola *et al*., 1974*a,b*). However, a determination of the number of systems and their properties required a genetic approach. We isolated mutants defective in each system by using very mild selection conditions which produced a selective pressure on one system at a time. We first isolated strains which were derepressed for the high-affinity LIV-I and leucine-specific uptake systems. These strains were isolated as D-leucine utilizers (Rahmanian *et al*., 1973) and fell into two complementation groups (*livR* and *lstR*) (Anderson *et al*., 1976; Oxender *et al*., 1977). Strains possessing the *livR* allele were derepressed in both LIV-I and leucine-specific transport systems, while the *lstR* strains were derepressed in only leucine-specific transport and possessed normal levels of the LIV-I system.

Since the *livR* strain was hypersensitive to L-valine, which is a toxic amino acid for *E. coli* K-12 strains, we were able to obtain mutants in the LIV-I transport system by selecting for resistance to low levels (0.5 μg ml^{-1}) of L-valine (Anderson and Oxender, 1977). In order to obtain mutants in which the genes were completely inactive we used the mutagenic phage Mu, which causes mutations by randomly inserting into the chromosome and thereby disrupting the primary sequence of genes (Howe and Bade, 1975). Using these genetic techniques we obtained a large number of transport defective mutants which fell into three separate classes (see Table 1). The mutant loci *livH, livJ*, and *livK* mapped in close proximity to each other at minute 74 on the *E. coli* chromosome. Each mutant had a different component of the high-affinity LIV-I system inactivated. In *livJ*, absence of the LIV-binding protein (LIV-BP) resulted in the loss of LIV-I transport without affecting the leucine specific uptake system. The *livH* mutant was defective in LIV-I and leucine-specific transport even though it had normal levels of the respective binding proteins (LIV-BP and LS-BP). Presumably the *livH* gene codes for an unidentified component necessary for the function of binding proteins.

Mutations (*livK*) that abolished the leucine-specific binding protein (LS-BP) (Furlong and Weiner, 1970) were found to be defective in both high-affinity transport systems (LIV-I and leucine-specific). It is possible that the Mu-induced *livK* mutations thus far obtained exert a polar effect on the *livH* allele. The remaining leucine transport activity in *livH* or *livK* mutants represents the LIV-II system which is recognized by having a 10-fold higher K_m for leucine (2–4 μM). The LIV-II system is completely resistant to L-threonine inhibition but is completely inhibited by the model amino acid, (−)*b*-2-aminonorbornane-2-carboxylic acid (BCH) (Tager and Christensen, 1971). This low-affinity LIV-II system is also the only one that is detected when wild-type cells are grown on L-leucine, which represses all high-affinity leucine uptake systems.

We have not detected a low-affinity system which is sensitive to L-threonine as reported by Guardiola *et al*. (1974*a,b*). These authors, and

Table 1 Summary of mutations in the branched-chain amino acid uptake systems of *E. coli*

		Uptake systems			Binding proteins	
Mutant locus	Selected phenotype	LIV-I	Leu-specific	LIV-II	LIV-BP	LS-BP
Wild type	–	+	+	+	+	+
livR	D-Leucine utilization[a]	Derepressed	Derepressed	+	Derepressed	Derepressed
lstR	D-Leucine utilization[b]	+	Derepressed	Elevated 2X	+	Derepressed
livH	L-Valine resistance[c]	–	–	+	+	+
livJ	L-Valine resistance	–	+	+	–	+
livK	L-Valine resistance	–	–	+	+	–
livH livP	Non-growth on low leucine[d]	–	–	–	+	+
livP	—	+	+	–	+	+

[a]Able to utilize 200 μg ml^{-1} D-leucine as source of L-leucine.
[b]Able to utilize 100 μg ml^{-1} D-leucine as source of L-leucine.
[c]Resistant to 0.05 μg ml^{-1} L-valine.
[d]Unable to utilize 20 μg ml^{-1} L-leucine.

Wood (1975), have also reported a very low affinity system with a K_m value around 200 μM but we have not investigated leucine transport in that concentration range.

We used a mutant strain (*livH*) which lacks all high-affinity uptake to obtain mutants in the remaining LIV-II system. Mutants with defects in the LIV-II system were unable to grow on low levels of leucine. These double mutants *livH livP* were shown to lack both the high- and low-affinity uptake systems (Anderson and Oxender, 1978). The new mutant locus *livP* mapped near minute 76 on the *E. coli* map. When this strain was converted to *livH*$^+$ by transduction the high-affinity uptake systems were restored but the LIV-II system was still missing. These results indicate that the high-affinity LIV-I and leucine-specific systems can operate independently of the LIV-II system. However, we have some evidence (D. L. Oxender, unpublished observation), and others have reported similarly (Guardiola *et al.*, 1974*a*; Anraku, as quoted in Yamoto and Anraku, 1977), that mutations exist that abolish both high and low affinity uptake, suggesting that a common component may be shared by the two systems.

Our present understanding of the components of the branched-chain amino acid transport systems is that they fall into two classes; one class is the high-affinity, binding protein-mediated systems (LIV-I and leucine specific) which require the products of the *livH, livJ*, and *livK* genes and which are under the control of the genes *livR* and *lstR*. The remaining class is the low-affinity LIV-II system which requires the *livP* gene product, and the level of which is raised somewhat by the *lstR* gene.

2 Regulation of the LIV-II System in *E. Coli*

Most of the studies on leucine transport regulation which will be reported in the succeeding sections concern the regulation of the high-affinity transport (LIV-I and leucine-specific systems combined), since the transport measurements were carried out using low concentrations of leucine that effectively eliminate transport by the lower affinity LIV-II system. However, the regulation of the LIV-II system can be observed in isolation, for, in the case of the *livH* mutants, the LIV-I and leucine-specific systems are non-functional. Table 2 shows the effect of growth medium supplementation and the *lstR* allele upon the level of the LIV-II system. High levels of L-leucine added to minimal medium repress the LIV-II system up to threefold over that found for wild-type strains grown in minimal medium. L-Leucine produces less than a twofold effect upon LIV-II transport in the strain carrying the *lstR* allele. These data support similar conclusions reached previously with *livH*$^+$ strains (Anderson *et al.*, 1976; Quay and Oxender, 1976). When the medium of the wild-type strain was supplemented with isoleucine and valine, or

Table 2 Regulation of the LIV-II uptake system

Strain	Genotype	L-Leucine[a] Supplement	V_{max}[b] (leu nmol min^{-1} (mg dry wt)$^{-1}$)
AE173	*livH*::Mu *lstR*	+	4.1 (2.8–5.2)
		–	7.7 (5–9)
AE89	*livH*::Mu *lstR*$^+$	+	2.1 (1.5–2.4)
		–	7 (5–8.5)

[a]Cells grown in 0.2 per cent glucose minimal medium. Supplementation is with 50 μg L-leucine ml^{-1}.
[b]Median estimate of V_{max} of uptake. Range in parentheses is 70 per cent confidence interval.

methionine, no repression was seen nor did these amino acids increase the level of repression by L-leucine. In summary, regulation of the LIV-II system is qualitatively similar to that of the LIV-I system, but differs markedly in sensitivity.

3 Regulation of the High-Affinity Systems in *E. Coli* and *S. Typhimurium*: Leucine Repression of Transport and Biosynthetic Activity

Since both the biosynthetic enzymes and the transport activity are repressed by growth on medium containing leucine, it was first important to examine if these two distinct processes were regulated in a concerted manner. For these studies mutants of *E. coli* and *S. typhimurium* defective in either transport regulation or in the regulation of the biosynthetic enzymes were examined (Quay *et al*., 1975*b*). Thus, reciprocal experiments were made in which transport regulation was measured in strains with operator-promotor mutations in the *leu* operon or with mutations causing alterations in the regulation of the *ilv* operon. In all such cases studied, transport was found to be regulated in a normal fashion. These experiments showed that the regulation of transport was separate from the regulation of the biosynthetic enzymes. However, the *livR* mutation in *E. coli* does cause changes in the absolute level of the biosynthetic enzymes.

In order for a cell to grow at a typical exponential rate, both the transport and the biosynthetic systems must be regulated so as to maintain a relatively constant level of amino acids. When a mutation causes the transport capacity of a cell to be raised the biosynthetic activities must be lowered in a compensating manner. When this can be achieved, the net result is an unaltered branched-chain amino acid pool with a shift upward in transport capacity and downward in biosynthetic capability. These predictions have been confirmed (Quay *et al*., 1977).

Even though the LIV-I system transports a variety of amino acids, only the presence of leucine causes repression of transport capacity. Methionine and alanine have also been implicated in the regulation of the LIV-I system (Guardiola *et al*., 1974*a,b*), however, we were able to show that high levels of methionine can cause an increase in the internal level of leucine and thereby produce an indirect effect of repression in the LIV-I system (Quay and Oxender, 1976). Since the LIV-I and LIV-II transport systems serve for leucine, isoleucine, and valine the regulation by leucine alone seems somewhat paradoxical. At least two hypotheses could explain this situation. (a) The regulatory pattern of the enteric organism *E. coli* has been adapted to the feast–famine conditions of the gut where coordinate changes in the levels of the three branched-chain amino acids would be expected to occur. In such a situation the selection of leucine as a regulator of transport may have been fortuitous. (b) The cell must regulate the leucine level independent of isoleucine and valine because of the central role played by leucine in various metabolic processes. The latter hypothesis is supported by an examination of some of the leucine-linked metabolic interactions in *E.coli*. Table 3 summarizes a number of the metabolic steps where growth of the cells on leucine causes marked changes.

Table 3 The role of leucine as a regulator of metabolic activities

Effector	Action	Reference
Leucyl-tRNA	Represses $NADP^+$:NADPH oxidoreductase	R. L. Hanson, personal communication
Leucine	Represses membrane-bound proline oxidase	Deutch and Soffer, 1975
Leucine	Represses cystathionine synthetase	Greene and Radovich, 1975
Leucine	Represses serine hydroxymethyltransferase	Greene and Radovich, 1975
Leucine	Inhibits aspartokinase III	Richaud *et al*., 1974
Leucine	Inhibits growth following nutritional shiftdown	Alfoldi and Kerekes, 1964
Leucine	Stimulates accumulation of unique tRNA species of leucine, histidine, arginine, valine and phenylalanine	Fournier and Peterkofsky, 1975; Yegian and Stent, 1969
Leucyl-tRNA	Stimulates lysyl-, methionyl-, and arginyl-tRNA synthetases	Hirschfield *et al*., 1975
Leucine	Stimulates *S*-adenosylmethionine synthetase	Greene and Radovich, 1975
Leucine	Stimulates threonine and serine deaminases via exchange transport	Pardee and Prestidge, 1955; Quay *et al*., 1977
Leucine	Stimulates D-alanine and glycine (dag) transport	Robbins and Oxender, 1973
Leucine	Spares glycine utilization as a nitrogen source	Fraser and Newman, 1975
$tRNA_1^{leu}$	Most abundant $tRNA^{leu}$ isoaccepting species. (In addition, $tRNA^{leu}$ species taken together are the most abundant tRNA.)	Von Ehrenstein, 1970
$tRNA_1^{leu}$	Transcription of the genes for $tRNA_1^{leu}$ is regulated in a unique manner	Ikemura and Dahlberg, 1973
$tRNA_1^{leu}$	Although the most abundant $tRNA^{leu}$, this species is used for protein synthesis to a much lower extent that other $tRNA^{leu}$ species (indicating a possible regulatory role)	Wettstein and Stent, 1968; Kano-Sueoka and Sueoka, 1969; Kitchingman and Fournier, 1975

4 Role of Leucyl tRNA Synthetase and Leucyl tRNA

Since there were extensive parallelisms in the factors that regulate transport and biosynthesis of the branched-chain amino acids, we drew on the previous studies on the regulation of biosynthesis by Umbarger (1973). These studies indicated that aminoacyl-tRNA synthetases serve as part of the regulatory factors in branched-chain amino acid biosynthesis. To determine if leucine must be activated to an appropriate $tRNA^{leu}$ to participate in the repression of transport we measured transport activity in a temperature-sensitive leucyl-tRNA synthetase mutant of *E. coli* (Quay *et al*., 1975). We used a strain with a deletion of the leucine biosynthetic operon to eliminate any possible complications that might result from changes in the levels of the biosynthetic enzymes. Isogenic strains containing a deletion of the leucine biosynthetic operon and a temperature-sensitive leucyl-tRNA synthetase ($leuS^{ts}$) were grown in minimal medium with repressing levels of leucine at the permissive temperature (36 °C) and at a marginally non-permissive temperature (41 °C). At the permissive temperature the two strains had nearly identical levels of leucine, isoleucine, valine, histidine, and proline transport. Histidine and proline transport served as controls in these experiments. When the cultures were grown at 41 °C, the growth of the $leuS^{ts}$ mutant slowed and ceased after about one-half generation. During this period of marginal growth the differential rate of synthesis of transport components for leucine, isoleucine, and valine increased five- to 10-fold in the $leuS^{ts}$ mutant strain. The temperature change had no effect on the transport of proline or histidine in either strain. Threonine deaminase was derepressed in the $leuS^{ts}$ mutant indicating that the loss of charging ability of the mutant was growth limiting. Both RNA polymerase initiation and protein synthesis were found to be necessary for the transport derepression. Measurements of the LIV-binding protein showed four- to six fold increases during growth at 41 °C, which is consistent with the proposed role of this protein in the rate-limiting step for LIV-I transport (Penrose *et al*., 1968; Rahmanian and Oxender, 1972). These results showed that the transport activity is not regulated simply by the cellular level of free leucine but rather by some interaction of leucine with a functional leucyl-tRNA synthetase.

In the regulation of the enzymes for valine and isoleucine biosynthesis the first enzyme is threonine deaminase, which has been shown to be under multivalent regulation by leucine, isoleucine and valine (Freundlich *et al*., 1962). In order to test if the LIV-I transport regulation was under multivalent control we examined leucine transport regulation in a temperature-sensitive valyl-tRNA synthetase mutant and a mutant with a defect in the isoleucyl-tRNA synthetase. The results of these studies indicated that LIV-I transport regulation was solely regulated by the level of leucine and, therefore, not under multivalent control. These results indicated that the leucine transport activity in *E. coli* was regulated by the level of modified leucyl-tRNA or the leucyl-tRNA synthetase enzyme. To distinguish between these two possibilities we obtained a mutant strain with a defect in the maturation of leucyl-tRNA, the *hisT* locus.

5 Role of *hisT* Gene Product

The *hisT* gene codes for an enzyme that converts uridine to pseudouridine (ψ) in the tRNAs for leucine, histidine, and tyrosine. The regulation of the biosynthetic enzymes for these three amino acids is defective in a *hisT* mutant *S. typhimurium* strain (Cortese *et al*., 1974). There is a loss in the repressibility of these biosynthetic operons in a *hisT* mutant. Since these experiments were not performed in auxotrophic strains, a possible role of the *hisT* gene product in the maximum level of derepression had not been examined, even though the biosynthetic enzymes were significantly derepressed in prototrophic strains grown in minimal medium. We examined the regulation of leucine transport in *hisT* strains of both *S. typhimurium* (Oxender and Quay, 1976*a*) and *E. coli* (Quay *et al*., 1978), and found that strains carrying a *hisT* mutation were defective in both repression and derepression of LIV-I transport activity. As in previous cases where transport activity was altered this change in activity was the result of a primary change in the level of the binding proteins. These data could be taken to support a positive role of mature uncharged $tRNA^{leu}$ in the expression of the binding proteins and, therefore, branched-chain amino acid transport activity. Alternatively, it could be that the different pattern of isoaccepting species of leucyl tRNAs present in a *hisT* strain produces an increased level of repression of transport over that observed in a wild-type strain at a given level of external leucine.

The biosynthesis pathway for leucine measured in *hisT* mutants of *S. typhimurium* also did not show a repression or derepression regulation (Oxender and Quay, 1976*a*). Bresalier *et al*. (1975) have similarly shown that when *hisT* mutants of *S. typhimurium* are grown in a chemostat they show reduced levels of derepression in the isoleucine, valine, and leucine enzymes, contrary however to conclusions made from other studies using prototrophic strains of *S. typhimurium* (Cortese *et al*., 1974) and *E. coli* (Lawther and Hatfield, 1977). Thus, the general conclusion from these data is that the regulation of both biosynthesis and transport of the branched-chain amino acids in *hisT* mutants is similar.

6 Role of rho Factor

In addition to regulation by the classical mechanism of inhibition of RNA polymerase initiation by a corepressor–aporepressor complex (Jacob and

Monod, 1961) it now appears that mechanisms exist for changing the frequency of RNA transcriptional termination at an attenuator site in the 'leader region' of the DNA proximal to the structural genes (Bertrand *et al*., 1975). The protein factor rho (Beckwith, 1963) seems to be involved in transcriptional termination at the attenuator site in the tryptophan leader region (Korn and Yanofsky, 1976). We examined branched-chain amino acid transport regulation in a rho mutant to test for an attenuator site in the leucine transport operon (Quay and Oxender, 1977). The uptake of a number of amino acids was examined in a rho mutant (*rho* 120) and in the parental strain, each being grown on rich or minimal medium. In cells grown in minimal medium, the rho mutant showed increased uptake of arginine, histidine, leucine, isoleucine, and valine varying from 30 to 90 per cent; glutamine transport, on the other hand, was decreased to one-third that of the parental strain, while no change in the transport of proline or tryptophan was observed in the rho mutant. Growth in rich medium had two effects: a tendency to negate the transport differences between the parental strain and the *rho* 120 strain, and a generalized decrease in all transport capacities. In additional experiments, with leucine limitations in an auxotroph, we found that while the rho mutant shows a 15-fold derepression upon shifting the cells from 0.6 to 0.05 mM leucine, the parental strain undergoes only a three- to fourfold increase in leucine transport under similar conditions. This derepression primarily affects the LIV-I transport system.

In analogous experiments, we measured the specific activities of the various periplasmic binding proteins in these strains. The binding protein activities for leucine, arginine, glutamine, and histidine were all regulated in a manner related to their transport activities. The two membrane-bound transport systems (proline and tryptophan) were not affected by the rho mutation, whereas, all binding protein systems were altered. As an alternative to explaining these effects as due to reduced attenuation, the levels of the binding-protein transport systems might reflect a pleiotropic effect of the *rho* 120 mutation on the energy balance of the cell, as it is known that the β subunit of the Ca^{2+}–Mg^{2+} ATPase is altered by *rho* mutations (Adhya and Gottesman, 1978).

When *hisT* and *rho* 120 were combined in a single strain of *E. coli* the *hisT* effect was seen to be dominant in that leucine transport could not be derepressed (Quay *et al*., 1978). This result is consistent with a model in which *hisT* primarily affects initiation of mRNA synthesis and rho affects a subsequent event such as attenuation.

7 Role of *relA* Product

The finding that growth on rich medium negates the increased transport observed in *rho* mutants indicates that additional regulatory controls exist under these growth conditions. A recent model of regulation, suggested by

Travers (1976), in which RNA polymerase can exist in two distinct forms depending on the presence or absence of ppGpp, is attractive as a means of explaining some aspects of promotor specificity during nutritional shifts. Since the *relA* locus is required for ppGpp synthesis during nutritional step-down conditions we examined transport regulation in *relA* mutants. Cells were grown on rich medium to a steady state of growth and then shifted to minimal medium, allowed to grow for several hours and then harvested. The results of these experiments indicate that the shift-down growth condition prevented the derepression of the transport activity.

An attractive model for the effect of ppGpp on transport regulation is that it produces a modification of the specificity of the RNA polymerase, since studies by Travers (1976) indicate that binding of ppGpp to the RNA polymerase induces a conformational change which alters the specificity of DNA sequences which are recognized as initiation codons. However, this cannot be the only action of ppGpp since its production by other conditions such as threonine or arginine limitation does not lead to derepression of leucine transport.

Recently, the maturation of $tRNA^{leu}$ has been shown to be incomplete in a *relA* strain starved for leucine or phenylalanine (Kitchingman and Fournier, 1977). In this strain, leucine starvation leads to the accumulation of a form of the major leucyl tRNA species, $tRNA_1^{leu}$ which is undermodified in the uridine at position 41; this is the same deficiency as that found for the $tRNA_1^{leu}$ in *hisT* mutants (Rizzino *et al*., 1974). Therefore, the finding that both *relA* and *hisT* mutants are unable to produce $tRNA^{leu}$ with pseudouridine at position 41, and that they are both also unable to derepress leucine transport fully suggests a regulatory role for this key nucleotide, and further implies that the modified uncharged $tRNA^{leu}$ may act as a positive effector in the regulation of leucine transport.

The regulatory scheme shown in Figure 1 summarizes our current hypothesis of the regulation of leucine transport. As shown by these studies, the regulation of transport is extremely complex and requires the intricate interaction of leucine, $tRNA^{leu}$, rho factor, ppGpp, the *livR* product (X) and the *hisT* gene product.

During a number of these studies we examined the transport of a variety of amino acids to determine the uniqueness of the regulatory scheme for transport. We were able to conclude that other transport systems that require amino acid binding proteins such as those for histidine and arginine appear to be regulated in a manner similar to that of leucine.

5 CONCLUDING REMARKS

In summary, it is clear that the bacterial cell has evolved a regulatory apparatus for leucine transport which rivals that for amino acid

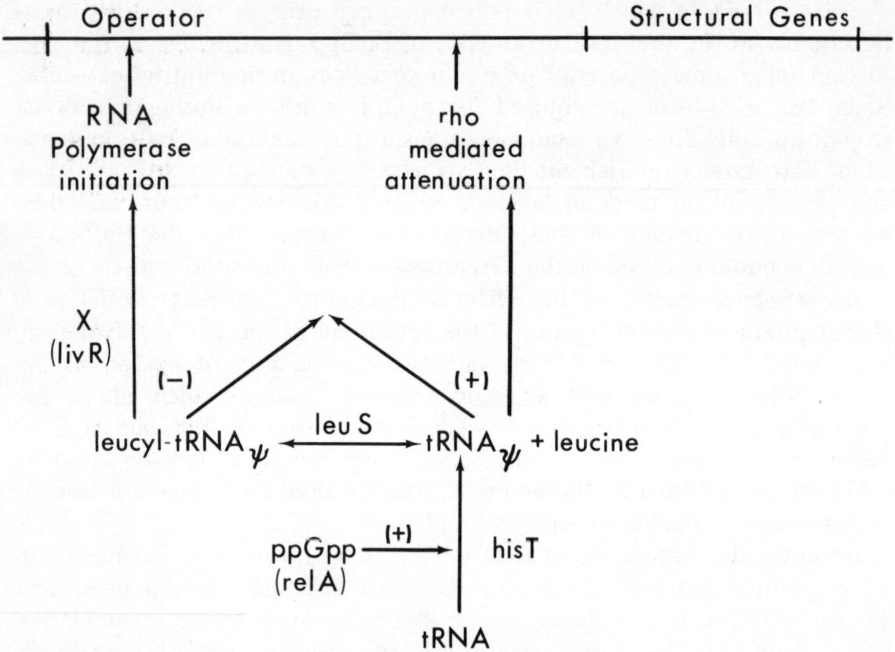

Figure 1 Hypothetical scheme showing effectors and sites of action (arrowheads) for the transcriptional regulation of leucine transport genes. Genes shown to be involved in regulation are *livR, leuS, relA*, and *hisT* (see text). The symbols ψ, ppGpp, and X refer, respectively, to pseudouridine, guanosine tetraphosphate, and the unknown product of the *livR* gene. Positive and negative effects are labelled (+) and (−)

biosynthetic enzymes. Both transport and biosynthesis act to control the level of the branched-chain amino acid pool, biosynthesis by metabolic input and transport by environmental input. However, an additional role for transport is that of retaining the biosynthetic pool. We have shown (Anderson and Oxender, 1978) that cells lacking both high- and low-affinity transport systems (*livH livP* double mutants) excrete leucine when grown in minimal medium. Ample evidence in other transport systems (cf. Anderson and Oxender, 1978) suggest this is a common phenomenon. That the high-affinity LIV-I system can influence the biosynthetic leucine pool is suggested by our findings (Quay *et al*., 1975*b*) that the branched-chain amino acid biosynthetic enzyme levels are maintained at lower levels in strains derepressed for LIV-I (*livR* strains), presumably due to a greater efficiency in maintaining a repressing pool of branched-chain amino acids. The significance of multiple transport systems and their regulatory fine-tuning for *E. coli* cells in the wild can be suggested from their repression patterns and behaviour under certain artificial conditions. The high-affinity, binding protein-mediated leucine transport systems are clearly superior for recovering the last traces of leucine from the environment, and this role is consistent with their derepression in minimal medium. However, if this system is artificially elevated by mutation, the cell scavenges so well that it faces a sudden loss of leucine when the medium is abruptly exhausted by transport; this is evidenced by a protracted lag in derepressing its biosynthetic enzymes and resuming growth (Quay *et al*., 1975*b*). Since LIV-I transport promotes the exchange of one branched-chain amino acid for another (external leucine for internal isoleucine), the derepressed LIV-I transport mutation also sensitizes the cell to growth inhibition by leucine when this amino acid is abruptly introduced into the medium (Quay *et al*., 1977). The need for strict control of this system is evident. The LIV-II system, of lower affinity, is also regulated but over a narrow range; this quasi-constitutive, high-capacity system is of primary importance to the cell when leucine is in abundance. In nature this would probably accompany excess valine and isoleucine as well, presenting few of the problems associated with excess transport capacity.

Finally, one may ask are other amino acid transport systems as complex as the leucine transport systems? Certainly some of the other amino acid transport systems such as that for proline do not have derepressible scavenging systems. If the leucine system proves to be unusually complex, part of the answer may be in the diverse roles which leucine apparently plays in the economy of the cell (Table 3), which would require the pool of this amino acid to be very carefully regulated.

REFERENCES

Adhya, S. and Gottesman, M. (1978). *Annu. Rev. Biochem*., **47**, 967–996

Alfoldi, L. and Kerekes, E. (1964). *Biochim. Biophys. Acta*, **91**, 155–157

Anderson, J. J., Quay, S. C., and Oxender, D. L. (1976). *J. Bacteriol*., **126**, 80–90

Anderson, J. J. and Oxender, D. L. (1977). *J. Bacteriol*., **130**, 384–392

Anderson, J. J. and Oxender, D. L. (1978). *J. Bacteriol*., **136**, 168–174

Beckwith, J. (1963). *Biochim. Biophys. Acta*, **76**, 162–164

Benko, V. P., Wood, T. C., and Segel, I. H. (1967). *Arch. Biochem. Biophys*., **122**, 783–804

Berger, E. A. and Heppel, L. A. (1974). *J. Biol. Chem*., **249**, 7747–7755

Bertrand, K., Korn, L., Lee, F., Platt, T., Squires, C. L., Squires, C., and Yanofsky, C. (1975). *Science, N.Y*., **189**, 22–26

Betteridge, P. R. and Ayling, P. D. (1976). *J. Gen. Microbiol*., **95**, 324–334

Boezi, J. A. and DeMoss, R. D. (1961). *Biochim. Biophys. Acta*, **49**, 471–484

Bresalier, R. A., Rizzino, A. A., and Freundlich, M. (1975). *Nature, Lond*., **253**, 279–280

Burrous, S. E. and DeMoss, R. D. (1963). *Biochim. Biophys. Acta*, **73**, 623–637

Celis, T. F. R., Rosenfeld, H. J., and Mass, W. K. (1973). *J. Bacteriol*., **116**, 619–626

Cortese, R., Lamdsberg, R. R., Vonder Haar, R. A., Umbarger, H. E. and Ames, B. N. (1974). *Proc. Natl. Acad. Sci. U.S.A*., **71**, 1857–1861

Crabeel, M. and Grenson, M. (1970). *Eur. J. Biochem*., **14**, 197–204

Deutch, C. E. and Soffer, R. L. (1975). *Proc. Natl. Acad. Sci. U.S.A*., **72**, 405–408

Fournier, M. J. and Peterkofsky, A. (1975). *J. Bacteriol*., **122**, 538–548

Fraser, J. and Newman, E. M. (1975). *J. Bacteriol*., **122**, 810–817

Freundlich, M., Burns, R. O., and Umbarger, H. E. (1962). *Proc. Natl. Acad. Sci. U.S.A.*, **48**, 1804–1808
Furlong, C. E. and Weiner, J. H. (1970). *Biochem. Biophys. Res. Commun.*, **38**, 1076–1083
Gale, E. F. (1943). *Bacteriol. Rev.*, **7**, 139–173
Greene, R. C. and Radovich, C. (1975). *J. Bacteriol.*, **124**, 269–278
Guardiola, J., DeFelice, M., Klopotowski, T., and Iaccarino, M. (1974*a*). *J. Bacteriol.*, **117**, 383–392
Guardiola, J., DeFelice, M., Klopotowski, T., and Iccarino, M. (1974*b*). *J. Bacteriol.*, **117**, 393–405
Harrison, L. I., Christensen, H. N., Handlogten, M. N., Oxender, D. L., and Quay, S. C. (1975). *J. Bacteriol.*, **122**, 957–965
Hirschfield, I. N., Yeh, F. M., and Sawyer, L. E. (1975). *Proc. Natl. Acad. Sci. U.S.A.*, **72**, 1364–1367
Howe, M. M. and Bade, E. G. (1975). *Science, N.Y.*, **190**, 624–632
Ikemura, T. and Dahlberg, J. E. (1973). *J. Biol. Chem.*, **248**, 5033–5041
Inui, Y. and Akedo, H. (1965). *Biochim. Biophys. Acta*, **94**, 143–152
Jacob, F. and Monod, J. (1961). *J. Mol. Biol.*, **3**, 318–356
Kadner, R. J. (1975). *J. Bacteriol.*, **122**, 110–119
Kano-Sueoka, T. and Sueoka, N. (1969). *Proc. Natl. Acad. Sci. U.S.A.*, **62**, 1229–1236
Kay, W. W. and Kornberg, H. L. (1971). *Eur. J. Biochem.*, **18**, 274–281
Kitchingman, G. R. and Fournier, M. J. (1975). *J. Bacteriol.*, **124**, 1382–1394
Kitchingman, G. R. and Fournier, M. J. (1977). *Biochemistry*, **16**, 2013–2022
Korn, L. J. and Yanofsky, C. (1976). *J. Mol. Biol.*, **106**, 231–241
Lawther, R. P. and Hatfield, G. W. (1977). *J. Bacteriol.*, **130**, 552–557
Monod, J. (1947). *Growth*, **11**, 223–289
Moore, P. A., Jayme, D. W., and Oxender, D. L. (1977). *J. Biol. Chem.*, **252**, 7427–7430
Oxender, D. L. (1972). *Annu. Rev. Biochem.*, **41**, 777–814
Oxender, D. L. and Quay, S. C. (1976*a*), *J. Cell. Physiol.*, **89**, 517–522
Oxender, D. L. and Quay, S. C. (1976*b*). In *Methods in Membrane Biology*, Vol. 6 (E. D. Korn, ed.), Plenum Press, New York, pp. 183–242
Oxender, D. L., Mayo, M. M., Anderson, J. J., and Quay, S. C. (1977). *J. Supramol. Struct.*, **6**, 419–431
Pall, M. L. (1971). *Biochim. Biophys. Acta*, **233**, 201–208
Pardee, A. B. and Prestidge, L. S. (1955). *J. Bacteriol.*, **70**, 667–674
Penrose, W. R., Nichoalds, G. E., Piperno, J. R., and Oxender, D. L. (1968). *J. Biol. Chem.*, **243**, 5921–5928
Quay, S. C. and Christensen, H. N. (1974). *J. Biol. Chem.*, **249**, 7011–7017
Quay, S. C. and Oxender, D. L. (1976). *J. Bacteriol.*, **127**, 1225–1238
Quay, S. C. and Oxender, D. L. (1977). *J. Bacteriol.*, **130**, 1024–1029
Quay, S. C., Dick, T. E., and Oxender, D. .L. (1977). *J. Bacteriol.*, **129**, 1257–1265
Quay, S. C., Kline, E. L., and Oxender, D. L. (1975*a*). *Proc. Natl. Acad. Sci. U.S.A.*, **72**, 3921–3924
Quay, S. C., Lawther, R. P., Hatfield, F. W., and Oxender, D. L. (1978). *J. Bacteriol.*, **134**, 683–686
Quay, S. C., Oxender, D. L., Tsuyumu, S., and Umbarger, H. E. (1975*b*). *J. Bacteriol.*, **122**, 994–1000
Rahmanian, M. and Oxender, D. L. (1972). *J. Supramol. Struct.*, 1, 55–59
Rahmanian, M., Claus, D. R., and Oxender, D. L. (1973). *J. Bacteriol.*, **116**, 1258–1266
Richaud, C., Mazat, J. P., Felenbok, B., and Patte, J.C. (1974). *Eur. J. Biochem.*, **48**, 147–156
Ring, K., Gross, W., and Heinz, E. (1970). *Arch. Biochem. Biophys.*, **137**, 243–252
Rizzino, A. A., Bresalier, R. S., and Freundlich, M. (1974). *J. Bacteriol.*, **117**, 449–455
Robbins, J. C. and Oxender, D. L. (1973). *J. Bacteriol.*, **116**, 12–18
Rosen, B. P. (1973). *J. Bacteriol.*, **116**, 627–635
Tager, H. S. and Christensen, H. N. (1971). *J. Biol. Chem.*, **246**, 7572–7580
Templeton, B. A. and Savageau, M. A. (1974). *J. Bacteriol.*, **120**, 114–120
Travers, A. (1976). *FEBS Lett.*, **69**, 195–198
Umbarger, H. E. (1973). In *Genetics of Industrial Organisms* (Z. Vanek, Z. Hostalek, and J. Culdin, eds), Academic Publishing House, Prague, pp. 195–218

Von Ehrenstein, (1970). In *Aspects of Protein Biosynthesis*, Part A (C. B. Anfinsen Jr, ed.), Academic Press, New York, p. 139
Wettstein, F. O. and Stent, G. (1968). *J. Mol. Biol.*, **38**, 25–40
Whipp, M. J. and Pittard, A. J. (1977). *J. Bacteriol.*, **132**, 453–461
Wiley, W. R. and Matchett, N. H. (1966). *J. Bacteriol.*, **92**, 1698–1705
Willis, R. C., Iwata, K. K., and Furlong, C. E. (1975). *J. Bacteriol.*, **122**, 1032–1037
Wood, J. M. (1975). *J. Biol. Chem.*, **250**, 4477–4485
Wood, J. M., Peycha, K. L. and Zadworny, D. (1978). *Abstr. Am. Soc. Microbiol. Conf. Microbial Membranes*
Yamoto, I. and Anraku, Y. (1977). *J. Biochem., Tokyo*, **81**, 1517–1523
Yegian, C. D. and Stent, G. (1969). *J. Mol. Biol.*, **39**, 45–58

Microorganisms and Nitrogen Sources
Edited by J. W. Payne

Section 1.7

Energetics of Bacterial Amino Acid Transport

I. R. BOOTH AND W. A. HAMILTON

Department of Microbiology, University of Aberdeen, Marischal College, Aberdeen AB9 1AS, UK

1 INTRODUCTION

Bacteria possess the ability to accumulate a wide range of sugars, amino acids, and ions against a concentration gradient. Energy derived from metabolism is required for this process and the mechanisms by which energy coupling to transport is achieved have been extensively studied.

Particularly lucid accounts of our knowledge and of the problems associated with energy coupling have been presented by Harold (1972, 1976), Simoni and Postma (1975), and Hamilton (1975, 1977). It is our aim in this chapter to develop further the concepts enunciated by these authors, and to attempt to extend them in order to accommodate more recent experimental data.

It is now clear that there are two distinct modes of coupling energy to amino acid transport. These are as follows:

(a) *ATP-dependent systems*. This method of energization was discovered by Berger (1973), and utilizes ATP, or a metabolite derived from ATP, as a direct source of energy. In general, these systems have several characteristics in common; they have as a component a periplasmic binding protein, are absent from membrane vesicle systems, are insensitive to uncouplers, but are very sensitive to inhibition by arsenate.

(b) *Proton-motive force-dependent systems*. These systems are dependent upon gradients of ions and of charge which are generated during either respiration or hydrolysis of ATP. In general terms they have been found to involve cotransport of protons, or, in some systems, of sodium ions.

There are many problems associated with experiments designed to determine the method of energy coupling to the uptake of a substrate. In particular, the use of amino acid transport for such investigations has been hampered by the multiplicity of such systems. Through overlapping specificities, several systems may be responsible for the uptake of a single compound. This often leads to one amino acid being transported by more than one mechanism, each of which may involve separate modes of energy coupling. Experimentation is further complicated by the possible involvement of the membrane potential in the functioning of the membrane-bound component of the ATP-dependent systems. The part played by ATP itself in these systems is uncertain and the evidence for its involvement is largely circumstantial. Consequently considerable care must be exercised when interpreting the experimental data available for amino acid transport.

It is clear that transport systems are an essential component of cellular structure, and thus it is evident that their energization must be an integral function of cell metabolism. Therefore, it is essential to consider energy coupling to transport systems in terms both of the mechanisms involved and of the relationship to cell metabolism as a whole.

2 PHOSPHATE BOND ENERGY-DEPENDENT TRANSPORT SYSTEMS

1 The Role of the Binding Protein

When *Escherichia coli* cells were subjected to an hypo-osmotic shock procedure (see Section 1.4.1.), Piperno and Oxender (1966) observed a marked reduction in the activity of the branched-chain amino acid transport system. The loss of transport activity was correlated with the release from the periplasm of a binding protein specific for the substrates of this uptake system. Subsequently workers have shown that osmotic shock treatment of *E. coli* causes the release of binding proteins for many amino acids, sugars, and ions (Boos, 1974) (see Section 1.4).

The involvement of a binding protein in a transport system may be established by the following criteria: (a) osmotic shock treatment results in a marked reduction in transport activity when compared with the effect on an osmotic shock-resistant uptake system; (b) concentrated, and possibly partially purified, shock fluid contains a protein which has a similar affinity for the substrates of the transport system as was demonstrated by whole cells; (c) loss of binding-protein activity subsequent to mutagenesis leads to impaired ability to accumulate the amino acid by that system; (d) during preparation of membrane vesicles, transport activity is lost due to removal of the binding protein. Many systems have been found to satisfy all of these criteria (Boos, 1974). However, there are at least two notable exceptions.

First, the uptake of dicarboxylic acids into *E. coli* is achieved by two systems, *dct* and *ct*. Osmotically-shocked cells lose the capacity for succinate uptake by either system and mutants lacking a binding protein are unable to grow on a range of carboxylic acids (Lo *et al*., 1972). However, membrane vesicles possess transport activity toward succinate, as do vesicles prepared from mutants lacking the binding protein (Rayman *et al*., 1972). More recently, Lo and Sanwal (1975) have identified three genetic loci involved in succinate transport, *dctA* and *dctB* code for membrane proteins, and the third locus, *cbt*, codes for a binding protein which is required for transport by whole cells but not for that by vesicles (see also Section 1.4.2.5).

Second, Willis and Furlong (1975) presented data which suggested the involvement of both a binding protein and sodium ions in glutamate uptake into *E. coli* K-12 cells. Subsequent work has shown that the sodium ion-dependent system is active in membrane vesicles and involves sodium ion cotransport (Kahane *et al*., 1975). Binding of glutamate to the released binding protein on the other hand is not sodium ion-dependent (Miner and Frank, 1974). Recently it has been demonstrated that cells grown on rich media lack the binding protein and although glutamate uptake capacity is not affected, the initial rate of transport may be reduced (Kahane *et al.*, 1976*a,b*). It has been suggested that the glutamate-binding protein and the membrane carrier are components of the same systems which in vesicles, and in whole cells under certain conditions, can also operate with the membrane carrier alone (Kahane *et al.*, 1976*b*). An alternative solution is that proposed recently for *E. coli* W, in which the binding protein-dependent and sodium ion-dependent systems for glutamate uptake are separate, with the latter being the major route of transport (Schellenberg and Furlong, 1977). Thus, loss of the binding protein in this instance would cause only a small reduction in transport activity (see also Section 1.4.2.5).

Clearly in the case of succinate uptake, and less unambiguously with glutamate transport, the binding protein is not an essential component of transport in vesicles but its presence is either obligatory or beneficial in whole cells.

Relatively little is known about the membrane-bound components of the ATP-dependent systems. Their involvement in transport has been detected by the isolation of mutants which have lost uptake activity but retain the binding protein. Mutants have been isolated which affect histidine transport in *Salmonella typhimurium*. Two systems, each with its own binding protein, interact with a common membrane protein, loss of which results in a lack of transport activity (Ames, 1976). Similar studies with sugar transport in *E. coli* have identified essential genetic loci other than that for the binding protein. However, the products of these genes have not been identified (Postma, 1976; Hazelbauer, 1975*a,b*; Szmelcman *et al.*, 1976). Thus, this is an area in which much work needs to be done in order that the mechanism of energy coupling should become clear.

Throughout this section considerable emphasis has been placed on the involvement of binding proteins in ATP-dependent systems. It has also been demonstrated, however, that the participation of a binding protein in transport does not necessarily imply energy coupling via ATP. On the other hand, Harold and Spitz (1975) demonstrated that the uptake of glutamate and of phosphate into *Streptococcus faecalis* occurred by a mechanism similar to that of the ATP-dependent systems of *E. coli*. However, no binding protein was detected in these cells nor were they susceptible to osmotic shock. Thus, there is no obligatory requirement for a binding protein in this mode of energy coupling.

Since the binding protein does not appear to be essential for energy coupling from phosphate bond energy, it is pertinent to inquire as to the role of the binding protein (see also Section 1.1.4.1 and 1.6.4.). The following observations have been made:

(a) The binding proteins are present in relatively large amounts (10^{-6} or 10^{-5} M) in the periplasm (Oxender and Quay, 1975).
(b) They do not themselves appear to be involved in solute membrane translocation steps, although they contain a second binding site that could interact with membrane components (Ames and Spudich, 1976).
(c) Mutant strains which produce enhanced amounts of binding protein have similarly increased levels of transport and this has led to the suggestion that the binding proteins are the limiting component of the transport complex (Oxender and Quay, 1976).
(d) The binding proteins appear to increase the affinity of the transport system for the solute by interacting with a membrane component. This may result from the possibility that the solute-binding protein complex is the actual substrate for the membrane-bound transport components (Oxender and Quay, 1975).

Thus the role of the binding protein would appear to be that of increasing the availability of the substrate to the membrane carrier. In this regard it is interesting to note that the binding proteins have been shown to play an integral role in certain chemotactic responses (Ordal and Adler, 1974*a,b*). This too would indirectly enhance the supply of solute to the transport system.

2 Energy Coupling in Binding Protein Systems

In something of a classic paper, Berger (1973) examined energy coupling to the glutamine and proline transport systems of *E. coli*. These systems are, respectively, sensitive and resistant to osmotic shock. He was able to demonstrate that glutamine transport was driven directly by phosphate bond energy, formed either by oxidative phosphorylation or by glycolysis. Proline uptake, on the other hand, was found to be coupled to a high energy membrane state derived either by respiration or from the action of the adenosine triphosphatase (ATPase) on glycolytically-produced ATP. There is some evidence, however, contradictory to this clear-cut description of the binding protein transport systems being energized solely by phosphate bond energy. Several studies have since concluded that there is a direct role for the proton-motive force (p.m.f.) in these transport systems (Plate *et al*., 1974; Singh and Bragg, 1976; Lieberman and Hong, 1976; Lieberman *et al*., 1977; Plate, 1976). A summary of the available evidence for the involvement of phosphate bond energy and of the proton-motive force is given below.

(i) Role of phosphate bond energy

These studies (Berger, 1973) were carried out using *E. coli* ML308–225 and a derivative mutant, DL-54, which is impaired in the ATPase and cannot synthesize ATP via oxidative phosphorylation. Using cells of the parent strain depleted of their energy reserves, Berger demonstrated that either glucose or lactate stimulated both glutamine and proline uptake to an equivalent extent. In the mutant strain, DL-54, glucose energized glutamine uptake into starved cells to a much greater extent than did lactate. Proline uptake was energized by either carbon source to a similar level. Therefore, in the absence of an active ATPase in the mutant, energy derived from respiration could not be coupled to glutamine accumulation. It was suggested that ATP, or a metabolite derived from ATP, acted directly to energize glutamine uptake.

A schematic presentation of these results is given in Figure 1. In the parent, glucose or lactate metabolism leads to the generation of a proton-motive force and of ATP, which is synthesized via both substrate and oxidative phosphorylation. The mutant, DL-54, is unable to synthesize ATP via oxidative phosphorylation and thus lactate respiration, whilst generating a p.m.f., does not cause significant ATP synthesis. Consequently only proline uptake is strongly stimulated by lactate. In contrast, glucose generates both ATP (via glycolysis) and a p.m.f. (via respiration), and thus both glutamine and proline are taken up (Figure 1).

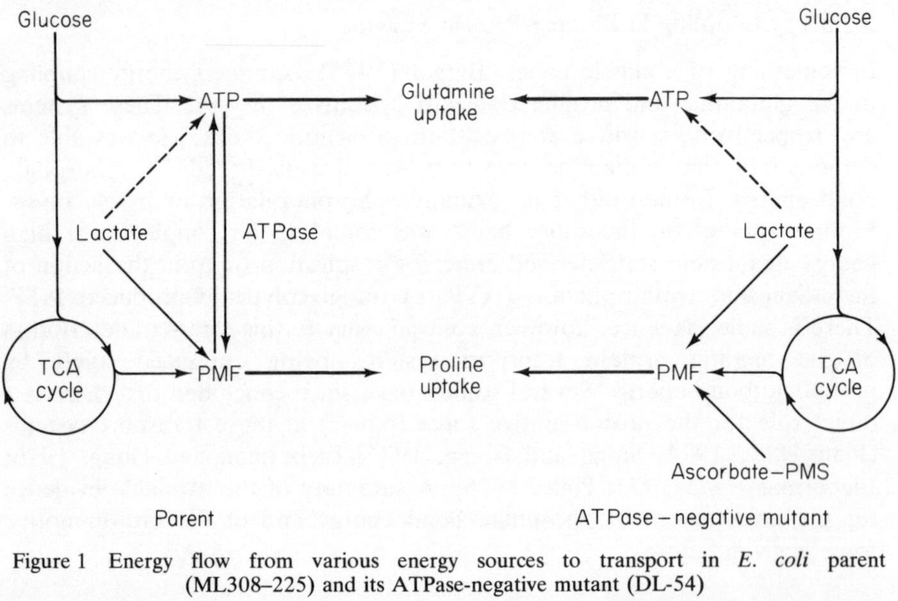

Figure 1 Energy flow from various energy sources to transport in *E. coli* parent (ML308–225) and its ATPase-negative mutant (DL-54)

Support for the role of ATP in glutamine uptake was obtained from the inhibitory effect of arsenate. Arsenate completely inhibited glutamine uptake into energy-depleted cells of both ML308–225 and DL-54 whether glucose or lactate was the energy source. However, proline uptake was only significantly inhibited by arsenate in ML308–225 when glucose was the energy source. Thus it was suggested that ATP, or some closely related metabolite, is closely involved in energy coupling to glutamine uptake.

Subsequently to Berger's study there have been several complementary investigations of energy coupling in *E. coli* transport systems possessing a binding protein. The following substrates appear to have a similar dependence on phosphate bond energy: glutamine, α,ε-diaminopimelic acid, arginine, histidine, ornithine (Berger and Heppel, 1974; Gutowski and Rosenberg, 1976*a*), methionine (Kadner and Winkler, 1975), leucine, isoleucine, and valine (Oxender, 1972; Wood, 1975), glycylglycine (Cowell, 1974), ribose (Curtis, 1974), and potassium ions (Rhoads and Epstein, 1977). As was mentioned above, similar systems, though lacking a binding protein, have been reported in *Strep. faecalis* (Harold and Spitz, 1975). Whilst the exact nature of the energy coupling is not known, there is in each case good evidence that uptake is not obligatorily coupled to the proton-motive force.

(ii) Role of the proton-motive force

In a subsequent investigation of energy coupling to shock-sensitive transport systems, Berger and Heppel (1974) demonstrated that a p.m.f. alone would not energize transport. When ascorbate-phenazine methosulphate (ascorbate-PMS) was provided as energy source in DL-54 no amino acid uptake occurred via any of these systems. Shock-resistant systems, on the other hand, did respond to this energy donor. This is in contrast to lactate as an energy source which stimulated both shock-resistant and shock-sensitive systems, although the latter to a much lesser extent. These authors demonstrated that the difference was due to a small amount of ATP synthesis which occurs with lactate but not with ascorbate-phenazine methosulphate (Berger and Heppel, 1974). Thus, generation of a p.m.f. alone was not sufficient for activation of systems involving a binding protein.

However, there are several lines of evidence which suggest a possible direct involvement for the p.m.f. in shock-sensitive transport systems.

First, in all studies to date using mutants with defective ATPase and thus incapable of oxidative phosphorylation, it has been noted that shock-sensitive systems remain sensitive to agents which collapse the proton-motive force. It has been shown that inhibitors of respiration and proton ionophores cause significant reductions in the activity of these systems. Similarly anaerobiosis which prevents the generation of a p.m.f. in these mutants, also inhibits transport by 30–60 per cent (Berger, 1973; Berger and Heppel, 1974; Cowell, 1974; Rhoads and Epstein, 1977). Therefore, the p.m.f. appears to increase the efficiency of the shock-sensitive systems.

Second, further evidence for the involvement of the p.m.f. comes from investigations of the effect of colicin K on glutamine transport. The lethal effect of colicin K is thought to come about by abolition of the proton-motive force and the consequent reduction in ATP. In an ATPase-negative mutant the loss of ATP does not occur. When cells of such an ATPase-negative mutant were treated with colicin K the activity of the glutamine transport system declined rapidly, although the ATP content rose to a level higher than that prior to treatment. Assuming that ATP content reflected that of other high energy phosphates, the conclusion was drawn that both a p.m.f. and phosphate bond energy were required for shock-sensitive transport systems (Plate *et al*., 1974). When the rate of loss of glutamine transport was compared with that of proline, it was discovered that the decline of the latter system occurred more rapidly. The two rates only approached equality when the parent strain was treated with colicin K and there was a simultaneous reduction in both the p.m.f. and cellular ATP content. Thus it appears that the glutamine transport system may in fact be coupled solely to phosphate bond energy but with less efficiency than in the presence of a proton-motive force.

Third, it has recently been suggested that the membrane potential is the component of the p.m.f. necessary for the activity of the shock-sensitive transport systems. Two groups of workers have isolated mutants which are defective in energy coupling to both ATP synthesis and active transport (Lieberman and Hong, 1974, 1976; Lieberman *et al*., 1977; Plate, 1976). Both groups have isolated temperature-sensitive variants of their mutants and have shown that at the non-permissive temperature glutamine transport is inhibited although the ATP content of the cells is unaltered. Designated *ecf*^{ts}, the mutant isolated by Lieberman and Hong (1976) and Lieberman *et al*. (1977) is best understood. The mutant possesses an altered inner membrane protein which at the non-permissive temperature allows the dissolution of the membrane potential while the pH gradient remains constant. Thus, it would appear that the membrane potential has a function in the activity of shock-sensitive transport systems. It is interesting to note that the mutants isolated by both groups of workers are also colicin K resistant and it has been suggested that the altered protein is the site of action of colicin K (Plate, 1976).

The energization of the shock-sensitive transport systems thus appears to involve two components: phosphate bond energy and a membrane potential. Whilst the former is essential, the role of the membrane potential may be less so. The mechanism of these systems therefore requires the interplay of the two major energy currencies of the cell.

3 The Mechanism of Energy Coupling in Binding Protein Systems

The mechanisms of both solute translocation and energy coupling in shock-sensitive systems are unknown. It is clear that solute translocation involves interaction between the periplasmic binding protein and other protein components more firmly embedded in the inner membrane. Mutants of *S. typhimurium* have been isolated in which the histidine-binding protein has lost the ability to interact with the membrane protein while retaining solute-binding activity. One class of revertants which have regained transport activity is characterized by the possession of an altered membrane protein which now accommodates the change in the binding protein (Ames and Spudich, 1976). Thus, it is apparent that the solute binding site and membrane attachment site of the binding protein are separate, and these results raise the possibility that the requirements for interaction are relatively simple.

Energy coupling is obviously complex and the roles of phosphate bond energy and the membrane potential are not known at the present time. It is plausible that the membrane potential may be necessary for the interaction between the membrane protein and the binding protein–solute complex. The p.m.f. has recently been suggested to act as a regulator of the activity of various membrane-bound proteins. In particular the activities of the chloroplast ATPase, certain transport proteins and some colicins have been reported to be influenced by the p.m.f. (Schuldiner and Kaback, 1977). Thus, the involvement of the membrane potential in the osmotic shock-sensitive transport systems may be one more example of this phenomenon.

No information is available on the method by which phosphate bond energy is coupled to translocation. Early workers suggested that a conformational change consequent upon phosphorylation of the membrane protein might be involved (Berger and Heppel, 1974). In this regard it is interesting to note the recent demonstration of exchange of amino acids across a shock-sensitive transport system. The *E. coli* LIV-I transport system has been reported to catalyse the exchange of extracellular leucine for intracellular isoleucine and valine (Quay *et al*., 1977). This is contrary to the previously held belief that shock-sensitive transport systems function irreversibly (Harold, 1976). Clearly exchange transport creates some problems in energy coupling to uptake if futile cycles of transport at the expense of ATP are to be avoided. It has recently been suggested that entry of maltose into *E. coli* occurs via a shock-sensitive system with energy coupling through phosphate bond energy. However, exit of maltose from the cell is mediated by a different route and is controlled by the proton-motive force (Ferenci *et al*., 1977). Clearly such systems must be stringently controlled to avoid non-essential dissipation of energy (see also Section 2.6).

In as much as the immediate energy donor for shock-sensitive systems remains unknown, speculation about the mechanism of energy coupling seems premature. However, an attempt should be made to discriminate between two possibilities; one is that a common phosphorylation mechanism operates for all such systems, while the alternative is that each carrier is independently coupled. If the former is the case it should be possible to isolate a class of mutants which are pleiotropically affected in their ability to transport via shock-sensitive systems. Such a mutant would be analogous to the Enzyme I and HPr mutants of the phosphoenol pyruvate:sugar phosphotransferase (PT) system (Kundig *et al*., 1964). Indeed there may be other parallels between the two systems. Although phosphorylation of the substrate does not occur in the shock-sensitive systems, both classes of transport show very high uptake capacity compared with chemiosmotic systems. Further, in neither case does rapid leakage of the transported substrate occur when the energy supply becomes limiting, in marked contrast to systems driven by the p.m.f. Therefore, both PT and binding protein systems appear to be more efficient than chemiosmotic systems, and thus may be evolutionary more advanced. This may also explain why in a single organism there may exist two (three for sugars) different mechanisms for uptake of a single substrate. Since each mechanism has its benefits as well as its costs, none has achieved precedence over the others (see also Section 1.5.6).

3 PROTON-MOTIVE FORCE-DEPENDENT TRANSPORT SYSTEMS

In recent years the chemiosmotic model of active transport in substrate translocation systems not involving binding proteins or phosphate bond energy has attracted almost universal support (Mitchell, 1966; Harold, 1976). The model proposes that uptake of many metabolites is coupled to transmembrane gradients of ions and of charge. Oxidative and photosynthetic electron transport chains, the bacteriorhodopsin of *Halobacterium halobium*, and the proton-translocating ATPase are considered to be organized anisotropically within the cell membrane such that they generate gradients of protons and of charge (Harold, 1976). This electrochemical gradient is termed the proton-motive force (p.m.f.); until recently this was often referred to as a 'high-energy membrane state', a much less explicit term. The proton-motive force, Δp, is composed of a membrane potential, $\Delta\Psi$, and a pH gradient, ΔpH. Thus

$$\Delta p = \Delta\Psi - Z\Delta\text{pH}, \tag{1}$$

where Z is the factor converting pH into millivolts, and is approximately 59 at 25 °C (Mitchell, 1966). The bulk of this chapter is thus taken up with a consideration of energy coupling to the p.m.f.-dependent systems.

1 Properties of Membrane Vesicle Systems

Membrane vesicle systems have been extensively used in studies of energy coupling to transport. This has often influenced the results obtained and it is therefore pertinent to consider the properties of these systems.

(i) Membrane orientation in vesicles

It has been claimed that vesicles retain the same membrane orientation as in the parent cell membrane, and similar selective permeability properties to the whole cell (Kaback, 1974*b*; 1976; Kaback and Hong, 1973). However, inner membrane vesicles prepared by osmotic lysis by the method of Kaback (1974*a*) have been shown by several laboratories to display enzyme activities which were cryptic in spheroplasts (Futai, 1974; Weiner, 1974). These results suggested that the vesicle preparations were either a mixture of right side out and everted membranes, or that certain membrane proteins were translocated from the inner to the outer surface of the membrane during vesicle preparation. Confirmation that the latter alternative is correct came in recently published results from Rosen's laboratory (Adler and Rosen, 1977). Three lines of evidence suggest that certain segments of the membrane containing NADH oxidase and ATPase activities are inverted during vesicle preparation.

First, after vesicle preparation, NADH oxidase and ATPase activities were found to be distributed between the inner and outer faces of the membrane. In whole cells all of the enzyme activities were located on the inner face of the membrane, while in vesicles 60 per cent of NADH oxidase and ATPase activities were on the outer face. All of the D-lactate dehydrogenase, on the other hand, was retained on the membrane inner face. This evidence supports that of Futai (1974) and Weiner (1974), who reported loss of crypticity for certain enzyme activities but not for others during vesicle preparation. It also agrees with immunological studies which found that all the D-lactate dehydrogenase activity was located wholly inside the vesicle (Futai and Tanaka, 1975; Short *et al*., 1975).

Second, antibodies prepared to the membrane ATPase agglutinated vesicle preparations. Adler and Rosen (1977) found that both the oxidase activities and the proline and calcium ion transport activities precipitated with the agglutinated vesicles. This suggests a uniform vesicle population rather than a mixture of inverted and right-side-out vesicles.

Third, the criterion chosen to demonstrate conclusively the inversion of certain sections of the membrane during vesicle preparation was taken from work on calcium ion transport in *E. coli*. Whole cells exclude calcium ions by an antiport mechanism involving exchange for protons. Everted vesicles, prepared by cell lysis in a French pressure cell, accumulate calcium ions with either D-lactate or NADH as energy source. This is due to the reversal of polarity of the p.m.f. caused by inversion of the membrane (Figure 2(a)). Thus calcium ions are accumulated in response to a p.m.f., internal acid and positive (Rosen and McClees, 1974). In vesicles prepared by osmotic lysis, on the other hand (Kaback, 1974*a*), D-lactate did not act as an energy donor for calcium ion uptake, whereas NADH and ATP were both very efficient energy donors. Further, when D-lactate and either NADH or ATP were supplied simultaneously not only was calcium uptake inhibited but also that of the amino acid, proline. In control experiments only D-lactate energized proline uptake (Adler and Rosen, 1977; B. P. Rosen, personal communication).

These results, presented schematically in Figure 2(b) can only be explained by inversion of specific segments of the membrane. Clearly, in view of the low membrane permeability to the substrates, NADH

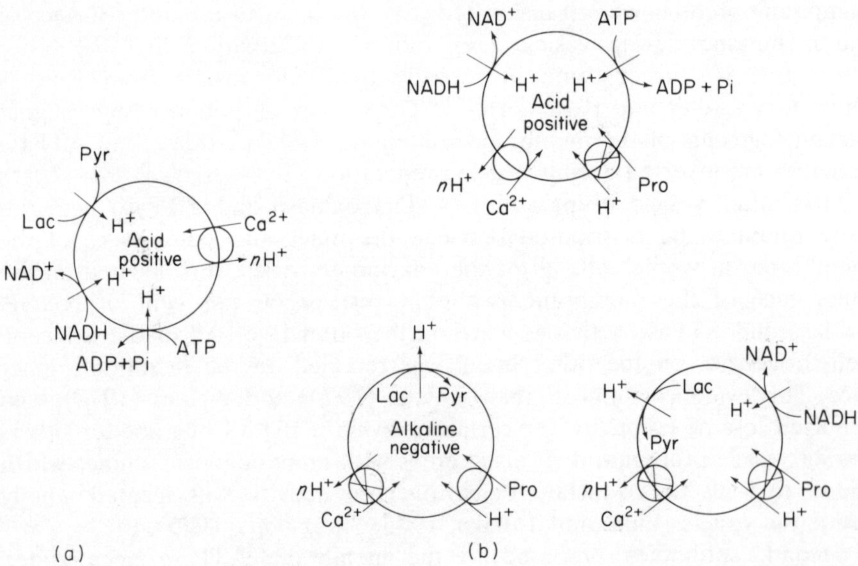

Figure 2 Energy coupling to calcium ion (Ca^{2+}) and proline (Pro) transport in (a) inverted vesicles and (b) mosaic vesicles of *E. coli*. (Lac, lactate; Pyr, pyruvate)

oxidation and ATP hydrolysis must occur at the outside of the vesicle, with consequent injection of protons into the intravesicular space and generation of a p.m.f., inside acid and positive. Conversely, vesicles are permeable to lactate and thus oxidation could occur either internally or externally. However, from the uptake data it is evident that most D-lactate oxidation occurred intravesicularly with generation of a p.m.f., inside alkaline and negative. Calcium ion and proline transport, therefore, respond to the appropriately orientated p.m.f. Inhibition of uptake of either solute by simultaneous utilization of two energy sources suggests that the associated proton movements are vectorially opposed with no resultant p.m.f. Thus, the inverted NADH oxidase and ATPase must occur in the same vesicle membrane as the D-lactate dehydrogenase.

In summary, the vesicle membrane appears to contain both inverted and right-side-out segments. The process of inversion is non-random, affecting the NADH oxidase and ATPase but not the D-lactate dehydrogenase. Any conclusions regarding the anisotropy of other proteins, such as transport carriers, must be tempered by the possibility that these too may be inverted.

(ii) Energy source for uptake

One major experimental difference between whole cells and vesicles is that while the former generally contain large pools of respiratory substrates, vesicles are empty of such reserves. Thus, generation of a p.m.f. in vesicles requires the addition of a suitable membrane-permeable energy source. This has given rise to many anomalous results, but principally to the impression that oxidation of certain compounds is more readily coupled to transport processes. As was indicated above, NADH is oxidized by vesicles but the released energy is not coupled to transport of amino acids. However, intravesicular NADH, generated by an alcohol dehydrogenase/NAD^+ couple, was equally effective as D-lactate as an energy source for amino acid uptake (Futai, 1974). Similarly the high efficiency of NADH as an energy source was recorded in *Bacillus subtilis* once account was taken of the low membrane permeability to this compound (Hampton and Freese, 1974).

Succinate is also a relatively inefficient energy donor for transport, especially for neutral and acidic amino acid uptake into vesicles (Kaback, 1974*b*). The reason for this is again tied to entry into vesicles of the energy source. During respiration succinate is converted stoicheiometrically into fumarate with the extrusion of two protons via a single energy coupling site (Kaback and Milner, 1970, Haddock and Jones, 1977). As two protons are taken up in symport with succinate (Gutowski and Rosenberg, 1976*b*) there is therefore no net proton translocation and hence no pH gradient

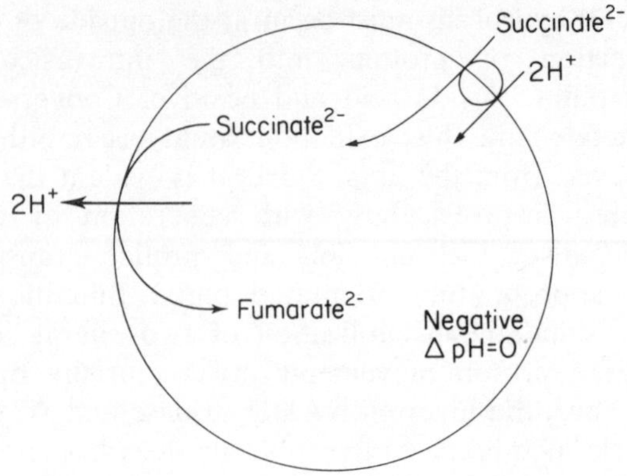

Figure 3 Succinate uptake and oxidation in *E. coli* membrane vesicles

generated. A membrane potential should be formed, however, since proton extrusion is electrogenic while proton re-entry is electroneutral (Figure 3). Recently it has been confirmed that under conditions in which D-lactate oxidation generates a pH gradient of around 102 mV, no pH gradient is developed during succinate oxidation. However, oxidation of either energy source results in a membrane potential of approximately 70 mV (Ramos *et al*., 1976). Thus, succinate oxidation is a more efficient energy donor for cation uptake, driven solely by $\Delta\Psi$, than for neutral or acidic substrates which respond to both the pH gradient and the membrane potential.

Proline is often used as a typical neutral amino acid in transport assays. In experiments with vesicles oxidizing succinate it is in fact atypical. As indicated by Kaback (1974*b*), succinate is 4 per cent as effective in stimulating proline uptake as is D-lactate. Succinate is, however, invariably present as the sodium salt and, as described by Lombardi and Kaback (1972), sodium ions inhibit proline uptake into *E. coli* vesicles. Thus if one assays the effectiveness of succinate as an energy donor in *E. coli* by studying proline uptake and ignores considerations of either the transport of succinate itself or the presence of its neutralizing counter-ion, one invariably reaches the wrong conclusion.

The efficiency of D-lactate as an energy donor in vesicle transport reactions is almost certainly the result of its own rapid translocation through the membrane (Matin and Konings, 1973; Harold and Levin, 1974; Nichols and Hamilton, 1976). In whole cells of *E. coli* the respiration rate is unaffected by pH below pH 8.5 (Padan *et al*., 1976) whereas in vesicles the rate of oxygen consumption falls by 50 per cent over the range of pH 5.5–7.5 (Ramos and Kaback, 1977*a*). In both cells and vesicles the pH gradient diminishes from two pH units at pH 5.5 to zero at pH 7.8 (Padan *et al*., 1976; Ramos and Kaback, 1977*a*). The uptake of D-lactate occurs in response to the pH gradient (Ramos and Kaback,

1977*b*) and therefore, lowering ΔpH results in a reduced internal D-lactate concentration. Thus, an apparent effect of pH on the rate of respiration may be explained by the need to transport an oxidizable substrate which itself responds to ΔpH, a pH-dependent variable.

(iii) Intravesicular phase

When vesicles are prepared the cell contents are replaced by buffer. This is one of the advantages of vesicle systems since it leads to the separation of transport from the influence of intracellular control mechanisms and from substrate metabolism subsequent to uptake. Whilst this facilitates study of the mechanism of uptake, it prevents consideration of uptake as an integral part of cell metabolism. It has the further disadvantage that different factors to those which operate in whole cells, control the intravesicular space. When vesicles, prepared in phosphate buffer, were respiring reduced-PMS the intravesicular pH was approximately pH 7.8. Substitution of Tris (Tris(hydroxymethyl)aminomethane) or MES (2-(*N*-morpholino)-ethane sulphonic acid) buffers for phosphate, changed the internal pH to pH 8 and pH 7, respectively (Ramos and Kaback, 1977*a*). Thus the intravesicular pH was affected by the buffer used to prepare the vesicles.

Similarly the apparent simplification of the intravesicular space may account for the markedly lower values of the membrane potential measured by lipid-soluble cations in these systems. These are usually 50–60 mV lower than similar estimates in whole cells. As will be discussed below, this is felt to be due to the limited availability of a suitable counterion in the buffers routinely used in vesicle preparation.

These examples indicate that one cannot move from cells to vesicles without some loss of confidence in the system. Whilst it is true that vesicles have a simplified internal phase, this, far from being an advantage, may prove to be their most severe limitation.

In general, vesicles have proved extremely useful in studying the mechanism of some transport systems, and indeed the contribution of Kaback's group (Kaback 1974*a*, 1976) has been extremely valuable in this respect. However, it remains that transport is a cellular function and vesicles are of little use as an aid to understanding uptake at this level. Further, it has yet to be demonstrated that the quantitative measurement of ΔpH and $\Delta\Psi$ have the same meaning in vesicles as they have in cells. Consequently whilst vesicles have played a significant role in studies of the mechanism of membrane transport, they should only be used with a clear vision of their shortcomings.

2 Proton-linked Transport Systems

As was stated in equation (1) the proton-motive force is composed of two elements, the membrane potential and the pH gradient. Mitchell (1973) has derived the following relationship for equating the accumulation of a translocated species to the magnitude of the two components of the p.m.f:

$$Z \log \frac{[\mathrm{S}]_i}{[\mathrm{S}]_o} = (m + n)\,\Delta\Psi - nZ\Delta\mathrm{pH}, \tag{2}$$

where m is the charge on the substrate transported, n, is the number of protons cotransported, $[\mathrm{S}]_i$ and $[\mathrm{S}]_o$ are the intracellular and extracellular concentrations of substrate, respectively. Z, $\Delta\Psi$, and ΔpH have the same meanings as in equation (1). Whilst in most cases the magnitudes of m, ΔpH, $[\mathrm{S}]_i$ and $[\mathrm{S}]_o$ can be obtained unequivocally, there are considerable differences in the reported values of $\Delta\Psi$ and, as a result, in the predicted values of n. A more detailed discussion of the quantitative aspects of the chemiosmotic model in relation to metabolizing cells will be presented below.

(i) Function of the proton-motive force in metabolizing and non-metabolizing cells

Many early studies were carried out in non-metabolizing cells in which artificial gradients of pH and potential were generated and which provided qualitative evidence for the involvement of the p.m.f. in amino acid transport. A model to relate uptake of amino acids into starved cells of *Staphylococcus aureus* to the components of the p.m.f. has been derived from equation (2) (Table 1; Niven *et al*., 1973). The model assumes that protons are taken up with neutral and acidic amino acids but not with cationic amino acids, and that electroneutral uptake of anions occurs. The following predictions were made: (a) that transport of lysine, a cationic amino acid, should respond to the membrane potential; (b) that the uptake of glutamate, an anionic amino acid, should be driven solely by the pH gradient; (c) that the uptake of neutral amino acids, e.g. isoleucine and glycine, should respond to both the membrane potential and pH gradient. Energy-depleted cells of *Staph. aureus* can be treated with ionophores and manipulated in such a way that at the final steady state they have a pH gradient vectorially opposed by a membrane potential, but no net p.m.f. Thus, in contrast to metabolizing cells, a pH gradient, interior alkaline, is

Table 1 Model relating amino acid uptake to the imposed driving force in energy-depleted cells (n, H^+: amino acid stoicheiometry)

Amino acid	n	Transport complex	Net driving force
Cationic	0	$(aa^+)^+$	$\Delta\Psi$
Neutral	1	$(aa^0H^+)^+$	$\Delta\Psi - Z\Delta$pH
Acidic	1	$(aa^-H^+)^0$	$- Z\Delta$pH

balanced by a membrane potential, interior positive, and vice versa. This system was used to test the above predictions. It was found that lysine was transported only in response to a membrane potential (interior negative), and that glutamate was taken up only in response to an imposed pH gradient (inside alkaline). Isoleucine accumulation did not occur at the steady state since $\Delta p = 0$. Transient uptake of isoleucine and glycine could be induced in similar experiments in which $\Delta p \neq 0$ due either to enhanced ΔpH or to enhanced $\Delta\Psi$. These results give unequivocal support for the chemiosmotic model of transport (Niven *et al*., 1973; Niven and Hamilton, 1973, 1974).

Such studies are qualitative only, since it is difficult to quantify the imposed gradients. Furthermore, by virtue of the experimental manipulations, the results do not provide any evidence concerning the validity of the assumed values of n, the number of protons taken up per molecule of substrate (Table 1). Since the experiments were designed such that the membrane potential and pH gradient were equal and vectorially opposed, the net p.m.f. was zero. From equation (2), using the example of glutamate ($m = -1$), the driving force is derived as follows:

if $n = 1$, $m + n = 0$, and

$$Z \log \frac{[S]_i}{[S]_o} = -Z\Delta pH;$$

if $n = 2$, $m + n = 1$, and

$$Z \log \frac{[S]_i}{[S]_o} = \Delta\Psi - 2Z\Delta pH = -Z\Delta pH;$$

if $n = 3$, $m + n = 2$, and

$$Z \log \frac{[S]_i}{[S]_o} = 2\Delta\Psi - 3Z\Delta pH = -Z\Delta pH.$$

Consequently for glutamate the net driving force for uptake into these non-metabolizing cells is the pH gradient regardless of the value of n. Thus, if a membrane potential was generated (inside negative) in starved *Staph. aureus* cells, the pH gradient (internal acid) would oppose any uptake in response to that potential. While the data obtained in these experiments therefore provide evidence for the involvement of the pH gradient in glutamate uptake, they do not exclude the possibility that electrogenic uptake of this amino acid in response to both the pH and the potential gradients may in fact be the mechanism of uptake in metabolizing cells. Similar considerations apply to lysine uptake. Even if n is greater than zero, the net driving force for lysine uptake in these non-metabolizing cells is always the membrane potential. Thus, uptake of lysine in response to a pH gradient, inside alkaline, would not be observed because the membrane potential is internal positive under these conditions and thus opposes uptake. However, it is possible that in metabolizing cells lysine uptake involves both the membrane potential and the pH gradient. Therefore, while this work with non-metabolizing cells demonstrates the ability of the components of the p.m.f. to drive amino acid uptake in *Staph. aureus*, the data obtained is equally consistent with proton:amino acid stoicheiometries other than those originally proposed in the model.

Similar qualitative demonstrations of the ability of the p.m.f. to drive amino acid transport have been made using both bacterial cells and vesicles. The uptake of neutral amino acids into *Strep. faecalis* cells in response to an imposed membrane potential has been reported (Asghar *et al*., 1973). In vesicle preparations from *E. coli*, potassium diffusion potentials (inside negative) have been shown to drive uptake of proline, glycine, tyrosine, glutamate, and various sugars (Hirata *et al*., 1974; Schuldiner and Kaback, 1975). Similarly, leucine uptake into *H. halobium* vesicles in response to a potassium diffusion potential has been reported (MacDonald and Lanyi, 1975). Thus, it can be seen that there is good qualitative evidence that the proton-motive force is involved in amino acid transport in non-metabolizing cells and in non-respiring vesicles. Further evidence of its adequacy to drive transport in metabolizing cells and vesicles requires quantitation both of the levels of substrate accumulation achieved and of the available driving force.

(ii) Quantitation of the proton-motive force

Many methods have been developed which are claimed to measure the magnitude of the components of the proton-motive force. Sufficient space is not available for a detailed discussion of all the methods, and this will be found elsewhere (Rottenberg, 1975; Maloney *et al*., 1975).

Measurement of the pH gradient This is now routinely carried out by estimating the distribution of weak acids and bases across the cell membrane. While the membrane is very permeable to the undissociated acid, it is essentially impermeable to the anion. Conditions are chosen such that the concentration of the undissociated acid is very low; thus the pK of the acid is usually below the pH range under study. The degree of dissociation of the acid is determined by the pH and thus, the transmembrane concentration ratio of the acid is a measure of the pH gradient. Consequently when the internal pH is more alkaline than that of the cell's environment the acid will accumulate within the cell. This is illustrated in Figure 4. When the converse is true and the internal pH is

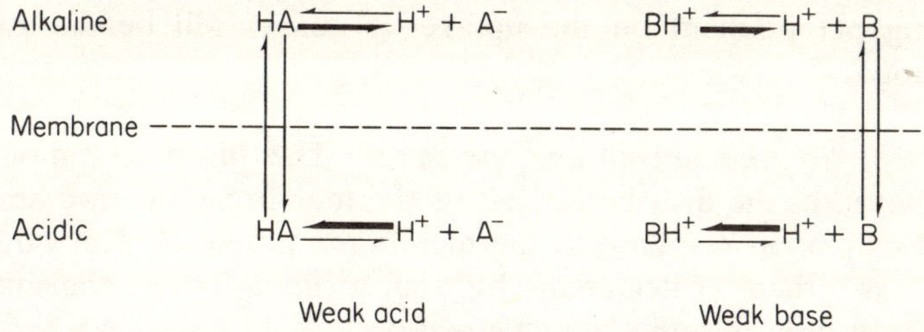

Figure 4 Distribution of weak acids and weak bases across a membrane in response to a pH gradient

lower than that outside the cell a weak base is used to measure ΔpH. The principle involved is the same.

Dimethyloxazolidinedione (DMO, p*K* 6.3), a weak acid which has been routinely used in many laboratories with a range of organisms, has the virtue of not being metabolized (Padan *et al*., 1976; Michel and Oesterhelt, 1976; Harold and Papineau, 1972*a*; Mitchell and Hamilton, 1977). Other metabolizable acids such as acetate (p*K* 4.76) and propionate (p*K* 4.87) have also been used in *E. coli* membrane vesicles with much success (Ramos and Kaback, 1977*a*). Weak bases have been less frequently used, but uptake of methylamine (p*K* 10.16–10.79) in response to a pH gradient, inside acid, was reported in *E. coli* cells (Padan *et al*., 1976).

Quantitative measurements of the pH gradient in several laboratories suggest that bacteria maintain their intracellular pH (pH_i) constant over a wide range of extracellular pH (pH_e). In *E. coli* the internal pH is approximately pH 7.8 over the range of external pH 5.5–8.0, but may rise to a maximum pH_i 8.5 when the medium is at pH 9 (Padan *et al*., 1976; Lagarde, 1977). Similar data are obtained with *Staph. aureus* in which a pH_i 7.8 is also observed over a similar range of external pH (Mitchell and Hamilton, 1977; W. J. Mitchell and W. A. Hamilton, unpublished observations).

Whilst equivalent data have been obtained with membrane vesicles from both *E. coli* and *S. typhimurium*, the limitations on these systems have already been discussed.

In the fermentative organisms, *Strep. faecalis* and *Clostridium pasteurianum*, and in the halophile *H. halobium*, a slightly different situation exists. Although in general the pH gradient is larger at acid pH_e than at neutrality, the internal pH appears not to be so well regulated as in respiring bacteria (Harold *et al*., 1970; Harold and Papineau, 1972*a*; Riebeling *et al*., 1975; Michel and Oesterhelt, 1976).

Thus, whole cells have been shown to possess a pH-dependent component of the p.m.f., in consequence of the maintenance of the intracellular pH near or slightly above neutrality. The effect of the diminishing pH gradient on the uptake of anions will be discussed in a later section.

Measurement of the membrane potential The most commonly used method measures the distribution across the membrane of trace amounts of permeant cations in response to the membrane potential. The value of the potential is then calculated by substituting the transmembrane concentration ratio into the Nernst equation:

$$\Delta\Psi = \frac{RT}{F} \ln \frac{[\text{cation}]_i}{[\text{cation}]_o}, \tag{3}$$

where R is the gas constant, F the Faraday constant, and T the absolute temperature. For this equation to be applied in a valid manner the permeant cation must be entirely free to respond to the potential. Any constraint on ion movement immediately invalidates the estimate of $\Delta\Psi$ obtained.

Lipid-soluble cations have been increasingly favoured as agents to measure the membrane potential. They have two advantages. First, ionophores need not be present in the membrane as is required with measurement of the potassium ion distribution. Second, incorporation of ^{14}C or ^{3}H labels renders their assay within the cell easy compared with methods available for natural ions. However, their use in quantitative studies is debatable for at least three reasons.

First, both the cations used to date, dibenzyldimethylammonium (DDA^+) and triphenylmethylphosphonium ($TPMP^+$), adhere both to cells and to membrane filters. With whole cells of *Strep. faecalis* it has been demonstrated that DDA^+ and $TPMP^+$ exchange with the natural cations of the cell wall (Harold and Papineau, 1972*b*). Similar observations have been made with *H. halobium* and *Staph. aureus* (Michel and Oesterhelt, 1976; W. J. Mitchell and W. A. Hamilton, unpublished observations). Thus, there is a high background of uptake that is not related to the transmembrane potential.

Second, it has been found that the uptake of these cations is accelerated by a lipid-soluble anion, e.g. tetraphenylboron (TPB^-). Furthermore, a new level of uptake is achieved in the presence of this ion, suggesting that the cations do not normally achieve thermodynamic equilibrium with the membrane potential (Altendorf *et al*., 1975; Szmelcman and Adler, 1976; W. J. Mitchell and W. A. Hamilton, unpublished observations).

Third, a limitation on the use of lipid-soluble cations is that their uptake appears to require a motile cation which is expelled from the cells. Thus, DDA^+ uptake into *Strep. faecalis* cells or into *E. coli* membrane vesicles only occurred when sodium ions were present inside the cell or vesicle.

Potassium-containing cells and vesicles did not accumulate DDA^+ (Harold and Papineau, 1972*b*; Altendorf *et al*., 1975). However, uptake of DDA^+ into potassium-containing *E. coli* vesicles did occur when nigericin, an ionophore catalysing potassium ion–proton exchange, was added (Altendorf *et al*., 1975). The explanation would seem to be that whilst sodium ion-containing vesicles expel sodium ions, and thus allow DDA^+ to accumulate, no equivalent process occurs in potassium ion-loaded vesicles in the absence of the ionophore, nigericin. Support for this conclusion comes from the observation that whilst potassium ion-loaded vesicles respiring D-lactate did not take up DDA^+, when a potassium diffusion potential was generated by treatment with valinomycin in the absence of respiration, DDA^+ was rapidly accumulated (Altendorf *et al*., 1975). Thus, the absence of an appropriate counter-ion limits the ability of DDA^+ to respond to the membrane potential and therefore, the extent of uptake of this cation cannot be used as a measure of $\Delta\Psi$.

The use of $TPMP^+$, though subject to some of these limitations, has been applied to the study of the membrane potential in *E. coli* cells and vesicles (Schuldiner and Kaback, 1975). Although the values obtained were somewhat low (approximately 70 mV) internal consistency was observed in vesicle systems by comparison with the distribution of Rb^+ in the presence of valinomycin. However, examination of the experimental techniques employed by various groups shows that in each case $TPMP^+$ was added prior to the generation of the membrane potential and therefore, that protons may act as counter-ions (Lombardi *et al*., 1973; Griniuviene *et al*., 1974; Michel and Oesterhelt, 1976; W. J. Mitchell and W. A. Hamilton, unpublished observations). If $TPMP^+$ were to act as an indicator of the membrane potential it should be taken up in response to a preformed potential and not just during the generation of the potential. When this point was investigated in cells of *Staph. aureus* it was found that $TPMP^+$ was only taken up to a significant extent during generation of the p.m.f. Addition of $TPMP^+$ to cells with a preformed membrane potential resulted in little or no uptake. Treatment of such cells with valinomycin, however, in the presence of low external potassium ion concentration, increased the uptake of $TPMP^+$ by at least fourfold (W. J. Mitchell, I. R. Booth and W. A. Hamilton, unpublished observations). Thus, the lipophilic cation did not respond to a preformed potential, unless a suitable counter-ion flux was available. In consequence it is clear that lipophilic cations do not fulfil the obligatory criterion of the Nernst equation of responding passively to the membrane potential. This method cannot, therefore, be used to assess the membrane potential quantitatively, although qualitatively it may indicate the presence and polarity of one.

By far the most reliable method of measuring the membrane potential at the present time uses the distribution of potassium ions in the presence of valinomycin. This method has been used in mitochondria with much success provided adequate safeguards are taken (Rottenberg, 1975). In bacteria the method is subject to some limitations, most of which can be overcome. First, the majority of Gram-negative species of bacteria are insensitive to valinomycin as a result of the impermeability of the outer membrane. Whilst this can be overcome by EDTA treatment of the cells, some inner membrane damage may also occur. Mutants of *E. coli* which are sensitive to deoxycholate have been reported to be also sensitive to valinomycin (Collins *et al*., 1976). However, it has been suggested that such mutants may also be affected in respiratory function (Holland, 1975).

In the presence of valinomycin the membrane potential resulting from metabolic activity is a function of the external potassium ion concentration. At concentrations higher than 0.1 mM, potassium ion influx cannot be compensated by proton extrusion and thus the potential falls. In *E. coli* cells the potential appears to fall from 120 mV to 85 mV over the concentration range 0.1–1 mM (Padan *et al*., 1976).

Many workers feel that because potassium is a naturally occurring ion intracellular binding may lead to overestimation of the membrane potential (V. P. Skulachev, personal communication). Whilst intracellular potassium ion binding is thought to be significant in the Halobacteria, the concentration of the ion in these species is also significantly higher than in the non-halophilic species (Cope and Damadian, 1970; Eisenbach and Caplan, 1977). In *Staph. aureus* no intracellular binding of potassium ions was found to occur (W. J. Mitchell and W. A. Hamilton, unpublished observations).

One simplification of the method cannot be avoided and introduces errors into the measurement. In essence one measures the concentration of potassium ions in the two phases, whereas strict application of the Nernst equation requires the measurement of ion activities. Since it is difficult to estimate the activity of ions inside the cell, concentration terms are substituted. Further, all methods of potassium ion measurement treat the cell as a single homogenous phase, which it is not. Due to the ion-exchange properties of the cell wall, it is almost certain that locally high concentrations of potassium ions occur close to the membrane. Thus, the measured transmembrane concentration gradient is probably an overestimate, and consequently so is the membrane potential. Recent evidence suggests that the Donnan potential of the cell wall may result in a three- to fivefold accumulation of potassium ions close to the inner membrane of the Gram-negative cell (Stock *et al*., 1977). This type of effect would lower the potential measured in *Staph. aureus* by approximately 40 mV from 165 mV to 125 mV (W. J. Mitchell and W. A. Hamilton, unpublished observations). A further limitation on the method may be that the organism possesses natural potassium ion porters.

However, it is obvious that since the rate of potassium ion uptake in the presence of valinomycin is much greater than in its absence, the natural porters contribute little to the equilibrium achieved.

The membrane potential in *Staph. aureus* has recently been measured using the valinomycin–K^+ method and estimating whole cell potassium ion content by flame photometry. As was mentioned above, values of 160–170 mV were obtained, but these may be lowered by up to 40 mV when a correction is made for the contribution of cell wall potassium ions (W. J. Mitchell and W. A. Hamilton, unpublished observations). Estimates of the membrane potential of 120 mV have been recorded for valinomycin-treated cells of *E. coli* using radioactive rubidium as a marker of potassium ion distribution (Padan *et al*., 1976). Lipophilic cations when applied to whole cells have registered potentials of a similar magnitude but in general they appear to be less reliable than the valinomycin–potassium ion method. In vesicle systems, lipophilic cations have been used to demonstrate membrane potentials of approximately 70 mV (Kaback, 1976) but it is felt that this figure represents the limitations of the method rather than an accurate estimate of the potential (Altendorf *et al*., 1975).

Estimates of the membrane potential in valinomycin-treated cells of *E. coli* and of *Staph. aureus* have been made by measuring potassium flux with ion-specific electrodes. By measuring potassium ion flux one avoids the problems discussed above and can therefore be more confident of the estimate obtained. In both *E. coli* and *Staph. aureus* the membrane potential developed on energization was approximately 135 mV (Collins and Hamilton, 1976).

Thus, it can be seen that, as indicated above, measurement of the membrane potential is a contentious issue. Whilst it is acknowledged that no convenient method of measurement is considered accurate, it is felt that the values obtained by the potassium ion flux method (Collins and Hamilton, 1976) are the most reasonable estimates. In further considerations of the relationship of active transport to the p.m.f. it is these values which will be adhered to.

The membrane potential does not appear to be affected by the extracellular pH, in the manner of the pH gradient. In cells of both *E. coli* and *Staph. aureus* the membrane potential was found to be constant across the pH range pH 5.5–8.5 (Padan *et al*., 1976; W. J. Mitchell and W. A. Hamilton, unpublished observations). In *Staph. aureus* it is possible that the membrane potential is somewhat diminished below pH 5.5 and that the pH gradient makes the most significant contribution to the p.m.f. under these conditions (W. J. Mitchell and W. A. Hamilton, unpublished observations). The effect of the constancy of the membrane potential whilst the pH gradient diminishes, is that the p.m.f. as a whole decreases as the extracellular pH is raised. This point is clearly illustrated by the data

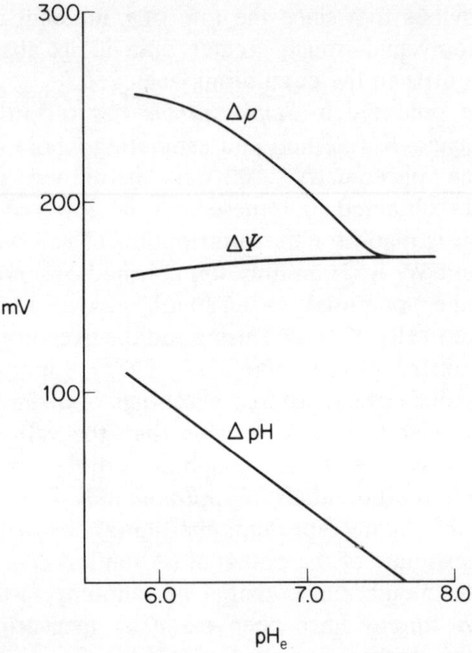

Figure 5 The magnitude of the pH gradient, ΔpH, membrane potential, ΔΨ, and proton-motive force, Δ*p*, in *Staph. aureus* cells over a range of external pH_e

for *Staph. aureus* presented in Figure 5. At pH 5.8 the p.m.f. is 260 mV and decays to approximately 160 mV at pH 8. A subsequent decline in the p.m.f. as the pH is further raised may occur if the pH gradient becomes inverted, as in the case in *E. coli* (Padan *et al*., 1976). Thus, the p.m.f. is pH-dependent and this should affect substrate transport, according to the model proposed in Table 1, in the following manner: (a) the transport of lysine or other cationic amino acids should be substantially unaffected by pH; (b) the uptake of neutral amino acids should follow a curve similar to that of the p.m.f., and thus, be reduced at alkaline pH; (c) the transport of glutamate, and similar anions entering the cell by electroneutral symport, should be markedly sensitive to pH and should not occur at all above pH 7.8.

The relevance of this model, derived from studies on non-metabolizing cells, to amino acid transport in metabolizing cells has already been alluded to and will be further discussed in the next section.

(iii) The relationship between the proton-motive force and transport of amino acids

The relationship developed by Mitchell (1973) to relate the accumulation of a solute to the p.m.f. is derived from the Nernst equation. This is most easily appreciated by considering the uptake of cationic amino acids, in which no protons are involved.

Equation (2) simplifies to

$$Z \log \frac{[S]_i}{[S]_o} = m\Delta\Psi,$$

all the terms having the same meaning as in equation (3) above, with Z being equal to $2.303RT/F$. The accumulation of the amino acid relates to the membrane potential through the charge, m, on the amino acid. A precondition for the valid application of the Nernst equation is that the cation must be entirely free to respond to the membrane potential.

This condition becomes even more significant in situations in which protons are also involved. Because the substrates are either neutral or negatively charged they are only transported by virtue of cotransport with protons. Coupling to energy is through the protons but in order to gain information about this coupling it is necessary to measure the accumulation of the substrate. The techniques used to measure the p.m.f. are special cases of substrate transport in which m and n are known and therefore, $\Delta\Psi$ and ΔpH can be calculated. As it was important for these special substrates to achieve thermodynamic equilibrium in order to measure the p.m.f., so it is essential that amino acids achieve thermodynamic equilibrium with the p.m.f. before n may be calculated using the derivative Nernst equations. A comparison of the theoretical value of n obtained in this manner can then be made with the value obtained by experimental determination of the proton to substrate stoichiometry. However, measurements of the stoichiometry of amino acid transport systems are themselves very difficult due to the low activity of these systems (W. J. Mitchell, I. R. Booth, and W. A. Hamilton, unpublished observations). Until such comparisions can be made for a number of systems the true quantitative relationship between the p.m.f. as driving force and the levels of accumulation attained must remain in doubt.

Attempts to obtain such data are only just beginning. Using vesicles from *E. coli*, Ramos and Kaback (1977*b*) have attempted to correlate the magnitude of the p.m.f. with the accumulation of a range of sugars, amino acids, and carboxylic acids, and thus to predict n. The surprising conclusion was reached that in each case the p.m.f. was insufficient to account for uptake if $n = 1$. However, this conclusion was based on a membrane potential of 70–75 mV, estimated by the distribution of $TPMP^+$ across the membrane. The limitations of this method for estimating $\Delta\Psi$ have been discussed above (Section 1.7.3.2(ii)) and it seems likely that the value obtained by Ramos and Kaback (1977*b*) is an underestimate, possibly by as much as 60 mV (Collins and Hamilton, 1976). A higher value of the membrane potential (135 mV, Collins and Hamilton, 1976) would remove the requirement for $n > 1$, at least in the case of neutral substrates. However, it was also observed that the anions, glutamate, gluconate, succinate, glucuronate, and glucose-6-phosphate, were all accumulated under conditions where no pH gradient existed, and this clearly contradicts the concept of electroneutral symport of these anions driven solely by the pH gradient. It was suggested that n was a variable parameter, being one at low pH_e where the pH gradient is large, and increasing to two or more as the pH gradient diminished at higher pH_e (Ramos and Kaback, 1977*b*). Apparent confirmation of this hypothesis was obtained from the effects on anion uptake of collapsing the pH gradient and membrane potential at pH 5.5 ($n = 1$) and at pH 7.6 ($n \geqslant 2$). Thus at pH 5.5, uptake was sensitive to collapse of the pH gradient by nigericin, but apparently not to reduction of the membrane potential by valinomycin. On the other hand, at pH 7.5 all the systems were sensitive to collapse of the membrane potential (Ramos and Kaback, 1977*b*). However, collapse of the pH gradient at pH 5.5 did not result in complete inhibition of anion uptake; the result was qualitative rather than quantitative. Considerable accumulation of the anions occurred even in the absence of a pH gradient at pH 5.5. Thus, it seems more likely that *n is not variable* but that for *univalent anions* it is *greater than one* and that the membrane potential is involved in uptake at all values of external pH. Consequently, electroneutral symport of anions is unlikely to occur, except in the case of, for example, D-lactate, which was found to respond solely to the pH gradient (Ramos and Kaback, 1977*b*).

There is a strong teleological argument for the evolution of electrogenic transport systems for anionic substrates. Electroneutral transport is only efficient at those pH values at which a large pH gradient exists. However, the selective pressure is more likely to have been kinetic rather than thermodynamic (I. R. Booth and W. A. Hamilton, unpublished observations), since the membrane potential has been suggested to change significantly the kinetics of transport when electrogenic transport is occurring (Kaback, 1976). This is supported by the relative effects on the uptake of neutral molecules of collapsing the pH gradient or the membrane potential. In most cases collapse of the membrane potential results in a much greater inhibition of transport than does the dissipation of the pH gradient (Ramos and Kaback, 1977*b*). In this respect it has recently been suggested that different transport systems respond to the membrane potential and pH gradient in contrasting ways. Thus, it was concluded that at pH 5.5 some systems are coupled to the p.m.f., others primarily to the membrane potential, and third that some are driven by the pH gradient (Ramos and Kaback, 1977*b*). These conclusions were based on the observed effects on substrate accumulation of ionophores which diminish either the membrane potential or the pH gradient. However, implicit in the suggestion is the assumption that all the transported substrates achieve thermodynamic equilibrium with the driving force. Whilst we would not challenge the observations on a qualitative basis, we believe that they are not quantitatively valid because the assumption that thermodynamic equilibrium occurs in all cases is incorrect. In those situations where a kinetic steady state is achieved rather than thermodynamic equilibrium with the p.m.f., the effects of the ionophores cannot be interpreted wholly in terms of their effects on the components of the p.m.f. The principal reason for this will become evident from consideration of the effects of energy coupling on the kinetics of transport.

(iv) Role of energy coupling in transport

Whilst it is not yet possible to identify conclusively the mechanism of energy coupling, there is general acceptance of the idea that it occurs through alteration of the kinetics of carrier movement such that net uptake occurs. Recent evidence suggests that the carrier protein is both functionally and structurally symmetrical (Burnell *et al*., 1975*a,b*; Tsuchiya and Rosen, 1976; Lancaster and Hinkle, 1977*a,b*). Energy coupling is thought to impose asymmetry on the carrier function. Thus, Winkler and Wilson (1966) suggested that energy coupling to lactose transport in *E. coli* caused a decrease in the affinity of the system for lactose at the inner face of the membrane. On the other hand, Lancaster and his colleagues studying partial de-energization of the same uptake system, concluded that both entry and exit reactions were affected by energy coupling (Lancaster *et al*., 1975). The apparent disparity between these two reports may be removed by recent evidence concerning the nature of the lactose carrier. In an elegant series of experiments Kaback and his colleagues have produced a considerable body of evidence to suggest that the carrier protein bears a net negative charge at its active site (Schuldiner *et al*., 1975*a,b*; 1976). Further, these authors suggested that the role of the membrane potential in energy coupling was to reorientate the free carrier from the inner face of the membrane to the outer face (Kaback, 1976). The membrane potential might thus affect the rate of lactose uptake by increasing the rate of translocation outwards of the free carrier. Similarly, by decreasing the lifetime of free carrier at the inner face of the membrane, efflux of the substrate could be reduced. Thus, energy coupling through the effect of the membrane potential on the negatively charged carrier would affect both influx and efflux of the substrate.

There is good evidence to suggest that the affinity of the proton-linked transport systems for their substrate is affected by pH. Thus, for alanine uptake into *E. coli* cells by the *dagA* system (see Sections 1.1.3.1.i and 1.5.2.4) the K_m varies from 6 μM at pH 5.5 to 22 μM at pH 7.8 (I. R.

Booth and W. A. Hamilton, unpublished observations). Similar data have been obtained for lactose (I. R. Booth and W. A. Hamilton, unpublished observations) and gluconate (Robin and Kepes, 1973) uptake into *E. coli*. Whilst this is a relatively small effect, clearly if the internal pH is higher than that outside the cell, this will contribute to the accumulation of the substrate. Collapse of the pH gradient may also accelerate efflux of the substrate through an increase in affinity of the carrier at the inner surface.

If a transported substrate achieves thermodynamic equilibrium with the p.m.f., changes in its accumulation may be predicted from changes in either $\Delta\Psi$ or ΔpH. Such a system should then respond to extracellular pH in a manner identical to the p.m.f. However, if the transported substrate achieves a steady state governed by factors other than the p.m.f., then its response to changes in $\Delta\Psi$ and ΔpH are not quantitatively predictable. Although the carrier itself will respond to changes in the potential and pH gradients, leak pathways and metabolism are usually independent of the p.m.f. The response of a kinetic steady state to a reduction in the rate of influx or efflux, caused by changes in the p.m.f., will be dependent upon the relative contributions of the carrier and of leak (or metabolism) to efflux. Thus, the equation derived by Mitchell (1973; equation (2)) cannot be applied. Furthermore, the effects of ionophores which reduce the p.m.f. cannot be interpreted in the simple manner that has been used to date.

(v) Components of a kinetic steady state

Under conditions which do not allow the further metabolism of a translocated substrate, it has been found that passive leak through the membrane is the most significant component which determines the kinetic steady state. Thus, for uptake of non-metabolizable β-galactosides into *E. coli*, it was observed that a leak pathway controlled the level of accumulation and therefore that the transmembrane concentration ratio was a function of the external substrate concentration (Cohen and Rickenberg, 1956). Similar findings have been made with regard to α-aminoisobutyric acid uptake into *Streptococcus lactis* (Thompson, 1976) and 2-keto-3-deoxygluconate uptake into *E. coli* (Lagarde and Stoeber, 1975). Thus, the kinetic steady state for a non-metabolizable substrate is controlled by passive leakage from the cells. Consequently, the accumulation achieved does not relate directly to the available driving force in the form of the proton-motive force.

With metabolizable substrates the factors contributing to the steady state include the rates of transport, of diffusion-dependent leak pathways and of metabolism of the substrate. Clearly this situation will be more complex than with non-metabolizable substrates. In *E. coli* K-12 uptake of D-alanine occurs via the *dagA* transport system, and the amino acid is metabolized by the D-alanine oxidase (Raunio and Jenkins, 1973). In merodiploid strains, possessing double the complement of the *dagA* gene, the transport of D-alanine is increased twofold both in rate and in the steady-state level. Thus, the rate of entry is the determining factor in the steady state (Lee *et al*., 1975). This is the opposite of the situation in which a leak pathway governs the steady-state accumulation of non-metabolizable β-galactosides in *E. coli*. Doubling of the rate of transport has little effect on the final accumulation achieved (Maloney and Wilson, 1974).

The level of glutamate uptake into the soluble pool of *Staph. aureus* has been shown, in this laboratory, to be a kinetic steady state governed by metabolism of the amino acid (W. J. Mitchell and W. A. Hamilton, unpublished observations). The amino acid is thought to be taken up by electrogenic proton symport ($2H^+$/glutamate$^-$) and its uptake is sensitive to inhibitors both of the membrane potential and of the pH gradient. This obviously contradicts the model proposed from the study of glutamate uptake into non-metabolizing cells (Section 1.7.3.2(i)). However, under appropriate conditions, glutamate uptake into non-metabolizing cells in response to an imposed membrane potential can be demonstrated. The system requires a pH gradient to be orientated alkaline inside at the same time as the potential is generated, whereas in the earlier experiments the membrane potential was opposed by a pH gradient (inside acid) (W. J. Mitchell and W. A. Hamilton, unpublished observations; Niven and Hamilton, 1974).

In summary, it is clear that protons are involved in energy coupling to some amino acid transport systems. However, there is a paucity of data relating the p.m.f. to substrate accumulation and this is reflected by our ignorance of the mechanism of energy coupling. From the studies that have been made two general conclusions can be drawn.

First, it is likely that the uptake of only a few substrates achieves thermodynamic equilibrium with the p.m.f. Kinetic steady states, controlled either by leakage of the substrate from the cell or by metabolism of the accumulated solute, are more frequently observed. Thus, the effects of agents which collapse part of the p.m.f. must be interpreted with this possibility in mind. Further, the substitution of accumulation data from a kinetic steady state into equation (2) above will yield only minimal values of n with little meaning for the transport mechanism.

Second, whilst some aspects of the chemiosmotic model of active transport have been modified, the principal elements are unchanged. However, it has become clear that anions are rarely taken up on membrane carriers by electroneutral symport. Mechanisms have been evolved which diminish the role of the pH gradient in the transport of anions. Discussed above were the electrogenic proton symports for glucose-6-phosphate, gluconate, glucuronate, glutamate, and possibly succinate. A further mechanism for overcoming the problem of the diminishing pH gradient, that of sodium ion cotransport, has recently been discovered and will be discussed in the next section.

3 Sodium Ion-Linked Transport Systems

The involvement of sodium ions in bacterial transport was first reported for marine bacteria many years ago (Drapeau and MacLeod, 1963; Stevenson, 1966). Subsequently amino acid transport in another halophile, *H. halobium*, has been extensively studied (MacDonald and Lanyi, 1975; Lanyi *et al*., 1976*a,b*; MacDonald *et al*., 1977*a*). All the uptake systems require sodium ions, but they exhibit considerable variability in their sensitivity to the absence of a transmembrane sodium ion gradient. Uptake of all the amino acids studied is energized by a sodium-ion gradient generated by the external addition of sodium chloride. With the exception of glutamate all the systems respond to a membrane potential generated by potassium efflux in the presence of valinomycin (MacDonald *et al*., 1977*a*). Whilst glutamate uptake is dependent solely on the sodium ion gradient, that of glutamine, although sodium ion-dependent, does not appear to require a sodium ion gradient across the membrane. Serine uptake is dependent on both the membrane potential and the sodium ion gradient. Aspartate transport, however, demonstrates properties intermediate between those of serine and glutamate (MacDonald *et al*., 1977a). These four amino acids are representatives of four apparently different methods of coupling to the ion and charge gradients in *H. halobium*. It is interesting to speculate that the different requirements for a sodium ion gradient might reflect the different affinities or stoicheiometries of the transport proteins for this ion. The potential of sodium ion coupling in this respect will be discussed below.

Among non-halophilic bacteria there are also now many reports of sodium ion-dependent transport systems. The involvement of sodium ions in α-aminoisobutyric acid uptake by a marine pseudomonad was reported by MacLeod and his colleagues (Wong *et al*., 1969). In subsequent studies however they considered that sodium and potassium gradients provided none of the energy required for accumulation of the α-aminoisobutyric acid (Thompson and MacLeod, 1971, 1973, 1974). While this view was maintained in a later paper, it was suggested, however, that the absence of any evidence of sodium cotransport could result from the rapid extrusion of sodium ions from the cells under normal physiological conditions (Sprott *et al*., 1975).

Similarly, glutamate transport in *E. coli* has been variously reported either to require or not to require sodium ions (Miner and Frank, 1974; Willis and Furlong, 1975; Kahane *et al*., 1975). Recently the situation in *E. coli* has been clarified to some extent by the characterization of multiple transport systems for glutamate, one of which involves sodium ion cotransport (Schellenberg and Furlong, 1977) (see also Section 1.4.2.5). Several studies have shown that the K_m for glutamate uptake by this system is decreased several-fold as the external sodium ion concentration is raised (Frank and Hopkins, 1969; Miner and Frank, 1974; Halpern *et al.*, 1973; Kahane *et al.*, 1975). The affinity constant of the transport protein for sodium ions is 10 mM, substantially higher than the K_m for glutamate (Kahane *et al.*, 1975). On the other hand, the affinity for sodium ions is much lower than that of proton-dependent systems for protons (Robin and Kepes, 1973; I. C. West, personal communication, P. J. F. Henderson, personal communication). The effect of this difference on energy coupling will be discussed below.

Recent evidence suggests that glutamate uptake into membrane vesicles from *E. coli* B can be driven by either a membrane potential or a sodium ion gradient. The effects of the two gradients were demonstrated to be additive, in a manner similar to that of proton-dependent systems (Hasan and Tsuchiya, 1977). Similar experiments with whole cells of *E. coli* B have supported these observations (Tsuchiya *et al.*, 1977), but other work with vesicles from another strain have suggested that the membrane potential is not involved in transport (MacDonald *et al.*, 1977*b*). However, this latter conclusion rests on the inability of a valinomycin-induced potassium diffusion potential to stimulate glutamate uptake into sodium ion-loaded vesicles (MacDonald *et al.*, 1977*b*). Since this is equivalent to the effect of a pH gradient, internal acid, in similar experiments with proton-linked systems, the inability to demonstrate uptake linked to a membrane potential would not necessarily preclude a role for $\Delta\Psi$ in transport.

In addition to the above examples, sodium ion-dependent transport of threonine by *Brevibacterium flavum* and of proline by *Mycobacterium phlei* have been reported (Shiio and Miyajima, 1972; Hirata *et al.*, 1974). Although this latter system has been claimed to be coupled directly to phosphate bond energy, there is now evidence that sodium ion cotransport occurs in response to a sodium ion gradient (Prasad *et al.*, 1976, MacDonald *et al.*, 1977*b*). The involvement of sodium ions in transport of amino acids into an alkalophilic species of *Bacillus* has also recently been reported (Koyama *et al.*, 1976; Kitada and Horikoshi, 1977). Thus, sodium ion-dependent transport systems are widespread and are not just a feature of halophilic organisms.

(i) Energy coupling to sodium ion gradients

In view of the widespread occurrence of sodium ion-dependent systems, it is relevant to enquire of the advantages of such a mechanism. Some of the benefits which accrue from coupling to sodium ions rather than to protons can be seen from brief discussion of the mechanism of generation of sodium ion gradients. Sodium ions are extruded from the cell in exchange for protons by an antiport mechanism (West and Mitchell, 1974; Harold and Papineau, 1972*a*; Lanyi and MacDonald, 1976). In *E. coli* and *Strep. faecalis* the antiports have been suggested to be electroneutral whilst that in *H. halobium* may be electrogenic (West and Mitchell, 1974; Harold and Papineau, 1972*a*; Lanyi and MacDonald, 1976). In *S. typhimurium* membrane vesicles, sodium extrusion has been suggested to be electrogenic since it occurred under conditions where there was no pH gradient (Tokuda and Kaback, 1977). However, although there is no pH gradient, proton circulation is still occurring in order to generate the membrane potential. Proton re-entry into the cell is thought to occur via the sodium ion-proton antiport (Mitchell, 1973; West and Mitchell, 1974; Padan *et al*., 1976). Thus, the observations of sodium ion extrusion in the absence of a pH gradient does not necessarily imply that the antiport is electrogenic, and therefore driven by the membrane potential. Whether the antiport is electrogenic or not, it is clear that while the pH gradient diminishes with increasing external pH, the sodium ion gradient is relatively unaffected (Tokuda and Kaback, 1977). Thus, the driving force for the accumulation of substrates by sodium-linked systems is largely independent of the external pH.

One possible advantage arising out of rapid ejection of the coupling ion is that it can lead to effectively unidirectional transport of the substrate. Carrier-mediated exchange of the transported substrate is regarded as an essential feature of proton-linked systems. In sodium ion-linked systems, however, essentially unidirectional uptake with little exchange has been reported in several instances (Halpern *et al*., 1973; Kahane *et al*., 1975; MacDonald *et al*., 1977*b*). But, it was recently demonstrated that unidirectional uptake of amino acids via sodium-linked systems in both *E. coli* and *H. halobium* only occurred when the internal concentration of Na^+ was severely depleted. When the intracellular (or intravesicular) sodium ion concentration was high, rapid exchange of the amino acids occurred via the carrier protein (MacDonald *et al*., 1977*a,b*). Therefore, unidirectional uptake arises from the maintenance of an intracellular concentration of the coupling ion that is low compared with the affinity of that system for the ion. Whilst this is possible for sodium ions, the ubiquity of protons in cellular processes presumably prevents similar effects for proton-linked systems.

Thus, it may be that the ubiquity of protons has limited their efficiency as coupling ions in active transport. Clearly though, both coupling ions have their advantages; that of sodium ions appears to be its capacity for introducing unidirectional uptake. However, this is also its shortcoming since it requires constant sodium ejection, there being no cellular buffering power for sodium ions. On the other hand, the buffering power of the cell contents for protons is one of the advantages of this coupling ion, but exchange transport is the price paid. Thus, while benefits accrue from coupling to sodium ion these are not sufficient for this method of energy coupling to supplant the central role of protons in the majority of bacterial transport systems.

4 CONCLUSIONS

The central position of the chemiosmotic hypothesis (Mitchell, 1966, 1973) in energy coupling to amino acid transport has been clearly demonstrated. Thus, proton circulation across the cell membrane can now be seen to generate a membrane potential and to regulate the intracellular pH. Consequently, ΔpH is a function of the extracellular pH and decreases as the external pH becomes more alkaline. The relative constancy of the membrane potential across a wide range of external pH must certainly have influenced the evolution of energy coupling to transport. Thus the membrane potential has been found to play a major role in all three classes of energy coupling.

In the binding protein systems of *E. coli* it has been proposed that the membrane potential increases the efficiency of coupling to phosphate bond energy. Whilst ATP or a similar high energy phosphate compound will energize these systems alone, the membrane potential will not (Berger and Heppel, 1974). However, it has been observed that in the presence of a sufficient supply of ATP, the absence of a membrane potential lowers the activity of these systems (Lieberman *et al*., 1977; Plate *et al*., 1974; Plate, 1976). Thus, the role of the membrane potential in the binding protein systems appears to be ancillary to that of ATP.

Many proton- and sodium ion-linked transport systems involve net positive charge translocation across the membrane in response to the transmembrane potential. However, it has been suggested that in lactose transport the membrane potential acts on a negatively charged active site to orientate the empty carrier at the outer face of the membrane. The loaded carrier protein will bear no net charge and its movement will thus be independent of the potential. Clearly this may not be the case with every transport system. In some instances the carrier itself may be uncharged and the loaded carrier bear a net positive charge. Here the potential would act to accelerate influx. The net effect in both instances is to accelerate influx and diminish efflux of the translocated solute. Thus, the membrane potential is essential to these systems, whereas the binding protein-dependent systems have a less stringent requirement for it.

A role for the membrane potential in the Na^+/H^+ antiport in *H. halobium* has been demonstrated and a similar electrogenic antiport activity was suggested to operate in *S. typhimurium* (Lanyi and MacDonald, 1976; Tokuda and Kaback, 1977). Therefore, the membrane potential has the additional function of enhancing the driving force for sodium ion-linked systems in these organisms. The total possible driving force for a 1 : 1 sodium ion–amino acid symport would be equivalent to $2\Delta p$ ($\Delta\Psi + \Delta p$Na; where $\Delta p\text{Na} = \Delta\Psi - 2Z\Delta\text{pH}$), i.e. equivalent to a stoicheiometry of $2H^+$ per substrate molecule. Consequently, electrogenic Na^+/H^+ antiport confers a secondary advantage on sodium ion symports.

Thus, it can be seen that the membrane potential occupies a central position in energy coupling to amino acid transport.

It was stressed at the outset that transport is a cellular function. One consequence of this is the multiplicity of transport systems with overlapping specificities and different mechanisms which obscure studies of energy coupling in whole cells. These difficulties can only be partially resolved by resorting to vesicle preparations. In particular, while they have proved useful in studying ion-linked systems, vesicles can be of little service in studying the energy coupling to binding protein systems. This is due in part to the loss of the binding protein during vesicle preparation, but there may be the additional problem of providing phosphate bond energy as an energy supply for uptake. Thus, whole cells with their intact metabolic machinery may prove more useful in such studies.

Quantitative studies of ion-linked transport systems are at an early stage of development. In *E. coli* membrane vesicles two general observations have been made: (a) the p.m.f. is insufficient to account for the observed substrate accumulation, and (b) the accumulation ratios of many substrates do not respond in the predicted manner to changes in the p.m.f.

This has led to the hypothesis that energy coupling is variable, being modulated by pH-dependent proton–substrate stoicheiometries (Ramos and Kaback, 1977*b*). In part this hypothesis arises out of what are probably underestimates of the p.m.f., particularly of $\Delta\Psi$. However, the experiments have revealed that many anions are transported electrogenically rather than by the previously assumed electroneutral symport. We believe that the failure of many substrates to respond in the predicted way to the driving force is a reflection of their participation in a kinetic steady state rather than a thermodynamic equilibrium. Thus, the accumulation observed is a function of two conflicting forces. The p.m.f. (or sodium ion motive force) which causes uptake is opposed by leak pathways which dissipate the accumulated substrate. In whole cells the leak pathway may involve either exit or metabolism of the transported substrate. Thus, the thermodynamically feasible accumulation of substrate is modulated in whole cells not through the energy coupling mechanism but by the subsequent 'leak' of the accumulated molecules.

ACKNOWLEDGEMENT

We would like to thank our colleague, W. J. Mitchell, for his useful contribution to discussions on the subject material of this chapter.

REFERENCES

Adler, L. W. and Rosen, B. P. (1977). *J. Bacteriol.*, **229**, 959–966
Altendorf, K., Hirata, H., and Harold, F. M. (1975). *J. Biol. Chem.*, **250**, 1405–1412
Ames, G. F.-L. (1976). *J. Cell. Physiol.*, **89**, 543–544
Ames, G. F.-L. and Spudich, E. N. (1976). *Proc. Natl. Acad. Sci. U.S.A.*, **73**, 1877–1881
Asghar, S. S., Levin, E., and Harold, F. M. (1973). *J. Biol. Chem.*, **248**, 5225–5233
Berger, E. A. (1973). *Proc. Natl. Acad. Sci. U.S.A.*, **70**, 1514–1518
Berger, E. A. and Heppel, L. A. (1974). *J. Biol. Chem.*, **249**, 7747–7755
Boos, W. (1974). *Annu. Rev. Biochem.*, **43**, 123–146
Burnell, J. N., John, P., and Whatley, F. R. (1975*a*). *Biochem. J.*, **150**, 527–536
Burnell, J. N., John, P., and Whatley, F. R., (1975*b*). *FEBS Lett.*, **58**, 215–218
Cohen, G. N. and Rickenberg, H. V. (1956). *Ann. Inst. Pasteur, Paris*, **91**, 693–720
Collins, S. H. and Hamilton, W. A. (1976). *J. Bacteriol.*, **126**, 1224–1231
Collins, S. H., Jarvis, A. W., Lindsay, R. J., and Hamilton, W. A., (1976). *J. Bacteriol.*, **126**, 1232–1244
Cope, F. W. and Damadian, R. (1970). *Nature, Lond.*, **228**, 76–77
Cowell, J. L. (1974). *J. Bacteriol.*, **120**, 139–146
Curtis, S. J. (1974). *J. Bacteriol.*, **120**, 295–303
Drapeau, B. R. and MacLeod, R. A. (1963). *Biochim. Biophys. Acta*, **12**, 111–115
Eisenbach, M. and Caplan, S. R. (1977). *Trends Biochem. Sci.*, **2**, 245–247
Ferenci, T., Boos, W., Schwarz, M., and Szmelcman, S. (1977). *Eur. J. Biochem.*, **75**, 187–193
Frank, L. and Hopkins, I. (1969). *J. Bacteriol.*, **100**, 329–336
Futai, M. (1974). *J. Membr. Biol.*, **15**, 15–28
Futai, M. and Tanaka, Y. (1975). *J. Bacteriol.*, **124**, 470–475
Griniuviene, B., Chmieliauskaite, V., and Grinius, L. (1974). *Biochem. Biophys. Res. Comm.*, **56**, 206–213
Gutowski, S. J. and Rosenberg, H. (1976*a*). *Biochem. J.*, **154**, 731–734
Gutowski, S. J. and Rosenberg, H. (1976*b*). *Biochem. J.*, **152**, 647–654
Haddock, B. A. and Jones, C. W. (1977). *Bacteriol. Rev.*, **41**, 47–99
Halpern, Y. S., Barash, H., Dover, S., and Druck, K., (1973). *J. Bacteriol.*, **114**, 53–58
Hamilton, W. A. (1975). *Adv. Microbial Physiol.*, **12**, 1–53
Hamilton, W. A., (1977). *Symp. Soc. Gen. Microbiol.*, **27**, 185–216
Hampton, M. L. and Freese, E. (1974). *J. Bacteriol.*, **118**, 497–504
Harold, F. M. (1972). *Bacteriol. Rev.*, **36**, 172–230
Harold, F. M. (1976). *Curr. Top. Bioenerg.*, **6**, 83–149
Harold, F. M. and Levin, E. (1974). *J. Bacteriol.*, **117**, 1141–1148
Harold, F. M. and Papineau, D. J. (1972*a*). *J. Membr. Biol.*, **8**, 27–44
Harold, F. M. and Papineau, D. J. (1972*b*). *J. Membr. Biol.*, **8**, 45–62
Harold, F. M. and Spitz, E. (1975). *J. Bacteriol.*, **122**, 266–277
Harold, F. M., Pavlasova, E., and Baarda, J. R., (1970). *Biochim. Biophys. Acta*, **196**, 235–244
Hasan, S. M. and Tsuchiya, T. (1977). *Biochem. Biophys. Res. Commun.*, **78**, 122–128
Hazelbauer, G. L. (1975*a*). *J. Bacteriol.*, **122**, 206–214
Hazelbauer, G. L. (1975*b*). *J. Bacteriol.*, **124**, 119–126
Hirata, H., Altendorf, K., and Harold, F. M., (1974). *J. Biol. Chem.*, **249**, 2939–2945
Hirata, H., Kosmakos, F. C., and Brodie, A. F. (1974). *J. Biol. Chem.*, **249**, 6965–6970
Holland, I. B. (1975). *Adv. Microbial Physiol.*, **12**, 56–140
Kaback, H. R. (1974*a*). *Methods Enzymol.*, **31**, 698–709

Kaback, H. R. (1974*b*). *Science, N.Y.*, **186**, 882–892
Kaback, H. R. (1976). *J. Cell. Physiol.*, **89**, 575–594
Kaback, H. R. and Hong, J.-S. (1973). *CRC Crit. Rev. Microbiol.*, **2**, 333–376
Kaback, H. R. and Milner, L. S. (1970). *Proc. Natl. Acad. Sci. U.S.A.*, **66**, 1008–1015
Kadner, R. J. and Winkler, H. M., (1975). *J. Bacteriol.*, **123**, 985–991
Kahane, S., Marcus, M., Metzer, E., and Hapern, Y. S. (1976*a*). *J. Bacteriol.*, **125**, 762–769
Kahane, S., Marcus, M., Metzer, E., and Halpern, Y. S. (1976*b*). *J. Bacteriol.*, **125**, 770–775
Kahane, S., Marcus, M., Barash, H., Halpern, Y. S., and Kaback, H. R. (1975). *FEBS Lett.*, **56**, 235–239
Kitada, M. and Horikoshi, K. (1977). *J. Bacteriol.*, **131**, 784–788
Koyama, N., Kiyomiya, A., and Nosoh, Y. (1976). *FEBS Lett.*, **72**, 77–79
Kundig, W., Ghosh, S., and Roseman, S. (1964). *Proc. Natl. Acad. Sci. U.S.A.*, **52**, 1067–1074
Lagarde, A. E. (1977). *Biochem. J.*, **168**, 211–221
Lagarde, A. E. and Stoeber, F. R. (1975). *Eur. J. Biochem.*, **55**, 343–354
Lancaster, J. R. and Hinkle, P. C. (1977*a*). *J. Biol. Chem.*, **252**, 7657–7662
Lancaster, J. R. and Hinkle, P. C. (1977*b*). *J. Biol. Chem.*, **252**, 7663–7667
Lancaster, J. R., Hill, R. J., and Struve, N. G. (1975). *Biochim. Biophys. Acta*, **401**, 285–298
Lanyi, J. K. and MacDonald, R. E. (1976). *Biochemistry*, **15**, 4608–4614
Lanyi, J. K., Yearwood-Drayton, V., and MacDonald, R. E. (1976*a*). *Biochemistry*, **15**, 1595–1602
Lanyi, J. K., Renthal, R., and MacDonald, R. E. (1976*b*). *Biochemistry*, **15**, 1603–1609
Lee, M., Robins, J. C., and Oxender, D. L. (1975). *J. Bacteriol.*, **122**, 1001–1005
Lieberman, M. A. and Hong, J.-S. (1974). *Proc. Natl. Acad. Sci. U.S.A.*, **81**, 4395–4399
Lieberman, M. A. and Hong, J.-S. (1976). *Arch. Biochem. Biophys.*, **172**, 312–315
Lieberman, M. A., Simon, M., and Hong, J.-S. (1977). *J. Biol. Chem.*, **252**, 4056–4067
Lo, T. C. Y., and Sanwal, B. D. (1975). *Biochem. Biophys. Res. Commun.*, **63**, 278–285
Lo, T. C. Y., Rayman, M. K., and Sanwal, B. D. (1972). *J. Biol. Chem.*, **247**, 6323–6331
Lombardi, F. J. and Kaback, H. R. (1972). *J. Biol. Chem.*, **247**, 7844–7857
Lombardi, F. J., Reeves, J. P., and Kaback, H. R. (1973). *J. Biol. Chem.*, **248**, 3551–3565
MacDonald, R. E. and Lanyi, J. K. (1975). *Biochemistry*, **14**, 2882–2889
MacDonald, R. E., Greene, R. V., and Lanyi, J. K. (1977*a*). *Biochemistry*, **16**, 3227–3235
MacDonald, R. E., Lanyi, J. K., and Greene, R. V. (1977*b*). *Proc. Natl. Acad. Sci. U.S.A.*, **74**, 3167–3170
Maloney, P. C. and Wilson, T. H. (1974). *Biochim. Biophys. Acta*, **330**, 196–205
Maloney, P. C., Kashket, E. R., and Wilson, T. H. (1975). *Methods Membr. Biol.*, **5**, 1–49
Matin, A. and Konings, W. N. (1973). *Eur. J. Biochem.*, **34**, 58–67
Michel, H. and Oesterhelt, D. (1976). *FEBS Lett.*, **65**, 175–178
Miner, K. M. and Frank, L. (1974). *J. Bacteriol.*, **117**, 1093–1098
Mitchell, P. (1966). *Biol. Rev.*, **41**, 445–502
Mitchell, P., (1973). *J. Bioenerg.*, **4**, 65–91
Mitchell, W. J. and Hamilton, W. A. (1977). *Proc. Soc. Gen. Microbiol.*, **4**, 89–90
Nichols, W. W. and Hamilton, W. A. (1976). *FEBS Lett.*, **65**, 107–110
Niven, D. F. and Hamilton, W. A. (1973). *FEBS Lett.*, **37**, 244–248
Niven, D. F. and Hamilton, W. A. (1974). *Eur. J. Biochem.*, **44**, 517–522
Niven, D. F., Jeacocke, R. E., and Hamilton, W. A. (1973). *FEBS Lett.*, **29**, 248–252
Ordal, G. W. and Adler, J. (1974*a*). *J. Bacteriol.*, **117**, 509–516
Ordal, G. W. and Adler, J. (1974*b*). *J. Bacteriol.*, **117**, 517–526
Oxender, D. L. (1972). *Annu. Rev. Biochem.*, **41**, 777–814
Oxender, D. L. and Quay, S. C. (1975). *Ann. N.Y. Acad. Sci.*, **264**,, 358–372
Oxender, D. L. and Quay, S. C. (1976). *J. Cell. Physiol.*, **89**, 517–522
Padan, E., Zilberstein, D., and Rottenberg, H. (1976). *Eur. J. Biochem.*, **63**, 533–541
Piperno, J. R. and Oxender, D. L. (1966). *J. Biol. Chem.*, **241**, 5732–5734
Plate, C. A. (1976). *J. Bacteriol.*, **125**, 467–474
Plate, C. A., Suit, J. L., Jetten, A. M., and Luria, S. E. (1974). *J. Biol. Chem.*, **249**, 6138–6143
Postma, P. W. (1976). *FEBS Lett.*, **61**, 49–53
Prasad, R., Kalra, V. K., and Brodie, A. F. (1976). *J. Biol. Chem.*, **251**, 2493–2498

Quay, S. C., Dick, T. E., and Oxender, D L. (1977). *J. Bacteriol.*, **129**, 1257–1265

Ramos, S. and Kaback, H. R. (1977*a*). *Biochemistry*, **16**, 848–853

Ramos, S. and Kaback, H. R. (1977*b*). *Biochemistry*, **16**, 855–859

Ramos, S., Schuldiner, S., and Kaback, H. R. (1976). *Proc. Natl. Acad. Sci. U.S.A.*, **73**, 1892–1896

Raunio, R. P. and Jenkins, W. T. (1973). *J. Bacteriol.*, **115**, 560–566

Rayman, M. K., Lo, T. C. Y., and Sanwal, B. D. (1972). *J. Biol. Chem.*, **247**, 6332–6339

Rhoads, D. B. and Epstein, W. (1977). *J. Biol. Chem.*, **252**, 1394–1401

Riebeling, V., Thauer, R. K., and Jungermann, K. (1975). *Eur. J. Biochem.*, **55**, 445–453

Robin, A. and Kepes, A. (1973). *FEBS Lett.*, **36**, 133–136

Rosen, B. P. and McClees, J. S. (1974). *Proc. Natl. Acad. Sci. U.S.A.*, **71**, 5042–5046

Rottenberg, H. (1975). *Bioenergetics*, **7**, 61–74

Schellenberg, G. D. and Furlong, C. E. (1977). *J. Biol. Chem.*, **252**, 9055–9064

Schuldiner, S. and Kaback, H. R. (1975). *Biochemistry*, **14**, 5451–5461

Schuldiner, S. and Kaback, H. R. (1977). *Biochim. Biophys. Acta*, **472**, 401–418

Schuldiner, S., Kerwar, G. K., Weil, R., and Kaback, H. R. (1975*a*). *J. Biol. Chem.*, **250**, 1361–1370

Schuldiner, S., Kung, H., Weil, R., and Kaback, H. R. (1975*b*). *J. Biol. Chem.*, **250**, 3679–3682

Schuldiner, S., Rudnick, G., Weil, R., and Kaback, H. R. (1976). *Trends Biochem. Sci.*, **1**, 41–45

Shiio, J. and Miyajima, R. (1972). *J. Biochem., Tokyo*, **72**, 773–775

Short, S. A., Kaback, H. R., and Kohn, L. D. (1975). *J. Biol. Chem.*, **250**, 4291–4296

Simoni, R. D. and Postma, P. W. (1975). *Annu. Rev. Biochem.*, **43**, 423–454

Singh, A. P. and Bragg, P. D. (1976). *Biochim. Biophys. Acta*, **432**, 450–461

Sprott, G. D., Drozdowski, J. P., Martin, E. L., and MacLeod, R. A. (1975). *Can. J. Microbiol.*, **21**, 43–50

Stevenson, J. (1966). *Biochem. J.*, **99**, 257–260

Stock, J. B., Rauch, B., and Roseman, S. (1977). *J. Biol. Chem.*, **252**, 7850–7862

Szmelcman, S. and Adler, J. (1976). *Proc. Natl. Acad. Sci. U.S.A.*, **73**, 4387–4391

Szmelcman, S., Schwarz, M., Silhavy, T. J., and Boos, W. (1976). *Eur. J. Biochem.*, **65**, 13–19

Thompson, J. (1976). *J. Bacteriol.*, **127**, 719–730

Thompson, J. and MacLeod, R. A. (1971). *J. Biol. Chem.*, **246**, 4066–4074

Thompson, J. and MacLeod, R. A. (1973). *J. Biol. Chem.*, **248**, 7106–7111

Thompson, J. and MacLeod, R. A. (1974). *J. Bacteriol.*, **120**, 598–603

Tokuda, H. and Kaback, H. R. (1977). *Biochemistry*, **16**, 2130–2136

Tsuchiya, T. and Rosen, B. P. (1976). *J. Biol. Chem.*, **251**, 962–967

Tsuchiya, T., Hasan, S. M., and Raven, J. (1977). *J. Bacteriol.*, **131**, 848–853

Weiner, J. H. (1974). *J. Membr. Biol.*, **15**, 1–14

West, I. C. and Mitchell, P. (1974). *Biochem. J.*, **144**, 87–90

Willis, R. C. and Furlong, C. E. (1975). *J. Biol. Chem.*, **250**, 2581–2586

Winkler, H. H. and Wilson, T. H. (1966). *J. Biol. Chem.*, **241**, 2200–2211

Wong, P. T. S., Thompson, J., and MacLeod, R. A. (1969). *J. Biol. Chem.*, **244**, 1016–1025

Wood, J. M. (1975). *J. Biol. Chem.*, **250**, 4477–4485

Part 2

Peptides

Microorganisms and Nitrogen Sources
Edited by J. W. Payne

Section 2.1

Transport and Utilization of Peptides by Bacteria

J. W. PAYNE

Department of Botany,
University of Durham, Science Laboratories,
South Road, Durham DH1 3LE, UK

1 INTRODUCTION

Peptide transport systems are widely distributed in nature, occurring in prokaryotic bacteria, eukaryotic microorganisms, plants, and animals. In all these life forms peptides are of considerable nutritional importance, in certain cases apparently surpassing amino acids. These two important conclusions are based on a considerable body of experimental evidence obtained mainly during the last decade.

The broad aspects of peptide transport and utilization by microorganisms and mammals have been discussed at two CIBA Foundation Symposia (1972, 1977), and in a number of reviews. These reviews have dealt mainly with microorganisms (Sussman and Gilvarg, 1971; Barak and Gilvarg, 1975*a*; Matthews and Payne 1975*a,b*, 1980*a,b*; Payne 1972*a*, 1975*a*, 1976, 1977, 1979; Payne and Gilvarg, 1971, 1978); the mammalian intestine, and other animal organs and cells (Burston *et al.*, 1977; Matthews, 1975*a,b*; Matthews and Adibi, 1976; Matthews and Payne, 1975*a,b* 1980*a,b*); and plants (Matthews and Payne, 1979*a*). In this section coverage is confined to the prokaryotic bacteria (excluding the cyanobacteria) and concentrated upon recent developments. Consequently, for a fuller bibliography, especially of the early literature, the above references should be consulted.

2 NATURE OF THE BACTERIAL SURFACE AND ITS INFLUENCE ON PEPTIDE UPTAKE

The variety of forms encompassed within the Kingdom Prokaryotae is enormous (see, for example, *Bergey's Manual*). However, with few exceptions (Mollicutes) the cells are contained within a fairly rigid cell wall that is responsible for cell shape. Almost all species contain peptidoglycans amongst their cell wall polymers, but they differ enormously in their complement of other cell-envelope components, e.g. slime layers, capsules, outer membranes, etc. To date, peptide transport has been studied in relatively few species (Payne and Gilvarg, 1978) many of which are fairly closely related taxonomically, consequently, it is the surface structures of these species that will be mainly considered. For the remainder, it seems that their abilities to transport peptides may vary with the nature of the particular cell-envelope structures surrounding their cytoplasmic membrane, which is the location of peptide permeases.

Detailed investigations of peptide transport have been carried out with both Gram-negative and Gram-positive bacteria. In the former category, studies have been concentrated upon members of the Enterobacteriaceae, Pseudomonadaceae, and Bacteroidaceae, while in the latter group only the Streptococcaceae and Lactobacillaceae have been looked at in any detail. It is important to be aware of this rather narrow spectrum of experimentation during discussions that refer to peptide transport, peptide utilization, etc., in bacteria.

The structure and function of the cell envelope of Gram-negative bacteria has been reviewed (Costerton *et al.*, 1974; Braun and Hantke, 1974; Di Rienzo *et al.*, 1978). The cell envelopes of *E. coli* and *S. typhimurium* comprise, at their outermost surface a distinct membrane, the outer membrane, (OM), underlying this is the peptidoglycan layer, and between this and the cytoplasmic membrane (inner membrane, IM) lies the periplasmic space. The outer membrane and the peptidoglycan layer together constitute the cell wall.

Some time ago, the cell wall of *E. coli* was shown to have a molecular-sieving property that dramatically affected peptide uptake (Payne and Gilvarg, 1968*a*). Thus, the ability of various amino acid auxotrophs to grow on different members of a series of homologous oligopeptides ceased abruptly at a particular point in the series (Gilvarg and Katchalski, 1965; Payne and Gilvarg, 1968*a*). The nutritional cut-off did not always occur at the same homologue, and therefore sieving was not directly dependent on a particular number of amino acid residues in the peptide chain. Measurements of peptide size, using Sephadex gel filtration, indicated that overall volume (Stokes' radius) was the feature that determined whether or not a peptide could be utilized, the cut-off occurring at about the size of the model substrate pentalysine (molecular mass *c*. 650 daltons). Failure of the 'larger' peptides to inhibit utilization of the 'smaller' peptides, led to the suggestion that the steric exclusion was not effected by the peptide permease (location presumed to be at the inner membrane) but arose through a molecular sieving function of the cell wall. Other workers have since presented evidence for a size limit on peptide uptake by *E. coli* (Smith *et al.*, 1970), *Pseudomonas putida* (Cascieri and Mallette, 1976*a*), by *Strep. lactis* (Rice *et al.*, 1978), and *Saccharomyces cerevisiae* (Naider *et al.*, 1974; Marder *et al.*, 1977; see Section 2.2.9). In previous discussions (Payne and Gilvarg, 1971; Matthews and Payne, 1975*b*) of the size limitation for peptide uptake it was concluded that the speculation on the sieve-like nature of the cell wall would only be resolved if it could be shown that other molecules were subject to the same size-dependent exclusion as peptides. Since that time, this feature has been investigated extensively, and the molecular-sieving property of the Gram-negative cell wall clearly established. Thus, Nikaido and colleagues measured the permeability of the outer membrane of *Salmonella typhimurium* to saccharides and polyethyleneglycols of various sizes, and concluded that an exclusion limit of about 600 daltons operates for these molecules (Nakae and Nikaido, 1975; Decad and Nikaido, 1976). The question arises as to the structure responsible for the sieving. The above authors suggested that the effect arose because the penetration of these small molecules through the outer membrane occurred via water-filled channels or pores, and these were of limited size. The outer membranes of *E. coli* and *S. typhimurium* are very similar, comprising phospholipids, lipopolysaccharide and a limited range of proteins (Di Rienzo *et al.*, 1978); the proteins being considered the most likely components for pore-formation. The best characterized and most abundant of the proteins in *E. coli* and *S. typhimurium* is a lipoprotein of molecular mass *c.* 7000 daltons (Braun, 1975), which initially seemed a likely candidate for pore formation, and indeed on the basis of the primary sequence reported by Braun and Bosch (1972), Inouye (1974) proposed that it might create 'diffusion pores'. However, in a mutant lacking the Braun lipoprotein, permeability of the outer membrane to the test substrate 6-aminopenicillanic acid was unaltered, a result indicating that the lipoprotein does not form diffusion pores (Nikaido *et al.*, 1977*b*). It is possible that the outer membrane in the mutant could be altered in permeability to hydrophobic substrates for this was not tested, although it has been suggested that hydrophobic molecules penetrate the outer membrane 'by dissolving into the hydrocarbon interior regions' rather than through pores (Nikaido, 1976). In fact, it is now clear that the 'sieve' is another major protein which has been variously called protein I (Garten and Henning, 1974), protein 1 (Schnaitman, 1974), protein 0–8, 9 (Uemura and Mizushima, 1975), protein b, c (Lutgenberg *et al.*, 1975), matrix protein (Rosenbusch, 1974), protein A_1,A_2 (Bragg and Hou, 1972), and porin (Nakae, 1975, 1976*a*). Porin covers a large proportion of the cell surface and is present in about 10^5 copies per cell (Rosenbusch, 1974; Steven *et al.*, 1977). It spans the outer membrane, being tightly associated with the underlying peptidoglycan layer (Rosenbusch, 1974; Steven *et al.*, 1977; Lutgenberg *et al.*, 1977) and also accessible from the outside acting as a phage receptor (Datta *et al.*, 1977; Bassford *et al.*, 1977). In spite of its apparently tight association with the cell envelope, porin is released by osmotic shock (Benz *et al.*, 1978*a*). Porin has been shown in reconstitution experiments with vesicles to function as a hydrophilic pore with penetrability for small molecules and ions comparable to the intact cell *in vivo* (Nakae, 1975, 1976*a*, *b*; Benz *et al.*, 1978*b*), although differences have been noted between detergent solubilized and shock-fluid porin (Benz *et al.*, 1978*a*). In addition, porin-deficient mutants show pleiotropic transport defects for these small molecules (Bavoil *et al.*, 1977; Nikaido *et al.*, 1977*a*). Lutkenhaus (1977) recently presented evidence for the pore function of several outer membrane proteins and suggested that they may differ in facilitating transport of substances at high or low concentrations (von Meyenburg, 1971; Bavoil *et al.*, 1977). Benz *et al.* (1978*a*) also noted that shock fluid from porin-negative mutants exhibited pore activity which was different from that for porin. In fact, in addition to the apparently non-specific penetration via porin channels, specific penetration systems in the outer membrane have been reported (reviewed in Kadner and Bassford, 1978) for nucleosides (Hantke, 1976), maltose and malto-dextrins (Szmelcman *et al.*, 1976; Palva, 1978), vitamin B_{12} (Di Masi *et al.*, 1973; Kadner, 1978); and iron–siderophore complexes (Hantke and Braun, 1975; Ichihara and Mizushima, 1978).

It may reasonably be assumed therefore that peptides cross the outer membrane via pores, possibly porins. Whether porin-deficient mutants are defective in oligopeptide uptake has not been established. On the other hand, whether any of the oligopeptide-transport-deficient mutants (see Section 2.1.5.6) isolated on the basis of resistance to triornithine (TOR mutants) or other toxic peptides and designated *opp*$^-$ (oligopeptide permease deficient) are actually pore-deficient rather than permease-deficient is also not established. This seems unlikely however, as the map position of genes involved in the appearance of porin proteins b and c (Verhoef *et al.*, 1979) does not correspond to the position for *opp* (see Section 2.1.5.6).

If the occurrence of pores is widespread amongst bacteria and they are indeed used by peptides then, contrary to the suggestion made earlier, broad differences in cell-wall composition may have little effect on the ability of peptides to reach their specific permeases in the cytoplasmic membrane. This of course depends upon the depth to which pores penetrate the envelope and the influence of the periplasmic space on the approach of substrates to the inner membrane. The periplasmic space, which may under 'typical' experimental conditions comprise 20–40 per cent of the total cell volume (Stock *et al.*, 1977), is defined empirically as a region that contains a unique complement of proteins, predominently hydrolases and binding proteins, that are selectively released by osmotic shock (Heppel, 1971; Heppel *et al.*, 1972; Rosen and Heppel, 1973). It is implicit in most considerations of this space that it is a homogeneous region uniformly distributed around the cytoplasmic membrane, but there appears to be no real evidence to support this assumption. In fact, individual periplasmic enzymes show various degrees of release by osmotic shock, lysozyme-EDTA, penicillin-EDTA and polymyxin-B treatments, perhaps indicating some compartmentalization (Ananthaswamy, 1977). The outer limit of the periplasmic space is poorly defined and it is possible some periplasmic proteins may interact with the outer membrane (Garrard, 1972; Costerton *et al.*, 1974; McAllister *et al.*, 1977; Benz *et al.*, 1978*a*). Several circumstances may be envisaged in regard to peptides crossing this region (see also Barak and Gilvarg, 1975*a*). First, peptides reaching the cell at all points on its surface may simply diffuse through the cell wall (no pores being present), enter the periplasmic space from all directions, and either diffuse across it and/or be carried to the inner membrane permease by binding proteins. Second, peptides traverse the outer membrane via pores at a limited number of sites which extend inwards only as far as the inner regions of the cell wall; peptides thus enter the periplasmic space at specific locations and are then carried across and/or diffuse as in the first case above. Third, specific outer membrane pores either penetrate inwards as far as the cytoplasmic membrane (comparable with pili or flagella) or, at the interface of the cell wall and periplasmic space they are associated with further channels that link the inner membrane and cell wall. In either instance it is imagined that the peptide will be delivered into the more immediate environment of its permease, in effect increasing its local concentration, rather than being dispersed throughout the periplasmic space. Reports of regions of adhesion between the wall and inner membrane may have relevance here (Bayer, 1968, 1974).

The cell envelopes of Gram-positive and Gram-negative bacteria differ significantly. The former lack an outer membrane, have a thicker peptidoglycan layer, and contain teichoic/teichuronic acid, they also lack a periplasmic space although an apparently analogous membrane-wall-interlayer has been described (Giesbrecht *et al.*, 1977). In early studies it was claimed that isolated Gram-positive walls showed sieve-like properties (Scherrer and Gerhardt, 1971; Scherrer *et al.*, 1971) although there is now doubt as to whether the exclusion limits determined can be applied to whole cells (Scherrer *et al.*, 1977).

Finally, it should be noted that different growth conditions can markedly affect the composition of cell envelopes. Inhibition of protein synthesis can cause gross compositional differences (Rothfield and Pearlman-Kothencz, 1969; Shockman *et al.*, 1974, reviewed in Payne, 1975*a*). Variations in the supply of specific nutrients may cause particular membrane changes (Tempest, 1970; Robinson and Tempest, 1973; Lutgenberg *et al.*, 1976; reviewed in Braun, 1978). In many instances such changes may affect permeability of the envelope for peptides (Levine and Simmonds, 1962*a*, *b*; Kepes and Cohen, 1962) and other nutrients (Braun, 1978). These features should be should be kept in mind when designing experiments to study peptide uptake and utilization. For example, as the typical condition of *E. coli* in the gut is one of nutrient limitation with an average generation time of 20 h (Savage, 1977), the validity of using batch grown cells with a division time of 20–60 min for studies on the relevance of peptide utilization to bacterial nutrition *in vivo* is questionable.

3 UTILIZATION OF PEPTIDES IN BACTERIAL NUTRITION

Previous reviews of this topic, covering the early literature in particular, are in Matthews and Payne (1975*b*) and Payne (1976).

It has long been recognized that peptides are of great nutritional value for many types of bacteria, and as a result commercial peptones are routinely incorporated into a variety of growth media. The composition of peptones may vary somewhat from batch to batch and each one is different with respect to its peptide components, (e.g. size range, amino acid content), and its other nutrients (e.g. carbohydrates, vitamins). This diversity has practical value in facilitating preparation of selective peptone-containing media for the isolation and cultivation of individual bacterial species (Matthews and Payne, 1975*b*), but it makes it virtually impossible to define uniquely the nutritional role of the peptide components. To study this it is necessary to use both peptides of known sequence and culture media of defined composition.

At present there is no evidence that any peptide acts as a vitamin-like, or hormonal-growth factor for any bacterium, although this suggestion has been made with respect to mammalian cells (Pickart and Thaler, 1973). Bacteria usually hydrolyse peptides and use the amino acids either directly for protein synthesis or, after further breakdown, as sources of nitrogen, carbon, or sulphur.

When amino acids are utilized predominently for protein synthesis it is not unusual for a free and a peptide-bound residue to produce different growth responses. The difference may be in the rate of growth and/or the growth yield. These effects are most readily observed with amino acid auxotrophs. With auxotrophs of *E. coli* the responses to peptides and free amino acids are identical for virtually all amino acids tested, a feature that makes it particularly suitable as an assay organism (see Section 4.9.6.3). This identity is presumably a consequence of its non-fastidious nature and consequent simplicity of its growth media (see below), the presence of highly efficient processes for transporting and for hydrolysing peptides (see Sections 2.1.5 and 2.4), and the strict regulation of its amino acid pool levels (see Section 2.1.5.2) which in turn control the amounts and activities of enzymes involved in the biosynthesis and degradation of the amino acids. Related studies with other enteric organisms have not been carried out although results with *Salmonella typhimurium* suggest its responses may be identical (Miller and Mackinnon, 1974; McHugh and Miller, 1974; Jackson *et al.*, 1976; Kirsh *et al.*, 1978). With *Pseudomonas* species various responses have been observed. Amino acid auxotrophs (*his*, *trp*, *thr*, or *met*) of *P. putida* grew on a variety of di- and tripeptides up to twice as slowly as with free amino acids. Final growth yields were 5–10 times greater when *P. putida* his^- was grown on histidyl di- or tripeptides rather than on free His because the histidyl residues were protected from catabolism by L-histidine ammonia lyase (Cascieri and Mallette, 1976*a*). In *P. maltophilia* (Cascieri and Mallette, 1976*a*) and *P. aeruginosa* (Miller and Becker, 1978) methionyl peptides could satisfy the methionine requirement, and although growth rates were all broadly similar a twofold variation in growth yield was noted amongst a range of methionyl peptides. These effects occur because peptide uptake and/or hydrolysis can be the rate-limiting process for overall growth. With multiple auxotrophic species such as the *Streptococci* and *Lactobacilli* growth responses to peptides and amino acids are commonly quite different. Peptides may be superior to amino acids in stimulating growth rate when the medium contains many required amino acids and competition for uptake between them causes a growth limitation; growth yields are often greater on peptides because the residues are better protected from catabolism. These different responses may further be modified by variations in the content of other nutrients in the medium. It was these complex interactions between the multi-nutrient requirements of these organisms and the composition of their growth media that gave rise to their apparent requirement for specific peptide growth factors (strepogenin) (Woolley and Merrifield, 1958, 1963; Kihara and Snell, 1960*a*, *b*; reviewed in Matthews and Payne, 1975*b*, and Payne, 1976). *Bacteroides* species are reportedly able to grow upon peptides but not upon free amino acids (Pittman and Bryant, 1964; Pittman *et al.*, 1967) although other reports indicate that free amino acids can stimulate the response to peptides (Wright, 1967; Lev, 1977).

Systematic studies on the relative nutritional value of amino acids and peptides as sole sources of carbon or nitrogen have not been reported. Conditions of nitrogen limitation can produce derepressed synthesis of low-affinity transport systems for certain amino acids (see Sections 1.1.4.3 and 2.5.4), which presumably enhances utilization of the free amino acids; it is not known if peptide-transport systems respond similarly. Enhanced peptidase synthesis and secretion has been noted for some species (Litchfield and Prescott, 1970; Wagner *et al*., 1972; reviewed in Payne, 1975*b*) growing on peptides (see Section 2.1.5.4) although there is no evidence that added peptides produce changes in intracellular peptidase activities (see Section 2.4.7). It has been suggested for *Salmonella typhimurium* that carbon-limited growth may produce enhanced activity of certain transport systems under cyclic AMP control, both amino acid and peptide transport systems were suggested as being in this category (Alper and Ames, 1978). Direct studies are needed on amino acid and peptide transport under these carbon-limited conditions in order to substantiate this suggestion and to assess its possible nutritional relevance. If equal energy is needed to transport an amino acid molecule and a peptide molecule then under carbon-limited conditions it is likely that peptides will be nutritionally superior. For such studies, nutrient-limited cultures grown in the chemostat are most suitable, and these should also be useful for evaluating the likely nutritional values of amino acids and peptides for bacteria in nature where they will mostly exist under conditions of nutrient limitation, e.g. it has been estimated for *E. coli* that its average generation time in the gut is 20 h (Savage, 1977) while it may be only one-fiftieth of this using minimal-media, batch culture in the laboratory.

4 METHODS FOR STUDYING PEPTIDE TRANSPORT AND UTILIZATION

Those who study transport processes in bacteria are accustomed to using radioactively labelled substrates. Unfortunately for those inclined to study the transport of peptides the radioactive substrates are not generally available, only about six labelled dipeptides and no higher peptides ever having been commercially available. Some investigators have themselves synthesized particular labelled peptides and useful results have ensued, although this has led to the situation where general conclusions have had to be drawn from results with a limited range of substrates. Given this situation, alternative methods have been used. However, these have been far from satisfactory with the result that progress has been relatively slow and no doubt many have been deterred from entering the field. Fortunately, a leap forward in methodology has come with the recent development of direct assays based on fluorescent labelling of substrates. The various methods are summarized below.

1 Use of Amino Acid Auxotrophs

(i) Monitoring Growth Response

It has already been mentioned that peptides of appropriate amino acid composition can usually substitute for free amino acids in satisfying the growth requirements of amino acid auxotrophs. If an auxotroph has intracellular peptidases but lacks extracellular activity, so that peptides cannot be cleaved before they undergo transport, then the transport of these nutrients can be monitored by studying the auxotrophic growth response to different peptides containing the required amino acid. However, assessment of transport is only indirect for growth response is not only dependent on the transport step. It is a relatively insensitive technique, each test needing about 1 μmol of peptide at a concentration of about 2 mM. Incubations typically take from 12 to 48 h with growth being measured by change in turbidity of the culture, although plate techniques have been used (see Section 4.9.6.2). This approach has formed part of almost all bacterial studies and is particularly useful when interest centres not simply on transport but on relative utilizability of different peptides, comparisons with amino acids, or when peptide toxicity is of concern. When the method is extended to include competition studies, in which peptides devoid of the essential amino acid(s) are used, then much fundamental information on transport can be obtained, e.g. *N*- and *C*-terminal requirements, stereospecificity, size restriction, etc. However, transport can only be studied satisfactorily under conditions permitting adequate growth and experiments into energy coupling, pH and temperature dependence etc., cannot be performed. The methodology and the types of results it can yield have been considered in greater detail in previous reviews (Sussman and Gilvarg, 1971; Payne and Gilvarg, 1971, 1978; Barak and Gilvarg, 1975*a*; Payne 1975*a*, 1976, 1977).

(ii) Monitoring Enzyme Synthesis

As an alternative to measuring peptide-dependent growth, i.e. cell multiplication over many generations, this technique involves assaying the peptide-dependent synthesis of a single enzyme (Bell *et al.*, 1977). Conditions for the derepressed or induced synthesis of an enzyme, e.g. alkaline phosphatase, β-galactosidase, are established in the auxotrophic cells just before adding the peptide as sole source of the essential amino acid, transport of the peptide then supplies the required nutrient and permits synthesis of the enzyme, the rate and extent of which can be measured using substrates that yield coloured or fluorescent products (see also Section 4.9.6.3). This approach shares the disadvantage of the previous method in requiring incubation conditions that permit growth; however, it is much more sensitive and a few minutes incubation is all that is needed. The method has been used to characterize peptide-transport mutants and to study competition between peptides, and the results endorsed previous conclusions from growth tests on the structural specificities for peptide transport (Bell *et al.*, 1977). This method, and the previous ones, gave similar types of results but with this technique it is convenient to decrease the peptide concentration progressively until the rate at which it supplies the essential amino acid just begins to slow enzyme synthesis. This concentration is analogous to the 'limit concentration', defined for amino acids by Ames (1964) as the external concentration that just permits normal growth in the absence of endogenous biosynthesis of the specified nutrient.

A further method in which determination of enzyme activity was used as an indirect measure of peptide transport was described by Cascieri and Mallette (1974).

(iii) Incorporation of Radioactive Amino Acids into Protein

In this method an auxotroph requiring at least two amino acids is used (Payne and Bell, 1977*a*; Payne, 1977). One required amino acid is supplied free, in excess, and radioactively labelled, the other is supplied in a peptide. Protein synthesis is dependent on the simultaneous availability of both amino acids. Thus, transport of the peptide can be monitored from the rate and extent of incorporation of labelled amino acid into protein. The method is very quick and sensitive but suffers from the same disadvantages as described in Section 2.1.4.1 (ii) above.

2 Use of Radioactively Labelled Peptides

The usual transport assay involves incubation of a bacterial suspension with a radioactively labelled substrate, periodic removal of samples, followed by rapid membrane filtration and measurement of the radioactivity within the bacteria retained on the filters. The method is direct, sensitive, and fast, amino acid auxotrophs are not required, and incubations can be performed under varied conditions. All these features should apply to peptide transport assays providing an investigator prepares his own labelled peptides.

References to the early use of radioactively labelled peptides were given in Payne and Gilvarg (1971) and since that report the following have been described: *E. coli*, [^{14}C]-Gly-Gly (De Felice *et al.*, 1973; Cowell, 1974; Neuhaus *et al.*, 1977; Staros and Knowles, 1978) and [^{14}C]-D-Ala-[^{14}C]-D-Ala (Neuhaus *et al.*, 1977); *S. typhimurium*, Ala-Pro-[^{14}C]-Gly (Jackson *et al.*, 1976) and [^{14}C]-Gly-Pro (Yang *et al.*, 1977); *P. putida*, Gly-Gly-[^{14}C]-Ala (Cascieri and Mallette, 1976*a*); *Streptococcus* spp., Gly-[^{14}C]-Leu (Law, 1978), [^{14}C]-Leu-Gly, Gly-[^{14}C]-Leu-Gly, Gly-Gly-[^{14}C]-Leu-Gly, and Gly-Gly-Gly-[^{14}C]-Leu-Gly (Rice *et al.*, 1978). In most studies, chloramphenicol is added to prevent labelled amino acid being continuously incorporated into newly synthesized protein. When protein synthesis is inhibited in this way, and incubations are carried out for up to about 10 min, it is assumed that peptide uptake into the intracellular pool can be measured. However, recent studies (Payne and Bell, 1977*b,c*, 1979) have shown that this assumption may not always hold because with *E. coli* (and *S. typhimurium*, J. W. Payne and G. Bell, unpublished results) the peptides may be very rapidly hydrolysed by cytoplasmic peptidases and the released amino acid residues (and peptide fragments, Sections 2.1.5.2 and 2.1.5.3) may undergo immediate, rapid exodus concomitant with further uptake of the peptide substrate. In addition, after peptide scission certain of the released amino acids may be rapidly metabolized with only a fraction undergoing exodus (see Section 2.1.5.2). Given these possibilities, measurement of counts within the bacteria may be a misleading technique liable to underestimate peptide transport. When Cowell (1974) attempted to measure uptake of counts from [^{14}C]-Gly-Gly into chloramphenicol-treated *E. coli* he noted that uptake failed to plateau out and suggested that some form of metabolism might be responsible, and in consequence had to base his transport measurements on single 15 s readings.

3 Use of Fluorescent Labelled Peptides

Because fluorescent substances can be determined with such great sensitivity it is in principle possible to measure directly the transport of fluorescent labelled peptides. Unfortunately, with a fluorescent group attached (to the *N*-terminal α-amino group) the peptides are unlikely to be substrates for the transport systems. Consequently, the approach taken involves labelling and assaying the peptide *after* it has been incubated with the bacteria.

(i) *Dansyl Method*

Bacteria are incubated with one or more peptides and periodically samples are removed with a syringe and passed through membrane filters to remove the cells. The peptides in the filtrates are fluorescently labelled with dansyl chloride (dimethylaminonaphthalene sulphonyl chloride) and the derivatives separated by two-dimensional thin-layer chromatography. The peptides are quantitated from the intensities of their fluorescent spots measured *in situ* using a scanner spectrofluorimeter. In this way the transport rate for each peptide substrate can be determined by measuring its removal from the incubation medium. In a complementary way, the pool contents of the bacteria retained on the membrane filter may be extracted, dansylated and chromatographed so that the absorbed substrates (and their cleavage products and metabolites) can be quantitated, although this is really useful only when intact peptide accumulates. This technique has been used successfully not only with bacteria (Payne and Bell, 1977*b,c*, 1979) but also with yeast (Nisbet and Payne, 1979*a,b*) and plants (Higgins and Payne, 1977; see also Section 4.6).

Frequently, additional information accrues from use of this 'visual' analysis. For example, the exodus of peptidase-released amino acid residues, their metabolites (Payne and Bell, 1977*b,c* 1979; see also Section 2.1.5.2), and peptide fragments (Sections 2.1.5.3 and 2.1.5.4) was first noted in this way.

(ii) *Fluorescamine*

At pH 8–9 fluorescamine (fluoram) reacts in milliseconds with the unprotonated form of primary amino groups to give intensely fluorescent derivatives. Furthermore, there is a sufficient difference between the pK_b values of the *N*-terminal α-amino groups of typical small peptides and most protein amino acids to effect selective labelling of peptides when the pH is lowered to about 6.2 (Perrett *et al*., 1975). In addition, and in contrast to dansyl chloride, neither fluorescamine itself nor the products it forms on reaction with water or ammonia are fluorescent. These useful characteristics have been exploited in developing a solution assay for peptide transport (Nisbet and Payne, 1979*a,b*). Thus, as in the dansyl procedure, samples of incubation solutions are removed periodically and filtered to remove bacteria. After reaction with fluorescamine t... fluorescent intensity of the filtrates give a measure of their peptide content, from which the extent of peptide transport during the incubation period can be determined. Recently, a continuous, automated assay for peptide transport has been developed by coupling a pumping system that samples the incubation mixture, filters it to remove bacteria and reacts the filtrate with fluorescamine, to a fluorimeter equipped with a flow-cell and recorder (J. W. Payne, unpublished). If wanted, the reaction mixture can be collected using a fraction collector and, similar to the dansyl method, the fluorescamine-labelled products (peptide and amino acids) characterized using thin layer chromatography on silica gel.

(iii) Other Reagents

2-Methoxy-2,4-diphenyl-3-(2H)-furanone (MDPF) is closely related to fluorescamine and may have additional advantages in the assay of peptides (Udenfriend, 1978). Others with requisite sensitivity are *o*-phthaldehyde and dabsyl chloride (4-*N,N*-dimethylaminoazobenzene-4′-sulphonyl chloride) (Lin and Chang, 1975) which yield fluorescent and coloured derivatives respectively.

5 CHARACTERISTICS OF PEPTIDE UPTAKE

1 General

Early studies on bacterial nutrition indicated that amino acids and peptides had different routes of uptake, and this distinction was further emphasized by the isolation of mutants from several species that failed to transport a particular amino acid but responded normally to the same amino acid in peptide form (results reviewed in Payne and Gilvarg, 1971; Payne, 1976). Payne (1968) described the isolation of a mutant of *E. coli* that failed to utilize oligopeptides but responded normally to amino acids and to dipeptides. It became evident, therefore, that in this organism separate systems are used for di- and oligopeptide uptake, and the same is true in *S. typhimurium*. This feature is less clear for other species. In *E. coli*, mutants also defective in dipeptide uptake have been described. (Details of peptide transport mutants are given in Section 2.1.5.6.) Thus, when the characteristics of peptide uptake are considered it is best to regard di- and oligopeptides as separate classes of substrate.

In previous reviews (Payne and Gilvarg, 1971; Barak and Gilvarg, 1975*a*; Payne 1975*a*, 1976), classification of the structural specificities of peptide permeases has rested largely on evidence obtained from auxotrophic growth tests (Section 2.1.4.1). Although reservations were always placed on the interpretation of such results, it nevertheless seems from recent studies that a re-evaluation of the earlier classification may be needed. Consequently, these recent studies are described in the next sections, before the substrate specificities of the permeases are considered.

2 Peptide Uptake and Amino Acid Exodus

When the dansyl procedure was first used with *E. coli* it was immediately apparent that peptides undergo transport at a steady rate, followed by rapid intracellular hydrolysis and a concomitant exodus of their constituent amino acid residues (Payne and Bell, 1979). However, exodus is not identical for all amino acid residues. Under typical incubation conditions (peptide 50–200 μM), there is very little exodus of Glu, Asp, Asn, Gln, or Ser, although their peptides are well absorbed, hydrolysed, and other residues of the peptides undergo exodus. These missing residues are not present in corresponding amounts in the free amino acid pool, which stays fairly constant during the incubation, and it seems that they are rapidly metabolized, probably involving deamination products which would not be detected by dansylation. It is clear that metabolism can rapidly affect the amino acid residues released from peptides for in some instances exodus of these metabolic products is seen, e.g. trithreonine uptake is accompanied by exodus of Thr and Gly (presumably formed by action of threonine aldolase), valine peptide uptake produces exodus of proline also (Payne and Bell, 1977*b*), and with ornithine peptides some ornithine but mainly putrescine (produced by ornithine decarboxylase) efflux occurs (J. W. Payne, unpublished results).

The exodus of amino acids occurring when *E. coli* (and *S. typhimurium*) is incubated with one or several defined peptides is presumably an atypical phenomenon. In the mammalian gut, bacteria are more likely presented with a heterogeneous mixture of peptides, thereby achieving uptake of a range of amino acids that balances their nutritional needs and makes selective amino acid exodus unnecessary.

How might the observed amino acid exodus affect the different transport assays? In a typical auxotrophic growth test, an incubation medium (10 ml) containing a defined peptide (100 μM) and inocula (10^7 cells ml^{-1}) that divide about every hour is used; calculation based on representative rates of peptide uptake and amino acid exodus (Payne and Bell, 1979) indicate that after about 4 h incubation the peptide will have gone and the auxotroph will be growing on reabsorbed amino acid residues. This conclusion has been confirmed for an *E. coli* lysine auxotroph growing on trilysine (Payne and Bell, 1979), and for several other peptides (J. W. Payne, unpublished results). It is for this reason that virtually equal growth rates are observed for auxotrophs growing on glycine- (Payne, 1968) and lysine-homopeptides (Gilvarg and Katchalski, 1965) even though the homologues have significantly different rates of uptake and cleavage (Payne and Bell, 1979). A further consequence of amino acid exodus is that if an auxotroph has a defective permease for its required amino acid it would be expected to grow less well on peptides than the same auxotroph with the permease intact. One instance has been found in the literature in which this may have occurred with an *E. coli* auxotroph and the authors did indeed suggest amino acid exit as one explanation for their results (Levine and Simmonds 1960, 1962*a,b*). Other evidence compatible with this type of amino acid exodus was presented by Brock and Wolley (1964) for Gly-Gly uptake in *Streptococcus faecalis*, and for Gly-Val uptake in *Leuconostoc mesenteroides* (Shelton and Nutter, 1964). Reports that peptides can influence amino acid uptake (see Payne and Bell, 1977*c*), which are difficult to explain by direct interactions, are possibly explicable as competition from amino acid residues that have undergone exodus.

With assays based on radioactively labelled peptides, unless true initial rates of uptake are determined then measurement of the amount of radioactivity present in the cells at any time may underestimate both the rate and the extent of peptide transport. (Instances may be found in the literature that seem to bear this out, e.g. Young *et al*. (1964).) After a short period of incubation it is possible that a constant amount of radioactivity might be associated with the cells and that this could be interpreted as a constant accumulation level of the peptide; however, it is more likely to represent the steady-state value for the dynamic equilibrium between peptide uptake and amino acid exodus (see Section 2.1.5.4). Another expectation would be that for a peptide A-B in which efflux of A and B was unequal then transport of A-B might appear to be different depending on which residue carried the radioactive label. Examples of this may be present in the literature, such as the difference in uptake recorded for Gly-[^{14}C]-Ala and [^{14}C]-Gly-Ala in *Lactobacillus* spp. (Leach and Snell, 1960).

With the dansyl method, measurements of peptide uptake based on thin layer chromatography of incubation media will not usually be affected by amino acids present as a result of exodus. In the fluorescamine method, the fluorescent yield from amino acids is low (about 4 per cent) relative to that from peptides so exodus will not significantly affect measurements of peptide uptake during the early phase of an incubation.

3 Peptide Uptake and Exodus of Cleaved Peptide Residues

When the dansyl procedure was first used with wild-type *E. coli* to follow uptake of peptides, exodus of only free amino acids was seen (Payne and Bell, 1977*b,c*, 1979). However, with further use (J. W. Payne, unpublished results) we have now found two situations in which peptide uptake is accompanied by exodus of cleaved peptide residues (plus free amino acid residues).

First, when studying stereospecificity of transport by *E. coli* it was found that L-Ala-L-Ala-D-Ala, L-Val-L-Val-D-Val, and Gly-Gly-D-Leu were all taken up but they were not fully cleaved and exodus of L-Ala-D-Ala, L-Val-D-Val, and Gly-D-Leu occurred. Second, certain peptidase-deficient mutants of *Salmonella typhimurium* (Miller and MacKinnon, 1974) are able to take in tripeptides but some of them can be only partly cleaved and the unhydrolysed dipeptide fragments undergo exodus (J. W. Payne, unpublished). Thus, with strain TN 273 uptake of Met-Leu-Gly, Gly-Pro-Ala, and Gly-Leu-Gly leads to exodus of Leu-Gly, Pro-Ala, and Leu-Gly respectively, and with strain TN 272 uptake of Ala-Gly-Ala, Gly-Pro-Ala, and Leu-Gly-Gly leads to exodus of Gly-Ala, Pro-Ala, and Gly-Gly respectively.

How might this type of peptide exodus affect the various transport assays? Its occurrence will presumably be so infrequent as to be generally unimportant (although it may possibly occur with some peptides during protein turnover). However, in some auxotrophic growth tests peptide exodus may have occurred but not been considered, e.g. with *E. coli*, oligopeptides containing β-alanine (Payne, 1973), sarcosine (Payne, 1972*c*), or D-amino acids (Becker and Naider, 1974) are only partly utilized, indicating that they are taken up though the fate of non-utilizable fragments is unknown. With assays using radioactively labelled peptides similar confusion may arise as discussed above with amino acid exodus and only true initial rates will provide adequate transport data. Providing the peptide substrate and its cleaved fragment are separated on chromatography, the dansyl method should not be affected by this exodus, and on the contrary it is the only procedure that will readily reveal this effect. The fluorescamine assay cannot be applied under these circumstances because the increasing fluorescence from the fragment undergoing exodus will compensate for the decreasing fluorescence of the peptide being taken up and the net result may appear to be no transport. In fact, this was the initial conclusion reached for the L,L,D stereoisomers described above, and it was only following dansylation that this was found to be erroneous.

The way in which these initial observations on peptide exodus may lead to a better understanding of the process of peptide transport is considered in the next section.

4 Role of Peptidases in Uptake and Utilization

In previous reviews (Payne and Gilvarg, 1971; Barak and Gilvarg, 1975*a*; Payne, 1975*a*, 1976) consideration was given to the relation between the cellular locations of peptidases and their possible functions in peptide transport and utilization. *E. coli* and *S. typhimurium* were reported to possess only intracellular peptidases, precluding the possibility of extracytoplasmic peptide hydrolysis, while one or two less well studied species were reported to produce extracellular and membrane-bound peptidases. Since those reviews were written many detailed studies on bacterial peptidases have appeared; these are discussed in depth by Hermsdorf and Simmonds (Section 2.4) and only certain aspects are considered here.

In a series of papers, Miller and colleagues described the structural specificities and genetic loci of a range of peptidases in *S. typhimurium* (Miller and MacKinnon, 1974; McHugh and Miller, 1974; Miller, 1975*a,b*; Kirsh *et al*., 1978) and in *E. coli* (Miller and Schwartz, 1978). Their results led these authors to conclude that these two species possess an exactly similar complement of peptidases. In several cases it is possible to relate peptidase activities described by Miller to those described by other workers. Thus, peptidase A (Miller and MacKinnon, 1974; Miller and Schwartz, 1978) is probably the aminopeptidase I purified from *E. coli* K-12 by Vogt (1970). Peptidase N (Miller and MacKinnon, 1974; Miller, 1975*a*; Miller and Schwartz, 1978) is almost certainly the same enzyme as that purified from *E. coli* K-12 by Yang and Sommerville (1976) and from *E. coli* K-10 by Lazdunski and coworkers (Lazdunski *et al*., 1975*a,b,c*; Latil *et al*., 1976; Murgier *et al*., 1976*a,b*, 1977; Chappelet-Tordo *et al*., 1977). Good correspondence can be made between the several enzymes characterized by Simmonds and her colleagues (Simmonds *et al*., 1976; Hermsdorf, 1978; Hermsdorf *et al*., 1979; see also Section 2.4) and those described by Miller and others. Thus, the following probably correspond: peptidase D (Miller) and dipeptidase DP (Simmonds), this is also probably analogous to the dipeptidase studied by Brown (1973), but different from that described by Patterson and colleagues (Patterson *et al*., 1973; Hayman *et al*., 1974); peptidase B (Miller) and aminopeptidase AP (Simmonds); peptidase N (Miller) and oligopeptidase OP (Simmonds); peptidase with an R_f value of 0.47 (Miller and MacKinnon, 1974; Miller and Schwartz, 1978) and tripeptidase TP (Hermsdorf, 1978); peptidase with an R_f value of 0.28 (Miller and MacKinnon, 1974; Miller and Schwartz, 1978) and peptidase L (Simmonds *et al*., 1976; Hermsdorf *et al*., 1979). The dipeptidyl carboxypeptidase of *E. coli* described by Yaron *et al*. (1972) does not appear to correspond to any of the above enzymes, although it could not have been detectable by the gel-staining techniques of Miller which depend on the peptidases producing free amino acids. Recently, Deutch and Soffer (1978) further characterized this peptidase and isolated mutants lacking the activity (Dcp$^-$). Specific peptidases are involved in cleavage of proline peptides and two have been described in *E. coli* and *S. typhimurium*. Aminopeptidase P of *E. coli* (Yaron and Berger, 1970) apparently corresponds to peptidase P of *S. typhimurium* (McHugh and Miller, 1974) and *E. coli* (Miller and Schwartz, 1978) which cleaves *N*-terminal amino acids adjacent to proline from di- and oligopeptides. Peptidase Q, which occurs in both *S. typhimurium* (McHugh and Miller, 1974) and *E. coli* (Miller and Schwartz, 1978), cleaves the X–Pro bond in only dipeptides.

These recent studies have in two instances modified the earlier impression that peptidases are exclusively intracellular in these bacteria. This clearly has implications for the process of peptide uptake for no longer can it be generally assumed that peptides can only be hydrolysed within the bacteria. Thus, about 10 per cent of the activity of dipeptidyl carboxypeptidase apparently occurs in the periplasmic space with the remainder being in the cytoplasm (Deutch and Soffer, 1978). This exopeptidase cleaves dipeptidyl residues from the unblocked *C*-terminal of various peptides but it does not act on a peptide bond containing the imino group of a prolyl residue or on free tripeptides, although it does hydrolyse α-*N*-acyl tripeptides. This last activity, acting in the periplasmic space, is presumably responsible for the slight growth observed on certain α-*N*-acetyl tripeptides (Losick and Gilvarg, 1966; Deutch and Soffer, 1978) which are not substrates of the oligopeptide permease (Gilvarg and Katchalski, 1965; Payne, 1971*a*; see also Section 2.1.5.6). The possible action of this periplasmic enzyme seems unlikely to have affected many studies of peptide transport because of its limited specificity. However, if it were able to cleave free tetra- and higher peptides, and α-*N*-acyl oligopeptides, with release of dipeptidyl residues into the periplasmic space, then these might compete with true dipeptide substrates for uptake via the dipeptide permease, a situation which superficially could be interpreted as direct competition between di- and oligopeptides. Potentially a more significant source of difficulty is peptidase N which, although originally reported as inactive towards small peptides (Lazdunski *et al.*, 1975*b*), has subsequently been shown to cleave a range of such substrates with preference for amino acid residues with hydrophobic side-chains (Chappelet-Tordo *et al.*, 1977). The manner of its association with the cytoplasmic membrane is not completely established. The observation that intact cells hydrolyse β-naphthylamides (specific substrates of peptidase N) was interpreted as evidence for the enzyme being located on the outer surface of the membrane (Murgier *et al.*, 1976*a,b*, 1977) but it is not certain whether a process of substrate uptake and product exodus (see Section 2.1.5.2) was rigorously excluded here and in a related report (Cerny, 1976) (see also Watson, 1976; Payne, 1979). If the enzyme were to span the membrane, as is the case for a brush border intestinal aminopeptidase (Louvard *et al.*, 1976; see also discussion to Noren *et al.*, 1977), then it might conceivable constitute a transport system by which peptides are bound from the outer surface and amino acids vectorially discharged to the inner surface, as envisaged in one theoretical model (Payne, 1977); there is, however, no evidence for this, and on the contrary, a mutant of *E. coli* lacking peptidase N activity was not defective in peptide uptake (Payne and Bell, 1979). In those strains in which a membrane-bound form of the enzyme functions it seems that the ensuing 'membrane-digestion' type of peptide absorption (Ugolev *et al*., 1977) will normally be quantitatively unimportant relative to peptide transport through the di- and oligopeptide permeases. However, in peptide permease mutants it could take on some importance. Indeed, it seems possible that the ability of opp^- mutants of *E. coli* to utilize certain oligopeptides (Payne, 1968; Barak and Gilvarg, 1975*b*; Naider and Becker, 1975) may perhaps arise from peptidase N action rather than from the presence of several specialized oligopeptide transport systems (see Section 2.1.5.6), about which reservations have been expressed previously (Payne, 1977). It should be possible to resolve this point using appropriate opp^- strains with or without peptidase N activity. It is noteworthy that, although the periplasmic peptidase N is constitutively produced, its differential rate of synthesis is increased fourfold when growth is limited by phosphate, conditions which cause simultaneous derepression of periplasmic alkaline phosphatase (Lazdunski *et al*., 1975a,c; Murgier *et al*., 1976*b*). Neither carbon nor nitrogen limitation derepresses peptidase N synthesis. Could it be that peptidase N plays a role in removing the leader sequence from the periplasmic alkaline phosphatase (Inouye and Beckwith, 1977) and/or hydrolysing the leader sequence to smaller peptides and amino acids for reabsorption? It would be of interest to determine whether induction or derepression of other periplasmic enzymes synthesised with leader sequences might also be accompanied by increased synthesis of aminopeptidase N.

Evidence for membrane-bound peptidases which may function in peptide utilization has been presented for several other species. Miller and Becker (1978) made such a proposal for *Pseudomonas aeruginosa*, though two other strains, *Ps. maltophilia* and *Ps. putida* reportedly lack this activity (Cascieri and Mallette, 1976*a,b*). Streptococci that utilize milk products may possess both membrane peptidases and peptide transport systems (Law, 1977, 1978; Rice *et al*., 1978; Exterkate, 1977). Peptidase activity has been found associated with the membrane and cytoplasmic fractions of *Mycoplasma laidlawii* (Choules and Gray, 1971; Pecht *et al*., 1972). Extracellular aminopeptidases have been characterized from *Aeromonas proteolytica* (Wilkes *et al*., 1973), *Bacillus subtilis* (Wagner *et al*., 1972), and *Bacillus licheniformis* (Rodriguez-Absi and Prescott, 1978).

It is relatively easy to envisage the ways in which the above peptidases, located in the membrane, the periplasmic space, or the extracellular fluid, could affect the uptake of peptides, but how might intracellular peptidases influence measurements of peptide transport? Two alternatives may be imagined: (A) a peptide is taken in more slowly than the rate at which it can be hydrolysed in the cytoplasm, or, in contrast, (B) peptide uptake is faster than intracellular hydrolysis. These two circumstances are considered in turn to see how they could affect transport assays based on measurement of auxotrophic-growth, radioactivity, or fluorescence (see Section 2.1.4).

Consider case A. First, the auxotrophic growth rate will be at a maximum when the *absolute* rate at which a peptide provides the required amino acid is not limiting (Payne, 1972*b*), the *relative* rate of hydrolysis to uptake being of secondary importance. However, if the growth rate were below the maximum, one would not know if uptake or cleavage were the rate limiting step. Second, using a radioactively-labelled peptide in a kinetic assay the situation in Figure 1A might be observed. During phase (a), radioactivity would steadily accumulate within the cells, providing the amino acid residue released by hydrolysis was retained; a rate of peptide uptake could be determined from the plot but it would be an underestimate of the true rate if exodus of the amino acid residue were to occur to any extent during this period. Phase (b) would represent the steady-state level of radioactivity associated with the cells, arising from a dynamic equilibrium between peptide uptake and amino acid exodus. The plot (here and in comparable cases below) would be horizontal providing that none of the labelled amino acid was directly incorporated into protein or metabolized to products that were retained by the cell, either of which circumstance would produce a steady rise in cell-associated counts. The time taken to reach the plateau and its actual level would depend on the capacity of the cells for the hydrolysed amino acids. Although phase (b) seemingly appears to be a static situation, this could only be so in the unlikely event that the released amino acid residues within the cell were preventing further peptide uptake; more likely, peptide uptake would still be proceeding at its maximum velocity, and analysis of the incubation medium would reveal a progressive change from peptide to amino acids. Phase (c) would be reached when all peptide was eventually converted to amino acids which progressively underwent exodus until an equilibrium was reached that represented the fractional dilution with unlabelled amino acid from the cytoplasmic pool. Third, a fluorescence assay (dansyl, fluorescamine) of the incubation medium would show steady rates of peptide transport and amino acid exodus until peptide uptake and hydrolysis was complete Figure 1B (Payne and Bell, 1979). In case A, examination of cytoplasmic extracts would show only the free amino acid residues, as is the normal observation when using wild-type bacteria that contain a range of highly active peptidases.

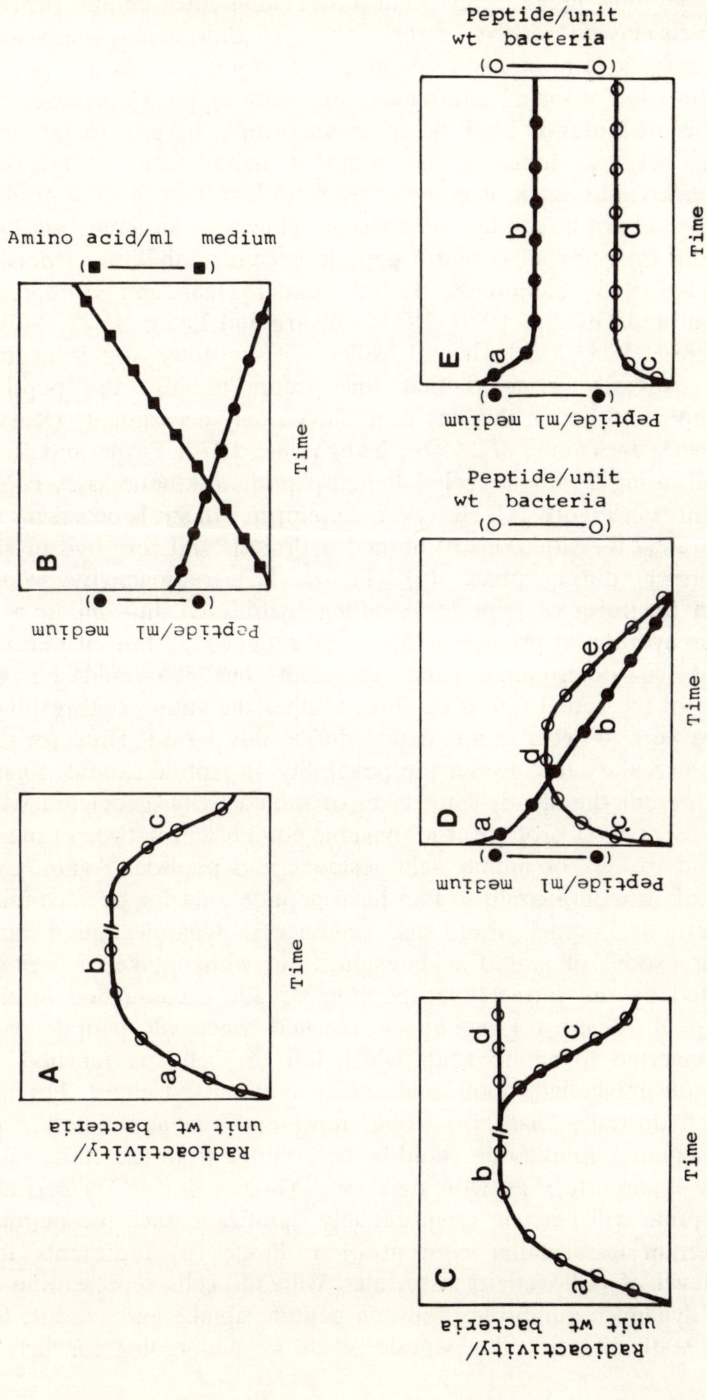

Figure 1 Speculations on the influence of intracellular peptidase action plus amino acid and peptide exodus during measurement of peptide uptake using radioactive (A,C) or fluorescent labelling (B,D,E) techniques. Uptake of peptide is slower (A,B) or faster (C,D,E) than the rate of intracellular hydrolysis. (See text for details)

In case B, peptide uptake can occur faster than intracellular peptide hydrolysis which may be positive or zero. These conditions may apply with particular substrates (e.g. sarcosine peptides, triornithine) that are poorly hydrolysed by the normal peptidases, or with typical peptides in peptidase-deficient mutants. First, using an auxotroph, its growth rate can be maximum here so long as the required amino acid is released sufficiently quickly, but again, if growth rate were less than this the growth limiting step would not be directly identifiable. However, an indication that these conditions may prevail is that a peptide becomes inhibitory (Meisler and Simmonds, 1963; Simmonds, 1970; Vonder Haar and Umbarger, 1972; Sussman and Gilvarg, 1970, 1971; Gilvarg and Levin, 1972; Miller and MacKinnon, 1974; McHugh and Miller, 1974; Miller and Schwartz, 1978), it is generally assumed that this occurs because the peptide accumulates intact, although this has been shown only occasionally (Kessel and Lubin, 1963; Jackson *et al*., 1976; Yang *et al*., 1977; Payne and Bell, 1979). Second, using a radioactively-labelled peptide, a kinetic assay could show the features in Figure 1C. However, to interpret this it is necessary to consider separately the situations of limited hydrolysis and zero hydrolysis. With the former, during phase (a) (Figure 1C), radioactivity would accumulate in the form of peptide (awaiting hydrolysis) and amino acid residues (after hydrolysis) providing they were retained. A rate of peptide uptake could be determined from the plot but it would be an underestimate of the actual rate if exodus of either the amino acid residues or the peptide were to occur to any extent during this period. Thus, for the first time it is necessary to consider the possibility of peptide exodus. Phase (b) would represent the steady-state level of radioactivity associated with the cells, representing (I propose) the dynamic equilibrium between uptake of peptide and exodus of amino acid residues and peptide. Again, this apparently static situation could in fact have peptide entering at maximum rate. An alternative model would also envisage a dynamic equilibrium, with a similar exodus of amino acid residues but with uptake of peptide being regulated by the intracellular peptide already accumulated in the cytoplasmic pool. Phase (c) would be reached when all peptide was eventually converted to amino acids which left the cell and reached an equilibrium with unlabelled, pool amino acids as discussed above. For the case of zero hydrolysis, phase (a) would represent true accumulation of peptide. A maximum uptake rate could be determined from the initial rate. The rate will apparently slow with time (see Yang *et al*., 1977) because exodus of peptide will become quantitatively significant once the peptide reaches a certain intracellular concentration. Phase (b) represents the steady-state level of radioactivity associated with the cells representing (I propose) the dynamic equilibrium between peptide uptake and exodus; far from being a static situation the peptide would be undergoing concurrent uptake and exodus, both possibly at maximum rates. An alternative model would be a static situation in which peptide uptake was shut off by the accumulated peptide. The dynamic and static models could readily be distinguished by adding a different peptide under steady-state conditions; a dynamic-exchange situation would allow immediate, competitive uptake of the second peptide and establishment of a new equilibrium involving both peptides, in a static, 'shut off' situation the second peptide would be unable to enter. Phase (d) would simply be a prolongation of the steady-state phase (c). Finally, with the fluorescence assays how will the circumstances of case B influence the results? In the first instance in which uptake can occur with slight intracellular hydrolysis, the phases in Figure 1D could occur. By monitoring the medium in phase (a), a true V_{max} for uptake could be determined from the intitial rate (say within 15 s), however, the average rate of peptide removal from the medium throughout this phase (say up to 1 min) is likely to be an underestimate of the actual rate of uptake to an extent dependent on the level of peptide exodus. In phase (b) a linear rate of peptide removal occurs; however, this is likely to conceal a dynamic equilibrium and to represent the *net* rate for peptide uptake minus peptide exodus. The concentration of amino acid residues would increase in the medium during these phases. Using the complementary approach of monitoring cell extracts would reveal during phase (c) a rapid increase in the concentration of peptide in the pool, reaching a steady-state level during phase (d), and eventually declining during stage (e). In the second instance, in which hydrolysis is zero, the stages shown in Figure 1E could occur. By monitoring the medium in phase (a), the initial rate (say within 15 s) would give V_{max} for uptake, but again the average rate throughout the period, say to 1 min, is likely to be below the V_{max} because of some peptide exodus. In phase (b) there is a steady-state, dynamic equilibrium between peptide uptake and exodus, and in principle this could be maintained for long periods. Assay of cell extracts would show a rapid rise in peptide concentration during phase (c), reaching a steady-state level during phase (d). The alternative static model of the steady state, in which peptide uptake is shut off, could be distinguished from the proposed dynamic equilibrium by adding a different peptide, as discussed above for Figure 1D.

The transport events depicted in Figure 1B and D were reported for peptide uptake in *E. coli* by Payne and Bell (1979). However, in view of the subsequent observations of peptide exodus (see Section 2.1.5.3) which has led to the above interpretation of these plots, it now seems likely that many of the transport rates reported in that study using the dansyl method were net uptake rates rather than maximum uptake rates. Furthermore, the reported variations in rates with peptide structure could reflect specificity of the peptidases rather than specificities of the permeases as was suggested. The situations represented in Figure 1B, D, and E have been observed (J. W. Payne, unpublished) using several of Miller's peptidase-deficient strains of *S. typhimurium*. The results have been obtained using a rapid, kinetic assay with fluorescamine in conjunction with dansyl-TLC. To obtain results illustrated in Figure 1E it is necessary to decrease the peptide concentration to 20–50 μM and to use cell cultures at *c*. 5×10^9 cells ml^{-1} in order to maintain a significant proportion (about 30 per cent) of the peptide within the bacteria, under these conditions concentration gradients of about 250-fold are typically observed and these can be increased several-fold by decreasing the initial concentration of the peptide. It seems likely that the presence in these cells of intact peptide at about 10 mM concentration is responsible for the inhibitory effects frequently observed.

5 Substrate Specificities for Peptide Transport and Utilization

This topic has received great attention in earlier reviews (Payne, 1975*a*, 1976; Barak and Gilvarg, 1975*a*; Payne and Gilvarg, 1978; Matthews and Payne, 1980*a*) which should be consulted for detailed results and references. However, the cited conclusions from these earlier studies may need re-evaluation in light of the above observations on amino acid and peptide exodus, and the possible influence of peptidases; such a re-evaluation is attempted below. Second, most of the conclusions derive from growth tests, and the following calculation is important when assessing the results from this test system. Consider the rate at which a required amino acid must be supplied to an auxotroph to sustain steady growth. Assume the amino acid is utilized for protein synthesis only. In a culture containing 1 mg dry weight bacteria, let 50 per cent (500 μg) be protein of which 5 per cent is Ala (approx. molecular mass 100 daltons). Thus, total Ala in culture is $(5/100) \times (1/100) \times 500 = 0.25$ μmol, and this amount would be needed for the culture to double in mass in (say) 60 min. Throughout this period therefore, an average rate of uptake of *c*. 4 nmol Ala min^{-1} (mg dry weight)$^{-1}$ (2 nmol Ala-Ala, 1.3 nmol Ala-Ala-Ala) is required. Now typical rates of peptide uptake cover the range 4–40 nmol min^{-1} (mg dry weight)$^{-1}$ (De Felice *et al*., 1973; Cowell, 1974; Jackson *et al*., 1976; Yang *et al*., 1977; Payne and Bell, 1979) and these remain effectively constant until uptake is complete (Payne and Bell, 1979). Thus, for those peptides taken in most slowly, reducing their rate of uptake by 1 per cent could decrease growth rate, while for the fastest peptides, lowering their rate of uptake by 90 per cent could be without effect on growth rate. Thus, a particular structural change in a 'slow' peptide might apparently have a dramatic effect on transport and utilization, while the same modification to a 'fast' peptide might appear to be without effect.

(i) N-terminal α-amino Group

Early studies, showing that derivatization of this group impaired the utilization of peptides, did not distinguish between an effect primarily on transport or intracellular hydrolysis. Subsequently it was shown by Gilvarg and Katchalski (1965) that α-*N*-acyl peptides could be cleaved by the intracellular peptidases of *E. coli* although they could not be utilized by auxotrophs, and they concluded that the transport systems had a requirement for the unsubstituted α-amino group. (However, with certain peptides this conclusion could be influenced by action of the periplasmic dipeptidyl carboxypeptidase (Deutch and Soffer, 1978; see Section 2.1.5.4).) In latter studies (Payne, 1971*a*, 1974) it was established that *E. coli* utilizes mono-α-*N*-substituted peptides in which the positive charge is retained, e.g. *N*-monoalkyl derivatives, but not *N*-dialkyl or *N*-acyl derivatives. This specificity applies to both di- and oligopeptides. Similar conclusions have been reached with various strains of *E. coli* (Becker and Naider, 1974), *Pseudomonas* (Cascieri and Mallette, 1976*a*; Miller and Becker, 1978), *Streptococci* (Law, 1977, 1978), and *S. typhimurium* (Jackson *et al*., 1976). The fluorescence assays have not been applied to this type of derivatived peptide so a more critical appraisal is not yet possible.

(ii) C-terminal α-carboxyl Group

The peptide permeases do not have an absolute requirement for this group. Various species can utilize peptides devoid of the carboxyl group (Payne and Gilvarg, 1968*b*; Payne, 1973) or with it variously derivatized (Ames *et al*., 1973; Fickel and Gilvarg, 1973; Becker and Naider, 1974; Cascieri and Mallette, 1976*a*; Allen *et al*., 1978). However, these modifications are not without effect for the derivatives are less good competitors than the parent compounds (see above references, and Mostillo and Gilvarg, 1977). The specificity of the dipeptide permease appears to be stricter than that of the oligopeptide permease. Thus, uptake of dipeptides with this group absent or derivatized seems to be via the oligopeptide system, for it is competitively affected by oligopeptides and does not occur in mutants lacking the oligopeptide system (Payne and Gilvarg, 1968*a*; Payne, 1973; Ames *et al*., 1973; Cascieri and Mallette, 1976*a*; Hirshfield and Price, 1975; Law, 1978). However, alaphosphin (Ala-Ala in which the carboxyl group is replaced by a phosphoryl group) (see Section 4.7.2.1(iii)) still inhibits oligopeptide permease deficient mutants implying it can be handled by the dipeptide system (Atherton *et al.*, 1979*b*), in contrast, dipeptide tetrazole analogues use the oligopeptide system (Franklin and Snow, personal communication). Fluorescence assays have not yet been carried out on peptides with modified *C*-termini.

(iii) Peptide Bonds

Growth tests have indicated that α-peptide linkages are required for peptide uptake in *E. coli* (see Payne, 1976, for references), *Pseudomonas* (Cascieri and Mallette, 1976*a*), *Salmonella* (Yang *et al.*, 1977), and *Streptococci* (Law, 1978). Certain β-alanyl peptides are utilized although it may be that they are transported by a separate system (Payne, 1973; Kirsh *et al.*, 1978). Peptides in which the peptide-bond nitrogen is methylated are transported by *E. coli* (Payne, 1972*c*) although using dansyl-TLC has shown that the rates are slower than for the parent peptides (Payne, 1977). These substituted bonds are very resistant to enzymic hydrolysis and the peptides accumulate intact in *E. coli* (Payne and Bell, 1979), *Streptococcus faecalis* (T. M. Nisbet and J. W. Payne, unpublished) and *S. cerevisiae* (Nisbet and Payne, 1979*a*).

(iv) Stereochemical Specificity

Results indicate that the peptide transport systems show specificity for L-residues. Auxotrophic growth tests are of little use for such studies because even if the substrates were to be transported the L-residue specificity of the intracellular peptidases makes it unlikely that they would be hydrolysed.

Early studies showed that uptake of radioactively-labelled L-Leu-Gly or Gly-Gly by *E. coli* was not inhibited by a variety of dipeptides containing D-residues although all the L-isomers did inhibit (Levine and Simmonds, 1962*b*; Kessel and Lubin, 1963). Similar conclusions have been reached for *Pseudomonas* (Cascieri and Mallette, 1976*a*), *Streptococci* (Law, 1977, 1978), *Lactobacilli* (Leach and Snell, 1960; Kihara *et al.*, 1961; Shelton and Nutter, 1964; Yoder *et al.*, 1965*a,b*), and *Salmonella* (Yang *et al.*, 1977), although these studies have usually involved indirect ways of judging uptake of the D-isomers. The rate reported for the direct uptake of [^{14}C]-D-Ala-[^{14}C]-D-Ala by *E. coli* (Neuhaus *et al.*, 1977) is less than 0.1 per cent of the typical rate for all L-dipeptides (Payne and Bell, 1979; J. W. Payne and G. Bell, unpublished). Uptake of the following peptides was undetectable using dansyl-TLC assays of media or cell extracts; or using a kinetic fluorescamine assay under the conditions of Figure 1E: the L,D, D,L, and D,D forms of Ala-Ala, Val-Val, and Leu-Leu; D-Ala-Gly, Gly-D-Ala, D-Leu-Gly, and Gly-D-Leu (J. W. Payne, unpublished).

For oligopeptides, Shankman *et al.* (1962) reported that L-Val-L-Val-D-Val was taken up by *Lactobacillus* spp. although other positional substitutions of the D-residue prevented uptake. A similar conclusion was reached for uptake of the stereoisomers of Met-Met-Met by *E. coli* (Becker and Naider, 1974). This result was anticipated from the observed lack of specificity towards the *C*-terminus (Payne and Gilvarg, 1971). Effects of the positional substitution of D-residues on the antibacterial action of some tri, tetra-, and pentaphosphonopeptides has been reported (Allen *et al.*, 1978; Atherton *et al.*, 1979*a*), and interpreted in terms of specificity of transport, although it is possible that a contributing factor to the apparent specificity of the higher homologues could be the selective action of the periplasmic dipeptidyl carboxypeptidase (Deutch and Soffer, 1978, and see Section 2.1.5.4). Using the dansyl-TLC assay, rates of peptide removal from the medium by *E. coli* were determined for the following peptides (J. W. Payne, unpublished); uptake of D-Leu-Gly-Gly and D-Ala-Gly-Gly was undetectable although the L-isomers were rapidly transported; the rate of uptake of L-Ala-L-Ala-D-Ala was *c.* 40 per cent that for the L,L,L form, but no uptake of the L,D,L and D,D,D was detectable; L-Val-L-Val-D-Val was taken up at *c.* 20 per cent the rate of the L,L,L form, but uptake of D,D,D was not detectable; Gly-Gly-D-Leu was removed at *c.* 25 per cent the rate of the L-isomer. With the above three peptides carrying *C*-terminal D-residues, intact L-Ala-D-Ala, L-Val-D-Val, and Gly-D-Leu respectively were detected in cell extracts, and as tripeptide uptake continued these dipeptides and the respective free *N*-terminal residues underwent selective exodus (see Section 2.1.5.2). It is noteworthy that when *E. coli* is incubated with these dipeptides they cannot be detected inside the cell. How the dipeptides leave the cell is a matter for future study. Finally, it should be emphasized that inability to detect uptake of certain of these tripeptides using this assay arises from a defect in transport and not hydrolysis because other non-hydrolysable peptides (e.g. Gly-Sar, Orn-Orn-Orn) are readily detected in cell extracts (Payne and Bell, 1979).

(v) Side-chain Specificity

Evidence from diverse studies indicate that the transport systems are of very broad specificity able to handle peptides carrying all types of side-chains (see reviews by Barak and Gilvarg, 1975*a*; Payne, 1975*a*; Payne and Gilvarg, 1971, 1978), although the extent to which the different structures influence the kinetics of transport is still not fully clear.

With dipeptides, competitive inhibitions of auxotrophic growth have been reported for many species (Payne, 1968; Cascieri and Mallette, 1976*a*; Jackson *et al.*, 1976; Miller and Becker, 1978; Law, 1977, 1978).

Similarly, diverse dipeptides inhibit uptake of radioactively-labelled ones (Leach and Snell, 1960; Levine and Simmonds, 1962*b*; Kessel and Lubin, 1963; Yang *et al*., 1977; Law, 1978; Rice *et al*., 1978), and the same conclusion has been reached for fluorescence assays (Payne and Bell, 1979). In addition, 'unnatural' and derivatized side-chains are recognized (Ames *et al*., 1973; Kenig and Abraham, 1976; Alper and Ames, 1978; and see Section 4.7.2). Finally, in several species, putative Dpp^- mutants (selected as resistant to a particular toxic dipeptide) fail to utilize or transport other dipeptides (Kessel and Lubin, 1963; De Felice *et al*., 1973; Kenig *et al*., 1976; J. W. Payne and R. A. Alves, unpublished).

Similar conclusions have been reached for oligopeptides. With auxotrophs, competitive inhibition of the utilization of nutrient peptides by other oligopeptides has been reported for various species (Payne, 1968; Barak and Gilvarg, 1974; Ames *et al*., 1973; Jackson *et al*., 1976; Cascieri and Mallette, 1976*a*; Diddens *et al*., 1976; Miller and Becker, 1978; Rice *et al*., 1978). With radioactive- (Smith *et al*., 1970; Jackson *et al*., 1976; Payne and Bell, 1977*a*; Rice *et al*., 1978) and fluorescence- (Payne and Bell, 1977*b,c*, 1979) assays competition for uptake between many varied oligopeptides has been noted. In addition, oligopeptides with 'unnatural' and derivatized side chains are absorbed via the general oligopeptide permease (Payne, 1968; 1972*a*; Ames *et al*., 1973; Diddens *et al*., 1976; Alper and Ames, 1978; see also Section 4.7). Finally, Opp^- mutants (selected as resistant to a specific peptide, e.g. triornithine) lose ability to utilize and transport all tested oligopeptides (Payne, 1968; De Felice *et al*., 1973; Jackson *et al*., 1976; Barak and Gilvarg, 1974, 1975*b*; Becker and Naider, 1974; Naider and Becker, 1975; Payne and Bell, 1977*a,b*, 1979). The residual ability of Opp^- strains to utilize certain oligopeptides (Payne, 1968; Barak and Gilvarg, 1975*b*; Naider and Becker, 1975) exemplifies a broad, but much diminished ability to transport oligopeptides (Payne and Bell, 1979; J. W. Payne, unpublished), although there is as yet no clear interpretation of these observations (see Section 2.1.5.6).

The general conclusion that a vast spectrum of small peptides can be transported by one or two peptide permeases should not conceal the fact that the side chains are likely to affect binding affinities (V_t) and perhaps rates of uptake (V_{max}), and the very limited available data support this (Leach and Snell, 1959, 1960; Kessel and Lubin, 1963; De Felice *et al*., 1973; Cowell, 1974; Jackson *et al*., 1976; Yang *et al*., 1977; Mostillo and Gilvarg, 1977; Payne and Bell, 1977*a,b*, 1979). However, it is possible that the results from many of these studies refer to net uptake rates (see Section 2.1.5.4). Thus, final conclusions must await direct measurement of initial rates of uptake for diverse peptides, ideally under the conditions of Figure 1E (Section 2.1.5.4); preliminary results from such a study using kinetic fluorescence assays of initial rates in *S. typhimurium* show V_{max} is very similar for different peptides.

6 Di- and Oligopeptide Permeases and Transport Mutants

The existence of separate di- and oligopeptide permeases is indicated in several bacterial species from the results of competition studies showing minimal interactions between the substrates during uptake, and from the isolation of transport mutants, but how widespread this duality is amongst bacteria remains to be established. With the following species, using auxotrophic growth tests, radioactive and fluorescence assays, it has been shown that oligopeptides compete with each other for uptake but only to a very limited extent with dipeptides: *E. coli* (Kessel and Lubin, 1963; Payne, 1968, 1971*a,b*; Smith and Dunn, 1970; Barak and Gilvarg 1974, 1975*b*; De Felice *et al*., 1973; Naider and Becker, 1975; Diddens *et al*., 1976; Payne and Bell, 1979), *S. typhimurium* (Ames *et al*., 1973; Jackson *et al*., 1976; Yang *et al*., 1977); *Pseudomonas* spp. (Cascieri and Mallette, 1976*a*; Miller and Becker, 1978); *Streptococcus* spp. (Law, 1978; Rice *et al*., 1978), and *Lactobacillus* spp. (Dunn *et al*., 1957; Shelton and Nutter, 1964). Staros and Knowles (1978) recently reported that dipeptide transport in *E. coli* was specifically blocked after attachment to the cells of a photoaffinity-labelled dipeptide. On the other hand, other results indicate that in these species dipeptides can affect uptake of oligopeptides, leading to the conclusion that dipeptides can use the oligopeptide permease to a slight extent. Many of the investigations on which these conclusions have been based would pay re-examination, bearing in mind the possible ways in which peptidases may influence the results (see Section 2.1.5.4).

Mutants defective in dipeptide uptake (*dpp*$^-$) are more difficult to isolate on the basis of resistance to toxic dipeptides, presumably because they can also enter via the oligopeptide permease. Starting with an Opp$^-$ strain of *E. coli*, De Felice *et al*. (1973) readily selected an *opp*$^-$ *dpp*$^-$ strain from it. Possible specific *dpp*$^-$ mutants of *E. coli* were reported by Kessel and Lubin (1953) and Vonder Haar and Umbarger (1972), and we have recently isolated such a mutant in *E. coli* W (R. A. Alves and J. W. Payne, unpublished). With regard to the map position some confusion exists for the *dpp*$^-$ gene. De Felice *et al*. (1973) reported its map position as between *opp* (27 min) and the *proC* marker (9 min). The 5 min locus reported in Bachman *et al*. (1976) has been found to be incorrect (P. E. Hartman, personal communication).

Mutants defective in oligopeptide uptake (*opp*$^-$) have been isolated from *E. coli* (Payne, 1968; De Felice *et al*., 1973; Barak and Gilvarg, 1974; Becker and Naider, 1974; Diddens *et al*., 1976) and *S. typhimurium* (Ames *et al*., 1973; Jackson *et al*., 1976), using as selection procedure resistance to toxic oligopeptides (usually triornithine). All tested oligopeptides use the system associated with the above mutation and yet uptake of certain peptides can still be detected in *opp*$^-$ strains; it has been suggested that this may indicate the operation of specialized systems (Barak and Gilvarg,

1975*b*; Naider and Becker, 1975; Jackson *et al.*, 1976).

It should be noted, however, that our own studies using fluorescence assays (J. W. Payne and R. A. Alves, unpublished results) indicate that it is misleading to consider the additional system(s) present in Opp^- strains as being 'specialized' with a very narrow range of peptide substrates e.g. Met_3, Leu_3 and Thr_3, although this was the impression obtained from auxotrophic growth tests (Barak and Gilvarg, 1975*b*; Naider and Becker, 1975). In contrast, Opp^- Dpp^- strains tested to date retain ability to absorb a wide range of di- and oligopeptides at rates varying from about 5 to 40 per cent that of the parent. Substrates perhaps regarded as definitive for the Opp, e.g. Lys_3, are still absorbed to a small degree by opp^- mutants. Frequently, however, these residual rates are inadequate to maintain growth of the auxotroph utilizing the peptide as its sole source of required amino acid. At this time it has not yet been clarified whether the additional uptake mode represents the existence of additional peptide permease(s) mediating active, intact uptake (see below) and/or a membrane digestion mode (Ugolev *et al.*, 1977) perhaps involving peptidase N (as discussed in Section 2.1.5.4).

With respect to the genetics of the oligopeptide permease there are a number of features which need clarification; amongst these are the following. (a) When spontaneous opp^- mutants are selected by their resistance to toxic oligopeptides they are isolated at an unusually high frequency. Barak and Gilvarg (1974) reported frequencies for transport mutants of about 10^{-4} to 10^{-5} for a variety of *E. coli* strains, using resistance to triornithine as selection procedure; and we have found similar values. In other studies (J. W. Payne and R. A. Alves, unpublished results), we found mutation rates of between 2×10^{-5} and 4×10^{-5} in an *E. coli* W strain (M.26.26) using selection for resistance to the toxic peptides triornithine, norVal-Gly-Gly, Gly-Gly-norLeu, and Ala-Ala-aminoxyAla. In all these studies, using this type of selection, no other mutations conferring resistance (e.g. altered peptidases, modified intracellular target site, etc.) were picked up. In one instance, several hundred isolates were examined and all were transport defective, but it was not established that the mutational defect was identical in each case. In short, the spontaneous loss of oligopeptide transport activity in a normal population of exponential cells not subjected to any selective pressure during growth, occurs at a frequency considerably higher than is normal when asking for a specific kind of mutation. (b) Examination of a random selection of about thirty laboratory strains of *E. coli*, unselected for any property relating to peptide transport, reveals that many of the strains are already oligopeptide transport mutants. It can be argued that there is no reason why different strains should have the same behaviour, and that if their previous history were known it might indicate that some strains could have been selected to become Opp^-; but nevertheless, the frequency at which unselected strains display the Opp^- phenotype is unusually high. Certainly, a similar proportion (about 20 per cent) would not be found to be defective, e.g. in a glycine transport system. (c) Mapping of the *opp* gene by two groups (Barak and Gilvarg, 1974; De Felice *et al.*, 1973) showed an approximate locus at 27 min, but large and unexplained differences were reported in the frequency of linkage to a variety of adjacent markers. (d) Inability to obtain revertants of opp^- mutants is invariably found. No spontaneous opp^+ revertants have been isolated from any of a number of strains (J. W. Payne, unpublished results, and discussion to Payne, 1972*a*; Barak and Gilvarg, 1975*a*), or by nitrosoguanidine mutagenesis (Barak and Gilvarg, 1974, 1975*a*). These authors failed to isolate revertants and reported that the reversion frequency must be less than 2×10^{-8}, and our own studies indicate it is lower than this. This property is a characteristic of all *opp* mutants examined, although they have not all been shown to arise by the same mutation. Thus, these results are characteristic of the Opp^- phenotype arising by deletion (or possibly a non-revertible point mutation). (e) opp^- mutants are sometimes produced when strains of *E. coli* are cultivated in a chemostat (in minimal media devoid of peptides), under particular nutrient limitations (J. W. Payne, unpublished results, and quoted in CIBA Foundation Symposium, 1972). (f) Different strains of *E. coli* can show general differences in rates of oligopeptide uptake (Payne and Bell, 1979). It may be that the above characteristics are not necessarily related to one another; on the other hand, the possibility that the oligopeptide permease occurs in a region of the chromosome particularly susceptible to gene amplification–deletion could have implications for several of the above features; this possibility is considered below. (It was reported by Gratia (1966) that deletions in the region of 27 min on the *E. coli* chromosome occur frequently (see also, Postle and Reznikoff, 1978).)

It is clearly established that duplications may be very frequent events (see review by Anderson and Roth, 1977). Several authors have calculated the frequency of spontaneously generated duplications to be of the order of $5 \times 10^{-4} - 7 \times 10^{-5}$ (Glansdorf and Sand, 1968; Horiuchi *et al.*, 1963; Straus and Straus, 1976; Folk and Berg, 1971; Straus, 1974; see also above review). Thus, the common case is: *extremely* frequent appearance of duplications (when a selection exists, usually if not always, for increased function) such as 1 in 1000; and also *extremely* frequent loss of the duplication. Therefore, the outstanding characteristic of duplications is their instability. Anderson and Roth (1977) quote many examples illustrating the instability of a duplication, and indicate how they will be subject to frequent loss or further amplification. Normal recombination between two copies should lead either to loss of the duplication (deletion) or to further amplification (i.e. triplication) of the chromosomal segment. Both these events involve legitimate recombination between homologous sequences and might be expected to occur frequently. Nevertheless, some duplications are stably maintained (see below), and Glandsdorf and Sand (1968) indicated that while a particular duplication mutant was found at a frequency of about 10^{-5}, the frequency of its reversion to wild-type (non-duplicated) was 10^{-10}.

How then might these considerations relate to the observed properties, (a)–(f) of the Opp$^+$ and Opp$^-$ strains? With (a), although duplications occur at the required frequency, a simple duplication of the entire gene (Figure 2(a)) cannot cause loss of transport activity, unless gene dosage causes a negative phenotype (doubtful), or dosage of a regulatory element causes a negative phenotype (although unlikely, it should be remembered that transport activity is not *completely* lost for any oligopeptide in Opp$^-$ strains). Similarly, precise duplication starting within the gene (Figure 2(b)) will not cause a negative phenotype unless one invokes a regulatory element as above; an imprecise duplication could lead to loss of wild type activity, but the molecular mechanism of this is not clear. A duplication with an inversion, within *opp* (not shown), would produce the observed

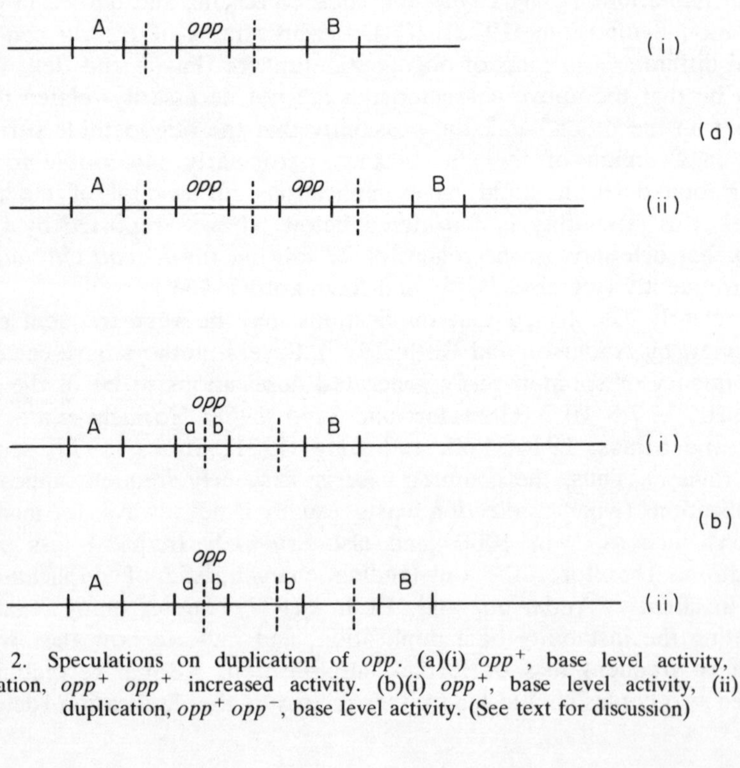

Figure 2. Speculations on duplication of *opp*. (a)(i) *opp*$^+$, base level activity, (ii) full duplication, *opp*$^+$ *opp*$^+$ increased activity. (b)(i) *opp*$^+$, base level activity, (ii) partial duplication, *opp*$^+$ *opp*$^-$, base level activity. (See text for discussion)

result, but again the molecular mechanism is difficult to picture. On the other hand, if the Opp^+ state were to represent a stably maintained replicated form (e.g. Figure 2(a)(ii) or 2(b)(ii)), then a deletion (of varying length) could yield a partial or possibly complete negative phenotype. However, it appears that a deletion in a duplicated region most commonly involves excision of the duplicated material only, and not of neighbouring genes; thus Opp^- would represent the beginning phenotype. With (b), the observation of 'natural' *opp*$^-$ mutants would require the same arguments as for (a) above. It would be of interest to check in all the 'natural' Opp^+ and Opp^- strains whether any correlation can be made with them harbouring a *rec*A$^-$ gene; because amplification requires normal recombination process, *rec*A$^-$ mutants give much lower frequencies of gene duplications (Anderson and Roth, 1977). For (c), if the reported variation in nearness of linked markers is true and not just an error, then it could be explicable if gene amplification–deletion could include variable regions of adjacent chromosome. In *E. coli*, duplicated regions may typically be up to 4 min in chromosome units (Anderson and Roth, 1977), although very large, stably maintained regions have been reported in *S. typhimurium* (Straus and Straus, 1976). With (d), if Opp^- arises from deletion, its high frequency of isolation could perhaps arise if multiple gene copies of *opp* exist. The failure then to obtain *opp*$^+$ revertants might indicate that certain DNA sequences initially present in Opp^+, and which facilitate duplication, had been lost. With (e), if *opp*$^+$ normally carries reasonably stable amplified genes, then when circumstances which favour the presence of the duplicated genes no longer prevail, as with prolonged growth in the chemostat under certain nutrient limitations (Horiuchi *et al*., 1963; Rigby *et al*., 1974), the cell population would no longer remain committed to the amplification genotype, and the inherent instability of amplifications would permit return to the transport diminished phenotype. With (f), the fact that different strains show different capacities for uptake could represent the presence of a single *opp*$^+$ gene that is stably replicated to different extents. However, it is perhaps more likely that multiple oligopeptide transport systems are present in these species (probably arising through earlier duplication and subsequent separate evolution), which have somewhat different properties (high and low affinity, etc.) or are under different regulatory control (e.g. N-control, see Section 2.6). A complete transport negative phenotype will, therefore, require more than one mutation to occur; mapping will be difficult, etc.

Alternative explanations may be invoked for some of the above observations on the Opp^+ and Opp^- strains. For example, *opp* could be a hot spot for insertion of an unknown IS sequence, phage, or transposon. This could lead to the Opp^- phenotype at the observed frequency; Opp^+

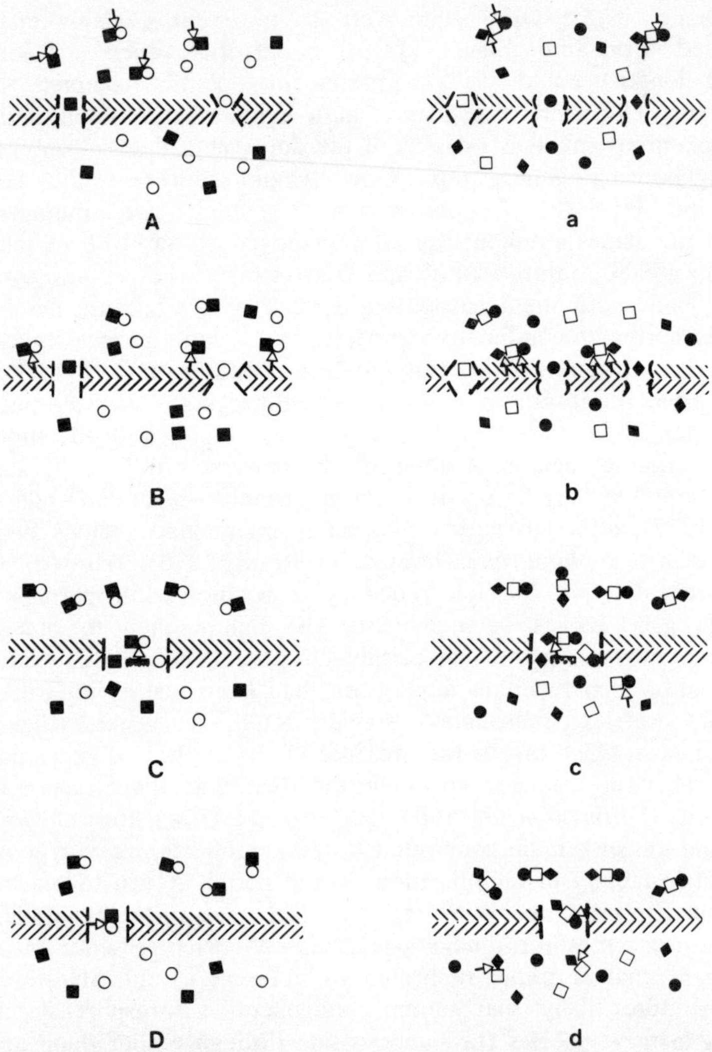

Figure 3 Theoretical models for hydrolysis-dependent peptide uptake in bacteria. The cytoplasmic bilayer membrane is represented by crosshatched section and contains several specific transport systems. The cell exterior, cell wall, and periplasmic space are above, and the cell interior is below the membrane. Arrows indicate peptidase action. A–D represent dipeptide uptake and a–d oligopeptide uptake. (A,a) Hydrolysis by free peptidases either outside the cell or within the cell wall or periplasmic space, with release of free amino acids and uptake through amino acid permeases. (B,b) Hydrolysis by envelope-bound peptidases with release of free amino acids and uptake through amino acid permeases. (C,c) Hydrolysis by membrane-bound peptidases with uptake of cleavage products through associated sites not accessible to the free substrates. (D,d) Group translocation in which membrane-bound peptidase-carrier serves both to hydrolyse and translocate peptides. (See text for discussion)

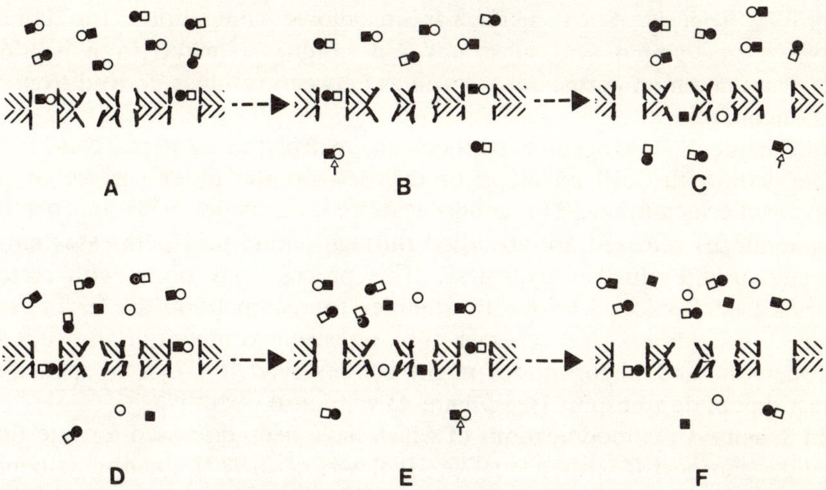

Figure 4 Theoretical model for uptake and exodus of intact peptides in bacteria. The cytoplasmic bilayer membrane is represented by crosshatched section. Outside the cell is to the top and inside to the bottom. The membrane is shown with two different amino acid permeases (in centre) and two identical dipeptide permease sites (at each end). Each permease is represented with an interior (bottom) and exterior (top) binding face involved initially in uptake and exodus respectively. Arrows indicate peptidase action. Two peptide substrates are present, one (■○) is sensitive to hydrolysis, the other (□●) is resistant. A–F represent a sequence of transport and hydrolysis events. (See text for discussion)

could only be regenerated by precise excision of the insertion element and this might be a very rare event (as in Tn10).

It remains therefore, to establish whether there is any relationship between duplication/deletion and Opp$^-$ and Opp$^+$ expression. Measurement of K_m values and initial rates of uptake for a range of peptides in the wild type and mutants is needed, as well as further precise mapping of the *opp* gene. Finally, the residual system present in *opp*$^-$ mutants has to be mapped, and mutants lacking both systems (double *opp*$^-$) assayed for any residual transport capability.

7 Mechanisms of Transport: Current Models

Figure 3 is a schematic summary of possible peptide uptake systems in which hydrolysis preceeds appearance of the substrate in the cytoplasm. Certain aspects of these models have been discussed earlier (Payne and Gilvarg, 1978).

In scheme A, a, exogenous peptides are hydrolysed by free peptidases either outside the cell or within the cell wall or periplasmic space (see Section 2.1.5.4 for examples). The released amino acid residues are transported by their specific permeases. This process occurs in several microbial species. Such a mechanism allows competition for uptake between free amino acid substrates and residues released from peptides that share common permeases, as well as between residues derived from di- and oligopeptides.

In scheme B, b exogenous peptides are hydrolysed by peptidases bound either within the cell envelope or exposed on the outer surface of the cytoplasmic membrane. The amino acids (B), or amino acids and peptide fragments (b) released are absorbed through amino acid permeases either directly or after further hydrolysis. This process may occur with certain species (see Section 2.1.5.4). Possibilities for competition are as in A, a above. In addition, for schemes a,b, apparent competition for uptake between di- and oligopeptides might be observed if a system mediating intact dipeptide transport (see Figure 4) were also operative.

In Scheme C, c, modifications of which have been discussed for intestinal peptide transport (Matthews, 1975; Ugolev *et al*., 1977), membrane-bound peptidases produce cleavage products (without outward release) that are transported by associated carriers inaccessible to the same substrates supplied free in solution. Hydrolysis and translocation are considered to be distinct events mediated sequentially by separate components. A simple form of this model might be a pore lined with the peptidase and carrier in series. With a general dipeptidase (C) one or more amino acid carriers of broad specificity might be involved. With oligopeptides, an exopeptidase of broad specificity might produce amino acid and dipeptide (subsequently cleaved inside) (c), or amino acids only might be produced and passed to a broad specificity carrier (not shown). Uptake specificity would primarily be a property of the peptidases, one or more of which may function for each of the di- and oligopeptide systems. No competition between amino acids, di- and oligopeptides would occur. The model can accommodate most of the information on bacterial peptide uptake except for the observation of intact peptide accumulation (see Section 2.1.5.4).

The final scheme D, d, is one of group translocation in which hydrolysis and vectorial discharge of cleavage products are obligatorily coupled and performed by a single molecular species. This model also fits most of the available information apart from the crucial observation of the uptake of intact peptides.

The situation of intact peptide accumulation is illustrated in Figure 4A–F. In A, two different peptides bind at two identical peptide permease sites. In B, the peptides are transported into the cell where one (at left) is rapidly hydrolysed; this part of the model represents transport of peptide intact and is the scheme considered in earlier reviews (see CIBA Foundation Symposia, 1972, 1977; Payne, 1975*a*, 1976; Payne and Gilvarg, 1978). The discussion in these reviews generally implied that the released amino acids are retained by the cells, but it is now known (see

Section 2.1.5.2) that these residues may undergo exodus concomitant with further peptide uptake, as shown in C–F. B,C also show uptake of a non-hydrolyzable peptide (see Sections 2.1.5.3, 2.1.5.4, and Figure 1). When this substrate reaches a certain intracellular concentration (it is possible an initial overshoot may occur, C) its rate of exodus will become significant and with uptake of the peptide continuing a steady state distribution across the membrane will be reached (D–F). Throughout this period, C–F, uptake of the other peptide is continuing and its released residues are undergoing exodus. The two dipeptides represent extreme cases with respect to hydrolysis but it is possible (see Section 2.1.5.4 and Figure 1) that a slowly cleaved peptide might undergo several passages of uptake and exodus before being hydrolysed. For convenience, the scheme here shows that dipeptide uptake and exodus are mediated by the same system, for which there is some preliminary evidence from studies with transport mutants (J. W. Payne and R. A. Alves, unpublished). The scheme can be appropriately modified to accommodate uptake of intact oligopeptides with various types of intracellular hydrolysis and exodus (see Section 2.1.5.3).

6 CONCLUSIONS

In this as in previous reviews the results of studies on peptide uptake and utilization are summarized. However, for the first time, detailed consideration is given to the various assay procedures used and to a critical evaluation of the type of result provided by each method. Recent findings are presented on the exodus of amino acid residues released by intracellular hydrolysis of transported peptides and on the exodus of peptides, and it is concluded that the results of earlier experiments may need reinterpretation in light of these findings. The occurrence of exodus has clear implications for the design of antimicrobial smugglins (Matthews and Payne, 1975; Payne, 1977, see also Section 4.7.2.1). (A smugglin is a derivatized peptide that can be transported via a peptide permease, the derivative being a potentially toxic but normally impermeant moiety.) Providing the structure of the smugglin satisifies the structural specificities of the peptide permease, it is now clear that in order to trap the toxic moiety within the cell, efficient intracellular hydrolysis is important. Thus, it becomes crucial to determine the specificities of mammalian and bacterial peptidases, and in particular, to identify any differences that may exist. The development of fluorescence labelling techniques, in particular the continuous monitoring system based on an automated fluorescamine assay, have at last provided the means by which the molecular mechanism of peptide transport can be determined conveniently. The next several years should be an exciting and fruitful period in this area.

ACKNOWLEDGEMENTS

Financial support for the author's work was provided by the Science Research Council, the Royal Society, and Smith, Kline and French Foundation. I thank Drs. G. F. L. Ames and M. D. Watson for their helpful comments on parts of the manuscript.

REFERENCES

Allen, J. G., Atherton, F. R., Hall, M. J. Hassall, C. H., Holmes, S. W., Lambert, R. W., Nisbet, L. J., and Ringrose, P. S. (1978). Phosphonopeptides, a new class of synthetic antibacterial agents. *Nature, Lond.,* **272**, 56–58

Alper, M. D. and Ames, B. N. (1978). Transport of antibiotics and metabolite analogs by systems under cyclic AMP control: positive selection of *Salmonella typhimurium cya* and *crp* mutants. *J. Bacteriol.,* **133**, 149–157

Ames, G. F. L. (1964). Uptake of amino acids by *Salmonella typhimurium. Arch. Biochem. Biophys.,* **104**, 1–18

Ames, B. N., Ames, G. F. L., Young, J. D., Isuchiya, D., and Lecocq, J. (1973). Illicit transport, the oligopeptide permease. *Proc. Natl. Acad. Sci. U.S.A.,* **70**, 456–458.

Ananthaswamy, H. N. (1977). Release of periplasmic enzymes from *Escherichia coli* by penicillin ethylenediaminetetra-acetate treatment. *J. Bacteriol.,* **131**, 710–712

Anderson, R. P. and Roth, J. R. (1977). Tandem genetic duplications in phage and bacteria. *Annu. Rev. Microbiol.,* **31**, 473–505

Atherton, F. R., Hall, M. J., Hassall, C. H., Lambert, R. W., and Ringrose, P. S. (1979*a*). Phosphonopeptides as antibacterial agents: rationale, chemistry, and structure–activity relationships. *Antimicrobial Agents Chemother,* **15**, 677–683

Atherton, F. R., Hall, M. J., Hassall, C. H., Lambert, R. W., Lloyd, W. J., and Ringrose, P. S. (1979*b*). Phosphonopeptides as antibacterial agents: mechanism of action of alaphosphin. *Antimicrobial Agents Chemother.,* **15**, 696–705

Bachman, B. J., Low, K. B., and Taylor, A. L. (1976). Recalibrated linkage map of *Escherichia coli* K-12. *Bacteriol. Rev.,* **40**, 116–167

Barak, Z. and Gilvarg, C. (1974). Triornithine resistant strains of *E. coli*: Isolation, definition and genetic studies, *J. Biol. Chem.,* **249**, 143–148

Barak, Z. and Gilvarg, C. (1975*a*). Peptide transport. In *Biomembranes*, Vol. 7 (H. Eisenberg, E. Katchalski-Katzir, and L. A. Manson, eds), Plenum, New York, pp. 167–218

Barak, Z. and Gilvarg, C. (1975*b*). Specialized peptide transport system in *Escherichia coli. J. Bacteriol.,* **122**, 1200–1207

Bassford, P. J., Deidrich, D. L., Schnaitman, C. L., and Reeves, P. (1977). Outer membranes of *Escherichia coli*. VI. Protein alteration in bacteriophage-resistant mutants. *J. Bacteriol.,* **131**, 608–622

Bavoil, P., Nikaido, H., and von Meyenberg, K. (1977). Pleiotropic transport mutants of *Escherichia coli* lack porin, a major outer membrane protein. *Mol. Gen. Genet.,* **158**, 23–33

Bayer, M. E. (1968). Areas of adhesion between wall and membrane of *Escherichia coli. J. Gen. Microbiol.,* **53**, 395–404

Bayer, M. E. (1974). Ultrastructure and organization of the bacterial envelope. *Ann. N.Y. Acad. Sci.,* **235**, 6–28

Becker, J. M. and Naider, F. (1974). Stereospecificity of tripeptide utilization in a methionine auxotroph of *Escherichia coli* K-12. *J. Bacteriol.,* **120**, 191–196.

Bell, G., Payne, G. M., and Payne, J. W. (1977). Monitoring enzyme synthesis as a means of studying peptide transport and utilization in *Escherichia coli. J. Gen. Microbiol.,* **98**, 485–491

Benz, R., Boehler-Kohler, B. A., Dieterle, R., and Boos, W. (1978*a*). Porin activity in the osmotic shock fluid of *Escherichia coli. J. Bacteriol.,* **135**, 1080–1090

Benz, R., Janko, K., Boos, W., and Lauger, P. (1978*b*). Formation of large, ion-permeable membrane channels by the matrix protein (porin) of *Escherichia coli*. *Biochim. Biophys. Acta*, **511**, 305–319.

Bragg, P. D. and Hou, C. (1972). Organization of proteins in the native and reformed outer membrane of *Escherichia coli*. *Biochim. Biophys. Acta*, **274**, 478–488

Braun, V. (1975). Covalent lipoprotein from outer membrane of *Escherichia coli*. *Biochim. Biophys. Acta*, **415**, 335–377.

Braun, V. (1978). Structure–function relationships of the Gram-negative bacterial cell envelope. *Symp. Soc. Gen. Microbiol.*, **28**, 111–138

Braun, V. and Bosch, V. (1972). Repetitive sequences in the murein lipoprotein of the cell wall of *Escherichia coli*. *Proc. Natl. Acad. Sci. U.S.A.*, **69**, 970–974.

Braun, V. and Hantke, K. (1974). Biochemistry of bacterial cell envelopes. *Annu. Rev. Biochem.*, **43**, 89–121

Brock, T. D. and Wooley, S. O. (1964). Glycylglycine uptake in Streptococci and a possible role of peptides in amino acid transport. *Arch. Biochem. Biophys.*, **105**, 51–57.

Brown, J. L. (1973). Purification and properties of dipeptidase M from *Escherichia coli* B. *J. Biol. Chem.*, **248**, 409–416

Burston, D., Marrs, T. C., Sleisenger, M. H., Sopanen, T., and Matthews, D. M. (1977). Mechanisms of peptide transport. In *Peptide Transport and Hydrolysis*, CIBA Foundation Symposium, Vol. 50, Associated Scientific Publishers, Amsterdam, pp. 79–108

Cascieri, T. and Mallette, M. F. (1974). New method for study of peptide transport in bacteria. *Appl. Microbiol.*, **27**, 457–463

Cascieri, T. and Mallette, M. F. (1976*a*). Peptide utilization by *Pseudomonas putida* and *Pseudomonas maltophilia*. *J. Gen. Microbiol.*, **92**, 283–295.

Cascieri, T. and Mallette, M. F. (1976*b*). Intracellular peptide hydrolysis by *Pseudomonas putida* and *Pseudomonas maltophilia*. *J. Gen. Microbiol.*, **92**, 296–303.

Cerny, G. (1976). Method for distinction of Gram-negative and Gram-positive bacteria. *Eur. J. Appl. Microbiol.*, **3**, 223–225

Chappelet-Tordo, D., Lazdunski, C., Murgier, M., and Lazdunski, A. (1977). Aminopeptidase N from *Escherichia coli*: ionizable active-center groups and substrate specificity. *Eur. J. Biochem.*, **81**, 299–305

Choules, G. L. and Gray, W. L. (1971). Peptidase activity in the membranes of *Mycoplasma laidlawii*. *Biochem. Biophys. Res. Commun.*, **45**, 849–855

CIBA Foundation Symposium, Vol. 4 (1972). *Peptide Transport in Bacteria and Mammalian Gut*, Associated Scientific Publishers, Amsterdam

CIBA Foundation Symposium, Vol. 50 (1977). *Peptide Transport and Hydrolysis*, Associated Scientific Publishers, Amsterdam

Costerton, J. W., Ingram, J. M., and Cheng, K. J. (1974). Structure and function of the cell envelope of Gram-negative bacteria. *Bacteriol. Rev.*, **38**, 87–110

Cowell, J. L. (1974). Energetics of glycylglycine transport in *Escherichia coli*. *J. Bacteriol.*, **120**, 139–146

Datta, D. R., Arden, B., and Henning, U. (1977). Major proteins of the *Escherichia coli* outer cell envelope membrane as bacteriophage receptors. *J. Bacteriol.*, **131**, 821–829.

Decad, G. M. and Nikaido, H. (1976). Outer membrane of Gram-negative bacteria. XII. Molecular-sieving function of cell wall. *J. Bacteriol.*, **128**, 325–336

De Felice, M., Guardiola, L., Lamberti, A., and Iaccarino, M. (1973). *Escherichia coli* K-12 mutants altered in the transport systems for oligo- and dipeptides. *J. Bacteriol.*, **116**, 751–756

Deutch, C. E. and Soffer, R. L. (1978). *Escherichia coli* mutants defective in dipeptidyl carboxypeptidase. *Proc. Natl. Acad. Sci. U.S.A.*, **75**, 5998–6001

Diddens, H., Zahner, H., Kraas, E., Gohring, W., and Jung, G. (1976). On the transport of tripeptide antibiotics in bacteria. *Eur. J. Biochem.*, **66**, 11–23

Di Masi, D. R., White, J. C., Schnaitman, C., and Bradbeer, C. (1973). Transport of vitamin B_{12} in *Escherichia coli*: common receptor sites for vitamin B_{12} and the E colicins on the outer membrane of the cell envelope. *J. Bacteriol.*, **115**, 506–513

Di Rienzo, J. M., Nakamura, K., and Inouye, M. (1978). The outer membrane proteins of Gram-negative bacteria; biosynthesis, assembly, and functions. *Annu. Rev. Biochem.*, **47**, 481–532

Dunn, F. W., Humphreys, J., and Shive, W. (1957). Utilization of tripeptides. *Arch. Biochem. Biophys.*, **71**, 475–476

Exterkate, F. A. (1977). Pyrrolidone carboxylyl peptidase in *Streptococcus cremoris*: dependence on interaction with membrane components. *J. Bacteriol.*, **129**, 1281–1288

Fickel, T. E. and Gilvarg, C. (1973). Transport of impermeant substances in *E. coli* by way of the oligopeptide permease. *Nature New Biol.*, **241**, 161–163

Folk, W. R. and Berg, P. (1971). Duplication of the structural gene for glycyl-transfer RNA synthetase in *Escherichia coli*. *J. Mol. Biol.*, **58**, 595–610

Garrard, W. T. (1972). Synthesis, assembly and localization of periplasmic cytochrome *c*. *J. Biol. Chem.*, **247**, 5935–5943

Garten, W. and Henning, U. (1974). Cell envelope and shape of *Escherichia coli* K-12: isolation and preliminary characterization of the major ghost-membrane proteins. *Eur. J. Biochem.*, **47**, 343–352

Giesbrecht, P., Wecke, J., Reinicke, B., and Tesche, B. (1977). The demonstration of the existence of an interlayer between the cytoplasmic membrane and the cell wall proper of *Staphylococci*. *Arch. Microbiol.*, **115**, 25–35

Gilvarg, C. and Katchalski, E. (1965). Peptide utilization in *Escherichia coli*. *J. Biol. Chem.*, **240**, 3093–3098

Gilvarg, C. and Levin, Y. (1972). Response of *Escherichia coli* to ornithyl peptides. *J. Biol. Chem.*, **247**, 543–549

Glansdorf, N. and Sand, G. (1968). Duplication of a gene belonging to an arginine operon of *Escherichia coli* K-12. *Genetics*, **60**, 257–268.

Gratia, J. P. (1966). Studies on defective lysogeny due to chromosomal deletion in *Escherichia coli* I. Single lysogens. *Biken. J.*, **9**, 77–78.

Hantke, K. (1976). Phage T6-colicin K receptor and nucleoside transport in *Escherichia coli*. *FEBS Lett.*, **70**, 109–111

Hantke, K. and Braun, V. (1975). Membrane receptor dependent iron transport in *Escherichia coli*. *FEBS Lett.*, **49**, 301–315

Hayman, S., Gatmaitan, J. S., and Patterson, E. K. (1974). The relationship of extrinsic and intrinsic metal ions to the specificity of a dipeptidase from *Escherichia coli* B. *Biochemistry*, **13**, 4486–4494

Heppel, L. A. (1971). The concept of periplasmic enzymes. In *Structure and Function of Biological Membranes* (L. I. Rothfield, ed.), Academic Press, New York and London, pp. 223–247

Heppel, L. A., Rosen, B. P., Friedberg, I., Berger, E. A., and Weiner, J. H. (1972). Studies on binding proteins, periplasmic enzymes and active transport in *Escherichia coli*. In *The Molecular Basis of Biological Transport*, Miami Winter Symposia, Vol. 3 (F. J. Woessner and F. Huiging, eds), Academic Press, New York and London, pp. 133–156

Hermsdorf, C. L. (1978). Tripeptide-specific aminopeptidase from *Escherichia coli* AJ005. *Biochemistry*, **17**, 3370–3376

Hermsdorf, C. L., Simmonds, S., and Saunders, A. (1979). Soluble di- and aminopeptidases in *Escherichia coli* K-12: dispensible enzymes. *Int. J. Peptide Protein Chem.*, **13**, 146–151

Higgins, C. F. and Payne, J. W. (1977). Characterization of active dipeptide transport by germinating barley embryos: effects of pH and metabolic inhibitors. *Planta*, **136**, 71–76

Hirshfield, I. N. and Price, M. B. (1975). Utilization of selected leucine peptide amides by *Escherichia coli*. *J. Bacteriol.*, **122**, 966–975

Horiuchi, T., Horiuchi, S., and Novick, A. (1963). The genetic basis of hyper-synthesis of β-galactosidase. *Genetics*, **48**, 157–169

Ichihara, S. and Mizushima, S. (1978). Identification of an outer membrane protein responsible for the binding of the Fe–enterochelin complex to *Escherichia coli* cells. *J. Biochem.*, **83**, 137–140

Inouye, H. and Beckwith, J. (1977). Synthesis and processing of an *Escherichia coli* alkaline phosphatase precursor *in vitro*. *Proc. Natl. Acad. Sci. U.S.A.*, **74**, 1440–1444

Inouye, M. (1974). A three-dimensional molecular assembly model of a lipoprotein from *Escherichia coli* outer membrane. *Proc. Natl. Acad. Sci. U.S.A.*, **71**, 2396–2400

Jackson, M. B., Becker, J. M., Steinfield, A. S., and Naider, F. (1976). Oligopeptide transport in proline peptidase mutants of *Salmonella typhimurium*. *J. Biol. Chem.*, **251**, 5300–5309

Kadner, R. J. (1978). Repression of synthesis of the vitamin B_{12} receptor in *Escherichia coli*. *J. Bacteriol.*, **136**, 1050–1057

Kadner, R. J. and Bassford, P. J. (1978). The role of the outer membrane in active transport.

In *Bacterial Transport* (B. P. Rosen, ed.), Marcel Dekker, New York and Basel, pp. 413–463

Kenig, M. and Abraham, E. P. (1976). Antimicrobial activities and antagonists of bacilysin and anticapsin. *J. Gen. Microbiol.*, **94**, 37–45

Kenig, M., Vandamme, E., and Abraham, E. P. (1976). The mode of action of bacilysin and anticapsin and biochemical properties of bacilysin-resistant mutants. *J. Gen. Microbiol.*, **94**, 46–54

Kepes, A. and Cohen, G. N. (1962). Permeation. In *The Bacteria*, Vol. V (I. C. Gunsalus and R. Y. Stanier, eds), Academic Press, New York, pp. 179–221

Kessel, D. and Lubin, M. (1963). On the distinction between peptidase activity and peptide transport. *Biochim. Biophys. Acta,* **71**, 656–663

Kihara, H. and Snell, E. E. (1960*a*). Peptides and bacterial growth. VIII. The nature of strepogenin. *J. Biol. Chem.*, **235**, 1409–1414

Kihara, H. and Snell, E. E. (1960*b*). Peptides and bacterial growth. IX. Release of double inhibitions by single peptides. *J. Biol. Chem.*, **235**, 1415–1418

Kihara, H., Ikawa, M., and Snell, E. E. (1961). Peptides and bacterial growth. X. Relation of uptake and hydrolysis to utilization of D-alanine peptides for growth of *Streptococcus faecalis*. *J. Biol. Chem.*, **236**, 172–178

Kirsh, M., Dembinski, D. R., Hartman, P. E., and Miller, C. G. (1978). *Salmonella typhimurium* peptidase active on carnosine. *J. Bacteriol.*, **134**, 361–374

Latil, M., Murgier, M., Lazdunski, A., and Lazdunski, C. (1976). Isolation and genetic mapping of *Escherichia coli* aminopeptidase mutants. *Mol. Gen. Genet.*, **148**, 43–47

Law, B. A. (1977). Dipeptide utilization by starter Streptococci. J. Dairy Res., **44**, 309–317

Law, B. A. (1978). Peptide utilization by Group N Streptococci. *J. Gen. Microbiol.*, **105**, 113–118

Lazdunski, A., Murgier, M., and Lazdunski, C. (1975*a*). Evidence for an aminoendopeptidase localized near the cell surface of *Escherichia coli*: regulation of synthesis by inorganic phosphate. *Eur. J. Biochem.*, **60**, 349–355

Lazdunski, C., Basuttil, J., and Lazdunski, A. (1975*b*). Purification and properties of a periplasmic aminoendopeptidase from *Escherichia coli*. *Eur. J. Biochem.*, **60**, 363–369

Lazdunski, A., Pellissier, C., and Lazdunski, C. (1975*c*). Regulation of *Escherichia coli* K-10 aminoendopeptidase activity: effects of mutations involved in the regulation of alkaline phosphatase. *Eur. J. Biochem.*, **60**, 357–362

Leach, F. R. and Snell, E. E. (1959). Occurrence of independent uptake mechanisms for glycine and glycine peptides in *Lactobacillus casei*. *Biochim. Biophys. Acta,* **34**, 292–293

Leach, F. R. and Snell, E. E. (1960). The absorption of glycine and alanine and their peptides by *Lactobacillus casei*. *J. Biol. Chem.*, **235**, 3523–3531

Lev, M. (1977). Casamino acids enhance growth of *Bacteroides melaninogenicus*. *J. Bacteriol.* **129**, 562–563

Levine, E. M. and Simmonds, S. (1960). Metabolite uptake by serine-glycine auxotrophs of *Escherichia coli*. *J. Biol. Chem.*, **235**, 2902–2909

Levine, E. M. and Simmonds, S. (1962*a*). Effect of cultural conditions on growth and metabolite uptake by serine–glycine auxotrophs of *Escherichia coli*. *J. Bacteriol.*, **84**, 683–693

Levine, E. M. and Simmonds, S. (1962*b*). Further studies on metabolite uptake by serine–glycine auxotrophs of *Escherichia coli*. *J. Biol. Chem*., **237**, 3718–3724

Lin, J. K. and Chang, J. Y. (1975). Chromophoric labeling of amino acids with 4-dimethylaminoazobenzene-4′-sulfonyl chloride *Analyt. Chem.*, **47**, 1634–1638

Litchfield, C. D. and Prescott, J. M. (1970). Regulation of proteolytic enzyme production by *Aeromonas proteolytica*. II. Extracellular aminopeptidase. *Can. J. Microbiol.*, **16**, 23–28

Losick, R. and Gilvarg, C. (1966). Effect of α-acetylation on utilization of lysine oligopeptides in *Escherichia coli*. *J. Biol. Chem.*, **241**, 2340–2346

Louvard, D., Semeriva, M., and Maroux, S. (1976). The brush border intestinal aminopeptidase, a transmembrane protein as probed by macromolecular photolabelling. *J. Mol. Biol.*, **106**, 1023–1035.

Lutgenberg, B., Bronstein, H., van Selm, N., and Peters, R. (1977). Peptidoglycan associated outer membrane proteins in Gram-negative bacteria. *Biochim. Biophys. Acta,* **465**, 571–578

Lutgenberg, B., Peters, R., Bernheimer, H., and Berendsen, W. (1976). Influence of cultural conditions and mutations on the composition of the outer membrane proteins of *Escherichia coli. Mol. Gen. Genet.*, **147**, 251–262

Lutgenberg, B., Meijers, J., Peters, R., van der Hoek, P., and van Alphen, L. (1975). Electrophoretic resolution of the major outer membrane protein of *Escherichia coli* K-12 into four bands. *FEBS Lett.,* **58**, 254–258

Lutkenhaus, J. F. (1977). Role of a major outer membrane protein in *Escherichia coli. J. Bacteriol.*, **131**, 631–637

McAllister, T. J., Irvin, R. T., and Costerton, J. W. (1977). Cell surface-localized alkaline phosphatase of *Escherichia coli* as visualized by reaction product deposition and ferritin-labeled antibodies. *J. Bacteriol.,* **130**, 318–328

McHugh, G. L. and Miller, C. G. (1974). Isolation and characterization of proline peptidase mutants of *Salmonella typhimurium. J. Bacteriol.,* **120**, 364–371

Marder, R., Becker, J. M., and Naider, F. (1977). Peptide transport in yeast. Utilization of leucine- and lysine-containing peptides by *Saccharomyces cerevisiae. J. Bacteriol.*, **131**, 906–916

Matthews, D. M. (1975*a*). Absorption of peptides by mammalian intestine. In *Peptide Transport in Protein Nutrition* (D. M. Matthews and J. W. Payne, eds), North Holland and American Elsevier, Amsterdam, Oxford, New York, pp. 61–146

Matthews, D. M. (1975*b*). Intestinal absorption of peptides. *Physiol. Rev.,* **55**, 537–608

Matthews, D. M. and Adibi, S. A. (1976). Peptide absorption. *Gastroenterology,* **71**, 151–161

Matthews, D. M. and Payne, J. W. (eds) (1975*a*). *Peptide Transport in Protein Nutrition*, North-Holland and American Elsevier, Amsterdam, Oxford, New York

Matthews, D. M. and Payne, J. W. (1975*b*). Peptides in the nutrition of microorganisms and peptides in relation to animal nutrition. In *Peptide Transport in Protein Nutrition* (D. M. Matthews and J. W. Payne, eds), North Holland and American Elsevier, Amsterdam, Oxford, New York, pp. 1–60

Matthews, D. M. and Payne, J. W. (1980*a*). Membrane transport of small peptides. *Curr. Topics Membranes Transport*, **14**, 331–425.

Matthews, D. M. and Payne, J. W. (1980*b*). Peptide transport. In *CRC Handbook of Nutrition and Food*, in press.

Meisler, N. and Simmonds, S. (1963). The metabolism of glycyl-L-leucine by *Escherichia coli. J. Gen. Microbiol.,* **31**, 109–117

Miller, C. G. (1975*a*). Genetic mapping of *Salmonella typhimurium* peptidase mutations. *J. Bacteriol.*, **122**, 171–176

Miller, C. G. (1975*b*). Peptidases and proteases of *Escherichia coli* and *Salmonella typhimurium. Annu. Rev. Microbiol.,* **29**, 485–504

Miller, C. G. and Mackinnon, K. (1974). Peptidase mutants of *Salmonella typhimurium. J. Bacteriol.*, **120**, 355–363

Miller, C. G. and Schwartz, G. (1978). Peptidase-deficient mutants of *Escherichia coli. J. Bacteriol.*, **135**, 603–611

Miller, R. V. and Becker, J. M. (1978). Peptide utilization in *Pseudomonas aeruginosa*: evidence for membrane-associated peptidase. *J. Bacteriol.,* **133**, 165–171

Mostillo, R. and Gilvarg, C. (1977). Determinants of peptide binding to the oligopeptide transport system of *Escherichia coli*. *Fed. Proc. Fed. Am. Soc. Exp. Biol*., **36**, 826

Murgier, M., Lazdunski, A., and Lazdunski, C. (1976*b*). Control of aminopeptidase and alkaline phosphatase synthesis in *Escherichia coli*: evidence for a regulation by inorganic phosphate independent of *phoR* and *phoB* phosphatase regulatory genes, *FEBS Lett.,* **64**, 130–134

Murgier, M., Pelissier, C., Lazdunski, A., and Lazdunski, C. (1976*a*). Existence, localization and regulation of the biosynthesis of aminoendopeptidases in Gram-negative bacteria. *Eur. J. Biochem.,* **65**, 517–520

Murgier, M., Pelissier, C., Lazdunski, A., Bernadac, A., and Lazdunski, C. (1977). Aminopeptidase N from *Escherichia coli*: unusual interactions with the cell surface. *Eur. J. Biochem.,* **74**, 425–433

Naider, F. and Becker, J. M. (1975). Multiplicity of oligopeptide transport systems in *Escherichia coli. J. Bacteriol.,* **122**, 1208–1215

Naider, F., Becker, J. M. and Katzir-Katchalski, E. (1974). Utilization of methionine-containing peptides and their derivatives by a methionine-requiring auxotroph of *Saccharomyces cerevisiae. J. Biol. Chem.,* **249**, 9–20

Nakae, T. (1975). Outer membrane of *Salmonella typhimurium*: reconstitution of sucrose-permeable membrane vesicles. *Biochem. Biophys. Res. Commun.*, **64**, 1224–1230

Nakae, T. (1976*a*). Identification of the outer membrane protein of *E. coli* that produces transmembrane channels in reconstituted vesicle membranes. *Biochem. Biophys. Res. Commun.*, **71**, 877–884

Nakae, T. (1976*b*). Outer membrane of *Salmonella*. Isolation of protein complex that produces transmembrane channels. *J. Biol. Chem.*, **251**, 2176–2178

Nakae, T. and Nikaido, H. (1975). Outer membrane as a diffusion barrier in *Salmonella typhimurium*. Penetration of oligo- and polysaccharides into isolated outer membrane and cells with degraded peptidoglycan layer. *J. Biol. Chem.*, **250**, 7359–7365

Neuhaus, F. C., Goyer, S., and Neuhaus, D. W. (1977). Growth inhibition of *Escherichia coli* W by D-norvalyl-D-alanine: an analogue of D-alanine in position 4 of the peptide subunit of peptidoglycan. *Antimicrobial Agents Chemother.*, **11**, 638–644

Nikaido, H. (1976). Outer membrane of *Salmonella typhimurium*: transmembrane diffusion of some hydrophobic substances. *Biochim. Biophys. Acta,* **433**, 118–132

Nikaido, H., Bovoil, P., and Hirota, Y. (1977*b*). Outer membranes of Gram-negative bacteria. XV. Transmembrane diffusion rates in lipoprotein-deficient mutants of *Escherichia coli. J. Bacteriol.,* **132**, 1045–1047

Nikaido, H., Song, S. A., Shaltiel, L., and Nurminen, M. (1977*a*). Outer membrane of *Salmonella*. XIV. Reduced transmembrane diffusion rates in porin-deficient mutants. *Biochem. Biophys. Res. Commun.,* **76**, 324–330

Nisbet, T. M. and Payne, J. W. (1979*a*). Peptide uptake in *Saccharomyces cerevisiae*: characteristics of transport system shared by di- and tripeptides. *J. Gen. Microbiol.*, **115**, 127–133

Nisbet, T. M. and Payne, J. W. (1979*b*). Specificity of peptide uptake in *Saccharomyces cerevisiae*, and isolation of a bacilysin-resistant, peptide-transport-deficient mutant. *FEMS Lett.*, **6**, 193–196

Noren, O., Sjostrom, H., Svenson, B., Jeppesen, L., Staun, M., and Josefsson, L. (1977). Intestinal brush border peptidases. In *Peptide Transport and Hydrolysis*, CIBA Foundation Symposium Vol. 50, Associated Scientific Publishers, Amsterdam, pp. 177–197

Palva, E. T. (1978). Major outer membrane protein in *Salmonella typhimurium* induced by maltose. *J. Bacteriol.,* **136**, 286–294

Patterson, E. K., Gatmaitan, J. S., and Hayman, S. (1973). Substrate specificity and pH dependence of dipeptidases purified from *Escherichia coli* and mouse ascites tumor cells. *Biochemistry,* **12**, 3701–3709

Payne, J. W. (1968). Oligopeptide transport in *Escherichia coli*: specificity with respect to side chain and distinction from dipeptide transport. *J. Biol. Chem.*, **243**, 3395–3403

Payne, J. W. (1971*a*). The requirement for the protonated α-amino group for the transport of peptides in *Escherichia coli. Biochem. J.*, **123**, 245–253

Payne, J. W. (1971*b*). The utilization of prolyl peptides by *Escherichia coli. Biochem. J.,* **123**, 255–260

Payne, J. W. (1972*a*). Mechanisms of bacterial peptide transport. In *Peptide Transport in Bacteria and Mammalian Gut*, CIBA Foundation Symposium, Vol. 4, Associated Scientific Publishers, Amsterdam, pp. 17–32

Payne, J. W. (1972*b*). Variations in the peptidase activities of *Escherichia coli* in response to environmental changes. *J. Gen. Microbiol.*, **71**, 281–289

Payne, J. W. (1972*c*). Effects of *N*-methyl peptide bonds on peptide utilization by *Escherichia coli. J. Gen. Microbiol.,* **71**, 259–265

Payne, J. W. (1973). Peptide utilization in *Escherichia coli*: studies with peptides containing β-alanyl residues. *Biochim. Biophys. Acta,* **298**, 469–478

Payne, J. W. (1974). Peptide transport in *E. coli*: permease specificity towards terminal amino group substituents. *J. Gen. Microbiol.,* **80**, 269–276

Payne, J. W. (1975*a*). Transport of peptides in microorganisms. In *Peptide Transport in Protein Nutrition* (D. M. Matthews and J. W. Payne, eds), North Holland and American Elsevier, Amsterdam, Oxford, New York, pp. 283–364

Payne, J. W. (1975*b*). Microbial peptidohydrolases. In *Peptide Transport in Protein Nutrition* (D. M. Matthews and J. W. Payne, eds), North Holland and American Elsevier, Amsterdam, Oxford, New York, pp. 365–391.

Payne, J. W. (1976). Peptides and microorganisms. *Adv. Microbial. Physiol.*, **13**, 55–113

Payne, J. W. (1977). Transport and hydrolysis of peptides by microorganisms. In *Peptide Transport and Hydrolysis*, CIBA Foundation Symposia, Vol. 50, Associated Scientific Publishers, Amsterdam, pp. 305–334

Payne, J. W. (1979). Microbial peptidases. In *CRC Handbook of Nutrition and Food*, in press.

Payne, J. W. and Bell, G. (1977*a*). A radioisotope incorporation method for studying the transport and utilization of peptides by *Escherichia coli. FEMS Lett.,* **1**, 91–94

Payne, J. W. and Bell, G. (1977*b*). The transport and utilization of peptides by *Escherichia coli*: interrelations between peptide uptake and amino acid exodus and biosynthesis. *FEMS Lett.,* **2**, 259–262

Payne, J. W. and Bell, G. (1977*c*). Substrate interactions during transport of amino acids and peptides by *Escherichia coli*: relevance to regulation of peptide uptake. *FEMS Lett.,* **2**, 301–304

Payne, J. W. and Bell, G. (1979). Direct determination of the properties of peptide transport systems in *Escherichia coli* using a fluorescent labelling procedure. *J. Bacteriol.,* **137**, 447–455

Payne, J. W. and Gilvarg, C. (1968*a*). Size restriction on peptide utilization in *Escherichia coli. J. Biol. Chem.,* **243**, 6291–6299

Payne, J. W. and Gilvarg, C. (1968*b*). Role of the terminal carboxyl group in peptide transport in *Escherichia coli. J. Biol. Chem.,* **243**, 335–340

Payne, J. W. and Gilvarg, C. (1971). Peptide transport. *Adv. Enzymol. Relat. Areas Molec. Biol.,* **35**, 187–244

Payne, J. W. and Gilvarg, C. (1978).Transport of peptides in bacteria. In *Bacterial Transport* (B. P. Rosen, ed.), Marcel Dekker, New York, pp. 325–383

Pecht, E., Giberman, E., Keysary, A., Yariv, J., and Katchalski, E. (1972). Hydrolysis of alanine oligopeptides by an enzyme located in the membrane of *Mycoplasma laidlawii. Biochim. Biophys. Acta,* **290**, 267–273

Perrett, D., Webb, J. P. W., Silk, D. B. A., and Clark, M. L. (1975). The assay of dipeptides using fluorescamine and its application to determining dipeptidase activity. *Analyt. Biochem.,* **68**, 161–166

Pickart, L. and Thaler, M. M. (1973). Tripeptide in human serum which prolongs survival of normal liver cells and stimulates growth in neoplastic liver. *Nature New Biol.,* **243**, 85

Pittman, K. A. and Bryant, M. P. (1964). Peptides and other nitrogen sources for growth of *Bacteroides ruminicola. J. Bacteriol.,* **88**, 401–410.

Pittman, K. A., Lakshmanan, S., and Bryant, M. P. (1967). Oligopeptide uptake by *Bacteroides ruminicola. J. Bacteriol.,* **93**, 1499–1508

Postle, K. and Reznikoff, W. S. (1978). Hind II and Hind III restriction maps of the attΦ80-tonB-trp region of the *Escherichia coli* genome and location of the *tonB* gene. *J. Bacteriol.,* **136**, 1165–1173

Rice, G. H., Stewart, F. H. C., Hillier, A. J., and Jago, G. R. (1978). The uptake of amino acids and peptides by *Streptococcus lactis. J. Dairy Res.,* **45**, 93–107

Rigby, P. W. J., Burleigh, B. D., and Hartley, B. S. (1974). Gene duplication in experimental enzyme evolution. *Nature, Lond.,* **251**, 200–204

Robinson, A. and Tempest, D. W. (1973). Phenotypic variability of the envelope proteins of *Klebsiella aerogenes. J. Gen. Microbiol.,* **78**, 361–370

Rodriguez-Absi, J. and Prescott, J. M. (1978). Isolation and properties of an aminopeptidase from *Bacillus licheniformis. Arch. Biochem. Biophys.,* **186**, 383–391

Rosen, B. P. and Heppel, L. A. (1973). In *Bacterial Membranes and Walls*, Vol. 1 (L. Leive, ed.), Marcel Dekker, New York, pp. 131–208

Rosenbusch, J. (1974). Characterization of the major envelope protein from *E. coli*. Regular arrangement on the peptidoglycan and unusual dodecylsulphate binding. *J. Biol. Chem.,* **249**, 8019–8029

Rothfield, L. and Pearlman-Kothencz, M. (1969). Synthesis and assembly of bacterial membrane components: a lipopolysaccharide–phospholipid–protein complex excreted by living bacteria. *J. Mol. Biol.,* **44**, 477–492

Savage, D. C. (1977). Microbial ecology of the gastrointestinal tract. *Annu. Rev. Microbiol.,* **31**, 107–133

Scherrer, R. and Gerhardt, P. (1971). Molecular sieving by the *Bacillus megaterium* cell wall and protoplast. *J. Bacteriol.,* **107**, 718–735

Scherrer, R., Beaman, T. C., and Gerhardt, P. (1971). Macromolecular sieving by the dormant spore of *Bacillus cereus*. *J. Bacteriol.*, **108**, 868–873

Scherrer, R., Berlin, E., and Gerhardt, P. (1977). Density, porosity and structure of dried cell walls isolated from *Bacillus megaterium* and *Saccharomyces cerevisiae*. *J. Bacteriol.*, **129**, 1162–1164

Schnaitman, C. (1974). Outer membrane proteins of *Escherichia coli*. III. Evidence that the major protein of *Escherichia coli* O111 outer membrane consists of four distinct polypeptide species. *J. Bacteriol.*, **118**, 442–453

Shankman, S., Gold, V., Higa, S., and Squires, R. (1962). On the mode of action of a peptide inhibitor of growth of *P. cerevisiae*. *Biochem. Biophys. Res. Commun.*, **9**, 25–31

Shelton, D. C. and Nutter, W. E. (1964). Uptake of valine and glycylvaline by *Leuconostoc mesenteroides*. *J. Bacteriol.*, **88**, 1175–1184

Shockman, G. D., Daneo-Moore, L., and Higgins, M. L. (1974). Problems of cell wall and membrane growth, enlargement, and division. *Ann. N.Y. Acad. Sci.*, **235**, 161–197

Simmonds, S. (1970). Peptidase activity and peptide metabolism in *Escherichia coli* K-12. *Biochemistry*, **9**, 1–9

Simmonds, S., Szeto, K. S., and Fletterick, C. G. (1976). Soluble tri- and dipeptidases in *Escherichia coli* K-12. *Biochemistry*, **15**, 261–271

Smith, R. L., Archer, E. G., and Dunn, F. W. (1970). Uptake of ^{14}C-labelled tri-, tetra-, and pentapeptides of phenylalanine and glycine by *Escherichia coli*. *J. Biol. Chem.*, **245**, 2967–2971

Smith, R. L. and Dunn, F. W. (1970). Bacterial utilization of oligopeptides containing β-2-thienylalanine and phenylalanine. *J. Biol. Chem.*, **245**, 2962–2966

Staros, J. V. and Knowles, J. R. (1978). Photoaffinity inhibition of dipeptide transport in *Escherichia coli*. *Biochemistry*, **17**, 3321–3325

Steven, A. C., ten Heggeler, B., Kistler, J., and Rosenbusch, J. P. (1977). Ultrastructure of a periodic protein layer in the outer membrane of *Escherichia coli*. *J. Cell Biol.*, **72**, 292–301.

Stock, J. B., Rauch, B., and Roseman, S. (1977). Periplasmic space in *Salmonella typhimurium* and *Escherichia coli*. *J. Biol. Chem.*, **252**, 7850–7861

Straus, D. S. (1974). Induction of mutagens of tandem gene duplications in *gly*S region of the *Escherichia coli* chromosome. *Genetics*, **78**, 823–830

Straus, D. S. and Straus, L. D. (1976). Large overlapping tandem genetic duplications in *Salmonella typhimurium*, *J. Mol. Biol.*, **103**, 143–153

Sussman, A. J. and Gilvarg, C. (1970). Peptidases in *Escherichia coli* K-12 capable of hydrolyzing lysine homopeptides. *J. Biol. Chem.*, **245**, 6518–6524

Sussman, A. J. and Gilvarg, C. (1971). Peptide transport and metabolism in bacteria. *Annu. Rev. Biochem.*, **40**, 397–408

Szmelcman, S., Schwartz, M., Silhavy, T. J., and Boos, W. (1976). Maltose transport in *Escherichia coli* K-12. *Eur. J. Biochem.*, **65**, 13–19

Tempest, D. W. (1970). The place of continuous culture in microbiological research. *Adv. Microbial. Physiol.*, **4**, 223–250

Udenfriend, S. (1978). Fluorescent methods for isolation, characterization and assay of peptides and proteins. In *Versatility of Proteins*, (C. H. Li, ed.), Academic Press, New York and London, pp. 23–27

Uemura, J. and Mizushima, S. (1975). Isolation of outer membrane proteins of *Escherichia coli* and their characterization on polyacrylamide gels. *Biochim. Biophys. Acta*, **413**, 163–175

Ugolev, A. M., Timofeeva, N. M., Smirnova, L. F., De-Laey, P., Gruzdkov, A. A., Iezuitova. N. N., Mityushova, N. M., Roschina, G. M., Gurman, E. G., Gusev, V. M., Tsevetkova, V. A., and Shcherbakov, G. G. (1977). Membrane and intracellular hydrolysis of peptides: differentiation, role and interrelations with transport. In *Peptide Transport and Hydrolysis*, CIBA Foundation Symposia, Vol. 50, Associated Scientific Publishers, Amsterdam, pp. 221–245

Verhoef, C., Lugtenberg, B., van Boxtel, R., de Graff, P., and Verheij, H. (1979). Genetics and biochemistry of the peptidoglycan-associated proteins b and c in *Escherichia coli K-12*. *Mol. Gen. Genet*., **169**, 137–146

Vogt, V. M. (1970). Purification and properties of an aminopeptidase from *Escherichia coli*. *J. Biol. Chem.*, **245**, 4760–4769 von Meyenburg, K. (1971). Transport limited growth rates in a mutant of *Escherichia coli*. *J. Bacteriol.*, **107**, 878–888

Vonder Haar, R. A. and Umbarger, H. E. (1972). Isoleucine and valine metabolism in *Escherichia coli*. XIX. Inhibition of isoleucine biosynthesis by glycylleucine. *J. Bacteriol.*, **112**, 142–147

Wagner, F. W., Chung, A., and Ray, L. E. (1972). Characterization of the aminopeptidase from *Bacillus subtilis* as an extracellular enzyme. *Can. J. Microbiol.*, **18**, 1883–1887

Watson, R. R. (1976). Substrate specificities of aminopeptidases. *Methods Microbiol.*, **9**, 1–15

Wilkes, S. H., Bayliss, M. E., and Prescott, J. M. (1973). Specificity of *Aeromonas* aminopeptidase towards oligopeptides and polypeptides. *Eur. J. Biochem.*, **34**, 459–466

Woolley, D. W. and Merrifield, R. B. (1958). Specificity of peptides. *Science, N.Y.*, **128**, 238–240

Woolley, D. W. and Merrifield, R. B. (1963). Anomalies in the structural specificity of peptides. *Ann. N.Y. Acad. Sci.*, **104**, 161–171

Wright, D. E. (1967). Metabolism of peptides by rumen microorganisms. *Appl. Microbiol.*, **15**, 547–550

Yang, L. M. and Sommerville, R. (1976). Purification and properties of a new aminopeptidase from *Escherichia coli* K-12. *Biochim. Biophys. Acta*, **445**, 406–419

Yang, S. L., Becker, J. M., and Naider, F. (1977). Transport of [^{14}C]-Gly-Pro in a proline peptidase mutant of *Salmonella typhimurium*. *Biochim. Biophys. Acta*, **471**, 135–144

Yaron, A. and Berger, A. (1970). Aminopeptidase P. *Methods Enzymol.*, **19**, 521–534

Yaron, A., Mlynar, D., and Berger, A. (1972). A dipeptidocarboxypeptidase from *Escherichia coli*. *Biochem. Biophys. Res. Commun.*, **47**, 897–902

Yoder, O. C., Beamer, K. C., Cipollono, P. B., and Shelton, D. C. (1965*a*). Kinetic study of L-valine and glycyl-L-valine uptake by *Leuconostoc mesenteroides*. *Arch. Biochem. Biophys.*, **110**, 336–343

Yoder, O. C., Beamer, K. C., and Shelton, D. C. (1965*b*). Structural and stereochemical specificity in transport systems for glycine, valine and their dipeptides in *Leuconostoc mesenteroides*. *Fed. Proc. Fed. Am. Soc. Exp. Biol.*, **24**, 352

Young, E. A., Bowen, D. O., and Diehl, J. F. (1964). Transport studies with peptides containing unnatural amino acids. *Biochem. Biophys. Res. Commun.*, **14**, 250–255

Microorganisms and Nitrogen Sources
Edited by J. W. Payne

Section 2.2

Transport and Utilization of Peptides by Yeast

JEFFREY M. BECKER

Department of Microbiology, University of Tennessee, Knoxville, Tennessee 37916, USA

and

Department of Biophysics, Weizmann Institute of Science, Rehovot, Israel

AND

FRED NAIDER

Chemistry Department, The College of Staten Island of the City University of New York, Staten Island, New York 10301, USA

and

Department of Biophysics, Weizmann Institute of Science, Rehovot, Israel

1 INTRODUCTION

The research which will be discussed in this chapter was conceived during the summer of 1971. By that time a considerable number of studies had been conducted on peptide transport and utilization in *Escherichia coli* (Payne and Gilvarg, 1971). In addition, there was also a growing body of information on peptide absorption through the gut in mammals (Elliot and O'Connor, 1972; Payne and Gilvarg, 1971). Several statements appeared in the literature which implied that peptide transport was a general phenomenon which would be characterized by similar features in a variety of cell types. We felt, however, that the difference between a simple prokaryote—*E. coli*—and a complex mammalian organ—the intestine—was large and that studies in a well-defined unicellular eukaryotic organism could provide valuable insights into peptide transport in higher cell types.

Yeasts were chosen as our model primarily because they were well characterized in many laboratories, could be grown on defined synthetic media, could be genetically manipulated, and were available as different amino acid auxotrophs. This final feature was especially important in preliminary studies as it allowed us rapidly to screen peptide transport by yeast using an indirect growth assay (Barak and Gilvarg, 1975) (see also Section 2.1). The stimulation of yeast growth by peptones, tryptones, and other plant and animal hydrolysates had been known for many years and gave credence to the idea that yeast could utilize peptides for growth.

The term 'yeasts' applies to a heterogeneous, ill-defined group of microorganisms. The diversity of yeast is illustrated by the fact that over 35 genera and over 350 species are recognized (Lodder, 1970). The features by which the species are distinguished are, for the most part, morphological and physiological. In general, yeasts are recognized by overall similarities and affinities rather than by constant definitive characteristics. By tradition, yeasts are limited to fungi in which the unicellular form is predominate, and vegetative reproduction is usually by budding. The ability to ferment many sugars and assimilate various nitrogen-containing compounds is typical of yeast.

Fortunately, within the heterogeneous collection of organisms regarded as yeasts, one species, *Saccharomyces cerevisiae*, has been the focus of most of the biochemical and physiological work, in part because of its widespread occurrence (Lund, 1958) and its industrial importance. *S. cerevisiae* fulfills all of the criteria listed above for the study of peptide transport in a unicellular, eukaryotic microorganism. A list of the various strains used in our studies appears in Table 1.

Our original goals were to characterize thoroughly the structural requirements for peptide uptake in yeast and to use that information to aid us in the isolation of components of the peptide transport system. After

Table 1 List of organisms used for peptide transport in yeasts

Organism and strain designation	Relevant phenotypic requirements	Source
Saccharomyces cerevisiae G1333	Methionine	G. Simchen, Hebrew University
Saccharomyces cerevisiae ATCC 9896 (Fleischmann strain 139)	Wild type	H. Lichstein, University of Cincinnati
Saccharomyces cerevisiae ZI-2D	Lysine and leucine	N. Khan, Brooklyn College
Saccharomyces cerevisiae XJB1-1C	Methionine	R. Mortimer, University of California, Berkley
Candida albicans WD 18-4	Methionine and lysine	A. Sarachak, Wichita State University

commencement of our studies the work of Fickel and Gilvarg (1973) and of Ames *et al.* (1973) opened up the area of drug design via 'illicit transport' through the peptide transport system (see Section 4.7). We therefore expanded our horizons and initiated studies on peptide uptake in a pathogenic yeast—*Candida albicans*. The scope of this chapter for the most part will be limited to a review of our progress in both of these areas. Since numerous reviews of peptide transport in bacteria (Barak and Gilvarg, 1975; Elliot and O'Connor, 1972; Payne, 1976; Payne and Gilvarg, 1971) and the gut (Barak and Gilvarg, 1975; Elliot and O'Connor, 1972; Matthews, 1975; Payne and Gilvarg, 1971) have appeared recently, no attempt will be made to discuss these topics systematically. After the initiation of our studies with yeast, a number of reports on peptide transport in another eukaryote—the fungus *Neurospora crassa*—have appeared (Wolfinbarger and Marzluf, 1974, 1975*a*,*b*). These studies have revealed a number of the structural factors involved in peptide uptake by *N. crassa*, and have provided evidence for a genetic basis for peptide transport in this fungus. The investigations with *N. crassa* are the subject of another chapter in this volume (Section 2.3) and will not be discussed further in our review.

2 YEAST CYTOLOGY

The cell envelope of yeast is composed of the protoplasmic membrane (also called cytoplasmic membrane, cell membrane, and plasmalemma), the cell wall, and the periplasmic space—the intervening region between the membrane and the wall (Matile *et al.*, 1969). For *Saccharomyces* species it appears that some enzymes are situated within the periplasmic space (Lampen, 1968; Matile *et al.*, 1969). Later in this review (Section 2.2.3) we will provide evidence that, in the strains we have examined, peptidases are not present in this region of the cell. The thick cell wall of yeast is responsible for the cell's resistance to mechanical stress; the cell membrane is the site of specific solute exchange with the medium.

Although the cell wall of *S. cerevisiae* has been studied for many years, a general agreement on the structure of this cell envelope component has still not been reached. This is due in part to the very complex and heterogeneous mixture of polymers which compose the cell wall, and to the different procedures and variety of sources used to isolate cell walls. The major cell wall components are believed to include glucan, mannan, lipid, and protein (Northcote and Horne, 1952; Phaff, 1971). Evidence exists which suggests that molecules of certain dimensions can pass through the cell wall of yeast. In a careful study, Scherrer *et al.* (1974) found that the cell wall of *S. cerevisiae* permitted the passive diffusion of a number of oligomers of ethylene oxide. Recent data of Arnold and Lacy (1977) has shown that sugars such as L-arabinose, lactose, mannitol, and pentaerythritol are able to permeate through cell walls. Furthermore, basic proteins such as cytochrome *c* (molecular mass approx. 12 000 daltons) somehow penetrate the cell wall and disrupt the plasma membrane (Svihla *et al.*, 1969). Treatment of *S. cerevisiae* with such proteins leads to extensive cell damage and causes the release into the surrounding medium of cellular constituents of molecular mass ≤10 000 daltons. However, the studies of Gerhardt also led to the conclusion that molecules larger than 620 daltons were excluded by a filtering effect of the cell wall (Scherrer *et al.*, 1974), and it was shown in another investigation that *S. cerevisiae* was unable to utilize proteins as the sole source of nitrogen. The studies cited clearly illustrate the uncertainty which exists concerning the action of the yeast cell wall as a barrier to the free diffusion of molecules. This confusion makes it difficult to interpret the results obtained from investigations of the chain length dependence of peptide transport. This dependence will be discussed in detail later (Section 2.2.9).

In contrast to the uncertainty concerning the sieving action of the yeast cell wall there appears to be little doubt that the yeast cell membrane functions as a protective barrier for the protoplast and serves to control solute entry and departure to and from the cell. The cell membrane contains many proteins of known enzymic activity (Matile, 1970; Nurminen and Suomalainen, 1973; Nurminen *et al.*, 1970) such as ATPases and enzymes involved in cell wall synthesis, and it is assumed to be the site of transport components. We have found compelling evidence that the system mediating peptide uptake in yeast has features characteristic of those found for both amino acid and sugar transport systems. Peptide transport in yeast shows saturation kinetics (K_m for $(Met)_3$ uptake equals 7.7×10^{-5} M), substrate specificity and stereospecificity, and the velocity of transport is dependent on both pH and temperature (Becker and Naider, 1977). Since most transport systems are assumed to be located in the plasma membrane it is reasonable to assume that this is also true of the components involved in peptide translocation. As discussed below, however, yeasts contain a number of intracellular organelles which are enclosed by membranes. It is quite possible, therefore, that peptide transport is also affected by proteins localized on intracellular membranes.

It is beyond the scope of this review to discuss yeast cytology in full. However, one yeast cellular organelle, the vacuole, should be mentioned because it may have bearing on peptide metabolism in yeast. The vacuole, visible by light microscopy, forms the most conspicuous organelle of the yeast. This organelle is surrounded by a membrane and contains a significant fraction of the cellular amino acid pool (Wiemken and Nurse, 1973). Isolated vacuoles which have been identified as the lysosome of the yeast cell (Matile and Wiemken, 1967), have been shown to contain high amounts of several hydrolytic enzymes including proteases, ribonucleases, and esterases. An aminopeptidase has also been localised in the vacuole, although a significant amount of aminopeptidase-like activity has been shown outside the cell membrane (Matile *et al*., 1971). The possibility of a extracellular aminopeptidase is especially significant in connection with peptide utilization in yeasts and will be discussed in more detail later (Section 2.2.3).

Isolated yeast vacuoles were shown to contain a specific transport system for the amino acid arginine (Boller *et al*., 1975). The authors hypothesize that the uptake of amino acids into yeast cells involves transport from the medium into the cytoplasm and also from the cytoplasm into the vacuole. Isolated vacuoles were also shown to accumulate *S*-adenosylmethionine by means of a highly specific transport system (Schwenke and deRobichon-Szulmajster, 1975). The accumulation of various amino acids within the vacuoles under different physiological conditions was investigated by Indge *et al*. (1977). Large amounts of glycine were found rapidly to enter the vacuole (see also Section 1.2). Although these studies are in their early stages they suggest the possibility that peptide transport could be influenced at both the cell and vacuolar membranes.

3 LOCATION OF PEPTIDASES IN *S. CEREVISIAE*

Peptide uptake by cells can occur via a number of mechanisms (see Barak and Gilvarg, 1975; and Payne, 1976, for a review and discussion of mechanisms of peptide transport). In order to claim confidently that one is investigating the uptake of the intact peptide rather than its component amino acids, the absence of extracellular or periplasmic peptidase activity must be established. In the case of *E. coli* and other Gram-negative bacteria, numerous studies have shown that the peptidase activities of these microorganisms toward natural peptide substrates are located intracellularly (Matheson and Murayama, 1966; Payne, 1974; Simmonds and Toye, 1966; Sussman and Gilvarg, 1971; Van Lenten and Simmonds, 1967) (see also Section 2.4.6). The case in yeast is not as clear. A number of investigations have demonstrated that proteinases A, B, and C are located in the vacuole of yeast (Cabib *et al*., 1973; Hasilik *et al*., 1974; Holzer *et al*., 1975; Lenney *et al*., 1974; Matern *et al*., 1974). Matile and coworkers (Matile, 1966; Matile *et al*., 1971), however, have provided evidence that an aminopeptidase (similar to leucine aminopeptidase (LAP) and active against leucyl-β-naphthylamide) is localized between the plasmalemma and the cell wall. However, this aminopeptidase activity does not constitute the total aminopeptidase activity of the cell. In a later study it was observed that 81 per cent of the leucine aminopeptidase activity in *S. cerevisiae* remains in spheroplasts and that the LAP activity is enriched almost fivefold in the vacuolar fraction compared to that in whole cells (Lenney *et al*., 1974). Since identical strains were used in both studies, the variance in results is not clear (see also Section 3.3.3).

A recent examination of the action and utilization of small peptide fragments from the yeast mating hormone (α-factor) has suggested that there is an extracellular aminopeptidase activity that is associated with the cell wall or periplasmic space of *S. cerevisiae* (Stotzler *et al*., 1977). We believe, however, that the fact that this activity only cleaved Trp from Trp-Leu-Gln-Leu and did not further degrade the resulting Leu-Gln-Leu, argues for a very specific aminopeptidase activity. It is possible that the hydrolytic activity reported by Stotzler *et al*. (1977) is integrally associated with the receptor of the α-factor and may be responsible for the recovery of *S. cerevisiae* from G1 arrest by the mating hormone (Chan, 1977).

Our studies on peptide utilization in yeast point to the absence of extracellular or periplasmic peptidase activities capable of hydrolysing the substrates we tested. No activities toward $(Met)_3$ were found using whole cells or a supernatant solution from cells which were osmotically shocked (Becker *et al*., 1973). Furthermore, equivalent peptidase activity toward $(Met)_3$ was found in extracts of whole cells and osmotically-shocked cells. There is also indirect evidence which suggests that the peptidase activity in yeast is located intracellularly. In *S. cerevisiae* amino acids to not compete with the utilization of peptides, and peptides do not compete with the utilization of amino acids (Marder *et al*., 1977). Although many tripeptides competed with the uptake of Met-Met-[^{14}C]-Met by *S. cerevisiae*, amino acids were not effective competitors (Becker and Naider, 1977). Finally we have found that many hydrolysable peptides which are similar in structure to utilized peptides do not serve as growth substrates for yeast (Becker *et al*., 1973; Marder *et al*., 1977; Naider *et al*., 1974). These facts are strong evidence against extracellular peptidase activity and we conclude that the peptidase activity in *S. cerevisiae* (and in fact for *C. albicans*) is located intracellularly and that peptides are not cleaved prior to transport.

The question of the exact intracellular location of the peptidase activity is at present less certain (see also Section 3.3.3). As mentioned above a LAP-like activity appears enriched in the vacuole as compared with the cytoplasm (Lenney *et al.*, 1974). Matile *et al.* (1971) have shown that aminopeptidase III activity is isolated in vacuoles whereas the three other aminopeptidase isozymes are extra-vacuolar. Using acrylamide gel electrophoresis we have found four independent peptidase activities toward $(Met)_3$, $(Leu)_3$, and similar tri- and tetrapeptides (B. Rose, unpublished results). At least one independent dipeptidase activity also is present in yeast (B. Rose, unpublished data; Rohm, 1974). The presence of peptidase activity in both the vacuole and cytoplasm makes unclear what path a peptide follows after entering the cytoplasm. In certain instances it may be totally hydrolysed by cytoplasmic peptidases. In others, final metabolism to the constituent amino acids may necessitate passage of the peptide across the vacuolar membrane. Since little is known about the relative permeabilities of peptides to the cytoplasmic or vacuolar membrane, any results on the structural specificity of peptide transport in yeast may apply to either or both of these carriers. Future discussions of peptide transport in this review are made with this qualification in mind.

4 EVIDENCE FOR SEPARATE AMINO ACID AND PEPTIDE TRANSPORT SYSTEMS

There are many examples in the early fermentation literature of the ability of yeast to use peptides as sources of nitrogen (Harris, 1958; Suomalainen and Oura, 1971). Although most of these studies were not systematic in nature, they do represent the first indication of separate amino acid and peptide transport systems, because differences were observed in the relative effectiveness of amino acids and peptides to serve as nitrogen precursors. Our initial studies with *S. cerevisiae* showed that several lysine auxotrophs of yeast grew readily on lysine but not on dilysine, trilysine, or higher oligomers of this amino acid (Becker *et al.*, 1973). Since all of the lysine homopeptides examined were hydrolysed to lysine by enzymes in a cell extract of the various mutants, the failure of the lysine auxotrophs to utilize lysine-containing peptides was ascribed to the absence of a peptide transport system.

Recently, more definitive evidence has appeared for separate amino acid and peptide transport systems in yeast. Methionine was not able to inhibit the uptake of Met-Met-[^{14}C]-Met by *S. cerevisiae* 139 (Table 2) (Becker and Naider, 1977). Furthermore, although phenylalanine effectively prevented a leucine–lysine auxotroph from growing on leucine it had no significant effect on the utilization of leucine-containing peptides (Figure 1). Since phenylalanine is thought to prevent leucine from entering the cell (Wolfinbarger and DeBusk, 1971) this observation is strong evidence for

Table 2 Competition for transport of trimethionine by various structurally related compounds[a]

Competitor	Concentration relative to trimethionine	Inhibition of trimethionine transport (%)
None		—
Methionine	10-fold	14
Dimethionine	10-fold	30
Leu-Leu-Leu	10-fold	70
Lys-Lys-Lys	20-fold	0
Ala-Ala-Ala	10-fold	71
Gly-Met-Gly	10-fold	0
Met-Met-Ile	10-fold	86
L-Met-L-Met-D-Met	10-fold	40
L-Met-D-Met-L-Met-OMe	10-fold	30
L-Met-L-Met-L-Met-OMe	10-fold	93
D-Met-D-Met-D-Met-OMe	10-fold	14
Acetyltrimethionine	20-fold	0

[a]Cells suspended at 0.5 mg dry weight ml^{-1} of water were added to an equal volume of reaction mixtures containing 0.5 per cent glucose, 0.1M phosphate buffer (pH 5.5), radioactive trimethionine (0.024 mM, 0.5 μCi μmol^{-1}) and methionine or peptide at concentrations (molarity) indicated. Initial rates of transport of $(Met)_3$ were determined. (Reproduced from Becker and Naider (1977) with the permission of Academic Press Inc.)

separate amino acid and peptide transport systems. Finally we have recently generated a mutant of *S. cerevisiae* Z1-2D which can grow on leucine but not on leucine peptides (R. Marder, unpublished results). Since this mutant contains enzymes which can hydrolyse $(Leu)_2$ and $(Leu)_3$, its failure to grow on these peptides provides conclusive evidence for separate amino acid and peptide transport systems in this strain of yeast.

5 EVIDENCE FOR ONE PEPTIDE TRANSPORT SYSTEM IN YEAST

Studies of peptide transport in *E. coli* have shown that separate di- and oligopeptide transport systems are present in this microorganism (Payne and Gilvarg, 1971). Our investigations with *S. cerevisiae* suggest that this may not be the case with yeast. Competition studies have shown that both dipeptides and tripeptides can compete with the utilization of tripeptides and vice versa. For example, we found that the utilization of $(Leu)_2$ by *S. cerevisiae* Z1-2D was greatly inhibited by the presence of $(Met)_2$ or $(Met)_3$ in the growth medium (Marder *et al.*, 1977). Similar results were observed with the growth of this mutant on $(Leu)_3$ when either $(Met)_2$ or $(Met)_3$ was also present. In addition, high concentrations of $(Met)_2$ were found to

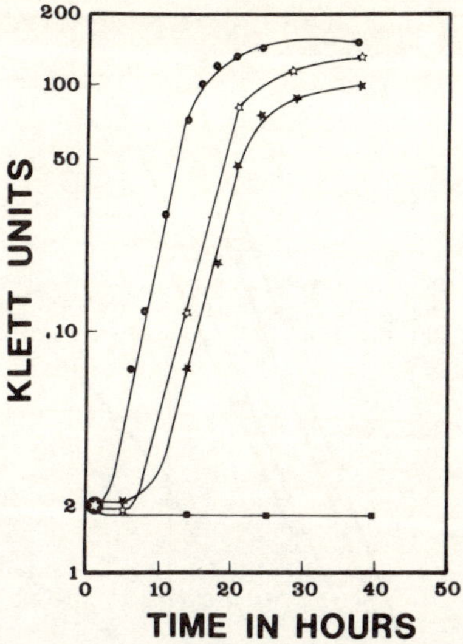

Figure 1 Growth of *S. cerevisiae* Z1-2D on Leu and $(Leu)_2$ in the presence and absence of Phe. Each growth flask contained lysine (0.11 mM). Leu and $(Leu)_2$ were supplied in the appropriate growth flasks so that each flask contained a leucine content equivalent to 0.23 mM of free amino acid. Turbidity, in Klett units, was measured for growth on Leu (○), Leu + Phe at a 46-fold molar excess to Leu (■), $(Leu)_2$ (★), and $(Leu)_2$ + Phe at a 92-fold molar excess to $(Leu)_2$ (☆). (Reproduced from Marder *et al.* (1977) with the permission of the American Society for Microbiology).

reduce the initial rate of uptake of Met-Met-[^{14}C]-Met into *S. cerevisiae* 139 (Becker and Naider, 1977). Both of these results can be interpreted in terms of separate transport systems which have some overlapping affinity. However, using a one-step selection procedure we isolated a mutant of *S. cerevisiae* which simultaneously lost the ability to take up di- and trileucine (R. Marder, unpublished results). This latter result makes it likely that the peptide transport system in at least one strain of yeast can take in both di- and tripeptides.

6 SIDE-CHAIN SPECIFICITY OF THE PEPTIDE TRANSPORT SYSTEM

To date, studies of peptide utilization in yeast have been carried out using only methionine, leucine, and lysine auxotrophs. Nevertheless peptides containing imino acids (Pro), or hydrophobic (Leu, Ile, Val, Ala, Phe), aromatic (Phe), sulphur (Met), basic (Lys), and neutral (Gly) residues have been examined via either growth response or competition experiments. For the most part, yeast can utilize di- and oligopeptides containing all of these amino acids. Unlike *E. coli*, however, the growth curve exhibited by yeast

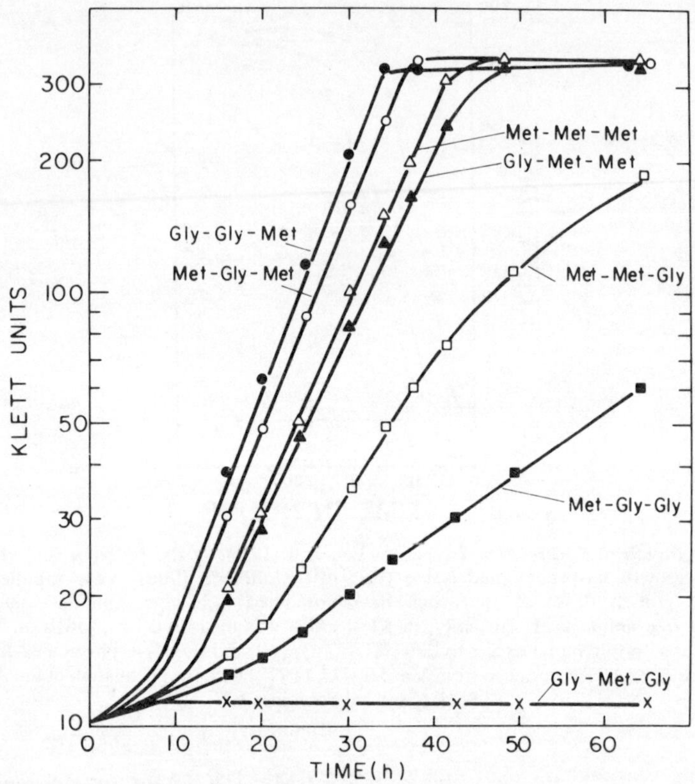

Figure 2 Growth of *S. cerevisiae* G1333 on various methionine-containing tripeptides. All growth flasks contained the equivalent of 0.067 mM in methionine residues. (Reproduced from Naider *et al*. (1974) with the permission of the American Society of Biological Chemists, Inc.)

appears to be sensitive to both the composition and sequence of the peptide substrate. Figure 2, for example, depicts the growth response of the methionine auxotroph *S. cerevisiae* G1333 to seven different tripeptides containing methionine and glycine (Naider *et al*., 1974). Although all of these peptides were readily cleaved by enzymes from a cell extract of strain G1333, they exhibited different abilities to serve the nutritional needs of the mutant for methionine. In all cases the tripeptides with methionine at the carboxyl terminus proved to be superior growth substrates. Similar results were observed for growth of *S. cerevisiae* G1333 on dipeptides and tetrapeptides containing methionine. Another methionine-requiring mutant of *S. cerevisiae* (strain XJB1-1C) (J. M. Becker and F. Naider, unpublished observations) and a leucine–lysine auxotroph (strain Z1-2D) (Marder *et al*., 1977) were also observed to grow differently on similar peptides of varied sequence. However, although strain XJB1-1C exhibits a variable growth response to seven tripeptides composed of methionine and glycine, the trends were not similar to those found in strain G1333 (Figure 2). In addition, whereas strain G1333 utilized peptides containing *C*-terminal methionine more efficiently, the leucine auxotroph strain Z1-2D exhibits no corresponding preference for leucine at this position. We can conclude that whereas sequence does influence peptide utilization in yeast, no simple relationships can be perceived by comparing one strain to another.

The results of competition studies in both growth and uptake experiments indicate that hydrophobic peptides are preferred by the peptide transport system in yeast. In extensive studies with strain Z1-2D, peptides composed of methionine competed strongly with the utilization of $(Leu)_2$ and $(Leu)_3$ whereas alanine peptides were poor competitors and glycine or lysine peptides did not compete (Marder *et al.*, 1977). In addition, whereas $(Leu)_3$, $(Ala)_3$, and Met-Met-Ile effectively prevented the uptake of Met-Met-[^{14}C]-Met by strain 139, neither Gly-Met-Gly nor $(Lys)_3$ competed with $(Met)_3$ transport (Becker and Naider, 1977). The finding with Gly-Met-Gly is consistent with the failure of strain G1333 to utilize this tripeptide and is comparable to the lack of growth of *S. cerevisiae* Z1-2D on Gly-Leu-Gly (Becker *et al.*, 1973; Marder *et al.*, 1977). The inability of $(Lys)_3$ to compete with $(Met)_3$ uptake is in agreement with the failure of four independent lysine auxotrophs of *S. cerevisiae* and one of *C. albicans* to utilize $(Lys)_3$ as a lysine source (Becker *et al.*, 1973; Lichliter *et al.*, 1976). Thus, the failure of the peptide transport system in yeast to recognize lysine homopeptides and its preference for peptides containing amino acids with hydrophobic residues is a characteristic which applies to all strains so far examined. Further work is necessary to establish whether this finding is in fact general to yeast and will apply to peptides containing other hydrophobic and ionic amino acid residues.

7 IMPORTANCE OF PEPTIDE TERMINI FOR PEPTIDE UTILIZATION

A major characteristic of oligopeptide transport in *E. coli* is the importance of the amine terminus and the lack of importance of the carboxyl terminus for permeation into the cell (Barak and Gilvarg, 1975; Payne, 1976; Payne and Gilvarg, 1971). In numerous studies acylation of the α-amino nitrogen resulted in a derivative which was not recognized by the oligopeptide permease. The case in *S. cerevisiae* appears to be more complex. In a preliminary study, peptides containing a toxic amino acid were less inhibitory when acylated on the amine terminus (Dunn and Dittmer, 1951). However, in this study a high concentration of the free dipeptide was necessary for growth inhibition and DL-dipeptides were employed in all investigations. Recently it was reported that benzyloxycarbonyl-leucylleucine supported the growth of approximately 15 strains of *S. cerevisiae* (Jones, 1977). It should be emphasized, however, that this latter study was primarily a genetic analysis of peptidase mutants and no attempt was made to determine if an extracellular or periplasmic peptidase was involved in Z-Leu-Leu utilization. We found that *S. cerevisiae* G1333 was able to utilize certain di- and tripeptides in which the α-NH_2 groups carried an acetyl or tert-butyloxycarbonyl group (Naider *et al.*, 1974). However, various growth-supporting peptides, particularly those containing glycine at the carboxyl terminus, were not utilized after acetylation. In a more recent study *S. cerevisiae* Z1-2D was not able to grow on Ac-Gly-Leu or Ac-Leu-Leu-Leu although di- and trileucine are excellent growth substrates (Marder *et al.*, 1977). These variations in results make it impossible to generalize on the importance of the α-amine for peptide transport. What is definite, however, is that various acylated di- and oligopeptides do enter certain yeasts. It was found that [^{3}H]-Ac-Met-Met-Met was taken up by *S. cerevisiae* 139 by a saturable uptake system having a K_m of 1.6×10^{-4} M (Becker and Naider, 1977). Although Ac-(Met)$_3$ did not prevent Met-Met-[^{14}C]-Met uptake, (Met)$_3$ did reduce the uptake of [^{3}H]-Ac-Met-Met-Met. It is possible therefore that acylated peptides enter certain yeast through a system other than a peptide transport system that mediates uptake of free peptides. Studies on peptide transport deficient mutants should resolve this question.

In all studies to date, modification of the α-carboxyl group by esterification has not prevented utilization of the peptide derivative by *S. cerevisiae* (Marder *et al.*, 1977; Naider *et al.*, 1974). Strain G1333 grew on (Met)$_3$-OMe, (Met)$_4$-OMe and (Met)$_5$-OMe as methionine sources although the lag phase was significantly longer than that observed with the analogous free peptide. Strain Z1-2D grew on Leu-Leu-OBzl as well as it grew on dileucine. Finally, trimethionine methyl esters were as effective as the free peptides in preventing uptake of radioactive (Met)$_3$ into *S. cerevisiae* 139. It is apparent that for a number of independent mutants of *S. cerevisiae* a free α-carboxyl is not necessary for peptide uptake. This finding conflicts markedly with the need for a free carboxyl for dipeptide uptake in *E. coli* and the absolute failure of *C. albicans* to take up the methyl ester forms of normally used tetra- and pentapeptides (Lichliter *et al.*, 1976; Payne and Gilvarg, 1971).

8 STEREOSPECIFICITY OF PEPTIDE UTILIZATION IN YEAST

A knowledge of the stereospecificity of the peptide transport system in microorganisms is essential for the rational design of non-hydrolysable transport substrates, affinity labels, and peptide carriers which might be useful materials to deliver toxic agents into microorganisms. Although a great deal of information is available on the structural requirements for peptide uptake in bacteria, very few studies have explored the importance of the configuration at the chiral centres for peptide transport. In one attempt at a systematic study, we observed that *E. coli* was able to transport L,L,L, L,L,D, and D,L,L-trimethionine methyl ester but was unable to utilize any other diastereomer of this tripeptide (Becker and Naider, 1974). Furthermore, the non-utilized diastereomers were not competitors of trimethionine uptake even when present in large molar excess in the growth medium. In yeast, the stereospecificity of peptide uptake has at present been studied in *S. cerevisiae* in strains G1333, Z1-2D and 139 and in *C. albicans* WD18-4. The most complete study found that *S. cerevisiae* G1333 utilized only two (L,L,L-OMe and L,L,D-OMe) of the eight stereoisomers of trimethionine methyl ester (Figure 3). This was despite

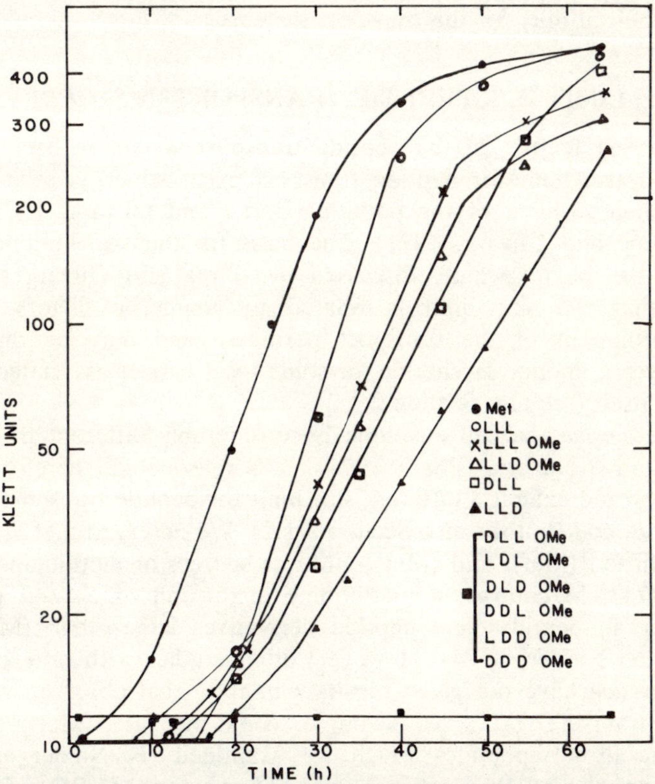

Figure 3 Growth of *S. cerevisiae* G1333 on diastereomers of trimethionine and trimethionine methyl ester. The configurations of the substrates examined are indicated in the figure. All growth flasks contained the equivalent of 0.067 mM in methionine residues

the fact that the cell contained peptidases which liberated methionine from all diastereomers except D-Met-D-Met-D-Met-OMe and suggests that the other diastereomers are not utilized because they are not transported. Interestingly, although D-Met-L-Met-L-Met-OMe is not utilized by strain G1333, the corresponding free acid is a good growth substrate. This finding is reasonable in view of the generally poor growth response of this mutant to peptide esters and suggests that the peptide transport system cannot tolerate simultaneous alteration of both peptide termini. In direct competition studies we also observed that L-Met-L-Met-D-Met competed with the uptake of radioactive $(Met)_3$ by *S. cerevisiae* 139, albeit much less effectively than the all L-trimethionine (Becker and Naider, 1977). In addition L-Met-D-Met-L-Met-OMe and D-Met-D-Met-D-Met-OMe were found to be much less effective competitors than L-Met-L-Met-L-Met-OME with $(Met)_3$ transport. These results suggest that the peptide transport system in yeast can apparently accommodate certain peptides containing D-residues. The incorporation of D-residues in a peptide, however, appears to lower their affinity for the transport system.

9 SIZE LIMIT FOR PEPTIDE TRANSPORT IN *S. CEREVISIAE*

An interesting feature of the peptide transport system in bacteria is the fact that a size limit for peptide transport exists which is related to the hydrodynamic volume of the peptide (Barak and Gilvarg, 1975; Payne, 1976; Payne and Gilvarg, 1971). The basis for the size limit to peptide transport has been recently discussed by Barak and Gilvarg (1975). It appears that the size limit is related to geometric factors near the microenvironment of the transport permease and may be due to the peptidoglycan, lipopolysaccharide, or some local barrier associated with the permease itself (see also Section 2.1.2).

In *S. cerevisiae* the cell envelope is considerably different from that in Gram negative bacteria (Phaff, 1971). It is reasonable, therefore, that *S. cerevisiae* would exhibit a different size limit for peptide transport than that found for *E. coli* (but see also Section 2.1.2). We observed that *S. cerevisiae* G1333 utilized $(Met)_5$ and $(Met)_5$-OMe as sources of methionine (Naider, *et al*., 1974), In contrast a methionine auxotroph of *E. coli* showed a cut-off for all peptides and peptide derivatives larger than $(Met)_4$-OMe (Becker and Naider, 1974). However, other studies with different strains of *S. cerevisiae* have not given results similar to that observed with strain G1333. Strain XJB1-1C, for example, could not use $(Met)_5$ and grew poorly on all tetrapeptides which we examined (F. Naider and J. M. Becker, unpublished observations). In addition, strain Z1-2D did not grow on the three tetrapeptides or the one pentapeptide containing leucine which we tested. Furthermore these tetrapeptides did not compete with the uptake of di- or trileucine and Gly-Leu-Gly-Gly was not utilized even when supplied in the growth medium at very high concentration (Marder *et al*., 1977). As mentioned previously the growth response of strain Z1-2D depends on peptide sequence. It is, therefore, impossible to definitely conclude that we have found a size limit in strain Z1-2D. Currently we are attempting to prepare an homologous series of oligomers with the sequence: Lys-Leu$_n$ (n = 1–3). Since strain Z1-2D grows on Lys-Leu, a positive growth response when n = 2 and a negative response when n = 3 will be strong evidence for a size limit in this mutant. At present we can only note that the ability of *S. cerevisiae* to utilize peptides of different chain lengths is not equivalent to that of *E. coli* and may be a strain-specific property. It is important to point out, however, that *S. cerevisiae* is an organism whose sexual conjugation is controlled by peptides with 10–13 residue chain lengths (Stotzler *et al*., 1976). The presence of a size limit for peptide transport may, therefore, reflect on the mechanism of action of the mating factors and their ultimate metabolism by the yeast.

10 REGULATION OF PEPTIDE TRANSPORT AND UTILIZATION

The ability of yeast to transport and utilize peptides appears to be a non-inducible characteristic of *S. cerevisiae*. In response to peptides added to the growth medium, we have found no evidence for the induced synthesis of peptidases or peptide transport systems in the strains we have examined. The constitutivity of peptidases and peptide transport has also been established for bacteria (Barak and Gilvarg, 1975; Miller, 1975).

On the other hand, growth in the presence of ammonium ion appears to repress the transport of peptides in *S. cerevisiae* (Becker and Naider, 1977). Cells grown in medium with ammonium nitrate as the nitrogen source take up radioactivity from trimethionine at a reduced rate compared to cells grown on proline or glutamic acid as the nitrogen source. This phenomenon of nitrogen catabolite repression has been shown also for general amino acid permeases in yeasts (Grenson *et al*., 1970; Woodward and Cirillo, 1977). Nitrogen metabolism appears to play a role in the regulation of synthesis and activity of cellular peptidases. Aminopeptidase III activity was derepressed twofold when protoplasts were incubated without a nitrogen source (Matile *et al*., 1971). In a recent paper Hansen *et al*. (1977) have shown that the availability of nitrogen affects the level of proteinases and peptidases in yeasts. It has been suggested that the inhibition of amino acid transport by ammonium ion results from the uptake of this metabolite into the cell and its subsequent incorporation into the α-amino groups of glutamic acid and other amino acids (Roon *et al*., 1975). However, additional mechanisms of nitrogen catabolite repression have been proposed (Dubois *et al*., 1977). It is not clear at this time whether any of these mechanisms are operative for peptide transport.

11 ENERGETICS OF PEPTIDE TRANSPORT

Much information is now known about coupling of energy and transport in bacteria, especially *E. coli* (Simoni and Postma, 1975) (see also Section 1.7). Although fewer studies have been carried out in eukaryotic cells, a series of papers from the laboratory of Eddy (Cockburn *et al*., 1975; Seaston *et al*., 1973, 1976) have shown that the mechanism of amino acid transport in *S. cerevisiae* may be comparable to the mechanism of amino acid uptake in bacteria (see also Section 1.2). The results of these studies are consistent with the notion that amino acid uptake is driven by a proton symport mechanism according to the chemiosmotic hypothesis of Mitchell (1977). Our findings on peptide transport in yeast demonstrate an energy requirement for peptide uptake (Becker and Naider, 1977). Energy uncouplers such as sodium azide, dinitrophenol, and potassium cyanide all cause total inhibition of trimethionine uptake. Transport is also pH-dependent with the optimum pH at 5.5 for both free trimethionine and acetylated trimethionine. It is unlikely that the pH effect is related solely to the concentration of the proper ionic form of the transported substrate as acetylated trimethionine is completely ionized by pH 4–5 and does not change ionization at higher values. Possibly, hydrogen ion is affecting some component of the transport system. On the other hand, the peptide transport carrier may have binding sites for peptides and protons. This would be consistent with the finding of Eddy on proton symport mechanisms in yeast amino acid transport (Cockburn *et al*., 1975; Seaston *et al*., 1973). Nevertheless, at present we cannot conclude whether transport is an active process because we have no evidence that peptides are accumulated against a concentration gradient. In our only detailed analysis, trimethionine was rapidly hydrolysed either in the process of transport or immediately after entering *S. cerevisiae* (Becker and Naider, 1977). Studies with peptidase deficient mutants or gratuitous substrates are necessary to prove whether yeast actively transport oligopeptides.

12 CONCLUSIONS ON STUDIES WITH *S. CEREVISIAE*

Our studies with *S. cerevisiae* have shown that, unlike in *E. coli*, the utilization of peptides varies to some extent from strain to strain. Nevertheless, it appears that the following generalizations can be made. Systems do exist in yeast which can transport peptides from the extracellular environment into the cell. The peptide transport system is capable of taking in peptides composed of all amino acids examined to date. Results of competition studies seem to indicate that peptide transport systems in yeast have a higher affinity for peptides containing hydrophobic amino acids. There are other differences between peptide utilization in yeast and *E. coli*. For example, the growth of *S. cerevisiae* is markedly affected by the sequence of a peptide substrate, whilst this does not appear to be the case in *E. coli*. In the few cases examined, *S. cerevisiae* was found to have a different size limit for transport than that found for *E. coli*. Preliminary evidence also points to the existence of one transport system in *S. cerevisiae* which handles both di- and tripeptides while in *E. coli* genetically distinct di- and oligopeptide transport systems have been found in all strains examined. Finally, acylation of the amine terminus of a peptide does not generally prevent the resulting derivative from entering *S. cerevisiae*. In contrast, in all cases noted in the literature, *E. coli* was not able to transport α-*N*-acylated di- or oligopeptides. We believe that additional studies with yeast are necessary before these conclusions can be viewed with certainty. At present we are using these preliminary indications as guides to design affinity labels and peptide-drug conjugates which may be toxic to yeast. Our rationale in both of these areas is outlined in the following sections.

13 DESIGN OF AFFINITY LABELS FOR THE PEPTIDE TRANSPORT SYSTEM

A major emphasis of our current studies on peptide transport in yeast is the isolation of components of the peptide transport system. To aid in these studies we are synthesizing peptides with reactive groupings that we hope will form covalent bonds to the permease molecules. Such molecules should specifically react with the transport proteins and can serve as a readily identifiable label during purification. Based on the information from the structural specificity studies we have chosen to synthesize affinity labels having the following structures:

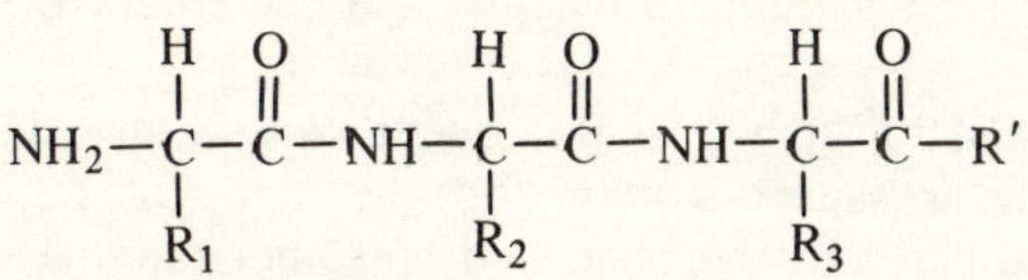

where

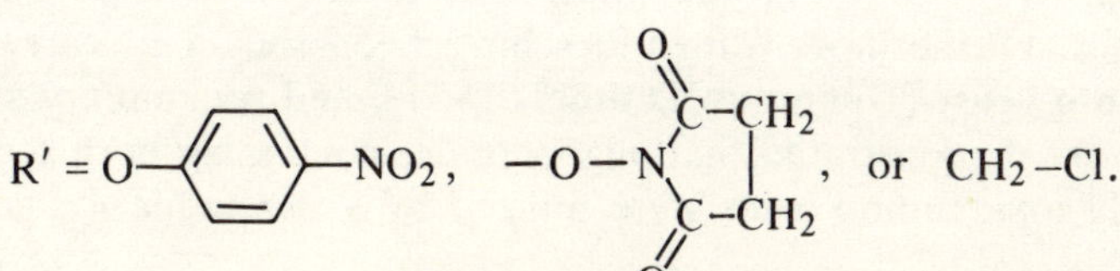

We intend to use peptides composed primarily of hydrophobic residues since these seem to have the highest affinity for the peptide transport system in yeast. It should be pointed out that kinetic studies on *E. coli* and *S. typhimurium* indicate that hydrophobic peptides also have high affinity for the oligopeptide and dipeptide transport systems in these microorganisms (Payne and Bell, 1977; Yang *et al*., 1977). Thus, the peptides we propose to synthesize may also label components of the peptide transport systems in bacteria. It is obvious that our knowledge of peptide transport phenomena will be greatly increased by physical-chemical analysis of molecular components of the transport system.

14 PEPTIDE DRUGS FOR *CANDIDA ALBICANS*

The structural specificity of the peptide transport system in *E. coli* permits peptides containing virtually any amino acid side-chain to enter the bacterium. Taking advantage of this broad side-chain specificity, Ames *et al*. (1973) and Fickel and Gilvarg (1973) simultaneously demonstrated that normally impermeant molecules could be brought into Gram-negative bacteria in the form of peptides. It is obvious that such 'illicit transport' (Ames *et al*., 1973) could prove very useful in the design of peptide–drug conjugates which would be toxic to certain pathogenic microorganisms (see also Section 4.7.2.1). The concept of illicit transport in the design of therapeutically useful drugs requires that: (a) The peptide conjugate must enter the pathogenic cell in the intact form. (b) After entering the cell either the conjugate must elicit a lethal action or it must be enzymatically degraded to toxic components. (c) The peptide conjugate must not be degraded by serum enzymes. (d) The peptide conjugate must not be rapidly cleared and must have a half-life inside the host which is long enough to permit it to reach, enter and kill the pathogen. (e) Ideally the peptide conjugate would not be taken up by host tissue nor by the natural microbial flora of the host, or at the very least the peptide conjugate should enter the pathogen much more efficiently than it enters the host tissue.

It is clear that the design of such a 'magic bullet' warrants a detailed knowledge of the peptide transport system and the array of peptidase activities in both the pathogen and the host, and an understanding of the mechanism of action of the toxic agent. Our studies with *S. cerevisiae* have uncovered many differences between the structural requirements of the peptide transport system(s) in this yeast and those in bacteria and the mammalian gut. Furthermore, dipeptides have been shown to carry toxic amino acids into yeast (Dunn and Dittmer, 1951), and preliminary studies in our laboratory demonstrated that oligopeptides containing both required and potentially toxic amino acids were utilized by *S. cerevisiae* for growth.

Met-Met-*p*-F-Phe, Met-Met-thienylalanine, and Met-Met-ethionine are all sources of methionine in a methionine auxotroph resistant to the toxic amino acid moiety of the oligopeptide (J. M. Becker and F. Naider, unpublished results). We decided, therefore, to investigate peptide utilization in the pathogenic yeast *Candida albicans* and to determine whether any trends could be observed which might aid us in the design of a drug against this microorganism.

The data in Table 3 indicate a comparison of the utilization of peptides by *E. coli*, *S. cerevisiae*, and *C. albicans*. In many respects the structural requirements for peptide utilization in *C. albicans* WD18-4 mirror those found in *S. cerevisiae*. Specifically, lysine-containing peptides for the most part are not transported, *N*-α-acylated peptides are utilized, and the size limit for transport is different from that in *E. coli*. Differences found between *C. albicans* and *S. cerevisiae* include the insensitivity of the growth response of *C. albicans* to peptides of similar composition but different sequence and the inability of the pathogen to utilize any peptide esters. We emphasize that our work with *C. albicans* is limited to one strain and that additional studies will have to be carried out to generalize these findings. Nevertheless, similarities between *S. cerevisiae* and *C. albicans* are evident,

Table 3 Utilization of peptides by *E. coli*, *S. cerevisiae*, and *C. albicans*

	Growth response[a]				
Peptide	*E. coli* (Met⁻)	*S. cerevisiae* (Met⁻)	*C. albicans* (Met⁻ Lys⁻)	*E. coli* (Lys⁻)	*S. cerevisiae* (Lys⁻)
Met-Met	+	+	+		
Ac-Met-Met	−	+	+		
L-Met-L-Met-L-Met	+	+	+		
D-Met-L-Met-L-Met	+	+	+		
L-Met-L-Met-D-Met	+	+	−		
Ac-Met-Met-Met	−	+	+		
Boc-Gly-Met-Met	−	+	+		
Met-Met-Met-OMe	+	+	−		
Gly-Met-Gly	+	−	+		
Met-Met-Gly-Met	+	+	+		
Met-Met-Gly-Gly	+	−	+		
Met-Met-Met-Gly	+	−	+		
Met-Met-Met-Met-OMe	+	+	−		
Met-Met-Met-Met-Met	−	+	+		
Lys-Lys			−	+	−
Lys-Gly			+	ND	+
Gly-Lys			−	ND	−
Lys-Lys-Lys			−	+	−

[a](+) indicates that the peptide substitutes for the required amino acid; (−) indicates that the peptide is not a growth substrate. ND, not done. (Reproduced from Lichliter *et al.* (1976) with the permission of the American Society for Microbiology.)

and we are currently incorporating our knowledge of peptide utilization in *C. albicans* into the design of conjugates of peptides and fluorinated pyrimidines which may prove toxic to the yeast, such as those indicated below:

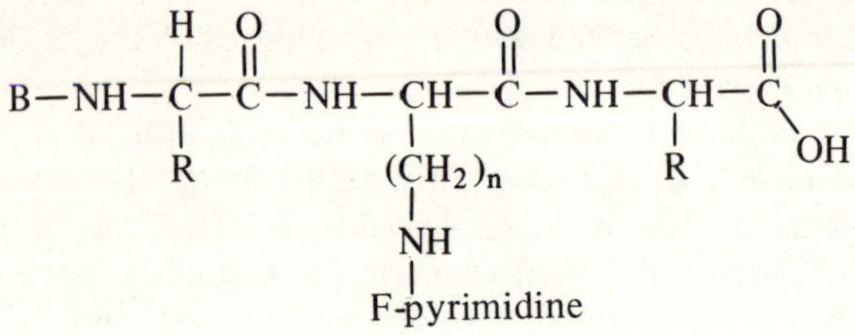

I

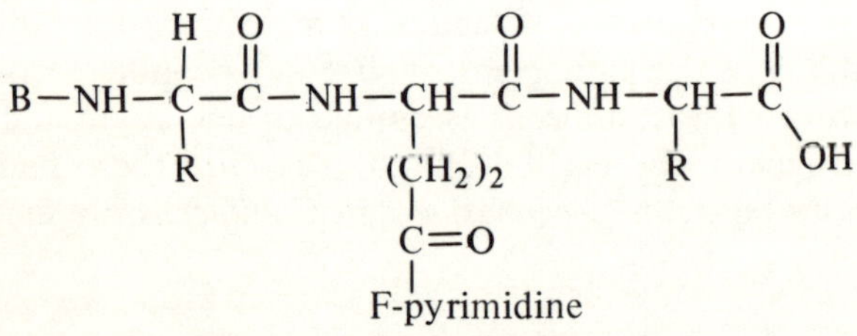

II

Our initial synthesis is devised to place the fluorinated pyrimidine in the side chain of a di- or oligopeptide using a bond which should be susceptible to enzymatic hydrolysis (compounds **I** and **II**). Thus, 5-fluorocytosine can be put on a peptide via the carboxyl side-chain of glutamic acid (compound **II**), and fluoro-orotic acid can be incorporated into the peptide via the δ- or ε-amino group of peptides containing ornithine or lysine (compound **I**). Alternative routes use the α-amino group of the peptide as the point of attachment for the fluorinated pyrimidine. In our initial evaluations, however, it is desirable to examine conjugates containing both free and blocked α-amines to evaluate their relative efficacy against bacteria and yeast. All of our conjugates at this stage are designed with a free carboxyl terminus in accord with the structural requirements of peptide utilization in *C. albicans* WD18-4.

Our goals in this project are first to establish that peptide-fluorinated pyrimidine conjugates can enter yeast. Thus, we have begun working with 5-fluorocytosine which is known to be toxic to *C. albicans*. The second phase of our work will be to use peptides to bring in a normally impermeant pyrimidine such as 5-fluoro-orotic acid. If we are successful in demonstrating that such ‘illicit transport’ is possible in *Candida* we believe that an effective drug can be designed against this pathogen.

ACKNOWLEDGEMENTS

Work reported in this review has been supported by grants from the Research Corporation, the National Science Foundation, the American Cancer Society, the National Institutes of General Medical Sciences, and the Faculty Research Award Program of the City University of New York. We are grateful to these institutions for supplying the funds to carry out our research. Several graduate students, research assistants, and research associates have provided invaluable assistance in the performance and development of studies on peptide transport in yeast. We are especially indebted to Yael Nuchamowitz, Bruce Rose, Robert Marder, Phillip Roberts, Wayne Lichliter, David Parker, Susan Winslow, and Dr Alvin Steinfeld for participation in these studies. Both authors acknowledge the support of Career Development Awards from the National Institutes of General Medical Sciences. We thank Dr Alvin S. Steinfeld for critically reading this manuscript and for his many helpful suggestions. Significant portions of this review were completed while the authors were on sabbatical at the Weizmann Institute of Science.

REFERENCES

Ames, B. N., Ames, G. F., Young, J. D., Tsuchiya, D., and Lecocq, J. (1973). *Proc. Natl. Acad. Sci. U.S.A.*, **70**, 782

Arnold, W. N. and Lacy, J. S. (1977). *J. Bacteriol.,* **131**, 564

Barak, Z. and Gilvarg, C. (1975). In *Biomembranes* (H. Eisenberg, E. Katchalski-Katzir, and L. A. Manson, eds), Plenum Press, New York, p. 167

Becker, J. M., Naider, F., and Katchalski, E. (1973). *Biochim. Biophys. Acta,* **291**, 388

Becker, J. M. and Naider, F. (1974). *J. Bacteriol.,* **120**, 191

Becker, J. M. and Naider, F. (1977). *Arch. Biochem. Biophys*., **178**, 245

Boller, T., Durr, M., and Wiemken, A. (1975). *Eur. J. Biochem.,* **54**, 81

Cabib, E., Ulane, R., and Bowers, B. (1973). *J. Biol. Chem.,* **248**, 1451

Chan, R. K. (1977). *J. Bacteriol.,* **130**, 766

Cockburn, M., Earnshaw, P., and Eddy, A. A. (1975). *Biochem. J.,* **146**, 705

Dunn, F. W. and Dittmer, K. (1951). *J. Biol. Chem.,* **188**, 263

Dubois, E., Vissers, S., Grenson, M., and Wiame, J. M. (1977). *Biochem. Biophys. Res. Commun.,* **75**, 233

Elliot, K. and O'Connor, M. (eds) (1972). *Peptide Transport in Bacteria and Mammalian Gut*, Ciba Foundation Symposium, Elsevier, Amsterdam

Fickel, T. E. and Gilvarg, C. (1973). *Nature New Biol.,* **241**, 161

Grenson, M., Hou, C., and Crabeel, M. (1970). *J. Bacteriol.,* **103**, 770

Hansen, R. J., Switzer, R. L., Hinze, H., and Holzer, H. (1977). *Biochim. Biophys. Acta,* **496**, 103

Harris, G. (1958). In *The Chemistry and Biology of Yeasts* (A. H. Cook, ed.), Academic Press, New York, p. 437

Hasilik, A., Muller, H., and Holzer, H. (1974). *Eur. J. Biochem.,* **48**, 111

Holzer, H., Betz, H., and Ebner, E. (1975). *Curr. Top. Cell. Regul*., **9**, 103

Indge, K., Seaston, A., and Eddy, A. A. (1977). *J. Gen. Microbiol.,* **99**, 243

Jones, E. W. (1977). *Genetics*, **85**, 23
Lampen, J. O. (1968). *Antonie Van Leeuwenhock,* **34**, 1
Lenney, J. F., Matile, Ph., Wiemken, A., Schellenberg, M., and Meyer, J. (1974). *Biochem. Biophys. Res. Commun*., **60**, 1378
Lichliter, W. D., Naider, F., and Becker, J. M. (1976). *Antimicrob. Agents Chemother*., **10**, 483
Lodder, J. (1970). In *The Yeasts*, 2nd edn (J. Lodder, ed.), North-Holland, Amsterdam, p. 1
Lund, A. (1958). In *The Chemistry and Biology of Yeasts* (A. H. Cook, ed.), Academic Press, New York, p. 63
Marder, R. M., Becker, J. M., and Naider, F. (1977). *J. Bacteriol.,* **131**, 906
Matern, H., Betz, H., and Holzer, H. (1974). *Biochem. Biophys. Res. Commun.,* **60**, 1051
Matheson, A. T. and Murayuma, T. (1966). *Can. J. Biochem.,* **44**, 1407
Matile, Ph. (1966). In *Proceedings of the 2nd International Special Symposium on Yeasts, Bratislava*, Publishing House of the Slovak Academy of Science, Bratislava, p. 503
Matile, Ph. (1970). *Fed. Eur. Biochem. Soc. Symp*., **20**, 39
Matile, Ph., Moor, H., and Robinow, C. F. (1969). In *The Yeasts*, Vol. I (A. H. Rose and J. S. Harrison, eds), Academic Press, New York, p. 219
Matile, Ph. and Wiemken, A. (1967). *Arch. Mikrobiol.,* **56**, 148
Matile, Ph., Wiemken, A., and Guyer, W. (1971). *Planta,* **96**, 43
Matthews, D. M. (1975). In *Intestinal Absorption and Malabsorption* (T. Z. Csaky, ed.), Raven Press, New York, p. 95
Miller, C. G. (1975). *Annu. Rev. Microbiol.,* **29**, 485
Mitchell, P. (1977). *Symp. Soc. Gen. Microbiol.,* **27**, 383
Naider, F., Becker, J. M., and Katchalski-Katzir, E. (1974). *J. Biol. Chem.,* **249**, 9
Northcote, D. H. and Horne, R. W. (1952). *Biochem. J.,* **51**, 232
Nurminen, T. and Suomalainen, H. (1973). In *Proceedings of the Third International Special Symposium on Yeasts* (H. Suomalainen and C. Waller, eds), Oy Press, p. 169
Nurminen, T., Oura, E., and Suomalainen, H. (1970). *Biochem. J.,* **116**, 61
Payne, J. W. (1974). In *Peptide Transport in Protein Nutrition* (D. M. Matthews and J. W. Payne, eds), North Holland, Amsterdam, p. 17
Payne, J. W. (1976). *Adv. Microbial. Physiol*., **13**, 55
Payne, J. W. and Bell, G. (1977). *FEMS Lett.,* **1**, 92
Payne, J. W. and Gilvarg, C. (1971). *Adv. Enzymol. Relat. Areas Mol. Biol*., **35**, 187
Phaff, H. (1971). In *The Yeasts*, Vol. II (A. H. Rose and J. S. Harrison, eds), Academic Press, New York, p. 135
Rohm, K. H. (1974). *Hoppe-Seyler's Z. Physiol. Chem*., **355**, 675
Roon, R. J., Larimore, F., and Levy, J. S. (1975). *J. Bacteriol.,* **124**, 325
Scherrer, R., Louden, R. L., and Gerhardt, P. (1974). *J. Bacteriol.,* **118**, 534
Schwenke, J. and DeRobichon-Szulmajster, H. (1976). *Eur. J. Biochem.,* **65**, 49
Seaston, A., Carr, G., and Eddy, A. A. (1976). *Biochem. J.,* **154**, 669
Seaston, A., Inkson, C., and Eddy, A. A. (1973). *Biochem. J.,* **134**, 1031
Simmonds, S. and Toye, N. O. (1966). *J. Biol. Chem.,* **241**, 3852
Simoni, R. D. and Postma, P. W. (1975). *Annu. Rev. Biochem.,* **44**, 523
Stotzler, D., Betz, R., and Duntze, W. (1977). *J. Bacteriol.,* **132**, 28
Stotzler, D., Klitz, H. H., and Duntze, W. (1976). *Eur. J. Biochem*., **118**, 534
Suomalainen, H. and Oura, E. (1971). In *The Yeasts*, Vol. II (A. H. Rose and J. S. Harrison, eds), Academic Press, New York, p. 3
Sussman, A. J. and Gilvarg, C. (1971). *Annu. Rev. Biochem.,* **40**, 397
Svihla, G., Dainko, J. L., and Schlenk, F. (1969). *J. Bacteriol.,* **100**, 498
Van Lenten, E. J. and Simmonds, S. (1967). *J. Biol. Chem.,* **242**, 1439
Wiemken, A. and Nurse, P. (1973). In *Proceedings of the Third International Special Symposium on Yeasts* (H. Suomalainen and C. Waller, eds), Oy Press, p. 331
Wolfinbarger, L., Jr and Debusk, A. G. (1971). *Arch. Biochem. Biophys*., **14**, 503
Wolfinbarger, L., Jr and Marzluf, G. A. (1974). *J. Bacteriol.,* **119**, 371
Wolfinbarger, L., Jr and Marzluf, G. (1975*a*). *J. Bacteriol.,* **122**, 949

Wolfinbarger, L., Jr and Marzluf, G. A. (1975*b*). *Arch. Biochem. Biophys.*, **171**, 637
Woodward, J. R. and Cirillo, V. P. (1977). *J. Bacteriol.*, **130**, 714
Yang, S. L., Becker, J. M., and Naider, F. (1977). *Biochim. Biophys. Acta*, **471**, 135

Microorganisms and Nitrogen Sources
Edited by J. W. Payne

Section 2.3

Transport and Utilization of Peptides by Fungi

L. WOLFINBARGER, JR
Department of Biological Sciences, Old Dominion University, Norfolk, Virginia 23508, USA

1 INTRODUCTION

Studies of peptide transport and utilization in fungi have, with the exception of the yeasts, been limited to the Ascomycete fungus *Neurospora crassa*. Although not as extensive as the reports of peptide transport in bacteria, the studies in *Neurospora* have permitted a clearer understanding of the overall capability of this fungal organism to grow on proteins in the environment as sources of required metabolites. Thus, this chapter will attempt not only to discuss peptide transport and utilization, but the broader picture of extracellular digestion of proteins to peptides and amino acids followed by their transport and utilization. The increasing interdependence of intracellular pools of amino acids and extracellular protease production (Cohen and Drucker, 1977) as well as peptide transport (Wolfinbarger and Marzluf, 1975*a*), have made it essential that we study all aspects of transport and utilization of proteins, peptides, and amino acids if we ever hope to acquire an understanding of individual systems for transport of given metabolites.

Transport of amino acids in fungi has been reviewed in some detail (see Section 1.3 and Scarborough, 1973; Payne, 1975). In *Neurospora*, specifically, there are at least five reported transport systems for amino acids. In the following discussion of peptide transport it is important to appreciate that in order to preclude any significant involvement of extracellular hydrolysis of peptides to free amino acids, it is necessary to demonstrate growth of an amino acid auxotroph on a given peptide in the presence of an excess of an amino acid which competes with the required free amino acid for transport to an extent that would normally limit growth. Growth of the auxotroph under such conditions is an indication of a functional peptide transport system, independent of amino acid transport. As will be seen, *Neurospora* does possess the ability to hydrolyse certain extracellular peptides, and so the capacity to be able to block amino acid transport was a prerequisite to a study of peptide transport by conventional growth studies (see Section 2.1). In addition, there has been some suggestion (Magill *et al*., 1973; Wolfinbarger and Marzluf, 1974) that certain dipeptides may be transported by a general amino acid transport system (Pall, 1969; Wolfinbarger and DeBusk, 1971; Rao *et al*., 1975).

2 PEPTIDE UTILIZATION IN *NEUROSPORA*

In an early attempt to evaluate the overall capability of *Neurospora* to grow on peptides as a source of some required amino acids, mutants auxotrophic for four different amino acids were screened for their abilities to grow on a wide spectrum of peptides (0.33 mM) (Wolfinbarger and Marzluf, 1974). The growth assays involved the collection of mycelial pads which formed after 3 days growth at 30 °C.

1 Lysine Auxotrophs

A single lysine auxotroph (*lys–1*) was tested for growth on peptides containing lysine (Wolfinbarger and Marzluf, 1974). As a control, similar growth assays were run in the presence of excess arginine so as to effectively prevent transport of any free lysine which had been liberated extracellularly as a result of peptidase action. Such experiments provided two types of information. First, they revealed which lysine-containing peptides would support growth of *lys-1* and second, which peptides were hydrolysed extracellularly prior to transport of the constituent amino acids. *Lys-1* grew well on the tripeptide Lys-Lys-Lys, but poorly on the dipeptides tested. Inclusion of excess arginine precluded growth on all peptides except

Table 1 Growth[a] of lysine auxotroph of *Neurospora crassa* (*lys-1*) on free lysine and peptides containing lysine in the presence or absence of arginine

	Growth	
Source of lysine[b]	Control	Plus Arg[c]
Lys	31	5
Lys-Gly	1	0
Gly-Lys	4	0
Lys-Lys	1	0
Lys-Lys-Lys	52	40
Lys-Leu	2	1

[a]Growth as milligrams dry weight mycelial pad following 3 days growth at 30 °C. Values represent averages of from two to five separate determinations.
[b]Source of lysine supplements at 0.33 mM.
[c]Arginine concentration at 3.3 mM.
Reproduced from Wolfinbarger and Marzluf (1974) with the permission of the American Society for Microbiology.

for trilysine (Table 1). It is of some interest that, in contrast, lysine auxotrophs of *Saccharomyces cerevisiae* will not grow on trilysine as a source of lysine (see Section 2.2.6 and Becker *et al*., 1973; Naider *et al*., 1973).

2 Methionine Auxotrophs

A total of three methionine auxotrophs, *me-1*, *me-2*, and *me-3* were screened for their ability to grow on methionine-containing peptides (Wolfinbarger and Marzluf, 1974). All three auxotrophs grew, to various extents, on all of the methionine-containing peptides tested. Some differences in total growth yields were detected, for example, *me-3* grew to a total pad weight of 31 mg on Met-Gly whereas *me-1* and *me-2* attained pad weights of only 4 and 7 mg, respectively (Table 2). At present it is not known if such differences are due to strain backgrounds or location of the biochemical lesion(s) in the metabolic pathway.

Testing for growth of methionine auxotrophs on methionine-containing peptides is difficult because of the presence of at least three known methionine transport systems (Wolfinbarger and DeBusk, 1971; Pall, 1971). Methionine transport by the neutral and general amino acid transport systems can be blocked by the inclusion of excess phenylalanine (Wolfinbarger and Marzluf, 1974), but methionine transport by the methionine-specific transport system is not blocked by phenylalanine. However, an auxotrophic mutant designated *cys-3* is deficient in many methionine biosynthetic enzymes, and in the methionine-specific transport

Table 2 Growth[a] of methionine auxotrophs on free methionine and peptides containing methionine

	Strain		
Supplement[b]	*me-1*	*me-2*	*me-3*
None	0	0	0
Met	48	96	58
Met-Met	80	90	48
Met-Met-Met	38	51	38
Met-Gly-Met-Met	13	38	11
Met-Ala-Ser	34	68	51
Gly-D,L-Met	8	8	40
Met-Gly	4	7	31
Met-His	2	4	2
His-Met	3	7	5
Met-Leu	35	52	33

[a]Growth data as milligrams dry weight mycelial pad following 3 days growth at 30 °C. Values represent averages of two to seven separate determinations.
[b]Supplements at concentrations of 0.33 mM.
Reproduced from Wolfinbarger and Marzluf (1974) with the permission of the American Society for Microbiology.

Table 3 Growth[a] of methionine auxotrophs *me-1* and *cys-3* on free methionine or methionine-containing peptides in the presence or absence of competing amino acids

	Inhibiting amino acid[b]			
Source of methionine[c]	None	Phe	Arg	Phe + Arg
me-1				
Met	60	34	70	62
Met-Met	105	28	99	41
Met-Ala-Ser	36	28	42	33
Gly-D,L-Met	11	2	9	2
Met-Gly	4	1	12	4
cys-3				
None	0	—	—	—
Met	69	1	73	—
Met-Met	66	4	85	—
Met-His	0	0	0	—
Met-Gly	60	0	1	—
Met-Met-Met	68	49	74	—
Met-Ala-Ser	61	50	64	—

[a]Growth as milligrams dry weight mycelial pad following 3 days growth at 30 °C. Values represent averages of two or more separate determinations.
[b]Inhibiting amino acid at concentration of 3.3 mM.
[c]Source of methionine supplement at 0.33 mM.
Reproduced from Wolfinbarger and Marzluf (1974) with the permission of the American Society for Microbiology.

system (Marzluf and Metzenberg, 1968). Growth of *cys-3* on various methionine-containing peptides in the presence of excess phenylalanine permits the blocking of extracellularly-liberated methionine entry via amino acid transport systems. Thus, growth of *cys-3* under such conditions presumably requires that a given peptide be transported intact and hydrolysed intracellularly before it can support any increase in cellular mass. Referal to Table 3 reveals that, of those peptides tested, only tripeptides supported growth of *cys-3* when excess phenylalanine was included in the growth medium.

3 Histidine Auxotrophs

Only one mutant strain, *his-6*, was tested for growth on various histidine-containing peptides and the results obtained did not permit any generalizations to be made regarding peptide transport and utilization (Table 4). Histidine is transported by combinations of three amino acid transport systems (Wolfinbarger and DeBusk, 1971; Woodward *et al*., 1967; Magill *et al*., 1972), and to block growth of *his-6* on histidine requires the inclusion of excess neutral and basic amino acids, for example phenylalanine and arginine (Wolfinbarger and Marzluf, 1974). However, *his-6* grows readily on the dipeptide histidylglycine, as a source of histidine, and requires only phenylalanine or arginine, not both, to block growth (Table 4). In their early paper, Wolfinbarger and Marzluf (1974) interpreted these results to mean that histidine-containing dipeptides were

Table 4 Growth[a] of histidine auxotroph *his-6* on histidine or peptides containing histidine in the presence or absence of competing amino acids

	Inhibiting amino acid[b]			
Source of histidine[c]	None	Phe	Arg	Arg + Phe
His	81.5	86.3	98.0	0.0
His-Gly	39.7	1.0	1.3	0.0
Gly-His	1.5	0.0	0.0	0.0
His-Leu	64.7	3.3	34.7	5.0
His-Met	16	2	7	4
Met-His	20	1	2	1
β-Ala-His	3.0	—	—	—
Ala-His	5.5	—	—	—
His-Gly-Gly	3.3	—	—	—

[a]Growth as milligrams dry weight mycelial pad following 3 days growth at 30 °C. Values represent averages of from three to five separate determinations.
[b]Inhibiting amino acid at concentrations of 3.3 mM.
[c]Sources of histidine at concentrations at 0.33 mM.
Reproduced from Wolfinbarger and Marzluf (1974) with the permission of the American Society for Microbiology.

transported exclusively by a general amino acid transport system and not by the specific neutral and basic amino acid transport systems. This conclusion seemed to be supported by the findings of Magill *et al*. (1973) in that His-Gly apparently inhibited transport of histidine by the general system. In addition, Wolfinbarger and Marzluf reported the isolation of a mutant strain that was unable to grow on His-Gly as a source of histidine, *his-6*; *hgu-4*, and found that segregants, presumably the *hgu-4* lesion alone, were resistant to the growth inhibitory properties of D-glutamate. This latter amino acid has been reported to be transported exclusively by the general amino acid transport system (Wolfinbarger and Kay, 1973). In short, there appeared at that time to be a strong case for the involvement of an amino acid transport system in the accumulation of certain dipeptides. Later studies have modified this viewpoint and although the involvement of the general amino acid transport system in the transport of histidine-containing dipeptides has not been ruled out, subsequent work has shown that the 'system' is more complex than thought. Transport assays with [^{3}H]-histidylglycine were unsuccessful, in that no transport activity could be detected under a variety of conditions, conditions under which it was known that extensive general amino acid transport activity existed (Wolfinbarger and Kay, 1973). In addition, intracellular peptidase activity for His-Gly was variable, and not reproducible, even though the ability of *his-6* to grow on His-Gly required the presence of such activity. Finally, further biochemical analysis of the *hgu-4* phenotype revealed a generalized transport deficiency for all amino acids (Wolfinbarger, 1976) and not a deficiency exclusively in the general system.

In conclusion therefore, the growth assays of a histidine auxotroph on various histidine-containing peptides have provided more questions than answers, and further research in this area will be required before any meaningful conclusions can be made.

4 Leucine Auxotrophs

A total of four leucine auxotrophic strains, *leu-1*, *leu-2*, *leu-3*, and *leu-4*, have been tested for their growth responses on a variety of leucine-containing peptides (Wolfinbarger and Marzluf, 1974). This series of growth assays was more extensive, involving a larger number and variety of peptides than with the other auxotrophic strains and provided a better estimate of the relative involvement of amino acid and peptide transport systems in growth of *Neurospora* on peptides as a sole source of required amino acids. The data in Table 5 reveal that, in general, leucine auxotrophs do not grow on leucine-containing dipeptides, except for those that are hydrolyzed extracellularly. However, tripeptides are readily transported and utilized for growth, even in the presence of excess

Table 5 Growth[a] of leucine auxotrophs and wild-type strain on peptides containing leucine

	Strain				
Supplement[b]	Wild-type	*leu-1*	*leu-2*	*leu-3*	*leu-4*
None	82.8	0	0	0	0
Leu	74.0	42.0	56.7	28.0	37.0
Leu-Leu	83.0	43.7	14.3	2.0	2.0
Leu-Leu-Leu	—	73.0	68.0	20.4	55.0
Leu-Gly-Gly	70.0	0.0	0.0	0.0	0.0
Leu-Gly	68.0	0.0	0.0	0.0	0.0
Gly-Leu	75.0	0.0	0.0	0.0	0.0
Gly-Leu-Tyr	—	33.5	52.0	19.8	26.0
Leu-Phe	68	2.0	0.0	0.0	0.0
Phe-Leu	56	6.5	3.5	2.0	3.0
His-Leu	72	0.0	0.0	0.0	0.0
Leu-β-Ala	78	0.0	0.0	0.0	0.0
Gly-D-Leu	84	—	—	0.0	—
Gly-Ala-Leu	—	31.0	49.0	14.0	23.0
Met-Leu	74	0.0	0.0	0.0	2.0
Lys-Leu	—	0.0	0.0	0.0	0.0

[a]Growth as milligrams dry weight following 3 days growth at 30 °C. Values represent the averages of from three to ten separate determinations.
[b]Supplements at 0.33 mM.
Reproduced from Wolfinbarger and Marzluf (1974) with the permission of the American Society for Microbiology.

competing amino acids, e.g. phenylalanine (Wolfinbarger and Marzluf, 1974).

The general conclusions from the above growth assays of *Neurospora* amino acid auxotrophs on peptides is that dipeptides do not support growth whereas tripeptides do permit growth, even in the presence of excess competing amino acids. These data suggest the presence of an oligopeptide transport system in this fungus that does not recognize or transport dipeptides or amino acids. This system would appear to be analogous to the oligopeptide transport system in yeast in that dipeptides do not compete well with tripeptides for transport (see Section 2.2 and Becker and Naider, 1977 but see addendum, 2.2). It is not, however, consistent with the bacterial oligopeptide transport system which does appear capable of recognizing dipeptides to a certain extent (see Section 2.1 and addendum, 2.1). With some exceptions, *Neurospora* possesses intracellular peptidase activity for most of the peptides tested, even those which will not support growth (Wolfinbarger and Kay, 1973).

3 PEPTIDASE ACTIVITY IN PEPTIDE UTILIZATION

Studies of peptidase enzymes in fungi have been undertaken in a variety of microorganisms, for example, *Endothia parasitica* (Whitaker and Caldwell,

1973), *Saccharomyces cerevisiae* (Becker *et al*., 1973; Naider *et al*., 1973; Hansen *et al*., 1977), *Neurospora crassa* (Tan, 1976; Johnson and Brown, 1974), *S. carslbergensis* (Felix and Brouillett, 1966; Maddox and Hough 1970), *Aspergillus* (Nakadai *et al*., 1972,1973; Johnson and Peterson, 1936), etc. A general conclusion arising from the majority of these studies is that peptidase enzymes are localized intracellularly and are not secreted into the medium. There are, however, exceptions to this generalization (Becker *et al*., 1973; Tan, 1976).

With apologies to those who have carried out the many fine studies of peptidase enzymes in other fungal microorganisms, coverage of peptidase activities in this paper will be restricted to those studies directly related to utilization of peptides supplied from an extracellular source. The reader should not, however, get the idea that intracellular peptidases are involved solely in hydrolysis of peptides obtained from the cellular medium, for they are in all probability also involved in a variety of cellular processes ranging from trimming of nascent polypeptide chains (Johnson and Brown, 1974) to protein turnover (Hansen *et al*., 1977) and destruction of toxic peptides (Meisler and Simmonds, 1963) (see also Sections 2.4 and 3.4).

In studies of peptide utilization employing growth assays, it is advantageous to use a microorganism which does not secrete peptidohydrolytic enzymes into the medium. The presence of such enzymes in the medium makes it difficult to decide from a simple growth response whether the peptides themselves, or their hydrolytic products, are being transported. In studies with *Neurospora crassa*, the presence of a variety of extracellular peptidohydrolytic enzymes has made it necessary to prevent transport of the free form of the required amino acid before any conclusions could be made regarding transport of peptides. Wolfinbarger and Marzluf (1975*b*) were able to demonstrate that peptides were transported intact via a peptide transport system (as opposed to extracellular hydrolysis and transport of amino acids) yet they still have not been able to discern whether the peptide is hydrolysed during or after the transport step. Unlike studies in bacterial systems, where the transported peptide can be detected intact in the cell following the transport step (see Section 2.1, and Levine and Simmonds, 1962; Simmonds, 1966; Payne, 1972) they were able to find only the hydrolysis products of a transported tripeptide (Wolfinbarger and Marzluf, 1975*b*). Hence, the direct involvement of a plasma membrane-localized peptidase enzyme in the transport step in *Neurospora* remains a viable possibility.

A large number, at least 11 (Tan, 1976), of intracellular peptidases have been described in *Neurospora* and they reportedly exhibit considerable overlap in their specificities for peptide substrates (Table 6). Johnson and Brown (1974), and Tan (1976) studied intracellular peptidases in *Neurospora* as potential 'trimming' enzymes in protein maturation. The

Table 6 Peptidase activities on polyacrylamide gels[a]

Substrate	I	II	IIIA	IIIB	IV	V	VI	VII	VIII	IX	X
Met-Phe, Leu-Gly											
Met-Leu, Met-Ala		+	+++	+++		+	+		+	+	
Gly-Leu-Tyr		++			+		+	+			
Met-Ala-Ser	+	+++	+		+		+		+		
Met-Pro					+	+	+				
Lys-Trp, Trp-Glu, Leu_3		+++	+								
Glu-Trp											+
Gly-Leu-Gly-Leu		++	+								
Gly-Met	+?			+			+		+	+	
Gly-Ala-Leu		++	+		+						
Met-His		+		++							
Leucyl-*p*-nitroanilide		++									
R_f	0.79	0.69	0.65	0.63	0.59	0.57	0.56	0.53	0.52	0.44	0.27

[a]The peptidase activities were stained for bands of peptide hydrolysis in the gels by the method described by Tan (1976). The relative intensities of the peptidase activities are indicated by the number of plus signs (+) where +++ indicates intense peptidase activity and +? indicates faint, weak activity (S. T. Tan, unpublished data).

more extensive study by Tan, however, attempted also to assess the involvement of these peptidases in protein and peptide utilization by *Neurospora*. He reported 11 different peptidases, as discernible on disc-gel electrophoresis using a limited number of peptide substrates, and concluded that these represented a minimum estimate for the total number of such enzymes. A minimum of three enzymes, peptidases II, IIIB, and IV, were reported capable of hydrolysing leucine-β-naphthylamide (the traditional substrate for a leucine aminopeptidase). However, the relationship of these aminopeptidases to those described by others (Siepen *et al*., 1975; Yu *et al*., 1976) is not known at present. The peptidases IIIB, V, IX, and X hydrolyse dipeptides only. Of the peptide substrates studied, glutamyltryptophan appeared to be the only peptide hydrolysed by only one peptidase (peptidase X). Peptidase activities were constitutive, and varied little in either a qualitative or quantitative manner with developmental stage. It should be noted that all peptidase activity studied was soluble in the cytoplasm, and the particulate membrane fraction was not examined for peptidase activity (Tan, 1976). The implication of Tan's work to a study of peptide transport and utilization is that it may be difficult to separate transport from hydrolysis because of the overlap in peptidase specificities (note that Leu-Gly is hydrolysed by seven of the 11 peptidases), unless one chooses to study transport of a peptide such as Glu-Trp. Such studies might however be complicated by the apparent absence, or low activity, of a dipeptide transport mechanism in *Neurospora* (Wolfinbarger and Marzluf, 1974).

Analysis of crude cell-free extracts of *Neurospora* revealed that the peptides lysylleucine, leucylphenylalanine, lysylglycine, and histidylglycylglycine were not hydrolysed by intracellular peptidase activities (under the conditions employed) and that the dipeptides histidylglycine and glycylhistidine were hydrolysed only poorly. Peptidase activity for histidylglycine was found to be variable, even though the *his-6* auxotroph can grow on this peptide as a source of histidine (Wolfinbarger and Marzluf, 1974). The potential involvement of peptidase inhibitors, such as those described by Kaehn and Kula (1977), in peptide utilization by *Neurospora* has not been determined.

There is little doubt that peptidase activity is involved in peptide utilization by *Neurospora crassa*. The sum total of intracellular peptidase activity is both qualitatively and quantitatively different from the extracellular peptidase activity but it is not at present known whether the latter peptidases are specifically secreted or are primarily liberated due to cell death or damage. The inability of various auxotrophs of *Neurospora* to grow on certain peptides, while possessing intracellular peptidase activity capable of hydrolysing these same peptides, tends to support the idea that peptidase activity is not located within the cell wall or membrane. This observation holds true only as a generality however, and does not preclude the possible localization of specific peptidases in the plasma membrane.

4 SIZE RESTRICTION ON PEPTIDE UTILIZATION

Because of the potential involvement of a peptide transport mechanism in the utilization of extracellular proteins it becomes of interest to know what is the upper size limit for a given peptide transport system(s). The molecular sieving studies with *Neurospora crassa* cell walls by Trevithick and Metzenberg (1966 *a,b*) suggested that, unlike bacterial cells (Payne and Gilvarg, 1968), this structure in fungi might not restrict access by peptides to a plasma membrane-localized transport mechanism. Thus, if this conclusion were correct, then the inability of an amino acid auxotroph of *Neurospora* to grow on large peptides e.g. penta- or hexapeptides, might be taken as evidence for an upper size limit for transport by a peptide transport system. Investigating the same topic in bacteria, Payne and Gilvarg (1968) had earlier fractionated an enzymic digest of peptone (Neopeptone, Difco) into mixtures of peptides of similar hydrodynamic volumes by passage of the Neopeptone through a Sephadex G-15 column. Employing such mixtures in bacterial growth assays, they showed that *E. coli* could not utilize peptides larger than about a pentapeptide and that this size restriction was mediated not by the peptide transport system itself but by the bacterial cell wall acting as a permeability barrier. Using similar ideas and techniques, Wolfinbarger and Marzluf (1975*a*) showed that amino acid auxotrophs of *Neurospora* were also unable to grow on peptides larger than pentapeptides (Figure 1). The *leu-2* auxotroph employed in these growth assays was presumed to have the same cell wall porosity as reported for *Neurospora* by Trevithick and Metzenberg (1966 *a,b*). In order to reveal whether or not it was the cell wall that was restricting access of larger peptides, a mutant (*os-1*) was crossed into the *leu-2* background and the double mutant, *leu-2*; *os-1* was used in growth assays on peptides. The *os-1* mutation results in larger ‘pores’ in the cell wall, such that molecules approaching a molecular mass of 18 500 daltons can permeate the cell wall, as opposed to 4750 daltons for the wild type (Trevithick and Metzenberg, 1966*b*), and presumably for the parent *leu-2* strains. The *leu-2*; *os-1* double mutant was also unable to grow on peptides larger than a pentapeptide (Wolfinbarger and Marzluf, 1975*a*). Thus, it was suggested that, unlike bacterial cells, fungal cells (*Neurospora* specifically) might be restricted in growth on larger peptides by an inability of the transport mechanism itself to accommodate peptides exceeding a hydrodynamic volume similar to that of trileucine. Such a hydrodynamic volume could presumably apply to a pentapeptide of mixed amino acid residues.

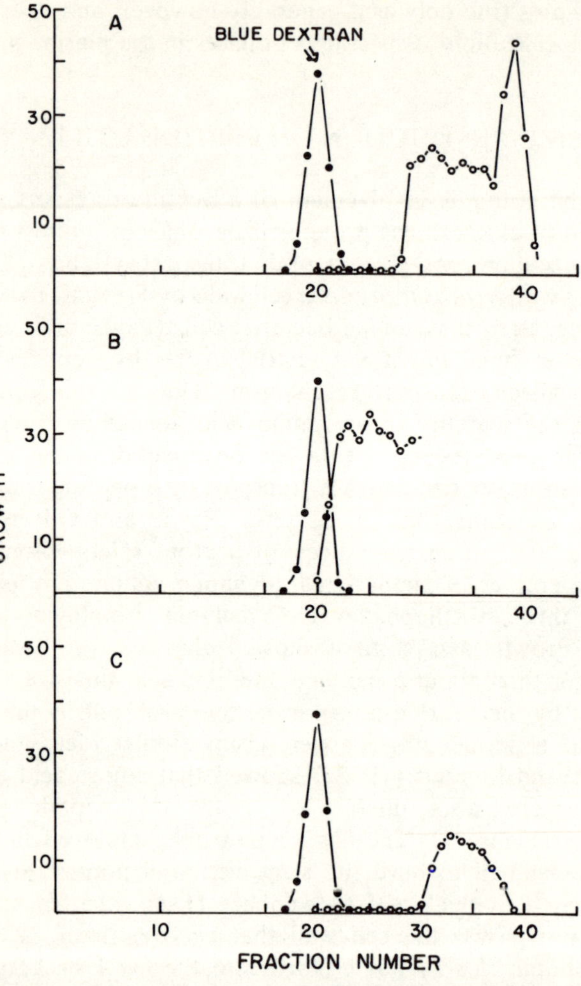

Figure 1 Growth (○——○) (milligrams dry weight of mycelial pad) of *Neurospora crassa* leucine auxotroph (*leu-2*) on 1-ml aliquots of the various fractions of Neopeptone (A), on acid hydrolysed fractions of Neopeptone (B), and on the fractions of Neopeptone plus 3.3 mM phenylalanine (C). The 'peaks' of blue dextran (●——●) indicate the void volume as it related to subsequent fractions of Neopeptone. (Reproduced from Wolfinbarger and Marzluf (1975*a*) with permission of the American Society for Microbiology.)

Marder *et al*. (1977), studying yeast, have reported that the restriction on upper size limit of peptide transport is different for different strains and have not been able to show peptide uptake is curtailed at a particular hydrodynamic volume of peptide (see also Section 2.2.9).

Additional evidence in support of the presence of a single oligopeptide transport system in *Neurospora* came from study of a mutant strain unable to grow on any leucine-containing peptide as a source of leucine. The mutation was designated as *glt-r* (Wolfinbarger and Marzluf, 1975*a*) and was crossed into the leu-2 background to yield the double mutant *leu-2*; *glt-r*. The *glt-r* mutation was originally picked up in a tyrosine-sensitive mutant (*tyr-s*) (Wolfinbarger and Marzluf, 1976) as conferring resistance to the tyrosine-containing tripeptide, glycylleucyltyrosine, hence the *glt-r* locus designation. Employing the *leu-2*; *glt-r* strain in growth assays on the peptide fractions obtained by Sephadex G-15 chromatography of Neopeptone, it was found that this double mutant was unable to grow on any of the peptides as a source of leucine (Figure 2).

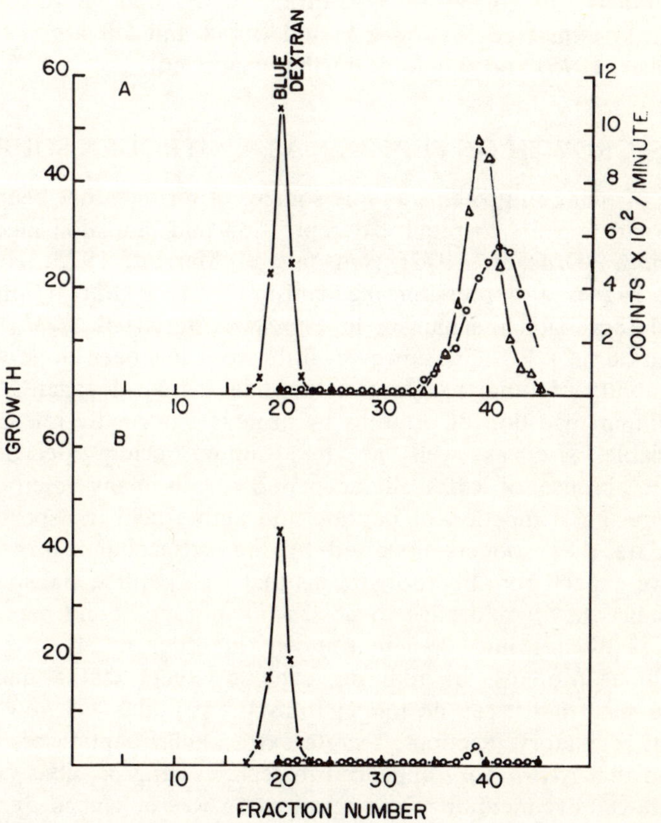

Figure 2 Growth (○——○) (milligrams dry weight of mycelial pad) of *Neurospora crassa* leucine auxotroph (*leu-2*; *glt-r*) on 1-ml aliquots of the various fractions of Neopeptone (A) and on 1-ml aliquots of fractionated Neopeptone in the presence of 3.3 mM phenylalanine (B). The 'peak' of radioactive leucine occurred in fractions of Neopeptone to which [^{14}C]-leucine was added prior to fractionation over the Sephadex G-15 column (not used in growth tests) (△——△). (Reproduced from Wolfinbarger and Marzluf (1975*a*) with permission of the American Society for Microbiology.)

Growth occurred only on those fractions shown to contain free leucine (Wolfinbarger and Marzluf, 1975*a*). The *leu-2*; *glt-r* strain was found to be equivalent to *leu-2* with respect to the peptides hydrolysed by intracellular peptidase activity. Thus, the presumptive removal, via mutation of the oligopeptide transport system in *Neurospora* resulted in the inability to grow on peptides ranging in size from a pentapeptide down to a tripeptide (remember that, in general, dipeptides are not handled well if at all by the oligopeptide transport system). Furthermore, since the peptides present in the fractionated Neopeptone should be very heterogeneous in amino acid content and sequence, it may tentatively be concluded that there is only one oligopeptide transport mechanism, at least for leucine-containing oligopeptides. This conclusion is contrary to the findings in certain other microbial systems (see Section 2.1, and Barak and Gilvarg, 1975; Naider and Becker, 1975) and should be further examined.

5 GROWTH ON PEPTIDES AS A NITROGEN SOURCE

Growth of fungi on proteins as sole sources of nitrogen has been fairly well studied and appears to require derepression and induction mechanisms in *Neurospora* (Drucker, 1972; Hanson and Marzluf, 1973; Heiniger and Matile, 1974), a depression mechanism in *Aspergillus* (Cohen, 1973), and a derepression mechanism in *Candida* (Ogrydziak *et al*., 1977) (see also Section 3.2). To date however, little work has been done with respect to the ability of fungi to grow on peptides as a sole agent of nitrogen. Extracellular digestion of proteins by proteases normally releases peptides of variable size as well as free amino acids. Because of the apparent absence of extracellular peptidases in many microbial cells it is the presumed function of peptide and amino acid transport systems to accumulate the products liberated by the extracellular proteases. What might we expect from microorganisms that lack peptide transport systems? Presumably they would have to possess membrane-bound peptidases or to secrete peptidases into the environment in order to effectively grow on extracellular proteins. In addition, can we expect that a microorganism mutated such that it can no longer transport peptides will now, because of different regulatory reactions, secrete extracellular peptidases? It has been reported that *Neurospora* appeared to secrete some peptidase activity into the extra-cellular medium when germination was enhanced by inclusion of small amounts of a required amino acid (Wolfinbarger and Marzluf, 1975*a*). Using the extracellular protease regulatory system as an analogy (see Section 3.2), it is of interest to know if *Neurospora* can be induced to secrete peptidases by first starving it for a nitrogen source and then adding peptides into the medium. Now, if conidia of *leu-2* are inoculated into a nitrogen-free medium (i.e. Vogel's minimal medium N with the NH_4NO_3

omitted (Vogel, 1956)) supplemented with 2 per cent sucrose and the peptide fractions obtained by column separation of Neopeptone, it is found that growth occurs on *all* sizes of peptides (L. Wolfinbarger, Jr. unpublished data). Indeed, both strains *leu-2*, and *leu-2*; *glt-r* exhibit comparable growth on peptides ranging in size from 30 down to two amino acid residues when they are used as a source of both nitrogen and leucine. Of particular interest however, is the observation that *leu-2* secretes peptidohydrolytic activity (as assayed using hydrolysis of casein yellow (Calbiochem, San Diego, Calif.)) only in response to peptides larger than a pentapeptide, i.e. those peptides which were unable to support growth of *leu-2* when used solely as a source of leucine (see Figure 1). The *leu-2*; *glt-r* mutant however, is found to secrete peptidohydrolytic activity in response to all sizes of peptides in the Neopeptone fractions. Only those fractions earlier shown to contain free leucine (i.e. presumably not containing peptides) (see Figure 2) failed to induce the secretion of extracellular peptidohydrolytic activity by strain *leu-2*; *glt-r* (L. Wolfinbarger, Jr, unpublished data). We are presently examining this peptide-induced extracellular peptidohydrolytic activity to determine whether peptidases and/or proteases are present. Dr Harvey Drucker at Battelle Northwest has shown that the total peptidohydrolytic activity produced in response to peptides of three to five amino acid residues exhibits an immunological cross reactivity with antibody prepared against the purified alkaline protease synthesized by *Neurospora*.

Cohen and Drucker (1977) have recently suggested an involvement of amino acid pools in the regulation of extracellular protease production. Their studies reveal that under nitrogen-starvation conditions the total nitrogen content of cells declines from a value of 9 per cent, immediately after transfer to a nitrogen-free medium, to a value of 4 per cent, 4 h after transfer. Addition of bovine serum albumin results in production of substantial extracellular protease activity (after approximately 2.5 h induction) and the cell nitrogen again rises towards its original level. In addition, the amino acids arginine, tryptophan, and threonine cause a repression (analogous to ammonium ion) of protease synthesis under conditions of nitrogen starvation. The amino acids cysteine, methionine, and cystine repress protease synthesis under sulphur-starvation conditions, and all of the above amino acids plus histidine, isoleucine, leucine, lysine, phenylalanine, and valine repress protease synthesis in carbon-starved cells. In short, intracellular amino acids may effect regulation of extracellular protease production in *Neurospora* (Cohen and Drucker, 1977) (see also Section 3.2).

The obvious implications of these results lie in the potential involvement of peptide transport systems in the utilization of both proteins and peptides as sources of needed metabolites. Thus, one may ask the question: how does a fungal organism recognize the presence of an exogenous source of a needed metabolite such as an amino acid, or nitrogen, when these components are tied up, chemically, in a molecule that is too large to permeate the cell wall, something it would presumably have to do if it is to interact with a plasma membrane-localized receptor? Does a 'common' receptor mechanism recognize both proteins and small peptides? Is the presumed receptor located in or on the exterior face of the cell wall, and if it is, what is the secondary and even tertiary signal(s) that 'turns on' the appropriate genetic response? Presumably, since a tripeptide can apparently 'induce' the appearance of extracellular proteases in the *glt-r* mutant, peptides are not the 'secondary signals'. Is the cell surface 'receptor' a protease, such that in the presence of an exogenous protein, low levels of peptides and/or amino acids are released and subsequently transported to act as the 'secondary signal'? If such were to be the case, and peptide and amino acid transport were to be blocked using appropriate mutants, could a cell population be induced to produce and/or secrete extracellular proteolytic activity? If the answer to this last question were to be yes, would that imply a second messenger system analogous to cyclic AMP production by adenyl cyclase? It should be obvious from such speculations that studies of peptide transport and utilization in any cell system have far reaching implications for numerous other systems under study.

6 KINETICS OF PEPTIDE TRANSPORT

Studies of the kinetics of peptide transport in fungi, and all other organisms, have been hampered by the scarcity of commercially radiolabelled peptides (see also Section 2.1). As discussed elsewhere in this volume the kinetics of peptide transport have been examined in bacteria (Section 2.1) and yeast (Section 2.2). In the fungus *Neurospora* transport studies to date have been restricted to a single publication (Wolfinbarger and Marzluf, 1975*b*) and to transport of a single tripeptide. However, in view of the strong probability that *Neurospora* possesses only a single oligopeptide transport system it is anticipated that conclusions derived from the single study will be applicable to transport of other peptides.

The radiolabelled tripeptide glycyl-D,L-leucyl-[^{3}H]-tyrosine was synthesized by the dicyclohexylcarbodiimide condensation of the carbobenzoxy (CBZ)-derivative of glycylleucine and the methylester of [^{3}H]-tyrosine. Although the synthesis of a peptide using these reagents normally leads to the presence of both stereoisomers the kinetics of inhibition by glycyl-L-leucyl-L-tyrosine suggested that either both the D- and L-stereoisomers were transported equally or that the D-stereoisomer was not transported to any great degree. It also remains a possibility that little of

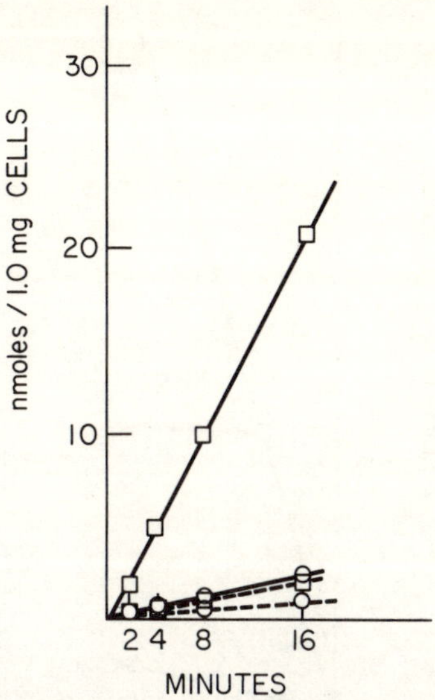

Figure 3 Transport (pH 5.8) of glycyl-D,L-leucyl-[^{3}H]-tyrosine (0.1 mM) by *Neurospora crassa* leucine auxotroph (*leu-2*) in the presence (□----□) and absence (□——□) of 1.0 mM glycylleucyltyrosine, and by strain *leu-2*; *glt-r* in the presence (○----○) and absence (○——○) of 1.0 mM glycyleucyltyrosine. Cells grown to mycelial stage (17–18 h) in acetate-minimal medium prior to transport assays. (Reproduced from Wolfinbarger and Marzluf (1975*b*) with permission of Academic Press Inc.)

the D-stereoisomer was present from the synthesis. Kinetic transport assays using this tripeptide in mycelia of *Neurospora* suggested that there was indeed an oligopeptide transport system and that mutation in a single gene, the *glt-r* locus, resulted in a 90 per cent reduction in transport velocity (Figure 3). Transport activity for the tripeptide exhibited typical saturation for substrate with an apparent K_m of 3.4×10^{-5} M and a V_{max} of 2.07 nmol min^{-1} (mg dry weight cells)$^{-1}$. Comparable values have since been reported for peptide transport in yeast (Becker and Naider, 1977). It was not determined whether the transport was active in *Neurospora* because it was not possible to effectively separate the transport step from the hydrolysis of the tripeptide. Sampling of the intracellular radioactivity after as little as 16 min revealed only radioactive tyrosine. In yeast, sampling times as short as 30 s has indicated intracellular hydrolysis of transported peptides (Becker and Naider, 1977). However, sensitivity to energy inhibitors such as sodium azide have traditionally been employed to help determine whether or not a transport system is active, and analysis of the transport activity in *Neurospora* and in yeast (Becker and Naider, 1977) revealed a rapid inhibition of peptide transport by azide when added either prior to, or after, addition of radioactive tripeptide (Figure 4). The intracellular

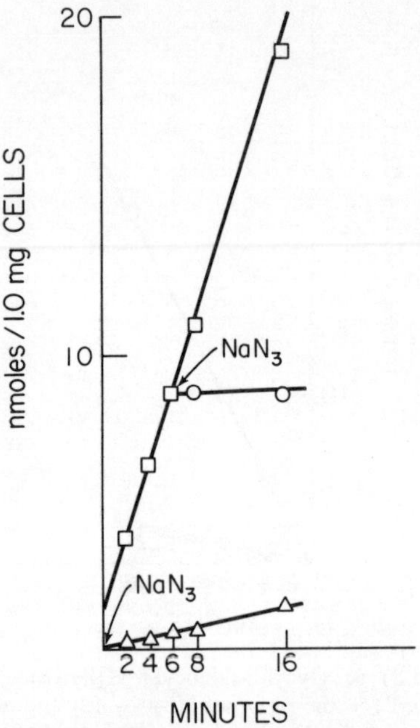

Figure 4 Transport (pH 5.8) of glycyl-D,L-leucyl-[^{3}H]-tyrosine (0.1 mM) by *Neurospora crassa* leucine auxotroph (*leu-2*) cells (17–18 h mycelia) in the presence (○——○) and (△——△) and absence (□——□) of 5 mM NaN_3 added at the times indicated. (Reproduced from Wolfinbarger and Marzluf (1975*b*) with permission of Academic Press Inc.)

peptidase activity, as measured *in vitro* (Tan, 1976; Wolfinbarger and Marzluf, 1975*b*), is not sensitive to inhibition by azide, and accumulation of the tripeptide by metabolic drag can be excluded as a possibility for transport energization. Hence, it seems reasonable to suggest that the transport of tripeptides by *Neurospora* and *Saccharomyces cerevisiae* is active at least in the sense that it is dependent upon metabolic energy. It is anticipated that the availability of peptidase-deficient mutants will permit demonstration of the accumulation of intact oligopeptides.

Although of a somewhat cursory nature, preliminary data (Wolfinbarger and Marzluf, 1975*b*) suggest that the oligopeptide transport system in *Neurospora* is constitutive in character. Deprivation of readily useable carbon or nitrogen sources did not dramatically change the level of peptide transport activity. This observation appears to be analogous to that observed for intracellular peptidase activity (Tan, 1976), yet markedly different from the situation with extracellular peptidohydrolytic activity (Drucker, 1972; Hanson and Marzluf 1973; Heiniger and Matile, 1974). A comparable oligopeptide transport system in yeast appears to be regulated to a certain extent by the available nitrogen source (Becker and Naider, 1977).

7 SUMMARY

The major difficulty in reviewing peptide transport and utilization in fungi at the present time is the lack of extensive information. The yeasts represent the only fungal system other than *Neurospora* for which a significant amount of information has been published. Consequently, the major thrust of this review has been to present an overview of the interrelationships between extracellular peptidohydrolases, peptide and amino acid transport systems, and intracellular peptidases. Since comparable systems in yeast are being covered elsewhere in this text (Section 2.2) this review of fungi has been confined almost exclusively to *Neurospora*.

Considerable differences are becoming apparent in the properties of different fungal systems for protein and peptide transport and utilization. Differences occur in the regulatory mechanism(s) for production and secretion of extracellular peptidohydrolases; yeast may have more than one oligopeptide transport system (Marder *et al*., 1977) (but see also Section 2.2.5) and grows on dipeptides (Becker *et al*., 1973; Naider *et al*., 1973; Marder *et al*., 1977), and the upper size limit for peptide transport in yeast may be larger than for *Neurospora* (Naider *et al*., 1973).

We may summarize our current understanding of the systems in *Neurospora* by suggesting that under starvation conditions extracellular proteases are secreted which hydrolyse proteins down to monomer amino acids and peptides. Those peptides smaller than a hexapeptide (with the exception of dipeptides) are transported by the single oligopeptide transport system. The transported peptides are hydrolysed either during, or immediately following, the transport step and the products derived from the liberated amino acids, or perhaps the amino acids themselves (Cohen and Drucker, 1977), dictate continued synthesis and/or release of extracellular peptidohydrolytic activity.

ACKNOWLEDGEMENTS

During the writing of this manuscript the author was supported in part by a grant from the Old Dominion Universtity Research Foundation. Appreciation is extended to Sai Tee Tan for permission to quote some of his unpublished data. The excellent technical assistance of Nadean Salalila in preparation of the manuscript is also acknowledged.

REFERENCES

Barak, Z. and Gilvarg, C. (1975). *J. Bacteriol*., **122**, 1200–1207

Becker, J. M. and Naider, F. (1977). *Arch. Biochem. Biophys*., **178**, 245–255

Becker, J. M., Naider, F., and Katzir-Katchalski, E. (1973). *J. Biol. Chem.*, **249**, 9–20
Cohen, B. L. (1973). *J. Gen. Microbiol.*, **79**, 311–320
Cohen, B. L. and Drucker, H. (1977). *Arch. Biochem. Biophys.*, **182**, 601–613
Drucker, H. (1972). *J. Bacteriol.*, **110**, 1041–1049
Felix, F. and Brouillet, N. (1966). *Biochim. Biophys. Acta*, **122**, 127–133
Hansen, R. J., Switzer, R. L. Hinze, H., and Holzer, H. (1977). *Biochim. Biophys. Acta*, **496**, 103–114
Hanson, M. and Marzluf, G. A. (1973). *J. Bacteriol.*, **116**, 785–789
Heininger, U. and Matile, Ph. (1974). *Biochem. Biophys. Res. Commun.*, **60**, 1425–1432
Johnson, G. L. and Brown, J. L. (1974). *Biochim. Biophys. Acta*, **370**, 530–540.
Johnson, M. J. and Peterson, W. H. (1936). *J. Biol. Chem.*, **112**, 25–31.
Kaehn, K. and Kula, M. R. (1977). *FEBS Lett.*, **79**, 85–90
Levine, E. M. and Simmonds, S. (1962). *J. Bacteriol.*, **84**, 683–689
Maddox, I. S. and Hough, J. S. (1970). *Biochem. J.*, **117**, 843–847
Magill, C. W., Sweeney, H., and Woodward, V. W. (1972). *J. Bacteriol.*, **110**, 313–320
Magill, C. W., Nelson, S. D., D'Ambrosio, S. M., and Glover, G. I. (1973). *J. Bacteriol.*, **113**, 1320–1325
Marder, R., Becker, J. M., and Naider, F. (1977). *J. Bacteriol.*, **131**, 906–916
Marzluf, G. A. and Metzenberg, R. L. (1968). *J. Mol. Biol.*, **33**, 423–427
Meisler, N. and Simmonds, S. (1963). *J. Gen. Microbiol.*, **31**, 109–114.
Naider, F. and Becker, J. M. (1975). *J. Bacteriol.*, **122**, 1208–1215
Naider, F., Becker, J. M. and Katzir-Katchalski, E. (1973). *J. Biol. Chem.*, **249**, 9–20
Nakadai, T., Nasumo, S., and Iguchi, N. (1972). *Agric. Biol. Chem., Tokyo*, **36**, 28–32
Nakadai, T., Nasumo, S., and Iguchi, N. (1973). *Agric. Biol. Chem., Tokyo*, **37**, 757–765
Ogrydziak, D. M., Demain, A. L., and Tannenbaum, S. R. (1977). *Biochim. Biophys. Acta*, **497**, 525–538
Pall, M. L. (1969). *Biochim. Biophys. Acta*, **173**, 113–119
Pall, M. L. (1971). *Biochim. Biophys. Acta*, **233**, 201–207
Payne, J. W. (1972). *J. Gen. Microbiol.*, **71**, 259–263
Payne, J. W. (1975). In *Peptide Transport in Protein Nutrition* (D. M. Matthews and J. W. Payne, eds), North-Holland and American Elsevier, Amsterdam and New York, pp. 293–312
Payne, J. W. and Gilvarg, C. (1968). *J. Biol. Chem.*, **243**, 6291–6299
Rao, E. Y. T., Rao, T. K., and DeBusk, A. G. (1975). *Biochim. Biophys. Acta*, **413**, 45–51
Scarborough, G. A. (1973). *Int. Rev. Cytol.*, **34**, 103–122
Siepen, D., Yu, P. H., and Kula, M. R. (1975). *Eur. J. Biochem.*, **56**, 271–281
Simmonds, S. (1966). *J. Biol. Chem.* **241**, 2502–2509
Tan, S. T. (1976). Ph. D. Dissertation, Ohio State University
Trevithick, J. R. and Metzenberg, R. L. (1966*a*). *J. Bacteriol.*, **92**, 1010–1015
Trevithick, J. R. and Metzenberg, R. L. (1966*b*). *J. Bacteriol.*, **92**, 1016–1021
Vogel, H. J. (1956). *Microbial Genet. Bull.*, **13**, 42–43
Whitaker, J. R. and Caldwell, P. V. (1973). *Arch. Biochem. Biophys.*, **159**, 188–200
Wolfinbarger, L., Jr (1976). *Biochim. Biophys. Acta,* **436**, 774–788
Wolfinbarger, L., Jr and DeBusk, A. G. (1971). *Arch. Biochem. Biophys.*, **144**, 503–511
Wolfinbarger, L., Jr and Kay, W. W. (1973). *Biochim. Biophys. Acta*, **330**, 335–343
Wolfinbarger, L., Jr and Marzluf, G. A. (1974). *J. Bacteriol.*, **119**, 371–378
Wolfinbarger, L., Jr and Marzluf, G. A. (1975*a*). *J. Bacteriol.*, **122**, 949–956
Wolfinbarger, L., Jr and Marzluf, G. A. (1975*b*). *Arch. Biochem. Biophys.*, **171**, 637–644
Wolfinbarger, L., Jr and Marzluf, G. A. (1976). *J. Gen. Microbiol.*, **93**, 189–193
Woodward, C. K., Read, C. P., and Woodward, V. W. (1967). *Genetics*, **56**, 598
Yu, P. H., Kula, M. R., and Tsai, H. (1976). *Can. J. Biochem.*, **54**, 699–703

Microorganisms and Nitrogen Sources
Edited by J. W. Payne

Section 2.4

Role of Peptidases in Utilization and Transport of Peptides by Bacteria

CAROL L. HERMSDORF AND SOFIA SIMMONDS

Department of Molecular Biophysics and Biochemistry, Yale University, New Haven, Connecticut, USA

1 INTRODUCTION

That many bacteria are capable of transporting and hydrolysing simple peptides has long been evident from the results of growth tests and uptake experiments, especially tests with amino acid auxotrophs of *Escherichia coli*. The data initially obtained with *E. coli* auxotrophs (summarized by Fruton and Simmonds, 1950) led to the hypothesis that hydrolysis of peptides precedes the utilization of the constituent amino acids for growth, and results subsequently obtained in several laboratories attest to the fact that peptidases must participate not only in the use of exogenous peptides as growth fractors and in the destruction of toxic peptides composed solely of L-amino acids, but in other aspects of peptide and protein metabolism including cell wall synthesis and degradation (Blumberg and Strominger, 1974; Goldberg and St. John, 1976; Miller, 1975; Payne, 1975, 1976, 1977). This presentation will emphasize the recent developments. More detailed descriptions of the earlier work may be found in Payne (1975, 1976).

The ease with which *E. coli* may be cultured in chemically defined media and converted to mutants that are readily characterized both biochemically and genetically has made this bacterium highly useful for the study of all aspects of peptide metabolism. Although much of the work directed toward the identification of *E. coli* enzymes active toward peptide bonds linking L-amino acids was initiated in order to find the biochemical bases for the ability of small peptides to act as growth factors (see for example Stone, 1953; Simmonds and Griffith, 1962; Van Lenten and Simmonds, 1967) and as growth inhibitors (e.g., Simmonds *et al.*, 1951; Meisler and Simmonds, 1963; Sussman and Gilvarg, 1970), other work in this field was prompted mainly by an interest in comparing bacterial peptidases with those of mammalian tissues (e.g. Sarid *et al.*, 1962; Matheson, 1963; Yaron, 1974; Patterson, 1976) or, more recently, by a desire to identify the 'methionine aminopeptidase' specifically concerned with the removal of amino-terminal Met residues from nascent polypeptide chains during protein biosynthesis (summarized by Simmonds, 1970; see also Matheson *et al.*, 1971; Brown and Krall, 1971) and to determine the role of peptidases in protein degradation *in vivo* (summarized by Goldberg and St. John, 1976; see also Kowit *et al.*, 1976) (and Section 3.4). The results of such studies on *E. coli* K-12, the most thoroughly investigated strain, are outlined in Table 1.

2 DIPEPTIDASES

While only one peptidase (DP) limited in its action to the hydrolysis of dipeptides has been obtained from extracts of *E. coli* K-12 (Table 1), two broad specificity dipeptidases have been isolated from *E. coli* B. One, termed dipeptidase M, resembles DP in its high activity toward Met-X substrates, activation by Mn^{2+} (but not by Zn^{2+}), stabilization by β-mercaptoethanol, and dimeric structure (subunit molecular weight *c.* 47 000 daltons) (Brown, 1973). These properties distinguish dipeptidase M from the second dipeptidase which was purified by Patterson and her collaborators (Patterson, 1976; Hayman *et al.*, 1974); it also is a dimer (subunit molecular weight 53 000 daltons), but, unlike dipeptidases M and DP, it contains two atoms of Zn per molecule and is not an 'SH-enzyme', The specificity of the Zn-dipeptidase is altered in different ways by the addition of Mn^{2+} or Co^{2+}: substrates rapidly hydrolysed by the Zn-enzyme *per se* are generally hydrolysed at slower rates in the presence of added Mn^{2+} or Co^{2+}, whilst with other substrates the addition of Mn^{2+} or Co^{2+} has an activating effect. The presence of a Zn-dependent dipeptidase in strain K-12 seems unlikely in view of the marked inhibition of dipeptide

Table 1 Peptidases shown to be present in Escherichia coli K-12[a]

Enzyme	Properties and occurrence	Known synthetic substrates
Dipeptidases[b]		
Dipeptidase DP (peptidase D[c])[d]	Mn(or Co)- and BME-dependent; inhibited by Zn. Probably a dimer, subunit molecular weight *c*. 55 000 daltons. Activity absent in *pepD*$^-$ mutants.	Dipeptides X-X and X-Y, where X and Y are Ala, Gly, Leu, Lys, Met, or Phe. Highest activity to Met-Ala(or Met). Very low activity to Leu-X and Lys-X.
Iminodipeptidase[e]	(No information available.)	Dipeptides with *N*-terminal Pro or Hyp. (Inactive to other substrates.)
Peptidase Q[c]	Activity absent in *pepQ*$^-$ mutants.[c]	Leu-Pro (and probably other A-Pro dipeptides).
Aminopeptidases[b]		
Tripeptidase TP	Probably Mn-dependent; inhibited by Zn. Not affected by SH-reagents or serine protease inhibitors. Particle molecular weight *c*. 80 000 daltons. Active only to unsubstituted tripeptides; inactive to *N*- or *C*-terminal substituted dipeptides. Activity present in *pepN*$^-$ *pepB*$^-$ *pepA*$^-$ *pepD*$^-$ mutant.	Leu (or Met or Lys or Phe)-Gly-Gly; Leu (or Met)-Gly-Leu (or Met); Leu (or Met)-Leu-Gly; [Lys_3 and Lys-A-B substrates]; Met-Ala-Met (or Ser). Prefers *N*-terminal Met. (Inactive to Gly_3.)
Aminopeptidase I[f] (peptidase A[c])	Mn(or Mg)-dependent, heat-stable enzyme with weak dipeptidase activity; inhibited by Zn. Hexamer, subunit molecular weight 52 000 daltons. Found ribosome-bound after disruption of cells in buffers of low ionic strength. Activity absent in *pepA*$^-$ mutants.	Di- and oligo-peptides with unsubstituted *N*-terminal NH_2-groups including Leu (or Met)-Gly-Gly and Met-Gly-Met-Met. Met-A-B substrates hydrolyzed most rapidly; Met-Ala-Ser-OMe hydrolysed more slowly than Met-Ala-Ser.
Aminopeptidase AP (peptidase B[c])[d]	Probably Mn-dependent; inhibited by Co and Zn; stimulated by BME. Particle molecular weight *c*. 230 000 daltons. Shows high dipeptidase activity; A-A and A-B dipeptides hydrolysed slower than A-A- and A-B- tripeptides. Activity absent in *pepB*$^-$ mutants.	[Leu (or Met or Phe)-Gly]; Leu (or Met)-Leu; Met-Met; [Leu (or Met or Phe)-Gly-Gly; Leu (or Met)-Gly-Leu(or Met)]; Leu (or Met)-Leu-Gly; Met-Ala-Met (or Ser); [Met-Phe-Gly(or Met); Met-Gly-Met-Met; Lys-Ala-Ala; Lys-Trp-Lys]. (Inactive to Lys-Lys and Lys_3).

Table 1 Peptidases shown to be present in Escherichia coli K-12[a] (*continued*)

Enzyme	Properties and occurrence	Known synthetic substrates
'Aminopeptidase L'	Mn-stimulated enzyme with high dipeptidase activity. BME inhibitory with substrates A-X and A-X-Y where A is Met or Leu and X is Met, Leu, or Tyr; BME stimulatory with all other known substrates. Particle molecular weight *c*. 241 000 daltons. Activity absent in *pepN*⁻ *pepA*⁻ *pepB*⁻ *pepD*⁻ *pepQ*⁻ mutant.	[Ala (or Phe)-Gly]; [Leu (or Met)-Gly; Leu (or Met)-Ala; Ala-Met][g]; Leu (or Met)-Leu; Leu-Tyr; [Ala-Gly-Gly]; [Leu (or Met)-Gly-Leu(or Met)][g]; Leu (or Met)-Leu-Gly; Leu-Tyr-Leu; Met-Ala-Ser; [Leu-*p*-nitroanilide]. (Inactive to Gly-Gly, Lys-Lys, and Lys_3.)
Oligopeptidase (aminopeptidase II;[h] 'peptidase N'[c])[d]	Possibly divalent metal-dependent; inhibited by Zn; not affected by SH-reagents or serine protease inhibitors. Monomer, particle weight *c*. 100 000 daltons. Shows high dipeptidase activity. Activity absent in strain AJ005 and known '*pepN*⁻ mutants'.	[Met-Ala; Leu (or Phe)-Gly]; Leu-Leu; Lys-Lys; [Leu (or Phe)-Gly-Gly]; Lys_3; Met-Gly (or Leu)-Gly; Lys_4; β-naphthylamides of Ala, Arg, Lys [Ser, Leu, and Met], and of Ala-Ala; Leu-*p*-nitroanilide.
Proline iminopeptidase[e]	Mn-dependent; inhibited by other metals and SH-reagents. Activity restricted to substrates with *N*-terminal Pro.	Pro-Gly; Pro-Gly-Gly; oligopeptides of structure Pro-A-B-; poly-L-proline
Peptidase P[c]	(No information available.)	(Assumed responsible for ability of *pepQ*⁻ mutants to hydrolyse Leu-Pro slowly and Gly-Pro-Ala rapidly.)
Endopeptidases		
Tetralysine endopeptidase[i]	Probably divalent metal-dependent. Particle molecular weight less than 100 000 daltons.	Lys_4 (cleaved to two equivalents of Lys-Lys).
Aminoendopeptidase (aminopeptidase N)[j]	Divalent metal-independent. Monomeric, particle weight 45 000 daltons; contains one essential SH-group. Associated with outer surface of cytoplasmic membrane. Constitutive but derepressed under phosphate limitation. Absent in '*pepN*⁻ mutant'.	*p*-Nitroanilides of Ala and Leu (but not of Phe); β-naphthylamides of Ala, [Leu, and Lys]; [^{125}I]-labelled casein; serum albumin, [F_c fragment of rabbit IgG]. Active to Ala-A, [Leu-A] dipeptides, Ala_3, Ala_5.
Acetylphenylalanine naphthyl ester-hydrolysing enzyme[k]	Activity localized in periplasm; relatively heat stable. Absent in Pr⁻ mutant. (Two other enzymes with similar activity believed to be present in cytoplasm.)	*N*-Ac-D,L-Phe naphthyl ester. (Inactive to iodinized casein.)

Benzoylarginine β-naphthylamide-hydrolysing enzyme[k]	Absent in Tr^- mutant.	α-*N*-Bz-D,L-Arg β-naphthylamide.

[a]Unless otherwise noted, enzymes are cytoplasmic and substrates are composed of L-amino acid residues; relatively poor substrates are listed in brackets. Abbreviations for metals refer to divalent cations; BME is β-mercaptoethanol.
[b]Data are those of Hermsdorf (1978), Simmonds (1972), S. Simmonds (unpublished data), and Simmonds *et al*. (1976) except where otherwise indicated.
[c]Name, gene designation, and data for enzyme activity in mutant cell-extract are taken from Miller (1975) and Miller and Schwartz (1978).
[d]The identity of the two enzymes named were established by Hermsdorf *et al*. (1979) and S. Simmonds (unpublished results) using mutant strains obtained from C. G. Miller.
[e]Sarid *et al*. (1962).
[f]Vogt (1970); see also Simmonds (1972).
[g]Although poor substrates when tested in the absence of BME, peptides are relatively good substrates (compared to the reference substrate, Leu-Leu) in the presence of BME.
[h]Yang and Somerville (1976). The identity of aminopeptidase II and peptidase N was suggested by Miller (1975) on the basis of activity to amino acid β-naphthylamides. Similarity of properties of purified preparations suggest the identity of aminopeptidase II and oligopeptidase (Hermsdorf, 1978; Hermsdorf *et al*., 1979).
[i]Sussman and Gilvarg (1970).
[j]Enzyme purified from strain K-10 is said also to be present in strain K-12 *per se* and absent in its *pepN*$^-$ mutants (Lazdunski *et al*., 1975*a*; Latil *et al*., 1976; Murgier *et al*., 1976*a*, 1977; Chappelet-Tordo *et al*., 1977).
[k]Kowit *et al*. (1976).

cleavage observed both *in vivo* and *in vitro* upon the addition of Zn^{2+} (Simmonds, 1970).

Salmonella typhimurium also has been reported to contain a broad specificity dipeptidase, which was identified after polyacrylamide gel electrophoresis of crude cell extracts as a band showing activity toward Leu-Gly but not toward Leu-Gly-Gly (Miller and MacKinnon, 1974). Because an enzyme with similar mobility in the gel and similar activity toward leucine peptides is present in extracts of *E. coli* K-12, and because *pepD* mutations map near *proAB* in both organisms (Miller 1975), *S. typhimurium* peptidase D may be equated with *E. coli* K-12 DP (Table 1), (see Addendum).

The polyacrylamide gel electrophoresis and gene mapping techniques have established the presence in extracts of *Salmonella* and *E. coli* K-12 of a second dipeptidase (peptidase Q) apparently specific for substrates of structure X-Pro (McHugh and Miller, 1974; Miller, 1975; Miller and Schwartz, 1978). However, *Salmonella* is thought to lack an enzyme equivalent in activity to the strain K-12 iminodipeptidase (Miller, 1975). Nor has *Salmonella* yet been reported to contain the *E. coli* B enzyme acting specifically at the peptide bond linking the β-carboxyl group of aspartic acid to the α-NH_2 group of a second amino acid (Haley, 1970) although such activity has been observed in extracts of *E. coli* K-12 (Yang and Somerville, 1976).

3 AMINOPEPTIDASES

Many enzymes displaying aminopeptidase activity to oligopeptides act also on dipeptides. Consequently, the restricted action of tripeptidase (TP) (Table 1) makes it apparently unique among bacterial enzymes characterized to date. This enzyme, discovered during a study of lysine-homopolymer metabolism (Sussman and Gilvarg, 1970) and later shown to act on a variety of tripeptides, may be equated (Simmonds *et al*., 1976; Hermsdorf, 1978) with the Leu-Gly-Gly-cleaving tripeptidase observed as a 'variable' band (R_f c. 0.4) on polyacrylamide gel electrophoretic analysis of extracts of *S. typhimurium* (Miller and MacKinnon, 1974) and *E. coli* K-12 (Miller and Schwartz, 1978). It may be of significance with respect to the physiological functions of peptidases that no mutants lacking 'band 4' activity have been detected in either organism (Miller and Schwartz, 1978), although inability of leucine auxotrophs to grow on exogenous Leu-Gly-Gly is characteristic of *Salmonella* and *E. coli* mutants lacking all the activities ascribed to aminopeptidases I and AP and the oligopeptidase (Table 1), i.e. *pepA*$^-$ *pepB*$^-$ *pepN*$^-$ mutants (Miller, 1975; Miller and Schwartz, 1978).

Of the aminopeptidases showing dipeptidase activity that have been obtained in at least partially purified form from *E. coli* K-12 (Table 1), only aminopeptidase I has been crystallized (Vogt, 1970). A similar hexameric enzyme (subunit molecular mass c. 60 000 daltons) has been purified from the ribosomal fraction of *E. coli* B cells disrupted in buffer of low ionic strength (Matheson and Dick, 1970; Matheson *et al*., 1970; Dick *et al*., 1970); this 'r-peptidase', like aminopeptidase I, is heat-stable, especially in the presence of SH-compounds, is activated by Mn^{2+} or Mg^{2+}, and shows a preference for substrates containing Met or Leu at the amino-terminal position. Although insensitive to Zn^{2+} and displaying relatively high activity toward some dipeptides (cf. Table 1), r-peptidase is believed to be the strain B counterpart of aminopeptidase I (Vogt, 1970; Matheson *et al*., 1971). Data available for *S. typhimurium* indicate that it too contains a large, heat-stable enzyme (peptidase A) capable of releasing methionine from Met-Ala-Ser and leucine from Leu-NH_2, Leu-Gly, Leu-Ala-NH_2, Leu-Gly-Gly, and Leu-Leu-Leu; this enzyme is assumed to be the *Salmonella* equivalent of aminopeptidase I (Miller and MacKinnon, 1974).

Three other enzyme preparations showing 'Met/Leu-aminopeptidase' activity have been separated from *E. coli* K-12 extracts: AP, L, and oligopeptidase (Table 1). Aminopeptidase AP has been equated with Miller's peptidase B, identified in polyacrylamide gels as a band showing activity toward Leu-Gly and Leu-Gly-Gly (Miller and Schwartz, 1978). Whether this homology may be extended to the peptidase B of *S. typhimurium* (active toward the tripeptide substrates of peptidase A but having relatively less ability to hydrolyse Leu-NH_2 and Leu-Ala-NH_2 (Miller, 1975) is stated to be uncertain by Miller and Schwartz, 1978.

The preparation designated aminopeptidase L contains a protein of approximately the size of AP, but the two are readily separated by ion-exchange chromatography (Simmonds *et al*., 1976). Although the anomalous effects of β-mercaptoethanol (BME) on L activity suggest the presence of at least two peptidases, repeated efforts further to fractionate preparations of L have been unsuccessful and parallel losses in BME-stimulated and BME-inhibited activity occur during various treatments. Thus, such preparations have been assumed to contain a single enzyme which undergoes BME-induced changes in conformation and, therefore, in apparent di- and aminopeptidase activity; that the enzyme appears from its behaviour on gel filtration to be a polymeric protein adds support to this assumption (S. Simmonds, unpublished data). It also seems probable that L is responsible for the 'variable' bands of Leu-Gly-Gly-cleaving activity observed in polyacrylamide gel electrophoresis patterns of *E. coli* K-12 (Miller and Schwartz, 1978) and *S. typhimurium* (Miller and MacKinnon, 1974); the failure of such bands

(R_f 0.28) consistently to display activity toward Leu-Gly and the high activity seen toward Met-Ala-Ser (for *Salmonella*) can readily be explained by the absence of BME in the test medium (see Table 1 with special reference to footnote *g*).

The oligopeptidase was initially identified by Sussman and Gilvarg (1970) as the only *E. coli* K-12 enzyme acting on di-, tri-, and tetralysine, and was later shown to be an aminopeptidase active toward a variety of di- and tripeptides and toward leucine *p*-nitroanilide (Simmonds *et al.*, 1976; Hermsdorf *et al.*, 1979). Similarity of behaviour during gel electrophoresis and of ability to act as a 'leucine aminopeptidase' suggests a marked resemblance between oligopeptidase and the so-called peptidase N observed in extracts of *E. coli* K-12 (Miller and Schwartz, 1978) and of *Salmonella* (Miller and MacKinnon, 1974); moreover, a mutant of strain K-12 characterized as *pepN*$^-$ by Miller and Schwartz, 1978 has been found to lack the enzymic activity attributed to oligopeptidase (Table 1). In both *E. coli* and *S. typhimurium*, 'peptidase N' is said to be the only enzyme capable of hydrolysing alanine β-naphthylamide, an observation forming the basis for the proposed identity of *E. coli* 'peptidase N' with aminopeptidase II isolated from strain K-12 by Yang and Somerville (1976); this proposal receives additional support from the close similarity in the physical properties of purified preparations of aminopeptidase II (Yang and Somerville, 1976) and oligopeptidase (Hermsdorf and Simmonds, 1978).

Lazdunski and coworkers have presented evidence that *E. coli* contains only one enzyme catalysing the hydrolysis of alanine *p*-nitroanilide (APNA) and have purified it from strain K10 (Lazdunski *et al.*, 1975*a*,*b*; Chappelet-Tordo *et al.*, 1977); the purified enzyme appears to be responsible for the alanine β-naphthylamide (ABN)-cleaving activity ascribed to *E. coli* K-12 'peptidase N'. Strain K-12 mutants lacking hydrolytic activity to APNA ('*pepN*$^-$ mutants') also lack the band of APNA-cleaving activity that is found in polyacrylamide gel electrophoresis patterns of wild type cell-extracts (Latil *et al.*, 1976). These investigators also report they 'have found the aminopeptidase in [*S. typhimurium* active to APNA] shows the same electrophoretic migration on polyacrylamide gel as in *E. coli* and cross-reacts with antibodies raised against purified *E. coli* aminopeptidase;' moreover, the *pepN* gene of strain K-12, like that of *Salmonella* (Miller, 1975), maps close to *pyrD*. Despite this impressive evidence, it is not entirely certain that the enzyme purified from *E. coli* K10 is identical with the oligopeptidase/aminopeptidase II purified from strain K-12. The former, originally termed 'aminoendopeptidase' (Lazdunski *et al.*, 1975*a*,*b*) and subsequently renamed 'aminopeptidase N' (Murgier *et al.*, 1977), has physical properties and catalytic activity (see Table 1 under Endopeptidases) differing significantly from those of the strain K-12 preparations. Nor is it clear from the published reports whether the enzyme responsible for the absence of APNA-hydrolysing activity of '*pepN*$^-$' mutants of strain K-12 is capable of hydrolysing iodinized casein and simple di- and tripeptides, albeit all the mutants apparently contain material cross-reacting with 'an anti-wild-type aminopeptidase antiserum' (Latil *et al*., 1976). Additional confusion arises from the fact that 30–40 per cent of the activity attributed to the strain K10 'aminopeptidase N' (i.e. APNA- and ABN-hydrolysing activity) is released when K10 cells are subjected to osmotic shock or are converted to lysozyme spheroplasts, whereas the activity of other *E. coli* strains, including K-12, is characteristic of truly cytoplasmic enzymes (Murgier *et al*., 1977). It may be suggested, therefore, that the strain K10 enzyme, which was purified from the supernatant solution from lysozyme spheroplasts and which represents only about one-third of the total APNA-hydrolysing activity of the cells (Lazdunski *et al*., 1975*a*), probably is an endopeptidase devoid of activity toward typical substrates of trypsin (benzoylarginine ethyl ester) and chymotrypsin (acetyltyrosine ethyl ester) but capable of hydrolysing some amino acid *p*-nitroanilides and β-naphthylamides and, though at much slower rates, some di- and oligopeptides. It may be noted that Régnier and Thang (1975) have prepared from *E. coli* ML304 an enzyme which appears to exhibit substrate specificity not markedly different from that of the K10 'periplasmic enzyme'; however, the physical properties of these two endopeptidases differ significantly.

The ability to hydrolyse *p*-nitroanilides and naphthylamides of alanine and other amino acids is common among Gram-negative bacteria, having been detected in a variety of Enterobacteriacea as well as in *Pseudomonas* strains; whether it is equally widespread among Gram-positive organisms is still uncertain (Sheahan *et al*., 1975; Murgier *et al*., 1976*b*). However, three intracellular enzymes (AP I, II, and III) hydrolysing amino acid β-naphthylamides have been partially purified from a strain of *Bacillus subtilis* (Desmond *et al*., 1975). These Co^{2+}-stimulated enzymes, which also readily hydrolyse dialanine and degrade tetraalanine, apparently in a stepwise manner, to alanine, may be distinguished on the basis of their physical properties and catalytic activity toward the β-naphthylamides. Moreover, only AP I (molecular mass *c*. 210 000 daltons; most active to Ala-naphthylamide) and AP II (molecular mass *c*. 70 000 daltons; most active to Lys-naphthylamide) are produced by cells during the early part of the growth cycle whereas AP III (molecular mass *c*. 230 000 daltons; most active to α-Asp-naphthylamide) is produced mainly in early stationary-phase cells. The bacilli also produce an enzyme active toward D-Ala-β-naphthylamide, D-Ala-D-Ala-β-naphthylamide, and tri-D-alanine. This 'D-aminopeptidase' is found mainly in the periplasm and cell wall fraction of early stationary-phase cells. Whilst this enzyme may play a role in cell wall metabolism, its mode of action clearly differentiates it from the other cell wall-metabolizing peptidases to be considered below.

Proline iminopeptidase and peptidase P differ from the other known *E. coli* K-12 aminopeptidases that show dipeptidase activity in that each acts only on proline-containing substrates (Table 1). In the case of the well characterized iminopeptidase, proline provides the carboxyl carbon of the sensitive peptide bond in susceptible substrates. Activity attributable to a proline iminopeptidase has been observed also in *E. coli* B (Sarid *et al.*, 1962) but is said to be absent from *S. typhimurium* (Miller, 1975). The latter conclusion is based on the inability of $pepD^-$ $pepA^-$ $pepB^-$ $pepN^-$ mutants of amino acid auxotrophs to utilize Pro-Leu or Pro-Gly-Gly. While this would suggest that one or more of the missing aminopeptidases (i.e. the enzymes homologous to *E. coli* K-12 aminopeptidases I and AP and the oligopeptidase listed in Table 1) has iminopeptidase activity, this conclusion is not completely clear cut because the failure of amino acid auxotrophs to grow on small peptides, despite the ability of the bacteria readily to transport and to hydrolyse the peptides is amply documented in the literature (e.g. Meisler and Simmonds, 1963; Sussman and Gilvarg, 1970). However, if it were to be correct then it would not be without precedent, for non-specific aminopeptidases are known to be produced by bacteria. One of the best characterized is the aminopeptidase isolated from the culture filtrate of *Clostridium histolyticum* by Kessler and Yaron (1976*a*,*b*) in the course of a search for 'proline-specific enzymes' (Yaron, 1974). A very large hexameric enzyme (subunit molecular mass *c.* 53 000 daltons), clostridial aminopeptidase, catalyses the removal of all types of amino-terminal amino acids but is inactive toward substrates of structure X-Pro, X-Pro-Y, and Gly-Gly-Y. It requires Mn^{2+} for activity, and, although several other divalent cations also activate, Zn^{2+} is inhibitory. The complete absence of endopeptidase activity distinguishes the specifictity of this aminopeptidase from two other *C. histolyticum* peptidases, clostridiopeptidases A and B, i.e. collagenase (Seifter and Harper, 1971) and clostripain (Mitchell and Harrington, 1971), respectively (see also Section 3.1.3.2). Another non-specific aminopeptidase capable of effecting removal of terminal proline residues has been purified from the marine bacterium *Aeromonas proteolytica* by Prescott and his collaborators (Wagner *et al.*, 1972*b*; Wilkes *et al.*, 1973; Prescott and Wilkes, 1976). This, too, is an extracellular enzyme, but is a small protein (molecular mass *c.* 29 000 daltons) containing two atoms of Zn.

In a formal sense, the activity of proline iminopeptidase resembles that of pyrrolidonyl peptidases (pyrrolidone carboxylyl peptidases) which effect the hydrolysis of the CO–NH linking the α-carboxyl carbon of pyrrolidone carboxylic acid to the α-amino nitrogen atom of an α-amino acid or a polypeptide chain (as in some fibrinopeptides). First detected in a strain of *Pseudomonas fluorescens* (Doolittle and Armentrout, 1968), enzymes with this activity have been purified from *Pseudomonas* spp. and from *B. subtilis* (Doolittle, 1970; see also DeLange and Smith, 1971) and have been demonstrated to be present in a variety of bacteria. In *Streptococcus cremoris*, at least, the enzyme appears to be associated with the spheroplast membrane (Exterkate, 1977).

While neither the *E. coli* K-12 peptidase P (Table 1) nor the apparently homologous *S. typhimurium* peptidase P has been fully characterized, these enzymes appear to be essential for the growth of proline auxotrophs on Gly-Pro-Ala and other X-Pro-Y tripeptides in lieu of proline (McHugh and Miller, 1974; Jackson *et al*., 1976; Miller and Schwartz, 1978). From the data available for *Salmonella* (McHugh and Miller, 1974), peptidase P appears strikingly similar to the *E. coli* B aminopeptidase P which acts on substrates of structure X-Pro-Y (where Y may be OH, a free amino acid, an amino acid methyl ester, or even a long polypeptide as in polyproline or carboxymethylated papain) (Yaron and Berger, 1970). Like so many *E. coli* peptidases, aminopeptidase P is Mn^{2+}-dependent and polymeric, being composed of subunits (molecular weight *c*. 60 000 daltons) that aggregate to yield dimers and tetramers. The dimer is believed to be the active form and available evidence suggests that intermolecular interactions, as in tetramer formation, lead to loss of enzymic activity (Yaron, 1974).

The genus *Bacillus* has been shown to contain a number of aminopeptidases. For example, in addition to the *B. subtilis* intracellular enzymes AP I, II, and III (Desmond *et al*., 1975) noted above, an extracellular aminopeptidase also has been obtained from that organism (Wagner *et al*., 1972*a*). *B. stearothermophilus* is the source of three enzymes, which include: a highly heat-stable enzyme (particle molecular weight *c*. 400 000 daltons) containing twelve subunits of two types, α and β, which have the same molecular weight and are present in different ratios in aminopeptidase I types A, B, and C (α:β ratios of 10:2, 8:4, and 6:6, respectively); and two other thermolabile and dimeric enzymes, of which aminopeptidase II (subunit molecular weight *c*. 46 000 daltons) shows activity toward peptides of various chain lengths, and aminopeptidase III (subunit molecular weight *c*. 47 500 daltons) acts specifically on tripeptides (Roncari *et al*., 1976; Deranleau and Zuber, 1977). The thermostable aminopeptidase I is of special interest because of its response to changes in activating divalent cations, temperature, and pH; e.g. replacement of Co^{2+} by Zn^{2+} results in marked loss of catalytic activity but the Zn-apoenzyme complex has greater heat-stability; maximal activity toward leucine *p*-nitroanilide is seen at pH 7.5–8.0 but Gly-Leu-Tyr hydrolysis occurs best at about pH 9.3; while high pH promotes the conversation of the $\alpha_8\beta_4$ form to the $\alpha_{10}\beta_2$ and α_{12} forms. These, and other, observations led to the assumption 'that both types of subunits have hydrolytic activity but different specificity [and] aminopeptidase I hybrids can equally well be called multienzyme complexes' (Roncari *et al*., 1976).

4 ENDOPEPTIDASES

Tetralysine endopeptidase is the only *E. coli* K-12 enzyme known to catalyse the hydrolysis of the central bond in a tetrapeptide (Table 1). However, tetralysine is one of the substrates for dipeptidyl carboxypeptidase (DCP) purified from *E. coli* B (Yaron, 1974, 1976). This atypical carboxypeptidase effects removal of an X-Y dipeptide from A-B-X-Y and longer peptides and also from α-amino-blocked tripeptides (e.g. Z-Ala_3). DCP is reported to have a particle molecular weight of 97 000 daltons and, at least after dialysis against EDTA, its activity is markedly stimulated by Co^{2+}, properties not unlike those of tetralysine endopeptidase. During purification, DCP is accompanied by a Mn^{2+}-dependent dipeptidase whose properties resemble those of the strain K-12 DP (Table 1) and the strain B peptidase M (Brown, 1973). Thus, in test mixtures containing partially purified preparations of DCP plus Mn^{2+}, the cleavage of the penultimate peptide bond of appropriate substrates is coupled to dipeptide cleavage so that, as noted by Yaron (1976), the 'complementary action of the two enzymes . . . represents an alternative mechanism for carboxypeptidase action by which a polypeptide chain is successively degraded from the carboxyl end to amino acids.' That this type of sequential degradation may be of physiological significance is suggested both by the observation that dipeptides released by the action of DCP are competitive inhibitors of DCP activity and by the apparent absence from *E. coli* of activity ascribed to typical carboxypeptidases.

Indeed, until recently few carboxypeptidases had been identified in bacteria. One, carboxypeptidase G from *Pseudomonas stutzeri*, is a Zn^{2+}-activated enzyme catalysing the hydrolytic removal of carboxyl-terminal glutamic (and aspartic) acid residues from polyglutamates (e.g. folic acid derivatives) and from other peptides and *N*-benzyloxycarbonyl amino acids (e.g. Tyr-Glu, Z-Glu, Z-Asp) (McCullough *et al*., 1971). Two others have been partially purified from the growth medium of *B. cereus* (Twardowski *et al*., 1977): a 'carboxypeptidase A' that, like the analogous mammalian enzyme (Hartsuck and Lipscomb, 1971), hydrolyses hippuryl-Phe (Bz-Gly-Phe) and is inactivated by *o*-phenanthroline; and a 'carboxypeptidase B' that is analogous to the mammalian carboxypeptidase B (Folk, 1971) in showing activity toward hippuryl-Lys. The activity of the *B. cereus* carboxypeptidase

A is not inhibited by phenylmethane sulphonyl fluoride (a serine-protease inhibitor), a property distinguishing it from serine proteases of microbial origin that exhibit both carboxypeptidase and endopeptidase activity (Morihara, 1974).

It is not unusual for bacterial enzymes classified as proteases also to act as carboxypeptidases and/or aminopeptidases (e.g. streptococcal proteinase which shows both types of activity) (Morihara, 1974). This property appears to be absent from the two serine protease-like enzymes isolated from *E. coli* B: 'protease I' is a periplasmic enzyme, is active toward the chymotrypsin substrate *N*-acetyl-D,L-phenylalanine β-naphthylester (NAPE), but not to trypsin substrates (Pacaud and Uriel, 1971) which, in turn are substrates for the cytoplasmic 'protease II' (Pacaud and Richard, 1975). Protease I is apparently responsible for most, but not all, of the bacterial activity to NAPE. Thus, polyacrylamide gel electrophoretic analysis indicates the presence of three bands of activity toward NAPE in extracts of strain B (Pacaud and Uriel, 1971), and similar gel patterns have been observed in *E. coli* K-12 and *S. typhimurium* (Miller, 1975; Miller *et al*., 1976). The strongest activity band has been assigned to protease I, and mutants lacking this band have been obtained from *E. coli* (Kowit *et al*., 1976). While the *E. coli* mutant (Pr^- strain) was devoid of all periplasmic activity toward NAPE, its ability to degrade protein was unimpaired, and Kowit *et al*. (1976) (see also Goldberg and St. John, 1976) have expressed doubts as to the advisability of classifying protease I as a protease, suggesting instead that a term like NAPE-hydrolysing enzyme (as in Table 1) or acetylphenylalanine esterase is more appropriate. For similar reasons, the name benzoyl-arginine β-naphthylamide-hydrolysing enzyme is used in Table 1 to denote the enzyme inferred to be the *E. coli* K-12 homologue of the strain B protease II. Mutant strains lacking protease I or protease II activity (as judged by tests using as substrates NAPE and *N*-methyl-*N*-tosyl-L-lysine β-naphthyl ester, respectively) also are known in *Salmonella*, and they, too, appear normal with respect to ability to degrade protein (Miller *et al*., 1976; Miller and Zipser, 1977).

5 MUREIN PEPTIDASES

In view of their complex structure it is only to be expected that cell-wall peptidoglycans (mureins) (Reaveley and Burge, 1972; Schleifer and Kandler, 1972) would be hydrolysed by enzymes specifically adapted for that purpose. As would be expected, therefore, the D-amino acid-containing peptides found in mureins are substrates for specific D-amino acid carboxypeptidases and endopeptidases present in Gram-negative and Gram-positive bacteria (Blumberg and Strominger,

1974). *E. coli* contains at least three enzymes with D-alanine carboxypeptidase (ACP) activity: ACP IA and ACP IB effect the removal of the terminal D-alanine from the murein precursor UDP-*N*-acetylmuramyl-L-alanyl-D-glutamyl-γ-*meso*-diaminopimelyl-D-alanyl-D-alanine (UDP-MurNAc-L-Ala-D-Glu-*m*-Dap-D-Ala-D-Ala), whilst ACP II releases D-alanine from UDP-MurNAc-L-Ala-D-Glu-*m*-Dap-D-Ala (Nguyen-Distèche *et al.*, 1974; Blumberg and Strominger, 1974). The membrane-bound ACP IA and IB differ in several respects. Thus, ACP IA is obtained as a 'doublet' (subunits of molecular mass *c.* 32 000 and 34 000 daltons), is competitively inhibited by penicillins, and displays both carboxypeptidase and transpeptidase activity (with hydroxylamine or glycine as the acceptor) not only toward UDP-MurNAc-L-Ala-D-Glu-*m*-Dap-D-Ala-D-Ala but also toward the corresponding compound containing lysine in place of diaminopimelic acid and toward diacetyl-L-Lys-D-Ala-D-Ala; ACP IB (and the probably identical ACP IC isolated from the cytoplasm) consists of a single polypeptide (molecular mass *c.* 41 000 daltons), does not bind penicillin, shows carboxypeptidase but only weak transpeptidase activity, acts only on the diaminopimelate-containing UDP-MurNAc-pentapeptide, and also possesses endopeptidase activity toward the cell-wall disaccharide–tetrapeptide dimer (i.e. cleaves the CO–NH bond cross-linking the carboxyl group of a terminal D-alanine of one L-Ala-D-Glu-*m*-Dap-D-Ala chain to the amino group at the D-configuration carbon atom of the *meso* acid in the second chain) (Tamura *et al.*, 1976; Spratt and Strominger, 1976).

Streptococcus faecalis likewise contains a membrane-bound ACP which, like *E. coli* ACP IA, catalyses both the hydrolysis of the peptide bond in the terminal dialanine group of substrates like diacetyl-L-Lys-D-Ala-D-Ala, and the transfer of the diacetyl-Lys-Ala group from the tripeptide to glycine and also to diglycine (Coyette *et al.*, 1974). Other membrane-bound enzymes similar to the *E. coli* ACP IA and ACP II have been purified from bacilli, e.g. *B. subtilis* D-alanine carboxypeptidase and *B. sphaericus* L,D-carboxypeptidase. The former, like *E. coli* ACP IA, catalyses the hydrolytic removal of the terminal D-Ala from the UDP-MurNAc-pentapeptide precursor of the cell-wall peptidoglycan and binds penicillins covalently; it has been used to study the site of penicillin attachment by Georgopapadakou *et al.* (1977), who conclude that the antibiotic must be covalently bound to a specific serine residue in the enzyme. The other (*B. sphaericus*) enzyme, the substrate specificity of which has been extensively investigated by Arminjon *et al.* (1977), more nearly resembles *E. coli* ACP II, splitting the bond in the L-Lys-D-Ala group at the end of MurNAc-tetrapeptides; it acts also on substrates containing ω-amidated *meso*-diaminopimelate residues in place of lysine.

As noted above, several of the known carboxypeptidases exhibit endopeptidase activity, but specific endopeptidases also have been identified. Several have been found in *E. coli* which cleave the cross-linking peptide bond in the cell-wall disaccharide–tetrapeptide dimer (Blumberg and Strominger, 1974). In contrast, the 'γ-D-glutamyl-(L)-*meso*-diaminopimelate endopeptidases' of *B. subtilis* (Guinand *et al*., 1976) and *B. sphaericus* (Arminjon *et al*., 1977) display activity toward free peptides such as L-Ala-D-Glu-*m*-Dap-D-Ala and L-Ala-D-Glu-Dap, cleaving the bond linking the γ-carboxyl group of D-glutamic acid to the amino group at the L-configuration carbon atom of *meso*-diaminopimelic acid; however, the preferred substrates are the *N*-acetylmuramyl peptides representing the monomers of the spore cortex peptidoglycans. These endopeptidases, and also the L,D-carboxypeptidases, are found in particulate preparations from sporulating *Bacillus* cells and, at least in the case of *B. sphaericus*, in the media of sporulating cultures.

The bonds linking *N*-acylmuramic acid groups to the L-alanine of peptidoglycans are cleaved to release peptides (Blumberg and Strominger, 1974), and it is noteworthy that the *E. coli* K-12 amidase is active toward MurNAc-L-Ala-D-Glu-*m*-Dap but not toward the nucleotide-containing murein precursor (Beck and Park, 1976), a property apparently not shared by the ACP Is and II of this strain. However, like the activity of ACP IA and at least one of the *E. coli* endopeptidases, the activity of the amidase is known to be penicillin-sensitive.

6 LOCALIZATION OF PEPTIDASES

While many bacteria secrete into their growth media enzymes capable of attacking the peptide bonds of peptides and/or proteins, *E. coli* apparently does not (Sussman and Gilvarg, 1971; Miller, 1975; Payne, 1976). All the evidence available from studies carried out with strains K-12 and B indicates that the hydrolysis of small peptides composed of L-amino acids (and/or glycine) results from the action of peptidases localized within the bacterial cell and having the properties generally ascribed to constitutive enzymes.

Much of the work on the localization of *E. coli* peptidases has involved the use of a limited number of di- or tripeptides as substrates to test for the release of enzymic activity when the bacteria were subjected to osmotic shock or converted to lysozyme spheroplasts (Matheson and Murayama, 1966; Matheson *et al*., 1971; Van Lenten and Simmonds, 1967; Simmonds, 1970). Because a given dipeptide generally serves as a substrate for several peptidases (Table 1, cf. dipeptidase DP, aminopeptidases I, AP, and L, and oligopeptidase), the release of a single enzyme may not be readily demonstrable, especially if the released enzyme is responsible for only a very small amount of the total cellular activity toward the test substrate. However, when *E. coli* K-12 cells were converted to spheroplasts or subjected to osmotic shock, the release of enzymic activity toward Leu-Gly (known to be hydrolysed by DP, I, AP, L, and oligopeptidase) and toward Phe-Gly (hydrolysed rapidly only by DP) was almost identical with the extent of β-galactosidase leakage (β-galactosidase being the usual model for a truly cytoplasmic enzyme) (Van Lenten and Simmonds, 1967). Osmotically shocked *E. coli* K-12 cells likewise retain essentially all their activity toward Met-Ala-Ser (the hydrolysis of which is effected in large part by aminopeptidase I but which is also a known substrate for aminopeptidases AP and L and the tripeptidase) (Vogt, 1970). In addition, the strict substrate specificity of β-aspartyl dipeptidase made it possible clearly to establish the cytoplasmic localization of that strain B enzyme (Haley, 1970).

When *E. coli* cells are disrupted in buffers of low ionic strength, and the cell-free extracts are separated by centrifugation into ribosomal and soluble fractions, both fractions display dipeptidase and aminopeptidase activity, with the soluble fraction accounting for most of the total cell-extract activity toward each of a large number of di- and tripeptides tested (Matheson, 1963; Matheson *et al*., 1971; Simmonds, 1970, 1972; S. Simmonds, unpublished data). Under these conditions, much of the activity associated with the ribosomes of strains K-12 and B is due to the presence of aminopeptidase I (Vogt, 1970) and 'r-peptidase' (Dick *et al*., 1970; Matheson *et al*. 1970), respectively. Such apparent binding of aminopeptidase I to ribosomes is prevented by disrupting the cells in a buffer of high ionic strength (containing 0.2 M KCl) (Vogt, 1970; Simmonds, 1972). Since r-peptidase can be selectively removed from the ribosomal surface with 1 M ammonium chloride, a treatment known to remove protein loosely attached to ribosomes (Dick *et al*., 1970), and the reconstitution of a ribosome–aminopeptidase I complex could not be demonstrated (Vogt, 1970), the presence of such a complex *in vivo* is questionable. However, it should be noted that the apparent subcellular distribution of this aminopeptidase activity (in cells disrupted in low ionic strength buffer) appears to vary with the strain, and to be influenced by the temperature of growth and stage of the growth cycle at which the cells are harvested, with 'a tendency [in both strains K-12 and B] toward more ribosomal-bound enzyme in early log cultures grown at 37 °' (Matheson *et al*., 1971). It has also been observed that ribosomal fractions from strain K-12, which show little of the Leu-Gly-hydrolysing activity attributed to aminopeptidase I, may still exhibit significant activity toward peptides not readily hydrolysed by that aminopeptidase (Simmonds, 1970). Moreover, highly purified r-peptidase is characterized by lower specific activity toward dipeptides in the Met-X series (relative to its Leu-Leu cleaving activity) than is seen with sucrose-washed ribosomes (Matheson and Dick, 1970; Matheson *et al.*, 1970). Thus, the possibility that ribosome-bound peptidases may exist *in vivo* cannot be ignored.

Whilst the results summarized above may be used as evidence that di-, tri-, and aminopeptidases are cytoplasmic enzymes in *E. coli*, the situation with respect to the endopeptidases and proteases is more difficult to evaluate. It has been evident since the initial demonstration by Chaloupka (1961) that *E. coli* cell-extracts release trichloroacetic acid-soluble material from cellular protein and from casein, with enzymic activity responsible for such proteolysis being present in ribosomal, soluble, and plasma-membrane fractions made from penicillin spheroplasts. The presence of proteolytic activity toward casein in the periplasmic, cytoplasmic, and membrane fractions of osmotically-shocked cells of *E. coli* has also been observed (Kowit *et al*., 1976; see also Régnier and Thang, 1975), but the precise nature of the enzymes involved is still uncertain.

As noted earlier, the endopeptidase designated 'aminoendopeptidase, aminopeptidase N' in Table 1 was originally described as a periplasmic enzyme because significant amounts of activity toward the model substrate L-alanine *p*-nitroanilide (APNA) were released when *E. coli* K10 cells were subjected to osmotic shock or converted to lysozyme spheroplasts (Lazdunski *et al*., 1975*b*). However, subsequent studies revealed that strain K10 was unique since other *E. coli* strains exhibited APNA-hydrolysing activity that was typical of a cytoplasmic enzyme. Nor may the APNA-hydrolysing enzyme of most of the *E. coli* strains tested (including K-12) be considered membrane-bound since it is released upon lysis of spheroplasts. While there is a considerable body of evidence indicating that the APNA-hydrolysing enzyme of *E. coli* K10 has 'mainly an electrostatic interaction with the outer surface [of the cytoplasmic membrane], probably mediated by magnesium bridges [but involving other] interactions . . .' (Murgier *et al*., 1977), there is a paucity of evidence that the enzyme can in fact catalyse the hydrolysis of CO–NH bonds in oligo- or polypeptide chains.

A not dissimilar lack of clarity applies to the cellular localization of the endopeptidase (i.e. protein-hydrolysing) activity ascribed to proteases I and II of *E. coli* B. Assays for enzymic activity using *N*-acetyl-L-phenylalanine β-naphthylester (NAPE) and *N*-benzoyl-D,L-arginine *p*-nitroanilide (BANA) as test substrates clearly demonstrated that activity toward NAPE is primarily localized in the periplasm (being distributed between the periplasmic and cytoplasmic fractions of osmotically-shocked cells in a manner similar to the activity of the model enzyme alkaline phosphatase) and that the activity toward BANA is restricted to the cytoplasm (being distributed like β-galactosidase activity) (Pacaud and Richaud, 1975). The assignment of similar sites for the localization of endopeptidase activities is inferred from the fact that NAPE and BANA are specifically hydrolysed by purified proteases I and II, respectively (Pacaud and Uriel, 1971; Pacaud and Richaud, 1975). However, as noted in the section on endopeptidases (Section 2.4.4), there is reason to question the use of the term 'protease' as a meaningful name for the NAPE-cleaving enzyme and, possibly, for the BANA-cleaving enzyme also (Kowit *et al*., 1976; see also Goldberg and St. John, 1976).

Attempts unequivocally to show the presence of membrane-associated enzymes are hampered by the technical difficulties of preparing bacterial membranes without causing dissociation of enzyme from the membrane, and/or inactivation of the enzyme; the possibility of the artifactual association of cytoplasmic enzymes with the membrane is another hazard (for further discussion, see Payne, 1975, 1976). The relative ease with which membranes may be prepared from *Mycoplasma laidlawii* probably accounts for the use of that organism in the early work on membrane-bound peptidases, and the presence of peptidase activity in such membrane preparations has been reported by Choules and Gray (1971) and by Pecht *et al*. (1972). However, as mentioned earlier, proteolytic activity has been observed in membrane fractions from *E. coli* (Kowit *et al*., 1976; Régnier and Thang, 1975). At least two other bacterial species also are said to contain membrane-associated peptidase activity, and in both species the activity resides in somewhat atypical enzymes, i.e. the highly thermostable, very large polymeric (and possibly 'multienzyme') aminopeptidase I of *B. stearothermophilus* (Roncari *et al*., 1976; Deranleu and Zuber, 1977) and the pyrrolidone carboxylyl peptidase of *Strep. cremoris* (Exterkate, 1977). The work on the *Strep. cremoris* enzyme illustrates the difficulty encountered in this type of study, for it proved necessary to avoid the use of mechanical methods to disrupt lysozyme spheroplasts and to localize the enzyme activity in a particulate fraction mainly consisting of membrane material.

In addition to the membrane-associated enzymes just described, most of the murein carboxy- and endopeptidases discussed in an earlier section (Section 2.4.5) also are classified as membrane-bound enzymes. An exception is the D-alanine carboxypeptidase II of *E. coli* K-12, which appears to be a periplasmic enzyme (Beck and Park, 1976). The binding of murein peptidases to the cell membrane may be relatively weak or very strong. For example, D-alanine carboxypeptidase IB could be solubilized by treatment of an *E. coli* particulate preparation with a buffer containing LiCl (and the possibly identical alanine carboxypeptidase IC was isolated from the soluble fraction of disrupted cells) whereas solubilization of alanine carboxypeptidase IA required use of a buffer containing Triton X-100 (Tamura *et al*., 1976).

Secretion of extracellular enzymes exhibiting protease activity is not uncommon among bacteria (Morihara, 1974), but relatively few examples of extracellular peptidases are known (Payne, 1975, 1976) (see also Section 3.1.6). However, a given bacterium may produce both types of enzymes. For example, *B. subtilis* secretes not only an alkaline serine protease and a neutral metalloprotease (Dancer and Mandelstam, 1975) but also at least one aminopeptidase (Wagner *et al*., 1972*a*). Production of extracellular aminopeptidases has been observed in species as diverse as *Clostridium histolyticum* (i.e. the clostridial aminopeptidase characterized by Kessler and Yaron (1976*a*,*b*)) and the marine bacterium *Aeromonas proteolytica* (i.e. the *Aeromonas* aminopeptidase characterized by Prescott and Wilkes (1976)); both of these bacteria also produce extracellular proteases (Seifter and Harper, 1971; Mitchell and Harrington, 1971; Wilkes and Prescott, 1976). Interestingly enough, certain bacterial enzymes known to act as typical carboxypeptidases (i.e. the *B. cereus* carboxypeptidases A and B) were obtained from the bacterial culture fluid and their respective activities were in fact first detected in partially purified preparations of the extracellular phospholipase C (Twardowski *et al*., 1977).

7 PRODUCTION AND REGULATION OF ACTIVITY

While it is generally agreed that in bacteria the intracellular peptidases catalysing the hydrolysis of small peptides are constitutive enzymes, their synthesis and activity *in vivo* may be subject to several types of regulation and may vary with the phase of the bacterial growth cycle. The production and/or secretion of extracellular peptidases, as well as the production of murein peptidases, also is influenced by environmental factors (Payne, 1975, 1976).

Evidence for the constitutive nature of *E. coli* peptidases has come from studies which, for the most part, involved assays for peptidase activity on crude extracts made from cells cultured in peptide-free media; where tested for comparison, growth on peptide-containing media had no significant effect on the level of peptidase activity demonstrable in the resultant cell extracts (Payne, 1972*a*,*b*; Sarid *et al*., 1962; Simmonds, 1970, 1972; Sussman and Gilvarg, 1970; Vogt, 1970). Due to the presence in *E. coli* of so many enzymes displaying overlapping substrate specificity (Table 1), activity assays on cell extracts *per se* lack the precision necessary to detect the presence of a specific enzyme and, especially, to determine whether its biosynthesis can be repressed or derepressed *in vivo*. This difficulty is illustrated by the inability to differentiate between the *E. coli* oligopeptidase-deficient mutant (strain AJ005) and its parent on the basis of crude cell-extract activity toward Leu-Leu, a substrate not only for the oligopeptidase but also for aminopeptidases I, AP, and L (Simmonds *et al*.,

1976). Moreover, misleading data may result from the use of activity-assay systems that lack a required peptidase activator or from the use of inadequate analytical methods to measure peptide hydrolysis. The former problem is illustrated by the failure of the strain AJ005 cell extract to exhibit activity toward Lys-Lys when tested in the absence of a divalent cation essential for dipeptidase DP activity (Sussman and Gilvarg, 1970); the latter difficulty was encountered when estimation of the liberation of free amino acids from glycine-containing peptides was found to be complicated by the presence of glycine-deaminating enzymes in the cell extract (Meisler and Simmonds, 1963; cf. Simmonds and Griffith, 1962, and Stone and Hoberman, 1953). It therefore was reassuring to have the constitutive nature of *E. coli* peptidase activity confirmed by a quite different experimental procedure: the induced synthesis *in vivo* of β-galactosidase by lysine and glycine auxotrophs of *E. coli* W cultured on amino acid homopeptides (the di-, tri-, and tetralysines and glycines) was indistinguishable from the induced synthesis observed when each auxotroph was cultured in medium containing only the free amino acid (Bell *et al.*, 1977). However, even this procedure does not completely overcome some of the uncertainties associated with peptidase activity assays on cell extracts. For example, if, as seems probable (Sussman and Gilvarg, 1970), strain B contains the full complement of peptidases found in strain K-12, then each of the exogenous lysine homopeptides could serve as a substrate for more than one enzyme (Table 1). To take the simplest case, dilysine can be hydrolysed by both dipeptidase DP and oligopeptidase. Thus, the activity *in vivo* of only one of these peptidases might suffice to provide the lysine needed for β-galactosidase production, thereby masking the possible absence of the other peptidase.

Despite the limitations of the procedure, measurements of the ability of *E. coli* extracts to hydrolyse small peptides have led to the recognition that peptidase activity *per se* is not invariant. The extensive studies by Payne (1972*b*) involving chemostat-grown cultures and activity tests carried out in several assay systems have shown clearly that demonstrable dipeptidase activity varies in response to environmental changes such as carbon-source limitation, which yields cells with low activity (toward the dipeptides Gly-Gly, Pro-Gly, and Val-Val), and oxygen depletion, which yields cells with high activity. Whilst ammonium ion-limited media resulted in cells showing an intermediate level of dipeptidase activity, these cells contained the least activity toward triglycine. The additional observation that phosphate limitation produced both high dipeptidase and high tripeptidase activity is of special interest since phosphate limitation is also said to derepress the biosynthesis of the so-called aminoendopeptidase (i.e. alanine *p*-nitroanilide-hydrolysing activity) found in *E. coli* (Table 1) and a variety of Gram-negative organisms (Lazdunski *et al.*, 1975*c*; Murgier *et al.*,

1976*b*). Unlike the di- and tripeptide-cleaving activity, alanine *p*-nitroanilide-hydrolysing activity apparently is not altered by growth of *E. coli* in nitrogen- or carbon-limited media (Lazdunski *et al*., 1975*b*).

Among the results of Payne's studies was the observation that all the peptidase activities of the strain B cell extracts were independent of the pH of the growth medium in the range pH 5.5–7.0 (Payne, 1972*b*). Somewhat different results have been obtained with batch-grown cultures of strain K-12 where cell-extract dipeptidase activity (toward Gly-Leu, Leu-Gly, and Phe-Gly) decreased after the cessation of culture growth and the pH of the culture fell below *c*. pH 5.5 (Simmonds, 1970). The same result was immediately evident after exponential-phase cultures were acidified to pH 5 by the addition of HCl, but only if the growth medium was deficient in 'trace metals'. Exposure of cells grown in a metal-deficient culture medium to relatively high concentrations of Zn^{2+} likewise caused a fall in the dipeptidase activity of the cell extracts. These effects may be attributed to the inactivation of peptidases that were present in the bacterial cells before acid or Zn^{2+} treatment. Certainly, the fact that many strain K-12 peptidases appear to be metalloenzymes the activity of which *in vitro* is inhibited by Zn^{2+} suggests that divalent cations may play a role in the regulation of enzymic activity *in vivo*. This suggestion receives additional support from the changes observed in the substrate specificity of the purified Zn-containing dipeptidase isolated from *E. coli* B in the presence of added Co^{2+} or Mn^{2+} (see Section 2.4.2 on dipeptidases), as well as from the variations in the dipeptidase activity of strain B cell extracts observed when a phosphate buffer is replaced by one containing Tris and/or augmented with EDTA (Payne, 1972*a*,*b*).

Replacement of glycerol by glucose in a peptone-containing medium has also been found to alter the apparent dipeptidase activity of strain K-12 (to Leu-Gly and Phe-Gly). In experiments using the soluble and ribosomal fractions of cells disrupted in a low ionic-strength buffer (i.e. when aminopeptidase I and at least one other, undefined, peptidase are ribosome-bound), the ribosomal fraction from glucose-grown cells showed less Leu-Gly hydrolysing activity than was seen with glycerol-grown cells; the ribosomal fraction also accounted for a lower percentage of the total cell-extract activity. Furthermore, only in glucose-grown cells did the activity of each subcellular fraction (not only to Leu-Gly but also to Phe-Gly) drop sharply after the cessation of exponential culture growth. These changes may be related to the acid-inactivation effect mentioned above, since cultures growing on glucose showed a significant decrease in pH at the end of the exponential growth phase (Simmonds, 1970). The glucose-mediated changes in extract activity, especially activity toward Leu-Gly, apparently is not due to inactivation of aminopeptidase I (Table 1) since Vogt (1970) has shown that the heat-stable activity of extracts toward Met-Ala-Ser was not altered when cultures were grown in a peptide-free medium containing glucose rather than glycerol. Nor would it appear to be due to changes in the activity of the oligopeptidase (Table 1), the activity of which (as measured toward alanine β-naphthylamide) was found by Yang and Somerville (1976) to be significantly greater in extracts from stationary-phase cells than in extracts from exponential-phase cells. However, as pointed out by these investigators, the activity assays on the crude extracts may be unreliable 'because of possible complications in the assay [ascribable] to the presence of inhibitors', and they have in fact shown that extracts contain an inhibitor, possibly a low molecular weight, heat-stable protein. Thus, not only metal-ions but also organic compounds must be considered in relation to the regulation of *E. coli* peptidase activity *in vivo*.

Still another type of inhibition may be operative, i.e. the binding of peptides *per se* to enzymes other than peptidases. This is illustrated by the discovery that the bacteriostatic effect of many leucine-containing peptides on the growth *E. coli* K-12 (Simmonds *et al*., 1951) is prevented by addition of isoleucine because, at least in the case of Gly-Leu, the inhibitory peptide binds to threonine deaminase thereby preventing catalysis of the first step in isoleucine biosynthesis (Vonder Haar and Umbarger, 1972). Such binding to the deaminase would appear selectively to protect Gly-Leu (and even leucine-containing tripeptides (Simmonds, 1977)) from hydrolysis *in vivo* since growth inhibition is reversed not only by free isoleucine but also by isoleucine-containing peptides; and glycine auxotrophs whose growth is inhibited in media containing Gly-Leu readily grow on exogeneous Gly-Phe (Meisler and Simmonds, 1963; Simmonds, 1970). Moreover, although the bacteriostatic effects of leucine peptides is especially pronounced when the inocula are composed of stationary-phase organisms, extracts prepared from such cells show 'normal' levels of dipeptidase activity.

Although product inhibition is commonly observed to influence the rate of many enzyme-catalysed reactions, little is known about the effects of reaction products on the activities of bacterial peptidases. As noted in Section 2.4.4 on endopeptidases, this type of inhibition has been observed for the dipeptidyl carboxypeptidase from *E. coli* B (Yaron, 1974, 1976), and it has also been reported for the extracellular *Aeromonas proteolytica* aminopeptidase (Prescott and Wilkes, 1976). Results obtained with *E. coli* K-12 subcellular fractions, and with aminopeptidase AP preparations contaminated with dipeptidase DP, suggest that aminopeptidases may also be subject to product inhibition, since the aminopeptidase-catalysed degradation of tripeptides to dipeptides was accelerated when the latter were being further degraded by dipeptidases (Simmonds, 1970, 1972). In this connection, it is noteworthy that some of the hydrolytic activities of the membrane-bound *B. stearothermophilus* aminopeptidase I are inhibited by dipeptides, whilst other activities are not demonstrable unless both types of subunits (see Section 2.4.3. on aminopeptidases) are present to provide a 'multienzyme complex' (Roncari *et al*., 1976). Additional work on this metalloenzyme has led to the suggestion that its activity may be regulated by one or more structural components of the membrane (Deranleau and Zuber, 1977). The potential importance of interactions between membrane components and peptidases as determining factors in enzymic activity has also been emphasized by Exterkate (1977) as a result of his studies on the membrane-bound pyrrolidone carboxylyl peptidase of *Strep. cremoris*.

The initial work on the dipeptidase activity of *E. coli* K-12 (reviewed by Simmonds, 1970) indicated that in resting cells, intracellular enzymes behave like cryptic enzymes (i.e. when cells were tested directly for ability to hydrolyse dipeptides they exhibited much less activity than that seen with broken-cell preparations). The extent of such crypticity increased markedly when cell cultures were held in the stationary phase of the growth cycle for prolonged periods, although these cells could still actively concentrate dipeptides *per se* and their extracts were fully active. Furthermore, intact spheroplasts and osmotically-shocked cells exhibited the low levels of dipeptidase activity characteristic of intact bacteria, and toluene treatment of freshly-harvested bacteria did not permit full expression of the activity. This may be contrasted with the behaviour of the strain K-12 aminoendopeptidase (Table 1) the activity of which (assayed against alanine-β-naphthylamide and -*p*-nitroanilide) was fully demonstrable in spheroplasts; some 75–85 per cent of the total activity was demonstrable even in assays on intact cells (Murgier *et al*., 1977). While the physiological significance of the cryptic behaviour of intracellular peptidases has been questioned, especially as a factor in the metabolism of 'normal log phase growth of *E. coli* in its natural environment' (Sussman and Gilvarg, 1971), it has been pointed out by Payne (1975) that 'in [*E. coli's*] natural environment log phase growth is certainly not the normal condition and prolonged periods in stationary phase or under growth-limiting conditions is more common.' That enzyme production as well as enzyme activity *in vivo* may be limited under such conditions is indicated by the results described above for the cell-extract activity observed when chemostat cultures were grown under conditions of carbon and, especially, nitrogen limitation (Payne 1972*b*).

As noted earlier, the *E. coli* intracellular peptidases appear to be constitutive enzymes the biosynthesis of which is not stimulated by the presence of exogenous peptides. In contrast, the synthesis and secretion by *Aeromonas proteolytica* of both the *Aeromonas* aminopeptidase and endopeptidase ('neutral protease') is stimulated by exogenous protein or by complex peptide mixtures (e.g. enzymic digests of soybean protein or casein) (Prescott and Wilkes, 1976; Wilkes and Prescott, 1976). The amount of each enzyme produced and the ratio between them differ markedly depending on the peptide composition of the medium, with minimal production being observed in a medium containing acid-hydrolysed casein; this medium also results in a high value for the ratio of aminopeptidase to protease activity. Moreover, maximal production of the two enzymes occurs at different times during the growth cycle. Likewise, the production by *B. subtilis* of aminopeptidases I, II, and III, and D-aminopeptidase has been shown to vary during the growth cycle (see Section 2.4.3 on aminopeptidases) (Desmond *et al*., 1975). The synthesis of aminopeptidase III and D-aminopeptidase occurs mainly in early stationary-phase cultures, i.e. at a time just prior to sporulation when *B. subtilis* also produces two extracellular proteases (Dancer and Mandelstam, 1975). Whether any of these enzymes are specifically induced by exogenous peptides is uncertain, but such induction of an extracellular aminopeptidase, presumably different from the aminopeptidases mentioned above, has been observed (Wagner *et al*., 1972*a*).

In *B. stearothermophilus*, peptidase production is dependent upon the temperature at which the organism is cultured (Roncari *et al*., 1976). Thus, of the three enzymes with aminopeptidase (AP) activity that have been identified (see Section 2.4.3 on aminopeptidases), AP I and II are found in large amounts in cells from cultures grown at 55° C. Cells grown at 37° C contain little AP I or II, but show increased amounts of AP III and such cells contain still another aminopeptidase (I_m) that is strongly particle-bound. These variations would hardly be surprising in a thermophilic bacterium except for the fact that only AP I is highly thermostable and that the activity of AP III is restricted to tripeptide substrates. It would appear, therefore, that in cells cultured at 55° C the combined activities of AP I and AP II are sufficient to compensate for the absence of AP I_m and the low level of AP III activity.

The highly specific nature of murein-peptidase activity would in itself suggest that these activities might be maximal in cells at the time of most active peptidoglycan metabolism, i.e. during the growth and division of cells and in appropriate bacteria, during sporulation. That this probably is true of at least some of the murein peptidases has become evident from recent work on the carboxypeptidases I and II (CPI and CPII) and cell wall amidase of *E. coli* K-12 (see Section 2.4.5 on murein peptidases). Thus, measurement of each enzymic activity in toluene-treated cells showed only CPII activity varied as a function of the 'division cycle', with maximal activity being observed at the time of cell division (Beck and Park, 1976). However, the increase in CPII activity is believed to represent the activation of latent CPII present in non-dividing cells rather than synthesis *de novo* of this periplasmic enzyme (Beck and Park, 1977). Whether the murein peptidases isolated from sporulating bacilli (i.e. the carboxypeptidases, endopeptidase, and amidase noted in Section 2.4.5 on murein peptidases) also are present in latent form in vegative cells is not known.

8 FUNCTIONS OF BACTERIAL PEPTIDASES

As detailed knowledge of the bacterial metabolism of amino acids, peptides, and proteins has increased, so has the number of physiological functions assigned to bacterial enzymes catalysing the hydrolysis of peptide bonds in naturally occurring compounds. In view of the extensive discussions that have recently appeared concerning the roles of peptidases in the bacterial metabolism of small and large peptides (Goldberg and St. John, 1976; Miller, 1975; Payne, 1975, 1976) and of peptidoglycans (Blumberg and Strominger, 1974; Tamura *et al*., 1976), these aspects of peptidase function will be reviewed only briefly (see also Section 3.4.4).

The early work on the growth response of *E. coli* amino acid auxotrophs to small peptides led to the realization that peptides must be hydrolysed before the constituent amino acids become available for growth (Fruton and Simmonds, 1950); the fact that more extensive cell multiplication might occur in media containing a peptide in lieu of free amino acids was shown (in certain instances) to be the result of the protection of the peptide-bound amino acid (e.g. proline) from degradation *in vivo* (Stone and Hoberman, 1953; see also Simmonds 1970, 1972). The importance of 'uptake mechanisms' (i.e. transport systems) in the utilization of exogenous peptides for growth (Simmonds and Griffith, 1962; Kessel and Lubin, 1963) and the role of *E. coli* peptidases in relieving the bacteriostatic effects of growth-inhibitory peptides (Simmonds *et al*., 1951; see also Meisler and Simmonds, 1963; Simmonds, 1970) also were recognized. During this same early period, the work of E. E. Snell and his collaborators (see Introduction) (reviewed in detail by Payne, 1976) provided evidence for similar phenomena in the more fastidious Gram-positive Lactobacillaceae and, in addition, demonstrated that the stimulation of growth by particular peptides or ill-defined mixtures of peptides (as in strepogenin) could variously be attributed to the relative ease with which they may be 'absorbed' by cells unable to take up at an adequate rate all the required free amino acids from the growth medium, or to their use in correcting amino acid imbalances in the medium.

The nutritional activity of small peptides has been used to good effect in the isolation of peptidase-deficient mutants of *E. coli* (Kessel and Lubin, 1963; Sussman and Gilvarg, 1970) and of *S. typhimurium* (McHugh and Miller, 1974; Miller and MacKinnon, 1974), mutant detection being based on the inability of a specific amino acid auxotroph to grow on peptide(s) containing the required amino acid. In the case of *E. coli* K-12, which is sensitive to valine, the isolation of peptidase-deficient mutants has also been based on the development of resistance to valine-containing peptides (Miller, 1975; Miller and Schwartz, 1978). Use also has been made of potentially chromogenic enzyme substrates (i.e. β-naphthylamides of free or *N*-acylated amino acids, the hydrolysis products of which can be visualized by diazotization) to detect aminopeptidase- and 'protease'-deficient mutants of *E. coli* (Kowit *et al*., 1976; Latil *et al*., 1976; Miller, 1975) and *S. typhimurium* (Miller, 1975; Miller *et al*., 1976).

While the results of the early work with amino acid auxotrophs were consistent with the view that growth on exogenous peptides required the active uptake and hydrolysis of peptides, the unequivocal differentiation between these two processes depended on the production of mutant strains either specifically unable to transport peptides or specifically unable to hydrolyse them. The first genetic evidence of this kind was presented by Kessel and Lubin (1963), who isolated two mutants of an *E. coli* W glycine auxotroph unable to grow on diglycine: one a diglycine transport-deficient mutant, the other a 'dipeptidase-deficient' mutant. The authors' conclusion that a mutation had resulted in the specific loss of diglycine-cleaving activity, although warranted at the time, no longer is fully justified in view of current knowledge about the overlapping specificity of *E. coli* enzymes capable of hydrolysing dipeptides. (For additional discussion of this point, see Payne (1976); possible explanations of the failure of Kessel and Lubin to observe hydrolysis of diglycine are presented by Simmonds *et al*. (1976)). Since Kessel and Lubin's report, a large amount of evidence for the occurrence in bacteria of peptide uptake in the absence of peptide hydrolysis has been presented, and the existence of several peptide-transport systems which differ in their 'substrate specificity' was established (Payne, 1973, 1976; Sussman and Gilvarg, 1971). Of special interest is the recent report that the uptake and hydrolysis of small defined peptides can lead to an unequal cellular accumulation of the hydrolysis products because of the selective release from the cells of specific amino acids (Payne, 1977). This could suggest perhaps that some of the amino acids liberated by the hydrolysis of intracellular proteins (Goldberg and St. John, 1976) may become unavailable for re-utilization in the synthesis of new protein molecules under conditions of nitrogen-limitation (but see also Section 2.1). In a similar manner, cells might be deprived of amino acids liberated as the result of proteolytic modification of the large molecules serving as precursors of bacteriophage proteins (Miller, 1975). Although such loss would be of little consequence for the survival of organisms when there is an adequate exogenous source of amino acids or, in the case of bacteria capable of synthesizing all their amino acids, an adequate source of ammonia-nitrogen plus a utilizable carbon source, it might prove lethal to more fastidious bacteria.

The discovery that the bacterial synthesis of proteins involves the formation of nascent polypeptide chains containing amino-terminal methionine residues prompted a search for the 'methionine aminopeptidase' specifically concerned with the removal of *N*-terminal methionine during the biosynthesis of proteins that in their final form lack that residue. A number of investigators reported the presence in bacterial extracts of methionine aminopeptidase activity toward di- and oligopeptides (summarized by Simmonds, 1970), and the isolation and characterization of the *E. coli* K-12 aminopeptidase I (Vogt, 1970) was a direct outcome of an attempt to identify such an enzyme, as was Matheson's initial demonstration that *E. coli* B ribosomes (containing 'r-peptidase') possessed methionine aminopeptidase activity (Matheson and Murayama, 1966). Whether these enzymes (the ability of which to bind to ribosomes places them at the site of protein synthesis) actually play a role in protein biosynthesis is uncertain: whilst Matheson *et al.* (1970) believe r-peptidase may play such a role, Vogt's study of the substrate specificity of aminopeptidase I has led him to the opposite conclusion (Vogt, 1970). Interest in this problem led also to the isolation of the strain B dipeptidase M (Brown and Krall, 1971; Brown, 1973), and despite the inability of highly purified preparations to hydrolyse substrates larger than dipeptides, Brown (1973) has not ruled out a functional role for dipeptidase M in protein biosynthesis if only because the enzyme 'has not been tested on what appears to be the proper substrate . . . i.e. a polypeptidyl-tRNA . . . attached to a ribosome.' The same argument might of course be advanced in favour of such a function for any of the known *E. coli* di- or amino-peptidases that can rapidly hydrolyse bonds linking methionine to alanine or serine residues (alanine and serine being commonly found at the amino-termini of *E. coli* cellular protein); and the possibility exists that in fact several aminopeptidases might participate in this phase of protein biosynthesis, their action being determined not so much by the primary structure of the nascent polypeptide chain but more by the secondary and tertiary structure, i.e. only amino-termini at the surface of a specific convoluted structure might be accessible to attack by a specific peptidase.

The possible indirect participation of peptidases in protein synthesis is suggested by the observation that action of *B. cereus* carboxypeptidase A (see Section 2.4.4 on endopeptidases) on the aggregated form of elongation factor 1 (from rabbit reticulocytes) effects disaggregation, yielding the monomeric species that catalyses the binding of aminoacyl-tRNAs to ribosomes during peptide chain elongation (Twardowski *et al*., 1977). Disaggregation appears to result from the removal of only a small fragment from the carboxyl-terminal end of a monomer polypeptide, and it is tempting to postulate the involvement of similar limited degradation (from either the carboxyl- or amino-terminus) in the production of catalytically active forms of enzymes, like the aminopeptidase P of *E. coli* B, that are composed of subunits having a tendency to form inactive polymers (see Section 2.4.3 on aminopeptidases) (Yaron, 1974). Indeed, Twardowski *et al*. (1977) have suggested that 'proteolytic modification of [elongation factor 1] might represent a general mechanism by which disaggregation occurs.'

The ability of bacteria to degrade proteins, and especially aberrant, non-functional, or experimentally-induced incomplete proteins, is well documented (see Goldberg and St. John, 1976; Miller, 1975; Payne, 1976) (see also Section 3.4). Such degradation is evident in cells in growing cultures, and often is increased after cultures enter the stationary phase or when growth is artificially limited by the use of culture media deficient in nitrogen or carbon sources. While the initial steps in proteolysis undoubtedly are catalysed by proteases *per se*, the complete degradation of proteins to free amino acids must also require the participation of peptidases. That the amino acids so liberated may be re-utilized for the synthesis of new proteins, and that amino- and dipeptidases (like those listed in Table 1) play an important role in such 'protein turnover' is hardly a new concept (Simmonds, 1970). Nevertheless, relatively little information is available about the details of the degradative process, and mutant strains of bacteria that lack one or more of the known peptidases (Table 1) have only recently become available for use in studies of protein degradation *in vivo*. Doubtless part of the difficulty can be attributed to inadequate knowledge of the catalytic properties of individual enzymes and of the mechanisms controlling enzymic activity.

Among the recent work relevant to this aspect of peptidase function are studies with peptidase-deficient mutant strains of *E. coli* and *S. typhimurium*. Goldberg and his collaborators have used *E. coli* mutants lacking either the acetylphenylalanine β-naphthylester (NAPE)-hydrolysing enzyme or the benzoylarginine β-naphthylamide (BANA)-hydrolysing enzyme (i.e. mutants described as lacking 'protease I' and 'protease II', respectively) to study the effects of such deficiencies on protein degradation during growth and during glucose-starvation (Goldberg and St. John, 1976; Kowit *et al*., 1976). Comparisons of the rates at which isotopically-labelled amino acids were released from labelled cell protein showed no significant difference between mutant and wild-type strains; nor did the mutants appear to differ from their parents in the ability rapidly to degrade abnormal (puromycin-containing) polypeptides. Similar results were obtained by Miller *et al*. (1976) when a presumably analogous 'protease I' mutant of *S. typhimurium* was compared to the parent strain with respect to ability to degrade intracellular protein either labelled with an isotopic amino acid or made aberrant by incorporation of canavanine in lieu of arginine. A different technique, involving study of the rates at which mutationally-generated fragments of β-galactosidase disappeared from *Salmonella* cells (as measured either by polyacrylamide gel electrophoretic analysis or by complementation tests for β-galactosidase activity), again failed to show a significant difference between peptidase mutants, lacking either 'protease I' or 'protease II' and the parent strain (Miller and Zipser, 1977). Since (as discussed in Section 2.4.4 on endopeptidases) there is some doubt about the ability of the NAPE- and BANA-hydrolysing enzymes rapidly to hydrolyse proteins *per se* (Kowit *et al*., 1976), these results only provide evidence that enzymes active toward NAPE and BANA play no major part in protein degradation *in vivo*.

Studies with mutant strains lacking activity attributable to enzymes known to act on small peptides have provided more meaningful data. Thus, Miller and Zipser (1977) also reported that the rates at which the β-galactosidase fragments disappeared were not signifigacantly altered by mutations leading to the loss of one or another of the *Salmonella* enzymes controlled by the (presumably) structural genes *pepN*, *pepA*, *pepB*, *pepP*, and *pepQ* (i.e. *Salmonella* enzymes assumed to be analogous to the *E. coli* K-12 enzymes listed in Table 1 as the oligopeptidase, aminopeptidase I, aminopeptidase AP, peptidase P, and peptidase Q, respectively). Of more interest is the statement that the quadruple mutant *pepN*$^-$ *pepA*$^-$ *pepB*$^-$ *pepD*$^-$ (i.e. lacking enzymes analogous to *E. coli* oligopeptidase, aminopeptidases I and AP, and dipeptidase DP) showed a reduction in both the rate and extent of intracellular protein degradation *in vivo* during starvation, and produced significant amounts of (unidentified) small molecular weight peptides; in contrast, the more rapid protein degradation in the parent wild-type *Salmonella* gave rise only to free amino acids (Miller, 1975). The importance of enzymes with aminopeptidase activity is illustrated by the ability of triple mutants containing wild-type alleles of *pepN*, *pepA*, or *pepB* (e.g. a *pepN*$^+$ *pepA*$^-$ *pepB*$^-$ *pepD*$^-$ strain) to degrade protein somewhat faster than did the quadruple mutant, and by the further finding that presence of both *pepN*$^+$ and *pepA*$^+$ genes sufficed to restore the protein degradation rate to the wild-type level (i.e. a double mutant of genotype *pepN*$^+$ *pepA*$^+$ *pepB*$^-$ *pepD*$^-$ is said to release trichloroacetic acid-soluble material from isotopically-labelled protein at the wild-type rate). It should be noted that even the quadruple mutant may be assumed still to contain the several peptidases analogous to the *E. coli* tripeptidase TP, aminopeptidase L, and proline-peptide specific peptidases P and Q (Table 1), as well as 'proteases I and II'. Yet the combined activities of these enzymes appears insufficient to permit 'normal' rates of intracellular protein degradation to free amino acids. Moreover, the observation that both the rate and the extent of protein hydrolysis were decreased implies that one or more of the missing aminopeptidases may play an important role in the initial degradative reactions and/or that enzymes catalysing the initial reactions may be subject to product inhibition.

Extracellular aminopeptidases might logically be expected to continue the degradation of exogenous proteins initiated by extracellular endopeptidases (proteases) and thereby produce peptides of sizes suitable for transport into cells. It therefore is difficult to explain why, in *Aeromonas proteolytica*, maximal endopeptidase production follows, rather than precedes, the peak of aminopeptidase activity during the growth cycle (Wilkes and Prescott, 1976); such progression suggests that these enzymes, the production and secretion of which are stimulated to different extents in media containing complex mixtures of peptides, are also of importance in the metabolism of intracellular proteins. This certainly would appear to be true for the alkaline serine protease which, together with a neutral metalloprotease, is produced and secreted by *B. subtilis* 168. Dancer and Mandelstam (1975), who have shown that the serine protease is necessary for sequential sporulation, believe it unlikely that the extracellular form of the enzyme is involved in the events leading to sporulation. They postulate that some serine protease also is present and active within the cells, in part because mutants unable to produce extracellular serine protease neither sporulate nor show intracellular protein turnover when starved, and in part because serine protease appears to be responsible for the degradation *in vivo* of the neutral metalloprotease, an enzyme 'dissociated' from sporulation.

Two other *B. subtilis* enzymes, aminopeptidase III and D-aminopeptidase, also are produced mainly in early stationary phase by both sporulating and non-sporulating strains. It has been suggested that aminopeptidase III may function in a selective degradation of peptides which probably occurs in the early stationary phase and which might be related to the cessation of culture growth or the early events leading to sporulation. It has also been suggested that the D-aminopeptidase (which is localized mainly in the cell-wall and periplasmic fractions of cells from early stationary phase cultures) may play a role either in the metabolism of peptide antibiotics or in the degradation of cell-wall peptides during the release of spores (Desmond *et al*., 1975).

At least one strain of bacillus, *B. sphaericus*, secretes some murein peptidases during sporulation, but this apparently is not true of *B. subtilis* (Arminjon *et al*., 1977; Guinand *et al*., 1976). Two *B. subtilis* murein peptidases, the endopeptidase and amidase (see Section 2.4.5 on murein peptidases), are thought to function in the synthesis of the specific cortex peptidoglycan, since their presence is highest at the time of cortex formation and their activity in vegetative cells is low (Guinand *et al*., 1976). The D-alanine carboxypeptidase of *B. subtilis* likewise is not essential for either the growth or the division of cells; its most potent catalytic action (i.e. removal of the terminal D-alanine residue from the UDP-*N*-acetyl-muramyl-pentapeptide intermediate in peptidoglycan synthesis) has led to the suggestion that it may play a role in regulating the trans-peptidase-catalysed cross-linking of peptide chains during cell-wall synthesis (Georgopapadakou *et al.*, 1977). The same role may be ascribed to the D-alanine carboxypeptidases IA and IB of *E. coli*, whose catalytic properties are analogous to those of the *B. subtilis* enzyme (see Section 2.4.5 on murein peptidases) and the activity of which also appears to be dispensible during normal growth and cell division (Tamura *et al.*, 1976).

As mentioned earlier, the D-carboxypeptidase I activity of *E. coli* remains constant during the division cycle, as does the activity of the *E. coli* amidase, phenomena which have led to the suggestion that neither type of activity has a special function in relation to septum formation at the time of cell division (Beck and Park, 1976). On the other hand, the direct correlation between the marked increase in demonstrable carboxypeptidase II activity and cell division, as well as the increased facility of dividing cells to release this periplasmic enzyme (following treatment with EDTA or toluene) are indicative of a specific role in relation to processes associated with cell division. These observations by Beck and Park (1977) have led them to postulate that 'Carboxypeptidase II may prevent the linkage of diaminopimelic acid in the murein layer to the lipoprotein by removing the subterminal D-alanine residue before it can be utilized to drive the hypothetical murein-lipoprotein cross-linking reactions, or its role may be . . . [that of preventing] the association of the outer membrane and the murein layer at the time of septum formation.' (For further speculations and discussion of the possible functions of murein peptidases, see Blumberg and Strominger, 1974.)

Finally, mention may be made of some bacterial peptidases the activity of which may have significance not only to the organism in which they are synthesized but to others as well. Two quite dissimilar examples may be cited. One concerns the extracellular enzymes secreted by *C. histolyticum*, collagenase and clostripain (i.e. clostridiopeptidases A and B) and the clostridial aminopeptidase. By definition, collagenases are endopeptidases acting on native collagen (Seifter and Harper, 1971), and products of collagenase action will include peptides that can serve as substrates for the broad-spectrum aminopeptidase (Kessler and Yaron, 1976*a,b*) and the more selective clostripain (Mitchell and Harrington, 1971). Whilst such degradative activities would be of nutritional advantage to the *Clostridia*, they are obviously highly deleterious to the collagen-containing tissues of higher animals (see also Section 3.1.3.2).

The second example is taken from the *S. cremoris* strains used as starter cultures in the dairy industry and again involves the concerted action of both proteases and peptidases, including the pyrrolidonyl peptidase mentioned earlier among the membrane-bound 'aminopeptidases'. Pyrrolidonyl peptidase (pyrrolidone carboxylyl peptidase) is considered to be one of several cellular peptidases participating in the degradation of oligopeptides presumed to arise via the action of cell wall-associated proteases on the casein in milk. Whilst such degradation again would yield easily transportable peptides of nutritional value to the bacteria, the overall results also are of 'importance to growth and acid production in milk and to the nature of the endproducts in cheese' (Exterkate, 1977) (see also Section 3.1.5).

9 CONCLUSIONS

Much progress has been made in recent years on the enzymology of bacterial peptidases and a promising start has been made in the study of their functions *in vivo*. The newly developed methods, permitting the isolation of peptidase-deficient mutants, have already been of significant help in both these aspects of peptidase biochemistry and should continue to be so, the availability of such mutants greatly facilitating the isolation of specific enzymes from cells containing several peptidases active toward the same substrates, as well as providing bacterial strains useful in the elucidation of the role played by specific peptidases in the bacterial metabolism of peptides, proteins, and peptidoglycans. The recent work, and not excluding that done with mutants, has also emphasized the incomplete state of currently available knowledge concerning the full complement of peptidases in even so popular a bacterium as *E. coli*, and the confusion surrounding the identity or non-identity of enzymes described by different investigators prompts a reiteration of the 'plea [made by J. W. Payne, 1977] that only *true* peptides be used as substrates in studies of peptidase activity, and that the use of artificial substrates such as aminoacyl-β-naphthylamides, aminoacyl-*p*-nitroanilides, etc. be curtailed.'

ACKNOWLEDGEMENTS

Grateful acknowledgement is made for support by U.S. Public Health Service Research Grant GM 23202 from the National Institute of General Medical Sciences for the work described by Hermsdorf (1978), Hermsdorf *et al*. (1979), and S. Simmonds (unpublished results).

REFERENCES

Arminjon, F., Guinand, M., Vacheron, M.-J., and Michel, G., (1977). *Eur. J. Biochem*., **73**, 557

Beck, B. D. and Park, J. T. (1976). *J. Bacteriol*., **126**, 1250

Beck, B. D. and Park, J. T. (1977). *J. Bacteriol.*, **130**, 1292
Bell, G., Payne, G. M., and Payne, J. W. (1977). *J. Gen. Microbiol.*, **98**, 485
Brown, J. L. (1973). *J. Biol. Chem.*, **248**, 409
Brown, J. L. and Krall, J. F. (1971). *Biochem. Biophys. Res. Commun.*, **42**, 390
Blumberg, P. and Strominger, J. L. (1974). *Bacteriol. Rev.* **38**, 291
Chaloupka, J. (1961), *Nature, Lond.*, **189**, 512; *Folia Microbiol., Praha*, **6**, 231
Chappelet-Tordo, D., Lazdunski, C., Murgier, M. and Lazdunski, A., (1977). *Eur. J. Biochem.*, **81**, 299–307
Choules, G. L. and Gray, W. R. (1971). *Biochem. Biophys. Res. Commun.*, **45**, 849
Coyette, J., Perkins, H. R., Polacek, I., Shockman, G. D., and Ghuysen, J. M. (1974). *Eur. J. Biochem.*, **44**, 459.
Dancer, B. N. and Mandelstam, J. (1975). *J. Bacteriol.*, **121**, 406
DeLange, R. J. and Smith, E. L. (1971). In *The Enzymes*, Vol. 3, Academic Press, New York, p. 81
Deranleau, D. A. and Zuber, H. (1977). *Int. J. Peptide Protein Res.*, **9**, 258
Desmond, E. P., Starnes, W. L., and Behal, F. J. (1975). *J. Bacteriol.*, **124**, 353
Dick, A. J., Matheson, A. T., and Wang J. H. (1970). *Can. J. Biochem.*, **48**, 1181
Doolittle, R. F. (1970). *Methods Enzymol.*, **19**, 555
Doolittle, R. F. and Armentrout, R. W. (1968). *Biochemistry* **7**, 516
Exterkate, F. A. (1977). *J. Bacteriol.*, **129**, 1281
Folk, J. E. (1971). In *The Enzymes*, Vol. 3, Academic Press, New York, p. 57
Frére, J.-M., Leyh-Bouille, M., Ghuysen, J.-M., Nieto, M., and Perkins, H. R. (1976), *Methods Enzymol.*, **45** Part B, 610
Fruton, J. S., and Simmonds, S. (1950). *Cold Spring Harbor Symp. Quant. Biol.*, **14**, 55
Georgopapadakou, N., Hammarström, S. and Strominger, J. L. (1977). *Proc. Natl. Acad. Sci. U.S.A.*, **74**, 1009
Goldberg, A. L. and St. John, A. C. (1976). *Annu. Rev. Biochem.*, **45**, 747
Guinand, M., Michel, G., and Balassa, G. (1976). *Biochem. Biophys. Res. Commun.*, **68**, 1287
Haley, E. E. (1970). *Methods Enzymol.*, **19**, 730
Hartsuck, J. A. and Lipscomb, W. N. (1971). In *The Enzymes*, Vol. 3, Academic Press, New York, p. 1
Hayman, S., Gatmaitan, J. S., and Patterson, E. K. (1974). *Biochemistry*, **13**, 4486
Hermsdorf, C. L. (1978). *Biochemistry*, **17**, 3370
Hermsdorf, C. L., Simmonds, S. and Saunders, A. (1979). *Int. J. Peptide Protein Res.* **13**, 146
Jackson, M. B., Becker, J. M., Steinfeld, A. S., and Naider, F. (1976). *J. Biol. Chem.*, **251**, 5300
Kessel, D. and Lubin, M. (1963). *Biochim. Biophys. Acta*, **71**, 656
Kessler, E. and Yaron, A. (1976*a*). *Eur. J. Biochem.*, **63**, 271
Kessler, E. and Yaron, A. (1976*b*). *Methods Enzymol.*, **45** Part B, 544
Kowit, J. D., Choy, W.-N., Champe, S. P., and Goldberg, A. L. (1976). *J. Bacteriol.*, **128**, 776
Latil, M., Murgier, M., Lazdunski, A., and Lazdunski, C. (1976). *Mol. Gen. Genet.*, **148**, 43
Lazdunski, C., Busuttil, J., and Lazdunski, A. (1975*a*). *Eur. J. Biochem.*, **60**, 363
Lazdunski, A., Murgier, M., and Lazdunski, C. (1975*b*). *Eur. J. Biochem.*, **60**, 349
Lazdunski, A., Pellisier, C., and Lazdunski, C. (1975c). *Eur. J. Biochem.*, **60**, 357
McCullough, J. L., Chabner, B. A., and Bertino, J. R. (1971). *J. Biol. Chem.*, **246**, 7207
McHugh, G. L. and Miller, C. G. (1974). *J. Bacteriol.*, **120**, 364
Matheson, A. T. (1963). *Can. J. Biochem. Physiol.*, **41**, 9
Matheson, A. T. and Dick, A. J. (1970). *FEBS Lett.*, **6**, 235
Matheson, A. T. and Murayama, T. (1966). *Can. J. Biochem.*, **44**, 1407
Matheson, A. T., Dick, A. J., and Rollin, F. (1970). *Can. J. Biochem.*, **48**, 1292
Matheson, A. T., Visentin, L. P., Boutet, A., and Rollin, C. F. (1971). *Can. J. Biochem.*, **49**, 1340
Meisler, N. and Simmonds, S. (1963). *J. Gen. Microbiol.*, **31**, 109; see also Meisler, N. (1963). Ph.D. Thesis, Yale University
Miller, C. G. (1975). *Annu. Rev. Microbiol.*, **29**, 485
Miller, C. G. and MacKinnon, K. (1974). *J. Bacteriol.*, **120**, 355
Miller, C. G. and Schwartz, G. (1978). *J. Bacteriol.*, **135**, 603
Miller, C. G. and Zipser, D. (1977). *J. Bacteriol.*, **130**, 347

Miller, C. G., Heiman, C., and Yen, C. (1976). *J. Bacteriol.*, **127**, 490
Mitchell, W. A. and Harrington, W. F. (1971). In *The Enzymes*, Vol. 3, Academic Press, New York, p. 699
Morihara, K. (1974). *Adv. Enzymol. Relat. Areas Mol. Biol.*, **41**, 179
Murgier, M., Lazdunski, A., and Lazdunski, C. (1976*a*). *FEBS Lett.*, **64**, 130
Murgier, M., Pelissier, C., Lazdunski, A., and Lazdunski, C. (1976*b*). *Eur. J. Biochem.*, **65**, 517
Murgier, M., Pelissier, C., Lazdunski, A., Bernadac, A., and Lazdunski, C. (1977). *Eur. J. Biochem.*, **74**, 425
Nguyen-Distèche, M., Ghuysen, J. M., Pollock, J. J., Reynolds, P., Perkins, H. R., Coyette, J., and Salton, M. R. J. (1974). *Eur. J. Biochem.*, **41**, 457
Pacaud, M. and Richaud, C. (1975). *J. Biol. Chem.*, **250**, 7771
Pacaud, M. and Uriel, J. (1971). *Eur. J. Biochem.*, **23**, 435
Patterson, E. K. (1976). *Methods Enzymol.*, **45** Part B, 377
Payne, J. W. (1972*a*). *J. Gen. Microbiol.*, **71**, 267
Payne, J. W. (1972*b*). *J. Gen Microbiol.*, **71**, 281
Payne, J. W. (1973). *Biochim. Biophys. Acta*, **28**, 469
Payne, J. W. (1975). In *Peptide Transport in Protein Nutrition* (D. M. Matthews and J. W. Payne, eds) North Holland, Amsterdam, Chap. 10
Payne, J. W. (1976). *Adv. Microbial Physiol.*, **13**, 55
Payne, J. W. (1977). In *Peptide Transport and Hydrolysis*, Ciba Foundation Symposium Vol. 50, Associated Science Publishers, Amsterdam, p. 305
Pecht, M., Giberman, E., Keysary, A., Yariv, J., and Katchalski, E. (1972). *Biochem. Biophys. Res. Commun.*, **290**, 267
Prescott, J. M. and Wilkes, S. H. (1976). *Methods Enzymol.*, **45** Part B, 530
Reaveley, D. A. and Burge, R. E. (1972). *Adv. Microbial Physiol.*, **7**, 1
Régnier, P. and Thang, M.-N. (1975). *Eur. J. Biochem.*, **54**, 445
Roncari, G., Stoll, E., and Zuber, H. (1976). *Methods Enzymol.*, **45** Part B, 522
Sarid, S., Berger, A., and Katchalski, E. (1962). *J. Biol. Chem.*, **237**, 2207
Schleifer, K. H. and Kandler, O. (1972). *Bacteriol. Rev.*, **36**, 407
Seifter, S. and Harper, E. (1971). In *The Enzymes*, Vol. 3, Academic Press, New York, p. 649
Sheahan, J., Eitenmiller, R. R., and Carpenter, J. A. (1975). *Appl. Microbiol.*, **29**, 726
Simmonds, S. (1970). *Biochemistry*, **9**, 1
Simmonds, S. (1972). In *Peptide Transport in Bacteria and Mammalian Gut*, Ciba Foundation Symposium Vol. 4, Associated Science Publishers, Amsterdam, p. 44
Simmonds, S. and Griffith, D. (1962). *J. Bacteriol.*, **83**, 256
Simmonds, S., Harris, J. I., and Fruton, J. S. (1951). *J. Biol. Chem.*, **188**, 251
Simmonds, S., Szeto, K. S., and Fletterick, C. G. (1976). *Biochemistry*, **15**, 261
Spratt, B. G. and Strominger, J. L. (1976). *J. Bacteriol.*, **127**, 660
Stone, D. (1953). *J. Biol. Chem.*, **202**, 821
Stone, D. and Hoberman, H. D. (1953). *J. Biol. Chem.*, **202**, 203
Sussman, A. J. and Gilvarg, C. (1970). *J. Biol. Chem.*, **245**, 6518
Sussman, A. J. and Gilvarg, C. (1971). *Annu. Rev. Biochem.*, **40**, 397
Tamura, T., Imae, Y., and Strominger, J. L. (1976). *J. Biol. Chem.*, **251**, 414
Twardowski, T., Hill, J. M., and Weissbach, H. (1977). *Arc. Biochem. Biophys.*, **180**, 444
Van Lenten, E. J. and Simmonds, S. (1967). *J. Biol. Chem.*, **242**, 1439
Vonder Haar, R. A. and Umbarger, H. E. (1972). *J. Bacteriol.*, **112**, 142
Vogt, V. M. (1970). *J. Biol. Chem.*, **245**, 4760
Wagner, F. W., Chung, A., and Ray, L. E. (1972*a*). *Can. J. Microbiol.*, **18**, 1883
Wagner, F. W., Wilkes, S. H., and Prescott, J. M. (1972*b*). *J. Biol. Chem.*, **247**, 1208
Wilkes, S. H., Baylis, M. E., and Prescott, J. M. (1973). *Eur. J. Biochem.*, **34**, 459
Wilkes, S. H. and Prescott, J. M. (1976). *Methods Enzymol.*, **45** Part B, 404
Yang, L. M., and Somerville, R. L. (1976). *Biochim. Biophys. Acta*, **445**, 406
Yaron, A. (1974). *Israel J. Chem.*, **12**, 651
Yaron, A. (1976). *Methods Enzymol.*, **45** Part B, 599
Yaron, A. and Berger, A. (1970). *Methods Enzymol.*, **19**, 521

Microorganisms and Nitrogen Sources
Edited by J. W. Payne

Section 2.5

Regulation of Peptide Transport in Bacteria

J. W. PAYNE

Department of Botany, University of Durham, Science Laboratories, South Road, Durham DH1 3LE, UK

1 INTRODUCTION

Regulation is a somewhat neglected facet of studies of solute transport systems in bacteria. The regulation of carbohydrate transport is best understood (Saier and Moczydlowski, 1978). Relatively little is known about regulation of amino acid transport, especially with regard to general control (see Section 1.6), and even less is known concerning regulation of peptide transport.

This chapter will consider the limited information that there is relevant to regulation of peptide uptake, but more particularly it will focus upon those aspects which are overdue for investigation and which, with recent improvements in methodology (see Section 2.1.4.3), are now amenable to study.

2 INDUCTION AND REPRESSION OF TRANSPORT PROTEIN SYNTHESIS

Regulation here is concerned with modification of the concentration of transport sites by changes in the synthesis and turnover rates of their components. This process, acting at the genetic level, does not permit an 'immediate' change in transport activity. Commonly, presence of a substrate in the medium leads to induction of its permease. Continued presence of a substrate at high concentration may lead to repression of its permease, while derepression of a high-affinity system may occur when substrate levels approach zero.

There is no evidence for differences in the rate of peptide uptake by bacteria whether cells have been grown with or without peptides in the medium. However, the studies of this feature are scant (see Payne and Gilvarg, 1971) and the topic would certainly pay a thorough investigation using peptides that cannot be cleaved in peptidase-deficient mutants (see Section 2.1.5.4). It should be noted, however, that examples of inducible amino acid transport systems in bacteria are uncommon, occurring for particular amino acids acting mainly as carbon or nitrogen sources. Kepes (1964) pointed out the energetic rationale for this situation and for the contrasting one with sugar permeases, which are generally inducible. Peptide uptake in *E. coli* and *S. typhimurium* is demonstrably very efficient, and as peptides are able to supply the full range of amino acids, certain of which can be used not only for protein synthesis but also as carbon and nitrogen sources (Gutnick *et al*., 1969), it might reasonably be anticipated that peptide permeases would be inducible. Inducible amino acid permeases have been demonstrated in the following instances, and it might be of interest therefore to investigate the possible inducibility of peptide permeases using peptide substrates containing these residues. An inducible system for transporting tryptophan has been described in *E. coli* (Boezi and De Moss, 1961; Burrous and De Moss, 1963; Whipp and Pittard, 1977), and in *Pseudomonas acidovorans* (Rosenfeld and Feigelson, 1969). A major proline permease in *S. typhimurium* is induced by proline in the medium (Ratzkin *et al*., 1978). An inducible uptake system for proline was also reported in *Pseudomonas aeruginosa* (Kay and Gronlund, 1969*a,b*) and one for lysine in *Pseudomonas putida* (Fan *et al*., 1972; Miller and Rodwell, 1971). However, it is likely that certain of the conclusions from these early studies regarding inducibility may need re-evaluation in light of the demonstrated effect of nitrogen limitation in transport systems (see Section 2.5.4).

No investigation into the possibility of repression of peptide permeases has been reported. It is an aspect deserving study. Many amino acid transport systems are subject to repression. In *E. coli*, a decreased level of a general aromatic permease (for Tyr, Trp, Phe) occurs in cells grown with tyrosine or phenylalanine, growth with tyrosine also results in a decreased level of the tyrosine-specific system (Kuhn and Somerville, 1974; Whipp and Pittard, 1977). The four transport systems for the basic amino acids arginine, lysine, and ornithine are repressible; formation of the specific systems for arginine and ornithine is repressible by either arginine or ornithine, formation of the specific lysine system, and the general system is repressible only by lysine (Celis, 1977*a,b*). Repression of the uptake system for threonine and serine (Templeton and Savageau, 1974), and of isoleucine, valine, and leucine uptake has been reported (Quay and Oxender, 1976; Kiritani and Ohnishi, 1977; see also Section 1.6).

Coordination between the control of transport and the biosynthetic pathway for an amino acid is varied. Whipp and Pittard (1977) showed that the *tyrR* gene product, known to be involved in regulation of aromatic biosynthetic enzymes, is also involved in the regulation of the aromatic amino acid transport systems. On the other hand, independent regulation of transport and biosynthesis occurs in *E. coli* for arginine (Celis, 1977*b*), and for branched-chain amino acids (Quay and Oxender, 1976; Quay *et al*., 1975; see also Section 1.6), and in *S. typhimurium* for proline (Ratzkin *et al*., 1978; Ratzkin and Roth, 1978).

Although adequate investigations have yet to be carried out into whether induction–repression of peptide permeases can occur, induction of the synthesis and secretion of extracellular peptidases by peptides in the medium has been demonstrated for *Aeromonas proteolytica* (Litchfield and Prescott, 1970), *Bacillus subtilis* (Ray and Wagner, 1972: Wagner *et al*., 1972) and *Bacillus licheniformis* (Absi and Prescott, 1978). It would be of some interest to know whether peptide permease activity changed in parallel with synthesis of these peptidases.

3 REGULATION OF THE ACTIVITY OF TRANSPORT SYSTEMS

In contrast to the above type of regulation one is concerned here with a more immediate modification of transport activity in response to presence of substrate. Two situations may be envisaged: uptake of substrate may be competitively inhibited by presence of a structural analogue in the medium; uptake of substrate may be inhibited or stimulated by presence of the substrate, an analogue, or a metabolic derivative within the cell; the mechanism thus allows an analogous form of control to that of end-product regulation of enzymic reactions.

Inhibition of the uptake of one peptide by another present in the medium has been demonstrated with many forms of bacteria, the inhibition being monitored using effects on growth or by direct assays for peptide uptake; rarely however, are the data good enough to demonstrate true competition (see Sections 2.1.4 and 2.1.5.5). In all cases to date, the peptide competitors are themselves taken up by the permeases, their relative inhibitory abilities being judged an index of their affinities for the system. Staros and Knowles (1978) in effect produced a specific irreversible inhibitor of uptake by photoaffinity-labelling a dipeptide analogue to the dipeptide permease of *E. coli*; their results are, however, open to various interpretations. Regulation of this type (*cis*-inhibition) for amino acid uptake is considered in Section 1.6.2.

The second, 'feedback'-type regulation has been noted with amino acids in various bacteria; both inhibition (trans-inhibition) and stimulation (trans-stimulation) of uptake being demonstrated (see Section 1.6.2.1). In bacteria there appears to be only one report of trans-inhibition, occuring for methionine transport in *E. coli*; methionine itself is apparently the inhibitory effector (Kadner, 1975). It has been suggested that such an effect should be an expected feature of carrier-mediated transport systems (Cuppoletti and Segel, 1974), and it is therefore of some interest that it has not been observed more often. Could it be that it is a feature only to be found with shock-sensitive, phosphate-bond-driven transport systems (e.g. Met), the shock-resistant systems in general being primarily under genetic control? The apparent trans-stimulation effect in which external leucine can exchange for internal isoleucine via the LIV-I system would accord with this idea (Quay *et al*., 1977). Trans-inhibition of amino acid uptake has been reported in a variety of other microorganisms: in the yeasts *Saccharomyces cerevisiae* (Crabell and Grenson, 1970; Kotyk and Rihova, 1972) and *Rhodotorula glutinis* (Saiyid and Kotyk, 1972), and fungus *Neurospora crassa* (Wiley and Matchett, 1968; Pall, 1971; Pall and Kelly, 1971). Detailed investigations have been performed with *Streptomyces hydrogenans* and the varied effects of amino acid pool composition during the different phases of the growth cycle demonstrated (Gross *et al*., 1970; Ring *et al*., 1970; Alim and Ring, 1976; Langheinrich and Ring, 1976).

It is not known whether such feedback mechanisms might operate for peptide uptake. It is possible that under typical culture conditions used in the laboratory, peptides are cleaved so rapidly upon entering the cytoplasm that no effective increase in peptide pool (it is assumed that a finite pool of peptides exists at any instant as necessary intermediates in protein turnover) can occur, and so direct feedback regulation of transport by intact peptides seems unlikely. In fact, continuous peptide uptake and cleavage is normally observed (Payne and Bell, 1979). On the other hand, it is arguable that, under conditions in which protein synthesis is severely depressed (by nutrient deprivation other than a depletion of the amino acid pool) or completely inhibited, it would be useful to the organism to be able to control peptide uptake rather than having to perform both peptide uptake and efflux of the cleaved amino acid residues (see Section 2.1.5.2). Thus, it may be pertinent that inhibition of protein synthesis by chloramphenicol decreased peptide uptake about 15 per cent and incubation in media containing most of the protein amino acids (but not just one or two at the same total concentration) also decreased peptide uptake (Payne and Bell, 1977*a*, 1979). No evidence was obtained relating to the possible mechanisms of these effects (but see next section). The possibility that intact peptides might feedback-inhibit peptide permeases can be examined by using peptidase-deficient mutants that have accumulated one peptide to a maximum steady-state concentration (i.e. preloading) and then determining whether a second peptide, added under these conditions, will undergo rapid exchange with the first (suggesting the steady state is one of dynamic exchange, see Section 2.1.5.4) or its uptake will be inhibited (suggesting feedback regulation). The continuous monitoring, automated-fluorescamine system is ideally suited for such studies together with visualization of the media components using dansyl-TLC (see Section 2.1.4.3).

4 CARBON AND NITROGEN CONTROL

It is well known that inducible transport systems (for sugars) are subject to catabolite repression in the presence of glucose. In *E. coli* and *S. typhimurium*, levels of cyclic AMP (cAMP) respond to the availability of carbon sources, the greater the limitation, the greater the concentration of cAMP; cell metabolism is in turn regulated by the levels of cAMP so as to tune it for the most efficient response to the prevailing limitation (see reviews by Rickenberg, 1974; Pastan and Adhya, 1976).

While discussing the general aromatic amino acid permease of *S. typhimurium*, Ames (1972) suggested that cAMP may be the regulator involved in derepression of this system. A similar suggestion was made for glutamine uptake in *E. coli* (Willis *et al*., 1975). Ring and colleagues have recently presented evidence that *de novo* synthesis of protein components of certain amino acid transport systems in *Streptomyces hydrogenans* is stimulated by addition of exogenous cAMP and in consequence rates of uptake and efflux are increased (Ring *et al*., 1977*a*,*b*). The effect of cAMP was time dependent, reaching a maximum after about 60 min incubation, and there was no indication that the activity of the transport systems was influenced directly.

Alper and Ames (1978) have suggested that as the cellular concentration of cAMP increases in response to carbon limitation it yields a CAMP–CRP complex (CRP is cAMP receptor protein) which in turn interacts to a graded extent (thereby causing activation) with a hierarchy of operons involved in the utilization and transport of fuel sources. They presented evidence for such activation of a variety of transport systems, including ones for amino acids that might function to supply their substrates primarily as carbon sources rather than simply for protein synthesis. In addition, they presented indirect evidence, based on increased sensitivity to peptide-analogue antibiotics, that peptide permeases might be subject to this regulation.

Relevant to these considerations is a recent suggestion that in *E. coli* the adenylate cyclase complex, which is mainly responsible for controlling levels of cAMP, is regulated by the proton electrochemical gradient (Peterkofsky and Gazdar, 1979). It was previously demonstrated that the phosphoenolpyruvate: sugar phosphotransferase system (PTS) (Postma and Roseman, 1976) which catalyses the uptake of glucose and some other sugars by a mechanism involving a change in the state of phosphorylation of PTS proteins, is involved in controlling adenylate cyclase activity, and suggestions have been made that the fraction of some PTS protein in the phospho form is proportional to adenylate cyclase activity (Peterkofsky and Gazdar, 1975; Saier and Feucht, 1975). However, such a model does not account for the fact that uptake of sugars via the lactose system, which occurs without modification and is a proton-symport mechanism driven by a proton-motive force (p.m.f.), also lowers cAMP levels (Peterkofsky and Gazdar, 1979). Furthermore, these authors showed that in the absence of sugars, a protonophore (CCCP) which decreases p.m.f. also inhibits adenylate cyclase activity. Thus, it was suggested that the steady-state electrochemical proton gradient (maintained in the absence of a substrate for a proton-symport system) regulates (in part) the activity of adenylate cyclase and proton symport reduces this steady-state p.m.f.; regulation was proposed to be mediated by a protein factor which could exist in two forms—energized (under steady-state p.m.f.) and de-energized (as a result of proton-symport), the latter state leading to inhibition of adenylate cyclase activity (and thus decreased levels of cAMP). It was further suggested that certain sugars, e.g. lactose, might cause twofold inhibition, first by lowering p.m.f. during initial proton-symport uptake, and second, following intracellular β-galactosidase action, glucose could leak from the cell (no mechanism suggested or associated ion-fluxes considered) and then be reaccumulated via the PTS.

Alper and Ames (1978) divided the transport systems they studied into several classes based on the extent of inhibition relative to the cAMP level, and interpreted '. . .these complex patterns as indicating that each specific transport system is induced at a particular cAMP level . . .'. However, it does seem that a mechanism of control based mainly on the level of cAMP is efficient only up to a point; for example, it is hardly efficient to activate synthesis of components of an amino acid transport system in the absence of its substrates in the medium. Thus, other regulatory parameters need consideration, and those identified by Oxender and colleagues (see Section 1.6) for controlling leucine transport, viz. $tRNA^{leu}$, *rho* factor, ppGpp (*relA*), and *hisT* gene product, stand out in this respect.

The protein factor *rho* seems to be involved in changing the frequency of RNA transcriptional termination at an attenuator site in the DNA proximal to the structural genes (leader region). In a *rho* mutant, transport via shock-sensitive, phosphate-bond dependent systems was stimulated as a result of increased synthesis (i.e. decreased attenuation) of their binding proteins (normally the rate-limiting component of the transport system); shock-resistant, p.m.f.-driven systems were unaffected (Quay and Oxender, 1977). A possible reservation on the effects arising solely from direct attentuation arises from the observation that the β-subunit of the Ca,Mg-ATPase is altered by *rho* mutations. The *relA*, and *hisT* genes, and the requirement for tRNA are interrelated. Thus, the *relA* locus is required for ppGpp synthesis during nutritional step down. In *relA* mutants starved for Leu, maturation of the main $tRNA^{leu}$ is incomplete, being undermodified with respect to pseudouridine (Ψ); under such conditions derepression of leucine transport is prevented. Similarly, the *hisT* gene codes for the enzyme (pseudouridine synthetase) that converts uridine to pseudouridine in $tRNA^{leu}$ (and His, Tyr) and *hisT* mutants are defective in derepression of synthesis of the binding protein of the LIV-I transport system. Together these results indicate that the uncharged, fully mature form of $tRNA^{leu}$ may act as a positive effector in regulation, and the above nucleotide is also shown to play a regulatory role (see Section 1.6). On the other hand, Gross *et al*. (1970) suggested that, in *Streptomyces hydrogenans*, specific tRNA molecules loaded with amino acid seem to provide positive feedback signals (they suggested a form of trans-stimulation) counteracting the negative feedback effects (trans-inhibition) of the free amino acid pool components.

Thomale and Nass (1978) have presented results that seem particularly relevant to the regulatory phenomenon being considered here although they did not touch upon this aspect in their paper. These workers showed that in a stringent *E. coli* strain growing exponentially under specific amino acid limitation, not only were the concentrations of the relevant isoaccepting tRNAs altered but drastic changes also occurred in the pattern of *all* isoaccepting tRNAs. They pointed out that previously this effect has been observed only in the following instances during unbalanced growth (i.e. RNA synthesis continuing but protein synthesis stopped): in relaxed cells after withdrawal of amino acids, or in wild-type cells after treatment with chloramphenicol or puromycin. In both instances the modifications involved substitution of uracil for pseudouridine (and dihydrouridine) (this being equivalent to failure to complete maturation of the tRNA) (Kitchingman and Fournier, 1976; Huang and Mann, 1974), and Thomale and Nass (1978) claimed that the changes they observed in isoaccepting tRNAs arose from the same structural deficiencies. It was suggested by Thomale and Nass (1978) that not only amino acid deficiency but also continuation of net tRNA synthesis is necessary to produce changes in isoaccepting tRNAs and the reason that previous workers had not observed such changes in stringent cells was because the amino acids were completely removed from the medium leading to a total stoppage of tRNA net synthesis. They suggested the following model for a sequence of events accompanying amino acid deficiency: binding of uncharged tRNA to ribosome will produce an effector molecule (not ppGpp) which directly or indirectly inhibits the action of certain enzymes that modify tRNAs, which in turn leads to correspondingly undermodified molecules (i.e. containing U in place of Ψ).

Thus, we have seemingly opposed observations. The regulatory picture presented by Oxender and colleagues proposes: (a) leucine limitation is a necessary and sufficient stimulus for derepression of transport, which results from enhanced synthesis of LIV-I binding proteins; (b) mutants in *hisT* cannot be fully derepressed for transport (or for biosynthesis, Bresalier *et al*., 1975); thus, the fully mature, uncharged $tRNA^{leu}$ with Ψ residues is proposed as necessary for full transport derepression, i.e. uncharged $tRNA\Psi^{leu}$ is a positive effector of transport (Quay *et al*., 1978). On the other hand, the conclusions of Thomale and Nass (1978) are that as a result of amino acid limitation undermodified tRNA with U residues are produced. There may be many explanations for this situation. It is possible that the modifications to the tRNAs have been wrongly assigned by Thomale and Nass, or perhaps enzymic interconversions of Ψ and U can occur in both directions, etc.

Changes in isoaccepting tRNA species were found to be fairly specific for amino acid limitation, and do not occur with carbon limitation (Thomale and Nass, 1978). It would be of some interest however, to know if a general nitrogen limitation caused similar changes.

It seems likely, therefore, that control elements like those above will also be operating for peptide transport (see next section). The inhibitory effects of treatment with chloramphenicol or amino acid mixtures (Payne and Bell, 1977*a,b*) may in part arise from such causes.

Finally, we must consider the effects of nitrogen limitation in modifying transport processes in bacteria. It has been known for some time that in fungi the synthesis of amino acid transport systems, especially general permeases, is under nitrogen control (Benko *et al*., 1969; Grenson and Hou, 1972; Roon *et al*., 1975), and the peptide transport system of *S. cerevisiae* is similarly regulated (Becker and Naider, 1977; Nisbet and Payne, 1979*a,b*), (see also Sections 1.3.3 and 3.2.3). However, only recently has this topic been investigated in bacteria (Willis *et al*., 1975; Betteridge and Ayling, 1975; Funanage *et al*., 1978; Kahane *et al*., 1978; Kustu *et al*., 1979).

Many bacteria produce elevated levels of glutamine synthetase (GS) in response to limitation of nitrogen in the growth medium, reflecting its role, together with glutamate synthase (GOGAT), in the assimilation of NH_4^+ into glutamate under nitrogen-limiting conditions (see Section 4.2). Some bacteria, e.g. *Klebsiella aerogenes*, also show nitrogen control of the synthesis of other enzymes, e.g. histidine (*hut*) and proline (*put*) degradative enzymes, involved in glutamate synthesis. Although this has not been demonstrated in *S. typhimurium* there is evidence that it has a measure of nitrogen control of degradative enzymes (Funanage *et al*., 1978). Furthermore, *K. aerogenes* (Prival and Magasanik, 1971), *S. typhimurium* (Funanage *et al*., 1978), and *E. coli* (Kahane *et al*., 1978) have mechanisms whereby the catabolite repression normally exerted over these degradative enzymes is released when nitrogen is also limiting.

Zaboura and Halpern (1978) showed that the enzymes of γ-aminobutyric acid (GAB) degradation in *E. coli*, although a catabolite-sensitive system, escape from this repression under nitrogen-limiting conditions and they suggested this relief was mediated by GS. Subsequently, Kahane *et al*. (1978) presented evidence that a specific, membrane-bound, GAB transport system is similarly subject to the dual physiological control catabolite repression and nitrogen availability, and that GS is the effector in the regulation of GAB carrier. In addition, Funanage *et al*. (1978) reported that *S. typhimurium* mutants with high GS levels also displayed increased transport rates for Gln, Arg, Asp and Lys, but a decreased rate for Pro. Similarly, in mutants with decreased GS levels, uptake rates for Gln and Lys were decreased. Their results therefore point to a regulatory correlation between nitrogen limitation, GS synthesis, and derepression of shock-sensitive, but not shock-resistant (Pro, contrast with GAB above) amino acid transport systems. Recently, Kustu *et al*. (1979) have presented detailed evidence that nitrogen limitation (under conditions of catabolite repression, glucose as carbon source) in *S. typhimurium* leads to derepression of the synthesis of GS and also of the binding proteins for the shock-sensitive transport systems for His, Gln, Lys/Arg/Orn, and Glu/Asp. Circumstantial evidence indicated that synthesis of transport systems without periplasmic components might also be under nitrogen control. Furthermore, since the structural genes for the binding proteins are not clustered with *glnA*, the structural gene for GS, and are in fact scattered on the chromosome, nitrogen control clearly affects the expression of several unlinked genes. Kustu *et al*. (1979) question the view of Magasanik and colleagues (see reviews, Magasanik, 1976; Tyler, 1978) that GS controls its own synthesis (autogenous regulation), and suggest the presence of a regulatory gene, *glnR*, closely linked to *glnA*.

It seems likely, therefore, that nitrogen limitation could play an important role in regulation of peptide transport. Unfortunately, no investigations have yet been made into this point, but it is clearly a feature that needs early attention.

5 INTERRELATIONS IN THE REGULATION OF AMINO ACID AND PEPTIDE TRANSPORT

So far in this chapter entitled 'Regulation of Peptide Transport' much detail has been given on the regulation of amino acid systems and relatively little for the peptide permeases. There are two reasons for this. First, peptides and amino acids are intimately related in that each directly supplies free amino acid residues to the cytoplasmic pool, because of this unique interrelation I believe it is important to consider their regulation together. The second reason therefore, follows from this, for there is much more relevant information available concerning amino acid regulation.

When it was first noted that specific exodus of the hydrolytically released amino acid residues occurred concomitantly with peptide uptake, and continued until all peptide had been accumulated (Payne and Bell, 1977*b*, 1979) (see also Section 2.1.5.2), it was suggested that regulation of peptide uptake might not occur and that instead any imbalances in amino acid pool composition arising from peptide uptake could be corrected by a mechanism of selective amino acid exodus (as had been postulated earlier (Payne, 1975)). In contrast to the above situation *in vitro*, in which E. coli was supplied with single defined peptides, when the bacteria are growing in a natural environment like the gut, which contains heterogeneous peptides, a more balanced amino acid accumulation should occur, making the need for exodus minimal. Of course, what was implied by regulation in the above case was a homotropic mechanism of direct feedback, or trans-inhibition, which is an unlikely mechanism when the full complement of intracellular peptidases is functioning. But regulation with respect to general control mechanisms, i.e. mediated by nitrogen or carbon limitation, is likely to be even more important for peptide transport systems (which supply the entire amino acid pool) than for individual amino acid transport systems.

The efficiency of peptide transport as a sole means of supplying required amino acids for growth of amino acid auxotrophs was mentioned earlier (Section 2.1.5). Consider now the case in which peptides act as a nitrogen source. In a culture containing 1 mg dry weight of cells it will have (approx.) 0.5 mg protein. Assume for simplicity that this protein is the only nitrogen component and it contains 16 per cent nitrogen; thus, total nitrogen content = $0.5 \times 10^{-3} \times 16 \times 10^{-2}$g nitrogen (i.e. 8×10^{-5}g N) and this amount will need to be supplied for the culture to double in mass in, say, 1 h. Now typical rates of net peptide uptake in *E. coli* and *S. typhimurium* are 20 nmol min^{-1} (mg dry wt)$^{-1}$ (see Section 2.1.5.4), thus, the nitrogen transported in a representative tripeptide with an amino acid residue mass weight of 100 daltons is (approx.) $20 \times 3 \times 10^{-9} \times 10^{2} \times 16 \times 10^{-2} \times 60$ g nitrogen h^{-1} (i.e. 5.8×10^{-5} g nitrogen h^{-1}). But not all the protein amino acids can act as sources of nitrogen (Gutnick *et al*., 1969; Tyler, 1978); assume 25 per cent can, therefore, utilizable nitrogen transported in peptide = $25 \times 10^{-2} \times 5.8 \times 10^{-5}$ g nitrogen h^{-1} (i.e. 1.5×10^{-5} g nitrogen h^{-1}). But utilizable residues will not occur in all peptides, and for the general case in which every peptide that can be formed from the twenty protein amino acids occurs in equal amount, then any one residue will occur in 9.5, 13.5, and 17.1 per cent of all di-, tri-, and tetrapeptides respectively; and similarly, the 25 per cent utilizable residues (i.e. 5/20) will occur in $3(19^2 + 18^2 + 17^2 + 16^2 + 15^2)/20^3 = 54$ per cent of all tripeptides. In other words, just over half of all tripeptides will contain residues that are sources of nitrogen. Thus, total utilizable nitrogen transported in peptide = $54 \times 10^{-2} \times 1.5 \times 10^{-5}$ g nitrogen h^{-1} (i.e. 8×10^{-6} g nitrogen h^{-1}). From this calculation it may be concluded that, if *mixed* peptides are to maintain growth rate when acting as sole nitrogen source their average rate of uptake must increase about 10-fold over that observed in NH_4^+/glucose media. This increase might not, of course, be necessary for *specific* peptides that are especially good sources of nitrogen; the observation (Kustu *et al*., 1979) that Arg-Arg (in contrast to Arg) is an excellent nitrogen source for *S. typhimurium*, and supports the same growth rate in wild type and a nitrogen-regulatory mutant, accords with this idea. Interestingly, about a 10-fold increase in peptide transport capacity has been observed in *S. cerevisiae* grown under nitrogen limitation (Becker and Naider, 1977; Nisbet and Payne, 1979*a,b*) (see Section 2.5.4), but no relevant studies have been carried out with bacteria.

It was reasoned above that, because peptides and free amino acids both supply the same pool constituents, it is to be expected that their transport will be potentially susceptible to similar forms of regulation under nitrogen limitation. Such limitation will affect the steady-state composition of the amino acid pool; some of these changes may be rectified by shifts in the productivity of biosynthetic pathways, but some may persist and become transmitted to tRNA species, which in turn may mediate derepression of transport/biosynthetic systems by the mechanisms discussed above (see Section 2.5.4). Now peptide transport systems are driven by phosphate bond energy (see Section 2.6.4) and (although not yet demonstrated, see Section 2.6.5) would be expected to possess periplasmic binding proteins (BP). BP is the component that usually limits the rate of transport for the overall system (Wilson and Smith, 1978) and under nitrogen limitation their synthesis is derepressed and transport rates are increased (see Section 2.6.4). In part, the derepressed synthesis of BP may be a device to improve recapture of amino acids that leak from the cytoplasm into the periplasmic space (and this may explain why apparently several are coordinately derepressed by nitrogen limitation (Kustu *et al*., 1979)) but clearly they also mediate uptake of substrate from the medium. It is the latter aspect that leads to the question, how might a specific response be abstracted from the total complement of available responses that are primed by nitrogen limitation? For it is expected that only by operation of an extra level of control will regulation be made really efficient, i.e. it is not efficient to derepress synthesis of (say) peptide binding protein in the absence of its substrates in the medium (see also Section 2.5.4 for similar argument with respect to carbon limitation). The bacterium requires the means to sense the levels of nutrients in its environment and to communicate this information to the inside of the cell, in order that a 'rational' selection can be made of which genes to trigger from amongst the range which are subject to the relevant, general regulatory parameters functioning under nitrogen and/or carbon limitation. If more than one specific response were possible, because of the availability of several potentially suitable nutrients in the medium, it would be advantageous for a hierarchical control to operate in which response to the best nitrogen source was preferentially triggered. Thus, the requirement is for chemoreception. But it is for more than that. The requirement is for an integrated control over chemoreception and consequent taxis, expression of transport capacity, and the levels of biosynthetic/degradative enzymes within the cell. Relevant aspects of these three elements are considered first before a model is presented for integrated control for the cases of substrates acting either primarily in a direct manner for biosynthesis, e.g. amino acids in protein synthesis, or mainly as carbon and/or nitrogen source.

Chemotaxis in bacteria has been reviewed recently by Adler (1975). Aspects essential to the present argument are the following. Periplasmic binding proteins are the chemoreceptors, i.e. the part of the chemosensor system that recognizes the substrates and initially binds them. However, although various lines of evidence (e.g. similar K_m values in wild type and certain mutants) indicate that the binding proteins are unambiguously involved in both transport and chemotaxis, their roles in these are separable. Thus, galactose-binding protein mutants have been isolated (Ordal and Adler, 1974) that show normal galactose chemotaxis but defective transport, or alternatively normal galactose transport and defective chemotaxis. These features led to the proposal (Adler, 1975) that: 'The binding protein thus appears to have three sites—one for binding the ligand, one for interacting with the next transport components, and one for interacting with the next chemotaxis components'. Although this model is entirely reasonable, for simplicity and symmetry in the following model it is proposed that instead the binding protein has two separate sites that bind to different regions of a common second component: one site is involved in binding that directly leads to transport, and after this binding the other site allows communication with a chemotaxis-specific component.

In the absence of a chemical stimulant (i.e. no attractant or repellent present) bacteria move in straight runs interrupted by a tumble, after which the cell sets off in a new random direction. Tumbling results from abrupt reversal of the direction of rotation of the flagella (from counter-clockwise to clockwise) and it is the frequency of tumbling that is altered by chemical stimulants. An increasing gradient of an attractant suppresses tumbling with the result that cells tend to swim more in a straight line and more up the gradient, as the concentration decreases they tumble more frequently. In other words, as a bacterium swims up a spatial gradient the fractional amount of chemoreceptor bound increases, and the runs are long; moving down a gradient, the fractional amount of chemoreceptor bound decreases, and the runs are about the same length as in isotropic solution. For repellants the opposite is true: in an increasing concentration they tumble more often, while a decreasing concentration suppresses tumbling (Alder, 1975).

These effects occur not only with a spatial gradient (e.g. a higher concentration of stimulant above than below) but also with a temporal gradient (e.g. a higher concentration of stimulant earlier than now) (Brown and Berg, 1974). These authors explained their results in positive gradients by a model in which the suppression of directional changes is proportional to the time rate of change of the fractional amount of chemoreceptor bound. The bacteria can be thought of as having a 'memory with a decay time'.

A further important observation (Tsang *et al*., 1973) is that there is an integration of multiple sensory data, apparently by algebraically adding the stimuli. The 'decision' by the bacteria whether to be attracted or repelled in a situation where both attractant and repellent occur depends on the relative effective concentrations of the two, i.e. how far each is above its threshold concentration. Similarly, for two attractants (e.g. ribose and galactose), the 'decision' to choose one over the other depends on the concentration of each in the environment, relative to the binding affinities of a specific galactose-binding protein and ribose-binding protein that interact with a common relay component (presume transport protein) (Strange and Koshland, 1976). The mechanism(s) for summing and relaying these stumuli is unknown; it is an analogous mechanism that is needed for the specific transport regulation postulated earlier. Specific for the chemotactic case, it has been shown that, for the usual transient response to a change in the concentration of a stimulant, the initial sensory adaptation specifically has an absolute requirement for methionine (Springer *et al*., 1977), a requirement earlier demonstrated to involve methylation of a membrane protein (Kort *et al*., 1975). A change in membrane potential also occurs during the chemotactic response (Szmelcman and Adler, 1976).

A recent study on chemotaxis of *S. typhimurium* towards amino acids and sugars (Melton *et al*., 1978) showed that those amino acids which were the best attractants were those that can serve as sole carbon or nitrogen source or as sole carbon source. Thus, one is impressed by the coordinate efficiency of the process, a premise that initiated development of an entirely hypothetical model for optimal coordination in the utilization of amino acids and peptides.

In this model (Figure 1) a binding protein (receptor) is considered in the absence of substrate to exist in a conformation that has little or no affinity for its cognate, membrane-bound component. Binding of substrate induces a conformational change leading to enhanced binding of the receptor–substrate complex. (This proposal has been made previously by Strange and Koshland, 1976.) It is, of course, not currently possible to test this proposal, not least because reconstitution experiments using purified receptor and osmotically-shocked cells are not generally successful. Evidence for such a conformational change has been presented for *E. coli* binding proteins for: galactose (Boos, 1972; Boos *et al*., 1972; McGowan *et al*., 1974; Strange and Koshland, 1976), arabinose (Parsons and Hogg, 1974), maltose (Szmelcman et al., 1976), leucine (Penrose *et al*., 1970), and glutamine (Weiner and Heppel, 1971), and a similar report appeared for an amino acid-binding protein from *S. cerevisiae* (Vorisek, 1973).

Clear evidence for the binding of a receptor (and thus by extrapolation a receptor–substrate complex) to a membrane-bound component of a transport system comes from the elegant studies of Ames and her colleagues. Thus, in the high-affinity histidine transport system of *S. typhimurium*, the binding protein J (*hisJ* gene product) has two sites, one at which the substrate binds and another required for it to bind to protein P (*hisP* gene product) (Kustu and Ames, 1974; Ames and Spudich, 1976); protein P has recently been shown to be membrane bound (Ames and Nikaido, 1978). Protein P, which has the properties of an integral, or intrinsic membrane protein, seems the most obvious candidate for receiving the substrate from the receptor and achieving its passage through the membrane; *hisP* mutants lose transport entirely (Ames and Nikaido, 1978; Ames *et al.*, 1977; Kustu *et al.*, 1979).

A novel aspect of the present model is the postulate that there is an additional peripheral, or extrinsic membrane component (protein) located at the inner surface of the membrane; this will be referred to as m-protein for messenger protein. m-Protein binds to the cytoplasmic face of the integral membrane transport proteins (e.g. protein P above); for each specific integral transport protein there will be a correspondingly specific m-protein. However, the integral proteins have variable affinity for m-proteins; in the absence of substrate in the medium m-proteins have strong affinity and remain bound at the membrane, but attachment of a binding protein–substrate complex to the integral protein causes the latter to undergo a conformational change which concomitantly decreases its affinity for its attached m-protein, causing it to be released into the cytoplasm. In this way a response can be triggered that is proportional to the fractional amount of binding protein–receptor complex (see above) and, importantly, has yielded a specific diffusible signal of the binding event. In the model (Figure 1) it is postulated that separate m-proteins exist, functioning primarily in regulation of transport and metabolism (tm-protein), or for chemotaxis (cm-protein); for simplicity it is envisaged that both can bind to a single integral protein (subunits not considered) each being released through conformational perturbations induced by interaction of separate regions of the binding protein with the integral protein (see above). An alternative model (not shown) would have an integral protein still with two sites for binding cm- and tm-proteins but with it being possible to attach only one of them at any time; separate cm-, tm-complexes would then exist in the membrane. Not all systems need have cm- *and* tm-proteins; a transport system could lack a chemotaxis component (not all nutrient substrates will be attractants), or a system could exist for chemotaxis alone (functional for toxic repellent molecules). In no case is it postulated that transport is necessary to trigger m-protein release, attachment of the binding protein–substrate complex is alone sufficient.

The fate and function of the released m-proteins is postulated to be as follows. The cm-proteins will be involved in a set of reactions linked to flagellar motion; these are not considered further here. The tm-proteins will help regulate the operons for their associated transport systems and biosynthetic/degradative pathways. Thus, on release of the tm-protein from the membrane it will diffuse and bind to a specific attachment site in the operator region of the operon for the biosynthetic pathway enzymes of the transport substrate; the binding will repress synthesis of the enzymes (during uptake of substrate its continued biosynthesis is wasteful). The affinity of the operator site for the tm-protein, although normally less than the membrane site, is greater when the latter is modified by binding protein–substrate attachment. This repression could of course be further modified under the influence of the transported substrate. In addition, the tm-protein will bind to a specific attachment site in the operator region of the operon for the proteins of the transport system: this binding will derepress synthesis of these proteins. Similar considerations apply here regarding the affinity of binding as were mentioned above for the biosynthetic pathway operon. Conceivably tm-protein might also derepress synthesis of enzymes required for substrate metabolism. The structural

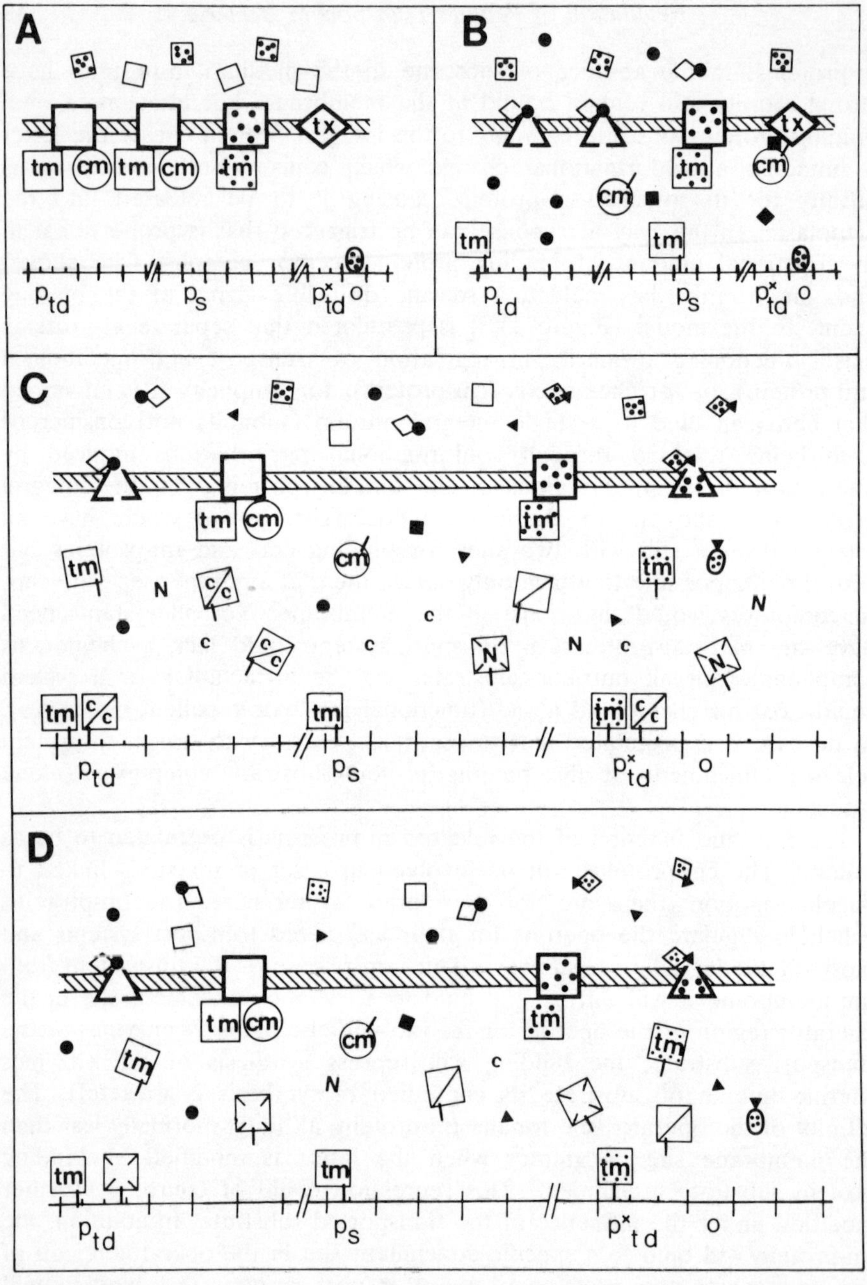

Figure 1 Hypothetical scheme for coordinating nutrient availability, transport, and metabolism.

A. Steady-state situation in which constitutive protein synthesis occurs. No exogenous substrates are available for either of two periplasmic binding proteins , , which consequently have little affinity for their respective integral proteins, and . Messenger proteins for transport (tm) and chemotaxis (cm) remain attached at the cytoplasmic face to appropriate integral proteins □, , of permeases which mediate transport of the substrates, ●, and ▲, respectively (see B–D); only the first system can elicit a chemotactic response. tx is a membrane-bound system, linked in some way to the flagellar, that evokes appropriate taxis. p_{td} is a promoter in an operon which contains structural genes for both the transport system (p_t) and degradative enzymes (p_d) for the substrate, ●, (p_t and p_d are shown linked for simplicity but could be separate); the substrate, ●, (e.g. an amino acid) can function directly as a metabolic substrate, and as either a carbon or a nitrogen source, consequently p_{td} contains binding sites for tm, cAMP-CRC (, see C), glutamine synthetase, GS (, see D), and RNA polymerase, RP (not shown). p_s is a promoter region in the operon for the biosynthetic enzymes for the substrate, ●; it contains binding sites only for tm and RP. p_{td}^x is a promoter region in an operon which contains structural genes for both the transport system and the degradative enzymes for the substrate, ▲, (see C,D); ▲ can function directly as a metabolic substrate and as a carbon source, consequently p_{td}^x contains binding sites for tm, cAMP–CRC, and RP. o is an operator that is protected by a repressor, .

B. Single exogenous nutrient. ● attaches to its binding protein, □, causing conformational change and giving complex . This complex has enhanced affinity for integral protein □, and on binding causes it to undergo conformational change to , this in effect opens a pore allowing substrate translocation. In addition, the conformational change releases cm and tm. Free cm can bind an effector, ■, which allows generalized cm binding to tx; different cm can have different affinities for ■ related to 'desirability' of exogenous substrate. tm binds to p_{td} and to p_s causing transcription of the former operon to be stimulated through enhanced initiation and the latter to be repressed; conceivably tm may bind to promoters of other genes whose products are metabolically related to ●. As exogenous ● is depleted,  will revert to □ and tm will rebind; the order of binding affinities for ● is □ $> p_{td} \approx p_s >$ .

C. Two exogenous nutrients and carbon limitation. Carbon limitation increases cAMP (c) which binds to cAMP-receptor protein (CRC) , to form complex ; attachment of this complex to p_{td} and to p_{td}^x exerts positive control as normal. Released tms also bind to p_{td} and to p_{td}^x, thus focusing the general  effect upon relevant nutrients available exogenously; in addition, cooperativity in binding to p_{td} or to p_{td}^x may occur and activation may be greater with both regulators than with either one alone. The biosynthetic operon is again repressed by tm binding to p_s. The repressor is inactivated on binding the transported substrate, ▲ (or metabolite, not shown), leaving the operator empty.

D. Two exogenous nutrients and nitrogen limitation. It is speculated that in cells growing on ammonia, GS exists in the form , but nitrogen depletion limits availability of nitrogen, leading to increased  which has enhanced affinity for p_{td}. Released tm also binds to p_{td} thus focusing the general  effect upon relevant nutrients available exogenously; in addition, cooperativity in binding to p_{td} may occur and activation may be greater with both regulators than with either one alone. Similar conditions obtain for p_s and o as described in C above.

The model considers tm only to be involved in regulating transcription initiation; a related model could involve tm in regulating transcription termination.

gene for the tm-protein would be expected to be found in the operon for the transport system (or conceivably that for metabolism of substrate) so that its synthesis could be coordinated with that of other components of the permease. In this regard it will be of some interest to establish the function of the product of the recently discovered *hisQ* gene which has been shown to map between *hisJ* and *hisP* (Ames and Nikaido, 1978; Ames *et al.*, 1978). (Thc possible relation of β-galactoside transacetylase to this model is also of interest.)

Relative to the above picture, the needs of the cell may change under conditions of carbon and/or nitrogen limitation. Carbon limitation will lead to increased levels of cAMP and production of cAMP–CRC. It was mentioned above how operation of the PTS and changes in p.m.f. resulting from substrate movements could affect cAMP levels. cAMP–CRC provides positive control by stimulating transcription of many operons useful under carbon limitation. It was suggested by Alper and Ames (1978) (see above) that a hierarchy of such operons might occur, this being possible if their operators have different affinities for cAMP–CRC. Based on the present model, it is postulated that, in addition, fine tuning of this control will occur in which attachment of a tm-protein together with cAMP–CRC can enhance derepression over that occuring with either activator alone (see Figure 1). The two effectors could bind independently of each other but it could be advantageous if it were cooperative and attachment of tm-protein increased the affinity for cAMP–CRC binding. Given that any differences in the affinities of the operons for cAMP–CRC reflect the intrinsic value of the relevant substrates as carbon sources, then the tm-proteins would serve to focus cellular activity on those particular ones which were available.

Similar arguments may be made for nitrogen limitation. However, a component (analogous to cAMP–CRC) that is selectively produced under nitrogen limitation has not (yet) been identified. It has been suggested (see above and review by Tyler, 1978) that GS may be such a substance, but Ames and colleagues (Kustu *et al.*, 1979) have questioned this view and suggested that control might be mediated by the as yet unidentified product of the *glnR* gene. Such a hypothetical molecule could be expected to act in an exactly analogous way to cAMP–CRC, showing different affinities for operons relevant to nitrogen utilization and stimulating synthesis. tm-Proteins would, as above, exercise derepression over the operons for the transport and biosynthesis/degradation of nitrogen compounds. The hypothetical nitrogen effector and the tm-protein acting in concert would serve to specify the particular operons which should undergo enhanced derepression for optimal utilization of environmental nutrient supplies. It was mentioned above that conditions of nitrogen starvation cause derepression of the proteins for several transport systems (Funanage *et al.*, 1978; Kustu *et al.*, 1979). Interestingly, when arginine is used as nitrogen source it commandeers a high-affinity histidine transport system; it will be of interest to know how general is the ability to take over a high-affinity system under such conditions. Under nitrogen limitation, certain tm-proteins for amino acids might (together with other regulatory agents, e.g. tRNA) repress synthesis of the binding proteins for other amino acids.

Some substances, e.g. peptides and amino acids, may act both as nitrogen and carbon sources and so regulation by cAMP–CRC and the hypothetical nitrogen effector might be expected, with operator binding sites for both, and perhaps joint binding if a poor substrate were acting both as nitrogen and carbon source. It was mentioned above that the effects of catabolite repression can be overcome under conditions of nitrogen limitation. This could readily be achieved if operon binding of hypothetical nitrogen effector (in the absence of cAMP–CRC) caused derepression.

There is of course little or no support for much of this model. However, Anderson *et al*. (1976) presented evidence that the product of the *livR* gene is '. . . a diffusible, negative controlling element for high-affinity leucine transport systems'. It may also be relevant to the proposed tm-protein effect on transport and metabolism that the *gal* operon of *E. coli* (coding for synthesis of a number of enzymes concerned with galactose metabolism) is located at map position 17 min, but the map position of a regulatory factor occurs at 61 min; 61 min is the location of an L-Ara permease that may form part of a transport operon involved also in Gal transport. A methylgalactoside regulatory gene is located at 17 min (Bachman *et al*., 1976).

6 CONCLUSIONS

It has been emphasized that little detail is yet available concerning regulation of peptide uptake, although new methods for studying peptide transport now make study of this feature feasible. The discussion integrated considerations of the regulation of amino acid and peptide transport because of their obvious interrelatedness via intracellular pool levels. The recent demonstration of the occurrence of peptide efflux, and more particularly, the process of selective amino acid efflux of residues hydrolytically released from transported peptides are considered as perhaps more efficient 'regulatory' mechanisms than controlling peptide uptake via *trans*-inhibition. Attention is given to the likely effects of carbon and/ or nitrogen limitation on the regulation of peptide uptake, a potentially most important aspect as peptides can act not only as sources of amino acids for protein synthesis but also as efficient sources of carbon and nitrogen. Derepression of peptide transport in yeast has been noted under nitrogen limitation, but no study with bacteria has yet been carried out. It is postulated that nitrogen and carbon control will be found for peptide uptake in bacteria. Based on the premise that a highly efficient coordination will be needed between specific transport capacity and particular biosynthetic/degradative enzyme systems on the one hand, and the nutrients available in the environment on the other hand, especially under carbon and/or nitrogen limitation, an entirely hypothetical scheme is presented for the types of control that might be useful.

Finally, it is the author's opinion that, in studying transport regulation, especially the effects of carbon or nitrogen limitation where limited derepression is apparently commonly observed, serious attention should be given to the possibility that the observed changes may arise not from a regulatory effect in the normal sense but as a result of gene duplication. Duplicated genes are very frequently produced and although they are not generally stably preserved (Anderson and Roth, 1977) it is quite clear that potentially this could affect interpretation of changes in transport capacity. It is well documented (see review by Anderson and Roth, 1977) that subjecting bacteria to particular limitations, e.g., carbon- or nitrogen-limited growth in chemostat culture, may select for strains carrying amplification of the gene that is rate limiting for growth under the conditions. Not only does this fortunate event occur at high frequency (thus arguing for the maintenance within the chromosome of the information that achieves it), but in this way populations presumably adapt efficiently to short-term environmental stress without stably altering the genetic material that prolonged evolution has optimized for them. Because functioning of transport systems may in many instances constitute a rate-limiting process, it is reasonable to expect amplification of transport genes.

REFERENCES

Absi, J. R. and Prescott, J. M. (1978). Isolation and properties of an aminopeptidase from *Bacillis licheniformis*. *Arch. Biochem. Biophys.*, **186**, 383–391

Adler, J. (1975). Chemotaxis in bacteria. *Annu. Rev. Biochem.*, **44**, 341–356

Alim, S. and Ring, K. (1976). Regulation of amino acid transport in growing cells of *Streptomyces hydrogenans*.II. Correlation between transport capacity and growth rate in chemostat cultures. *Arch. Microbiol.*, **111**, 105–110

Alper, M. D. and Ames, B. N. (1978). Transport of antibiotics and metabolite analogues by systems under cyclic AMP control: Positive selection of *Salmonella typhimurium cya* and *crp* mutants. *J. Bacteriol.*, **138**, 149–157

Ames, G. F. L. (1972). Components of histidine transport. In *Membrane Research*, 1st ICN-UCLA Symposium on Molecular Biology (C. F. Fox, ed.), Academic Press, New York and London, pp. 409–426

Ames, G. F. L. and Nikaido, H., (1978). Identification of a membrane protein as a histidine transport component in *Salmonella typhimurium*. *Proc. Natl. Acad. Sci. U.S.A.*, **75**, 5447–5451

Ames, G. F. L. and Spudich, E. N., (1976). Protein–protein interaction in transport: periplasmic histidine-binding protein J interacts with P protein. *Proc. Natl. Acad. Sci. U.S.A.*, **73**, 1877–1881

Ames, G. F. L., Biek, D. P., and Spudich, E. N. (1978). Duplications of histidine transport genes in *Salmonella typimurium* and their use for the selection of deletion mutants. *J. Bacteriol.*, **136**, 1094–1108

Ames, G. F. L., Noel, K. D., Taber, H., Spudich, E. N., Nikaido, K., Afong, J., and Ardeshir, F., (1977). Fine structure map of the histidine transport genes in *Salmonella typhimurium. J. Bacteriol.*, **129**, 1289–1297

Anderson, J. J., Quay, S. C., and Oxender, D. L. (1976). Mapping of two loci affecting the regulation of branched chain amino acid transport in *Escherichia coli* K-12. *J. Bacteriol.*, **126**, 80–90

Anderson, R. P. and Roth, J. R. (1977). Tandem genetic duplications in phage and bacteria. *Annu. Rev. Microbiol.*, **31**, 473–505

Bachman, B. J., Low, K. B., and Taylor, A. L. (1976). Recalibrated linkage map of Escherichia coli K-12. *Bacteriol. Rev.*, **40**, 116–167

Becker, J. M. and Naider, F. (1977). Peptide transport in yeast: uptake of radioactive trimethionine in *Saccharomyces cerevisiae. Arch. Biochem. Biophys.*, **178**, 245–255

Benko, P. V., Wood, T. C., and Segel, I. H. (1969). Multiplicity and regulation of amino acid transport in *Penicillium chrysogenum. Arch. Biochem. Biophys.*, **129**, 498–508

Betteridge, P. R. and Ayling, P.D. (1975). The regulation of glutamine transport and glutamine synthetase in *Salmonella typhimurium. J. Gen. Microbiol.*, **95**, 324–334

Boezi, J. A. and De Moss, R. D. (1961). Properties of a tryptophan transport system in *Escherichia coli. Biochim. Biophys. Acta*, **49**, 471–484

Boos, W. (1972). Structurally defective galactose-binding protein isolated from a mutant negative in β-methylgalactoside transport system of *Escherichia coli. J. Biol. Chem.*, **247**, 5414–5424

Boos, W., Gordon, A. S., Hall, R. E., and Price, H. D. (1972). Transport properties of the galactose-binding protein of *Escherichia coli*: substrate induced conformational change. *J. Biol. Chem.*, **247**, 917–923

Bresalier, R. A., Rizzino, A. A., and Freundlich, M. (1975). Reduced maximal levels of derepression of the isoleucine, valine and leucine enzymes in *hisT* mutants of *Salmonella typhimurium. Nature, Lond.*, **253**, 279–280

Brown, D. A. and Berg, H. C. (1974). Temporal stimulation of chemotaxis in *Escherichia coli. Proc. Natl. Acad. Sci. U.S.A.*, **71**, 1388–1392

Burrous, S. E. and De Moss, R. D. (1963). Studies on tryptophan permease in *Escherichia coli. Biochim. Biophys. Acta*, **73**, 623–637

Celis, T. F. R. (1977*a*). Properties of an *Escherichia coli* K-12 mutant defective in the transport of arginine and ornithine. *J. Bacteriol.*, **130**, 1234–1243

Celis, T. F. R. (1977*b*). Independent regulation of transport and biosynthesis of arginine in *Escherichia coli* K-12. *J. Bacteriol.*, **130**, 1244–1252

Crabeel, M. and Grenson, M. (1970). Regulation of histidine uptake by specific feedback inhibition of two histidine permeases in *Saccharomyces cerevisiae. Eur. J. Biochem.*, **14**, 197–203

Cuppoletti, J. and Segel, I. H. (1974). Trans inhibition kinetics of sulphate transport system of *Penicillium notatum*: analysis based on an iso uni univelocity equation. *J. Membrane Biol.*, **17**, 239–252.

Fan, C. L., Miller, D. L., and Rodwell, V. W. (1972). Metabolism of basic amino acids in *Pseudomonas putida*. Transport of lysine, ornithine and arginine. *J. Biol. Chem.*, **247**, 2283–2290

Funanage, V. L., Ayling, P. D., Dendinger, S. M., and Brenchley, J. E. (1978). *Salmonella typhimurium* LT-2 mutants with altered glutamine synthetase levels and amino acid uptake activities. *J. Bacteriol.*, **136**, 588–596

Grenson, M. and Hou, C. (1972). Ammonia inhibition of the general amino acid permease and its suppression in NADPH-specific glutamate dehydrogenaseless mutants of *Saccharomyces cerevisiae. Biochem. Biophys. Res. Commun.*, **48**, 749–756

Gross, W., Ring, K., and Heinz, E. (1970). Positive feedback regulation of amino acid transport in Streptomyces hydrogenans. *Archs Biochem. Biophys.*, **137**, 253–261

Gutnick, D., Calvo, J. M., Klopotowski, P., and Ames, B. N. (1969). Compounds which serve as the sole source of carbon or nitrogen for *Salmonella*. *J. Bacteriol*., **100**, 215–219

Huang, P. C. and Mann, M. B. (1974). Comparative fingerprint and composition analysis of the three forms of ^{32}P-labeled phenylalanine tRNA from chloramphenicol treated *Escherichia coli*. *Biochemistry*, **13**, 4704–4710

Kadner, R. J. (1975). Regulation of methionine transport activity in *Escherichia coli*. *J. Bacteriol*., **122**, 110–119

Kahane, S., Levitz, R., and Halpern, Y. S. (1978). Specificity and regulation of γ-aminobutyrate transport in *Escherichia coli*. *J. Bacteriol*., **135**, 295–299

Kay, W. W. and Gronlund, A. F. (1969*a*). Proline transport in *Pseudomonas aeruginosa*. *Biochim. Biophys. Acta*, **193**, 444–450

Kay, W. W. and Gronlund, A. F. (1969*b*). Influence of carbon or nitrogen starvation in amino acid transport in *Pseudomonas aeruginosa*. *J. Bacteriol*., **100**, 276–286

Kepes, A. (1964). The place of permeases in cellular organization. In *The Cellular Functions of Membrane Transport* (J. F. Hoffman, ed.) Prentice Hall, Englewood Cliffs, N.J., pp. 155–169

Kiritani, K. and Ohnishi, K. (1977). Repression and inhibition of transport systems for branched-chain amino acids in *Salmonella typhimurium*. *J. Bacteriol*., **129**, 589–598

Kitchingman, G. R. and Fournier, M. J. (1976). *In vivo* maturation of an undermodified *Escherichia coli* leucine transfer RNA. *Biochem. Biophys. Res. Commun*., **73**, 314–322

Kort, E. N., Goy, M. F., Larsen, S. H., and Adler, J. (1975). Methylation of a membrane protein involved in bacterial chemotaxis, *Proc. Natl. Acad. Sci. U.S.A*., **72**, 3939–3943

Kotyk, A. and Rihova, L. (1972). Transport of α-aminoisobutyric acid in *Saccharomyces cerevisiae*: feedback control. *Biochim. Biophys. Acta*, **288**, 380–387

Kuhn, J. and Somerville, R. L. (1974). Uptake and utilization of aromatic D-amino acids in *Escherichia coli* K-12. *Biochim. Biophys. Acta*, **332**, 298–312

Kustu, S. G. and Ames, G. F. L. (1974). The histidine-binding protein J, a histidine transport component has two different functional sites. *J. Biol. Chem*., **249**, 6976–6983

Kustu, S. G., McFarland, N. C., Hui, S. P., Esmon, B., and Ames, G. F. L. (1979). Nitrogen control in *Salmonella typhimurium* co-regulation of synthesis of glutamine synthetase and amino acid transport systems. *J. Bacteriol*., **138**, 218–234

Langheinrich, W. and Ring, K. (1976). Regulation of amino acid transport in growing cells of *Streptomyces hydrogenans*. I. Modulation of transport capacity and amino acid pool composition during the growth cycle. *Arch. Microbiol*., **109**, 227–235

Litchfield, C. D. and Prescott, J. M. (1970). Regulation of proteolytic enzyme production by *Aeromonas proteolytica*. II. Extracellular aminopeptidase. *Can. J. Microbiol*., **16**, 23–28

McGowan, E. B., Silhary, T. J., and Boos, W. (1974). Involvement of a tryptophan residue in the binding site of *Escherichia coli* galactose-binding protein. *Biochemistry*, **13**, 993–999

Magasanik, B. (1976). Classical and postclassical modes of regulation of the synthesis of degradative bacterial enzymes. *Prog. Nucleic Acid Res. Mol. Biol*., **17**, 99–115

Melton, T., Hartman, P. E., Stratis, J. P., Lee, T. L., and Davis, A. T. (1978). Chemotaxis of *Salmonella typhimurium* to amino acids and some sugars. *J. Bacteriol*., **133**, 708–716

Miller, D. L. and Rodwell, V. W. (1971). Metabolism of basic amino acids in *Pseudomonas putida*. Properties of the inducible lysine transport system. *J. Biol. Chem*., **246**, 1765–1772

Nisbet, T. M. and Payne, J. W. (1979*a*). Peptide uptake in *Saccharomyces cerevisiae*: characteristics of transport system shared by di- and tripeptides. *J. Gen. Microbiol*., **115**, 127–133

Nisbet, T. M. and Payne, J. W. (1979*b*). Specificity of peptide uptake in *Saccharomyces cerevisiae*, and isolation of a bacilysin-resistant, peptide transport-deficient mutant *FEMS Lett*., **6**, 193–196

Ordal, G. W. and Adler, I. (1974). Properties of mutants in galactose taxis and transport. *J. Bacteriol*., **117**, 517–526

Pall, M. L. (1971). Amino acid transport in *Neurospora crassa*. IV. Properties and regulation of a methionine transport system. *Biochim. Biophys. Acta*, **233**, 201–207

Pall, M. L. and Kelly, K. A. (1971). Specificity of transinhibition of amino acid transport in *Neurospora*. *Biochem. Biophys. Res. Commun*., **42**, 940–946

Parsons, R. G. and Hogg, R. W. (1974). Crystallization and characterization of the L-arabinose-binding protein of *Escherichia coli* B/r. *J. Biol. Chem*., **249**, 3602–3607

Pastan, I. and Adhya, S. (1976). Cyclic adenosine 5′-monophosphate in *Escherichia coli. Bacteriol. Rev*., **40**, 527–551

Payne, J. W. (1975). Transport of pertides in microorganisms In *Peptide Transport in Protein Nutrition* (D. M. Matthews and J. W. Payne, eds) North Holland, Amsterdam, p. 343

Payne, J. W. and Bell, G. (1977*a*). Substrate interactions during transport of amino acids and peptides by *Escherichia coli*: relevance to regulation of peptide uptake. *FEMS Lett*., **2**, 301–304

Payne, J. W. and Bell, G. (1977*b*). The transport and utilization of peptides by *Escherichia coli*: interrelations between peptide uptake and amino acid exodus and biosynthesis. *FEMS Lett*., **2**, 259–262

Payne, J. W. and Bell, G. (1979). Direct determination of the properties of peptide transport systems in *Escherichia coli* using a fluorescent labelling procedure *J. Bacteriol*., **137**, 447–455

Payne, J. W. and Gilvarg, C. (1971). Peptide transport. *Adv. Enzymol. Relat. Areas Mol. Biol*., **35**, 187–244

Penrose, W. R., Zand, R., and Oxender, D. L. (1970). Reversible conformational changes in a leucine-binding protein from *Escherichia coli*. *J. Biol. Chem*., **245**, 1432–1439

Peterkofsky, A. and Gazdar, C. (1975). Interaction of Enzyme I of the phosphoenolpyruvate:sugar phosphotransferase system with adenylate cyclase of *Escherichia coli. Proc. Natl. Acad. Sci. U.S.A*., **72**, 2920–2924

Peterkofsky, A. and Gazdar, C. (1979). *Escherichia coli* adenylate cyclase complex: regulation by the proton electrochemical gradient. *Proc. Natl. Acad. Sci. U.S.A*., **76**, 1099–1103

Postma, P. W. and Roseman, S. (1976). The bacterial phosphoenolpyruvate:sugar phosphotransferase system. *Biochim. Biophys. Acta*, **457**, 213–257

Prival, M. and Magasanik, B. (1971). Resistance to catabolite repression of histidase and proline oxidase during nitrogen limited growth of *Klebsiella aerogenes*. *J. Biol. Chem*., **246**, 6288–6296

Quay, S. C. and Oxender, D. L. (1976). Regulation of branched-chain amino acid transport in *Escherichia coli*. *J. Bacteriol*., **127**, 1225–1238

Quay, S. C. and Oxender, D. L. (1977). Regulation of amino acid transport in *Escherichia coli* by transcription termination factor rho. *J. Bacteriol*., **130**, 1024–1029

Quay, S. C., Dick, T. E., and Oxender, D. L. (1977). Role of transport systems in amino acid metabolism: leucine toxicity and the branched-chain amino acid transport systems. *J. Bacteriol*., **129**, 1257–1265

Quay, S. C., Oxender, D. L., Tsuyumu, S., and Unbarger, H. E. (1975). Separate regulation of transport and biosynthesis of leucine, isoleucine, and valine in bacteria. *J. Bacteriol* **122**, 994–1000

Quay, S. C., Lawther, R. P., Hatfield, G. W., and Oxender, D. L. (1978). Branched-chain amino acid transport regulation in mutants blocked in tRNA maturation and transcriptional termination. *J. Bacteriol*., **134**, 683–686

Ratzkin, B. and Roth, J. (1978). Cluster of genes controlling proline degradation in *Salmonella typhimurium*. *J. Bacteriol*., **133**, 744–754

Ratzkin, B., Grabnar, M., and Roth, J. (1978). Regulation of the major proline permease gene of *Salmonella typhimurium J. Bacteriol*., **133**, 737–743

Ray, L. E. and Wagner, F. W. (1972). Characteristics of an aminopeptidase activity from the cultural fluid of *Bacillus subtilis*. *Can. J. Microbiol*., **18**, 853–859

Rickenberg, H. V. (1974). Cyclic AMP in procaryotes. *Annu. Rev. Microbiol*., **28**, 353–369

Ring, K., Foit, B., and Ehle, H. (1977*a*). Stimulation of uptake of 2-aminoisobutyric acid in *Streptomyces hydrogenans* by exogenous dibutyryl cyclic AMP. *FEMS Lett*. **2**, 27–30

Ring, K., Gross, W., and Heinz, E. (1970). Negative feedback regulation of amino acid transport in *Streptomyces hydrogenans*. *Archs. Biochem. Biophys*., **137**, 243–252

Ring, K., Langheinrich, W., Ehle, H., and Foit, B. (1977*b*). Effect of dibutyryl cyclic adenosine monophosphate on active amino acid transport in *Streptomyces hydrogenans*. *Arch. Microbiol*., **115**, 199–205

Roon, R. J., Larimore, F., and Levy, T. F. (1975). Inhibition of amino acid transport by ammonium ion in *Saccharomyces cerevisiae*. *J. Bacteriol*., **124**, 325–331

Rosenfeld, H. and Feigelson, P. (1969). Product induction in *Pseudomonas acidovorans* of a permease which transports L-tryptophan. *J. Bacteriol.*, **97**, 705–709

Saier, M. H. and Feucht, B. U. (1975). Coordinate regulation of adenylate cyclase and carbohydrate permeases by the phosphoenolpyruvate:sugar phosphotransferase system in *Salmonella typhimurium. J. Biol. Chem.*, **250**, 7078–7080

Saier, M. J. and Moczydlowski, E. G. (1978). The regulation of carbohydrate transport in *Escherichia coli* and *Salmonella typhimurium*. In *Bacterial Transport* (B. P. Rosen, ed.), Marcel Dekker, New York and Basel, pp. 103–125

Saiyid, N. H. and Kotyk, A. (1972). Amino acid transport in *Rhodotorula glutinis. Folia Microbiol., Praha*, **17**, 261–271

Springer, M. S., Goy, M. F., and Adler, J. (1977). Sensory transduction in *Escherichia coli*: a requirement for methionine in sensory adaptation. *Proc. Natl. Acad. Sci. U.S.A.*, **74**, 183–187

Staros, J. V. and Knowles, J. R. (1978). Photoaffinity inhibition of dipeptide transport in *Escherichia coli. Biochemistry*, **17**, 3321–3325

Strange, P. G. and Koshland, D. E. (1976). Receptor interactions in a signalling system: competition between ribose receptor and galactose receptor in the chemotaxis response. *Proc. Natl. Acad. Sci. U.S.A.*, **73**, 762–766

Szmelcman, S. and Adler, J. (1976). Change in membrane potential during bacterial chemotaxis. *Proc. Natl. Acad. Sci. U.S.A.*, **73**, 4387–4391

Szmelcman, S., Schwartz, M., Silhavy, T. J., and Boos, W. (1976). Maltose transport in *Escherichia coli* K-12. *Eur. J. Biochem.*, **65**, 13–19

Templeton, B. A. and Savageau, M. A. (1974). Transport of biosynthetic intermediates: regulation of homoserine and threonine uptake in *Escherichia coli. J. Bacteriol.*, **120**, 114–120

Thomale, J. and Nass, G. (1978). Alteration of the intracellular concentration of aminoacyl-tRNA synthetases and isoaccepting tRNAs during amino-acid-limited growth in *Escherichia coli. Eur. J. Biochem.*, **85**, 407–418

Tsang, N., Macnab, R., and Koshland, D. E. (1973). Common mechanism for repellents and attractants in bacterial chemotaxis. *Science, N.Y.*, **181**, 60–63

Tyler, B. (1978). Regulation of the assimilation of nitrogen compounds. *Annu. Rev. Biochem.*, **47**, 1127–1162

Vorisek, J. (1973). Study of the specificity of amino acid binding protein from *Saccharomyces cerevisiae* membranes. *Folia Microbiol., Praha*, **18**, 17–29

Wagner, F. W., Chung, A., and Ray, L. E. (1972). Characterization of the aminopeptidase from *Bacillus subtilis* as an extracellular enzyme. *Can. J. Microbiol.*, **18**, 1883–1888

Weiner, J. H. and Heppel, L. A. (1971). A binding protein for glutamine and its relation to active transport in *Escherichia coli. J. Biol. Chem.*, **246**, 6933–6970

Whipp, M. J. and Pittard, A. J. (1977). Regulation of aromatic amino acid transport systems in *Escherichia coli* K-12. *J. Bacteriol.*, **132**, 453–461

Wiley, W. R. and Matchett, W. H. (1968). Tryptophan transport in *Neurospora crassa*. II metabolic control. *J. Bacteriol.*, **95**, 959–967

Willis, R. C., Iwata, K. K., and Furlong, C. E. (1975). Regulation of glutamine transport in *Escherichia coli*. *J. Bacteriol.*, **122**, 1032–1037

Wilson, D. B. and Smith, J. B. (1978). Bacterial transport proteins. In, *Bacterial Transport* (B. P. Rosen, ed.), Marcel Dekker, New York and Basel, pp. 495–557

Winkler, H. H. (1971). Kinetics of exogenous induction of the hexose-6-phosphate transport system of *Escherichia coli J. Bacteriol.*, **107**, 74–78

Zaboura, M. and Halpern, Y. S. (1978). Regulation of γ-aminobutyric acid degradation in *Escherichia coli* by nitrogen metabolism enzymes. *J. Bacteriol.*, **133**, 447–451

Microorganisms and Nitrogen Sources
Edited by J. W. Payne

Section 2.6

Energetics of Peptide Transport in Bacteria

J. W. PAYNE
Department of Botany, University of Durham, Science Laboratories, South Road, Durham DH1 3LE, UK

1 INTRODUCTION

Extensive and detailed information is available concerning the energetics of amino acid transport in microorganisms and this is presented in Sections 1.2 and 1.7. Much less is known about the energetics of peptide transport. Nevertheless, it is appropriate to consider this as a separate topic, not only for ease of comparison with the amino acid work but also to focus upon those features which, being unique to peptides, will require special attention in future studies.

2 MECHANISMS OF PEPTIDE TRANSPORT IN RELATION TO ENERGY COUPLING

In principle, peptides may enter bacteria in several ways (Figure 1). At physiological concentrations passive diffusion of peptides (Figure 1A) is unlikely to be significant. However, at high concentrations it may be that uptake of small lipophilic peptides may occur, in a similar manner to that suggested for the passage of small hydrophobic molecules through the outer membrane of Gram-negative bacteria (Nikaido, 1976). Uptake can only occur down a concentration gradient, although accumulated peptides could be quickly hydrolysed by intracellular peptidases. Kessel and Lubin (1973) reported that a strain of *Escherichia coli* apparently defective in dipeptide transport could still accumulate Gly-Gly at high external concentration, which they suggested might occur by diffusion. However, the transport defect was uncharacterized and it is possible that mediated uptake occurred through a system with a poor K_m. Cowell (1974) calculated that not more than 10 per cent of the total uptake of Gly-Gly into *E. coli* was by diffusion; in fact the additional Gly-Gly uptake mode revealed in his experiments (and equated with diffusion) was most probably transport via the oligopeptide permease (see Section 2.1.5.6). Payne (1968, 1971) showed that at high concentrations certain hydrophobic tripeptides could enter an oligopeptide transport-deficient mutant of *E. coli*, and the uptake was apparently not susceptible to competition by other oligopeptides (as judged by growth tests), but later studies (Barak and Gilvarg, 1975; Naider and Becker, 1975) indicate that this possibly occurred by an additional transport mechanism rather than by diffusion.

Facilitated diffusion (Figure 1B) is a mediated uptake process that is not directly energy linked; energy may be required, however, to maintain the

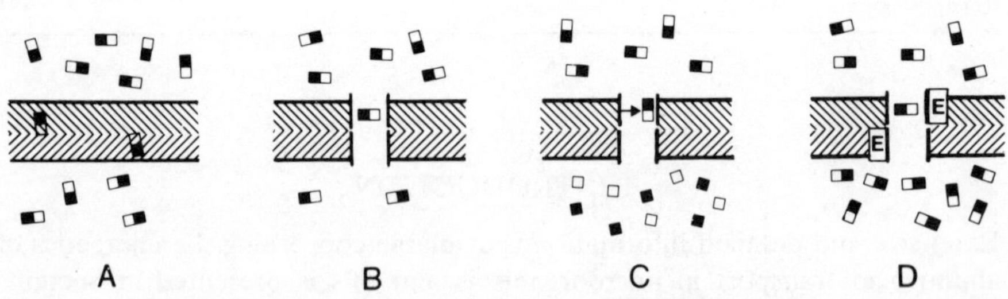

Figure 1 Mechanisms of peptide uptake in relation to energy coupling. The cytoplasmic bilayer membrane is represented by cross-hatched section and peptide transport systems by open section. The cell exterior is above and cell interior below. (A) Passive diffusion; (B) facilitated diffusion; (C) group translocation, the arrow represents peptidase action; (D) active transport, section E represents energy coupling component. (See text for discussion)

system in a functional conformation. Uptake can only be down a concentration gradient, although rapid intracellular hydrolysis would allow uptake to proceed continuously. This mechanism has not been demonstrated in bacteria but could be expected to occur in energy-depleted organisms that could not carry out active transport.

Group translocation (Figure 1C) could be linked to any of a variety of covalent modifications. The most obvious one for peptides would be hydrolytic (see Section 2.1.5.7) in which case the energy released by peptide bond cleavage might be linked to uptake in some way (see below). More commonly the mechanism requires a separate source of energy (Christensen, 1975). The problems in establishing whether this mechanism (in addition to active transport) operates in bacteria were mentioned earlier (Section 2.1.5.7). The γ-glutamyl cycle is a classic example of a group-translocation mechanism and the possibility of it functioning to transport peptides in microorganisms is considered in Section 4.1.

Active transport, in which energy is used to concentrate the native substrate against a concentration gradient (Figure 1D), has been clearly demonstrated for peptide uptake in several microorganisms. These results are considered in the next section.

3 EVIDENCE FOR ACTIVE TRANSPORT OF PEPTIDES

1 Indirect Evidence

Various studies have shown that in the following organisms peptide accumulation requires a metabolizable source of energy: for dipeptides, in *E. coli* (Kessel and Lubin, 1963; Meisler and Simmonds, 1963; Simmonds, 1966; De Felice *et al.*, 1973; Cowell, 1974; Payne and Bell, 1977*a*, 1979), *L. casei* (Leach and Snell, 1959, 1960), *Leuconostoc mesenteroides* (Mayshak *et al.*, 1966, Yoder *et al.*, 1965), *Pseudomonas putida* (Cascieri and Mallette, 1976), *Streptococcus lactis* (Rice *et al.*, 1978), and *Saccharomyces cerevisiae* (J. W. Payne and T. M. Nisbet, unpublished results); and for oligopeptides, in *E. coli* (Smith *et al.*, 1970; Payne, 1972*a*; De Felice *et al.*, 1973; Cowell, 1974), *S. typhimurium* (Jackson *et al.*, 1976), *Pseudomonas putida* (Cascieri and Mallette, 1976), *Bacteroides ruminicola* (Pittman *et al.*, 1967), *L. casei* (Young *et al.*, 1964), *Strep. lactis* (Rice *et al.*, 1978), *Saccharomyces cerevisiae* and *Strep. faecalis* (J. W. Payne and T. M. Nisbet, unpublished results; and see Section 2.2), and *Neurospora crass* (Wolfinbarger and Marzluf, 1976; and see Section 2.3).

Most of these reports presented additional evidence for an energy requirement by showing that poisons such as dinitriphenol, azide, etc., inhibited peptide uptake.

2 Direct Evidence

To provide direct evidence for active transport it is necessary to demonstrate the intracellular accumulation of unmodified peptide against a concentration gradient. The presence of highly active intracellular peptidases normally makes it impossible to demonstrate this feature. However, this difficulty can be overcome with varying degrees of success in the following ways: (a) by modifying incubation conditions to minimize intracellular peptidase activity, (b) by using peptides resistant to intracellular hydrolysis, (c) by using mutants deficient in intracellular peptidases.

In the first case (which is the least useful) it has been shown that peptidase activity in *E. coli* varies with growth phase and with particular nutrient limitations (Simmonds, 1970; Payne, 1972*b*; Miller, 1975*a*; see also Sections 2.4.7 and 3.4). In stationary-phase cells, in which peptidase activity is low, accumulation of intact, radioactively-labelled Gly-Leu was demonstrated (Levine and Simmonds, 1962*a*; Simmonds, 1966).

For case (b), examples of peptides which are sufficiently resistant to intracellular hydrolysis in *E. coli* for their intact accumulation to be demonstrated are: ε-substituted lysine oligopeptides (Payne, 1972*a*), sarcosine peptides (Payne, 1972*c*; Payne and Bell, 1977*b*, 1979), and triornithine (Payne and Bell, 1977*b*, 1979). All these peptides share the transport system used by normal physiological peptides that are readily cleaved. In *S. cerevisiae*, sarcosine peptides, which are transported by the same system as other peptides (Nisbet and Payne, 1979*a,b*) are accumulated intact against a concentration gradient (J. W. Payne and T. M. Nisbet, unpublished results), providing the clear evidence for active transport that to date has been lacking for yeast (see Section 2.2).

By far the best procedure is (c). Early on, Kessel and Lubin (1963) reported active transport of [^{14}C]-Gly-Gly by a 'diglycine-peptidase'-deficient mutant of *E. coli.* Since then various peptidase-deficient mutants have been described in *E. coli* (Sussman and Gilvarg, 1970; Latil *et al*., 1976; Miller and Schwartz, 1978), *S. typhimurium* (McHugh and Miller, 1974; Miller, 1975*a,b*; Miller and MacKinnon, 1974; Kirsch *et al*., 1978) (see also Sections 2.1.5.4, and 2.4), and *S. cerevisiae* (Jones, 1977; Frey and Rohm, 1979). [^{14}C]-Gly-Pro was shown to be accumulated more than 100-fold in the intact form by the proline-peptidase mutant TN 87 of *S. typhimurium* (Yang *et al*., 1977), and Ala-Pro-[^{14}C]-Gly was also transported intact by this strain (Jackson *et al*., 1976). Using fluorescent assay methods (see Section 2.1.4.3) with a variety of Miller's peptidase-deficient mutants of *S. typhimurium* a great many peptides have been shown to be accumulated intact to concentration ratios of several hundred-fold (calculated on dry weight basis and assuming no change in cell volume) (J. W. Payne, unpublished results).

4 MODE OF ENERGY COUPLING

In bacteria, many active transport systems for amino acids (see reviews by Simoni and Postma, 1975; Konings, 1977; Harold, 1977; Haddock and Hamilton, 1977; also Sections 1.1.4.2 and 1.7) are coupled to the circulation of protons across the cytoplasmic membrane (the proton motive force, p.m.f.). It has also been shown for *E. coli* that ATP or some form of phosphate bond energy may be used to drive transport. The latter occurs with osmotic shock-sensitive systems, permeases that require periplasmic binding proteins for activity; the former, p.m.f.-linked systems do not involve binding proteins and are functional in osmotically-shocked cells and membrane vesicles (Berger, 1973; Berger and Heppel, 1974). Berger and Heppel reached these conclusions from the following observations: (a) arsenate (which drastically depletes intracellular ATP levels) almost completely abolishes shock-sensitive but not shock-resistant transport activity; (b) mutants defective in Ca, Mg-ATPase, and thus unable to perform oxidative phosphorylation, cannot drive shock-sensitive transport with the respiratory substrate D-lactate or the artificial electron donor ascorbate/phenazine methosulphate as energy source, whereas these electron donors support shock-resistant transport; (c) acting by substrate-level phosphorylation (glycolysis), glucose stimulates shock-sensitive transport in these mutants, and this stimulation is little affected by anaerobiosis or the uncoupler 2,4-dinitrophenol.

The p.m.f., Δp, arising from primary active extrusion of H^+, is an electrochemical gradient comprising a membrane potential component $\Delta\Psi$ and a pH gradient ΔpH, thus

$$\Delta p = \Delta\Psi - Z\Delta\text{pH};$$

either component may supply the main driving force for transport depending on the substrate and the conditions, particularly the external pH (Ramos and Kaback, 1977*a,b*). It has been suggested on the basis of studies with mutants defective in the energy coupling factor (ECF) protein that the $\Delta\Psi$ component may also contribute to the efficiency of energy coupling in the shock-sensitive systems, which reportedly required more than ATP alone (Lieberman and Hong, 1976). However, this view has recently been retracted (Hong *et al*., 1979). These authors presented evidence implicating acetyl phosphate as energy donor for the binding-protein-dependent transport systems, and claimed that neither the p.m.f. nor the *ecf* gene product is required for the shock-sensitive transport systems.

These concepts have been applied to peptide uptake in bacteria by Cowell (1974) and Payne and Bell (1979). Using [^{14}C]-Gly-Gly, Cowell showed that its uptake corresponded with a binding-protein-dependent mechanism: initial rate of uptake being decreased 85 per cent after osmotic shock treatment, and no uptake occurring with membrane vesicles. Cowell presented the following evidence that the active transport of Gly-Gly, like the substrates (e.g. glutamine) of other shock-sensitive systems, required phosphate bond energy, but not respiration or the energized membrane state (as, for example, in proline uptake). (a) Gly-Gly transport was strongly inhibited by arsenate; (b) oxidizable energy sources such as D-lactate, succinate, and ascorbate could not serve as energy sources for Gly-Gly transport in strain DL-54, which lacks a functional Ca, Mg-ATPase; (c) Gly-Gly transport occurred when energy was supplied only from ATP produced by glycolysis (i.e. anaerobic transport assays with glucose as energy source in strain DL-54); and (d) when substrate-level phosphorylations and electron transport were functioning but Ca, Mg-ATPase activity was absent (i.e. mutant DL-54 using glucose), Gly-Gly transport showed significant resistance to the uncouplers 2,4-dinitrophenol and carbonylcyanide-*p*-trifluoromethoxyphenylhydrazone.

Using fluorescence dansyl methods, Payne and Bell (1979) endorsed many of Cowell's experiments and extended his conclusions to additional dipeptides, and also to oligopeptides which are transported by a separate system in *E. coli* (see Section 2.1.5.6). Thus, arsenate decreased uptake of di- and oligopeptides by over 80 per cent with glucose, lactate, or succinate as energy sources. 2,4-Dinitrophenol and carbonylcyanide chlorophenylhydrazone (CCCP) had relatively little effect on peptide uptake with glucose as energy source (i.e. substrate-level phosphorylation), but greater inhibition occurred using lactate (i.e. ATP via oxidative phosphorylation). DNP was generally more inhibitory than CCCP, perhaps indicating it has secondary effects; e.g. Berger and Heppel (1974) reported that DNP could lower intracellular ATP by about 40 per cent. Anaerobic conditions only slightly inhibited peptide uptake (although, interestingly, it caused specific exodus of certain amino acids, the significance of which remains to be determined). Results with KCN were variable and inhibition was generally greater than expected simply from inhibiting oxidative phosphorylation in an ATP dependent system; it seems likely that KCN also depletes ATP pools (as occurs with DNP), presumably during attempts by the cells to maintain a proton gradient. The unknown pool component, which was shown to undergo massive selective exodus in the presence of KCN, has since been identified as putrescine (J. W. Payne, unpublished results). I am not aware of any reports of this phenomenon in the literature and its explanation must await further study, but cognizance should be taken of the extensive exodus of this polycation when interpreting transport studies using cyanide. Observations apparently compatible with the suggestion (see above) that the coupling of phosphate bond energy to transport may be less effective in the absence of the p.m.f. were (a) the marked inhibitory effects of DNP and KCN (but see above), (b) the inhibitory effect of anaerobiosis, (c) the decreased uptake seen in strain NR 70 (which is not only defective in Ca, Mg-ATPase but is also unable to maintain a proton gradient) compared with its parent strain 7.

No relevant studies with any other bacteria have been reported. In *S. cerevisiae* (Becker and Naider, 1977), azide, KCN, and DNP inhibit uptake of radioactively-labelled trimethionine but the apparent failure of arsenate to inhibit is in doubt because the incubations were performed in phosphate buffer which competitively inhibits arsenate transport (see also Section 2.2.11).

5 ROLE OF PERIPLASMIC BINDING PROTEINS

Shock-sensitive transport systems for amino acids and sugars almost invariably involve periplasmic binding proteins. However, although it has been reported that peptide transport is markedly diminished in *E. coli* following osmotic shock no binding proteins have been detected for either dipeptides (Cowell, 1974; B. P. Rosen, personal communication; Z. Barak and C. Gilvarg, unpublished observations) or oligopeptides (Z. Barak and C. Gilvarg, unpublished observations cited in Payne and Gilvarg, 1978). Cowell suggested that failure to detect binding activity for [^{14}C]-Gly-Gly in the shock fluid might be caused by enzymic hydrolysis of the peptide, a view which this author would share (see also Section 2.1.5.4.). Before perhaps concluding prematurely that peptide uptake is atypical in having no requirement for binding proteins it is necessary to assay shock fluids using peptidase resistant substrates, or preferably to perform the study using shock fluids from peptidase-deficient strains. Furthermore, a range of peptides should be tested for glycine peptides may well have the lowest binding affinities.

Consistent with the effects of osmotic shock, Cowell (1974) stated that membrane vesicles of *E. coli* failed to transport [^{14}C]-Gly-Gly and Barak and Gilvarg found the same effect with [^{14}C]-Gly-Gly-Gly (Z. Barak and C. Gilvarg, unpublished observations cited in Payne and Gilvarg, 1978).

6 ASPECTS FOR FUTURE STUDY

1 Energy Requirements for Intact Peptide Uptake

It is important for the economy of the bacterial cell that it has efficient transport systems for small nutrients such as amino acids, since the energy required to transport one molecule of an amino acid is likely to be much less than that needed for the biosynthesis of the same molecule from glucose. In fact, it has been calculated that for the synthesis of one molecule of histidine, the potential energy corresponding to 41 ATP molecules is expended (quoted in Ames, 1972). But what of the energy consumption for the uptake of an intact peptide; could it be that the same amount of energy is needed for uptake intact of an amino acid, di-, tri-, and tetrapeptide molecule? If so, peptide uptake would in energy terms represent a very superior mechanism. In principle, it should be possible to determine the amount of energy consumed in the uptake of a given amount of intact peptide, by using energy-starved, peptidase-deficient strains (see Section 2.6.3.2) together with appropriate peptide substrates that cannot be cleaved, and supplying known, limited amounts of an energy source, e.g. glucose. Uptake would ideally be measured by continuous sampling using the automated fluorescamine procedure (see Section 2.1.4.3).

2 Relation of Substrate Charge to Energy Requirement for Intact Peptide Uptake

In the above discussion, a 'model' peptide molecule is being considered. Real peptides, under physiological conditions, can be neutral, or can carry one or more positive or negative charges. This inherent diversity of charge would presumably pose problems for the design of a general peptide permease energized by a symport or antiport mechanism, for varying stoicheiometries would be required (see also Sections 1.2.7 and 1.7). Nevertheless, it is necessary that charge equilibration be exercised across the membrane during transport processes and so uptake of charged peptides must involve associated (but not co-) fluxes of counterions. This is illustrated in Figure 2A which shows, from top to bottom: uptake intact of a neutral peptide, formed from uncharged residues, with no associated ion fluxes; uptake intact of a neutral peptide, formed from equal numbers of positive (e.g. Lys, Arg) and negative (e.g. Glu, Asp) residues, with no associated ion fluxes; uptake intact of peptide with net negative charge, with either associated influx of cation or efflux of anion; uptake intact of peptide with net positive charge with either associated influx of anion or efflux of cation. Multiple charged peptides would require equivalently greater counterion fluxes. OH^- and K^+ are used only for illustration and other ions, e.g. H^+, Na^+, may conceivably be involved. In view of these considerations, during attempts to measure the energy needed for the uptake of an intact peptide (using automated fluorescamine method, see above), cognizance must be made of the possible energy consumption/yield

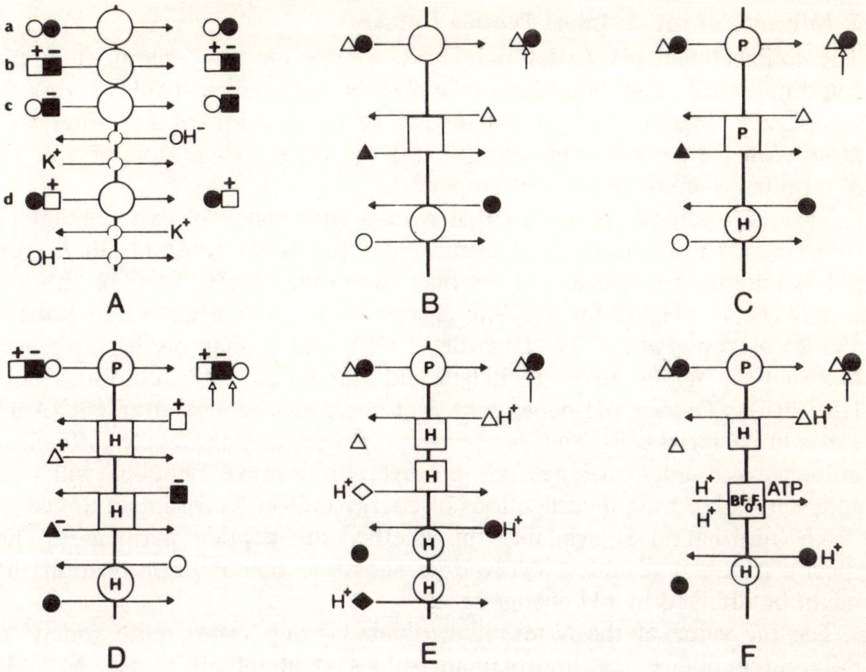

Figure 2 Possible ion fluxes and countertransport events associated with peptide uptake. Membrane, with exterior to the left, is represented by vertical line and contains various carrier systems. P-carriers represent phosphate bond-linked, shock-sensitive systems; H-carriers represent electrochemical-proton-gradient-linked, shock-resistant systems. Peptide uptake intact is shown in A; peptide uptake with intracellular hydrolysis (open arrows represents peptidase action) is in B–F. (A) Intact uptake with counterion fluxes: (a) neutral amino acid residues, neutral peptide, no associated ion fluxes; (b) equal and opposite charged amino acid residues, net peptide charge zero, no associated ion fluxes; (c) neutral and negative amino acids, net peptide charge negative, associated anion efflux or cation influx; (d) neutral and positive amino acids, net peptide charge positive, associated cation efflux or anion influx. (B) simple mixed homologous countertransport: efflux of peptidase-released amino acids drives uptake of different amino acids sharing a common permease. (C) Mixed homologous countertransport through variously energized permeases: efflux of amino acids drives uptake of different amino acids through permeases energized similarly (top) or differently (bottom) to peptide uptake. (D) Mixed homologous countertransport through carriers driven by various components of the electrochemical proton gradient: neutral peptide yields positive, negative, and neutral amino acids, any associated ion fluxes are not considered. (E) Mixed heterologous countertransport: efflux of amino acids with associated H^+ ions, creates H^+ gradient that drives substrate influx by H^+ symport via different permease. (F) Recycled energy: efflux of amino acids with associated H^+ ions creates H^+ gradient which, in absence of extracellular substrate, re-enters via Ca, Mg-ATPase (BF_0F_1) driving it in reverse to yield ATP. (See text for discussion)

of these counterion fluxes (by simultaneously monitoring the system using, for example, H^+-selective electrodes). Thus, net energy consumption may differ for uptake intact of neutral, positive, or negative peptide molecules.

3 Influence of pH on Intact Peptide Uptake

The extracellular pH (pH_e) may have effects on three parameters that could influence intact peptide uptake: (a) the pH gradient (ΔpH) across the cytoplasmic membrane, (b) the degree of protonation of any interactive groups on the peptide permease(s), (c) the degree of protonation of the *N*-terminal α-amino group on the peptide.

Various studies have shown that over a wide range of extracellular pH (pH_e) bacteria maintain their intracellular pH (pH_i) constant. In *E. coli*, pH_i is maintained at near pH 7.8 over the range of pH_e 5.5–8.0, although it may rise to pH_i 8.5 for pH_e 9.0, (Padan *et al.*, 1976; Ramos and Kaback, 1977*c*; Navon *et al.*, 1977; Lagarde, 1977), and similar results have been reported for *Staph. aureus* (Mitchell and Hamilton, 1977; see also Section 1.7.3.2(ii)). Thus, a pH-dependent proton electrochemical gradient (ΔpH) exists in bacterial cells, and its operation, magnitude, and likely differential influence on uptake of neutral, positive, or negative peptides will need consideration during investigations of energy utilization for intact uptake.

No information is available on whether the peptide permease(s) has specific charged residues involved in substrate binding/translocation that might be affected by pH changes.

The p*K* value of the *N*-terminal α-amino group varies quite widely for different peptides; e.g. approximate values at about 30 °C are, Met-Met (7.5), Lys-Lys (7.5), Leu-Leu (8.0), Gly-Gly (8.2), Ala-Ala (8.4), Gly-Asp (8.6), Asp-Gly (9.1). Thus, in the range pH_e 6.5–8, which is used for most uptake studies, quite large variations may occur in the proportions of protonated and non-protonated species present in the medium. This may influence to different degrees the uptake and competitive abilities of various peptides, especially if the permease has absolute specificity for one or other species. Furthermore, because of the general range of the p*K* values, if incubations are performed at the typical value of pH_e 6.8 then, after intact uptake (pH_i 7.8), the intracellular peptide will generally be present with a markedly lower proportion of protonated to non-protonated species than exists in the outside medium. Wide-ranging discussions of hydrogen ion dissociations in relation to amino acid transport have been presented by Christensen and colleagues (Christensen and Handlogten, 1975; Garcia-Sancho *et al.*, 1977; Christensen, 1977).

Yang *et al.* (1977) reported that uptake of [^{14}C]-Gly-Pro by the peptidase-deficient strain TN 87 of *S. typhimurium* was an optimum at pH 6.

4 Relation of Peptide Bond Hydrolysis to Energy Requirement for Peptide Uptake

Only the case of intracellular hydrolysis is considered here. It has been suggested by a number of authors that the free energy released on peptide bond hydrolysis might be used to help fuel transport. However, this possibility has never been formally quantified and related to the likely energy requirement for transport, and neither has any proposal been made as to a mechanism for coupling the released energy. In practice, any energy gain may not always be as large as expected because proteolysis by certain enzymes is actually ATP-dependent (Goldberg and St. John, 1976; see also Section 3.4.5.8).

The p*K* of the α-amino group in the *N*-terminal residue of a typical peptide is about 1–2 pH units lower (about pH 8) than the p*K* for the same residue as the free amino acid (about pH 9.6). Thus, intracellular aminopeptidase action at pH 7.8 will liberate amino acids with a relatively greater capacity than the peptide substrate to bind protons; *N*-terminal cleavage may thus have secondary effects on ion fluxes. Similarly, hydrolysis of neutral peptide bonds produces species with amino and carboxylate groups which at pH_i 7.8 may (in effect) carry a partial positive and a negative charge, respectively; again this may have secondary effects on ion fluxes. These considerations lead to the speculation that these associated ion fluxes might generate energy that could be utilized in peptide uptake.

5 Relation of Amino Acid and Peptide Exodus to Energy Requirement for Peptide Uptake

It has been demonstrated in strains of *E. coli* (Payne and Bell, 1977*a,b*, 1979) and *S. typhimurium* (J. W. Payne, unpublished results) having full peptidase activity that, concomitant with peptide uptake, efflux of the hydrolytically released amino acid residues can occur (see also Section 2.1.5.4). This raises the question as to whether uptake of other substrates (amino acid, peptide), can be driven by efflux of the intracellularly released amino acids. A variety of transport-related mechanisms may be operating under the conditions of peptide uptake–hydrolysis–amino acid efflux, and their possible interrelations are complex; it is helpful to consider other situations first.

According to the chemiosmotic hypothesis, as related to active transport in bacteria, oxidation of electron donors by a membrane-bound respiratory chain or hydrolysis of ATP catalysed by a membrane-bound Ca, Mg-ATPase is accompanied by electrogenic expulsion of protons, producing an electrochemical proton gradient (Δp) made up of an electrical potential component and a chemical component, the pH gradient (ΔpH). Uptake of neutral substances is considered to occur via symport with protons (i.e. cotransport). Cation transport is purportedly linked to the electrical component $\Delta\Psi$, and it is postulated that transport of organic acids is dependent upon the chemical gradient of protons ΔpH, i.e. the undissociated acid is taken through the membrane and considered to accumulate in the ionized form as a result of the relative intracellular alkalinity. Now it has been shown that the membrane can be energized, not only as described above but also by artificially generating a transmembrane potential (a K^+ diffusion potential induced by valinomycin) or an artificially generated pH difference (inside alkaline) (Altendorf *et al*., 1975; Schuldiner and Kaback, 1975; Lagarde, 1977). It has also been demonstrated that H^+ ion fluxes accompany downhill (West and Mitchell, 1973) or uphill (Lagarde, 1977) substrate movements via their specific carrier. Furthermore, the downhill movement of one solute can drive uphill in the opposite direction another solute that shares the same transport system (Winkler and Wilson, 1966; Bentaboulet and Kepes, 1977). This homologous countertransport is specific and depends on the integrity of a functional carrier system (Kepes, 1969; Wong and Wilson, 1970); in practice it will prove difficult to detect such countertransport unless the energized state of the membrane is first eliminated. Now, the suggestion has been made (Bentaboulet *et al*., 1979) that if the downhill flux of a substrate is accompanied by a flux of H^+, it should be possible to establish a transient electrochemical gradient of H^+ at the expense of the dissipation of the initial chemical gradient of the solute, and that this transient H^+ gradient could be coupled to the uphill transport of a different solute via a *different* carrier, provided that it also is a H^+-cotransport system. In fact, these authors demonstrated this heterologous transport, driving amino acid uptake by sugar efflux. They further demonstrated that, unlike homologous countertransport, heterologous countertransport is completely inhibited by uncouplers.

The relationship of the foregoing to the peptide uptake–amino acid exodus system is as follows. Details of the carrier-mediated exodus of several amino acids from *E. coli* have been presented (Brown, 1970; Halpern *et al*., 1973; Halpern and Ennis, 1975; Quay *et al*., 1977). Results fitting a process of homologous countertransport were also presented in several cases. It seems certain therefore, that efflux of amino acids derived from a transported peptide will be able to drive uptake of other amino acids that share a common permease, e.g. Gly/Ala, Thr/Ser, Phe/Tyr/Trp, Leu/Ile/Val; Lys/Arg/Orn (see Section 1.1.3); overall this process comprises a form of 'mixed-homologous countertransport' (see Figure 2B). Although never demonstrated in *E. coli*, evidence compatible with this process has been presented for a *Streptococcus* (Brock and Wooley, 1964). Simple counterflow was probably observed with valine and Val peptides in *Leuconostoc mesenteroides* (Yoder *et al*., 1965), and with glycine and in Gly-Gly in *E. coli* (Levine and Simmonds, 1962*b*).

Uptake of a peptide, being a shock-sensitive mechanism driven by phosphate-bond energy (see Section 2.6.4), can, depending on the composition of the peptide, give rise to the accumulation of amino acids that are substrates of shock-sensitive systems (e.g. Gln, Asn, Met) and/or shock-resistant systems (e.g. Pro, Ala, Phe, Glu, Asp, Lys) (see Section 1.1.3). If the latter amino acids took part in 'mixed homologous countertransport' it would in effect produce an interconversion of energy currency from a phosphate bond-driven form to an electrochemical proton gradient form (see Figure 2C). Considering this feature in more detail for those amino acids which enter via shock-resistant systems then, depending on its composition, uptake of a peptide could lead to the presence in the cytoplasm of amino acids, able to undergo homologous countertransport, that are neutral (e.g. Ala), and/or acidic (e.g. Asp), and/or basic (e.g. Lys), and the uptake of which is driven normally by different components of the electrochemical proton gradient, (see Figure 2D) as discussed above.

Figure 2D is intended primarily to illustrate the implications of the intracellular generation of differently charged amino acids and not to indicate possible transport mechanisms or associated ion fluxes. In fact, Hamilton and coworkers (Niven and Hamilton, 1974; Niven *et al.*, 1973) (see also Section 1.7) have studied the uptake of amino acids in response to Δp by energy-depleted cells of *S. aureus*, and have proposed that the carriers for the various amino acids function thus: neutral amino acids are transported by H^+ symport mechanisms in response to Δp, acidic amino acids by H^+ symports in response to ΔpH, and basic amino acids by uniports in response to $\Delta\Psi$. The availability of these various energy components will differ depending on incubation conditions, e.g. pH, while intracellular peptide hydrolysis may remain constant. Therefore, other possibilities might need consideration: for example, uniports can be utilized for accumulation of cations, extrusion of anions, or facilitated diffusion of uncharged substrates, while an antiport mechanism could involve simultaneous translocation by a carrier in opposite directions of two co-substrates with opposite charges (Rosen and Kashket, 1978).

The uptake of peptides containing glutamate will lead to a further variation, for it has been shown that Glu uptake can be driven by Na^+ ions (Tsuchiya *et al.*, 1977; MacDonald *et al.*, 1977; Hasan and Tsuchiya, 1977), and a Na^+/H^+ antiport process has been described in *E. coli* (West and Mitchell, 1974).

Mixed heterologous countertransport also seems feasible in this system. For example, after uptake and cleavage of a peptide containing a neutral amino acid, e.g. Phe, Pro, Ala, efflux of the amino acid through its specific permease should produce an electrochemical gradient of H^+ that could be used either for the uphill transport of another amino acid through a different permease or, alternatively, for the uptake of a different species such as a sugar, e.g. gluconate or methyl thiogalactoside (see Figure 2E). However, unlike the transient proton gradient generated by efflux of a fixed and limited amount of preloaded amino acid (Bentaboulet *et al*., 1979) the continuing uptake of peptide provides a replenishing source of substrate for efflux and thus the means to sustain an electrochemical gradient of H^+. In principle therefore, this could be a convenient system in which to study heterologous countertransport. For uptake and cleavage of peptides containing amino acids with net negative or positive charges, or Glu (Na^+), related discussions could be made concerning their possible functions in heterologous countertransport.

A different situation exists when cells take up peptides under aerobic conditions from a medium in which there are no substrates which can undergo countertransport. If we again assume that efflux of certain amino acids will produce a proton gradient then, in the absence of extracellular cosubstrates, we can further assume that no H^+-symport carrier will facilitate the uptake of the protons alone, for such a property would result in an uncoupled system; thus, it is feasible that dissipation of this gradient through re-entry of the H^+ ions via the Ca, Mg-ATPase could lead to ATP synthesis, as required by chemiosmotic theory (Mitchell, 1976) (see Figure 2F). In fact, coupling between H^+ entry and ATP formation has been demonstrated in whole cells of *E. coli* (Maloney *et al*., 1974; Grinius *et al*., 1975; Maloney, 1978) and *Strep. lactis* (Maloney *et al*., 1974; Maloney, 1977), and in membrane vesicles from *E. coli* (Tsuchiya and Rosen, 1976*a,b*; Tsuchiya, 1977). The possibility that ATP synthesis may follow the efflux of amino acids raises speculation as to the fate of the ATP. An intriguing thought is that it might be used to help drive further peptide uptake. The stoicheiometry of energy required: oligopeptide molecule transported is unknown (see Section 2.6.6.1). For *E. coli*, an apparent stoicheiometry of $10H^+$ per ATP molecule synthesized has been suggested (Maloney, 1978), but this may be a significant overestimate (Maloney, 1977) and a value of 2:1 might be anticipated. Consequently, it is not useful at this time to speculate on what proportion this (postulated) 'recycled energy' (efflux–influx-dependent ATP synthesis, Figure 2F) might contribute to the total energy required for uptake of a peptide molecule.

The above considerations have focused upon those amino acids which use shock-resistant p.m.f.-linked permeases. However, following peptide uptake, efflux of amino acids that enter via ATP-linked shock-sensitive systems may occur (Figure 2C). Presumably the amino acid efflux will occur via phosphate bond-driven permeases, the energetic implications of which is unknown; a similar conclusion applies to the efflux of peptides observed under certain conditions (see Section 2.1.5.4).

It would be helpful here to know more about the process of efflux but few studies have been carried out. It seems there may be multiple channels for efflux. Some may be the carriers used for influx but some reports claim that separate carrier systems may occur (Parnes and Boos, 1973; Halpern *et al.*, 1973; Israeli *et al.*, 1977). Leak pathways also exist. Leakage of a substance is often noted in mutants defective in the transport system(s) for the compound (Anderson and Oxender, 1978). Perhaps binding proteins are patrolling in the periplasmic space to recapture substrates lost by leakage.

Finally, it is noteworthy that in studies designed to determine Δp in whole cells Gly-Gly buffer is frequently used. In the light of its likely uptake, hydrolysis, and possible involvement in the above postulated mechanisms this buffer seems an inappropriate choice.

REFERENCES

Anderson, J. J. and Oxender, D. L. (1978). Genetic separation of high- and low-affinity transport systems for branched-chain amino acids in *Escherichia coli* K-12. *J. Bacteriol.*, **136**, 168–174

Altendorf, K., Hirata, H., and Harold, F. M. (1975). Accumulation of lipid soluble ions and of rubidium as indicators of the electrical potential in membrane vesicles of *Escherichia coli. J. Biol. Chem.*, **250**, 1405–1412

Ames, G. F. L. (1972). Components of histidine transport. In *Membrane Research*, First ICN-UCLA Symposium on Molecular Biology (C. F. Fox, ed.), Academic Press, New York and London, p. 411

Barak, Z. and Gilvarg, C. (1975). Specialized peptide transport system in *Escherichia coli. J. Bacteriol.*, **122**, 1200–1207

Becker, J. M. and Naider, F. (1977). Peptide transport in yeast: uptake of radioactive trimethionine in *Saccharomyces cerevisiae. Arch. Biochem. Biophys.*, **178**, 245–255

Bentaboulet, M. and Kepes, A. (1977). Counter transport mediated by the lactose permease of *Escherichia coli. Biochim. Biophys. Acta*, **471**, 125–134

Bentaboulet, M., Robin, A., and Kepes, A. (1979). Artificially induced active transport of amino acid driven by the efflux of a sugar via a heterologous transport system in de-energized *Escherichia coli. Biochem. J.*, **178**, 103–107

Berger, E. A. (1973). Different mechanisms of energy coupling for the active transport of proline and glutamine in *Escherichia coli. Proc. Natl. Acad. Sci. U.S.A.*, **70**, 1514–1518

Berger, E. A. and Heppel, L. A. (1974). Different mechanisms of energy coupling for the shock-sensitive and shock-resistant amino acid permeases of *Escherichia coli. J. Biol. Chem.*, **249**, 7747–7755

Brock, T. D. and Wooley, S. O. (1964). Glycylglycine uptake in *Streptococci* and a possible role of peptides in amino acid transport. *Arch. Biochem. Biophys.*, **105**, 51–57

Brown, K. D. (1970). Maintenance and exchange of the aromatic amino acid pool in *Escherichia coli. J. Bacteriol.*, **106**, 70–81

Cascieri, T. and Mallette, M. F. (1976). Peptide utilization by *Pseudomonas putida* and *Pseudomonas maltophilia. J. Gen. Microbiol.*, **92**, 283–285

Christensen, H. N. (1975). *Biological Transport*, 2nd edn, Benjamin, London and New York, pp. 247–271

Christensen, H. N. (1977). Hydrogen-ion dissociation as a factor in amino acid transport. *FEBS Symp*. **42**, 222–235

Christensen, H. N. and Handlogten, M. E. (1975). A cycle of deprotonation and reprotonation energizing amino-acid transport? *Proc. Natl. Acad. Sci. U.S.A.*, **72**, 23–27

Cowell, J. L. (1974). Energetics of glycylglycine transport in *Escherichia coli. J. Bacteriol.*, **120**, 139–146

De Felice, M., Guardiola, J., Lamberti, A., and Iaccarino, M. (1973). *Escherichia coli* K-12 mutants altered in the transport systems for oligo- and dipeptides. *J. Bacteriol.*, **116**, 751–756

Frey, J. and Rohm, K. H. (1979). The glucose induced inactivation of aminopeptidase I in *Saccharomyces cerevisiae*. *FEBS Lett.*, **100**, 261–265

Garcia-Sancho, J., Sanchez, A., and Christensen, H. N. (1977). Role of proton dissociation in the transport of acidic amino acids by the Ehrlich ascites tumor cell. *Biochim. Biophys. Acta*, **464**, 295–312

Goldberg, A. L. and St. John, A. C. (1976). Intracellular protein degradation in mammalian and bacterial cells. *Annu. Rev. Biochem.*, **45**, 747–803

Grinius, L., Slusnyte, R., and Griniuviene, B. (1975). ATP synthesis driven by a proton motive force imposed across *Escherichia coli* cell membranes. *FEBS Lett.*, **57**, 290–293

Haddock, B. A. and Hamilton, W. A. (eds.) (1977). *Symp. Soc. Gen. Microbiol.*, **27**

Halpern, Y. S., Barash, H., and Druck, K. (1973). Glutamate transport in *Escherichia coli* K-12: non-identity of carriers mediating entry and exit. *J. Bacteriol.*, **113**, 51–57

Halpern, Y. S. and Ennis, H. L. (1975). Analysis of glutamate exit in *Escherichia coli*. *J. Bacteriol.*, **122**, 332–334

Harold, F. M. (1977). Membranes and energy transduction in bacteria. *Curr. Topics Bioenerg.*, **6**, 83–149

Hasan, S. M. and Tsuchiya, T. (1977). Glutamate transport driven by an electrochemical gradient of sodium ion in membrane vesicles of *Escherichia coli* B. *Biochem. Biophys. Res. Commun.*, **78**, 122–128

Hong, J. S., Hunt, A. G., Masters, P. S., and Lieberman, M. A. (1979). Requirement of acetyl phosphate for the binding protein-dependent transport systems in *Escherichia coli*. *Proc. Natl. Acad. Sci. U.S.A.*, **76**, 1213–1217

Israeli, E., Kalra, V. K., and Brodie, A. (1977). Different binding sites for entry and exit of amino acids in whole cells of *Mycobacterium phlei*. *J. Bacteriol.*, **130**, 729–735

Jackson, M. B., Becker, J. M., Steinfeld, A. S., and Naider, F. (1976). Oligopeptide transport in proline peptidase mutants of *Salmonella typhimurium*. *J. Biol. Chem.*, **251**, 5300–5309

Jones, E. W. (1977). Proteinase mutants of *Saccharomyces cerevisiae*. *Genetics*, **85**, 23–33

Kepes, A. (1969). Carrier properties of β-galactoside permease. The role of permease in the leak of β-galactosides from *Escherichia coli*. In *The Molecular Basis of Membrane Function* (D. C. Tosteson, ed.), Prentice Hall, Englewood Cliffs, N.J.

Kessel, D. and Lubin, M. (1963). On the distinction between peptidase activity and peptide transport. *Biochim. Biophys. Acta*, **71**, 656–663

Kirsh, M., Dembinski, D. R., Hartman, P. E., and Miller, C. G. (1978). *Salmonella typhimurium* peptidase active on carnosine. *J. Bacteriol.*, **134**, 361–374

Konings, W. N. (1977). Active transport of solutes in bacterial membrane vesicles. *Adv. Microbial Physiol.*, **15**, 175–251

Lagarde, A. E. (1977). Evidence for an electrogenic 3-deoxy-2-oxo-D-gluconate-proton co-transport driven by the proton motive force in *Escherichia coli* K-12. *Biochem. J.*, **168**, 211–221

Latil, M., Murgier, M., Lazdunski, A., and Lazdunski, C. (1976). Isolation and genetic mapping of *Escherichia coli* aminopeptidase mutants. *Mol. Gen. Genet.*, **148**, 43–47

Leach, F. R. and Snell, E. E. (1959). Occurrence of independent uptake mechanisms for glycine and glycine peptides in *Lactobacillus casei*. *Biochim. Biophys. Acta*, **34**, 292–293

Leach, F. R. and Snell, E. E. (1960). The absorption of glycine and alanine and their peptides by *Lactobacillus casei*. *J. Biol. Chem.*, **235**, 3523–3531

Levine, E. M. and Simmonds, S. (1962*a*). Further studies on metabolite uptake by serine–glycine auxotrophs of *Escherichia coli*. *J. Biol. Chem.*, **237**, 3718–3724

Levine, E. M. and Simmonds, S. (1962*b*). Effects of cultural conditions on growth and metabolite uptake by serine–glycine auxotrophs of *Escherichia coli*. *J. Bacteriol.*, **84**, 683–693

Lieberman, M. A. and Hong, J. S. (1976). Energization of osmotic shock-sensitive transport systems in *Escherichia coli* requires more than ATP. *Arch. Biochem. Biophys.*, **172**, 312–315

MacDonald, R. E., Lanyi, J. K., and Greene, R. V. (1977). Sodium-stimulated glutamate uptake in membrane vesicles of *Escherichia coli*: the role of ion gradients. *Proc. Natl. Acad. Sci. U.S.A.*, **74**, 3167–3170

McHugh, G. L. and Miller, C. G. (1974). Isolation and characterization of proline peptidase mutants of *Salmonella typhimurium*. *J. Bacteriol.*, **120**, 364–371

Maloney, P. C. (1977). Obligatory coupling between proton entry and the synthesis of adenosine-5′-triphosphate in *Streptococcus lactis*. *J. Bacteriol.*, **132**, 564–575

Maloney, P. C. (1978). Coupling between H^+ entry and ATP formation in *Escherichia coli*. *Biochem. Biophys. Res. Commun.*, **83**, 1496–1501

Maloney, P. C., Kashket, E. R., and Wilson, T. H. (1974). A proton motive force drives ATP synthesis in bacteria *Proc. Natl. Acad. Sci. U.S.A.* **71**, 3896–3900

Mayshak, J., Yoder, O. C., Beamer, K. C., and Shelton, D. C. (1966). Inhibition and transport kinetic studies involving L-leucine, L-valine, and their dipeptides in *Leuconostoc mesenteroides*. *Arch. Biochem. Biophys.*, **113**, 189–194

Meisler, N. and Simmonds, S. (1963). The metabolism of glycyl-L-leucine by *Escherichia coli*. *J. Gen. Microbiol.*, **31**, 109–123

Miller, C. G. (1975*a*). Peptidases and proteases of *Escherichia coli* and *Salmonella typhimurium*. *Annu. Rev. Microbiol.*, **29**, 485–504

Miller, C. G. (1975*b*). Genetic mapping of *Salmonella typhimurium* peptidase mutations. *J. Bacteriol.*, **122**, 171–176

Miller, C. G. and Mackinnon, K. (1974). Peptidase mutants of *Salmonella typhimurium*. *J. Bacteriol.*, **120**, 355–363

Miller, C. G. and Schwartz, G. (1978). Peptidase-deficient mutants of *Escherichia coli*. *J. Bacteriol.*, **135**, 603–611

Mitchell, P. (1976). Vectorial chemistry and the molecular mechanics of chemiosmotic coupling: power transimission by proticity. *Biochem. Soc. Trans.*, **4**, 399–430

Mitchell, W. J. and Hamilton, W. A. (1977). Energisation of amino-acid uptake in *Staphylococcus aureus*. *Proc. Soc. Gen. Microbiol.*, **4**, 89–90

Naider, F. and Becker, J. M. (1975). Multiplicity of oligopeptide transport systems in *Escherichia coli*. *J. Bacteriol.*, **122**, 1208–1215

Navon, G., Ogawa, S., Shulman, R. G., and Yamane, T. (1977). High resolution ^{31}P nuclear magnetic resonance studies of metabolism in aerobic *Escherichia coli* cells. *Proc. Natl. Acad. Sci. U.S.A.*, **74**, 888–891

Nikaido, H. (1976). Outer membrane of *Salmonella typhimurium*, transmembrane diffusion of some hydrophobic substances. *Biochim. Biophys. Acta*, **433**, 118–132

Nisbet, T. M. and Payne, J. W. (1979*a*). Peptide uptake in *Saccharomyces cerevisiae*: characteristics of transport system shared by di- and tripeptides, *J. Gen. Microbiol.*, **115**, 127–133

Nisbet, T. M. and Payne, J. W. (1979*b*). Specificity of peptide uptake in *Saccharomyces cerevisiae*, and isolation of a bacilysin-resistant, peptide-transport-deficient mutant. *FEMS Lett.*, **6**, 193–196

Niven, D. F. and Hamilton, W. A. (1974). Mechanisms of energy coupling to the transport of amino acids by *Staphylococcus aureus*. *Eur. J. Biochem.*, **44**, 517–522

Niven, D. F., Jeacocke, R. E., and Hamilton, W. A. (1973). The membrane potential as the driving force for the accumulation of lysine by *Staphylococcus aureus*. *FEBS Lett.*, **29**, 248–252

Padan, E., Zilberstein, D., and Rottenberg, H. (1976). The proton electrochemical gradient in *Escherichia coli* cells. *Eur. J. Biochem.*, **63**, 533–541

Parnes, J. R. and Boos, W. (1973). Unidirectional transport activity mediated by the galactose-binding protein of *Escherichia coli*. *J. Biol. Chem.*, **248**, 4436–4445

Payne, J. W. (1968). Oligopeptide transport in *Escherichia coli*: specificity with respect to side chain and distinction from dipeptide transport. *J. Biol. Chem.*, **243**, 3395–3403

Payne, J. W. (1971). The requirement for the protonated α-amino group for the transport of peptides in *Escherichia coli*. *Biochem. J.*, **123**, 245–253

Payne, J. W. (1972*a*). Mechanisms of bacterial peptide transport. *CIBA Found. Symp.*, **4**, 17–32

Payne, J. W. (1972*b*). Variations in the peptidase activities of *Escherichia coli* in response to environmental changes. *J. Gen. Microbiol.*, **71**, 281–289

Payne, J. W. (1972*c*). Effects of *N*-methyl peptide bonds on peptide utilization by *Escherichia coli*. *J. Gen. Microbiol.*, **71**, 259–265

Payne, J. W. and Bell, G. (1977*a*). The transport and utilization of peptides by *Escherichia coli*: interrelations between peptide uptake and amino acid exodus and biosynthesis. *FEMS Lett.*, **2**, 259–262

Payne, J. W. and Bell, G. (1977*b*). Substrate interactions during transport of amino acids and peptides by *Escherichia coli*: relevance to regulation of peptide uptake. *FEMS Lett.*, **2**, 301–304

Payne, J. W. and Bell, G. (1979). Direct determination of the properties of peptide transport systems in *Escherichia coli* using a fluorescent labeling procedure. *J. Bacteriol.*, **137**, 447–455

Payne, J. W. and Gilvarg, C. (1978). Transport of peptides in bacteria. In *Bacterial Transport* (B. P. Rosen, ed.) Marcel Dekker, New York and Basel, pp. 325–383

Pittman, K. A., Lakshmanan, S., and Bryant, M. P. (1967). Oligopeptide uptake by *Bacteroides ruminicola. J. Bacteriol.*, **93**, 1499–1508

Quay, S. C., Dick, T. E., and Oxender, D. L. (1977). Role of transport systems in amino acid metabolism: leucine toxicity and the branched-chain amino acid transport systems. *J. Bacteriol.*, **129**, 1257–1265

Ramos, S. and Kaback, H. R. (1977*a*). The relationship between the electrochemical proton gradient and active transport in *Escherichia coli* membrane vesicles. *Biochemistry*, **16**, 854–859

Ramos, S. and Kaback, H. R. (1977*b*). pH-dependent changes in proton:substrate stoichiometries during active transport in *Escherichia coli* membrane vesicles. *Biochemistry*, **16**, 4271–4275

Ramos, S. and Kaback, H. R. (1977*c*). The electrochemical proton gradient in *Escherichia coli* membrane vesicles. *Biochemistry*, **16**, 848–854

Rice, G. H., Stewart, F. H. C., Hillier, A. J., and Jago, G. R. (1978). The uptake of amino acids and peptides by *Streptococcus lactis. J. Dairy Res.*, **45**, 93–107

Rosen, B. P. and Kashket, E. R. (1978). Energetics of active transport. In *Bacterial Transport* (B. P. Rosen, ed.) Marcel Dekker, New York and Basel, pp. 559–620

Schuldiner, S. and Kaback, H. R. (1975). Membrane potential and active transport in membrane vesicles from *Escherichia coli. Biochemistry*, **14**, 5451–5461

Simmonds, S. (1966). The role of dipeptidases in cells of *Escherichia coli* K-12. *J. Biol. Chem.*, **241**, 2502–2508

Simmonds, S. (1970). Peptidase activity and peptide metabolism in *Escherichia coli* K-12. *Biochemistry*, **9**, 1–9

Simoni, R. D. and Postma, P. W. (1975). The energetics of bacterial active transport. *Annu. Rev. Biochem.*, **44**, 523–554

Smith, R. L., Archer, E. G., and Dunn, F. W. (1970). Uptake of [^{14}C]-labeled tri, tetra-, and pentapeptides of phenylalanine and glycine by *Escherichia coli. J. Biol. Chem.*, **245**, 2967–2971

Sussman, A. J. and Gilvarg, C. (1970). Peptidases in *Escherichia coli* K-12 capable of hydrolyzing lysine homopeptides. *J. Biol. Chem.*, **245**, 6518–6524

Tsuchiya, T. (1977). Adenosine-5′-triphosphate synthesis driven by a proton motive force in membrane vesicles of *Escherichia coli. J. Bacteriol.*, **129**, 763–769

Tsuchiya, T. and Rosen, B. P. (1976*a*). ATP synthesis by an artificial proton gradient in right-side-out membrane vesicles of *Escherichia coli. Biochem. Biophys. Res. Commun.*, **68**, 497–502

Tsuchiya, T. and Rosen, B. P. (1976*b*). Adenosine 5′-triphosphate synthesis energized by an artificially imposed membrane potential in membrane vesicles of *Escherichia coli. J. Bacteriol.*, **127**, 154–161

Tsuchiya, T., Hasan, S. M., and Raven, J. (1977). Glutamate transport driven by an electrochemical gradient of sodium ions in *Escherichia coli. J. Bacteriol.*, **131**, 848–853

West, I. C. and Mitchell, P. (1973). Stoicheiometry of lactose-H^+ symport across the plasma membrane of *Escherichia coli. Biochem. J.*, **132**, 587–592

West, I. C. and Mitchell, P. (1974). Proton/sodium ion antiport in *Escherichia coli. Biochem. J.*, **144**, 87–90

Winkler, H. H. and Wilson, T. H. (1966). The role of energy coupling in transport of β-galactosides by *Escherichia coli. J. Biol. Chem.*, **241**, 2200–2211

Wolfinbarger, L., Jr and Marzluf, G. A. (1976). Specificity and regulation of peptide transport in *Neurospora crassa*. *Arch. Biochem. Biophys.*, **171**, 637–644

Wong, P. T. S. and Wilson, T. H. (1970). Counterflow of galactosides in *Escherichia coli*. *Biochim. Biophys. Acta*, **196**, 336–350

Yang, S. L., Becker, J. M., and Naider, F. (1977). Transport of [^{14}C]-Gly-Pro in a proline peptidase mutant of *Salmonella typhimurium*. *Biochim. Biophys. Acta*, **471**, 135–144

Yoder, O. C., Beamer, K. C., and Shelton, D. C. (1965). Kinetic study of L-valine and glycyl-L-valine uptake by *Leuconostoc mesenteroides*. *Arch. Biochem. Biophys.*, **110**, 336–342

Young, E. A., Bowen, D. O., and Diehl, J. F. (1964). Transport studies with peptides containing unnatural amino acids. *Biochem. Biophys. Res. Commun.*, **14**, 250–255

Part 3

Proteins

Microorganisms and Nitrogen Sources
Edited by J. W. Payne

Section 3.1

Transport and Utilization of Proteins by Bacteria

BARRY A. LAW

National Institute for Research in Dairying, Shinfield, Reading, UK

1 INTRODUCTION

The utilization of exogenous proteins by bacteria requires the degradation of the extracellular proteins to peptides and amino acids before cellular uptake, and the direct uptake of utilizable proteins probably does not occur as such in bacteria. However, protein transport is relevant to utilization in that extracellular protein-degrading enzymes, which are produced inside the cell, must be moved outside to act on their substrates. Proteinase-producing bacteria form a vital part of natural ecosystems in which they not only make protein carbon and nitrogen available to themselves by degradation, but also interact with non-proteolytic microorganisms by supplying them with nutrients. In the course of their existence, proteinase-producing bacteria are likely to encounter proteins as diverse in structure as serum albumins, plant proteins, and connective tissue proteins such as collagen and keratin, as part of the bodies of dead animals and plants littering their environment. This multiplicity of potential nutrient sources is matched by the enormous diversity of bacterial proteinases. The role of extracellular proteinases in protein utilization by bacteria is clearly seen in the context of advances in our understanding of peptide and amino acid transport, dealt with at length elsewhere in this volume. In summary, the scheme of utilization by bacteria producing extracellular proteinases would be expected to follow a pattern of extracellular protein degradation to amino acids and to peptides whose maximum transportable size appears to be limited to about seven amino acid residues. Before further utilization, transported peptides are hydrolysed to their constituent amino acids by intracellular peptidases of wide specificity. Most detailed information on peptide uptake is available for Gram-negative bacteria (see Section 2.1) but recent reports show that lactic acid bacteria, particularly the streptococci, possess very efficient systems of cell-bound extracellular proteinases (Thomas *et al*., 1974; Exterkate, 1975, 1976*b*) which make peptides available near the cell surface for peptide-transporting systems (Law, 1977, 1978) of similar specificity to those of Gram-negative bacteria. This Section sets out to describe and consider the characteristics, locations, and secretion of bacterial extracellular proteinases in relation to their function in the utilization of proteins as sources of nutrients.

2 BACTERIA WHICH PRODUCE EXTRACELLULAR PROTEINASES

Representative examples of bacteria which have been reported to produce extracellular proteinases are listed in Table 1. In the published data concerning extracellular proteolytic enzymes derived from various bacteria it is not always clear whether the enzymes described are truly extracellular (actively secreted as a cellular function of viable, undamaged organisms) or are released mainly by moribund or lysing cells. This problem was emphasized by Pollock (1962) in his review of extracellular bacterial enzymes, yet even quite recent reports describing new extracellular enzymes are often lacking in definitive evidence for their extracellular nature. However, for the purpose of this review it will be useful to include many instances in which proteinases have been isolated from the supernatant solutions of bacterial cultures and more or less uncritically described as extracellular since, as far as viable bacterial populations are concerned, the proteinases will allow utilization of proteins in their environment, however released.

Glenn (1976) suggested that only Gram-positive bacteria produce truly extracellular enzymes, and it is certainly true that there are more reports and data available about proteinases from Gram-positive than Gram-negative bacteria. Well-characterized enzymes from the former group are the subtilisins from strains of *Bacillus subtilis*, the pronase enzymes from *Streptomyces griseus*, collagenases and clostripain from *Clostridium histolyticum*, and streptococcal proteinase from Group A streptococci (Table 1A). The best known proteinases of Gram-positive bacteria are the subtilisins, a group of alkaline serine proteinases (Hartley, 1960) produced by strains of *B. subtilis* (Kraut, 1971; Markland and Smith, 1971). This organism, and other species of *Bacillus* also produce other types of proteinases (Priest, 1977), such as EDTA-sensitive alkaline serine proteinase (*B. Sphaericus*) and neutral, non-serine proteinases (*B. thermoproteolyticus*, *B. megaterium*), as well as an acid, chymosin-like (rennin) proteinase (*B. cereus*), so rarely produced by bacteria (most are mould enzymes). Species of *Streptomyces* and *Arthrobacter* also produce examples of alkaline serine proteinases but, in contrast, the serine proteinases of *Staphylococcus aureus* are reported to have acid/neutral pH optima (Rydên *et al*., 1973; Drapeau, 1976). The best known bacterial thiol proteinase is that of the Group A streptococci, but *Cl. histolyticum* also produces such a proteinase, though it is unlike the streptococcal enzyme in being metal-activated. The collagenases and pseudocollagenase of *Cl. histolyticum* are non-thiol, EDTA-sensitive enzymes.

Many types of enzyme which are extracellular and cell-free in Gram-positive bacteria are found in the periplasmic space of Gram-negative bacterial cells, and it has been suggested that the more complex cell envelope of the latter prevents the release of secreted enzymes which would be cell-free in Gram-positive cultures (Glenn, 1976). However, Heppel (1971) excludes both proteinases and peptidases from the list of known periplasmic enzymes (see also Section 2.4.6) and up to the time of writing there have been reports of only two such proteinases, both concerning *Escherichia coli* (Regnier and Thang, 1972; Lazdunski *et al*., 1975). These apparently periplasmic proteinases are considered in more detail in Section 3.1.5, but for the present it is difficult to justify the extension of the parallel between cell-free extracellular enzymes in Gram-positive bacteria and periplasmic (extracellular) enzymes in Gram-negative bacteria to the particular instance of proteinases, with so few known examples of periplasmic proteinases in the latter group. In contrast, there have been increasing numbers of reports of proteinases isolated from the culture supernatant solutions of Gram-negative species representing the genera *Pseudomonas*, *Aeromonas*, *Proteus*, *Escherichia*, *Flavobacterium*, *Serratia*, *Bacteroides*, *Rhodopseudomonas*, and *Vibrio*. In many cases the enzymes have been described as extracellular chiefly on the

Table 1 Bacteria representative of those reported to produce extracellular proteinases†

Organism	Protein substrates	Proteinase type (Hartley system)‡	Metal ion activation	Ca^{2+} stabilization	Molecular mass (daltons)	Reference
A Gram-positive bacteria						
Arthrobacter sp.	Wide range (e.g. casein, haemoglobin, gelatin)	Serine, alkaline	No	No	22 000	Hofsten and Reinhammer, (1965), Hofsten and Tjeder (1965)
Bacillus cereus	Casein	Acid, chymosin-like				Melachouris and Tuckey (1968)
B. cereus	Casein	Neutral, EDTA-sensitive	Zn^{2+}	Yes	50 000	Islam and Blanshard (1973)
B. licheniformis	Haemoglobin, casein	Serine, alkaline, and neutral EDTA-sensitive	Ca^{2+}			Hall *et al*. (1966)
B. megaterium	Casein	Neutral, EDTA-sensitive	Ca^{2+}	Yes		Millet (1969)
B. mesentericus	Casein, haemoglobin	Alkaline?				Stoeva and Mesrob (1977)
B. sphaericus	Casein	Serine, alkaline, EDTA-sensitive	Bound Ca^{2+}	Yes	27–32 000	Yoshida *et al*. (1977)
B. subtilis	Wide range (e.g. elastin, ribonuclease, casein)	Serine, alkaline (subtilisin)	No	Yes	26–28 000	Markland and Smith (1971)
		Neutral, EDTA-sensitive	Zn^{2+}, Ca^{2+}	Yes	35–40 000	Matsubara and Feder (1971)
B. thermoproteolyticus	Casein	Neutral, EDTA-sensitive	Zn^{2+}, Ca^{2+}	Yes	35–40 000	Matsubara and Feder (1971)
Clostridium histolyticum	Collagen (collagenases A and B)	EDTA-sensitive	Ca^{2+}, Zn^{2+}, (Co^{2+}, Mn^{2+}, Fe^{2+}, ?)		105 000 and 57 400	Seifter and Harper (1971)
	Gelatin (pseudo-collagenase)					

	Wide range (e.g. casein, lysozyme, insulin, etc.)	Thiol, EDTA-sensitive	Ca^{2+}		50 000	Mitchell and Harrington (1971)
Micrococcus caseolyticus	Casein	Neutral				Matsubara and Feder (1971)
Micrococcus freudenreichii	Casein	Acid(?)				McDonald (1965)
Mycobacterium tuberculosis	Collagen (two enzymes)			Possibly	77 000	Takahashi (1967)
Myxobacter AL-1	Cell wall peptides, albumin, gelatin, casein	Alkaline, EDTA-sensitive	None known		8700–14 300	Matsubara and Feder (1971)
Staphylococcus aureus	Haemoglobin, casein	Serine, acid/neutral	No		29 000	Ryden *et al*. (1973)
	Haemoglobin, casein	Serine, acid/neutral	No	No	12 000	Drapeau (1976)
Group A streptococci	Casein, fibrin M protein, gelatin	Thiol	No	No	32 000	Liu and Elliot (1971)
Streptococcus lactis	Casein, serum, albumin, haemoglobin	Thiol				Williamson *et al*. (1964)
Streptomyces fradiae	Native keratin, elastin casein (five proteinases)	Serine, alkaline	No	Yes	14 000 and 16 500 (major proteinases)	Nickerson *et al*. (1963), Morihara *et al*. (1967)
Streptomyces griseus	Multiple proteinase (Pronase)	Serine, alkaline (trypsin-like) and EDTA-sensitive				Matsubara and Feder (1971), Siegel and Awad (1973), Vosbeck *et al*. (1975)
Streptomyces madura	Collagen	EDTA-sensitive acid (?)	Ca^{2+}		35 000	Rippon (1968)
B Gram-negative bacteria						
Aeromonas proteolytica	Casein, haemoglobin, gelatin	Neutral, EDTA-sensitive	Zn^{2+}	Yes	38 000	Wilkes and Prescott (1976)
Bacteroides amylophilus	Casein	Serine, alkaline (and neutral?)				Blackburn (1968), Blackburn and Hullah (1974)

Table 1 (*continued*)

Organism	Protein substrates	Proteinase type (Hartley system)‡	Metal ion activation	Ca^{2+} stabilization	Molecular mass (daltons)	Reference
Bact. nodosus	Casein, gelatin, fibrin	Alkaline, EDTA-sensitive	Co^{2+}, Mn^{2+}, Ca^{2+}, Mg^{2+}			Broad and Skerman (1976)
Escherichia freundii	Casein	Alkaline, EDTA-sensitive	Zn^{2+}	No	45 400	Nakajima *et al*. (1974)
Flavobacterium sp.	Elastin	Neutral (?)				Mandl and Cohen (1960)
Proteus mirabilis		Neutral, EDTA-sensitive	None known (inhibited by Mg^{2+})	No		Hampson *et al*. (1963)
Pseudomonas aeruginosa	Casein, egg albumin, haemoglobin	Neutral, EDTA-sensitive	Zn^{2+}	Yes	48 400	Morihara (1963)
	Elastin, casein	Neutral, EDTA-sensitive	Zn^{2+}		39 500	Morihara (1964), Morihara *et al*. (1965)
	Casein, elastin	Neutral, EDTA-sensitive		Yes	23 000	Scharman and Balke (1974), Wretlind and Wadström (1977)
	Collagen	Alkaline, EDTA-sensitive			Possible subunits	Schoellmann and Fisher (1966)
	Collagen, azocoll, casein	Neutral, EDTA-sensitive	Co^{2+}, Zn^{2+}		17 500	Carrick and Berk (1975)
Ps. fluorescens	Casein				23 000	Mayerhofer *et al*. (1973)
	Casein	Thiol(?)	No		38 400	Law *et al*. (1977), Alichanidis and Andrews (1977)
Ps. fragi	Casein	Neutral, EDTA-sensitive	$Zn^{2+}Co^{2+}$, Mn^{2+}	Yes	40–50 000	Porzio and Pearson (1975)
Ps. lachrymans	Casein, plant proteins	Neutral/alkaline (unclassified by normal criteria)				Keen *et al*. (1966)

Ps. maltophila	Casein, azocoll	Serine, alkaline, EDTA-sensitive	Ca^{2+}, Sr^{2+}, Ba^{2+}, Co^{2+}, Cu^{2+}, Mg^{2+}, Zn^{2+}		35 000	Boethling (1975*a*)
Pseudomonas sp.	Casein, muscle albumin	Neutral, EDTA-sensitive				Li and Jordan (1968)
	Casein	EDTA-sensitive	Zn^{2+}, Ca^{2+}	Yes	49 000	Barach *et al*. (1976), Barach and Adams (1977)
Rhodopseudomonas gelatinosa	Gelatin, casein	Neutral, EDTA-sensitive	None known	Yes		Klemme and Pfleiderer (1977)
Serratia marcescens	Casein, collagen-derived peptides	Neutral/alkaline, EDTA-sensitive	Fe^{2+}	Yes	30–39 000	McQuade and Crewther (1969), Aiyappa and Harris (1976)
Serratia sp.	Casein, haemoglobin, insulin	Alkaline, EDTA-sensitive	Zn^{2+} (Co^{2+}, Mn^{2+})	No		Miyata *et al*. (1970*a*,*b*)
Vibrio sp.	Haemoglobin	Serine(?), neutral, EDTA-sensitive	Ca^{2+}	Yes	49 000	Merkel and Sipos (1971)

†Includes truly extracellular proteinases and those merely reported as recoverable from culture filtrates. Cell-bound extracellular proteinases are excluded (see Section 3.1.5).

‡Hartley (1960); see Section 3.1.3.1.

basis of their recovery from culture fluids, usually after prolonged incubation and at unspecified stages of growth. However, the proteinase activities of a *Pseudomonas* sp. causing proteolysis in skin lesions of fish (Li and Fleming, 1967), and the *Ps. myxogenes* proteinase cited by Pollock (1962) are probably truly extracellular enzymes since they are not found intracellularly. In other instances, the release of the proteinases into the growth medium has been shown to be under several types of metabolic control (see Section 3.1.7) suggesting that their presence in the medium is not due to uncontrolled cell lysis. Generally speaking, Gram-negative bacteria are noted for producing neutral and alkaline metal chelator-sensitive proteinases (Table 1B); in particular, *Ps. aeruginosa* produces a large number of such proteinases, all active against casein but differing in their activities with collagen and elastin.

In addition to the cell-free proteinases already mentioned, some bacteria have firmly cell-bound proteinases accessible to external substrates (e.g. Group N streptococci, *Bacteroides amylophilus*); these will be discussed in Section 3.1.5.

3 PROPERTIES OF BACTERIAL EXTRACELLULAR PROTEINASES

1 Classification

Microbial proteinases can be classified by the system of Hartley (1960) into the familiar four groups based on their mechanism of action: serine proteinases, thiol proteinases, metalloproteinases (sensitive to metal chelating agents such as EDTA), and acid proteinases. However, a more comprehensive system has been developed by Morihara (1974) and is based on primary specificitics (determined on synthetic substrates and on oxidized insulin B-chain) and on effects of secondary (i.e. more long-range) interactions on primary specificity. The detailed experimental data by which the enzyme specificities were determined as a basis for this classification are reviewed by Morihara (1974) and are merely summarized for the purpose of the present discussion.

Primary specificity is dependent on the type of amino acid residues comprising the peptide bond being hydrolysed. Serine and thiol proteinases show specificity for basic, acidic, aromatic (or hydrophobic), and small aliphatic amino acid residues at the carboxyl side of the splitting point. Metal chelator-sensitive neutral proteinases act where there is a basic and bulky or hydrophobic amino acid residue at the amino side. Where the splitting point has aromatic or hydrophobic residues on both sides, the peptide bond will be sensitive to acid proteinases. There are some difficulties with this classification as applied to bacteria because the streptococcal thiol proteinase has a much broader specificity than clostridial thiol proteinase (clostripain), but studies with synthetic substrates establish that amino acid residues on the carboxyl side of the splitting point are also important in the streptococcal proteinase. Similarly, metal chelator-sensitive alkaline proteinases have much broader specificities than the neutral enzyme but the activities of both are affected by residues on the amino side. The effects of secondary interactions from amino acid residues not involved in the susceptible peptide bond vary from small in trypsin-like serine proteinases and metal chelator-sensitive neutral proteinases, to quite marked in serine alkaline types, and more important than primary specificity in streptococcal proteinase and metal chelator-sensitive alkaline proteinases. Morihara (1974) has speculated that it is the requirement for secondary interactions from both sides of peptide bonds, as well as the absence of a charged α-amino or carboxyl group which renders terminal peptide bonds insensitive to proteinases (i.e. endopeptidases). It will be apparent from this consideration of proteinase specificity that the hydrolysis of protein substrates by bacterial proteinases is therefore partially dependent on the interactions between substrate and enzyme at sites other than the catalytic centre, the binding of elastin by elastase being an extreme example of such dependence (see below; degradation of elastin). Svendsen (1976) has reviewed the various chemical modification studies which have led to a better understanding of the surface properties of enzymes which are involved in protein substrate binding by the subtilisins; for example, nitration, iodination, glutarylation, or succinylation of the enzyme increase the rate of hydrolysis of gelatin, but carboxyl group modification decreases it. In all these instances the hydrolysis of small ester substrates was unaffected. Comparisons between results obtained by chemical modification studies and those obtained by X-ray diffraction and kinetic studies on synthetic polypeptides suggest that complex secondary interactions can produce more than one mode of productive binding between enzymes and large substrates.

2 Biological Activities and Action on Protein Substrates

Non-structural proteins such as albumin, casein, insulin, or haemoglobin are degraded by most bacterial proteinases, but to variable extents dependent on the specificity of the particular enzyme. Casein degradation by microbial acid proteinases, with their high activity on κ-casein, has proved economically useful since they can be used as substitutes for chymosin in cheese manufacture (Green, 1977); however, the only such bacterial enzyme characterized so far is that of *B. cereus* (e.g. Islam and Blanshard, 1973) and this has undesirable milk-clotting activities in ultra-high-temperature sterilized (UHT) milk. A heat-resistant proteinase from a psychrotrophic strain of *Ps. fluorescens* also degrades κ-casein, as well as β- and αs_1-casein and causes the gelation of UHT milk (Law *et al.*, 1977) but, unlike the chymosin-like enzymes, it is optimally active at near-neutral pH (Alichanidis and Andrews, 1977). However, despite extensive studies of casein-degrading enzymes from these and other Gram-negative bacteria (Malik, 1976; Adams *et al.*, 1975, 1976; Barach *et al.*, 1976, Barach and Adams, 1977; Porzio and Pearson, 1975; Mayerhofer *et al.*, 1973; Cousin and Marth, 1977) little is known about their role in protein utilization. It is known that some species of *Pseudomonas* can transport and utilize peptides (Cascieri and Mallette, 1976) so it is possible to envisage a nutritional function for the extracellular proteinases produced by the bacteria when growing in milk; pseudomonads generally do not have an absolute requirement for preformed amino acids but can use them as sources of carbon for energy as well as for their nitrogen content.

Only a relatively small range of bacteria are known to produce enzymes which can hydrolyse structural and connective tissue proteins. These are often bacteria associated with pathological lesions, and their ability to degrade such proteins may confer a dual advantage on the producer organisms. Obviously the secretion of enzymes, close to the cell, which degrade proteins resistant to attack by other bacterial species may allow the producer to dominate an ecosystem through its nutritional advantage. Alternatively, or in addition, the penetration of protective barriers (made up of keratin, collagen, or elastin) by specific enzymes may allow the exploitation of more easily digestible underlying protein sources, inaccessible to other bacteria.

(i) Degradation of collagen

Collagen is a connective tissue protein of vertebrates and invertebrates that has a highly-organized helical structure (Ramachandran, 1967; von Hippel, 1967; Piez, 1967) conferred by regular repeating triad units containing glycine, proline, or hydroxyproline, and one of many other types of amino acid (X; particularly alanine). The terminal regions of collagen chains do not have the characteristic repeating units (Gly-Pro-X- and Gly-Pro-Hyp-) and are excluded from the helical folding. They are susceptible to hydrolysis by general peptidases but peptide bonds in the helical chain are only attacked by true collagenases. Enzymes cleaving denatured collagen (gelatin) are not classified as true collagenases (e.g. the pseudocollagenase of *Cl. histolyticum*; Mitchell, 1968).

The two collagenases, A and B, produced by *Cl. histolyticum* have been described in detail by Seifter and Harper (1971) as EDTA-sensitive metalloenzymes probably containing Zn^{2+} and activated by Ca^{2+}. Collagenase A degrades collagen extensively, catalysing about 200 cleavages per chain. Collagenase B appears to disrupt the structural organization of collagen without extensive peptide bond hydrolysis. Seifter and Harper (1971) suggested that it is the ability of collagenases to bind the substrate and cause structural changes, as well as their bond specificity, which enables them to degrade such a complex substrate. Digestion of collagen by collagenase A yields *N*-terminal glycine peptides as small as tripeptides but the number average molecular mass of the peptide mixture is about 500 daltons (Seifter *et al*., 1959). Seifter and Harper (1971) considered that further breakdown by other enzymes (e.g. the pseudocollagenase) would be required to render these products of proteolysis nutritionally useful. However, Hauschild (1965) demonstrated that *Cl. perfringens* utilized peptide fractions containing up to 12 amino acid residues. This species also produces a collagenase, so clostridial collagen degradation probably has direct nutritional significance. *Streptomyces madurae* produces a collagenase with properties similar to those of clostridial collagenase A, though its molecular mass is lower (35 000 daltons; see Table 1A).

Corneal collagen is degraded by a collagenase from *Ps. aeruginosa* which appears to exist in active subunits or aggregates (Schoellman and Fisher, 1966; Carrick and Berk, 1975). The enzyme described by Schoellman and Fisher (1966) was inferred to be a collagenase because it hydrolysed Cbz-Gly-Pro-Gly-Gly-Pro-Ala at the Gly-Gly bond; it was not tested directly against collagen. The enzyme purified by Carrick and Berk (1975) from a different strain, degraded native collagen but not the collagenase-specific hexapeptide, indicating a different bond specificity from clostridial collagenase. *Pseudomonas* enzymes cause extensive destruction of collagen-based connective tissue *in vivo* (Diener *et al*., 1973; Kreger and Griffin, 1974), although the nature of the products is not known. Other bacteria producing collagenase are listed in Table 1.

(ii) Degradation of elastin

Elastin is another example of a connective tissue fibrous protein resistant to attack by most proteinases. Elastases have side-chain specificities towards non-aromatic, uncharged amino acid residues of which elastin is largely composed (e.g. alanine, glycine, valine). However, other proteinases have specificities which apparently should allow them to degrade elastin, yet they are not elastolytic. Morihara and Tsuzuki (1967) showed that the elastases of *B. subtilis*, *B. thermoproteolyticus*, *Ps. aeruginosa*, *Streptomyces griseus*, and *Streptomyces fradiae* were able to bind elastin to a much greater extent than were other non-elastolytic proteinases. Also, the ability of subtilisin elastase to degrade this substrate (but not globular proteins or ester substrates) requires free ε-amino groups in the enzyme (Gertler,

1971). This could suggest that close contact with elastin, achieved by electrostatic binding between enzyme and substrate, is necessary for catalysis. Morihara and Tsuzuki (1967) envisaged a mechanism of action of elastases involving two stages; adsorption of enzyme onto the substrate in such a way that the steric hindrance of bulky elastin cross-links is overcome; peptide bond hydrolysis to release fragments from between the cross-links. These fragments are presumed to be susceptible to general proteinase attack.

Elastase production, like that of collagenase, has been linked with the pathogenicity of *Ps. aeruginosa* (Margaretten *et al.*, 1961; Mull and Callahan, 1965; Wretlind and Wadstrom, 1977). Several enzymes have been isolated from culture supernatant solutions and described as neutral, EDTA-sensitive proteinases (e.g. Morihara and Tsuzuki, 1975; Wretlind and Wadstrom, 1977). A species of *Flavobacterium* isolated from peridontal disease pockets produces an elastase (Mandl and Cohen, 1960) which appears to differ from the *B. subtilis* and *Ps. aeruginosa* enzymes in being insensitive to metal-chelating agents, and not inhibited by high NaCl concentrations (like the *Streptomyces fradiae* elastase; Morihara *et al.*, 1967).

(iii) Degradation of keratin

Keratin owes its structural properties to a high content of disulphide cross-links. It is very resistant to enzymic hydrolysis, but *Streptomyces fradiae* produces an enzyme system which appears to degrade native keratin by first reducing the disulphide bridges and then acting as a proteinase, similar to a conjugated keratinase described by Noval and Nickerson (1959), Nickerson *et al.* (1963), and Nickerson and Durand (1963). This latter enzyme produced peptides (but no free amino acids) from wool keratin; only poly-L-lysine served as a synthetic substrate.

(iv) Degradation of other proteins

Muscle albumin degradation by a truly extracellular proteinase from a pseudomonad isolated from skin lesions in rainbow trout has been demonstrated by Li and Jordan (1968). Utilization of myofibrillar pig muscle proteins was reported in *Ps. fragi* (Dutson *et al.*, 1971; Tarrant *et al.*, 1971); protein breakdown led to structural changes in the muscle as well as the release of non-protein N. A number of proteolytic Gram-negative species also break down beef muscle proteins (specific sarcoplasmic and myofibrillar proteins) (Dainty *et al.*, 1975). The trout muscle-degrading enzyme was characterized as a metal chelator-sensitive proteinase, but the nature of the pig and beef muscle degrading enzymes has not yet been reported.

3 Other Properties of Bacterial Extracellular Proteinases

Most of these enzymes have relatively low molecular mass (20 000–40 000 daltons), and many require divalent metal ions for their activity (often Zn^{2+}) or for stability (usually Ca^{2+}) (Table 1). Many, though not all, do not contain cysteine/cystine, a property which Pollock and Richmond (1962) suggested might be important in conferring flexibility to aid the secretion of proteinase molecules through the bacterial cell membranes. As a group, the bacterial extracellular proteins (exoproteins) do not have any obviously distinctive amino acid composition besides the low incidence of cysteine, and their polarities do not differ markedly from those of intracellular proteins (Glenn, 1976). Equally, the fact that extracellular proteinases have to be secreted through membranes and walls does not seem to signify any unique secondary structural properties, as evidenced by the export of proteinase protein fragments by mutants of *B. subtilis* (Shoer and Rappaport, 1972). If comparisons can be drawn with other extracellular enzymes, then the tertiary structure has as little significance as secondary structure, e.g. studies with inactive phosphatase, containing amino acid analogues (Attias *et al*., 1969; Morris and Schlessinger, 1972). In any case, recent ideas on protein secretion from cells suggest that pre-folded, complete molecules are not transported as such (see Section 3.1.6).

The general properties of the extracellular proteinases produced by Gram-positive bacteria have been reviewed elsewhere (Kraut, 1971; Markland and Smith, 1971; Liu and Elliot, 1971; Seifter and Harper, 1971; Mitchell and Harrington, 1971; Matsubara and Feder, 1971; Kelly and Fogarty, 1976; Drapeau, 1976) and the reader is referred to these papers because the recording of this detailed information is beyond the scope of this Section. However, because the properties of the Gram-negative bacterial proteinases have not been reviewed recently, some of their characteristics will be considered below (also see Table 1).

Elastin and collagen degradation by strains of *Ps. aeruginosa* has already been discussed in relation to the biological activity of bacterial proteinases. The elastase activity has been attributed to enzymes with molecular mass of 39 000 (Morihara *et al*., 1965) and 23 000 daltons (Wretlind and Wadstrom, 1977; Scharmann and Balke, 1974); presumably there are several elastases produced by different strains. All the enzymes appear to be neutral, EDTA-sensitive proteinases, but the elastase described by Scharmann and Balke (1974) also showed characteristics of a serine enzyme and had a slightly more alkaline (pH 8.3) optimum activity against casein. Morihara and Tsuzuki (1975) showed that the elastase of their strain contained Zn^{2+}, yet this ion inhibited the elastase purified by Wretlind and Wadstrom (1977). Despite these differences, the elastases of *Ps. aeruginosa* have many properties in common, and generally (with the exception of the serine enzymes) are typical of other neutral metal chelator-sensitive proteinases (cf. Matsubara and Feder, 1971).

Carrick and Berk (1975) purified and partially characterized the collagenase of *Ps. aeruginosa* as a distinct metal chelator-sensitive neutral proteinase, which appeared to exist either as monomers (molecular mass 17 500 daltons) or dimers (molecular mass 34 000 daltons). The metal content of the enzyme was not reported, but it was activated by Ca^{2+} and to a lesser extent by Zn^{2+}. According to its amino acid analysis, the collagenase contained cysteine, unlike many other extracellular enzymes.

Psychrotrophic and psychrophilic species of *Pseudomonas* produce a variety of very heat-stable proteinases whose action on casein in milk has already been discussed. For example, the *Ps. fluorescens* proteinase which gels UHT-milk (Law *et al*., 1977) is insensitive to EDTA, inhibited by Co^{2+} and Zn^{2+}, and may be a thiol proteinase (Alichanidis and Andrews, 1977). It has a near-neutral pH optimum with casein as substrate and its molecular mass is 38 400 daltons. Another strain of the same species produces a proteinase of similar molecular mass (37 000 daltons) but it is inhibited by EDTA (re-activated by Zn^{2+} and Co^{2+}) and is insensitive to thiol-blocking reagents (Juan and Cazzulo, 1976). Barach *et al*. (1976) isolated an EDTA-sensitive Zn^{2+}-containing neutral proteinase (molecular mass 48 000 daltons) from a raw milk pseudomonad. Its heat resistance in milk was dependent on Ca^{2+}, and Barach and Adams (1977) compared its properties to those of thermolysin, the *B. thermoproteolyticus* heat-resistant proteinase. They speculated that the heat resistance of both enzymes could be due to their high content of hydrophobic side chains, their flexibility (conferred by an absence of cysteine/cystine), and their metal ion content, the metal ions probably acting to ensure accurate re-folding of the enzymes after treatment at very high temperatures, as previously suggested by Dahlquist *et al*. (1976). *Ps. fragi* produces a typical neutral, EDTA-sensitive proteinase (molecular mass 40 000–50 000 daltons) activated by Zn^{2+} and stabilized by Ca^{2+}, though the enzyme was not stable above 50–60 °C even in the presence of the latter (Porzio and Pearson, 1975). An alkaline serine proteinase is produced by *Ps. maltophila* (Boethling, 1975*a*); its molecular mass is 35 000 daltons and it is inhibited by EDTA and re-activated by a wide variety of divalent metal ions.

The properties of the *Serratia* proteinase (Miyata *et al*., 1970*a*, *b*) have been reported in some detail. It is an alkaline metalloenzyme containing Zn^{2+} but is not a serine enzyme and does not contain cysteine. Compared with other bacterial extracellular enzymes its molecular mass (60 000 daltons) is high. Details of an *Aeromonas* neutral proteinase were reviewed by Wilkes and Prescott (1976). This is a cysteine-containing metalloenzyme that contains Zn^{2+} and is possibly stabilized by Ca^{2+}. Some of the properties of less well-studied proteinases produced by other Gram-negative bacteria are listed in Table 1B.

4 EXTRACELLULAR EXOPEPTIDASES PRODUCED BY BACTERIA

There are fewer examples of extracellular bacterial exopeptidases than there are of extracellular endopeptidases. Exopeptidases recovered from the growth media of bacterial cultures are usually aminopeptidases (Table 2), with alkaline pH optima and (like almost all exopeptidases) a requirement for divalent metal ions. Most reported examples are produced by Gram-positive bacteria (*B. subtilis*, *B. licheniformis*, *Cl. histolyticum*, and *Streptomyces fradiae*), although there is one particularly well-documented enzyme produced by the Gram-negative organism *Aeromonas proteolytica* (Prescott and Wilkes, 1966, 1976). *Streptomyces fradiae* also produces a carboxypeptidase, according to a report by Morihara *et al*. (1967). The exopeptidases presumably contribute to the protein-utilizing abilities of bacteria by producing free amino acids directly from proteins, or from peptides that are too large for direct transport into the organism. Excluded from this category are unusual exopeptidases that act on peptides containing D-amino acid residues, such as the extracellular L,D-carboxypeptidase of *B. sphaericus* involved in sporulation (Vacheron *et al*., 1977), and other cell wall degrading enzymes.

5 SURFACE-BOUND EXTRACELLULAR PROTEINASES

In addition to the bacterial extracellular proteinases released as free enzymes into the growth medium, there appear to be several instances of cell-wall-bound, periplasmic, and membrane-bound proteinases and peptidases accessible to substrates from outside the cell (in the Group N streptococci, *Lactobacillus bulgaricus*, *E. coli*, and *Bact. amylophilus*). Such an arrangement appears to offer advantages over the release of free enzymes. Thus, retention of proteolytic activity on the producer cell is less likely to distribute the benefits of proteinase producing ability at random on other species competing for nutrients in the same environment; amino acids and transportable peptides are produced as close as possible to the appropriate uptake systems within the cell envelope.

Cell-wall-bound proteinases appear to be involved in exogenous protein utilization by lactic acid bacteria growing in milk. Thomas *et al*. (1974) first reported the cell wall location of a proteinase in *Streptococcus lactis*, which

Table 2 Bacteria which produce extracellular exopeptidases

Organism	Peptidase type	Metal ion activation	Molecular weight (daltons)	Reference
Bacillus subtilis	Alkaline, EDTA-sensitive aminopeptidase	Co^{2+} and/or Zn^{2+}		Ray and Wagner (1972), Wagner *et al*. (1972)
B. licheniformis	Alkaline, EDTA-sensitive aminopeptidase	Co^{2+}		Hall *et al*. (1966)
Clostridium histolyticum	Alkaline, EDTA-sensitive thiol aminopeptidase	Mn^{2+}, Co^{2+}	340 000 (6 × 56,000 subunits)	Kessler and Yaron (1976)
Streptomyces fradiae	Alkaline, EDTA-sensitive leucine aminopeptidase	Ca^{2+}, Sr^{2+}, Co^{2+}		Morihara *et al*. (1967)
	Neutral carboxypeptidase	Co^{2+}		Morihara *et al*. (1967)
Aeromonas proteolytica	Alkaline, EDTA-sensitive aminopeptidase	Zn^{2+} (Co^{2+}, Mn^{2+})	29 000	Prescott and Wilkes (1966, 1976)

was released from the cells on stable protoplast formation by bacteriophage lysin, and unlike the cell-free proteinase of *Streptococcus lactis* (Williamson *et al*., 1964) the cell-bound enzyme was EDTA-sensitive and unaffected by thiol-blocking reagents. Exterkate (1975) described a casein-utilizing system in *Streptococcus cremoris* strain HP that consisted of acid and neutral cell-wall-bound proteinases, as well as membrane-bound endopeptidases (active against low molecular weight synthetic peptide derivatives), two aminopeptidases, and a pyrrolidone carboxylyl peptidase. Other strains of *Streptococcus cremoris* differ widely in the numbers and types of proteinases in their cell walls (Exterkate, 1976*b*). Recent studies on a third type of proteinase, in cell walls of *Streptococcus cremoris* strain AM1, indicate that it is only stable and active in the presence of Ca^{2+} in the growth medium. However, the organism may also produce a cell-free extracellular proteinase whose stability is not dependent on Ca^{2+} (F. A. Exterkate, personal communication). Strains of *Streptococcus lactis* (but not *Streptococcus cremoris*) also produce a surface-bound peptidase released from stable protoplasts by lysozyme (B. A. Law, unpublished results); its activity in whole cells overcomes competition between dipeptides for the uptake system (Law, 1977, 1978) and it may therefore have a valid function in the later stages of protein utilization by these organisms. The cell-bound proteinase activity of *Streptococcus lactis* strain C2 may be coded on plasmid DNA, rather than chromosomal DNA. Evidence for this is based on the finding that loss of specific plasmid DNA coincides with proteinase loss (Efstathiou and McKay, 1976), but definitive proof awaits the establishment of chromosomal markers for streptococcal structural genes. Exterkate (1976*a*) has also reported proteinase-deficient variants of *Streptococcus cremoris* and, because these strains grow only slowly in milk, the characterization of the genetic basis of proteinase loss is of considerable interest to those concerned with the manufacture of cultured dairy products which depend on the acid-forming abilities of lactic streptococci.

Cell wall proteinases may also be formed by other types of lactic acid bacteria, some of which are used in dairy product manufacture. For example, Argyle *et al*. (1976) showed that a metal chelator-sensitive proteinase active at pH 5.2–5.8 was released from *Lb. bulgaricus* by lysozyme treatment, although the possible release of intracellular enzymes was not monitored. Recent studies (Shankar, 1978) with *Lb. bulgaricus* and *Streptococcus thermophilus*, which are used as combined cultures in yoghurt manufacture, suggests a complementary relationship between the casein-utilizing systems of these organisms; the lactobacillus has surface-bound proteinase activity, and the streptococcus apparently produces cell-free and cell-bound extracellular peptidases capable of degrading the products of the lactobacillus enzyme. However, the evidence for cell-bound peptidase was not unambiguous as it was only based on observations of peptide hydrolysis with whole cells and requires further investigation (see also Section 2.1.5.2).

In Gram-negative bacteria many degradative enzymes are located in the periplasmic space between the inner cytoplasmic membrane and the cell wall (Heppel, 1971), from where they may be prevented from escaping into the surrounding medium (i.e. being truly extracellular) by the outer membrane which is absent in Gram-positive bacteria (Greenawalt and Whiteside, 1975). Peptide hydrolases are not usually included in the list of periplasmic enzymes. However, Regnier and Thang (1972) reported that 10 per cent of the proteinase activity of *E. coli* strain ML 3046 was released from cells by an osmotic-shock treatment used to release typical periplasmic enzymes but not an intracellular marker enzyme. Murgier *et al.* (1976, 1977) have provided evidence that an amino-endopeptidase (naphthylamidase) is located near the cell surface of *E. coli* K10 and other Gram-negative bacteria. They used the criteria of substrate hydrolysis by whole cells, accessibility of the enzyme to fluorescamine, and enzyme release by osmotic shock or spheroplast formation. The enzyme from *E. coli* K10 has been purified, and characterized as having aminopeptidase and endopeptidase activity. Its molecular mass is 45 000 daltons and unlike most aminopeptidases it is not inhibited by EDTA but appears to be a thiol proteinase of restricted specificity (Lazdunski *et al.*, 1975). However, the significance of this enzyme to protein utilization is unclear and, in fact, it is an intracellular enzyme in most strains of *E. coli* (see also Section 2.4.6).

Another Gram-negative species, *Bact. amylophilus* (a rumen anaerobe) produces a cell-bound serine proteinase accessible to protein substrates but not released by osmotic shock or spheroplast formation; *n*-butanol treatment is required to release the enzyme from membrane fragments of ruptured spheroplasts (Blackburn and Hullah, 1974). This variation on the theme of surface-binding of proteinases emphasizes the need for more information on the nature of the physical and/or chemical binding of these enzymes to surface structures in bacteria.

6 SYNTHESIS AND SECRETION OF EXTRACELLULAR PROTEINASES

In order to secrete proteinases (or any other extracellular protein) a mechanism is required which allows large, highly-charged protein molecules to be transported through the lipophilic membrane of the cell from interior to exterior. Pollock and Richmond (1962) envisaged that extracellular enzymes would pass through relatively easily because they are likely to be flexible due to the absence of disulphide cross-links. However, it has already been noted that some extracellular enzymes do contain cysteine so the potential to form disulphide bridges does not necessarily prevent secretion. Some properties of secreted proteins in eukaryotes which have been considered as possible aids to transmembrane passage were discussed by Lampen (1974) but they do not provide the basis of a unifying hypothetical mechanism for transport in either the eukaryotic or the prokaryotic cell (see also Section 3.2.2.6). Many of the difficulties of explaining the passage of proteins through membranes can be overcome by adopting the idea of translation and extrusion of newly-formed polypeptides at membrane sites. This concept of direct simultaneous translation and extrusion of extracellular proteins was developed from studies with membrane-bound ribosomes of eukaryotic cells (Redman and Sabatini, 1966); its extension to prokaryotes also helps to explain how bacteria produce the proteinases they need for utilization of external proteins without the danger of intracellular accumulation of potentially destructive enzymes. The site of synthesis of proteins destined for extracellular secretion is therefore of particular interest to this discussion.

In a recent review Glenn (1976) considered the evidence relating to the site of exoprotein synthesis in bacterial cells and concluded that there was little experimental data to support cytoplasmic synthesis, even though the concept of inactive zymogen molecules becoming active on secretion may be superficially attractive, e.g. *Streptococcus faecalis* proteinase forms (Liu *et al*., 1963; Liu and Elliot, 1965). In any case, it is generally true that active extracellular proteinases are never found in the cytoplasm, despite some controversy over the sporulation-associated serine proteinase of *B. subtilis* (Leighton *et al*., 1973; Prestidge *et al*., 1971; Reysset and Millet, 1972). The alternative site of exoprotein synthesis is the cell membrane, but according to Glenn (1976) there is little reliable direct evidence for membrane-associated ribosomes on which the proteins would be produced. However, some of the objections to the evidence for these membrane-bound ribosomes have been overcome in a recent study by Randall and Hardy (1977). Without using lysozyme, they isolated a membrane-bound fraction of *E. coli* ribosomes which preferentially synthesized membrane proteins and a secreted periplasmic protein *in vitro*. In contrast, the major product of free polysomes was a soluble cytoplasmic protein. The existence of membrane-associated ribosomes or polysomes is implicit in the translation-extrusion model for the synthesis and secretion of exoproteins in bacteria (May and Elliot, 1968; Lampen, 1974). It has been proposed that specific exoprotein mRNA carries a common 5′ sequence preventing binding to free ribosomes, but allowing recognition by membrane-bound ribosomes (Lampen, 1974). In this way the mRNA could traverse the space from its transcription site (chromosomal or plasmid DNA) to translation site without cytoplasmic translation. Glenn

(1976) has reviewed the evidence for such specific exoprotein synthesis and also discussed the apparent problem of how the normally short-lived mRNA can survive long enough to reach the site of exoprotein synthesis. Relevant to this topic are the reports (Both *et al.*, 1972; Semets *et al.*, 1973) that the synthesis of extracellular proteinase continues *de novo* in *B. amyloliquefaciens* and *B. subtilis* in the presence of inhibitors of mRNA synthesis, a result indicating the existence of stable mRNA, although direct measurements show that mRNA was generally unstable. Recently, however, Randall and Hardy (1977) reported that mRNA associated with membrane-bound polysomes in *E. coli* was more stable than the bulk of cellular mRNA. Also, comparisons of the continuation times and the rates of surface-bound proteinase synthesis in the absence of mRNA synthesis in *Streptococcus cremoris* strongly suggest that the proteinase is produced from proteinase-specific long-lived mRNA rather than from a large pool of short-lived mRNA (F. A. Exterkate, personal communication). Furthermore, attempts to demonstrate a large pool of preformed mRNA sufficient to account for the continued synthesis of exoproteins in the absence of new mRNA synthesis have been unsuccessful (Brown and Coleman, 1975*a*, *b*). Perhaps the observations with both *E. coli* and *Streptococcus cremoris* can be explained if the exoprotein mRNAs are not inherently more stable, but rather they are protected from the action of cellular nucleases by the proximity of the cell membrane associated with the polysomes.

A mechanism ('the signal hypothesis') originally proposed to account for exoprotein synthesis and secretion in eukaryotic cells (Blobel and Sabatini, 1971; Blobel and Dobberstein, 1975) may be extended to the situation in prokaryotes (Both *et al.*, 1972; Priest, 1977), in combination with the translation–extrusion model. It is proposed that a signal region occurs at the 5′ end of the exoprotein-specific mRNA, and that translation of this portion produces a signal *N*-terminal amino acid sequence on the nascent polypeptide chain which, as it emerges from the ribosome, is recognized by specific membrane proteins. These proteins interact with the translating ribosome and form a 'tunnel' through which the growing polypeptide chain is extruded by the adoption of a hydrophobic conformation compatible with the hydrophobic groups in the lipid bilayer (Lampen, 1974). Hydrophobic amino-terminal groups may also aid passage (Yamamoto and Lampen, 1975). The newly-formed protein presumably does not take up its native conformation until it is outside the membrane and the signal sequence has been removed by a specific endopeptidase. Detailed discussions of the experimental evidence for these proposals have been published and the reader is referred to reviews by Lampen (1974), Glenn (1976), and Priest (1977). Also, a more recent report (Smith *et al.*, 1977) claims to provide direct evidence that secreted proteins thread through the membrane as growing chains and that the membrane-associated polysomes in *E. coli* are functionally attached to the membrane and not trapped artificially when cells are disrupted.

7 REGULATION OF EXTRACELLULAR PROTEINASE PRODUCTION

Factors affecting the synthesis of extracellular proteinases (and other exoproteins) have been reviewed recently (Glenn, 1976) and the genus *Bacillus* has been discussed at great length (Priest, 1977). Both reviews consider a model that has been proposed to try to explain the control mechanism for exoprotein synthesis operating in bacteria which produce extracellular enzymes maximally during the post-exponential growth phase in batch culture (Coleman *et al*., 1975). Control at the transcription level is envisaged such that exoprotein mRNA synthesis competes unfavourably with the synthesis of non-translatable RNA and mRNA specific for intracellular proteins. Only when the latter two processes are reduced (during post-exponential phase) can exoprotein mRNA synthesis commence due to higher availability of RNA polymerase and RNA precursors. More recent studies of cellular protein:exoprotein synthesis ratios in *Staphylococcus aureus* (Abbas-Ali and Coleman, 1977) indicate that the competition model may be valid in this organism as well as in *Bacillus*, for which it was originally proposed. It is implicit in this model that the 'switching on' of exoprotein synthesis corresponds to conditions of nutrient limitation, characteristic of post-exponential batch cultures. One would therefore expect chemostat cultures, grown under specific nutrient limitations, to produce extracellular enzymes maximally. Indeed, it has been shown that proteinase production by *B. licheniformis* is maximal under conditions of nitrogen limitation (Wouters and Bieysman, 1977), but the observed specific association between proteinase production with nitrogen limitation on the one hand and alkaline phosphatase production with phosphorus limitation on the other indicates that specific control mechanisms additional to that proposed by Coleman *et al*. (1975) probably operate for particular extracellular enzymes. There is ample evidence for general catabolite repression of proteinase synthesis by easily-utilizable sources of carbon (e.g. McDonald and Chambers, 1966; Litchfield and Prescott, 1970*a*; Boethling, 1975*b*; Juffs, 1976), as well as more specific, end-product repression by amino acids (Glenn, 1976). The mechanism of end-product control is unclear at present but repression of cell wall proteinase synthesis in *Streptococcus cremoris* appears to be mediated *via* a transcriptional inhibitor of specific mRNA; the concentration of the inhibitor is directly related to intracellular amino acid concentrations (F. A. Exterkate, personal communication). However, inhibition by glucogenic amino acids may operate in the same way as catabolite repression (e.g. via the growth rate / RNA transcription model).

In some bacteria, amino acids and/or peptides appear to induce extracellular proteinase formation. This presumably accounts for those proteinases produced during exponential growth. Synthesis of the extracellular proteinase of *Micrococcus freudenreichii* was increased by certain amino acids, but repressed by compounds regarded only as sources of carbon (McDonald and Chambers, 1966). These authors considered that their observations indicated that amino acids were themselves utilized as carbon sources rather than as units for protein synthesis. Other workers have reported proteinase induction by proteins, e.g. *Serratia* proteinase (Murakami *et al*., 1969), but a recent study by Loriya *et al*. (1977) has shown that the true inducer of *Serratia marcescens* proteinase is leucine, produced by the action of basal enzyme on albumin. *Aeromonas proteolytica* produces extracellular aminopeptidase and endopeptidase. The latter is induced by peptides and repressed by some individual amino acids and by easily metabolized sources of carbon (Litchfield and Prescott, 1970*a*). Control of the aminopeptidase appeared to be distinct and more complex, although peptide induction is also involved (Litchfield and Prescott, 1970*b*). Other examples of extracellular proteinases or aminopeptidases induced by amino acids and/or peptides have been reported by Hofsten and Tjeder (1965) and Daatselaar and Harder (1974).

In addition to the effects of nutrients on proteinase production, physical factors such as pH, temperature, and even hydrostatic pressure can affect amounts of proteinase secreted by some bacteria. For example, *Ps. aeruginosa* produces increasing amounts per unit of growth with increasing incubation temperature up to 30 °C, but *Ps. fluorescens* strains showed increased production only up to 20 °C, and produced very little proteinase at 30 °C (Juffs, 1976). In a marine species of *Vibrio*, gelatinase production increased with increasing growth temperature but decreased with increasing hydrostatic pressure (Weimer and Morita, 1974). However, definitive reports of such phenomena are lacking and the mechanisms by which physical factors exert their effects, whether directly or indirectly, remain unknown.

8 SUMMARY AND CONCLUSIONS

Bacteria appear unable to take up proteins directly and instead they utilize them nutritionally by secreting proteinases outside the cell membrane either as free enzymes or as enzymes bound at sites within the cell envelope; the products of such proteinase action are transportable amino acids and peptides. The specificities of bacterial proteinases as a group are such that even the most resistant proteins (e.g. keratin) can be degraded by some of the enzymes. Proteinases representative of the serine, thiol, metal chelator-sensitive, and acid types are produced by bacteria, although the last group is not well represented. Secretion of the enzymes does not appear to involve the transport of completed proteins but rather their simultaneous synthesis and extrusion, with folding into the active form only occurring outside the cell envelope. Control mechanisms for exoprotein (including proteinase) synthesis and secretion during the stationary phase of growth have been proposed but secretion during exoponential growth has received less attention. It is not known specifically how proteinase induction operates, especially since amino acids are reported to act as inducers of proteinases in some bacteria, rather than acting in their more predictable role as end-product inhibitors. The function of many bacterial extracellular proteinases can be seen as simply nutritional in that they act as scavenger enzymes, providing easily transported carbon and nitrogen sources from proteins, but they can also aid the penetration of physical barriers by invasive, or potentially invasive bacteria such as clostridia or pseudomonads, respectively. Additional functions for *Bacillus* proteinases relate to sporulation.

REFERENCES

Abbas-Ali, B. and Coleman, G. (1977). Nutritional shifts and their effect on the secretion of extracellular proteins by *Staphylococcus aureus* (Wood 46). *Biochem. Soc. Trans.*, **5**, 420–422.

Adams, D. M., Barach, J. T. and Speck, M. L. (1975). Heat resistant proteases produced in milk by psychrotrophic bactéria of dairy origin. *J. Dairy Sci.*, **58**, 828–838.

Adams, D. M., Barach, J. T. and Speck, M. L. (1976). Effect of psychrotrophic bacteria from raw milk on milk proteins and stability of milk proteins to ultrahigh temperature treatment. *J. Dairy Sci.*, **59**, 823–827.

Aiyappa, P. S. and Harris, J. O. (1976). The extracellular metalloprotease of *Serratia marcescens*. I. Purification and characterization. *Mol. Cell Biochem.*, **13**, 95–100.

Alichanidis, E. and Andrews, A. T. (1977). Some properties of the extracellular protease produced by the psychrotrophic bacterium *Pseudomonas fluorescens* strain AR-11. *Biochim. Biophys. Acta*, **485**, 424–433.

Argyle, P. J., Mathison, G. E. and Chandan, R. C. (1976). Production of cell-bound proteinase by *Lactobacillus bulgaricus* and its location in the bacterial cell. *J. Appl. Bacteriol.*, **41**, 175–184.

Attias, J., Schlessinger, M. J. and Schlessinger, S. (1969). The effect of amino acid analogues on alkaline phosphatase formation in *E. coli* K-12. *J. Biol. Chem.*, **244**, 3810–3817.

Barach, J. T. and Adams, D. M. (1977). Thermostability at ultrahigh temperatures of thermolysin and a protease from a psychrotrophic Pseudomonas. *Biochim. Biophys. Acta*, **485**, 417–423.

Barach, J. T., Adams, D. M. and Speck, M. L. (1976). Stabilization of a psychrotrophic *Pseudomonas* protease by calcium against thermal inactivation in milk at ultrahigh temperature. *Appl. Environ. Microbiol.*, **31**, 875–879.

Blackburn, T. H. (1968). Protease production by *Bacteroides amylophilus* strain H18. *J. Gen. Microbiol.*, **53**, 27–36.

Blackburn, T. H. and Hullah, W. A. (1974). The cell-bound protease of *Bacteroides amylophilus* H18. *Can. J. Microbiol.*, **20**, 435–441.
Blobel, G. and Dobberstein, B. (1975). Transfer of proteins across membranes. I. Presence of proteolytically processed and unprocessed nascent immunoglobulin light chains on membrane-bound ribosomes of murine myeloma. *J. Cell Biol.*, **67**, 835–851.
Blobel, G. and Sabatini, D. D. (1971). Ribosome-membrane interactions in eukaryotic cells. In *Biomembranes*, Vol. 2 (L. A. Manson, ed.), Plenum Publishing, New York, pp. 193–195.
Boethling, R. S. (1975*a*). Purification and properties of a serine protease from *Pseudomonas maltophilia*. *J. Bacteriol.*, **121**, 933–941.
Boethling, R. S. (1975*b*). Regulation of extracellular protease secretion in *Pseudomonas maltophilia*. *J. Bacteriol.*, **123**, 954–961.
Both, G. W., McInnes, J. L., Hanlon, J. E., May, B. K. and Elliott, W. H. (1972). Evidence for an accumulation of messenger RNA specific for extracellular protease and its relevance to the mechanism of enzyme secretion in bacteria. *J. Mol. Biol.*, **67**, 199–207.
Broad, T. E. and Skerman, T. M. (1976). Partial purification and properties of extracellular proteolytic activity of *Bacteroides nodosus*. *N.Z. J. Agric. Res.*, **19**, 317–322.
Brown, S. and Coleman, G. (1975*a*). Stability of labelled messenger ribonucleic acid in *Bacillus amyloliquefaciens* during the phases of minimal and maximal extracellular enzyme formation. *J. Mol. Biol.*, **96**, 335–344.
Brown, S. and Coleman, G. (1975*b*). Messenger ribonucleic acid content of *Bacillus amyloliquefaciens* throughout its growth cycle compared with *Bacillus subtilis* 168. *J. Mol. Biol.*, **96**, 345–352.
Carrick, L., Jr and Berk, R. S. (1975). Purification and partial characterization of a collagenolytic enzyme from *Pseudomonas aeruginosa*. *Biochim. Biophys. Acta*, **391**, 422–434.
Cascieri, T., Jr and Mallette, M. F. (1976). Peptide utilization by *Pseudomonas putida* and *Pseudomonas maltophilia*. *J. Gen. Microbiol.*, **92**, 283–295.
Coleman, G., Brown, S. and Stormonth, D. A. (1975). A model for the regulation of bacterial extracellular enzyme and toxin biosynthesis. *J. Theor. Biol.*, **52**, 143–148.
Cousin, M. A. and Marth, E. H. (1977). Cheddar cheese made from milk that was precultured with psychrotrophic bacteria. *J. Dairy Sci.*, **60**, 1048–1056.
Daatselaar, M. C. C. and Harder, W. (1974). Some aspects of the regulation of the production of extracellular proteolytic enzymes by a marine bacterium. *Arch. Microbiol.*, **101**, 21–34.
Dahlquist, F. W., Long, J. W. and Bigbee, W. L. (1976). Role of calcium in the thermal stability of thermolysin. *Biochemistry*, **15**, 1103–1111.
Dainty, R. H., Shaw, B. G., de Boer, K. A. and Scheps, E. S. J. (1975). Protein changes caused by bacterial growth on beef. *J. Appl. Bacteriol.*, **39**, 73–81.
Diener, B., Carrick, L., Jr and Berk, R. S. (1973). *In vivo* studies with collagenase from *Pseudomonas aeruginosa*. *Infect. Immun.*, **7**, 212–217.
Drapeau, G. R. (1976). Protease from Staphylococcus aureus. *Methods Enzymol.*, **45**, (Part B) 469–476.
Dutson, T. R., Pearson, A. M., Price, J. F., Spink, G. C. and Tarrant, P. J. V. (1971). Observations by electron microscopy on pig muscle inoculated and incubated with *Pseudomonas fragi*. *Appl. Microbiol.*, **22**, 1152–1158.
Efstathiou, J. D. and McKay, L. L. (1976). Plasmids in *Streptococcus lactis*: evidence that lactose metabolism and proteinase activity are plasmid linked. *Appl. Environ. Microbiol.*, **32**, 38–44.
Exterkate, F. A. (1975). An introductory study of the proteolytic system of *Streptococcus cremoris* strain HP, *Neth. Milk Dairy J.*, **29**, 303–318.
Exterkate, F. A. (1976*a*). The proteolytic system of a slow lactic acid-producing variant of *Streptococcus cremoris* HP. *Neth. Milk Dairy J.*, **30**, 3–8.
Exterkate, F. A. (1976*b*). Comparison of strains of *Streptococcus cremoris* for proteolytic activities associated with the cell wall. *Neth. Milk Dairy J.*, **30**, 95–105.
Gertler, A. (1971). Selective, reversible loss of elastolytic activity of elastase and subtilisin resulting from electrostatic changes due to maleylation. *Eur. J. Biochem.*, **23**, 36–40.

Glenn, A. R. (1976). Production of extracellular proteins by bacteria. *Annu. Rev. Microbiol.*, **30**, 41–62.

Green, M. L. (1977). Reviews in the progress of Dairy Science: milk coagulants. *J. Dairy Sci.*, **44**, 159–188.

Greenawalt, J. W. and Whiteside, T. L. (1975). Mesosomes: membranous bacterial organelles. *Bacteriol. Rev.*, **39**, 405–463.

Hall, F. F., Kunkel, H. O. and Prescott, J. M. (1966). Multiple proteolytic enzymes of *Bacillus licheniformis. Arch. Biochem. Biophys.*, **114**, 145–153.

Hampson, S. E., Mills, G. L. and Spencer, T. (1963). Observations on a protease from *Proteus mirabilis. Biochim. Biophys. Acta*, **73**, 476–487.

Hartley, B. S. (1960). Proteolytic enzymes. *Annu. Rev. Biochem.*, **29**, 45–71.

Hauschild, A. H. W. (1965). Incorporation of ^{14}C from amino acids and peptides into protein by *Clostridium perfringens* type D. *J. Bacteriol.*, **90**, 1569–1574.

Heppel, L. A. (1971). The concept of periplasmic enzymes. In *Structure and Functions of Biological Membranes* (L. I. Rothfield, ed.) Academic Press, New York and London, pp. 223–247.

Hofsten, B. V. and Reinhammar, B. (1965). An extracellular proteolytic enzyme from a strain of *Arthrobacter*. III. Stability and substrate specificity. *Biochim. Biophys. Acta*, **110**, 599–607.

Hofsten, B. V. and Tjeder, C. (1965). An extracellular proteolytic enzyme from a strain of *Arthrobacter*. I. Formation of the enzymes and isolation of mutant strains without proteolytic activity. *Biochim. Biophys. Acta*, **110**, 576–584.

Islam, M. A. and Blanshard, J. M. V. (1973). Purification and properties of an extracellular proteolytic enzyme from *Bacillus cereus*. *J. Dairy Res.*, **40**, 427–440.

Juan, S. M. and Cazzulo, J. J. (1976). The extracellular protease from *Pseudomonas fluorescens. Experientia*, **32**, 1120–1122.

Juffs, H. S. (1976). Effects of temperature and nutrients on proteinase production by *Pseudomonas fluorescens* and *Ps. aeruginosa* in broth and milk. *J. Appl. Bact.*, **40**, 23–32.

Keen, N. T., Williams, P. H. and Walker, J. C. (1966). Characterization of a protease produced by *Pseudomonas lachrymans. Phytopathology*, **57**, 257–262.

Kelly, C. T. and Fogarty, W. M. (1976). Microbial alkaline enzymes. *Process Biochemistry*, **11**, 3–9.

Kessler, E. and Yaron, A. (1976). An extracellular aminopeptidase from *Clostridium histolyticum. Eur. J. Biochem.*, **63**, 271–287.

Klemme, J.-H. and Pfleiderer, C. (1977). Production of extracellular proteolytic enzymes by phototrophic bacteria. *FEMS Lett.*, **1**, 297–299.

Kraut, J. (1971). Subtilisin: X-ray structure. In *The Enzymes*, Vol. 3 (P. D. Boyer, ed.), Academic Press, New York and London, pp. 547–560.

Kreger, A. S. and Griffin, O. K. (1974). Physicochemical fractionation of extracellular cornea-damaging proteases of *Pseudomonas aeruginosa. Infect. Immuno.*, **9**, 828–834.

Lampen, J. O. (1974). Movement of extracellular enzymes across membranes. *Symp. Soc. Exp. Biol.*, **28**, 351–374.

Law, B. A. (1977). Dipeptide utilization by starter streptococci. *J. Dairy Res.*, **44**, 309–317.

Law, B. A. (1978). Peptide utilization by Group N streptococci. *J. Gen. Microbiol.*, **105**, 113–118.

Law, B. A., Andrews, A. T. and Sharpe, M. E. (1977). Gelation of ultra-high-temperature-sterilized milk by proteases from a strain of *Pseudomonas fluorescens* isolated from raw milk. *J. Dairy Res.*, **44**, 145–148.

Lazdunski, A., Murgier, M. and Lazdunski, C. (1975). Evidence for an aminoendopeptidase localized near the cell surface of *Escherichia coli. Eur. J. Biochem.*, **60**, 349–355.

Leighton, T. J., Doi, R. H., Warren, R. A. J. and Kelin, R. A. (1973). The relationship of serum protease activity to RNA polymerase modification and sporulation in *Bacillus subtilis. J. Mol. Biol.*, **76**, 103–122.

Li, M. F. and Flemming, C. (1967). A proteolytic pseudomonad from skin lesions of rainbow trout (*Salmo gairdnerii*). I. Characteristics of the pathogenic effects and the extracellular proteinase. *Can. J. Microbiol.*, **13**, 405–416.

Li, M. F. and Jordan, C. (1968). A proteolytic pseudomonad from skin lesions of rainbow trout (*Salmo gairdnerii*). II. Some properties of the proteinase. *Can. J. Microbiol.*, **14**, 875–880.

Litchfield, C. D. and Prescott, J. M. (1970*a*). Regulation of proteolytic enzyme production by *Aeromonas proteolytica*. I. Extracellular endopeptidase. *Can. J. Microbiol.*, **16**, 17–22.

Litchfield, C. D. and Prescott, J. M. (1970*b*). Regulation of proteolytic enzyme production by *Aeromonas proteolytica*. II. Extracellular aminopeptidase. *Can. J. Microbiol.*, **16**, 23–27.

Liu, T. Y. and Elliot, S. D. (1965). Activation of Streptococcal proteinase and its zymogen by bacterial cell walls, *Nature, Lond.*, **206**, 33–34.

Liu, T. Y. and Elliott, S. D. (1971). Streptococcal proteinase. In *The Enzymes*, Vol. 3 (P. D. Boyer, ed.), Academic Press, New York and London, pp. 609–647.

Liu, T. Y., Neumann, N. P., Elliot, S. D., Moore, S. and Stein, W. H. (1963). Chemical properties of streptococcal proteinase and its zymogen. *J. Biol. Chem.*, **238**, 251–256.

Loriya, Zh. K., Bryukner, B. and Egorov, N. S. (1977). Effect of amino acids on synthesis of exocellular protease in *Serratia marcescens*, *Microbiology*, *U.S.S.R.*, **6**, 31–35.

McDonald, I. J. (1965). Distribution of proteinase in cultures of a species of *Micrococcus* in synthetic medium. *Can. J. Microbiol.*, **11**, 693–701.

McDonald, I. J. and Chambers, A. K. (1966). Regulation of proteinase formation in a species of *Micrococcus*. *Can. J. Microbiol.*, **12**, 1175–1185.

McQuade, A. B. and Crewther, W. G. (1969). Activity against a synthetic substrate by a preparation of extracellular proteinase from *Serratia marcescens*. *Biochim. Biophys. Acta*, **191**, 762–764.

Malik, A. C. (1976). Heat stable proteases from strains of *Pseudomonas fluorescens* (in milk). *Diss. Abstr. Int. B*, **11**, 5499–5500.

Mandl, I. and Cohen, B. B. (1960). Bacterial elastase. I. Isolation, purification and properties. *Arch. Biochem. Biophys.*, **91**, 47–53.

Margaretten, W., Nakai, H. and Landing, B. H. (1961). Significance of selective vasculitis and the 'bone marrow' syndrome in *Pseudomonas* septicemia. *New Engl. J. Med.*, **265**, 773–776.

Markland, F. S., Jr and Smith, E. L. (1971). Subtilisins: primary structure, chemical and physical properties. In *The Enzymes*, Vol. 3 (P. D. Boyer, ed.), Academic Press, New York and London, pp. 561–608.

Matsubara, H. and Feder, J. (1971). Other bacterial, mold, and yeast proteases. In *The Enzymes*, Vol. 3 (P. D. Boyer, ed.), Academic Press, New York and London, pp. 721–795.

May, B. K. and Elliott, W. H. (1968). Characteristics of extracellular protease formation by *Bacillus subtilis* and its control by amino acid repression. *Biochim. Biophys. Acta*, **157**, 607–615.

Mayerhofer, H. J., Marshall, R. T., White, C. H. and Lu, M. (1973). Characterization of a heat-stable protease of *Pseudomonas fluorescens* P26, *Appl. Microbiol.*, **25**, 44–48.

Melachouris, N. and Tuckey, S. L. (1968). Properties of a milk-clotting microbial enzyme, *J. Dairy Sci.*, **51**, 650–655.

Merkel, J. R. and Sipos, T. (1971). Marine bacterial proteases. I. Characterization of an endopeptidase produced by *Vibrio* B-30. *Arch. Biochem. Biophys.*, **145**, 121–136.

Millet, J. (1969). Étude de la mégatériopeptidase prótease exocellulaire de *Bacillus megaterium*. I. Purification et propriétés générales. *Bull. Soc. Chim. Biol.*, **51**, 61–68.

Mitchell, W. M. (1968). Pseudocollagenase: a protease from *Clostridium histolyticum*. *Biochim. Biophys. Acta*, **159**, 554–557.

Mitchell, W. M. and Harrington, W. F. (1971). Clostripain. In *The Enzymes*, Vol. 3 (P. D. Boyer, ed.), Academic Press, New York and London, pp. 699–719.

Miyata, K., Maejima, K., Tomoda, K. and Isono, M. (1970*a*). *Serratia* protease. Part I. Purification and general properties of the enzyme. *Agric. Biol. Chem.*, *Tokyo*, **34**, 310–318.

Miyata, K., Tomoda, K. and Isono, M. (1970*b*). *Serratia* protease. Part II. Substrate specificity of the enzyme. *Agric. Biol. Chem.*, *Tokyo*, **34**, 1457–1462.

Morihara, K. (1963). *Pseudomonas aeruginosa* proteinase. I. Purification and general properties. *Biochim. Biophys. Acta*, **73**, 105–112.

Morihara, K. (1964). Production of elastase and proteinase by *Pseudomonas aeruginosa*. *J. Bacteriol.*, **88**, 745–757.

Morihara, K. (1974). Comparative specificity of microbial proteinases. *Adv. Enzymol. Relat. Areas Mol. Biol.*, **41**, 179–243.

Morihara, K., Oka, T. and Tsuzuki, H. (1967). Multiple proteolytic enzymes of *Streptomyces fradiae*. Production, isolation and preliminary characterization. *Biochim. Biophys. Acta*, **139**, 382–397.

Morihara, K. and Tsuzuki, H. (1967). Elastolytic properties of various proteinases from microbial origin. *Arch. Biochem. Biophys.*, **120**, 68–78.

Morihara, K. and Tsuzuki, H. (1975). *Pseudomonas aeruginosa* elastase: affinity chromatography and some properties as a metallo-neutral proteinase. *Agric. Biol. Chem., Tokyo*, **39**, 1123–1128.

Morihara, K., Tsuzuki, H., Oka, T., Inoue, H. and Ebata, M. (1965). *Pseudomonas aeruginosa* elastase. Isolation, crystallization, and preliminary characterization. *J. Biol. Chem.*, **240**, 3295–3304.

Morris, H. and Schlessinger, M. J. (1972). Effects of proline analogues on the formation of alkaline phosphatase in *E. coli. J. Bacteriol.*, **111**, 203–210.

Mull, J. D. and Callahan, W. S. (1965). The role of the elastase of *Pseudomonas aeruginosa* in experimental infection. *Exp. Mol. Pathol.*, **4**, 567–575.

Murakami, M., Fukunaga, K., Matsuhashi, M. and Ono, M. (1969). Stimulative effect of proteins on protease formation by *Serratia* sp. *Biochim. Biophys. Acta*, **192**, 378–380.

Murgier, M., Pelissier, C., Lazdunski, A. and Lazdunski, C. (1976). Existence, location and regulation of the biosynthesis of amino-endopeptidase in Gram-negative bacteria. *Eur. J. Biochem.*, **65**, 517–520.

Murgier, M., Pelissier, C., Lazdunski, A., Bernadac, A. and Lazdunski, C. (1977). Aminopeptidase from *Escherichia coli.* Unusual interactions with the cell surface. *Eur. J. Biochem.*, **74**, 425–433.

Nakajima, M., Mizusawa, K. and Yoshida, F. (1974). Purification and properties of an extracellular proteinase of psychrophilic *Escherichia freundii. Eur. J. Biochem.*, **44**, 87–96.

Nickerson, W. J. and Durand, S. C. (1963). Keratinase. II. Properties of the crystalline enzyme. *Biochim. Biophys. Acta*, **77**, 87–99.

Nickerson, W. J., Noval, J. J. and Robinson, R. S. (1963). Keratinase I. Properties of the enzyme conjugate elaborated by *Streptomyces fradiae. Biochim. Biophys. Acta*, **77**, 73–86.

Noval, J. J. and Nickerson, W. J. (1959). Decomposition of native keratin by *Streptomyces fradiae. J. Bacteriol.*, **77**. 251–263.

Piez, K. A. (1967). In *Treatise on Collagen*, Vol. I. (G. N. Ramachandran, ed.). Academic Press, New York, p. 207.

Pollock, M. R. (1962). Exoenzymes. In *The Bacteria*, Vol. IV (I. C. Gunsalus and R. Y. Stanier, eds.), Academic Press, New York and London, pp. 121–178.

Pollock, M. R. and Richmond, M. H. (1962). Low cyst(e)ine content of bacterial extracellular proteins: its possible physiological significance, *Nature, Lond.*, **194**, 446–449.

Porzio, M. A. and Pearson, A. M. (1975). Isolation of an extracellular neutral proteinase from *Pseudomonas fragi. Biochim. Biophys. Acta*, **384**, 235–241.

Prescott, J. M. and Wilkes, S. H. (1966). *Aeromonas* aminopeptidase: purification and some general properties. *Arch*. Biochem. Biophys., **117**, 328–336.

Prescott, J. M. and Wilkes, S. H. (1976). *Aeromonas* aminopeptidase. *Methods Enzymol.*, **45** (Part B), 530–543.

Prestidge, L., Gage, V. and Spizizen, J. (1971). Protease activity during the course of sporulation in *Bacillus subtilis. J. Bacteriol.*, **107**, 815–823.

Priest, F. G. (1977). Extracellular enzyme synthesis in the genus *Bacillus. Bacteriol. Rev.*, **41**, 711–753.

Ramachandran, G. N. (1967). In *Treatise on Collagen*, Vol. I (G. N. Ramachandran, ed.), Academic Press, New York, p. 103.

Randall, L. L. and Hardy, S. J. S. (1977). Synthesis of exported proteins by membrane-bound polysomes from *Escherichia coli. Eur. J. Biochem.*, **75**, 43–53.

Ray, L. E. and Wagner, F. W. (1972). Characteristics of an aminopeptidase activity from the cultural fluid of *Bacillus subtilis*. *Can. J. Microbiol.*, **18**, 853–859.

Redman, C. M. and Sabatini, D. D. (1966). Vectorial discharge of peptides released by puromycin from attached ribosomes. *Proc. Natl. Acad. Sci., U.S.A.*, **56**, 608–615.

Regnier, Ph. and Thang, M. N. (1972). Subcellular distribution and characterization of endo- and exo-cellular proteases in *E. coli. Biochimie*, **54**, 1227–1236.

Reysett, G. and Millet, J. (1972). Characterization of an intracellular protease in *Bacillus subtilis* during sporulation. *Biochem. Biophys. Res. Commun.*, **49**, 328–334.

Rippon, J. W. (1968). Extracellular collagenase produced by *Streptomyces madurae. Biochim. Biophys. Acta*, **159**, 147–152.

Rydên, A.-C., Lindberg, M. and Philipson, L. (1973). Isolation and characterization of two protease-producing mutants from Staphylococcus aureus, *J. Bacteriol.*, **116**, 25–32.

Scharmann, W. and Balke, E. (1974). Untersuchungen über eine protease (Elastase) von *Pseudomonas aeruginosa* I. *Hoppe-Seyler's Z. Physiol. Chem.*, **355**, 443–450.

Schoellmann, G. and Fisher, E., Jr (1966). A collagenase from *Pseudomonas aeruginosa. Biochim. Biophys. Acta*, **122**, 557–559.

Seifter, S. and Harper, E. (1971). The collagenases. In *The Enzymes*, Vol. 3 (P. D. Boyer, ed.), Academic Press, New York and London, pp. 649–697.

Seifter, S., Gallop, P. M., Klein, L. and Meilman, E. (1959). Studies on collagen II. Properties of purified collagenase and its inhibition. *J. Biol. Chem.*, **234**, 285–293.

Semets, E. V., Glenn, A. R., May, B. K. and Elliott, W. H. (1973). Accumulation of messenger ribonucleic acid specific for extracellular protease in *Bacillus subtilis* 168. *J. Bacteriol.*, **116**, 531–534.

Shankar, P. A. (1978). Interrelationships of *Streptococcus thermophilus* and *Lactobacillus bulgaricus* in yoghurt culture. Thesis, University of Reading.

Shoer, R. and Rappaport, H. P. (1972). Analysis of a *Bacillus subtilis proteinase mutant. J. Bacteriol.*, **109**, 575–583.

Siegel, S. and Awad, W. M. (1973). The proteolytic enzymes of the K-1 strain of *Streptomyces griseus* obtained from a commercial preparation (Pronase). IV. Structure–function studies of the two smallest serine endopeptidases; stabilization by glycerol during reaction with acetic anhydride, *J. Biol. Chem.*, **248**, 3233–3240.

Smith, W. P., Tai, P.-C., Thompson, R. C. and Davis, B. D. (1977). Extracellular labelling of nascent polypeptides traversing the membrane of *Escherichia coli. Proc. Natl. Acad. Sci. U.S.A.*, **74**, 2830–2834.

Stoeva, St. P. and Mesrob, B. K. (1977). Proteolytic and esterolytic activity of a milk-clotting protease (MCP) from *Bacillus mesentericus* strain 76. *FEBS Lett.*, **80**, 86–88.

Svendsen, I. (1976). Chemical modifications of the subtilisins with special reference to the binding of large substrates. A review. *Carlsberg Res. Commun.*, **41**, 237–291.

Takahashi, S. (1967). Isolation of a collagenolytic enzyme from *Mycobacterium tuberculosis. J. Biochem., Tokyo*, **61**, 258–259.

Thomas, T. D., Jarvis, B. D. W. and Skipper, N. A. (1974). Localization of proteinase(s) near the cell surface of *Streptococcus lactis. J. Bacteriol.*, **118**, 329–333.

Tarrant, P. J. V., Pearson, A. M., Price, J. F. and Lechowich, R. V. (1971). Action of *Pseudomonas fragi* on the proteins of pig muscle. *Appl. Microbiol.*, **22**, 224–228.

Vacheron, M.-J., Guinand, M., Arminjon, F. and Michel, G. (1977). Mise en evidence et separation d'une γ-D-glutamyl-(L)-*meso*-diaminopimelate endopeptidase et d'une L,D-carboxypeptidase exocellulaires a partir de *Bacillus sphaericus* 9602. *Biochimie*, **59**, 15–21.

Von Hippel, P. H. (1967). In *Treatise on Collagen*, Vol. I (G. N. Ramachandran, ed.), Academic Press, New York, p. 253.

Vosbeck, K. D., Greenberg, B. D. and Awad, W. M. (1975). The proteolytic enzymes of the K-1 strain of *Streptomyces griseus* obtained from a commercial preparation (Pronase). Specificity and immobilisation of aminopeptidase. *J. Biol. Chem.*, **250**, 3981–3987.

Wagner, F. W., Chung, A. and Ray, L. E. (1972). Characterization of the aminopeptidase from *Bacillus subtilis* as an extracellular enzyme. *Can. J. Microbiol.*, **18**, 1883–1891.

Weimer, M. S. and Morita, R. Y. (1974). Temperature and hydrostatic pressure effects on gelatinase activity of a *Vibrio sp.* and partially purified gelatinase. *Z. Allgem. Mikrobiol.*, **14**, 719–725.

Wilkes, S. H. and Prescott, J. M. (1976). *Aeromonas* neutral protease. *Methods Enzymol.*, **45**, (Part B), 404–415.

Williamson, W. T., Tove, S. B. and Speck, M. L. (1964). Extracellular proteinase of *Streptococcus lactis. J. Bacteriol.*, **87**, 49–53.

Wouters, J. T. M. and Bieysman, P. J. (1977). Production of some exocellular enzymes by *Bacillus licheniformis* 749/C in chemostat cultures, *FEMS Lett.*, **1**, 109–112.

Wretlind, B. and Wadström, T. (1977). Purification and properties of a protease with elastase activity from *Pseudomonas aeruginosa*. *J. Gen. Microbiol.*, **103**, 319–327.

Yamamoto, S. and Lampen, J. O. (1975). Membrane penicillinase of *Bacillus licheniformis* 749/C, a phospholipoprotein, *J. Biol. Chem.*, **250**, 3212–3213.

Yoshida, K., Hidaka, H., Miyado, S., Shibata, U., Saito, K. and Yamada, Y. (1977). Purification and some properties of *Bacillus sphaericus* protease. *Agric. Biol. Chem., Tokyo*, **41**, 745–754.

Microorganisms and Nitrogen Sources
Edited by J. W. Payne

Section 3.2

Transport and Utilization of Proteins by Fungi

BERNARD L. COHEN

Department of Genetics, University of Glasgow, Glasgow G11 5JS, UK

1 INTRODUCTION

Fungi are celebrated for their degradative capacities and extracellular proteases are among the most potent weapons in their armamentarium. Indeed there is probably no protein—from the sulphydryl-tanned keratin of skin, fur, and feather to the phenol-tanned sclerotin of arthropod cuticle—which is totally immune to fungal hydrolysis. This widespread capacity for extracellular proteolysis, which may be both useful and harmful to Man, accounts for the general interest of my topic. Scientific interest centres on the desire to understand the many aspects of cell physiology that are involved (Walsh, 1976). However, as will become apparent, very little is known about most aspects of protease production in fungi although much is known about a few aspects, especially the relatively uninteresting (!) enzymology of proteases.

Moreover it is unfortunately true that many contributions to the fungal protease literature suffer from methodological limitations that result in them being rather uninformative. A number of references to such works will be found listed in an appendix to the references.

2 THE EXTRACELLULAR PROTEASES OF FUNGI

Fungal extracellular proteases are important in industrial and food processing industries (Keay, 1971; Perlman, 1976). Largely for this reason there has been a very large number of publications describing the properties of individual enzymes or of the enzymes produced by particular organisms. This literature has been well reviewed (Matsubara and Feder, 1971) and was recently summarized for the Aspergilli in particular (Cohen, 1977).

1 The Enzymes Themselves

The classification of proteolytic enzymes is difficult, but the present system, based on type of action and catalytic mechanism, is both practical and probably recognizes natural groupings (see also Section 3.1.3.1). Two different types of proteolytic action have been recognized amongst fungal extracellular enzymes (Commission on Biochemical Nomenclature, 1972). These are: peptide exohydrolases such as aminopeptidase or carboxypeptidase (E.C.3.4.11–15) which remove the terminal amino acids of proteins or peptides, and peptide endohydrolases (E.C.3.4.21–24) which, with varying specificity, cleave internal peptide bonds.

The exohydrolases participate in, but cannot themselves rapidly complete, the digestion of protein. They therefore seem to be of only secondary importance as proteolytic enzymes. It should, however, be noted that they offer technical advantages in that sensitive and specific chromogenic substrates exist by which these enzymes may be detected or assayed and which may form the basis of mutant selection methods (Wolf and Fink, 1975; Jones, 1977). I have not pursued in the literature the interesting question of whether or not exohydrolases are secreted *only* by fungi which also secrete endohydrolases.

The endopeptidases, commonly called proteinases or proteases (Barrett, 1977), are recognized by their ability to cleave peptide bonds distant from the termini of polypeptide chains and thus to solubilize insoluble proteins. They are further classified by catalytic mechanism, usually determined indirectly through reactivity towards inhibitors that react with particular residues in the active site region. This approach differentiates four broad classes of proteases. One group (E.C.3.4.21) have a reactive serine residue in the active site and may be inhibited by diisopropyl-fluorophosphate or phenylmethylsulphonyl fluoride. The thiol proteases (E.C.3.4.22) have a reactive sulphydryl inhibitable by *p*-chloromercuribenzoate, *N*-ethyl maleimide or iodoacetamide. The carboxylic proteases (E.C.3.4.23) are inhibited by various bacterial polypeptides such as pepstatin, whilst the metallo-proteases (E.C.3.4.24) require a zinc or other metal ion and may be inhibitable by chelating agents. In general the serine enzymes are most active under alkaline conditions with an optimum around pH 8 but with appreciable activity above pH 9 and below pH 5. The thiol and metallo-enzymes are generally most active from pH 4 to about pH 8 and are often known as 'neutral' proteases. The carboxylic or acid proteases are active below pH 4. Thus between them, these enzymes facilitate polypeptide hydrolysis between pH values of 10 and 2.

2 Which Fungi Produce Extracellular Proteases?

Representatives of each of the major taxa have been shown to be capable of protease production during a non-autolytic phase of growth. Amongst Basidiomycetes, *Trametes sanguinea* releases an acid protease; amongst Phycomycetes, *Mucor* and *Sorangium* release acid and alkaline proteases respectively, whilst a large number of Ascomycetes and Fungi Imperfecti (especially Moniliales) release enzymes of all functional classes (Matsubara and Feder, 1971).

Within major taxa, only one comprehensive survey for protease production has been reported (Ahearn *et al.*, 1968) and even this is rather uninformative since it was designed without regard to the (then unknown) existence of induction and repression. In this study *c.* 800 yeast isolates were cultured on media containing beef extract and tryptophan or Yeast Nitrogen Base, both with added casein, gelatin, or albumin. Thus a considerable amount of low molecular weight organic and inorganic nitrogen was present and the many reported instances of absence of protease production (e.g. by four unspecified isolates of *Saccharomyces carlsbergensis*) may only be due to sustained nitrogen metabolite repression. Support for this interpretation is provided by Maddox and Hough (1970) who cultured a defined strain of *S. carlsbergensis* on casein or *N*-acetyltyrosine ethyl ester as sole nitrogen source and observed substantial extracellular protease production, presumably via derepression (see Section 3.2.3 on regulation).

Despite the methodological limitations which may be seen with hindsight to affect the yeast survey, 31 out of 76 species tested yielded at least some proteolytic isolates and 21 species were uniformly proteolytic.

3 Proteases and Fungal Taxonomy

Two approaches are possible: one via comparison of protein sequence studies, the other by much less sensitive indicators of homology such as kinetic, electrophoretic and chromatographic properties. The only results so far published relate to a few Aspergilli. Thus, partial sequence analyses (i.e. by peptide 'fingerprinting') have been made for the serine proteases of the four closely related species: *A. flavus, A. oryzae, A. sulphureus,* and *A. sojae.* The pretext for these studies was that five independent groups of workers had each been studying these enzymes and comparisons using uniform conditions were needed to clarify the results. The comparisons demonstrated that the enzymes of *A. flavus* and *A. oryzae* were probably identical; the serine protease of *A. sojae* was similar but not identical, whilst that of *A. sulphureus* was homologous but less related (Bretschneider *et al.,* 1971; Turkova *et al.,* 1972). It seems unfortunate that most of the strains used in these studies are 'private' isolates rather than from international culture collections.

An electrophoretic comparison within the *flavus* group using mobility of alkaline proteases in polyacrylamide gels has been reported by Nasuno (1971, 1972). The proteases of nine species (Raper and Fennell, 1965) were divided into four groups on electrophoretic criteria and the results were considered generally to support the relationship suggested by other taxonomic criteria.

Because of the diversity of experimental materials and methods used by students of *Aspergillus* proteases, two questions which cannot yet be answered are: (a) whether all members of the genus produce proteases of each of the functional classes, and (b) whether in all members of the genus the extracellular proteases are regulated (as in *A. nidulans*) by repression alone. My current studies are directed towards obtaining partial answers to these questions, using internationally available strains representative of most of the main subdivisions of the genus. Results so far indicate that the answers to both questions are generally affirmative (Cohen, 1977).

The widespread distribution of protease-producing ability coupled with the generally saprophytic way of life of fungi suggests that protease production is an ancestral function rather than one which has evolved independently several times.

4 What Classes of Protease do Producers Produce?

Although the literature on fungal proteases is vast it is curiously difficult to give a secure answer to this question. The reasons are not difficult to discern. Many studies antedate the relatively recent clarification of protease classification; few investigators (including myself, for instance) simultaneously investigate both acid and neutral/alkaline proteases; many investigators have failed to adopt defined culture media, growth conditions, and internationally available strains; a regrettable number of studies have taken as their starting material some partially purified industrial preparation. These failings are variously evident in many otherwise excellent studies of extracellular proteases of the industrially important *oryzae–flavus–sojae* group of Aspergilli (for example, Kundu *et al.,* 1968; Nordwig and Jahn, 1968; Nasuno, 1972; Klapper *et al.,* 1973; Turkova *et al.,* 1976), but they are quite general. In consequence the most that can be said is that probably most protease-producing fungi secrete enzymes of each main type under appropriate conditions but that there may be some specialization, e.g. conditions under which members of the *Aspergillus niger* group release alkaline protease have not yet been found (B. L. Cohen, unpublished observations). The apparently sharp dichotomy between those species which do and those which do not secrete acid proteases may be real in some cases, but it may often have an artefactual basis since typical enzymes of this class are irreversibly inactivated above pH 6 whilst the neutral/alkaline proteases are inactivated below about pH 4 (Matsubara and Feder, 1971). Thus fungi such as *A. niger* which acidify their culture medium will not normally be found to produce alkaline proteases, whilst investigators who buffer their culture media above pH 6 will never detect an acid protease even if they subsequently run assays at an appropriate pH.

5 How Clear is the Distinction between Intracellular and Extracellular Proteases?

What we presently understand of the nature, distribution, and functions of the intracellular endoproteases is discussed elsewhere in this volume (see Sections 3.3 and 3.4). Here it may be useful to point out that equivalent studies of intra- and extracellular proteases have not yet been made in any single species. Thus relations between these enzymes remain to be explored. The principal difficulty of this subject is that of determining the intracellular distribution and activity of the extracellular enzymes between the moments of synthesis and export, whilst distinguishing them from any coexisting intracellular enzymes. Moreover, there is a difficulty in determining whether an apparently intracellular enzyme is genuinely so located or may also or instead be periplasmic or intramural. As has been indicated elsewhere (Cohen, 1977), protoplast studies may help here. In this whole field there is a need for better, preferably electron-dense specific histochemical reagents which would permit the localization with high certainty of any endoprotease in the presence of others. However, the nature of fungal endoprotease activity makes the discovery or development of such reagents rather improbable: the enzymes have rather lower bond cleavage specificities than comparable mammalian enzymes.

6 Synthesis and Secretion of Extracellular Proteases

A reasonably complete knowledge of the ultrastructural organization of extracellular protease synthesis and secretion exists only for the pancreatic enzymes such as trypsinogen (Bloebel, 1977; Jameson and Palade, 1977) (see also Section 3.1.6). Whilst one may anticipate that comparable processes in fungi will involve similar forms of organization, it must be admitted that the evidence is scanty. Nor does this conclusion apply exclusively to fungal proteases; it is true of any fungal macromolecular product. In particular I know of no publication which establishes even such basic facts as whether extracellular protease synthesis takes place on the ribosomes of the endoplasmic reticulum or whether it is restricted to other particular locations in the cell. There is, again, no evidence on the possible role of the Golgi apparatus (or its equivalent) in packaging and processing fungal extracellular proteases. Evidence indicating protease secretion in *Neurospora crassa* by reverse micropinocytosis was put forward (Matile, 1965; Matile *et al.,* 1965) but later qualified on the grounds that the enzyme studied was probably an intracellular protease (Heiniger and Matile, 1974). Although Chang and Trevithick (1974) appear to have recognized the same activity as an extracellular enzyme in the same organism, the lack of specific assays and electron-dense histochemical reagents and the absence of consistent use of electrophoresis preclude reinterpretation of these studies sufficiently to accept Matile's original, attractive proposition. Further evidence suggestive of reverse pinocytosis has been presented for secretion of non-proteolytic hydrolases of *Sclerotinia fructigena* (Hislop *et al.,* 1974) but the question remains open at present.

It is not even certain where in a mycelial colony active protease secretion occurs. Yanagita and Nomarchi (1967), in an interesting and neglected paper, argue that acid protease of *Aspergillus niger* is secreted by a region of approximately constant length not closer than 40 μm to the hyphal tip, whilst Chang and Trevithick (1974) provide evidence that secretion is confined largely to the region ahead of the region of wall rigidification and thus (in *Neurospora crassa*) within 2 mm of the hyphal tip. It remains possible, however, that relatively small and unglycosylated extracellular proteases could be secreted and pass through even rigidified walls. An indication that protease synthesis, if not secretion, may not be confined to the immediately subapical region is provided by the observations of Gealt and Axelrod (1974) who found that specific activity for protease production declines with the onset of conidiation. With the reservation that their calculation does not take account of the mass of the conidial apparatus and thus underestimates the specific protease-producing capacity of the remaining mycelium, this too points towards an extensive region behind the apex as being the most active secretion zone. Conidiation commences several millimetres behind the apex, corresponding to the extent of hyphal elongation in *c*. 22h.

3 REGULATION OF EXTRACELLULAR PROTEASE PRODUCTION: REPRESSION RULES—OK!

1 Four Patterns of Protease Regulation

The development of an understanding of the regulation of protease production (i.e. of synthesis and secretion) has been retarded by delayed application of modern concepts. Thus, although results showing sulphur- and nitrogen-repression of protease synthesis were obtained by Tomonaga *et al.,* (1964) and by various other workers subsequently (e.g. Matile, 1965; Maddox and Hough, 1970; Imshenetskii and Popova, 1970; Shinmyo *et al.,* 1971; Imshenetskii *et al.,* 1971), it was not until 1972 that these effects began to be related to catabolite repression as a general phenomenon, and it became clear that in *A. nidulans* protease production could be regulated by repression alone without need of any inducer (Cohen, 1972, 1973*a,b*). Meanwhile, in *Neurospora crassa* a different pattern of regulation was being explored in which both repression by carbon, nitrogen, and sulphur catabolites and also induction by substrate were involved (Drucker, 1972, 1973, 1975; Hanson and Marzluf, 1973, 1975; Cohen *et al.,* 1975; Cohen and Drucker, 1977). As the discussion below will make clear, many publications which mention protease regulation and frequently claim induction, invariably fail to give the crucial experimental results which would allow one to distinguish genuine induction from catabolite derepression (e.g. Ahearn *et al.,* 1968; Maddox and Hough, 1970; O'Sullivan and Mathison, 1971; Abdelal *et al.,* 1977; Meyers and Ahearn, 1977).

Despite recent advances in analysis of protease regulation in fungi, there is still far to go before any comprehensive understanding of the regulatory mechanisms or their genetic basis is achieved and further developments must await other advances in the field. Nevertheless it may be helpful to attempt a somewhat theoretical discussion of the possible patterns of protease regulation.

There are, hypothetically, four different mechanisms by which the production (i.e. synthesis *and* secretion) of an extracellular enzyme could be controlled. These differ importantly in their requirements for the recognition of potential substrate molecules and in their energy cost to the producing organism.

(i) Constitutive synthesis with regulated secretion

In one conceivable pattern of protease regulation, enzyme *synthesis* would be constitutive but *secretion* would be regulated by induction and/or repression according to metabolic need. There is no evidence for such a mechanism and this is not surprising since it would seem to be energetically costly. Indeed what little evidence there is indicates *de novo* synthesis of protease in response to the onset of conditions favourable to protease production (Tomonaga, 1966; Heiniger and Matile, 1974) and the absence of any store of unsecreted extracellular proteases in repressed (non-secreting) mycelium (Cohen, 1973*b*).

Accordingly the discussion which follows will assume tight coupling between synthesis and secretion, i.e. either that synthesis is regulated but secretion is constitutive, or that synthesis and secretion are coordinately controlled. Thus, the operative steps in protease regulation may be imagined to affect protease structural gene transcription or mRNA translation.

(ii) Constitutive synthesis and secretion

In the case of constitutive synthesis, control and the ability to recognize substrate are absent: extracellular enzyme is synthesized at all times and there is concomitant secretion. So far as I am aware such a system has never been unequivocally described in any free-living, wild-type fungus, nor would such an energetically unfavourable plan be expected to evolve.

Constitutive protease production might, however, be anticipated to typify obligate 'proteotrophs' such as dermatophytes (if these *are* obligate proteotrophs) and other fungi highly adapted as pathogens of both plants and animals, especially if restricted to reliably protein-rich environments. The evidence for constitutivity is not at present clear in any case because it requires that the organism be capable of unrestricted growth and protease production on defined media in the absence of proteins or polypeptides, a requirement that could not be met by any truly obligate proteotroph. In the few reported studies on protease production by dermatophytes or other pathogens, this test has apparently not been made (O'Sullivan and Mathison, 1971; Minocha *et al.*, 1972; Cheung and Maniotis, 1973).

(iii) Induced synthesis with constitutive secretion

Induction of an enzyme implies the existence of a molecular mechanism whereby the subtrate, or some other 'effector' molecule metabolically related to substrate, serves to initiate *de novo* synthesis of the induced enzyme (Jacob and Monod, 1961; Yagil, 1975). Note, however, that in cells which have no provision to store newly-synthesized extracellular enzymes, feedback inhibition of synthesis by accumulation of such enzymes might occur. In such a situation, control of *secretion* by induction would be indistinguishable from control of *synthesis.* Exoenzyme control by induction is efficient, but more complex than control by repression. Presumably, therefore, it occurs only where it confers (or at some time conferred) an advantage.

Hypothetically, two mechanisms of exoenzyme induction could exist: distant or local. In *distant induction,* the presence of a potentially digestible but non-diffusible source of (substrate) protein in the environment would be detected by the receipt of a reporter molecule such as a peptide or amino acid released by interaction of a putative explorer molecule, such as an exoprotease, with substrate. The reporter could be a fragment either of the substrate or of the explorer, release of which would probably be constitutive. If such a mechanism exists it could be recognized experimentally as induced protease synthesis and/or release in response to an external inducer that is separated from the cells by some barrier sufficient to ensure spatial separation of substrate and fungus whilst permitting diffusion of explorer and reporter molecules.

So far as I am aware the possible existence of such a mechanism has not been tested experimentally.

In *local induction* there would be recognition of the presence of freely diffusable substrate protein only when it is in contact with the cell membrane, implying freedom to penetrate the cell wall layers. This form of induction has been postulated to be involved in the induction of protease production by bovine serum albumin and other soluble proteins in *Neurospora crassa.* The reporter-group or effector in this induction may be a peptide fragment resulting from activation of a membrane-bound zymogen form of extracellular protease, and investigation of this system is continuing (Drucker, 1975; H. Drucker, personal communication).

Several other investigations claim to have identified induction of protease production, but in each case the evidence consists of the production of protease when cultured in the presence of protein as sole carbon and/or nitrogen source and an answer to the essential question whether protease is also produced by carbon- or nitrogen-starved cells in the *absence* of substrate protein is not reported. In a true induction system the carbon, nitrogen, or sulphur starvation would *not* induce protease production (as in *Neurospora crassa,* see Cohen *et al.,* 1975). Those examples where induction is claimed but not demonstrated should be treated conservatively and be considered *pro tem* to fall into the following category, from which they have not been experimentally distinguished.

(iv) Derepressed synthesis with constitutive secretion

In those few fungi (apart from *N. crassa*) in which the proper experimental work has been done, regulation of protease production solely by derepression has proved to be the rule. Even in *N. crassa*, induction of protease synthesis is balanced against repression.

Derepression of protease synthesis and secretion entails the lifting by nutrilite limitation of carbon, nitrogen, or sulphur catabolite repression. It is the simplest conceivable system of regulation for extracellular catabolic enzymes. The genetic and biochemical bases of carbon, nitrogen, and sulphur catabolite repression have been the subject of considerable study and are reviewed elsewhere (Metzenberg, 1972; Arst and Bailey, 1977; Pateman and Kinghorn, 1977). In those cases where catabolite repression of protease production has been established, the evidence—while not watertight—indicates control of synthesis rather than control of secretion.

Proteases may be unique amongst catabolite-repressed molecules in being subject to all three forms of repression, and therefore studies of protease regulation may possibly be the most favourable to throw light upon the complex interrelations of these mechanisms. In both *A. nidulans* and *N. crassa,* derepression for any *one* of the three nutrilites is sufficient to permit protease synthesis, so there must hypothetically exist a single point at which all three systems of repression converge. The possibility that the *areA* regulatory gene product of *A. nidulans* is involved in such a triple function has not been properly investigated, although at least one allele (*xprD1*), later shown to be at this locus (Arst and Cove, 1973), is insensitive to carbon, nitrogen, and sulphur repression of protease synthesis (Cohen, 1973*b*).

Studies of catabolite repression have also been undertaken in *Saccharomyces cerevisiae* (e.g. Grenson *et al.,* 1974), but unfortunately this organism is not an active producer of extracellular proteases and we are unable to make a potentially interesting intergeneric comparison (see also Section 3.3).

There have been two cases in which cell-derived molecules with apparent regulatory effects on fungal protease production have been recognized, but unfortunately neither has been followed up. Shinmyo *et al*. (1971) extracted an 'internal inducer' of acid protease synthesis from *A. niger*. Although their report speaks of induction of this protease, it is in fact another case where the cultural conditions used do not allow a definite discrimination to be made between induction and derepression of protease synthesis. In the second case (Drucker, 1972), an inducing factor was obtained from culture filtrates of induced *N. crassa*. The implications of such autogenous inducers for protease regulation generally would be interesting, but remain obscure.

2 Speculations on the Evolution of Protease Regulation

Assuming that synthesis of an unnecessary enzyme is an expensive, i.e. non-adaptive, luxury, some very gross and naive simplifications can be used

Table 1 Predicted adaptive values‡ of different extracellular protease regulatory mechanisms under different conditions of nutrient supply

Regulatory mechanism	Quantity and distribution of substrate proteins in typical habitat			
	Little† Uniform	Much Uniform	Little Non-uniform	Much Non-uniform
A. Non-protein carbon, nitrogen and sulphur source concentrations not limiting for growth				
Constitutive (neither inducible nor repressible)	–	–	–	–
Inducible (not repressible)	–	–	–	–
Inducible and repressible	±	±	±	±
Repressible (not inducible)	±	±	±	±
B. Non-protein carbon and/or nitrogen and/or sulphur concentrations limiting for growth				
Constitutive (neither inducible nor repressible)	– to ±	++	– to ±	+
Inducible (not repressible)	± to +	++	± to +	+
Inducible and repressible	± to +	++	± to +	+
Repressible (not inducible)	±	++	+ (locally)	+

†Not enough when hydrolysed to support normal growth.
‡The criterion adopted is that of normal asexual growth and differentiation. A different and possibly more realistic criterion would be the ability to complete asexual or sexual sporulation despite starvation for low molecular weight nutrilites. ++ indicates very advantageous; + indicates advantageous; ± indicates more or less neutral; – indicates disadvantageous.

to derive predictions which relate patterns of protease regulation to fungal habitat (Table 1). It is of some slight comfort that this 'thought experiment' gives results broadly consistent with what is so far known. Only two systems of regulation—those so far characteristic of *N. crassa* and *A. nidulans*—seem likely to be of adaptive value over a range of possible habitats, and the difference between these two is probably only noticeable where very little protein is available. It would probably be stretching the plausibility of these hypotheses (and the circularity of the arguments) too far to equate burnt wood (habitat of *N. crassa*) with such a uniformly protein-poor habitat.

4 TRANSPORT AND UPTAKE OF SUBSTRATE PROTEINS AND POLYPEPTIDES

1 Transport of Substrate Protein? Closing the Gap

Despite the wide spread of the mycelial world, any individual mycelium may well find itself in nature to be in need of nutrients yet spatially separated from the nearest source of protein substrate. There may be a gap and we may ask whether and how this gap can be closed. The following three classes of mechanism for bringing fungal mycelium and substrate protein into close proximity may be worth some discussion.

(i) Vectorial growth

Can a mycelium grow towards a depot of *insoluble* protein? Is there chemotaxis? Presumably in nature, insoluble proteins rarely occur except in association with diffusable nutrient molecules, so that the possibility of vectorial growth towards *insoluble* protein may be dismissed as a purely academic question.

(ii) Gross substrate transport

It is difficult to conceive of any simple mechanism whereby distant particles—or even molecules—of protein could be physically transported towards itself by a fungal cell. Any such mechanism would entail cytoplasmic extensions or the assembly of extracellular contractile structures for which there is no precedent in fungi and which would probably be prohibitively expensive in metabolic energy for their potential substrate yield. Not surprisingly, no such mechanism has been reported. Note, however, that the raptorial fungi, with their hyphal traps, achieve a similar end by different means.

(iii) Creation of a concentration gradient

The possibility of creating net transport by manufacturing a concentration gradient does not seem to have been previously considered as a possible mechanism for substrate transport. Surprisingly, it probably occurs. Both H. Drucker (personal communication) and I have observed that under certain conditions of starvation, both *N. crassa* and *A. nidulans* precipitate otherwise soluble proteins onto the mycelial surfaces (without substantial pH change). For example, in stationary thin layer cultures of carbon-starved *A. nidulans* (strains *xprC1*, Cohen, 1973*a*), virtually all of the protein from one per cent milk medium was precipitated *on* the mycelium even when growth was quite sparse. A suspension of ferritin behaved similarly (Figure 1). Under related conditions, particles of the insoluble protein elastin and other proteins may become quite firmly attached to the surface of *N. crassa* mycelia (H. Drucker, personal communication). The physicochemical basis for these observations is unknown, although phenol-based tanning of the proteins seems a strong possibility. Whatever the mechanism, the effect is to precipitate a layer of very concentrated insoluble substrate protein in juxtaposition with the cell wall and thus to set up a concentration gradient which will lead to a diffusion-powered flow of soluble protein from distant sites towards the mycelium. This is a most remarkable phenomenon which might provide a novel adaptive *raison d'être* for some of the hitherto mysterious 'secondary' metabolic activities of fungi. How many other natural substrates are precipitated locally by fungal metabolites?

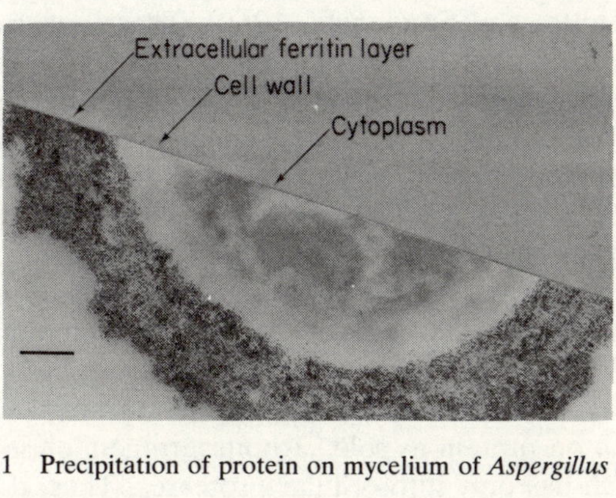

Figure 1 Precipitation of protein on mycelium of *Aspergillus nidulans*

The strain of *bi*A1 *w*A5 *ino*B2 *xpr*C1, was grown in stationary thin liquid layer culture for 48 h at 37 °C (Cohen, 1973*c*). The medium was minimal medium modified by providing ammonium dihydrogen phosphate (50 mM) as sole nitrogen source. Carbon sources were casein (10 mg ml^{-1} skim milk), ferritin (Pentex, from horse spleen, 10 mg ml^{-1}), and residual sugars, etc., in the skim milk. Mycelium was harvested by gentle filtration and prepared for electron microscopy. At time of harvest virtually all the ferritin was adherent to hyphae. The electron micrograph was made by Dr P. T. P. Oliver. Part of a transverse hyphal section is shown. The bar indicates 0.1 μm.

2 Uptake of Substrate Proteins and of the Products of Proteolysis

Having discussed ways in which potential substrate proteins may be concentrated in the vicinity of the fungal plasmalemma, the obvious next question to consider is whether such proteins (assuming they can penetrate the wall layers) are able to pass without digestion into any intracellular compartment. Here there are two possibilities to be discussed.

(i) Protein permeases

Elsewhere in this volume, Wolfinbarger considers the evidence relating to peptide uptake mechanisms in fungi (*N. crassa*) and concludes that only small peptides are taken up (Section 2.3.4). Similarly, *Saccharomyces cerevisiae* will not grow on pentapeptides (Marder *et al.*, 1977) (see Section 2.2.9). It is therefore rather unlikely that any sort of 'permease' channel exists for the uptake of intact protein molecules.

(ii) Micropinocytosis

However unlikely protein permeases may be, the occurrence of micropinocytosis seems to have been observed in at least one fungal cell (Wilson, 1973) and this might be thought to provide a more plausible mechanism for the internalization of undigested protein molecules. It has been sought once, in *S. carlsbergensis,* using fluorescein-labelled protein, but not found (Holter and Ottolenghi, 1960). These negative results, both the non-utilization of pentapeptides and the absence of micropinocytotic uptake of protein, might be explained away if polypeptide uptake and micropinocytosis of proteins were each regulated by catabolite repression. Since the experiments referred to were not conducted under derepressing conditions, neither of the mechanisms would have been active. Unlikely though the coexistence of protease production and protein imbibition may seem, it should also be noted that no-one's experiments on regulation of protease production would have discovered a limited amount of imbibition: careful substrate-balance experiments have not been done.

One simple way of testing for uptake of intact protein molecules would be to seek growth on protein as sole carbon, nitrogen, or sulphur source of an organism which has only intracellular proteases. Thus, if a yeast such as *S. cereviseae* were to prove able to grow—however poorly—there would be reason to suspect whole-protein uptake. I have recently made a similar test by constructing an amino acid auxotrophic strain of *A. nidulans* that also carries the *xprC* mutation, abolishing all extracellular protease production. The double mutant strain *argA1 xprC1* proved to be unable to use protamine as sole arginine source. However, even this failure to grow remains inconclusive as evidence against protein uptake for several reasons, including the usual weakness of negative results.

5 MISCELLANY

A number of topics are gathered under this heading because, although they do not obviously relate extracellular proteases to protein utilization, they may turn out to do so or are otherwise interesting.

1 Living Organisms as Sources of Protein. Role of Extracellular Proteases in Parasitism

The particular role of proteases in fungal pathogenesis has been dealt with in surprisingly few publications. For example, Kucera (1971) considered that the extracellular protease of the entomophagous fungus *Beauveria bassiana* was toxic, whereas Sikura and Bevzenko (1974) apparently found that virulence of the same organism was not related to the amount of protease produced. However, Samsinakova and Misikova (1973) concluded that protease and/or lipase were essential for pathogenesis of *B. bassiana* and that their role was to prepare the integument for chitinase attack. Budtz-Jorgensen (1974) found no relationship between proteolytic activity of *Candida albicans* and *C. tropicalis* isolates and the pathogenesis of denture stomatitis. The possibility of a correlation between proteolytic activity of dermatophytic fungi and skin pathogenesis has been investigated experimentally by Minocha *et al.* (1972) but unfortunately the proteases were obtained by extracting fungal mycelia after growth in Sabouraud's medium and the authors do not distinguish protease content from protease production. (A more informative study would have considered the extracellular rather than the intracellular proteases.)

That the hydrolytic enzymes of animal cells are typically confined within the lysosomal system has been well established, and Wilson (1973), in an extensive and stimulating review, has concluded that similar lysosomal systems do indeed exist in fungi. In the absence of direct evidence it may be useful to assume, until proved otherwise, that not only the intracellular proteases but also the extracellular proteases of fungi are organized in lysosomal systems. It thus becomes of little sense to consider proteases in isolation, for they represent only one of many hydrolytic enzymes which must be coordinately involved in pathogenic processes.

In passing we may note with surprise that there seems to be little evidence that haustoria or penetration pegs are specialized structures for extracellular enzyme secretion, and none that they actually produce proteases. It is possible, however, that it is more efficient for parasitic fungi to release toxins which activate host lysosomes, thus killing the host cell and solubilizing its macromolecules in advance of hyphal invasion.

Finally, one may point to those fungi which form hyphal traps for the capture of animal prey and predict that they, like fungal plant and animal pathogens, may yet be discovered to secrete extracellular proteases (and other hydrolases) after penetration. Speculation on the regulatory control of protease production in such parasitical situations suggests that it will prove either to involve induction by substrate and to be insensitive to catabolite repression or—more likely—will be a constitutive part of a developmental sequence of gene activity initiated by penetration or activation of the hyphal trap.

2 Morphogenesis

Little of value is known about possible interrelationships between extracellular proteases and morphogenesis. The proteaseless mutant *xprC1* of *A. nidulans* grows and differentiates normally, suggesting that the enzymes are dispensable and countering the suggestion by Gealt and Axelrod (1974) of an association between protease production and conidiation. A. J. Clutterbuck (personal communication) has also failed to fing any change of protease production in non-conidiating mutants of *A. nidulans*. Similarly, Wood and Goodenough (1977) found no changes in protease related to fruiting in *Agaricus bisporus,* although the enzymes assayed might possibly have been intracellular proteases released by autolysis.

The paucity of information under this heading may well be thought to reflect a genuine independence of extracellular proteases and morphogenesis. They are, after all, catabolic enzymes which may not be synthesized at detectable levels under many cultural conditions: it would be inexplicable if they also had essential morphogenetic functions.

3 Hyphal Exudate

In a recent paper, McPhee and Colotelo (1977) have drawn attention to the possibility that exudation of fluid from hyphae may have developmental and physiological significance. It may be of interest to note here that exudate droplets on cleistothecia of *A. nidulans* contain enough of the extracellular proteases to be analysed by electrophoresis (B. L. Cohen, unpublished observations).

4 For the Relief of Dyspepsia in Ants

The most unusual connection in which fungal proteases have been recognized is as faecal contents of the fungus-growing ant *Atta texana*

(Boyd and Martin, 1975). These were, apparently, intracellular rather than extracellular enzymes. The ability of proteases to survive extreme physical and chemical conditions has been noted elsewhere (Pringle, 1975).

ACKNOWLEDGEMENTS

I am grateful to Dr A. J. Clutterbuck for several valuable criticisms and for suggesting the term 'proteotroph' to describe organisms which are especially adapted to the use of proteins as nitrogen source.

REFERENCES

Abdelal, A. T. H., Kennedy, E. H. and Ahearn, D. G. (1977). Purification and characterisation of a neutral protease from *Saccharomycopsis lipolytica*. *J. Bacteriol.*, **130**, 1125–1129

Ahearn, D. G., Meyers, S. P. and Nichols, R. A. (1968). Extracellular proteases of yeasts and yeast-like fungi. *Appl. Microbiol.*, **16**, 1370–1374.

Arst, H. N. and Bailey, C. R. (1977). The regulation of carbon metabolism in *Aspergillus nidulans*. In *Genetics and Physiology of Aspergillus* (J. E. Smith and J. A. Pateman, eds), Academic Press, New York and London, pp. 131–146

Arst, H. N. and Cove, D. J. (1973). Nitrogen metabolite repression in *Aspergillus nidulans*. *Mol. Gen. Genet.*, **126**, 111–141.

Barrett, A. J. (1977). In *Proteinases in Mammalian Cells and Tissues*, North-Holland, Amsterdam, New York and Oxford, pp. 1–56.

Bloebel, G. (1977). Synthesis and segregation of secretory proteins: the signal hypothesis. In *International Cell Biology 1976–77* (B. R. Brinkley and G. E. Palade, eds), Rockefeller University Press, New York, pp. 318–325

Boyd, N. D. and Martin, M. M. (1975). Faecal proteinases of the fungus-growing ant *Atta texana*: their fungal origin and ecological significance. *J. Insect Physiol.*, **21**, 1815–1820.

Bretschneider, G., Nordwig, A., Mikes, O. and Turkova, J. (1971). Properties of proteases from *Aspergillus oryzae* and *Aspergillus flavus*: a comparative study. *Hoppe-Seyler's Z. Physiol. Chem.*, **352**, 1372–1376

Budtz-Jorgenson, E. (1974). Proteolytic activity of *Candida* spp. as related to the pathogenesis of denture stomatitis. *Sabouraudia*, **12**, 266–271

Chang, P. L. Y. and Trevithick, J. R. (1974). How important is secretion of exoenzymes through apical cell walls of fungi? *Arch. Microbiol.*, **101**, 281–293

Cheung, S.-S. C. and Maniotis, J. (1973). A genetic study of an extracellular elastin-hydrolysing protease in the ringworm fungus *Arthroderma benhamiae*. *J. Gen. Microbiol.*, **74**, 299–304

Cohen, B. L. (1972). Ammonium repression of extracellular protease in *Aspergillus nidulans*. *J. Gen. Microbiol.*, **71**, 293–299

Cohen, B. L. (1973*a*). The neutral and alkaline proteases of *Aspergillus nidulans*. *J. Gen. Microbiol.*, **77**, 521–528

Cohen, B. L. (1973*b*). Regulation of intracellular and extracellular neutral and alkaline proteases in *Aspergillus nidulans*. *J. Gen. Microbiol.*, **79**, 311–320

Cohen, B. L. (1973*c*). Growth of *Aspergillus nidulans* in a thin liquid layer. *J. Gen. Microbiol.*, **76**, 277–282

Cohen, B. L. (1977). The proteases of Aspergilli. In *Genetics and Physiology of Aspergillus* (J. E. Smith and J. A. Pateman, eds), Academic Press, New York and London, pp. 281–292

Cohen, B. L. and Drucker, H. (1977). Regulation of exocellular protease in *Neurospora crassa*: induction and repression under conditions of nitrogen starvation. *Arch. Biochem. Biophys.*, **182**, 601–613

Cohen, B. L., Morris, J. E. and Drucker, H. (1975). Regulation of two extracellular proteases of *Neurospora crassa* by carbon-, nitrogen- and sulphur-metabolite repression. *Arch. Biochem. Biophys.*, **169**, 324–330

Commission on Biochemical Nomenclature (1972). *Enzyme Nomenclature,* Elsevier, Amsterdam, pp. 228–250

Drucker, H. (1972). Regulation of exocellular proteases in *Neurospora crassa*: induction and repression of enzyme synthesis. *J. Bacteriol.*, **110**, 1041–1049

Drucker, H. (1973). Regulation of exocellular proteases in *Neurospora crassa*: role of *Neurospora* proteases in induction. *J. Bacteriol.*, **116**, 593–599

Drucker, H. (1975). Regulation of exocellular proteases in *Neurospora crassa*: metabolic requirements of the process. *J. Bacteriol.*, **122**, 1117–1125

Gealt, M. A. and Axelrod, D. E. (1974). Co-ordinate regulation of enzyme inducibility and developmental competence in *Aspergillus nidulans. Dev. Biol.*, **41**, 224–232

Grenson, M., Dubois, E., Piotrowska, M., Drillien, R. and Aigle, M. (1974). Ammonia assimilation in *Saccharomyces cerevisiae* as mediated by two glutamate dehydrogenases. *Mol. Gen. Genet.*, 128, 73–85

Hanson, M. and Marzluf, G. A. (1973). Regulation of a sulfur-controlled protease in *Neurospora crassa. J. Bacteriol.*, **116**, 785–789

Hanson, M. and Marzluf, G. A. (1975). Control of the synthesis of a single enzyme by multiple regulatory circuits in *Neurospora crassa. Proc. Natl. Acad. Sci., U.S.A.*, **72**, 1240–1244

Heiniger, U. and Matile, P. (1974). Protease secretion in *Neurospora crassa. Biochem. Biophys. Res. Commun.*, **60**, 1425–1432

Hislop, E. C., Barnaby, V. M., Shellis, C. and Laborda, F. (1974). Localization of α-L-arabinofuranosidase and acid phosphatase in mycelium of *Sclerotinia fructigena. J. Gen. Microbiol.*, **81**, 79–99

Holter, H. and Ottolenghi, P. (1960). Observations on yeast protoplasts. *C. R. Trav. Lab. Carlsberg,* **31**, 409–422

Imshenetskii, A. A. and Popova, N. V. (1970). Biosynthesis of proteinase by *Aspergillus terricola* on media with organic nitrogen compounds. Translated from *Mikrobiologiya,* **39**, 421–427

Imshenetskii, A. A., Kasatkina, I. D. and Zheltova, E. T. (1971). Repression of protease synthesis in *Aspergillus terricola* by exogenic amino acids. Translated from *Mikrobiologiya,* **40**, 439–444

Jacob, F. and Monod, J. (1961). Genetic regulatory mechanisms in the synthesis of proteins. *J. Mol. Biol.*, **3**, 318–356

Jameson, J. D. and Palade, G. E. (1977). Production of secretory proteins in animal cells. In *International Cell Biology, 1976–77* (B. R. Brinkley and K. R. Porter, eds), Rockefeller University Press. New York, pp. 308–317

Jones, E. W. (1977). Proteinase mutants of *Saccharomyces cerevisiae. Genetics*, **85**, 23–33

Keay, L. (1971). Microbial proteases. *Process Biochem.,* **6**, 17–21

Klapper, B. F., Jamieson, D. M. and Mayer, R. M. (1973). Factors affecting the synthesis and release of the extracellular protease of *Aspergillus oryzae* NRRL 2160. *Biochim. Biophys. Acta*, **304**, 513–519

Kucera, M. (1971). Toxins of the entomophagous fungus *Beauveria bassiana.* II. Effect of nitrogen sources on formation of toxic protease in submerged culture. *J. Invert. Pathol.,* **17**, 211–215

Kundu, A. K., Das, S., Manna, S. and Pal, N. (1968). Extracellular proteases of *Aspergillus oryzae. Appl. Microbiol.,* **16**, 1799–1801

McPhee, W. J. and Colotelo, N. (1977). Fungal exudates. I. Characteristics of hyphal exudates in *Fusarium culmorum. Can. J. Bot.,* **55**, 358–365

Maddox, I. S. and Hough, J. S. (1970). Proteolytic enzymes of *Saccharomyces carlsbergensis. Biochem. J.,* **117**, 843–852

Marder, R., Becker, J. M. and Naider, F. (1977). Peptide transport in yeast: utilization of leucine- and lysine-containing peptides by *Saccharomyces cerevisiae. J. Bacteriol.,* **131**, 906–916

Matile, P. (1965). Intrazelluläre lokalisation proteolytischer enzyme von *Neurospora crassa. Z. Zellforsch.*, **65**, 884–896

Matile, P., Jost, M. and Moor, H. (1965). Intrazelluläre lokalisation proteolytischer enzyme von *Neurospora crassa. Z. Zellforsch.*, **68**, 205–216

Matsubara, H. and Feder, J. (1971). Other bacterial, mold and yeast proteases. In *The Enzymes*, 3rd edn (P. D. Boyer, ed.), Academic Press, New York and London, pp. 721–795

Metzenberg, R. L. (1972). Genetic regulatory systems in *Neurospora. Annu. Rev. Genet.*, **6**, 111–132

Meyers, S. P. and Ahearn, D. G. (1977). Extracellular proteolysis by *Candida lipolytica. Mycologia,* **69**, 646–651

Minocha, Y., Pasricha, J. S., Mohapatra, L. N. and Kandhari, K. C. (1972). Proteolytic activity of dermatophytes and its role in the pathogenesis of skin lesions. *Sabouraudia,* **10**, 79–85

Nasuno, S. (1971). Polyacrylamide gel disc electrophoresis of alkaline proteinases from *Aspergillus* species. *Agric. Biol. Chem., Tokyo,* **35**, 1147–1150

Nasuno, S. (1972). Electrophoretic studies of alkaline proteinases from strains of *Aspergillus flavus* group. *Agric. Biol. Chem., Tokyo,* **36**, 684–689

Nordwig, A. and Jahn, W. F. (1968). A collagenolytic enzyme from *Aspergillus oryzae. Eur. J. Biochem.,* **3**, 519–529

O'Sullivan, J. and Mathison, G. E. (1971). The localisation and secretion of a proteolytic enzyme complex by the dermatophytic fungus *Microsporum canis. J. Gen. Microbiol.,* **68**, 319–326

Pateman, J. A. and Kinghorn, J. (1977). The regulation of nitrogen metabolism in *Aspergillus nidulans.* In *Genetics and Physiology of Aspergillus* (J. E. Smith and J. A. Pateman, eds), Academic Press, New York and London, pp. 203–242

Perlman, D. (1976). The fermentation industries. *Am. Soc. Microbiol. News,* **43**, 82–89

Pringle, J. R. (1975). Methods for avoiding proteolytic artefacts in studies of enzyme and other proteins from yeasts. In *Methods in Cell Biology,* Volume XII (D. M. Prescott, ed.), Academic Press, New York and London, Chap. IX

Raper, K. B. and Fennell, D. I. (1965). *The Genus Aspergillus,* The Williams and Wilkins Company, Baltimore

Samsinakova, A. and Misikova, S. (1973). Enzyme activities in certain entomophagous representatives of Deuteromycetes (Moniliales) in relationship to their virulence. *Ceska Mykologie,* **27**, 55–60

Shinmyo, A., Tahara, T., Matsumoto, H. and Terui, G. (1971). Kinetic studies on enzyme production by microbes. VIII. Autoinduction of acid protease in *Aspergillus niger. J. Ferment. Technol.,* **49**, 535–543

Sikura, A. N. and Bevzenko, T. M. (1974). Activity of some enzymes of the fungus *Beauveria bassiana* Vuill. under deep cultivation conditions. Abstracted from *Mikologiya i Fitopatologiya,* **8**, 65–67

Tomonaga, G. (1966). Preferential synthesis of extracellular protease by *Aspergillus niger* in sulfur deficiency. *J. Gen. Appl. Microbiol.,* **12**, 267–276

Tomonaga, G., Ohama, H. and Yanagita, T. (1964). Effect of sulfur compounds on the protease formation by *Aspergillus niger. J. Gen. Appl. Microbiol.,* **10**, 373–386

Turkova, J., Valentova, O. and Coupek, J. (1976). Isolation of aminopeptidase from *Aspergillus flavus. Biochim. Biophys. Acta,* **420**, 309–315

Turkova, J., Mikes, O., Hayashi, K., Danno, G. and Polgar, L. (1972). Alkaline proteinases of the genus *Aspergillus. Biochim. Biophys. Acta,* **257**, 257–263

Walsh, K. A. (1976). Unifying concepts among proteases. In *Proteases and Biological Control* (E. Reich, D. B. Rifkin and E. Shaw, eds), Cold Spring Harbor Laboratories, Cold Spring Harbor, Maine, pp. 1–11

Wilson, C. L. (1973). A lysosomal concept for plant pathology. *Annu. Rev. Phytopathol.,* **11**, 247–272

Wolf, D. H. and Fink, G. R. (1975). Proteinase C (carboxypeptidase) mutant of yeast. *J. Bacteriol.,* **123**, 1150–1156

Wood, D. A. and Goodenough, P. W. (1977). Fruiting of *Agaricus bisporus.* Changes in extracellular enzyme activity during growth and fruiting. *Arch. Microbiol*., **114**, 161–165

Yagil, G. (1975). Quantitative aspects of protein induction. *Curr. Topics Cell Regul.,* **9**, 183–236

Yanagita, T. and Nomarchi, Y. (1967). Kinetic analysis of the region of protease formation in the hypha of *Aspergillus niger. J. Gen. Appl. Microbiol.,* **13**, 227–235

APPENDIX TO REFERENCES

This list, whilst not exhaustive, identifies most recent reports in which one or more of the following three criteria apply and from which clear information about protease synthesis and regulation can not be obtained: (a) strains were from a 'private' collection or are not adequately identified; (b) organisms were grown on complex media or were cultured well into stationary phase; (c) starting material for enzyme studies was a crude commercial product. Investigations concerned primarily with enzyme purification and characterization are not included here.

Aleeva, V. D. and Aravina, L. A. (1973). *Prikl. Biokhim. Mikrobiol.,* **9**, 554–558

Aleeva, V. D., Fedotova, A. I. and Aravina, L. A. (1973). *Mikrobiologiya,* **42**, 428–433

Aravina, L. A., Ponomareva, V. D., Tereshin, I. M., Kasatkina, I. D. and Grekova, V. K. (1976). *Mikrobiologiya,* **45**, 770–776

Bossman, H. B. (1973). *Biochim. Biophys. Acta,* **293**, 476–489

Bretschneider, G. and Nordwig, A. (1971). *Hoppe-Seyler's Z. Physiol. Chim.,* **352**, 1039–1042

Burgum, A. and Prescott, J. M. (1965). *Arch. Biochem. Biophys.,* **111**, 391–398

Dion, W. M. (1950). *Can. J. Res.,* **28**, 586–599

Dworschack, R. G., Koepsell, H. J. and Lagoda, A. A. (1952). *Arch. Biochem. Biophys.,* **41**, 48–60

Ebeling, W., Hennrich, N., Klockow, M., Metz, H., Orth, H. D. and Lang, H. (1974). *Eur. J. Biochem.,* **47**, 91–97

Fromentin, H. (1974). *Bull. Soc. Fr. Mycol. Med.,* **4**, 215–219

Gendina, S. B. and Popova, N. V. (1973). *Mikrobiologiya,* **42**, 505–509

Ikeda, Y., Nakamura, K., Uchida, K. and Ishitani, C. (1957). *J. Gen. Appl. Microbiol.,* **3**, 93–101

Imshenetskii, A. A., Kasatkina, I. D. and Zheltova, E. T. (1971). *Mikrobiologiya,* **40**, 18–22

Kundu, A. K. and Manna, S. (1975). *Appl. Microbiol.,* **30**, 507–513

Martin, S. M. and Jönsson, A. G. (1965). *Can. J. Biochem.* **43**, 1745–1753

Misaki, T., Yamada, M., Okazaki, T. and Sawada, J. (1970). *Agric. Biol. Chem., Tokyo,* **34**, 1383–1392

Ong, P. S. and Gaucher, G. M. (1972). *Proc. IV IFS: Ferment. Technol. Today,* 271–278

Prudlov, B., Ushakova, V. I. and Egorov, N. S. (1972). *Mikrobiologiya,* **41**, 791–797

Rao, K. S. N., Sreekantiah, K. R. and Rao, T. N. R. (1970). *J. Food Sci. Technol.,* **7**, 26–30

Remold, H., Fasold, H. and Staib, F. (1968). *Biochim. Biophys. Acta,* **167**, 399–406

Roschlau, W. H. E. and Ives, D. A. J. (1974). *Folia Haematol., Leipzig,* **101**, S22–37

Sekine, H. (1972). Agric. Biol. Chem., Tokyo, **36**, 198

Singh, K. and Vezina, C. (1972). *Can. J. Microbiol.,* **18**, 1165–1167

Skiba, J. and Stachowicz, K. J. (1975). *Prace Inst. Lab. Bad. Przem. Spoz.,* **25**, 41–46

Suzuki, M., Kuno, M., Asai, T. and Nakao, Y. (1974). *Agric. Biol. Chem., Tokyo,* **38**, 657–661

Zakirov, M. Z., Schelokova, S. S. and Karavaeva, N. N. (1975). *Prikl. Biokhim. Mikrobiol.,* **11**, 686–690

Microorganisma and Nitrogen Sources
Edited by J. W. Payne

Section 3.3

Proteolysis in Yeast

DIETER H. WOLF AND HELMUT HOLZER
Biochemisches Institut der Universität Freiburg im Breisgau and Institut für Toxikologie und Biochemie der Gesellschaft für Strahlenund Umweltforschung, München D-7800 Freiburg im Breisgau, West Germany

1 INTRODUCTION

Proteases were amongst the first enzymes to be investigated biochemically. In many instances, as with the secretory enzymes trypsin and chymotrypsin of the intestine, their physiological function was obvious: they were needed to digest exogenous food-proteins so as to supply the organism with the amino acids and nitrogen necessary for life. This article is concerned mainly with the proteases of the simple eukaryotic yeast *Saccharomyces cerevisiae,* from which, as yet, no secreted proteolytic enzymes which might obviously serve such a digestive function have been detected (Suomalainen and Oura, 1971). However, intracellular proteolytic activities were found in yeast as early as 1898 by Hahn and Geret, and the first characterization of the enzymes was reported by Dernby (1917) and by Willstätter and Grassman (1926). What then are the intracellular proteolytic enzymes needed for? In 1942 Schoenheimer established the concept of the 'dynamic state of body constitutents', indicating that cellular constituents are continuously being degraded and resynthesized, and from this has developed the specific concept of protein turnover which explains the apparently constant concentrations of intracellular proteins as being the result of a balanced interplay between their synthesis and degradation. One part of this concept, the regulation of protein levels by alteration of their biosynthetic rates, has subsequently become a well-established phenomenon and is presently an area under intensive investigation (Epstein and Beckwith, 1968; O'Malley *et al.*, 1977). Protein degradation catalysed by proteases, the second regulatory mechanism in Schoenheimer's concept, has attracted less attention but nevertheless its general occurrence in prokaryotic and eukaryotic cells has been shown by many authors and its importance in cellular regulation is now well recognized (Schimke and Doyle, 1970; Pine, 1972; Goldberg and Dice, 1974; Goldberg and St. John, 1976; Switzer, 1977).

In addition to their role, outlined above, in the regulation of protein levels, the following functions have either been demonstrated or postulated for intracellular proteases in general (see also Section 3.4). (a) Degradation of unneeded cellular proteins in order to supply amino acids and energy during cell differentiation and under starvation conditions (Mandelstam, 1960; Kornberg *et al.,* 1968). (b) Specific action on individual proteins resulting in inactivation or modification of a particular enzyme activity or modification of enzyme specificity (Switzer, 1977; Holzer, 1976*a*; Rouget and Chapeville, 1971). (c) Maturation of proteins: by cleaving the translation product of monocistronic messenger RNA coding for several distinct polypeptides (Fink, 1971); by removing *N*-formylmethionine or methionine from newly synthesized polypeptide chains (Stewart *et al.,* 1971); by removing the 'signal sequence' from nascent polypeptides traversing membranes (Blobel and Dobberstein, 1975; Lingappa *et al.,* 1977); by converting inactive pro-proteins (zymogens) into the biologically active products (Neurath and Walsh, 1976). (d) Selective elimination of defective protein molecules

(Goldschmidt, 1970; Platt *et al.,* 1970; Etlinger and Goldberg, 1977). (e) Utilization of intracellularly transported exogenous peptides as sources of amino acids and nitrogen (Payne, 1976).

The yeast *Saccharomyces cerevisiae* serves as an extremely useful tool for studying proteolysis in the eukaryotic cell, for it is a simple eukaryote that can be grown quickly and under defined nutritional conditions very like bacteria, and it is thus readily available in large quantities, thereby facilitating biochemical studies. In addition, the yeast cell is easily accessible to genetic manipulation and analysis, thereby making it feasible to obtain mutants of the proteolytic system which might be useful in helping to uncover the links between the events catalysed *in vitro* by the proteolytic system and its biological function(s) *in vivo*.

2 THE PROTEOLYTIC SYSTEM

1 The Proteases Themselves

Several intracellular proteases have been found and characterized in *Saccharomyces cerevisiae*: two endopeptidases, protease A and B (Lenney and Dalbec, 1967; Hata *et al.,* 1967*a,b*); two carboxypeptidases, called carboxypeptidase Y (Hata *et al*., 1967*b*; Doi *et al*., 1967) (formerly called protease C) and carboxypeptidase S (Wolf and Weiser, 1977); and, in addition, at least three aminopeptidases (Masuda *et al*., 1975; Frey and Röhm, 1977; K. H. Röhm, personal communication). Table 1 summarizes some of their properties.

(i) Protease A

Protease A is an acid endoprotease with a molecular mass of about 60 000 daltons, exhibiting an isoelectric point of pH 3.8, and containing about 10 per cent hexose (Hata *et al*., 1967*b*). It hydrolyses casein and haemoglobin rapidly between pH 2 and 4 but is less active at higher pH values (Lenney and Dalbec, 1967; Hata *et al.,* 1967*a,b*). It does not act on esters of tyrosine and arginine, nor does it attack any small peptide substrate that has been tested (Hata *et al*., 1976*a*). It is not inhibited by EDTA, *p*-chloromercuribenzoate, diisopropylfluorophosphate, or phenylmethylsulphonylfluoride, thus indicating, respectively, that a divalent cation, an —SH group, or a serine residue of the enzyme are not essential for catalysis (Lenney and Dalbec, 1967; Hata *et al.,* 1967*a*). The finding that pepstatin is a highly effective inhibitor of protease A (Saheki *et al*., 1974) allows protease A to be easily differentiated from other proteases. The inhibition by pepstatin indicates the presence of a carboxyl group of an aspartic acid or glutamic acid residue at the active site of

Table 1 Intracellular proteolytic enzymes from *Saccharomyces cerevisiae*

Enzyme	Characteristics	Molecular weight (daltons)	Inhibitory agents†‡	References
Protease A	Acid endopeptidase	60 000	Pepstatin	Lenney and Dalbec (1967); Hata *et al*. (1967*a,b*); Doi *et al*. (1967); Saheki *et al*. (1974)
Protease B	Serine endopeptidase	32–44 000	DIFP, PMSF, PCMB chymostatin antipain	Lenney and Dalbec (1967); Hata *et al*. (1967*a*); Ulane and Cabib (1976); Bünning (1977)
Carboxypeptidase Y	Serine exopeptidase	61 000	DIFP, PMSF, TPCK, PCMB	Hata *et al*. (1967*a*); Doi *et al*. (1967); Hayashi *et al*. (1970, 1973*a,b*); Kuhn *et al*. (1974, 1976)
Carboxypeptidase S	Exopeptidase (Zn^{2+})	n.d.	EDTA	Wolf and Weiser (1977)
Aminopeptidase I (probably identical with aminopeptidase II isolated by Frey and Röhm (1977))	Exopeptidase (Zn^{2+})	200 000	EDTA, PCMB	Masuda *et al*. (1975); Frey and Röhm (1977); K. H. Röhm (personal communication)
Aminopeptidase II (probably identical with the dipeptidase isolated by Frey and Röhm (1977))	Exopeptidase (Me^{2+})	approx. 34 000; 140 000	n.d.	Masuda *et al*. (1975); Frey and Röhm (1977); K. H. Röhm (personal communication)
Vacuolar aminopeptidase (identical with aminopeptidase I isolated by Frey and Röhm (1977))	Exopeptidase (Zn^{2+})	640 000	EDTA	Frey and Röhm (1977); K. H. Röhm (personal communication); Matile *et al*. (1971)

†For the proteinaceous intracellular inhibitors from yeast see Section 3.3.2.2.

‡Abbreviations: DIFP, diisopropylfluorophosphate; PMSF, phenylmethylsulphonylfluoride; PCMB, *p*-chloromercuribenzoate; TPCK, *N*-tosyl-L-phenylalanine chloromethyl ketone; EDTA, ethylene diaminetetraacetic acid; n.d., not determined.

proteinase A (Tang, 1976). The enzyme activity is also sensitive to the intracellular inhibitor protein I^A (Saheki *et al*., 1974; Holzer, 1975; Nuñez de Castro and Holzer, 1976; Bünning *et al*., 1977) (see also Section 3.3.2.2).

(ii) Protease B

Protease B was first partially purified and characterized by Lenney and Dalbec (1967) and by Hata *et al.,* (1967*a*). Highly purified enzyme was later obtained by Ulane and Cabib (1976) and by Bünning (1977) with the aid of affinity chromatography using the immobilized, protease B-specific inhibitor from yeast; to achieve better and more gentle elution from the affinity column, Bünning used nitrated protease B-specific inhibitor (Bünning, 1977; Bünning and Holzer, 1977). The enzyme was shown to consist of a single polypeptide chain with molecular mass of 32 000 (P. Bünning, unpublished results) to 44 000 daltons (Ulane and Cabib, 1976); the reason for this discrepancy might be due to the different treatment of the yeast crude extracts. The isoelectric point of protease B was found to be 5.8 (Bünning, 1977). The enzyme hydrolyses acid-denatured haemoglobin and casein with high activity between pH 6 and 10; with casein the pH optimum for activity was shown to be around 9 (Hata *et al.,* 1967*a*). It attacks azocoll (Ulane and Cabib, 1976; Bünning, 1977; Cabib and Ulane, 1973; Saheki and Holzer, 1974) and hide powder azure (Ulane and Cabib, 1976) in the neutral pH range, and has some activity toward the ester substrates *N*-benzoyl-L-arginine ethylester, acetyl-L-tyrosine ethylester (Hata *et al*., 1967*a*; Ulane and Cabib, 1976), the histochemically-used acetyl-D,L-phenylalanine-β-naphthylester (D. H. Wolf and C. Ehmann, unpublished results) and several amino acid *p*-nitrophenylesters (Ulane and Cabib, 1976), but it does not hydrolyse a variety of benzyloxycarbonyl (Cbz) dipeptides such as Cbz-Phe-Leu, Cbz-Tyr-Leu, Cbz-Glu-Tyr, or Cbz-Gly-Leu (Lenney and Dalbec, 1967; D. H. Wolf, C. Ehmann, and P. Bünning, unpublished results).

The enzyme is not affected by EDTA, indicating that divalent cations are not required for activity. However, its activity is strongly inhibited by diisopropylfluorophosphate and phenylmethylsulphonylfluoride (Lenney and Dalbec, 1967; Hata *et al.,* 1967*a*), and by *p*-chloromercuribenzoate and $HgCl_2$ (Lenney and Dalbec, 1967; Hata *et al.,* 1967*a*; Ulane and Cabib, 1976; Bünning, 1977). Furthermore, inhibition is also caused by the actinomycete metabolites chymostatin and antipain, chymostatin being the more powerful inhibitor (Ulane and Cabib, 1976; Lenney, 1975); activity is also strongly inhibited by the specific intracellular inhibitor protein I^B (Bünning and Holzer, 1977; Ulane and Cabib, 1974; Betz *et al.,* 1974) (see Section 3.3.2.2). Taken together, these studies classify protease B as an endopeptidase, its activity being dependent upon one serine residue and the presence of sulphydryl residue(s) in the enzyme protein.

(iii) Carboxypeptidase Y

Carboxypeptidase Y was first described by Hata *et al.* (1967*a*). It was originally called protease C because of its activity against the protein substrate casein (Doi *et al.,* 1967). Further studies, however, showed it to have strong carboxypeptidase activity (Hayashi *et al.,* 1970). The enzyme has been purified to homogeneity (Doi *et al.,* 1967), and shown to be composed of one polypeptide chain with a molecular mass of around 61 000 daltons (Hayashi *et al*., 1973*a*; Kuhn *et al*., 1974). Carboxypeptidase Y is an acidic protein, exhibiting an isoelectric point of around 3.6 (Hata *et al*., 1967*b*). It is a glycoprotein, each molecule having about 16 residues of glucosamine and about 15 per cent hexose (Hayashi *et al*., 1973*a*; Kuhn *et al*., 1974). It removes most amino acid residues, including proline, from the *C*-termini of proteins and peptides at pH 5.5–6.5 (Hayashi *et al*., 1970, 1973*a*; Kuhn *et al.,* 1974) (for detailed information on substrate specificities see Hayashi, 1976). In addition, the enzyme hydrolyses ester substrates, such as acetyl-L-phenylalanine-ethylester, acetyl-L-tyrosine-ethyl ester (Doi, *et al.,* 1967; Hayashi, 1976) and the histochemical reagent acetyl-D,L-phenylalanine-β-naphthyl ester (Wolf and Fink, 1975) with high activity, and it exhibits amidase activity against benzoyl-L-tyrosine-*p*-nitroanilide (Aibara *et al*., 1971). Inhibition of carboxypeptidase Y activity is brought about by its intracellular inhibitor protein I^c (Matern *et al*., 1974*a*). The enzyme is not affected by chelating agents (Doi *et al.,* 1967) but its activity towards peptides and esters is highly sensitive to the action of the group-specific inhibitors diisopropylfluorophosphate, phenylmethylsulphonylfluoride, and Cbz-L-phenylalanine chloromethyl ketone and tosyl-L-phenylalanine chloromethyl ketone (Doi *et al*., 1967; Hayashi *et al*., 1975; Kuhn *et al*., 1976). Quantitative analysis using diisopropylfluorophosphate revealed one serine residue per enzyme molecule at the active site, and using Cbz-L-phenylalanine chloromethyl ketone one histidine residue that was essential for activity was detected (Hayashi *et al*., 1973*b*, 1975; Kuhn *et al.,* 1976). Modification studies using iodoacetamide and phenylglyoxal uncovered a single methionine residue and a single arginine residue that were necessary for peptidase activity but not for esterase activity (Kuhn *et al.,* 1976), suggesting that both these amino acid residues are located in a region of the active site that is involved in the binding of peptide but not ester substrates. The extent of inhibition of the enzyme by *p*-hydroxymercuribenzoate varied with the structure of the substrate tested, indicating that the particular cysteine residue involved although not essential for enzymic activity is presumably located at or near the substrate binding site (Bai and Hayashi, 1975).

(iv) Carboxypeptidase S

A second enzyme exhibiting carboxypeptidase activity was found in a mutant that lacked carboxypeptidase Y activity (Wolf and Weiser, 1977). This enzyme, carboxypeptidase S, cleaves the carboxypeptidase Y substrate Cbz-Gly-Leu but is inactive against Cbz-Phe-Leu and Cbz-Leu-Leu (Wolf and Weiser, 1977). The activity against Cbz-Gly-Leu is inhibited by the isomer containing D-Leu, Cbz-Gly-D-Leu (P. Bünning and D. H. Wolf, unpublished results). In contrast to carboxypeptidase Y the enzyme is a metal ion-dependent carboxypeptidase strongly inhibited by chelating agents like EDTA and *o*-phenanthroline; reactivation of the enzyme is achieved by Zn^{2+}, Co^{2+}, and Mn^{2+}, Zn^{2+} being the most effective (Wolf and Weiser, 1977; P. Bünning and D. H. Wolf, unpublished results). The enzyme is not inhibited by phenylmethylsulphonylfluoride (Wolf and Weiser, 1977).

(v) Aminopeptidases

The situation concerning the aminopeptidases is less clear. Matile and coworkers (Matile and Wiemken, 1967; Matile *et al.*, 1971) reported four aminopeptidase activities, but did not characterize them biochemically in detail. Masuda *et al.* (1975) purified one enzyme to homogeneity and called it aminopeptidase I. They estimated the molecular mass of the protein to be 200 000 daltons, and showed that it was strongly inhibited by chelating agents like EDTA and *o*-phenanthroline, indicating that its activity was dependent on divalent cations of which Zn^{2+} was the most effective; the enzyme was also inhibited by sulphydryl reagents such as *p*-chloromercuribenzoate, but diisopropylfluorophosphate had no effect. Masuda *et al.* (1975) reported that aminopeptidase I cleaved a variety of dipeptides and amino acid amides in the pH range 7.5–8.0. The synthetic substrate leucine-*p*-nitroanilide was also cleaved. The dipeptide Ile-Phe acted as an inhibitor although in contrast Leu-Phe was rapidly hydrolysed. The enzyme hydrolysed glucagon, releasing only the amino acids which were expected to be cleaved from the known NH_2-terminal sequence.

Masuda *et al.* (1975) also purified a second aminopeptidase that they called aminopeptidase II and which differed in several respects from their aminopeptidase I. Its molecular mass was estimated to be around 34 000 daltons. It cleaved the same dipeptides as did aminopeptidase I, and in addition it also cleaved Ile-Phe, which inhibited aminopeptidase I. In contrast to aminopeptidase I, aminopeptidase II was not able to cleave leucine-*p*-nitroanilide and leucine amide. Its pH optimum for activity is around 7.6. Its activity was inhibited by EDTA and by *o*-phenanthroline, indicating its dependence on divalent cations.

Frey and Röhm (1977) purified an aminopeptidase having a high molecular mass (640 000 daltons) which they also named aminopeptidase I. It is composed of 12 subunits. Enzyme activity specifically depends on Zn^{2+} ions and is activated by Cl^- anions; inactivation is brought about by metal chelating agents. A variety of L-amino acids and peptide derivatives containing a free α-*N*-terminal amino group are attacked, although substrates with *N*-terminal lysine are not hydrolysed (K. H. Röhm, personal communication). Because of differences in substrate specificity and molecular weight, Frey and Röhm (1977) claim that this enzyme is different from aminopeptidase I and II described by Masuda *et al.* (1975). However, they assume it to be identical with the only vacuolar aminopeptidase ('lysosomal aminopeptidase') so far described (Matile *et al.,* 1971) as they found the enzyme to be localized in the vacuole (see Section 3.3.3). To distinguish this enzyme from the aminopeptidases I and II of Masuda *et al.* (1975), and to indicate its vacuolar localization, we propose, for the purpose of this review, to call this enzyme 'vacuolar aminopeptidase'.

J. Frey and K. H. Röhm (unpublished results) also purified a second aminopeptidase, which they called aminopeptidase II. This enzyme has a molecular mass around 200 000 daltons, cleaves a variety of *N*-terminal amino acids including lysine, and cleaves lysine-*p*-nitroanilide. Because of the broad similarity in their properties they assume this enzyme to be identical with aminopeptidase I described by Masuda *et al*. (1975).

In addition, Frey and Röhm (1977) characterized a specific dipeptidase. The enzyme showed no activity on amino acid amides or dipeptide amides. Its molecular mass was estimated to be around 140 000 daltons. These authors claim (personal communication) that aminopeptidase II described by Masuda *et al*. (1975) is identical with this dipeptidase. However, as Masuda *et al*. only reported on the cleavage of dipeptides by aminopeptidase II, a conclusive judgement about the identity of the two is not possible. Furthermore, it should be noted that the reported molecular weight of 34 000 daltons for aminopeptidase II (Masuda *et al*., 1975) is very different from the molecular mass of 140 000 daltons reported for the dipeptidase (J. Frey and K. H. Röhm, unpublished results).

2 Protease Inhibitors

The proteases described above are only a part of the proteolytic system of yeast. Thus, proteases A and B and carboxypeptidase Y had been found to occur in an inactive state in freshly prepared crude extracts, and it had been suspected that they might exist as inactive zymogens which could undergo activation by similar mechanisms to those found for trypsin and chymotrypsin (Lenney and Dalbec, 1967; Hayashi *et al.,* 1967). Evidence that protease inhibitors might be responsible for the inactivity of proteases A and B and carboxypeptidase Y in fresh cell homogenates was first presented by Lenney and Dalbec (1969) and Lenney (1973). Subsequently, three classes of proteins that specifically inhibit protease A, protease B, and carboxypeptidase Y have been isolated and characterized in the laboratories of E. Cabib (Ulane and Cabib, 1974), H. Holzer (Saheki *et al*., 1974; Nuñez de Castro and Holzer, 1976; Betz *et al*., 1974; Matern *et al.,* 1974), and J. F. Lenney (Lenney, 1975). Table 2 summarizes some of the characteristics of the inhibitors. Evidence that the protease inhibitors are independently synthesized proteins rather than being the cleavage products of protease pro-forms came from the studies of Beck *et al.* (1976), and Beck and Wolf (1977), who showed that a mutant not able to derepress protease A inhibitor is still able to derepress protease A, indicating the independent synthesis of protease and inhibitor.

(i) Protease A Inhibitor

Using baker's yeast as starting material, Saheki *et al.* (1974) isolated two protease A-inhibiting proteins, I^A2 and I^A3. Both inhibitors are stable to heat and to trichloroacetic acid. The two proteins exhibit the same molecular weight on SDS-electrophoresis (Saheki *et al.,* 1974; Bünning *et al.,* 1977) but show different isoelectric points of 5.7 for I^A2 and 6.3 for I^A3, respectively (Saheki *et al.,* 1974). Both inhibitory proteins specifically block protease A activity; neither protease B nor carboxypeptidase Y, nor an acid protease from another source (pepsin), are inhibited. The kinetics of I^A3 inhibition of protease A revealed an apparent K_i value of 5.5×10^{-8} M and was concluded to be a 'non-stoicheiometric inhibition' of the 'pseudo-irreversible' type (Nuñez de Castro and Holzer, 1976). I^A3 was analysed in more detail and shown to consist of 68 amino acids per molecule, these included a large number of the polar amino acids glutamate, glutamine, aspartate, asparagine, and lysine, while proline, arginine, cysteine, and tryptophan were absent; from the amino acid analysis the molecular mass was estimated to be 7676 daltons (Nuñez de Castro and Holzer, 1976). Recent experiments, using a haploid strain of *Saccharomyces cerevisiae,* suggest that only I^A3 occurs in this organism whereas I^A2 was found in *Saccharomyces carlsbergensis* (Bünning *et al.,* 1977). The existence of iso-inhibitors in baker's yeast has also been found for protease B inhibitor (see next section).

Table 2 Intracellular protease-inhibiting polypeptides from baker's yeast

Inhibitor	Specific substrate	Molecular weight (daltons)	Isoelectric point	Stability	References
$I^{A}2$ $I^{A}3$	Protease A	7700	5.7 6.3	Heat and acid resistant	Saheki *et al.* (1974); Nuñez de Castro and Holzer (1976); Bünning *et al.* (1977)
$I^{B}1$ $I^{B}2$	Protease B	8500	8.0 7.0	Heat and acid resistant	Bünning and Holzer (1977); Ulane and Cabib (1974); Betz *et al.* (1974); Maier (1977)
I^{C}	Carboxypeptidase Y	25 000	6.6	Heat and acid labile	Matern *et al.* (1974*a*); Matern (1975)

(ii) Protease B Inhibitor

Lenney and Dalbec (1969) and Ferguson *et al.* (1973) found evidence for heat-stable material from yeast that inhibited protease B. Starting from baker's yeast, Betz *et al.* (1974) were able to purify to homogeneity two protease B-inhibiting proteins, I^B1 and I^B2, which are equivalent in molecular size, but differ substantially in charge. Ulane and Cabib (1974) found only one protease B inhibitor protein from a haploid strain of *Saccharomyces cerevisiae*. The discrepancy between these findings was resolved by Bünning and Holzer (1977), who showed that I^B1 occurs exclusively in *Saccharomyces carlsbergensis* whereas I^B2 is found exclusively in *Saccharomyces cerevisiae*. As one finds both proteins I^B1 and I^B2, not only in commercially produced baker's yeast but also in single-cell cultures of baker's yeast, with a constant ratio of 1:3, one may suggest the origin of both proteins is in the genomes of a polyploid baker's yeast cell, consisting of one haploid *Saccharomyces carlsbergensis* genome and three haploid *Saccharomyces cerevisiae* genomes (Bünning and Holzer, 1977; Kreger van-Rij, 1969). The protease B inhibitor proteins did not inhibit protease A or carboxypeptidase Y, thus showing a high degree of specificity for protease B; enzyme–inhibitor association constants were calculated as $3.7 \times 10^9\ \mathrm{M}^{-1}$ for I^B1 and $3.9 \times 10^9\ \mathrm{M}^{-1}$ for I^B2 (Bünning and Holzer, 1977) with inhibition being maximal between pH 6 and 9 and of the non-competitive type when purified inhibitor was used (Betz *et al.*, 1974). The molecular mass of the protease B inhibitor from *Saccharomyces cerevisiae* was calculated from amino acid analysis and SDS-electrophoresis to be about 8500 daltons (Ulane and Cabib, 1974; Maier, 1977). The inhibitor protein has an isoelectric point of 7.0–7.1 and is heat- and acid-stable (Ulane and Cabib, 1974; Betz *et al.*, 1974). It consists of 74 amino acids, a large proportion of which are polar amino acids; it contains one tyrosine residue, but arginine, tryptophan, methionine, and cysteine are absent (Ulane and Cabib, 1974; Maier, 1977). The amino acid sequence has been determined (Maier, 1977); it has an *N*-terminal threonine, and the *C*-terminus has the sequence -Val-His-Thr-Asn-OH. This *C*-terminus is essential for inhibitor action, for cleavage of the penultimate threonine is paralleled by a drastic decrease in inhibitory activity (Maier, 1977).

(iii) Carboxypeptidase Y Inhibitor

The carboxypeptidase Y inhibitor I^C was purified to homogeneity by Matern *et al.* (1974*a*) and shown to be one polypeptide chain composed of 217 amino acids with a molecular mass of about 25 000 daltons (Matern *et al.*, 1974*a*; Barth, 1977). An earlier calculation by Hayashi and Hata

(1972), using the difference between the molecular weight of inhibited carboxypeptidase Y and the active enzyme, had given a molecular mass of about 19 000 daltons for I^C. However, as these authors prepared their inhibited enzyme from a yeast autolysate, their lower molecular weight might have arisen from proteolytic attack. The inhibitor protein has an isoelectric point of 6.6 and shows maximum inhibition of carboxypeptidase Y between pH 5 and 9 (Matern, 1975). It specifically inhibits carboxypeptidase Y, and shows no action on proteases A or B, carboxypeptidase S, or carboxypeptidases A and B from bovine pancreas (Matern *et al.,* 1974*a*; D. H. Wolf, unpublished results). Inhibition was shown to be competitive, with a K_i of 2.5×10^{-9} M (Matern, 1975).

3 SUBCELLULAR LOCALIZATION OF PROTEASES AND PROTEASE INHIBITORS

Matile and Wiemken (1967) and Wiemken (1969) reported that proteases A and B and carboxypeptidase Y were localized in the vacuole of the yeast cell, an organelle functionally related to the lysosome of the animal cell. In contrast, Cabib *et al.* (1973) were unable to find the acid protease A located in the vacuoles, although they did confirm the location of protease B. Experiments of Lenney *et al.* (1974) and Matern *et al.* (1974*b*), however, support the findings of Matile and Wiemken in that they also found proteases A and B and carboxypeptidase Y in the vacuolar fraction. A question that remains unanswered, however, is whether proteases A and B and carboxypeptidase Y are located together in one category of vacuole or if different categories of vacuoles exist, each category having one protease activity, or one set of proteases with a similar pH optimum for activity. There is also no answer yet to the question of whether these proteases are always contained in the vacuoles or if conditions exist *in vivo* under which protease molecules also appear in the cytoplasm of the cell. In contrast to the vacuolar localization of proteases A and B and carboxypeptidase Y, their specific inhibitory proteins were found to reside in the cytoplasm of the cell (Lenney *et al.,* 1974; Matern *et al.,* 1974*b*; Hasilik *et al.,* 1974). The situation concerning the location of carboxypeptidase S is less clear; centrifugation experiments using metabolically-lysed cells show it occurs exclusively in the pellet, indicating it is associated with a membranous fraction and absent from the cytosol (P. Bünning and D. H. Wolf, unpublished results). The high molecular weight, vacuolar aminopeptidase was found to be localized in the vacuole of the cell (K. H. Röhm, personal communication; Matile *et al.*, 1971). Frey and Röhm (1977) and K. H. Röhm (personal communication) reported their dipeptidase and aminopeptidase II to be localized in the cytosol. No experiments to localize the aminopeptidases I and II of Masuda *et al.* (1975) have been reported as yet.

4 REGULATION OF THE SYNTHESIS OF PROTEASES AND PROTEASE INHIBITORS

Dependence of protease and inhibitor activities on growth conditions has been documented in several publications (Saheki and Holzer, 1975; Betz, 1975; Hansen *et al.,* 1977). Saheki and Holzer (1975) reported a several-fold increase of proteases A and B, carboxypeptidase Y, and protease B inhibitor activities during transition of cells from logarithmic- to stationary-phase growth. A more detailed analysis was performed by Hansen *et al.* (1977), who showed that the increase of about three- to 10-fold in the protease and inhibitor activities arose from glucose exhaustion, and that a similar increase in these activities was observed when cells growing logarithmically on glucose were shifted to a medium containing acetate as sole carbon source; the presence of glucose in acetate medium prevented the increase in the protease and inhibitor activities. Addition of glucose to acetate-grown cells did not inactivate the proteases or their inhibitors, as is the case for several enzymes sensitive to 'catabolite inactivation', such as cytoplasmic malate dehydrogenase (Hansen *et al.,* 1977). The results of Hansen *et al.* (1977) therefore indicate that proteases A and B, carboxypeptidase Y, and their respective inhibitors are catabolite-repressible proteins. The aminopeptidase I and II activities described by Masuda *et al.* (1975) show quite small glucose effects, generally about twofold (Hansen *et al.,* 1977). Studies by Frey and Röhm (1977) demonstrated a several-fold increase of the vacuolar aminopeptidase activity during transition from logarithmic- to stationary-phase growth, a result which is in accord with studies of Matile *et al.* (1971) on the 'lysosomal' aminopeptidase. The derepression of proteases upon glucose exhaustion suggests their involvement in the degradation of particular enzymes which have become superfluous under those conditions, thereby making amino acids available for synthesis of new enzymes necessary for growth on non-glucose carbon sources, such as acetate or ethanol. Thus, participation in the adaptation of the cell machinery to the utilization of different growth substrates may be an important function of the intracellular proteases in yeast.

Carboxypeptidases Y and S seem to serve an additional function. Their activities were shown to increase six- to 10-fold upon cell growth on a peptide such as N-Cbz-glycylleucine as sole nitrogen source (Wolf and Ehmann, 1978*a*; Wolf *et al*., 1979). Thus their involvement in food supply seems obvious.

5 CONTROL OF PROTEOLYSIS

Just as enzymic activity is usually dependent on the amount of enzyme present, it seems obvious from studies on protease and inhibitor synthesis that changes in the levels of the proteases at least partly control proteolysis, but to give the proteolytic events the necessary specificity seems to require more than just changes in activity of this type. One further possible way to control proteolysis lies in alteration of the susceptibility of the protein substrate to the protease. This susceptibility may be brought about by changes in the tertiary structure of the protein substrate, such a change perhaps being triggered by some newly appearing metabolite functioning as a signal, or by removing a metabolite or a coenzyme that is bound to the protein in its active state (Schimke and Doyle, 1970; Grisolia, 1964). One example of this kind studied in *S. cerevisiae in vitro* is tryptophan synthase, which was shown to undergo a two- to threefold enhanced inactivation upon removal of the coenzyme, pyridoxal-5′-phosphate (Manney, 1968*b*; Schött and Holzer, 1974).

As the proteases and their substrates are located in different compartments of the eukaryotic yeast cell, additional possibilities for control of proteolysis may exist. One possibility lies in the limited uptake, probably by pinocytosis, of protein substrates into the vacuole. Another possibility might involve a selective release of proteolytic enzymes from the vacuole. In addition, the function of the specific protease inhibitors can usefully be considered in this context. Their function was hypothesized as a safeguard against unwanted proteolytic cleavage following vacuolar leakage, or during transport of newly synthesized proteases from the cytoplasmic polysomes to the vacuoles (Saheki and Holzer, 1974; Holzer and Saheki, 1976). One prerequisite for such a function would appear to be a stringently coupled regulation of protease and inhibitor activity. In fact, Saheki and Holzer (1975) and Hansen *et al.* (1977) did find a parallel derepression of protease and inhibitor activities upon glucose exhaustion of *S. cerevisiae*. In addition, however, Beck *et al.* (1976) and Beck and Wolf (1977), isolated a temperature-sensitive (and simultaneously pH-sensitive) mutant with altered regulation of protease A-inhibitor activity, with which they were able to show that regulation of the protease A inhibitor was dissociated from regulation of proteases A and B. Thus, at least one element in the regulation of protease A and protease A inhibitor is different, and the question arises as to whether the yeast cell uses this type of element under certain circumstances *in vivo* to regulate protease and inhibitor levels independently, if so, this would add another facet to the cellular control of proteolysis: viz. regulating protease activity by inhibitor levels.

Mechanisms based on the following findings *in vitro* might also play a role in the control of proteolysis. Hayashi *et al.* (1967*a,b*) reported the activation of the latent proteases B and C through the action of protease A *in vitro*. Identification of the latent proteases as protease–inhibitor complexes (Lenney and Dalbec, 1969; Lenney, 1973), which had formed during disruption of the yeast cell, and thus also of the vacuolar compartment (Lenney *et al.*, 1974; Matern *et al.*, 1974*b*) led to the finding that activation of the proteases in crude extracts is due to digestion of the protease inhibitors (Saheki *et al.*, 1974; Holzer, 1975; Lenney, 1973; Saheki and Holzer, 1975). This cleavage of inhibitor molecules in crude extracts is brought about in an autocatalytic cycle by 'cross-hydrolysis' of protease B inhibitor by protease A, protease A inhibitor by protease B, and carboxypeptidase Y inhibitor by proteases A and B (Saheki *et al.*, 1974; Holzer, 1975). This autocatalytic cycle can be started in crude extracts by adjusting the pH to around 5, a procedure which dissociates small amounts of protease A–inhibitor complex, liberating free protease A (Saheki *et al.*, 1974; Holzer, 1975). Studies on the carboxypeptidase Y–inhibitor complex indicate that prior dissociation of the inhibitor from the carboxypeptidase must occur for the inhibitor to undergo hydrolysis (Barth *et al.*, 1978). Whether this holds true also for protease A inhibitor and protease B inhibitor has yet to be established.

The finding that the proteases can be activated *in vitro* by cross-hydrolysis of their specific inhibitors may have its equivalence *in vivo* and provide another means of controlling proteolytic events. Proteases may be allowed to leave the vacuolar compartment, and then, forming inactive protease–inhibitor complexes in the cytosol, they could become activated at any time and at any place in the cell upon a specific signal causing dissociation of protease–inhibitor complex and leading to digestion of the inhibitors and formation of free proteases, the action of which could in turn trigger a specific cellular process. Such a model has been presented for the activation of chitin synthetase (Holzer, 1975, 1978; Holzer and Saheki, 1976) (see Section 3.3.6.1). Studies *in vitro* on the carboxypeptidase Y–inhibitor complex showed that inhibition of carboxypeptidase Y was lost to variable degrees upon incubation with different substrates, and evidence was presented that the inhibitor had been dissociated from the substrate binding site of carboxypeptidase Y under those conditions (Barth *et al.*, 1978). Extrapolating these observations to the situation of protease action and control in the cytosol could mean that the substrate itself might control proteolysis at any time and at any place in the cell. The difference between this model and that outlined at the beginning of this section, in which it was the structure of the substrate that was responsible for controlling proteolysis, is the involvement here of the protease inhibitor as a second element enhancing the 'specificity' by lowering the apparent 'sensitivity' of the protease against its substrates. In other words, only those proteins which are or become good substrates as a result of some metabolic signal will be cleaved, whereas other proteins—which because of the broad substrate specificity of the proteases might erroneously be hydrolysed too—will remain unaffected. In addition, as already outlined above, dissociation of the protease–inhibitor complex might be facilitated locally by some signal. Thus, both effects, enhancing the binding constant of the substrate enabling it to dissociate the protease–inhibitor complex, and weakening the protease–inhibitor complex by some metabolically controlled signal, might provide sensitive control over protease action.

6 PROTEASE FUNCTION

As the only proteases so far found in *Saccharomyces cerevisiae* are intracellular enzymes, it may be assumed that they function in intracellular processes. One might divide these processes into 'specific protease function' and 'general protease function'. However, in several cases, such as the specific protease function in catabolite inactivation (see Section 3.3.6.1(i)) or in the degradation of aberrant proteins (see Section 3.3.6.1(iv)), the distinction is hazy. Furthermore, the 'general' process of protein turnover may well conceal 'specific' proteolytic processes until their 'specificity' can be detected by direct correlation of an altered turnover of a specific protein or a class of proteins with an environmental or metabolic signal.

1 Specific Protease Function

A variety of protease effects on native yeast proteins *in vitro* has been reported and these have been summarized by Pringle (1975). Some of these effects could be related to possible specific events occurring *in vivo*, whereas others appear to be due to a general accessibility of proteins to enzymic hydrolysis necessary for the 'general' process of protein turnover *in vivo*.

(i) Inactivation and Modification of Biologically Active Proteins

The autocatalytic, 'cross' inactivation of the inhibitors of proteases A and B and carboxypeptidase Y has been discussed already in Section 3.3.5.

Several enzymes have been shown to be irreversibly inactivated or modified after addition of glucose to cells, for example, components of the galactose uptake system (Spiegelman and Reiner, 1947; Matern and Holzer, 1977) and α-glucoside-(maltose) permease (Robertson and

Halvorson, 1957), cytoplasmic malate dehydrogenase (Witt *et al.,* 1966), fructose-1,6-disphosphatase (Gancedo, 1971), phosphoenolpyruvate carboxykinase (Haarasilta and Oura, 1975; Gancedo *et al.,* 1974), α-isopropyl malate synthase (Brown *et al.,* 1975), and uridine nucleosidase (Magni *et al.,* 1977). The working hypothesis has been put forward that the mechanism of 'catabolite inactivation' of these enzymes involves proteolysis (Holzer, 1976*a*; Holzer and Saheki, 1976). The enzymes tested *in vitro* (fructose-1,6-disphosphatase and cytoplasmic malate dehydrogenase) were shown to be hydrolysed by protease B (Holzer and Saheki, 1976; Neeff and Mecke, 1976). In the case of cytoplasmic malate dehydrogenase further evidence has appeared to support the idea of proteolysis as a possible inactivation mechanism: thus, using immunochemical techniques, an increased rate of disappearance of precipitable, radioactively-labelled enzyme was observed in crude extracts of cells which had been grown on a non-fermentable carbon source and subsequently treated with glucose (Neeff and Mecke, 1976; Neeff *et al*., 1978).

However, studies on recently isolated proteinase B mutants (Wolf *et al*., 1979; Wolf and Ehmann, 1978*b*; Zubenko *et al*., 1979) make the hypothesized involvement of proteinase B in the primary inactivation event of fructose-1,6-*bis*phosphatase and cytoplasmic malate dehydrogenase unlikely. Even though inactivation *in vitro* of these enzymes by proteinase B does not occur any more in mutant extracts (Wolf and Ehmann, 1979), inactivation *in vivo* of both enzymes takes place in the same way as in the wild type (Wolf and Ehmann, 1978*b*; Zubenko *et al*., 1979; Wolf and Ehmann, 1979). This finding does not generally exclude a proteolytic inactivation mechanism. Another proteinase might exist, which is responsible for the inactivations *in vivo*.

The general involvement of proteolysis during the adjustment of the cell machinery to the new conditions brought about by a rapid transition from growth on a non-fermentable carbon source to growth on glucose was shown by Nauhaus *et al.* (1977). They reported that about 10 per cent of the total cellular protein was found as amino acids or peptides in the surrounding liquid, when glucose was added to derepressed cells suspended in buffer, indicating a high rate of proteolysis. When glucose was added to derepressed cells suspended in growth medium no extracellular appearance of peptides and amino acids was detected but the intracellular pool was found to be enhanced (D. Mecke and J. Neeff, personal communication). The different behaviour of cells suspended in buffer or in growth medium under those conditions may be due to the fact that cells suspended in buffer are not able to grow and thus the intracellular accumulation of amino acids and peptides is not desirable and may even be harmful, whereas under conditions supporting growth the accumulated amino acids may be used up quickly for new protein synthesis.

(ii) Activation of biologically inactive proteins

Studies on the mechanism and control of yeast budding (Cabib *et al.,* 1974; Cabib, 1976) have implicated a protease in the conversion of chitin synthetase from the inactive zymogen to the active enzyme (Cabib and Ulane, 1973; Cabib *et al.,* 1973; Cabib and Farkas, 1971). Chitin is a characteristic constituent of the bud scar. The activating protease *in vitro* ('activating factor') was identified as protease B (Cabib and Ulane, 1973; Hasilik and Holzer, 1973). The fact that chitin synthesis takes place at a precise location in the cell can partly be explained by the finding that the chitin synthetase zymogen is attached to the plasma membrane (Duran *et al.,* 1975), the location of septum formation and thus necessarily of chitin synthesis. However, an explanation for the exact timing of initiation of chitin synthesis and how activation of the zymogen, which is scattered over the whole plasma membrane (Duran *et al.,* 1975), is achieved at a precise location, is not yet available. However, based on the findings of Cabib and coworkers and their proposal that activation of chitin synthetase was by protease B, the following model was put forward in an attempt to explain this sophisticated process (Holzer, 1975, 1978; Holzer and Saheki, 1976): proteases A and B are released from a vacuole near the site where chitin synthesis begins and form inactive complexes with their specific cytosol inhibitors. However, in response to a timed signal, for instance a decrease of pH, traces of protease A are liberated from the protease–inhibitor complex, which cross-activate (see Section 3.3.5) protease B, which subsequently activates chitin synthetase. When it becomes necessary to stop chitin synthesis this might be brought about by subsequent proteolytic inactivation of chitin synthetase by protease B, which was shown to occur *in vitro* (Hasilik, 1974).

Whether this model has to undergo modification, or even needs replacement, rests on future studies. Recent experiments on mutants lacking proteinase B activity indicate that the activation of chitin synthetase *in vivo* by action of proteinase B is questionable. Growth and chitin insertion proceed normally in the mutants (Wolf *et al*., 1979; Wolf and Ehmann, 1979; Zubenko *et al*., 1979). Another proteolytic enzyme might be responsible for *in vivo* activation of chitin synthetase (Wolf and Ehmann, 1978*b*).

(iii) Protein Maturation

Nascent, cytoplasmically-synthesized polypeptide chains in eukaryotes start with the amino acid methionine (Haselkorn and Rothman-Denes, 1973; Lucas-Lenard and Lipman, 1971). However the *N*-terminal methionine is subsequently removed from the majority of polypeptide chains. Evidence indicating that cleavage of the methionine was by an aminopeptidase came from analysis of iso-1-cytochrome *c* mutants (Stewart *et al.,* 1971). The methionyl codon, AUG, is the beginning of the genetic message of iso-1-cytochrome *c*. However, wild-type isocytochrome *c* carries threonine at the *N*-terminus. Sequence analysis of iso-1-cytochrome *c* variants from revertants of chain-initiation mutants led to the conclusion that the aminopeptidase could cleave methionine efficiently from threonyl and alanyl residues but not at all from leucyl, isoleucyl, and arginine residues.

Another process of protein maturation was assumed in the case of cytochrome *c* oxidase. This mitochondrial enzyme is composed of seven subunits, four of which are coded by nuclear DNA and three by mitochondrial DNA (Schatz and Mason, 1974). Poyton and McKemmie (1979) reported that the four nuclear-coded subunits are synthesized in the cytoplasm as one single polypeptide chain, which has to be cleaved into the four subunits found in the mature cytochrome oxidase. The proteolytic steps and the cleavage sites are unknown as yet.

Hasilik and Tanner (1976) showed that a precursor-protein form of carboxypeptidase Y occurred, having a molecular mass 5000 daltons higher than the authentic vacuolar enzyme. The precursor protein occurs in association with a membranous fraction. Incubation of the precursor protein in a crude extract at pH 5 converted it to the usual final enzyme form, possibly as a result of proteolysis. As carboxypeptidase Y is a vacuolar enzyme, the authors suggested that its mode of synthesis might be similar to that observed for secretory proteins in higher eukaryotes, which in their nascent form carry a peptide extension at the *N*-terminus of the polypeptide chain ('signal peptide') that is cleaved off during membrane passage (Blobel and Dobberstein, 1975; Lingappa *et al.,* 1977).

The question as to whether proteolysis plays an important functional role in the maturation or regulation of the *his4* and *trp5* gene products has yet to be answered (Fink, 1971). From genetic evidence, Duntze and Manney (1968) proposed that the product of the *trp5* region, tryptophan synthase, was a dimer of a single multifunctional polypeptide chain. However, dissociation of the purified enzyme yielded four monomers, each of molecular mass 35 000 daltons (Wolf and Hoffmann, 1974). Taking this finding, as well as genetic data on strains with nonsense mutations near the middle of the *trp5* region that yield only a monofunctional, active protein fragment of 35 000 daltons molecular mass (Manney, 1968*a*), and also considering the analogous bacterial enzyme (Yanofsky and Crawford, 1972), one might be tempted to propose that tryptophan synthase is composed of two different monofunctional polypeptide chains. However, if one looks at the analogous enzyme from another much studied eukaryote, *Neurospora crassa* (Matchett and DeMoss, 1975), one would have to assume that a dimer of one bifunctional polypeptide chain also exists for *S. cerevisiae*. Similar problems are connected with the *his4* gene product, which specifies three enzymic activities in the histidine pathway (Fink, 1971). Until recently, genetic and biochemical studies could not clearly answer the question as to whether the *his4* gene codes for three different polypeptide chains or one polypeptide chain with three discrete enzyme functions. As both tryptophan synthase and the *his4* gene product are highly sensitive to protease B (Schött and Holzer, 1974; Wolf and Hoffmann, 1974) the idea was put forward that, instead of being synthesized as different polypeptide chains, each mature enzyme complex may be formed by proteolytic cleavage after synthesis as one single polypeptide chain (Fink, 1971). After having purified the *his4* gene product several times as a complex of three polypeptide chains, Bigelis *et al.* (1977) were recently able to isolate the *his4* gene product as one single polypeptide chain containing three discrete enzymic activities, thereby eliminating the possibility that three distinct polypeptides are synthesized at the messenger RNA level. However, it is still an open question whether proteolytic cleavage of the single polypeptide chain occurs *in vivo* and is a prerequisite to some regulatory event.

(iv) Degradation of Nonsense Proteins

In bacteria, the growth-inhibiting effect of various di- and tripeptides is well established (Sussman and Gilvarg, 1971) (see also Sections 2.1 and 4.7). Thus, to protect the cell against accumulation of such deleterious peptides, as well as against the accumulation of aberrant proteins, proteases and peptidases are necessary to hydrolyse such components. Such a hydrolytic system has been found in bacteria (Goldberg and Dice, 1974; Goldberg and St. John, 1976) (see also Section 3.4). However, very little is known about the occurrence of an analogous process in yeast. Evidence for the proteolysis of subunits of the fatty acid synthetase complex, which as a result of mutation were unable to assemble to the authentic complex, was given by Dietlein and Schweizer (1975). However, it is not known if this degradation is brought about by some or all of the proteases already characterized or if there exist additional proteolytic enzymes apparently specific for digesting aberrant proteins, as has been reported for bacteria (Pine, 1972; Goldberg and Dice, 1974; Goldberg and St. John, 1976) (but see also Section 3.4).

2 General Protease Function

(i) Protein Turnover during Growth and Starvation

Even though protein turnover may seem to be a very wasteful event many studies have indicated its importance for cellular processes (Schimke and

Doyle, 1970; Pine, 1972; Goldberg and Dice, 1974; Goldberg and St. John, 1976). One of its functions, the catabolism of abnormal proteins, has already been mentioned above. This process should be most efficient during rapid growth of the cells, the time of highest rates of protein synthesis. Furthermore, adaptation to changes in the environment, and most importantly for microorganisms, adaptation to poor growth conditions, depends on the ability of the organism to degrade temporarily unneeded proteins so as to provide amino acids for the synthesis of enzymes essential for such conditions. Furthermore, protein degradation may be an important means of mobilizing energy in times of low caloric intake by providing amino acids as carbon sources, thus enabling the cell to survive.

During logarithmic growth of yeast in complete medium a rate of protein degradation of 1 per cent per hour was observed (Betz, 1976). It was also shown that protein degradation is enhanced about twofold when protein synthesis is inhibited by using cycloheximide, and in mutants conditionally defective in protein initiation or elongation. From these observations it was concluded that the activity of the pre-existing proteolytic apparatus in growing yeast cells was stimulated upon cessation of protein synthesis. It was speculated that this stimulation might be triggered by accumulation of GTP (or a metabolite), which is involved in initiation, aminoacyl-tRNA binding, and translocation in protein synthesis (Betz, 1976).

The first indication of a significant stimulation of protein turnover in resting cells compared to growing cells came from studies by Halvorson (1958*a,b*). In these studies Halvorson also noted the apparent energy dependence of the turnover process, thus, 4×10^{-4} M dinitrophenol inhibited protein breakdown more than 90 per cent. Using valinomycin, sodium cyanide, and sodium fluoride as energy blockers, Betz and Weiser (1976*a*) confirmed this energy dependence of protein breakdown in resting cells. As the hydrolysis of peptide bonds is an exergonic reaction the apparent energy requirement of the breakdown process is a puzzling observation (see also Section 3.4). However, one may hypothesize that the observed energy requirement is not due to the hydrolytic reaction *per se* but rather to events indirectly connected with protein degradation such as activating proteolytic enzymes, removing protease inhibitors, or facilitating the entry of protein substrates into the vacuoles through some uptake process (Goldberg and St. John, 1976; Holzer, 1976*a,b*).

As has already been discussed in Section 3.3.4, starvation for glucose and, to some extent, for nitrogen leads to a drastic increase of proteases A and B, carboxypeptidase Y, and the vacuolar aminopeptidase. Thus, it seems likely that these enzymes participate in the enhanced turnover accompanying adaptation of the cell to poorer growth conditions.

(ii) Protein Turnover and Proteinase during Differentiation (Sporulation)

The differentiation process of sporogenesis in yeast is triggered by transition of the cells to extremely poor conditions, such as when only acetate is supplied as nutrient (Croes, 1967*a,b*). Thus, to adapt the protein machinery to this differentiation process the cell has only its internal amino acid pool and its vegetative protein molecules as sources of new building blocks. Several investigators found intensive protein turnover in sporulating yeast cells (Betz and Weiser, 1976*a*; Hopper *et al*., 1974; Magee and Hopper, 1974; Klar and Halvorson, 1975) with 30 to 70 per cent of the pre-existing vegetative protein reported as being broken down up until the time mature asci appeared. Protein breakdown in haploid cells (Betz and Weiser, 1976*a*) or in cells homozygous for mating type (Hopper *et al*., 1974) and thus not able to sporulate, was found to be between 10 and 50 per cent respectively, under similar sporulation conditions. This indicates an enhanced proteolysis specifically associated with sporulating cells.

However, turnover was not specific for vegetative cell proteins, and proteins synthesized at different times during sporulation had approximately the same degradation rates (Betz and Weiser, 1976*a*). The breakdown of several enzymes has been followed during the sporulation process, although none of the measured enzymic changes proved to be sporulation specific, occurring also in haploid, non-sporulating cells (Betz and Weiser, 1976*b*). Protein breakdown during sporulation was also shown to be inhibited by uncouplers and inhibitors of energy metabolism (Betz and Weiser, 1976*a*). The possible role of energy in this degradative process has already been discussed in connection with similar inhibition during starvation (see previous section).

In contrast to growing cells (see Section 3.3.6.2(i)) interruption of protein synthesis by cycloheximide greatly reduced protein degradation in sporulating cells (Betz and Weiser, 1976*a*; Magee and Hopper, 1974). This effect becomes clearer when one analyses the behaviour of the protease activities during sporulation. Thus, although no new proteases have been detected specificially during sporulation, Klar and Halvorson (1975) and Betz and Weiser (1976*a*), measured a drastic increase of proteases A (two- to 11-fold) and B (six- to 12-fold) during sporulation. There is disagreement, however, concerning increase of activity of carboxypeptidase Y: Klar and Halvorson (1975) found about a five-fold increase, but no such increase was found by Betz and Weiser (1976*a*). In contrast to sporulating cells, no increase of protease activities was observed in non-sporogenic cultures (haploid and α/α strains), indicating the sporulation specificity of the process. The increase of protease activities in sporulating cells was shown to arise from *de novo* synthesis, as inhibition of protein synthesis by cycloheximide completely blocked the increase (Klar and Halvorson, 1975). The parallel inhibition of sporulation itself by cycloheximide is not surprising as this drug inhibits generally the synthesis of new proteins necessary for sporulation. The sporulation-specific increase of known protease activities and the parallel sporulation-specific increase in the amount of protein broken down during this differentiation process points to the crucial role of these events for the adapting cell. However, final proof of whether only some or all of the known proteases are needed for sporulation, or even if other, as yet undetected, proteases are involved in the sporulation process, should come from studies of mutants lacking the known proteolytic enzymes. Wolf and Fink (1975) isolated a mutant devoid of carboxypeptidase Y activity. The lesion was shown to be in the structural gene of carboxypeptidase Y (Wolf and Weiser, 1977), ensuring that any possible phenotypic change was due to lack of the enzyme activity. Diploids homozygous for the carboxypeptidase Y lesion were still able to sporulate, apparently normally, indicating that either carboxypeptidase Y is not involved in the sporulation process or that the remaining activities can adequately substitute for the enzyme. Diploid cells homozygous for a defect in protease B activity were found to exhibit reduced or no sporulation (Wolf and Ehmann, 1979; Zubenko *et al.*, 1979). As the lesion was found to be in the structural gene of protease B (Zubenko *et al.*, 1979), this strongly suggests a necessary role for proteinase B in this differentiation process. A mutant obtained by Hartwell (personal communication) was shown to exhibit constitutively low protease A activity, and diploids homozygous for this alteration were unable to sporulate (Betz, 1977). This finding may point to an essential role of protease A in the sporulation process; however, as the mutation does not seem to be in the structural gene of protease A one must be cautious of this interpretation for the mutation may be pleiotropic, affecting different functions at the same time which are unlinked to each other.

(iii) Peptide Utilization

As this topic is discussed in detail by Becker and Naider in Section 2.2 it will only be briefly mentioned here. The ability of *S. cerevisiae* to utilize a variety of peptides as a nitrogen source was reported early on by Nielsen (1943). More detailed studies have since been performed by Becker, Naider, and their coworkers (Becker *et al.,* 1973; Naider *et al.,* 1974; Becker and Naider, 1977; Marder *et al.,* 1977). Using amino acid auxotrophs, these investigators showed that peptides are taken up and cleaved intracellularly. It seems reasonable to believe that the peptidases characterized to date (see Section 3.3.2.1) are involved in supplying the auxotrophs with their required amino acids from the extracellularly supplied peptides. This view is substantiated by the finding that carboxypeptidase Y and S activities increase several-fold when cells are grown on a dipeptide like *N*-Cbz-Gly-Leu as sole nitrogen source (Wolf *et al*., 1979; Wolf and Ehmann, 1978*a*). Studies on a triple mutant, auxotrophic for leucine and lacking carboxypeptidase Y and carboxypeptidase S activity show that, in contrast to the wild type, it is unable to grow on the carboxypeptidase substrate *N*-Cbz-Gly-Leu (Wolf *et al*., 1979). Depending on their relative activities during growth phase and stationary phase, the know peptidases might serve the dual functions of supplying the cell with nitrogen from extracellular peptides and of being involved in intracellular turnover by supplying the cell with amino acids for new protein synthesis derived from intracellularly-released peptides.

7 CONCLUDING REMARKS

During the last ten years, there has been substantial progress in characterizing the proteolytic system in *Saccharomyces cerevisiae* and its great importance in fundamental cellular processes has been recognized. However, there is still very limited knowledge about the detailed functions of the individual proteases and the various processes in which they are involved. One powerful tool to fill in these gaps in our knowledge is a combined genetic and biochemical approach. The availability of mutants lacking one or more of the proteases or their inhibitors and analysis of the physiological alterations brought about by their absence should uncover their functions, and indeed, such an approach has successfully been started (Wolf and Fink, 1975; Jones, 1977; Wolf *et al*., 1979; Wolf and Ehmann, 1978*b*; Zubenko *et al*., 1979).

Understanding of another aspect of proteolysis is still very rudimentary: How is regulation of the proteolytic processes achieved? As several proteolytic enzymes normally occur in separate locations (organelles) from their substrates in the eukaryotic yeast cell, studies have not only to take into account regulation at the levels of protease activity and substrate accessibility but in addition the interplay between enzymes and organelles. The problem of regulation at these different levels of complexity is a challenge that will require the close collaboration of biochemists and cytologists.

ACKNOWLEDGEMENTS

The experimental work of the authors cited in this article was supported by the Deutsche Forschungsgemeinschaft, Bonn-Bad Godesberg, grants No. Wo 210/1, Wo 210/2, SFB 46 ('Molekulare Grundlagen der Entwicklung'), and Verband der Chemischen Industrie e.V. (Fonds der Chemischen Industrie). Thanks are due to Mrs D. Montfort for her expert assistance in preparing this manuscript.

REFERENCES

Aibara, S., Hayashi, R. and Hata, T. (1971). *Agric. Biol. Chem., Tokyo*, **35**, 658–666

Bai, Y. and Hayashi, R. (1975). *FEBS Lett.*, **56**, 43–45

Barth, R. (1977) Doctoral dissertation, Faculty of Biology, University of Freiburg

Barth, R., Wolf, D. H. and Holzer, H. (1978). *Biochim. Biophys. Acta*, **527**, 63–69

Beck, I. and Wolf, D. H. (1977). *11th Meet. Fedn. Eur. Biochem. Soc., Abstr.* A 6–5–853

Beck, I., Fink, G. R. and Wolf, D. H. (1976). *Abstr. Tenth Int. Congr. Biochem.*, p. 420

Becker, J. M. and Naider, F. (1977). *Arch. Biochem. Biophys.*, **178**, 245–255

Becker, J. M., Naider, F. and Katchalski, E. (1973). *Biochim. Biophys. Acta,* **291**, 388–397

Betz, H. (1975). *Biochim. Biophys. Acta*, **404**, 142–151

Betz, H. (1976). *Biochem. Biophys. Res. Commun.,* **72**, 121–129

Betz, H. (1977). In *Proceedings of the Symposium on Cell Differentiation in Microorganisms, Plants and Animals* (L. Nover and K. Mothes, eds), VEB Gustav Fischer, Jena, and Elsevier, North Holland, pp. 243–255

Betz, H. and Weiser, U. (1976*a*). *Eur. J. Biochem*., **62**, 65–76

Betz, H. and Weiser, U. (1976*b*). *Eur. J. Biochem*., **70**, 385–395

Betz, H., Hinze. and Holzer, H. (1974). *J. Biol. Chem.,* **249**, 4515–4521

Bigelis, R., Keesey, J. and Fink, G. R. (1977). In *Eucaryotic Genetics Systems*, ICN–UCLA Symposia on Molecular and Cellular Biology, Vol. VIII, Academic Press, New York, pp. 179–187

Blobel, G. and Dobberstein, B. (1975). *J. Cell Biol.,* **67**, 835–851

Brown, H. D., Satyanarayana, T. and Umbarger, H. E. (1975). *J. Bacteriol.,* **121**, 959–969

Bünning, P. (1977). Doctoral dissertation, Faculty of Biology, University of Freiburg

Bünning, P. and Holzer, H. (1977). *J. Biol. Chem.,* **252**, 5316–5323

Bünning, P., Meussdoeffer, F. and Holzer, H. (1977). *Hoppe-Seyler's Z. Physiol. Chem*., **358**, 1186–1187

Cabib, E. (1976). *Trends Biochem. Sci.,* **1**, 275–277

Cabib, E. and Farkas, V. (1971). *Proc. Natl. Acad. Sci. U.S.A*., **68**, 2052–2056

Cabib, E. and Ulane, R. (1973). *Biochem. Biophys. Res. Commun.,* **50**, 186–191

Cabib, E., Ulane, R. and Bowers, B. (1973). *J. Biol. Chem.,* **248**, 1451–1458

Cabib, E., Ulane, R. and Bowers, B. (1974). *Curr. Topics Cell. Regul.,* **8**, 1–32

Croes, A. F. (1967*a*). *Planta,* **76**, 209–226

Croes, A. F. (1976*b*). *Planta,* **76**, 227–237

Dernby, K. G. (1917). *Biochem. Z.,* **81**, 107–208

Dietlein, G. and Schweizer, E. (1975). *Eur. J. Biochem.,* **58**, 177–184

Doi, E., Hayashi, R. and Hata, T. (1967). *Agric. Biol. Chem., Tokyo,* **31**, 160–169

Duntze, W. and Manney, T. R. (1968). *J. Bacteriol.,* **96**, 2085–2093

Duran, A., Bowers, B. and Cabib, E. (1975). *Proc. Natl. Acad. Sci. U.S.A*., **72**, 3952–3955

Epstein, W. and Beckwith, J. R. (1968). *Annu. Rev. Biochem.,* **37**, 411–436

Etlinger, J. D. and Goldberg, A. L. (1977). *Proc. Natl. Acad. Sci. U.S.A*., **74**, 54–58

Ferguson, R. A., Katsunuma, T., Betz, H. and Holzer, H. (1973). *Eur. J. Biochem.,* **32**, 444–450

Fink, G. R. (1971). In *Metabolic Pathways,* Vol. 5, *Metabolic Regulation* (H. J. Vogel, ed.), Academic Press, New York and London, pp. 199–223

Frey, J. and Röhm, K. H. (1977). *Hoppe-Seyler's Z. Physiol. Chem.,* **358**, 1202–1203

Gancedo, C. (1971). *J. Bacteriol.,* **107**, 401–405

Gancedo, C. Schwerzmann, L. and Malano, J. (1974). In *Proceedings of the Fourth International Symposium on Yeasts, Vienna, 1974,* Part 1, p. 9, Abstr. 5

Goldberg, A. L. and Dice, J. F. (1974). *Annu. Rev. Biochem.,* **43**, 835–869

Goldberg, A. L. and St. John, A. C. (1976). *Annu. Rev. Biochem.,* **45**, 747–803
Goldschmidt, R. (1970). *Nature, Lond.,* **228**, 1151–1154
Grisolia, S. (1964). *Physiol. Rev.,* **44**, 657–712
Haarasilta, S. and Oura, E. (1975). *Eur. J. Biochem.,* **52**, 1–7
Hahn, M. and Geret, L. (1898). *Chem. Ber.,* **31**, 200–202
Halvorson, H. (1958*a*). *Biochim. Biophys. Acta,* **27**, 255–266
Halvorson, H. (1958*b*). *Biochim. Biophys. Acta,* **27**, 267–276
Hansen, R. J., Switzer, R. L., Hinze, H. and Holzer, H. (1977). *Biochim. Biophys. Acta,* **496**, 103–114
Haselkorn, R. and Rothman-Denes, L. B. (1973). *Annu. Rev. Biochem.,* **42**, 397–438
Hasilik, A. (1974). *Arch. Microbiol.,* **101**, 295–301
Hasilik, A. and Holzer, H. (1973). *Biochem. Biophys. Res. Commun.,* **53**, 552–559
Hasilik, A. and Tanner, W. (1976). *Biochem. Biophys. Res. Commun.,* **72**, 1430–1436
Hasilik, A., Müller, H. and Holzer, H. (1974). *Eur. J. Biochem.,* **48**, 111–117
Hata, T., Hayashi, R. and Doi, E. (1967*a*). *Agric. Biol. Chem., Tokyo,* **31**, 150–159
Hata, T., Hayashi, R. and Doi, E. (1967*b*). *Agric. Biol. Chem., Tokyo,* **31**, 357–367
Hayashi, R. (1976). *Methods Enzymol.,* **45** (Part B), 568–587
Hayashi, R. and Hata, T. (1972). *Agric. Biol. Chem., Tokyo,* **36**, 630–638
Hayashi, R., Aibara, S. and Hata, T. (1970). *Biochim. Biophys. Acta,* **212**, 359–361
Hayashi, R., Bai, Y. and Hata, T. (1975). *J. Biol. Chem.,* **250**, 5221–5226
Hayashi, R., Moore, S. and Stein, W. H. (1973*a*). *J. Biol. Chem.,* **248**, 2296–2302
Hayashi, R., Moore, S. and Stein, W. H. (1973*b*). *J. Biol. Chem.,* **248**, 8366–8369
Hayashi, R., Oka, Y., Doi, E. and Hata, T. (1967). *Agric. Biol. Chem., Tokyo,* **31**, 1102–1104
Holzer, H. (1975). *Adv. Enzyme Regul.,* **13**, 125–134
Holzer, H. (1976*a*). *Trends Biochem. Sci.,* **1**, 178–181
Holzer, H. (1976*b*). In *Metabolic Interconversion of Enzymes 1975* (S. Shaltiel, ed.), Springer-Verlag, Berlin and Heidelberg, pp. 168–174
Holzer, H. (1978). In *Protein Turnover and Lysosome Function* (H. L. Segal and D. J. Doyle, eds), Academic Press, New York, pp. 305–314
Holzer, H. and Saheki, T. (1976). *Tokai J. Exp. Clin. Med.,* **1**, 115–125
Holzer, H., Betz, H. and Ebner, E. (1975). *Curr. Topics Cell Regul.,* **9**, 103–156
Hopper, A. K., Magee, P. T., Welch, S. K., Friedman, N., and Hall, B. D. (1974). *J. Bacteriol.,* **119**, 619–628
Jones, E. W. (1977). *Genetics,* **85**, 23–33
Klar, A. J. S. and Halvorson, H. O. (1975). *J. Bacteriol.,* **124**, 863–869
Kornberg, A., Spudich, J. A., Nelson, D. L. and Deutscher, M. P. (1968). *Annu. Rev. Biochem.,* **37**, 51–78
Kreger-van Rij, N. J. W. (1969). In *The Yeasts,* Vol. 1 (A. H. Rose and J. S. Harrison, eds) Academic Press, New York, pp. 5–78
Kuhn, R. W., Walsh, K. A. and Neurath, H. (1974). *Biochemistry,* **13**, 3871–3877
Kuhn, R. W., Walsh, K. A. and Neurath, H. (1976). *Biochemistry,* **15**, 4881–4885
Lenney, J. F. (1973). *Fed. Proc. Fed. Am. Soc. Exp. Biol.,* **32**, 659, 2510 Abs
Lenney, J. F. (1975). *J. Bacteriol.,* **122**, 1265–1273
Lenney, J. F. and Dalbec, J. M. (1967). *Arch. Biochem. Biophys.,* **120**, 42–48
Lenney, J. F. and Dalbec, J. M. (1969). *Arch. Biochem. Biophys.,* **129**, 407–409
Lenney, J. F., Matile, P., Wiemken, A., Schellenberg, M. and Meyer, J. (1974). *Biochem. Biophys. Res. Commun.,* **60**, 1378–1383
Lingappa, V. R., Devilliers-Thiery, A. and Blobel, G. (1977). *Proc. Natl. Acad. Sci. U.S.A.,* **74**, 2432–2436
Lucas-Lenard, J. and Lipman, F. (1971). *Annu. Rev. Biochem.,* **40**, 409–448
Magee, P. T. and Hopper, A. K. (1974). *J. Bacteriol.,* **119**, 952–960
Magni, G., Santarelli, J., Natalini, P., Ruggieri, S. and Vita, A. (1977). *Eur. J. Biochem.,* **75**, 77–82
Maier, K. (1977). Doctoral dissertation, Faculty of Biology, University of Freiburg
Mandelstam, J. (1960). *Bacteriol. Rev.,* **24**, 289–309
Manney, T. R. (1968*a*). *Genetics,* **60**, 719–733
Manney, T. R. (1968*b*). *J. Bacteriol.,* **96**, 403–408

Marder, R., Becker, J. M. and Naider, F. (1977). *J. Bacteriol.,* **131**, 906–916
Masuda, T., Hayashi, R. and Hata, T. (1975). *Agric. Biol. Chem., Tokyo,* **39**, 499–505
Matchett, W. H. and DeMoss, J. A. (1975). *J. Biol. Chem.,* **250**, 2941–2946
Matern, H. (1975). Doctoral dissertation, Faculty of Biology University of Freiburg
Matern, H. and Holzer, H. (1977). *J. Biol. Chem.,* **252**, 6399–6402
Matern, H., Betz, H. and Holzer, H. (1974*b*). *Biochem. Biophys. Res. Commun.,* **60**, 1051–1057
Matern, H., Hoffman, M. and Holzer, H. (1974*a*). *Proc. Natl. Acad. Sci. U.S.A.,* **71**, 4874–4878
Matile, Ph. and Wiemken, A. (1967). *Arch. Mikrobiol.,* **56**, 148–155
Matile, Ph., Wiemken, A. and Guyer, W. (1971). *Planta,* **96**, 43–53
Naider, F., Becker, J. M. and Katzir-Katchalski, E. (1974). *J. Biol. Chem.,* **249**, 9–20
Nauhaus, J., Essig, U., Neeff, J. and Mecke, D. (1977). *Hoppe-Seyler's Z. Physiol. Chem.,* **358**, 281–282
Neeff, J. and Mecke, D. (1976). *Abstr. Tenth Int. Congr. Biochem.,* p. 419.
Neeff, J., Hĕgele, E., Nauhaus, J., Heer, U. and Mecke, D. (1978). *Eur. J. Biochem.,* **87**, 489–495
Nielsen, N. (1943). *Engebn. Bio.,* **19**, 375–408
Neurath, H. and Walsh, K. A. (1976). *Proc. Natl. Acad. Sci. U.S.A.,* **73**, 3825–3832
Nuñez de Castro, J. and Holzer, H. (1976). *Hoppe-Seyler's Z. Physiol. Chem.,* **357**, 727–734
O'Malley, B. W., Towle, H. C. and Schwartz, R. J. (1977). *Annu. Rev. Genet.,* **11**, 239–275
Payne, J. W. (1976). *Adv. Microbial Physiol.,* **13**, 55–113
Pine, J. M. (1972). *Annu. Rev. Microbiol.,* **26**, 103–126
Platt, T., Miller, J. H. and Weber, K. (1970). *Nature, Lond.,* **228**, 1154–1156
Poyton, R. O. and McKemmie, E. (1979). In *Limited Proteolysis in Microorganisms* (G. N. Cohen and H. Holzer, eds), DHEW Publication No. (NIH) 79–1591. U.S. Printing Office, Washington, pp. 145–149
Pringle, J. R. (1975). In *Methods in Cell Biology,* Vol. 12 (D. M. Prescott, ed.) Academic Press, New York, San Francisco and London, pp. 149–184
Robertson, J. J. and Halvorson, H. O. (1957). *J. Bacteriol.,* **73**, 186–198
Rouget, P. and Chapeville, F. (1971). *Eur. J. Biochem.,* **23**, 459–467
Saheki, T. and Holzer, H. (1974). *Eur. J. Biochem.,* **42**, 621–626
Saheki, T. and Holzer, H. (1975). *Biochim Biophys. Acta,* **384**, 203–214
Saheki, T., Matsuda, Y. and Holzer, H. (1974). *Eur. J. Biochem.,* **47**, 325–332
Schatz, G. and Mason, T. L. (1974). *Annu. Rev. Biochem.,* **43**, 51–87
Schimke, R. T. and Doyle, D. (1970). *Annu. Rev. Biochem.,* **39**, 929–976
Schoenheimer, R. (1942). *The Dynamic State of the Body Constituents,* Harvard University Press, Cambridge, Mass.
Schott, E. H. and Holzer, H. (1974). *Eur. J. Biochem.,* **42**, 61–66
Suomalainen, H. and Oura, E. (1971). In *The Yeasts* Vol. 2 (A. H. Rose and J. S. Harrison, eds), Academic Press, New York and London, pp. 3–74
Spiegelman, S. and Reiner, J. M. (1947). *J. Gen. Physiol.,* **31**, 175–193
Stewart, J. W., Sherman, F., Shipman, N. A. and Jackson, M. (1971). *J. Biol. Chem.,* **246**, 7429–7445
Sussman, A. J. and Gilvarg, C. (1971). *Annu. Rev. Biochem.,* **40**, 397–408
Switzer, R. L. (1977). *Annu. Rev. Microbiol.,* **31**, 135–157
Tang, J. (1976). *Trends Biochem. Sci.,* **1**, 205–208
Ulane, R. E. and Cabib, E. (1974). *J. Biol. Chem.,* **249**, 3418–3422
Ulane, R. E. and Cabib, E. (1976). *J. Biol. Chem.,* **251**, 3367–3374
Wiemken, A. (1969). Doctoral Dissertation, Swiss Federal Institute of Technology, Zürich
Willstatter, R. and Grassmann, W. (1926). *Hoppe-Seyler's Z. Physiol. Chem.,* **153**, 250–282
Witt, I., Kronau, R. and Holzer, H. (1966). *Biochim. Biophys. Acta,* **118**, 522–537
Wolf, D. H. (1977). *Hoppe-Seyler's Z. Physiol. Chem.,* **358**, 1301–1302
Wolf, D. H. and Ehmann, C. (1978*a*). *FEBS Lett.,* **91**, 59–62
Wolf, D. H. and Ehmann, C. (1978*b*). *FEBS Lett.,* **92**, 121–124
Wolf, D. H. and Ehmann, C. (1979). *Eur. J. Biochem.,* **98**, 375–384
Wolf, D. H. and Fink, G. R. (1975). *J. Bacteriol.,* **123**, 1150–1156

Wolf, D. H. and Hoffman, M. (1974). *Eur. J. Biochem.*, **45**, 269–276
Wolf, D. H. and Weiser, U. (1977). *Eur. J. Biochem.*, **73**, 553–556
Wolf, D. H., Ehmann, C. and Beck, I. (1979). In *Biological Functions of Proteinases*, (H. Holzer and H. Tschesche, eds), Springer Verlag, Berlin–Heidelberg–New York, pp. 55–72
Yanofsky, C. and Crawford, I. P. (1972). In *The Enzymes*, 3rd edn, Vol. 7 (P. D. Boyer, ed.), Academic Press, New York, pp. 1–31
Zubenko, G. S., Mitchell, A. P. and Jones, E. W. (1979). In *Limited Proteolysis in Microorganisms*, (G. N. Cohen and H. Holzer, eds), DHEW Publication No. (NIH) 79-1591, U.S. Government Printing Office, Washington, pp. 49–53

Microorganisms and Nitrogen Sources
Edited by J. W. Payne

Section 3.4

Turnover of Microbial Cell Proteins and their Peptide Residues

MARTIN J. PINE

Department of Experimental Therapeutics, Grace Cancer Drug Center, Roswell Park Memorial Institute, Buffalo, New York 14263, USA

1 INTRODUCTION

In its usual context protein turnover refers to the replacement of any internal proteins in metazoans, or more generally as the complete breakdown and resynthesis of intracellular proteins above and beyond the bounds of net growth or of lysis and replacement of old cells. The same proteins may be recycled during steady-state conditions or the profile of cellular protein may be transformed during environmental change. Alternatively cell proteins may not be recycled but catabolized for energy, or they may be sloughed off through wear and tear or voided externally rather than reutilized. Thus, non-proliferating *Escherichia coli* continues to produce the unit component of protein-lipopolysaccharide that is normally assembled into the microcapsule, but it is secreted into the medium rather than used to replace or extend old microcapsule (Rothfield and Pearlman-Kothencz, 1969). Modern considerations of the fates of individual proteins may allow loose and overlapping connotations of any recycling of peptide bonds or of any commonly occurring hydrolytic step during protein assembly or processing, which will also be discussed. Thus, proteolytic cleavages may be involved at different stages of translation, maturation, or secretion of viral and microbial cell proteins, sometimes involving breakdown to relatively small polypeptides. Turnover of some microbial cell proteins, in contrast, may involve the accumulation of large metastable protein moieties. Finally, protein turnover may sometimes denote the exchange and transfer of proteins between cell cytoplasm, nucleus, and other organelles in eukaryotes.

The magnitude of the 'dynamic equilibrium' of cellular proteins was first indicated in animals in the early experiments of Schoenheimer (1942) with stable isotopes. A large body of subsequent evidence was required to show that much of this isotopic equilibration and rapid change in enzyme activity represented total hydrolysis and resynthesis of cell proteins without cell growth. Of particular importance initially were experiments with the cellular proteins of uniform cultures of animal cells (Eagle *et al.*, 1959) and *E. coli* (Mandelstam, 1958*a*, *b*), and with individual proteins that can be alternately stable or labile under different conditions (Schimke, 1964). Various proteins of higher animals display a variety of catabolic behaviours. Thus, the extracellular stromal elements are metabolically stable except in metamorphosis. Specialized cell proteins like the myofibrillar elements of skeletal muscle may serve as major nutritional reservoirs. Enzyme proteolysis, investigated most extensively in the liver, acts as a major mechanism regulating enzyme activity. Protein turnover underlies adaptation of fairly long-lived cells to many demands: starvation, developmental and functional changes brought about by hormones or by changes in diet or temperature, the challenge of toxicity and infection, and perhaps certain needs for supramolecular assembly (Schimke, 1973; Goldberg and St. John, 1976). Probably the same general principles operate in the simplest microbial metazoans. For example, most of the cells of nematodes are fixed and non-dividing even at the beginning of the animal's life. Correspondingly, the half-life of the soluble cell proteins of *Turbatrix aceti*, estimated at 33 h, is even shorter than that of the proteins of the liver cell (Prasanna and Lane, 1977). The ciliate protozoan *Tetrahymena pyriformis* would differ from this model by its capacity to respond to environmental change by cellular outgrowth rather than stasis. Correspondingly, the breakdown of cellular protein is relatively limited during active growth (generation time 6 h), but during amino acid starvation it can increase fourfold in rate and encompass half of the cellular protein overnight, or almost all within 7 days (M. J. Pine, unpublished data). The prokaryote *E. coli* which has been most often studied can also increase the rate and extent of its cellular protein breakdown during growth restriction. Prokaryotes with extremely slow generation times may, however, show considerable protein turnover during growth (Bond and Bradley, 1978). Even these distinctions are to some extent circumstantial. Thus, protein turnover in the liver, including even the breakdown of abnormal proteins, can be strongly suppressed so as to allow the organ mass to increase during growth or regeneration (Scornik *et al*., 1978). Cell protein breakdown rates in animal cell cultures also can vary twofold, depending on hormonal and nutritional status (Hershko *et al*., 1971). These factors, and the fact that prokaryote cellular organization is simpler and far different from that of the eukaryote, should dictate many differences in the nature and function of intracellular proteolysis in the biological systems studied. The similarities that do exist may represent retained evolutionary characters or functional coincidences. Past reviews and treatises on protein turnover that will complement the present coverage are those of Rechcigl (1971), Pine (1972), Schimke (1973), Goldberg and Dice (1974), Holzer *et al.* (1975), and Goldberg and St. John (1976).

2 METHODS OF MEASURING PROTEIN TURNOVER

1 Measurement of Enzyme Activity

During relatively slow adaptation in eukaryote and prokaryote cells, for example in sporulation of yeast (Betz and Weiser, 1976*b*) or in growth adaptation of *Nocardia erythropolis* (Bond *et al*., 1977), proteolysis affects significantly the net level of individual cell proteins. In the kinetic model used in describing the dynamics of mammalian cell proteins, the level of any protein (P) under steady-state conditions may be considered, at least in a first approximation, as the balance of a constant or zero-order rate of synthesis (k_s) and a random or first-order rate of degradation (k_d):

$$\frac{dP}{dt} = k_s - k_d P = 0$$

at equilibrium.

During any condition of enzyme induction or deinduction, assuming that synthesis and proteolysis are maintained at steady levels, the enzyme's half-life should be equal to the time required for activity to reach midway in its final equilibrium level (Berlin and Schimke, 1965). Alternately, an individual enzyme such as catalase can be specifically inactivated or its assembly prevented by administering reagents that modify or prevent assembly of its functional group. The respective rates of regeneration of new enzymes or decay of the remaining enzymes have thus been determined directly in animal systems from measurement of enzyme activity (Rechcigl, 1971). Breakdown has also been estimated from the fall in enzyme activity when synthesis is blocked by ribosomal inhibitors. This blockade can in itself, however, alter proteolysis in a general manner, usually inhibiting it moderately in animal systems (Goldberg and St. John, 1976) or partly in starving *E. coli* (Pine, 1973*a*), or stimulating proteolysis in yeast to maximal levels (Betz, 1976). There are many limitations inherent in deducing metabolic turnover from activity measurement. Conditions that are required for enzyme induction or deinduction may, by themselves, produce transient alterations in the basal rate of proteolysis (Chee and Swick, 1976). Moreover, a decrease in the activity of an enzyme may indicate other events than proteolysis—such as covalent modification. In addition, the remarkably short half-lives calculated for ornithine decarboxylase in animal systems have been subsequently found to be due to the synthesis of short-lived antizymes (Heller *et al*., 1977). Although synthesis and breakdown of an individual protein can operate in dynamic equilibrium in microbial systems, these dynamics may also be sequential. Thus in the slime mould *Dictyostelium discoideum*, glycogen phosphorylase is stable when actively synthesized during early sporulation and is then rapidly destroyed when synthesis stops (Thomas and Wright, 1976).

The actual destruction of a protein can be verified by the disappearance of its immunological reactivity. Cell proteins may thus be directly isolated by immunoprecipitation or merely selected by this method for further isolation, for example by electrophoresis (Roberts *et al*., 1977). Immunoprecipitation can also detect inactive or fragmented forms of the enzyme that retain some of its original conformation. Enzymes that are abundant in the cell and of distinctively low molecular weight, like the labile proteins of *Bacillus megatherium* (Setlow, 1975*b*), or of unusually high molecular weight, like β-galactosidase of *E. coli*, can be identified by electrophoresis alone, even when enzymically inactive. In addition, β-galactosidase mutants with defects in different loci can complement each other to generate activity *in vitro* or *in vivo*. (Lin and Zabin, 1972).

The overall intactness of a protein can be judged from quantitative comparisons of several criteria such as titration with antibody, electrophoretic resolution, and determination of optical absorption and functional activity (Lau *et al*., 1977). These criteria should distinguish the alternate fates of proteolysis versus non-destructive inactivation or modification.

Specific peptide residues in individual cell proteins are occasionally modified functionally after assembly, and these residues may not be recycled. Thus, the production of free 3-methylhistidine is a distinctive index of the breakdown of myosin (Funabuki *et al*., 1976).

2 Isotopic Methods

Isotopic labelling coupled with any procedure of isolating individual proteins provides the most versatile means of following protein turnover without directly altering the process being studied. Labelling with heavy isotopes may be used merely as a separatory device, for example to determine by isopycnic centrifugation whether newly appearing enzyme activity comes from new or old protein (Williams and Neidhardt, 1969). Labelling may be followed either in the peptide residues or in the prosthetic group of a protein. The peptide component of the acyl carrier protein of fatty acid biosynthesis in *E. coli* is quite stable, but the conjugated pantetheine residue turns over at about 4 per cent per minute, with no known functional or regulatory significance (Bauza *et al*., 1976). To determine the absolute or relative rates of protein biosynthesis, an ideal protein label should be freely accessible to the site of biosynthesis. Correspondingly, to determine the relative stability of a protein, the label should be efficiently dissipated without reutilization. In *E. coli* conventional amino acid labels generally appear empirically to fulfill these criteria. Thus, external leucine can largely supplant internal *de novo* sources as a precursor of cell protein, and in the most unstable cell proteins the incorporated leucine can be returned to the free form almost immediately (Pine, 1970). In *Neurospora* (Weiss, 1973), yeast (Wiemkin and Nourse, 1972), and animal cells (Van Venrooij *et al*., 1974), a large part of the intracellular pool of free amino acids is compartmentalized, probably in lysosomes or related structures such as osmotic vacuoles. These would appear to act sometimes as relative depots for much of the intracellular supply of recycled amino acids.

In some systems, incorporated amino acid label may be reutilized in preference to circulating supplies, adding significantly to estimates of protein half-lives (Poole, 1971). Reutilization can be minimized by the extremes of either flooding the biological system with a large excess of carrier (Mortimer *et al*., 1972) or by providing labels that are rapidly destroyed (Swick and Ip, 1974). Also, amino acid pools may be internally labelled with relatively general and multisite metabolic precursors such as $^{14}CO_2$ (Millward, 1970), $^{3}H_2O$ (Humphrey and Davies, 1976; Pine and Schimke, 1978), $^{2}H_2O$ (Borek *et al*., 1958) or $H_2{}^{18}O$ (Bernlohr, 1972). These labels may enter precursors either spontaneously or via enzyme-catalysed exchange reactions such as transamination, sometimes much faster than the precursors can be used up or can exchange with extracellular precursors (Pine and Schimke, 1978). Internal labelling with $^{3}H_2O$ has the added advantage that incorporations can be directly normalized with the activity of the cell water. As an alternative to any stratagem for equilibrating isotopic precursors, protein dynamics may be inferred from the activity of a supplied amino acid label in aminoacyl-tRNA (Kairallah, 1978).

Estimation of relative protein half-lives may be greatly simplified in steady-state systems by the technique of double labelling or isotope-ratio measurement (Arias *et al*., 1969). Pulsed labels of ^{3}H and ^{14}C in the same precursor are alternately given a long and a short time before harvesting the cell protein. The ratio of activities of old/new labelling should increase logarithmically in proportion to the half-life of the protein. Relative labilities can thus be compared between different proteins in the same labelling experiment. Absolute turnover rates can be inferred by interpolation from the isotope ratios of proteins with known half-lives. Even without knowledge of absolute breakdown rates the method is particularly useful in scanning large populations of proteins for their individual or collective differences in metabolic stability according to any criteria for separating or distinguishing them *en masse*. Several of the soluble cell proteins of *E. coli* can thus be shown to be unstable during growth (Larrabee and Larrabee, 1978). Although protein breakdown is generally followed from the disappearance of incorporated label the process may also be followed by the recycling of label into other molecules. Thus, the turnover of glycine label in *E. coli* cell proteins has been followed from its reappearance in the purine residues of the cellular nucleic acid (Koch and Levy, 1955). Also the metabolic recycling of *E. coli* cell proteins has been followed during growth by the reappearance of old ^{35}S label into newly synthesized β-galactosidase (Hogness *et al*., 1955). These latter studies showed that most of the cell protein of *E. coli* is metabolically stable during growth.

3 SUBSTRATES OF INTRACELLULAR PROTEOLYSIS

1 Growth-related Intracellular Proteolysis

A small but heterogeneous assortment of proteolytic events occur during active growth of *E. coli*. At least 5 per cent of the leucine that is incorporated into cell protein during late logarithmic growth is recycled back to free leucine within several minutes and could represent aberrant protein synthesis or maturative and processing events (Pine, 1970; Brunschede and Bremer, 1971). A slower proteolysis that may include true protein recycling has been variously estimated to involve 1–7 per cent of the cellular protein, with average half-lives of 13–60 min. (Willetts, 1967*b*; Pine, 1970; Nath and Koch, 1971; Brunschede and Bremer, 1971). Permeases such as that for β-galactosides (Koch, 1963) may fall into this category. Of the 200-odd major soluble components of *E. coli* cell protein that can be resolved on polyacrylamide gel by their size and charge, only about eight may be significantly unstable during growth (Larrabee and Larrabee, 1978). The bulk of the cell protein of *E. coli* is thus relatively inert metabolically (Willetts, 1967*b*), but the overall breakdown, which is detectable for the most part just after the protein is labelled, is significant, and can amount to from 2.5 to 3.5 per cent per hour, depending on growth rate (Pine, 1970). In contrast, the cell protein of *Bacillus cereus* has been reported to be broken down relatively uniformly during growth, at 5 per cent per hour (Urbá, 1959).

2 Cell Proteins Catabolized during Starvation

During adaptation to new nitrogen or carbon sources, or during starvation, cellular protein and RNA continue to provide precursors for the endogeneous turnover of macromolecular components. These sources are the only significant stores of nitrogen and sometimes of carbon and energy, for example in glycogen-less *E. coli* in early logarithmic growth. The small amount of cell protein that is normally broken down during growth is expanded to include the more stable cell proteins (Pine, 1966). During diauxie, the growth transition from one carbon source to another, the rise in the breakdown of the stable proteins can be dramatic (Willetts, 1965). There is also an increase in the total rate of breakdown of all cell proteins, including the least stable ones, but this may not occur under some conditions of nutrient depletion. Therefore growth arrestment or starvation *per se* is not a direct cause of the attack upon stable proteins (Pine, 1970; Nath and Koch, 1971). During a variety of starvation conditions, maximal breakdown rates of 2.5–5 per cent of cell protein per hour have been reported in *E. coli* (Mandelstam, 1958*a*; Willetts, 1967*a*; Schlessinger and Ben Hamida, 1966). *Bacillus cereus* (Urbá, 1959), *Pseudomonas* *saccharophila* (Young and Klein, 1967), and *Arthrobacter crystallopoietes* (Scherer and Boylen, 1977). Comparatively low breakdown rates of 1 per cent per hour in starving yeast (Halvorson, 1958) and 0.5 per cent per hour in starving *Arthrobacter atrocyaneus* (Scherer and Boylen, 1977) could represent limitations in the ability of the end-products of breakdown to exchange with extracellular or pool amino acids.

Over prolonged starvation periods of several days the maximum amount of cell protein that can be mobilized is influenced by the manner of starvation. In *E. coli* this amounts to about 20 per cent of the cell protein during abrupt starvation of NH_4^+ and carbon, perhaps representing the limit of uncompensated degradation of cell protein. During partial or complete restriction of NH_4^+ alone a breakdown limit of 30 per cent of the cell protein is reached (Pine, 1973*a*). The same maximum is also reached during alternating cycles of amino acid starvation and replenishment in a chemostat. However when growth is reduced to a low steady-state level ($k = 0.1\ h^{-1}$) by glucose or amino acid limitation, the otherwise inert bulk of the cell protein can break down at a uniform rate ($k = 0.007\ h^{-1}$). This may mean that all of the cell protein is eventually susceptible to catabolism during starvation, but that most of the catabolism can occur only during the slow intercalation of new cell material (M. J. Pine, unpublished data). Many organisms such as *Arthrobacter* are well-known to endure very long periods of starvation. However, their pattern of overall proteolysis of prelabelled cell protein does not greatly differ from that of *E. coli* and effectively stops when up to one-third of the cell protein has been used up (Scherer and Boylen, 1977). These limits do not, however, imply that proteolysis is completely stopped. Thus a fraction of the proteins that are resynthesized during starvation are themselves highly unstable, and therefore provide a continuing source for recycling that would not be apparent from the net breakdown of old label (Pine, 1965; Brunschede and Bremer, 1971).

Microbial cells usually do not contain high levels of any one specialized protein that can serve as a proteolytic store during starvation. In general a large spectrum of cell proteins must serve as the proteolytic supply. In the blue-green bacterium *Anacystis nidulans*, however, the accessory pigment phycocyanin comprises 10 per cent of the soluble cell protein and to some extent fulfils the function of a preferred store. It is selectively destroyed in quantity during starvation and is selectively resynthesized during recovery (Lau *et al*., 1977). The diverse examples of selective inactivation of individual proteins of microbial cells under different conditions of starvation have been summarized and discussed by Switzer (1977) from the standpoint of their possible mechanisms, including proteolysis. Irreversible inactivation has been suggested as an ultimate means of avoiding counterproductive cycling of metabolites by opposing pathways, and for some enzymes it may be an economic and slow-acting supplement to feedback control. Thus the NADP-linked glutamate dehydrogenase of yeast, which is specifically involved in the synthesis of glutamate is normally stable. It is destroyed rapidly in *Candida utilis* (Hemmings, 1978) and *Saccharomyces* (Mazon, 1978) when the medium is depleted of carbon source. The half-life for disappearance may be as short as 20 min, and inactivation is concomitant with the disappearance of immunologically recognizable protein (Hemmings, 1978). The proteins of yeast mitochondrial membrane and an assortment of enzymic activities may be lost in the process of catabolite inactivation during growth enrichment, for example following glucose addition (Hansen *et al.*, 1977). The induced mannitol dehydrogenase activity of *Nocardia erythropolis* as well as several electrophoretically distinguishable fractions of its cell proteins disappear variably during growth or adaptation to new substrates (Bond *et al.*, 1977; Bond and Bradley, 1978). The ribulose diphosphate carboxylase activity of autotrophically-grown *Hydrogenomonas* is stable during energy starvation, but after heterotrophic growth the enzyme can disappear by a process sensitive to 2,4-dinitrophenol and actinomycin D. The possible interpretation that the enzyme may be destroyed is strengthened by its instability in extracts only of heterotrophically-grown cells (Kuehn and McFadden, 1968). Aminoacyl-tRNA synthetases of *E. coli* are inactivated during amino acid starvation, particularly when it is intermittent (Williams and Neidhardt, 1969). Even though inactivation does not necessarily indicate breakdown, it may indicate some predisposition to do so.

3 Proteins Synthesized during Growth Arrest

The cell proteins synthesized during starvation and adaptation would of course include those enzymes whose synthesis is specifically derepressed by environmental change. Certain protein species appear to be produced in unusual amounts in amino acid-starved *E. coli*, but not in a mutant (rel^-) for stringent control of many biosynthetic processes during amino acid starvation. Synthesis may thus involve a constellation of proteins directly or indirectly derepressed by this locus (Chao, 1977). Of the particulate cell proteins of *E. coli*, two outer coat proteins continue to be produced in large amounts during amino acid starvation (Aguanno and Larrabee, 1976). In amino acid-starved *E. coli*, cell number continues to increase through the completion of initiated rounds of chromosome synthesis and septum formation. Nitrogen-starved yeast cells continue to bud and presumably produce the minimal protein ingredients necessary for the formation of considerably smaller daughter cells than the parent (Johnston *et al.*, 1977).

4 Sources of Protein Turnover in Sporulation and Spore Germination

In the bacilli, sporulation is accomplished by replication of the nuclear body, migration to the forespore area and its enclosure by septation. Within it are synthesized at separate stages spore enzymes, low molecular weight components such as antibiotics and Ca dipicolinate, and the several layers of the spore coat which eventually isolate the maturing spore from its environment. Although some species (e.g. *Bacillus brevis*) sporulate continuously and during active cell growth, sporulation in bacilli as in many other microbes is, however, typically induced by starvation, and may be endotrophic (without any exogenous supply of substrates). Protein turnover is then primarily unidirectional. The bulk of the cell proteins of the developing bacterial sporangium are eventually committed to net destruction at an overall breakdown rate of 20 per cent per hour (including all sporangial contents), to supply precursors for the developing forespore (Spudich and Kornberg, 1968; Bernlohr, 1972). During endotrophic ascospore formation in yeast, 60–70 per cent of the pre-existing cell protein is broken down (Betz and Weiser, 1976*a*).

For spore germination, which can also be endotrophic, the spores of *B. megatherium* contain the necessary enzymes and components for all macromolecular synthesis, but lack both the free pools of, and the biosynthetic enzymes for, several amino acids (Setlow and Primus, 1975). During the first hour of germination, amino acids and energy can be generated mostly by the breakdown of a select population of three proteins averaging 12 000 daltons in molecular mass which comprise 20 per cent of the total spore protein (Setlow, 1975*a*, *b*). Breakdown at first greatly exceeds protein synthesis, and amino acids are excreted to be taken up later by the outgrowing cell. These proteins do not have a balanced amino acid composition. Tryptophan must be supplied from another protein source and cystine may be utilized from the outer spore coat (Setlow and Primus, 1975). The breakdown of the three small internal spore proteins may have the additional function of activating cellular DNA, since these proteins have high isoelectric points, bind to DNA and confer ultraviolet light resistance to the spore specifically at the time of their synthesis (Setlow, 1975*b*).

Swarming of vegetative cells of the slime mould *Dictyostelium discoideum* to form fruiting bodies is the culmination of the combined processes of starvation, differentiation, and sporulation. Enzymes that are involved in the breakdown of glycogen in the cell and in the synthesis of carbohydrate elements of the fruiting body are synthesized and then broken down or discarded in separate phases (Thomas and Wright, 1976).

5 Proteolysis during Virus Production

Host proteins do not appear to be generally metabolized for the production of viral proteins. Thus when the lysogenic lambda prophage of

E. coli K-12 is induced by mitomycin C into vegetative growth, the overall rate of intracellular proteolysis proceeds normally up to the time of cell lysis (Pine, 1975). However, a large number of simple and complex viruses require the late imposition of proteolytic steps that range from minor polypeptide excision to extensive breakdown, and involve the assembly of coat, core and tail precursors (Hershko and Fry, 1975). The small RNA-containing picorna-, toga-, and oncornaviruses of animals are translated by very different schemes, but perhaps all of their intrinsic products require proteolytic processing. Thus the viral mRNAs are translated initially as oversize proteins or as a single polyprotein (Butterworth, 1977; Murphy *et al*., 1978). The cleavage can be more or less generally impaired in poliovirus by serine protease inhibitors, or by translation with amino acid analogues to produce denatured products (Jacobson *et al*., 1970). More specific protease cleavages can be inhibited in other RNA viruses by complexants of the translation product such as Zn^{2+}, or specific cleavage may not occur in viral mutants at non-permissive temperatures (Butterworth, 1977).

4 ROLES OF INTRACELLULAR PROTEOLYSIS

1 Breakdown of Altered Proteins

One important function of protein turnover is internally to remove or replace altered or impaired proteins or organelles, either molecule by molecule or as discrete structures (DeDuve and Wattiaux, 1966). Breakdown may be accomplished through a cognitive mechanism or by random selection.

(i) Replacement of Damaged Organelles

The outermost or coat proteins of animal cells are replaced at a normal rate by the normal process of membrane turnover if carbohydrate or protein components are removed by sialidase or trypsin treatments (Buck and Warren, 1977). More extensive membrane toxicity may induce a more dramatic repair. Thus adsorption of Alcian Blue to the plasma membrane of *Chaos chaos* rapidly produces clumps that are endocytosed, but are mostly sloughed off *en masse*. Within 35 min the membrane is entirely replaced, presumably from an internal store of proteins and perhaps by direct interpolation rather than by intermediate vesicle formation. A repeated dye treatment produces a much slower cycle of replacement (Nachmias, 1966). Chemical toxicity in higher animal cells frequently produces damaged mitochondria which are rapidly engulfed and digested within secondary lysosomes (DeDuve and Wattiaux, 1966).

(ii) Incomplete Proteins

Premature protein termination during translation can produce considerable overall conformational pertubations in the product. Thus about one-quarter of *E. coli* cell proteins synthesized in the presence of a high level of puromycin are found as denatured granules (Prouty *et al*., 1975) and their overall breakdown is increased clearly above the secondary proteolytic effects due to growth inhibition (Pine, 1967). Nonsense (termination) or restart mutations of enzymes fall into this class. Although such mutations of β-galactosidase produce immunologically cross-reactive material, all of them, including those of nearly normal molecular weight are enzymically inactive and can have initial half-lives as short as several minutes, depending on the region of the termination. This instability determines the steady-state level of the protein in the cell (Lin and Zabin, 1972).

(iii) Mis-sense Proteins

Although cell proteins of *E. coli* that contain a nonsense suppressor tRNA are proteolysed normally (Pine, 1967), those of the more active and non-specific ribosomal suppressor mutant *ram* are proteolysed more actively than normal (Goldberg, 1972). The same is true after relatively prolonged exposure of normal *E. coli* to dihydrostreptomycin (Hewitt and Kogut, 1977). These proteolytic effects are probably the results of manifold mis-sense errors, possibly in the order of 3 per cent or more in a fraction of the translated codons (Strigini and Brickman, 1973). Multiple protein errors are no doubt also introduced at the level of mRNA through 5-fluorouracil incorporation or through DNA damage with ultraviolet irradiation. These treatments correspondingly produce cell proteins with abnormally high metabolic lability (Pine, 1967). Single mis-sense mutations in an individual protein may potentiate proteolysis, but individual proteins may require significant alteration to become labile. Thus no metabolic instability was found (M. J. Pine, unpublished data) in any of 20 of the most heat-sensitive variants of β-galactosidase of Langridge (1968) that had been obtained by serine or tyrosine substitution for normal sequences. In an examination of 32 completely inactive mis-sense mutants of β-galactosidase, only seven proved to be unstable, and their genetic loci were at only three specific sites of the *lacZ* gene (Zipser and Bhasvar, 1976). The RNA polymerase of *E. coli* in contrast appears to be more easily disposed to destruction. Although this enzyme is normally stable, the β-subunits of temperature-sensitive mutants disappear even at permissive temperatures (Kirschbaum *et al*., 1975).

(iv) Analogue-containing Proteins

Rabinovitz and Fischer (1964) first noted that haemoglobin containing the valine analogue α-amino- β-chlorobutyric acid was rapidly destroyed almost immediately after it had been synthesized by rabbit reticulocytes whereas normal haemoglobin is catabolically inert. The same discrimination in the breakdown of analogue-containing cell proteins has also been noted in *E. coli* (Pine, 1967; Goldberg, 1972). Among comparisons of different analogues of phenylalanine, methionine, and tryptophan there is a good correlation between the degree of structural perturbation which an analogue introduces into cell protein and its subsequent rate of breakdown, which can be in excess of the normal rate by two- to 10-fold, depending on the conditions of measurement (Goldberg, 1977; Pine, 1978). The arginine analogue canavanine almost entirely produces large dense and amorphous intracellular granules of denatured material, including even the soluble cell proteins. As the granules do not contain structure, nor membranes, nor protease activity, they are not autophagic structures (Prouty *et al*., 1975). The granules are also not necessary for proteolytic recognition, for *m*-fluoro- and dihydrophenylalanine-containing cell proteins are labile but produce no granules. Selenomethionine, which can sometimes completely replace methionine for normal growth, produces normally stable cell proteins (Pine, 1978).

2 Senescence

Orgel (1963) originally proposed an error catastrophe theory to account for aging as the result of accumulated errors in transcription and translation that could not be remedied or reversed by the cell. Protein abnormalities could be created or minimized by proteolysis, and protein turnover may eventually be unable or insufficient to remedy translational errors. In support of an error catastrophe in ageing nematodes, several isolated enzymes such as malate dehydrogenase, isocitrate lyase, and enolase of *Turbatrix aceti* have lower activity and heat stability in senescent than in young animals (Sharma *et al*., 1976). Also, the breakdown of labelled cell protein is greatly reduced in senescence (Prasanna and Lane, 1977).

3 SOS Functions

In *E. coli*, DNA damage sets off a pattern of emergency, or SOS responses mediated genetically by the *recA* locus. The effects include an error-prone repair of DNA damage and the inhibition of septation so as to produce filamentous cell forms. The *recA* response has been postulated to be effected by an induced protease activity that may initiate new cell functions or modify existing ones. The induction of vegetative growth of the chromosome-integrated lambda phage of *E. coli* is the best explored. The phage can proliferate when its normal translation product, the C-repressor protein is inactivated by a single mid-cleavage (Roberts *et al*., 1977). Only a small part of the cell protein is proteolysed, for ultraviolet irradiation and mitomycin treatment do not increase proteolytic rates of normal cell proteins (Pine, 1967).

4 Membrane Transport

In all biological systems, in general, exported proteins appear to be assembled as oversized precursors (preproteins) containing signal peptides or amino-terminal extensions of 11–30 predominantly hydrophobic residues that attach the nascent peptides to membranes. The signal peptides and their cleavage have been postulated in some schemes to be instrumental in intercalation across the lipid bilayer, either to enable some membrane proteins to span both the inner and outer surfaces, or to effect secretion (Blobel and Dobberstein, 1975). Preprotein forms are generally demonstrable when their mRNA is translated in heterologous or membrane-free systems, and are distinguished from, and indeed are precursors to, the pro- or inactive forms of enzymes or hormones that are stored or packaged.

The alkaline phosphatase of *E. coli* is exemplary of an external but fixed enzyme, freely accessible to the low molecular weight components of the environment, but retained within an operational locus called the periplasmic space. An enzymically active precursor several thousand daltons heavier than the phosphatase is the immediate product of translation *in vitro* (Inouye and Beckwith, 1977). The dynamics of excretion can be followed by labelling the protruding terminus of the tRNA-linked nascent pre-enzyme with an impermeable radiolabelled alkylating reagent. Thus polypeptide elongation at the inner side of the cell membrane, transport via the amino terminus, and maturative cleavage at the outer side of the membrane appear to be simultaneous events (Smith *et al*., 1977).

The protein component of the *E. coli* endotoxin which forms the major and outermost protein of the cell envelope is initially assembled as a precursor with 20 partly hydrophobic residues more than its final 85 residues. *N*-Terminal cysteine exposed by the cleavage is then conjugated to a lipid moiety, in turn linked to carbohydrate (Inouye *et al*., 1977). The penicillinase of *B. licheniformis* is also initially produced with a leader sequence (G. Blobel, unpublished results). After excretion and processing it is bound as a metastable membrane form to the surface lipid at the serine-26 residue. A peptide cleavage *C*-terminal to this residue releases soluble and freely diffusing extracellular enzyme from the fixed lipopeptide linkage (Yamamoto and Lampen, 1976).

The ribulose diphosphate carboxylase of *Chlamydomonas reinhardtii*, a major component of the chloroplast, is assembled from two distinct polypeptides. A large subunit is synthesized *in situ*. A small, 16 500 dalton subunit under control of the nuclear gene is produced in the cytosome as an immunologically reactive precursor of 20 000 daltons. The precursor is synthesized on free rather than bound ribosomes (Dobberstein *et al*., 1977). In this case the signal peptide would effect transport of the protein after, rather than during, its assembly.

It seems likely that signal peptides would be continuously extruded from the cell during growth, and they may retain their initiating *N*-formylmethionyl termini. Synthetic *N*-formylmethionyl oligopeptides are in fact chemotactic stimuli for neutrophils, as are solvent extracts of culture supernates. It is thus possible that signal peptides are used by the body defences for the detection of prokaryote invasion (Day *et al*., 1977).

5 Viral Metabolism

(i) *Early Translational Cleavages*

In prokaryotes such as *E. coli* an RNA virus can serve as a polycistronic mRNA to produce several proteins independently at different initiating sites. Since animal cell proteins in contrast are translated on monocistronic mRNAs, the single-stranded RNA genome of poliovirus is initially assembled as a hypothetical 250 000 dalton polyprotein. It is cleaved into three intermediate fragments even before translation is completed and slower cleavages then produce the dozen-odd final viral polypeptides (Jacobson *et al*., 1970; Korant, 1975). Although all products are obligatory produced in equivalent amounts, further scission of the generated viral RNA polymerase to an inactive product may additionally serve as a means of regulating activity late in infection (Korant, 1975). In one mutant involving a deletion in the poliovirus capsid, the translation product is overproteolysed to its complete destruction (Cole and Baltimore, 1973). Biosynthetic control of individual viral components from a single mRNA may be effected by occasional read-through. Thus the 35 S RNA of Rauscher leukaemia virus normally translates in abundance an 80 000 dalton precursor of the viral structural protein. In about 5 per cent of the translations the single termination codon for this protein is also translated, producing a 200 000 dalton polyprotein from which the rarer reverse transcriptase is cleaved (Murphy *et al*., 1978).

(ii) *Late Maturative Cleavages*

The structural proteins synthesized by coliphage T4 assemble spontaneously to form an intermediate or prehead structure. Cleavages in two of the major essential proteins metamorphose the prehead periphery to its mature, expanded hexagonal shape (Hsiao and Black, 1977). Several core proteins are also proteolysed, at least one in its entirety, in accompaniment to DNA entry. Most internal protein is thus displaced except for basic protein, and several acidic polypeptide cleavage products which contribute to internal packing through repulsive collapse of the DNA strands. All of these events are interrelated so that any mutation that prevents prehead assembly also allows the several proteolysed members of the prehead proteins to accumulate as the intact precursors (Laemmli, 1975).

Murine leukaemia viruses are initially assembled with the RNA in a loose, concentrically-coiled configuration with a 70 000 dalton precursor of group-specific antigen. The latter is then cleaved to components as small as 10 000 daltons to produce the collapsed RNA configuration of mature virus (Yoshinaka and Luftig, 1977).

The DNA-containing filamentous F1 phage of *E. coli* is characteristic of the small viruses that can bud out of prokaryote cells without lysing them. The coat preprotein of this virus is initially assembled on the cell membrane. A 20-amino acid signal sequence initially attaches the nascent protein to the membrane (Chang *et al*., 1978).

The functions of maturative protein cleavage in other viruses remain to be explored. Core proteins of the large poxviruses are cleaved specifically during the later stages of virion assembly (Moss and Rosenblum, 1973). During the synthesis of adenovirus, five unstable proteins are initially included in a proviral component, persist in the early virion, and eventually are processed to a core and a capsid component and to a very small peptide (Ishibashi and Maizel, 1974).

5 PROTEASES

Evidence has thus far accumulated at different levels of detail on the existence of enzyme systems involved in different aspects of intracellular protein turnover. Thus, for eukaryotes there exists a relatively detailed enzymology of isolated lysosomal proteases. In the prokaryotes, there is various enzymological and genetic evidence of specific or general impairment of degradation of cell proteins in *recA*$^-$, *deg*$^-$, and *pep*$^-$ mutants of *E. coli*, a protease mutant of sporulating *B. subtilis*, ATP-dependent proteases in *E. coli*, and proteases the unique location, activity, or origin of which suggest roles in the breakdown and processing of such specific products as viral or transported proteins.

1 Protein Initiation

All polypeptide synthesis appears to be initiated with the codon AUG which codes for a special initiating isoacceptor tRNA of methionine. In prokaryotes, this methionyl-tRNA$_{fmet}$ is first *N*-formylated by a reaction that can be dispensed with under folateless conditions (Pine *et al*., 1969) but otherwise improves the efficiency of initiation. The *N*-formylmethionine termini are retained on polypeptide chains synthesized in cell-free translating systems (Capecchi, 1966). Their persistence is due mostly to the lability of the formylase to reducing agents. *N*-Formylmethionine peptides are also substrates of the enzyme, but not free *N*-formylmethionine (Adams, 1968). The identity of the enzyme that removes methionine is still not established (see also Section 2.4.3). An aminopeptidase I (Vogt, 1970), a dipeptidase M (Brown, 1973), and other peptidases that attack methionyl peptides either lack affinity for the putative nascent termini or are not sufficiently restrictive for the single limited cleavage that takes place on most *N*-termini. The overall retention of methionine on the *N*-termini of mature cell proteins varies remarkably even among the Enterobacteriaceae, in contrast to near identity in the profiles of the seven favoured amino acids that are exposed by the maturative removal of methionine (Sarimo and Pine, 1969).

2 Enzymes Involved in the Proteolysis of Normal Cell Proteins in the *Enterobacteriaceae*

The protease or proteases responsible for the breakdown of normal cell proteins in *E. coli* and *Salmonella typhimurium* do not correspond to any of the known major proteases of these organisms (Kowit *et al*., 1976; Miller *et al*., 1976; Miller and Zipser, 1977). Cell-free extracts of *E. coli* show rapid proteolysis only for the very labile cell proteins produced during growth arrestment (Pine, 1965) and for protein fragments below 29 000 daltons (Kemshead and Hipkiss, 1976).

Intracellular peptides produced by protein turnover as well as incoming extracellular peptides can be hydrolysed by a system of at least eight peptidases (see also Section 2.4). These have been identified primarily by chromatographic separation in *E. coli* (Simmonds *et al*., 1976) and by electrophoresis, histochemical staining, and mutant selection in *S. typhimurium* (Miller and MacKinnon, 1974; McHugh and Miller, 1974). Of broadest specificity is aminopeptidase N, a surface enzyme located at the cellular poles (but see also Section 2.4.3). It attacks naphthylamides, permitting the selection of non-chromogenic, non-hydrolysing mutants (Miller and MacKinnon, 1974; Murgier *et al*., 1977). Within the cell are three more aminopeptidases, one with broad specificity and heat-stable, and two with narrower specificities. A tripeptidase and dipeptidase occur; the latter in different sources is either Zn^{2+}-containing or stimulated by other divalent cations. Aminoacyl and internal proline residues are cleaved respectively by a separate dipeptidase and iminopeptidase. From a naphthalamidase-deficient *S. typhimurium*, mutants for the heat-stable aminopeptidase, dipeptidase, and one additional aminopeptidase were obtained upon stepwise pencillin counterselection against the utilization of different leucine peptides by a leucine auxotroph. The quadruple mutant is unable to recycle its cell proteins during starvation and instead accumulates large amounts of low molecular weight peptides that permeate Sephadex G-25 gels. This multiple mutant is also slow to adapt to step-down conditions of growth (Yen, 1974). The mutant can still utilize some leucine peptides for growth, indicating that the full functioning of so many peptidases is more essential for the recycling of intracellular protein than for the utilization of exogenous peptides.

3 The *Deg-lon* Gene in *E. coli*

Degradation-deficient (*deg*$^-$) mutants could be obtained from an *E. coli* merodiploid with two β-galactosidase fragments that could not generate stable enzyme activity by phenotypic complementation (Bukhari and Zipser, 1973). Mutants with stable activity were selected by lactose growth and were found to degrade, at least several times more slowly than normal, both termination and unstable mis-sense mutants of β-galactosidase. Other defective proteins could also be spared, as indicated by the improved burst size of temperature-sensitive mutants of lambda and T5 phages at high temperature (Gottesman and Zipser, 1978) and by failure to degrade puromycin-released cell proteins selectively (Goldberg *et al*., 1975). The *deg*$^-$ mutant, however, appears on closer examination to correspond to the previously described pleiotropic mutation *lon*$^-$, which also confers many other defects possibly of membrane origin. *lon*$^-$ defects relating to death and filament induction by mutagens and inability to integrate lysogenic phages are lost upon acquisition of the second-site mutation *sul*$^-$ (or *sfi*$^-$), but the Deg$^-$ phenotype with its residual defect of capsule overproduction are still retained. Even these two latter characters are not necessarily dependent, for the Deg$^-$ phenotype persists in capsuleless derivatives (Gottesman and Zipser, 1978). Thus the *lon* gene could function by regulating or affecting a protease activity which appears generally responsible for the selective destruction of abnormal cell proteins, and other characters affected by the *lon* gene may have no direct relationship with the protease. The immediate genetic provenance of Deg as well as the attribution of any specific protease activity remain to be established. Thus *deg*$^-$ mutants do not lack any of the presently known protease or peptidase activities of *E. coli* (Miller and Zipser, 1977).

4 Membrane Transport

The involvement of signal peptide cleavage in transmembrane transport of proteins is borne up by the finding of membrane-bound proteases that can carry out the maturative cleavages of pre-enzymes. Protease activity for the *E. coli* alkaline phosphatase precursor is found only in the outermost component of the fractions of the cell envelope (Inouye and Beckwith, 1977). The protease for maturative cleavage of the coat of filamentous DNA phage F1 can be demonstrated with everted vesicles (Chang *et al.*, 1978) but whether this reaction is effected by a different protease on the inner membrane remains to be established. A protease activity present in *C. reinhardtii* membranes can effect signal peptide cleavage from the small subunit precursor of ribulose diphosphate carboxylase. Its activity is inhibited by sulphydryl reagents (Dobberstein *et al.*, 1977).

5 Viral Development

The proteases of the host cell, and indeed non-specific digestive proteases, can perform at least the initial proteolytic cleavages that generate the precursors of viral products from the large polyprotein translated by the small RNA picornaviruses. Host cell proteases can also perform similar cleavages upon the products of oncornavirus translation (Korant, 1975). After infection, proteolytic activity upon the viral translation products increases and more sites are cleaved (Butterworth, 1977). The latter stages of cleavage indeed are associated with the virion products produced, as exemplified in a picorna- and an oncornavirus respectively. Thus the 23 000 dalton capsid component of the encephalomyelocarditis virus copurifies with cleavage activity for a site near the amino terminus of the capsid precursor. The site is not attacked by host cell extracts (Lawrence and Thach, 1975). Likewise, a 15 000 dalton component of the internal protein of the Rous sarcoma virus can cleave its own precursor protein. The cleavage is specific in that it cannot be duplicated by the Rauscher murine leukaemia virus protein (Helm, 1977). In coliphage T4, a 21 000 dalton component of the prehead has been postulated to be the protease, or its activator, responsible for the assembly cleavages of head and internal proteins. Since this component is missing from the mature head, it may, like the major internal protein, be completely cleaved to small polypeptides (Laemmli, 1975). Maturation of at least four of these head and internal proteins involves cleavages of very similar glutamylalanyl sequences close to the amino termini (Isobe *et al.*, 1977).

6 Neutral Proteases in Eukaryotes

Extracts of *Tetrahymena pyriformis* contain a very active protease activity that is mostly neutral and occurs in soluble and particulate fractions (Levy and McConkey, 1977). Two of the three protease-carboxypeptidase activities in the osmotic vacuole of yeast are neutral. Roles have been suggested for these proteases and their cytoplasmic inhibitors in protein turnover and activation (see also Section 3.3.6). *Neurospora crassa* also contains an intracellular aminopeptidase along with a specific inhibitor (Kaehn and Kula, 1977).

7 Metamorphosis and Sporulation

Early in sporulation most bacilli actively secrete hydrolases for substrates such as starch and casein. More than one kind of protease may be excreted. Of greater relevance, however, is an intracellular serine protease that is indispensable for sporulation. In *B. cereus* a temperature-conditional mutant for this enzyme has been obtained (Cheng and Aronson, 1977*a*, *b*). The mutant cannot sporulate adequately unless a full complement of amino acids is provided in the medium, and correspondingly it cannot maintain the characteristic heightened rate of cell protein turnover throughout sporulation. In addition, the activated spores of this mutant germinate very slowly because the outermost, keratin-like coat protein is almost absent. In normal spore coat assembly, an initial 65 000 dalton precursor is cleaved, perhaps with some peptide loss, into 12 000 dalton monomers that pack together with the accompaniment of disulphide interchange. The mutant cannot cleave the precursor properly even though an accompanying and dispensible extracellular metalloprotease is still excreted. Thus the internal protease appears to have specific and overlapping roles in both intracellular proteolysis in the sporangium and protein maturation in the spore coat. The protease is chymotrypsin-like and inhibited by phenylmethane sulphonylfluoride. In *B. subtilis* this enzyme is accompanied by an inhibitor that is specific for this enzyme and is without effect upon an accompanying extracellular form of phenylmethane sulphonylfluoride-sensitive protease (Millet, 1977).

In yeast, ascospore formation is accompanied by the disintegration of the osmotic vacuole which contains the three major protease activities, with massive breakdown of the sporangial proteins (Holzer *et al*., 1975).

In germinating *B. cereus* spores a new phenylmethane sulphonylfluoride-sensitive protease is activated, but unlike the sporangial protease, it can effectively break down only the special class of low molecular weight basic proteins that are consumed during early germination (Setlow, 1976). This breakdown appears less controlled than the breakdown of freshly synthesized cell protein in being insensitive to energy poisoning (Setlow and Primus, 1975).

The synthesis of antibiotics is closely linked with the sporulation process in the Bacillaceae. The antibiotic bacitracin contains a number of unusual amino acids. It is initially assembled via thioacyl esters, probably of pantetheine, by a system that is insensitive to phenylmethane sulphonylfluoride (Froyshov, 1977). Very substantial amounts of the antibiotic can be liberated, however, from ribosomal and cell extracts of presporulating *B. licheniformis* by the action of its sporulation-specific seryl protease. The role of sporulation protease in antibiotic production or storage remains to be defined (Vitkovic and Sadoff, 1977).

8 ATP-dependent Proteases

The SOS functions induced by activation of the *recA* gene in response to DNA damage in *E. coli* all appear to be events linked to the activity of a common protease. Thus, error-prone DNA repair, cell filamentation, and prophage lambda induction are all suppressed by the papain-inhibiting antibiotic antipain at 0.5 mM, a concentration that does not generally inhibit other cell functions (Mehn *et al*., 1977). The protease may be the product of the *recA* gene or be modified by it. Extracts of *recA* constitutive cells specifically cleave added C-repressor protein of the lambda phage, an event that initiates its vegetative growth. The repressor is attacked at only one site and presumably must be in fully native condition, as purified repressor is not attacked. The cleavage reaction requires added ATP and a divalent cation (Roberts *et al*., 1977).

In addition, a more generalized ATP-stimulated neutral protease activity has been demonstrated in *E. coli*. The system is soluble and attacks incomplete proteins such as small β-galactosidase termination fragments and free globin. The enzyme activity is sensitive to *N*-ethylmaleimide and EDTA but not to phenylmethane sulphonylfluoride. It retains its ATP dependence upon purification (Murakami *et al*., 1977).

The best-investigated model of an ATP-dependent protease is the soluble system that can break down analogue-containing haemoglobin in rabbit reticulocytes. This system requires intact ATP and its α, β- or β, γ-methylene analogues. The protease degrades haemoglobin with different degrees of molecular perturbation to different extents, producing small polypeptide fragments (Etlinger and Goldberg, 1977).

6 CONTROL OF INTRACELLULAR PROTEOLYSIS

The rate of proteolysis of *E. coli* cell proteins is subject to considerable variation, depending on the nutritional state of the cell. Although some nutritional depletions may stop growth without altering the proteolytic rate (Pine, 1970; Nath and Koch, 1971), the depletion of almost any essential element, energy source, amino acid, purine, or pyrimidine stimulates proteolysis of the more stable cell proteins. This stimulation is virtually immediate and can be detected within a few minutes after depletion, and restoration of nutrilite lowers proteolysis back to normal levels with corresponding speed (Pine, 1973*a*). Thus, intracellular proteolysis is most likely to be regulated by feedback effects or protease modification, at least within moderate time intervals, rather than by synthesis or breakdown of the protein itself.

1 Stringent Control

The increase in proteolytic rate during nutritional depletion appears to be one of the events involved in the regulatory regimen known as stringent control. Many additional cellular events such as the synthesis of stable RNA and phospholipid are correspondingly inhibited and the synthesis of many individual proteins appears to be favoured by this regulation (Taguchi *et al*., 1977). The stringent response can be signalled by the accumulation of uncharged tRNA, and is thought to be mediated by the nucleotide guanosine-3′-diphosphate-5′-diphosphate (ppGpp). This mediator is synthesized from ATP and GDP by the enzyme product of the *relA* gene, called stringent factor. Adaptive synthesis of ppGpp is complex, involving the additional participation of a complete, but amino acid-deficient ribosomal assembly. This would include a functional form of the 50s protein L11, primed with mRNA plus cognate but uncharged tRNA at the amino acid acceptor (A) site (Block and Haseltine, 1975). Relaxed (*relA*$^-$) mutants are unable to increase their basal ppGpp levels during amino acid starvation, and correspondingly intracellular proteolysis along with other stringently controlled events remains unchanged (Sussman and Gilvarg, 1969). *RelA*$^-$ mutants can still accumulate ppGpp during carbon starvation. This may be through a second regulatory mechanism, the inhibition of the phosphorylative pathway for ppGpp breakdown as controlled by the *spoT* locus (Lazzarini and Stamminger, 1975). Correspondingly, *relA*$^-$ mutants can still increase proteolysis during carbon starvation (Sussman and Gilvarg, 1969).

During stringent control, amino acid deficiencies increase intracellular proteolysis more or less in accordance with the abundance of the amino acid in the cell protein. The increases also correlate well in degree with the extent of stringent suppression of RNA synthesis (Pine, 1973*b*). Amino acyl-tRNA synthetase inactivation also stimulates intracellular proteolysis, and correspondingly these increases are not found in *relA*$^-$ mutants (St. John *et al*., 1978). When chloramphenicol stops protein synthesis, tRNAs can become charged with the amino acids generated by intracellular proteolysis. Correspondingly, chloramphenicol reverses the enhanced proteolysis due to amino acid starvation, but it cannot reverse the same event when it is due to inactivation of amino acyl-tRNA synthetase. Tetracycline, in contrast, also reverses the latter event. This antibiotic is distinctive in being able to suppress ppGpp synthesis directly (St. John *et al*., 1978). Starvation of cell precursors other than amino acids, such as

Mg^{2+} or PO_4^{3-}, also stimulates intracellular proteolysis, but these may all relate to increases again in ppGpp levels (Goldberg and St. John, 1976).

2 Energy Control

Carbon starvation also stimulates intracellular proteolysis in accordance with the fall in the level of free amino acid pools, and presumably of amino acyl-tRNA (Pine, 1975). A more important effect, however, is the preservation of the ppGpp level by its decreased rate of breakdown (St. John and Goldberg, 1978; Voellmy and Goldberg, 1977). Severe energy depletion by removal of energy source and poisoning of oxidation and fermentative pathways inhibits and may almost entirely eliminate intracellular proteolysis. Thus energy supply has two overlapping roles, in stimulating the stringent response under moderate restriction and in supporting intracellular proteolysis at very low levels (Pine, 1975; St. John and Goldberg, 1978; Olden and Goldberg, 1977). The energy requirement for intracellular proteolysis appears to be related directly to a minimal ATP requirement at about 15 per cent of the normal cellular ATP level. Proteolysis cannot be maintained by direct membrane transduction of energy, for example by membrane-associated oxidative enzymes (Olden and Goldberg, 1977). The partial inhibition of intracellular proteolysis by a number of protease inhibitors could in fact be related to energy metabolism too, since the inhibitions occur only when glucose-deficient, but not nitrogen-deficient, cells are treated (Rafaeli-Eshkol and Hershko, 1974). Under the specific condition of glucose deficiency, these inhibitors, in fact, become energy poisons and deplete the cell of glutathione (Shechter *et al*., 1973). It is possible that glutathione plays a special role in intracellular proteolysis, at least in higher animal cells (Yushok *et al*., 1977), but it is also apparent that such poisoning produces many additional effects with no proven relationship to intracellular proteolysis (Rossman *et al*., 1974). ATP may have a number of functions in intracellular proteolysis. A number of ATP- dependent proteases have already been described, and they may be the major effectors of protein turnover (Etlinger and Goldberg, 1977; Murakami *et al*., 1977; Roberts *et al*., 1977). In the eukaryotes, ATP may also be a requirement for the active concentration of protons into the lysosome beyond the amounts accumulated by Donnan equilibrium (Schneider and Cornell, 1978). Autophagic digestion must also require energy.

3 Cyclic Adenosine-3′,5′-Monophosphate

In animal cell cultures, intracellular proteolysis is altered in common with other effects such as those induced by hormones. These have been referred to as the pleiotypic response (Hershko *et al*., 1971). Some effects could be putatively mediated through cAMP. In *E. coli*, however, cAMP-less mutants have normal proteolytic responses to glucose starvation, and administered cAMP is without proteolytic effects (Pine, 1975).

4 Fidelity of Translation

The cell protein of *E. coli* is efficiently cycled during amino acid starvation, the rate of protein resynthesis equalling the rate of decomposition (Mandelstam, 1958*a*, *b*). Considering the diversity of ribosome-associated elements involved in ppGpp synthesis, it is likely that overall rates of protein synthesis and breakdown are both mutually adjusted, but they may not be optimal for proper translation. In one study, β-galactosidase synthesized by a *relA*$^{-}$ mutant during arginine starvation was more thermosensitive and may have had a lower activity than in stringent cells (Hall and Gallant, 1972). In another study, the antibody-reactive β-galactosidase appeared to have no enzyme activity when synthesized in histidine starvation even in stringent *E. coli*. Since β-galactosidase is an unusually long molecule, it may be correspondingly more likely to accumulate translational errors (Subrahmanyam and Das, 1977). Incomplete polypeptides may also be prematurely ejected from the ribosomes, and indeed this phenomenon has been suggested to account for the comparatively low molecular weight, metabolically unstable proteins produced during amino acid deficiency (Brunschede and Bremer, 1971). Direct evidence for the production of incomplete β-galactosidase during starvation is lacking (Subrahmanyam and Das, 1976). A mutant of *E. coli*, conditional for the hydrolysis of ribosomally ejected polypeptide-tRNAs, accumulates them at non-permissive temperatures (Menninger, 1976). The rate of such accumulation appears sufficient to account for at least part of the unstable protein populations produced during growth or starvation.

7 MECHANISMS OF INTRACELLULAR PROTEOLYSIS

1 Site of Protein Breakdown

Inherent in the structure of any cell protein is its predisposition to breakdown, as determined either by its direct vulnerability to protease action or by its accessibility to the proteolytic apparatus. In the cells of higher animals at least, it is clear on morphological grounds that lysosomes are involved in the autophagic digestion of organelles (Gordon, 1973). In the higher animals the lysosomes of the liver cell are transformed, under conditions that favour protein breakdown, into larger and denser forms, mostly for the purpose of generalized breakdown of cell components, including endoplasmic reticulum (Mortimore *et al*., 1978). The simplest model of membrane turnover found thus far is of the plasma membrane of cultured hepatoma where all proteins disappear at uniform rates (Doyle, *et al*., 1978), suggesting intact engulfment. Cell protein breakdown is partly inhibited by the antibiotic pepstatin, an inhibitor of pepsin-like enzymes such as the lysosomal protease cathepsin D (Dean, 1978). Also the bases chloroquine (Poole *et al*., 1978) and ammonia (Seglen and Reith, 1976) act as inhibitors of intracellular proteolysis of different animal cells by virtue of their ability to concentrate in and swell lysosomes and to inhibit their general functions. In some cells, however, chloroquine affects primarily the breakdown of exogenous cell proteins and of the additional cell proteins broken down during starvation (Poole *et al*., 1978). The labile cell proteins as well as abnormal amino acid analogue-containing proteins preferentially bind to lysosomal granules. This preferential attachment has been suggested to play a role in proteolytic selection where discrimination is evident in the half-life of individual proteins (Dean, 1978). An alternative older model of Haider and Segal (1972) proposed non-discriminating entry by lysosomal engulfment, followed by selective escape of the most resistant proteins. In other schemes, proteolysis may combine relatively uniform degradation in lysosomes with a second system that degrades individual proteins selectively (Dice *et al*., 1978). Although prokaryote cells have no defined digestive organelle such as the lysosome, the labile termination mutant X-90 of β-galactosidase tends to associate with membrane fractions of the cell (Kowit and Goldberg, 1977). The denatured granules of analogue-containing proteins are not enclosed by membrane and can eventually disappear without obvious participation of membrane structures (Prouty *et al*., 1975). The abnormal protein degrading system of the reticulocyte contains no lysosomes. The distinctive ATP-requiring protease activity in these cells appears to be completely soluble (Etlinger and Goldberg, 1977). Selective breakdown at least of such abnormal proteins or of specialized proteins (Roberts *et al*., 1977; Setlow, 1976) could be effected without compartmentation. It remains to be determined whether the sporulation-specific protease of the bacterial sporangium (Cheng and Aronson, 1977*a*, *b*) breaks down the enzymes that selectively and precipitously disappear at the commencement of sporulation (Waindle and Switzer, 1973).

2 Determinants of Proteolytic Susceptibility

The most labile cell proteins, identified by short-term labelling, are also inherently more susceptible to proteolysis by conventional proteases that have non-selective roles in digestion (Goldberg *et al*., 1975; Goldberg, 1977). Also, inherent proteolytic susceptibility may determine the relative priority of breakdown of individual cell proteins and their stability under different conditions. Thus, when growth stops the most labile cell proteins cannot be regenerated as quickly as before and hence would offer less substrate competition against the more inert cell proteins. It has been suggested that the lack of such competition could in itself explain in part the increased spectrum of cell proteins that are broken down in *E. coli* when growth is arrested, without the necessity of postulating separate proteolytic systems (Pine, 1970).

Dissociation has been implicated as a primary step in proteolytic attack upon supramolecular protein assemblies or proteins composed of subunits. This would account for different aspects of heterogeneity in the turnover of individual protein components of membranes (Schimke, 1975). In temperature-sensitive mutants of RNA polymerase in *E. coli*, the β_1 subunit decomposes as fast as or faster than the β subunit, even when the latter contains the mutation, and both subunits disappear rapidly when they are produced in excess (Kirschbaum *et al*., 1975). Thus the relaxed or dissociated—rather than mutated—state of the subunit may be the signal form for proteolysis.

Complete denaturation or a reversible unfolding to reveal susceptible peptide sequences has been postulated in a number of schemes in the events preceding proteolysis. Assuming that this susceptibility is on the average related to irreversible denaturation *in vitro*, approximate correlations have been noted between the stability of individual enzymes *in vivo* and their stability to heat and acid, provided that specific cofactors and a number of non-specific cell components are present (Bond, 1975, 1976; Haider and Segal, 1972). Binding of an enzyme to any ligand, either non-specifically or at specific sites for substrate or cofactors, generally stabilizes tertiary conformation and also decreases resistance to proteases *in vitro*. There are, however, few convincing demonstrations that stability *in vivo* is actually increased by binding with substrate (Schimke *et al*., 1965) or with a modifying factor of an enzyme (Dunaway and Segal, 1976). Other general protein features that have been associated with increased proteolysis *in vivo* are: the relative size of the whole molecule or of the polypeptide component (Dice *et al*., 1973); the protein's relative acidity, expressed as isoelectric point (Dice and Goldberg, 1975) or the ratio of charge to subunit surface area (Momany *et al*., 1976); or the presence of carbohydrate. Such features may not be characteristic of proteolysis in all biological systems or in all physiological states (Dice *et al*., 1978).

3 Intermediates in Breakdown

The simplest and most economical model of intracellular proteolysis would envision a rate-limiting cleavage of substrate at a specific or most susceptible recognition site, followed by rapid and complete recycling to free amino acids by secondary peptidase action as unprotected sites are increasingly exposed. Thus, decomposition would be strictly first-order, at least under steady-state conditions and without significant accumulation of intermediates. The efficient destruction of NADP-linked glutamate dehydrogenase of carbon-starved *Candida utilis* is indicated by the rapid disappearance of immunoreactive material concomitantly with enzyme activity (Hemmings, 1978). Because breakdown is examined most often during growth transition the overall kinetics may imply little mechanistically. The most discriminating studies of initial cleavage of intracellular proteins have been made with deletion and termination mutants of β-galactosidase. These can be examined for overall intactness of both the whole molecule and a limited auto-alpha peptide sequence near the *N*-terminus by means of complementation of activity in cell extracts (Lin and Zabin, 1972). The first-order decay rates of the auto-alpha activity in over one-third of the examined labile mutants show pronounced breaks (Lin and Zabin, 1972). This could indicate substrate heterogeneity, alternate sites of cleavage, or, as conjectured by Kowit and Goldberg (1977), alternate degradation systems. Two mutants of β-galactosidase produce relatively large molecular weight stable intermediates. Thus the nearly normal-sized 120 000 dalton termination mutant polypeptide X-90 is accompanied by a metastable product produced by an endoproteolytic cleavage 30 000 daltons from the *C*-terminus. Disappearance of total auto-alpha activity parallels the disappearance of this product, indicating a second cleavage of the molecule as the rate-limiting step in the final decomposition. In the *deg*$^-$ mutant, the steady-state level of the parent molecule is increased while that of the scission product is decreased (Kowit and Goldberg, 1977). The largest detectable product of the β-galactosidase deletion mutant U163 is a polypeptide of 65 000 daltons, 20 000 daltons smaller than expected from the size of the deletion. This is accompanied by a second derivative that is still associated with a 12 000 dalton fragment cleaved from the amino terminus (Villarejo and Zabin, 1974). In the *deg*$^-$ cell, the full transcript of U163 still fails to appear but, like the X-90 mutant, the heavier peptide species is favoured over the smaller during steady-state conditions. Thus at least for abnormal proteins a putative *deg*$^+$ protease may determine the initial recognition-cleavage step, but breakdown is not necessarily efficient afterwards. Even normal microbial proteins undergo a number of different fates in their disposal. Thus after cell aggregation of the starving slime mould *Dictyostelium discoideum*, galactose transferase is mostly extruded preferentially into the medium (Sussman, 1965). Enzymic deadaptation may occur by loss of a cofactor rather than the protein moiety. This is true for the energy-dependent loss of lysine decarboxylase activity in *Bacillus cadaveris* which occurs through destruction of the pyridoxal phosphate coenzyme only (Mandelstam, 1954). The enzymes of the *trp* locus of *E. coli* for tryptophan biosynthesis are stable to many nutritional deficiencies except for the TrpC protein which has dual activities for phosphoribosyl anthranylate isomerase and indole glycerophosphate synthetase. The latter activity is relatively low initially and disappears more readily, sometimes with the former activity still intact. Since these losses are also accompanied by lost immunoreactivity, it has been suggested that only a part of the TrpC protein may be degraded (Mosteller *et al.*, 1977). Likewise upon inactivation of the C repressor protein during induction of the lambda phage of *E. coli* K-12, only a fragment comprising about half of the polypeptide chain is actually destroyed (Roberts *et al.*, 1977). During deadaptation from galactose utilization in yeast, all of the galactose-metabolizing enzymes remain intact, while the galactose permease is altered to a form with low substrate affinity. A proteolytic model may yet exist for this alteration (Matern and Holzer, 1977).

REFERENCES

Adams, J. M. (1968). *J. Mol. Biol.* **33**, 571.
Aguanno, J. J. and Larrabee, A. R. (1976). *J. Bacteriol.*, **128**, 735.
Arias, I. M., Doyle, D. and Schimke, R. T. (1969). *J. Biol. Chem.*, **244**, 3303.
Bauza, M. T., DeLoach, J. R., Aguanno, J. J. and Larrabee, A. R. (1976). *Arch. Biochem. Biophys.*, **174**, 344
Berlin, C. M. and Schimke, R. T. (1965). *Mol. Pharmacol.*, **1**, 149
Bernlohr, R. (1972). *J. Biol. Chem.*, **247**, 4893
Betz, H. (1976). *Biochem. Biophys. Res. Commun.*, **72**, 121
Betz, H and Weiser, U. (1976*a*). *Eur. J. Biochem.*, **62**, 65
Betz, H. and Weiser, U. (1976*b*). *Eur. J. Biochem.*, **70**, 385
Blobel, G. and Dobberstein, B. (1975). *J. Cell Biol.*, **67**, 835
Block, R. and Haseltine, W. A. (1975). *J. Biol. Chem.*, **250**, 1212
Bond, J. S. (1975). In *Intracellular Protein Turnover* (R. T. Schimke and N. Katunma, eds.), Academic Press, New York, p. 281
Bond, J. S. (1976). *Biochim. Biophys. Acta*, **451**, 238
Bond, J. S. and Bradley, S. G. (1978). Personal communication
Bond, J. S., Pang, R. H. L. and Bradley, S. G. (1977). *Develop. Indust. Microbiol.*, **18**, 419
Borek, E., Ponticorvo, L. and Rittenberg, D. (1958). *Proc. Natl. Acad. Sci. U.S.A.*, **44**, 369
Brown, J. L. (1973). *J. Biol. Chem.*, **248**, 409
Brunschede, H. and Bremer, H. (1971). *J. Mol. Biol.*, **57**, 35
Buck, C. A. and Warren, L. (1977). *J. Cell Physiol.*, **89**, 187
Bukhari, A. I. and Zipser, D. (1973). *Nature New Biol.*, **243**, 238–241
Butterworth, B. E. (1977). *Curr. Topics Microbiol. Immunol.*, **77**, 141
Capecchi, M. R. (1966). *Proc. Natl. Acad. Sci. U.S.A.*, **55**, 1517
Chang, C. N., Blobel, G. and Model, P. (1978). *Proc. Natl. Acad. Sci. U.S.A.*, **75**, 361
Chao, L. (1977). *Biochem. Biophys. Res. Commun.*, **77**, 42
Chee, P. Y. and Swick, R. T. (1976). *J. Biol. Chem.*, **251**, 1029
Cheng, Y.-S. E. and Aronson, A. I. (1977*a*). *Proc. Natl. Acad. Sci. U.S.A.*, **74**, 1254
Cheng, Y.-S. and Aronson, A. I. (1977*b*). *Arch. Microbiol.*, **115**, 61
Cole, C. N. and Baltimore, D. (1973). *J. Mol. Biol.*, **76**, 325
Day, A. R., Radding, J. A. and Freer, R. J. (1977). *FEBS Lett.*, **77**, 291
Dean, R. T. (1978). In *Protein Turnover and Lysosomal Function* (H. L. Segal and D. Doyle, eds.), Academic Press, New York, p. 29
DeDuve, C. and Wattiaux, R. (1966). *Annu. Rev. Physiol.*, **28**, 435

Dice, J. F. and Goldberg, A. L. (1975). *Proc. Natl. Acad. Sci. U.S.A.*, **72**, 3893

Dice, J. F., Dehlinger, P. J. and Schimke, R. T. (1973). *J. Biol. Chem.*, **248**, 4220

Dice, J. F., Walker, C., Byrne, B. and Cardiel, A. (1978). *Proc. Natl. Acad. Sci. U.S.A.*, **75**, 2093

Dobberstein, B., Blobel, G. and Chua, N.-H. (1977). *Proc. Natl. Acad. Sci. U.S.A.*, **74**, 1082

Doyle, D., Baumann, H., England, B., Friedman, E., Hou, E. and Tweto, J. (1978). In *Protein Turnover and Lysosomal Function* (H. L. Segal and D. Doyle, ed.), Academic Press, New York, p. 689

Dunaway, G. A. and Segal, H. L. (1976). *J. Biol. Chem.*, **251**, 2323

Eagle, H., Piez, K. A., Fleishman, R. and Oyama, V. I. (1959). *J. Biol. Chem.*, **234**, 592

Etlinger, J. D. and Goldberg, A. L. (1977). *Proc. Natl. Acad. Sci. U.S.A.*, **74**, 54

Froyshov, O. (1977). *FEBS Lett.*, **81**, 315

Funabuki, R., Watanabe, Y., Nishigawa, N. and Hareyama, S. (1976). *Biochim. Biophys. Acta*, **451**, 143.

Goldberg, A. L. (1972). *Proc. Natl. Acad. Sci. U.S.A.*, **69**, 422

Goldberg, A. L. (1977). In *Intracellular Protein Catabolism*, Vol. II (V. Turk and N. Marks, eds), Plenum Publishing, New York, p. 49

Goldberg, A. L. and Dice, J. F. (1974). *Annu. Rev. Biochem.*, **43**, 835

Goldberg, A. L. and St. John, A. C. (1976). *Annu. Rev. Biochem.*, **45**, 747

Goldberg, A. L., Olden, K. and Prouty, W. F. (1975). In *Intracellular Protein Turnover* (R. T. Schimke and N. Katanuma, eds), Academic Press, New York.

Gordon, A. H. (1973). In *Lysosomes in Biology and Pathology*, Vol. 3 (J. T. Dingle, ed.), North Holland, Amsterdam, p. 89

Gottesman, S. and Zipser, D. (1978). *J. Bacteriol.*, **133**, 844

Haider, M. and Segal, H. L. (1972). *Arch. Biochem. Biophys.*, **148**, 228

Hall, B. and Gallant, J. (1972). *Nature New Biol.*, **237**, 131

Halvorson, H. O. (1958). *Biochim. Biophys. Acta*, **27**, 255

Hansen, R. J., Switzer, R. L., Hinze, H. and Holzer, H. (1977). *Biochim. Biophys. Acta*, **496**, 103

Heller, J. S., Fong, W. F. and Canellakis, E. S. (1977). *Proc. Natl Acad. Sci. U.S.A.*, **73**, 1858

Helm, K. v.d. (1977). *Proc. Natl. Acad. Sci. U.S.A.*, **74**, 911

Hemmings, B. A. (1978). *J. Bacteriol.*, **133**, 867

Hershko, A. and Fry, M. (1975). *Annu. Rev. Biochem.*, **44**, 775

Hershko, A., Mamont, P., Shields, R. and Tompkins, G. M. (1971). *Nature New Biology*, **232**, 206

Hewitt, J. and Kogut, M. (1977). *Eur. J. Biochem.*, **74**, 285

Hogness, D. S., Cohn, M. and Monod, J. (1955). *Biochim. Biophys. Acta*, **16**, 99

Holzer, H., Betz, H. and Ebner, E. (1975). *Curr. Topics Cell. Regul.*, **9**, 103

Hsiao, C. L. and Black, L. W. (1977). *Proc. Natl. Acad. Sci. U.S.A.*, **74**, 3652

Humphrey, T. J. and Davies, D. D. (1976). *Biochem. J.*, **156**, 561

Inouye, H. and Beckwith, J. (1977). *Proc. Natl. Acad. Sci. U.S.A.*, **74**, 1440

Inouye, S., Wang, S. S., Sekizawa, J., Halegoua, S. and Inouye, M. (1977). *Proc. Natl. Acad. Sci. U.S.A.*, **74**, 1004

Ishibashi, M. and Maizel, J. V. (1974). *Virology*, **57**, 409

Isobe, T., Black, L. W. and Tsugita, A. (1977). *J. Mol. Biol.*, **110**, 165

Jacobson, M. F., Asso, J. and Baltimore, D. (1970). *J. Mol. Biol.*, **49**, 657

Johnston, G. C., Singer, R. A. and McFarlane, E. S. (1977). *J. Bacteriol.*, **132**, 723

Kaehn, K. and Kula, M. R. (1977). *FEBS Lett.* **79**, 85

Kairallah, E. (1978). In *Protein Turnover and Lysosomal Function* (H. L. Segal and D. Doyle, eds), Academic Press, New York, p. 89

Kemshead, J. T. and Hipkiss, A. R. (1976). *Eur. J. Biochem.*, **71**, 185

Kirschbaum, J. B., Claeys, I. V., Nasi, S., Molhott, B. and Miller, J. H. (1975). *Proc. Natl. Acad. Sci. U.S.A.*, **72**, 2375

Koch, A. L. (1963). *Ann. N.Y. Acad. Sci.*, **102**, 602

Koch, A. L. and Levy, H. R. (1955). *J. Biol. Chem.*, **217**, 947

Korant, B. D. (1975). In *Proteases and Biological Control* (E. Reich, D. B. Rifkin and E. Shaw, eds), Cold Spring Harbor Laboratory, Cold Spring Harbor, N.Y., p. 621

Kowit, J. D. and Goldberg, A. L. (1977). *J. Biol. Chem.*, **252**, 8350

Kowit, J. D., Choy, W.-N., Champe, S. P. and Goldberg, A. L. (1976). *J. Bacteriol.*, **128**, 776
Kuehn, G. D. and McFadden, B. A. (1968). *J. Bacteriol.*, **95**, 937
Laemmli, U. K. (1975). In *Proteases and Biological Control* (E. Reich, D. B. Rifkin and E. Shaw, eds), Cold Spring Harbor Laboratory, Cold Spring Harbor, N.Y., p. 661
Langridge, J. (1968). *J. Bacteriol.*, **96**, 1711
Larrabee, K. L. and Larrabee, A. R. (1978). In *Protein Turnover and Lysosomal Function* (H. L. Segal and D. Doyle, eds), Academic Press, New York, p. 287
Lau, R. H., Mackenzie, M. M. and Doolittle, W. F. (1977). *J. Bacteriol.*, **132**, 771
Lawrence, C. and Thach, R. E. (1975). *J. Virol.*, **15**, 918
Lazzarini, R. A. and Stamminger, G. (1975). In *Control of Ribosome Synthesis* (N. C. Kjeldgaard and O. Maalöe, eds), Academic Press, New York, p. 393
Levy, M. R. and McConkey, C. L. (1977). *J. Cell. Physiol.*, **90**, 253
Lin, S. and Zabin, I. (1972). *J. Biol. Chem.*, **247**, 2205
Mandelstam, J. (1954). *J. Gen. Microbiol.*, **11**, 426
Mandelstam, J. (1958*a*). *Biochem. J.*, **69**, 103
Mandelstam, J. (1958*b*). *Biochem. J.*, **69**, 110
Matern, H. and Holzer, H. (1977). *J. Biol. Chem.*, **252**, 6399
Mazon, M. (1978). *J. Bacteriol.*, **133**, 780
McHugh, G. L. and Miller, C. G. (1974). *J. Bacteriol.*, **120**, 364
Mehn, M. S., Rossman, T. and Troll, W. (1977). *Proc. Natl. Acad. Sci. U.S.A.*, **74**, 1152
Menninger, J. R. (1976). *J. Biol. Chem.*, **251**, 3392
Miller, C. G. and Mackinnon, K. (1974). *J. Bacteriol.*, **120**, 355
Miller, C. G. and Zipser, D. (1977). *J. Bacteriol.*, **130**, 347
Miller, C. G., Heiman, C. and Yen, C. (1976). *J. Bacteriol.*, **127**, 490
Millet, J. (1977). *FEBS Lett.*, **74**, 59
Millward, D. J. (1970). *Clin. Sci.*, **39**, 577
Momany, F. A., Aguanno, J. J. and Larrabee, A. R. (1976). *Proc. Natl. Acad. Sci. U.S.A.*, **73**, 3073
Mortimore, G. E., Ward, W. F. and Schworer, C. M. (1978). In *Protein Turnover and Lysosomal Function* (H. L. Segal and D. Doyle, eds), Academic Press, New York, p. 67
Mortimore, G. E., Woodside, K. H. and Henry, J. E. (1972). *J. Biol. Chem.*, **247**, 2776
Moss, B. and Rosenblum, E. N. (1973). *J. Mol. Biol.*, **81**, 267
Mosteller, R. D., Nishimoto, K. R. and Goldstein, R. V. (1977). *J. Bacteriol.*, **131**, 153
Murakami, K., Voellmy, R. and Goldberg, A. L. (1977). *Fed. Proc. Fed. Am. Soc. Exp. Biol.*, **36**, 764
Murgier, M., Pelissier, C., Lazdunski, A., Bernadac, A. and Lazdunski, C. (1977). *Eur. J. Biochem.*, **74**, 425
Murphy, E. C., Kopchick, J. J., Watson, K. F. and Arlinghaus, R. B. (1978). *Cell*, **13**, 359
Nachmias, V. T. (1966). *Exp. Cell Res.*, **43**, 583
Nath, K. and Koch, A. L. (1971). *J. Biol. Chem.*, **246**, 6956
Olden, K. and Goldberg, A. L. (1977). *Proc. FEBS, Copenhagen*.
Orgel, L. E. (1963). *Proc. Natl. Acad. Sci. U.S.A.*, **49**, 517
Pine, M. J. (1965). *Biochim. Biophys. Acta*, **104**, 439
Pine, M. J. (1966). *J. Bacteriol.*, **92**, 847
Pine, M. J. (1967). *J. Bacteriol.*, **93**, 1527
Pine, M. J. (1970). *J. Bacteriol.*, **103**, 207
Pine, M. J. (1972). *Annu. Rev. Microbiol.*, **26**, 103
Pine, M. J. (1973*a*). *J. Bacteriol.*, **115**, 107
Pine, M. J. (1973*b*). *J. Bacteriol.*, **116**, 1253
Pine, M. J. (1975). In *Intracellular Protein Turnover* (R. T. Schimke and N. Katanuma, eds), Academic Press, New York, p. 65
Pine, M. J. (1978). *Antimicrobial Agents Chemother.*, **13**, 676
Pine, M. J. and Schimke, R. T. (1978). In *Protein Turnover and Lysosomal Function* (H. L. Segal and D. Doyle, eds), Academic Press, New York, p. 273
Pine, M. J., Gordon, B. and Sarimo, S. S. (1969). *Biochim. Biophys. Acta*, **179**, 439
Poole, B. (1971). *J. Biol. Chem.*, **216**, 6587

Poole, B., Ohkuma, S. and Warburton, M. (1978). In *Protein Turnover and Lysosomal Function* (H. L. Segal and D. Doyle, eds), Academic Press, New York, p. 43
Prasanna, H. R. and Lane, R. S. (1977). *Fed. Proc. Fed. Am. Soc. Exp. Biol.,* **36**, 292
Prouty, W. F., Karnovsky, M. J. and Goldberg, A. L. (1975). *J. Biol. Chem.*, **250**, 1112
Rabinovitz, M. and Fischer, J. M. (1964). *Biochim. Biophys. Acta*, **91**, 313
Rafaeli-Eshkol, D. and Hershko, A. (1974). *Cell*, **2**, 31
Rechcigl, M. (1971). *Enzyme Synthesis and Degradation in Mammalian Systems*, University Park Press, Baltimore, Md
Roberts, J. W., Roberts, C. W. and Mount, D. W. (1977). *Proc. Natl. Acad. Sci. U.S.A.*, **74**, 2283
Rossman, T., Norris, C. and Troll, W. (1974). *J. Biol. Chem.*, **249**, 3412
Rothfield, L. and Pearlman-Kothencz, M. (1969). *J. Mol. Biol.*, **44**, 477
St. John, A. C. and Goldberg, A. L. (1978). *J. Biol. Chem.*, **253**, 2705
St. John, A. C., Conklin, K., Rosenthal, E. and Goldberg, A. L. (1978). *J. Biol. Chem.*, **253**, 3945
Sarimo, S. S. and Pine, M. J. (1969). *J. Bacteriol.*, **98**, 368
Scherer, C. G. and Boylen, C. W. (1977). *J. Bacteriol.*, **132**, 584
Schimke, R. T. (1964). *J. Biol. Chem.*, **239**, 3808
Schimke, R. T. (1973). *Adv. Enzymol. Relat. Areas Mol. Biol.*, **37**, 135
Schimke, R. T. (1975). In *Intracellular Protein Turnover* (R. T. Schimke and N. Katanuma, eds), Academic Press, New York, p. 173
Schimke, R. T., Sweeney, E. W. and Berlin, C. M. (1965). *J. Biol. Chem.*, **241**, 322
Schlessinger, D. and Ben-Hamida, F. (1966). *Biochim. Biophys. Acta*, **119**, 171
Schneider, D. L. and Cornell, E. (1978). In *Protein Turnover and Lysosomal Function* (H. L. Segal and D. Doyle, eds), Academic Press, New York, p. 59
Schoenheimer, R. (1942). *Dynamic State of Body Constituents*, Harvard University Press, Cambridge, Mass.
Scornik, O. A., Botbol, V., Conde, R. and Amils, R. (1978). In *Protein Turnover and Lysosomal Function* (H. L. Segal and D. Doyle, eds), Academic Press, New York, p. 119
Seglen, P. O. and Reith, A. (1976). *Exp. Cell Res.*, **100**, 276
Setlow, P. (1975*a*). *J. Biol. Chem.*, **250**, 631
Setlow, P. (1975*b*). *J. Biol. Chem.*, **250**, 8168
Setlow, P. (1976). *J. Biol. Chem.*, **251**, 7853
Setlow, P. and Primus, G. (1975). *J. Biol. Chem.*, **250**, 623
Sharma, H. K., Gupta, S. K. and Rothstein, M. (1976). *Arch. Biochem. Biophys.*, **174**, 323
Shechter, Y., Rafaeli-Eshkol, D. and Hershko, A. (1973). *Biochem. Biophys. Res. Commun.*, **54**, 1518
Simmonds, S., Szeto, K. S. and Fletterick, C. G. (1976). *Biochemistry*, **15**, 261
Smith, W. P., Tai, P.-C., Thompson, R. C. and Davis, B. D. (1977). *Proc. Natl. Acad. Sci. U.S.A.*, **74**, 2830
Spudich, J. A. and Kornberg, A. (1968). *J. Biol. Chem.*, **243**, 4600
Strigini, P. and Brickman, E. (1973). *J. Mol. Biol.*, **75**, 659
Subrahmanyam, C. S. and Das, H. K. (1976). *J. Bacteriol.*, **126**, 1042
Subrahmanyam, C. S. and Das, H. K. (1977). *FEBS Lett.*, **81**, 299
Sussman, A. J. and Gilvarg, C. (1969). *J. Biol. Chem.*, **244**, 6304
Sussman, M. (1965). *Biochem. Biophys. Res. Commun.*, **18**, 763
Swick, R. W. and Ip, M. M. (1974). *J. Biol. Chem.*, **249**, 6836
Switzer, R. L. (1977). *Annu. Rev. Microbiol.*, **31**. 135
Taguchi, M., Izui, K. and Katsuki, H. (1977). *FEBS Lett.*, **77**, 270
Thomas, D. A. and Wright, B. E. (1976). *J. Biol. Chem.*, **251**, 1258
Tweto, J. and Doyle, D. (1975). *J. Biol. Chem.*, **251**, 872
Urbá, R. C. (1959). *Biochem. J.*, **71**, 513
Van Venrooij, W. J., Moonen, H. and Van Loon-Klassen, H. (1974). *Eur. J. Biochem.*, **50**, 207
Villarejo, M. R. and Zabin, I. (1974). *J. Bacteriol.*, **120**, 466
Vitkovic, L. and Sadoff, H. L. (1977). *J. Bacteriol.*, **131**, 897

Voellmy, R. and Goldberg, A. L. (1977). *Abstract* 11th FEBS meeting, A6-2-858, Copenhagen
Vogt, V. M. (1970). *J. Biol. Chem.*, **245**, 4769
Waindle, L. M. and Switzer, R. L. (1973). *J. Bacteriol.*, **114**, 517
Weiss, R. L. (1973). *J. Biol. Chem.*, **248**, 5409
Wiemkin, A. and Nourse, P. (1972). *Planta*, **109**, 293
Willetts, N. S. (1965). *Biochem. Biophys. Res. Commun.*, **20**, 692
Willetts, N. S. (1967*a*). *Biochem. J.*, **103**, 453
Willetts, N. S. (1967*b*). *Biochem. J.*, **103**, 462
Williams, L. S. and Neidhardt, F. C. (1969). *J. Mol. Biol.*, **43**, 529
Yamamoto, S. and Lampen, J. O. (1976). *Proc. Natl. Acad. Sci. U.S.A.*, **73**, 1457
Yen, C. (1974). Studies of the role of peptidases in intracellular protein degradation. Ph.D. thesis, Case Western Reserve University
Yoshinaka, Y. and Luftig, R. B. (1977). *Biochem. Biophys. Res. Commun.*, **79**, 319
Young, H. L. and Klein, H. P. (1967). *J. Bacteriol.*, **93**, 830
Yushok, W. D., Zabrenski, J. and Leary, R. (1977). *Proc. Am. Assoc. Cancer Res.*, **18**, 78
Zipser, D. and Bhavsar, P. (1976). *J. Bacteriol*, **127**, 1538

Part 4

Related Systems

Microorganisms and Nitrogen Sources
Edited by J. W. Payne

Section 4.1

Possible Relation of the γ-Glutamyl Cycle to Amino Acid and Peptide Transport In Microorganisms

ALTON MEISTER

Department of Biochemistry, Cornell University Medical College, 1300 York Avenue, New York, New York 10021, USA

1 INTRODUCTION

The γ-glutamyl cycle is the metabolic pathway that accounts for the biosynthesis of glutathione and its degradation. (The reader is referred to recent reviews (Meister, 1975; Meister and Tate, 1976), which cite the relevant literature in greater detail than is possible here.) There is substantial evidence that this series of enzyme-catalysed reactions occurs in a number of mammalian tissues, and it has been proposed that glutathione (or a γ-glutamyl compound derived from glutathione) functions in one of the mechanisms that mediates translocation of amino acids across mammalian cell membranes. Although virtually all of the information about the enzymes of the cycle and the function of the cycle *in vivo* have come from studies on mammalian tissues, it is reasonable to suppose that the cycle may also occur in other living forms including microorganisms. Indeed, there is evidence that a number of non-mammalian organisms contain enzymes and substrates of the γ-glutamyl cycle. This chapter was prepared at the request of the Editor for inclusion in a volume concerned with the transport of amino acids and related compounds by microorganisms in the hope that a brief summary of the information available about the γ-glutamyl cycle in mammals, together with a recital of the sparse data now available about these reactions in microorganisms, might serve to inform investigators and thus to stimulate research in this area. Thus, this brief chapter summarizes the salient features of the γ-glutamyl cycle as they have been revealed through studies on mammalian tissues and also cites published findings that seem relevant to the possibility that the cycle may function in microorganisms.

2 THE γ-GLUTAMYL CYCLE IN MAMMALS

Glutathione (L-γ-glutamyl-L-cysteinylglycine) is found in virtually all living cells; in mammalian tissues it occurs predominantly intracellularly in concentrations that range from about 0.5 to about 10 mM. Glutathione is usually the most prevalent thiol compound in tissues, and glutamine and glutathione are generally the most abundant γ-glutamyl compounds. The intracellular concentration of glutathione is much greater than that of glutathione disulphide. Glutathione participates in a number of reactions in which its sulphydryl group is oxidized, leading to formation of glutathione disulphide. Such reactions are believed to play a role in the protection of cell membranes against oxidative damage and also in the maintenance of the sulphydryl groups of proteins. Thus, glutathione peroxidase catalyses the interaction of glutathione with peroxides, including hydrogen peroxide, to form glutathione disulphide. Glutathione disulphide is also formed in reactions between glutathione and compounds containing disulphide linkages catalysed by transhydrogenases. The high intracellular concentration of glutathione is maintained by the highly active and widely distributed enzyme glutathione reductase. However, the reactions catalysed by glutathione peroxidase, transhydrogenases, and glutathione reductase do not account for the substantial turnover of glutathione that takes place in a number of mammalian tissues. The intracellular concentration of glutathione is much higher than those of γ-glutamylcysteine, cysteine, and cystine; it therefore appears that glutathione serves as a storage form of cysteine. Indeed, the concentrations of cysteine and cystine in mammalian tissues and body fluids are extremely low. Since free cysteine and cystine are rapidly metabolized, it would seem that under physiological conditions the release of cysteine from glutathione must occur in a regulated manner in order that cysteine may be used efficiently for protein synthesis.

The γ-glutamyl cycle (Figure 1), was elucidated by studies on the enzymes that catalyse the synthesis of glutathione and by investigations that demonstrated that 5-oxoproline is a quantitatively significant metabolite of glutathione (Meister, 1973; Orlowski and Meister 1970; Van Der Verf *et al*., 1971). The degradation of glutathione is catalysed by γ-glutamyl transpeptidase, a membrane-bound enzyme that transfers the

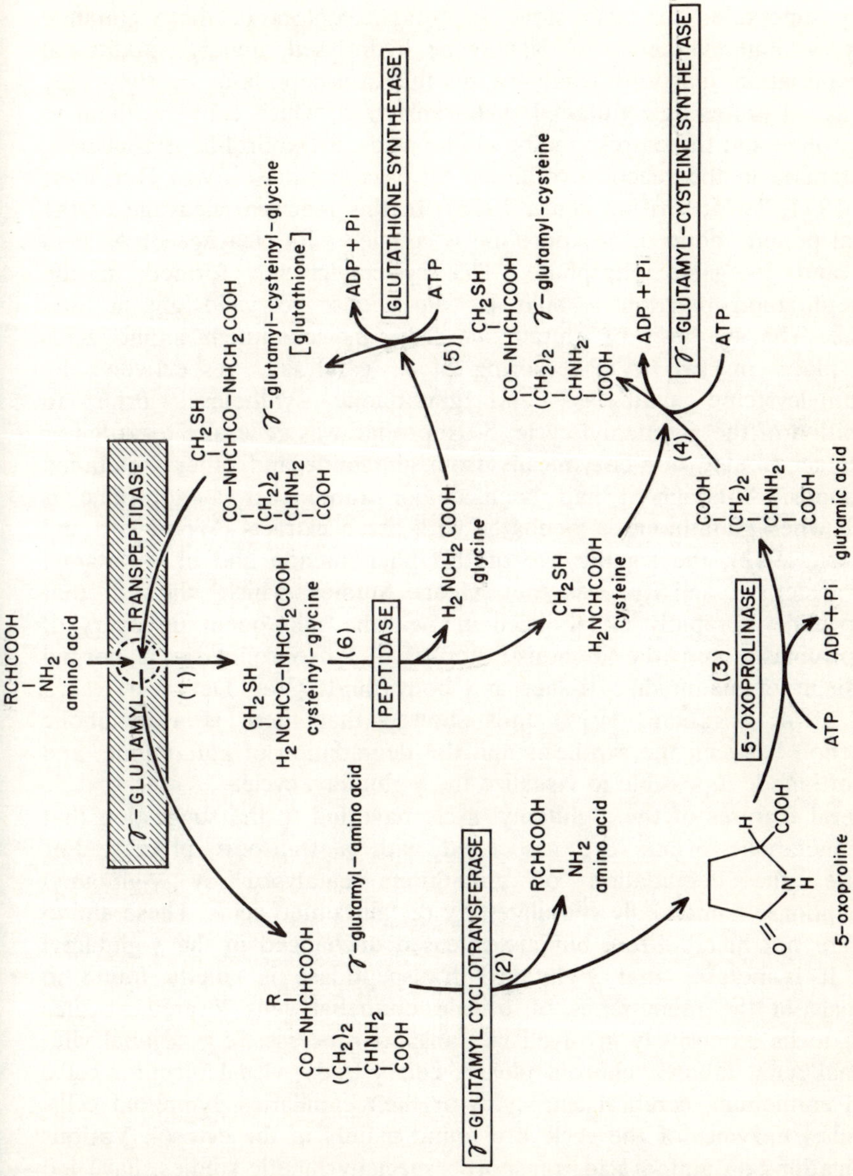

Figure 1 The γ-glutamyl cycle. (Reproduced from A. Meister (1973). *Science, N.Y.*, **180**, 33–39 with the permission of the Editor. Copyright © 1973 American Association for the Advancement of Science)

γ-glutamyl moiety of glutathione to an amino acid acceptor to form a γ-glutamyl amino acid and cysteinylglycine. The transpeptidase can also catalyse the hydrolysis of glutathione to glutamate and cysteinylglycine. Certain dipeptides can also serve as good acceptors of the γ-glutamyl moiety. γ-Glutamyl amino acids may be hydrolysed, undergo additional transpeptidation reactions catalysed by the transpeptidase, or they may serve as substrates of γ-glutamyl cyclotransferase, which converts them to 5-oxoproline and the corresponding amino acids. 5-Oxoproline is converted to glutamate in the reaction catalysed by 5-oxoprolinase (Van Der Werf *et al*., 1971, 1975; Griffith *et al*., 1976). In this reaction, cleavage of the internal peptide bond of 5-oxoproline is coupled with cleavage of ATP to ADP and inorganic phosphate. The cysteinylglycine formed in the transpeptidation reaction is split by dipeptidase to yield glycine and cysteine. The synthesis of glutathione from its constituent amino acids takes place in two ATP-requiring steps catalysed, respectively, by γ-glutamylcysteine synthetase and glutathione synthetase. Prior to recognition of the γ-glutamyl cycle, 5-oxoproline was generally regarded as an artefact formed non-enzymically from glutamine and other γ-glutamyl compounds. Although it had been demonstrated that 5-oxoproline is formed when glutathione is incubated with tissue extracts (Woodward and Reinhart, 1942), the significance of this phenomenon and of γ-glutamyl cyclotransferase activity was not clear. Studies which showed that 5-oxoproline is rapidly metabolized *in vivo*, the subsequent discovery of 5-oxoprolinase, and the demonstration that 5-oxoproline is a normal constituent of mammalian tissues and body fluids (Van Der Werf *et al*., 1971, 1974; Richman, 1975) thus showed that there is a metabolic connection between the synthesis and the degradation of glutathione, and therefore made it possible to visualize the γ-glutamyl cycle.

Several features of the γ-glutamyl cycle have led to the suggestion that these metabolic events are associated with a transport process. For example, the degradation of glutathione catalysed by γ-glutamyl transpeptidase is markedly stimulated by certain amino acids. These amino acids are not metabolized, but are released unchanged in the γ-glutamyl cycle. It is notable that γ-glutamyl transpeptidase is chiefly found in mammals in the membranes of certain epithelial cells located at sites known to be extensively involved in transport processes, e.g. jejunal villi, proximal renal tubules, choroid plexus, ciliary body, visual receptor cells, retinal epithelium, cerebral astrocytes or their capillaries, lymphoid cells. The other enzymes of the cycle are found mainly in the cytosol. Various investigations on amino acid transport, especially kinetic studies, have led to the conclusion that amino acid transport involves several steps such as binding of amino acids to sites on cell membranes, carrier-mediated translocation of carrier-amino acid complexes, release of the amino acid from the carrier intracellularly, and energy-dependent reactivation of the carrier ('recovery') (Pardee, 1968; Heinz, 1972). The γ-glutamyl cycle appears to have properties that fit such a general description of the steps in amino acid transport. The membrane-bound γ-glutamyl transpeptidase would be a key enzyme in this process by interacting with intracellular glutathione (or perhaps another γ-glutamyl donor derived from glutathione) and with extracellular amino acid. According to this proposal, the γ-glutamyl group is the carrier and the amino acid is translocated into the cell as a γ-glutamyl amino acid. The amino acid is released intracellularly from its γ-glutamyl carrier by the action of γ-glutamyl cyclotransferase or possibly also in some instances by the hydrolytic activity of γ-glutamyl transpeptidase. The energy-requiring steps of the cycle, i.e. those involved in the resynthesis of glutathione and in the conversion of 5-oxoproline to glutamate, would serve in the 'recovery' phase of the transport process.

It has not been proposed that the γ-glutamyl cycle mediates the transport of all amino acids or that it is the only amino acid translocation mechanism. The available data about amino acid transport in mammalian tissues indicate that energy is required for active transport, that there are probably a number of different translocation mechanisms, and that these systems (whose existence has been postulated largely on the basis of measurements of uptake rates and the observation of competitive effects) exhibit 'overlapping' specificity. The γ-glutamyl cycle is thus far the only amino acid translocation mechanism proposed in which specific enzymes have been implicated. Although there are probably other translocation mechanisms for amino acids, we have virtually no information about the chemical phenomena that may be involved. Although a direct and unequivocal demonstration that the γ-glutamyl cycle functions in amino acid transport is still required, the data that are available seem strongly to support the transport hypothesis.

A number of modifications of the γ-glutamyl cycle have been considered (Meister and Tate, 1976). Thus, a cycle involving cleavage of only two molecules of ATP would result if the formation of 5-oxoproline were bypassed by hydrolysis of γ-glutamyl amino acids or a γ-glutamyl acceptor such as a γ-glutamyl dipeptide. A cycle in which there is cleavage of only one molecule of ATP is also conceivable if successive transpeptidation reactions occur; one of these would involve cysteine, thus providing an alternative route to γ-glutamylcysteine. A model has also been proposed for exchange diffusion involving intermediate formation of γ-glutamyl amino acid–enzyme complexes (Meister and Tate, 1976). The amino acid specificity of γ-glutamyl transpeptidase suggests that certain neutral amino acids, especially glutamine and cystine, may be major substrates of the γ-glutamyl cycle. The idea that glutamine might play a special role is particularly attractive in view of the high concentrations of glutamine in mammalian body fluids and tissues. Thus, glutamine may be readily converted to γ-glutamylglutamine and there is, in fact, evidence that γ-glutamylglutamine is present in low concentrations in many mammalian tissues (Kanazawa and Sano, 1967). Not only is glutamine an excellent acceptor amino acid substrate of the transpeptidase, but γ-glutamylglutamine is the best γ-glutamyl amino acid substrate of γ-glutamyl cyclotransferase. That the substrate specificity of cyclotransferase is parallel to that of the amino acid acceptor specificity of the transpeptidase supports the view that the cyclotransferase and transpeptidase function in sequence as catalysts in the γ-glutamyl cycle. Glutamine may also serve as a γ-glutamyl donor for γ-glutamyl transpeptidase. Although the enzyme exhibits relatively low affinity *in vitro* for glutamine as a γ-glutamyl donor as compared to glutathione, it is conceivable that glutamine might serve as a γ-glutamyl donor under certain physiological conditions. The finding that cystine is an excellent acceptor substrate of the transpeptidase has led to the suggestion that the transport of this amino acid may be effected by the cycle and also that such transport may be integrated with other aspects of the metabolism of cystine. For example, the transport of cystine as γ-glutamylcystine might be followed by reduction of this molecule to cysteine and γ-glutamylcysteine; the latter compound would be converted directly to glutathione (Thompson and Meister, 1975).

That the reactions of the γ-glutamyl cycle occur *in vivo* was initially suggested by the observation that the enzymes and substrates of the cycle are widely distributed in mammalian tissues. Subsequently, the reactions of the cycle were demonstrated *in vivo* by labelling studies and through use of specific enzyme inhibitors. Studies in which the incorporation of labelled glutamate and labelled 5-oxoproline into glutathione was determined showed that the turnover of glutathione is much higher in the kidney than it is in the liver (Sekura and Meister, 1974), and this finding is consistent with the much higher activities of the cycle enzymes in kidney as compared to liver. When 5-oxoproline and glutamate are administered to animals, they are incorporated at about the same rate into glutathione. Administration of labelled glutamate leads to prompt appearance of the label in tissue 5-oxoproline (Sekura *et al*., 1976). When mice are treated with a competitive inhibitor of 5-oxoprolinase, they accumulate 5-oxoproline in their tissues (Van Der Werf and Meister, 1975; Van Der Werf *et al*., 1974). Animals who have received a 5-oxoprolinase inhibitor show a marked decrease in the usually rapid conversion of administered labelled 5-oxoproline to labelled respiratory carbon dioxide. Animals given a 5-oxoprolinase inhibitor together with amino acids exhibit increased 5-oxoproline accumulation. These findings can be explained in terms of the γ-glutamyl cycle, since administration of amino acids would be expected to increase transpeptidation, and the γ-glutamyl amino acids formed would become substrates of intracellular γ-glutamyl cyclotransferase. When an inhibitor of γ-glutamyl cyclotransferase is administered to mice, there is a significant depression of the steady-state concentration of 5-oxoproline in the kidney (Griffith *et al.*, 1978). Administration of such an inhibitor also prevents the accumulation of 5-oxoproline which is observed after administration of large amounts of amino acids. These studies therefore support the view that γ-glutamyl cyclotransferase and 5-oxoprolinase are major catalysts for the formation and utilization, respectively, of 5-oxoproline *in vivo*.

According to the γ-glutamyl cycle, the observed steady-state levels of glutathione in a tissue such as kidney result from a balance between the rate of glutathione synthesis and the rate of its utilization by γ-glutamyl transpeptidase. If this balance is disturbed by decreasing glutathione synthesis or by increasing glutathione utilization, one would expect a substantial decrease in the tissue glutathione level. When mice are injected with methionine sulphoximine (which inhibits glutamine and glutathione synthesis) or with prothionine sulphoximine (which inhibits only glutathione synthesis), there is a very rapid decline in the glutathione level of the kidney. The rate of this decline seems to reflect the normal rate of glutathione utilization by γ-glutamyl transpeptidase (Griffith *et al.*, 1979). The rate at which the glutathione level falls after injection of prothionine sulphoximine was found to be decreased when an inhibitor of γ-glutamyl transpeptidase was also administered (Griffith *et al.*, 1978). The rate and extent of the transpeptidase-mediated utilization of glutathione can be increased by administration of transpeptidase acceptor substrates such as glycylglycine and certain amino acids. Thus, when glycylglycine was given to rats, there was a marked decrease in the tissue levels of glutathione (Palekar *et al.*, 1975). Presumably, the very rapid transpeptidation reaction that occurs after administration of glycylglycine uses glutathione at a faster rate than it can be resynthesized and therefore the level of glutathione falls. When mice were given relatively large doses of amino acids, there were dramatic decreases in the level of glutathione in the kidney, and also a substantial increase in the 5-oxoproline level (Griffith *et al.*, 1978). The most probable explanation for these results is that the increased concentration of plasma amino acids leads to increased transpeptidation and that this reaction utilizes intracellular glutathione more rapidly than it can be resynthesized. According to the γ-glutamyl cycle, such utilization of glutathione would be expected to lead to increased levels of γ-glutamyl amino acids; indirect evidence that this occurs is the finding of increased levels of 5-oxoproline. Administration of an inhibitor of γ-glutamyl transpeptidase leads to a substantial decrease in the rate of decline of the glutathione levels and of 5-oxoproline formation after amino acid administration.

The findings summarized above are consistent with the conclusion that amino acids may be translocated into cells by the γ-glutamyl cycle pathway. Thus, there seems to be a close relationship between increased extracellular amino acid concentration on the one hand, and decreased intracellular glutathione levels and increased levels of 5-oxoproline, on the other. Clearly, a direct demonstration that γ-glutamyl amino acids are translocated would be of importance. Evidence that a translocation pathway for γ-glutamyl amino acids does indeed exist is suggested by recent studies on the uptake of methionine sulphoximine by kidney (O. W. Griffith and A. Meister, unpublished results). In these experiments it was found that more methionine sulphoximine was found in the kidney after administration of γ-glutamylmethionine sulphoximine to mice than after giving methionine sulphoximine itself. Furthermore, in these studies γ-glutamylmethionine sulphoximine was found in the kidney. The observation that γ-glutamylmethionine sulphoximine is well transported is consistent with the postulated function of the γ-glutamyl cycle.

We need to know more about the orientation of γ-glutamyl transpeptidase in the cell membrane. Although the available data indicate that a substantial fraction of the transpeptidase is located on the outer surface of the cell membrane (Meister and Tate, 1976), the findings also indicate that the transpeptidase is a key enzyme in the utilization of intracellular glutathione. This suggests that intracellular glutathione is transported to the membrane-bound transpeptidase. The mechanism of such translocation is not yet clear. It is possible that a protein carrier system is involved or that there is a channel of some sort within the membrane, or both. That intracellular glutathione is transported to the membrane-bound transpeptidase is consistent with findings on a patient who has an inborn error associated with marked reduction of γ-glutamyl transpeptidase activity (Schulman *et al.*, 1975). This patient has a marked glutathionaemia and glutathionuria and it has been suggested that the glutathione present in this patient's blood plasma is secreted by or leaks from renal cells into the glomerular filtrate. The transport of intracellular glutathione to the plasma in this patient may represent an aspect of a process that occurs normally by which substrate (glutathione or a γ-glutamyl compound derived from it) is provided to the membrane-bound enzyme. Normally there is very little glutathione present in the blood plasma, but evidently in the absence of the transpeptidase, glutathione passes out of the cells into the plasma. Other inborn errors of metabolism involving enzymes of the γ-glutamyl cycle have also been observed (Meister, 1974*a*, 1977; Wellner *et al.*, 1974; Konrad *et al.*, 1972; Richards *et al.*, 1974); the findings on these patients are consistent with the function of the γ-glutamyl cycle *in vivo* and also with the transport hypothesis.

The enzymes of the cycle have been purified from a number of sources and their properties and mechanisms of action have been extensively studied. Thus, γ-glutamyl transpeptidase has been obtained in highly purified form from mammalian kidney; the enzyme has also been isolated from liver, other tissues and tumours (Meister and Tate, 1976). A very highly purified preparation of the enzyme from rat kidney was found to have a molecular mass of about 70 000 daltons and to contain about 19 per cent carbohydrate (Tate and Meister, 1974*a,b*, 1975, 1976*a*). This enzyme may be dissociated by treatment with sodium dodecyl sulphate in the presence of 2-mercaptoethanol into two unequal glycopeptide subunits that have molecular masses of 23 000 and 46 000 daltons. The enzyme is rapidly inactivated by the γ-glutamyl analogues 6-diazo-5-oxo-L-norleucine and L-azaserine and such inactivation is prevented by γ-glutamyl substrates but not by acceptor substrates. 6-Diazo-5-oxo-L-norleucine reacts specifically, covalently, and stoicheiometrically at the γ-glutamyl site of the enzyme which is associated with the subunit of molecular weight 23 000 daltons (Tate and Meister, 1977). Kinetic and other investigations indicate a ping-pong mechanism involving two half-reactions, the first of which leads to formation of a γ-glutamyl enzyme. Since the γ-glutamyl enzyme can react with water or with another molecule of γ-glutamyl donor, the transpeptidase can catalyse hydrolysis and autotranspeptidation, as well as transpeptidation (Tate and Meister 1975; Thompson and Meister, 1976). D-γ-Glutamyl compounds are good γ-glutamyl substrates of the enzyme whereas D-amino acids are not acceptor substrates. Thus, use of a D-γ-glutamyl substrate eliminates the possibility of autotranspeptidation. The most active amino acid acceptors of the transpeptidase include the L-isomers of certain neutral amino acids such as glutamine, cystine, methionine, asparagine, and alanine. Glycylglycine is an excellent acceptor substrate (Hanes *et al*., 1952). Dipeptides of the X-Gly type are almost always much more active than dipeptides of the Gly-X type (Meister and Tate, 1977; Thompson and Meister, 1977). The accumulated data indicate that aminoacylglycines bind to the region of the glutathione binding site of the enzyme that attaches to cysteinylglycine. It is notable that the relative activity found with peptides of the X-Gly type fall in an order which is almost parallel to that found for the corresponding free amino acids. This finding suggests that free amino acids bind to the site on the enzyme that normally binds the cysteinyl moiety of cysteinylglycine. These considerations suggest that the high acceptor activity exhibited by the transpeptidase towards certain dipeptides is a reflection of the enzyme's affinity for glutathione. Although it is conceivable that the enzyme might function physiologically in the transport of certain dipeptides, e.g. those of the X-Gly type, the data do not support the view that the transpeptidase functions as a general transporter of dipeptides in mammalian systems.

Several highly purified and active preparations of γ-glutamyl cyclotransferase have been obtained from mammalian tissues. The specificity of highly purified rat kidney γ-glutamyl cyclotransferase (Taniguchi and Meister, 1978) has been extensively studied; the enzyme acts most effectively on γ-glutamylglutamine, but also exhibits high activity toward many di-γ-glutamyl amino acids. The latter may be formed by transpeptidation reactions. Thus, the preferred substrates of γ-glutamyl cyclotransferase have the structure, γ-Glu-γ-Glu-NHR, in which the nature of the R moiety seems to have relatively little effect on activity (Taniguchi and Meister, 1976). The findings suggest that the major physiological function of the enzyme is associated with its activity toward γ-glutamylglutamine and possibly also toward di-γ-glutamyl amino acids.

5-Oxoprolinase has been obtained in highly purified form from rat kidney (Van Der Werf *et al*., 1975). The reaction is of interest in that the enzyme couples the hydrolysis of the internal peptide bound of 5-oxoproline with the hydrolysis of ATP. The equilibrium between 5-oxoproline and glutamate lies distinctly in favour of cyclization (Wilson and Cannon, 1937; Meister *et al*., 1963), but in the reaction catalysed by 5-oxoprolinase, the hydrolysis of 5-oxo-L-proline is coupled with hydrolysis with ATP and the overall reaction goes essentially to completion. Uncoupling phenomena have been observed when either 5-oxoproline or ATP is replaced by analogues (Van Der Werf *et al*., 1975; Griffith and Meister, 1976). The mechanism of the reaction has been studied with 5-oxo-L-proline containing ^{18}O in the amide carbonyl group (Griffith *et al*., 1976). The findings are consistent with a mechanism involving phosphorylation of the amide carbonyl group of 5-oxoproline to give an enzyme-bound intermediate that can react with water to form enzyme-bound γ-glutamyl phosphate, which is hydrolysed to glutamate.

Highly purified preparations of γ-glutamylcysteine synthetase and glutathione synthetase have been obtained. These reactions have been shown to involve acyl phosphate intermediates, i.e. γ-glutamyl phosphate and γ-glutamylcysteinyl phosphate, respectively (Nishimura *et al*., 1963, 1964; Meister, 1974*b*). Extensive studies have been carried out on the specificity of the synthetases and these have provided an approach to the mapping of the active sites of these enzymes.

Studies on a variety of substrate analogues show that suitable modification of the glutamyl moiety of the substrates can lead to specific inhibitors or non-metabolizable analogues at each of the steps of the γ-glutamyl cycle (Griffith and Meister, 1977). Although most of the reactions of the cycle involve the γ-carboxyl group of glutamate, the reaction pathways of the enzymes are quite different and are presumably associated with differences in the enzyme-bound conformations of the glutamate-carbon chain at the active sites of the several enzymes. The data obtained in these studies open the way to a number of experimental approaches. For example, in one of these β-glutamyl-L-α-aminobutyrate was found to be a highly active competitive inhibitor of γ-glutamyl cyclotransferase; this compound is effective in inhibiting γ-glutamyl cyclotransferase *in vivo* (Griffith *et al.*, 1978). Thus, blocks of the γ-glutamyl cycle have been observed *in vivo* after administration of specific inhibitors and in patients with inborn errors of metabolism associated with specific enzyme deficiencies (Meister 1974*a*, 1977).

3 ACTIVITY OF γ-GLUTAMYL CYCLE ENZYMES IN MICROORGANISMS AND OTHER NON-MAMMALIAN SPECIES

Although enzyme activity apparently comparable to that of mammalian γ-glutamyl transpeptidase has been found in a number of microorganisms (Samuels–Talalay, 1953, 1954; Milbauer and Grossowicz, 1965; Szewczuk and Mulczyk, 1970, R. V. Krishna, unpublished results), the only microorganism thus far reported to contain all of the enzymes of the γ-glutamyl cycle is yeast (Mooz and Wigglesworth, 1976). It has long been known that yeast contains glutathione and, in fact, glutathione was first isolated from this source (De Rey Pailhade, 1888; Hopkins, 1921). Snoke and Bloch (1955) and Snoke (1955) showed that yeast contains γ-glutamylcysteine synthetase and glutathione synthetase, and a very highly purified preparation of the latter enzyme has been isolated from yeast (Mooz and Meister, 1967). Mooz and Wigglesworth (1976) found that baker's yeast has substantial activities of γ-glutamyl transpeptidase, γ-glutamyl cyclotransferase, 5-oxoprolinase, and dipeptidase, and that the activities of these enzymes are comparable to that of glutathione synthetase. They succeeded in partially purifying the transpeptidase and observed about a 60 per cent increase in activity during purification, suggesting the presence of an inhibitor in the initial yeast autolysate. In their studies on γ-glutamyl transpeptidase, Mooz and Wigglesworth (1976) used γ-glutamyl-*p*-nitroanilide, γ-glutamyl-α-naphthylamide, and a variety of amino acid acceptors. Although they found substantial activity in the absence of added acceptor, this may have been associated with the presence of free amino acids in the enzyme preparation. They found increased transpeptidase activity in the presence of glycylglycine, glycine, and the L-isomers of methionine, alanine, glutamine, proline, cysteine, and glutamate. It is of interest that the best amino acid acceptors were methionine, glycine, and cysteine, and that glycylglycine was somewhat more active than the free amino acids. In studies in which 0.5 mM acceptor was used, L-cystine and glycylglycine were about equally active. These findings on the specificity of yeast γ-glutamyl transpeptidase closely resemble those obtained with transpeptidase preparations isolated from rat kidney. Mooz and Wigglesworth (1976) also found that suspensions of intact yeast cells also exhibit transpeptidase activity, suggesting that the enzyme may be located on the cell surface, as it seems to be in many mammalian cells. The finding of the γ-glutamyl cycle enzyme activities in yeast is of significance because it indicates that this metabolic pathway is not restricted in its distribution to mammalian tissues, and also because studies on the function of the cycle may be facilitated by experiments with a relatively simple free cell system.

The first demonstration that γ-glutamyl transpeptidase occurs in microorganisms came from the studies of Samuels-Talalay (1953, 1954), who examined cell-free extracts of *Escherichia coli* and *Proteus vulgaris* for their ability to catalyse glutathione hydrolysis and transpeptidation reactions of the type that had been elucidated earlier by Hanes *et al*. (1952) in studies on animal tissues. It was found that although cell-free extracts of *E. coli* synthesize glutathione from its component amino acids, such extracts did not hydrolyse glutathione (Samuels-Talalay, 1953, 1954). On the other hand, whole cell suspensions and cell-free extracts of *P. vulgaris* were found to catalyse glutathione breakdown effectively. Furthermore, these studies showed that *P. vulgaris* extracts catalysed transpeptidation reactions leading to the formation of new γ-glutamyl compounds (e.g. the γ-glutamyl derivatives of methionine, valine, phenylalanine, leucine, and glutamine).

Milbauer and Grossowicz (1965) found that γ-glutamyl transfer reactions are catalysed by a large number of bacterial species. These workers surveyed 16 different bacterial species for their ability to hydrolyse glutathione and they also determined the effects of adding glycylglycine and L-methionine on this reaction. The organisms capable of hydrolysing glutathione included *Aerobacter aerogenes, A. aerogenes* 8303, *Achromobacter delicatulum, Alcaligenes bookeri, Al. faecalis, Bacillus mycoides, B. subtilis, Klebsiella ozaenae, Proteus vulgaris* (IIX19), *P. morganii, Pseudomonas sp., Ps. fluorescens. Ps. non liquefaciens.* Three of the organisms examined (*E. coli, E. freundii*, and *Staphylococus aureus*) showed only weak activity. The most active organisms were *Ach. delicatulum, P. morganii*, and *P. vulgaris.* In the studies on the more active microorganisms, addition of an acceptor increased γ-glutamyl transfer. However, in some instances no such activation was observed, possibly because of the presence of amino acids in the preparations employed. In this work, Milbauer and Grossowicz (1965) demonstrated the formation of new γ-glutamyl products (including γ-glutamylglutathione). Studies on a cell-free preparation of *P. morganii* demonstrated γ-glutamyl transfer to a number of amino acid and dipeptide acceptors; although the quantitative values obtained differ somewhat from those obtained with various mammalian preparations, the findings were, in general, similar to the results that have been obtained with mammalian γ-glutamyl transpeptidase. Some of the observed differences might be explained by the presence of amino acids and other enzyme activities in the bacterial extracts employed. Szewcuk and Mulczyk (1970) purified a γ-glutamyl peptidase from *Pseudomonas aeruginosa*, which hydrolysed L-γ-glutamyl-α-naphthylamide, glutamine, and other γ-glutamyl compounds. The purified preparation was found to lack transpeptidase activity; the authors concluded that another enzyme was responsible for the transpeptidase activity found in the initial homogenates. The findings, considered together, provide substantial evidence for the belief that certain microorganisms contain an enzyme activity or activities which closely resemble the mammalian γ-glutamyl transpeptidase.

It would clearly be of considerable interest to learn whether those microorganisms that have γ-glutamyl transpeptidase activity also have γ-glutamyl cyclotransferase and 5-oxoprolinase activities. At present, only a small amount of information is available on this point. Preliminary studies in the author's laboratory (R. V. Krishna, unpublished results) indicate that *P. vulgaris, Ps. aeruginosa*, and *Ach. delicatulus* (ATCC 19103) have γ-glutamyl cyclotransferase and 5-oxoprolinase activities as well as glutathione synthetase activity. 5-Oxoprolinase activity has also been found in a pseudomonad isolated from soil by enrichment culture (Van Der Werf and Meister, 1974). This enzyme was purified about 26-fold in high yield and shown to be similar in its properties to rat kidney 5-oxoprolinase. However, in contrast to the kidney enzyme, 5-oxoprolinase from pseudomonas is only slightly inhibited by L-2-imidazolidone-4-carboxylate.

Finally, a recent paper describes the distribution of glutathione in bacteria (Fahey *et al*., 1978). This information, together with that obtained using mutants deficient in glutathione synthesis (Apontoweil and Berends, 1975; Fuchs and Warner, 1975) should be of considerable use in this area.

The investigations reviewed above seem to provide a reasonable basis on which to pursue further studies on the enzymes of the γ-glutamyl cycle in microorganisms. It seems relevant to observe, however, that a number of other non-mammalian forms also contain these activities. Thus, the enzymes of the cycle have been found in hydra (Tate and Meister, 1976*b*), certain insects (Bodnaryk, 1970, 1971, 1974; Bodnaryk *et al*., 1974), phytoplanktons (Kurelec *et al*., 1977), and plants (Goore and Thompson, 1967). Bodnaryk and his colleagues found that, in housefly larvae, γ-glutamyl transpeptidase is located on the brush border of the proximal portion of the Malpighian tubules (which are analogous to the mammalian kidney), and the striated border of the epithelial cells of the midgut. It is known that the latter site is involved in the uptake of dietary amino acids in insects. γ-Glutamyl transpeptidase was purified about 100-fold from an acetone powder of housefly larvae, and found to exhibit properties similar to those of the mammalian enzyme. γ-Glutamylphenylalanine was found to be a major constituent of housefly larvae; the concentration of this dipeptide, apparently formed by transpeptidation between glutathione and phenylalanine, increases continually during larval growth. The γ-glutamylphenylalanine levels decline rapidly when hardening and darkening of the puparium occurs, and at this stage γ-glutamyl transpeptidase and γ-glutamyl cyclotransferase activities are at their highest levels. The process of puparium formation is associated with abrupt ecdysone-induced formation of γ-glutamyl transpeptidase on the epidermal cell membrane at the epidermis–cuticle interface, and with rapid liberation of phenylalanine. It was concluded that γ-glutamylphenylalanine and γ-glutamyl transpeptidase are constituents of a γ-glutamyl cycle for the transport of phenylalanine (utilized for formation of quinones required for tanning cuticle proteins in housefly larvae) across certain cell membranes in this insect. In support of this view, it was found that 5-oxoproline is formed during γ-glutamyl phenylalanine utilization; at this stage of development the rate of formation of 5-oxoproline exceeds the rate of its utilization.

The enzymes of the γ-glutamyl cycle have also been found in the coelenterate *Hydra littoralis* (Tate and Meister, 1976*b*). Thus, this organism was found to exhibit a relatively high γ-glutamyl transpeptidase activity, i.e. about 12 per cent of that found in rat kidney. Histochemical studies showed that the enzyme is located mainly in the gastric and sub-hypostome regions. It is also found in the tentacles and basal disc. Homogenates of hydra exhibit significant activities of γ-glutamyl cyclotransferase, dipeptidase, and γ-glutamylcysteine synthetase. It thus appears that the cycle may play a role in the metabolism of glutathione in hydra. Possibly γ-glutamyl transpeptidase may function in digestive and absorptive processes in this organism. It is of interest that glutathione activates the feeding response of hydras and also affects or elicits a number of other behavioural responses. Thus, Loomis (1955) concluded that glutathione is the specific chemical stimulator that evokes the feeding response. Glutathione disulphide was found to be inactive, but Cliffe and Waley (1958) found that norophthalmic acid and ophthalmic acid (tripeptides in which the sulphydryl group of glutathione is replaced by a hydrogen atom and a methyl group, respectively) are also very efficient activators of the feeding response. *S*-Methylglutathione also activates the feeding response (Lenhoff 1959). Since the remarkable effect of glutathione is evidently not associated directly with its sulphydryl group, one must consider the possibility that the γ-glutamyl moiety of this tripeptide functions in the feeding reaction. The finding of transpeptidase activity in the tentacles of hydra suggests that this enzyme might be involved in the behavioural responses to glutathione. Thus, interaction of external glutathione with membrane-bound γ-glutamyl transpeptidase in the region of the head or tentacles might trigger the feeding response, possibly by leading to the formation of a γ-glutamyl derivative which might function as a specific neurotransmitter.

Kurelec *et al.* (1977) have shown the presence of the γ-glutamyl cycle enzymes in natural phytoplankton populations. Crude homogenates of these organisms and partially purified preparations obtained from homogenates were found to exhibit the activities of γ-glutamyl transpeptidase, γ-glutamyl cyclotransferase, 5-oxoprolinase, and glutathione synthesis. It is known that many phytoplanktons have transport systems which enable them to accumulate amino acids from sea water which contains extremely low concentrations of amino acids. The finding by Kurelec *et al.* (1977) of the γ-glutamyl cycle enzymes in phytoplankton populations has led these workers to postulate that the cycle might play an important role in these organisms by facilitating rapid accumulation of amino acids from the environment. Kurelec *et al.* (1977) did not find γ-glutamyl transpeptidase activity in zooplankton, organisms that do not exhibit transepidermal uptake of amino acids.

A γ-glutamyl transpeptidase has been purified from kidney fruit by Goore and Thompson (1967). The properties of this enzyme are similar, though not identical, to the transpeptidase found in mammalian kidney. These workers suggested that the enzyme functions physiologically to break down glutathione, and also that it may account for the biosynthesis of a number of γ-glutamyl amino acids in plants. It is notable that a wide variety of γ-glutamyl amino acids has been found in plants; the occurrence, distribution, and possible physiological function of these compounds has been reviewed (Meister 1965, 1975, Waley 1966).

REFERENCES

Apontoweil, P. and Berends, W. (1975). *Biochim. Biophys. Acta*, **399**, 10–22

Bodnaryk, R. P. (1970). *J. Insect Physiol.*, **16**, 919–929

Bodnaryk, R. P. (1971). *Can. J. Biochem.*, **50**, 524–528

Bodnaryk, R. P. (1974). *Insect. Biochem.*, **4**, 439–454

Bodnaryk, R. P., Bronskill, J. F., and Fetterly, J. R. (1974). *J. Insect. Physiol.*, **20**, 167–181

Cliffe, E. E. and Waley, S. G. (1958). *Nature, Lond.*, **182**, 804–805

De Rey Pailhade, J. (1888). *C.R. Acad. Sci.*, **106**, 1683

Fahey, R. C., Brown, W. C., Adams, W. B. and Worsham, M. B., (1978). *J. Bacteriol.*, **133**, 1126–1129

Fuchs, J. A. and Warner, H. R. (1975). *J. Bacteriol.*, **124**, 140–148

Goore, M. Y. and Thompson, J. F. (1967). *Biochim. Biophys. Acta*, **132**, 15–26

Griffith, O. W. and Meister, A. (1976). *Biochem. Biophys. Res. Comm.*, **70**, 759–765

Griffith, O. W. and Meister, A. (1977). *Proc. Natl. Acad. Sci. U.S.A.*, **74**, 3330–3334

Griffith, O. W., Anderson, M. E., and Meister, A. (1979). *J. Biol. Chem.*, **254**, 1205–1210

Griffith, O. W., Bridges, R. J., and Meister, A. (1978). *Proc. Natl. Acad. Sci. U.S.A.*, **75**, 5405–5408

Griffith, O. W., Van Der Werf, P., and Meister, A. (1976). In *Krok Foundation Symposium, Santa Barbara, California, June 1—3, 1975*, Raven Press, New York, pp. 63–69

Hanes, C. S., Hird, F. J. R., and Isherwood, F. A. (1952). *Biochem. J.*, **51**, 25–35

Heinz, E. (1972). In *Metabolic Pathways*, Vol. 6, Academic Press, New York, pp. 455–501

Hopkins, F. G. (1921). *Biochem. J.*, **15**, 286

Kanazawa, A. and Sano, I. (1967). *J. Neurochem.*, **14**, 596–598

Konrad, P. N., Richards, F., Valentine, W. N., and Paglia, D. E. (1972). *New Engl. J. Med.*, **286**, 557–561

Kurelec, B., Rijavec, M., Britvic, S., Muller, W. E. G., and Zahn, R. K. (1977). *Comp. Biochem. Physiol.*, **56B**, 415–419

Lenhoff, H. M. (1959). *Exp. Cell. Res.*, **17**, 570–573

Loomis, W. F. (1955). *Ann. N.Y. Acad. Sci.*, **62**, 209–227

Meister, A. (1965). *Biochemistry of the Amino Acids*, 2nd edn, Academic Press, New York

Meister, A. (1973). *Science, N.Y.*, **180**, 33–39

Meister, A. (1974*a*). *Ann. Int. Med.*, **81**, 247–253

Meister, A. (1974*b*) In *The Enzymes*, 3rd edn, Vol. **10**, (P. D. Boyer, ed.), Academic Press, New York, pp. 671–697

Meister, A. (1975). In *Metabolic Pathways*, 3rd edn, Vol. 7; *Biochemistry of Glutathione*, Academic Press, New York, pp. 101–188

Meister, A. (1977). In *The Metabolic Basis of Inherited Diseases*, 4th edn (J. B. Stanbury, J. B. Wyngaarden and D. S. Frederickson, eds), McGraw-Hill, New York, pp. 328–336

Meister, A. and Tate, S. S. (1976). *Annu. Rev. Biochem.*, **45**, 559–604

Meister, A., Bukenberger, M. W., and Strassburger, M. (1963). *Biochem. Z.* (Otto Warburg Festband). **338**, 217–229

Meister, A., Tate, S. S., and Thompson, G. A. (1977). *CIBA Found. Symp.*, **50**, 123–144

Milbauer, R. and Grossowicz, N. (1965). *J. Gen. Microbiol.*, **41**, 185–194

Mooz, E. D. and Meister, A. (1967). *Biochemistry*, **6**, 1722–1734

Mooz, E. D. and Wigglesworth, L. (1976). *Biochem. Biophys. Res. Commun.*, **68**, 1066–1072

Nishimura, J. S., Dodd, E. A., and Meister, A. (1963). *J. Biol. Chem.*, **238**, 1179–1180

Nishimura, J. S., Dodd, E. A., and Meister, A. (1964). *J. Biol. Chem.*, **239**, 2553–2558

Orlowski, M. and Meister, A. (1970). *Proc. Natl. Acad. Sci. U.S.A.*, **67**, 1248–1255

Palekar, A. G., Tate, S. S., and Meister, A. (1975). *Biochem. Biophys. Res. Commun.*, **62**, 651–657

Pardee, A. B. (1968). *Science, N.Y.*, **162**, 632–637

Richards, F., Cooper, M. R., Pearce, L. A., Cowan, R. J., and Spurr, C. L. (1974). *Arch. Intern. Med.*, **134**, 534–537

Richman, P. G. (1975). Doctoral dissertation, Cornell University Medical College, New York

Samuels-Talalay, P. (1953). *Biochem. J.*, **55**, 441–444

Samuels-Talalay, P. (1954). *Nature, Lond.*, **174**, 516–517

Schulman, J. D., Goodman, S. I., Mace, J. W., Patrick, A. D., Tietze, F., and Butler, E. J. (1975). *Biochem. Biophys. Res. Commun.*, **65**, 68–74

Sekura, R. and Meister, A. (1974). *Proc. Natl. Acad. Sci. U.S.A.*, **71**, 2404–2409

Sekura, R., Van Der Werf, P., and Meister, A. (1976). *Biochem. Biophys. Res. Commun.*, **71**, 11–18

Snoke, J. E. (1955). *J. Biol. Chem.*, **213**, 813

Snoke, J. E. and Bloch, K. (1955). *J. Biol. Chem.*, **213**, 825

Szewczuk, A. and Mulczyk, M. (1970). *Arch. Immun. Therap. Exp.*, **18**, 515–526

Taniguchi, N. and Meister, A. (1978). *J. Biol. Chem.*, **253**, 1799–1806

Tate, S. S. and Meister, A. (1974*a*). *Proc. Natl. Acad. Sci. U.S.A.*, **71**, 3329–333

Tate, S. S. and Meister, A. (1974*b*). *J. Biol. Chem.*, **249**, 7593–7602

Tate, S. S. and Meister, A. (1975). *J. Biol. Chem.*, **250**, 4619–4624

Tate, S. S. and Meister, A. (1976*a*). *Proc. Natl. Acad. Sci. U.S.A.*, **73**, 2599–2603

Tate, S. S. and Meister, A. (1976*b*). *Biochem. Biophys. Res. Commun.*, **70**, 500–505

Tate, S. S. and Meister, A. (1977). *Proc. Natl. Acad. Sci. U.S.A.*, **74**, 931–935

Thompson, G. A. and Meister, A. (1975). *Proc. Natl. Acad. Sci. U.S.A.*, **72**, 1985–1988

Thompson, G. A. and Meister, A. (1976). *Biochem. Biophys. Res. Comm.*, **71**, 32–36

Thompson, G. A. and Meister, A. (1977). *J. Biol. Chem.*, **252**, 6792–6797

Van Der Werf, P. and Meister, A. (1974). *Biochem. Biophys. Res. Commun.*, **56**, 90–96

Van Der Werf, P. and Meister, A. (1975). *Adv. Enzymol. Relat. Areas Mol. Biol.*, **43**, 519–556
Van Der Werf, P., Orlowski, M., and Meister, A. (1971). *Proc. Natl. Acad. Sci. U.S.A.*, **68**, 2982–2985
Van Der Werf, P., Griffith, O., and Meister, A. (1975). *J. Biol. Chem.*, **250**, 6686–6692
Van Der Werf, P., Stephani, R. A., and Meister, A. (1974). *Proc. Natl. Acad. Sci. U.S.A.*, **71**, 1026–1029
Waley, S. G. (1966). *Adv. Protein Chem.*, **21**, 2–112
Wellner, V. P., Sekura, R., Meister, A., and Larsson, A. (1974). *Proc. Natl. Acad. Sci. U.S.A.*, **71**, 2505–2509
Wilson, H. and Cannon, R. K. (1937). *J. Biol. Chem.*, **119**, 309–331
Woodward, G. E. and Reinhart, F. E. (1942). *J. Biol. Chem.*, **145**, 471–480

Microorganisms and Nitrogen Sources
Edited by J. W. Payne

Section 4.2

Ammonia Assimilation and Utilization in Bacteria and Fungi

C. M. BROWN

Department of Brewing and Biological Sciences, Heriot-Watt University, Chambers Street, Edinburgh EH1 1HX, UK

1 INTRODUCTION

Ammonia is situated at a branch point in nitrogen metabolism being the product of nitrogen fixation and assimilatory (and in some instances dissimilatory) nitrate reduction and the form of inorganic nitrogen converted to organic nitrogen (ammonia assimilation). Ammonia may also be produced intracellularly by the breakdown of urea, arginine, histidine, glutamate, etc. Ammonia and/or early products of ammonia assimilation are involved in the control of these pathways and in the uptake and utilization of other nitrogen sources such as amino acids. This chapter reviews some aspects of these processes. The role of ammonia as an electron donor in the metabolism of nitrifying bacteria is not pursued.

The words ammonia and ammonium are often used synonymously since there is no clear cut answer as to which form of the substrate is transported across membranes and assimilated. Since the p*K* of the dissociation $NH_4^+ \leftrightharpoons NH_3 + H^+$ is 9.25 at 24 °C it is evident that at physiological pH values the vast bulk of substrate is present as ammonium (ion), a fact rightly presented in favour of ammonium as being the usual substrate. Cole (1976), in discussing this problem, concludes that it would be difficult to design experiments to discriminate between active transport systems for ammonium and ammonia and free diffusion of ammonia across membranes, and the same is true in relation to the substrate involved in assimilation reactions. That free ammonia may diffuse across membranes is not in doubt, although as an uptake mechanism such a process requires a diffusion gradient (see Henderson, 1971) and as outlined later there is evidence for concentration against a diffusion gradient in bacteria and fungi. Alternatively, it is known that some ionophores such as gramicidin and valinomycin are able to translocate ammonium ion and some such mechanism may be envisaged as part of an active uptake system. Evidence for and against ammonium as assimilatory substrate is similarly difficult to evaluate. Some indication may be gained by studying the effects of pH on the K_m of the enzymes concerned. Brown and Stanley (1972) quoted data for the NADP-glutamate dehydrogenase from the yeast *Candida utilis* which indicated that a pH increase of 1 unit resulted in a 10-fold decrease in the apparent K_m of the enzyme, data consistent with ammonia as substrate. Unfortunately, enzyme activity decreased markedly at about pH 8, making a more detailed study impracticable. This type of evidence is at best circumstantial and it is not known in which form assimilation occurs, although ammonium ion is clearly the preferred candidate.

For the purpose of this chapter I propose to use the word 'ammonia' to denote the substrate transported and assimilated without specifying its form. This is in keeping with previous reviews from this and other laboratories (see Brown *et al.*, 1974; Miflin and Lea, 1976; Brown and Johnson, 1977).

2 AMMONIA UPTAKE IN MICROORGANISMS

There is clear evidence that bacteria are able to retain ammonia intracellularly at concentrations higher than in the culture medium. For example, nitrogen-fixing cultures of *Klebsiella pneumoniae* and *Azotobacter chroococcum* contain 3 mM and 2 mM ammonia respectively (Yoch and Pengra, 1966; Drozd *et al.*, 1972). Kleiner (1975) reported that nitrogen-fixing cultures of *Az. vinelandii* contained 0.6 mM ammonia intracellularly when the extracellular concentration was 2–8 μM i.e. a 100-fold difference in concentration across the cell envelope. He suggested that the membrane of this organism was impermeable to ammonia and that uptake would require a transport system. Further experiments indicated that the rate of translocation was dependent upon the nature of the anion present (to the ammonium ion) and this was explained by ammonium uptake via a symport system, a highly speculative interpretation. In nitrogen-fixing *K. pneumoniae* cultures the internal ammonia concentration was also about 100 times higher than the external concentration but leakage occurred on addition of dinitrophenol (Kleiner, 1976); the mechanism of this effect is unknown. Broach *et al.* (1976) described mutants of *Salmonella typhimurium* which were unable to grow on ammonia at concentrations less than 1 mM but, unlike mutants of *K. aerogenes* of similar phenotype (discussed below), were not defective in ammonia assimilation. It seems likely that these mutants had some defect in ammonia uptake and an inability to maintain ammonia intracellularly. A useful experimental tool for studying ammonia uptake is the use of [^{14}C]-methylammonium (methylamine) as a substrate analogue. Stevenson and Silver (1977) reported that three strains of *Escherichia coli* concentrated methylammonium some 100-fold intracellularly while cultures of *Bacillus subtilis* did not take up the analogue. In *E. coli*, methylammonium was not toxic (in the presence of ammonia or glutamate) and was not metabolized, making it an ideal substrate for uptake studies. That the methylammonium uptake system was equivalent to that for ammonia was indicated by the fact that 2 mM ammonia but not 2 mM glutamate completely inhibited methylammonium accumulation. Variations in uptake activity with pH suggested that two distinct systems may be present with pH optima at 7 and 9. Methylammonium accumulated intracellularly could be chased out with ammonia or higher alkylamines. This is the only report to my knowledge of the partial characterization of an uptake system for ammonia in bacteria. Stevenson and Silver (1977) did report some problems with the system since K_m values for methylammonium uptake varied from about 10 μM to 1 mM amongst samples.

Methylammonium has been widely used, however, in uptake studies in fungi. Hackette *et al.* (1970) reported that *Penicillium chrysogenum* would grow on methylammonium as sole nitrogen source. The K_m for uptake was about 10 μM and ammonia was a potent inhibitor of the process with a K_i of 27 μM. Uptake rate was low during growth in a nitrogen-sufficient medium but increased 800-fold during nitrogen starvation.

The most complete study reported is that of Pateman *et al.* (1974) for cultures of *Aspergillus nidulans* in which methylammonium uptake was compared with the uptake of ^{15}N-enriched ammonia. Experiments using ^{15}N-ammonia indicated that uptake rates were equal for all substrate concentrations in the range 0.5–50 mM but that the ^{15}N method was not sufficiently sensitive at initial concentrations of less than 0.5 mM. At that concentration, however, uptake proceeded at a constant rate for approximately 10 min and then ceased due to lack of substrate. At this time the extracellular ammonia concentration was estimated to be 50 μM, indicating a K_m for uptake of less than this value. The ammonia uptake system was affected by mutation, thus *amrA1* mutants had uptake rates of only 15 per cent of the wild type while *DER3* mutants grew only slowly on ammonia and had a K_m for uptake of 2 mM. Pateman *et al.* (1974) studied the system in more detail with [^{14}C]-methylammonium, reporting a K_m of 20 μM and 120-fold concentration of substrate. Uptake rates were high in organisms grown on alanine, proline, nitrate, and 0.15 per cent (w/v) Casamino acids, lower on 10 mM urea or 10 mM ammonia, and significantly lower still on 50 mM urea, 50 mM ammonia, or 0.6 per cent (w/v) Casamino acids. These last three conditions are known to produce high intracellular concentrations of ammonia. Development of the uptake system during nitrogen starvation following growth on 50 mM urea was prevented by cycloheximide. The methylammonium uptake system appears to be the physiological ammonia uptake process since, as with the *E. coli* and *N. crassa* uptake of methylammonium, it is inhibited by ammonia but not by urea, nitrate, glutamate, alanine, or Casamino acids. These results indicate that the intracellular ammonia concentration controls ammonia uptake by repression of the synthesis of some protein component.

Methylammonium uptake in the yeast *Saccharomyces cerevisiae* shares many features with the process in filamentous fungi, with 100-fold concentration of substrate being observed and ammonia inhibition of uptake occurring even at low concentrations (Roon *et al.*, 1975*a*). Mutants lacking methylammonium transport simultaneously loose the ability to take up and grow on ammonia while revertants regain both processes. Unlike the filamentous fungi, however, ammonia uptake in this yeast is not increased on nitrogen starvation.

3 AMMONIA ASSIMILATION AND THE CONTROL OF NITROGEN METABOLISM IN BACTERIA

The mechanisms of ammonia assimilation in bacteria have been reviewed in detail previously (Brown *et al.*, 1974; Brown and Johnson, 1977). In brief, two major mechanisms are known both resulting in the amination of 2-oxoglutarate to form glutamate. One route is catalysed by glutamate dehydrogenase (E.C.1.4.13, abbreviated GDH) and is a freely reversible reaction converting 2-oxoglutarate directly to glutamate with ammonia as nitrogen donor and NADPH or NADH as electron donor:

$$\text{2-Oxoglutarate} + NH_3 + NAD(P)H + H^+ \xrightleftharpoons{\text{GDH}} \text{Glutamate} + NAD(P)^+ + H_2O$$

In general, NADP-GDH is assumed to have a biosynthetic function, although the K_m for ammonia is usually high, indicating that it functions efficiently only at high ammonia concentrations. NAD-GDH is known to occur in some bacteria, principally in pseudomonads where it has a catabolic function associated with growth on amino acids as sources of carbon (see Brown *et al.*, 1973*a*). Other amino acid dehydrogenases specific to alanine and leucine are also known to occur especially in *Bacillus* spp., but again have catabolic functions (Meers and Kjaergaard-Pedersen, 1972; Porella, 1971).

An alternative mechanism present in most bacteria studied involves two enzymes, glutamine synthetase (E.C.6.3.1.2, abbreviation GS) and glutamate synthase (E.C.2.6.1.53, glutamine: 2-oxoglutarate aminotransferase, abbreviation GOGAT). These enzymes catalyse the overall amination of 2-oxoglutarate to glutamate with glutamine as intermediate:

$$NH_3 + \text{L-Glutamate} + \text{ATP} \xrightarrow[\text{GS}]{Me^{2+}} \text{L-Glutamine} + \text{ADP} + P_i$$

where Me^{2+} is a divalent metal ion;

$$\text{L-Glutamine} + \text{2-Oxoglutarate} + \text{NAD(P)H} + H^+ \xrightarrow[\text{GOGAT}]{} \text{L-Glutamate} + \text{L-Glutamate} + \text{NAD(P)}^+ + H_2O$$

This system is physiologically irreversible and has the same overall characteristics as GDH with the additional expenditure of one mole of ATP per mole of ammonia assimilated. This expenditure of energy is compensated for by the fact that GS has a much higher affinity for ammonia than has GDH and therefore enables organisms to 'scavenge' ammonia when this is present at low intracellular concentrations. GOGAT was first discovered in ammonia-limited cultures of *Klebsiella aerogenes* (Tempest *et al.*, 1970, 1973) but is now known to occur widely in bacteria, as discussed below.

The position of glutamine as an intermediate in the GS/GOGAT pathway is of particular note since the amide group of glutamine may serve as nitrogen donor not only to 2-oxoglutarate but also in the synthesis of carbamyl phosphate, glucosamine-6-phosphate, asparagine, histidine, tryptophan, *p*-aminobenzoate, AMP, CTP, and NAD (Ginsburg and Stadtman, 1973). Therefore, in organisms containing a GS/GOGAT pathway, the vast bulk of ammonia is converted first to glutamine, placing GS in a unique position in nitrogen metabolism. Not surprising then GS is subject both to control of synthesis and of activity, the complexity of the control system varying amongst bacteria. In *Bacillus subtilis*, GS activity is controlled by glutamine and by end-products of glutamine metabolism such as histidine, tryptophan, and AMP. These inhibitors have greatest effect when present together; for example, 5 mM glutamine alone is not inhibitory but together with 2.5 mM AMP lowers enzyme activity by 85 per cent (Deuel and Prusiner, 1974). In *E. coli*, however, control is much more complex. GS synthesis is repressed in organisms growing on good nitrogen sources such as glutamine or at high ammonia concentrations in which case ammonia assimilation proceeds via GDH and ammonia can substitute for glutamine in many biosynthetic reactions. Under conditions of ammonia limitation or growth on poor nitrogen sources a 20-fold derepression of GS synthesis occurs (Ginsburg and Stadtman, 1973) accounting then for about 1 per cent total cell protein. GS activity in *E. coli* is controlled by three different but related mechanisms, namely activation by divalent cations, feedback inhibition by products of glutamine metabolism, and covalent modification by adenylylation. The native enzyme has 12 identical subunits arranged in two hexagons, the whole aggregate behaving as a spherical particle of 600 000 daltons molecular mass which is stabilized by divalent cations. Each subunit contains a specific tyrosyl residue which can be adenylylated enzymically to form one 5′-adenylyl-*O*-tyrosyl derivative. Each enzyme molecule may contain between 0 and 12 adenyl residues and adenylylation and deadenylylation are controlled by intracellular metabolite concentrations, particularly those of glutamine and 2-oxoglutarate. Fully adenylylated GS is biosynthetically inactive and partially adenylylated enzyme is more sensitive to feedback inhibition than is the non-adenylylated enzyme. Each subunit is thought to have specific binding sites for the feedback inhibitors (which are products of glutamine metabolism) CTP, glucosamine-6-phosphate, carbamyl phosphate, tryptophan, histidine, AMP, and a single site for amino acids including serine, glycine, and alanine. *E. coli* GS catalyses a transferase reaction in addition to the biosynthetic reaction detailed above:

$$\text{L-Glutamine} + \text{Hydroxylamine} \xrightarrow[\text{ADP, arsenate}]{\text{Me}^{2+}} \gamma\text{-Glutamyl hydroxamate} + NH_3$$

The extent of adenylylation measured as an adenylylation number ($\bar{n}$ from 0 to 12) is usually estimated by the transferase reaction in which the adenylylated and non-adenylylated forms show different cation specificities. For example, while at pH 7.15 Mn^{2+} activates both forms of the enzyme equally, Mg^{2+} activates only the non-adenylylated form and, moreover, inhibits the activity of the adenylylated form in the presence of Mn^{2+}. The average state of adenylylation is therefore calculated using the equation $\bar{n} = 12 - 12(b/a)$, where a is a measure of the total transferase activity of both adenylylated and non-adenylylated subunits assayed in the presence of 0.3 mM $MnCl_2$ at pH 7.15 while b is a measure of the transferase activity of only non-adenylylated subunits assayed in the presence of 0.3 mM $MnCl_2$ at pH 7.15. For more details the review of Ginsburg and Stadtman (1973) is recommended.

The model for metabolite control of *E. coli* GS is illustrated in Figure 1 and summarized in the following paragraphs. Both adenylylation and de-adenylylation are catalysed by adenylyl transferase (ATase), the activity

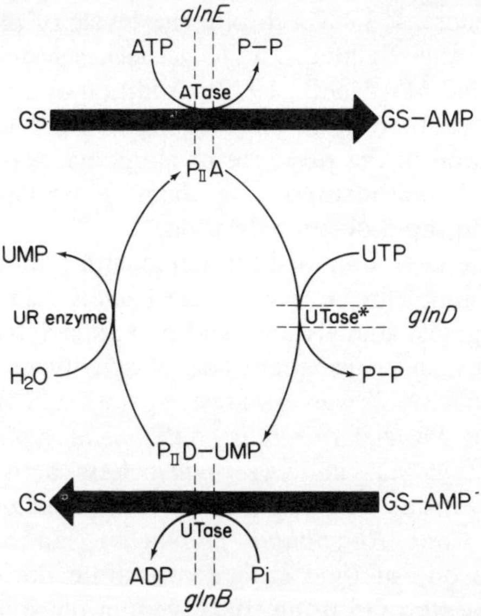

Figure 1 Control of glutamine synthetase (GS) activity in some Gram-negative bacteria. GS in non-modified and biosynthetically active while GS-AMP is adenylylated. *glnA* and *glnC* mutations influence the synthesis of GS protein. *UTase is activated by 2-oxoglutarate, ATP, and divalent cations and inhibited by glutamine and P_i. (Modified from Ginsburg and Stadtman (1973) by courtesy of the authors and with permission of Academic Press, Inc.)

of which is controlled by a small regulatory protein P_{II} which can exist in two forms. The non-modified regulator P_{IIA} interacts with ATase to stimulate GS adenylylation. Modification of P_{IIA} by uridylylation is catalysed by uridylyl transferase (UTase) which is activated by 2-oxoglutarate, ATP, and either Mg^{2+} or Mn^{2+} and potently inhibited by glutamine and inorganic phosphate. Uridylylated P_{II} ($P_{IID.UMP}$) stimulates the ATase-catalysed de-adenylylation of GS. $P_{IID.UMP}$ is de-uridylylated to form P_{IIA} by the Mn^{2+}-activated uridyl-removing (UR) enzyme. The overall effect of this system is that in the presence of high concentrations of glutamine and low concentrations of 2-oxoglutarate and ATP the P_{II} protein is maintained in non-uridylylated form which activates ATase to adenylylate GS, rendering it less active biosynthetically and more sensitive to feedback inhibition. Alternatively, with a low concentration of glutamine and inorganic phosphate and high levels of 2-oxoglutarate and ATP the P_{II} protein is uridylylated, resulting in ATase-catalysed de-adenylylation and biosynthetically active GS. Schutt and Holzer (1972) presented striking evidence of the effectiveness of this control system. When ammonia was added to a culture of *E. coli* in medium containing a growth-limiting nitrogen source then within a few seconds the intracellular level of glutamine increased several-fold, the levels of glutamate and ATP fell, and GS became adenylylated. The *K. aerogenes* system is similar to that of *E. coli* (Janssen and Magasanik, 1977). Addition of ammonia to a culture growing on proline resulted in an increase of $\bar{n}$ from 1 to 11 within 4 min. After 8 min incubation in the presence of ammonia, washing the cells in 1 per cent (w/v) KCl and resuspending them in proline medium (minus ammonia) resulted in rapid de-adenylylation.

Two methods have been employed to demonstrate that the adenylylation control of GS is common in Gram-negative bacteria. Tronick *et al.* (1973) prepared antisera against adenylylated and non-adenylylated *E. coli* GS and found cross-reactivity and consequent loss of activity in extracts and other organisms. Cross-reactivity was greatest against GS from a group of organisms including *Proteus mirabilis*, *Salmonella typhimurium*, *Serratia marcescens*, *K. pneumonia*, and *Az. agilis*. Less activity was found in *Pseudomonas aeruginosa*, *Ps. putida*, and *Az. vinelandii*, little against *Bacillus polymyxa* and *Rhizobium japonicum*, and none against *B. megaterium*. The second method is to demonstrate the removal of adenyl groups from adenylylated GS using snake venom phosphodiesterase with a concomitant increase in transferase activity in the presence of Mg^{2+}. This method and that using anti-*E. coli* GS serum give equivalent results. While control of ammonia assimilation by modulation of GS synthesis and activity is important in its own right, it has added significance in being involved in the control of a number of ammonia-repressible systems in bacteria, most notably histidine utilization (*hut* system) and nitrogen fixation, as described below. A number of mutations influencing the adenylylation system in *K. aerogenes* have been characterized, mainly in Magasanik's laboratory. *glnA* is the GS structural gene and *glnA* mutants do not synthesize GS. *glnC* mutants synthesize GS constitutively although it remains subject to control by adenylylation. *glnB* mutants are glutamine auxotrophs containing repressed levels of highly adenylylated enzyme. These mutants have an altered P_{II} protein which is unable to activate ATase-catalysed de-adenylylation. *glnE* mutants, in contrast, are unable to adenylylate GS and contain constitutively high levels of biosynthetically active enzyme. Both *glnB* and *glnE* mutants are insensitive to nitrogen metabolite control (Prival *et al.*, 1973; Magasanik *et al.*, 1974; Foor *et al.*, 1975; Janssen and Magasanik, 1977). The regulatory significance of some of these mutations is discussed below.

The conversion of glutamine and 2-oxoglutarate to two molecules of glutamate is catalysed by GOGAT. The physiological electron donor is either NADH or NADPH in heterotrophic bacteria while in blue-green algae, green algae, and higher plants reduced ferredoxin is involved (see Miflin and Lea, 1976; Brown and Johnson, 1977). It is interesting to note that in both sulphur and non-sulphur photosynthetic bacteria GOGAT is linked to the pyridine coenzymes rather than ferredoxin (Brown and Herbert, 1977*a*, *b*). *E. coli* GOGAT was purified by Miller (1973) and found to be an iron–sulphur protein made up of eight subunits, four of molecular mass 135 000 daltons and four of molecular mass 53 000 daltons and containing eight moles of flavin per aggregate. The *K. aerogenes* enzyme is also a multimeric iron–sulphur protein with two types of subunit, the heavier type being of higher molecular weight than that of *E. coli* (Trotta *et al.*, 1974). Both iron and sulphur are contained in the heavy subunit which also binds glutamine (the light subunit does not). Purified *K. aerogenes* GOGAT shows some activity with ammonia. The partially purified GOGAT from *Thiobacillus thioparus* is inhibited by atebrin and has an absorption spectrum consistent with the presence of flavin. The *T. thioparus* enzyme also shows some ammonia-linked activity (Adachi and Suzuki, 1977). All GOGATs studied show no activity with asparagine or other nitrogen donors (other than low activity with ammonia) and are specific to 2-oxoglutarate.

Thus, many bacteria contain two distinct routes of ammonia assimilation with the GS/GOGAT route best able to function at low ammonia concentrations. In *E. coli* mutation to glutamate auxotrophy requires the loss of both GOGAT and GDH and Berberich (1972) has presented evidence that each route can by itself synthesize sufficient glutamate for growth on ammonia. There was no evidence for genetic linkage of GS, GOGAT, and GDH. In *K. aerogenes* growing in the presence of an excess of ammonia the synthesis of both GS and GOGAT was repressed and the GS produced was adenylylated, while GDH synthesis was derepressed (Meers *et al.*, 1970; Brenchley *et al.*, 1973; Brenchley and Magasanik, 1974; Brown, 1976). In contrast, during ammonia-limited growth or growth on a poor nitrogen source, GS and GOGAT synthesis is derepressed, GS is non-adenylylated, and GDH synthesis repressed. These results strongly suggested a role for GS/GOGAT at low ammonia concentrations and reciprocal control of GS/GOGAT and GDH synthesis. Magasanik's group confirmed the former results when they reported the production of a mutant lacking GOGAT which was unable to grow with ammonia as nitrogen source if the medium concentration was less than 1 mM. A mutant with neither GOGAT nor GDH but able to synthesise GS (strain MK 261) was a glutamate auxotroph. Transductants prepared by treating strain MK 261 with phage grown on wild-type organisms and selecting on a medium containing low ammonia concentrations (1 mM) regained GOGAT but not GDH activity. These transductants grew well at high ammonia concentrations, which can be interpreted as a lack of physiological function for the GDH, although a more detailed investigation must preceed any such conclusion. Another way of demonstrating the significance of the GS/GOGAT route of ammonia assimilation is the use of the inhibitors methionine sulphoximine (MSX) and methionine sulphone (MSO). Brenchley (1973) reported that growth of wild-type *K. aerogenes* on ammonia was sensitive to 0.1 M concentrations of either inhibitor. She isolated a GDH-minus (*gltD*) mutant, however, which grew well on ammonia but which was sensitive to 1 mM inhibitor. Both GS and GOGAT are inhibited by MSX and MSO but inhibition can be relieved by glutamate and glutamine. Cultures of *S. typhimurium* were inhibited by 50 μM MSX, and glutamine (but not glutamate) relieved this inhibition. MSX at 2–4 μM concentration causes 50 per cent inhibition of GS activity in extracts, while 2–3 mM was required to inhibit GOGAT by an equivalent amount. Up to 50 mM MSX had no effect on the *Salmonella* GDH (Steimer-Veale and Brenchley 1974).

Some *Clostridium* spp. do not synthesize GDH and synthesis of GS/GOGAT is constitutive (Dainty, 1972; Brown, 1976), while some streptococci do not synthesize GOGAT and growth at low ammonia concentrations is characterized by a derepression of GDH synthesis (Griffith and Carlsson, 1974) as found in yeasts and some filamentous fungi (see below). Between these extremes most bacteria studied modulate the synthesis of GS (and its activity), GOGAT, and GDH to suit the availability of nitrogen in the culture medium (see Brown *et al.*, 1973*a*, 1974; Senior, 1975; Vandecasteele *et al.*, 1975; Brenchley *et al.*, 1975; Brown, 1976; Brown and Johnson, 1977).

While in neither *E. coli* (Berberich, 1972) nor *K. aerogenes* (Brenchley and Magasanik, 1974) are GS, GOGAT, and GDH linked genetically, in *K. aerogenes* the synthesis of GS has a marked influence on GDH synthesis while the presence or absence of GOGAT influences GS synthesis and activity. Additionally, GS has a central role in the control of the synthesis of enzymes subject to control by ammonia. In *K. aerogenes* the synthesis of histidase and proline oxidase is repressed by ammonia and by repressing carbon metabolites such as glucose. Under ammonia-limited conditions or growth on poor nitrogen sources, however, both enzymes are synthesized even in the presence of glucose (Prival and Magasanik, 1971) and this is related to the content and activity of GS (Prival *et al.*, 1973). In both *glnA* and *glnB* mutants which lack biosynthetically active GS and require glutamine for growth the synthesis of histidase and proline oxidase is repressed under nitrogen limitation. In *glnC* revertants of *glnB* which synthesize GS constitutively both enzymes are insensitive to repression both by glucose and ammonia. This evidence suggests that in *K. aerogenes* production of histidase and proline oxidase is activated by biosynthetically active (non-adenylylated) GS. Prival *et al.* (1973) also reported that *glnA* mutants possessed high levels of GDH and low histidase while *glnC* mutants possessed low GDH levels (and high histidase). Thus, while GS activates histidase synthesis it inhibits or represses the synthesis of GDH. The

GOGAT-less mutant of Brenchley *et al.* (1973) when placed in a glucose–histidine medium (derepressed for histidase and GS in wild type) would not degrade histidine. Revertants (constitutive for GS but lacking GOGAT) degraded histidine but would not grow on amonia. The probable explanation for this is that in a GOGAT-less mutant glutamine or some products of glutamine metabolism interferes with the GS-directed activation of histidine utilization, probably due to adenylylation of GS. The GOGAT-less, GS-constitutive mutants would not grow on ammonia since the alternative assimilatory route via GDH is repressed by the GS. While GS controls the synthesis of GDH in *K. aerogenes* the loss of GDH by mutation does not influence GS synthesis, or activity. In *S. typhimurium* the control of GS synthesis and activity is similar to that in *K. aerogenes*. Tyler *et al.* (1974) isolated non-adenylylated GS from a histidase-constitutive mutant of *K. aerogenes* and showed that it stimulated the transcription of the histidine utilization (*hut*) system in extracts of *S. typhimurium.* This stimulation was inversely proportional to the $\bar{n}$ of the GS. A *S. typhimurium* strain with a mutation in the *hut* promotor region showed only limited GS activation of *hut* transcription. Thus non-adenylylated GS activates transcription of *hut* DNA. Foor *et al.* (1975) additionally demonstrated that the synthesis of GS itself, i.e. the expression of *glnA*, requires non-adenylylated GS and is repressed when the enzyme is adenylylated.

Ammonia regulation of *hut*, proline oxidase, GDH, and GS is therefore effectively mediated through GS in *K. aerogenes*. In addition to these enzymes there is good evidence for the involvement of GS in the control of other ammonia-repressible systems such as urease, asparaginase, arginine degradation, tryptophan aminotransferase in *K. aerogenes*, and of nitrogen fixation in *K. pneumoniae* (Magasanik *et al*., 1974; Tubb, 1974; Streicher *et al*., 1974; Friedrich and Magasanik, 1977).

In *E. coli* the synthesis and activity of GS is controlled in an analogous manner to that of *K. aerogenes* but GS controls neither GDH nor proline oxidase in this organism (which does not synthesize *hut* enzymes) (Streicher *et al.*, 1976; Newman and Cole, 1977). Streicher *et al.* (1976), however, reported that when the GS of *K. aerogenes* with a *glnA* deletion was restored after conjugation with *E. coli* carrying a F′ *glnA* plasmid then the hybrid (*K. aerogenes* with *E. coli* GS) showed GS regulation of GDH and *hut*, i.e. the *E. coli* GS represses GDH synthesis in *K. aerogenes* but not in *E. coli. S. typhimurium* (like *E. coli*) does not show GS control of GDH, *hut*, or proline oxidase (Brenchley *et al.*, 1975). When the *Salmonella hut* genes were transferred to *K. aerogenes* by transduction then they were controlled by the *Klebsiella* GS, this suggests some difference between the *Klebsiella* and *Salmonella* transcription systems. In either *E. coli* or *S. typhimurium* the *Salmonella* GS activates both *Salmonella hut* and *Klebsiella hut.* Thus, within *S. typhimurium* the presence of GS is necessary but not sufficient to activate *Salmonella hut* transcription. Some other factor (present in *E. coli* and *K. aerogenes*) must be involved but is missing or is altered in *S. typhimurium*. Nitrogen-limited *K. aerogenes* and *S. typhimurium* contain similar amounts of non-adenylated GS and the additional factor required for GS control in *S. typhimurium* is unknown (Bloom *et al.* 1977).

The ammonia-repressible system receiving most attention in recent years is that of nitrogen fixation, and the reviews of Brill (1975) and Shanmugam and Valentine (1975*a*) are recommended for a background to this topic. The study of ammonia control of nitrogen fixation in *K. pneumoniae* has derived much benefit from a knowledge of the *hut* system of *K. aerogenes*. In *K. pneumoniae* ammonia represses the synthesis of the three proteins of the nitrogenase complex. As with the *hut* system it is likely that products of ammonia assimilation rather than ammonia *per se* are involved in regulation. Nagatani *et al.* (1971) isolated *asm* mutants lacking GOGAT which no longer fixed nitrogen and moreover were unable to utilize a number of fixed-nitrogen sources. This may be due to a similar cause to that proposed for the GOGAT-less *K. aerogenes* which did not express the *hut* genes. MSX and MSO which inhibit GS and GOGAT and therefore prevent the conversion of ammonia to glutamine and glutamate enable wild-type *K. pneumoniae*, *Az. vinelandii*, and *Rhodospirillum rubrum* to fix nitrogen in the presence of ammonia (Gordon and Brill, 1974; Weare and Shanmugam, 1976). As early as 1972 Streicher *et al.* proposed that GS was involved in control of nitrogen fixation, and this was confirmed later by Tubb (1974) and Streicher *et al.* (1974). For example, *glnA* mutants do not fix nitrogen whereas *glnC* mutants do so constitutively. When *glnA* mutants reaquire GS on conjugation with *E. coli* carrying a F′ *gln* plasmid they also reaquire the ability to fix nitrogen. One class of mutant of particular interest was derived from an *asm* (GOGAT-less) strain and mutated to glutamine auxotrophy. These were found to be *asm glnC* (Shanmugam and Valentine, 1975*b*) and formed nitrogenase constitutively. Since, however, ammonia assimilation was blocked by mutation and repression of GDH then, as in the presence of MSX, ammonia produced on nitrogen fixation was excreted. These results suggest a role for GS similar to that in the *hut* system, i.e. activation of the nitrogenase genes (*nif*) by non-adenylylated GS. Shanmugam *et al.* (1975), however, isolated a series of *asm* mutants which were glutamate or glutamine auxotrophs and all of which fixed nitrogen in the presence of ammonia. One class were *asm glnC*, as above (for example strain SK 24 mentioned later), whereas another class which were glutamine auxotrophs contained catalytically inactive GS which was detected immunologically. A third class contained no biosynthetically active GS nor cross-reacting protein. All three types of *gln* mutation mapped in the *glnA* region and their physiology may be explained if GS or alternatively another protein (with no GS activity nor cross-reactivity) activate *nif* transcription or if the *nif* promotor is also modified. While ammonia did not repress nitrogen fixation in the *asm glnC* strain SK 24, the presence of amino acids did (Shanmugam and Morandi, 1976), but only at high concentrations. Thus, 1 mg ml^{-1} each of glutamine and aspartate repressed nitrogen fixation completely, while with 1 mg ml^{-1} glutamine alone ammonia also acted as repressor. Shanmugam and Morandi (1976) suggested that regulation in this organism involved conversion of ammonia to amino acids which in turn act as effectors of GS protein. In support of this, back mutation resulting in ammonia utilization also results in sensitivity to ammonia repression. *K. pneumoniae* is a facultative anaerobe which fixes nitrogen only under anaerobiosis, and the results outlined above were obtained from anaerobic cultures. Eady *et al.* (1978) have reported that oxygen represses nitrogenase synthesis in the mutant strain SK 24 and propose that regulation by oxygen is independent of the GS-mediated ammonia control loop.

In the *Rhizobium*/legume symbiotic nitrogen-fixing system, there is good evidence that nitrogen fixed by the bacteroids is exported as ammonia and then assimilated by plant enzymes (Brown and Dilworth, 1975; Robertson *et al.*, 1975; Planque *et al.*, 1977). The bacteroids usually contain only low levels of ammonia-assimilatory enzymes, and physiologically are similar to the *Klebsiella* mutants and to organisms treated with MSX which fix nitrogen and excrete ammonia. A recent advance has been the ability to induce free-living rhizobia to fix nitrogen, thereby making the system amenable to conventional bacteriological experimentation (Pagan *et al.*, 1975; Kurz and La Rue, 1975; McComb *et al.*, 1975). In such a study, O'Gara and Shanmugam (1976) grew cowpea Rhizobium (strain 32HI) and *R. japonicum* in the presence of $^{15}N_2$ in a medium containing mannitol and glutamate. Nitrogenase activity was associated with a higher enrichment of ^{15}N into the culture supernatant solution than into cell material. Up to 94 per cent of the nitrogen fixed (ammonia) was excreted by *R. japonicum.* A number of *Rhizobium* strains grown in the presence of histidine also excrete ammonia and in experiments with *R. trifolii* the rate of histidine uptake was closely related to the rate of ammonia excretion. Ammonia excretion may therefore be a common feature of *Rhizobium* metabolism. Clearly, control of *nif* expression in rhizobia must differ in some respects from that in *K. aerogenes.* Kondorosi *et al.* (1977) found no indication of GS adenylylation in *R. meliloti.* These authors isolated GS-less and GOGAT-less mutants of this organism which did not synthesize GDH. GOGAT-less mutants were glutamate auxotrophs but were effective in forming nitrogen-fixing root nodules in *Medicago sativa.* GS-less mutants were glutamine auxotrophs and were ineffective. This suggests that GS is required for *nif* expression in *R. meliloti* bacterioids while GOGAT is not, and that assimilation of ammonia via GOGAT is not essential for bacteroid development nor in the nodule system.

Nitrogen fixation has also been studied in some detail in photosynthetic bacteria. In *Rhodospirillum rubrum* ammonia repression is reversed on addition of MSX with the excretion of up to 95 per cent nitrogen fixed as ammonia (Weare and Shanmugam, 1976). The site of MSX action appeared to be GS and glutamine was thought to be the controlling metabolite produced from ammonia. Mutants of *Rhodopseudomonas capsulata* unable to fix nitrogen simultaneously lost the ability to utilize glutamate, glutamine, alanine, arginine, proline, uracil, and xanthine but would utilize ammonia present in high concentrations (Wall *et al.* 1977). Such mutants produced normal levels of nitrogenase, GS (with intact adenylylation system), GOGAT, and GDH, and were therefore unlike any of the mutants of the other bacteria mentioned above. A unifying control model is therefore not possible at the present time.

4 AMMONIA ASSIMILATION AND THE CONTROL OF NITROGEN METABOLISM IN YEASTS AND FILAMENTOUS FUNGI

Ammonia assimilation in fungi differs markedly from the pathways and control loops illustrated above for bacteria. In essence the glutamine pathway appears to be of limited significance and GS does not play a central role in the control of nitrogen metabolism. There is no evidence for control of GS activity by adenylylation in fungi. The best understood mechanisms of assimilation are in *Candida utilis*, while control is best illustrated in *Aspergillus nidulans* and to a lesser extent in *Saccharomyces cerevisiae.*

Sims *et al.* (1968) demonstrated that when cultures of *C. utilis* were growing on $^{15}NH_3$ then almost all the ammonia assimilated was converted to glutamate and glutamine. About 75 per cent of the total was incorporated into the α-amino group of glutamate and much of the remainder into the amide group of glutamine. Sims and Folkes pointed out that controls must operate to ensure an appropriate flow of ammonia into glutamate and glutamine and to ensure that the rate of glutamate synthesis was sufficient to meet demands for glutamine synthesis and transamination to form other amino acids. The enzymes involved in the synthesis of glutamate and glutamine are NADP-GDH and GS. Many yeasts and filamentous fungi also synthesize NAD-GDH, but the main function of this enzyme appears to be the catabolism of glutamate (see Brown *et al.*, 1974; Pateman and Kinghorn, 1976). Yeast NADP-GDH has a high K_m for ammonia (Brown *et al.*, 1974), and in the absence of a high-affinity system (such as GS/GOGAT) in most of these organisms, mechanisms have been developed to enable yeasts to grow in the presence of low ammonia concentrations and on poor nitrogen sources such as nitrate (Brown and Stanley, 1972; Burn *et al.*, 1974). One such mechanism is the ability to accumulate ammonia intracellularly as suggested by the methylammonium uptake results of Roon *et al.* (1975*a*), but the principle method appears to be the derepression of NADP-GDH synthesis. For example, in *C. utilis* and *S. cerevisiae* the NADP-GDH level in ammonia-limited cultures was 20–30 times higher than in carbon-limited (ammonia excess) cultures (Brown, 1976), thus yeasts assimilate ammonia present at low concentrations by synthesizing large amounts of a rather inefficient enzyme. Folkes and Sims (1974) studied the kinetics of [^{15}N]-ammonia incorporation into glutamate in *C. utilis* and reported that the rate of glutamate synthesis was inversely related to the size of the intracellular amino acid pool while in the experiments described no variation in rate of enzyme synthesis was recorded. This suggested that NADP-GDH activity was controlled by amino acids and this was confirmed using cell extracts. In *S. cerevisiae* the synthesis of NADP-GDH was inversely related to the size of the amino acid pool (Brown and Stanley, 1972) and it may be that glutamate synthesis is controlled by adjusting both the synthesis and activity of this enzyme. The control of glutamine synthesis in *C. utilis* is better understood. Sims and Ferguson (1974) measured the rate of glutamine synthesis using [^{15}N]-ammonia and showed that it was closely related to the specific activity of GS, which in turn was inversely related to the intracellular concentration of glutamine (Ferguson and Sims, 1974*a*). A sudden increase in the supply of ammonia or decrease in the glucose supply resulted in a rapid decrease in GS activity, faster than could be accounted for by repression of enzyme synthesis followed by dilution during growth. This was shown subsequently to be due to irreversible enzyme inactivation. Reappearance of GS activity after removal of ammonia was inhibited by cycloheximide, indicating a requirement for *de novo* enzyme synthesis (Ferguson and Sims 1974*b*). Regulation of GS activity in *C. utilis* is therefore by inactivation followed by resynthesis. Candida GS is an octameric globular protein of 390 000 daltons aggregate molecular mass. Inactivation *in vivo* appears to involve dissociation into tetramers and then into dimers and monomers. While reaggregation to form octamers occurs in the presence of Mg^{2+} and glutamate the aggregate has no biosynthetic activity (Sims *et al.*, 1974).

The ammonia-repressible systems of *S. cerevisiae* include the general amino acid uptake system, the uptake systems for proline and ureidosuccinate, and the enzymes arginase, urea amidolyase, and allantoinase. Ammonia repression is relieved in a number of mutants, although opinion is divided on the relative importance of different regulators. In *S. cerevisiae*, NADP-GDH is the most important enzyme in ammonia assimilation and mutants lacking the structural gene for this enzyme (*gdhA*) grow only very slowly on ammonia. A further mutation (*gdhCR*) accelerates growth on ammonia (in the absence of NADP-GDH) apparently by derepressing the synthesis of NAD-GDH (Grenson and Hou, 1972; Dubois *et al.*, 1973; Dubois and Grenson, 1974; Dubois *et al.*, 1974; Grenson *et al.*, 1974). A report of GOGAT in this organism (Roon *et al.*,

1974) has not been substantiated and while GOGAT-like activity may be detected at low levels it is doubtful whether this enzyme (even if present, which itself is in doubt) plays any part in ammonia assimilation (Brown, 1976). *gdhA* mutants show a lack of ammonia repression, suggesting a role for NADP-GDH in ammonia control of nitrogen metabolism. The *gdhCR* mutation in organisms containing an intact *gdhA* gene results in only low levels of NADP-GDH activity and again a release of ammonia repression. A GS-less mutant, in contrast, while being a glutamine auxotroph, retained ammonia control of those activities listed above, thus GS is not involved in ammonia repression. The effect of the *gdhA* mutations was not lost in the presence of glutamate, glutamine, or asparagine in *S. cerevisiae* (Dubois *et al.*, 1974), although van de Poll (1973) reported that glutamate plus ammonia and glutamine repressed allantoinase in a *gdhA* mutant of *S. carlsbergensis*. Roon *et al.* (1975*b*) reported that in *S. cerevisiae* ammonia repression of amino acid uptake was lost in *gdhA* mutants but regained in a double mutant containing high levels of NAD-GDH (*gdhA gdhCR* phenotype) and proposed that some product of ammonia assimilation (mediated by NAD-GDH) was functional in ammonia regulation. These results are clearly at variance with those discussed above. Middelhoven (1977) has described another class of mutants showing relief of arginase, ornithine transaminase, and allantoinase in *S. cerevisiae* which are independent of the *gdhA* and *gdhCR* loci. These organisms were selected as methylammonium resistant and designated *argNmr* and *allNmr* and were not methylammonium transport mutants. Repression of allantoinase was not influenced by *gdhA* or *argNmr* but was by *allNmr*, suggesting that this latter locus had a specific role in allantoinase control. Ammonia repression of arginase was lost in *gdhA* mutants but that of ornithine transaminase was not, while both were synthesized in the presence of ammonia in *argNmr* mutants.

In many yeasts, NADP-GDH is partly repressed at high ammonia concentrations but markedly derepressed when ammonia is the growth-limiting substrate (Burn *et al.* 1974). A strain of *Schizosaccharomyces pombe* which did not show this derepression was found to contain a GS/GOGAT system (Brown *et al.*, 1973*b*). The occurrence of GOGAT is apparently restricted to a small number of yeasts and in these GS/GOGAT has a function similar to that in bacteria (Johnson and Brown, 1974). The GS of *Schiz. pombe* is not controlled by adenylylation but appears to be regulated by inactivation and resynthesis as in *C. utilis* but with glutamate as the most likely regulator molecule (van Andel and Brown, 1977).

There is no report to my knowledge of GOGAT in a filamentous fungus, although relatively few of these organisms have been studied intensively. Both NAD-GDH and NADP-GDH are known to occur in many filamentous fungi, with NADP-GDH functioning in ammonia assimilation as it does in yeasts (Pateman and Kinghorn, 1976). An exception is *Coprinus lagopus* (Fawole and Casselton, 1972) in which NAD-GDH has both aminating and deaminating functions. In *Neurospora crassa* and *Aspergillus nidulans*, mutants lacking NADP-GDH grow poorly on ammonia. In *N. crassa* the structural gene for NADP-GDH is *am-1* and a large number of *am* mutants have been characterized and used in the determination of the structure of the enzyme. It is a hexamer of molecular mass 280 000 daltons made up of identical subunits. The complete amino acid sequence of the wild-type enzyme is known (Wooton *et al.*, 1973). In *A. nidulans gdhA* is the structural gene for NADP-GDH (Kinghorn and Pateman, 1975). The activity of NADP-GDH is low during growth on glutamate in the presence of glucose. Replacement of glucose with amino acids, however, results in loss of activity which is prevented by cycloheximide, i.e. synthesis of a protease may be involved in activation in the presence of amino acids. *gdhB* is the structural gene for NAD-GDH and mutants lacking this enzyme but containing NADP-GDH grow normally on nitrate or ammonia but not with glutamate as carbon source.

Ammonia regulates a number of transformations of nitrogenous compounds in *A. nidulans.* Pateman and Kinghorn (1976) listed ammonia-repressible system as follows: nitrate reduction, asparaginase, purine degradation, amide degradation, amino acid-, urea-, purine- and ammonia-uptake, the degradation of arginine and histidine, and the synthesis of extracellular protease. Mutations in a number of genes influence ammonia regulation. *meaA* and *meaB* mutants are resistant to 100 mM methylammonium (Arst and Cove, 1969), and show derepression of thiourea- and glutamate-uptake, which are ammonia-repressible in wild-type organisms. *amrA* mutants grow poorly with ammonia as sole nitrogen source and have low levels of ammonia and methylammonium uptake. Glutamate uptake is not ammonia-repressed in these organisms, which, when grown on 10 mM ammonia, contain high intracellular ammonia levels (Pateman *et al.*, 1973). *xprD* mutants have a derepressed synthesis of extracellular protease and uptake of glutamate and a low rate of ammonia uptake but also contain high intracellular ammonia concentrations when grown on 10 mM ammonia. Chlorate is toxic to wild-type *A. nidulans* due to metabolism via nitrate reductase and in the wild type chlorate sensitivity is relieved by the presence of ammonia, which represses nitrate reductase. Arst and Cove (1973) described *areA* mutants which are chlorate-resistant in the absence of ammonia and grow poorly on nitrate, amino acids, amides, and purines. *areA* mutants and the *xprD* mutants are probably allelic and both are thought to be involved in regulation of ammonia repression. Thiourea and aspartate hydroxamate are toxic analogues of urea and asparagine but (as with chlorate sensitivity) toxicity is relieved by ammonia in wild-type organisms. Kinghorn and Pateman (1973, 1975) isolated mutants resistant simultaneously to chlorate, thiourea, and aspartate hydroxamate in the absence of ammonia. They reported that *tanA* organisms have normal levels of NADP-GDH and grow normally on all nitrogen sources, and are resistant with nitrate, but not amino acids, as nitrogen source, while *tamA* mutants conversely are resistant in alanine (but not in nitrate) medium and have low levels of thiourea and methylammonium uptake. *tamA* mutants have only 10 per cent NADP-GDH activity of wild-type organisms and ammonia efflux occurs when they are grown on amino acids. *tanA*, *tamA*, and *areA* loci are thought to specify components of the ammonia regulatory system, and *tanA* and *areA* may be at the same locus.

gdhA mutants of *A. nidulans* are derepressed for a number of ammonia-regulated systems but are not resistant to high concentrations of methylammonium. Pateman and Kinghorn (1976) suggest two hypotheses to account for the involvement of *gdhA* and the NADP-GDH in ammonia regulation. First, ammonia is assumed not to be the effector molecule but rather some product, such as glutamate or glutamine, which would be present at low intracellular concentrations in NADP-GDH-less mutants. Against this model are the facts that glutamate and glutamine are taken up readily by the wild type but have no direct repressive effects and that (as outlined in Section 4.2.2) the intracellular rather than the extracellular ammonia concentration is implicated in control of ammonia uptake and hence in ammonia regulation. The second hypothesis is that NADP-GDH has a regulatory function in addition to its biosynthetic function. This must be in addition to the products of the putative regulatory genes discussed above. Pateman and Kinghorn (1976) have proposed a model for ammonia regulation taking into account many of the genetical and biochemical data outlined above. They suggest that any model must take four main points into consideration: (a) The differential effects of internal and external ammonia concentrations. For example wild-type organisms when nitrogen-starved had derepressed thiourea- and glutamate-uptake systems which were repressed during growth on 10 mM ammonia. The intracellular ammonia levels (100–200 μM), however, did not vary, indicating that glutamate and thiourea uptake were controlled by the extracellular ammonia concentration. Conversely, ammonia uptake itself is controlled by the intracellular ammonia concentration. *gdhA* mutants do not respond to the intracellular concentrations of ammonia. (b) The role of NADP-GDH as a regulator protein must be included. (c) The roles of the putative regulatory genes *areA* and *tamA* must also be included. (d) Derepressed mutants other than *gdhA* (e.g. *mea* and *amr*) must be explained.

The model is detailed in Figure 2 and has the following features:

(a) Normal organisms monitor intracellular and extracellular ammonia levels

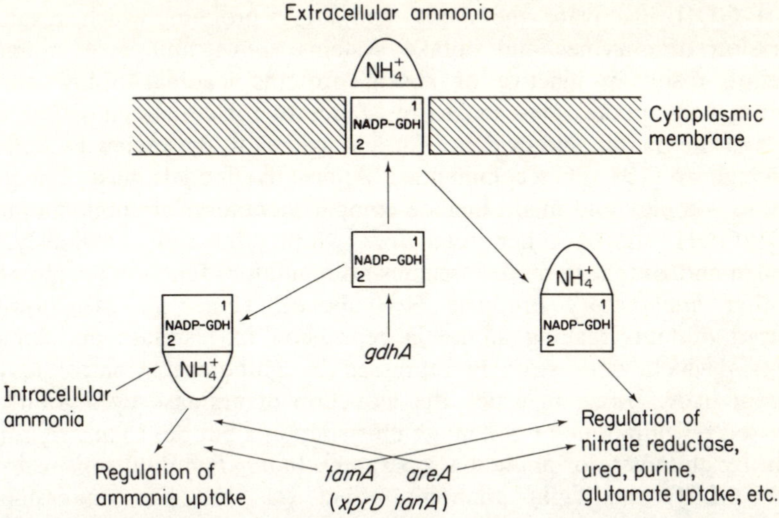

Figure 2 Ammonia regulation in *Aspergillus nidulans*. The NADP-GDH specified by *glnA* monitors extracellular (site 1) and intracellular (site 2) ammonia concentrations. *tamA* and *areA* gene products are involved in control of ammonia repressible activities. (Modified from Pateman and Kinghorn (1976) by courtesy of the authors and with permission of Edward Arnold (Publishers) Ltd.)

(b) The intracellular ammonia concentration determines the level of the ammonia-uptake system.
(c) Extracellular ammonia determines the levels of nitrate reductase, urea uptake and glutamate uptake.
(d) NADP-GDH is located in an ammonia monitoring site in the membrane and can complex with extra- but not intracellular ammonia. It therefore monitors the extracellular ammonia concentration and indirectly determines the levels of nitrate reductase, etc.
(e) A number of membrane components (specified by *mea*, *amr*, and *xpr* loci) are essential for uptake systems and monitoring of extracellular ammonia concentration.
(f) NADP-GDH can also complex with intracellular ammonia and determines ammonia uptake.
(g) A 'control pathway' converts information relating to intra- and extracellular ammonia concentrations (as NADP-GDH complexes) into signals for the control of the ammonia repressible systems. The regulatory genes *areA* and *tamA* specify components of this 'pathway'.

The role suggested for NADP-GDH explains why the *gdhA* mutation results in derepression since with an altered protein or no protein at all the organism is insensitive to the ammonia concentration. Ammonia plus

NADP-GDH inactivate the *tamA* and *areA* proteins which results in repression of enzymes and uptake systems. *tamA* and *areA* mutations therefore result in inactive or absent proteins leading to low enzyme activities and uptake rates and to resistance to toxic analogues. For more details of the model, and evidence for and against it, the review by Pateman and Kinghorn (1976) is recommended. Alternative models must of necessity be more complex and might include compartmentation of ammonia and/or NADP-GDH and/or other regulators within the cell. Obviously, the Pateman and Kinghorn model assumes no regulatory function for glutamine or other assimilatory products (see above). Hynes (1974), however, reported mutants lacking ammonia repression of histidase and amidase activities which were strongly repressed by glutamine. Control may be different in *N. crassa* in which the induction of arginase by arginine was repressed by glutamine or glutamate plus ammonia but not to any significant extent by glutamate or ammonia separately. In *am-1* mutant organisms (no NADP-GDH) the slight ammonia effect was lost but repression by glutamate and ammonia and by glutamine were still expressed (Vaca and Mora 1977).

5 CONCLUSIONS

The last ten years has seen a marked increase in interest in ammonia assimilation and ammonia control of nitrogen metabolism. The fact that GS in bacteria and NADP-GDH in fungi are central in current models of ammonia control is due to the significance of these enzymes in catalysing the first steps in ammonia assimilation in these different groups of microorganisms. While assimilation of ammonia in fungi is a relatively simple process, that in bacteria is more complex, involving the two-enzyme–glutamine pathway. The discovery of GOGAT gave new impetus to the study of bacterial nitrogen metabolism by emphasizing the almost all-embracing role of glutamine as nitrogen donor.

It is premature to extend the current information based on a handful of organisms to produce a unifying model. In bacteria. for example, control of nitrogen metabolism in the Gram-positive bacteria, apparently lacking a GS adenylylation loop, is an area ripe for research, while the rhizobium system is only just becoming amenable to experimentation with the ability to use nitrogen-fixing cultures of these organisms.

In fungi a major query must be, why no widespread GOGAT system?, with a secondary question about the relative economics of synthesizing a lot of an inefficient enzyme versus a smaller amount of enzymic pathway with a higher substrate affinity. It is probably only a matter of time before GOGAT is found in filamentous fungi, especially since both pyridine coenzyme- and ferredoxin-linked enzymes appear to be widespread in algae and higher plants (see Miflin and Lea, 1976, 1977).

The major pathways of ammonia assimilation are probably known, although variations may still be found, and work will now proceed with studies of control in a wider range of organisms. It is pertinent to record just how successful in this context has been the use of defined cultural conditions and the joint application of genetics and biochemistry in studies so far, techniques which will no doubt be exploited even more fully in the future.

ACKNOWLEDGEMENTS

I am grateful to many colleagues for discussions on nitrogen assimilation and control and especially wish to record thanks to Professor D. W. Tempest and Dr J. A. Cole. Work quoted from my own laboratory has been financed by the Science Research Council and the Natural Environment Research Council.

REFERENCES

Adachi, K. and Suzuki, I. (1977). Purification and properties of glutamate synthase from *Thiobacillus thioparus*. *J. Bacteriol.*, **129**, 1173–1182

van Andel, J. G. and Brown, C. M. (1977). Ammonia assimilation in the fission yeast *Schizosaccharomyces pombe* 972. *Arch. Microbiol.*, **111**, 265–270

Arst, H. M. and Cove, D. J. (1969). Methylammonium resistance in *Aspergillus nidulans*. *J. Bacteriol.*, **98**, 1284–1293

Berberich, M.-A. (1972). A glutamate-dependent phenotype of *E. coli* K12: the result of two mutations. *Biochem. Biophys. Res. Commun.*, **47**, 1498–1503

Bloom, F. R., Streicher, S. L., and Tyler, B. (1977). Regulation of enzyme synthesis of *Salmonella typhimurium*: a factor in addition to glutamine synthetase is required for activation of enzyme formation. *J. Bacteriol.*, **130**, 983–990

Brenchley, J. E. (1973). Effects of methionine sulfoxamine and methionine sulphone on glutamate synthesis in *Klebsiella aerogenes*. *J. Bacteriol.*, **114**, 666–673

Brenchley, J. E. and Magasanik, B. (1974). Mutants of *Klebsiella aerogenes* lacking glutamate dehydrogenase. *J. Bacteriol.*, **117**, 544–550

Brenchley, J. E., Baker, C. A., and Patil, L. G. (1975). Regulation of the ammonia assimilatory enzymes in *Salmonella typhimurium*. *J. Bacteriol.*, **124**, 182–189

Brenchley, J. E., Prival, M. J., and Magasanik, B. (1973). Regulation of the synthesis of enzymes responsible for glutamate formation in *Klebsiella aerogenes*. *J. Biol. Chem.*, **248**, 6122–6128

Broach, J., Neumann, C., and Kustu, S. (1976). Mutant strains (*nit*) of *Salmonella typhimurium* with a pleiotropic defect in nitrogen metabolism. *J. Bacteriol.*, **128**, 86–98

Brill, W. J. (1975). Regulation and genetics of bacterial nitrogen fixation. *Annu. Rev. Microbiol.*, **29**, 109–129

Brown, C. M. (1976). Nitrogen metabolism in bacteria and fungi. In *Continuous Culture 6: Applications and New Fields* (A. C. R. Dean, D. C. Ellwood, C. G. T. Evans, and J. Melling, eds), Ellis Horwood Ltd., Chichester, pp. 170–183

Brown, C. M. and Herbert, R. A. (1977*a*). Ammonia assimilation in purple and green sulphur bacteria. *FEMS Lett.*, **1**, 39–42

Brown, C. M. and Herbert, R. A. (1977*b*). Ammonia assimilation in members of the Rhodospirillaceae. *FEMS Lett.*, **1**, 43–45

Brown, C. M. and Dilworth, M. J. (1975). Ammonia assimilation by Rhizobium cultures and bacteroids. *J. Gen. Microbiol.*, **86**, 39–48

Brown, C. M. and Johnson, B. (1977). Inorganic nitrogen assimilation in aquatic microorganisms. *Adv. Aqua. Microbiol.*, **1**, 49–114

Brown, C. M. and Stanley, S. O. (1972). Environment-mediated changes in the cellular content of the 'pool' constituents and their associated changes in cell physiology. In *Environmental*

Control of Cell Synthesis and Function (A. C. R. Dean, S. J. Pirt, and D. W. Tempest, eds), Academic Press, London and New York, pp 363–389

Brown, C. M., Burn, V. J., and Johnson, B. (1973*b*). Presence of glutamate synthase in fission yeasts and its possible role in ammonia assimilation. *Nature, Lond.*, **246**, 115–116

Brown, C. M., Macdonald-Brown, D. S., and Meers, J. L. (1974). Physiological aspects of microbial inorganic nitrogen metabolism. *Adv. Microbial Physiol.*, **11**, 1–52

Brown, C. M., Macdonald-Brown, D. S., and Stanley, S. O. (1973*a*). The mechanisms of nitrogen assimilation in pseudomonads. *Antonie van Leeuwenhoek*, **39**, 89–98

Burn, V. J., Turner, P. R., and Brown, C. M. (1974). Aspects of inorganic nitrogen assimilation in yeasts. *Antonie van Leeuwenhoek*, **40**, 93–102

Cole, J. A. (1976). Microbial gas metabolism. *Adv. Microbial Physiol.*, **14**, 1–92

Dainty, R. H. (1972). Glutamate biosynthesis in *Clostridium pasteurianum* and its significance in nitrogen metabolism. *Biochem. J.*, **126**, 1055–1056

Deuel, T. F. and Prusiner, S. (1974). Regulation of glutamine synthetase from *Bacillus subtilis* by divalent cations, feedback inhibitors and L-glutamine. *J. Biol. Chem.*, **249**, 257–264

Drozd, J. W., Tubb, R. S., and Postgate, J. R. (1972). A chemostat study of the effect of fixed nitrogen sources on nitrogen fixation, membranes and free amino acids in *Azotobacter chroococcum. J. Gen. Microbiol.*, **73**, 221–232

Dubois, E. L. and Grenson, M. (1974). Absence of involvement of glutamine synthetase and of NAD-linked glutamate dehydrogenase in the nitrogen catabolite repression of arginase and other enzymes in *Saccharomyces cerevisiae. Biochem. Biophys. Res. Commun.*, **60**, 150–157

Dubois, E., Grenson, M., and Wiame, J.-M. (1973). Release of the ammonia effect on three catabolic enzymes by NADP-specific glutamate dehydrogenase-less mutations in *Saccharomyces cerevisiae. Biochem. Biophys. Res. Commun.*, **50**, 967–972

Dubois, E., Grenson, M., and Wiame, J.-M. (1974). The participation of the anabolic glutamate dehydrogenase in the nitrogen catabolite repression of arginase in *Saccharomyces cerevisiae. Eur. J. Biochem.*, **48**, 603–616

Eady, R. R., Issak, R., Kennedy, C., Postgate, J. R., and Ratcliffe, H. D. (1978). Nitrogenase synthesis in *Klebsiella pneumoniae* comparison of ammonium and oxygen regulation. *J. Gen. Microbiol.*, **104**, 277–285

Fawole, M. O. and Casselton, P. J. (1972). Observations on the regulation of glutamate dehydrogenase activity in *Coprinus lagopus. J. Exp. Bot.*, **23**, 530–551

Ferguson, A. R. and Sims, A. P. (1974*a*). The regulation of glutamine metabolism in *Candida utilis*. The role of glutamine in the control of glutamine synthetase. *J. Gen. Microbiol.*, **80**, 159–171

Ferguson, A. R. and Sims, A. P. (1974*b*). The regulation of glutamine metabolism in *Candida utilis*. The inactivation of glutamine synthetase. *J. Gen. Microbiol.*, **80**, 173–185

Folkes, B. F. and Sims, A. P. (1974). The significance of amino acid inhibition of NADP-linked glutamate dehydrogenase in the physiological control of glutamate synthesis in *Candida utilis. J. Gen. Microbiol.*, **82**, 77–95

Foor, F., Jannsen, K. A., and Magasanik, B. (1975). Regulation of synthesis of glutamine synthetase by adenylylated glutamine synthetase. *Proc. Natl. Acad. Sci. U.S.A.*, **72**, 4844–4848

Friedrich, B. and Magasanik, B. (1977). Urease of *Klebsiella aerogenes:* control of its synthesis by glutamine synthetase. J. Bacteriol., **131**, 446–452

Ginsburg, A. and Stadtman, E. R. (1973). Regulation of glutamine synthetase in *Escherichia coli*. In *The Enzymes of Glutamine Metabolism* (S. Prusiner and E. R. Stadtman, eds), Academic Press, London and New York, pp. 9–43

Gordon, J. K. and Brill, W. J. (1974). Derepression of nitrogenase synthesis in the presence of excess NH_4^+. *Biochem. Biophys. Res. Commun.*, **59**, 967–971

Grenson, M. and Hou, C. (1972). Ammonia inhibition of the general amino acid permease and its suppression in NADP specific glutamate dehydrogenase mutants of *Saccharomyces cerevisiae. Biochem. Biophys. Res. Commun.*, **48**, 749–756

Grenson, M., Dubois, E., and Piotrowska, M. (1974). Ammonia assimilation in *Saccharomyces cerevisiae* as mediated by the two glutamate dehydrogenases. *Mol. Gen. Genet.*, **128**, 73–85

Griffith, C. J. and Carlsson, J. (1974). Mechanism of ammonia assimilation in Streptococci. *J. Gen. Microbiol.*, **82**, 253–260

Hackette, S. L., Skye, G. E., Burton, C., and Segel, I. H. (1970). Characterisation of an ammonium transport system in filamentous fungi with methylammonium ^{14}C as the substrate. *J. Biol. Chem.*, **245**, 4241–4250

Henderson, P. J. F. (1971). Ion transport by energy conserving biological membranes. *Annu. Rev. Microbiol.*, **25**, 393–428

Hynes, M. J. (1974). Effects of ammonium, L-glutamate and L-glutamine on nitrogen catabolism in *Aspergillus nidulans. J. Bacteriol.*, **120**, 1116–1123

Janssen, K. A. and Magasanik, B. (1977). Glutamine synthetase of *Klebsiella aerogenes:* genetic and physiological properties of mutants in the adenylylation system. *J. Bacteriol.*, **129**, 993–1000

Johnson, B. and Brown, C. M. (1974). The enzymes of ammonia assimilation in *Schizosaccharomyces* spp. and in *Saccharomycodes ludwigii. J. Gen. Microbiol.*, **85**, 169–172

Kondorosi, A., Svab, Z., Kiss, G. B., and Dixon, R. A. (1977). Ammmonia assimilation and nitrogen fixation in *Rhizobium meliloti. Mol. Gen. Genet.*, **151**, 221–226

Kinghorn, J. R. and Pateman, J. A. (1973). A new class of mutants affecting ammonium regulation and methyl ammonium resistance in *Aspergillus nidulans. Heredity,* **31**, 427–428

Kinghorn, J. R. and Pateman, J. A. (1975). The structural gene for NADP L-glutamate dehydrogenase in *Aspergillus nidulans. J. Gen. Microbiol.*, **86**, 290–300

Kleiner, D. (1975). Ammonium uptake by nitrogen fixing bacteria. I. *Azotobacter vinelandii. Arch. Microbiol.*, **104**, 163–169

Kleiner, D. (1976). Ammonium uptake and metabolism by nitrogen fixing bacteria. II. *Klebsiella pneumoniae. Arch. Microbiol.*, **111**, 85–91

Kurz, W. G. W. and La Rue, T. A. (1975). Nitrogenase activity in rhizobia in absence of host plant. *Nature, Lond.*, **256**, 407–408

Magasanik, B., Prival, M. J., Brenchley, J. E., Tyler, B. M., De Leo, A. B., Streicher, S. L., Bender, R. A., and Paris, C. G. (1974). Glutamine synthetase as a regulator of enzyme synthesis. *Curr. Topics Cell. Regul.*, **8**, 119–138

McComb, J. A., Elliott, J., and Dilworth, M. J. (1975). Acetylene reduction by *Rhizobium* in pure culture. *Nature, Lond.* **256**, 409–410

Meers, J. L. and Kjaergaard-Pedersen, L. K. (1972). Nitrogen assimilation by *Bacillus licheniformis* organisms growing in chemostat cultures. J. Gen. Microbiol., **70**, 277–286

Meers, J. L., Tempest, D. W., and Brown, C. M. (1970). Glutamine (amide):2-oxo-glutarate amino transferase oxidoreductase (NADP), an enzyme involved in the synthesis of glutamate by some bacteria. *J. Gen. Microbiol.*, **64**, 187–194

Middelhoven, W. J. (1977). Isolation and characterization of methylammonium-resistant mutants of *Saccharomyces cerevisiae* with relieved nitrogen metabolite repression of allantoinase, arginase and ornithine transaminase synthesis. *J. Gen. Microbiol.*, **100**, 257–269

Miflin, B. J. and Lea, P. J. (1976). The pathway of nitrogen assimilation in plants. *Phytochemistry*, **15**, 873–885

Miflin, B. J. and Lea, P. J. (1977). Amino acid metabolism. *Annu. Rev. Plant. Physiol.*, **28**, 299–329

Miller, R. E. (1973). Glutamate synthase from *Escherichia coli:* an iron-sulphide flavoprotein. In *The Enzymes of Glutamine Metabolism* (S. Prusiner and E. R. Stadtman, eds), Academic Press, London and New York, pp. 183–203

Nagatani, H., Shimizu, M., and Valentine, R. C. (1971). The mechanism of ammonia assimilation in nitrogen fixing bacteria. *Arch. Mikrobiol.*, **79**, 164–175

Newman, B. M. and Cole, J. A. (1977). Lack of a regulatory function for glutamine synthetase protein in the synthesis of glutamate dehydrogenase and nitrite reductase in *Escherichia coli* K12. *J. Gen. Microbiol.*, **98**, 369–377

O'Gara, F. and Shanmugam, K. T. (1976). Regulation of nitrogen fixation by Rhizobia. Export of fixed N_2 as NH_4^+. *Biochim. Biophys. Acta*, **437**, 313–321

Pagan, J. D., Child, J. J., Scowcroft, W. R., and Gibson, A. H. (1975). Nitrogen fixation by *Rhizobium* cultured on a defined medium. *Nature, Lond.*, **256**, 406–407

Pateman, J. A. and Kinghorn, J. R. (1976). Nitrogen metabolism. In *Filamentous Fungi,* Vol. 2 (J. E. Smith and D. Berry, eds), Edward Arnold, London, pp. 159–237

Pateman, J. A., Dunn, E., Kinghorn, J. R., and Forbes, E. (1974). Ammonium and methylammonium transport in wild type and mutant cells of *Aspergillus nidulans. Mol. Gen. Genet.*, **133**, 225–236

Pateman, J. A., Kinghorn, J. R., Dunn, E., and Forbes, E. (1973). Ammonium regulation in *Aspergillus nidulans. J. Bacteriol.*, **114**, 943–950

Planque, K., Kennedy, I. R., de Vries, G. E., Quispel, A., and van Brussel, A. A. N. (1977). Location of nitrogenase and ammonia-assimilatory enzymes in bacteroids of *Rhizobium leguminosarum* and *Rhizobium lupini. J. Gen. Microbiol.*, **102**, 95–104

van de Poll, K. W. (1973). Ammonium repression in a mutant of *Saccharomyces carlsbergensis* lacking NADP dependent glutamate dehydrogenase activity. *FEBS Lett.*, **32**, 265–266

Porella, K. (1971). The induction of dehydrogenase activity for branched chain amino acids in *Bacillus subtilis. Arch. Mikrobiol.*, **77**, 339–343

Prival, M. J. and Magasanik, B. (1971). Resistance of catabolite repression of histidase and proline oxidase during nitrogen-limited growth of *Klebsiella aerogenes. J. Biol. Chem.*, **246**, 6288–6296

Prival, M. J., Brenchley, J. E., and Magasanik, B. (1973). Glutamine synthetase and the regulation of histidase formation in *Klebsiella aerogenes. J. Biol. Chem.*, **248**, 4334–4344

Robertson, J. G., Warburton, M. P., and Farnden, W. J. F. (1975). Induction of glutamate synthetase during nodule development in Lupin. *FEBS Lett.*, **55**, 33–37

Roon, R. J., Even, H. L., and Larimore, F. (1974). Glutamate synthase: properties of the reduced nicotinamide adenine dinucleotide-dependent enzyme from *Saccharomyces cerevisiae. J. Bacteriol.*, **118**, 89–95

Roon, R. J., Larimore, F., and Levy, J. S. (1975*b*). Inhibition of amino acid transport by ammonium ion in *Saccharomyces cerevisiae. J. Bacteriol.*, **124**, 325–331

Roon, R. J., Even, H. L., Dunlop, P., and Larimore, F. L. (1975*a*). Methylamine and ammonia transport in *Saccharomyces cerevisiae. J. Bacteriol.*, **122**, 502–509

Schutt, H. and Holzer, H. (1972). Biological inactivation of glutamine synthetase in *Escherichia coli. Eur. J. Biochem.*, **26**, 68–72

Senior, P. J. (1974). Regulation of nitrogen metabolism in *Escherichia coli* and *Klebsiella aerogenes:* studies with the continuous culture technique. *J. Bacteriol.*, **123**, 407–418

Shanmugam, K. T. and Morandi, C. (1976). Amino acids as repressors of nitrogenase biosynthesis in *Klebsiella pneumoniae. Biochim. Biophys. Acta*, **437**, 322–332

Shanmugam, K. T. and Valentine, R. C. (1975*a*). Molecular biology of nitrogen fixation. *Science, N.Y.*, **187**, 919–924

Shanmugam, K. T. and Valentine, R. C. (1975*b*). Microbial production of ammonium ion from nitrogen. *Proc. Natl. Acad. Sci. U.S.A.*, **72**, 136–139

Shanmugam, K. T., Chan, I., and Morandi, C. (1975). Regulation of nitrogen fixation. Nitrogenase-derepressed mutants of *Klebsiella pneumoniae. Biochim. Biophys. Acta*, **408**, 101–111

Sims, A. P. and Ferguson, A. R. (1974). The regulation of glutamine metabolism in *Candida utilis*. Studies with $^{15}NH_3$ to measure *in vivo* rates of glutamine synthesis. *J. Gen. Microbiol.*, **80**, 143–158

Sims, A. P., Folkes, B. F., and Bussay, A. H. (1968). Mechanisms involved in the regulation of nitrogen assimilation in microorganisms and plants. In *Recent Aspects of Nitrogen Metabolism in Plants* (E. J. Hewitt and C. V. Cutting, eds), Academic Press, London and New York, pp. 15–63

Sims, A. P., Toone, J., and Box, V. (1974). The regulation of glutamine synthesis in the food yeast *Candida utilis:* the purification and subunit structure of glutamine synthetase and aspects of enzyme deactivation. *J. Gen. Microbiol.*, **80**, 485–499

Steimer-Veale, K. and Brenchley, J. E. (1974). Characterization of *Salmonella typhimurium* strains sensitive and resistant to methionine sulfoximine. *J. Bacteriol.*, **119**, 848–856

Stevenson, R. and Silver, S. (1977). Methylammonium uptake by *Escherichia coli:* evidence for a bacterial NH_4^+ transport system. *Biochem. Biophys. Res. Commun.*, **75**, 1133–1440

Streicher, S. L., Gurney, E. G., and Valentine, R. C. (1972). The nitrogen fixation genes. *Nature, Lond.*, **239**, 485–499

Streicher, S. L., Shanmugam, K. T., Ausubel, F., and Morandi, C. (1974). Regulation of nitrogen fixation in *Klebsiella pneumoniae:* evidence for a role of glutamine synthetase as a regulator of nitrogenase synthesis. *J. Bacteriol.*, **120**, 815–821

Streicher, S. L., De Leo, A. B., and Magasanik, B. (1976). Regulation of enzyme formation in *Klebsiella aerogenes* by episomal glutamine synthetase of *E. coli. J. Bacteriol.*, **127**, 185–192

Tempest, D. W., Meers, J. L., and Brown, C. M. (1970). Synthesis of glutamate in *Aerobacter aerogenes* by a hitherto unknown route. *Biochem. J.*, **117**, 405–407

Tempest, D. W., Meers, J. L., and Brown, C. M. (1973). Glutamate synthetase (GOGAT); a key enzyme in the assimilation of ammonia by prokaryotic organisms. In *The Enzymes of Glutamine Metabolism* (S. Prusiner and E. R. Stadtman, eds), Academic Press, London and New York, pp. 167–182

Tronick, S. R., Ciardi, J. E., and Stadtman, E. R. (1973). Comparative biochemical and immunological studies of bacterial glutamine synthetases. *J. Bacteriol.*, **115**, 858–868

Trotta, P. P., Platzer, K. E. B., Haschemeyer, R. H., and Meister, A. (1974). Glutamine-binding subunit of glutamate synthase and partial reactions catalysed by this glutamine amido transferase. *Proc. Natl. Acad. Sci. U.S.A.*, **71**, 4607–4611

Tubb, R. S. (1974). Glutamine synthetase and ammonium regulation of nitrogenase synthesis in *Klebsiella. Nature, Lond.*, **251**, 481–484

Tyler, B., De Leo, A. B., and Magasanik, B. (1974). Activation and transcription of *hut* DNA by glutamine synthetase. *Proc. Natl. Acad. Sci. U.S.A.*, **71**, 255–229

Vaca, G. and Mora, J. (1977). Nitrogen regulation of arginase in *Neurospora crassa. J. Bacteriol.* **131**, 719–725

Vandecasteele, J. P., Lemal, J., and Coudert, M. (1975). Pathways and regulation of glutamate synthesis in a *Corynebacterium* sp. overproducing glutamate. *J. Gen. Microbiol.*, **90**, 178–180

Wall, J. D., Johansson, B. C., and Gest, H. (1977). A pleiotropic mutant of *Rhodopseudomonas capsulata* defective in nitrogen metabolism. *Arch. Microbiol.*, **115**, 259–263

Weare, N. M. and Shanmugam, K. T. (1976). Photoproduction of ammonium ion from N_2 in *Rhodospirillum rubrum. Arch. Microbiol.*, **110**, 207–213

Wooton, J. C., Chambers, C. K., Taylor, J. G., and Fincham, J. R. S. (1973). Amino acid sequence homologies between the NADP-dependent glutamate dehydrogenase of *Neurospora* and the bovine enzyme. *Nature, Lond.*, **241**, 42–43

Yoch, D. C. and Pengra R. M. (1966). Effect of amino acids on the nitrogenase system of *Klebsiella pneumoniae. J. Bacteriol.*, **92**, 618–622

Microorganisms and Nitrogen Sources
Edited by J. W. Payne

Section 4.3

The Utilization of Amides by Microorganisms

PATRICIA H. CLARKE
Department of Biochemistry, University College, London, UK

1 INTRODUCTION

Amide groups are common in biological molecules. The processes of amidation and deamidation enable the storage of nitrogen groups that can be released when required for further metabolism and the nitrogen group for many syntheses is provided from an amide donor. Examples of these reactions are the involvement of glutamine in ammonia assimilation and as a participant in the synthesis of anthranilate from chorismate in the tryptophan biosynthetic pathway (Crawford, 1975). Glutamine is also involved in the synthesis of imidazoleglycerol phosphate in the histidine biosynthetic pathway (Martin *et al.*, 1971). The central role of glutamine in biosynthesis is reflected in the complexity of the regulation of its activity and of its synthesis. Glutamine is an essential participant in nitrogen assimilation in some microorganisms and is synthesized by reactions requiring an active glutamine synthetase. The synthesis of glutamine synthetase is under autogenous regulation and, in addition, this enzyme is considered to be a regulatory protein controlling the synthesis of several other enzymes concerned with the metabolism of nitrogenous compounds (Magasanik *et al.*, 1974) (see also Section 4.2). From another point of view glutamine can be considered as one of the common amino acids and as such is a normal component of proteins. These important properties of glutamine transcend its position as a simple amide available for many microorganisms as a growth substrate. Asparagine is another amide with a central role in metabolism. These two amides are considered in detail elsewhere in this volume (Section 4.4) and will be mentioned only briefly in this chapter.

Urea is also a rather special amide. As the end-product of nitrogen metabolism in many animal species it is to be expected that microorganisms would have evolved to utilize it for growth and thereby continue the nitrogen cycle. Urea is broken down to carbon dioxide and ammonia by the enzyme urease. This reaction is very easy to detect, either by using a colour reaction for ammonia, or by including a pH indicator in the growth medium so that a pH shift can be observed. The production of urease is widespread among microorganisms and is a reasonably constant character of microbial species. Several variants of tests for urease production are used in diagnostic bacteriology (Cowan and Steel, 1965; Holding and Collee, 1971). Nicotinamide is a normal cell metabolite and there are many reports of nicotinamidases in microbial species. In some cases urease and nicotinamidase activities have been studied at the same time as amidase activities for alphatic and aryl amides.

This chapter is concerned mainly with the enzymes and metabolic pathways for organic acid amides that have no special roles in central metabolism but enable amides to be used for growth as carbon or nitrogen sources or both. Glutamine, asparagine, nicotinamide, and urea will be considered only in so far as their metabolic reactions are related to those of the simple amides.

2 AMIDES AS GROWTH SUBSTRATES

The first general survey of amides as growth substrates was that carried out by Den Dooren de Jong (1926). This is not readily available but Marjory Stephenson recognized the importance of these findings and included a list of species and growth substrates in the chapter on nutrition in *Bacterial Metabolism* (1949). A limited amount of information on growth of bacteria on amide substrates has therefore been accessible to microbiologists for many years. In Den Dooren de Jong's survey, representative species of stock cultures of bacteria were tested for growth, with amides providing either the sole growth substrate or the nitrogen source only. The minimal medium consisted of tap water (*sic*) containing 0.1 per cent K_2HPO_4 with the amides added alone to give a concentration of 1.0 per cent, or the same medium to which 1.0 per cent glucose and 1.0 per cent $CaCO_3$ had also been added. This medium was unlikely to be optimal for all species and negative results were obtained with some species now known to be able to metabolize some of the amides tested. Asparagine, not unexpectedly, was able to support the growth of enterobacteria and pseudomonads while only a few of the simple amides were utilized as sole growth substrates. For utilization as nitrogen source the essential requirements are that the amide should be able to enter the cells and be hydrolysed to provide ammonia, and many more strains were positive in this test. A strain of *Erwinia* (listed as *B. herbicola*) grew with formamide, acetamide, propionamide, butyramide, isobutyramide, valeramide, and stearamide. As well as the aliphatic amides this strain could use succinamide and succinimide as nitrogen sources. Other species which used some of the amides as nitrogen sources were *Klebsiella* (*Bacterium*) *aerogenes, Bacillus polymyxa*, and *Mycobacterium phlei*. A strain of *Pseudomonas aminovorans*, that had been isolated from soil in media containing amines as carbon source, was also able to utilize some of the amides. These findings suggested that enzymes hydrolysing amides might be found in various bacterial species but that hydrolysis of amides might not be linked to further metabolism of the resultant organic acid.

The results of Den Dooren de Jong's survey are now of mainly historical interest, but although there have been subsequent investigations of amide utilization by particular groups of microorganisms there are no general surveys with which to compare them. Stanier *et al*. (1966) included a few amides in a nutritional survey of the pseudomonads and found that most strains of *Pseudomonas aeruginosa* could utilize acetamide as a carbon source. Kelly and Clarke (1960) described an aliphatic amidase, hydrolysing acetamide, in a strain of *P. aeruginosa* and the reaction catalysed by this enzyme has been used as the basis for simple tests for identification of bacterial isolates. Buhlman *et al.* (1961) used a minimal medium containing glucose, acetamide, and a pH indicator and found that only those strains which could be classed as *P. aeruginosa* were able to produce an alkaline reaction. A positive result in this test is presumptive evidence of acetamidase activity but negative results may occur if growth is poor, even though the organism is capable of producing an amidase. Mossel and van Zadelhoff (1971) were concerned with the isolation and identification of *P. aeruginosa* from water and the method they used was to incubate the samples at 42 °C on plates containing glycerol–mannitol–acetamide–cetrimide–phenol red (GMAC) agar. They reported that only nine of 500 isolates of *P. aeruginosa* failed to grow and produce an alkaline reaction. They stated that six of these appeared to be unable to produce an acetamidase. Oberhofer and Rowen (1974) used acetamide agar to discriminate between non-fermentative Gram-negative bacteria and in a more extended survey Oberhofer *et al.* (1977) included several other amides. The medium was made up from a minimal salt base, as used for the Simmons citrate test, and included 0.5 or 1.0 per cent of the test amide and was used for slope cultures. The results were read after 4 and 7 days incubation. Most of the *P. aeruginosa* strains gave an alkaline reaction with acetamide and some were also weakly positive with butyramide. The *P. acidovorans* strains were positive for both acetamide and butyramide. Only a few strains of *P. cepacia* were included and these were all positive for acetamide. Of the other non-fermentative organisms examined, *Achromobacter xylosoxidans* gave positive tests for acetamide and butyramide, as did strains identified as *Alcaligenes odorans*. It should be emphasized that these simple tests provide only limited information about amide metabolism. Low enzyme levels, combined with a poor growth rate, might give a negative result for a strain which would be found to produce an amidase under more favourable growth conditions. On the other hand, the prolonged incubation period could allow mutants to grow up and thereby give positive results for strains that produce little or no amidase activity over a shorter growth period.

Selective isolation in minimal salt media, with amides providing the sole carbon or nitrogen source for growth, has revealed many organisms capable of utilizing amides. These will be discussed in later sections but it should be emphasized that there may be many more still to be discovered. Kelly (1965) isolated bacteria which were able to grow with fluoroacetamide. This compound is used as a pesticide and may be expected to enter the natural environment in the normal course of events as well as by accidental spillage. Fluoroacetamide is growth inhibitory to *Pseudomonas aeruginosa* strains with an active amidase and can be used to isolate amidase-negative mutants (Clarke and Tata, 1973). The existence of bacteria that are able to utilize fluoroacetamide as a growth substrate indicates that it can be hydrolysed to fluoroacetate and metabolized further without building up a toxic level of fluorinated intermediates. Other organisms may be found that are able to metabolize synthetic and potentially toxic amides.

3 AMIDE UPTAKE

Very little attention has been paid to amide uptake by microorganisms although studies on amide utilization have led in some cases to extensive investigations of the enzyme systems and their regulation and to the elucidation of substrate and inducer specificities. It is known that some microorganisms can produce more than one amidase capable of attacking a particular amide and that these enzymes may be regulated in different ways. For example, Hynes (1975*a*) suggests that *Aspergillus nidulans* may have three amidases capable of hydrolysing hexanoamide. There was no indication that an amide uptake system was associated with any of these enzymes. One indication for specific uptake systems is the existence of strains which are unable to utilize potential growth substrates although they are capable of producing the necessary metabolic enzymes. There are no examples of organisms known to be cryptic for amides, that is to say organisms unable to grow on certain amides because of the absence of an uptake system rather than the lack of amidase activity.

Brammar *et al.* (1966) carried out some preliminary studies on amide uptake by *Pseudomonas aeruginosa* with the object of establishing whether or not catabolite repression by succinate or amide analogue repression by cyanoacetamide and other acetamide analogues could be due to inhibition of uptake of inducing amides. It was known that in *Escherichia coli* the effect of glucose in preventing the synthesis of enzymes required for the metabolism of certain sugars is due in part to inhibition of uptake of the alternative sugar into the bacterial cell (Paigen and Williams, 1970).

In *Pseudomonas* species the most potent of the cell metabolites causing catabolic repression are succinate and the other intermediates of the tricarboxylic acid cycle, while glucose produces much less severe repression. It had been shown previously that succinate produced severe repression of amidase synthesis in the presence of both substrate and non-substrate amide inducers and it was important to know whether this effect could be ascribed to inducer exclusion (Brammar and Clarke, 1964). With the wild-type strain it was possible to measure incorporation of the radioactive label from $[^{14}C]$-*N*-acetylacetamide into bacterial cells grown either in succinate medium or in acetamide medium. There was no difference in the apparent rates of uptake in the presence of succinate and no significant difference in the rates for bacteria grown on acetamide rather than succinate. Similar results were obtained with an amidase-negative mutant using $[^{14}C]$-acetamide, and the labelled amide was recovered unaltered from the bacterial cells. The acetamide analogue cyanoacetamide, which inhibits the induction of amidase by acetamide and *N*-acetylacetamide, was also shown to have no effect on the rates of amide uptake. It was clear that repression of amidase synthesis by cyanoacetamide and succinate was not due to exclusion of amide inducers. From these experiments it was concluded that the entry of acetamide and *N*-acetylacetamide into the bacteria was not an inducible character. It was suggested that the uptake of acetamide and related amides was carried out by a low activity constitutive transport system.

With the isolation of mutant strains producing amidases with altered substrate specificities, enabling them to grow on amides not available to the wild-type strain, the question of amide uptake arose again. Betz and Clarke (1972) described a group of mutants able to grow on phenylacetamide and compared the activities of whole cells and extracts for acetamide, butyramide, valeramide, and phenylacetamide. There were no cases of crypticity for any of these substrates and the activities of cell-free extracts corresponded with those of the intact bacteria. Some members of this group of mutants had lost the ability to grow on acetamide and it was shown that the purified enzyme from one of these mutants was devoid of acetamidase activity. This indicated that the bacteria were freely permeable to all the amides and that the amide growth phenotype was solely dependent on the substrate specificity of the enzyme. It appeared that there were no permeability problems hindering the selection of mutants producing amidases with new substrate specificities.

The question of the specificity of amide uptake systems in *Pseudomonas aeruginosa* was re-investigated by Farin (1976) who wished to establish how many different amide-binding proteins were to be expected in this organism. Farin (1976) examined a number of mutants with different amidase activities in order to establish the way in which the radioactive label became incorporated during incubation with [^{14}C]-acetamide. He found that with all acetamide-positive strains there was high rate ^{14}C uptake that was strictly coordinate with the induction of amidase activity in inducible strains and remained at a constant activity in constitutive strains. The altered enzyme mutant PhB3, that had lost the ability to hydrolyse acetamide (Betz and Clarke, 1972), did not show this high rate of acetamide uptake. The apparent uptake system had the same K_m for acetamide as the amidase and Farin (1976) concluded that the rate-limiting step in the uptake of acetamide by strains with an active acetamidase was the enzyme itself. Farin (1976) considered that the entry of amides into the bacteria was by passive diffusion which would necessarily be a constitutive character and that this could be at a sufficient rate in bacteria possessing active amidase enzymes to be no hindrance to amide utilization.

He commented that it has been shown that plant cells, *Chara ceratophylla*, were highly permeable to amides and that Collander and Barlund (1933) had found that the migration of non-electrolyte solutes across the cell membrane of these cells could be related to the lipophilic properties of the solutes. Later, Stein (1967) considered the passage of non-electrolytes across membrane barriers and suggested that a better interpretation of differences in permeation would be found in the numbers of hydrogen bonds that the solutes could form with water. The greater difficulties in interrupting the hydrogen bonds of the smaller and more hydrophilic molecules would slow down their passage through the cell membranes. Stein (1967) plotted the number of possible hydrogen bonds against the rates of penetration of cell barriers and found good correlations for a number of compounds. These included acetamide, propionamide, and butyramide, which were found to be good permeants for erythrocytes and *Arabacia* cells as well as the cells of *Chara ceratophylla.* Cohen and Bangham (1972) examined permeation of amides into artificial membrane vesicles and found that aliphatic amides penetrated these structures very well. Collander and Barland (1933) had calculated that the rate of uptake of acetamide was only slightly less than that of water for *Chara ceratophylla.*

In view of the wide range of amides that can be metabolized by various microbial species it would be of interest to determine whether passive diffusion can account for amide uptake in all cases or whether specific transport systems are required for particular amides.

4 ENZYMES ACTING ON AMIDES

Amide bonds can be hydrolysed by several different classes of enzymes. Some of these enzymes are also able to hydrolyse esters and the esterase activity may be more pronounced than the amidase activity. As well as catalysing reactions involving water as a reactant, some amide-hydrolysing enzymes may be able to transfer the acyl group to other acceptors to produce various acyl derivatives. Many proteolytic enzymes have been found to have esterase and amidase activities for low molecular weight substrates. For example, chymotrypsin hydrolyses peptide bonds adjacent to the carbonyl group of aromatic amino acids but also hydrolyses esters of *N*-acetylphenylalanine and *N*-acetyltryptophan and amides such as benzoylglycineamide. The action of proteolytic enzymes on simple amides is presumably incidental to their physiological activity of splitting peptide bonds but has been very useful in studying the mechanism of action of these enzymes. Some of the findings obtained from studies of the reactions of the proteolytic enzymes may be relevant to investigations of the mechanism of action of the amidases attacking simpler substrates. Chymotrypsin may be an important pointer in this respect. Chymotrypsin belongs to the class of serine proteases whose structure is known. The catalytic activity of these enzymes can be interpreted in terms of a serine residue at the active site and the participation of certain reactive residues around the substrate cleft (Hartley, 1974). It is suggested that the interaction of chymotrypsin with its substrates involves the formation of an acyl intermediate with the active serine residue. The first evidence for the importance of a serine residue for this enzyme came from the observation that its activity could be inhibited by diisopropyl fluorophosphate (DFP) and later confirmation came from isolation of peptides containing a phosphorylated serine (Balls and Jansen, 1952; Oosterbaan *et al.*, 1958).

Some of the enzymes to be described in later sections have been shown to be inhibited by DFP although the implications of these findings have yet to be confirmed. In one case only, that being the amidase of *Pseudomonas acidovorans* studied by Heyman and Alt (1974), a phosphorylated peptide has been isolated from the enzyme inactivated with [^{32}P]-DFP. There is therefore some evidence that for this amidase, attacking amides with an aromatic ring, the mechanism may be similar to that of the serine proteases.

Other proteolytic enzymes with the ability to hydrolyse esters and amides have a cysteine residue at the active site and the catalytic activity is thought to involve the formation of a thioester intermediate. Papain (Stockwell and Smith, 1957) is an example of this group of enzymes and has a broad specificity, hydrolysing peptides, amides, anilides, and esters. Papain is also able to catalyse transferase reactions with these substrates. Thiol enzymes are inhibited by oxidation and by alkylating reagents, such as iodoacetamide, and by organic mercurials such as mercuribenzoate. Inhibition by these compounds is frequently taken as evidence for the involvement of a thiol group at the active site whereas the reduced thiol may actually be essential for enzyme structure. Many of the amidases described in later sections are inhibited by these thiol reagents but there is no case in which there is conclusive evidence for a thioester intermediate.

Pepsin also hydrolyses certain amides and esters and it is thought that two aspartate carboxyl groups are involved in the catalytic reaction. It has been suggested that an acyl intermediate might be formed with pepsin during the reaction with its substrates and there is some evidence for this (Fruton, 1976). This is another possible model for the amidases but it has not been explored in any of the systems studied so far.

Esterases and lipases may have very low substrate specificities and hydrolyse a large range of alkyl esters or aryl esters of different chain lengths. The acetanilidase isolated by Alt *et al.* (1975*b*) was discovered during an investigation of esterases. Although most esterases are devoid of amide hydrolase activity they frequently have transferase activity. Fitch (1963, 1964) purified a choline esterase from *Pseudomonas fluorescens* which hydrolysed acetylcholine and transferred the acetyl group to amyl alcohol to give amyl acetate. Esterases may also transfer acyl groups to hydroxylamine to form hydroxamates and such reactions may be confused with amidase transfer reactions. The unequivocal demonstration of the existence of an amidase, and the delineation of its specificity, requires that it should be isolated as a single protein species.

5 ω-AMIDASES

An enzyme hydrolysing the monoamide of succinic acid was recognized by Ramaley *et al.* (1972) by its transferase reaction. *Bacillus subtilis* 168 was found to produce an enzyme catalysing the formation of succinyl hydroxamate from succinate and hydroxylamine. The enzyme hydrolyses succinamate and has transferase activity with succinate and hydroxylamine. It was the latter reaction that led to its discovery since succinylhydroxamate can also be produced by the reaction of succinyl-CoA transferase and this latter enzyme was being investigated in this organism. The *B. subtilis* ω-amidase was shown to hydrolyse succinamate but was unable to grow with this compound as a carbon source. This was probably due to the inability to utilize succinate as a sole carbon source, since this organism was also unable to grow in a minimal salt medium with succinate and ammonium salts as growth substrates. There was no evidence for induction of the ω-amidase and in a medium with glucose as the carbon source, and succinamate as the nitrogen source, the enzyme was synthesized throughout growth. There were no differences in enzyme levels of cultures grown with succinamate or with glutamate or aspartate as nitrogen sources. However, the enzyme levels were reduced when ammonium chloride was present in the medium. This suggests that the ω-amidase may have a role in nitrogen assimilation although this possibility was not investigated in more detail. The *B. subtilis* ω-amidase was membrane bound and could be released by sonication or by the action of detergents.

Some of the properties of the *B. subtilis* ω-amidase were found to be similar to those of the ω-amidase isolated from rat liver (Meister *et al.*, 1955; Hersh, 1971, 1972). An enzyme isolated from *Thermus aquaticus* YT-1 carries out similar reactions and also appears to be a constitutive enzyme. Neither of these ω-amidase enzymes catalysed the hydrolysis of acetamide or the hydroxylamine transferase reaction with this amide as substrate. Succinamate, succinimide, and succinamide were substrates in the transferase reaction with both ω-amidases but only the *B. subtilis* enzyme acted on asparagine. As with the *Pseudomonas aeruginosa* aliphatic amidase (Brammar and Clarke, 1964), the corresponding acids were also able to participate in the transferase reaction, but while fumarate and succinate were substrates for both enzymes only the *T. aquaticus* enzyme was able to catalyse the transferase reaction with fumarate. The natural substrates of these ω-amidases and their role in metabolism is uncertain (Fernald and Ramaley, 1972).

Hersh (1970) described an enzyme from *Pseudomonas* AM that hydrolyses *N*-methyl-α-glutaramate. This enzyme is induced by growth in medium containing methylamine and may be important in the assimilation of methylamine as a nitrogen source. The enzyme has no activity on the unsubstituted glutaramate or on succinamate or α-oxosuccinamate. It is suggested that the enzyme catalyses the first reaction of methylamine with α-oxoglutarate and that the second cyclization reaction is non-enzymic.

6 ALIPHATIC AMIDASE OF A THERMOPHILIC BACILLUS

A thermophilic bacillus isolated by Epstein and Grossowicz (1969) was found to be able to grow in a minimal salt medium with acetamide as the sole carbon and nitrogen source. This strain synthesizes an inducible aliphatic amidase and there are several points of similarity with the aliphatic amidase of *Pseudomonas aeruginosa.* A limited range of amides is able to act as substrates and inducers and the synthesis of the enzyme is subject to catabolite repression by glucose. Thalenfeld and Grossowicz (1976) reported that the relative activities of the amidase for the aliphatic amides were acetamide, 1.0; propionamide, 0.97; formamide, 0.35; butyramide, 0.01. The substituted amide fluoroacetamide was hydrolysed at 0.84 times the rate for acetamide. Although the relative activities are somewhat different, the range of amides that can be hydrolysed by this amidase is very similar to that reported for the *P. aeruginosa* enzyme (Kelly and Clarke, 1962). The induction specificity was markedly different and some of the amides that are inhibitors of amidase induction in *P. aeruginosa* were found to act as inducers for the amidase of the thermobacillus. This finding was in accord with the general observation that greater differences are to be expected between the specificities of regulatory systems than of metabolic enzymes of bacteria of different genera (Ornston and Parke, 1976).

Cyanoacetamide and thioacetamide are both amide analogue repressors of amidase synthesis in *P. aeruginosa* (Brammar and Clarke, 1964) but cyanoacetamide is able to induce the synthesis of the thermobacillus amidase. Other non-substrate inducers for the thermobacillus enzyme were dimethylacetamide, dimethylformamide, and urea, whereas thioacetamide and thiourea inhibited induction by urea.

Glucose and other carbon compounds repressed amidase synthesis, and diauxic growth was apparent in glucose–acetamide medium. This suggests that catabolite repression is important in the regulation of synthesis of this enzyme. Thalenfeld and Grossowicz (1976) isolated a catabolite-repression-resistant mutant by subculturing in glucose-formamide medium and plating out on glucose-formamide agar. Thalenfeld *et al.* (1977) showed that the rate of amidase synthesis of the wild-type strain could be related to the cell density of the culture and showed that, if cultures synthesizing amidase at a significant rate were diluted, then amidase synthesis ceased until the cell density had again increased to the original level as the result of growth. In these experiments cyanoacetamide was added as a gratuitous inducer to a minimal salt medium with glutamate as the growth substrate. In other experiments, cultures were induced and kept at constant cell densities and this led to steady-state levels of amidase activity. The catabolite-repression-resistant mutant did not show the dilution effects or the dependence of amidase activities on the cell density that had been observed with the wild-type strain. It was concluded that these effects were manifestations of catabolite repression and were dependent on the variations in the metabolic pools under the various states of growth. It was considered by Thalenfeld and Grossowicz (1976) that the lag before the maximum rate of amidase synthesis occurred, and the high inducer concentration required (*c.* 10^{-2} M), might be due to a permeability barrier, but they rejected this interpretation. They pointed out that above a minimal concentration the lag was unaffected by inducer concentration and concluded that there was no indication that the bacteria were not freely permeable to the amide substrates.

Thalenfeld *et al.* (1977) studied the possible effects of nitrogen sources on the synthesis of this amidase. They found that a high ammonium salt concentration had no effect on the rates of amidase synthesis and were unable to detect any relief from catabolite repression under conditions of nitrogen-limited growth. They concluded that the enzyme is regulated by induction by amide substrates and repression by carbon compounds but is not subject to regulation by nitrogen compounds.

7 AMIDASES OF *MYCOBACTERIUM* SPP.

Investigations by a number of workers have shown that *Mycobacterium* spp. produce a variety of amidases, although none has been well characterized. Some of these enzymes exhibit transferase as well as hydrolase activities but there is also evidence for two distinct enzymes, i.e. both a transferase and a hydrolase attacking the same amide. Halpern and Grossowicz (1957) found that extracts of *M. phlei* could hydrolyse formamide, acetamide, glycineamide, leucineamide, and valineamide. These activities could be distinguished from the nicotinamidase and asparaginase activities of this organism. Kimura (1959*a, b, c*) examined several species and showed that *M. avium*, *M. phlei*, and *M. smegmatis* all produced several amidases. He examined hydrolase and transferase activities for nicotinamide, benzamide, glutamine, asparagine, and acetamide in extracts of these bacteria. He found that although *M. avium* extracts had transferase activity for nicotinamide, this was not due to the enzyme hydrolysing nicotinamide. From the other amide reactions studied he concluded that amide transferase reactions were frequently not associated with hydrolase reactions. Grossowicz and Halpern (1957) had previously obtained evidence that the asparaginase of *M. phlei* was not the same enzyme as that catalysing the formation of aspartylhydroxamate from aspartate and hydroxylamine which was also present in the crude bacterial extracts. A transferase enzyme from *M. avium* was partially purified by Kimura (1959*d*) and shown to catalyse hydroxamate formation from butyrate and valerate.

Draper (1967) studied the amidases of *M. smegmatis* in more detail. He showed that synthesis of enzymes with aliphatic amidase activity was induced with growth on acetamide. With both intact bacteria and with

Table 1 Amidase activity of *Mycobacterium smegmatis*

Amide	Specific activities of partially purified preparation† (μmol NH_3 min^{-1} (mg protein)$^{-1}$)
Formamide	400
Acetamide	4.0
Propionamide	5.0
Butyramide	10.2
Valeramide	5.6
Glycollamide	3.8
Acrylamide	6.3
Phenylacetamide	2.1
Benzamide	0.11
Nicotinamide	0.15

†Recalculated from Draper (1967).

cell-free preparations the highest activities were obtained with formamide as substrate and this was hydrolysed at about 100 times the rate for the other amides. For the aliphatic amides other than formamide, the optimal substrate was butyramide (Table 1). The simplest explanation of these results was that there were at least two amidases present, a formamidase which acted only on formamide, and a second amidase with a broad substrate specificity. However, Draper (1967) found that formamidase and butyramidase activities were both induced by growth with acetamide and that these activities were not separated by a partial purification process (2.2-fold). He also concluded that the amide transferase and hydrolase activities were not due to the same enzyme since the partially purified amide hydrolase had lost all the transferase activity detected in the crude extracts. The *M. smegmatis* crude extracts also had transferase activity for propionate and some other aliphatic acids and this may be similar to the *M. avium* enzyme described by Kimura (1959*d*).

The multiplicity of amidases reported for the mycobacteria is rather confusing and further information is needed about enzyme specificity and the regulation of synthesis of these enzymes. However, the characteristic patterns of amide hydrolysis have been used for species classification by several groups of bacteriologists (Bonicke, 1960; Juhlin, 1960; Schneidau, 1963; Urabe *et al*., 1965; Helbeque *et al*., 1975). A multiplicity of enzymes acting on amides has also been reported for *Aspergillus nidulans* (Hynes, 1975*a*) (see Section 4.3.11).

8 AMIDASES OF YEASTS

Gorr and Wagner (1933) described the hydrolysis of acetamide, asparagine and urea by dried preparations of *Candida (Torula) utilis*. They found that

Table 2 Amidase activities of *Candida utilis*

Amide	Specific activities of partially purified preparation† (μmol NH_3 min^{-1} (mg protein)$^{-1}$)
Acetamide	36
Propionamide	21
Butyramide	12
Valeramide	21
Hexanoamide	24
Acrylamide	17
Nicotinamide	5

†Recalculated from Brady (1969).

the highest acetamidase activity occurred in cultures grown with acetamide and this is the first record of an inducible aliphatic amidase. Brady (1969) reinvestigated the *Candida utilis* amidases and found that preparations grown in a complex medium containing glucose were able to hydrolyse a wide range of amides. Asparaginase activity was lost during a 60-fold purification but the preparation contained both nicotinamidase activity as well as hydrolase activity for a wide range of aliphatic amides (Table 2). The nicotinamidase activity of other fractions was thought to be due to a second enzyme. The aliphatic amidase had both amide hydrolase and amide transferase activities with acetamide as substrate but the other amide substrates were not tested. Brady (1969) prepared the amidase from cultures grown in glucose–yeast extract–peptone medium and did not investigate whether the activity could be increased by growth with amides. The specific activity of the cell-free extract used for enzyme purification was 0.59 (μmol NH_3 produced min^{-1} (mg protein)$^{-1}$. Table 2 gives the activities for various amide substrates with the 60-fold purified preparation. The K_m values for the aliphatic amides ranged from 2 to 5 mM. Formamide was not hydrolysed, the specific activities were lower for propionamide and butyramide than for acetamide, and the higher values for valeramide and hexanoamide could indicate that more than one amidase was present in these preparations.

9 ALIPHATIC AMIDASES OF *PSEUDOMONAS* SPP.

The early studies on the aliphatic amidases produced by *Pseudomonas* spp were reviewed by Clarke (1970) and will not be discussed in detail here. The induction by acetamide of an enzyme hydrolysing a limited range of low molecular weight aliphatic amides was reported for *P. aeruginosa* by Kelly and Clarke (1960) and similar enzymes were described for *P. putida, P. acidovorans*, and *P. cepacia* by Clarke (1972). These enzymes exhibit both amide hydrolase and amide transferase activities together with very low esterase activities.

1 *P. aeruginosa* Amidases

(i) Growth of *P. aeruginosa* on amides

The range of amides that can be used by *P. aeruginosa* as growth substrates is dependent on the substrate and inducer specificities of the amidases, on the general regulatory systems and on the existence of enzymes for the further metabolism of the carbon skeletons. The wild-type strain grows in minimal salt medium with acetamide and propionamide. None of the other aliphatic amides is utilized by a few substituted amides such as glycollamide and lactamide can support growth. The limited amide growth range contrasts with that of the *Mycobacterium* spp. and *Candida utilis*. The range of amide growth substrates can be extended by mutations in the structural gene of the enzyme leading to enzymes with altered substrate specificities, and by mutations in the amidase regulatory gene that alter the specific regulation of enzyme synthesis. In addition, mutations which have more general effects in derepressing enzyme synthesis may lead to growth on amide media that do not support growth of the wild-type strain. Table 3 lists some of the amides that have been used in studying this system and indicates the ways in which they have been employed in selective media for the isolation of mutants. There is a marked difference in the substrate and inducer specificities and this can probably be related to the amide-binding sites of the enzyme and regulator proteins respectively. Acetamide, propionamide, and glycollamide are good substrates and inducers and support growth. Fluoroacetamide and acrylamide are also good substrates and inducers but cause growth inhibition. Formamide is neither a good substrate nor a good inducer and, since 1-carbon compounds cannot be utilized by *P. aeruginosa*, it is merely a potential nitrogen source. Brammar *et al*. (1967) found that succinate–formamide medium allowed only very

Table 3 Amides as substrates and inducers for amidase of *Pseudomonas aeruginosa*

Amide	Characteristics†
Formamide	Poor substrate, poor inducer. Used to isolate regulator mutants.
Acetamide	Substrate, inducer, supports growth.
Propionamide	Substrate, inducer, supports growth.
Butyramide	Very poor substrate, amide analogue repressor, no growth. Used to isolate regulator and altered enzyme mutants.
Lactamide	Very poor substrate, good inducer, supports growth. Used to isolate catabolite repression-resistant mutants.
Fluoroacetamide	Substrate, inducer, inhibits growth due to lethal synthesis. Used to isolate amidase-negative mutants.
Valeramide	Not a substrate, amide analogue repressor, no growth.
Phenylacetamide	Not a substrate, amide analogue repressor, no growth.

†These effects are observed with wild-type strains. Mutants can be isolated, using various amide media, which are able to utilize amides that do not support growth of the wild type.

poor growth of the wild-type strain and they used this medium to isolate constitutive mutants and mutants altered in inducer specificity. The role of succinate in this medium is probably not merely to act as a carbon source; amidase synthesis is severely repressed by succinate, and growth of the wild-type strain on succinate–formamide medium is restricted both by the poor inducing capacity of formamide and by the catabolite repression exerted by succinate. Lactamide and butyramide are very poor substrates and no amides larger than these are hydrolysed at all. However, lactamide, unlike butyramide, is a very good inducer and as a sole carbon source can support growth of the wild-type strain. With succinate as carbon source and lactamide as nitrogen source the catabolite repression is severe enough to restrict growth, and this medium therefore allows the selection of mutants altered in catabolite repression.

Butyramide is a poor substrate, has no inducer activity, but can affect the regulatory process. Kelly and Clarke (1962) had found that some *N*-substituted amides could be used as gratuitous inducers and that cyanoacetamide, thioacetamide and β-hydroxypropionamide prevented amidase induction. Later, Brown and Clarke (1970) showed that butyramide behaved in the same manner and this effect was termed amide-analogue repression. This effect is thought to occur because butyramide is able to bind to the amidase regulatory protein at the amide-binding site but does this without producing the change of conformation necessary for gene expression. Butyramide, used as a carbon source, allows the selection of constitutive mutants altered in this response. These are constitutive butyramide-resistant mutants but some of the constitutive mutants isolated from succinate–formamide medium retain their sensitivity to butyramide repression. When the butyramide-sensitive constitutive mutant C11 (PAC111) was plated on butyramide medium it gave rise to both butyramide-resistant constitutive mutants and also to mutants producing an altered amidase with higher activity towards butyramide (Brown *et al.*, 1969). Butyramide as a nitrogen source, and succinate as a carbon source, provides a medium that demands two mutations for growth to occur. Smyth and Clarke (1975*a,b*) isolated several classes of mutants including altered enzyme mutants, butyramide-resistant mutants and catabolite-repression resistant mutants.

Betz and Clarke (1972) described the isolation of a family of altered enzyme mutants. These were derived by various routes, and five different classes of mutants were obtained that produced amidases with novel activities allowing them to grow on phenylacetamide. This amide could provide only the nitrogen source since phenylacetate is not utilized by *P. aeruginosa*. This study showed that the substrate specificity of an enzyme could be altered by stepwise mutation.

Brown and Clarke (1972) described another class of altered-enzyme mutants derived from a catabolite-repression-resistant strain. These were able to utilize acetanilide as the carbon source for growth. This was an interesting finding since enzymes attacking acetanilide are known for other *Pseudomonas* species (see Section 4.3.10.). The mutation in the amidase structural gene giving rise to an acetanilidase from an acetamidase is the replacement of a threonine residue of the wild-type enzyme by an isoleucine residue in the mutant AI3 enzyme (Brown and Clarke, 1972). The mutation resulting in the production of the butyramide-hydrolysing enzyme of mutant B6 is also a single site mutation resulting in the replacement of a serine residue of the wild-type enzyme by a phenylalanine residue (Paterson and Clarke, 1979).

Fluoroacetamide is potentially a toxic compound since it can be hydrolysed to produce fluoracetate. Clarke and Tata (1973) used this as the basis of a method for isolating acetamide-negative mutants and it was shown later (Farin and Clarke, 1978) that these included mutations in the regulator gene *amiR* as well as in the amidase structural gene *amiE*.

Urea had been shown by Kelly and Clarke (1962) to be an inhibitor of amidase activity and was used by Gregoriou *et al.* (1977) to isolate altered enzyme mutants. They started with a mutant able to grow on acetanilide as carbon source. Growth was inhibited by urea, and mutants were isolated which grew on acetanilide + urea medium and produced amidases with altered properties.

Goto *et al.* (1975) used urea inhibition to discriminate between amidases produced by different species of *Pseudomonas* and other non-fermentative Gram-negative bacteria.

(ii) Enzyme properties

The *P. aeruginosa* amidase is a hexameric protein composed of identical subunits each of about 35 000 daltons. The mutant amidases with altered substrate specificities are similar in general structure to the wild-type enzyme but several are thermolabile and it is thought that some of them dissociate more readily into their subunits (Betz and Clarke, 1972; Brown *et al.*, 1973). Most of the detailed studies have been carried out on the wild-type enzyme (A amidase) prepared from derepressed constitutive mutants of strain PAC1. Comparisons with enzymes prepared from the mutant strains have shown that mutations conferring new hydrolase activities also give new transferase and esterase activities, so that the differences between wild-type and mutant enzymes are related to differences in substrate binding and not to differences in catalytic activities.

The A amidase is inactivated by oxidation and is maintained in the active form by adding mercaptoethanol or dithiothreitol to the buffer solutions. It is inhibited by thiol reagents such as PCMB (*p*-chloromercuribenzoate) and DTNB (5,5′-dithio-*bis*-(2-nitrobenzoate)). One of the thiol groups of each subunit reacts very rapidly with iodoacetamide or iodoacetate and it was suggested that it might be involved at the active site. However, there is no protection from this inhibition by the enzyme substrates and the importance of maintaining the thiol groups in the reduced form is now thought to be related to subunit interaction and enzyme conformation. The enzyme is only active in the oligomeric form and the exact conformation may be critical for enzyme activity. Hollaway *et al*. (1980) have found that chloroacetone is a very potent inhibitor and that the enzyme is completely inhibited when only two molecules of chloroacetone have become bound to each molecule of the hexamer. Chloroacetone is thought to bind at the active site of the enzyme and these results suggest that when two of the sites have become blocked with chloroacetone then the others are inaccessible to acetamide because of conformational changes or steric hindrance. Findlater and Orsi (1973) found that acetaldehyde, propionaldehyde, and acetaldehyde-ammonia were also potent competitive inhibitors of the amidase from the *P. aeruginosa* PAC strain.

It was expected that if the amidase possessed a serine residue at the active site then it would be inhibited by DFP (diisopropylfluorophosphate). Although Findlater (1974) found some inhibition by 2mM DFP he thought this was probably due to impurities. However, the A amidase can accommodate only amides of low molecular weight, or presumably molecules of similar size, at the active site and the possibility of an active serine residue cannot be ruled out (see Sections 4.3.5 and 4.3.10).

The early studies on amidase specificity indicated that the *N*-substituted amides could not be attacked by the enzyme and Kelly and Clarke (1962) were able to use *N*-methylacetamide as gratuitous inducer, and later *N*-acetylacetamide was used for this purpose. Although hydrolysis of *N*-methylacetamide by whole cells, and cell-free extracts, was not detected, it was found later by Findlater and Orsi (1973) that the purified enzyme had both hydrolase and transferase activity for this amide. The K_m for *N*-methylacetamide was high (>200 mM for the hydrolase reaction) and since methylamine is not utilized by *P. aeruginosa* the reaction is unlikely to have physiological significance. It does, however, help to explain how a single amino acid change allows the mutant AI3 enzyme to accept *N*-phenylacetamide (acetanilide) as a growth substrate (Brown and Clarke, 1972). If the A amidase can accept an amide with a small group substituted on the amide nitrogen then a slight change in the conformation of the enzyme protein might be enough to allow the more bulky molecule of *N*-phenylacetamide to reach the active site and undergo hydolysis at a reasonable rate.

Findlater (1974) found that hydroxamates of acetic and propionic acids could be hydrolysed and that the corresponding hydrazides were rather poor substrates for both the hydrolase and transferase reactions.

(iii) Regulation of Amidase Synthesis

The earlier work indicated that a regulator gene *amiR* defined a protein with an amide-binding site. Some amides were found to be inducers while others had a repressing effect. Farin and Clarke (1978) suggest that the regulator protein exerts positive control of amidase synthesis. Systems under positive regulation can give rise to mutants that are non-inducible with a negative phenotype for the normal substrate inducer. These were identified for the amidase system by reversion studies. The amidase structural gene can be marked, by introducing a mutation specifying an altered enzyme, and a series of acetamide-negative mutants were isolated from a strain carrying the *amiE* mutation defining the B amidase and a wild-type inducible *amiR* gene. Some of these acetamide-negative mutants gave rise to revertants, with a range of different regulatory phenotypes but all producing B amidase. Other *amiR* mutants were isolated that were temperature-sensitive for amidase synthesis and it was shown that these had mutations in the regulator gene and not in the structural gene. These results indicate that, like the serine deaminase of *Escherichia coli* (Bloom *et al*., 1975), the synthesis of *P. aeruginosa* amidase requires the *amiR* product for gene expression.

Smyth and Clarke (1975*a*, *b*) re-examined catabolite repression of amidase synthesis and obtained evidence that cyclic AMP (cAMP) may be involved. Amidase synthesis by inducible and constitutive strains is stimulated by cAMP which also gives partial relief of catabolite repression. Promoter mutants were isolated that were insensitive to stimulation by cAMP and relatively insensitive to catabolite repression.

An unusual observation on amidase regulation was that amidase synthesis occurs in acetate medium as the wild-type culture reaches the end of the exponential growth phase (Kelly, 1961; Kelly and Kornberg, 1962). The reason for this is not known but it occurs also in *Aspergillus nidulans* (Hynes, 1977) (see Section 4.3.11). Mutants unable to metabolize acetate can be induced normally by acetamide and other amides so that for *P. aeruginosa* there is no evidence that a later metabolite of acetate is the true inducer. Potts and Clarke (1976) showed that catabolite repression was unaffected by nitrogen limitation and confirmed that this amidase is not regulated by nitrogen compounds.

(iv) Genetic Studies

The characteristics of many mutants have already been discussed. Genetic studies using transduction were carried out to investigate gene linkage

(Brammar *et al.* 1967; Smyth and Clarke, 1975*b*), and to map some of the *amiE* mutations (Betz *et al.*, 1974). Day *et al.* (1975) showed that the *ami* genes of strain PAO were located in the late region of the chromosome in experiments based on conjugal transfer by the sex factor FP2.

2 *P. putida* and *P. fluorescens* Amidases

Stanier *et al.* (1966) found that some strains of *P. putida* biotype A were able to grow on acetamide, and Clarke (1972) showed that these strains produced an aliphatic amidase which was closely related to that of *P. aeruginosa*. Some of the *P. putida* strains could also grow on butyramide but this was due to a difference in inducer specificity not in enzyme specificity. In these strains butyramide acted as an inducer and not a repressor. Previously, Jakoby and Fredericks (1964) had described an acetamidase produced by a strain identified as *P. fluorescens* isolated from soil by enrichment with acetamide. The enzyme had both hydrolase and transferase activity with acetamide, acrylamide, and propionamide but they were unable to detect any activity with formamide or butyramide. Preincubation with 1 mM DFP for 5 min resulted in the complete loss of activity but they also found that inhibition by DFP could be reversed by mercaptoethanol and concluded that the inhibition could have been due to an unknown contaminant reacting with the thiol groups. They concluded from the kinetic data that substrate binding induced conformational changes in the enzyme protein. Betz and Clarke (1973) found that *P. putida* biotype A strains grew on phenylacetamide and that this activity did not correspond to whether or not the strains produced an acetamidase. *P. putida* can utilize phenylacetate as a carbon source and phenylacetamide could be used as the sole growth substrate. The enzyme also hydrolysed mandelamide and isobutyramide and was induced by phenylacetamide and mandelamide. The properties of the enzyme have not been investigated in detail but it seems to be quite unrelated to the aliphatic amidases of the *Pseudomonas* species (Clarke, 1974; Laverack and Clarke, 1979).

10 ACETANILIDASE FROM *P. ACIDOVORANS*

A strain of *P. acidovorans* was isolated from soil by enrichment culture with acetanilide and the acetanilidase involved was characterized by Alt *et al.* (1975*b*). The enzyme was induced to high levels by acetanilide or by phenacetin. Alt *et al.* (1975*a*) established that the purified enzyme was a stable monomer with a molecular mass of about 57 000 daltons. They pointed out that this was similar in size to the subunits of esterases from mammalian sources but that these enzymes are oligomeric proteins. The enzyme from *P. acidovorans* also hydrolysed phenacetin, 4-nitroacetanilide,

4-nitropropylanilide, and 2-nitroacetanilide. It had no activity for acetamide, or phenylacetamide although they found that the crude extract could hydrolyse phenylacetamide, suggesting that another amidase was present. When the strain had been grown with acetamide it was possible to detect amidase activity in the crude extracts but all acetamidase activity was lost during the purification of the acetanilidase. The acetamidase activity of the crude extracts may be due to a third amidase.

The purified acetanilidase also hydrolysed phenyl acetate, 4-nitrophenyl acetate, and 2-nitrophenyl acetate. The esterase activity was of a similar order to the amide hydrolase activity and the authors class this enzyme as an amidase/esterase. All the substrates that could be hydrolysed were also effective as acyl donors with hydroxylamine as the acceptor. One interesting finding was that phenetidine acted as a very good acceptor for the acyl group donated by 4-nitrophenylacetanilide and this reaction could be followed by the release of 4-nitrophenol. This enzyme was strongly inhibited by thiol reagents, including PCMB, and since this enzyme appears to be a monomer there could be no subunit interaction to consider. Thus, thiol groups, of which three are known to be present, may be required for the catalytic site or may be required in the reduced state for maintenance of the correct active enzyme conformation. Inhibition by PCMB was reversed by cysteine and was not affected by the presence or absence of the substrate 4-nitroacetanilide. However, inhibition by the thiol reagents was delayed by the competitive inhibitor 2-nitrophenol. The three thiol groups were equally reactive to PCMB but with ω-bromoacetophenone there was total inactivation after modification of about 50 per cent of the SH groups (Heyman and Alt, 1974). These findings might suggest that this amidase is a thiol enzyme but it is also inhibited by organophosphorous compounds. Alt *et al.* (1975*a*) found that DFP, *bis*(4-nitrophenyl) phosphate, and diethyl-4-nitrophenyl phosphate were all very strong inhibitors and that inactivation was delayed in the presence of the substrate 4–nitroacetanilide. The enzyme protein was labelled with [^{32}P]-DFP, which was shown to be bound at only one of the serine residues. A labelled peptide was isolated and it was found that the amino acid sequence around the active serine was Gly-Ser-Ala (Heyman and Alt, 1974). This would indicate that the active site residue is serine and that the thiol groups have a secondary role.

11 AMIDASES OF *ASPERGILLUS NIDULANS*

The utilization of amides as growth substrates by *Aspergillus nidulans* was studied by Hynes and Pateman (1970*a*) who found that a large number of amides could be used as nitrogen sources in glucose minimal salt medium (Table 4). These included the aliphatic amides, from formamide to hexanoamide, a few amides with substituent groups, such as lactamide and glycollamide, and both benzamide and phenylacetamide. It seemed unlikely

Table 4 Amidases of *Aspergillus nidulans*

	Enzymes†			
	1 Formamidase	2 Acetamidase	3 General amidase	4 ?
Gene	*fmdS*	*amdS*	*gmdS*	
Amide induction	Non-inducible	Inducible	Non-inducible	
Catabolite repression	Weak	Severe	None	
Nitrogen repression	Present	Present	Present	
Major substrates	Formamide	Acetamide, propionamide, butyramide, valeramide, hexanoamide	Benzamide, phenylacetamide, butyramide, valeramide, hexanoamide	Valeramide, hexanoamide

†Data from Hynes (1975*a*)

that all these amides would be the substrates of a single amidase and attempts were made to identify the number of enzymes involved in the catabolism of these amides.

Mutants were obtained that were able to grow on acetamide, but not on formamide and these were also able to grow on a number of other amides. Other mutants were isolated that were able to grow on formamide but not on acetamide. Genetic analysis confirmed that at least two gene loci were concerned, *fmdS* specifying the formamidase and *amdS* specifying the acetamidase (Hynes and Pateman, 1970*a,b,c*). At a later time, Hynes (1975*a*) obtained evidence for a third amidase with a fairly broad specificity able to hydrolyse benzamide, phenylacetamide, butyramide, valeramide, hexanoamide, and also nicotinamide, indoleacetamide, naphthalene-acetamide, anthranilamide, leucineamide, prolineamide, and phenylalanine-amide. From the residual activities of mutants with defects in the genes for these three amidases (*fmdS*, *amdS*, and *gmdS*) Hynes (1975*a*) suggested that there is at least one more enzyme able to hydrolyse hexanoamide. The amidases and their principal substrates are listed in Table 4.

Investigations of the *A. nidulans* amidases have been mostly directed towards understanding the genetics of these systems and exploring the regulation of synthesis, and there is relatively little information on the properties of the enzymes themselves. Hynes and Pateman (1970*a*) showed that amidase activities were higher when cultures had been grown with amide substrates rather than with urea as the nitrogen source. The inducer specificities for the formamidase and the acetamidase were not clear-cut and many of the amides were able to induce both activities while β-hydroxypropionamide induced the acetamidase but not the formamidase. Acrylamide was shown to be a good substrate but not an inducer of the acetamidase and it was used by Hynes and Pateman (1970*b*) to isolate regulatory mutants which could utilize acrylamide producing high uninduced levels of acetamidase. Most of the acrylamide mutants mapped at a single gene locus, unlinked to the *amdS* gene, and were partially constitutive for acetamidase synthesis but were still subject to catabolite repression by glucose and also by ammonium salts. This regulatory gene was designated *amdR* and was shown to be semi-dominant to the wild-type inducible allele, indicating that the acetamidase was probably under positive control. Dunsmuir and Hynes (1973) examined several temperature-sensitive mutants which had reduced growth on acetamide at higher temperatures. Three of these had mutations in the *amdS* gene, so that their growth defects could be ascribed to defective enzyme proteins. One mutant had a defect in gene *amdR* and produced an acetamidase that was similar to the wild-type enzyme in the rate of inactivation during heating at 50 °C over a period of 3 min. The rate of decrease in amidase activity of a culture transferred from 25 °C to 40 °C was also consistent with a loss of ability to synthesize the enzyme rather than any defect in the enzyme itself. They concluded that the acetamidase gene *amdS* is under positive regulation exerted by the specific regulatory gene *amdR*.

The *A. nidulans* enzymes are also subject to general regulatory controls and many are regulated by repression by both carbon and nitrogen compounds. Hynes and Pateman (1970*b*) had isolated among the acrylamide mutants one that was originally designated *amdT*102. This appeared at first to be a fairly simple case of a mutation conferring resistance to catabolite repression but the regulation of enzymes for the metabolism of nitrogen compounds in *A. nidulans* is rather complex and appears to involve a number of gene loci (see also Section 4.2.4). This is reviewed by Arst and Cove (1973) who described a mutation in a gene *areA* that results in the inability to utilize a variety of nitrogen compounds, including amino acids, purines, amides, nitrite, and nitrate, while not affecting growth on ammonium salts. Other mutations in *areA* led to derepression of one or more activities that are repressed by ammonium ions in the wild-type strain. They suggested that the *areA* locus specifies a protein that is concerned with the synthesis of enzymes of nitrogen metabolism either as a positive or negative control element and that this regulator protein is normally inactive in the presence of ammonium ions. Hynes (1973) obtained other mutants with lesions in the same gene locus as the original *amdT*102 and these had pleiotropic effects and were later considered to belong to the same group as the *areA* mutations (Hynes, 1975*b*). The *areA* and *amdR* mutations have independent effects on the synthesis of amidase and it was suggested that the wild-type *areA* protein is involved in derepression of acetamidase synthesis upon relief from ammonium repression. Hynes (1974) extended these observations and suggested that the regulation of the synthesis of various enzymes required for the metabolism of nitrogen compounds was affected by glutamate and glutamine and that this was not necessarily the same system as that concerned in regulation by ammonium ions.

Arst (1976) suggested that the *amdR* gene may not be solely concerned with the regulation of acetamidase synthesis but may also control the regulation of synthesis of two other structural genes for enzymes required for nitrogen metabolism. He suggests that the *amdR* gene, exerting positive control over acetamidase synthesis, should be regarded as a integrator gene (Britten and Davidson, 1969), and that this gene, which he designates as *intA*, also controls the synthesis of γ-aminobutyrate transaminase and γ-aminobutyrate permease.

Hynes (1975*c*) showed that a *cis*-dominant mutation, closely linked to *amdS*, resulted in high levels of acetamidases when cultures were grown in acetamide medium and suggested that this mutation *amdI*9 is at the site of action of a regulatory molecule involved in induction by acetamide. This mutation also leads to increased acetamidase synthesis in acetate medium. Hynes (1977) suggests that the actual acetamidase inducer may not be acetamide but may be a product of acetate metabolism since various acetate mutants were not induced by acetamide. The acetamidase of *Pseudomonas aeruginosa* is synthesized at the end of the exponential growth phase in acetate medium (see Section 4.3.9.1(iii)) and similar effects involving catabolite repression may also affect the synthesis of the acetamidase of *A. nidulans*. The complex regulation of the *A. nidulans* acetamidase is discussed by Hynes (1978).

Acetamidase-negative mutants of *A. nidulans* were isolated from minimal salt medium containing acetate and urea (Hynes and Pateman, 1970*c*). Mutants were isolated from constitutive *amdR* mutants and some were revertants to the wild-type *amdR*$^{+}$, while others were unable to grow on either acetate or acetamide. The *amdS* mutants lacked almost all acetamidase activity but were able to grow with formamide. One class of mutants isolated was considered to be *amdR*$^{-}$ and this is particularly interesting in view of the positive regulation suggested for this system. Fluoroacetamide is only inhibitory to growth when the acetamidase level of the culture reaches relatively high levels and negative regulatory mutations would be as effective as structural gene mutations in reducing amidase activity as is the case for *P. aeruginosa* (Farin and Clarke, 1978).

12 FUTURE PROSPECTS

Amides, and the enzymes that attack them, offer many advantages for study to biochemists, microbiologists, and geneticists. Many amides are commercially available and precursors can be obtained for the synthesis of many others. This gives a wealth of chemically related compounds that can be tested for biological activity. The homologous series of aliphatic amides can be used to study substrate and inducer specificities and modifications can be introduced by the substitution of hydrogen atoms by hydroxyl, carbonyl, halogen, and other groups. This has been found to be very valuable in studying several of the amidases discussed in this chapter. The amidase activity of physiological significance is the release of ammonia by hydrolysis. Ammonia can be estimated by a variety of methods and many amidases can also participate in transferase and esterase reactions. This gives an opportunity to use several substrates and several different reactions to study the mechanism of action of the amidase enzymes.

Amides are potential growth substrates, providing either nitrogen or carbon or sometimes acting as sole growth substrates. The amide can be incorporated into suitable media for use in genetic studies. Selective media can be devised for mutant isolation and can provide the basis for examining the types of mutation that lead to amidases with novel activities.

Utilization of amides has been investigated in relatively few species to date but this is likely to change and the number of microbial amidases described in the literature will increase.

REFERENCES

Alt, J., Heymann, E., and Krisch, K. (1975*a*). *Eur. J. Biochem*., **53**, 357

Alt, J., Krisch, K., and Hirsch, P. (1975*b*). *J. Gen. Microbiol.*, **87**, 260

Arst, H. N. (1976). *Nature*, *Lond*., **262**, 231

Arst, H. N. and Cove, D. J. (1973). *Mol. Gen. Genet*., **126**, 111

Balls, A. K. and Jansen, E. F. (1952). *Adv. Enzymol. Relat. Areas Mol. Biol*., **13**, 321

Betz, J. L. and Clarke, P. H. (1972). *J. Gen. Microbiol.*, **73**, 161

Betz, J. L. and Clarke, P. H. (1973). *J. Gen. Microbiol.* , **75**, 167

Betz, J. L., Brown, J. E., Clarke, P. H., and Day, M. (1974). *Genet. Res., Cambridge*, **23**, 335

Bloom, F. R., McFall, E., Young, M. C., and Carothers, A. M. (1975). *J. Bacteriol.,* **121**, 1092

Bonicke, R. (1960). *Zentralbl. Backteriol*. Abt. 1, Orig., **179**, 209

Brady, B. L. (1969). *J. Gen. Microbiol.,* **59**, 47

Brammar, W. J. and Clarke, P. H. (1964). J. Gen. Microbiol., **37**, 307

Brammar, W. J., Clarke, P. H., and Skinner, A. J. (1967). *J. Gen. Microbiol.,* **47**, 87

Brammar, W. J., McFarlane, N. D., and Clarke, P. H. (1966). *J. Gen. Microbiol.,* **44**, 303

Britten, R. J. and Davidson, E. H. (1969). *Science, N.Y.,* **165**, 349

Brown, J. E. and Clarke, P. H. (1970). *J. Gen. Microbiol.,* **64**, 329

Brown, J. E., Brown, P. R., and Clarke, P. H. (1969). *J. Gen. Microbiol.,* **57**, 273

Brown, P. R. and Clarke, P. H. (1972). *J. Gen. Microbiol.,* **70**, 287

Brown, P. R., Smyth, M. J., Clark, P. H., and Rosemeyer, M. A. (1973). *Eur. J. Biochem.,* **34**, 177

Buhlman, X., Visher, W. A., and Bruhlin, H. (1961). *J. Bacteriol.,* **82**, 787

Clarke, P. H. (1970). *Adv. Microbial Physiol.,* **4**, 179

Clarke, P. H. (1972). *J. Gen. Microbiol.,* **71**, 241

Clarke, P. H. (1974). *Biochem. Soc. Trans.,* **2**, 831

Clarke, P. H. and Tata, R. (1973). *J. Gen. Microbiol.,* **75**, 231

Cohen, B. E. and Bangham, A. D. (1972). *Nature, Lond.,* **236**, 173

Collander, R. and Barlund, H. (1933). *Acta Bot., Fenn.,* **11**, 1

Cowan, S. T. and Steel, K. J. (1965). *Manual for the Identification of Medical Bacteria*, Cambridge University Press, Cambridge

Crawford, I. P. (1975). *Bacteriol. Rev.,* **39**, 87

Day, M., Potts, J. R., and Clarke, P. H. (1975). *Genet. Res., Cambridge,* **25**, 71

Den Dooren de Jong, L. E. (1926). *Bijdrage Tot de Kennis van het Mineralsatieproces*, Nijgh and Van Ditmar, Rotterdam

Draper, P. (1967). *J. Gen. Microbiol.,* **46**, 111

Dunsmuir, P. and Hynes, M. J. (1973). *Molec. Gen. Genet.,* **123**, 333

Epstein, I. and Grossowicz, N. (1969). *J. Bacteriol.,* **99**, 414

Farin, F. (1976). Studies on the mechanism of regulation of amidase synthesis. Ph.D. thesis, University of London

Farin, F. and Clarke, P. H. (1978). *J. Bacteriol*., **135**, 379

Fernald, N. J. and Ramaley, R. F. (1972). *Arch. Biochem. Biophys*., **153**, 95

Findlater, J. D. (1974). The aliphatic amidase of *Pseudomonas aeruginosa*. Ph.D. thesis, University of Dublin

Findlater, J. D. and Orsi, B. A. (1973). *FEBS Lett*., **35**, 109

Fitch, W. M. (1963). *Biochemistry,* **2**, 1217

Fitch, W. M. (1964). *J. Biol. Chem.,* **239**, 1328

Fruton, J. S. (1976). *Adv. Enzymol. Relat. Areas Mol. Biol*., **44**, 1

Gorr, G. and Wagner, J. (1933). *Botanische Z.,* **266**, 96

Goto, S., Tokida, H., Kaneko, Y., Ogawo, M., Dogosaki, I., Kuwabara, S., and Tada, S. (1975). *Jap. J. Bacteriol.,* **30**, 131

Gregoriou, M., Brown, P. R., and Tata, R. (1977). *J. Bacteriol.,* **132**, 377

Grossowicz, N. and Halpern, Y. S. (1957). *J. Biol. Chem.,* **228**, 643

Halpern, Y. S. and Grossowicz, N. (1957). *Biochem. J.,* **65**, 716

Hartley, B. S. (1974). *Symp. Soc. Gen. Microbiol.,* **24**, 151

Helbecque, D. M., Hanzel, V., and Eidus, L. (1975). *J. Clin. Microbiol.,* **1**, 50

Hersh, L. B. (1970). *J. Biol. Chem.,* **245**, 3526

Hersh, L. B. (1971). *Biochemistry,* **10**, 2884
Hersh, L. B. (1972). *Biochemistry,* **11**, 2251
Heymann, E. and Alt, J. (1974). *Hoppe-Zeyler's Z. Physiol. Chem.,* **355**, 1205
Holding, A. J. and Collee, J. G. (1971). *Methods Microbiol.,* **6** (Part A), 1–32
Hollaway, M. R., Clarke, P. H. and Ticho, T. (1980). In press
Hynes, M. J. (1973). *Mol. Gen. Genet.,* **125**, 99
Hynes, M. J. (1974). *J. Bacteriol.,* **120**, 1116
Hynes, M. J. (1975*a*). *J. Gen. Microbiol.,* **91**, 99
Hynes, M. J. (1975*b*). *Aust. J. Biol. Sci.,* **28**, 301
Hynes, M. J. (1975*c*). *Nature, Lond.,* **253**, 210
Hynes, M. J. (1977). *J. Bacteriol.,* **131**, 770
Hynes, M. J. (1978). *Mol. Gen. Genet.,* **161**, 59
Hynes, M. J. and Pateman, J. A. J. (1970*a*). *J. Gen. Microbiol.,* **63**, 317
Hynes, M. J. and Pateman, J. A. J. (1970*b*). *Mol. Gen. Genet.,* **108**, 97
Hynes, M. J. and Pateman, J. A. J. (1970*c*) *Mol. Gen. Genet.,* **108**, 107
Jakoby, W. B. and Fredericks, J. (1964). *J. Biol. Chem.,* **239**, 1978
Juhlin, I. (1960). *Acta Pathol. Microbiol. Scand.,* **50**, 195
Kelly, M. (1961). An investigation of the action of *Pseudomonas aeruginosa* on amides and related compounds. Ph.D. Thesis, University of London
Kelly, M. (1965). *Nature, Lond.,* **208**, 809
Kelly, M. and Clarke, P. H. (1960). *Biochem. J.,* **74**, 21P
Kelly, M. and Clarke, P. H. (1962). *J. Gen. Microbiol.,* **27**, 305
Kelly, M. and Kornberg, H. L. (1962). *Biochim. Biophys. Acta,* **64**, 190
Kimura, T. (1959*a*). *J. Biochem., Tokyo,* **46**, 973
Kimura, T. (1959*b*). *J. Biochem., Tokyo,* **46**, 1133
Kimura, T. (1959*c*). *J. Biochem., Tokyo,* **46**, 1271
Kimura, T. (1959*d*). *J. Biochem., Tokyo,* **46**, 1399
Laverack, P. D. and Clarke, P. H. (1979). Biotechnol. Lett., **1**, 353
Magasanik, B., Prival, M. J., Brenchley, J. E., Tyler, B. M., Deleo, A. B., Streicher, S. L., Bender, R. A., and Paris, C. G. (1974). *Curr. Topics Cell. Regul.,* **8**, 119
Martin, R. G., Berberich, M. A., Ames, B. N., Davis, W. W., Goldberger, R. F., and Yourno, J. D. (1971). *Methods Enzymol.,* **17** (Part B), 3
Meister, A., Levintow, L., Greenfield, R. E., and Abendschein, P. A. (1955). *J. Biol. Chem.,* **215**, 441
Mossel, D. A. A. and van Zadelhoff, C. (1971). *J. Gen. Microbiol.,* **69**, xiv
Oberhofer, T. R. and Rowen, J. W. (1974). *Appl. Microbiol.,* **28**, 720
Oberhofer, T. R., Rowen, J. W., and Cunningham, G. F. (1977). *J. Clin. Microbiol.,* **5**, 208
Oosterbaan, R. A., Kunst, P., van Rotterdam, J., and Cohen, J. A. (1958). *Biochim. Biophys. Acta,* **27**, 556
Ornston, L. N. and Parke, D. (1976). *Curr. Topics Cell. Regul.,* **12**, 210
Paigen, K. and Williams, B. (1970). *Adv. Microbial Physiol.,* **4**, 252
Paterson, A. and Clarke, P. H. (1979). *J. Gen. Microbiol.,* **114**, 75
Potts, J. R. and Clarke, P. H. (1976). *J. Gen. Microbiol.,* **93**, 377
Ramaley, R. F., Fernald, N., and DeVries, T. (1972). *Arch. Biochem. Biophys.,* **153**, 88
Schneidau, J. D. (1963). *Am. Rev. Infect. Dis.,* **88**, 563
Smyth, P. F. and Clarke, P. H. (1975*a*). *J. Gen. Microbiol.,* **90**, 81
Smyth, P. F. and Clarke, P. H. (1975*b*). *J. Gen. Microbiol.,* **90**, 91
Stanier, R. Y., Palleroni, N. J., and Douderoff, M. (1966). *J. Gen. Microbiol.,* **43**, 159
Stein, W. D. (1967). In *The Movement of Molecules across Cell Membranes*, Academic Press, New York
Stephenson, M. (1949). In *Bacterial Metabolism*, 3rd edn, Longmans, Green and Co., London, pp. 184–192
Stockell, A. and Smith, E. L. (1957). *J. Biol. Chem.,* **227**, 1
Thalenfeld, B. and Grossowicz, N. (1976). *J. Gen. Microbiol.,* **94**, 131
Thalenfeld, B., Epstein, I., and Grossowicz, N. (1977). *Biochim. Biophys. Acta,* **497**, 112
Urabe, K., Takei, N., and Saito, H. (1965). *Am. Rev. Infect. Dis.,* **91**, 120

Microorganisms and Nitrogen Sources
Edited by J. W. Payne

Section 4.4

Synthesis and Functions of Microbial Asparaginases and Glutaminases

H. E. WADE

Centre for Applied Microbiology and Research, Porton Down, Wiltshire, UK

1 INTRODUCTION

Although the structures and free energies of glutamine and asparagine are similar, evolution has favoured glutamine as amino-nitrogen donor for important steps in the synthesis of nucleic acids, proteins, and certain structural polysaccharides. In advanced life-forms, these amides have acquired specialized roles, glutamine providing a non-toxic vehicle for translocating readily available amino-nitrogen for synthesis and detoxification and asparagine playing a similar role in certain higher plants. It is only in the context of these specialized functions that glutaminase and asparaginase can be regarded as essential. In spite of their very wide distribution among microorganisms (Zanin *et al*., 1969; Wade *et al*., 1971*b*; Arima *et al*., 1972; Imada *et al*., 1973), there is little evidence that they serve a vital role under conditions of unrestricted growth. Under unfavourable conditions, however, their presence may have an important bearing upon survival.

From the time that the anti-cancer property of asparaginase from *Escherichia coli* became known (Mashburn and Wriston, 1964) and the necessity for high substrate affinity became recognized, much effort has gone into the characterization of microbial amidases with particular attention to substrate specificity and affinity. The data produced brought to light some taxonomic conformities and are consistent with the suggestion that asparaginases and glutaminases may benefit bacteria under certain nutritionally-deprived conditions (Wade *et al*., 1971*a*).

The emergence of glutamine rather than the simpler asparagine for the more vital metabolic roles may reflect the greater prominence of glutamate in the 'primordial soup' (Sagan and Khare, 1971). Ammonia probably served as the first amino-nitrogen donor but as the pH of the 'soup' fell with evolutionary time, this became less efficient, the protonated NH_4^+ ion being inefficient. At pH values nearer neutrality, glutamine emerged as the more effective alternative (Hartman, 1973). The energy requirements for most aminations by glutamine are such as to require the activation of the receptor by ATP. This requirement, and the need to avoid deamidation by free glutaminase, have been satisfied by locating these activities in close proximity to each other in multifunctional enzymes loosely termed glutamine transamidases. Many of these can use ammonia as a less effective alternative to glutamine and contain subunits which, in isolation, express exclusively glutaminase activity (Prusiner, 1973). Whether independent glutaminases evolved from transamidases or not, they appear in modern life-forms to have taken on less fundamental roles.

2 DISTRIBUTION OF GLUTAMINASES AND ASPARAGINASES

1 Animals and Higher Plants

It has long been known that in mammals a phosphate-dependent glutaminase in the kidney plays a vital role in controlling the pH of the blood. Although asparaginase is known to be present in the liver of many animals (Steensholt, 1944) its function is obscure and it has attracted little interest. Its inexplicable occurrence in the sera of the guinea pig and its close relatives (Old *et al*., 1963) and in the sera of New World monkeys (Peters *et al*., 1970) is indicative of its general lack of status in the biochemistry of higher animals. The enzyme maintains normally undetectable levels of asparagine in the circulation of these animals to no obvious purpose or disadvantage. Even in the majority of warm-blooded animals, the concentration of this amino acid in the circulation is only about 0.05 mM, about one-tenth that of glutamine.

In certain higher plants however, asparagine has assumed some importance (see also Section 4.6.5). A concentration of 30mM is present in the phloem of *Lupus albus* (Atkins *et al*., 1975) and about 60 per cent of the seed proteins originate from the amide nitrogen (Miflin and Lea, 1977). It is not surprising therefore that asparaginase, which is optimally present during the deposition of the storage proteins, should be implicated in this process (Lea *et al*., 1978; Sodek *et al*., 1980).

2 Microorganisms

The glutaminases and asparaginases of microorganisms have received far more attention, not only because of their ease of study but also because of their potential clinical use which has progressed to the point where two asparaginases are in routine use for the treatment of acute lymphatic leukaemia and are currently a part of the drug protocol in all the major clinical trials on this disease in the Western World.

(i) Escherichia Coli

The initial demonstration by Broome (1961) that the anti-tumour properties of guinea pig serum (Kidd, 1953) were due to the asparaginase present in it, led to a search for other sources of the enzyme and the demonstration of a similar enzyme in *Escherichia coli* (Mashburn and Wriston, 1964). Not surprisingly therefore, the asparaginases and glutaminases of this bacterium received much attention.

Two specific intracellular glutaminases have been described (Table 1) and interesting regulatory functions attributed to them.

Glutaminase A, first described by Meister *et al*. (1955), operates optimally at pH 5 and is induced when energy reserves fall, such as towards the end of rapid growth. Hartman (1973) postulated that, by depleting glutamine under these circumstances, this enzyme switches off major biosynthetic pathways, thereby conserving energy.

To glutaminase B, in collaboration with glutamine synthetase, has been attributed the function of maintaining a balance between glutamine and glutamate that avoids a lethal deprivation of ATP when an NH_4^+-limited bacterium is suddenly presented with an abundance of this ion (Prusiner, 1975). However, the evidence for such a vital function is not very convincing. The activity of glutamine synthetase is very adequately controlled by adenylation and many feedback systems (Stadtman, 1970; Schutt and Holzer, 1972), and glutamate synthase (Tempest *et al*., 1970), which is generally distributed among bacteria (Brown, 1976), offers a more attractive route for converting glutamine into glutamate by avoiding the liberation of NH_4^+ and lessening the chance of establishing a futile metabolic cycle (see also Section 4.2.3).

Following the clinical interest aroused in asparaginase, the closer study

Table 1 Glutaminases and asparaginases from *Escherichia coli*

	Strain	Inducibility	Location	Relative activities		K_m (mM)		Reference
				Gln	Asn	Gln	Asn	
Glutaminase A		Inductive	Cytoplasmic	100	0	0.5	—	Hartman, 1971
Glutaminase B		Constitutive	Cytoplasmic	100	0	*c*. 3	—	Prusiner, 1975; Prusiner *et al*., 1976
Asparaginase EC1	B	Constitutive	Cytoplasmic	—	—	—	—	Ho *et al*., 1970
I	K12	Constitutive	Cytoplasmic	—	—	—	*c*. 2	Schwartz, 1971
Asparaginase EC2	B	Inductive	Periplasmic	2	98	6.25	0.0115	Ho *et al*., 1970
II	K12	Inductive	Periplasmic	—	—	—	0.005	Schwartz, 1971

of *E. coli* revealed the coexistence of two very different asparaginases in both strain K12 (Schwartz *et al*., 1966) and strain B (Ho *et al*., 1970) (Table 1). Only the periplasmic, inducible enzyme has received much attention because it alone shows anti-tumour activity as a result of its longer persistence in the blood and higher affinity for asparagine.

(ii) Other Microorganisms

The high-affinity asparaginases from several other members of the Enterobacteriaceae display similar properties to the periplasmic enzyme from *E. coli* (Table 2). These enzymes have similar molecular mass of about 100 000–140 000 daltons, show a minor activity towards glutamine, are fully expressed in the intact cell, and in some cases are demonstrably periplasmic. In spite of these similarities, there are marked serological differences. Antisera raised to the periplasmic asparaginase from *E. coli* does not produce a precipitin line on Ouchterloney plates against asparaginases from *Erwinia carotovora* (MacLennan *et al*., 1971), *Serratia marcescens* (Boyd and Phillips, 1971), *Proteus vulgaris* (Tosa *et al*., 1971), or *Vibrio succinogenes* (Distasio *et al*., 1977). For patients who have become immunologically sensitive to the asparaginase from *E. coli*, this serological distinction is of considerable benefit, allowing their treatment to be continued with the enzyme from *E. carotovora* (Ohnuma *et al*., 1973).

Several Gram-negative species outside the Enterobacteriaceae have amidases that display a very high affinity for glutamine or asparagine or both (Table 2). Some of the exceptionally high and contradictory K_m values shown may reflect the difficulties of obtaining reliable measurements of K_m at such low substrate concentrations using conventional methods of assay. Among the non-specific amidases, there is a tendency for the relative activities to fall in the range 50–70 per cent glutamine to 30–50 per cent asparagine. What may appear to be different enzymes from substrate specificity studies may in fact be related proteins having small differences in a common structure. Holcenberg *et al*. (1978) have shown that an eight-residue sequence of amino acids at the active centre is common to the amidases from *Pseudomonas* 7A, *Acinetobacter glutaminasificans* and *E. coli*. Chemical modifications can also bring about changes in enzyme specificity. The relatively minor activity of *E. carotovora* asparaginase towards glutamine (see Table 2) can be reduced from 10 to 2 per cent, relative to activity towards asparagine, by succinylation (H. E. Wade, unpublished data). The same reaction has no effect on the substrate specificity of the amidase from *Acinetobacter glutaminasificans* (Holcenberg *et al*., 1975).

The failure of asparaginases and glutaminases from Gram-positive bacteria, yeasts, and lower fungi to have much effect upon the growth of

Table 2 Microbial amidases

	Relative activities		K_m (mM)		
	Gln	Asn	Gln	Asn	Reference
Gram-negative bacteria					
Pseudomonadaceae					
Pseudomonas aeruginosa	66	34	0.0027	0.0031	Roberts *et al*., 1972
Pseudomonas aeruginosa Amidase A	66	34	0.11	—	Soda *et al*., 1972
Amidase B	100	0	0.18	—	Soda *et al*., 1972
Pseudomonas putrifaciens	100	0	0.035	—	Roberts *et al*., 1972
Pseudomonas acidovorans	40	60	0.022	0.015	Davidson *et al*., 1977*b*
Pseudomonas fluorescens	0	100	—	0.40	Mardashev *et al*., 1975
Pseudomonas species 7A	66	34	0.0046	0.0044	Roberts, 1976
Vibrio succinogenes	0	100	—	0.017	Kaftewitz and Goodman, 1974
Vibrio succinogenes	0.015	100	—	0.048	Distasio *et al*., 1977
Azotobacteraceae					
Azotobacter vinelandii	—	—	—	0.11	Gaffar and Shetha, 1977
Azotobacter agilis	—	—	—	0.10	Ehrenfeld *et al*., 1963
Achromobacteraceae					
Alcaligenes marshallii	57	43	0.0031	0.0025	H. E. Wade, unpublished data
Acinetobacter glutaminasificans	56	44	0.0058	0.0048	Roberts *et al*., 1972
Acinetobacter calcoaceticus Amidase A	10	90	—	2.1	Joner *et al*., 1973
Amidase B	50	50	—	3	Joner, 1976
Achromobacter strain 257	56	44	0.0019	0.0025	Wade, 1977

Enterobacteriaceae					
Escherichia coli B	2	98	6.25	0.0115	Ho *et al*., 1970
Citrobacter freundii	5	95	—	0.026	Bascombe *et al*., 1975
	3	97	—	0.029	Davidson *et al*., 1977*a*
Klebsiella aerogenes	5	95	1.1	0.01	Resnick and Magasanik, 1976
Erwinia carotovora	10	90	1.4	0.015	Wade *et al*., 1968; Wade, 1977
Serratia marcescens	—	—	3	0.012	Rowley and Wriston, 1967
	5	95	0.87	0.0554	Novak and Phillips, 1974
Proteus vulgaris	2	98	0.43	0.026	Tosa *et al*., 1972
Gram-positive bacteria					
Sarcina species	100	0	0.51	—	Holcenberg *et al*., 1973
Bacillus coagulans	0	100	—	4.7	Law and Wriston, 1971
Mycobacterium coagulans	—	100	—	2–3.1	Jayaram *et al*., 1968
Mycobacterium bovis	—	—	—	2.6	Soru and Zaharia, 1973
Mycobacterium BCG	—	—	—	2.6	Turcotte and Quevillon, 1975
Yeasts					
Candida guilliermondii	—	—	—	3.5	Davtyan *et al*., 1977
Candida utilis	2	98	—	0.11–0.13	Imada *et al*., 1972
Hansenula jadini	1	99	—	0.11–0.13	Imada *et al*., 1972
Cryptococcus albidus	0	100	—	0.11	Imada *et al*., 1972
Sporoblomyces roseus	3	97	—	0.11	Imada *et al*., 1972
Rhodotorula rubra	0	100	—	0.55	Imada *et al*., 1972
Saccharomyces cerevisiae	0	100	—	0.74	Dunlop *et al*., 1978
Saccharomyces margini	0	100	—	0.25–0.3	Imada *et al*., 1972; Jones, 1977
Other fungi					
Aspergillus sojae	—	—	50	—	Yamamoto and Hirovka, 1974
Fusarium oxysporum	0	100	—	0.04–0.08	Igarasi *et al*., 1974
Fusarium tricinctum	—	100	—	0.05	Scheetz *et al*., 1971
Tilachlidium tumicola	47	53	0.6	0.3	Imada *et al*., 1973

tumours has lessened interest in their biochemical properties. Those that have been characterized have shown poor affinity for substrate (Table 2). Although the substrate affinities of those from yeasts and lower fungi are generally higher, they do not match the very high affinities of amidases from many Gram-negative bacteria (Table 2). Generally, these latter asparaginases and glutaminases are periplasmic and do not have access to the cytoplasm, where they would be in competition with vital anabolic enzymes such as the appropriate aminoacyl tRNA synthetases, glutamine transamidases, and glutamate synthase, which have much higher K_m^{Gln} values in the range 0.1–2.0 mM.

Bacterial amidases are generally intracellular and usually freely accessible to exogenous substrate (see also Section 4.3). By contrast, yeast and other fungal amidases are generally extracellular.

3 SYNTHESIS OF GLUTAMINASES AND ASPARAGINASES BY MICROORGANISMS

The synthesis of microbial glutaminases and asparaginases that are expressed extracellularly appear to be influenced by conditions of growth. In general, in batch cultures, these enzymes are maximally produced towards the end of the log phase or in the stationary phase. The contrasting and sometimes contradictory reports on the effects of growth conditions on asparaginase production leave the strong impression of many different mechanisms of induction and control. There have been few comparative studies on the subject.

Early in the study of the *E. coli* periplasmic asparaginase, Schwartz *et al.* (1966) demonstrated the induction of the enzyme from strain K12 by anaerobiosis (Cedar and Schwartz, 1968); similar reports appeared later for strains B (Boeck and Ho, 1973), W (Resnick and Magasanik, 1976), 316/66 (Netrval, 1977), and A-1 (Barnes *et al.*, 1977) but anaerobiosis had no effect upon production from strain HAP (Roberts *et al.*, 1968). Anaerobiosis also stimulated asparaginase synthesis by *Serratia marcescens* (Heinemann and Howard, 1969) and *Bacillus cadaveris* (Evseer *et al.*, 1967) but not by *Proteus vulgaris* (Tosa *et al.*, 1971) and *Citrobacterium freundii* (Bascombe *et al.*, 1975).

As with many other periplasmic hydrolytic enzymes, the synthesis of asparaginase and glutaminase is generally subject to catabolic repression. Thus, repression by glucose was observed in *Serratia marcesens* (Heinemann and Howard, 1969), *E. coli* K12 (Cedar and Schwartz, 1968), and strain W (Garaev and Golub, 1977), in *Erwinia carotovora* (Callow *et al.*, 1971), *Proteus vulgaris* (Atalay *et al.*, 1976), *Pseudomonas boreapolis* (Smirnova *et al.*, 1976), and *Staphylococcus* sp. (Mikucki *et al.*, 1977). In contrast, this sugar had no effect upon the synthesis of asparaginase by

Klebsiella aerogenes (Resnick and Magasanik, 1976) and actually appeared to be necessary for the optimal production of the enzyme from *E. coli* strain B (Boeck and Ho, 1973).

Similar contrasting effects have been observed with different sources of nitrogen. The synthesis of the periplasmic, high-affinity asparaginase by *E. coli* K12 was depressed on a medium containing asparagine as the sole source of nitrogen, when growth became dependent upon a low-affinity cytoplasmic asparaginase (Willis and Woolfolk, 1974). In contrast, growth of *K. aerogenes* under these circumstances is dependent upon the high-affinity periplasmic asparaginase (Resnick and Magasanik, 1976). Aspartate and glutamate induce the synthesis of asparaginase by *Ps. ovalis* (Foda *et al*., 1974) and *E. coli* strain 316/66 (Netrval, 1977) but not by *P. vulgaris* (Tosa *et al*., 1971). Glutamate stimulates the synthesis of this enzyme in *Pseudomonas* sp. (Mardashev, 1967), *Erwinia carotovora* (Callow *et al*., 1971), and in *E. coli* strains HAP (Roberts *et al*., 1968) and A-1 (Barnes *et al*., 1977).

Ammonium ions apparently have varied effects, inducing the synthesis of asparaginase in *Azotobacter vinlandii* grown in a defined medium (Gaffar and Shetha, 1977), while in a complex medium the enzyme is induced in *Vibrio succinogenes* (Albanese and Kafkewitz, 1978) and repressed in *P. vulgaris* (Tosa *et al*., 1971). Even in a direct comparison between species, contrasting effects have been observed. The synthesis of the enzyme by *E. coli* strain K12 grown on a glutamate–glucose medium was stimulated by ammonium ions whereas its synthesis by *K. aerogenes* was repressed (Resnick and Magasanik, 1976). Only extensive fundamental studies are likely to throw light on this very confusing picture. Such an approach has been made by Magasanik and his colleagues, who have attempted to correlate the synthesis of ammonia-producing enzymes such as histidinase, asparaginase, and urease with the synthesis of glutamine synthetase, which activates the transcription of the structural genes of these hydrolytic enzymes with what appears to be the common purpose of securing a nitrogen source in an ammonium-deficient medium (Resnick and Magasanik, 1976; Friedrich and Magasanik, 1977).

4 UPTAKE OF ASPARAGINE

The presence of a wide variety of amidases in bacteria has complicated investigations into the active transport of asparagine. It is not surprising, therefore, that some disagreement should exist about the existence of a specific transport system (see also Section 1.1.3). Cedar and Schwartz (1968, 1969) concluded from their studies that asparagine is not actively accumulated in *E. coli*. However, Willis and Woolfolk (1974, 1975) reported evidence for the existence of two transport systems; a

Table 3 Uptake of asparagine and aspartate by bacteria

Substrate	Organism	Kinetically distinguished processes of uptake expressed as K_m (mM) 1	2	Reference
Asparagine	*E. coli*	0.0035	0.08	Willis and Woolfolk, 1975
	Streptococcus faecalis	0.16	2.1	Holden and Bunch, 1973
	Lactobacillus plantarum	0.17	2.1	Holden and Bunch, 1973
Aspartate	*E. coli*	0.0035	0.03	Kay, 1971

high-affinity, energy-requiring system and a low-affinity system (Table 3), and this had the attraction of placing asparagine in the fast-expanding group of substances that command dual systems of transport (Tempest and Neijssel, 1976), although the reliability of their data is lessened, with respect to the high-affinity system, by the uncertainty of the presence or otherwise of periplasmic asparaginase in the suspensions they studied. The kinetics of systems that bear on the uptake and utilization of asparagine are such (Table 3) that, even if a high-affinity transport system does exist in this bacterium, its function in taking up and utilizing low concentrations of this amino acid from the medium could be fulfilled as efficiently by the combined activities of periplasmic asparaginase (K_m 0.005 mM) and the high-affinity aspartate transport system described by Kay (1971). Some support for this view comes from the absence of a similar high-affinity asparagine transport system in *Lactobacillus plantarum* and *Streptococcus faecalis* (Table 3), two Gram-positive bacteria that are unlikely to contain a high-affinity asparaginase (Table 2).

5 AMIDASES AND BACTERIAL ADAPTATION

If the dependence of some microorganisms upon high-affinity asparaginase for the uptake and utilization of low concentrations of asparagine is an acceptable possibility, it is a short step to speculate on the relevance of these enzymes and of high-affinity glutaminases upon survival. The possession of such an enzyme by the majority of Gram-negative bacteria may confer upon this group the advantage of competing successfully against other microorganisms for a valuable source of energy and ammonia. Is it possible that Gram-positive bacteria and other microorganisms deficient in this enzyme evolved sporulation as an alternative method of surviving in the face of this competition?

Gram-negative bacteria have the ability to utilize very low

concentrations of ammonia through the possession of a high-affinity glutamine synthetase, the efficiency of which is such that its activity has to be suppressed by a complex system of adenylation when a sudden increase in the concentration of ammonia threatens to depress the concentration of ATP in the cell (Stadtman, 1970). The confinement of this system to Gram-negative bacteria is compatible with their possession of a high-affinity amidase. In *K. aerogenes* the close functional association of the two enzymes extends to the stimulation of asparaginase synthesis by glutamine synthetase under conditions of ammonia starvation. Other ammonia-producing enzymes are also affected (Resnick and Magasanik, 1976). This leaves the strong impression that, in the course of evolution, the primordial dependence upon ammonia as nitrogen source may have progressed to substances that readily produce ammonia and to the development in Gram-negative bacteria of amidases with extremely high affinities for asparagine and glutamine. It is hard to avoid the conclusion that this development was the consequence of advantages bestowed on the microorganisms by their possession of these enzymes.

REFERENCES

Albanese, E. and Kafkewitz, D. (1978). *Appl. Environ. Microbiol.*, **36**, 25

Arima, K., Sakamoto, T., Araki, C., and Tamura, G. (1972). *Agric. Biol. Chem., Tokyo*, **36**, 356

Atalay, A., Kolankaya, N., Hosrik, E., and Sumer, S. (1976). *Hacettepe Bull. Sci. Eng.*, **5**, 32

Atkins, C. A., Pate, J. S., and Sharkey, P. J. (1975). *Plant Physiol.*, **56**, 807

Barnes, W. R., Dorn, G. L., and Vela, G. R. (1977). *Appl. Environ. Microbiol.*, **33**, 257

Bascombe, S., Banks, G. T., Skarstedt, M. T., Fleming, A., Bettelheim, K. A., and Connors, T. A. (1975). *J. Gen. Microbiol.*, **91**, 1

Boeck, L. D. and Ho, P. P. K. (1973). *Can. J. Microbiol.*, **19**, 1251

Boyd, J. W. and Phillips, A. W. (1971). *J. Bacteriol.*, **106**, 578

Broome, J. D. (1961). *Nature, Lond.*, **191**, 1114

Brown, C. M. (1976). In *Continuous Culture : Applications and New Fields* (A. C. R. Dean, D. C. Ellwood, C. G. T. Evans, and J. Melling, eds), Ellis Horwood, Chichester, pp. 170–183

Callow, D. S., Capell, B. J., and Evans, C. G. T. (1971). *J. Gen. Microbiol.*, **65**, ii

Cedar, H. and Schwartz, J. H. (1968). *J. Bacteriol.*, **96**, 2043

Cedar, H. and Schwartz, J. H. (1969). *J. Biol. Chem.*, **244**, 4112

Davidson, L., Brear, D. R., Wingard, P., Hawkins, J. and Kitto, G. B. (1977*a*). *J. Bacteriol.*, **129**, 1379

Davidson, L., Burkom, M., Ahn, S., Chang, L., and Kitto, G. B. (1977*b*). *Biochim. Biophys. Acta*, **480**, 282

Davtyan, M. A., Stepanyan, K. R., and Oganesyam, S. P. (1977). *Biol. Zh. Arm.*, **30**, 14

Distasio, J. A., Niederman, R. A., and Fafkewitz, D. (1977). *Proc. Soc. Exp. Biol. Med.*, **155**, 528

Dunlop, P. C., Meyer, G. M., Bau, D., and Roon, R. J. (1978). *J. Biol. Chem.*, **253**, 1297

Ehrenfeld, E., Marble, S. J. and Meister, A. (1963). *J. Biol. Chem.*, **238**, 3711

Evseer, L. P., Nikolaev, A. Y., Eremenko, V. V., and Mardashev, S. R. (1967). *Biokhimiya*, **32**, 873

Foda, M. S., Khafagy, E. Z., and Badr-Eldin, S. M. (1974). *ZBPIA*, **129**, 525

Friedrich, B. and Magasanik, B. (1977). *J. Bacteriol.*, **131**, 446

Gaffar, S. A. and Shetha, Y. I. (1977). *Appl. Environ. Microbiol.*, **33**, 508

Garaev, M. M. and Golub, E. I. (1977). *Mikrobiologiya*, **46**, 433
Hartman, S. C. (1971). In *The Enzymes*, 3rd edn, Vol. 4 (P. D. Boyer, ed.), Academic Press, New York, pp. 79–100
Hartman, S. C. (1973). In *Enzymes of Glutamine Metabolism* (S. Prusiner and E. R. Stadtman, eds), Academic Press, New York, pp. 319–330
Heinemann, B. and Howard, A. J. (1969). *Appl. Microbiol.*, **18**, 550
Ho, P. P. K., Milikin, E. B., Bobbitt, J. L., Grinnan, E. L., Burck, P. J., Frank, B. H., Boeck, L. D., and Squires, R. W. (1970). *J. Biol. Chem.*, **245**, 3708
Holcenberg, J. S., Ericsson, L., and Roberts, J. (1978). *Biochemistry*, **17**, 411
Holcenberg, J. S., Roberts, J., and Dolowy, W. C. (1973) In *Enzymes of Glutamine Metabolism* (S. Prusiner and E. R. Stadtman, eds), Academic Press, New York, pp. 277-292
Holcenberg, J. S., Roberts, J., Schmid, F., and Schmer, G. (1975). *J. Biol. Chem.*, **250**, 4165
Holden, J. T. and Bunch, J. M. (1973). *Biochim. Biophys. Acta*, **307**, 640
Igarasi, S., Imada, A., Nakahama, K., Matsumoto, T., and Ootsu, K. (1974). *Experientia*, **30**, 814
Imada, A. and Igarasi, S. (1973). *J. Takeda Res. Lab.*, **32**, 140
Imada, A., Nakahama, K., and Igarasi, S. (1972). *J. Takeda Res. Lab.*, **31**, 460
Imada, A., Igarasi, S., Nakahama, K., and Isono, M. (1973). *J. Gen. Microbiol.*, **76**, 85
Jayaram, H. N., Ramakrishnan, T., and Vaidynathan, C. S. (1968). *Arch. Biochem. Biophys.*, **126**, 165
Joner, P. E. (1976). *Biochim. Biophys. Acta*, **438**, 287
Joner, P. E., Tristiansen, T., and Einarsson, M. (1973). *Biochim. Biophys. Acta*, **327**, 146
Jones, G. E. (1977). *J. Bacteriol.*, **130**, 128
Kafkewitz, D. and Goodman, D. (1974). *Appl. Microbiol.*, **27**, 206
Kay, W. W. (1971). *J. Biol. Chem.*, **246**, 7373
Kidd, J. G. (1953). *J. Exp. Med.*, **98**, 565
Lea, P. J., Fowden, L., and Miflin, B. J. (1978). *Phytochemistry*, **17**, 217
Law, A. S. and Wriston, J. C. (1971). *Arch. Biochem. Biophys.*, **147**, 744
MacLennan, A. P., Wade, H. E., and Hawkins, D. C. (1971). *Col. Int. CNRS*, Report No. 197, p. 409
Mardashev, S. R. (1967). *Biokhimiya*, **32**, 1093
Mardashev, S. R., Nikolaev, A. Ya., Sokolov, N. N., Kozlov, E. A., and Kutsman, M. E. (1975). *Biokhimiya*, **40**, 984
Mashburn, L. T. and Wriston, J. C. (1964). *Arch. Biochem. Biophys.*, **105**, 450
Meister, A., Levintow, L., Greenfield, R. E., and Adsendschein, P. A. (1955). *J. Biol. Chem.*, **215**, 441
Miflin B. J. and Lea, P. J. (1977). *Annu. Rev. Plant Physiol.*, **28**, 299
Mikucki, J., Szarapinska-Kwaszewska, J., and Krzeminski, Z. (1977). *Zentbl. Bakt. ParasitKde*, **132**, 135
Netrval, J. (1977). *Folia Microbiol., Praha*, **22**, 106
Novak, E. K. and Phillips, A. W. (1974). *J. Bacteriol.*, **117**, 593
Ohnuma, T., Holland, J. F., and Meyer, P. A. (1973). *Cancer*, **30**, 376
Old, L. J., Boyse, E. A., Campbell, H. A., and Daria, G. M. (1963). *Nature, Lond.*, **198**, 801
Peters, J. H., Lin, S. C., Berridge, B. J., Chao, W. R., and Cummings, J. G. (1970). *Life Sci. II*, **9**, 431
Prusiner, S. (1973). In *Enzymes of Glutamine Metabolism* (S. Prusiner and E. R. Stadtman, eds), Academic Press, New York, pp. 293–316
Prusiner, S. (1975). *J. Bacteriol.*, **123**, 992
Prusiner, J., Davis, J. N. and Stadtman, E. R. (1976). *J. Biol. Chem.*, **251**, 3447
Resnick, A. D. and Magasanik, B. (1976). *J. Biol. Chem.*, **251**, 2722
Roberts, J., Burson, G., and Hill, J. M. (1968). *J. Bacteriol.*, **95**, 2117
Roberts, J. (1976). *J. Biol. Chem.*, **251**, 2119
Roberts, J., Holcenberg, J. S., and Dolowy, W. C. (1972). *Proc. Am. Assoc. Cancer Res.*, Abstract 98, p. 25
Rowley, B. and Wriston, J. C. (1967). *Biochem. Biophys. Res. Commun.*, **28**, 160
Sagan, C. and Khare, B. N. (1971). *Science, N.Y.*, **173**, 417

Scheetz, R. W., Whelan, H. A., and Wriston, J. C. (1971). *Arch. Biochem. Biophys.*, 142, 184
Schutt, H. and Holzer, H. (1972). *Eur. J. Biochem.*, **26**, 68
Schwartz, J. H., Reeves, J. Y., and Broome, J. D. (1966). *Proc. Natl. Acad. Sci. U.S.A.*, **56**, 1516
Schwartz, J. H. (1971). *Col. Int. CNRS*, Report No. 197, p. 79
Smirnova, I. P., Zanin, V. A., and Berezov, T. T. (1976). *Prikl. Biokhim. Mikrobiol.*, **12**, 14
Soda, K., Ohshima, M., and Yamamoto, T. Y. (1972). *Biochem. Biophys. Res. Commun.*, **46**, 1278
Sodek, L., Lea, P. J., and Miflin, B. J. (1980). *Plant Physiol.*, **65**, 22
Soru, E. and Zaharia, O. (1973). *Rev. Roumaine Biochim.*, **10**, 145
Stadtman, E. R. (1970). In *The Enzymes*, 3rd edn, Vol 1 (P. D. Boyer, ed.), Academic Press, New York, pp. 397–459
Steensholt, G. (1944). *Acta Physiol. Scand.*, **8**, 342
Tempest, D. W., Meers, J. L., and Brown, C. M. (1970). *Biochem. J.*, **117**, 405
Tempest, D. W. and Neijssel, O. M. (1976). In *Continuous Culture 6: Applications and New Fields* (A. C. R. Dean, D. C. Ellwood, C. G. T. Evans, and J. Melling, eds), Ellis Horwood, Chichester, pp. 283–296
Tosa, T., Sano, R., Yamamoto, K., Nakamura, M., and Chibata, I. (1972). *Biochemistry*, **11**, 217
Tosa, T., Sano, R., Yamamoto, K., Nakamura, M., Ando, K., and Chibata, I. (1971). *Appl. Microbiol.*, **22**, 387
Turcotte, R. and Quevillon, M. (1975). *Ann. Microbiol.*, **126B**, 181
Wade, H. E. (1977). *Develop. Biol. Standard.*, **38**, 73
Wade, H. E., Robinson, H. K. and Phillips, B. W. (1971*a*). *J. Gen. Microbiol.*, **65**, 1
Wade, H. E., Robinson, H. K., and Phillips, B. W. (1971*b*). *J. Gen. Microbiol.*, **69**, 299
Wade, H. E., Elsworth, R., Herbert, D., Keppie, J., and Sargeant, K. (1968). *Lancet*, **II**, 776
Willis, R. C. and Woolfolk, C. A. (1974). *J. Bacteriol.*, **118**, 231
Willis, R. C. and Woolfolk, C. A. (1975). *J. Bacteriol.*, **123**, 937
Yamamoto, S. and Hirovka, H. (1974). *J. Ferment. Technol.*, **52**, 564
Zanin, V. A., Berezov, T. T., Nikolaev, A. Ya., and Abdumalikov, A. Kh. (1969). *Microbiology*, **38**, 432

Microorganisms and Nitrogen Sources
Edited by J. W. Payne

Section 4.5

Transport and Utilization of Nitrogen Sources by Algae

W. D. P. STEWART

Department of Biological Sciences, University of Dundee, Dundee, UK

1 INTRODUCTION

The algae are a diverse assemblage of organisms, most of which evolve O_2 by a higher plant type of photosynthesis, and which range in size from microscopic unicellular forms a few micrometers in diameter to large thallose marine forms such as *Macrocystis*, which may be 60 m long (see Chapman and Chapman, 1973; Round, 1973; Bold and Wynne, 1977). There is some evidence that the prokaryotic blue-green algae (see Fogg *et al*., 1973) may be better classified as bacteria (cyanobacteria) (Stanier and Cohen-Bazire, 1977), although there is controversy on this point (see for example Lewin, 1976). The use of algae, particularly unicellular and coenocytic forms, has led to substantial advances in our knowledge of transport processes in plants (see Lüttge and Pitman, 1976) although compared with the information on the exchange of ions such as Na^+ and K^+ (see MacRobbie, 1974; Hope and Walker, 1975; Raven, 1976*a*) and organic compounds such as hexoses (see for example Tanner *et al*., 1974, 1977), less is known about the mechanisms of transport of nitrogenous compounds in algae. Nitrogen metabolism on the other hand, has been studied extensively. Recent reviews covering parts of the subject area considered in this chapter are available (Healey, 1973; Hellebust, 1974; Morris, 1974; Miflin, and Lea, 1976, 1977; Raven, 1976*a*; Barber, 1979).

2 SOURCES OF NITROGEN FOR GROWTH

Algae, depending on type, may assimilate nitrogen sources ranging from N_2 to amino acids, amides, and peptides. Those algae which utilize N_2 and inorganic nitrogen ions can use these as sole nitrogen sources for growth, but although various forms assimilate organic nitrogen, many cannot use this as sole nitrogen source for growth.

Only the blue-green algae fix N_2 and four main groups can be distinguished: (1) unicellular forms which reproduce by binary fission, only a few of which have nitrogenase—all which fix N_2 do so anaerobically, a few strains also fix aerobically; (2) the pleurocapsalean algae which reproduce by multiple fission, about half of which fix N_2, but under anaerobic conditions only; (3) the non-heterocystous filamentous forms where again over half fix N_2, but anaerobically only, and (4) the heterocystous forms, virtually all of which fix N_2 aerobically and anaerobically. Growth on N_2 is usually as good as on combined nitrogen. Various reviews on N_2-fixation by algae are available (Stewart, 1973; Fogg, 1974; Rippka and Waterbury, 1977; Haselkorn, 1978; Stewart *et al.*, 1978, 1979). There is no evidence of N_2-fixation by eukaryotic algae and early unconfirmed reports (see Fogg, 1953) can be attributed to the effects of contaminating bacteria and/or poor methodology.

Most algae utilize NO_3^- or NH_4^+ equally well as sole nitrogen source for growth. NO_2^- can be assimilated but is often (see Lui and Roels, 1972) but not always (see Morris, 1974) extremely toxic. When provided with NO_3^- and NH_4^+ simultaneously most algae assimilate NH_4^+ more rapidly initially, and in the longer term they assimilate NH_4^+ rather than NO_3^- because NH_4^+ represses synthesis of nitrate reductase. There are a few reports of algae growing better on NO_3^- (see for example Proctor, 1957; Stross, 1963; Cain, 1965) but this is unusual. A strain of *Cyanidium*, unable to utilize NO_3^- has also been reported (Rigano *et al.*, 1975). Lui and Roels (1972) observed with the centric diatom *Biddulphia aurita* that growth was equally good on NO_3^-, NO_2^- or NH_4^+ and that irrespective of the nitrogen source the algae showed the same intracellular levels of soluble free amino nitrogen, combined amino nitrogen, and NH_4^+ nitrogen, although soluble amide nitrogen was higher in the NH_4^+-grown cells. Uptake of NO_3^- results in a pH increase in the medium; uptake of NH_4^+ results in a pH drop (see below). Care must be taken to ensure that the effects on growth are due directly to the type of nitrogen supplied and not attributable solely to pH changes resulting from its assimilation. Studies on the modelling of inorganic nitrogen utilization by algae include those of Strickland *et al.* (1969), Malone *et al.* (1975), Macisaac and Dugdale (1976), Mur *et al.* (1978), Paelinck and De Maeseneer (1978), Zevenboom and Mur (1978).

The extent to which organic nitrogen supports growth depends largely on the algal group and often upon the environment from which the organism was isolated. Early work by Allen (1952) and more recently by Wyatt *et al.* (1971) suggested that amino acids and other organic nitrogenous compounds supported good growth of various blue-green algae; this has not been substantiated (see Neilson and Doudoroff, 1973; Smith, 1973). It seems from ^{14}C tracer studies that in general blue-green algae assimilate organic nitrogenous compounds poorly and that even when supplied under otherwise optimum growth conditions these cannot serve as sole nitrogen sources for rapid algal growth. This is believed to be a permeability problem (Smith, 1973). Compounds assimilated include arginine, aspartate, glutamate, glycine, leucine, threonine (see Smith, 1973), glutamine (Rowell *et al*., 1977), and the nucleic acid precursors uracil, thymine, orotic acid, adenine, guanine, thymidine, guanosine, and uridine-5-phosphate (Pigott and Carr, 1971).

The ability of eukaryotic algae to grow on organic nitrogen as sole nitrogen source varies greatly. For example, Bollard (1966) noted that *Chlorella* grew on most protein amino acids except cysteine, histidine, hydroxyproline, lysine, methionine, phenylalanine, and tryptophan and on various peptides and amides, but that, in general, amines and amino alcohols did not support growth. Some other *Chlorella* strains and *Ankistrodesmus* do not assimilate amino acids (den Dooren de Jong, 1967, 1969). Lynch and Gillmor (1966) obtained no evidence of glutamic acid utilization by *Chlorella pyrenoidosa*. Kirk and Kirk (1978*a,b*) observed uptake of arginine, but not of other amino acids, by strains of *Volvox* and *Chlamydomonas* although neither alga could use arginine as sole nitrogen source. Various amino acids and amides are assimilated by species of marine phytoplankton (see Sloan and Strickland, 1966; Hellebust and Guillard, 1967; North and Stephens, 1967, 1969, 1971, 1972; Hellebust, 1970; Stephens and North, 1971; McCarthy, 1972; Carpenter *et al*., 1972; Liu and Hellebust, 1974*a,b*, 1976*a,b*; Williams and Hodson, 1977) and such algae seem, in general, to use amino compounds as nitrogen sources in the nutrient-poor waters where they occur. Urea is a good nitrogen source for many eukaryotic microalgae (Guillard, 1963; McCarthy, 1972; Syrett and Leftley, 1976; Antia *et al*., 1975, 1977; Kirk and Kirk, 1978*a,b,c*). It may be noted that growth of an alga on an organic compound does not necessarily mean that this compound is being assimilated directly since some algae produce extracellular enzymes which degrade organic compounds prior to assimilation (Pringsheim, 1952; Belmont and Miller, 1965). Contaminating bacteria may do likewise (see, for example, Section 3.1).

3 MOVEMENT ACROSS MEMBRANES

Nitrogenous compounds may move to and from the external medium via the plasmalemma; in eukaryotes they also move to and from the vacuole via the tonoplast, and to and from subcellular organelles such as the chloroplast. Transport across membranes depends on many factors, for example, the membrane itself (e.g. the extent to which lipid bilayers act as barriers and the proteinaceous layer catalyses transmembrane movement); on the nature of the transported compounds (e.g. whether they are undissociated, ionized, hydrophilic, etc.); on other transportable components present (e.g. is there competition for uptake); on external and internal concentrations; on the metabolic state of the cell; on whether the process is active or passive, etc. (see Lüttge and Pitman, 1976; Marre and Ciferri, 1977).

Dinitrogen, which constitutes about 80 per cent of the atmosphere, permeates lipid bilayers readily and enters vegetative cells of N_2-fixing blue-green algae by passive diffusion. Entry into heterocysts may be via the pores connecting the vegetative cells to heterocysts, rather than via the heterocyst wall.

Ammonia (i.e. NH_3 or NH_4^+) is assimilated rapidly by algae (see for example Cooper and Osterhout, 1930; Jacques, 1939*a,b*; Stewart, 1964; Kesseler, 1966; Eppley and Rogers, 1970; Pribil and Kotyk, 1970; Abeliovich and Azov, 1976; Meeks *et al*., 1977; Smith and Walker, 1978; Smith *et al*., 1978). The uptake mechanisms are difficult to study because of the presence in solution of both undissociated NH_3 and of the cation, NH_4^+, with the proportion of each depending on the pH thus:

$$\log \frac{[NH_3]}{[NH_4^+]} = pH - pK_a.$$

The pK_a is approximately 9.25 at 25 °C (see also Section 4.2.1). It is generally accepted that NH_3 may freely permeate the plasmalemma by passive diffusion and that at high pH values (above 9.25) this is probably the main mechanism of uptake. However, at lower pH values where the cationic form is more abundant other mechanisms are also involved. Much of the evidence for this comes from using the NH_4^+ analogue, [^{14}C]-methylamine which is relatively non-toxic, non-metabolizable in short-term experiments and which, according to studies using $^{15}NH_4^+$ and ^{14}C-labelled methylamine, is assimilated at rather similar rates to NH_4^+

(Pateman *et al.*, 1974; Pateman and Kinghorn, 1976) (see also Section 4.2.2). In *Hydrodictyon* (Smith *et al.*, 1978), for example, influx of methylamine is light-dependent, temperature-sensitive, occurs against a concentration gradient, and is decreased by NH_4^+, with prolonged influx resulting in an increase in $CH_3NH_3^+$ and an efflux of K^+, Na^+, and H^+. Smith *et al.* (1978) argue that it is not necessary to postulate active uptake of methylamine, and conclude for *Hydrodictyon*, like Smith and Walker (1978) for *Chara corallina*, that at pH values below about 9 methylamine penetrates the cells mainly as $CH_3NH_3^+$ and that a passive electrogenic uniport of this ion is the main mechanism of uptake.

The assimilation of NO_3^- and NO_2^- by algae are both active processes. Thus, uptake rates are higher in the light than dark (Morris and Ahmed, 1969; Haines and Wheeler, 1978), uncouplers inhibit NO_2^- assimilation (Ahmed and Morris, 1967; Kessler *et al.*, 1970), on illumination in the presence of NO_3^- there is a transitory release of NO_2^- (Strotmann and Reid, 1969), uptake is linked to photosystem II activity (Thacker and Syrett, 1972), the stimulatory effects of light on uptake are too rapid to be attributable to effects on enzyme synthesis and turnover, calculations suggest that the rates of NO_3^- uptake cannot be attributed solely to rates of utilization by nitrate reductase (Raven, 1976*a*) and algae accumulate intracellular NO_3^- concentrations much in excess of those present in the external medium (Chapman and Craigie, 1977). Uptake of NO_3^- and NO_2^- by higher plant chloroplasts is also active (see for example Magalhaes *et al.*, 1974; Canvin and Atkins, 1974). The overall results for algae suggest that ATP, rather than free energy across the plasmalemma, is the energy source for active transport. In plants NO_3^- transport and metabolism are usually tightly coupled (see Butz and Jackson, 1977) although NO_3^- uptake systems which are independent of nitrate reductase have occasionally been reported (Heimer and Filner, 1971; Schloemer and Garrett, 1974). Butz and Jackson (1977) propose a model which they consider to be applicable to algae and to higher plants in which tight coupling of the two processes is envisaged. They suggest that in algae like *Chlorella* the dimeric nitrate reductase molecule spans the algal plasmalemma, with one monomer exposed to the outside and the other to the inside (the *Chlorella* molecule is 8.9 nm long and the plasmalemma 9 nm thick, they point out), that nitrate reductase is catalytically active only when NO_3^- is being transported and that transport and activity are coupled in a 1 : 1 ratio. It is suggested that regulation involves a NO_3^--activated ATP-ase associated with the nitrate reductase monomer and to which nitrate reductase is tightly coupled. This is adjacent to the NAD(P)H binding site and thus ADP binding prevents NAD(P)H binding and thus NO_3^- reduction. Light-generated thiol, on the other hand, is thought to alleviate ADP inhibition allowing reduction to continue. This theoretical model can be fitted to various experimental facts, for example, the known ATP requirement for nitrate reductase, the reactivation of deactivated nitrate reductase by thiol (Eaglesham and Hewitt, 1971), and the fact that illumination increases the thiol concentration in chloroplasts (Ueda, 1966) and affects dithiols in algae (Anderson and Avron, 1976; Anderson and Duggan, 1976; Wolosiuk and Buchanan, 1977). There are various difficulties, however, in equating what may go on in chloroplasts with what goes on in whole algae which themselves have chloroplasts. Solomonson and Spehar (1977) propose that NO_3^- assimilation and CO_2 fixation are coordinately controlled and that cyanide produced from hydroxylamine (from NO_2^- reduction) and glyoxylate (from glycollate oxidation) act as regulators of NO_3^- reduction to NH_4^+. Okabe *et al.* (1979) have evidence that *in vitro* hydroxylamine inhibits ribulose-1,5-bisphosphate oxygenase activity and stimulates ribulose-1,5-bisphosphate carboxylase activity and their data indicate that it may not be necessary to implicate cyanide in the regulation of nitrate reductase. The enhanced O_2 evolution noted by various other workers in the presence of NO_3^- and attributed to NO_3^- acting as a terminal electron acceptor may be attributable, in part at least, to NO_3^- leading to an inhibition of ribulose-1,5-bisphosphate oxygenase activity.

Losada, on the basis of studies on *Chlorella fusca* and *Chlamydomonas*, suggests that NH_4^+ uncouples non-cyclic photophosphorylation which increases intracellular reductant supply (Chaparro *et al.*, 1976) and ADP (Losada, 1974) and that these inhibit nitrate reductase. Diez *et al.* (1977), studying *Ankistrodesmus braunii* nitrate reductase, reach a rather similar conclusion finding that ADP in the presence of Mg^{2+} shows a cooperative effect with NADH in causing inactivation of the enzyme *in vitro*.

Topinka (1978) found that *Fucus spiralis* utilized NO_3^-, NH_4^+, and NO_2^- with uptake being higher in the apices and young thalli and lower in older parts and plants. At comparable nitrogen concentrations uptake of NH_4^+ was more rapid than NO_3^- uptake. K_s values were: NH_4^+ (6–10 μM), NO_3^- (7–13 μM), and NO_2^- (26–28 μM). Haines and Wheeler (1978), in a detailed study, found NO_3^- uptake by *Hypnea musciformis* had a K_s of 4.9 μg atom N l^{-1} with a V_{max} of 2.85 μg atom N (g wet weight)$^{-1}$ h^{-1}, whereas NH_4^+ saturation did not occur at concentrations up to 20 μg atom N l^{-1}. NO_3^- uptake was reduced by 50 per cent when NH_4^+ was also available but NO_3^- had no effect on NH_4^+ uptake. Uptake rates of both ions were reduced in the dark by over 50 per cent. NO_3^- uptake by *Macrocystis pyrifera* saturated at 13.1 μg atom N l^{-1} and showed a V_{max} of 3.05 μg atom N (g wet weight)$^{-1}$ h^{-1}. Uptake of NH_4^+ was particularly interesting in that it showed saturatable kinetics at concentrations below 22 μg atom N l^{-1} but that at higher concentrations NH_4^+ was taken up linearly. The presence of NO_3^- did not affect NH_4^+ uptake and *vice versa*. Similarly with *Gracilaria* and *Neoagarrdiella* uptake did not saturate at concentrations as high as

40 μg atom N l^{-1} (see De Boer *et al*., 1978; D'Elia and De Boer, 1978). These workers suggest that uptake of NH_4^+ at high concentrations is mainly by diffusion whereas at low concentrations a high-affinity component is involved in uptake.

Sources of ATP which support NO_3^- uptake by algae are non-cyclic photophosphorylation, cyclic photophosphorylation alone, oxidative phosphorylation (Kessler, 1957; Czygan, 1963), and, when H_2 is utilized (Kessler, 1957), ATP can be generated by an oxyhydrogen reaction (see also Raven, 1976*b*). In *Ankistrodesmus braunii* (Ullrich, 1971, 1972, 1974), *Hydrodictyon africanum* (Raven and Glidewell, 1975; Raven, 1977), and *Chlorella* (Thomas *et al*., 1976) NO_3^- may also act as a terminal electron acceptor for non-cyclic electron flow. Raven (1976*b*, 1977) concludes that when algae such as *Chlorella* or *Hydrodictyon* are growing photolithotrophically up to 20 per cent of the net non-cyclic electron flow could be associated with NO_3^- rather than CO_2 as electron acceptor, and that in *Hydrodictyon* all the ATP required for photolithotrophic growth could be supplied by non-cyclic electron flow to CO_2 and NO_3^-.

Amino acid uptake in most (not all, see Wallen, 1973) algae is active. Thus, amino acids, when assimilated, may accumulate against a concentration gradient, the uptake rates obey Michaelis–Menten kinetics (which vary with the amino acid), metabolic inhibitors such as cyanide and dinitrophenol inhibit uptake and uptake is temperature-dependent (see North and Stephens, 1972; Liu and Hellebust, 1974*a,b*). In *Cyclotella cryptica* (Liu and Hellebust, 1974*a,b*) uptake of leucine, unlike uptake of most other amino acids, appears to be by diffusion (Michaelis–Menten kinetics are not obeyed and uptake rates increase linearly with increasing exogenous concentration) although the fact that uptake does not occur in the presence of metabolic inhibitors also suggests the possible involvement of an active process. The early findings that various diatoms and dinoflagellates assimilated lysine better than alanine (Hellebust, 1970) and that *Platymonas* assimilated arginine better than glycine, which in turn was assimilated better than alanine (North and Stephens, 1971), led to the suggestion that different amino acid transport systems occurred in algae. This has been verified by detailed studies, particularly on *Melosira nummuloides*, *Nitzschia ovalis*, and *Cyclotella cryptica*. In *M. nummuloides*, Hellebust (1970) detected at least three amino acid transporting systems—one for basic amino acids, one (or possibly more) for neutral amino acids, and one for dicarboxylic acids. North and Stephens (1972) using *N. ovalis* concluded rather similarly. With *C. cryptica* (Liu and Hellebust, 1974*a*) data obtained using pairs of amino acids suggest four transport systems: one for acidic amino acids, one for basics, one for neutral amino acids, and another general one for alanine, glycine, proline, aspartate, and glutamine.

In contrast to the above findings, the freshwater *Volvox carteri* f. *hagariensis* (Kirk and Kirk, 1978*a*) and *Chlamydomonas reinhardtii* strain 89 (Kirk and Kirk, 1978*b*) both transport arginine only, although arginine cannot serve as the sole nitrogen source for either alga and neither is it assimilated by the same carrier as urea which can serve as sole nitrogen source. In *Volvox* there is a temporal variation in the rates of arginine and urea uptake during the synchronous growth of the asexual life-cycle. Kirk and Kirk (1978*c*) conclude from studies on eight other unicellular and multicellular algae belonging to the Volvocales and Chlorococcales that all algae studied have arginine carriers and that some also possess carriers for neutral and/or acidic amino acids. They consider that while the capacity to transport all amino acids may be characteristic of certain oceanic and neritic algae which occur in nutrient-poor environments where amino acids could play an important role as nitrogen sources, there may have been a relaxation of the selective pressure to retain a full complement of amino acid scavenging systems, and that the arginine carrier found in organisms such as *Volvox* today may be the only system not yet lost through genetic drift.

In the macroalga *Nitella flexilis*, the mechanism of amino acid uptake is as complex as in microalgae. Uptake of neutral amino acids (glycine, alanine, and valine) appears to be by diffusion, but the kinetics of uptake, the strong inhibition by CCCP, dinitrophenol, and iodoacetamide, and the temperature dependency of aspartic acid uptake suggest that it is an active process. Arginine uptake involves active and possibly also passive mechanisms (Wallen, 1973).

Sullivan (1979) suggests that Na^+-dependent cotransport of organic acids may be involved in the assimilation of organic compounds by diatoms and there is much detailed evidence that in bacteria a close coupling of amino acid and inorganic ion transport may occur in certain instances (Thompson and Macleod, 1973; Heinz, 1974 Kahane *et al*., 1975) (see also Sections 1.1.3.1.xii, 1.2.4, 1.4.2.5, and 1.7.3). Hellebust (1978) using *Cyclotella cryptica* has shown that the uptake of alanine (neutral), glutamic acid (acidic) and arginine (basic) all occur only in the presence of NaCl concentrations which are higher than those required for growth or photosynthesis. Uptake shows a hyperbolic dependence on NaCl concentration and is inhibited by the Na^+-ionophore monensin, and by increasing concentrations of KCl. This study, together with the fact that the proton symport system which allows active transport in freshwater algae like *Chlorella* (see Komor *et al*., 1972; Komor and Tanner, 1974; Tanner *et al*., 1974) is unlikely to be involved in the marine *Cyclotella*, because of the high pH of sea water, suggests that Na^+ may be involved in organic compound transport in *Cyclotella*. Hellebust (1978), however, is cautious in interpreting his data in terms of cotransport, pointing out that it is also possible, for example, that the requirement for high external

Na^+ may simply refect a Na^+ requirement for effective binding of the organic solute to carrier molecules (see Thompson and MacCleod, 1971).

The possibility that the γ-glutamyl cycle (see Section 4.1), which has been proposed as a likely mechanism of amino acid uptake by rat kidney (Meister, 1973) and by higher plants (Mazelis, 1979, Mazelis and Creveling, 1978), may be involved in amino acid uptake by algae comes from the preliminary studies of Kurelec *et al*. (1977). Using natural populations of freshwater phytoplankton (mainly diatoms and dinoflagellates), these workers have demonstrated the presence in their samples of certain of the key enzymes of the cycle.

Amino acid uptake may be affected both by the nitrogen status of the alga and by the available sources of combined nitrogen. Thus, nitrogen-starved algae take up amino acids more rapidly than do nitrogen-sufficient ones and cells provided with exogenous NH_4^+ assimilate amino acids more slowly than when other sources of inorganic nitrogen are available (Liu and Hellebust, 1974*a*; North and Stephens, 1969, 1971; Wheeler *et al*., 1974; Wheeler, 1977). Also, in *Cyclotella cryptica* rapid salinity changes affect uptake with uptake and incorporation into proteins decreasing, and with proline accumulating intracellularly, with increase in salinity. Proline levels, however, are low in algae adapted to high salinity. The decreased uptake of amino acids may be related in part to the fact that on rapid salinity change the intracellular concentrations of amino acids derived from photoassimilated, or exogenous carbon, increase (Liu and Hellebust, 1976*a,b*).

Urea is an excellent nitrogen source for the growth of various algae (see above) and in the case of algae such as *Dunaliella* and *Phaeodactylum*, which have efficient uptake mechanisms which saturate at low external concentrations of urea, it may serve as an important nitrogen source in nature (Syrett and Leftley, 1976). Syrett and Bekheet (1977) who used *Chlorella fusca* var. *vacuolata* and the non-metabolizable analogue of urea, thiourea (see Pateman *et al*., 1973) conclude that uptake of urea involves both a metabolic process and passive diffusion, and McCarthy (1972) using *Ditylum brightwellii* found that organisms which do not metabolize urea assimilate it as efficiently as clones which do, suggesting that the mechanisms of uptake and metabolism of urea are separate events.

Urea metabolism in algae proceeds via one or other of the two following enzyme systems:

(a) urease, which catalyses

$$CO(NH_2)_2 + H_2O \longrightarrow CO_2 + 2NH_3;$$

(b) ATP: urea amidolyase (UALase), which catalyses the ATP-dependent degradation of urea to NH_4^+ and bicarbonate via two enzymes thus: (i) urea carboxylase (urea:CO_2-ligase) (ADP) which catalyses

$$\text{Urea} + \text{ATP} + HCO_3^- \xrightarrow[\text{Biotin}]{Mg^{2+},\ K^+} \text{Allophanate} + \text{ADP} + P_i$$

and (ii) allophanate hydrolyase (allophanate amidohydrolase), which catalyses

$$\text{Allophanate} + 3H_2O + H^+ \longrightarrow 2NH_4^+ + 2HCO_3^-$$

(see Roon and Levenberg, 1968; Thompson and Muenster, 1971; Whitney and Cooper, 1972). Leftley and Syrett (1973) and Syrett and Leftley (1976) have investigated the distribution of both enzymes in a variety of algae from the Chlorophyceae, Prasinophyceae, Xanthophyceae, Chrysophyceae, Bacillariophyceae, Cryptophyceae, Euglenophyceae, and Cyanophyceae. They conclude (Syrett and Leftley, 1976) that any one organism may contain urease or UALase but not both, with the Chlorophyceae having UALase only, and the others urease only (see also Thomas and Syrett, 1976).

4 ACCUMULATION AND STORAGE

Nitrogen constitutes about 1–3 per cent of the dry weight of eukaryotic algae (Round, 1973) and about 3–8 per cent of prokaryotic algae (Fogg *et al*., 1973). Under certain conditions, particularly when an excess of nitrogen is available, algae may store nitrogen which is not immediately required for metabolism in various storage 'pools'. In blue-green algae, where there are no discrete membrane-bound vacuoles into which excess nitrogen can be readily channelled, the major storage pools are the phycobiliproteins, and the structured granules which are copolymers of aspartic acid and arginine.

The phycobiliproteins are water-soluble pigments (phycocyanin, allophycocyanin, and sometimes phycoerythrin) which are aggregated in discrete structures (phycobilisomes) and which serve both as photosystem II light-harvesting and -transfer pigments and as nitrogen reserves (see Fogg *et al*., 1973). They are synthesized in the light and sometimes in the dark (Raboy *et al*., 1976), accumulate under conditions of nitrogen sufficiency, and are rapidly mobilized (within hours) on nitrogen starvation (see for example Stewart and Lex, 1970; Van Gorkom and Donze, 1971). Van Gorkom and Donze (1971) used the pattern of re-appearance of phycocyanin in *Anabaena* on re-admitting N_2 after nitrogen starvation to gauge the locus of nitrogenase activity. They found that in the presence of O_2 phycocyanin first reappeared in the vegetative cells adjacent to heterocysts, whereas under anaerobic conditions it developed equally in all vegetative cells. This accords with the generally accepted view that heterocystous algae fix N_2 in heterocysts under aerobic conditions, but that under anaerobic conditions the vegetative cells may also do so (see Stanier and Cohen-Bazire, 1977). Structured granules (cyanophycin granules) are densely osmiophilic structures composed of aspartic acid and arginine in a 1 : 1 molar ratio. The branched polypeptide has a polyaspartic acid core with arginyl residues attached to its free carboxyl groups (Simon, 1971; Simon and Weathers, 1976). Cyanophycin develops under conditions of nitrogen sufficiency, particularly in post-log phase cultures (Simon, 1971; Stewart *et al*., 1977), and is synthesized by a non-ribosomal process (Simon, 1973).

The amount of nitrogen stored by blue-green algae may vary with cell type. Akinetes, which develop adjacent to, or remote from heterocysts, depending on the species (see Fritsch, 1945) have a lower nitrogen content (4.8 per cent) than whole filaments (7.7 per cent) of *Anabaena cylindrica* (Fay, 1969). This decreased nitrogen content is largely attributable to the presence of the thick carbohydrate-rich wall of akinetes, but they are also depleted in phycocyanin (Fay, 1969) although they do contain numerous structured granules. Heterocysts, likewise, have a lower nitrogen content than vegetative cells, being depleted in various vegetative cell proteins, and in phycocyanin, and having a thick carbon-rich envelope (see Haselkorn, 1978). In algae such as *Tolypothrix* where the filaments terminate in hairs, the extent of hair formation can be altered by the availability of combined nitrogen. When an excess of nitrogen is available the nitrogen-depleted terminal hairs do not develop (Stein, 1963).

In eukaryotic algae excess nitrogen can be stored in the vacuoles which may occupy from about 10 per cent of the cell volume in algae like *Chlorella* to 80–90 per cent in certain giant-celled algae (Atkinson *et al*., 1974; Hope and Walker, 1975). Such vacuoles are bounded by a unit membrane, have a large surface area for interchange with the cytoplasm and in most algae, of which *Laminaria* is exemplary, accumulate nitrogen, particularly NO_3^-, much in excess of external concentrations, by what appears to be an active process. For example, *Valonia* and *Halicystis* show NO_3^- concentrations in their vacuoles which are 500–2000 times those of the surrounding water (Jacques and Osterhout, 1938) and *Laminaria* may have NO_3^- levels which are thousands of times those in sea water. Chapman *et al*. (1978) observed that the growth rate of *Laminaria saccharina* increased linearly with increase in NO_3^- concentration up to 10 μM and that above this external concentration intracellular NO_3^- increased markedly. Wheeler (1977) has shown that the rates of uptake of NO_3^- vary depending on the velocity of the water in which the alga grows. Such NO_3^- concentrations in *Laminaria* are used to sustain growth when NO_3^- levels in the sea are depleted, and when other environmental conditions for growth are good (Chapman and Craigie, 1977; Mann, 1978). There is evidence that the planktonic unicellular dinoflagellates which show vertical migration assimilate most nitrogen in the lower nutrient-rich waters at night, presumably store it in vacuoles, and use it up in daytime in the surface nutrient-poorer waters where they photosynthesise most actively (Eppley *et al*., 1968, 1969; Harrison, 1976; Bhovichitra and Swift, 1977). In other unicellular eukaryotic algae there is evidence of accumulation of amino acids in different pools. Thus Wheeler and Stephens (1977) have shown, using the chlorophyte *Platymonas cordiformis*, and measuring turnover rates by use of ^{14}C-labelled amino acids, that there are at least two functionally distinct amino acid pools, a 'metabolic' pool which is rapidly metabolized and in which alanine occurs, and a storage pool which contains lysine and arginine. Such studies do not localize the pools spatially, but the authors speculate that the vacuole may contain the storage pool and the cytoplasm the metabolic pool.

5 ENZYMES OF NH_4^+ ASSIMILATION

A variety of NH_4^+-assimilating enzymes has been reported in algae (see also Section 4.2) with most attention being paid to:

(1) *glutamic acid dehydrogenase*, which catalyses

$$\text{2-Oxoglutarate} + NH_4^+ + NAD(P)H \longrightarrow \text{L-Glutamate} + NAD(P)^+ + H_2O$$

(2) *the glutamine synthetase–glutamate synthase pathway*, which assimilates NH_4^+ via glutamine synthetase (GS) which catalyses

$$\text{Glutamate} + NH_4^+ + ATP \xrightarrow{Mg^{2+}} \text{Glutamine} + ADP + P_i$$

and glutamate synthase (glutamine (amide):2-oxoglutarate amino transferase) (GOGAT) which catalyses

$$\text{Glutamine} + \text{2-oxoglutarate} \xrightarrow{Fd_{red}/Fd_{ox}} \text{2 Glutamate}$$

Collectively NH_4^+ is assimilated into glutamate thus:

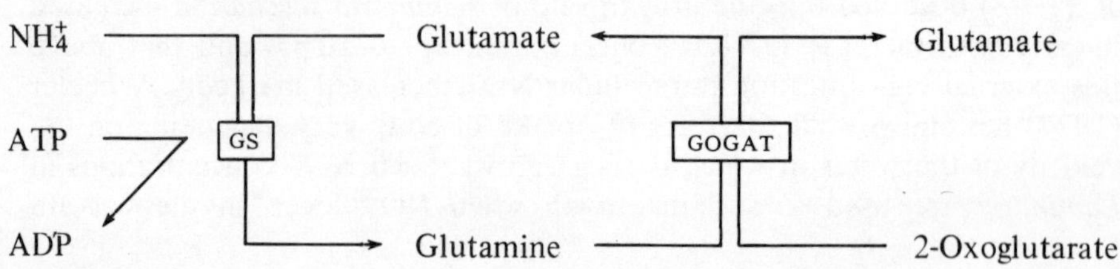

(3) *alanine dehydrogenase*, which catalyses

$$\text{L-Alanine} + NAD^+ + H_2O \longrightarrow \text{Pyruvate} + NADH + NH_4^+ + H^+$$

(4) *carbamoyl phosphate synthetase*, which catalyses

$$CO_2 + NH_4^+ + 2ATP + H_2O \longrightarrow \text{Carbamoyl phosphate} + 2ADP + P_i$$

Glutamic acid dehydrogenase activity is detectable in some (but not all) blue-green algae (see Neilson and Doudoroff, 1973). Even when detected its activity is usually very low and there is sometimes controversy over whether or not it is present in particular algae. For example, Hoare *et al.* (1967) failed to detect it in *Anacystis nidulans*, whereas Meeks *et al.* (1977) speculate that it is a major source of glutamate production in the same alga. However, activity is invariably detected in eukaryotic algae, for example in *Chlorella* (Morris and Syrett, 1965; Kretovich *et al.*, 1970; Talley *et al.*, 1972), *Ditylum brightwellii* (Eppley and Rogers, 1970), *Cyanidium caldarium* (Rigano *et al.*, 1975), *Skeletonema costatum* (Falkowski and Rivkin, 1976), and *Platymonas striata* (Edge and Ricketts, 1978). In *Chlorella* isoenzymes are found. Thus, Shatilov *et al.* (1970) propose on the basis that NADPH-dependent activity is induced by NH_4^+ while NADH-dependent activity remains constant that *C. pyrenoidosa* may have two forms of the enzyme. Talley *et al.* (1972), using a different *Chlorella* strain, isolated both NADH- and NADPH- dependent enzymes with molecular mass of 179 000 and 269 000 daltons respectively. The NADH-specific enzyme is produced in NO_3^--grown cultures and the NADPH-specific enzyme is induced by NH_4^+. The *Ditylum* enzyme is $NADP^+$-linked irrespective of nitrogen source (Eppley and Rogers, 1970). The eukaryotic algal enzymes usually have a high K_m for NH_4^+, for example 28 mM in *Skeletonema costatum* (Falkowski and Rivkin, 1976), although Morris and Syrett (1965) observed a very low K_m (0.5 mM) for a *Chlorella* enzyme.

Glutamine synthetase has been purified from several algae (Stacey *et al.*, 1977; Rowell *et al.*, 1978; Sampaio *et al.*, 1979). The *Anabaena* and *Nostoc* enzymes (Sampaio *et al.*, 1979) are both dodecameric enzymes, composed of 12 subunits, each of molecular mass 49 000–50 000 daltons, arranged in two superimposed hexagonal rings. Purification requires the presence of stabilizing ligands (Mg^{2+}, glutamate, and a thiol reagent). There is no evidence of regulation via adenylylation–deadenylylation, and in this respect the enzyme differs from that of other groups of N_2-fixing prokaryotes, e.g. free-living heterotrophic bacteria (Tronick *et al.*, 1973), *Rhizobium* (Rao *et al.*, 1978), and photosynthetic bacteria (Johansson and Gest, 1977). In *Anabaena cylindrica*, however, glutamine synthetase can be regulated by light and NH_4^+. Thus, on darkening the alga or adding NH_4^+ glutamine synthetase activity is reduced by about 50 per cent within 5–6 h and recovers within a few hours on re-exposure of the alga to light or removal of NH_4^+. Furthermore, so long as the alga remains metabolically active the deactivated enzyme can be immediately reactivated (within 5 min) by adding a thiol reagent such as dithiothreitol (Rowell *et al.*, 1979). This regulation of glutamine synthetase in response to light resembles the regulation of enzymes of the oxidative and reductive pentose phosphate pathways in response to light noted by others (Anderson and Avron, 1976; Anderson and Duggan, 1976; Duggan and Anderson, 1975; Wolosiuk and Buchanan, 1977). Light–dark regulation of higher plant glutamine synthetase has also been noted (Rhodes *et al.*, 1979). Glutamine synthetase of eukaryotic algae is smaller than that of prokaryotes and in some species at least is composed of eight monomers arranged in two superimposed tetramers (see Gromyko *et al.*, 1974; Miflin and Lea, 1977).

Glutamate synthase has been reported from various prokaryotic blue-green algae where it is ferredoxin-linked rather than NADP- or NAD-linked (Lea and Miflin, 1975; Meeks *et al.*, 1977) and although it was not detected in early enzymic studies on eukaryotic algae (Brown *et al.*, 1974) early tracer studies have provided evidence which fits the glutamine synthetase–glutamate synthase pathway of NH_4^+ assimilation by algae. Thus Baker and Thompson (1962) showed that nitrogen-depleted *Chlorella* first accumulated ^{15}N into the amide group of glutamine and more recently, Kanazawa *et al.*, (1972) showed, using ^{14}C as tracer, that glutamate was formed after glutamine when NH_4^+ was supplied to nitrogen-starved *Chlorella* cells. Enzymic evidence of ferredoxin-linked glutamate synthase in various eukaryotic algae including *Chlorella* (Lea and Miflin, 1975) and *Platymonas striata* (Edge and Ricketts, 1978) has now been obtained.

Alanine dehydrogenase activity has been reported in many algae (see for example Kates and Jones, 1964; Belmont and Miller, 1965; Romanov *et al.*, 1965; Shatilov *et al.*, 1968; Kretovich *et al.*, 1970; Haystead *et al.*, 1973; Scott and Fay, 1972; Batt and Brown, 1974; Neilson and Doudoroff, 1973; Rowell and Stewart, 1976). The enzyme from the prokaryotic *Anabaena cylindrica* has a molecular weight of 270 000 daltons and is composed of six identical subunits each of molecular mass 43 000 daltons. It can function both in aminating and deaminating directions, is NADH/NAD-specific and oxaloacetate can partially substitute for pyruvate. The K_m for NH_4^+ varies from 8 to 133 mM depending on the pH (Rowell and Stewart, 1976). Although carbamoyl phosphate synthase is present in various algae there is no evidence that it is of major importance in primary NH_4^+ assimilation (see Morris, 1974).

It is now accepted that in algae in general the glutamine synthetase–glutamate synthase pathway is usually the most important route of primary NH_4^+ assimilation. Evidence supporting this includes the following. First, the enzymes are invariably present and usually show sufficient activity to account for the rates of NH_4^+ assimilation, whereas glutamic acid dehydrogenase activities are not always detectable and are often low. Second, in various blue-green algae, like *Anabaena cylindrica*, the inhibition of glutamine synthetase by analogues such as L-methionine-D,L-sulphoximine (Stewart and Rowell, 1975; Ownby, 1977) or 5-hydroxylysine (Ladha *et al*., 1978) results in NH_4^+ liberation into the medium with the liberated NH_4^+ accounting for most of the nitrogen assimilated. Third, ^{15}N (Stewart *et al*., 1975) and ^{13}N (Wolk *et al*., 1976; Thomas *et al*., 1977) tracer studies show initial NH_4^+ incorporation into glutamine and then into glutamate. However, other enzymes may be important in NH_4^+ assimilation in some algae. For example, in certain blue-green algae and eukaryotic algae NH_4^+ is not liberated when glutamine synthetase activity is inhibited by analogues although the algae continue to assimilate nitrogen (W. D. P. Stewart, M. J. A. M. Sampaio, and P. Rowell, unpublished). Thus in these organisms, other NH_4^+-assimilating enzymes may be operative *in vivo*, although whether these are important when glutamine synthetase is functional is uncertain. Also, in *Anacystis nidulans* Meeks *et al*. (1977) conclude on the basis of ^{13}N kinetic studies that some NH_4^+ is assimilated by glutamic acid dehydrogenase and that this may be the most important source of glutamate for the biosynthesis of glutamine. In *Anabaena variabilis* Meeks *et al*. (1977) consider that alanine dehydrogenase may be important in primary NH_4^+ assimilation, although in *Anabaena cylindrica* Rowell and Stewart (1976) obtained no evidence for its role in primary NH_4^+ assimilation and considered it to be more important in the catabolism of alanine. In eukaryotic algae the glutamine synthetase–glutamate synthase pathway is the most important route of primary NH_4^+ assimilation (Miflin and Lea, 1976, 1977). This is particularly so in natural environments low in NH_4^+ where the low K_m of glutamine synthetase for NH_4^+ (for example, 1 mM in *Anabaena cylindrica* (Dharmawardene *et al*., 1973), 29 μM in *Skeletonema costatum* (Falkowski and Rivkin, 1976)) may be particularly beneficial to the algae. At high NH_4^+ concentrations glutamic acid dehydrogenase with its higher K_m for NH_4^+ (28 mM in *Skeletonema costatum* (Falkowski and Rivkin, 1976)) may sometimes be important (see Rigano *et al*., 1975) although Meeks *et al*. (1977) note that in *A. cylindrica* the glutamine synthetase–glutamate synthase pathway is the sole route of NH_4^+ assimilation of importance both in N_2- and NH_4^+-grown cultures. Morris and Syrett (1965) report that glutamic acid dehydrogenase from a *Chorella* species has a very low K_m for NH_4^+ and in this alga it could be important in primary NH_4^+ assimilation. It seems that in photosynthetic algae where the availability of ATP is less of a problem than in heterotrophs, the advantage of the glutamine synthetase–glutamate synthase pathway with its low K_m for NH_4^+ outweighs the disadvantage of the enzyme being ATP-requiring. Nevertheless, if uncoupling by high NH_4^+ leads to a shortage of ATP for glutamine synthetase activity (see Ohmori and Hattori, 1978) then perhaps only algae with alternative NH_4^+-assimilating enzymes such as glutamic acid dehydrogenase will be successful in environments high in NH_4^+.

In addition to a major role in NH_4^+ assimilation, Stewart and Rowell (1975) and Ownby (1977) found that the inhibition of glutamine synthetase by methionine sulphoximine relieved the inhibitory effect of NH_4^+ on nitrogenase synthesis and heterocyst production, suggesting that glutamine synthetase or a product is involved, possibly together with NH_4^+, in the regulation of both.

6 CELL TO CELL TRANSPORT

Algae range from microscopic unicells to large multicellular forms. In multicellular blue-green algae evidence of symplastic transport of organic nitrogen has been demonstrated (Wolk *et al*., 1976; Thomas *et al*., 1977). This transport is probably via the microplasmodesmata which link adjacent cells (see Lang and Fay, 1971). In colonial eukaryotes symplastic transport via plasmodesmata is probably common, when plasmodesmata are present. Their presence, however, varies even within a single group. Thus, in the Chlorophyceae plasmodesmata link adjacent cells of *Volvox* (Bisalputra and Stein, 1966; Pickett-Heaps, 1970), *Nitella* (Spanswick and Costerton, 1967), *Oedogonium* (Hill and Machlis, 1968), *Chara* (Pickett-Heaps, 1968), *Bulbochaete* (Fraser and Gunning, 1969), *Ulothrix*, and *Stigeoclonium* (Floyd *et al*., 1971; Floyd, 1972) but are apparently lacking in others like *Tribonema* (Falk and Kleinig, 1968) and *Microspora* (Pickett-Heaps, 1973). Endoplasmic reticulum has been observed in some plasmodesmata, e.g. those of *Volvox* (Bisalputra and Stein, 1966; Pickett-Heaps, 1970), *Nitella* (Spanswick and Costerton, 1967), and *Chara* (Pickett-Heaps, 1967*a,b*; 1968). In coenocytic forms such as *Acetabularia* and *Codium* where considerable differentiation occurs transport is mainly via cytoplasmic streaming. In thallose macroscopic forms such as kelps, specialized cells involved in transport over considerable distances are present. These are the 'sieve elements' or 'sieve tube equivalents' (see Schmitz and Srivastava, 1974*b*, Schmitz and Lobban, 1976). In *Alaria marginata* Schmitz and Srivastava (1975) report that the sieve elements, which are derived from the inner cortical cells, mature into cells more than 20 times their original length. Such cells, which maintain vascular continuity between the lamina and the stipe show cytoplasmic strands, vacuoles, a lobed nucleus, endoplasmic reticulum, ribosomes, dictyosomes, plastids, and mitochondria. There is no P protein. The longitudinal walls are considerably thicker than the cross-walls, and both have a secondary wall layer. The cross-walls show numerous sieve pores through which the protoplasts of adjacent sieve tubes connect. The pores are lined by plasmalemma and possibly a tubular endoplasmic reticulum. Old tubes become callose occluded, their protoplasm degenerates, and cellular organelles disappear. Details of the sieve elements of *Nereocystis*, which in contrast to *Alaria* has wide sieve elements and sieve pores, are presented by Zeigler (1963), Parker and Huber (1965), Nicholson (1976), and Schmitz and Srivastava (1976). Sideman and Scheirer (1977) consider sieve elements of *Laminaria saccharina* in detail. In the latter the sieve elements have small evenly spaced pores, traversed by protoplasmic strands surrounded by plasmalemma. These workers point out how there may be a phylogenetic trend in the development of sieve tube elements on moving from *Laminaria saccharina* with its small sieve pores and nucleated sieve elements, without P protein or callose, to *L. groenlandica* → *Alaria marginata* → *Nereocystis luetkeana* → *Macrocystis pyrifera* which has the largest pores and enucleated sieve elements, which contain P protein, and which may be callose occluded.

In the Rhodophyta, pit connections occur in the Florideae but only in some stages of the life-cycles of the Bangiodeae (Bold and Wynne, 1977). There is little convincing evidence of cytoplasmic connections through the pits and it appears that the pit material (which is not invariably present, see Hartmann and Eschrich, 1969) is wall-type material and is bounded on each side by the plasmalemma of the cells which it connects (Ramus, 1971). Pueschel (1977) however, suggests that only a single unit membrane delineates the plug at either end. This, if correct, is extremely unusual and requires verification. Cellular connections and transport cells in algae in general are considered by Esau (1969) and Marchant (1976).

Some of the most detailed biochemical studies on the transport of organic nitrogen in algae are those on the interchange of metabolites between heterocysts and vegetative cells of the blue-green alga *Anabaena cylindrica*. In aerobic cultures of this alga much, if not all, of the N_2 is fixed in the heterocysts (see Stewart, 1977) which are connected to adjacent vegetative cells via a pore with microplasmodesmata traversing the plasmalemma (see Lang and Fay, 1971). The primary route of NH_4^+ assimilation is the glutamine synthetase–glutamate synthase pathway. Glutamine synthetase is active in heterocysts, has a low K_m for NH_4^+, and scavenges newly fixed NH_4^+ (Dharmawardene *et al.*, 1973). Glutamate synthase is undetectable in heterocysts but occurs in vegetative cells (Meeks *et al.*, 1977). Results of ^{15}N and ^{13}N tracer studies suggest that NH_4^+ fixed into glutamine via glutamine synthetase in heterocysts is transported from heterocysts symplastically to vegetative cells where glutamate synthase is active and glutamate is produced. Some of the glutamate produced is transported back to heterocysts to provide the NH_4^+

acceptor, the remainder being used in the production of various amino acids, mainly by transamination, in the vegetative cells (Stewart *et al.*, 1975; Wolk *et al.*, 1976; Meeks *et al.*, 1977; Thomas *et al.*, 1977).

In eukaryotic algae there is extensive evidence of transport of organics, mainly photosynthate, with the early indirect evidence of transport in *Macrocystis pyrifera* (Sargent and Landtrip, 1952), being confirmed using ^{14}C in *Macrocystis* (Parker, 1965, 1966; Schmitz and Lobban, 1976; Lobban, 1978), *Nereocystis* (Nicholson and Briggs, 1972; Schmitz and Lobban, 1976), *Laminaria* (Lüning *et al.*, 1972; Schmitz *et al.*, 1972; Schmitz and Lobban, 1976), *Alaria* (Schmitz and Srivastava, 1975; Buggeln, 1976, 1977; Schmitz and Lobban, 1976) and *Cymathere, Costaria, Agarum, Hedophyllum, Postelsia, Lessoniopsis, Pterygophora, Egregia* and *Eisenia* (Schmitz and Lobban, 1976) as well as in *Sargassum pallidum*, a member of the Fucales (Titlyanov and Peshekhodko, 1973) and in the Rhodophyceae (*Delesseria sanguinea* and *Cystoclonium purpureum*) (Hartmann and Eschrich, 1969). Also various experiments show that development of the new blade in Laminariales is initially dependent on the presence of the old blade (Lüning, 1969; Chapman and Craigie, 1978). There is no evidence that the rates of growth of the apices of fucoids are dependent on the transport of carbohydrates to them since fucoid apices grow as fast as do whole thalli (see Moss, 1974; Stromgren, 1977*a,b,c*; Khailov, 1978; Khailov *et al.*, 1978). Transport of photosynthate from photosynthesizing hosts to heterotrophic parasites has also been reported (Turner and Evans, 1978).

It seems likely that if organic carbon is transported by algae so also is organic nitrogen. Indeed, mannitol, together with amino acids, are the major components found in the medullary exudates of *Macrocystis* (Parker, 1966; Jackson, 1977). Also, in *Delessaria sanguinea* (Hartmann and Eschrich, 1969) there is evidence for the translocation of the amino acids phenylalanine and leucine via the mid-rib. On exposing the apical part of the lamina of *Alaria marginata* to ^{14}C, Schmitz and Srivastava (1975) found labelled amino acids 10–16 cm from the site of exposure within 2 h. There were only minor differences in the distribution of ^{14}C in any of the amino acids in the two parts of the lamina but slightly higher labelling occurred in aspartic acid, serine, glycine, and cysteine and lower labelling in asparagine–glutamine, and glutamate in the translocate than in the exposed area. Such studies do not prove, of course, that ^{14}C-labelled amino acids *per se* are transported, rather than other ^{14}C-labelled compounds which are later converted to amino acids. There is also evidence of transfer of nitrogen assimilated in deeper nitrogen-rich waters to parts of *Macrocystis* thalli growing in upper nitrogen-depleted waters (Jackson, 1977) (see also Buggeln, 1977).

Transfer of organics in kelps is from sites of production to sites of utilization or storage, with various sinks, which may vary on a seasonal basis, being present in each plant. Thus in summer there is movement from non-growing, or slowly growing regions, to rapidly growing meristems and in autumn there is some evidence for movement to storage tissues (see for example Lüning, 1970; Lüning *et al.*, 1973). There must also be movement of fixed carbon from photosynthetic to non-photosynthetic parts of various algae. Evidence for transport via the medullary sieve elements has been obtained using fluorescein in *Macrocystis* (Parker, 1965), by removal of the cortex in *Nereocystis luetkeana* (Nicholson and Briggs, 1972), by autoradiography in *Laminaria hyperborea* (Steinbiss and Schmitz, 1973), and from the fact that the medullary exudate of *Macrocystis* comes mainly from the sieve element region (Crafts and Crisp, 1971; Schmitz and Srivastava, 1974*a*). Transport is usually strictly longitudinal along the axis of the plant but some lateral transport occurs in algae such as *Laminaria setchellii, Hedophyllum sessile*, and *Nereocystis luetkeana* where the longitudinally arranged sieve elements are connected to transverse connecting sieve elements (Schmitz and Srivastava, 1974*b*). Simultaneous bidirectional longitudinal transport has been demonstrated by establishing the necessary sources and sinks in *Laminaria sinclairii* and *Macrocystis* spp (Schmitz and Lobban, 1976).

Various rates of transfer have been calculated. Thus, in *Delesseria*, ^{14}C-labelled phenylalanine moves through the mid-rib at a rate of up to 63 cm h^{-1} (Hartmann and Eschrich, 1969). In the Laminariales rates of transfer of ^{14}C-label vary from 5 to 78 cm h^{-1}, depending on the alga (and on the type of sieve element it possesses) and also from region to region within the same plant (see Schmitz and Lobban, 1976). The rates are much higher than those solely attributable to diffusion and a pressure flow mechanism in which hydrostatic pressure differences between the source and sink ends of sieve tubes act as the major driving force is unlikely on anatomical grounds (Schmitz and Srivastava, 1974*b*; Penot *et al.*, 1976) and also because the data of Buggeln (1978) show no osmotic gradient between source and sink (34.6–36.1 $\times 10^2$ kPa). No cytoplasmic streaming occurs in sieve tube elements (Buggeln, 1978) and while the precise mechanism of movement is uncertain, it appears to be an active process both because of the high rates which occur and because translocation is reduced when the thalli are darkened (Buggeln, 1978).

7 THE PRODUCTION AND RELEASE OF EXTRACELLULAR NITROGEN

Fogg (1966) defines extracellular products as soluble organic substances liberated from healthy, as distinct from injured or decomposing, cells. Various nitrogenous compounds including amino acids, amides, peptides, polypeptides, and proteins released by algae fall into this category (Fogg, 1966, 1971).

Free amino acids have been reported in the filtrates of Euglenophyceae, Chlorophyceae, Chrysophyceae, Dinophyceae, and Bacillariophyceae (see Hellebust, 1974) but they usually represent only a small proportion of the intracellular organic nitrogen synthesized by eukaryotes. They are more abundant in filtrates of blue-green algae (see Fogg, 1966, 1971) and this may be related, in part, to the absence in prokaryotes of large vacuoles into which unwanted or reserve material may be channelled in most eukaryotes. Hood *et al*. (1969) speculate that it may also be attributable in part to a lack of repression of nitrogen-metabolizing enzymes in blue-green algae, but in fact the synthesis of most nitrogen-metabolizing enzymes appears to be repressible. Hellebust (1974) concludes that extracellular amino acids are released mainly by diffusion through the plasmalemma. Evidence supporting such a simple release process comes from the work of Gocke (1970) who found in *Scenedesmus quadricauda* that the extracellular and intracellular pools of amino acids were similar. In the case of prokaryotes on the other hand, the composition of the extracellular free amino acids and the intracellular pools are quite distinct indicating active or facilitated diffusion or differential membrane permeability to components of the intracellular pool. Certain marine phytoplankton deaminate assimilated amino acids intracellularly and subsequently release the carbon skeletons of such amino acids (Stephens and North, 1971).

The production of peptides and polypeptides by various Cyanophyceae, Chlorophyceae, Dinophyceae, and Bacillariophyceae has been reported (see Hellebust, 1974) but their role and significance has been most studied in the Cyanophyceae where, depending on the organism, its growth phase and environmental conditions, they may represent anything from a few per cent to most of the nitrogen assimilated into organic combination by the alga (Fogg, 1952; Jones and Stewart, 1969*a,b*). The extracellular nitrogen may be produced for short periods after the algae have been exposed to environmental stresses of temperature, salinity, desiccation, pH, etc. (Jones and Stewart, 1969*a,b*). There is also evidence that the type of extracellular peptides released may vary with the state of differentiation of the alga, and the nitrogen source. Thus David and Thomas (1972) report that the peptides released by *Anabaena* L-31 under N_2-fixing conditions when heterocysts are abundant are different in amino acid composition to those produced when the alga is growing on NO_3^- which inhibits heterocyst production. Similarly, Walsby (1974*b*) observed that certain extracellular peptides are produced by *Anabaena cylindrica* under N_2-fixing conditions only. It is possible, indeed probable, that just as extracellular peptides may be involved in bacterial differentiation (see Payne, 1976) they may play a similar role in blue-green algae. Ecologically, extracellular polypeptides of algae may be important as chelating agents (Fogg and Westlake, 1955) and Wetzel (1965) concludes that in a highly productive lake which he studied in Indiana the dissolved organic matter may act as chelating agents.

The production by algae of exotoxins as well as endotoxins of polypeptide nature has also been reported (for reviews see Schwimmer and Schwimmer, 1968; Gentile, 1971; Moore, 1977; Carmichael and Gorham, 1978). The commonest toxin producers are certain strains of *Anabaena flos-aquae, Aphanizomenon flos-aquae*, and *Microcystis aeruginosa*. In early studies Hughes *et al*. (1958) showed that a toxic strain of *Microcystis* (*M. aeruginosa* N.R.C.-1) produced two toxins which they distinguished on the basis of the survival times of mice injected interperitoneally with the toxins. One, SDF (slow death factor), caused death in 4–48 h, and the other, FDF (fast death factor), was released on cellular decomposition and caused death in 1–3 h. Subsequently the production of a toxin of fish and mammals by *Aphanizomenon flos-aquae* was reported (Sawyer *et al*., 1968; Gentile and Maloney, 1969; Gentile, 1971). This toxin shows many resemblances (Jakim and Gentile, 1968; Gentile, 1971; Alam *et al*., 1973) to saxitoxin (Schantz *et al*., 1975). More recently detailed studies of the toxins of *Anabaena flos-aquae* (anatoxins) have been reported by Gorham and coworkers. Expanding on early studies in which they obtained an exotoxin, VFDF (very fast death factor), from *Anabaena flos-aquae*, they have, by X-ray crystallography (Huber, 1972), synthesis (Campbell *et al*., 1977), etc, determined and confirmed its structure (Devlin *et al*., 1977) and renamed it anatoxin a. In structure it is rather similar to cocaine although it has no apparent effect on the central nervous system. Its main mode of action is as a post-synaptic depolarizing neuromuscular blocking agent (Carmichael *et al*., 1975; Biggs and Dryden, 1977). The LD_{min} of this stable toxin is 0.3 mg kg^{-1} with a 4–5 min survival time. Other anatoxins (b–d) may also exist and they may singly, or collectively, produce different symptoms in animals which ingest them (Carmichael and Gorham, 1978).

The rapid and substantial extracellular production of $^{15}NH_4^+$ and ^{15}N-labelled amides by *Westiellopsis prolifica* on exposure to $^{15}N_2$ has been reported by Fogg and Pattnaik (1966) who concluded that NH_4^+ produced by nitrogenase activity probably exceeded the capacity of NH_4^+-assimilating enzymes to utilize the available NH_4^+ which was liberated extracellularly. However, Walsby and Fogg (1975) who detected no extracellular NH_4^+ or amides in the filtrates of *Anabaena cylindrica* have re-interpreted the *Westiellopsis* data and conclude from the ^{15}N-labelling kinetics and other data that neither the NH_4^+ nor the amides were newly fixed, that the amides produced could not be glutamine or asparagine, and that amide production had perhaps been overestimated.

The production of amines by algae has long been suspected because of the nuisance odours which frequently occur in waters containing algal scums (Seydel, 1914; Liebmann, 1960) but it has been equivocal until recently whether these were produced directly by algae, or by bacterial breakdown of extracellular algal products. Evidence for their production by algae comes from the work of Rolle *et al.* (1971) who found methylamine and dimethylamine in gas bubbled through a mass culture of green algae, from that of Smith (1975) who also reported the algal production of two amines, and from that of Hartmann and Aufermann (1973) who found 3-methylbutylamine in the medium of the rhodophyte *Polysiphonia urceolata*. More recently Herrmann and Juttner (1977) using gas liquid chromatography–mass spectrometry obtained evidence that extracellular amines were produced by *Ankistrodesmus braunii, Microcystis aeruginosa, Ochromonas danica*, and *Nitzschia actinastroides*. Amines were also detected in natural waters containing a bloom of *Chlamydomonas*. 1-Aminopropane was the most widely distributed extracellular amine, being produced by *Ankistrodesmus, Nitzschia, Microcystis*, and *Ochromonas*. The amounts produced varied from 240 $\mu g\ l^{-1}$ by *Microcystis* to 300 $\mu g\ l^{-1}$ by *Ankistrodesmus*. *Ochromonas* medium also contained trimethylamine, 2-aminopropane, and ethanolamine, whereas *Chlamydomonas*-rich waters contained butanolamine. The finding of amines in pure cultures such as *Ochromonas danica* confirms that they may indeed be produced directly by algae and are not solely concomitant with bacterial breakdown of algal products.

Various algae liberate proteins extracellularly, for example, *Chlamydomonas, Volvox, Euglena, Ochromonas*, various Dinophyceae, Rhodophyceae, and Phaeophyceae (see Hellebust, 1974). Some such protein molecules at least have specific functions, e.g. the vitamin B_{12}-binding glycoprotein of *Euglena* (see Daisley, 1970), the growth inhibitors reported by Sieburth (1968) and Harris (1970, 1971), the sex attractants of Chlorophyceae (Föster *et al.*, 1956; Tsubo, 1961; Claes, 1971) and Phaeophyceae (Müller, 1968) etc.

Algae may grow symbiotically with other organisms and these may assimilate and/or liberate nitrogenous compounds. Plants involved include other algae, fungi, bryophytes, *Azolla*, certain cycads, and *Gunnera* (see Fogg *et al.*, 1973; Millbank, 1974; Becking, 1976; Peters, 1976; Silvester, 1976; Mague, 1977; Singh, 1977; Trench *et al.*, 1978). Animals involved include protozoa (see Karakashian, 1975), clams (see Cates and McLaughlin, 1976), coelenterates (see Muscatine *et al.*, 1975), corals (see Jones and Endean, 1973), and flatworms (see Holligan and Gooday, 1975). In the sacoglossan gastropod *Elysia viridis*, the chloroplasts of *Codium*, on which the animal feeds, remain partially functional when ingested (see Trench, 1975). The algal symbionts are usually simple types, for example Chlorophyceae or blue-green algae in lichens, dinoflagellates in some corals and sea anemones, Chlorophyceae in most flatworms, and zooxanthellae in some corals and coelenterates. Transport and interchange of carbon metabolites has been extensively studied in some of these but there are less data available on nitrogen (see Smith *et al*., 1969; Jennings and Lee, 1975).

The most detailed studies on nitrogen exchange are those on N_2-fixing algae. Nitrogen release or transfer from the prokaryote to eukaryote has been demonstrated in cycads (Bergersen *et al*., 1965), lichens (Millbank and Kershaw, 1969), *Gunnera* (Silvester and Smith, 1969), *Anthoceros* (Stewart and Rodgers, 1977), and *Azolla* (Peters *et al*., 1977). The form in which nitrogen is transferred from the alga appears to be mainly as amino acids and NH_4^+. Millbank (1974) studying the symbiotic phycobionts of five lichens concluded that the liberated products were partly polypeptidic in nature. More recent evidence-suggests that in symbioses between N_2-fixing algae and eukaryotic plants NH_4^+ is sometimes a major compound transferred from the prokaryote to eukaryote(s). Thus NH_4^+ can account for over 90 per cent of the nitrogen liberated when *Nostoc* colonies are isolated from *Blasia pusilla* and resuspended in fresh medium. This capacity to liberate NH_4^+ ceases within 48 h of resuspension (Stewart and Rodgers, 1977). NH_4^+ accounts for at least 50 per cent of the nitrogen liberated by the *Nostoc* phycobiont of *P. canina*, according to studies using ^{15}N as tracer, and digitonin to selectively inhibit metabolite uptake by the fungus (Stewart and Rowell, 1977). It is also a major nitrogenous compound liberated by the *Anabaena* which occurs symbiotically in *Azolla* (Peters, 1976). NH_4^+ liberation appears to be by diffusion, as happens with carbohydrate transfer in *Xanthoria parietina* (Collins and Farrar, 1978) and with NH_4^+ production by N_2-fixing free-living *Rhizobium* (Bergersen and Gibson, 1978). Various vesicles may be observed on the periphery of symbiotic algal cells (Peveling, 1973; Rodgers and Stewart, 1977) and these have been postulated as being involved in metabolite transfer between partners (Peveling, 1973). However, Collins and Farrar (1978) conclude that there are no convincing data that any special morphological features are involved in mass transfer in any lichen species.

The molecular events leading to NH_4^+ release have been investigated particularly in *Peltigera canina, P. polydactyla*, and *P. aphthosa* (Stewart and Rowell, 1977). In such lichens the blue-green algae show substantially reduced glutamine synthetase activity (Stewart and Rowell, 1977) and in the absence of major alternative routes of NH_4^+ assimilation, much of the N_2 fixed is liberated extracellularly by the algae. Studies using antiserum to purified glutamine synthetase from the *Nostoc* phycobiont of *P. canina* confirm that, although some algal glutamine synthetase is still present, the amount of glutamine synthetase protein synthesized is substantially reduced. It is not yet known whether that which is synthesized is as active as that in the free-living phycobiont, or whether it is also partially deactivated. Deactivation of glutamine synthetase in free-living blue-green algae, but not in the *Nostoc* from *P. canina* has been reported (Rowell *et al.*, 1979). The factors which regulate glutamine synthetase activity and synthesis *in vivo* are under active investigation in various laboratories. Reduced activity of glutamine synthetase has also been reported for the symbiotic *Anabaena* in *Azolla* (Ray *et al.*, 1978) and in a preliminary report it has been considered (Haselkorn, 1978) that the symbiotic *Anabaena* in *Azolla* does not synthesize glutamine synthetase. In lichens there is histochemical evidence that NADPH-dependent glutamic acid dehydrogenase which can assimilate liberated NH_4^+ is particularly active in those hyphae near to the algal cells (Stewart *et al.*, 1977). In *Azolla* and other photosynthetic symbionts the eukaryotic glutamine synthetase–glutamate synthase pathway may be the main route of assimilation of liberated NH_4^+ (Ray *et al.*, 1978). It appears that in symbioses the main function of the N_2-fixing alga, particularly in those where there is another photosynthetic partner, is to fix N_2 continuously and to release much of the nitrogen fixed for assimilation by the eukaryotic partner(s).

Nitrogen transfer in algal–animal associations has been less extensively studied, although it was suggested in early studies, e.g. Geddes (1882), Pütter (1911), Kawuguti (1953), that zooxanthellae associated with these functioned to remove nitrogenous animal wastes, particularly NH_4^+. Cates and McLaughlin (1976) have similar evidence using symbiotic and aposymbiotic *Condactylus* and *Cassiopea*. Since many such marine symbioses are characteristic of nitrogen-poor waters an obvious advantage of these is that they recycle and retain nitrogen within the ecosystem.

REFERENCES

Abeliovich, A. and Azov, Y. (1976). *Appl. Environ. Microbiol.*, **31**, 801

Ahmed, J. and Morris, I. (1967). *Arch. Mikrobiol.*, **56**, 219

Alam, M., Ikawa, M., Sassner, J. J., Jr, and Sawyer, P. J. (1973). *Toxicon*, **11**, 65

Allen, M. B. (1952). *Arch. Mikrobiol.*, **17**, 34

Anderson, L. E. and Avron, M. (1976). *Plant Physiol.*, **57**, 209

Anderson, L. E. and Duggan, J. X. (1976). In *Proceedings of the 2nd International Symposium on Photosynthetic Prokaryotes* (G. A. Codd and W. D. P. Stewart, eds), Dundee University, pp. 179–181

Antia, N. J., Berland, B. R., Bonin, D. J., and Maestrini, S. Y. (1975). *J. Mar. Biol. Assoc. U.K.*, **55**, 519

Antia, N. J., Berland, B. R., Bonin, D. J., and Maestrini, S. Y. (1977). *Phycologia*, **16**, 105

Atkinson, A. W., Jr, John, P. C. L., and Gunning, B. E. S. (1974). *Protoplasma*, **81**, 77

Baker, J. E. and Thompson, J. F. (1962). *Plant Physiol.*, **37**, 618

Barber, J. (1979). In *Handbook of Phycological Methods: Physiological and Biochemical Methods* (J. A. Hellebust and J. S. Craigie, eds), Cambridge University Press, Cambridge, pp. 449–462
Batt, T. and Brown, D. M. (1974). *Planta, Berl.*, **116**, 26
Becking, J. H. (1976). In *Proceedings of the 1st International Symposium on Nitrogen Fixation*, Vol. II (W. E. Newton and C. J. Nyman, eds), Washington State University Press, Seattle, pp. 556–580
Belmont, L. and Miller, J. D. A. (1965). *J. Exp. Bot.*, **16**, 318
Bergersen, F. J. and Gibson, A. H. (1978). In *Limitations and Potentials for Biological Nitrogen Fixation in the Tropics* (J. Dobereiner *et al.*, eds), Plenum Press, New York, pp. 263–274
Bergersen, F. J., Kennedy, G. S., and Wittman, W. (1965). *Aust. J. Biol. Sci.*, **18**, 1135
Bhovichitra, M. and Swift, E. (1977). *Limnol. Oceanogr.*, **22**, 73
Biggs, D. F. and Dryden, P. R. (1977). *Proc. West. Pharmacol Soc.*, **20**, 461
Bisalputra, T. and Stein, J. R. (1966). *Can. J. Bot.*, **44**, 1697
Bold, H. C. and Wynne, M. J. (1977). *Introduction to the Algae: Structure and Reproduction*, Prentice Hall, Englewood Cliffs, N. J.
Bollard, E. G. (1966). *Plant Soil*, **25**, 153
Brown, C. M., Macdonald Brown, D. S., and Meers, J. L. (1974). *Adv. Microbial Physiol.*, **11**, 1
Buggeln, R. G. (1976). *J. Phycol.*, **12**, 439
Buggeln, R. G. (1977). *J. Phycol.*, **13**, 212
Buggeln, R. G. (1978). *J. Phycol.*, **14**, 54
Butz, R. G. and Jackson, W. A. (1977). *Phytochemistry*, **16**, 409
Cain, J. (1965). *Can. J. Bot.*, **43**, 1367
Campbell, H. F., Edwards, O. E., and Kolt, R. (1977). *Can. J. Chem.*, **55**, 1372
Canvin, D. T. and Atkins, C. A. (1974). *Planta*, **116**, 207
Carmichael, W. W. and Gorham, P. R. (1978). *Mitt. Internat. Verein. Limnol.*, **21**, 285
Carmichael, W. W., Biggs, D. F., and Gorham, P. R. (1975). *Science, N.Y.*, **187**, 542
Carpenter, E. J., Remsen, C. C., and Watson, J. W. (1972). *Limnol. Oceanogr.*, **17**, 265
Cates, N. and McLaughlin, J. J. A. (1976). *J. Exp. Mar. Biol. Ecol.*, **21**, 1
Chaparro, A., Maldonado, J. M., Diez, J., Relimpio, A. M., and Losada, M. (1976). *Plant Sci. Lett.*, **6**, 335
Chapman, A. R. O. and Craigie, J. S. (1977). *Mar. Biol.*, **40**, 197
Chapman, A. R. O. and Craigie, J. S. (1978). *Mar. Biol.*, **46**, 209
Chapman, A. R. O., Markham, J. W., and Lüning, K. (1978). *J. Phycol.*, **14**, 195
Chapman, V. J. and Chapman, D. J. (1973). *The Algae*, Macmillan, London
Claes, H. (1971). *Arch. Mikrobiol.*, **78**, 180
Collins, C. R. and Farrar, J. F. (1978). *New Phytol.*, **81**, 71
Cooper, W. C. and Osterhout, W. J. V. (1930). *J. Gen. Physiol.*, **14**, 117
Crafts, A. S. and Crisp, C. E. (1971). *Phloem Transport in Plants*, Freeman, San Francisco
Czygan, F. C. (1963). *Planta*, **60**, 225
Daisley, K. W. (1970). *Int. J. Biochem.*, **1**, 561
David, K. A. V. and Thomas, J. (1972). In *Proceedings of the Symposium on Cellular Processes in Growth, Development and Differentiation*, Bhabha Atomic Research Centre, pp. 413–425
De Boer, J. A., Guigli, H. J., Israel, T. L., and D'Elia, C. F. (1978). *J. Physiol.*, **14**, 261
D'Elia, C. F. and De Boer, J. A. (1978). *J. Phycol.*, **14**, 266
Devlin, J. P., Edwards, O. E., Gorham, P. R., Hunter, N. R., Pike, R. K., and Stavric, B. (1977). *Can. J. Chem.*, **55**, 1367
Dharmawardene, M. W. N., Haystead, A., and Stewart, W. D. P. (1973). *Arch. Mikrobiol.*, **90**, 281
Diez, J., Chaparro, A., Vega, J. M., and Relimpio, A. (1977). *Planta*, **137**, 231
den Dooren de Jong, L. E. (1967). *Antonie van Leeuwenhoek J. Microbiol. Serol.*, **33**, 166
den Dooren de Jong, L. E. (1969). *Antonie van Leeuwenhoek J. Microbiol. Serol.*, **35**, 107
Duggan, J. X. and Anderson, L. E. (1975). *Planta*, **122**, 293
Eaglesham, A. R. and Hewitt, E. J. (1971). *FEBS Lett.*, **16**, 315

Edge, P. A. and Ricketts, T. R. (1978). *Planta*, **138**, 123

Eppley, R. W. and Rogers, J. N. (1970). *J. Phycol*., **6**, 344

Eppley, R. W., Holm-Hansen, O., and Strickland, J. D. H. (1968). *J. Phycol*., **4**, 333

Eppley, R. W., Rogers, J. N., and McCarthy, J. J. (1969). *Limnol. Oceanogr*., **14**, 912

Esau, K. (1969). *Handb. Pfl. Anat*., **5**, 1

Falk, H. and Kleinig, H. (1968). *Arch. Mikrobiol*., **61**, 347

Falkowski, P. G. and Rivkin, R. B. (1976). *J. Phycol*., **12**, 448

Fay, P. (1969). *Arch. Mikrobiol*., **67**, 62

Floyd, G. L. (1972). *J. Phycol*., **8**, 68

Floyd, G. L., Stewart, K. D., and Mattox, K. R. (1971). *J. Phycol*., **7**, 306

Fogg, G. E. (1952). *Proc. Roy. Soc., Lond., Ser. B*, **139**, 372

Fogg, G. E. (1953). *The Metabolism of Algae*, Methuen, London

Fogg, G. E. (1966). *Oceanogr. Mar. Biol. Rev*., **4**, 195

Fogg, G. E. (1971). *Arch. Hydrobiol. Beih. Ergebn. Limnol*., **5**, 1

Fogg, G. E. (1974). In *Algal Physiology and Biochemistry* (W. D. P. Stewart, ed.), Blackwell Scientific Publications, Oxford, pp. 560–582

Fogg, G. E. and Pattnaik, H. (1966). *Phykos*, **5**, 58

Fogg, G. E. and Westlake, D. E. (1955). *Verh. Int. Verein. Limnol*., **12**, 219

Fogg, G. E., Stewart, W. D. P., Fay, P., and Walsby, A. E. (1973). *The Blue-green Algae*, Academic Press, London and New York

Förster, H., Wiese, L., and Braunitzer, G. (1956). *Z. Naturforsch*., **11b**, 315

Fraser, T. W. and Gunning, B. E. S. (1969). *Planta*, **88**, 244

Fritsch, F. E. (1945). *The Structure and Reproduction of the Algae*, Vol. 2, Cambridge University Press, Cambridge

Geddes, P. (1882). *Proc. Roy. Soc. Edinburgh Ser. B*, **2**, 377

Gentile, J. H. (1971). In *Microbial Toxins* (S. Kadis, A. Ciegler, and S. J. Ajl, eds), Academic Press, New York and London, pp. 27–66

Gentile, J. H. and Maloney, T. E. (1969). *Can. J. Microbiol*., **15**, 165

Gocke, K. (1970). *Arch. Hydrobiol*., **67**, 285

Gromyko, E. A., Evstigneeva, Z. G., and Kretovich, W. L. (1974). *Biokhimia*, **34**, 1

Guillard, R. R. L. (1963). In *Marine Microbiology* (C. H. Oppenheimer, ed.), C. C. Thomas, Springfield, Ill., pp. 93–104

Haines, K. C. and Wheeler, P. A. (1978). *J. Phycol*., **14**, 319

Harris, D. O. (1970). *Plant Physiol*., **45**, 210

Harris, D. O. (1971). *Arch. Mikrobiol*., **76**, 47

Harrison, W. C. (1976). *J. Exp. Mar. Biol. Ecol*., **21**, 199

Hartmann, T. and Aufermann, B. (1973). *Mar. Biol*., **21**, 70

Hartmann, T. and Eschrich, W. (1969). *Planta*, **85**, 303

Haselkorn, R. (1978). *Annu. Rev. Plant Physiol*., **29**, 319

Haystead, A., Dharmawardene, M. W. N., and Stewart, W. D. P. (1973). *Plant Sci. Lett*., **1**, 439

Healey, F. P. (1973). *C.R.C. Crit. Rev. Microbiol*., **3**, 69

Heimer, Y. M. and Filner, P. (1971). *Biochim. Biophys. Acta*, **230**, 362

Heinz, E. (1974). *Curr. Top. Membr. Transp*., **9**, 137

Hellebust, J. A. (1970). In *Organic Matter in Natural Waters* (D. W. Hood, ed.), University of Alaska Press, Anchorage, pp. 225–256

Hellebust, J. A. (1974). In *Algal Physiology and Biochemistry* (W. D. P. Stewart, ed.), Blackwell Scientific Publications, Oxford, pp. 838–863

Hellebust, J. A. (1978). *J. Phycol*., **14**, 79

Hellebust, J. A. and Guillard, R. R. L. (1967). *J. Phycol*., **3**, 132

Herrmann, V. and Jüttner, F. (1977). *Analyt. Biochem*., **78**, 365

Hill, G. T. C. and Machlis, L. (1968). *J. Phycol*., **4**, 261

Hoare, D. S., Hoare, S. L., and Moore, R. B. (1967). *J. Gen. Microbiol*., **49**, 351

Holligan, P. M. and Gooday, G. W. (1975). *Symp. Soc. Exp. Biol*., **29**, 205

Hood, W., Leaver, A. G., and Carr, N. G. (1969). *Biochem. J*., **114**, 12p

Hope, A. B. and Walker, N. A. (1975). *The Physiology of Giant Algal Cells*, Cambridge University Press, Cambridge

Huber, C. S. (1972). *Acta Cryst. B*, **28**, 2577

Hughes, E. O., Gorham, P. R., and Zehnder, A. (1958). *Can. J. Microbiol.*, **4**, 225
Jackson, G. A. (1977). *Limnol. Oceanogr.*, **22**, 979
Jacques, A. G. (1939*a*). *J. Gen. Physiol.*, **22**, 501
Jacques, A. G. (1939*b*). *J. Gen. Physiol.*, **22**, 521
Jacques, A. G. and Osterhout, W. J. V. (1938). J. Gen. Physiol., **21**, 687
Jakim, E. and Gentile, J. (1968). *Science, N.Y.*, **162**, 915
Jennings, D. H. and Lee, D. L. (eds) (1975). *Symbiosis*, Symposia of the Society for Experimental Biology, Vol. 29, Cambridge University Press, Cambridge
Johansson, B. C. and Gest, H. (1977). *Eur. J. Biochem.*, **81**, 365
Jones, K. and Stewart, W. D. P. (1969*a*). *J. Mar. Biol. Assoc. U.K.*, **49**, 475
Jones, K. and Stewart, W. D. P. (1969*b*). *J. Mar. Biol. Assoc. U.K.*, **49**, 701
Jones, O. A. and Endean, R. (eds) (1973). *Biology and Geology of Coral Reefs*, Vol. II, Academic Press, New York
Kahane, S., Marcus, M., Barash, H., Halpern, Y. J., and Kaback, H. R. (1975). *FEBS Lett.*, **56**, 235
Kanazawa, T., Kanazawa, K., Kirk, M. R., and Bassham, J. A. (1972). *Biochim. Biophys. Acta*, **256**, 656
Karakashian, M. W. (1975). *Symp. Soc. Exp. Biol.*, **29**, 145
Kates, J. R. and Jones, R. J. (1964). *Biochim. Biophys. Acta*, **86**, 438
Kawaguti, S. (1953). *Biol. Okayama Univ.*, **1**, 171
Kesseler, H. (1966). *Verhöff. Inst. Meeresforsch. Bremerh.*, **2**, 357
Kessler, E. (1957). *Arch. Mikrobiol.*, **27**, 166
Kessler, E., Hofmann, A., and Zumft, W. G. (1970). *Arch. Mikrobiol.*, **72**, 23
Khailov, K. M. (1978). *Bot. Mar.*, **21**, 313
Khailov, K. M., Kholodov, V. I., Firsov, Y. K., and Prazukin, A. V. (1978). *Bot. Mar.*, **21**, 289
Kirk, M. M. and Kirk, D. L. (1978*a*). *Plant Physiol.*, **61**, 349
Kirk, D. L. and Kirk, M. M. (1978*b*). *Plant Physiol.*, **61**, 556
Kirk, D. L. and Kirk, M. M. (1978*c*). *J. Phycol.*, **14**, 198
Komor, E. and Tanner, W. (1974). In *Membrane Transport in Plants* (U. Zimmerman and J. Dainty, eds). Springer Verlag, Berlin, pp. 209–215
Komor, E., Haass, D., and Tanner, W. (1972). *Biochim. Biophys. Acta*, **266**, 649
Kretovich, W. L., Evstigneeva, Z. G., and Tomova, N. G. (1970). *Can. J. Bot.*, **48**, 1179
Kurelec, B., Rijavec, M., Britvic, S., Müller, W. E. G., and Zahn, R. K. (1977). *Comp. Biochem. Physiol.*, **56B**, 415
Ladha, J. K., Rowell, P., and Stewart, W. D. P. (1978). *Biochem. Biophys. Res. Commun.*, **83**, 688
Lang, N. J. and Fay, P. (1971). *Proc. Roy. Soc., Lond., Ser. B*, **178**, 193
Lea, P. J. and Miflin, B. J. (1975). *Biochem. Soc. Trans.*, **3**, 381
Leftley, J. W. and Syrett, P. J. (1973). *J. Gen. Microbiol.*, **77**, 109
Lewin, R. A. (1976). *Nature, Lond.*, **259**, 360
Liebmann, H. (1960). *Handbuch der Frischwasser—und Abwasser—Biologie*, Vol. 2, R. Oldenburg, München
Liu, M. S. and Hellebust, J. A. (1974*a*). *Can. J. Microbiol.*, **20**, 1109
Liu, M. S. and Hellebust, J. A. (1974*b*). *Can. J. Microbiol.*, **20**, 1119
Liu, M. S. and Hellebust, J. A. (1976*a*). *Can. J. Bot.*, **54**, 930
Liu, M. S. and Hellebust, J. A. (1976*b*). *Can. J. Bot.*, **54**, 938
Lobban, C. S. (1978). *J. Phycol.*, **14**, 178
Losada, M. (1974). In *Metabolic Interconversion of Enzymes*, Vol. III, (E. H. Fischer *et al.*, eds), Springer Verlag, Berlin, p. 257
Lui, N. S. T. and Roels, O. A. (1972). *J. Phycol.*, **8**, 259
Lüning, K. (1969). *Mar. Biol.*, **3**, 282
Lüning, K. (1970). *Helgolander wiss Meeresunters*, **20**, 79
Lüning, K., Schmitz, K., and Willenbrink, J. (1972). *Z. Pfl. Physiol.*, **67**, 418
Lüning, K., Schmitz, K., and Willenbrink, J. (1973). *Mar. Biol.*, **23**, 275
Lüttge, U. and Pitman, M. G. (1976). *Encyclopaedia of Plant Physiology, New Series*, Vols 2A and 2B, Springer Verlag, Berlin
Lynch, V. H. and Gillmor, G. G. (1966). *Biochim. Biophys. Acta*, **115**, 253

McCarthy, J. J. (1972). *J. Phycol.*, **8**, 216

Macisaac, J. J. and Dugdale, R. C. (1976). *Limnol. Oceanogr.*, **21**, 149

MacRobbie, E. A. C. (1974). In *Algal Physiology and Biochemistry* (W. D. P. Stewart, ed.), Blackwell Scientific Publications, Oxford, pp. 676–713

Magalhaes, A. C., Neyra, C. A., and Hageman, R. H. (1974). *Plant Physiol.*, **53**, 411

Mague, T. H. (1977). In *A Treatise on Dinitrogen Fixation*, Section IV, *Agronomy and Ecology* (R. W. F. Hardy and A. H. Gibson, eds), John Wiley and Sons, New York, pp. 85–140

Malone, T. C., Garside, C., Haines, H. C., and Roels, O. A. (1975). *Limnol. Oceanogr.*, **20**, 9

Mann, K. H. (1978). In *Proceedings of the 4th International Congress on Photosynthesis* (D. O. Hall, J. Coombs, and T. W. Goodwin, eds), The Biochemical Society, London, pp. 235–244

Marchant, H. J. (1976). In *Intracellular Communications in Plants: Studies in Plasmodesmata* (B. E. S. Gunning and A. W. Robards, eds), Springer Verlag, Berlin, pp. 59–80

Marre, E. and Ciferri, O. (1977). *Regulation of Cell Membrane Activities in Plants*, North Holland, Amsterdam

Mazelis, M. (1979). In *Nitrogen Assimilation of Plants* (E. J. Hewitt, and C. V. Cutting, eds), Academic Press, London and New York, pp. 407–417

Mazelis, M. and Creveling, R. K. (1978). *Plant Physiol.*, **62**, 798

Meister, A. (1973). *Science, N.Y.*, **180**, 33

Meeks, J. C., Wolk, C. P., Thomas, J., Lockau, W., Shaffer, P. W., Austin, S. M., Chien, W.-S., and Galonsky, A. (1977). *J. Biol. Chem.*, **252**, 7894

Miflin, B. J. and Lea, P. J. (1976). Phytochemistry, **15**, 873

Miflin, B. J. and Lea, P. J. (1977). *Annu. Rev. Plant Physiol.*, **28**, 299

Millbank, J. W. (1974). In *The Biology of Nitrogen Fixation* (A. Quispel, ed.), North Holland, Amsterdam, pp. 239–264

Millbank, J. W. and Kershaw, K. A. (1969). *New Phytol.*, **68**, 721

Moore, R. E. (1977). *Bioscience*, **27**, 797

Morris, I. (1974). In *Algal Physiology and Biochemistry* (W. D. P. Stewart, ed.), Blackwell Scientific Publications, Oxford, pp. 583–609

Morris, I. and Ahmed, J. (1969). *Physiol. Plant*, **22**, 1166

Morris, I. and Syrett, P. J. (1965). *J. Gen. Microbiol.*, **38**, 21

Moss, B. (1974). In *Algal Physiology and Biochemistry* (W. D. P. Stewart, ed.), Blackwell Scientific Publications, Oxford, pp. 788–813

Müller, D. G. (1968). *Naturwissenschaften*, **54**, 496

Mur, L. R., Gons, H. J., and Van Liere, L. (1978). *Mitt. Int. Verein. Limnol.*, **21**, 473

Muscatine, L., Cook, C. B., Pardy, R. L., and Pool, R. R. (1975). *Symp. Soc. Exp. Biol.*, **29**, 175

Neilson, A. H. and Doudoroff, M. (1973). *Arch. Mikrobiol.*, **89**, 15

Nicholson, N. L. (1976). *Bot. Mar.*, **7**, 23

Nicholson, N. L. and Briggs, W. R. (1972). *Amer. J. Bot.*, **59**, 97

North, B. B. and Stephens, G. C. (1967). *Biol. Bull.*, **133**, 391

North, B. B. and Stephens, G. C. (1969). In *Proceedings of the V International Seaweed Symposium* (R. Margalef, ed.), Subsecretaria Mercante Marine, Madrid, pp. 263–273

North, B. B. and Stephens, G. C. (1971). *Biol. Bull.*, **140**, 242

North, B. B. and Stephens, G. C. (1972). *J. Phycol.*, **8**, 64

Ohmori, M. and Hattori, A. (1978). *Arch. Microbiol.*, **117**, 17

Okabe, K., Codd, G. A. and Stewart, W. D. P. (1979). *Nature, Lond.*, **279**, 525

Ownby, J. D. (1977). *Planta*, **136**, 277

Paelinck, H. R. E. and de Maeseneer, J. (1978). *Mitt. Int. Verein. Limnol.*, **21**, 480

Parker, B. C. (1965). *J. Phycol.*, **1**, 41

Parker, B. C. (1966). *J. Phycol.*, **2**, 38

Parker, B. C. and Huber, J. (1965). *J. Phycol.*, **1**, 172

Pateman, J. A. and Kinghorn, J. R. (1976). In *Filamentous Fungi*, Vol. 2 (J. E. Smith and D. Berry, eds), Edward Arnold, London, pp. 159–239

Pateman, J. A., Kinghorn, J. R., Dunn, E., and Forbes, E. (1973). *J. Bacteriol.*, **114**, 943

Pateman, J. A., Dunn, E., Kinghorn, J. R., and Forbes, E. (1974). *Mol. Gen. Genetics*, **133**, 225

Payne, J. W. (1976). *Adv. Microbial Physiol.*, **13**, 56

Penot, M., Floc'h, J.-Y. and Penot, M. (1976). *Phycologia*, **15**, 299
Peters, G. A. (1976). In *Proceedings of the 1st International Symposium on Nitrogen Fixation*, Vol. 2 (W. E. Newton and C. J. Byman, eds) Washington State University Press, Seattle, pp. 592–610
Peters, G. A., Toia, R. E., Jr, and Lough, S. M. (1977). *Plant Physiol.*, **59**, 1021
Peveling, E. (1973). *New Phytol.*, **72**, 343
Pickett-Heaps, J. D. (1967*a*). *Aust. J. Biol. Sci.*, **20**, 539
Pickett-Heaps, J. D. (1967*b*). *Aust. J. Biol. Sci.*, **20**, 883
Pickett-Heaps, J. D. (1968). *Aust. J. Biol. Sci.*, **21**, 655
Pickett-Heaps, J. D. (1970). *Planta*, **90**, 174
Pickett-Heaps, J. D. (1973). *New Phytol.*, **72**, 347
Pigott, G. H. and Carr, N. G. (1971). *Arch. Mikrobiol.*, **79**, 1
Pribil, S. and Kotyk, A. (1970). *Biochim. Biophys. Acta*, **219**, 242
Pringsheim, E. G. (1952). *Quart. J. Microsc. Sci.* **93**, 71
Proctor, V. M. (1957). *Am. J. Bot.*, **44**, 141
Pueschel, C. M. (1977). *Protoplasma*, **91**, 15
Pütter, A. (1911). *Z. Allg. Physiol.*, **12**, 297
Raboy, B., Padan, B., and Shilo, M. (1976). *Arch. Mikrobiol.*, **110**, 77
Ramus, J. (1971). *Phycologia*, **10**, 99
Rao, V. R., Darrow, R. A., and Keister, D. L. (1978). *Biochem. Biophys. Res. Commun.*, **81**, 224
Raven, J. A. (1976*a*). In *Encyclopaedia of Plant Physiology, New Series*, Vol. 2A (U. Lüttge and M. G. Pitman, eds), Springer Verlag, Berlin, pp. 129–188
Raven, J. A. (1976*b*). In *The Intact Chloroplast* (J. Barber ed.), Elsevier, Amsterdam, p. 403
Raven, J. A. (1977). *J. Exp. Bot.*, **28**, 314
Raven, J. A. and Glidewell, S. M. (1975). *Photosynthetica*, **9**, 361
Ray, T. B., Peters, G. A., Toia, R. E., Jr, and Mayne, B. C. (1978). *Plant Physiol.*, **62**, 463
Rhodes, D., A. P. Sims and G. R. Stewart. (1979). In *Nitrogen Assimilation of Plants* (E. J. Hewitt, and C. V. Cutting, eds), Academic Press, London and New York, pp. 501–520
Rigano, C., Aliotta, G., and Rigano, V. M. (1975). *Arch. Microbiol.*, **104**, 297
Rippka, R. and Waterbury, J. B. (1977). *FEMS Lett.*, **2**, 83
Rodgers, G. A. and Stewart, W. D. P. (1977). *New Phytol.*, **78**, 441
Rolle, J., Payer, R., and Soeder, C. J. (1971). *Arch. Mikrobiol.*, **77**, 185
Romanov, V. I., Evstigneeva, Z. G., and Kretovich, W. L. (1965). *Prikl. Biokhim. Microbiol.*, **7**, 495
Roon, R. J. and Levenberg, B. (1968). *J. Biol. Chem.*, **243**, S213
Round, F. E. (1973). *The Biology of Algae*, Edward Arnold, London
Rowell, P. and Stewart, W. D. P. (1976). *Arch. Mikrobiol.*, **107**, 115
Rowell, P., Enticott, S., and Stewart, W. D. P. (1977). *New Phytol.*, **79**, 41
Rowell, P., Sampaio, M. J. A. M., and Stewart, W. D. P. (1978). *Proc. Soc. Gen. Microbiol.*, **5**, 103
Rowell, P., Sampaio, M. J. A. M., Ladha, J. K., and Stewart, W. D. P. (1979). *Arch. Microbiol.*, **120**, 195
Sampaio, M. J. A. M., Rowell, P., and Stewart, W. D. P. (1979). *J. Gen. Microbiol.*, **111**, 181
Sargent, M. C. and Landtrip, L. W. (1952). *Am. J. Bot.*, **39**, 99
Sawyer, P. J., Gentile, J. H., and Sassner, J. J., Jr (1968). *Can. J. Microbiol.*, **14**, 1199
Schantz, E. J., Ghazarossian, V. E., Schnoes, H. K., Strong, F. M., Springer, J. P., Pezzanite, J. O., and Clardy, J. (1975). *J. Am. Chem. Soc.*, **97**, 1238
Schloemer, R. H. and Garrett, R. H. (1974). *J. Bacteriol.*, **118**, 259
Schmitz, K. and Lobban, C. S. (1976). *Mar. Biol.*, **36**, 207
Schmitz, K. and Srivastava, L. M. (1974*a*). *Cytobiology.*, **10**, 66
Schmitz, K. and Srivastava, L. M. (1974*b*). *Planta*, **116**, 85
Schmitz, K. and Srivastava, L. M. (1975). *Can. J. Bot.*, **53**, 861
Schmitz, K. and Srivastava, L. M. (1976). *Am. J. Bot.*, **63**, 679
Schmitz, K., Lüning, K., and Willenbrink, J. (1972). *Z. Pfl. Physiol.*, **67**, 418
Schwimmer, M. and Schwimmer, D. (1968). In *Algae, Man and the Environment* (D. F. Jackson, ed.), Plenum Press, New York, p. 279
Scott, W. E. and Fay, P. (1972). *Br. Phycol. J.*, **7**, 283

Seydel, E. (1914). *Mitt. Fisch. Ver. Prov. Brandenburg*, **5**, 87
Shatilov, V. R., Evstigneeva, Z. G., and Kretovich, W. L. (1968). *Dokl. Akad. Nauk SSSR*, **178**, 482
Shatilov, V. R., Kaloshina, G. S., and Kretovich, W. L. (1970). *Proc. Acad. Sci. U.S.S.R.*, **194**, 964
Sideman, E. J. and Schierer, D. C. (1977). *Am. J. Bot.*, **64**, 649
Sieburth, J. McN. (1968). *Adv. Microbiol. Sea*, **1**, 63
Silvester, W. B. (1976). In *Symbiotic Nitrogen Fixation in Plants* (P. S. Nutman, ed.), Cambridge University Press, Cambridge, pp. 521–538
Silvester, W. B. and Smith, D. R. (1969). *Nature, Lond.*, **224**, 1231
Simon, R. D. (1971). *Proc. Natl. Acad. Sci. U.S.A.*, **68**, 265
Simon, R. D. (1973). *Arch. Mikrobiol.*, **92**, 115
Simon, R. D. and Weathers, P. (1976). *Biochim. Biophys. Acta*, **420**, 165
Singh, P. K. (1977). *Riv. I. R. Riso.*, **26**, 125
Sloan, P. R. and Strickland, J. D. H. (1966). *J. Phycol.*, **2**, 29
Smith, A. J. (1973). In *The Biology of Blue-Green Algae* (N. G. Carr and B. A. Whitton, eds), Blackwell Scientific Publications, Oxford, pp. 1–38
Smith, D. C., Muscatine, L., and Lewis, D. H. (1969). *Biol. Rev.*, **44**, 17
Smith, T. A. (1975). *Phytochemistry*, **14**, 865
Smith, F. A. and Walker, N. A. (1978). *J. Exp. Bot.*, **29**, 107
Smith, F. A., Raven, J. A., and Jayasuriya, H. D. (1978). *J. Exp. Bot.*, **29**, 121
Solomonson, L. P. and Spehar, A. M. (1977). *Nature, Lond.*, **265**, 373
Spanswick, R. M. and Costerton, J. W. F. (1967). *J. Cell. Sci.*, **2**, 451
Stacey, G., Tabita, F. R., and Van Baalen, C. (1977). *J. Bacteriol.*, **132**, 596
Stanier, R. Y. and Cohen-Bazire, G. (1977). *Annu. Rev. Microbiol.*, **31**, 225
Stein, J. R. (1963). *Br. Phycol. Bull.*, **2**, 206
Steinbiss, H. H. and Schmitz, K. (1973). *Planta*, **112**, 253
Stephens, G. C. and North, B. B. (1971). *Limnol. Oceanogr.*, **16**, 752
Stewart, W. D. P. (1964). *J. Exp. Bot.*, **15**, 138
Stewart, W. D. P. (1973). *Annu. Rev. Microbiol.*, **27**, 283
Stewart, W. D. P. (1977). *Br. Phycol. J.*, **12**, 89
Stewart, W. D. P. and Lex, M. (1970). *Arch. Mikrobiol.*, **73**, 250
Stewart, W. D. P. and Rodgers, G. A. (1977). *New Phytol.*, **78**, 459
Stewart, W. D. P. and Rowell, P. (1975). *Biochem. Biophys. Res., Commun.*, **65**, 846
Stewart, W. D. P. and Rowell. P. (1977). *Nature, Lond.*, **265**, 371
Stewart, W. D. P., Haystead, A. and Dharmawardene, M. W. N. (1975). In *Nitrogen Fixation by Free-living Micro-organisms*, I.B.P. Vol. 6 (W. D. P. Stewart ed.), Cambridge University Press, Cambridge, pp. 129–158
Stewart, W. D. P., Rowell, P., and Apte, S. K. (1977). In *Proceedings of the 2nd International Symposium on Nitrogen Fixation* (W. Newton, J. R. Postgate, and C. Rodriguez-Barrueco, eds), Academic Press, London and New York, pp. 287–307
Stewart, W. D. P., Rowell, P., Codd, G. A., and Apte, S. K. (1978). In *Proceedings of the 4th International Congress on Photosynthesis* (D. O. Hall, J. Coombs, and T. W. Goodwin, eds), The Biochemical Society, London, pp. 133–146
Stewart, W. D. P., Rowell, P., Ladha, J. K., and Sampaio, M. J. A. M. (1979). In *Nitrogen and Rice Symposium*, I.R.R.I., Manila, in press
Strickland, J. D. H., Holm-Hansen, O., Eppley, R. W., and Linn, A. J. (1969). *Limnol. Oceanogr.*, **14**, 23
Stromgren, T. (1977*a*). *Sarsia*, **63**, 39
Stromgren, T. (1977*b*). *Oikos*, **29**, 245
Stromgren, T. (1977*c*). *J. Exp. Mar. Biol. Ecol.*, **29**, 181
Stross, R. G. (1963). *Can. J. Microbiol.*, **9**, 33
Strotmann, H. and Reid, A. (1969). *Planta*, **88**, 250
Sullivan, C. W. (1979). In *Handbook of Phycological Methods: Physiological and Biochemical Methods* (J. A. Hellebust and J. W. Craigie, eds), Cambridge University Press, Cambridge, pp. 39–56

Syrett, P. J. and Leftley, J. W. (1976). In *Perspectives in Experimental Biology*, Vol. 2 (N. Sunderland, ed.), Pergamon Press, Oxford, pp. 221–234
Syrett, P. J. and Bekheet, I. A. (1977). *New Phytol.*, **79**, 291
Talley, D. J., White, L. H., and Schmidt, R. R. (1972). *J. Biol. Chem.*, **247**, 7927
Tanner, W., Haass, D., Decker, M., Loos, B., Komor, B. and Komor, E. (1974). In *Membrane Transport in Plants* (U. Zimmerman and J. Dainty, eds), Springer Verlag, Berlin, pp. 202–208
Tanner, W., Komor, E., Fenzl, F. and Decker, M. (1977). In *Regulation of Cell Membrane Activities in Plants* (E. Marre, and O. Ciferri, eds), North Holland, Amsterdam, pp. 79–90
Thacker, A. and Syrett, P. J. (1972). *New Phytol.*, **71**, 423
Thomas, E. M. and Syrett, P. J. (1976). *New Phytol.*, **76**, 409
Thomas, J., Meeks, J. C., Wolk, C. P., Shaffer, P. W., Austin, S. M., and Chien, W.-S. (1977). *J. Bacteriol.*, **129**, 1545
Thomas, R. J., Hipkin, C. R., and Syrett, P. J. (1976). *Planta*, **133**, 9
Thompson, J. and MacLeod, R. A. (1971). *J. Biol. Chem.*, **246**, 4066
Thompson, J. and MacLeod, R. A. (1973). *J. Biol. Chem.*, **248**, 7106
Thompson, J. F. and Muenster, A. E. (1971). *Biochem. Biophys Res. Commun.*, **43**, 1049
Titlyanov, E. A. and Peshekhodko, V. M. (1973). *Trudy Biol. Pochv. Inst. N.S.*, **20**, 137
Topinka, J. A. (1978). *J. Phycol.*, **14**, 241
Trench, R. K. (1975). *Symp. Soc. Exp. Biol.*, **29**, 229
Trench, R. K., Pool, R. R., Jr., Logan, M., and Engelland, A. (1978). *Proc. Roy. Soc. Lond., Ser. B.*, **202**, 423
Tronick, S. R., Ciardi, J. E., and Stadtman, E. R. (1973). *J. Bacteriol.*, **115**, 858
Tsubo, Y. (1961). *Bot. Mag. Tokyo* **74**, 519
Turner, C. H. C. and Evans, L. V. (1977). *New Phytol.*, **79**, 363
Turner, C. H. C. and Evans, L. V. (1978). *Br. Phycol. J.*, **13**, 51
Ueda, K. (1966). *Cytologia*, **31**, 461
Ullrich, W. R. (1971). Planta, **100**, 18
Ullrich, W. R. (1972) In *Proceedings of the 2nd International Conference on Photosynthesis* (G. Forti, M. Avron, and A. Melandri, eds), Dr W. Junk, N. V., Holland, pp. 861–871
Ullrich, W. R. (1974). *Planta*, **116**, 143
Van Gorkom, H. J. and Donze, M. (1971). *Nature, Lond.*, **234**, 231
Wallen, D. G. (1973). *J. Phycol.*, **9**, 148
Walsby, A. E. (1974*a*). *Br. Phycol. J.*, **9**, 371
Walsby, A. E. (1974*b*). *Br. Phycol. J.*, **9**, 383
Walsby, A. E. and Fogg. G. E. (1975). *Br. Phycol. J.*, **10**, 339
Wetzel, R. G. (1965). *Mem. 1st Ital. Idrobiol.*, **18**, Suppl. 137
Wheeler, P. A. (1977). *J. Phycol.*, **13**, 301
Wheeler, P. A. and Stephens, G. C. (1977). *J. Phycol.*, **13**, 193
Wheeler, P. A., North, B. B., and Stephens, G. C. (1974). *Limnol. Oceanogr.*, **19**, 249
Whitney, P. A. and Cooper, T. G. (1972). *J. Biol. Chem.*, **247**, 1349
Williams, S. K., II and Hodson, R. C. (1977). *J. Bacteriol.*, **130**, 266
Wolk, C. P., Thomas, J., Shaffer, P. W., Austin, S. M. and Galonsky, A. (1976). *J. Biol. Chem.*, **251**, 5027
Wolosiuk, R. A. and Buchanan, B. B. (1977). *Nature. Lond.*, **266**, 565
Wyatt, J. T., Lawley, G. G., and Barnes, R. D. (1971). *Naturwissenschaften*, **58**, 570
Zeigler, H. (1963). *Protoplasma*, **57**, 786
Zevenboom, W. and Mur, L. R. (1978). *Mitt. Int. Verin. Limnol.*, **21**, 261

Microorganisms and Nitrogen Sources
Edited by J. W. Payne

Section 4.6

Transport and Utilization of Amino Acids and Peptides by Higher Plants

C. F. HIGGINS and J. W. PAYNE

Department of Botany, Science Laboratories, University of Durham, South Road, Durham DH1 3LE, UK

1 INTRODUCTION

Unlike the microorganisms considered elsewhere in this volume, higher plants are generally autotrophic; thus, the utilization of external amino acids, peptides, and proteins is normally unimportant. However, several interesting examples may be envisaged, although these are of minor importance considering the number of species involved. One might expect some parasitic species to utilize organic nitrogenous compounds present in their hosts; the uptake of amino acids and/or peptides from symbiotic microorganisms in root nodules may also be considered in this category. Similarly, certain aquatic plants may be able to utilize organic compounds present in highly eutrophic waters. A third, and better established example is the utilization of captured insects as a nitrogen source by carnivorous plants. Finally, the uptake and utilization of the endosperm storage reserves by monocot embryos can also be considered as a system in which growth is dependent upon an external (although in this case controlled) supply of organic nitrogen.

Fortunately, this paucity of examples is adequately compensated for by two situations which rarely apply to microorganisms. First, being eukaryotic, the cells of higher plants contain a number of organelles, each of which may be capable of specific transport and utilization of amino acids. Second, and more importantly, higher plants are multicellular. Thus, we have a situation, almost unique amongst the organisms considered in this volume, in which amino acids are produced by some cells and utilized by others, thus necessitating their transport within the plant body; this transport may be over a short distance (cell to cell transfer), or a long distance (translocation).

As will be seen, repeatedly, our knowledge of amino acid transport processes in higher plants is limited, especially when compared with our understanding of similar systems in microorganisms and mammals. This seems to be due not simply to the difficulties in handling and complexities of the tissue, but also to a feeling amongst many biochemists, that plants are generally of lesser significance than other organisms.

The metabolism and interconversion of amino acids by plant cells is very similar to that of other organisms. Thus, in this chapter we will concentrate upon the uptake of amino acids and peptides, considering their metabolic fates only where specific to plants.

2 CARNIVOROUS PLANTS: TRANSPORT AND UTILIZATION OF AMINO ACIDS AND PEPTIDES

About 400 plant species, out of a world total of over a quarter million, are now known to be carnivorous, each being specially adapted to trap and digest insects, and to utilize nutrients derived from them. The traps and digestive structures are very varied, ranging from simple, sticky leaves (*Pinguicula*) (the common and Latin names of species cited in this text are listed in Appendix 1) and the cup-like leaves of pitcher plants (*Sarracenia* and *Nepenthes*) to the complex structures of the well known Venus flytrap (see Darwin, 1875; Lloyd, 1942; Heslop-Harrison, 1976, 1978; for general reviews).

The ability of such carnivorous plants to utilize organic compounds as a source of nitrogen has been known for over a hundred years, yet surprisingly little information is available concerning the nature of the absorption process. In his classic work *Insectivorous Plants*, Darwin (1875) showed that members of the genera *Utricularia* (bladderworts) and *Drosera* (sundews) exhibit enhanced growth and reproductive capacity when supplied with insects or with animal proteins. Furthermore, he concluded that members of two other genera, *Pinguicula* (butterworts) and *Dionaea* (Venus's flytrap), were also capable of specifically absorbing certain organic nitrogenous compounds, to which they responded by secreting digestive enzymes. Subsequent studies, on a wide variety of species, have confirmed that growth and reproductive capacity are promoted by a supply of insects, protein, peptones, or other forms or organic nitrogen, especially under conditions of nitrogen limitation (Darwin, 1878; Oudman, 1936; Lloyd, 1942; Pringsheim and Pringsheim, 1962, 1967; Harder and Zemlin, 1967, 1968; Sorenson and Jackson, 1968; Dore Swamy and Mohan Ram, 1969; Chandler and Anderson, 1976*a*). This, added to the fact that these species are commonly found in soils deficient in nitrogen, has led to the hypothesis that the carnivorous habit is an important means of obtaining nitrogen (Lloyd, 1942). However, digested insects may also supply phosphorus, sulphur, and other elements to the growing plant (Chandler and Anderson, 1976*a*).

It is well known that nutrients can pass from a trapped insect into the tissues of an insectivorous plant. Plummer and Kethley (1964) raised ants on a $^{32}P/^{35}S$-labelled diet. The isotopes became incorporated into insect protein and, when fed to a pitcher plant (*Sarracenia*), the label was found to be absorbed into the plant tissue. Similarly, ^{14}C-labelled nutrients are absorbed from *Daphnia* by *Drosera* (Fabian-Galan and Sălăgeanu, 1968) and from *Chlorella* by *Pinguicula* (Heslop-Harrison and Knox, 1971). However, in none of these cases were the forms in which the label was taken up investigated. No reports have appeared indicating the uptake of intact protein by pinocytosis. On the contrary, the secretion of proteases into the traps of carnivorous plants (*Nepenthes*; Steckelberg *et al*., 1967; Nakayama and Amagase, 1968; Amagase, 1972; Jentsch, 1972; Tökés *et al*., 1974. *Sarracenia*: Hepburn *et al*., 1927. *Drosera*: Amagase, 1972; Amagase *et al*., 1972. *Dionaea*: Scala *et al*., 1969), often as a specific response to insects or other forms of organic nitrogen (Darwin, 1875; Clancy and Coffey, 1977), indicates that absorption may well be in the form of amino acids and/or small peptides. However, the degree to which protein is digested is unknown. Total degradation to amino acids has been claimed in the pitcher plants, *Nepenthes* (Amagase *et al*., 1972) and *Sarracenia* (Plummer and Kethley, 1964), while in the Venus flytrap there is apparently no peptidase activity secreted into the traps (Scala *et al*., 1969), indicating that peptide uptake may be important.

Only a few reports of specific amino acid uptake have appeared. Hepburn *et al*. (1927) showed that asparagine and 'peptones' are absorbed more rapidly than water or inorganic ions from the pitchers of *Sarracenia*. Similarly, Lüttge (1965) found that L-alanine is absorbed by *Nepenthes* more rapidly than phosphate or sulphate ions. More recently it has been shown that *Drosera* will absorb [^{35}S]-methionine intact, although the kinetics of uptake were not investigated (Chandler and Anderson, 1976*b*). After feeding plants with ^{35}S-labelled insects, ^{35}S-labelled methionine and cysteine were found in the leaves, the results being compatible with the absorption of label from the insect in the amino acid form. In a more detailed study, Plummer and Kethley (1964) showed that 24 different amino acids could be absorbed from the pitcher of *Sarracenia*; characteristic differences in their rates of uptake implying the existence of specific absorption processes. Three dipeptides were also shown to be taken up from the pitcher fluid by the leaf; they appeared to be absorbed intact with subsequent intracellular hydrolysis, seemingly providing the first demonstration of peptide transport in a higher plant.

The weight of evidence clearly points to the existence of specific processes for the absorption of amino acids from the traps of insectivorous plants. Furthermore, when one considers the ubiquitous role of peptide transport systems in the nutritional utilization of proteins (Matthews and Payne, 1975), and their occurrence in other plant species (Higgins and Payne, 1977*a,b*), it seems probable that they will also be of importance in insectivores. This is a field which would seem to lend itself to further investigation, especially considering the dearth of systems suitable for studying the utilization of extracellular organic nitrogen in higher plants.

3 PLANT TISSUE IN CULTURE

Ever since the pioneering work of Haberlandt (1902), the culture of plant tissue *in vitro* has provided an important tool for studies of plant physiology and metabolism. Attempts to culture tissues from a wide variety of higher plant species on chemically defined media have made available a considerable body of information concerning the utilization of amino acids.

1 Amino Acid Utilization

Tissue cultures were originally established using inorganic nitrogen sources, although complex additives such as malt and yeast extracts were soon found to stimulate growth (White, 1934, 1939; Gautheret, 1935, 1939). Similarly, the development of cell suspension culture evolved using coconut milk, yeast extract, and other undefined growth media (Steward *et al*., 1952; Nickell, 1956; Steward *et al*., 1958*a,b*). However, in most cases the stimulatory properties of these complex media can be replaced by protein hydrolysates or defined amino acid mixtures (Spruce callus: Reinert and White, 1956; Steinhart *et al.*, 1961; Risser and White, 1964. Sugarcane cells: Nickell and Maretzki, 1969. Tobacco leaf cells: Sandstedt and Skoog, 1960; Filner, 1965. Maize endosperm: Straus, 1960. Broad bean cell suspensions: Mitchell and Gildow, 1975. Soybean root cells: Gamborg *et al.*, 1968. *Scorzonera* crown gall: Paris and Duhamet, 1953. Barley embryos: Cameron-Mills and Duffus, 1977). In many cases a complete amino acid mixture is not necessary, a mixture of a few, or even a single amino acid sufficing to stimulate growth (Galston, 1948; Sandstedt and Skoog, 1960; Maretzki and Nickell, 1967; Gamborg *et al.*, 1968; Nickell and Maretzki, 1969; Mitchell and Gildow, 1975; Sheridan, 1977). However, although a few amino acids do stimulate growth when added singly to a nitrate-based medium, most are actually powerful growth inhibitors (Riker and Gutsche, 1948; Nickell and Burkholder, 1950; Earle and Torrey, 1965; Filner, 1966; Nickell and Maretzki, 1969; Gamborg, 1970; Behrend and Mateles, 1975; Sheridan, 1977; Bright *et al.*, 1978). These inhibitory properties may be overcome by presenting the amino acid as part of a balanced mixture of amino acids, or in some cases by the addition of an additional specific single amino acid (Sandstedt and Skoog, 1960; Steinhart *et al.*, 1961; Earle and Torrey, 1965; Filner, 1966; Dunham and Bryan, 1969, 1971; Henke *et al.*, 1974; Behrend and Mateles, 1975; Bright *et al.*, 1978). These complex interactions are apparently due, at least in part, to interference with nitrate and/or ammonium assimilation; effects on nitrate uptake (Filner, 1966; Heimer and Filner, 1970, 1971), induction of nitrate reductase (Filner, 1966; Heimer and Filner, 1970, 1971; G. R. Stewart, 1972; Radin, 1975; Oaks *et al.*, 1977) and ammonium utilization (Behrend and Mateles, 1975) have all been reported. Other explanations involve feedback regulation of amino acid biosynthetic enzymes such as aspartate kinase and homoserine dehydrogenase (Miflin, 1969, 1971; Bright *et al.*, 1978).

Amino acids can also be utilized by plant tissue and cell cultures as the sole source of nitrogen (Sunflower: Riker and Gutsche, 1948. Sorrel tumors: Nickell and Burkholder, 1950. Tobacco leaf cells: Frank *et al.*, 1951. Groundsel roots: Skinner and Street, 1954. Jerusalem artichoke: Nitsch and Nitsch, 1957. Intact duckweed fronds: Hillman, 1961. Rice callus: Furuhashi and Yatazawa, 1970. Soybean cells: Thompson *et al.*, 1977. Alfalfa, tobacco, and wheat cells: Fukunaga *et al.*, 1978). Most species are only able to utilize a limited number of amino acids, most commonly alanine and aspartic and glutamic acids and their amides, although considerable variation is found in those amino acids which are able to satisfy the nitrogen requirements of different species. Again, complex inhibitions and interactions occur when mixtures of amino acids are supplied (Furuhashi and Yatazawa, 1970), the explanation of which awaits further investigation. One or two reports have also appeared indicating apparent requirements for certain essential amino acids, although these have not been substantiated (arginine in sugarcane: Maretzki and Nickell, 1967; tyrosine in spruce: Reinert and White, 1956).

Stereospecific preference for L-isomers is exhibited during amino acid utilization (Nitsch and Nitsch, 1957; Gamborg, 1970), although it is not known whether this is exerted at the transport level or during intracellular metabolism.

Recent work on the growth of isolated cell protoplasts has also shown stimulatory and inhibitory effects of amino acids similar to those on intact cells (Kao and Michayluk, 1975; Bourgin, 1976; von Arnold and Eriksson, 1977).

These, and similar results suggest that many tissues, from a wide variety of higher plant species, are capable of absorbing and utilizing amino acids. The use of cell suspension cultures as a rapid means of selecting mutant strains, offers great potential for the future, not only for plant breeding purposes, but also as an aid to the detailed characterization of plant metabolic and transport processes. Similarly, auxotrophic mutants obtained in this way could be used to study transport processes (as are many bacterial and fungal auxotrophs), by monitoring the growth of cells on different sources of a required nutrient. However this possibility, and the related idea that cell and tissue cultures would provide a simple model system with which to study amino acid metabolism in plants, have not yet been fully realized; this can probably be attributed to the complex and poorly understood controlling effects which amino acids exert on growth, and to the considerable variations in these properties that are found amongst different tissues and species.

2 Peptide Utilization

The study of peptide utilization by plant cells and tissues in culture has been neglected, although there are indications that they may affect growth in a variety of ways. Peptones (Riker and Gutsche, 1948) and acid-hydrolysed casein (no peptides) can enhance tissue growth, while enzymic casein hydrolysates (containing peptides) often show inhibitory effects (Nitsch and Nitsch, 1957; Staba, 1962; Furuhashi and Yatazawa, 1970). Several individual defined dipeptides have been found to be capable of providing the sole source of nitrogen for growth (Duckweed plants: Bollard, 1966. Deadly nightshade (*Atropa*) callus: Salonen and Simola, 1977) although in neither of these studies was any attempt made to determine whether the peptides were hydrolysed before or after absorption. However, the more extensive growth of *Atropa* obtained on diglycine (1.25 mM) rather than glycine (2.5 mM) suggests that rapid extracellular hydrolysis may not be important.

4 AMINO ACID TRANSPORT IN PLANT TISSUES

Our understanding of the mechanism(s) of amino acid transport across plant cell membranes is very fragmentary, although the absorption of amino acids seems to be a universal property of higher plant cells (see above). The use of radio-tracers and other detection methods have shown that a variety of tissues, from many different plant species, all have the ability to take up amino acids. (Leaves, stems and roots from a variety of species: Webster, 1954; Naylor *et al.*, 1958. Sunflower hypocotyls: Reinhold and Powell, 1956. Mustard roots: Wright, 1962. Soybean roots in cell suspension culture: King and Oleniuk, 1973. Cultured sugarcane cells: Maretzki *et al.*, 1969. Intact radish leaves: Streeter and Thompson, 1972. Broad bean leaves: C. R. Stewart, 1972, 1973. Barley leaves: Naylor and Tolbert, 1956. Tobacco leaf cells in culture: Francki *et al.*, 1971.) In fact, no tissue so far examined seems to lack amino acid transport capacity. Certain characteristics of these systems are given below although detailed studies are generally lacking, and as yet few generalizations can be made.

1 Kinetics

Birt and Hird (1958*b*), using excised carrot tissue, first showed that the uptake of certain amino acids exhibits saturation kinetics. Initial analysis of alanine uptake by soybean roots also indicated the operation of a single saturable system (King and Oleniuk, 1973), although further studies showed that uptake in fact exhibits biphasic kinetics (King, 1976). Apart from a few reports of multiphasic kinetics (Hancock, 1975; Shtarkshall and Reinhold, 1974; Lien and Rognes, 1977) most subsequent studies have shown biphasic kinetics (Epstein, 1966) to predominate. The apparent absence of biphasic kinetics in pea leaf fragments (Cheung and Nobel, 1973), mustard roots (Wright, 1962) and carrot tissue (Birt and Hird, 1958*b*) may be explained by the limited concentration ranges used (0.2–10 mM), and biphasic kinetics may prove to be a general characteristic of amino acid transport in all plant systems. Thus, over low concentration ranges (0–0.5 mM) uptake shows simple Michaelis–Menten kinetics; at higher concentrations (0.5–50 mM) uptake consists of two components; a second, saturable, energy-dependent system, and a non-saturable, energy-independent mode of uptake. Biphasic kinetics are generally explained by proposing the existence of two, probably independent, carriers acting in parallel, one of high affinity and low capacity, the second of low affinity and high capacity (see Table 1), although it has also been suggested

Table 1 Kinetic parameters for amino acid transport in a variety of plant systems

Plant	Tissue	Amino acid	System 1 (high affinity) K_m (mM)	System 1 (high affinity) V_{max}[a]	System 2 (low affinity) K_m (mM)	System 2 (low affinity) V_{max}[a]	Reference
Duckweed (*Spirodela polyrhiza*)	Leaf frond	Gly	0.077	1.4	4.2	8.4	Borstlap, 1977
		Ala	0.035	0.9	1.5	3.4	
		Leu	0.028	0.6	1.0	3.3	
		Ile	0.074	1.5	3.5	4.0	
		Val	0.063	1.5	3.9	4.4	
		D-Val	0.517	1.5	8.2	4.2	
Pea (*Pisum sativum*)	Leaf fragments	Gly			2.9	16.7	Cheung and Nobel, 1973
		Ala			1.8	20.0	
		Ser			2.0	14.9	
		Thr			3.7	17.0	
		Asp			17.0	10.1	
		Glu			15.0	12.3	
		AIB			1.0	1.7	
Soybean (*Glycine max*)	Root cells	Ala	0.046	3.1[b]	8.0	96[b]	King, 1976
Sugar cane (*Saccharum officinarum*)	Cell culture	Arg	0.103	6.6[b]			Maretzki and Thom, 1970
		Lys			2.45	30[b]	
Melon (*Cucumis melo*)	Root tips	Phe	0.020	36.0			Watson and Fowden, 1975
Caesalpinia tinctoria	Root tips	Phe	0.078	40.0			Watson and Fowden, 1975
Barley (*Hordeum sativum*)	Leaf slices	AIB[c]	0.3	0.8	10.5	4.8	Reinhold *et al.*, 1970
		AIB[d]	0.6	3.7	9.5	8.4	

[a]Calculated in micrograms per gram fresh weight per hour.
[b]Only determined on a dry weight basis.
[c]Before 'ageing'.
[d]After 'ageing'.

that biphasic kinetics are actually the result of a single carrier, an 'all-or-none' transformation occurring at a critical external concentration (Laites, 1969). However, Maretzki and Thom (1970) explained the biphasic kinetics obtained for arginine uptake by sugarcane cells, by proposing that uptake by the high-affinity (low K_m) system is due to an arginine specific system, while the low-affinity (high K_m) component of uptake is due to non-specific uptake of arginine by a lysine-specific transporter. Similarly, King (1976) proposed that the low-affinity component of alanine uptake by soybean root cells is due to non-specific uptake by a previously characterized, general, neutral amino acid transport system (King and Hirji, 1975).

2 Structural Specificity

Bacteria, fungi, and mammals all have a number of well-defined transport systems for amino acids; some are specific for individual amino acids while others have various overlapping specificities. In plants, however, the situation is still relatively confused. No general pattern has yet emerged, although the considerable variations observed may of course reflect actual differences between tissues and species. Several reports that the whole range of protein amino acids competitively inhibit the uptake of one another have appeared, indicating that a single transport system is able to handle most or all protein amino acids. (Carrot tissue: Birt and Hird, 1958*b*. Pea leaf fragments: Cheung and Nobel, 1973. Melon root tips: Watson and Fowden, 1975. Tobacco cells: Harrington and Smith, 1977. Barley leaves: Lien and Rognes, 1977.) Similarly, in the duckweed, Borstlap (1977) showed that all the neutral protein amino acids share a common system. This low specificity, in comparison with the situation in other organisms, might be expected if non-protein amino acids, often present in large quantities in plant tissue, are transported via the same system(s) as the protein amino acids. However, in sugarcane the basic amino acids, arginine and lysine, apparently use separate systems (Maretzki and Thom, 1970), while in maize embryos only alanine, cysteine, glycine, serine, and glutamic acid were able to inhibit glutamine uptake competitively (Stewart, 1971). In a detailed study on soybean cells, King and Hirji (1975) proposed the existence of three systems, one each for basic, acidic, and neutral amino acids, though with partially overlapping specificities. As yet, several anomalies prevent complete acceptance of this simple scheme. For example, methionine, tryptophan, and phenylalanine were able to inhibit uptake of amino acids from all three groups, while aspartic acid and asparagine showed no competitive ability whatsoever. As a general rule, plant tissue seems to handle the 'charged' amino acids less well than the 'neutral' ones (Birt and Hird, 1958*b*; Francki *et al*., 1971;

Cheung and Nobel, 1973; Watson and Fowden, 1975; Harrington and Smith, 1977).

Little is known regarding the structural requirements for amino acid transport. Watson and Fowden (1975) found that α-*N*-methylation barely alters the ability of amino acids to compete for transport, while α-*N*-acetylation has a more profound effect (cf. peptide transport in *E. coli*, Section 2.1.5.5). Similarly, α-*C*-methylation considerably reduces the competitive ability of an amino acid. In barley leaf tissue (Shtarkshall and Reinhold, 1974) and tobacco cells (Harrington and Smith, 1977) it seems that the ionic forms of both the amino and carboxyl groups are required. Amino acid transport also exhibits stereospecificity for L-residues. Although D-amino acids can be absorbed by a metabolically dependent process, especially when presented at high concentrations, and the maximal rates of uptake of the D- and L- forms may on occasions even be the same (Birt and Hird, 1956, 1958*a,b*; Borstlap, 1977), the affinity of the D-isomers for transport is generally very much lower than that of the L-isomers, as shown by their greatly reduced competitive ability (Birt and Hird, 1958*b*; Stewart, 1971; Cheung and Nobel, 1973; Borstlap, 1977; Harrington and Smith, 1977). No uptake of the D-isomers of alanine, valine, and leucine could be detected in germinating barley embryos (Higgins and Payne, 1978*c*).

3 Energetics

A wide variety of metabolic inhibitors (e.g. DNP, FCCP (abbreviations are listed in Appendix 2), azide, cyanide, iodoacetate) have been reported to inhibit amino acid uptake by plant tissue, although at high external amino acid concentrations there is also an energy-independent component to uptake (Webster, 1954; Birt and Hird, 1956, 1958*a,b*; Maretzki and Thom, 1970; Stewart, 1971; King and Oleniuk, 1973; Hancock, 1975; Watson and Fowden, 1975; Harrington and Smith, 1977). Oxygen accelerates amino acid uptake (Reinhold and Powell, 1956; Naylor *et al*., 1958) and uptake increases consumption of oxygen (Birt and Hird, 1958*a*). Amino acids can also be accumulated against a concentration gradient (Birt and Hird, 1956, 1958*a*; Shtarkshall *et al*., 1970; Cheung and Nobel, 1973; Hancock, 1975; Borstlap, 1977), leaving little doubt that there is an active component to uptake, although the mode of energization is still unclear. Light is important (Naylor *et al*., 1958) and uptake may be partially inhibited by DCMU (Francki *et al*., 1971; Cheung and Nobel, 1973; Lien and Rognes, 1977), although complete inhibition in the dark, or with DCMU, is only achieved if respiration is also inhibited, indicating that both photosynthesis and respiration may energize transport. Pre-lighting can energize transport when subsequently placed in the dark. ATP can partly replace this light requirement (Francki *et al*., 1971), and PMS also stimulates transport (Cheung and Nobel, 1973), indicating that ATP may be involved in the mechanism by which transport is energized.

Ion gradients are known to be one means by which amino acid transport is driven in bacteria and mammals (see Section 1.7.3) and the inhibitory effects of monovalent cations on amino acid transport (Birt and Hird, 1956; Stewart, 1971) may reflect the energization of transport by alkali metal gradients (as in animal tissues). Similarly, the apparent influx of H^+ ions accompanying amino acid uptake by tomato internode discs, has led to the suggestion that a proton gradient may drive amino acid transport in plants (van Bel and van Erven, 1976). This view is supported by recent evidence for transient depolarization of the cell membrane during amino acid uptake in duckweed fronds (Novacky *et al.*, 1978) and oat coleoptiles (Etherton and Rubinstein, 1978).

4 Effect of 'Shock' Procedures

The inhibition of amino acid transport by an osmotic or cold shock treatment is a common phenomenon in microorganisms (Heppel, 1969) (see Section 1.4.1). However, analogous studies with plant tissues are somewhat inconclusive. Stewart (1971) reported little or no effect of osmotic shock on amino acid transport in maize embryos. However, investigations of amino acid transport into leaves of *Phaseolus* (Amar and Reinhold, 1973) and *Avena* (Rubinstein *et al.*, 1977), showed that cold, osmotic, or EDTA shock can reduce uptake of AIB and leucine by over 80 per cent. Shock also inhibits transport in carrot cells (Wahlström and Eriksson, 1976). In *Phaseolus* these treatments caused 3.5 per cent of the cell protein to be released into the medium, which, when added back to the tissues could apparently partly restore transport, though no binding activity for AIB was detectable in the proteins of the shock fluid (Amar and Reinhold, 1973). However, in *Avena*, little protein was released by shock, and transport could not be restored by adding the protein back to the tissue (Rubinstein *et al.*, 1977). It seems that in neither case was inhibition of transport due to non-specific alteration of membrane permeability, for shock had no effect on AIB efflux in *Phaseolus*, and in *Avena* it did not greatly alter chloride uptake or protein synthesis. Under conditions of normal growth, shock inhibition is reversible within a few hours. Analogy with microorganisms would suggest the involvement of 'amino acid-binding proteins', at least in some plant tissues, but proof awaits their positive identification in shock fluids.

5 Regulation of Uptake

For many years 'ageing' of cut plant tissue has been known to increase respiration and ion transport, effects which can considerably complicate the interpretation of many results. Similar effects are seen for amino acid transport; preincubation of tissue in water (Birt and Hird, 1956; Dalgarno and Hird, 1960; Hancock, 1975) or in calcium sulphate (Shtarkshall *et al*., 1970; Wahlström and Eriksson, 1976) causes an increase in amino acid transport activity. The age of cell cultures also affects amino acid transport (King and Oleniuk, 1973; Dorée *et al*., 1972). This ageing process is poorly understood, although it apparently exerts its effect mainly on efflux rather than influx, (Shtarkshall *et al*., 1970), and has a preferential effect upon the high-affinity uptake system (Reinhold *et al*., 1970). It has been suggested that these effects may be due to CO_2 levels in the culture vessels (Dorée *et al*., 1972).

The linear rates of amino acid uptake obtained over long periods of time (up to 8 h) (Carrot slices: Birt and Hird, 1956. Mustard roots: Wright, 1962. Sugarcane cells: Maretzki and Thom, 1970. Barley leaves: Shtarkshall *et al*., 1970; Lien and Rognes, 1977. Tobacco leaf cells: Francki *et al*., 1971; Harrington and Smith, 1977. Duckweed fronds: Newton, 1974; Borstlap, 1977. *Egeria*: Petzold and Jacob, 1975), and their massive accumulation against a concentration gradient, imply that efflux of amino acids is generally negligible. This is certainly the case in the duckweed (Borstlap, 1977), although efflux may be a problem in unaged tissue (Shtarkshall *et al*., 1970; Rubin and Zaitlin, 1976). These results also indicate (for the concentrations studied) a lack of feedback inhibition of transport by accumulated amino acids and their metabolites (a situation distinct from that found with microorganisms; see Sections 1.2, 1.3, 1.5) although a few reports have suggested the contrary. For example, soybean and rapeseed cells grown under conditions of nitrogen starvation show greatly increased rates of amino acid uptake when compared with 'nitrogen-sufficient' cells. This is apparently due to inhibition by NH_4^+ ions, possibly via conversion to glutamine (King and Oleniuk, 1973; King and Khanna, 1978). Externally supplied NH_4^+ ions also inhibit phenylalanine uptake by melon root tips (Watson and Fowden, 1975). Francki *et al.* (1971) observed a positive correlation between rates of amino acid transport and rates of protein synthesis in tobacco leaf cells, and suggested that this might be due to feedback regulation of uptake.

Plant hormones may also exert a controlling influence on amino acid transport. The accumulation of amino acids caused by cytokinin is a well known, yet poorly understood phenomenon (Mothes *et al*., 1961). Reinhold and Powell (1956) have also suggested that IAA can stimulate amino acid transport (in sunflower hypocotyls) while ABA apparently has no effect on leucine uptake by duckweed (Newton, 1974*b*). This is an area which clearly warrants further investigation, particularly considering the possibility that the effects of hormones on transport could be a primary, rather than a secondary result of their action. Finally, plant parasites are also believed to affect membrane permeability of the host plant. This has been observed for amino acid uptake by squash (*Cucurbita*) infected with *Hypomyces*, although nothing is known about how this control is effected (Hancock, 1975).

It can be seen that although numerous reports of amino acid transport systems in plants have appeared, too little systematic work has been done, and further studies are needed before meaningful comparison with similar systems in other groups of organisms can be made.

6 Intracellular Transport by Organelles

In transport studies the assumption is very commonly made that the plasmalemma is the only rate-limiting membrane. Although in many cases this is justified, transport into various intracellular organelles may well be an important consideration; this being especially true in plant cells where the vacuole and other organelles can occupy a considerable proportion of the total cell volume.

Thus, in higher plants the vacuole is often a depository for excess absorbed substances, and observed concentration gradients may well represent gradients across the tonoplast, rather than the plasmalemma, the cytoplasm being more or less at equilibrium with the extracellular solution. Similarly, the rate-limiting step in absorption could be movement across the tonoplast rather than the plasmalemma. Due to the delicacy of the tonoplast, and the consequent difficulties of isolating vacuoles, studies on tonoplast permeability are lacking. However, the ability of the cell to concentrate specific compounds within the vacuole indicates that the tonoplast constitutes a selective barrier.

A whole range of metabolic processes requires the movement of amino acids from one organelle to another. The complete spectrum of amino acids is required by the mitochondria and chloroplasts for protein synthesis; in addition, glutamic and aspartic acid are important metabolites in the transfer of reducing equivalents between these organelles and the cytoplasm (Heber, 1975) and in the transport of amino groups to the mitochondria for deamidation. In addition to these two acids, glycine and serine are believed to shuttle between the mitochondria, peroxisomes, and chloroplasts during photo-respiration and operation of the glyoxylate cycle (Tolbert, 1973). Amino acids may be important products of photosynthesis, and as such must leave the chloroplast, and during C_4 photosynthesis (Hatch and Slack, 1970), aspartic acid and/or alanine are believed to shuttle between the chloroplasts of the mesophyll and bundle sheath cells. There must therefore, be a considerable flux of at least some amino acids across organellar membranes, yet, in most cases, it is not even known whether this process is carrier mediated.

In both mitochondria (Lehninger, 1971) and chloroplasts (Heldt and Rapley, 1970*a*) the outer of the two membranes seems to be freely permeable to small molecules, the inner membrane being the site of all specific transport processes.

The only study of amino acid transport in plant mitochondria indicates that glutamic acid enters via the general tricarboxylic acid carrier; there appears to be no specific glutamate/aspartate carrier as found in mammalian mitochondria (Day and Wiskich, 1977). In chloroplasts, however, glutamic and aspartic acids enter using the dicarboxylic acid carrier, displaying saturation kinetics and competitive inhibition of uptake with other dicarboxylic acids (Heldt and Rapley, 1970*b*; Heldt *et al*., 1975). Unlike mitochondria, there seems to be no relation between this dicarboxylate carrier and the movement of tricarboxylates or phosphate, and unlike plant mitochondria the tricarboxylic acid carrier will not handle glutamic and aspartic acids. Thus, there seem to be a number of important differences between glutamic and aspartic acid transport by mitochondria and chloroplasts from different sources.

Early studies showed that most, if not all, protein amino acids must be able to enter isolated chloroplasts (Ongun and Stocking, 1965; Aach and Heber, 1967; Roberts *et al*., 1970), although no detailed studies were made. Nobel and Wang (1970) using 'swelling' techniques found that isolated pea chloroplasts are freely permeable to a range of amino acids, more so than the plasmalemma. Permeability was stereospecific (Wang and Nobel, 1971) and they proposed the existence of two specific carriers, one for glycine, alanine, leucine, isoleucine, and valine, and a second for serine, threonine, and methionine (Nobel and Cheung, 1972). However, contrary to their evidence, Gimmler *et al*. (1974), using radiotracers and osmotic volume changes, found very low rates of amino acid uptake, and no evidence for carrier-mediated transport of neutral amino acids. This discrepancy has not yet been resolved. McLaren and Barber (1977), using radiotracers found similar rates of uptake to those reported by Gimmler *et al*., implying that the chloroplasts are not freely permeable to amino acids, although the observed rates of uptake would still be sufficient to support protein synthesis. They found that leucine could be accumulated against a concentration gradient, saturating at 5 mM. Uptake was competitively inhibited by isoleucine, although unlike Nobel and Cheung (1972) this leucine carrier did not handle glycine, and in fact they found no evidence for mediation of glycine uptake. These discrepancies may be accounted for by the high concentrations ($>$100 mM) of amino acid used by Nobel and his coworkers, but do serve to emphasize the need for further investigations of metabolite transport into plant organelles.

5 LONG DISTANCE TRANSPORT OF AMINO ACIDS

Long distance transport in plants is concerned with the movement of compounds from the sites of synthesis (e.g. leaves, roots) to the sites of utilization (e.g. meristems, reproductive organs), and involves a number of distinct processes: movement of compounds to the vascular tissue, loading of the xylem or phloem, translocation, and removal from the vascular elements. Although there is a considerable body of information concerning the long distance movement of ions and sugars, information specifically concerning amino acids is limited.

Movement of amino acids to and from the vascular tissue could follow one of two routes; the extracellular compartment (apoplast) or the cytoplasmic continuum, via plasmodesmata (symplast). Clearly, if the apoplast is important, cells must have the facility to specifically secrete and absorb amino acids across the plasmalemma, a process for which there is much evidence (see above). The relative importance of these two pathways is unknown, although it seems likely that both will be involved to some extent in the movements of amino acids within plant tissues.

1 Movement in the Phloem

The mechanism of phloem transport has been considered elsewhere (Zimmermann and Milburn 1975); here, specific consideration will be given only to the amino acids. Amino acids are found in the phloem at concentrations of about 1–10 mg ml^{-1} (Ziegler, 1974) although this depends upon the species and time of year (Mittler, 1958; Weatherley *et al*., 1959; Zimmermann, 1960; Tammes and van Die, 1964; Hall and Baker, 1972; Pate *et al*., 1975). For example, the total amino acid content of the phloem may increase by up to tenfold on leaf senescence, with all the protein amino acids becoming abundant (Peel and Weatherley, 1959; Weatherley *et al*., 1959; Zimmerman, 1960; Chopowick and Forward, 1974).

Although the whole spectrum of protein amino acids may be found in the phloem, two or three amino acids generally comprise 70–100 per cent of the total amino acid content, the others being present in trace amounts or absent. Qualitative, as well as quantitative, variations also occur, depending upon the time of year (Leckstein and Llewellyn, 1975). Normally, the primary amino acids are the 'acidics' and their amides, although occasionally others may predominate. Thus, the most abundant amino acids in the castor bean (*Ricinus*) are threonine, aspartic acid, and glutamic acid (Hall and Baker, 1972), while asparagine replaces threonine in willow (Peel and Weatherley, 1959) and the lupin (Pate *et al*., 1975). Oak (*Quercus*), false acacia (*Robina*) (Ziegler, 1956), and ash (*Fraxinus*) (Mittler, 1958) also have a similar phloem composition. However, in the palm (*Yucca*) (Tammes and van Die, 1964) and germinating castor bean (Stewart and Beevers, 1967) glutamine is the only abundant amino acid, while in soybeans (*Glycine*) serine predominates.

Although all amino acids are capable of movement in the phloem, (Crafts and Crisp, 1971) metabolic processes normally restrict those available for transport. This is borne out by following the fate of labelled amino acids supplied to leaves. Only certain applied amino acids enter the phloem directly, others are extensively metabolized. For example, in *Avena* and *Glycine*, leucine and lysine enter the phloem unaltered, while added serine is metabolized, the carbon skeleton being mainly converted to sucrose (Housley *et al.*, 1977; Peterson *et al.*, 1977). Similarly, aspartic acid, glycine, and alanine in corn (Hofstra and Nelson, 1969) and alanine in sunflower (Chopowick and Forward, 1974) are converted to sucrose prior to translocation, while other amino acids enter the phloem unaltered (Nelson and Gorham, 1959; Yamaguchi and Islam, 1967; Eschrich and Hartmann, 1969; Gardner and Peel, 1971), In the castor bean, added alanine, glutamic acid, and aspartic acid are converted to glutamine before entering the phloem, while valine is transported intact (Stewart and Beevers, 1967).

On supplying [^{14}C]CO_2 to leaves, similar results are obtained. Only certain labelled amino acids appear in the phloem (Fisher, 1970), for example, serine, glutamine, alanine, and glutamic acid in broom (*Spartium*) (Pate *et al.*, 1975), while in the soybean only labelled serine appears (Nelson *et al.*, 1961; Nelson, 1962), though later asparagine, alanine, aspartic acid, and glutamic acid are also found (Clauss *et al.*, 1964).

Little is known concerning the entry of amino acids into, and their removal from, the phloem, although it is believed to be accomplished by highly specialized 'transfer cells' (Gunning *et al.*, 1968; Pate and Gunning, 1972). Entry appears to be selective (Kursanov *et al.*, 1960; Nelson *et al.*, 1961; Brovchenko, 1964) as amino acids do not appear in the phloem in the same ratios as they occur in the surrounding tissues. Loading also apparently shows stereospecificity (Joy and Antcliff, 1966; Eschrich and Hartmann, 1969), saturation kinetics (Peterson *et al.*, 1977) and relies upon metabolic energy (Kursanov and Brovchenko, 1961; Kursanov, 1963; Brovchenko, 1964).

The possibility of an important role for peptide translocation does not seem to have been considered, although it might well be advantageous to protect transported amino acids from metabolic processes in the phloem by combining them in peptide linkages. There seems to be scope here for specific studies of transport across plant cell membranes, especially in species with large, easily isolated vascular tissues.

2 Movement in the Xylem

Free amino acids are found in the xylem, but generally at much lower concentrations than in the phloem (Pate *et al*., 1975) and, as in the phloem, only a small number of amino acids tend to predominate: aspartic acid, asparagine, glutamine, and homoserine in the field pea (Brennan *et al*., 1964; Pate *et al*., 1964); glutamine, asparagine, aspartic acid, and arginine in apple trees (Cooper *et al*., 1972; Tromp and Ovaa, 1976), glutaoine, glutamic acid, and γ-aminobutyric acid in the tomato (van Die, 1959), while glutamine alone predominates in the willow (Sauter, 1976). Other amino acids may be present in small quantities, sulphur often being transported as cysteine or methionine (Pate, 1965), and small peptides have also been reported (Pollard and Sproston, 1954; Fejer and Konya, 1958; Virtanen and Miettinen, 1963). However, the relative levels of each amino acid are considerably more variable than is found for the phloem, depending upon age, season, and nitrogen source (Kulayeva *et al*., 1957; Stoev *et al*., 1959; Pate, 1962; Pate and Wallace, 1964; Cooper *et al*., 1972; Tromp and Ovaa, 1976; Sauter, 1976). Thus, nodulated legumes have high levels of aspartic acid, glutamine, and asparagine in the xylem, while unnodulated strains contain only aspartic acid in any quantity (Wieringa and Bakhuis, 1975). This is in agreement with evidence suggesting that glutamine and asparagine are the most important forms of nitrogen secreted by symbiotic microorganisms (Watanabe, 1951; Zelitch *et al*., 1951).

Xylem is a dead tissue, thus, any active secretion into, or removal of compounds from, the xylem must be due to neighbouring parenchymatous cells, again probably 'transfer cells' (Gunning and Pate, 1969). Few details are known concerning the entry and removal of amino acids from the xylem, although the indications are that specific transport systems, rather than simple diffusion, may be involved. The relative concentrations of different amino acids vary as the sap passes up the stem suggesting a selective removal process (Brennan *et al*., 1964; Pate *et al*., 1964; van Die and Vonk, 1967; van Bel, 1974). The effects of pH upon amino acid removal from tomato xylem also indicates that the process is selective (van Bel and Hermans, 1977) as does the apparent stereospecificity of the process (Joy and Antcliff, 1966; van Die and Vonk, 1967).

6 GERMINATING SEEDLINGS

Many seeds contain substantial protein storage reserves, which are mobilized upon germination; the hydrolytic products and their metabolites being transported to the growing embryo (Mayer and Poljakoff-Mayber, 1975). Although in both monocots (Brown, 1946; Folkes *et al* 1952;

McConnell, 1957; Folkes and Yemm, 1958; Ingle *et al*., 1964) and dicots (Oota *et al*., 1953; Larson and Beevers, 1965; Stewart and Beevers, 1967), nitrogen is transferred from the endosperm and/or cotyledons to the growing embryo, the structure and metabolism of monocot seeds makes them particularly amenable to study (Figure 1).

The monocot storage material, contained in the endosperm and aleurone layer, is separated from the embryo by an absorptive organ called the scutellum (Sargant and Robertson, 1905; Avery, 1930). The scutellum is readily separated from the endosperm without significant damage to the tissue, exposing the scutellar epithelium (Zamski, 1973) across which all compounds must be transported before they can be utilized by the growing embryo. Thus, uptake by the scutellum can be studied *in vitro* without the presence of cut or damaged tissue. This is in marked contrast to most plant transport studies where excised or fragmented tissues must be used.

In many dicots the main storage reserves are contained within the cotyledons which are inseparable from the embryo without damage to the tissue. There is no epithelial connection or absorptive organ across which compounds must be transported. In fact, there is not even a necessity for transport across the plasmalemma, for movement could be entirely symplastic, via the plasmodesmata. One exception is the endospermic

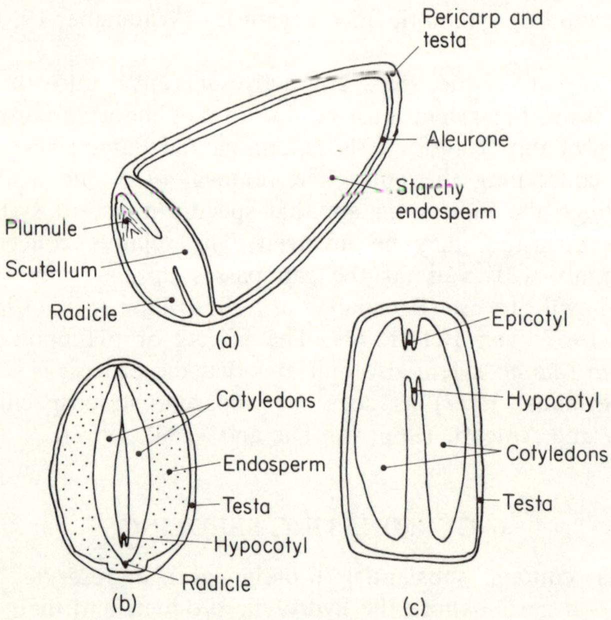

Figure 1 Diagram showing the structures of three typical seeds: (a) barley (monocot), (b) castor bean (endospermic dicot), and (c) broad bean (non-endospermic dicot)

dicots (e.g. castor bean) where a large part of the storage protein is contained in the endosperm. The hydrolytic products of these proteins must be absorbed through the cotyledonary or embryo surfaces, easily separable from the endosperm, and therefore providing a useful system for direct comparison with monocots (Figure 1).

In dicots, extensive metabolism of amino acids occurs in the cotyledons prior to transport to the embryo, especially in dark-grown seedlings. Amino acid nitrogen is generally transported as glutamine, although some other protein amino acids may also be transported intact (Stewart and Beevers, 1967; Mayer and Poljakoff-Mayber, 1975). In the pea, however, homoserine is the most abundant transported amino acid (Larson and Beevers, 1965), and in cotton, asparagine predominates (Capdevila and Dure, 1977). In the castor bean the formation of sugars from amino acids, prior to transport, also seems to play an important role (Stewart and Beevers, 1967). Thus, few generalizations concerning amino acid interconversions can yet be made. Similarly, the mechanism of nitrogen transfer to the embryo of dicots is poorly understood. It may be apoplastic, the young xylem and phloem elements may be involved, or it may be simple cell to cell transfer.

In monocots the situation seems to be less complex. Most amino acids are present in the storage proteins of the endosperm in approximately the quantities required for protein synthesis by the embryo, and may therefore be released by hydrolysis and transported to the embryo unchanged (Jones and Pierce, 1966, 1967). Some interconversions and transaminations do occur but, unlike dicots, these mostly occur after transport and within the growing embryo (Folkes and Yemm, 1958; Joy and Folkes, 1965; Oaks, 1965*a*; Pragnell *et al*., 1969). The free amino acid pools of both endosperm and embryo are reasonably balanced (Jones and Pierce, 1967; C. F. Higgins and J. W. Payne, unpublished results), although high levels of glutamine do become an important transport compound in shoots and roots of older seedlings, once they have begun to assimilate inorganic nitrogen (Yemm and Willis, 1956; Oji and Izawa, 1972; Margaris and Thanos, 1974).

1 Amino Acid Transport

Protein synthesis and growth are poor in isolated monocot embryos. However, the role of the endosperm can be replaced by a balanced mixture of amino acids (Kent and Brink, 1947; Harris, 1956; Folkes, 1959; Oaks and Beevers, 1964; Norstog, 1973), although when supplied individually amino acids are generally inhibitory (Brown, 1906; Harris, 1956), indicating that the scutellum can absorb amino acids. A number of studies have shown that exogenously supplied ^{14}C-labelled amino acids are taken up by isolated embryos, and incorporated into protein, although some interconversions may occur within the embryo prior to incorporation (Joy and Folkes, 1965; Pragnell *et al*., 1969). Exogenous amino acids prevent the *de novo* synthesis of amino acids from carbohydrates within the embryo (Joy and Folkes, 1965; Oaks, 1965*b*; Pragnell *et al*., 1969) although this is apparently not the case for proline (Jones and Pierce, 1967).

In the only detailed study on the scutellar absorption of amino acids, Stewart (1971) showed that glutamine was apparently taken up by a mediated process. Uptake had a broad, acidic pH optimum, exhibited stereospecificity, but was not 'shock' sensitive. It was inhibited by monovalent cations, and competitively by some, but not all, L-amino acids. We have also observed uptake of a variety of L-amino acids by isolated barley embryos, though D-isomers are not absorbed (Higgins and Payne, 1978*a*,*c*, and unpublished data).

Evidence concerning the control of amino acid movement from the endosperm is unclear. The initial activation of proteases and peptidases is hormonal (Jacobsen and Varner, 1967), but it is not yet clear whether there is further control of enzymic activity or amino acid transport by end-product inhibition. Using maize, Oaks (1965*c*) found that exogenously supplied amino acids reduced the rate of nitrogen loss from the endosperm, and proposed that proteolysis and/or transport might be controlled by a feedback mechanism. However, results obtained using barley do not entirely accord with this proposal. Large concentrations of amino acids can accumulate in the embryo, suggesting a lack of feedback inhibition of transport, and, similarly, no effect of accumulated amino acids on peptidase activity could be detected (Jones and Pierce, 1966, 1967).

2 Peptide Transport

Although it has generally been assumed that the transfer of nitrogenous compounds from the endosperm to the embryo of monocot seedlings occurs as amino acids, recent findings have shown that peptide transport may also be of considerable importance (Higgins and Payne, 1978*a*; Sopanen *et al*., 1978).

A number of earlier reports indicated that peptide transport might be involved. Peptides were detected in barley endosperm (Grant and Wang, 1972; C. F. Higgins and J. W. Payne, unpublished results) and embryo (Folkes and Yemm, 1958), and an endosperm leachate, containing a high proportion of peptides was shown to satisfy the nitrogen requirements of isolated maize embryos (Oaks and Beevers, 1964). The relatively high levels of peptidase activity in the embryo of germinating barley seedlings, compared with levels in the endosperm, also suggested that peptides may be transported to the embryo prior to hydrolysis (Mikola and Kolehmainen, 1972).

Using a sensitive, fluorescent detection technique we have now extensively characterized a peptide transport system in germinating barley embryos (Higgins and Payne, 1977*b*) and the general agreement between results obtained using this technique and radiotracers (Sopanen *et al.*, 1978) is encouraging. Demonstration of peptide transport was first achieved using the non-physiological peptides Gly-Sar and Gly-Sar-Sar. These peptides have *N*-methylated peptide bonds, making them exceptionally resistant to enzymic hydrolysis. Both peptides are accumulated intact by embryos against a concentration gradient (Higgins and Payne 1977*a*,*b*; Burston *et al.*, 1977; Sopanen *et al.*, 1977). The fact that intact Gly-Sar can be detected in the shoots of embryos which have been incubated such that only the scutellar surface has come into contact with the peptide solution necessarily requires that the dipeptide is first transported across the scutellum, and this feature together with the extremely high levels of peptide which accumulate, eliminates the possibility that absorption might arise from non-specific surface binding. Furthermore, uptake is prevented by anaerobic conditions and a range of metabolic inhibitors (Higgins and Payne, 1977*b*). Thus, although the strict criteria for an active process have not been demonstrated, in that it has not been shown that the peptide exists in free solution once absorbed by the embryo, or that it is accumulated against an electrochemical gradient, it appears that peptide transport is an active process.

The mode of energization for this process is, however, unclear. The absence of a sodium requirement (Higgins and Payne, 1977*b*; Sopanen *et al.*, 1978) implies that, unlike peptide transport in the mammalian gut (Matthews, 1975), a sodium-dependent mechanism does not operate. There are, however, indications that a proton gradient may be involved in the energization of transport. Thus, acetate inhibits peptide uptake, and causes exodus of amino acids at pH values below its pK_a. This is usually interpreted as being due to an 'uncoupling' effect, with the undissociated form of the acetate molecule acting as a proton shuttle, disrupting the H^+ ion distribution across the plasmalemma (Higgins and Payne, 1977*b*).

The rate of Gly-Sar transport is essentially linear for periods of up to 6 h, and uptake exhibits saturation kinetics (Higgins and Payne, 1977*b*; Sopanen *et al.*, 1977). Unlike amino acid transport systems in plants, biphasic kinetics have not (yet) been observed, although the limited concentration ranges used may not have revealed the true situation. Transport shows an acidic pH optimum at about pH 3.8, although considerable uptake occurs over the range of pH 3.5–7.0. The pH optimum for uptake corresponds with the pH optimum for the activity of several endosperm proteases (Burger *et al.*, 1970; Sundblom and Mikola, 1972), and also the reported pH of the endosperm (Kirsi and Mikola, 1971). The pH optimum for uptake of glycylisoleucine and trialanine is also about pH 4 (C. F. Higgins and J. W. Payne, unpublished results).

Sopanen *et al*. (1978) reported a slightly higher pH optimum for diglycine uptake, with maximum uptake occurring over the range pH 4–5. Amino acid uptake by maize embryos also shows an acidic pH optimum (Stewart, 1971). Buffer composition has little effect on uptake; interchange of phosphate, tris, and citrate buffers gives similar rates of uptake, as does substitution of K^+ for Na^+ (Higgins and Payne, 1977*b*). However, at high concentration Na^+ and K^+ may become somewhat inhibitory (Sopanen *et al*., 1978).

In addition to Gly-Sar, a range of physiological peptides is also taken up from the medium by barley embryos (Higgins and Payne, 1978*a,c*). Although intracellular peptidase activity is too high to allow detection of these peptides intact within the embryo, a number of results lead to the conclusion that they are indeed absorbed intact and subsequently hydrolysed intracellularly. Extracellular hydrolysis, followed by amino acid uptake, is precluded for two reasons. First, although peptide uptake is competitively inhibited by a wide range of other peptides, amino acids have no effect (Higgins and Payne, 1978*a*; Sopanen *et al*., 1977, 1978). Second, only low levels of free amino acid appear in the medium during incubation with a peptide. This eliminates the possibility of extracellular hydrolysis and amino acid uptake, unless hydrolysis were accompanied by rapid amino acid uptake, and the rates of amino acid uptake from free solution are demonstrably too slow to permit this explanation (Higgins and Payne, 1978*a*). Perhaps the most convincing argument for intact peptide transport concerns the uptake of certain D-amino acid residues; free D-amino acids are not absorbed by barley embryos, yet when they are presented in peptide form they may be taken up at a considerable rate (Higgins and Payne, 1978*c*). The possibility that vectorially orientated hydrolysis occurs during membrane translocation such that intact peptides never actually appear intracellularly has been considered in relation to other peptide transport systems (Matthews, 1975; Payne, 1977) and must also be considered here. However, the intact absorption of sarcosyl peptides, and some peptides containing D-residues (Higgins and Payne, 1978*c*), clearly shows that transport and hydrolysis are not necessarily linked. The fact that competitive inhibition is observed between these peptides and a range of physiological peptides implies that all peptides can be transported by this system, which does at least have the ability to mediate transport in the absence of hydrolysis (Higgins and Payne, 1978*b*).

Use of sarcosyl peptides and a wide range of physiological peptides has shown that the peptide transport system is capable of handling both di- and tripeptides. However, the utilization of larger peptides is still somewhat ambiguous. Only small amounts of Gly-Sar-Sar-Sar are absorbed, which could be accounted for by non-specific uptake. Although some tetra- and pentapeptides are removed quite rapidly from the medium, and also exert competitive effects, the possibility that this is a result of partial extracellular hydrolysis, although unlikely, has not been entirely eliminated (Sopanen *et al.*, 1977; Higgins and Payne, 1978*a*,*b*). Competition experiments have shown that di-, tri-, and higher oligopeptides are transported by the same system; the complete inhibition of dipeptide uptake by competing tripeptides, and *vice versa*, indicates that only one transport system exists (Sopanen *et al.*, 1977; Higgins and Payne, 1978*b*). The possibility that several systems exist, each capable of handling di- and tripeptides, or that there is a second system with specific structural requirements, still exists, although logically this would not be expected. The variety of peptides with very dissimilar amino acid side chains (e.g. proline, glutamic acid, lysine, and leucine) which compete for transport also argues against this.

Investigations of the structural requirements for peptide uptake are as yet incomplete. There seems to be no major discrimination against particular amino acid side-chains; all peptides so far examined appear to be handled by the same system and, as in all other organisms studied to date, *N*-methylation of the peptide bond reduces affinity but does not prevent transport.

Peptide transport shows stereospecificity (Higgins and Payne, 1978*a*,*c*). The presence of a D-amino acid residue in a peptide generally reduces its rate of transport, often to zero, although the location of the residue in the peptide is important. A D-residue at the *N*-terminus often has little or no effect, while at the *C*-terminus of di- or tripeptides it can totally inhibit uptake. This is particularly surprising in the case of tripeptides, and would seem to suggest that, unlike bacteria, (see Section 2.1) oligopeptide transport has a requirement for the *C*-terminus, but lacks specificity towards the *N*-terminus (see below). Some peptides containing D-residues are accumulated intact, showing the transport system and the hydrolases have different specificities.

The *N*- and *C*-terminal requirements for peptide transport have now been studied in some detail (C. F. Higgins and J. W. Payne, unpublished results). Although one would not expect a system which will accommodate both di- and oligopeptides to have requirements for both termini (as binding sites aligned for the α-amino and α-carboxyl groups of a dipeptide cannot also be aligned for the same groups in a tripeptide and *vice versa*) both seem to be important in barley. It is clear that substitution (as an ester or amide) or complete removal of the *C*-terminal carboxyl group of a peptide considerably impairs, although does not totally inhibit, its uptake. Substitution of the *N*-terminal amino group (by acylation or alkylation) also has a drastic effect on uptake. However, an imino group (proline) can be handled regardless of its position in the peptide.

A number of differences have been observed between the peptide transport system(s) of barley and those of other groups of organisms. It is pertinent to consider these in relation to the environment in which the system operates; specific variations between species would be expected to reflect specific biological functions. This is especially relevant in the case of barley embryos which differ from all other systems examined so far, in that they exist in a controlled environment. Thus, the size, amino acid composition, and concentration of peptides is relatively constant and predetermined, unlike the enormous fluctuations that may be presented to bacteria or yeasts in their natural environments. Thus, in barley one could envisage specific peptide transport systems with, say, a requirement for a particular *N*-terminal residue, reflecting the specificities of protease action within the endosperm. It is our hope that comparison of environmental differences may shed some light on the reasons for the considerable variability, and also the common factors, amongst peptide transport systems of different species.

As mentioned, differences have been found between the peptide transport system in barley and those in the varied microorganisms that have been studied. It is conceivable that in the future these differences may be put to use in designing systemic herbicides, based on the 'smugglin' principle (see Sections 2.1, 4.7). Thus, a toxic peptide could be designed such that it will enter and kill a pathogenic organism, yet, due to the differences in transport specificities, it would be unable to enter the host cells, and thus consequently be harmless.

These studies on peptide transport in barley have shown that peptides may be of considerable importance in certain plant systems, and it is our hope that they will emphasize the need for further investigations of the role of peptides in plants, a field which has been sorely neglected.

7 CONCLUSIONS

Throughout this chapter we have indicated fields where we believe further research may be profitable, and have shown that plant systems can be amenable to contemporary biochemical study. However, there is also a need for comparative studies, to determine whether information gained so far, with just a few species, is applicable throughout the higher plant kingdom, or, if differences arise, why? In particular, information gained in higher plants may well provide a new outlook on certain aspects of metabolite transport, and thus help to solve problems which have arisen as a result of studies on other groups of organisms.

ACKNOWLEDGEMENTS

Financial support for our own research described here was provided by the Science Research Council.

REFERENCES

Aach, H. G. and Heber, U. (1967). *Z. Pflanzenphysiol.*, **57**, 317–328
Amagase, S. (1972). *J. Biochem., Tokyo*, **72**, 73–81
Amagase, S., Mori, M., and Nakayama, S. (1972). *J. Biochem., Tokyo*, **72**, 765–767
Amar, L. and Reinhold, L. (1973). *Plant Physiol.*, **51**, 620–625
von Arnold, S. and Eriksson, T. (1977). *Physiol. Plant.*, **39**, 257–260
Avery, G. S. (1930). *Bot. Gaz.*, Chicago, **89**, 1–39
Behrend, J. and Mateles, R. I. (1975). *Plant Physiol.*, **56**, 584–589
van Bel, A. J. E. (1974). *Acta Bot. Neerl.*, **23**, 305–313
van Bel, A. J. E. and van Erven, A. (1976). *Z. Pflanzenphysiol.*, **80**, 74–76
van Bel, A. J. E. and Hermans, H. P. (1977). *Z. Pflanzenphysiol.*, **84**, 413–418
Birt, L. M. and Hird, F. J. R. (1956). *Biochem. J.*, **64**, 305–311
Birt, L. M. and Hird, F. J. R. (1958*a*). *Biochem. J.*, **70**, 277–286
Birt, L. M. and Hird, F. J. R. (1958*b*). *Biochem. J.*, **70**, 286–292
Bollard, E. G. (1966). *Plant Soil*, **25**, 153–166
Borstlap, A. C. (1977). *Acta Bot. Neerl.*, **26**, 115–128
Bourgin, J.–P. (1976). *Z. Naturforsch.*, **31c**, 337–338
Brennan, H., Pate, J. S., and Wallace, W. (1964). *Ann. Bot., Lond.*, **28**, 527–540
Bright, S. W. J., Wood, E. A., and Miflin, B. J. (1978). *Planta*, **139**, 113–117
Brovchenko, M. I. (1964). *Sov. Plant Physiol.*, **10**, 349–355
Brown, H. T. (1906). *Trans. Guiness Res. Lab.*, **1**, 288–299
Brown, R. (1946). *Ann. Bot., Lond.*, **10**, 73–96
Burger, W. C., Prentice, N., Moeller, M., and Robbins, G. S. (1970). *Phytochemistry*, **9**, 49–58
Burston, D., Marrs, T. C., Sleisenger, M. H., Sopanen, T., and Matthews, D. M. (1977). *Ciba Found. Symp.*, **50**, 79–108
Cameron-Mills, V. and Duffus, C. M. (1977). *Ann. Bot.*, Lond., **41**, 1117–1127
Capdevila, A. M. and Dure, L., III (1977). *Plant Physiol.*, **59**, 268–273
Chandler, G. E. and Anderson, J. W. (1976*a*). *New Phytol.*, **76**, 129–141
Chandler, G. E. and Anderson, J. W. (1976*b*). *New Phytol.*, **77**, 625-634
Cheung, Y.–N. S. and Nobel, P. S. (1973). *Plant Physiol.*, **52**, 633–637
Chopowick, R. E. and Forward, D. F. (1974). *Plant Physiol.*, **53**, 21–27
Clancy, F. G. and Coffey, M. D. (1977). *Can. J. Bot.*, **55**, 480–488
Clauss, H., Mortimer, D. C., and Gorham, P. R. (1964). *Plant Physiol.*, **39**, 269–273
Cooper, D. R., Hill-Cottingham, D. G., and Shorthill, M. J. (1972). *J. Exp. Bot.*, **23**, 247–254
Crafts, A. S. and Crisp, C. E. (1971). *Phloem Transport in Plants*, Freeman, San Francisco
Dalgarno, L. and Hird, F. J. R. (1960). *Biochem J.*, **76**, 209–215
Darwin, C. (1875). *Insectivorous Plants*, John Murray, London
Darwin, C. (1878). *J. Linn. Soc., Lond., Bot.*, **17**, 17–32
Day, D. A. and Wiskich, J. T. (1977). *Plant Sci. Lett.*, **9**, 33–36
van Die, J. (1959). *Proc. K. Ned. Wet. Ser. C*, **62**, 50–58
van Die, J. and Vonk, C. R. (1967). *Acta Bot. Neerl.*, **16**, 147–152
Dorée, M., Leguay, J. J. and Terrine, C. (1972). *Physiol. Veg.*, **10**, 115–131
Dore Swamy, R. and Mohan Ram, H. Y. (1969). *Phytomorphology*, **19**, 363–371
Dunham, V. L. and Bryan, J. K. (1969). *Plant Physiol.*, **44**, 1601–1608
Dunham, V. L. and Bryan, J. K. (1971). *Plant Physiol.*, **47**, 91–97
Earle, E. D. and Torrey, J. G. (1965). *Am. J. Bot.*, **52**, 891–899
Epstein, E. (1966). *Nature, Lond.*, **212**, 1324–1327
Eschrich, W. and Hartmann, T. (1969). *Planta*, **85**, 213–227
Etherton, B. and Rubinstein, B. (1978). *Plant Physiol.*, **61**, 933–937
Fabian-Galan, G. and Sălăgeanu, N. (1968). *Rev. Roum. Biol., Ser. Bot.*, **13**, 275–280
Fejer, D. and Konya, E. (1958). *Naturwissenschaften*, **45**, 387–388
Filner, P. (1965). *Exp. Cell Res.*, **39**, 33–39
Filner, P. (1966). *Biochim. Biophys. Acta*, **118**, 299–310
Fisher, D. B. (1970). *Plant Physiol.*, **45**, 107–113
Folkes, B. F. (1959). *Soc. Exp. Biol. Symp.*, **13**, 126–147

Folkes, B. F. and Yemm., E. W. (1958). *New Phytol.*, **57**, 106–131

Folkes, B. F., Willis, A. J., and Yemm, E. W. (1952). *New Phytol.*, **51**, 317–341

Frank, E. M., Riker, A. J., and Dye, S. L. (1951). *Plant Physiol.*, **26**, 258–267

Francki, R. I. B., Zaitlin, M., and Jensen, R. G. (1971). *Plant Physiol.*, **48**, 1418

Fukunaga, Y., King, J., and Child, J. J. (1978). *Planta*, **139**, 199–202

Furuhashi, K. and Yatazawa, M. (1970). *Plant Cell Physiol.*, **11**, 559–567

Galston, A. W. (1948). *Am. J. Bot.*, **35**, 281–287

Gamborg, O. L. (1970). *Plant Physiol.*, **45**, 372–375

Gamborg, O. L., Miller, R. A., and Ojima, K. (1968). *Exp. Cell Res.*, **50**, 151–158

Gardner, D. C. J. and Peel, A. J. (1971). *Phytochemistry*, **10**, 2385–2387

Gautheret, R. J. (1935). Thesis, University of Paris

Gautheret, R. J. (1939). *C. R. Acad. Sci.*, **208**, 118–120

Gimmler, H., Schäfer, G., Kraminer, H., and Heber, U. (1974). *Planta*, **120**, 47–61

Grant, D. R. and Wang, C. C. (1972). *Cereal Chem.*, **49**, 201–207

Gunning, B. E. S. and Pate, J. S. (1969). *Protoplasma*, **68**, 107–133

Gunning, B. E. S., Pate, J. S., and Briarty, L. G. (1968). *J. Cell Biol.*, **37**, C7–12

Haberlandt, G. (1902). *Sitzungsber. Akad. Wiss. Wien, Math.-Naturwiss. Kl., Abt. 1*, **111**, 69–92

Hall, S. M. and Baker, D. A. (1972). *Planta*, **106**, 131–140

Hancock, J. G. (1975). *Physiol. Plant Pathol.*, **6**, 65–73

Harder, R. and Zemlin, I. (1967). *Planta*, **73**, 181–193

Harder, R. and Zemlin, I. (1968). *Planta*, **78**, 72–78

Harrington, H. M. and Smith, I. K. (1977). *Plant Physiol.*, **60**, 807–811

Harris, G. P. (1956). *New Phytol.*, **55**, 253–268

Hatch, M. D. and Slack, C. R. (1970). *Annu. Rev. Plant Physiol.*, **21**, 141–162

Heber, U. (1975). In *Proceedings of the 3rd International Congress on Photosynthesis Research* (M. Avron, ed.), Elsevier, Amsterdam, pp. 1335–1348

Heimer, Y. M. and Filner, P. (1970). *Biochim. Biophys. Acta*, **215**, 152–165

Heimer, Y. M. and Filner, P. (1971). *Biochim. Biophys. Acta*, **230**, 362–372

Heldt, H. W. and Rapley, L. (1970*a*). *FEBS Lett.*, **7**, 139–142

Heldt, H. W. and Rapley, L. (1970*b*). *FEBS Lett.*, **10**, 143–148

Heldt, H. W., Sauer, F., and Rapley, L. (1972). In *Proceedings of the 2nd International Congress on Photosynthesis Research* (G. Forti, M. Avron, and A. Melandri, eds.), Elsevier, Amsterdam, pp. 1345–1355

Heldt, H. W., Fliege, R., Lehner, K., Milovancev, M., and Werdan, K. (1975). In *Proceedings of the 3rd International Congress on Photosynthesis Research* (M. Avron, ed.), Elsevier, Amsterdam, pp. 1369–1379

Henke, R. R., Wilson, K. G., McClure, J. W., and Treick, R. W. (1974). *Planta*, **116**, 333–345

Hepburn, J. S., Jones, F. M., and St. John, E. Q. (1927). *Trans. Wagner Free Inst. Sci. Philadelphia*, **11**, 1–50

Heppel, L. A. (1969). *J. Gen. Physiol.*, **54**, 95s–109s

Heslop-Harrison, Y. (1976). *Endeavour* **35**, 114–122

Heslop-Harrison, Y. (1978). *Scient. Am.*, **238**, 104–115

Heslop-Harrison, Y. and Knox, R. B. (1971). *Planta*, **96**, 183–211

Higgins, C. F. and Payne, J. W. (1977*a*). *Planta*, **134**, 205–206

Higgins, C. F. and Payne, J. W. (1977*b*). *Planta*, **136**, 71–76

Higgins, C. F. and Payne, J. W. (1978*a*). *Planta*, **138**, 211–216

Higgins, C. F. and Payne, J. W. (1978*b*). *Planta*, **138**, 217–221

Higgins, C. F. and Payne, J. W. (1978*c*). *Planta*, **142**, 299–305

Hillman, W. S. (1961). *Bot. Rev.*, **27**, 221–287

Hofstra, G. and Nelson, C. D. (1969). *Can. J. Bot.*, **47**, 1435–1442

Housley, T. L., Peterson, D. M., and Schrader, L. E. (1977). *Plant Physiol.*, **59**, 217–220

Ingle, J., Beevers, L., and Hageman, R. H. (1964). *Plant Physiol.*, **39**, 735–740

Jacobsen, J. V. and Varner, J. E. (1967). *Plant Physiol.*, **42**, 1596–1600

Jentsch J. (1972). *FEBS Lett.*, **21**, 273–276

Jones, M. and Pierce, J. S. (1966). *Proc. Am. Soc. Brew. Chem.*, **10**, 10–17

Jones, M., and Pierce, J. S. (1967) *J. Inst. Brew., Lond.*, **73**, 577–583

Joy, K. W. (1969). *Plant Physiol.*, **44**, 845–848
Joy, K. W. and Antcliff, A. J. (1966). *Nature, Lond.*, **211**, 210–211
Joy, K. W. and Folkes, B. F. (1965). *J. Exp. Bot.*, **16**, 646–666
Kao, K. N. and Michayluk, M. R. (1975). *Planta*, **126**, 105–110
Kent, N. F. and Brink, R. A. (1947). *Am. J. Bot.*, **34**, 600–601
King, J. (1976). *Can. J. Bot.*, **54**, 1316–1321
King, J. and Hirji, R. (1975). *Can. J. Bot.*, **53**, 2088–2091
King, J. and Khanna, V. (1978). *Planta*, **139**, 193–197
King, J. and Oleniuk, F. H. (1973). *Can. J. Bot.*, **51**, 1109–1114
Kirsi, M. and Mikola, J. (1971). *Planta*, **96**, 281–291
Kulayeva, O. N., Silina, E. I., and Kursanov, A. L. (1957). *Sov. Plant Physiol.*, **4**, 497–504
Kursanov, A. L. (1963). *Adv. Bot. Res.*, **1**, 209–278
Kursanov, A. L., and Brovchenko, M. I. (1961). *Sov. Plant Physiol.*, **8**, 211–217
Kursanov, A. L., Brovchenko, M. I., and Pariiskaya, A. N. (1960). *Sov. Plant Physiol.*, **6**, 544–552
Laites, G. C. (1969). *Annu. Rev. Plant Physiol.*, **20**, 89–116
Larson, L. A. and Beevers, H. (1965). *Plant Physiol.*, **40**, 424–432
Leckstein, P. M. and Llewellyn, M. (1975). *Planta*, **124**, 89–91
Lehninger, A. L. (1971). *Biomembranes*, **2**, 147–164
Lien, R. and Rognes, S. E. (1977). *Physiol. Plant.*, **41**, 175–183
Lloyd, F. E. (1942). *The Carnivorous Plants*, Chronica Botanica Co., Waltham, Mass.
Lüttge, U. (1965). *Planta*, **66**, 331–344
McConnell, W. B. (1957). *Can. J. Biochem. Physiol.*, **35**, 1259–1266
McLaren, J. S. and Barber, D. J. (1977). *Planta*, **136**, 147–151
Maretzki, A. and Nickell, L. G. (1967). *7th Int. Congr. Biochem. Asbstr.*, J-207
Maretzki, A. and Thom, M. (1970). *Biochemistry*, **9**, 2731–2736
Maretzki, A., Nickell, L. G., and Thom, M. (1969). *Physiol. Plant.*, **22**, 827–839
Margaris, N. S. and Thanos, C. A. (1974). *Phytochemistry*, **13**, 1645–1647
Matthews, D. M. (1975). *Physiol. Rev.*, **55**, 537–608
Matthews, D. M. and Payne, J. W. (1975). *Peptide Transport in Protein Nutrition*, North Holland/American Elsevier, Amsterdam/New York
Mayer, A. M. and Poljakoff-Mayber, A. (1975). *The Germination of Seeds*, 2nd edn, Pergamon Press, Oxford.
Miflin, B. J. (1969). *J. Exp. Bot.*, **20**, 810–819
Miflin, B. J. (1971). *Arch. Biochem. Biophys.*, **146**, 542–550
Mikola, J. and Kolehmainen, L. (1972). *Planta*, **104**, 167–177
Mitchell, J. P. and Gildow, F. E. (1975). *Physiol. Plant.*, **34**, 250–253
Mittler, T. C. (1958). *J. Exp. Biol.*, **35**, 74–84
Mothes, K., Englebrecht, L. and Schütte, H. R. (1961). *Physiol. Plant.*, **14**, 72–75
Nakayama, S. and Amagase, S. (1968). *Proc. Jpn. Acad.*, **44**, 358–362
Naylor, A. W., Rabson, R. and Tolbert, N. E. (1958). *Physiol. Plant.*, **11**, 537–547
Naylor, A. W. and Tolbert, N. E. (1956). *Physiol. Plant.*, **9**, 220–229
Nelson, C. D. (1962). *Can. J. Bot.*, **40**, 757–770
Nelson, C. D. and Gorham, P. R. (1959). *Can. J. Bot.*, **37**, 431–438
Nelson, C. D., Clauss, H., Mortimer, D. C. and Gorham, P. R. (1961). *Plant Physiol.*, **36**, 581–588
Newton, R. J. (1974*a*) *Plant Cell Physiol.*, **15**, 249–254
Newton, R. J. (1974*b*). *Physiol. Plant.*, **30**, 108–112
Nickell, L. G. (1956). *Proc. Natl. Acad. Sci. U.S.A.*, **42**, 848–850
Nickell, L. G. and Burkholder, P. R. (1950). *Am. J. Bot.*, **37**, 538–547
Nickell, L. G. and Maretzki, A. (1969). *Physiol. Plant.*, **22**, 117–125
Nitsch, J. P. and Nitsch, C. (1957). *Am. J. Bot.*, **44**, 555–564
Nobel, P. S. and Cheung, Y.-N. S. (1972). *Nature New Biol.*, **237**, 207–208
Nobel, P. S. and Wang, C. T. (1970). *Biochim. Biophys. Acta*, **211**, 79–87
Norstog, K. (1973). *In Vitro*, **8**, 307–308
Novacky, A., Fischer, E., Ullrich-Eberius, C. I., Lüttge, U., and Ullrich, W. R. (1978). *FEBS Lett.*, **88**, 264–267
Oaks, A. (1965*a*). *Plant Physiol.*, **40**, 142–149

Oaks, A. (1965*b*). *Plant Physiol.*, **40**, 149–155
Oaks, A. (1965*c*). *Can. J. Bot.*, **43**, 1077–1082
Oaks, A. and Beevers, H. (1964). *Plant Physiol.*, **39**, 37–43
Oaks, A., Aslam, M., and Boesel, I. (1977). *Plant Physiol.*, **59**, 391–394
Oji, Y. and Izawa, G. (1972). *Plant Cell Physiol.*, **13**, 249–259
Ongun, A. and Stocking, C. R. (1965). *Plant Physiol.*, **40**, 825–831
Oota, Y., Fujii, R., and Osawa, S. (1953). *J. Biochem., Tokyo*, **40**, 649–661
Oudman, J. (1936). *Extr. Rec. Trav. Bot. Neerl.*, **33**, 351–433
Paris, D., and Duhamet, L. (1953). *C. R. Acad. Sci.*, **236**, 1690–1692
Pate, J. S. (1962). *Plant Soil*, **17**, 333–356
Pate, J. S. (1965). *Science*, N.Y., **149**, 547–548
Pate, J. S. and Gunning, B. E. S. (1972). *Annu. Rev. Plant Physiol.*, **23**, 173–196
Pate, J. S., and Wallace, W. (1964). *Ann. Bot., Lond.*, **28**, 83–99
Pate, J. S., Sharkey, P. J., and Lewis, O. A. M. (1975). *Planta*, **122**, 11–26
Pate, J. S., Wallace, W., and van Die, J. (1964). *Nature, Lond.*, **204**, 1073–1074
Payne, J. W. (1977) *CIBA Found. Symp.*, **50**, 305–334
Peel, A. J. and Weatherley, P. E. (1959) *Nature, Lond.*, **184**, 1955–1956
Peterson, D. M., Housley, T. L., and Schrader, L. E. (1977). *Plant Physiol.*, **59**, 221–224
Petzold, U. and Jacob, F. (1975). *Biol. Zentralbl.*, **94**, 167–185
Plummer, G. L. and Kethley, J. B. (1964). *Bot. Gaz., Chicago*, **125**, 245–260
Pollard, J. K. and Sproston, T. (1954). *Plant Physiol.*, **29**, 360–364
Pragnell, M. J., Jones, M., Pierce, J. S., Mikola, J., and Enari, T.-M. (1969). *J. Inst. Brew., Lond.*, **75**, 511–515
Pringsheim, E. G. and Pringsheim, O. (1962). *Am. J. Bot.*, **49**, 898–901
Pringsheim, E. G. and Pringsheim, O. (1967). *Z. Pflanzenphysiol.*, **57**, 1–10
Radin, J. W. (1975). *Plant Physiol.*, **55**, 178–182
Reinert, J. and White, P. R. (1956). *Physiol. Plant.*, **9**, 177–189
Reinhold, L. and Powell, R. G. (1956). *Nature, Lond.*, **177**, 658–659
Reinhold, L., Shtarkshall, R. A., and Ganot, D. (1970). *J. Exp. Bot.*, **21**, 926–932
Riker, A. J. and Gutsche, A. E. (1948). *Am. J. Bot.*, **35**, 227–238
Risser, P. G. and White, P. R. (1964). *Physiol. Plant.*, **17**, 620–635
Roberts, G. R., Keys, A. J., and Whittingham, C. P. (1970). *J. Exp. Bot.*, **21**, 683–692
Rubin, G. and Zaitlin, M. (1976). *Planta*, **131**, 87–89
Rubinstein, B., Mahar, P., and Tattar, T. A. (1977). *Plant Physiol.*, **59**, 365–368
Salonen, M.-L. and Simola, L. K. (1977). *Physiol. Plant.*, **41**, 55–58
Sandstedt, R. and Skoog, F. (1960). *Physiol. Plant.*, **13**, 250–256
Sargant, E. and Robertson, A. (1905). *Ann. Bot., Lond.*, **19**, 115–124.
Sargent, P. A. and King, J. (1974). *Can. J. Bot.*, **52**, 1747–1755
Sauter, J. J. (1976). *Z. Pflanzenphysiol.*, **79**, 276–280
Scala, J., Iott, K., Schwab, D. W., and Semersky, F. E. (1969). *Plant Physiol.*, **44**, 367–371
Sheridan, W. F. (1977). *Physiol. Plant.*, **41**, 172–174
Shtarkshall, R. A., and Reinhold, L. (1974). In *Membrane Transport in Plants* (U. Zimmerman and J. Dainty, eds), Springer-Verlag: Berlin, Heidelberg, and New York
Shtarkshall, R. A., Reinhold, L., and Harel, H. (1970). *J. Exp. Bot.*, **21**, 915–925
Skinner, J. C. and Street, H. E. (1954). *New Phytol.*, **53**, 44–67
Sopanen, T., Burston, D., and Matthews, D. M. (1977). *FEBS Lett.*, **79**, 4–7
Sopanen, T., Burston, D., Taylor, E., and Matthews, D. M. (1978). *Plant Physiol.*, **61**, 630–633
Sorenson, D. R. and Jackson, W. T. (1968). *Planta*, **83**, 166–170
Staba, E. J. (1962). *J. Pharm. Sci.*, **51**, 249–254
Steckelberg, R., Lüttge, U., and Weigl, J. (1967). *Planta*, **76**, 238–241
Steinhart, C. E., Standifer, L. C., and Skoog, F. (1961). *Am. J. Bot.*, **48**, 465–472
Steward, F. C., Caplin, S. M., and Millar, F. K. (1952). *Ann. Bot., Lond.*, **16**, 57–77
Steward, F. C., Mapes, M. O., and Mears, K. (1958*a*). *Am. J. Bot.*, **45**, 705–708
Steward, F. C., Mapes, M. O., and Smith, J. (1958*b*). *Am. J. Bot.*, **45**, 693–703
Stewart, C. R. (1971). *Plant Physiol.*, **47**, 157–161
Stewart, C. R. (1972). *Plant Physiol.*, **50**, 551–555
Stewart, C. R. (1973). *Plant Physiol.*, **51**, 508–511

Stewart, C. R. and Beevers, H. (1967). *Plant Physiol.*, **42**, 1587–1595
Stewart, G. R. (1972). *L. Symp. Biol. Hung.*, **13**, 127–135
Stoev, K. D., Mamarov, P. T., and Benchev, I. V. (1959). *Sov. Plant Physiol.*, **6**, 424–430
Straus, J. (1960). *Am. J. Bot.*, **47**, 641–647
Streeter, J. G. and Thompson, J. F. (1972). *Plant Physiol.*, **49**, 572–578
Sundblom, N.-O. and Mikola, J. (1972). *Physiol. Plant.*, **27**, 281–284
Tammes, P. M. L. and van Die, J. (1964). *Acta Bot. Neerl.*, **13**, 76–83
Thompson, J. F., Madison, J. T., and Muenster, A.-M. E. (1977). *Ann. Bot., Lond.*, **41**, 29–39
Tökés, Z. A., Woon, W. C., and Chambers, S. M. (1974). *Planta*, **119**, 39–46
Tolbert, N. E. (1973). *Curr. Topics Cell. Regul.*, **7**, 21–50
Tromp, J. and Ovaa, J. C.: (1976). *Physiol. Plant.*, **37**, 29–34
Virtanen, A. I. and Miettinen, J. K. (1963). In *Plant Physiology*, Vol. 3 (F. S. Steward, ed.), Academic Press, New York, pp. 539–668
Wahlström, D. and Eriksson, T. (1976). *Physiol. Plant.*, **38**, 138–140
Wang, C.-T. and Nobel, P. S. (1971). *Biochim. Biophys. Acta*, **241**, 200–212
Watanabe, A. (1951). *Arch. Biochem. Biophys.*, **34**, 50–55
Watson, R. and Fowden, L. (1975). *Phytochemistry*, **14**, 1181–1186
Weatherley, P. E., Peel, A. J., and Hill, G. P. (1959). *J. Exp. Bot.*, **10**, 1–16
Webster, G. C. (1954). *Plant Physiol.*, **29**, 382–385
White, P. R. (1934). *Plant Physiol.*, **9**, 585–600
White, P. R. (1939). *Am. J. Bot.*, **26**, 59–64
Wieringa, K. T. and Bakhuis, J. A. (1957). *Plant Soil*, **8**, 254–262
Wright, D. E. (1962). *Arch. Biochem. Biophys.*, **97**, 174–180
Yamaguchi, S. and Islam, A. S. (1967). *Hilgardia*, **38**, 207–229
Yemm, E. W. and Willis, A. J. (1956). *New Phytol.*, **55**, 229–252
Zamski, E. (1973). *Israel J. Bot.*, **22**, 211–230
Zelitch, I., Rosenblum, E. D., Burris, R. H., and Wilson, P. W. (1951). *J. Biol. Chem.*, **191**, 295–298
Ziegler, H. (1956). *Planta*, **47**, 447–500
Ziegler, H. (1974). *Symp. Soc. Exp. Biol.*, **28**, 43–62
Zimmerman, M. H. (1960). *Annu. Rev. Plant Physiol.*, **11**, 167–190
Zimmermann, M. H. and Milburn, J. A. (eds) (1975). *Encylopaedia of Plant Physiology*, New Series, Vol. 1, Springer-Verlag, Berlin, Heidelberg, and New York

APPENDIX 1. SPECIES LIST

Alfalfa	*Medicago sativa*
Apple	*Malus* spp.
Ash	*Fraxinus* spp.
Barley	*Hordeum sativum* (Reinhold *et al.*, 1970; Shtarkshall *et al.*, 1970)
	Hordeum vulgare (all other references)
Bladderwort	*Utricularia* spp.
Broad bean	*Vicia faba*
Butterwort	*Pinguicula* spp.
Caesalpinia	*Caesalpinia tinctoria*
Carrot	*Daucus carota*
Castor bean	*Ricinus communis*
Cotton	*Gossypium hirsutum*
Deadly nightshade	*Atropa belladonna*

Duckweed	*Lemna minor* (Newton, 1974*a*)
	Spirodela oligorrhiza (Bollard, 1966)
	Spirodela polyrhiza (Borstlap, 1977)
Egeria	*Egeria densa*
False acacia	*Robina pseudoacacia*
Field pea	*Pisum arvense*
Groundsel	*Senecio vulgaris*
Haricot bean	*Phaseolus vulgaris*
Jerusalem artichoke	*Helianthus tuberosus*
Lupin	*Lupinus albus*
Maize	*Zea mays*
Melon	*Cucumis melo*
Mustard	*Brassica nigra*
Oak	*Quercus* spp.
Oat	*Avena sativa*
Palm	*Yucca flaccida*
Pea	*Pisum sativum*
Pitcher plant	*Nepenthes* spp.
	Sarracenia spp.
Radish	*Raphanus sativus*
Rapeseed	*Brassica napus*
Rice	*Oryza sativa*
Scorzonera	*Scorzonera* spp.
Sorrel	*Rumex acetosa*
Soybean	*Glycine max*
Spanish Broom	*Spartium junceum*
Spruce	*Picea* spp.
Squash	*Cucurbita maxima*
Sugarcane	*Saccharum officinarum*
Sundew	*Drosera* spp.
Sunflower	*Helianthus annus*
Tobacco	*Nicotiana tabacum*
Tomato	*Lycopersicum esculentum*
Venus flytrap	*Dionaea muscipula*
Wheat	*Triticum monococcum*
Willow	*Salix viminalis* (Peel and Weatherley, 1969)
	Salix caprea (Sauter, 1976)

APPENDIX 2. ABBREVIATIONS

ABA	abscisic acid
AIB	α-aminoisobutyric acid
ATP	adenosine 5′-triphosphate
DCMU	3-(3,4-dichlorophenyl)-1,1-dimethylurea
DNP	2,4-dinitrophenol
EDTA	ethylenediaminetetra-acetate
FCCP	*p*-trifluoromethoxy carbonyl cyanide phenylhydrazone
Gly-Sar	glycylsarcosine
Gly-Sar-Sar	glycylsarcosylsarcosine
Gly-Sar-Sar-Sar	glycylsarcosylsarcosylsarcosine
IAA	indole-3-acetic acid
PMS	*N*-methylphenazonium methosulphate
tris	2-amino-2-hydroxymethylpropane-1,3–diol

Microorganisms and Nitrogen Sources
Edited by J. W. Payne

Section 4.7

Peptides as Antimicrobial Agents

P. S. RINGROSE
*Roche Products Ltd, Welwyn Garden City, Hertfordshire, UK**

1 INTRODUCTION

In previous reviews of peptide antibiotics, much attention has focused on the biosynthesis, structural conformation, and mechanism of action of cyclic peptides. These antimicrobial peptides are structurally very different from normal peptides and, besides being cyclic, frequently contain ester links, D-configuration or unusual amino acid residues and other functional moieties, such as fatty acids, sugars, and heterocyclic residues. In recent years, however, as a result of more informed screening methods using antagonist-free media, interesting advances have occurred in the isolation and characterization of much simpler linear di- and tripeptide antibiotics. Detailed knowledge gained in the study of the mechanisms of action of

*Present address: Sandoz Forschungsinstitut, Vienna, Austria.

All amino acid moieties described in this chapter are of the L-configuration unless otherwise stated, this corresponds to the Cahn–Ingold–Prelog notation of (*S*) for natural amino acids. For other chiral centres and for amino acid mimetics, the (*S*/*R*) notation is used; N.B. L-aminoethylphosphonic acid has an *R*-configuration.

these short peptides has provided much useful information on structure–activity relationships and encouraged the rational design of novel synthetic peptides with antimicrobial activity. This review will largely concentrate on these recent advances, but appropriate mention will be made of the other more familiar types of antimicrobial peptides. The large number of peptide antibiotics now known (300–400) makes it necessary to categorize the various types of structure into the following groups:

- (I) Linear peptides
  - (a) Linear di- and tripeptides that act as amino acid mimetic delivery systems
  - (b) Linear oligopeptides that are active as the intact molecule
- (II) Cyclic peptides
- (III) Peptides attached to fatty acid moieties
- (IV) Peptides attached to sugar moieties
- (V) Peptides of undefined structure

The category numbers Ia, Ib, II, III, IV, and V will be used as a form of shorthand in this review.

An interesting feature, that will be developed is that within groups of structurally related peptides, the mechanism of action and nature of the target site are often very diverse (Table 1 and Figure 1). The mechanism of biosynthesis of the peptide antibiotics in non-ribosomal and generally involves specific multienzyme synthetases, with activation through a pantetheine 'transfer-arm'. This process has been described in detail elsewhere and would seem to be common even for structurally different peptides (Perlman and Bodanszky, 1971; Kurahashi, 1974; Corcoran, 1977; Katz and Demain, 1977).

2 LINEAR PEPTIDES

The linear peptides to be described in this section may be regarded in a pseudo-mechanistic sense as falling into two distinct categories. The first group are largely di- and tripeptides that act purely as disguised delivery systems for amino acid mimetics. After being transported into the cell, the peptides are enzymically cleaved, thus liberating intracellularly their active amino acid moiety. In contrast, the second group act either intracellularly or at membrane level as the intact molecule; consequently enzymic hydrolysis results in inactivation of the peptide.

1 Linear peptides that act as amino acid mimetic delivery systems

(i) The concept

Although the phenomenon of 'illicit transport' has only been described in detail comparatively recently as a way of enabling impermeant metabolites

Table 1 Antimicrobial peptides; structure/function categories

Mechanism	Structure Ia	Ib	II	III	IV	V
(1) Cell wall (a) Early intracellular stage (b) Membrane stage (c) Crosslinking stage	(a) Bacilysin (a) Alaphosphin (a) FCDP-Ala	(a) D-Norval-D-Ala	(a) D-cycloserine (a/b) Nisin (b) Bacitracin (c) Penicillin (c) Bicyclomycin	(b) Amphomycin	(a) Polyoxin (antifungal) (b) Vancomycin (b) Ristocetin	(b) Gardimycin (b) Stenothricin (b) Janiemycin
(2) Membrane (d) Disruption (e) Ionophore		(e) Gramicidin A	(d) Surfactin (d) Gramicidin S (d) Tyrocidine (d/e) Serratamolide (e) Valinomycin (e) Antamanide (d/e) Monamycin ((e) Alamethicin)	(d) Colistin (d) Polymixin (d) Octapeptin (d) Cerexin (d) Brevistin (d) Polypeptin (d) Circulin (d) Stendomycin		
(3) Nucleic acid (f) Precursor synthesis (g) Template (h) Polymerase	(f) Phosphinothricyl-alanylalanine (f) Alazopeptin (f) Tabtoxin		(g) Actinomycin D (g) Echinomycin (h) α-Amanitin (h) Gliotoxin (viral RNA)		(g) Bleomycin (g) Phleomycin	
(4) Protein (i) Precursor uptake and utilisation (j) Ribosome	(i) Ethionyl-Ala (i) Gly-thienylAla	(j) Negamycin (j) Triornithine (j) Edeine (i) Gly-Leu (i) Arg-D-Phe-Phe (antifungal)	(j) Osteogrycin B (j) Thiostrepton		(j) Gougerotin (j) Blasticidin S (j) Amicetin	

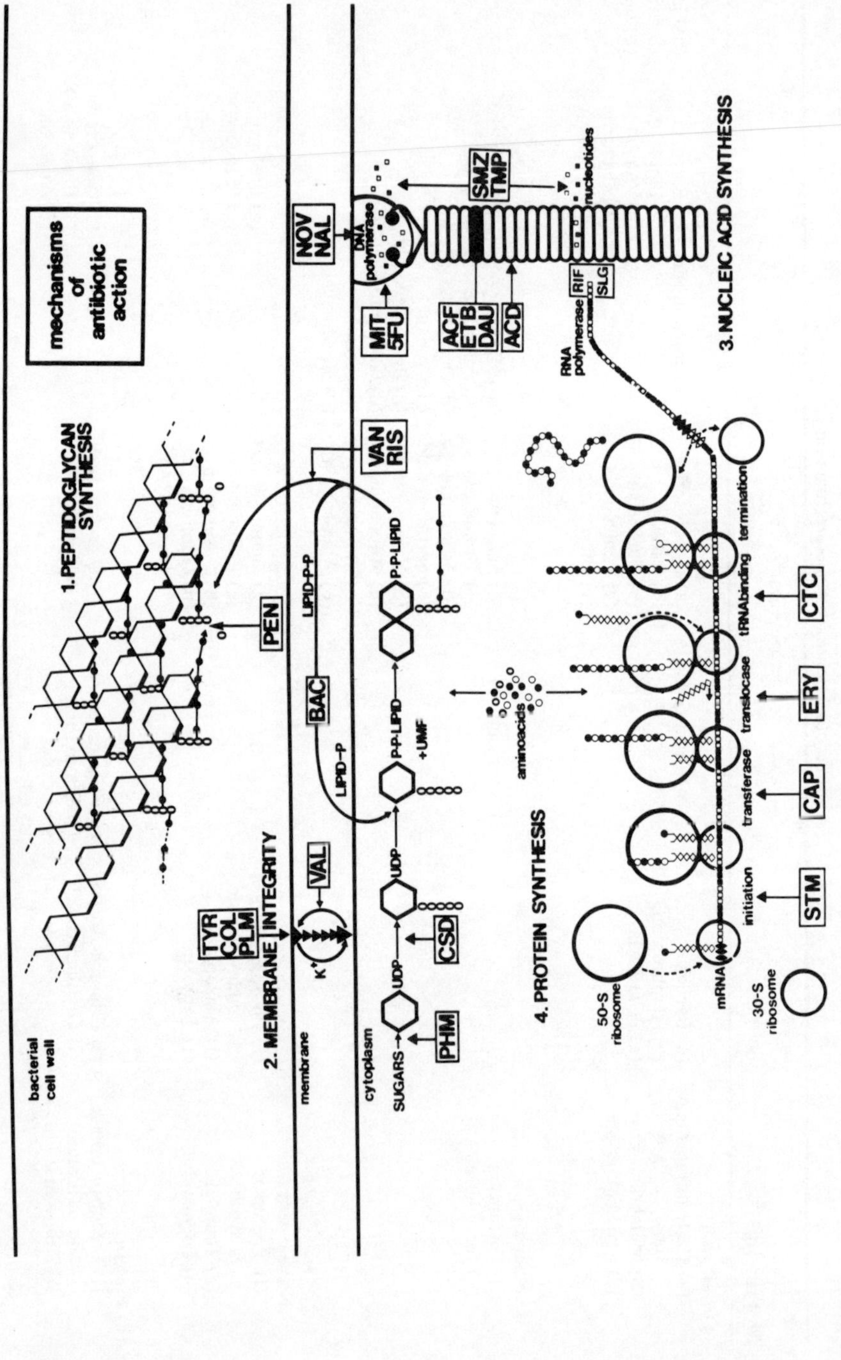
mechanisms of antibiotic action
bacterial cell wall
membrane
cytoplasm
1. PEPTIDOGLYCAN SYNTHESIS
2. MEMBRANE INTEGRITY
3. NUCLEIC ACID SYNTHESIS
4. PROTEIN SYNTHESIS
TYR COL PLM
VAL
K+
PEN
BAC
VAN RIS
LIPID-P-P
LIPID-P
P-P-LIPID
+UMF
SUGARS
UDP
PHM
CSD
NOV NAL
MIT 5FU
ACF ETB DAU
ACD
SMZ TMP
DNA polymerase
nucleotides
RNA polymerase
RIF
SLG
aminoacids
50-S ribosome
30-S ribosome
mRNA
initiation
transferase
translocase
tRNAbinding
termination
STM
CAP
ERY
CTC

Figure 1 Scheme for mechanisms of action of standard antibiotics

Abbreviations

1.	PHM	Phosphonomycin		5FU	5-Fluorouracil
	CSD	Cycloserine D*		SMZ	Sulfamethoxazole
	VAN	Vancomycin*		TMP	Trimethoprim
	RIS	Ristocetin*		ACF	Acriflavin
	BAC	Bacitracin*		ETB	Ethidium Br.
	PEN	Penicillin*		DAU	Daunomycin
				ACD	Actinomycin D*
2.	TYR	Tyrocidine*		RIF	Rifampicin
	COL	Colistin*		SLG	Streptolydigin
	PLM	Polymixin*			
	VAL	Valinomycin*			
			4.	STM	Streptomycin
3.	NOV	Novobiocin		CAP	Chloramphenicol
	NAL	Nalidixic acid		ERY	Erythromycin
	MIT	Mitomycin		CTC	Chlortetracycline

The antibiotics marked with an asterisk are peptides. For further details on antibiotic target sites see Gale *et al*. (1972)

(e.g. homoserine phosphate and histidinol phosphate) to enter the bacterial cell disguised as peptides (Ames *et al*., 1973; Fickel and Gilvarg, 1973), naturally-occurring antimicrobial peptides, that have subsequently been shown to use this mechanism, have been known for some time. Detailed studies on the molecular requirements and specificity of bacterial peptide transport in the late 1960s (Payne and Gilvarg, 1971) provided much of the early groundwork for proposing the 'smugglin' concept (Matthews and Payne, 1975).

Unlike the permeases involved in amino acid transport, peptide permeases are far less exacting in molecular terms, e.g. with respect to the nature of the component amino acid side groups, and to the *C*-terminal acidic function in oligopeptides (Payne, 1976). It is therefore possible for an amino acid mimetic which is not transported by the specific amino acid permeases, to be taken into the bacterial cell disguised as a peptide, i.e. 'illicit transport'. For antibacterial activity, however, the mimetic or 'warhead' must be rapidly released from its carrier peptide by intracellular peptidase action, so that it can act, for example, on a target enzyme. In order that a peptide may act in such a way as a carrier, it is necessary for its stereochemistry, sequence, chain length, etc., to conform to the structural specificities of bacterial peptide permeases and peptidases (see Sections 2.1 and 2.4). The optimal warhead is usually one that is poorly transported into the cell by itself, in this way once it is released intracellularly from its carrier, it is unlikely to take part in carrier-mediated efflux. As a consequence, very high intracellular concentrations of warhead can be achieved (see Figure 2).

Antimicrobial peptides of this type may therefore be regarded as being composed of a transport moiety, which may be one or more amino acids at either the amino- or carboxyl-terminus, and a warhead moiety, linked by a peptidase-labile bond. Peptides composed of L-amino acid residues are therefore normally required to meet these criteria.

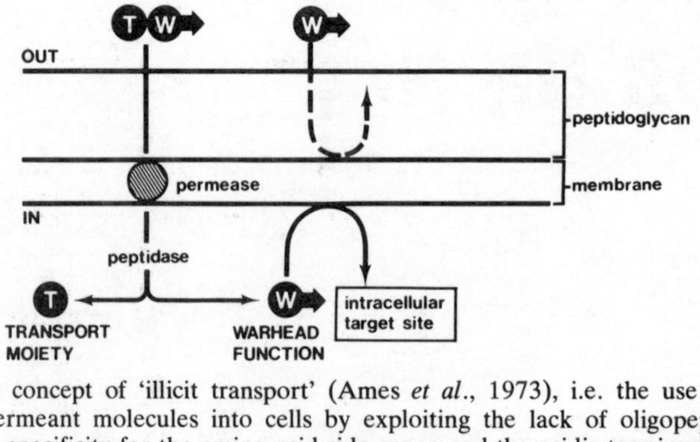

Figure 2 The concept of 'illicit transport' (Ames *et al*., 1973), i.e. the use of peptides to transport impermeant molecules into cells by exploiting the lack of oligopeptide permease specificity for the amino acid side group and the acidic terminus

(ii) Naturally-occurring examples of category Ia peptides

Bacilysin (**I**) is a well-characterized L,L-dipeptide antibiotic acting in the manner described above. It was first reported as an antibiotic substance, produced by a strain of *Bacillus subtilis*, which caused partial lysis of growing cultures of *Staphylococcus aureus* (Abraham *et al*., 1946). The isolation of bacilysin came much later (Rogers *et al*., 1965; Roscoe and Abraham, 1966) and its structure was unambiguously established in 1970 by Walker and Abraham. Earlier confusion with what appeared to be a related dipeptide, tetaine (Borowski and Konopa, 1957), was later resolved when it was shown that bacilysin and tetaine were identical compounds (Kaminski and Sokolowska, 1973). Furthermore, it would now appear that bacillin, first isolated by Woodruff and Foster in 1946, is also structurally identical with bacilysin (Atsumi *et al*., 1975).

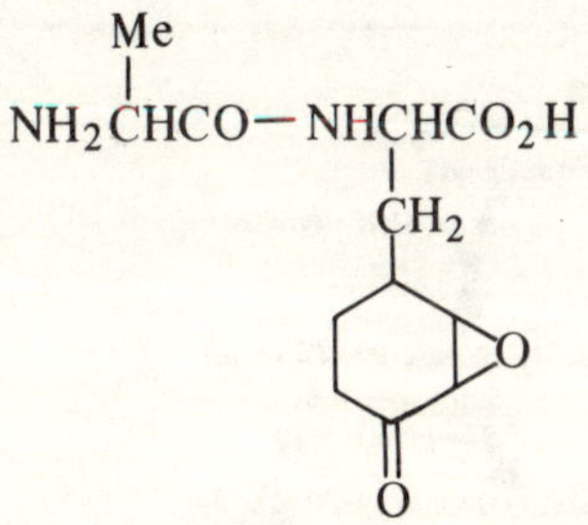

[Ala–X]

I

Bacilysin (tetaine, bacillin)

The unusual *C*-terminal L-amino acid moiety of bacilysin was independently isolated from the culture fluid of a strain of *Streptomyces griseoplanus* (Whitney *et al*., 1970, 1972; Shah *et al*., 1970; Neuss *et al*., 1970) and called 'anticapsin' because of its inhibitory action on the hyaluronic acid capsule of group A streptococci. Anticapsin was however without activity against staphylococci and thus by natural peptide derivatisation of this warhead amino acid, antibiotic activity was achieved against a wide range of bacteria and also against *Candida albicans* (Kenig and Abraham, 1976).

As with all antimicrobial peptides that use peptide permeases, the activity of bacilysin is strongly antagonized by many natural peptides but less so by amino acids. It is therefore necessary to use a defined, peptide-free medium for detecting activity, and ignorance of this fact is the reason why many examples of this type of agent have only been discovered comparatively recently (see Scannell and Pruess, 1974), for bacteriological testing media normally contain peptone and yeast extract supplements,

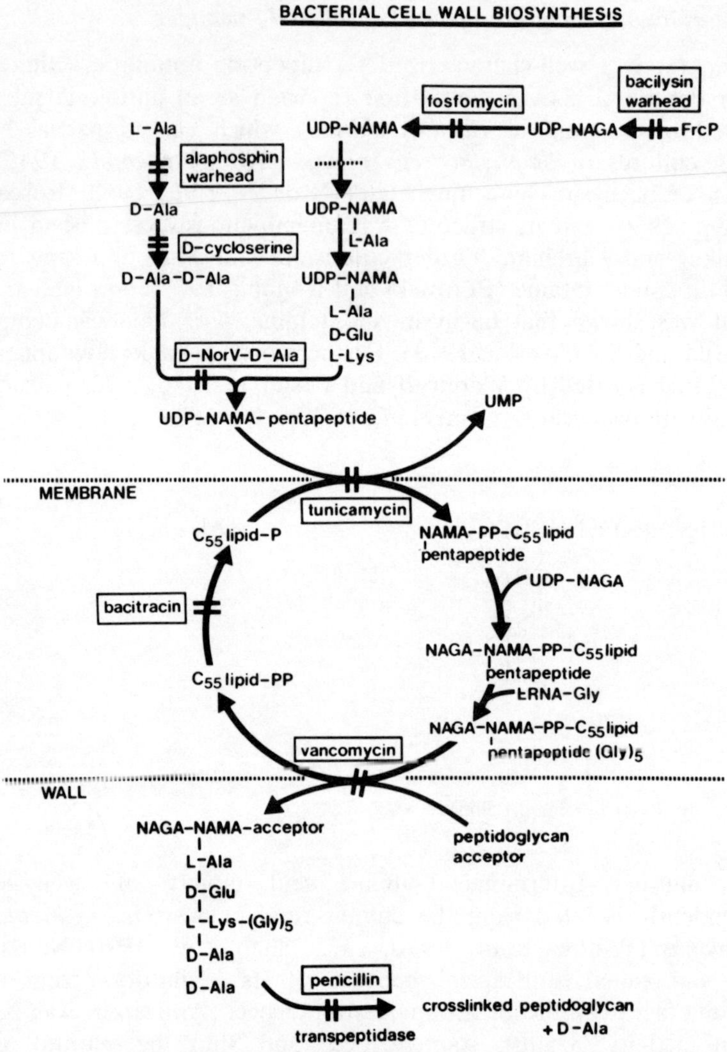

Figure 3 Bacterial cell wall biosynthesis

which are antagonistic towards all antimicrobial peptides in category Ia and some in category Ib.

Once the bacilysin molecule has penetrated the bacterial cell, its mechanism of action is due entirely to its anticapsin warhead. The target enzyme of anticapsin was shown to be L-glutamine:D-fructose-6-phosphate amino transferase (glucosamine synthetase, E.C.2.6.1.16) by Whitney and Funderburk (1970) using the isolated enzyme (see Figure 3.). Bacilysin *per se* is inactive as an inhibitor of this enzyme. Its activity against glucosamine synthetase in whole cells however is confirmed by the ability of glucosamine and *N*-acetylglucosamine to reverse bacilysin action (Vandamme *et al*., 1975; Kenig and Abraham, 1976). By specifically blocking this particular enzyme, bacterial cell wall biosynthesis is inhibited. The mechanism of enzyme inhibition may be irreversible, involving covalent interaction between the warhead epoxide and a cysteine thiol group in the enzyme (Kenig *et al*., 1976). This proposed mechanism of enzyme inhibition is, therefore, analogous to that reported for the other epoxide antibiotic, fosfomycin (Kahan *et al*., 1974), which acts on UDP-*N*-acetylglucosamine:pyruvyl transferase (see Figure 3).

Independent work on the mechanism of action of tetaine, before its structure was properly elucidated, indicated that its mechanism of action also involved inhibition of cell wall biosynthesis (Chmara and Borowski, 1973a,b). However, it was concluded, largely from work on isolated enzymes, that the target site was the enzyme that attached L-Ala to UDP-*N*-acetylmuramic acid. This may have been an artefact, since the crude enzyme preparation used could have hydrolysed tetaine to produce L-Ala and anticapsin.

Like bacilysin, the dipeptide antibiotic isolated from *Streptomyces collinus*, lindenbein (**II**), is antagonized by D-glucosamine and *N*-acetylglucosamine (Molloy *et al*., 1972). Fumarylcarboxyamido-L-2,3-diaminopropionyl-L-alanine would seem to be a 'warhead' delivery peptide, but the exact nature or mechanism of action of the 'warhead' has not been reported in detail. It is interesting to note, however, that the warhead moiety is at the *N*-terminus. It will be seen from other examples of category Ia peptides that the warhead can function effectively at either end of the carrier molecule.

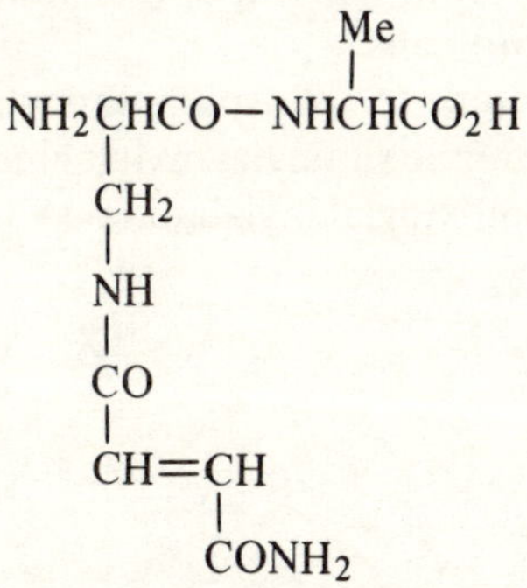

[X–Ala]

II

Fumarylcarboxyamido-L-2,3-diaminopropionyl-L-alanine (FCDP–Ala)

A surprisingly common property of most of the naturally occurring antimicrobial peptides considered in this section is that, although structurally diverse, their warheads interfere with either the synthesis or subsequent utilisation of glutamine. Unfortunately, this particular mechanism of action is generally undesirable in terms of clinical usefulness. Another feature is that L-alanine is often the other part of the peptide molecule (i.e. the transport moiety).

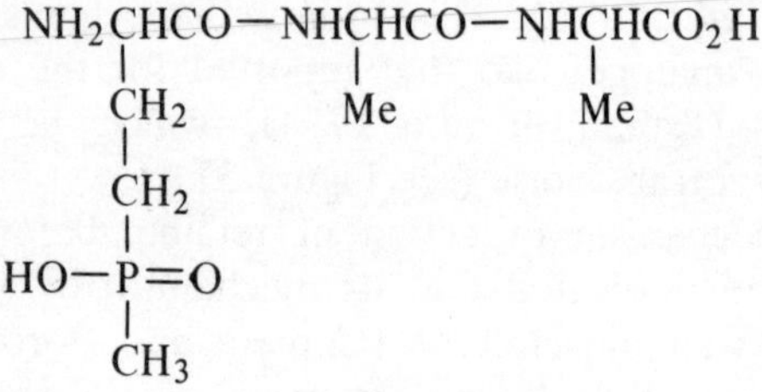

[X—Ala—Ala]

III

L-Phosphinothricylalanylalanine

Phosphinothricyl-alanyl-alanine (**III**) is a well-characterized example of a homo-L tripeptide delivery system which, in contrast to bacilysin, has its warhead at the *N*-terminus. Phosphinothricylalanylalanine is highly active against Gram-positive and Gram-negative bacteria and was first isolated from *Streptomyces viridiochromogenes* by Bayer *et al*. (1972). The warhead moiety, phosphinothricin, is a γ-phosphinate analogue of glutamic acid and consequently is a potent inhibitor of *E. coli* glutamine synthetase. In contrast, the intact tripeptide is inactive against the isolated enzyme (Bayer *et al*., 1972). Phosphinothricin is, however, only weakly antibacterial in its own right against a few organisms, and recent results of Diddens *et al*. (1976) indicate that the phosphinothricin tripeptide uses an oligopeptide permease to facilitate its entry into cells.

Two tripeptides, closely related to phosphinothricylalanylalanine are L-(N^5-phosphono)methionine-*S*-sulphoximinylalanylalanine (**IV**) (Pruess *et al*., 1973) and L-methionine-*S*-dioxidylalanylalanine (**V**) (Diddens *et al*.,

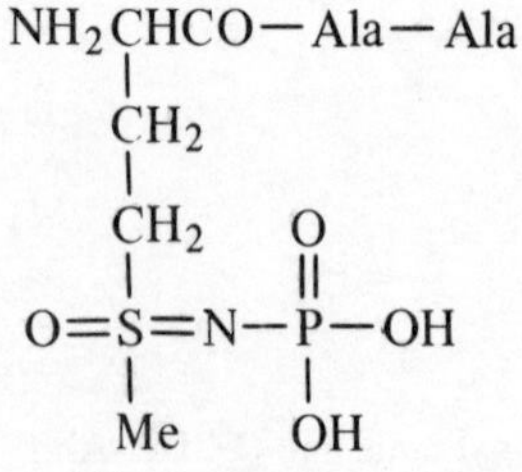

[X—Ala—Ala]

IV

L-(N^5-phosphono)methionine-*S*-sulphoximinylalanylalanine

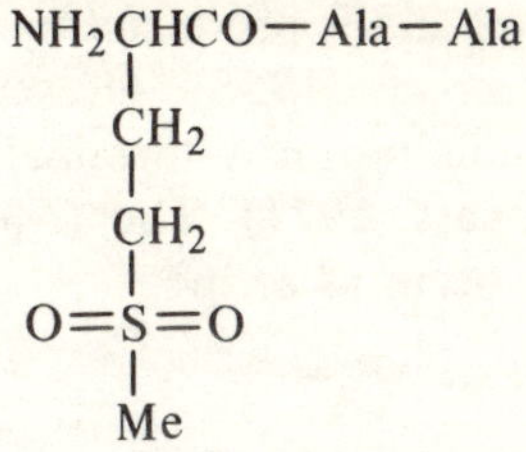

[X—Ala—Ala]

V

L-Methionine-*S*-dioxidylalanylalanine

1976). Both contain warheads that are potent inhibitors of glutamine synthetase. An interesting fact to emerge in the study of the latter tripeptide and its warhead was that synergy could be obtained by delivering the same warhead into cells simultaneously using two different transport systems; i.e. an oligopeptide-permease for the tripeptide mimetic and a methionine-permease for the warhead (Diddens *et al*., 1976).

Two tripeptides, alazopeptin (**VI**) (Patterson *et al*., 1966) and duazomycin B (Rao, 1962), have been characterized as delivery systems for the amino acid antimetabolite, 6-diazo-5-oxo-L-norleucine (DON). Both peptides are unusual in that they carry two molecules of the DON warhead, i.e. at the *C*-terminal and middle positions of the tripeptide. The *N*-terminal transport moiety is L-alanine in alazopeptin and γ-glutamic acid in duazomycin B. Both peptides are inhibitors of glutamine synthesis in whole cells and DON is believed to be an irreversible inhibitor of glucosamine synthetase (Bates and Handschumacher, 1969).

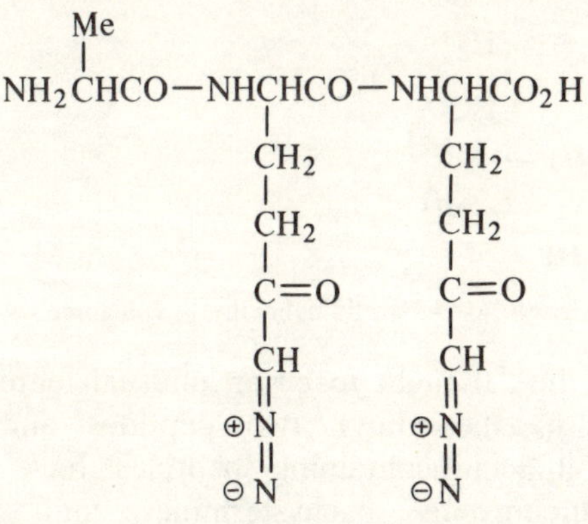

[Ala—X—X]

VI

Alazopeptin

Two closely related but unusual dipeptides with β-lactam rings have been described as having activity which is reversed by glutamine. Wildfire toxin or tabtoxin (**VII**) was the first to be investigated and is produced by a plant pathogenic strain of *Pseudomonas* (Woolley *et al*., 1955). However, its structure was only recently elucidated as containing an *N*-terminal amino acid mimetic with a β-lactam side-chain (Stewart, 1971). The *C*-terminal moiety is usually threonine but can sometimes occur as serine.

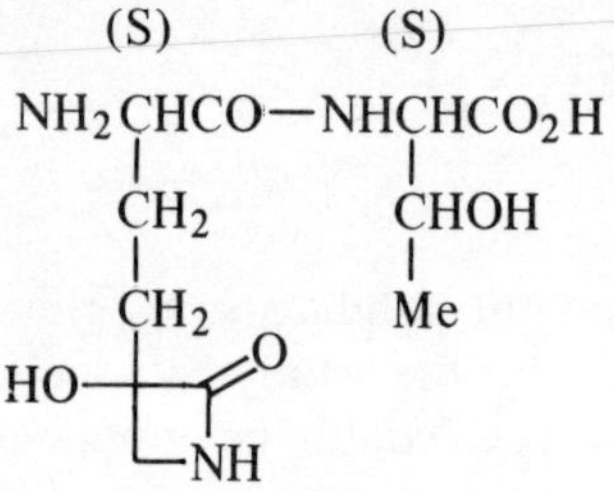

[X—Thr]

VII

Tabtoxin

In contrast, the dipeptide (**VIII**) isolated and characterized by Scannel *et al*. (1975) has its β-lactam warhead at the *C*-terminus and L-alanine as the transport moiety. An additional unusual feature is that the warhead amino acid contains a chlorine atom in the γ-position, as does the potent amino acid antagonist, armentomycin, i.e. L-2-amino-4,4-dichlorobutanoic acid (Argoudelis *et al*., 1967; Iwasaki *et al*., 1976).

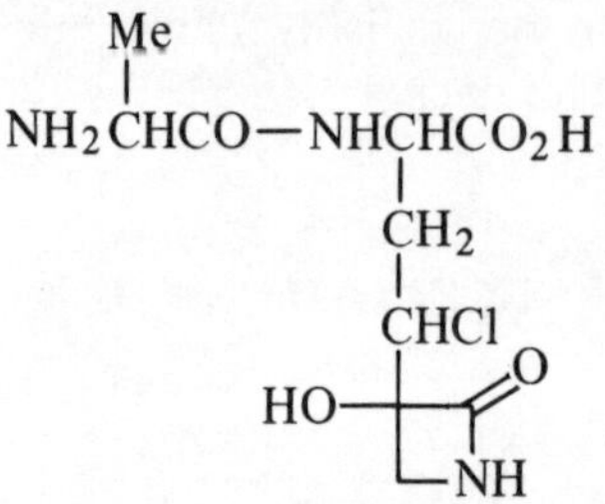

[Ala—X]

VIII

(*S*)-Alanyl-3-[α-(*S*)-chloro-3-(*S*)-hydroxy-2-oxo-3-azetidinylmethyl]-(*S*)-alanine

The β-lactam ring was at one time thought to be an unusual feature in nature but now, in addition to the above two peptides and the penicillins/cephalosporins, other β-lactam-containing 'peptides' have been described, e.g. bleomycins, phleomycins, pachystermines, and, more recently, nocardicin, thienamycin, and the olivanic acids (these will be briefly discussed in Sections 4.7.3 and 4.7.5).

Examples of natural peptides with warheads that do not act as glutamine antagonists are stravidin (**IX**) and valyl-1-hydroxy-2-amino-1-cyclobutane acetic acid (**X**).

Stravidin (MSD 235 s_3) is unusual in that its terminal amino group is *N*-methylated such that the transport moiety is *N*-methylisoleucine (Baggaley *et al.*, 1969). It is interesting to note that although monosubstituted *N*-alkyl peptides are transported adequately by bacteria, they may be cleaved more slowly (Payne, 1974). The warhead part of the dipeptide is the amino acid antibiotic amiclenomycin (Okami *et al.*, 1974). The antibacterial activity of both the dipeptide and its warhead are reversed by biotin (Stapley *et al.*, 1964; Okami *et al.*, 1974).

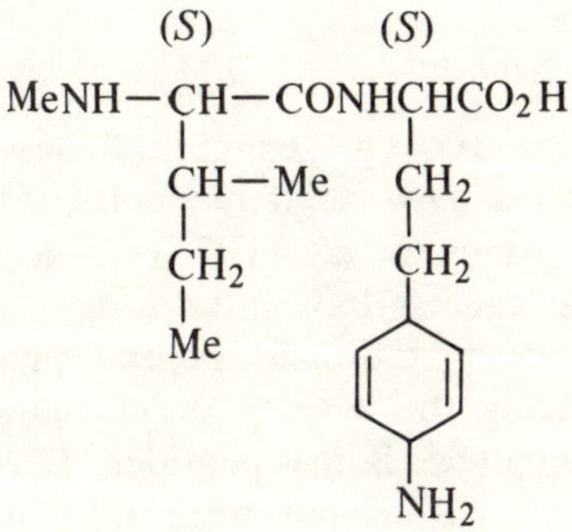

[MeIle—X]

IX

Stravidin (*N*-methylisoleucine-amiclenomycin)

The valyl-cyclobutane peptide (**X**) (Pruess *et al.*, 1974) is only active against Gram-positive bacteria and its activity is reversed by L-cysteine or L-methionine. The *C*-terminal γ-amino acid may act as the warhead but there is no direct proof that this is the case.

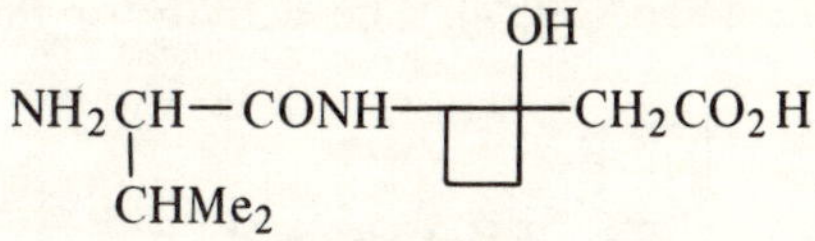

[Val—X]

X

1-(*S*)-Hydroxy-2-(*S*,*S*)-valylamido-cyclobutane-1-acetic acid

Feldamycin (**XI**) is a dipeptide analogue containing *N*-methylhistidine at its *N*-terminus (Arguodelis *et al.*, 1976). The mechanism of action of this peptide involves DNA synthesis and it would seem likely on the basis of analogous chemical structures that a warhead delivery system is involved. Its stereochemistry has not been reported.

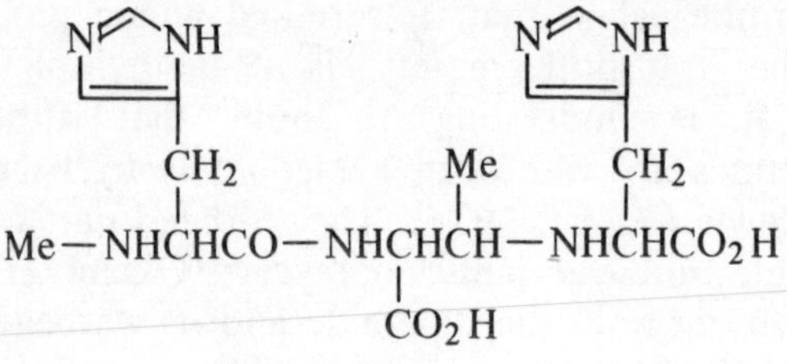

[N-MeHis—X—His]

XI

Feldamycin

(iii) Synthetic approaches for category Ia peptides

Partly as a result of the existence of the types of compound described above, various synthetic peptide analogues have been made in order to extend the 'illicit transport concept' (Ames *et al*., 1973) as a novel approach to designing chemotherapeutic agents. It is interesting to note, however, that even before there was any firm rationale, peptide mimetics were synthesized in the hope of enhancing the activity or changing the mechanism of action of amino acid mimetics. Some peptides, however, were made initially for entirely different reasons but were subsequently found to use a warhead delivery system, whilst other peptides, made for transport reasons, were found to have activity due to the intact form of the peptide and not because they supplied an amino acid metabolite (see Section 4.7.2.2. below).

It has been known for some time that certain peptides are biologically more active (i.e. nutritionally or inhibitory) than their component amino acids. Although some of these observations were based on the intracellular activities of the intact peptide, e.g. glycylleucine (Simmonds *et al*., 1947), much of the early synthetic work with amino acid mimetics was carried out with peptide derivatives of β-2-thienylalanine (**XII**) (Dunn, 1959; Dunn and Dittmer, 1951; Dunn *et al*., 1956, 1957). The activity of these latter

β-2-Thienylalanine

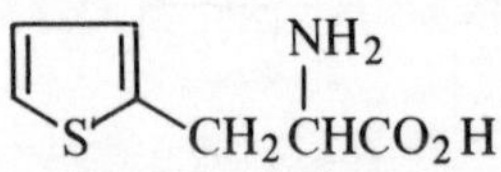

XII

Ethionine $EtSCH_2CH_2CH(NH_2)CO_2H$ **XIII**

Norleucine $CH_3CH_2CH_2CH_2CH(NH_2)CO_2H$ **XIV**

Examples of warheads used in synthetic peptides

peptides was, however, limited and depended on the presence of specific growth-stimulatory phenylalanine-containing peptides in the media.

Shankman and coworkers subsequently studied the growth inhibitory activity of peptides containing ethionine (**XIII**), norleucine (**XIV**), and sulphonamides in addition to trivaline peptides of mixed stereochemistry (Shankman *et al*., 1960, 1961, 1962*a,b*, 1963*a,b*). However, only the ethionine and norleucine peptides were shown to work via an apparent 'warhead delivery' mechanism. An interesting point to emerge from these early studies was that growth-inhibitory activity was optimal when the warhead moiety was at the carboxyl terminus. In addition to amino acid mimetics, certain natural amino acids have been shown to be growth inhibitory, e.g. valine (Leavitt and Umbarger, 1962), and even bactericidal, e.g. glycine (Hammes *et al*., 1973). These amino acids have also been used in peptide form (Val_3, Gly-Gly-Val, Gly-Gly) to show enhanced antibacterial activity. Unfortunately, this phenomenon is invariably very strain-specific (Payne and Gilvarg, 1971; De Felice *et al*., 1973; Yabu, 1969) and in some extreme cases (valine) is apparently limited to *E. coli* K12.

A further development came when Hill and Dunn (1969) applied the principle of 'double inhibition' (Kihara and Snell, 1960) and made double warhead dipeptides, containing both cyclopentane–glycine (an analogue of isoleucine) and β-2-thienylalanine (an analogue of phenylalanine).

More recently, others have investigated in greater detail, synthetic peptides containing ethionine, norvaline, norleucine, methionine dioxide, aza-adenylaminohexanoic acid, cysteine sulphonamide and fluorophenylalanine for antibacterial activity (Smith and Dunn, 1970; Ames *et al*., 1973; Payne, 1972*b*; Aleksiev *et al*., 1972; Barak and Gilvarg, 1974; Diddens *et al*., 1976) as well as for anticandidal activity (Lichliter *et al*., 1976).

Unfortunately, most peptides described in category Ia— of both natural and synthetic origin—are unsuitable for use as chemotherapeutic agents due to problems of instability in body fluids, toxicity, antagonism, spectrum of antimicrobial activity, and development of resistance (Scannell and Pruess, 1974; Barak and Gilvarg, 1974; Kenig *et al*., 1976).

An alternative approach to that described above (i.e. using peptides with aminocarboxylic acid warheads with unusual side groups), has been to replace the terminal carboxylic function of peptides with various other acidic functions, such as phosphonate, phosphinate, sulphonate, sulphonamide, hydroxamate, tetrazole, etc. (Hassal *et al*., 1977; Huber *et al*., 1975; Shiba *et al*., 1977; Smissman *et al*., 1976). The original rationale of our own approach was, however, to make analogues of the ubiquitous bacterial cell wall peptide fragment, D-Ala-D-Ala (Schleifer and Kandler, 1972), (see Section 4.7.2.2). Interest was concentrated on a series of phosphonopeptides which contained derivatives of α-aminoalkylphosphonic acids. The transport specificities of different bacterial species were investigated with phosphonopeptides of different chain length, stereochemistry, and amino acid sequence (Hassall *et al*., 1977). One of the most active peptides was selected from over 300 peptide analogues for further detailed investigation and this was the phosphonate analogue of L-alanyl-L-alanine, i.e. alaphosphin (**XV**) (Allen *et al*., 1978).

$$\underset{(S)}{\overset{CH_3}{\overset{|}{NH_2CH}}}CONH\underset{(R)}{\overset{CH_3}{\overset{|}{CH}}}{-}PO_3H_2 \qquad [Ala{-}X]$$

XV

Alaphosphin, (L-alanyl-L-1-aminoethylphosphonic acid)

This simple and yet novel dipeptide mimetic was shown to be a selective inhibitor of bacterial cell wall biosynthesis and, unlike most other inhibitors of cell walls, was more active against Gram-negative organisms than Gram-positive organisms (Ringrose *et al*., 1977). Alaphosphin is bactericidal and is more potent against a wide range of *E. coli* strains than any of the common clinically-used antibiotics. It is also particularly active against *Klebsiella, Serratia, Enterobacter*, and many other clinically important bacterial species. In addition, certain oligopeptide analogues of this series of phosphonopeptides show superior activity against certain Gram-positive organisms, in particular *Streptococcus faecalis*. A major advantage of the phosphonodipeptides relative to natural peptides is their increased stability to animal peptidases. Unlike most other peptides in category Ia, alaphosphin is active in animal infections, is absorbed across the intestine, and is well tolerated (Allen *et al*., 1977). Certain analogues of **XV** are, however, more potent.

The mechanism of action of alaphosphin involves active transport, via bacterial peptide permeases, followed by intracellular peptidase cleavage to yield the L-1-aminoethylphosphonic acid warhead (Ringrose *et al*., 1977). The warhead itself is antibacterially inactive and is not significantly transported into *E. coli* or other organisms. As a result of the inability of 1-aminoethylphosphonic acid to use effectively amino acid permeases, large concentrations of intracellular warhead are achieved after hydrolysis of alaphosphin inside bacterial cells. A concentration effect of 100- to 1000-fold for intracellular warhead is observed relative to extracellular alaphosphin. The enzyme target sites for the alaphosphin warhead are alanine racemase and, to a lesser extent, UDP-*N*-acetylmuramyl-L-alanine ligase (see Figure 3.) (Ringrose *et al*., 1977). As a result of this action, the intracellular levels of D-alanine decrease, peptidoglycan biosynthesis ceases, and cell lysis occurs within 20–40 min in sensitive organisms. In certain

Gram-negative organisms, the alaphosphin warhead is further metabolized to give a muramyl-1-aminoethylphosphonic acid derivative. An additional property of the cell wall-active phosphonopeptides is their ability to potentiate the action of certain other inhibitors of cell wall biosynthesis (e.g. D-cycloserine and the β-lactam group of antibiotics) by either double blockade or dual action on the intracellular pool of D-alanine.

Others have also explored the possibilities of changing the peptide acidic function (Franklin *et al.*, 1977). The tetrazole group has been shown to have a similar charge distribution and molecular size to the carboxyl group, (Morley, 1968). Peptides incorporating *C*-terminal aminotetrazole mimetics have shown some limited antibacterial activity, e.g. Phe-Gly-[Ile-tetrazole]. However, this particular tripeptide was significantly more active than its isoleucine-tetrazole warhead **(XVI)**. The only other tested aminotetrazole peptide to show antimicrobial activity was Phe-Gly-[Pro-tetrazole]. The activity of this tripeptide was limited to *S. aureus*, even though it was well transported and hydrolysed in *E. coli*. Its activity would, however, seem to be due to the intact dipeptide Gly-[Pro-tetrazole] **(XVII)** which was unexpectedly toxic to *S. aureus* (Franklin *et al.*, 1977). The tetrazole peptides may not, therefore, be interesting chemotherapeutically due to limited activity and potential lability to animal peptidases.

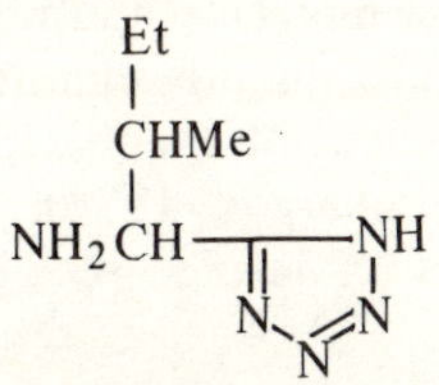

XVI

Ile-tetrazole

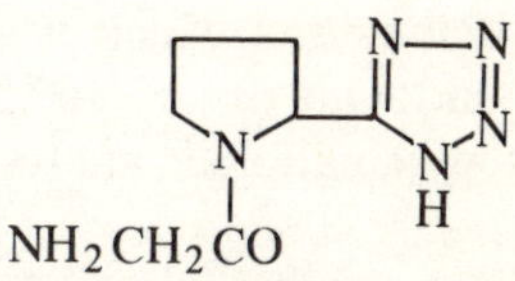

XVII

Gly-Pro-tetrazole

Dipeptides with the carboxyl replaced by sulphonate have also been examined. The four diastereoisomers of dialanine sulphonate (i.e. alanyl-1-aminoethylsulphonate) were, however, found to be without any significant antibacterial activity (Shiba *et al.*, 1977). Gly-Phe phosphonate was also found to be antibacterially inactive (Gilmore and McBride, 1974).

Peptide derivatization of aryl sulphonamides was first attempted by Shankman *et al.* (1963*a*) and will be discussed in the next section. More recently, sulphanilic acid has been linked to the dipeptide Phe-Gly, since sulphanilic acid, although active against isolated dihydrofolate synthetase, is not transported into cells (Franklin *et al.*, 1977). Activity was limited and very strain-specific. It was, however, shown that in *Escherichia coli* 198 (but not *E. coli* B), a 50-fold concentration of intracellular sulphanilic acid was achieved relative to the peptide in the medium. The antibacterial activity of Phe-Gly-sulphanilic acid was reversed by *p*-aminobenzoic acid, indicating direct action of the sulphanilic acid warhead (in contrast see Shankman *et al*., 1963*a*).

The possibilities of exploiting 'illicit transport' in chemotherapy still remain to be fully explored. However, the limitations of this concept, e.g. peptide stability to mammalian enzymes, resistance, potency, antimicrobial spectrum, antagonism, and toxicity, remain as difficulties to be resolved. The recent phosphonopeptide development is, however, an encouraging advance and is illustrative of how basic knowledge of transport mechanisms may facilitate the rational design of novel chemotherapeutic agents.

2 Linear peptides that are active as the intact molecule

One of the first small peptides to be studied in detail for its growth-inhibitory activity was glycylleucine (Simmonds *et al*., 1947, 1951). It became apparent from studies on the effect of growth conditions on intracellular peptidase levels in *E. coli*, that glycylleucine was most active when peptidase levels were lowest (Meisler and Simmonds, 1963; Simmonds, 1970). Other peptides of leucine and glycine were not inhibitory, neither were the component amino acids. The dipeptide was therefore assumed to be acting on an intracellular target site as the intact molecule and not as a mimetic-delivery system. This target site has been subsequently identified as threonine deaminase, which is also regulated by isoleucine feedback (Von der Haar and Umbarger, 1972; Wasmuth and Umbarger, 1974). For a detailed conformational analysis of the Gly-Leu molecule, see Hruby (1974).

In addition to the work described earlier with carrier-type peptides, Shankman and coworkers investigated other synthetic peptides which were believed to act as the intact molecule or at least as a peptide metabolite. L,L,D-trivaline was the only trivaline stereoisomer active against selected strains of *Lactobacillus* and *Pediococcus* (Shankman *et al*., 1960, 1961, 1962, 1963*b*). It was concluded from this work that stability to bacterial enzymic hydrolysis was necessary for antibacterial activity. It was also suggested that a tripeptide inhibitor should have an *N*-terminal L,L-dipeptide sequence (for transport) and should be made resistant to *C*-terminal attack by carboxypeptidases by having an L,L,D sequence. It was proposed that a dipeptide inhibitor should have L,L stereochemistry (for transport) and should be made resistant to aminopeptidases either by using 'symmetrical' dipeptides (i.e. dimers of the same amino acid) or by inserting natural amino acids or amino acid mimetics known to inhibit hydrolysis (Shankman *et al.*, 1962). The mixed stereochemistry tripeptide plumbemycin (L-alanyl-L-asparaginyl-D-2-amino-5-phosphono-3-*cis*-pentenoic acid) would seem to conform to Shankman's principles (Park *et al.*, 1977). It is antibacterial and a threonine antagonist.

L-Valyl-L-valyl-sulphanilamide was also synthesized and found to be antibacterial (Shankman *et al*., 1963*a*). Unlike the recent work with peptides of sulphanilic acid (Franklin *et al*., 1977), the activity of Val-Val-sulphanilamide was not reversed with *p*-aminobenzoic acid or folic acid. It would seem that the intact peptide was required for intracellular activity and the mechanism of action did not involve dihydrofolate synthetase but inhibition of uptake of specific essential amino acids (Shankman *et al*., 1963*a*).

The 'Shankman principle' of using peptidase-resistant tripeptides has been further exploited with triornithine (**XVIII**). Triornithine is transported into cells via an oligopeptide permease (Payne, 1968). It is resistant to enzymic hydrolysis (Sussman and Gilvarg, 1970) and exerts its action by inhibition of protein biosynthesis (Gilvarg and Levin, 1972; Barak *et al*., 1973*a,b*). No inhibitory effect, however, has yet been detected in cell-free protein synthesizing systems. Although the tripeptide is bactericidal, its action is very strain-specific and most studies have been limited to *E. coli* strains W, B, and K12 and to *Salmonella typhimurium*; the latter organism being the more sensitive.

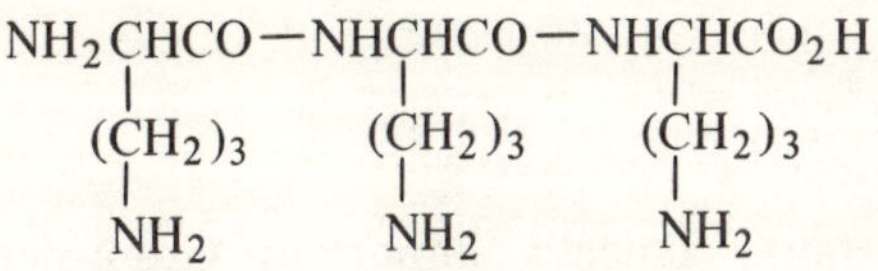

XVIII

Tri-L-ornithine

Triornithine has been a most useful 'tool' for characterizing oligopeptide permease (Opp) systems (Payne, 1968; Ames *et al*., 1973; Naider and Becker, 1975; Barak and Gilvarg, 1975). The emergence of resistance to triornithine is, however, very high, i.e. frequency of 10^{-4}–10^{-5} (Barak and Gilvarg, 1974), but it would now appear that the specific permease used by triornithine is only one of a number of Opp systems (Naider and Becker, 1975; Barak and Gilvarg, 1975; W. J. Lloyd and P. S. Ringrose, unpublished observations) (but see also Section 2.1.6).

Di- and tetraornithine were without activity, but certain analogues of triornithine containing other amino acids in various positions have been shown to have improved activity, e.g. Orn-Orn-Lys and Ala-Orn_3 (Gilvarg and Levin, 1972). The bactericidal or protein-biosynthesis inhibitory activity was inversely related to the susceptibility to peptidase cleavage for a large number of related peptides based on triornithine. Trithreonine has also been demonstrated as being toxic to *E. coli* K12 (Barak and Gilvarg, 1975); however, it should be remembered that some homo-tripeptides, e.g. trivaline (L,L,L) are active only as a result of intracellular hydrolysis (Payne and Gilvarg, 1971).

In attempts to use ornithine residues in peptides, linear copolymers of ornithine and leucine were made which were weakly antibacterial towards *S. aureus* (Kasymova *et al.*, 1973; Shenfeld *et al.*, 1974). Cyclo-L-Orn-D-Val and cyclodipeptides containing N^5-*p*-aminobenzenesulphonylornithine were, however, inactive as antibacterial agents (Ikeda *et al.*, 1972). Basic polypeptides of Orn, Lys, or Arg can be antibacterial, but in a completely different way to triornithine and they may cause an alteration in surface charge on bacteria leading to their aggregation (Keller *et al.*, 1968).

One of the few naturally-occurring dipeptide mimetics that could be placed in category Ib is negamycin (**XIX**). Negamycin is a 'dipeptide-like' bactericidal antibiotic consisting of δ-hydroxy-β-lysine linked to methylhydrazino acetic acid and is produced by a streptomycete (Hamada *et al.*, 1970; Kondo *et al.*, 1971).

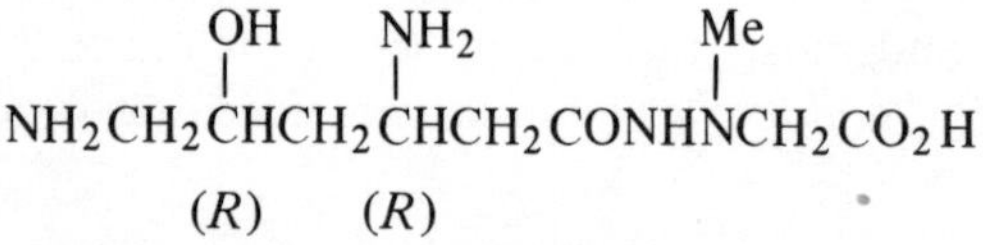

XIX

Negamycin

It is active against drug-resistant, clinically important Gram-negative bacteria including *Pseudomonas*; it inhibits protein biosynthesis by causing miscoding of the mRNA and by inhibition of chain-termination (Mizuno *et al.*, 1970*a,b*; Uehara *et al.*, 1972, 1976*a,b*). Although structurally unrelated to the streptamine and deoxystreptamine-containing aminoglycosides that cause miscoding (e.g. streptomycin, kanamycin), negamycin also binds to the ribosomal 30s subunit. Unlike the aminoglycosides, it will bind to polysomal ribosomes and its action is not inhibited by chloramphenicol (Uehara *et al.*, 1976*b*). However, a three-dimensional resemblance between negamycin and kanamycin has been drawn to try and explain some of the activity of negamycin (Uehara *et al.*, 1972). Its activity *in vivo* is greater than *in vitro* and its acute toxicity is reportedly low despite the presence of a hydrazino group (Hamada *et al.*, 1970).

Structural analogues of negamycin have been synthesized and studies of their activites indicate: (a) the natural (*R*) configuration of the β-amino carbon atom was essential for activity, (b) the δ-hydroxyl group was not necessary, and (c) acylation of the ε-amino group caused loss of activity (Uehara *et al.*, 1976*c*). The aza analogue, with an N atom replacing a C atom in the β-position, was also inactive (Streicher and Reinshagen, 1975).

A related hydrazino dipeptide antibiotic (**XX**) is linatine (1-[*N*-γ-L-Glu]amino-D-Pro), but in contrast to negamycin its mechanism of action involves interaction with pyridoxal phosphate (Klosterman *et al.*, 1967).

XX

Linatine

Another recently characterized peptidase-resistant antibiotic in category Ib is the L,D,L antifungal tripeptide L-Arg-D-alloThr-L-Phe (**XXI**), which is produced by the dermatophyte *Keratinophyton terreum* (König *et al.*, 1973).

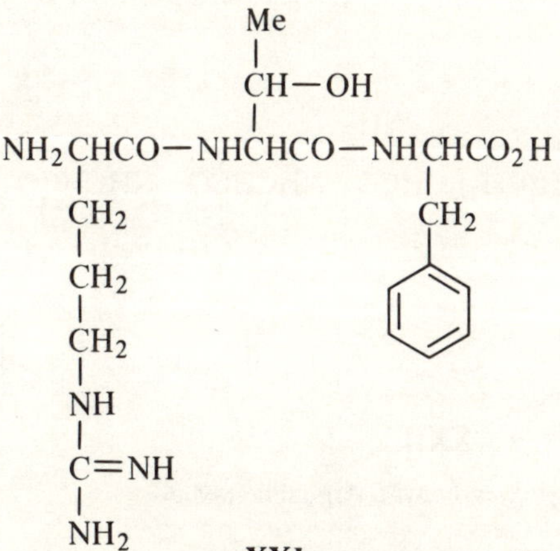

XXI

L-Arg-D-alloThr-L-Phe

This particular peptide was not active against bacteria. All work carried out so far on bacterial oligopeptide permease stereospecificity suggests that an L,D,L tripeptide would not be recognized (Becker and Naider, 1974; Payne, 1976). The activity of this mixed stereochemistry peptide against filamentous fungi was antagonized by both L-His and L-Thr. The synthetic tripeptide L-Arg-L-Thr-L-Phe was inactive, but certain other D-amino acids could substitute in the middle position, e.g. D-Phe (König *et al.*, 1973; Eisele, 1975*a*) to give active peptides. Substitution of D,L-4-chloroPhe and D,L-3-fluoroPhe as the central residue in the tripeptides showed some antifungal as well as antibacterial activity (Eisele, 1975*a*). The antibacterial activity was probably due to the L,L,L stereoisomer, which could act as a warhead delivery peptide. The *N*- and *C*-terminal amino acids (L-Arg and L-Phe) were, however, necessary for activity in all the cases tried.

The L,D,L sequence was subsequently shown only to be a requirement inasmuch as it prevented enzymic hydrolysis, since the D-amino acid could be replaced by *N*-α-methyl-L-amino acids, to give peptides which were also resistant to cleavage. The *N*-α-methyl amino acid substituents had to be lipophilic to retain activity. Trifunctional amino acids in this middle position abolished activity (Eisele and Eisele, 1975). It is interesting to note that it was the nature of the middle amino acid in the peptide that determined the L-amino acid that antagonized the activity of the tripeptide. The overall requirement for activity was therefore a peptidase-resistant, lipophilic L-Arg-X-L-Phe tripeptide.

N-Substituted benzyloxycarbonyl (Cbz) peptides were also examined for antifungal activity. The most interesting compound was N^{ω}-Cbz-L-Arg-L-Phe-L-Phe (**XXII**), although the middle substituent could be any amino acid; all N^{ω}-Cbz-Arg tripeptides were antagonized by L-Asp and L-Asn (Eisele, 1975*b*).

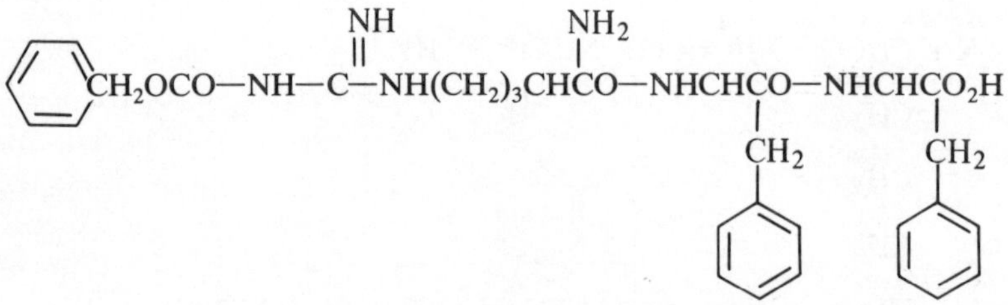

XXII

N^{ω}-Benzyloxycarbonyl-L-Arg-L-Phe-L-Phe

A closely related series of Cbz-protected peptides was examined by Nicolaides *et al.* (1968) and Norrby (1971) for antiviral activity. A number of di- and tripeptides were active against measles and herpes virus (but not against polio, rabies, or influenza), e.g. Cbz-D-Phe-L-Phe-L-(NO_2)Arg, Cbz-D-Phe-D-Phe, and Cbz-Tyr-(*S*-benzyl)Cys-OMe. It is of interest that the Cbz-protected free amino acids were inactive, regardless of stereochemistry.

Returning to bacterial peptides, an important dipeptide in the synthesis of bacterial cell wall peptidoglycan is D-Ala-D-Ala (Neuhaus, 1962). This dipeptide is resistant to the L,L-specific aminopeptidases inside the cell but is hydrolysed by D,D-carboxypeptidases in the cell wall (Ghuysen, 1977). A stable analogue of D-Ala-D-Ala could therefore be antibacterial by (a) inhibiting the enzyme that attaches D-Ala-D-Ala to the UDP-*N*-acetylmuramic acid tripeptide; (b) acting as a false substrate for the above enzyme and subsequently inhibiting transpeptidase; or (c) acting as an analogue of penicillin (Tipper and Strominger, 1965; Böhme *et al.*,

1971; Lee, 1971). This particular dipeptide has therefore attracted a lot of chemical attention, nearly all of which has failed to produce compounds with antibacterial activity, in part due to the inability of D,D-dipeptides to utilize bacterial L,L-specific dipeptide permeases, e.g. amino-isobutyryl-D-Ala, D-Ala-aminoisobutyric acid (Gale and Smith, 1973), D-Ala-D,L-1-aminoethyltetrazole, D-Ala-D-Ala-hydroxamic acid (Smissman *et al.*, 1976), D-Ala-D,L-1-aminoethylphosphonic acid (Huber *et al.*, 1975), D,L-Ala-D,L-1-aminoethylsulphonic acid (Shiba *et al.*, 1977) and D-Ala-D-Ala-hydrazide (Okada *et al.*, 1976*a*). Various diastereoisomers of di- and trialanine have also been examined for antimicrobial activity and for effects on spore germination in bacilli, without much success (Okada *et al.*, 1976*b*).

Neuhaus *et al.* (1977) have more recently synthesized a number of D-Ala-D-Ala mimetics; the most active being D-α-aminovaleryl-D-Ala, i.e. D-norVal-D-Ala. This dipeptide was utilized by the D-Ala-D-Ala adding enzyme to give a muramyl pentapeptide with a false amino acid in the 4 position which was subsequently inhibitory to cell wall transpeptidation (see Figure 3). Unfortunately D-norVal, when used as a substrate for D-Ala-D-Ala synthetase, was converted into D-Ala-D-norVal, which unlike D-norVal-D-Ala was not a substrate for the D-Ala-D-Ala adding enzyme (Neuhaus, 1962). The activity of D-norVal-D-Ala was only weakly antibacterial, presumably because it failed to meet the dipeptide permease specificity. This activity was however significantly enhanced by another cell wall inhibitor, D-cycloserine (Neuhaus *et al.*, 1977).

The most comprehensive study to date of dialanine mimetics has, however, identified the L,L-dialanine phosphonate analogue, alaphosphin, as being antibacterially the most active (Hassall *et al.*, 1977; Ringrose *et al.*, 1977; Allen *et al.*, 1978), but for mechanistic reasons other than those just described (see previous section). Other attempts to 'deliver' a D-Ala-D-Ala peptide mimetic as part of a larger peptide molecule have so far been unsuccessful in terms of antibacterial activity.

As an alternative to analogues of D-Ala-D-Ala, some have prepared more closely related peptide analogues of the penicillin nucleus, e.g. *N*-phenylacetyl-L-Cys-D-Val (Holly *et al.*, 1952). This acyl L,D-dipeptide had some weak antibacterial activity and exhibited a low degree of synergistic action with benzylpenicillin.

HCO—Val—Gly—Ala—D—Leu—Ala—D—Val—Val—D—Val

—Trp[D—Leu—Trp]$_3$NHCH$_2$CH$_2$OH

XXIII

Gramicidin A

Probably one of the best studied linear peptide antibiotics is gramicidin A (**XXIII**). It is composed of 15 amino acids arranged in an alternating L,D,L,D sequence with the *N*-terminal valine residue formylated and the *C*-terminal tryptophan masked with ethanolamine (Sarges and Witkop, 1965). Although linear in sequence gramicidin A is now believed to exist as a helix dimer in the membrane, thus forming an ion channel (Veatch *et al*., 1974) which allows exchange to occur between K^+ and H^+ ions (Henderson *et al*., 1969). Direct K^+ binding to gramicidin A may, however, play a role in its mechanism of action (Byrn, 1974) (see also Section 4.8.3.1(i)). Helical structures have also been described for much smaller linear peptides in solution e.g. Cbz-Gly-Pro-Leu-Gly (Ueki *et al*., 1967).

The structure of the gramicidin A molecule is such that it is extremely resistant to peptidase attack. Many researchers have modified the structure of gramicidin A without much success in terms of retaining activity; the formyl group is necessary for activity but can be replaced with acetyl, the ethanolamine group is responsible for the antibiotic's membrane-haemolytic activity, and the four indole rings in the tryptophan residues are also required for antibacterial action (Rambhav and Ramachandran, 1972). It is interesting to note that bacterially produced *N*-formyl di- and tripeptides have chemotactic activity for phagocytic leucocytes; the non-acylated peptides are, however, inactive (Schiffmann *et al*., 1975).

Edeine is another non-cyclic oligopeptide antibiotic that has been extensively studied as an inhibitor of DNA synthesis and of polypeptide chain initiation (Kurylo-Borowska, 1962; Szer and Kurylo-Borowska, 1970). Edeines A and B are basic pentapeptides linked to spermidine derivatives and their overall structure has been described by Hettinger *et al*. (1970). The peptaibophols are a group of larger linear *N*-acetyl peptide antibiotics (emerimicin, antiamoebin) with a C-terminal phenylalaninol moiety attached to 15 amino acids, six of which are aminoisobutyric acid (Pandey *et al*., 1977).

Attempts have been made to synthesize simple linear oligopeptides resembling the common amino acid sequences present in the cyclic peptide antibiotics, gramicidin S, tyrocidin B, polymixin B, and bacitracin (see Section 4.7.3). Tri- and hexapeptides containing Phe, Leu, and Lys were, however, shown to have only weak antimicrobial activity, e.g. D-Phe-L-Leu-L-Lys-D-Phe-L-Lys-L-Leu-amide (Paulay and Bajusz, 1972).

Of the various linear peptides described in Section 4.7.2.2, it would appear that some exert their action directly on the membrane, whereas others need to gain access to the cell interior in order to exert their effect. Whilst it is comparatively easy to design peptides that are resistant to peptidase cleavage, e.g. by using alternating D-amino acids and by adding *N*-terminal acyl groups, it is more difficult to incorporate these features into peptides designed to utilize microbial peptide permeases. The use of *N*-methylated peptide bonds does not in most cases prevent permease recognition, but does inhibit enzymic cleavage (Payne, 1972*a*). In addition, hydrophobic peptides may diffuse into cells without using energy-linked permeases; in this case, however, there is no net concentration of intracellular peptide unless intracellular metabolism or binding occurs.

An obvious corollary of the two types of peptides described in this section is that agents which specifically inhibit peptidase action, such as bestatin (**XXIV**) (Suda *et al*., 1976), should potentiate peptides described in Section 4.7.2.2 (category Ib) whilst antagonizing peptides described in Section 4.7.2.1 (category Ia). Peptidase inhibitors have already been studied for preventing degradation of biologically active peptides, e.g. bestatin and the peptide antibiotic bleomycin (Umezawa *et al*., 1976), also bacitracin and the peptide hormones TRF and LH (McKelvy *et al*., 1976).

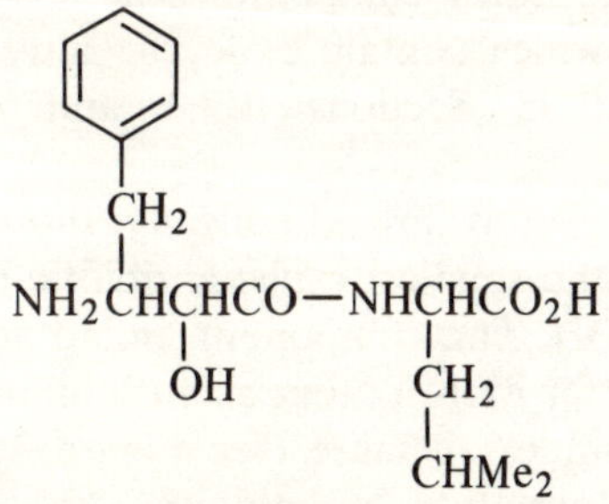

XXIV

Bestatin

A further chemotherapeutic extension of this use of peptidase inhibitors has been the recent findings that bestatin enhances delayed-type hypersensitivity and consequently increases immunity of mice to Ehrlich carcinoma cells (Umezawa *et al*., 1976). This may provide for the future a novel alternative use of peptides as indirect antimicrobial agents via immune stimulation.

3 CYCLIC PEPTIDES

The cyclic peptide antibiotics are the best characterized and most widely studied of all the antimicrobial peptides, largely because of the interest in relating conformation with the ability of certain members of this group to carry ions through membranes (Ovchinnikov, 1972; Pressman, 1976). Peptide ionophores will, however, be discussed in detail in the next chapter

(Section 4.8), but it will be seen from the following discussion that cyclic peptides, as a group, are extremely diverse in their mechanisms of action.

The structures of the cyclic peptides are such that they are all extremely resistant to enzymic degradation. There are usually no free α-amino or α-carboxyl groups. The component amino or hydroxy acids are often of D-configuration and are sometimes structurally very different from the amino acids found in proteins (Perlman and Bodanszky, 1971). The links within such peptide moieties can be ester or β- or γ-peptide bonds in addition to the normal α-peptide links. Hypercyclization is also a common phenomenon. Many of the larger cyclic peptides form pleated structures containing β-loops stabilized by intramolecular H-bonding (Hassall and Thomas, 1971), but cylindrical and α-helical structures are also found.

This section will describe various examples of homo-and heteromeric cyclic peptides in order of increasing ring size. Peptides such as valinomycin and serratamolide, which contain amino and hydroxy acids, and actinomycin D, which contains a heterocyclic ring system, will therefore be included, in addition to peptides containing only α-amino acids. Other heteromeric cyclic peptides which contain exocyclic fatty acid and sugar moieties will be discussed in Sections 4.7.4 and 4.7.5 respectively.

Some of the smallest peptide ring systems are chemically the most unusual. Although not strictly a peptide, the smallest cyclic antibiotic made from an amino acid is D-cycloserine (**XXV**). The component amino acid is aminoxy-D-alanine, but the cyclic form is far more potent as an inhibitor of bacterial D-Ala-D-Ala synthetase and alanine racemase (see Figure 3). As such D-cycloserine induces spheroplast formation by inhibiting the early stages of cell wall biosynthesis (Neuhaus, 1967; Strominger *et al.*, 1967).

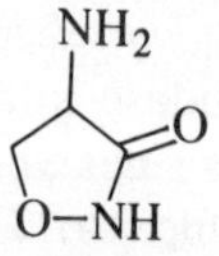

XXV

D-Cycloserine

The rigid planar ring of D-cycloserine is important for its activity, as has been shown by molecular models of various cyclic and ring-opened analogues (Neuhaus and Lynch, 1964; Roze and Strominger, 1966). Extensive biochemical work by Neuhaus and his coworkers has elucidated its mechanism of active transport, action on enzymes, and resistance development (Neuhaus, 1968; Reitz *et al.*, 1967; Wargel *et al.*, 1970, 1971; Lambert and Neuhaus, 1972). Although therapeutically useful in the past for the treatment of tuberculosis, other antibiotics have now superceded D-cycloserine owing to its limited neurological side effects. Recent work with a combination of deuterated D-fluoroalanine and D-cycloserine has indicated potent synergy which may be clinically useful (Kahan and Kropp, 1975). D-Cycloserine synergy with the peptide mimetic alaphosphin has also been described (Ringrose *et al.*, 1977; Allen *et al.*, 1978).

Cyclic dipeptides are most commonly of the diketopiperazine type (e.g. bicyclomycin, gliotoxin). However, penicillin and related β-lactam antibiotics, which are therapeutically the most important of all the antimicrobial peptides, can also be considered as cyclic dipeptides. It is interesting to note that these three antibiotics are, in fact, bicyclic peptides.

Gliotoxin (**XXVI**) is a potent antiviral epidithiapiperazine dione which also has antibacterial and antifungal activities (Rightsel *et al.*, 1964). It is structurally the simplest of a group of related diketopiperazines with disulphide bridges, e.g. sporidesmin and aranotin (Bell *et al.*, 1958; Hruby, 1974). A related group of simple diketopiperazines has also been recently investigated (Michel *et al.*, 1974).

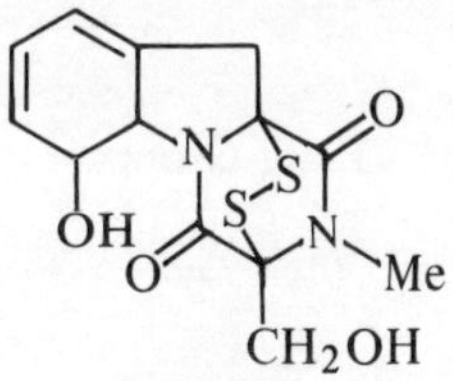

XXVI

Gliotoxin

Gliotoxin is active against most of the RNA viruses. It interferes specifically with viral RNA synthesis without having much effect on host cellular RNA synthesis (Miller *et al.*, 1968).

Bicyclomycin (**XXVII**) was isolated by Miyoshi *et al.* (1972) and its structure determined by Kamiya *et al.* (1972). It is a diketopiperazine with an oxytrimethylene bridge. Bicyclomycin is active against most Gram-negative bacteria with the exception of *Pseudomonas* and *Proteus*. Its mechanism of action is, however, unique in that it appears to interfere specifically with the linking of the Gram-negative outer-membrane lipoprotein to the ε-carboxyl of diaminopimelic acid in the peptidoglycan (Tanaka *et al.*, 1976; Braun *et al.*, 1974). A related diketopiperazine is cairomycin B, the mechanism of action of which has yet to be reported (Shimi *et al.*, 1977).

Diketopiperazines have also been suggested as being involved in epimerization of peptide L-amino acids *in situ* (Bycroft, 1969), and in the biosynthesis of larger microbial peptides (Mauger, 1968). Interestingly,

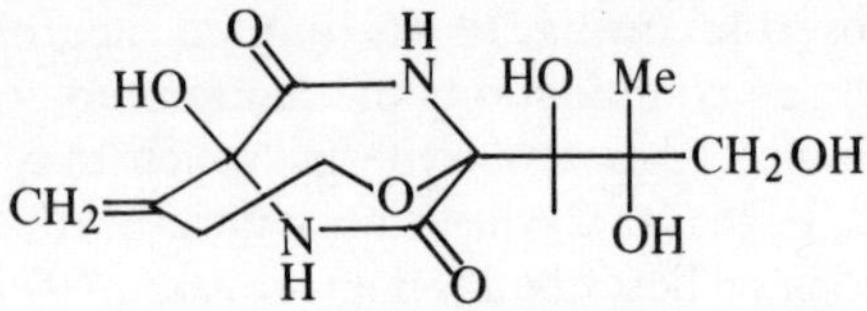

XXVII

Bicyclomycin

D-cycloserine will rearrange to give an inactive diketopiperazine if stored in aqueous solution (Neuhaus, 1967).

The penicillin group of antibiotics has been extensively reviewed elsewhere (Flynn, 1972), but recent studies have described a number of other related molecules with intriguing biological activities. The penicillin nucleus, 6-aminopenicillanic acid (APA) (**XXVIII**) is a bicyclic form of the dipeptide L-Cys-D-Val (**XXIX**) containing a β-lactam ring and a five-membered thiazolidene ring.

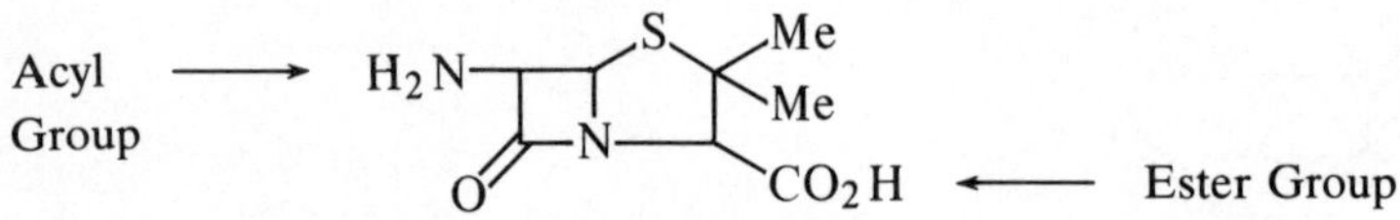

XXVIII

Aminopenicillanic acid

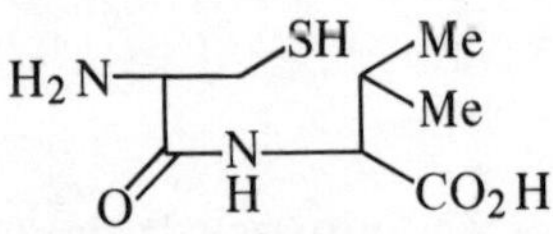

XXIX

Cys-Val

Indeed, the δ-(α-aminoadipyl)-L-cysteinyl-D-valine tripeptide, Arnstein's peptide, is a biosynthetic precursor of penicillin N (Arnstein and Morris, 1960; Adriaens *et al*., 1975). The close similarity of penicillin to the conformation of the terminal dipeptide D-Ala-D-Ala of bacterial peptidoglycan (Tipper and Strominger, 1965; Lee, 1971) has led many researchers to synthesize analogues (see Section 4.7.2.2).

The mechanism of action of the penicillins is believed to involve the transpeptidases and carboxypeptidases responsible for bacterial cell wall peptidoglycan cross-linking (Blumberg and Strominger, 1974; Ghuysen, 1977). Recent work with penicillin-binding proteins and on autolysin triggering has, however, made the picture less clear as to the precise mechanism(s) by which penicillins exert their bactericidal effect (Spratt,

1977; Tomasz and Waks, 1975). With the penicillins, the spectrum of antibacterial activity, resistance to acid conditions and β-lactamases, and the extent of intestinal absorption are greatly influenced by the nature of the acyl group on the 6-amino position and the ester group on the 3-carboxyl position (see **XXVIII**) (Flynn, 1972; Ferres *et al.*, 1974). It is interesting to note that, one of the most potent broad-spectrum penicillins, currently in clinical trial, i.e. piperacillin, has a 2,3-diketopiperazine substituent on the 6-amino acyl group (Ueo *et al.*, 1977).

The cephalosporins (**XXX**) are closely related to the penicillins, with a six-membered unsaturated thiazolidene ring and greater β-lactamase stability (Griffith, 1974). Other more recently isolated members of this class of cell-wall β-lactam antibiotics are the cephamycins (7-methoxycephalosporins) (**XXX**) (Stapley *et al.*, 1972; Miller *et al.*, 1972); mecillinam (amidinopenicillin) (Lund and Tybring, 1972), clavulanic acid (**XXXII**) (Brown *et al.*, 1976); olivanic acid (Brown *et al.*, 1977); thienamycin (**XXXI**) (Kropp *et al.*, 1976), and nocardicin (**XXXIII**) (Hashimoto *et al.*, 1976; Aoki *et al.*, 1976).

Each type of structure has its own characteristic antibacterial spectrum, mechanism of action and penicillin binding-protein profile (Spratt, 1977; Spratt *et al.*, 1977). Clavulanic acid is particularly interesting in that although it is weakly active antibacterially, it is a potent irreversible inhibitor of β-lactamase (Hunter and Reading, 1976).

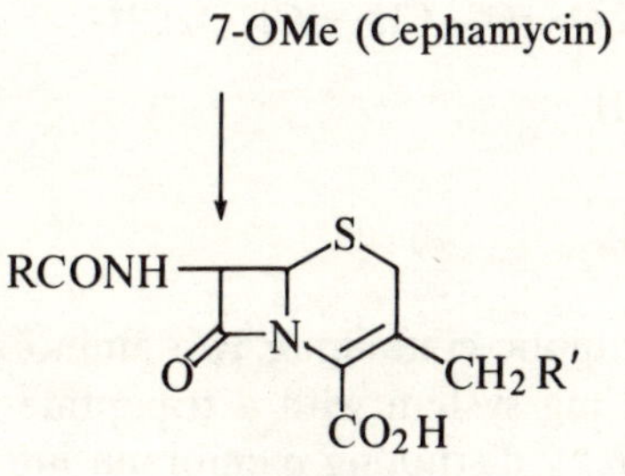

XXX

Cephalosporin

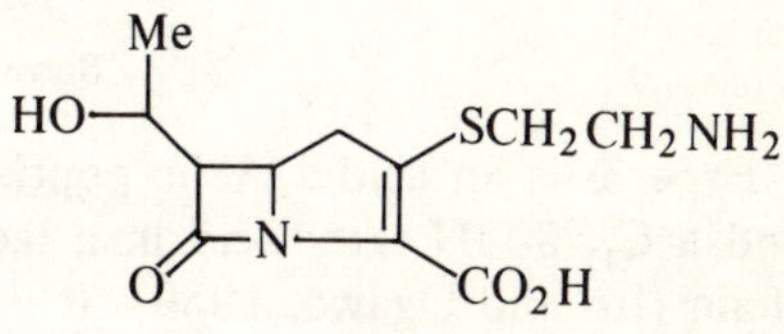

XXXI

Thienamycin

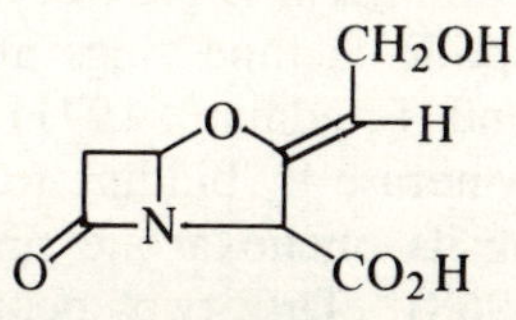

XXXII

Clavulanic acid

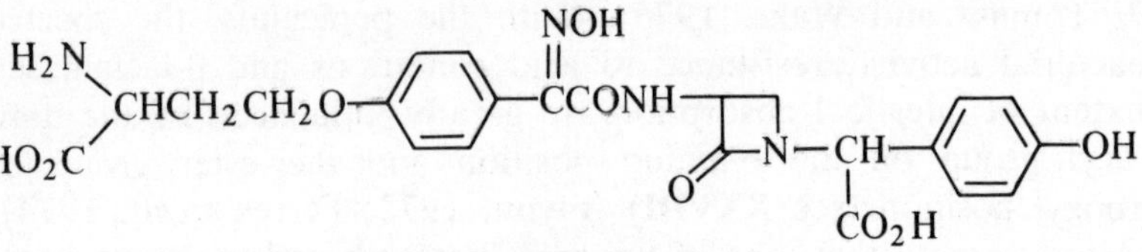

XXXIII

Nocardicin

Attempts at making synthetic cyclic dipeptides have not been successful in terms of antibacterial activity (Izumiya *et al*., 1964; Ikeda *et al*., 1972). One of the smallest cyclic depsipeptides containing only two amino acids is serratamolide (**XXXIV**). It is a cyclic dimer of L-Ser-β-hydroxydecanoic acid, produced by the bacterium *Serratia marcesens* and is weakly antibacterial, possibly by a K^+ ionophore effect (Wasserman *et al*., 1961, 1962; Deol *et al*., 1973). The lipophilic nature of the molecule, however, may mean that its action involves membrane disruption in a manner similar to that of surfactin or the polymixin fatty acyl peptides (Tsukagoshi *et al*., 1970; Storm *et al*., 1977). Related depsipeptides from *Serratia* have also been identified (Bermingham *et al*., 1970).

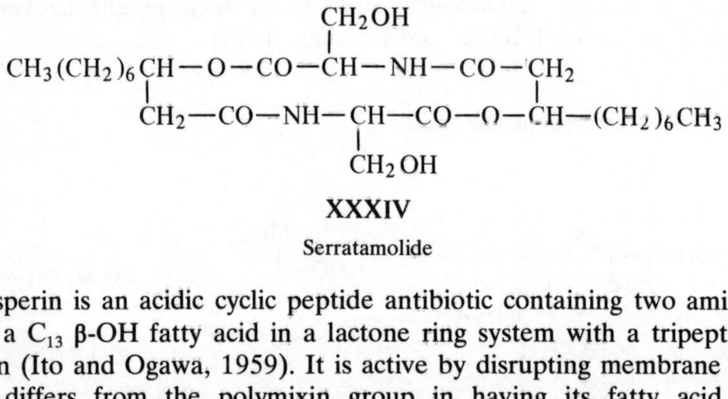

XXXIV

Serratamolide

Esperin is an acidic cyclic peptide antibiotic containing two amino acids and a C_{13} β-OH fatty acid in a lactone ring system with a tripeptide side-chain (Ito and Ogawa, 1959). It is active by disrupting membrane integrity but differs from the polymixin group in having its fatty acid directly incorporated into its small ring system (see Section 4.7.4).

There are a number of cyclic pentapeptides, one of the best characterized of which is actinomycin D (**XXXV**). This antibiotic consists of two identical cyclopentapeptide lactone rings attached to a phenoxazone chromophore (Goldberg and Friedman, 1971). Actinomycin D inhibits DNA-dependent RNA polymerase by binding to GC-rich regions of DNA. It does this by intercalating its phenoxazone ring between GC base pairs (Müller and Crothers, 1968). The two peptide rings are, however, necessary in the formation of this complex and form hydrogen bonds with deoxyguanosine residues as well as with the sugar–phosphate backbone (Sobell, 1972; Sobell *et al*., 1971). The most important intramolecular peptide hydrogen bonds are between the two ring D-valine residues, positioned above and below the chromophore plane. Intermolecular hydrogen bonding between the threonine residues and the deoxyguanosine bases plays an important role in complex formation. The peptide chains thus lie in the narrow groove of DNA and are positioned to give a twofold axis of symmetry (Sobell, 1974). A hybrid molecule containing the phenoxazone ring attached to a gramicidin S decapeptide was, not surprisingly, totally devoid of actinomycin D-like activity (Reich, 1966). An extensive review by Brockmann (1974) describes the various analogues of actinomycin D that have been made and their structure–activity relationships.

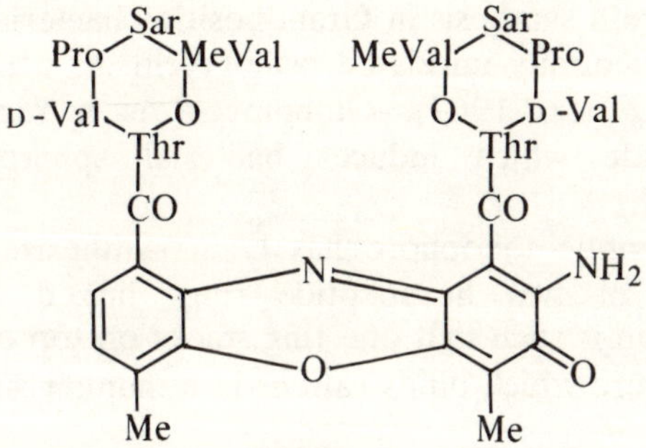

XXXV

Actinomycin D

Malformins are bicyclic pentapeptides, containing three D- and two L-amino acids (**XXXVI**) (Bodanszky and Stahl, 1974). They are produced by *Aspergillus niger* and are particularly active against Gram-negative organisms (Suda and Curtis, 1966). Other antibiotics with pentapeptide rings are the related antitubercular compounds: viomycin, capeomycin, and tuberactinomycin (Tsukamura, 1960). These molecules are in fact basic hexapeptides with a lysine or arginine-related moiety attached to a cyclic pentapeptide (Bycroft *et al*., 1971; Shiba *et al*., 1976). Viomycin has been studied extensively as an inhibitor of bacterial protein biosynthesis (Liou and Tanaka, 1976) and has been shown to contain a 10-membered loop structure (Bycroft, 1972).

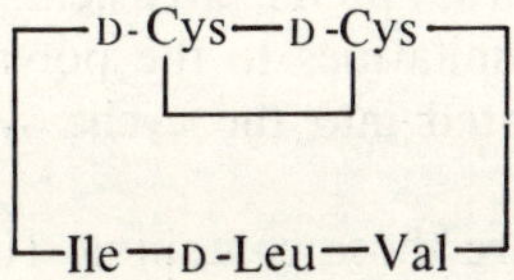

XXXVI

Malformin

Cyclic hexapeptides which have been particularly well studied in terms of structure are monamycin (Hall, 1970; Bevan *et al*., 1970) and enniatins A and B (Plattner *et al*., 1963; Quitt *et al*., 1963) (see also Section 4.8.3.2). They are both depsipeptides and can act as ionophores. Monamycin is particularly interesting in that it contains two unusual cyclic amino acids, L-hydroxypiperidiazine carboxylic acid and D-piperidiazine carboxylic acid (Hassall *et al*., 1969). Enniatin B is a regular cyclic molecule with a repeating L–D, ester–amide sequence of (L-MeVal-D-hydroxyisovaleric acid)$_3$. Monamycin is, however, a potent membrane lytic agent.

Osteogrycin B is also a cyclic hexapeptide and a member of the streptogramin group of antibiotics. Its ring is cyclized via the hydroxyl group of threonine and contains a characteristic 3-hydroxypicolinic acid. It is an inhibitor of ribosomal protein synthesis in Gram-positive bacteria and acts synergistically with the structurally unrelated osteogrycin A, which is classified as a macrolide (Vazquez, 1967). Globomycin is a recently characterized cyclic hexapeptide which induces bacterial spheroplast formation (Nakajima *et al*., 1978).

An interesting bicyclohexapeptide ionophore has been synthesized by Schwyzer (1970). It consists of two hexapeptide rings linked by a disulphide bond; its conformation is such that one ring stacks on top of the other to give a cylindrical structure which binds cations in a manner similar to that displayed by monactin.

Examples of cyclic heptapeptides are evolidine, patricin, and ilamycin (Studer and Lergier, 1965; Ondetti and Thomas, 1965; Cary *et al*., 1971). Surfactin comprises seven amino acids and one C_{15} β-OH fatty acid incorporated into the ring. It acts directly as a microbial anionic detergent and disrupts membranes (Tsukagoshi *et al*., 1970).

Octapeptide examples are mechanistically more diverse, e.g. α-amanitin and echinomycin. α-Amanitin is a bicyclic peptide which interacts directly and specifically with the RNA polymerase in the nucleoplasm which is responsible for synthesis of 'DNA-like' RNA. It is structurally related to the mushroom poison, phalloidin (Fiume and Wieland, 1970; Tata *et al*., 1970). Echinomycin is also a bicyclic octapeptide, but is pseudo-symmetrical containing two quinoxaline carboxylic acid chromophores. Its structure has obvious parallels with actinomycin D and it selectively binds to specific DNA nucleotide sequences in a bifunctional mode of intercalation (Waring and Wakelin, 1974).

Polypeptin is a nonapeptide cyclized through a C_7 β-OH fatty acid. It is, thus, related to surfactin and esperin and has similarities to the polymixin group, although the fatty acyl moiety is integrated into the cyclic system and there is no peptide side-chain (Sogn, 1976).

The most extensively studied cyclic peptides are those containing 10 and 12 amino acid/hydroxy acid moieties. Gramicidin S, tyrocidine, antamanide, bacitracin, and polymixin are decapeptides whilst valinomycin is a dodecadepsipeptide. All these peptides act in some way on membranes whether as ionophores, disrupters, or inhibitors of membrane-associated enzymes.

Gramicidin S (**XXXVII**) and tyrocidine are related cyclic homomeric peptides, produced by *Bacillus brevis*, that disrupt membrane integrity. Tyrocidine has, however, also been observed to interact directly with DNA (Schazschneider *et al.*, 1974). Gramicidin S is structurally and mechanistically unrelated to gramicidin A (see Section 4.7.2.2), which is an ion channel former. Gramicidin S (**XXXVII**) is composed of a repeated pentapeptide sequence (Val-Orn-Leu-D-Phe-Pro); tyrocidine A has the same pentapeptide but the other half of the molecule is composed of different amino acids (Kato and Izumiya, 1974). Both molecules contain a pleated, 10-membered β-loop structure. Many analogues of gramicidin S have been made with altered stereochemistry and amino acid sequences in larger and smaller ring structures. Although little significant antibacterial activity has been observed with many of these analogues, important structure–function relationships have been established (Snezhkova *et al.*, 1975).

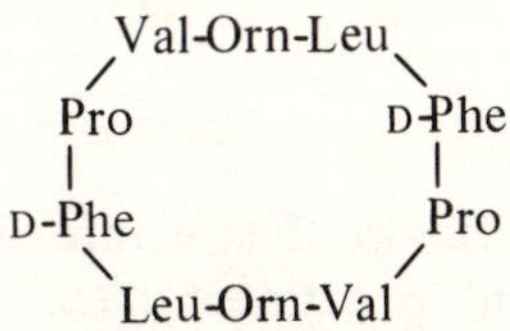

XXXVII
Gramicidin S

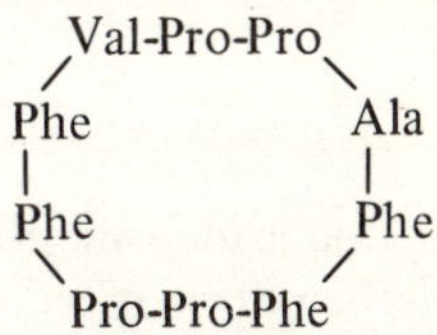

XXXVIII
Antamanide

Antamanide (**XXXVIII**) is a cyclic decapeptide formed from L-amino acids which, in contrast to gramicidin S, complexes strongly with Na^+ ions and acts as an antedote to the mushroom poison, phalloidin (Wieland *et al.*, 1968) (see also Section 4.8.3.1(iii)).

Polymixin, although a decapeptide, actually has a fatty acyl tripeptide attached to a heptapeptide ring and will be discussed later with related fatty acyl peptides in Section 4.7.4 (Storm *et al.*, 1977). Bacitracin (**XXXIX**) also has a heptapeptide ring but has a substituted thiazolidene ring attached to the tripeptide side-chain (Ressler and Kashelikar, 1966); its structure was, however, previously reported as having a hexapeptide ring with an exocyclic D-asparagine moiety. The mechanism of action of bacitracin has been studied in detail and involves inhibition of bacterial cell wall biosynthesis by interfering with the function of the lipid carrier molecule, undecaprenol phosphate (Siewert and

Strominger, 1967). Undecaprenol phosphate transfers disaccharide–peptide precursors of cell wall peptidoglycan from the inner to the outer surface of the membrane; in so doing, undecaprenol pyrophosphate is produced. In order that the prenol carrier is recycled, the pyrophosphate must be dephosphorylated to give the undecaprenol phosphate. It is this dephosphorylation step that bacitracin specifically blocks, not by inhibiting the lipid soluble phosphatase enzyme, but by directly binding to undecaprenol pyrophosphate in a ternary complex with a divalent metal ion (Stone and Strominger, 1971). The nitrogen atoms of the terminal amino group, the thiazolidene ring, and the ring histidine have been implicated as ion ligands (Cornell and Guiney, 1970).

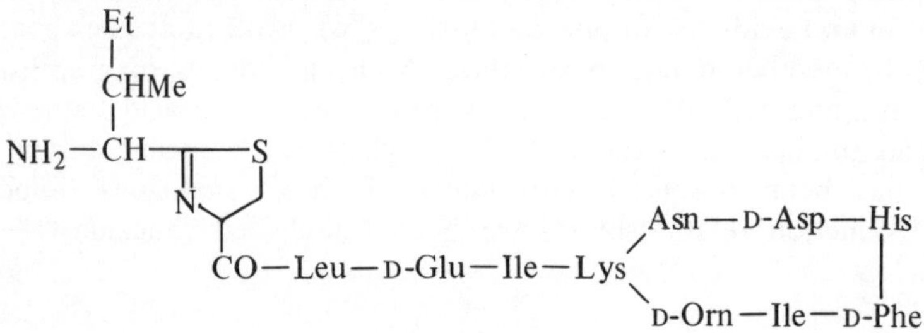

XXXIX

Bacitracin

The dodecadepsipeptide valinomycin has a cylindrical structure which changes conformation as it binds K^+ ions and penetrates membranes (Ovchinnikov *et al*., 1972; Shemyakin *et al*., 1969; Pinkerton *et al*., 1969). It is a trimer of [L-lactyl-L-Val-D-hydroxyisoval-D-Val] and will be discussed in the next chapter (Section 4.8). A recently characterized fungal undecapeptide is cyclosporin A which has a novel C_9-amino acid and exhibits immunosuppressive activities (Ruegger *et al*., 1976).

The larger cyclic peptides have been studied in less detail and will be discussed only briefly. Mycobacillin is a cyclic antifungal peptide with 13 residues of seven different amino acids, a number of which are of D-configuration (Sengupta and Bose, 1971). Mycobacillin forms complexes with phospholipids and sterols (Haldar and Bose, 1973) and has been evaluated as an agricultural fungicide (Nandi *et al*., 1975). Samarosporin and stendomycin both contain 14 amino acids. Samarosporin is a cyclic tetradecapeptide (Inoue *et al*., 1976) whereas stendomycin is a cyclic heptapeptide which has a fatty acyl heptapeptide side-chain and is believed to have some α- helical structure (Bodanszky *et al*., 1969; Urry and Ruiter, 1970). The antimicrobial peptide alamethicin contains 19 amino acids, is 40–60 per cent α-helical in structure, and forms ion-channels in membranes (Payne *et al*., 1970; McMullen *et al*., 1971). Alamethicin was originally believed to be cyclic but is now thought to be linear (Martin and Williams, 1976) (see also Section 4.8.3.1(ii)). A group of sulphur-containing cyclic peptide inhibitors of bacterial protein synthesis are thiostrepton, siomycin, and thiopeptin, also althiomycin and micrococcin (Pestka, 1971).

Other larger antibacterial polypeptides that have been investigated are nisin, duramycin, subtilin, and annamycin (Gross and Brown, 1976), also the much larger bacteriocins (Reeves, 1972). Nisin contains a number of small loops linked through the unusual amino acid lanthionine. The mechanism of action of the peptides nisin and subtilin may involve certain stages of bacterial cell wall biosynthesis (Linnett and Strominger, 1973), whilst individual members of the bacteriocins have very diverse cellular target sites.

4 PEPTIDES LINKED TO FATTY ACYL MOIETIES

The polymixins are the best characterized members of this category of antimicrobial peptides and were first isolated from *Bacillus polymyxa* in 1947 (Ainsworth *et al*., 1947). The selective activity of polymixin against *Pseudomonas* has proved invaluable in the topical treatment of infected burns. Other strains of *Bacillus* have been shown to produce related fatty acyl peptides and, during the last few years, the discovery and characterization of distinct new members of this series has enabled investigation of detailed structure–activity relationships (Storm *et al*., 1977). Peptides classed in this section include the polymixins, colistins, octapeptins, circulins, polypeptins, cerexin, amphomycin, brevistin, stendomycin, viscosin, the newly isolated but uncharacterized antibiotics B-43 and 339–29, and the totally synthetic fatty acyldilysines.

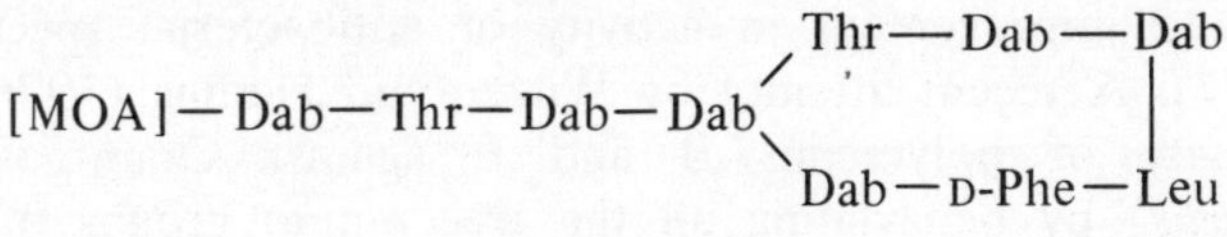

XL

Polymixin B_1

(MOA = 6-methyloctanoic acid: Dab = diaminobutyric acid)

Polymixins (**XL**) are basic decapeptides, containing a cyclic heptapeptide and a C_9 or C_8 *N*-terminal fatty acyl group on the exocyclic tripeptide (Vogler *et al*., 1965; Studer *et al*., 1965). They contain five to six diaminobutyric acid (Dab) residues; the cyclic heptapeptide being linked through the α-amino and carboxyl group of a diaminobutyric acid residue the γ-amino group of which is attached to the fatty acid tripeptide. Six different polymixins have been characterized; A(M), B_1, B_2, D_1, E_1, and E_2

(see review by Storm *et al*., 1977). More recently, two other polymixins, S_1 and T_1 have been added to the list (Shoji *et al*., 1977).

The primary site of action of the polymixins is on the bacterial membrane, in particular the Gram-negative outer membrane (Feingold *et al*., 1974). *Proteus* and *Neisseria* species are, however, particularly resistant to polymixin attack. Disruption of membrane integrity and leakage of intracellular contents are responsible for the bactericidal action of polymixins (Newton, 1956). This effect on membranes is, however, not entirely the result of the drug molecules acting as non-specific cationic detergents. Polymixins are far more potent than can be explained solely in terms of detergent action, and very small changes in their structure can result in significant changes in antimicrobial activity.

The selectivity of the polymixin group for Gram-negative membranes, as opposed to Gram-positive or animal cell membranes, has been explained in terms of the affinity of polymixin for certain phospholipids, in particular phosphatidyl ethanolamine (Feingold *et al*., 1974); methylation of this phospholipid to give lecithin reduces polymixin affinity. Thus, agents such as trimethoprim and sulphonamides, which inhibit the tetrahydrofolate one-carbon-carrier system, are synergistic with polymixin (Rosenblatt and Stewart, 1974). However, the precise mechanism of polymixin action may now have to be refined in order to be reconciled with the recent finding that polymixins retain some of their antibacterial activity after covalent binding to agarose beads (Laporte *et al*., 1977). This observation would seem to be consistent with the attack being restricted to the outer membranes of Gram-negative organisms (Brown and Tsang, 1977).

Many analogues of the polymixins have been synthesized, e.g. by changing the fatty acyl group, removal of the tripeptide side-chain, replacement of certain amino acids, expanding the heptapeptide ring, etc., without any marked improvement in activity or antibacterial spectrum (Storm *et al*., 1977). A recent attempt by Witzke and Heding (1976) to increase the activity of polymixins B and E against Gram-positive organisms (*S. aureus*) by benzylating all the free amino groups of the diaminobutyric acid residues resulted in a 100-fold increase in activity against *S. aureus* but a 10–100-fold decrease against *E. coli.*

Related fatty acyl decapeptides are colistin (polymixin E) and circulin (Suzuki *et al*., 1965; Studer *et al*., 1966). The structure of colistin A_1 is identical with that shown for polymixin B, except that L-leucine replaces D-phenylalanine. Various attempts to modify the acyl group and peptide sequence of deca-, nona-, penta-, and tetrapeptides based on colistin have met with little success in terms of retaining antibacterial potency (Chihara *et al*., 1974; Kurihara *et al*., 1974; Kurihara and Ito, 1974).

Cerexin (**XLI**) is also an acyl decapeptide but its structure is linear. It has a β-hydroxy C_{11} fatty acid, a preponderance of D-amino acids, and no Dab residues (Shoji and Hinoo, 1975; Shoji *et al*., 1976*e*).

[β-OH—C_{11}]—D-Asn—D-Val—D-Val—Asn—D-Asn—Hyl—

D-Thr—Ser—D-Trp—D-Ile

XLI

Cerexin

Other acylated straight-chain peptide antibiotics are amphomycin (Bodanszky *et al.*, 1973) and tsushimycin (Shoji *et al.*, 1968). Amphomycin is a linear undecapeptide acylated with a branched-chain unsaturated C_{13} fatty acid. It is active against Gram-positive organisms and selectively inhibits peptidoglycan biosynthesis (Tanaka *et al.*, 1977), at a site similar to that reported for tunicamycin, i.e., formation of *N*-acetylmuramylpentapeptide-P,P-lipid (see Figure 3.) (Ward, 1977).

Less closely related acyl peptides are brevistin (Shoji *et al.*, 1976*b*; Shoji and Kato, 1976), polypeptin (Sogn, 1976), and stendomycin (Bodanszky *et al.*, 1969). These peptides differ in the number and sequence of amino acids, the type of ring cyclization, and the nature of the fatty acid, e.g. stendomycin is a lactone-linked cyclic heptapeptide attached to an acyl heptapeptide side chain, whilst brevistin is a lactone-linked undecapeptide attached directly to a C_9 fatty acid.

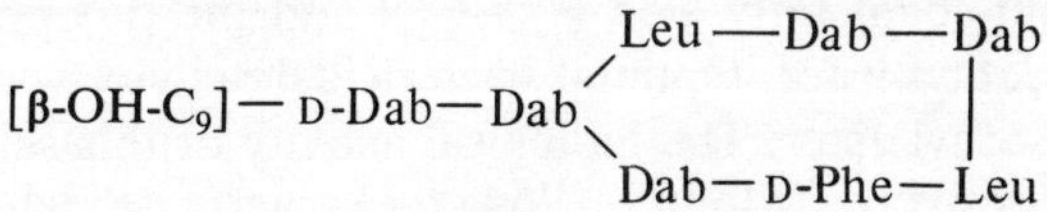

XLII

Octapeptin C_1 (333–25)

The octapeptins constitute a group of cyclic heptapeptides with a D-Dab side-group and a fatty-acyl moiety which differs from the polymixin acyl moiety in that it has a β-hydroxyl group and can be C_9, C_{10}, or C_{11} (Parker and Rathnum, 1975; Kato and Shoji, 1976). The A and B subgroups of octapeptins were previously classified as the antibiotic EM 49 (Meyers *et al.*, 1973), the mechanism of action of which was reported in detail (Meyers *et al.*, 1974). More recent evidence however, indicates that octapeptin not only alters membrane structure but also uncouples oxidative phosphorylation (Rosenthal *et al.*, 1976, 1977). Octapeptin C_1 (**XLII**) was also previously known as antibiotic 333–25 (Shoji *et al.*, 1976*a*). Other acyl-peptides, 339–29 and B-43, have yet to be categorized (Shoji *et al.*, 1976*c,d*).

Viscosin (**XLIII**) is a linear hydroxy-acyl hexapeptide, the structure of which was originally proposed by Ohno *et al.* (1953). Recent work by Hiramoto *et al.* (1971*a,b*) has, however, cast doubt on the proposed stereochemical sequence. Viscosin is reported to be active against tubercle bacillus and certain viruses (Kochi *et al.*, 1951).

[β-OH-C_{10}]—Leu—Gly—Ser—D-Val—Thr—Leu

XLIII

Viscosin

Esperin was mentioned earlier in Section 4.7.3 as a cyclic lipopeptide, but unlike other members of the fatty acyl–peptide group, the β-hydroxy fatty acid is incorporated into the ring structure (Ito and Ogawa, 1959). A more recently investigated antifungal lipopeptide is iturin A. It is a member of a related group of heptapeptides which contain a lipid moiety composed of C_{14} and C_{15} β-amino fatty acids (Peypoux *et al*., 1973). Iturin is also believed to be the active constituent of the antibiotics bacillomycins B and R, and eumycin (Besson *et al*., 1976). Echinocandin B is a recently characterized antifungal peptide which is particularly active against growing yeast cells. It is a cyclic hexapeptide with an unusually large acyl side-group, linoleic acid (Keller-Juslen *et al*., 1976).

As an alternative to modifying the structures of the above acyl-peptide antiobiotics chemically, small sequences of the 'polymixin-type' structure have been synthesized, in particular the acyl-peptide side-chain. [MOA]-Dab-Thr-Dab-Dab was found to be inactive (Moriér *et al*., 1975) but Vogler *et al*. (1964) synthesized a series of acyl-dipeptides in which the acyl group was a straight-chain fatty acid (C_{14}–C_{18}) and the dipeptide was lysyllysine or its methyl/ethyl ester. Optimal bactericidal activity was found for palmitoyl-lysyllysine ethyl ester. The biological activity of this substance has been studied extensively (Molin, 1964*a,b,c*), and its lability to hydrolysis in the alimentary tract to give innoxious fatty acid and lysine has been cited for its use as a food preservative. It should be noted, however, that certain fatty acids and their derivatives are weakly antimicrobial in their own right (Kabara *et al*., 1972; Freese *et al*., 1973). More recent chemical studies have pursued the detailed structure–activity relationships of this series of simple acyl-lysine peptides (Dabrowska *et al*., 1976; Moriér *et al*., 1975).

5 PEPTIDES LINKED TO SUGAR MOIETIES

Peptides attached to sugar moieties are structurally and mechanistically the most complex and diverse of all the categories of peptide antibiotics described in this review.

Vancomycin is probably one of the most studied antibiotics of this category in terms of its mechanism of action at a molecular level. It is a potent inhibitor of bacterial cell wall biosynthesis in Gram-positive organisms, causing the accumulation of UDP-*N*-acetylmuramylpentapeptide precursors of the cell wall peptidoglycan (Reynolds, 1961). The structure of vancomycin has very recently been elucidated in detail and this should prove most informative for interpreting its action (Sheldrick *et al.*, 1978). Vancomycin contains seven amino acids (aspartate, *N*-methylleucine, two β-hydroxychlorotyrosines, and three oxyphenylglycines) to give a compact tricyclic molecule which is linked to an aminodisacchardide, vancosamine. Vancomycin binds extremely specifically to the peptide acyl-D-Ala-D-Ala and close analogues (Perkins and Nieto, 1974). This binding can now be correlated with space-filling models of vancomycin, which indicate a narrow phenolic cleft suitable only for peptides terminating in D-Ala-D-Ala (Sheldrick *et al.*, 1978). The mechanism of action is thus thought to involve complexation of vancomycin with the cell wall precursor disaccharide-pentapeptide when it emerges from the membrane attached to the lipid carrier. In this way incorporation of precursor molecules into the peptidoglycan polymer in the cell wall is prevented (Gale *et al.*, 1972; Perkins and Nieto, 1974).

A closely related group of antibiotics are the ristomycins and ristocetins. They contain oxyphenylglycines arranged in a bicyclic system which is attached to six or seven sugar moieties (Lomakina, 1970). Unlike vancomycin, there are no chlorophenols. It is interesting to note that the sugar moieties in vancomycin and ristocetin do not seem to be required for activity. Vancomycin acts on bacterial cell wall biosynthesis at the step directly after the one that is inhibited by bacitracin, but like bacitracin, it inhibits an enzyme reaction by binding to the substrate and not the enzyme.

A number of peptide-like antibiotics have been described as interfering with the synthesis of peptidoglycan; however, recent work has shown that fragments of the muramyl-peptide structure can act as chemotherapeutic agents in their own right—not directly against microorganisms but indirectly by stimulating the host immune response. Detailed structure–activity relationships have established that the minimal structure for activity is the sugar dipeptide, *N*-acetylmuramyl-L-alanyl-D-isoglutamine **(XLIV)** (Jollès, 1976; Chedid *et al.*, 1976). This muramyl dipeptide is unusual in that it is active as an adjuvant both orally and parenterally. It has been shown to enhance non-specific immunity to *Klebsiella* infections even when administered after the bacterial challenge (Chedid *et al.*, 1977).

A group of sugar peptides that is specifically active against fungal cell walls is the polyoxins (Suzuki *et al.*, 1966). These antibiotics can be regarded as dipeptidyl pyrimidine nucleosides and probably act as mimetics of UDP-*N*-acetylglucosamine. Polyoxins B and D **(XLV)** have been studied in most detail and are potent inhibitors of fungal chitin synthetase (Hori *et al.*, 1974; Gooday, 1977). Despite the importance of chitin synthesis in the growth of the filamentous fungi, and to a lesser extent in the budding of yeasts, polyoxins have limited antifungal activity. This absence of activity is

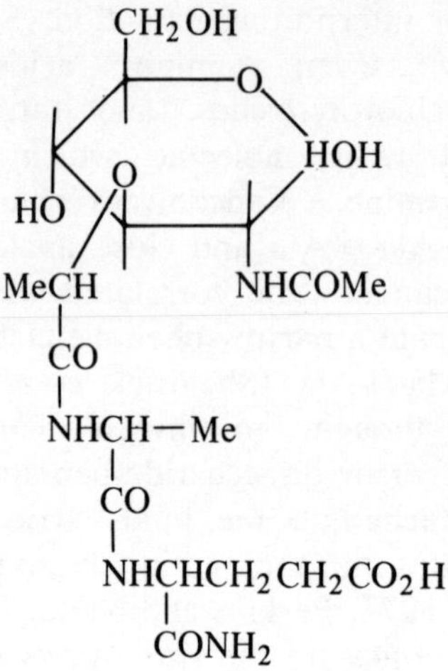

XLIV

Muramyldipeptide; (*N*-acetylmuramyl-L-alanyl-D-isoglutamine)

probably due to lack of penetration into certain organisms. It is, however, interesting to note that polyoxin activity is antagonized at whole cell level, but not enzyme level, by specific dipeptides, e.g. Gly-Ala, Gly-Val, Ala-Gly (Mitani and Inoue, 1968). This would suggest that polyoxins are transported into fungi via a dipeptide permease system and that consequently any environment containing small peptides will antagonize polyoxin action, (Bowers *et al*., 1974). Polyoxin could, therefore, be regarded as a disguised carrier peptide but with a difference, in that it does not require cleavage to release the enzyme inhibitor. It is also likely that the close analogue of polyoxin D, nikkomycin (Daehn *et al*., 1976) also uses a peptide permease.

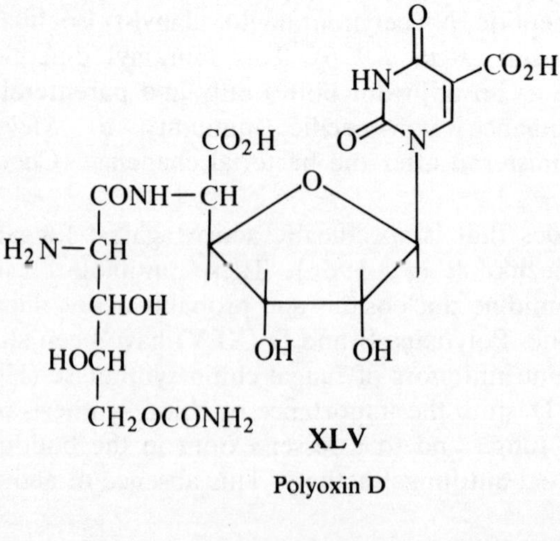

XLV

Polyoxin D

A series of antibiotics structurally related to the polyoxins is the aminoacyl-4-aminohexosyl cytosine group of protein synthesis inhibitors; e.g. gougerotin, blasticidin S, and amicetin (Suhadolnik, 1970). All are believed to inhibit peptidyl transferase activity of the larger ribosomal subunit. Gougerotin is the only true peptide in that it has the dipeptide sarcosylserine attached to the aminohexosylcytosine moiety **(XLVI)** (Clark, 1967). Blasticidin has only one amino acid (methyl-β-arginine) linked in the 4-position on the aminohexose moiety (Misato, 1967). Amicetin is more complex, with two hexose units and β-methylserine and *p*-aminobenzoic acid linked to the cytosine amino group.

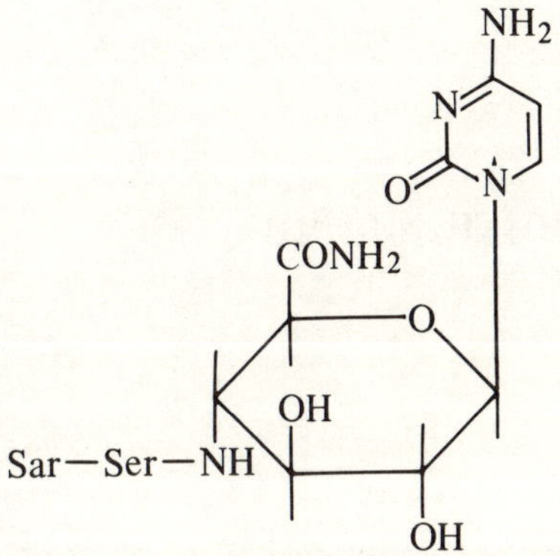

XLVI

Gougerotin

Many comparative structural studies have been carried out with this series of antibiotics and their synthetic analogues in order to relate structure with function in terms of their inhibitory activity on protein synthesis (Fox *et al*., 1966; Lichtenthaler and Trummlitz, 1974). A related aminoacyl pentose purine analogue is puromycin and this would seem to act as a direct structural analogue of aminoacyladenosine, i.e. the 3′-terminus of aminoacyl-tRNA (see Gale *et al*., 1972). Chloramphenicol, also a potent inhibitor of ribosomal peptidyl transferase, has been shown to resemble the acylamino pentose structure similar to uridylic acid (Jardetzky, 1963). Aminoacyl analogues of chloramphenicol have therefore been synthesized, but their activity has been poor on both whole cells and isolated ribosomes (Coutsogeorgopoulos, 1966; Ringrose and Lambert, 1973; Kraevskii *et al*., 1975).

Totally synthetic simple analogues of the nucleoside group of antibiotics have also been evaluated as chemotherapeutic agents (Sharma *et al*., 1975; Robins *et al*., 1971; Kawana *et al*., 1972*a,b*). Analogues with the greatest antibacterial activity were simple dipeptide (Phe-Gly, Leu-Gly, Lys-Gly) derivatives of 2′-amino-2′-deoxyuridine. Antiviral compounds have also been synthesized with peptides attached to the amino group of 5-aminouridine (Ivanovics *et al*., 1971).

Racemomycins (**XLVII**) belong to the basic streptothricin group of antibiotics and are aminoimidazole hexosamines linked to β-lysine peptides. Their antimicrobial and antiviral activities increase as their β-lysine content increases (Taniyama *et al*., 1971, 1972).

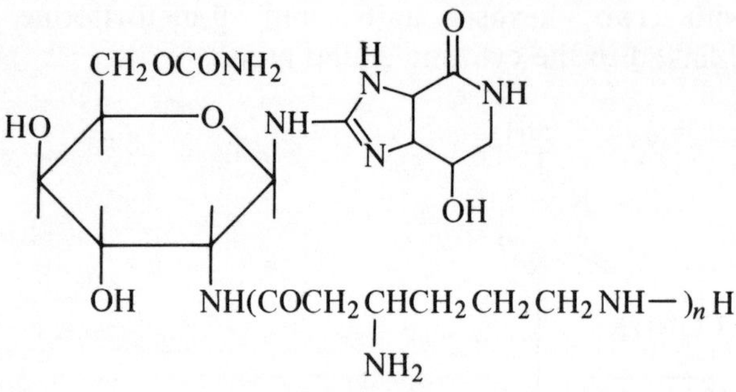

XLVII

Racemomycins
(n = 1–4)

A related compound to the racemomycins is myomycin which contains a β-lysine peptide linked to 3-amino-3-deoxy-D-mannose and myo-inositol (French *et al*., 1973).

A completely different category of peptide–sugar antibiotics containing sulphur is represented by bleomycin and phleomycin (Umezawa, 1977). Both contain identical disacchardies and a β-lactam ring, they also yield seven unusual amino acids on acid hydrolysis. The only difference between phleomycin and bleomycin is a double bond in one of the amino acid thiazole rings (Takita *et al*., 1972*a,b*). Both antibiotics complex copper and possess antibacterial and antitumour activity; their intracellular site of action being template DNA. The ability of copper-free bleomycin to induce single-strand scissions in DNA in whole cells and cell-free systems has been shown to be accelerated by thiol groups (Suzuki *et al*., 1969; Haidle, 1971; Takahashi *et al*., 1977). Bleomycin-induced nucleic acid degradation (Miyaki *et al*., 1974) would appear to be similar to that reported for the protein antibiotic, colicin E_2 (Ringrose, 1970, 1972, 1973). Phleomycin has also been observed to induce degradation of cellular DNA (Farrell and Reiter, 1973).

Various attempts to enhance the activity of known antibiotics in this category by attaching amino acids or peptides have been made but with little success, cf. earlier discussion on sulphonamides, chloramphenicol, etc.

Adriamycin, a planar, tetracyclic DNA-intercalator with an essential sugar moiety, has been derivatized with peptides, but no marked increase in activity was observed (Wilson *et al*., 1976). A recently characterized unusual aminoglycoside antibiotic is fortimycin. This molecule is composed of a disaccharide moiety bound to an amino acid (glycine) in such a way that removal of the amino acid results in a conformational change in the antibiotic and an almost complete loss of activity (Egan *et al*., 1976, 1977).

6 PEPTIDES OF UNDEFINED STRUCTURE

Three recently described peptide antibiotics have been identified as inhibitors of bacterial cell wall biosynthesis, but their structures are still not fully defined, i.e. gardimycin (Coronelli *et al*., 1976; Somma *et al*., 1977), stenothricin (Hasenböhler *et al*., 1974), and janiemycin (Linnett and Strominger, 1973; Meyers *et al*., 1970); the latter being related to enduracidin (Asai *et al*., 1968). All three antibiotics inhibited only the late stages of peptidoglycan biosynthesis, and for gardimycin and janiemycin a similar mechanism of action to that of vancomycin was indicated, i.e. inhibition of transfer of the disaccharide–peptide unit from the lipid intermediate to the cell wall acceptor.

Numerous reports of the discovery of other peptide antibiotics frequently appear in the literature with very little preliminary indication as to structure or mechanism of action, some are given names, e.g. the galantins (Shoji *et al*., 1975) whilst others are just given numbers, e.g. 61–26, S-520, TL-119, M-81, etc.

An antibacterial peptide of human origin has recently been described (Schlievert *et al*., 1976). It is found in amniotic fluid, has a molecular mass of less than 5000 daltons and has a specific requirement for zinc in order to exhibit its antibacterial properties.

7 CONCLUSIONS

An attempt has been made in this review to categorize both naturally-occurring and synthetic antimicrobial peptides into structural groups which in some cases have a common mechanism of action, e.g. the cyclic peptide ionophores, the lipopeptide membrane disrupters, the carrier peptides with warheads that inhibit glutamine synthetase. However, there are almost as many exceptions as there are rules, and Table 1 (see Section 4.7.1.) lists the various categories of peptides described in this chapter according to their mechanisms of action. The four basic types of antibacterial action are illustrated in Figure 1 (see Section 4.7.1) according to their macromolecular target sites, i.e. cell wall, membrane, nucleic acid, and ribosome.

Some peptides act as the intact molecules whereas others need to be cleaved in order to release the active constituent. Some act intracellularly and thus are required to penetrate the cytoplasmic membrane via a permease or by diffusion, whereas others act outside the membrane on the cell wall or even on the membrane itself by physical disruption or by destroying ion gradients. Some peptides inhibit enzymes directly whereas others bind to or destroy the substrate which may for example be DNA or a peptidoglycan precursor.

It can therefore be seen that antimicrobial peptides are an extremely diverse group of compounds with examples of many different mechanisms of action. In addition to being clinically useful as chemotherapeutic agents, a number of antimicrobial peptides have been extensively studied in terms of relating structure to function. A detailed understanding of peptide action at a molecular level in relation to transport systems, target sites, and peptidase susceptibilities is increasingly being used to facilitate the rational design of novel peptides and could lead to further useful developments in the future of microbial chemotherapy.

ACKNOWLEDGEMENTS

The author wishes to thank Drs C. H. Hassall, F. R. Atherton, R. W. Lambert, and W. J. Lloyd for their constructive comments during the writing of this review, also Drs T. Franklin, H. Zähner, and D. H. Williams for information prior to publication.

REFERENCES

Abraham, E. P., Callow, D., and Gilliver, K. (1946). *Nature, Lond.*, **158**, 818

Adriaens, P., Meesschaert, B., Wuyts, W., Vanderhaeghe, H., and Eyssen, H. (1975). *Antimicrobial Agents Chemother.*, **8**, 638

Ainsworth, G. C., Brown, A. M., and Brownlee, G. (1947). *Nature, Lond.*, **160**, 263

Aleksiev, B., Stoev, S., Spassov, A., Maneva, L., and Golovinsky, E. (1972). In *Proceedings of the 12th European Peptide Symposium* (H. Hanson and H. Jakubke, eds), North-Holland/Elsevier, New York, p. 245

Allen, J. G., Gilbert, D. A., Harris, G. S., Havas, L., Holmes, S. W., Jackson, M. R., Lenox-Smith, I., and Parkes, M. W. (1977). *10th Int. Congr. Chemother.*, Abstr. no. 366

Allen, J. G., Atherton, F. R., Hassall, C. H., Hall, M. J., Holmes, S. W., Lambert, R. W., Nisbet, L. J., and Ringrose, P. S. (1978). *Nature, Lond.*, **272**, 56

Ames, B. N., Ames, G. F., Young, J. D., Tsuchiya, D., and Lecocq, J. (1973). *Proc. Natl. Acad. Sci. U.S.A.*, **70**, 456

Aoki, H., Sakai, H., Kohsaka, M., Konomi, T., Hosada, J., Kubochi, Y., Iguchi, E., and Imanaka, H. (1976). *J. Antibiot.*, **29**, 492

Argoudelis, A. D., Reusser, F., Mizsak, S. A., and Baczynskyj, L. (1976). *J. Antibiot.*, **29**, 1007

Argoudelis, A. D., Herr, R. R., Mason, D. J., Pyke, I. R., and Zieseri, J. F. (1967). *Biochemistry*, **6**, 165

Arnstein, H. R. V. and Morris, D. (1960). *Biochem. J.*, **76**, 357

Asai, M., Muroi, M., Sugita, N., Kawashima, H., Mizuno, K. and Miyake, A. (1968). *J. Antibiot.*, **21**, 138

Atsumi, K., Oiwa, R., and Omura, S. (1975). *J. Antibiot.*, **28**, 77

Baggaley, K. H., Blessington, B., Falshaw, C. P., Ollis, W. D., Chaiet, L., and Wolf, F. J. (1969). *Chem. Commun.*, **1969**, 101

Barak, Z. and Gilvarg, C. (1974). *J. Biol. Chem.*, **249**, 143

Barak, Z. and Gilvarg, C. (1975). *J. Bacteriol.*, **122**, 1200

Barak, Z., Sarid, S., and Katchalski, E. (1973*a*). *Eur. J. Biochem.*, **34**, 317

Barak, Z., Sarid, S., and Katchalski, E. (1973*b*). *Eur. J. Biochem.*, **34**, 325

Bates, C. J. and Handschumacher, R. E. (1969). *Adv. Enzyme Regul.*, **7**, 183

Bayer, E., Gugel, K. H., Hägele, K., Hagenmaier, H., Jessipow, S., König, W. A., and Zähner, H. (1972). *Helv. Chim. Acta*, **55**, 224

Becker, J. M. and Naider, F. (1974). *J. Bacteriol.*, **120**, 191

Bell, M. R., Johnson, J. R., Wildi, B. S., and Woodward, R. B. (1958). *J. Am. Chem. Soc.*, **80**, 1001

Bermingham, M. A. C., Deol, B. S., and Still, J. L. (1970). *Biochem. J.*, **116**, 759

Besson, F., Peypoux, F., Michel, G., and Delcambe, L. (1976). *J. Antibiot.*, **29**, 1043

Bevan, K., Davies, J. S., Hall, M. J., Hassall, C. H., Morton, R. B., Phillips, D. A. S., Ogihara, Y., and Thomas, W. A. (1970). *Experientia*, **26**, 122

Blumberg, P. M. and Strominger, J. L. (1974). *Bacteriol. Rev.*, **38**, 291

Bodanszky, M. and Stahl, G. L. (1974). *Proc. Natl. Acad. Sci. U.S.A.*, **71**, 2791

Bodanszky, M., Izdebski, J., and Marumatsu, I. (1969). *J. Am. Chem. Soc.*, **91**, 2351

Bodanszky, M., Sigler, G. F., and Bodanszky, A. (1973). *J. Am. Chem. Soc.*, **95**, 2352

Böhme, E. H. W., Applegate, H. E., Toeplitz, B., Dolfini, J. E., and Gongoutas, J. Z. (1971). *J. Am. Chem. Soc.*, **93**, 4324

Borowski, E. and Konopa, J. (1957). *Acta Biochim. Polonica*, **4**, 241

Bowers, B., Levin, G., and Cabib, E. (1974). *J. Bacteriol.*, **119**, 564

Braun, V., Bosch, V., Hantke, K., and Schaller, K. (1974). *Ann. N.Y. Acad. Sci.*, **235**, 66

Brockmann, H. (1974). *Cancer Chemother. Rep.*, **58**, 9

Brown, A. G., Corbett, D. F., Eglington, A. J., and Howarth, T. T. (1977). *J. Chem. Soc. Chem. Commun.*, **1977**, 523

Brown, A. G., Butterworth, D., Cole, M., Hanscomb, G., Hood, J. D., Reading, C., and Rolinson, G. N. (1976). *J. Antibiot.*, **29**, 668

Brown, D. A. and Tsang, J. C. (1977). *J. Antibiot.*, **30**, 194

Bycroft, B. W. (1969). *Nature, Lond.*, **224**, 595

Bycroft, B. W. (1972). *J. Chem. Soc. Chem. Commun.*, **11**, 660

Bycroft, B. W., Cameron, D., Croft, L. R., Hassanali-Walji, A., Johnson, A. W., and Webb, T. (1971). *Nature, Lond.*, **231**, 301

Byrn, S. R. (1974). *Biochemistry*, **13**, 5186

Cary, L. W., Takita, T., and Ohnishi, M. (1971). *FEBS Lett.*, **17**, 145

Chedid, L., Audibert, F., Lefrancier, P., Choay, J., and Lederer, E. (1976). *Proc. Natl. Acad. Sci. U.S.A.*, **73**, 2472

Chedid, L., Parant, M., Parant, F., Lefrancier, P., Choay, J., and Lederer, E. (1977). *Proc. Natl. Acad. Sci. U.S.A.*, **74**, 2089

Chihara, S., Ito, A., Yahata, M., Tobita, T., and Koyama, Y. (1974). *Agric. Biol. Chem., Tokyo*, **38**, 521, 1767

Chmara, H. and Borowski, E. (1973*a*). *Biochem. Biophys. Res. Commun.*, **52**, 1381

Chmara, H. and Borowski, E. (1973*b*). *Biochem. Biophys. Res. Commun.*, **55**, 1147

Clark, J. M. (1967). In *Antibiotics*, Vol. 1 (D. Gottlieb and P. D. Shaw, eds), Springer-Verlag, New York, p. 278

Corcoran, J. W. (1977). *Annu. Rep. Med. Chem.*, **12**, 134

Cornell, N. W. and Guiney, D. G. (1970). *Biochem. Biophys, Res. Commun.*, **40**, 530

Coronelli, C., Tamoni, G., and Lancini, G. C. (1976). *J. Antibiot.*, **29**, 507

Coutsogeorgopoulos, C. (1966). *Biochim. Biophys. Acta*, **129**, 214

Dabrowska, M., Kupryszlewski, G., Muzalewski, F., and Kochanowski, J. (1976). *Pol. J. Pharmacol. Pharm.*, **28**, 77

Daehn, U., Hagenmaier, H., Hoehne, H., Koenig, W. A., Wolf, G., and Zaehner, H. (1976). *Arch. Microbiol.*, **107**, 143

De Felice, M., Guardiola, J., Lamberti, A., and Iaccarino, M. (1973). *J. Bacteriol.*, **116**, 751

Deol, B. S., Bermingham, M. A. C., Still, J. L., Haydon, D. A., and Gale, E. F. (1973). *Biochim. Biophys. Acta*, **330**, 192
Diddens, H., Zähner, H., Kraas, E., Göhring, W., and Jung, G. (1976). *Eur. J. Biochem.*, **66**, 11
Dunn, F. W. (1959). *J. Biol. Chem.*, **234**, 802
Dunn, F. W. and Dittmer, K. (1951). *J. Biol. Chem.*, **188**, 263
Dunn, F. W., Ravel, J. M. and Shive, W. (1956). *J. Biol. Chem.*, **219**, 809
Dunn, F. W., Humphreys, J., and Shive, W. (1957). *Arch. Biochem. Biophys.*, **71**, 475
Egan, R. S., Martin, J. R., Tadanier, J., Stanaszek, R. S., Mueller, S. L., Collum, P., Goldstein, A. W., De Vault, R. L., Sinclair, A. C., and Fager, E. E. (1976). *16th Intersci. Conf. Antimicrobial Agents Chemother.*, Abstr. no. 58
Egan, R. S., Stanaszek, R. S., Cirovic, M., Mueller, S. L., Tadanier, J., Martin, J. R., Collum, P., Goldstein, A. W., De Vault, R. L., Sinclair, A. C., Fager, E. E., and Mitscher, L. A. (1977). *J. Antibiot.*, **30**, 552
Eisele, K. (1975*a*). *Z. Naturforsch.*, **30**, 541
Eisele, K. (1975*b*). *Hoppe-Seyler's Z. Physiol. Chem.*, **356**, 1497
Eisele, A. and Eisele, K. (1975). *FEBS Lett.*, **55**, 153
Farrell, L. and Reiter, H. (1973). *Antimicrobial Agents Chemother.*, **4**, 320
Feingold, S., Hsu Chen, C. C., and Sud, I. J. (1974). *Ann. N.Y. Acad. Sci.*, **235**, 480
Ferres, H., Basker, M. J., and O'Hanlon, P. J. (1974). *J. Antibiot.*, **27**, 922
Fickel, T. E. and Gilvarg, C. (1973). *Nature, Lond.*, **241**, 161
Fiume, L. and Wieland, T. (1970). *FEBS Lett.*, **8**, 1
Flynn, E. H. (1972). *Cephalosporins and Penicillins; Chemistry and Biology*, Academic Press, New York and London
Fox, J. J., Watanabe, K. A., and Bloch, A. (1966). *Prog. Nucleic Acid Res. Mol. Biol.*, **5**, 290
Franklin, T. J., Snow, G. A., and White, A. J. (1977). Personal communication
Freese, E., Sheu, C. W., and Galliers, E. (1973). *Nature, Lond.*, **241**, 321
French, J. C., Bartz, Q. R., and Dion, H. W. (1973). *J. Antibiot.*, **26**, 272
Gale, E. F., Cundliffe, E., Reynolds, P. E. Richmond, M. H. and Waring, M. J. (1972). *The Molecular Basis of Antibiotic Action*, John Wiley, New York and London
Gale, G. R. and Smith, A. B. (1973). *J. Bacteriol.*, **114**, 460
Ghuysen, J. M. (1977). *J. Gen. Microbiol.*, **101**, 13
Gilmore, F. W. and McBride, H. A. (1974). *J. Pharm. Sci.*, **63**, 1087
Gilvarg, C. and Levin, Y. (1972). *J. Biol. Chem.*, **247**, 543
Goldberg, I. H. and Friedman, P. A. (1971). *Annu. Rev. Biochem.*, **40**, 775
Gooday, G. W. (1977). *J. Gen. Microbiol.*, **99**, 1
Griffith, R. S. (1974). *Adv. Clin. Pharmacol.*, **8**, 6
Gross, E. and Brown, J. H. (1976). In *Peptides 1976* (A. Loffet, ed.), Editions de l'Université de Bruxelles, Brussels, p. 183
Haidle, C. W. (1971). *Mol. Pharmacol.*, **7**, 645
Haldar, A. and Bose, S. K. (1973). *J. Antibiot.*, **26**, 358
Hall, M. J. (1970). *Biochem. Biophys. Res. Commun.*, **38**, 590
Hamada, M., Takeuchi, T., Kondo, S., Ikeda, Y., Naganawa, H., Maeda, K., Okami, Y., and Umezawa, H. (1970). *J. Antibiot.*, **23**, 170
Hammes, W., Schleifer, K. H., and Kandler, O. (1973). *J. Bacteriol.*, **116**, 1029
Harold, F. M. (1972). In *Metabolic Inhibitors*, Vol. III, Academic Press, New York, p. 305
Hasenböhler, A., Kneifel, H., König, A., Zähner, H., and Zeiler, H. J. (1974). *Arch. Microbiol.*, **99**, 307
Hashimoto, M., Komori, T., and Kamiya, T. (1976). *J. Am. Chem. Soc.*, **98**, 3023
Hassall, C. H. and Thomas, W. A. (1971). *Chem. Br.*, **7**, 145
Hassall, C. H., Morton, R. B., Ogihara, Y., and Thomas, W. A. (1969). *J. Chem. Soc., Chem. Commun.*, **1969**, 1079
Hassall, C. H., Allen, J. G., Atherton, F. R., Hall, M. J., Holmes, S. W., Lambert, R. W., Lenox-Smith, I., Nisbet, L. J., and Ringrose, P. S. (1977). *10th Int. Congr. Chemother.*, Abstr. no. 365
Henderson, P. J. F., McGivan, J. D., and Chappell, J. B. (1969). *Biochem. J.*, **111**, 521
Hettinger, T. P., Kurylo-Borowska, Z., and Craig, L. C. (1970). *Ann. N.Y. Acad. Sci.*, **170**, 1002

Hill, J. T. and Dunn, F. W. (1969). *J. Med. Chem.*, **12**, 737

Hiramoto, M., Okada, K., and Nagai, S. (1971*b*). *Chem. Pharm. Bull.*, **19**, 1315

Hiramoto, M., Okada, K., Nagai, S., and Kawamoto, H. (1971*a*). *Chem. Pharm. Bull.*, **19**, 1308

Holly, F. W., Peel, E. W., Luz, E. L., and Folkers, K. (1952). *J. Am. Chem. Soc.*, **74**, 4539

Hori, M., Eguchi, J., Kakiki, K., and Misato, T. (1974). *J. Antibiot.*, **27**, 260

Hruby, V. J. (1974). In *Chemistry and Biochemistry of Amino Acids, Peptides and Proteins*, Vol. 3 (B. Weinstein, ed.), Marcel Dekker, New York, p. 1

Huber, J. W., Gilmore, W. F., and Robertson, L. W. (1975). *J. Med. Chem.*, **18**, 106

Hunter, P. A. and Reading, C. (1976). *16th Intersci. Conf. Antimicrobial Agents Chemother.*, Abstr. no. 211

Iida, K., Hirata, S., Nakamuta, S., and Koike, M. (1977). *10th Int. Congr. Chemother.*, Abstr. no. 70

Ikeda, K., Kuromizu, K., Okamoto, K., and Izumiya, N. (1972). *Chem. Pharm. Bull.*, **20**, 1849

Inoue, N., Inoue, A., Furukawa, M., and Kanda, N. (1976). *J. Antibiot.*, **29**, 618

Ito, T. and Ogawa, H. (1959). *Bull. Agric. Chem. Soc. Japan*, **23**, 536

Ivanovics, G. A., Rousseau, R. J., and Robins, R. K. (1971). *J. Med. Chem.*, **14**, 1155

Iwasaki, T., Urabe, Y., Ozaki, Y., Miyoshi, M., and Matsumoto, K. (1976). *J. Chem. Soc. Perkin Trans 1*, **9**, 1019

Izumiya, N., Kato, T., Fujita, Y., Ohno, M., and Kondo, M. (1964). *Bull. Chem. Soc. Japan,* **37**, 1809

Jardetzky, O. (1963). *J. Biol. Chem.*, **238**, 2498

Jollès, P. (1976). *Experientia*, **32**, 677

Kabara, J. J., Swieczkowski, D. M., Conley, A. J., and Truant, J. P. (1972). *Antimicrobial Agents Chemother.*, **2**, 23

Kahan, F. M., Kanan, J. S., Cassidy, P. J., and Kropp, H. (1974). *Ann. N.Y. Acad. Sci.*, **235**, 364

Kahan, F. M. and Kropp, H. (1975). *15th Intersci. Conf. Antimicrobiol Agents Chemother.*, Abstr. no. 100

Kaminski, K. and Sokolowska, T. (1973). *J. Antibiot.*, **26**, 184

Kamiya, T., Maeno, S., Hashimoto, M., and Milne, Y. (1972). *J. Antibiot.*, **25**, 576

Kasymova, G. F., Burichenko, V. K., Meitus, E. E., and Poroshin, K.T. (1973). *Khim. Prir. Soedin, Tashkent*, **9**, 79

Kato, T. and Izumiya, N. (1974). In *Chemistry and Biochemistry of Amino Acids, Peptides and Proteins*, Vol. 2 (B. Weinstein, ed.), Marcel Dekker, New York, p. 1

Kato, T. and Shoji, J. (1976). *J. Antibiot.*, **29**, 1339

Katz, E. and Demain, A. L. (1977). *Bacteriol. Rev.*, **41**, 449

Kawana, M., Rousseau, R. J., and Robins, R. K. (1972*a*). *J. Org. Chem.*, **37**, 288

Kawana, M., Streeter, D. G., Rousseau, R. J., and Robins, R. K. (1972*b*). *J. Med. Chem.*, **15**, 561

Keller, R., Mueller-Eckhardt, C., Kayser, F.-H., and Keller, H. U. (1968). *Int. Arch. Allergy*, **33**, 239

Keller-Juslen, C., Kuhn, M., Loosli, H. R., Petcher, T. J., Weber, H. P., and Von Wartburg, A. (1976). *Tetrahedron Lett.*, **46**, 4147

Kenig, M. and Abraham, E. P. (1976). *J. Gen. Microbiol.*, **94**, 37

Kenig, M., Vandamme, E., and Abraham, E. P. (1976). *J. Gen. Microbiol.*, **94**, 46

Kihara, H. and Snell, E. E. (1960). *J. Biol. Chem.*, **235**, 115

Klosterman, H. J., Lamoureux, G. L., and Parsons, J. L. (1967). *Biochemistry*, **6**, 170

Kochi, M., Groupé, V., Pugh, L. H., and Weiss, D. (1951) *Bacteriol. Rev.*, **15**, 291

Kondo, S., Shibahara, S., Takahashi, S., Maeda, K., Umezawa, H., and Ohno, M. (1971). *J. Am. Chem. Soc.*, **93**, 6305

König, W. A., Loeffler, W., Meyer, W. H., and Uhmann, R. (1973). *Chem. Ber.*, **106**, 816

Kraevskii, A. A., Nikolaeva, L. V., Kukhanova, M. K., Gnuchev, N. V., and Gottikh, B. P. (1975). *Bioorg. Khim.*, 1, 368

Kropp, H., Kahan, J. S., Kahan, F. M., Sundelof, J., Darland, G., and Birnbaum, U. (1976). *16th Intersci. Conf. Antimicrobial Agents Chemother.*, Abstr. no. 228

Kurahashi, K. (1974). *Annu. Rev. Biochem.*, **43**, 445

Kurihara, T. and Ito, H. (1974). *Yakugaku Zasshi*, **94**, 1183

Kurihara, T., Takeda, H., and Ito, H. (1974). *Yakugaku Zasshi*, **94**, 1004
Kurylo-Borowska, Z. (1962). *Biochim. Biophys. Acta*, **61**, 897
Lambert, M. P. and Neuhaus, F. C. (1972). *J. Bacteriol.*, **110**, 978
Laporte, D. C., Rosenthal, K. F., and Storm, D. R. (1977). *Biochemistry*, **16**, 1642
Leavitt, R. I. and Umbarger, H. E. (1962). *J. Bacteriol.*, **83**, 642
Lee, B. (1971). *J. Mol. Biol.*, **61**, 463
Lichliter, W. D., Naider, F., and Becker, J. M. (1976). *Antimicrobial Agents Chemother.*, **10**, 483
Lichtenthaler, F. W. and Trummlitz, G. (1974). *FEBS Lett.*, **38**, 237
Linnett, P. E. and Strominger, J. L. (1973). *Antimicrobial Agents Chemother.*, **4**, 231
Liou, Y. and Tanaka, N. (1976). *Biochem. Biophys. Res. Commun.*, **71**, 477
Lomakina, N. N., Bognar, R., Brazhnikova, M. G., Sztaricskai, F., and Muravyeva, L. I. (1970). *Abstr. 7th Int. Symp. Chem. Natural Products, Riga*, p. 625
Lund, F. and Tybring, L. (1972). *Nature, Lond.*, **236**, 135
McKelvy, J. F., Leblanc, P., Laudes, C., Perrie, S., Jorgensen, Y., and Kordon, C. (1976). *Biochem. Biophys. Res. Commun.*, **73**, 507
McMullen, A. I., Marlborough, D. I., and Bayley, P. M. (1971). *FEBS Lett.*, **16**, 278
Martin, D. R. and Williams, R. J. P. (1976). *Biochem. J.*, **153**, 181
Matthews, D. M. and Payne, J. W. (1975). In *Peptide Transport and Protein Nutrition* (D. M. Matthews and J. W. Payne, eds), North-Holland/Elsevier, New York; Frontiers of Biology, Vol. 37, p. 428
Mauger, A. B. (1968). *Experientia*, **24**, 1068
Meisler, N. and Simmonds, S. (1963). *J. Gen. Microbiol.*, **31**, 109
Meyers, E., Parker, W. L. and Brown, W. E. (1974). *Ann. N.Y. Acad. Sci.*, **235**, 493
Meyers, E., Pansy, F. E., Basch, H. I., McRipley, R. J., Slusarchyk, D. S., Graham, S. F., and Trejo, W. H. (1973). *J. Antibiot.*, **26**, 457
Meyers, E., Weisenborn, F. L., Pansy, F. E., Slusarchyk, D. S., Von Saltza, M. H., Rathnum, M. L., and Parker, W. L. (1970). *J. Antibiot.*, **23**, 502
Michel, K. H., Chaney, M. O., Jones, N. D., Hoehn, M. M., and Nagarajan, R. (1974). *J. Antibiot.*, **27**, 57
Miller, P. A., Milstrey, K. P., and Trown, P. W. (1968). *Science, N.Y.* **159**, 431
Miller, T. W., Goegelman, R. T., Weston, R. G., Putter, I., and Wolf, F. J. (1972). *Antimicrobial Agents Chemother.*, **2**, 132
Misato, T. (1967). In *Antibiotics*, Vol. 1 (D. Gottlieb and P. D. Shaw, eds), Springer-Verlag, New York, p. 434
Mitani, M. and Inoue, Y. (1968). *J. Antibiot.*, **21**, 492
Miyaki, M., Kitayama, T., and Ono, T. (1974). *J. Antibiot.*, **27**, 647
Miyoshi, T., Miyairi, N., Aoki, M., Kosaka, M., Sakai, H. and Imanaka, H. (1972). *J. Antibiot.*, **25**, 569
Mizuno, S., Nitta, K., and Umezawa, H. (1970*a*). *J. Antibiot.*, **23**, 581
Mizuno, S., Nitta, K., and Umezawa, H. (1970*b*). *J. Antibiot.*, **23**, 589
Molin, N. (1964*a*). *Physiol. Plant.*, **17**, 443
Molin, N. (1964*b*). *Physiol. Plant.*, **17**, 456
Molin, N. (1964*c*). *Physiol. Plant.*, **17**, 468
Molloy, B. B., Lively, D. H., Gale, R. M., Gorman, M., Boeck, L. D., Higgens, C. E., Kastner, R. E., Huckstep, L. L., and Neuss, N. (1972). *J. Antibiot.*, **25**, 137
Moriér, E., Pacheco, H., Koeberle, J., and Yavordios, D. (1975). *Eur. J. Med. Chem.*, **10**, 221
Morley, J. S. (1968). *Fed. Proc. Fed. Am. Soc. Exp. Biol.*, **27**, 1314
Müller, W. and Crothers, D. M. (1968). *J. Mol. Biol.*, **35**, 251
Nakajima, M., Inukai, M., Haneishi, T., Terahara, A., Mamoru, A., Kinoshita, T., and Tamura, C. (1978). *J. Antibiot.*, **31**, 426
Nandi, J., Kumar De, B., and Bose, S. K. (1975). *J. Antibiot.*, **29**, 988
Neuhaus, F. C. (1962). *J. Biol. Chem.*, **237**, 778, 3128
Neuhaus, F. C. (1967). In *Antibiotics*, Vol. 1, *Mechanism of Action* (D. Gottlieb and P. D. Shaw, eds), Springer-Verlag, Berlin, p. 40
Neuhaus, F. C. (1968). *Antimicrobial Agents Chemother.*,**1967**, 304
Neuhaus, F. C. and Lynch, J. L. (1964). *Biochemistry*, **3**, 471

Neuhaus, F. C., Goyer, S. and Neuhaus, D. W. (1977). *Antimicrobial Agents Chemother.*, **11**, 638
Neuss, N., Molloy, B. B., Shah, R., and De La Higuera, N. (1970). *Biochem. J.*, **118**, 571
Newton, B. A. (1956). *Bacteriol. Rev.*, **20**, 14
Nicolaides, E., Dewald, H., Westland, R., Lipnik, M., and Posler, J. (1968). *J. Med. Chem.*, **11**, 74
Norrby, E. (1971). *Virology*, **44**, 599
Ohno, T., Tajima, S., and Toki, K. (1953). *Nippon Nogeikagaku Kaishi*, **27**, 665
Okada, Y., Tani, S., Yawatari, Y., and Yagyu, M. (1976*a*). *Chem. Pharm. Bull.*, **24**, 1925
Okada, Y., Okinaka, M., Yagyu, M., Watabe, K., Sano, K., and Kakiuchi, Y. (1976*b*). *Chem. Pharm. Bull.*, **24**, 3081
Okami, Y., Kitahara, T., Hamada, M., Naganawa, H., Kondo, S., Maeda, K., Takeuchi, T., and Umezawa, H. (1974). *J. Antibiot.*, **27**, 656
Ondetti, M. A. and Thomas, P. L. (1965). *J. Am. Chem. Soc.*, **87**, 4373
Ovchinnikov, Y. A. (1972). In *Peptides 1972* (H. Hanson and H. D. Jakubke, eds), North-Holland/Elsevier, Amsterdam, p. 3
Ovchinnikov, Y. A., Ivanov, V. T., and Shkrob, A. M. (1972). In *Granada Symposium on Molecular Mechanisms of Antibiotic Action on Protein Biosynthesis and Membranes* (E. Munoz, F. Garcia-Ferrandiz, and D. Vazquez, eds), Elsevier, Amsterdam, p. 459
Park, B. K., Hirota, A., and Sakai, H. (1977). *Agric. Biol. Chem., Tokyo*, **41**, 573
Pandey, R. C., Cook, J. C., and Rinehart, K. L. (1977). *J. Am. Chem. Soc.*, **99**, 5205
Parker, W. L. and Rathnum, M. L. (1975). *J. Antibiot.*, **28**, 379
Patterson, E. L., Johnson, B. L., De Voe, S. E., and Bohonos, N., (1966). *Antimicrobial Agents Chemother.*, **1965**, 115
Paulay, Z. and Bajusz, S. (1972). *Acta Chim. Acad. Sci. Hung.* **71**, 247
Payne, J. W. (1968). *J. Biol. Chem.*, **243**, 3395
Payne, J. W. (1972*a*). *J. Gen. Microbiol.*, **71**, 259
Payne, J. W. (1972*b*). In *Peptide Transport in Bacteria and Mammalian Gut*, Ciba Foundation Symposium 4 (K. Elliot and M. O'Connor, eds), Associated Scientific Publishers, Amsterdam, p. 17
Payne, J. W. (1974). *J. Gen. Microbiol.*, **80**, 269
Payne, J. W. (1976). *Adv. Microbial. Physiol.*, **13**, 55
Payne, J. W. and Gilvarg, C. (1971). *Adv. Enzymol. Relat. Areas Mol. Biol.*, **35**, 187
Payne, J. W., Jakes, R., and Hartley, B. S. (1970). *Biochem. J.*, **117**, 757
Perkins, H. R. and Nieto, M. (1974). *Ann. N.Y. Acad. Sci.*, **235**, 348
Perlman, D. and Bodanszky, M. (1971). *Annu. Rev. Biochem.*, **40**, 449
Pestka, S. (1971). *Annu. Rev. Microbiol.*, **25**, 487
Peypoux, F., Guinand, M., Michel, G., Delcambe, L., Das, B. C., Varenne, P., and Lederer, E. (1973). *Tetrahedron*, **29**, 3455
Pinkerton, M., Steinrauf, L. K., and Dawkins, P. (1969). *Biochem. Biophys. Res. Commun.*, **35**, 512
Plattner, P. A., Vogler, K., Studer, R. O., Quitt, P., and Keller-Schierlam, W. (1963). *Helv. Chim. Acta*, **46**, 927
Pressman, B. C. (1976). *Annu. Rev. Biochem.*, **45**, 501
Pruess, D. L., Scannell, J. P., Ax, H. A., Kellett, M., Weiss, F., Demny, T. C., and Stempel, A. (1973). *J. Antibiot.*, **26**, 261
Pruess, D. L., Scannell, J. P., Blount, J. F., Ax, H. A., Kellett, M., Williams, T. H. and Stempel, A. (1974). *J. Antibiot.*, **27**, 754
Quitt, P., Stander, R. O., and Vogler, K. (1963). *Helv. Chim. Acta*, **46**, 1715
Rambhav, S. and Ramachandran, L. K. (1972). *Ind. J. Biochem. Biophys.*, **9**, 225
Rao, K. V. (1962). *Antimicrobial Agents Chemother.*, **1961**, 178
Reeves, P. (1972). *The Bacteriocins*, Molecular Biology, Biochemistry and Biophysics, Vol. 11, Springer-Verlag, Berlin
Reich, E. (1966). *Symp. Soc. Gen. Microbiol.*, **16**, 266
Reitz, R. H., Slade, H. D., and Neuhaus, F. C. (1967). *Biochemistry*, **6**, 2561
Ressler, C. and Kashelikar, D. V. (1966). *J. Am. Chem. Soc.*, **88**, 2025
Reynolds, P. E. (1961). *Biochim. Biophys. Acta*, **52**, 403

Rightsel, W. A., Schneider, H. G., Sloan, B. J., Graf, P. R., Miller, F. A., Bartz, Q. R., Ehrlich, J., and Dixon, G. J. (1964). *Nature, Lond*., **204**, 1333
Ringrose, P. S. (1970). *Biochim. Biophys. Acta*, **213**, 320
Ringrose, P. S. (1972). *FEBS Lett*., **23**, 241
Ringrose, P. S. (1973). *Biochim. Biophys. Acta*, **312**, 656
Ringrose, P. S. and Lambert, R. W. (1973). *Biochim. Biophys. Acta*, **299**, 374
Ringrose, P. S., Atherton, F. R., Hall, M. J., Hassall, C. H., Lambert, R. W., Lloyd, W. J., and Nisbet, L. J. (1977). *10th Int. Cong. Chemother*., Abstr. no. 367
Robins, M. J., Simon, L. N., Stout, M. G., Ivanovics, M., Schweizer, P., Rousseau, R. J., and Robbins, R. K. (1971). *J. Am. Chem. Soc*., **93**, 1474
Rogers, H. J., Newton, G. G. F., and Abraham, E. P. (1965). *Biochem. J*., **97**, 573
Roscoe, J. and Abraham, E. P. (1966). *Biochem. J*., **99**, 793
Rosenblatt, J. E. and Stewart, P. R. (1974). *Antimicrobial Agents Chemother*., **6**, 84
Rosenthal, K. S., Ferguson, R. A., and Storm, D. R. (1977). *Antimicrobial Agents Chemother*., **12**, 665
Rosenthal, K. S., Swanson, P. E., and Storm, D. R. (1976). *Biochemistry*, **15**, 5783
Roze, U. and Strominger, J. L. (1966). *J. Mol. Pharmacol*., **2**, 92
Ruegger, A., Kuhn, M., Lichti, H., Loosli, H. R., Hugeunin, R., Quiquerez, C., and von Wartburg, (1976). *Helv. Chim. Acta*, **59**, 1075
Sarges, R. and Witkop, B. (1965). *J. Am. Chem. Soc*., **87**, 2011
Scannell, J. P. and Pruess, D. L. (1974). In *Chemistry and Biochemistry of Amino Acids, Peptides and Proteins* (B. Weinstein, ed.), Marcel Dekker, New York, p. 189
Scannell, J. P., Pruess, D. L., Blount, J. F., Ax, H. A., Kellett, M., Weiss, F., Demny, T. C., Williams, T. H., and Stempel, A. (1975). *J. Antibiot*., **28**, 1
Schazschneider, B., Ristow, H., and Kleinkauf, H. (1974). *Nature, Lond*., **249**, 757
Schiffmann, E., Corcoran, B. A., and Wahl, S. M. (1975). *Proc. Natl. Acad. Sci. U.S.A*., **72**, 1059
Schleifer, K. H. and Kandler, O. (1972). *Bacteriol. Rev*., **36**, 407
Schlievert, P., Johnson, W., and Galask, R. P. (1976). *Am. J. Obstet. Gynecol*., **125**, 906
Schwyzer, R. (1970). *Experientia*, **26**, 577
Sengupta, S. and Bose, S. K. (1971). *Biochem. J*., **121**, 839
Shah, R., Neuss, N., Gorman, M., and Boeck, L. D. (1970). *J. Antibiot*., **23**, 613
Shankman, S., Gold, V., and Higa, S. (1961). *Texas Rept. Biol. Med*., **19**, 358
Shankman, S., Makineni, S., and Gold, V. (1963*a*). *Arch. Biochem. Biophys*., **100**, 431
Shankman, S., Gold, V., Higa, S., and Squires, R. (1962*a*). *Biochem. Biophys. Res. Commun*., **9**, 25
Shankman, S., Higa, S., Lew, F., and Roberts, M. E. (1962*b*). *J. Med. Chem*., **5**, 42
Shankman, S., Higa, S., Makineni, S., and Noll, W. (1963*b*). *J. Med. Chem*., **6**, 746
Shankman, S., Higa, S., Florsheim, H. A., Schvo, Y. and Gold, V. (1960). *Arch. Biochem. Biophys*., **86**, 204
Sharma, R. A., Bobek, M., and Bloch, A. (1975). *J. Med. Chem*., **18**, 955
Sheldrick, G. M., Jones, P. G., Kennard, O., Williams, D. H. and Smith, G. A. (1978). *Nature, Lond*., **271**, 223
Shemyakin, M. M., Ovchinnikov, Y. A., Ivanov, V. T., Antonov, V. K., Vinogradova, E. I., Shkrob, A. M., Malenkov, G. G., Evstratov, A. V., Laine, I. A., Melnik, E. I., and Ryabova, I. D. (1969). *J. Membr. Biol*., **1**, 402
Shenfeld, A., Flowers, H. M., and Katchalski, E. (1974). *Antimicrobial Agents Chemother*., **5**, 179
Shiba, T., Nomoto, S., and Wakamiya, T. (1976). *Experientia*, **32**, 1109
Shiba, T., Miyoshi, K., and Kusumoto, S. (1977). *Bull. Chem. Soc., Japan*, **50**, 254
Shimi, I. R., Abedallah, N., and Fathy, S. (1977). *Antimicrobial Agents Chemother*., **11**, 373
Shoji, J. and Hinoo, H. (1975). *J. Antibiot*., **28**, 60
Shoji, J. and Kato, T. (1976). *J. Antibiot*., **29**, 380
Shoji, J., Hinoo, H. and Sakazaki, R. (1976*b*). *J. Antibiot*., **29**, 521
Shoji, J., Kato, T., and Hinoo, H. (1977). *J. Antibiot*., **30**, 427
Shoji, J., Kato, T., and Sakazaki, R. (1976*e*). *J. Antibiot*., **29**, 1268
Shoji, J., Hinoo, H., Wakisaka, Y., Koizumi, K., and Mayama, M. (1976*d*). *J. Antibiot*., **29**, 813
Shoji, J., Kozuki, S., Okamoto, S., Sakazaki, R., and Otsuka, H. (1968). *J. Antibiot*., **21**, 439

Shoji, J., Hinoo, H., Wakisaka, Y., Koizumi, K., Mayama, M., and Matsuura, S. (1976*c*). *J. Antibiot.*, **29**, 809

Shoji, J. Sakazaki, R., Wakisaka, Y., Koizumi, K., Mayama, M., and Matsuura, S. (1975). *J. Antibiot.*, **28**, 122

Shoji, J., Hinoo, H., Wakisaka, Y., Koizumi, K., Mayama, M., Matsuura, S., and Matsumoto, K. (1976*a*). *J. Antibiot.*, **29**, 516

Siewert, G. and Strominger, J. L. (1967). *Proc. Natl. Acad. Sci., U.S.A.*, **57**, 767

Simmonds, S. (1970). *Biochemistry*, **9**, 1

Simmonds, S., Harris, J., and Fruton, J. S. (1951). *J. Biol. Chem.*, **188**, 251

Simmonds, S., Tatum, E. L. and Fruton, J. S. (1947). *J. Biol. Chem.*, **170**, 483

Smissmann, E. E., Terada, A., and El-Antably, S. (1976). *J. Med. Chem.*, **19**, 165

Smith, R. L. and Dunn, F. W. (1970). *J. Biol. Chem.*, **11**, 2962

Snezhkova, L. G., Shepel, E. N., Ryabova, I. D., Miroshnikov, A. I., Ivanov, V. T., and Ovchinnikov, Y. U. (1975). *Bioorg. Khim.*, **1**, 347

Sobell, H. M. (1972). *J. Mol. Biol.*, **68**, 1

Sobell, H. M. (1974). *Scient. Am.*, **231**, 82

Sobell, H. M., Jain, S. C., Sakore, T. D., and Nordman, C. E. (1971). *Nature, Lond.*, **231**, 200

Sogn, J. (1976). *J. Med. Chem.*, **19**, 1228

Somma, S., Merati, W., and Parenti, F. (1977). *Antimicrobial Agents Chemother.*, **11**, 396

Spratt, B. G. (1977). *Eur. J. Biochem.*, **72**, 341

Spratt, B. G., Jobanputra, V., and Zimmerman, W. (1977). *Antimicrobial Agents Chemother.*, **12**, 406

Stapley, E. O., Mata, J. M., Miller, I. M., Demny, T. C., and Woodruff, H. (1964). *Antimicrobial Agents Chemother.*, **1963**, 20

Stapley, E. O., Jackson, M., Hernandez, S., Zimmerman, S. B., Currie, S. A., Mochales, S., Mata, J. M., Woodruff, H., and Hendlin, D. (1972). *Antimicrobial Agents Chemother.*, **2**, 122

Stewart, W. W. (1971). *Nature, Lond.*, **229**, 174

Stone, K. J. and Strominger, J. L. (1971). *Proc. Natl. Acad. Sci. U.S.A.*, **68**, 3223

Storm, D. R., Rosenthal, K. S., and Swanson, P. E. (1977). *Annu. Rev. Biochem.*, **46**, 723

Streicher, W. and Reinshagen, H. (1975). *Chem. Ber.*, **108**, 813

Strominger, J. L., Izaki, K., Matsuhashi, M. and Tipper, D. J. (1967). *Fed. Proc. Fed. Am. Soc. Exp. Biol.*, **26**, 9

Studer, R. O. and Lergier, W. (1965). *Helv. Chim. Acta*, **48**, 460

Studer, R. O., Lergier, W., and Vogler, K. (1966). *Helv. Chim. Acta*, **49**, 974

Studer, R. O., Lergier, W., Lanz, P., Boehni, E., and Vogler, K. (1965). *Helv. Chim. Acta*, **48**, 1371

Suda, S. and Curtis, R. W. (1966). *Appl. Microbiol.*, **14**, 475

Suda, H., Aoyagi, T., Takeuchi, T., and Umezawa, H. (1976). *Arch. Biochim. Biophys.*, **177**, 196

Suhadolnik, R. J. (1970). *Nucleoside Antibiotics*, John Wiley, New York

Sussman, A. J. and Gilvarg, C. (1970). *J. Biol. Chem.*, **245**, 6518

Suzuki, H., Nagai, K., Yamaki, H., Tanaka, N., and Umezawa, H. (1969). *J. Antibiot.*, **22**, 446

Suzuki, S., Isono, K., Nagatsu, J., Kawashima, Y., Yamagata, K., Sasaki, K., and Hashimoto, (1966). *Agric. Biol. Chem., Tokyo*, **30**, 817

Suzuki, T., Hayashi, K., Fujikawa, K., and Tsukamoto, K. (1965). *J. Biochem., Tokyo*, **57**, 226

Szer, W. and Kurylo-Borowska, Z. (1970). *Biochim. Biophys. Acta*, **224**, 477

Takahashi, K., Yoshioka, O., Matsuda, A., and Umezawa, H. (1977). *J. Antibiot.*, **30**, 861

Takita, T., Muraoka, Y., Fujii, A., Itoh, H., Maeda, K., and Umezawa, H. (1972*a*). *J. Antibiot.*, **25**, 197

Takita, T., Muraoka, Y., Yoshioka, T., Fujii, A., Maeda, K., and Umezawa, H. (1972*b*). *J. Antibiot.*, **12**, 755

Tanaka, N., Iseki, M., Miyoshi, T., Aoki, H., and Imanaka, H. (1976). *J. Antibiot.*, **29**, 155

Tanaka, H., Iwai, Y., Oiwa, R., Shinohara, S., Shimizu, S., Oka, T., and Omura, S. (1977). *Biochim. Biophys. Acta*, **497**, 633

Taniyama, H., Sawada, Y., and Kitagawa, T. (1971). *Chem. Pharm. Bull.*, **19**, 1627

Taniyama, H., Sawada, Y., Miyazeki, K., and Miyashi, F. (1972). *Chem. Pharm. Bull.*, **20**, 601
Tata, J. R., Hamilton, M. J., and Shields, D. (1972). *Nature, Lond.*, **238**, 161
Tipper, D. J. and Strominger, J. L. (1965). *Proc. Natl. Acad. Sci. U.S.A.*, **54**, 1133
Tomasz, A. and Waks, S. (1975). *Proc. Natl. Acad. Sci. U.S.A.*, **72**, 4162
Tsukagoshi, N., Tamura, G., and Arima, K. (1970). *Biochim. Biophys. Acta*, **196**, 204, 211
Tsukamura, M. (1960). *J. Biochem.*, **47**, 685
Uehara, Y., Kondo, S., Umezawa, H., Suzakake, K., and Hori, M. (1972). *J. Antibiot.*, **25**, 685
Uehara, Y., Hori, M., and Umezawa, H. (1976*a*). *Biochim. Biophys. Acta*, **442**, 251
Uehara, Y., Hori, M. and Umezawa, H. (1976*b*). *Biochim. Biophys. Acta*, **447**, 406
Uehara, Y., Hori, M., Kondo, S., Hamada, M., and Umezawa, H. (1976*c*). *J. Antibiot.*, **29**, 937
Ueki, T., Ashida, T., Kakudo, M., Sasada, Y. and Katsube, Y. (1967). *Nature, Lond.*, **216**, 1207
Ueo, K., Fukuoka, Y., Hayashi, T., Yusada, T., Taki, H., Tai, H., Tai, M., Watanabe, Y., Saikawa, I., and Mitsuhashi, S. (1977). *Antimirobial Agents Chemother.*, **12**, 455
Umezawa, H. (1977). *Lloydia*, **40**, 67
Umezawa, H., Ishizuka, M., Aoyagi, T., and Takeuchi, T. (1976). *J. Antibiot.*, **29**, 857
Urry, D. W. and Ruiter, A. L. (1970). *Biochem. Biophys. Res. Commun.*, **38**, 800
Vandamme, E. J., Kenig, M. D., and Abraham, E. P. (1975). *Abstr. Ann. Meeting Am. Soc. Microbiol.*, **75**, A11
Vazquez, D. (1967). In *Antibiotics*, Vol. 1 (D. Gottlieb and P. D. Shaw, eds), Springer-Verlag, New York, p. 387
Veatch, W. R., Fossel, E. T., and Blout, E. R. (1974). *Biochemistry*, **13**, 5249
Vogler, K., Studer, R. O., Lanz, P., Lergier, W., and Boehni, E. (1965). *Helv. Chim. Acta*, **48**, 1161
Vogler, K., Lanz, P., Quitt, P., Studer, R. O., Lergier, W., Boehni, E., and Fust, B. (1964). *Helv. Chim. Acta*, **47**, 526
Von der Haar, R. A. and Umbarger, H. E. (1972). *J. Bacteriol.*, **112**, 142
Walker, J. E. and Abraham, E. P. (1970). *Biochem. J.*, **118**, 563
Ward, J. B. (1977). *FEBS Lett.*, **78**, 151
Wargel, R. J., Shadur, C. A., and Neuhaus, F. C. (1970). *J. Bacteriol.*, **103**, 778
Wargel, R. J., Shadur, C. A., and Neuhaus, F. C. (1971). *J. Bacteriol.*, **105**, 1028
Waring, M. J. and Wakelin, L. P. G. (1974). *Nature, Lond.*, **252**, 653
Wasmuth, J. J. and Umbarger, H. E. (1974). *J. Bacteriol.*, **117**, 29
Wasserman, H. H., Keggi, J. J., and McKeon, J. E. (1961). *J. Am. Chem. Soc.*, **83**, 4107
Wasserman, H. H., Keggi, J. J., and McKeon, J. E. (1962). *J. Am. Chem. Soc.*, **84**, 2978
Whitney, J. G. and Funderburk, S. S. (1970). *Abstr. Xth Int. Congr. Microbiol., Mexico City*, p. 101
Whitney, J. G., Funderburk, S. S., Westhead, J. E., Lively, D. H., Solenberg, J. M., and Deeney, J. W. (1970). *Bacteriol. Proc.*, **7**
Whitney, J. G., Funderburk, S. S., Westhead, J. E., Lively, D. H., Solenberg, J. M. and Deeney, J. W. (1972). *Appl. Microbiol.*, **24**, 907
Wieland, T., Lüben, G., Ottenheym, H., Faesel, J., de Vriès, J. X., Konz, W., Prox, A., and Schmid, J. (1968). *Angew, Chem.*, **80**, 209
Wilson, D. W., Grier, D., Reimer, R., Bauman, J. D., Preston, J. F., and Gabbay, E. J. (1976). *J. Med. Chem.*, **19**, 381
Witzke, N. M. and Heding, H. (1976). *J. Antibiot.*, **29**, 1349
Woodruff, H. B. and Foster, J. W. (1946). *J. Bacteriol.*, **51**, 371
Woolley, D. W., Schaffner, G., and Braun, A. C. (1955). *J. Biol. Chem.*, **215**, 485
Yabu, K. (1969). *Biochim. Biophys. Acta*, **184**, 460

Microorganisms and Nitrogen Sources
Edited by J. W. Payne

Section 4.8

Microbial Peptide Ionophores

MARIS C. STEPHENSON AND COLIN RATLEDGE

Department of Biochemistry, University of Hull, Hull HU6 7RX, UK

1 INTRODUCTION

In this chapter we have chosen to adhere to Pressman's broad definition of ionophores as compounds which increase the cation permeability of lipid barriers (Pressman *et al.*, 1967). Ionophores can be divided into two main classes (a) monobasic acids and (b) neutral macrocyclic compounds. The former group, which does not include peptides, will not be considered here. Of the four distinct types of molecule which comprise the second group, that is, macrotetrolides, cyclic ethers, depsipeptides, and peptides, only the latter two qualify for discussion in this present framework.

We have divided the ionophores which fall within our purview into two classes which are mutually exclusive: first, those compounds which bind and transport mono- and/or divalent cations; and second, those molecules which are specific for trivalent cations. A great deal of work has been carried out with some of the representatives of the first group (e.g. valinomycin) as these ionophores have been invaluable in helping to illuminate some of the vexed questions of energy conservation during the coupling of electron transport to oxidative phosphorylation and the energy coupling mechanism of active transport systems. Much of this work, interesting though it may be, is strictly outside the objectives of this chapter and the reader is referred to the following reviews for greater details concerning the applications of ionophores in this area: Henderson, 1971; Harold, 1972, 1977; Harold *et al*., 1974; Skulachev, 1975 (see also Section 1.7).

2 GENERAL PROPERTIES OF PEPTIDE IONOPHORES

In choosing to adhere to Pressman's definition of an ionophore, caution must be exercised with respect to the specificity of action of the compound under consideration. It is not a prerequisite that an ionophore should exhibit ion selectivity, but where this is not the case it may be that the compound in question is simply acting as a general disrupter of membrane integrity. Thus, despite the fact that they increase membrane permeability to ions, gramicidin S, polymyxin, and the tyrocidins, for example, are not ionophores but simply fall into this latter category and do not therefore qualify for discussion here (but see Sections 4.7.3 and 4.7.4).

The importance of peptides in creating ionophores, unfortunately for this chapter, is somewhat limited. Since ionophores can be successfully formed without recourse to amino acids (e.g. the macrotetrolide actins, the monensin family, and nigericin) it may be that the use of amino acids by microorganisms to construct ionophores is fortuitous. Without doubt, the availability of amino acids has allowed them to be used as fundamental binding units either to produce true polypeptides, depsipeptides, or just substituted amino acids as in the case of some of the ferric ionophores (e.g. enterochelin). The conformational properties of the amino acids involved in a particular ionophore are probably vital to bring about the correct juxtapositioning of all the coordinating groups, although the amino acid residues *per se* can again be substituted by other amino acids without loss of chelating ability.

The overall strength of binding of a metal ion by a peptide ionophore is likely to be weak as the O atom is only weakly basic. Stability may be achieved by closure of a chelate ring with any adjacent $-NH_2$ group [the

N (peptide) atom will not participate in any metal binding activity (Freeman, 1973)] or —COO$^-$ group, which is the strongest of all ion binding groups likely to be found in an ionophore. The fact that none of the mono- or divalent cation ionophores chelate ions by means of a carboxyl group is probably an indication that the absolute stability constants for the complexes are unlikely to be exceptionally high. Release of the cation from the ionophore is therefore possible by a conformational change of structure requiring little input of energy. This is unlike the situation with the ferric ionophores where very high stability constants for the complexes exist, and thus disruption of the ionophore or reduction of the iron is required to achieve release of the metal. For a discussion of the more theoretical aspects of metal binding to amino acids and peptides, the reader is referred to the review by Freeman (1973).

3 PEPTIDE IONOPHORES FOR MONO- AND DIVALENT CATIONS

Although most of this group of ionophores were regarded as antibiotics upon isolation, their efficacy in disrupting mitochondrial energy conservation in higher organisms has limited their therapeutic applications. The antibiotic activity of the ionophores arises from their disruption of energy and ion metabolism of microorganisms as a direct consequence of increasing the ion permeability of the membrane. The ionphores in this section fall into two groups characterized by the mechanism by which this ion permeability is accomplished.

The first group of ionophores increases membrane permeability by carrying ions across lipid barriers as lipid-soluble complexes. Implicit in this definition is the movement of the metal–ionophore complex across the membrane; that is, the ionophore is acting as a true mobile carrier. To be an effective carrier, the compound should be capable of residing at an interface at which it should be able to bind rapidly and release cations; its ion complex should be lipophilic and small enough to exhibit sufficient mobility, and it should also exhibit ion selectivity. In order to bind cations effectively, such carriers generate a coordination sphere around the complexed ion; the dimensions and flexibility of the sphere will then govern ion selectivity and affinity. Characteristic of this group of ion carriers is the depsipeptide ionophore valinomycin.

The second group of ionophores increases permeability by altering the structure of the membrane. These compounds combine with each other and with certain lipids within the membrane, disrupting the continuity of the membrane by creating an aqueous channel or pore for the free diffusion of ions. Such channel formers are exemplified by gramicidin A (see also Section 4.7.2).

1 Peptide Ionophores

(i) Gramicidins

Producing species The same Dubos strain of *Bacillus brevis* produces all six linear gramicidins (Hotchkiss, 1944).

Structure and properties Gramicidins are linear polypeptides of molecular mass of about 1800. Their structures are given in Figure 1. Each gramicidin is a mixture of two molecules, one having the formylated amino acid as valine, as in Figure 1, the other having isoleucine (Sarges and Witkop, 1965*a,b,c,d*). They are of low solubility in water as there are no ionizable groups in the molecule, save for gramicidin C which contains tyrosine. The solubility of gramicidin A in water has been calculated as less than 10^{-10} M (Hladky and Haydon, 1972).

A. HCO-L-Val-Gly-L-Ala-D-Leu-L-Ala-D-Val-L-Val-
D-Val-L-Trp-D-Leu-L-Trp-D-Leu-L-Trp-D-Leu-L-Trp-NH.CH_2CH_2OH

B. HCO-L-Val-Gly-L-Ala-D-Leu-L-Ala-D-Val-L-Val-
D-Val-L-Trp-D-Leu-L-Phe-D-Leu-L-Trp-D-Leu-L-Trp-NH.CH_2CH_2OH

C. HCO-L-Val-Gly-L-Ala-D-Leu-L-Ala-D-Val-L-Val-
D-Val-L-Trp-D-Leu-L-Tyr-D-Leu-L-Trp-D-Leu-L-Trp-NH.CH_2CH_2OH

Figure 1 Gramicidins A, B, and C. (A similar series with L-isoleucine replacing the valine at the left-handside also occurs)

Nomenclature Gramicidins A, B, and C are the true gramicidins. Gramicidin D is the name of the mixture of gramicidins isolated from the Dubos strain of *B. brevis*. Most commercial preparations are type D. Gramicidin J is so named as it was isolated in Japan from the Nagano strain of *B. brevis*. This gramidicin, however, is identical with gramicidin S (= Soviet) which is a cyclic polypeptide and is therefore not a true gramicidin. Gramicidin J (= S) should be classed as a tyrocidin but neither it nor tyrocidin are ionophores (see Hunter and Schwartz, 1967*a,b*).

Biosynthesis Partial biosynthesis has been achieved using a soluble enzyme system from *B. brevis* (Bauer *et al*., 1972). This process appears to be independent of tRNA, i.e. is not synthesized in the manner of a conventional peptide chain. This probably applies to all the ionophores mentioned in this chapter.

Action in vitro The major studies on the action of the linear gramicidins have been made with gramicidin A. It may be assumed, however, that all the gramicidin molecules act similarly (Hladky, 1974). The literature concerning its mode of action has been reviewed by Bamberg *et al*. (1976) and by Urry *et al*. (1975).

Gramicidin A binds irreversibly to the surfaces of lipid membranes in such a way as to produce a pore which spans the membrane (Hladky and Haydon, 1970, 1972). The pore conducts Na^+ and K^+ with a selectivity of K^+ over Na^+ of about sixfold (Mueller and Rudin, 1967). Divalent ions are not conducted (Henderson *et al*., 1969; Bamberg and Lauger, 1977). The channel appears to be produced by a dimer of gramicidin (see Urry *et al*., 1975) and may be formed by head-to-head association of two monomers (Urry, 1971, 1972) or by the two peptides coiling about a common axis (Veatch *et al*., 1974). At the present time, which of these two models is correct cannot be determined (Apell *et al*., 1977). The simplistic view, of one ion penetrating the channel before the next one is attached, has been disturbed by the suggestion that the channel has two cationic binding sites one at each entrance to the channel (Sandblom *et al*., 1977). A complex method of uptake may result.

The increased cation permeability caused by gramicidin occurs not only with mitochondria, with which the effects were first elucidated (Chappell and Crofts, 1965; Pressman, 1965; Harold, 1970), but also with bacteria and therefore accounts for the inhibition of growth of *Streptococcus faecalis*. Here, the antibiotic at 10^{-7} M causes a rapid loss of K^+ from cells in exchange for external Na^+, Li^+, or NH_4^+ (Harold and Baarda, 1968).

(ii) Alamethicin Family

(a) Alamethicin

Producing species *Trichoderma viride* (Meyer and Reusser, 1967).

Structure There has been some confusion with this peptide which has been described as a cyclic molecule in four independent investigations (Payne *et al*., 1970; Meyer and Reusser, 1967; Ovchinnikov *et al*., 1971*b*; Jung *et al*., 1976). It is in fact a linear peptide of 19 residues, the difficulty having arisen because of an unusual *N*-terminal group (see below).

Alamethicin is a mixture of two components, F.30 (approx. 85 per cent) and F.50 (approx. 15 per cent) (McMullen, 1970; Melling and McMullen, 1974). The sequences of these molecules have been determined by Martin and Williams (1976); that of the F.30 component is as shown in Figure 2. The F.50 component differs only in that the terminal glutamine —COOH group is an amide (—$CONH_2$) group.

The unusual features of the molecule are its high content of

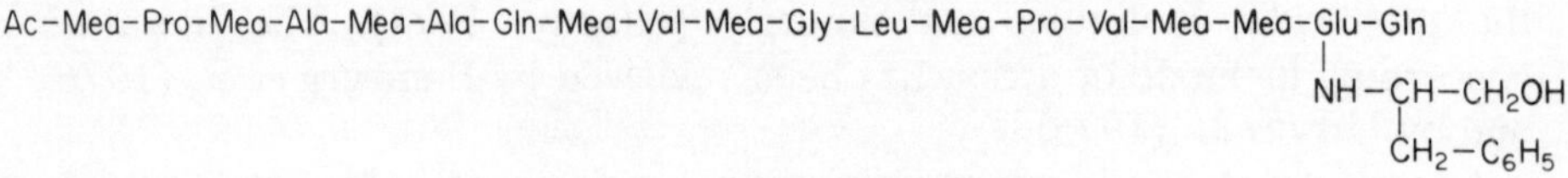

Figure 2 Alamethicin (F.30 component). Mea = 2-methylalanine, Ac = acetyl. (After Martin and Williams, 1976)

2-methylalanine residues and the presence of a phenylalaninol on the eighteenth residue. The biosynthesis of the 2-methylalanine residues is unknown and no pathway leading to it is immediately obvious as this compound will not undergo transamination. An isomerization of α-aminobutyric acid is perhaps the most likely route.

Biosynthesis Little is known of the biosynthesis of this antibiotic save that in common with other peptides it is not synthesized on the ribosomes, as synthesis was unaffected when cells were exposed to cycloheximide which blocks cell protein synthesis (Reusser, 1967; Bodanszky and Perlman, 1969). There is probably a stepwise addition of single amino acids on to the peptide chain.

Action in vitro Alamethicin forms cation complexes over a wide pH range but has little discrimination between alkali metals (Na^+, K^+, Li^+, and Cs^+). It has been considered to produce unit conducting channels through experimental lipid bilayers (Mueller and Rudin, 1968; 1970; Eisenberg *et al*., 1973; Gordon and Haydon, 1972; Gordon, 1974). The idea that these channels could be created by stacked rings of alamethicin was advanced before the linear structure became known. Models of the ionophore (Martin and Williams, 1976) show that it is long enough to stretch across a lipid bilayer so that the pores are formed by a number of molecules, probably six or more, laid side by side. The most recent suggestion is that there is one small channel surrounded by six larger ones (Gordon and Haydon, 1975, 1976). Both components of alamethicin behave similarly in lipid bilayer membranes (Boheim, 1974).

Alamethicin is toxic to Gram-positive bacteria, *Streptococcus haemolyticus, S. faecalis*, and *S. viridans* but not Gram-negatives (Meyer and Reusser, 1967). Its effects on bacterial membranes have not been extensively studied but it is probable that permeability barriers are selectively disrupted as indicated above with artificial membranes (see Harold, 1970).

(b) Suzukacillin

Producing species *Trichoderma viride*, strains 63 CL and 1037 (Ooka *et al*., 1966; Ooka and Takeda, 1972).

Structure and properties Suzukacillin, the structure of which is unknown, is a linear polypeptide of about 23 amino acids. The residues and their numbers are nine or ten 2-methylalanines, three glutamines, three valines, two alanines, two prolines, two leucines, one or two glycines, and one phenylalaninol (Jung *et al*., 1976). It is closely related to, but not identical with, alamethicin (see section immediately preceding). The crude material consists of two peptide components, A and B. All work on the structure and function has been with the major A component.

Biosynthesis Nothing is known (see Alamethicin).

Action in vitro There is a close similarity between alamethicin and suzukacillin in their membrane modifying and lysing properties. Suzukacillin, however, aggregates more strongly than alamethicin in aqueous media due to a longer α-helical part and a higher number of aliphatic residues (Jung *et al*., 1976). Part of the α-helix is exceptionally stable due to the 2-methylalanine residues shielding the peptide bonds from interactions with polar solvents. Thus, there is a higher probability that suzukacillin will form dimers which are favourable for the creation of pores through the lipid membranes (Boheim *et al*., 1976). Pore formation is not hindered if mixtures of suzukacillin and alamethicin are used (Boheim *et al*., 1976).

The peptide is an active antimicrobial agent inhibiting the growth of Gram-positive bacteria and some fungi at between 10 and 100 $\mu g\ ml^{-1}$ (Ooka *et al*., 1966).

(c) Trichotoxin A-40

Producing species *Trichoderma viride* NRRL 5242 (Hou *et al*., 1972). The compound described below does not correspond to the original description of the mycotoxin, given by Hou *et al*. (1972), though the same producing species was used in both cases.

Structure Trichotoxin is a linear polypeptide of 18–20 amino acid residues comprising one proline, one glycine, two alanines, two leucines, two glutamines, and ten 2-methylalanines, and which exhibits a right-handed, α-helical conformation (Irmscher *et al*., 1978). It is similar to alamethicin and suzukacillin but contains only one proline residue and lacks the phenylalaninol component.

Action in vitro Trichotoxin A-40 forms pores in bilayer lipid membranes comparable to those formed by alamethicin and suzukacillin (Boheim *et al*., 1978). The pores may be slightly less stable than those formed by the other two antibiotics. It exhibits antimicrobial activity at 1 $mg\ ml^{-1}$

methanol against *Brucella bronchiseptica*, *Mycobacterium phlei*, *Staphylococcus aureus*, and *Mucor ramannianus* but not *Candida albicans* (Hou *et al*., 1972). It is also toxic to mice if given intraperitoneally but not if administered orally.

(iii) Antamanide

Producing species It is produced from *Amanita phalloides*, which is a poisonous, green toadstool (Wieland *et al*., 1968).

Structure and properties The structure of antamanide is given in Figure 3. It is a cyclic peptide which has no antibiotic properties, though synthetic analogues have been prepared which do show antibiotic activity (Ovchinnikov *et al.*, 1972). It is an antitoxin used to counteract the accumulation of phallotoxins by the liver following ingestion of the toadstool.

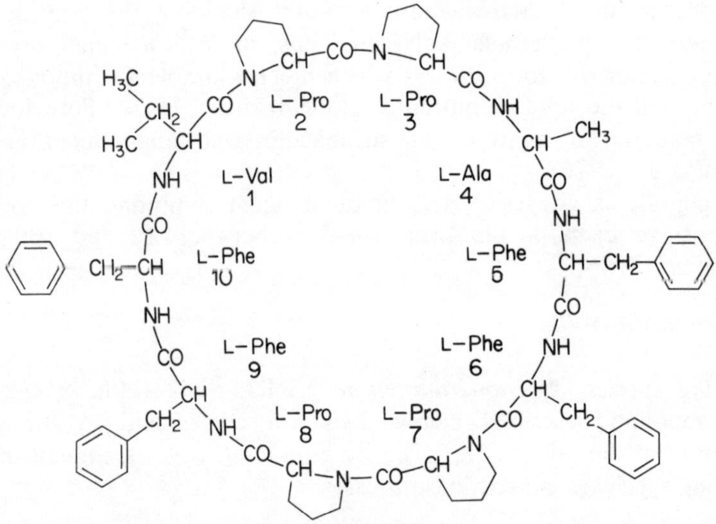

Figure 3 Antamanide

Biosynthesis Nothing is known.

Action in vitro The molecule possesses a high selectivity for Na^+ and Ca^{2+}. Other mono- and divalent cations show less than a tenth of the binding power of these two ions (Wieland, 1972). Conformationally, the equimolar complexes of antamanide with Li^+, K^+, Na^+, and Cs^+ are similar both in solution and the crystalline state (Patel and Tonelli, 1974). A saddle-like form occurs with the binding of Li^+ to four amide carbonyl oxygens situated approximately at the four corners of a square. With the binding of Ca^{2+}, a 'sandwich' complex of two antamanides to one ion has been proposed (Ivanov *et al.*, 1975). From this latter model, Ivanov *et al.* (1975) have suggested that the antitoxin action of this compound is due to its attachment to a membrane, via the polar group of a phospholipid and surface Na^{+} and Ca^{2+} ions. The cations are thus sandwiched between the membrane and antamanide. Antamanide, being a large planar molecule, then effectively blocks a considerable portion of the membrane surface which, it is argued, would then lead to impermeability to the toxins of *Amanita phalloides* amongst other modifications of membrane activity.

2 Depsipeptide Ionophores

Depsipeptides are compounds in which amino acids alternate with hydroxyacids and thus the connecting links are alternately peptide and ester bonds. Although they are, therefore, not true peptides, the depsipeptide ionophores have nevertheless been included in the scope of this chapter.

The naturally occurring depsipeptides were recognized as possessing antibiotic activity against a range of organisms (Brockmann and Schmidt-Kastner, 1955; Gaumann *et al.*, 1947). Due to the simplicity of the methods used for chemical synthesis of depsipeptides (Shemyakin *et al.*, 1966) a large number of analogues has been prepared and used in investigations of the relationship between structure and antibiotic activity. The antibiotic activity of the compounds synthesized has been investigated in a number of Gram-positive, Gram-negative, and acid-fast bacteria, yeasts, and phytopathogenic fungi (Shemyakin *et al.*, 1965) with the following conclusions:

(1) all linear analogues were completely inactive;
(2) activity of cyclic analogues was correlated to the ring size, which may vary within broad limits of 14–36 atom members, with an optimum of 18 members for the enniatin group and 36 members for the valinomycin group;
(3) the amino acid residues may tolerate considerable change without loss of activity;
(4) changes in the hydroxyacid composition caused inactivation;
(5) ester group linkages could be replaced by amides without loss of activity.

The antimicrobial activity of the depsipeptides was attributed to their ion-conducting properties by Moore and Pressman (1964), who showed that the ability of valinomycin to uncouple oxidative phosphorylation in mitochondria was strongly dependent on the presence of K^{+}, but not Na^{+}.

As a result of a cooperative study with Pressman (1965) of the antimicrobial and ion-translocating properties of depsipeptides, Shemyakin (1965) concluded that 'a quantitative parallelism exists between the antimicrobial activity of the depsipeptide antibiotics and their ion-transport inducing properties'.

The discovery of the mode of action of the depsipeptides as ionophores led to extensive investigations of the structures of the ion-carrier complexes. The general properties of the 1:1 complexes between carrier antibiotics and monovalent cations were summarized by Simon and Morf (1973) as follows:

(1) the metal ion is coordinated by 5–8 oxygen atoms;
(2) the external surface of the complex is lipophilic, enhancing the lipid-solubility of the carrier;
(3) the alkali metal cation complexes of the depsipeptide group are positively charged;
(4) the carrier molecules have sufficient flexibility to allow for a step-wise substitution of the solvent molecules, resulting in a low activation barrier and fast complexation reaction rates.

(i) Valinomycin

Producing species Valinomycin, an antibiotic produced by certain strains of *Streptomyces*, was first isolated by Brockmann and Schmidt-Kastner (1955) from the mycelium of cultures of *S. fulvissimus* strain Ihr 582. It has subsequently been isolated from the mycelium of *S. tsusimaeusis* strain N-329 by Nishimura *et al*. (1964).

Structure Brockmann *et al*. (1963) correctly identified the constituents of valinomycin as D- and L-valine, L-lactate, and D-hydroxyisovalerate, but concluded the sequence repeated only twice. Shemyakin *et al*. (1965) subsequently reported valinomycin to be a cyclododecadepsipeptide, with a molecular weight of 1111, consisting of three repetitions of the sequence D-valine, L-lactate, L-valine, and D-α-hydroxyisovalerate (Figure 4) and established this by synthesis (Shemyakin, 1965).

The crystal structure of the ion complex of valinomycin is known (Pinkerton *et al*., 1969). In the potassium complex the metal has six oxygen neighbours, all of which are carbonyls from ester groups. The carbonyl oxygen atoms from the amide groups are receptors in H-bonding which hold the valinomycin molecule in the correct conformation for forming a complex. The relevance of this work on the crystalline state to the behaviour of the molecule in solution has been checked by NMR, IR, and ORD spectroscopic studies in various solvents (Ivanov *et al*., 1969; Shemyakin *et al*., 1969; Ovchinnikov *et al*., 1972) which yielded data consistent with the results obtained by X-ray crystallography.

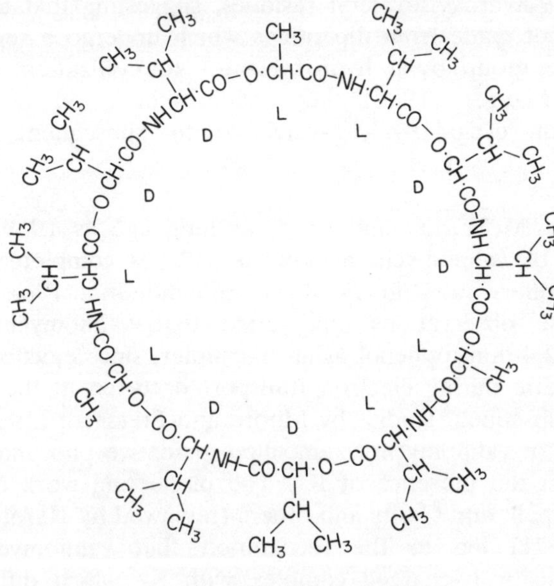

Figure 4 Valinomycin

For the uncomplexed molecule, there is agreement that more than one conformational state can exist. Conformation depends on the polarity of the solvent (Ivanov *et al.*, 1971; Patel and Tonelli, 1973) and changes in conformation upon complex formation do not involve *cis—trans* isomerization about the peptide bonds. The uncomplexed molecule possesses a hydrophilic and a hydrophobic side and thereby accumulates at the interface of a lipid membrane; on complex formation, as the ion enters the molecule, the ester carbonyl groups move towards the centre forming a rigid sphere which easily accommodates K^+ or Rb^+ but which cannot focus on ions as small as Na^+ and Li^+. As the carbonyl groups move inwards the lipid solubility of the complex is increased enabling it to sink into the central, non-polar core of the membrane (Ovchinnikov *et al.*, 1972).

Biosynthesis The biosynthesis of valinomycin has been investigated in a *Streptomyces* species PRL 1642 grown on a complex medium containing various labelled amino acids by following the incorporation of the label into the amino- and hydroxyacid constituents of valinomycin (MacDonald, 1960). Label from [1-^{14}C]-L-valine was incorporated to an equal extent into both D- and L-valine and to a lesser extent into D-α-hydroxyisovalerate, being present only in the carboxyl groups of these hydrolytic products. Since the label from [1-^{14}C]-D-valine was incorporated only very slightly, it was concluded that the carbon skeleton of L-valine was incorporated intact into the D-valyl,

L-valyl, and D-α-hydroxyisovaleryl residues, suggesting that the antibiotic was probably not made from dipeptides which undergo a replacement of the free amino group by a hydroxyl prior to cyclization. It has been suggested by Mauger (1968) that valinomycin could be formed by trimerization of cyclo-[D-valyl-L-valyl-] with subsequent hydroxyacid insertion.

Action in vitro McMurray and Begg, as long ago as 1959, found that concentrations of valinomycin as low as 10^{-7} M completely uncoupled oxidative phosphorylation in rat liver mitochondria. Pressman (1963) confirmed these observations and noted that valinomycin was quite different from 2,4-dinitrophenol as an uncoupler, since ejection of protons from mitochondria during electron transport occurred in the presence of valinomycin. Subsequent studies by Moore and Pressman (1964) indicated that the ability of valinomycin to uncouple oxidative phosphorylation was dependent upon the presence of K^+. The pioneering work of Pressman, Tosteson, Chappell, and Crofts and others (reviewed by Harold, 1970, and Henderson, 1971) led to the recognition that valinomycin forms a lipid-soluble, positively-charged complex with K^+ which diffuses rapidly across any lipid-phase including both natural and artificial membranes. Turnover numbers for valinomycin in mitochondria at room temperature are several hundred K^+ ions s^{-1} (Pressman *et al.*, 1967). The concentration dependence indicates that a single K^+ ion is carried by each valinomycin molecule, i.e. valinomycin acts as a mobile carrier, (Haynes *et al.*, 1969). Since K^+ ions equilibrate with the aqueous phase, the net effect of valinomycin is to catalyse the movement of K^+ across the membrane in a direction dictated by the electrochemical potential. Hence, Mitchell and Moyle (1965) considered uncoupling by valinomycin to result from the uptake of K^+ as a secondary electrochemical response to a metabolically-driven extrusion of protons.

The order of cation selectivity of valinomycin is $H^+ > Rb^+ > K^+ > Cs^+ > Na^+$. However, since the H^+ concentration is low at physiological pH values, valinomycin acts as a selective carrier for K^+ (Andreoli *et al.*, 1967). The apparent complexing preference for K^+ to Na^+ is 10 000:1 (Pressman *et al.*, 1967). Because of this extreme selectivity, and its ability to make membranes freely permeable to K^+, valinomycin has proved to be an extremely useful experimental tool for investigations of energy conservation and the coupling of energy to active transport in a wide range of biological systems. Valinomycin/K^+ have been used to monitor the magnitude of the potential across the membrane, to collapse such a potential, and even artificially to generate a transient potential across bacterial vesicle membranes. Citations of its usage are legion: the reader is referred to the reviews in this field mentioned in the Introduction to this chapter for further details.

(ii) Enniatins

Producing species Gaumann *et al*. (1947) isolated a new crystalline antibiotic substance called enniatin from the fungus *Fusarium orthoceras* var. *enniatin*. Enniatins have subsequently been isolated from other strains of Fusaria.

Structure Quitt *et al*. (1963) and Plattner *et al*. (1963) showed that the enniatins are a family of cyclohexadepsipeptides possessing alternating residues of D-hydroxyisovalerate and an *N*-methylamino acid (Figure 5). The *N*-methyl substituted amino acids of enniatins A, B, and C respectively are isoleucine, valine, and leucine. The structural requirements for antibiotic activity have been described in the introduction to this section.

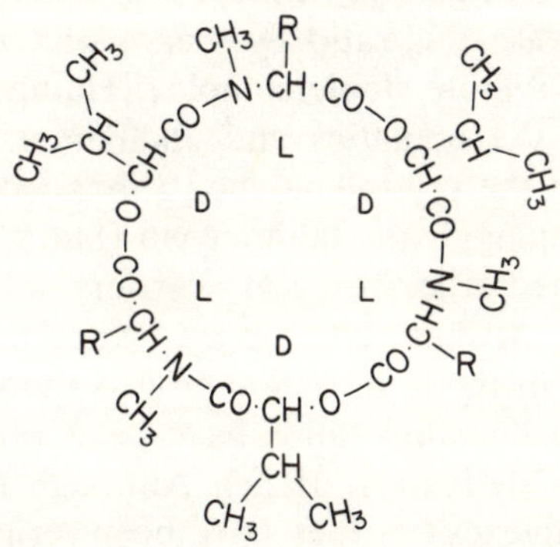

Figure 5 Enniatins and beauvericin. R = isoleucine, enniatin A; R = valine, enniatin B; R = leucine, enniatin C; R = benzyl group ($C_6H_5CH_2$—), beauvericin

A stoichiometric complex of enniatin B with K^+ was isolated (Shemyakin *et al*., 1967; Wipf *et al*., 1968) and the crystal structure of the enniatin B complex with KI was determined by X-ray analysis (Dobler *et al*., 1969). On the basis of ORD and CD, enniatins A, B, and C were shown to exist as equilibrium mixtures of two conformations P (a polar form) and N (a non-polar form); the composition of the mixture depending on the polarity of the solvent and the presence of cations (Shemyakin *et al*., 1969).

Action in vitro On its isolation, enniatin was recognized as an antibiotic being particularly active *in vitro* against several mycobacteria (Gaumann *et al*., 1947). It has subsequently been shown to inhibit the growth of other Gram-positive bacteria and fungi. Its antibiotic activity was attributed to its function as an ionophore when it was shown to resemble valinomycin in its structure and mode of action (Shemyakin *et al*., 1965).

Enniatin facilities K^+ diffusion across lipid bilayers (Mueller and Rudin, 1967) and induces K^+ uptake by mitochondria (Lardy *et al*., 1967;

Pressman, 1968). The ion selectivity of enniatin B is $Rb^+ > K^+ > Cs^+ > Na^+ > Li^+$ (Mueller and Rudin, 1967; Ovchinnikov, 1974) and of anniatin A is $K^+ > Na^+$, Rb^+, $Cs^+ > Li^+$, but, unlike valinomycin, there is only a difference of an order of magnitude in the stability constants for K^+ and Na^+. This comparatively low structural cation specificity is related to the difference in conformational flexibility between valinomycin, which has a rigid, H-bond-stabilized complex conformation, and the enniatins which display considerable conformational mobility (Ovchinnikov *et al.*, 1972). However, unlike the related enniatin-type compound, beauvericin (Section 4.8.3.2(iii)), the other enniatins do not form complexes with divalent metal cations such as Ba^{2+} and Ca^{2+}. An important consideration in the stability of complexes with polyvalent cations is the charge:surface area ratio. For the enniatins A, B, and C, with their aliphatic amino acid residues, this ratio is rather high. Beauvericin is, however, able to decrease this ratio by means of the large phenylalanine residues and thus form a more stable complex (Hamilton *et al.*, 1975).

Complexes between the enniatins and alkali metal ions are equimolar both in solution and in the crystalline state (Shemyakin *et al.*, 1969; Wipf *et al.*, 1968) and, by analogy with valinomycin (Stark and Benz, 1971), the enniatins were ascribed the role of carriers of metal ions across membranes. However, it has since been suggested that the enniatin ionophores may differ in mode of action from valinomycin since evidence of 2:1 and 3:2 ionophore:cation ratios have been obtained (Ivanov *et al.*, 1973; Ovchinnikov, 1974; Ivanov, 1975). Although the stability constants of these 'sandwich complexes', as they have been termed, are 1–2 orders of magnitude lower than those of the equimolar complexes, it is still thought possible that they may participate in ion transport because in these complexes the cation is more effectively screened from interaction with the solvent and anions.

(iii) Beauvericin

Producing species Hamill *et al.* (1969) isolated the biologically active metabolite beauvericin from the entomopathogenic fungus *Beauveria bassiana*, and identified it as a regular cyclohexadepsipeptide. It has subsequently been isolated from another entomopathogenic fungus *Paecilomyces fumosoroseus* (Bernardini *et al.*, 1975).

Structure Its structure of [D-α-hydroxyisovaleryl-L-N-methylphenylalanyl-]$_3$ places it as a member of the enniatin series of depsipeptides (Figure 5). The molecular structure of the barium picrate complex of beauvericin and of hydrated beauvericin have been determined by X-ray crystallography (Hamilton *et al.*, 1975; Geddes and Akrigg, 1976). The crystal structure of the hydrated beauvericin was similar to that found in the barium picrate complex (apart from possible differences in rotation in the side-chains) and also to the conformation which has been proposed to exist in polar solvents from spectroscopic studies and conformation analyses (Ovchinnikov *et al*., 1971*a*).

A general feature of the naturally-occurring ionophorous antibiotics is that they undergo conformational changes during cation binding which involve the breakage and formation of H-bonds. In beauvericin (where there is no possibility of intramolecular H-bond formation) the complex is formed not by adjusting the conformation to fit the ion but by two molecules surrounding the cation–anion complex. Obviously, the nature of the anion will affect the stability of the complex. Hamilton *et al*. (1975) pointed out that their proposed structure could also account for the observed second-order concentration dependence of beauvericin in lipid bilayers (Yafuso *et al*., 1974) and also for the apparent charge of +1 for calcium ions in beauvericin-mediated transport into bacterial chromatophores (Prince *et al*., 1974).

Biosynthesis Nothing is known.

Action in vitro West and Briggs (1968) demonstrated the existence of a toxic agent in the culture fluids of the insect pathogen *B. bassiana* as a result of injecting such fluids into larvae of *Galleria mellonella*. Hamill *et al*. (1969) isolated beauvericin from such culture fluids and demonstrated its toxicity to brine shrimp, Gram-positive bacteria, fungi, and mosquito larvae. In studies with insect cells cultured *in vitro*, Vey *et al*. (1973) showed that beauvericin inhibited the migration of *Leucophaea maderae* cardiac cells ane altered cell morphology, particularly that of the nucleus. Beauvericin synthesized by Ovchinnikov *et al.* (1971*a*) had the same specific rotation, IR-, UV-, NMR- and mass-spectra as the natural product and exhibited high antimicrobial activity. It also formed 1:1 complexes with alkali metal ions with stability constants, in ethanol, of 100 for Li^+, 300 for Na^+, 3100 for K^+, and 3500 for Rb^+ and Cs^+. As a result of extensive ORD and CD analysis Ovchinnikov *et al*. (1971*a*) concluded from a comparative study that beauvericin stands very close in conformational parameters and complexing properties to the thoroughly investigated enniatin antibiotics (see immediately previous section).

Roeske *et al*. (1974) found that, in addition to its affinity for the alkali metal cations, beauvericin has a high affinity for Ca^{2+}. They also showed the cation specificity in the presence of the anion picrate was $Rb^+ > Ba^{2+} > K^+ > Na^+ >> Ca^{2+} >>> Li^+$. It was suggested that the active form of beauvericin in ion transport might be a dimer or tetramer containing two (or four) cations and two (or four) beauvericin molecules.

Such a complex was proposed to possess binding sites for anions to explain the observation of Estrada-O *et al.* (1972) that the cation specificity of beauvericin in mitochondria depended on the anions present. However, in experiments investigating beauvericin-dependent cation transport into liposomes and chromatophores Prince *et al.* (1974) showed the selectivity of $Ca^{2+} > K^+ > Cs^+ > Li^+ > Na^+$ to be similar to that found by Dorschner and Lardy (1968) who demonstrated that, in the presence of K^+, 5 μM beauvericin could uncouple oxidative phosphorylation in rat liver mitochondria. In addition, Vallejos *et al.* (1975) demonstrated the inhibition of photophosphorylation in spinach chloroplasts where 25 μM beauvericin induced light-dependent uptake of Ca^{2+}.

Ivanov *et al.* (1973) have described the formation of 'sandwich complexes' of 2:1 and 3:2 ratios of ionophore:cation for the enniatin group ionophores, including beauvericin, and have suggested that, by stacking such complexes, the mode of action of this group of ionophores may be similar to the channel-forming ionophores rather than the depsipeptide carriers typified by valinomycin (Section 4.8.3.2(i)). Certainly beauvericin, with the ability to have its cation specificity altered after synthesis by the binding of different anions, represents a more advanced concept of function that that of valinomycin the cation specificity of which is predetermined by its structure and is not affected by the anions present. In a way, beauvericin may be considered to be intermediate between valinomycin and alamethicin (Section 4.8.3.1.(ii)). The second-order dependence on both cation and ionophore means that beauvericin will go from negligible transport to full saturation over a much smaller concentration range than would valinomycin, thus approaching the on–off behaviour of alamethicin which is sixth-order with respect to cation concentration.

(iv) Monamycins

Producing species Hassall and Magnus (1959) described the isolation from cultures of *Streptomyces jamaicensis* of a crystalline preparation which inhibited the growth of various Gram-positive bacteria.

Structure Monamycin is a mixture of at least 15 compounds, including six chlorinated derivatives (Bevan *et al.*, 1970). All are composed of one hydroxy- and five amino-acids, the configurations of which alternate between L and D around an 18-membered ring. The amino acid constituents include two unusual cyclic amino acids, L-hydroxypiperidazine carboxylic acid and D-piperidazine carboxylic acid (Bevan *et al.*, 1971; Hassall *et al.*, 1971*a,b*). The major and most active component, monamycin D_1, is shown in Figure 6.

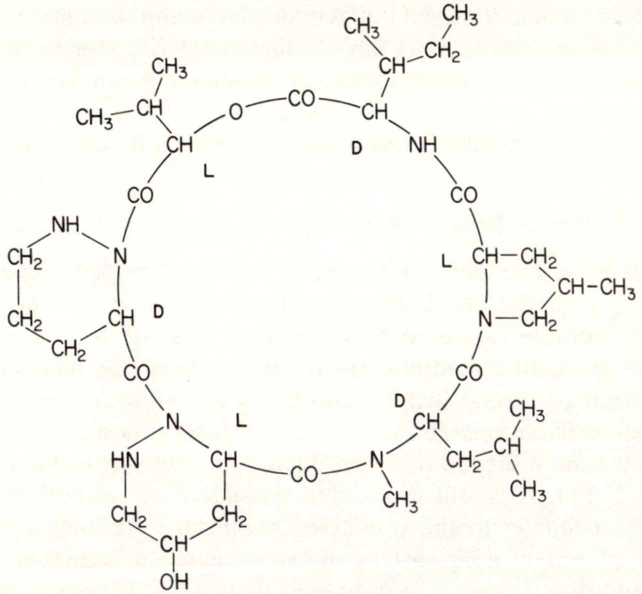

Figure 6 Monamycin D_1

Biosynthesis Bevan *et al*. (1969) isolated the unusual amino acid D-isoleucine from monamycins and, by following the fate of ^{14}C-labelled amino acids added to the culture medium, they suggested that L-isoleucine and L-alloisoleucine were precursors of these D-isoleucine residues (Davies *et al*., 1973). Thus, in contrast to the usual pattern of amino acid incorporation into cyclic peptide antibiotics, in which it has been suggested that the D residues arise from C_2 epimerization of the corresponding members of the L-series (Bycroft, 1969), Davies *et al*. (1973) suggested that D-amino acids are incorporated directly into monamycin. Hall *et al*. (1973) demonstrated that the naturally-occurring chlorinated monamycins were replaced by the corresponding brominated analogues if *S. jamaicensis* was grown in a low-chloride medium supplemented with NaBr.

Action in vitro Hall *et al*. (1973) demonstrated by diffusion and dilution assays that the antibiotic activity of the monamycins was inversely proportional to their unsolvated molecular weights, the bromo- analogues being the least active of the series. The bactericidal level of monamycin D against *Staphylococcus aureus* was 5 µg ml^{-1}. The monamycins form stable complexes with K^+, Rb^+, and Cs^+ but not readily with Na^+ and Li^+ (Hall, 1970). However, monamycin (<50 µg ml^{-1}) can cause haemolysis of human erythrocytes in a manner similar to that caused by tyrocidin-HCl and cetylpyridinium chloride which is in contrast to ionophores such as valinomycin, gramicidin A and nigericin (Hall and Sukan, 1972). This haemolysis was, however, markedly decreased by K^+, less so by Rb^+ and Cs^+ and only fractionally by Na^+ and Li^+. Whether the antibiotic activity of the monamycins is simply due to a lytic effect of the cell membrane or whether they act classically as ionophores remains to be established.

3 The Role of Peptide Ionophores *in vivo*

Despite extensive data concerning the functions of peptide ionophores *in vitro* there is a paucity of information regarding their role *in vivo*. The function of peptide antibiotics in general has been the subject of considerable speculation and discussion and previously popular ideas which have now been discarded include antibiotics as evolutionary relics, waste products of cellular metabolism, reserve food materials, spore coat components, and breakdown products of cellular macromolecules (Weinberg, 1971). It is still considered possible that antibiotics provide a competitive advantage to the producing organism by killing, or inhibiting the growth of other organisms in the immediate environment (Gottlieb, 1976). In addition, it has been suggested that antibiotic synthesis provides a means of avoiding cell death due to unbalanced growth (Weinberg, 1971) or occurs to eliminate certain toxic antimetabolites accumulated as exponential growth terminates (Woodruff, 1966). However, these last three suggestions require the antibiotic to be ineffective against the producing organism, and this is often not so.

In view of the fact that many of these antibiotics are seemingly secondary metabolites synthesized by the cells at the end of exponential growth, it is most unlikely that they are involved in the maintenance of balanced growth, but rather that they play a key role in the termination of vegetative growth and allow sporulation to take place. The role of peptide antibiotics in sporulation has been discussed extensively (see Katz and Demain, 1977 and references therein) and more specifically the role of ionophores in spore germination was investigated by Dring and Gould (1974). The precise role of peptide antibiotics during sporulation still remains obscure and includes suggestions that they may be involved in ion transport, in ensuring the commitment of the cell to sporulation, may be gratuitous inducers of certain proteases synthesized during sporulation (Hanson *et al.*, 1970) or may be required to participate in cell lysis allowing the release of free spores (Paulus, 1967). It has been suggested that valinomycin, together with high levels of K^+, may stimulate germination by facilitating K^+ influx into the spore (Dring and Gould, 1974). However, there is still no obvious known function *in vivo* of peptide ionophores and thus further discussion of this question is superfluous.

4 PEPTIDE IONOPHORES FOR TRIVALENT CATIONS

1 General Comments

This group of ionophores binds primarily with ferric ions though some may also chelate with other metals (Al, Cr, Ga, vanadyl, Co^{III}). They have been variously referred to as sideramines, siderochromes, and ironophores though the term *siderophore* (Lankford, 1973) seems more appropriate and will be used here. Their functions are almost entirely associated with the solubilization and subsequent uptake of iron into microorganisms. Iron, when in an aqueous environment, has a solubility of about 10^{-18} M (the solubility product of $Fe(OH)_3$ being approximately 10^{-39}). Even in an *in vivo* situation, where a pathogenic bacterium must acquire iron from the host tissues, iron is not freely available but is found as ferritin or transferrin. Indeed, there is some evidence to suggest that the virulence and even the degree of pathogenicity of a bacterium may be regulated by its ability to win iron from the host (Sussman, 1974; Hegenauer and Saltman, 1975; Weinberg, 1975). The physiological problems faced by microorganisms acquiring iron are, thus, unlike those they face with any other nutrient and these problems have been ably reviewed in recent years by Lankford (1973), Neilands (1974), and Emery (1978).

The principal feature which distinguishes this group of ionophores from those for mono- and divalent ions is that their synthesis is derepressed when the producing organism is grown with a suboptimal amount of iron. The need for iron drives the organism to produce often massive levels of the chelators in an attempt to sequester whatever iron may be available.

The mechanisms by which the ionophores transport ferric ions, after having solubilized them in the exterior environment, have not been studied in as great detail as some of the ionophores for mono- or divalent ions. It is imagined, however, that the whole complex moves physically across the membrane rather than the siderophore forming a pore or channel within the membrane to permit the facilitated passage of the iron.

Little is also known of how the iron, when transported to the interior surface of the membrane, is released from its carrier. The stability constants for most of the complexes are about 10^{30} and thus are higher than many synthetic chelators such as *o*-hydroxyquinoline, EDTA, etc. Release of iron thus cannot be by a change in the conformation of the ionophore upon reaching a lipid/water interface. Some input of energy is demanded.

Two means of release have so far been described. In the first, the iron is reduced to ferrous ion and, as this is only bound very weakly by the ionophore, it can then be picked up by the various proteins requiring iron in the cell (Brown and Ratledge, 1975). The second mechanism involves a partial degradation or hydrolysis of the carrier thereby destroying its integrity and thus its ability to bind the cation.

The list of ferric ionophores continues to increase as more and more organisms are examined. It may be anticipated that every organism which requires iron for its growth, and only the lactobacilli appear to be devoid of it (see Neilands, 1966, 1977), will be found to produce its own siderophore. A wide diversity of forms have arisen to achieve the creation of an appropriate hexadentate structure with which to bind ferric ions. The purpose of the diversity is probably to confer specificity on the process and prevent competing organisms acquiring iron without producing their own siderophores. Most of the structures employ amino acids, amines, or other nitrogen-containing compounds derived from amino acids. A complete list of the siderophores has been given by Neilands and Ratledge (1979).

Siderophores fall into two main categories according to the nature of their iron binding centre: those based on hydroxy- or dihydroxybenzoic acids and those using hydroxamic acid (—NOH.CO—) groupings. With both types of compound, the chelating centre is created by the provision of six oxygen atoms (see Anderson *et al*., 1976). (There is so far only one known exception to this and this is with mycobactin where a nitrogen atom of an oxazoline ring and five oxygens are used.)

Both types of chelate are exceptionally stable. The biosynthesis of the hydroxamate group has been discussed by Lankford (1973) and Emery (1974) with reference to the simplest naturally occurring hydroxamic acid, hadacidin (*N*-formyl-*N* hydroxyglycine), though there is no known role for this compound in iron transport.

In compiling the list of siderophores we have thought it artificial to exclude a compound on the basis of it being derived from an amine rather than an amino acid. Thus, if we had applied this principle we would have had to exclude arthrobactin

$$[CH_3.CO.NOH.(CH_2)_5.NH.CO.CH_2]_2C(OH)(COOH)$$

but include aerobactin

$$[CH_3.CO.NOH.(CH_2)_4.CH(COOH).NH.CO.CH_2]_2C(OH)(COOH)$$

Clearly, the iron-binding centres of both molecules are the same (i.e. two hydroxamate groups plus the free COOH and OH groups on the citrate moiety). As neither of these compounds is a peptide it will now be realized that we have widened the scope of the chapter to include those siderophores which have either an amino acid or an amine as part of their structure. To exclude compounds on the basis of the amino acid, or amine, not forming part of the chelating nucleus would again be artificial as this would have excluded enterochelin (the cyclic trimer of

2,3-dihydroxybenzoylserine) as the serine moieties serve to hold the three aromatic rings in the correct spatial configuration but do not participate in the chelating centre. Nevertheless the serine residues form an intrinsic and vital part of the structure. Having then included enterochelin we have felt obliged to include the other benzoyl amino acid derivatives for the sake of consistency.

2 Phenolate Siderophores

Structures and producing organisms The principal phenolic chelator is enterochelin, the cyclic triester of 2,3-dihydroxybenzoyl-L-serine (Figure 7) which is found in Enterobacteria (Pollack and Neilands, 1970; O'Brien and Gibson, 1970). 2-*N*-6-*N*-Di-(2,3-dihydroxybenzoyl)-L-lysine (Figure 8) has been isolated from *Azotobacter vinelandii* (Corbin and Bulen, 1969) and N^1N^8-*bis*-(2,3-dihydroxybenzoyl)-spermidine (Figure 9) and salicyloyl-

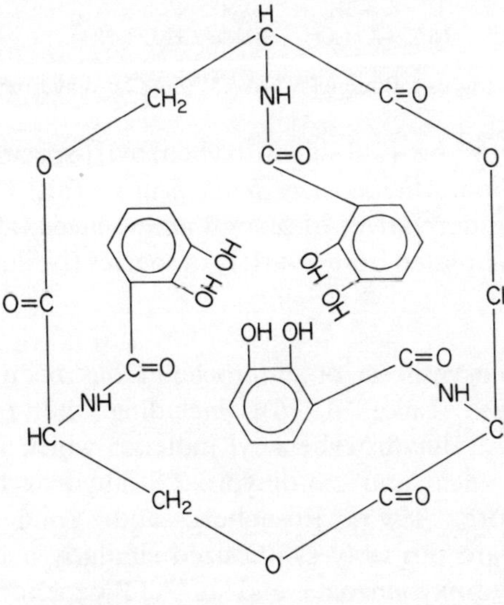

Figure 7 Cyclic triester of 2,3-dihydroxybenzoyl-L-serine (Enterochelin)

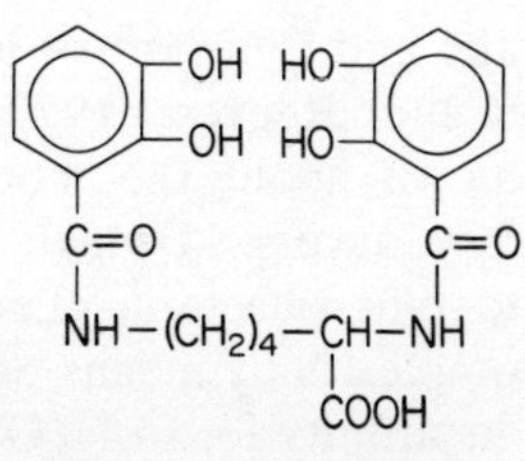

Figure 8 2-*N*,6-*N*-Di(2,3-dihydroxybenzoyl)-L-lysine

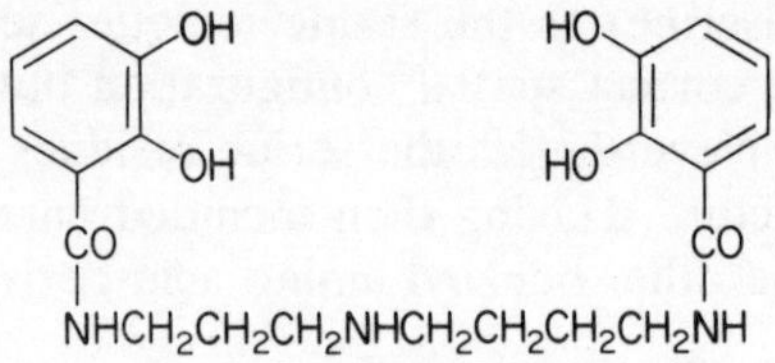

Figure 9 N^2,N^8-*bis*-(2,3-Dihydroxybenzoyl) spermidine

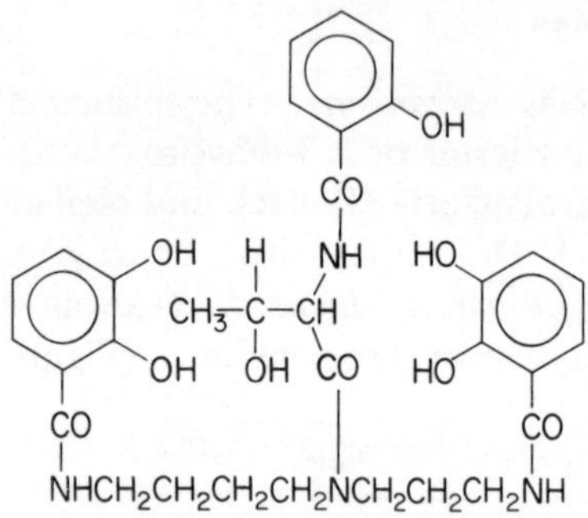

Figure 10 2-Hydroxybenzoyl-*N*-L-threonyl-N^4-[N^1,N^8-*bis*-(2,3-dihydroxybenzoyl)]spermidine

N-L-threonyl-N^4-[N^1,N^8-*bis*-(2,3-dihydroxybenzoyl)]-spermidine (Figure 10) have been isolated from *Micrococcus denitrificans* (Tait, 1975). Substituted 2,3-dihydroxybenzoyl derivatives of several amino acids (glycine, serine and threonine) have been isolated from a variety of species (Neilands and Ratledge, 1980).

Biosynthesis The biosynthesis of enterochelin has been fully elucidated (see Greenwood and Luke, 1976), including the pathway for the biosynthesis of the 2,3-dihydroxybenzoyl moieties which is from chorismic acid via isochorismic acid and 2,3-dihydro-2,3-dihydroxybenzoic acid (for reviews see Lankford, 1973; Rosenberg and Young, 1974). Other phenolate chelators are probably synthesized similarly using the particular amino acid, 2,3-dihydroxybenzoate and an ATP-dependent synthetase as the final step (*cf.* Brot and Goodwin, 1968).

Function in vivo The uptake and transport of iron by enterochelin is well documented (see Pugsley and Reeves, 1977, and references quoted therein) though little is known about the mode by which the complex traverses the cell boundary layers. Uptake of siderophores besides enterochelin can occur into enterobacteria (Leong and Neilands, 1976). Release of iron from the complex is by means of an esterase which cleaves the chelator and decreases its affinity for iron (O'Brien *et al.*, 1971).

Little is known of the functions of the other phenolates. Some of the substituted 2,3-dihydroxybenzoyl amino acids may be degradation products of larger molecules, as their ability to hold iron in solution against competing anions such as phosphate or citrate is probably minimal (see Ratledge *et al*., 1974). Their participation in iron transport processes have only been established where phosphate ions have been deliberately excluded.

3 Hydroxamate Siderophores

(i) Compounds Based on Ornithine, Putrescine, or Diaminopropane

Structures and producing species The two most abundant chelators in this category are the ferrichrome and fusarinine family and rhodotorulic acid.

Ferrichrome (Figure 11) is produced by Ascomycetes, Basidiomycetes, and many fungi imperfecti, including *Aspergillus, Neurospora, Paecilomyces, Penicillium, Spicaria, Ustilago*, and *Sphacelotheca* (see Neilands and Ratledge, 1980). Several minor variations of the molecule occur leading to a plethora of names: ferrichrome A, ferrichrome C, ferrichrysin, ferricrocin, ferrirubin, and ferrirhodin.

A non-cyclic variant of ferrichrome is coprogen (Figure 12) which occurs in *Neurospora crassa* and several species of *Penicillium* (Neilands, 1966). This structure contains a 5-hydroxy-3-methylpent-2-enoic acid moiety

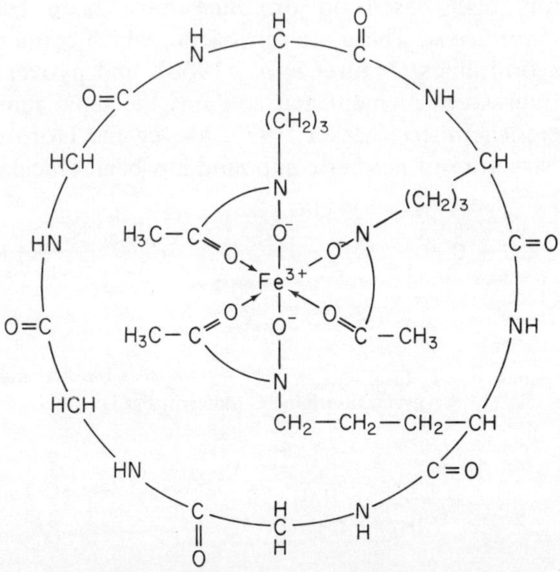

Figure 11 Ferrichrome

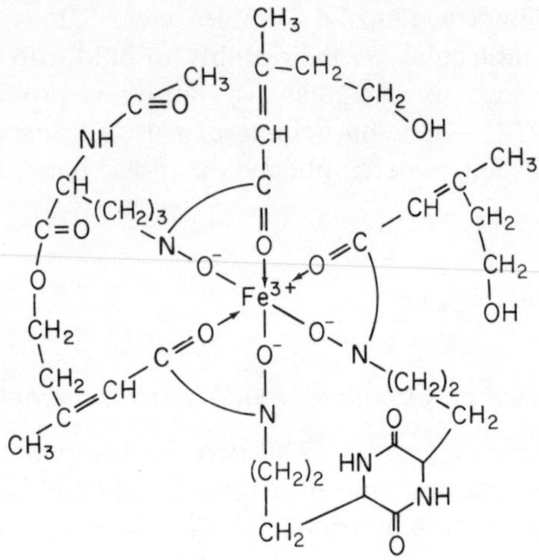

Figure 12 Coprogen

which also occurs, though as the *trans*-isomer, in dimerum acid (see below) and in the fusarinine (or fusigen) group of chelators. Fusarinines which occur in several fungi can be linear or cyclic dimers or trimers (see Figure 13) involving the above group acylated to the δ-*N*- of δ-*N*-hydroxyornithine (Diekmann, 1973). Triacetylfusarinines have been also isolated from an unknown *Penicillium* sp. (Moore and Emery, 1976) and from *Aspergillus* spp. (Diekmann and Krezdorn, 1975).

Two chelators also based on ornithine have been recovered from *Pseudomonas fluorescens*. These are ferribactin, which contains both L- and D- δ-*N*-hydroxyornithines (Maurer *et al.*, 1968), and pyoverdin which is a yellow-green fluorescent pigment and contains the same moiety though of unspecified stereochemistry (Meyer, 1977; Meyer and Hornsperger, 1978). The complete structure of neither compound has been elucidated.

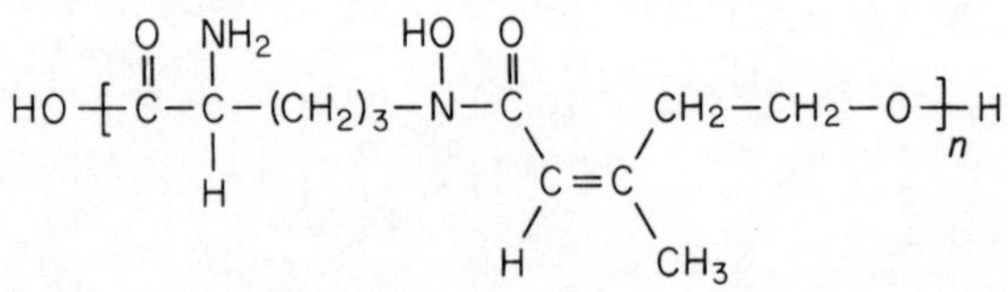

Figure 13 Fusarinines. $n = 1$, fusarinine; $n = 2$, fusarinine A; $n = 3$, fusarinine B; $n = 3$, cyclo, fusarinine C (deferrifusigen)

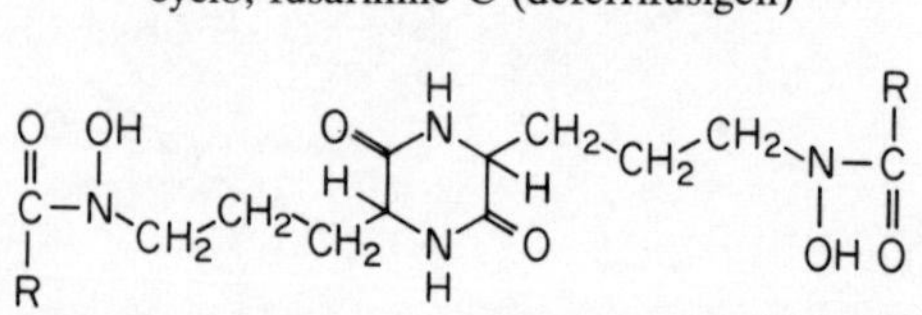

Figure 14 R = H, rhodotorulic acid; R = $-CH{=}CH(CH_3)CH_2CH_2OH$, dimerum acid

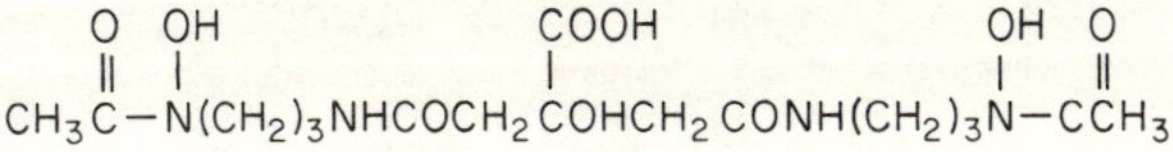

Figure 15 Schizokinen

Two chelators based on diketopiperazine, derived by cyclization of two ornithines, are known: rhodotorulic acid (Figure 14, R = H) which occurs in many yeasts, including species of *Rhodotorula, Sporidiobolus*, and *Sporobolomyces* (Atkin *et al*., 1970) and dimerum acid (Figure 14, R = $-CH{=}CH(CH_3)CH_2CH_2OH$) which is found in *Fusarium dimerum*.

Schizokinen (Figure 15) is the only known chelator to be based on the unusual diamine, diaminopropane (the next lowest homologue to putrescine). This siderophore was originally isolated from *Bacillus megaterium* (see Lankford, 1973; Byers, 1974) but has since been recovered from a number of blue-green algae (Simpson and Neilands, 1976; Murphy *et al*., 1976). It is essentially a substituted citric acid and is thus related to aerobactin and arthrobactin (see below).

Biosynthesis There has been a variety of studies on the biosyntheses of these compounds and the reader is referred to the review by Emery (1974) for details of these (see also Anke and Diekmann, 1974). More than one type of siderophore may be produced by a single species, as has been found by Diekmann and Krezdorn (1975) with several *Aspergillus* species producing fusarinines, triacetylfusarinines, and ferricrocin.

Function in vivo The documentation on iron uptake from these chelates is prolific (see Emery, 1974, 1978). The specificity of uptake can be variable (see Wiebe and Winkelmann, 1975; Arceneaux and Byers, 1976). There is an enigma with ferrioxamine A which though produced by *Ustilago sphaerogena* is not transported by it (Emery, 1971).

The mode of iron release may be by esterases acting on fusarinines as in *Fusarium roseum* or *Penicillium* sp. (Emery, 1976) or by reduction of the iron via an NADH:oxidoreductase as found in *Neurospora crassa* with coprogen and in *Aspergillus fumigatus* with ferrichrome (Ernst and Winkelmann, 1977).

(ii) Compounds Based on Lysine or Cadaverine

Structures Ferrioxamines are a family of compounds based on a trimer of *N*-hydroxy-*N*-succinylcadaverine and may be either cyclic (Figure 16) or linear (Figure 17).

Mycobactins (Figure 18) are complex lipophilic compounds involving two lysine moieties, one of which is ring closed upon itself, the other is

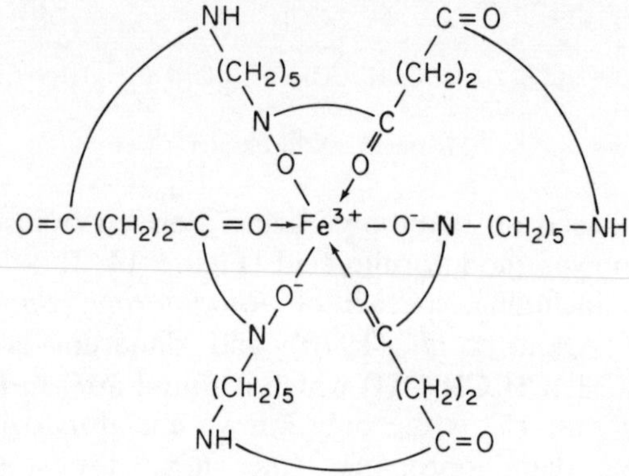

Figure 16 Ferrioxamine E

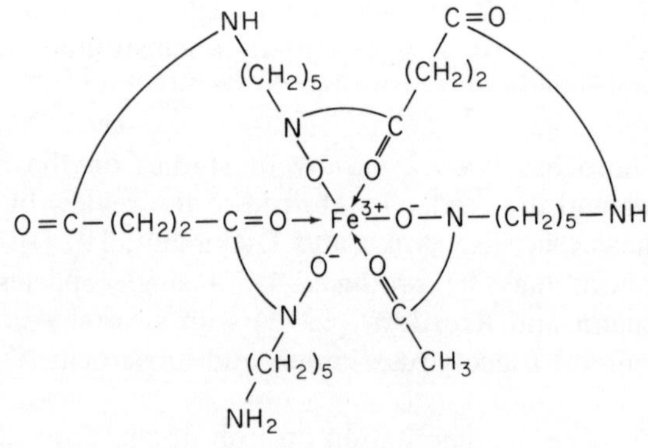

Figure 17 Ferrioxamine B

linked to an oxazoline ring derived from serine. There are numerous variations of the side-chains according to the producing species and consequently they have been used as chemotaxonomic markers (Snow, 1970; Ratledge and Patel, 1976).

Lysine and cadaverine, as their ε-*N*-hydroxy-ε-*N*-acetyl derivatives, occur condensed to the 1- and 6-carboxyls of citric acid, thus generating aerobactin (Gibson and Magrath, 1969) and arthrobactin, (Linke *et al.*, 1972) (see Section 4.8.4.1). They are, therefore, similar in structure to schizokinen (Figure 15).

Producing species Ferrioxamines, both linear and cyclic, are produced by various actinomycetes (*Micromonaspora, Streptomyces* spp., and *Nocardia* sp.) (Diekmann, 1973).

Mycobactins are produced by all mycobacteria so far examined except for *M. paratuberculosis* which requires it as an essential growth factor

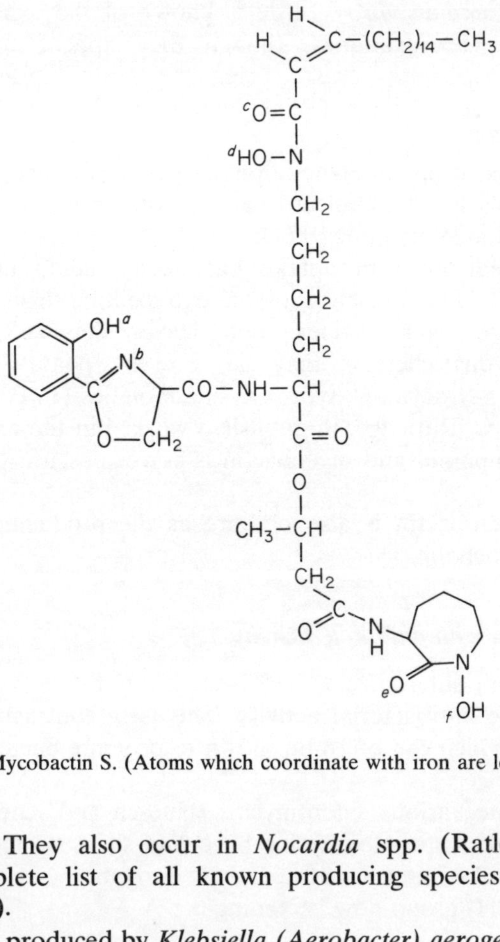

Figure 18 Mycobactin S. (Atoms which coordinate with iron are lettered *a* to *f*)

(Snow, 1970). They also occur in *Nocardia* spp. (Ratledge and Patel, 1976). A complete list of all known producing species is provided by Ratledge (1980).

Aerobactin is produced by *Klebsiella (Aerobacter) aerogenes* mutant 62/1 and arthrobactin by *Arthrobacter pascens.*

Biosynthesis There is little known about the biosynthesis of the ferrioxamines though the basic steps are probably similar to those occurring with the ferrichromes (Emery, 1974). The decarboxylation of lysine to give cadaverine may occur before or after its condensation with succinyl-CoA (see Schafft and Diekmann, 1978).

Mycobactins, too, are little studied biosynthetically. Tateson (1970) showed that there was at least one free intermediate between lysine and its incorporation into mycobactin. This was not ε-*N*-hydroxylysine. Studies using various derivatives of *N*-seryllysine and salicylic acid failed to elicit any synthetic activity in cell extracts (Brown, 1975). Salicylate though is readily incorporated into mycobactin using growing cells (Ratledge and Hall, 1970).

Function in vivo and in vitro Little is known of the exact mode of iron uptake with the ferrioxamines, though they appear to be taken up specifically in *Neurospora crassa* using the same binding sites as the ferrichrome/coprogen chelates (Winkelmann and Zahner, 1973; Ernst and Winkelmann, 1977).

Ferrioxamine B (as the methane sulphonate of the desferri form—Desferal) is used extensively for the clinical treatment of chronic or acute iron overloading (Jacobs and Worwood, 1974).

Mycobactin-mediated iron uptake has been studied in *M. smegmatis* (Ratledge, 1976). The presence of an extracellular iron chelator, other than mycobactin, which does not leave the cell, is required. The nature of this chelator may be a cyclic penta- or hexapeptide containing three ε-*N*-acyl-ε-*N*-hydroxylysine moieties (L. P. Macham, M. C. Stephenson, and C. Ratledge, unpublished work) but the exact relationship between this compound and mycobactin is as yet obscure (Stephenson and Ratledge, 1979).

The role of aerobactin is also obscure as the producing organism also synthesizes enterochelin.

(iii) Siderophores with antibiotic activity

The general term, sideromycin, had been given to substances containing iron and showing antibacterial activity. This is in contrast to the majority of siderophores which can often be shown to promote bacterial growth.

Structures Of the various sideromycins (Nüesch and Knüsel, 1967) only ferrimycin A, and some albomycins have had their structures elucidated. Ferrimycin is a substituted ferrioxamine B derivative (see Figure 17) with the terminal —NH_2 group now becoming

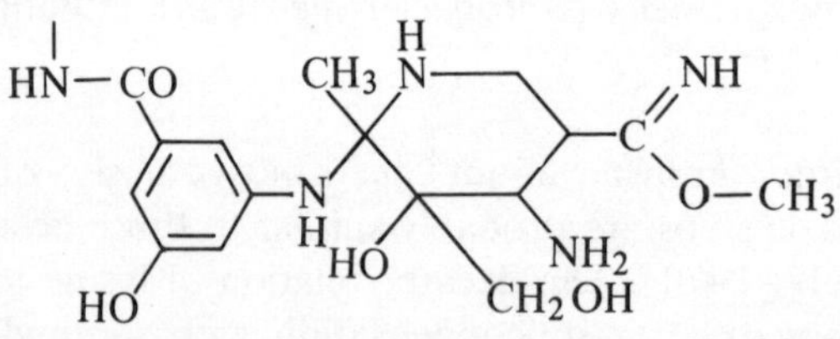

Albomycins are substituted ferrichromes and three types, δ_1, δ_2, and ε have been characterized. Each has a 3-methylcytosine derivative linked to the cyclohexapeptide via a —SO_2— bridge (see Neilands and Ratledge, 1979).

Sideromycins of unknown structures include grisein (which is probably an albomycin), succinimycin, danomycin, antibiotic A22765 and antibiotic LA5352 (Nüesch and Knüsel, 1967).

Producing organisms Ferrimycin is produced by *Streptomyces griseoflavus*, *S. lavendulae*, and *S. galilaeus* (Bickel *et al.*, 1960; Zahner, *et al.*, 1960). Albomycins are produced by *Actinomyces* (=*Streptomyces*) *griseus* (Gause and Brazhnikova, 1951).

Biosynthesis Nothing is known.

Function in vitro The antibiotics are active against Gram-positive and Gram-negative organisms with little mammalian toxicity. Bacteria, however, readily become resistant to them. It was originally thought that as the sideromycins and the siderophores were mutually antagonistic, the activity of the sideromycins was to deprive the susceptible bacteria of iron. It is now recognized, however, that the sideromycins merely use the siderophores' transport system to gain entry to the cell. Once inside, their antibacterial activity is solely due to the pendant group and not to the ferrioxamine or ferrichrome part (Knüsel *et al.*, 1969; Zimmerman and Knüsel, 1969).

There have been several recent reports of other iron-containing antibiotics, NRCS-15 (Haron, 1974), ferramidochloromycin (Shimi *et al.*, 1969), and ASK-753 (Shimi *et al.*, 1975) but as these are not antagonized by ferrioxamine (Shimi and Shoukry, 1976) they are probably gaining entry to the cell by other transport systems—perhaps those of other siderophores—and until their nature has been more fully examined they should not be regarded as true sideromycins.

REFERENCES

Anderson, B. F., Buckingham, D. A., Robertson, G. B., Webb, J., Murray, K. S., and Clarke, P. E. (1976). *Nature, Lond.*, **262**, 722

Andreoli, T. E., Tieffenberg, M., and Tosteson, D. C. (1967). *J. Gen. Physiol.*, **50**, 2527

Anke, T. and Diekmann, H. (1974). *Arch. Microbiol.*, **95**, 227

Apell, H. J., Bamberg, E., Alpes, H., and Läuger, P. (1977). *J. Memb. Biol.*, **31**, 171

Arceneaux, J. E. L. and Byers, B. R. (1976). *J. Bacteriol.*, **127**, 1324

Atkin, C. L., Neilands, J. B., and Phaff, H. J. (1970). *J. Bacteriol.*, **103**, 722

Bamberg, E., Kolb, H.-A., and Läuger, P. (1976). In *The Structural Basis of Membrane Function* (Y. Hatefi and L. Djavadi-Ohaniance, eds), Academic Press, New York and London, p. 143

Bamberg, E. and Läuger, P. (1977). *J. Memb. Biol.*, **35**, 351

Bauer, K., Roskoski, R., Kleinkauf, H., and Lipmann, F. (1972). *Biochemistry*, **11**, 3266

Bernardini, M., Carilli, A., Pacioni, G., and Santurbano, B. (1975). *Phytochemistry*, **14**, 1865

Bevan, K., Davies, J. S., Hall, M. J., Hassall, C. H., Morton, R. B., Phillips, D. A. S., Ogihara, Y., and Thomas, W. A. (1970). *Experientia*, **26**, 122

Bevan, K., Davies, J. S., Hassall, C. H., Morton, R. B., and Phillips, D. A. S. (1971). *J. Chem. Soc. C*, **1971**, 514

Bevan, K., Davies, J. S., Hassall, C. H., and Phillips, D. A. S. (1969). *J. Chem. Soc. Commun.*, **21**, 1246

Bickel, H., Gäumann, E., Nussberger, G., Reusser, P., Vischer, E., Voser, W., Wettstein, A., and Zahner, H. (1960). *Helv. Chim. Acta*, **43**, 2105

Bodanszky, M. and Perlman, D. (1969). *Science, N.Y.*, **163**, 352
Boheim, G. (1974). *J. Memb. Biol.*, **19**, 277
Boheim, G., Irmscher, G., and Jung, G. (1978). *Biochim. Biophys. Acta*, **507**, 485
Boheim, G., Janko, K., Leibfriz, D., Ooka, T., Konig, W. A., and Jung, E. (1976). *Biochim. Biophys. Acta*, **433**, 182
Brockmann, H. and Schmidt-Kastner, G. (1955). *Chem. Ber.*, **88**, 57
Brockmann, H., Springorum, M., Traxler, G., and Hofer, I. (1963). *Naturwissenschaften*, **50**, 689
Brot, N. and Goodwin, J. (1968). *J. Biol. Chem.*, **243**, 510
Brown, K. A. (1975). Ph.D. Thesis, University of Hull
Brown, K. A. and Ratledge, C. (1975). *FEBS Lett.*, **53**, 262
Bycroft, B. W. (1969). *Nature, Lond.*, **224**, 595
Byers, B. R. (1974). In *Microbial Iron Metabolism* (J. B. Neilands, ed.), Academic Press, New York, p. 83
Chappell, J. B. and Crofts, A. R. (1965). *Biochem. J.*, **95**, 393
Corbin, J. L. and Bulen, W. A. (1969). *Biochemistry*, **7**, 3734
Davies, J. S., Foley, M. H., Hassall, C. H., and Arroyo, V. (1973). *J. Chem. Soc. Commun.*, **20**, 782
Diekmann, H. (1973). In *Handbook of Microbiology*, Vol. 3 (A. I. Laskin and H. A. Lechevalier, eds), CRC Press, Cleveland, Ohio, p. 449
Diekmann, H. and Krezdorn, E. (1975). *Arch. Microbiol.*, **106**, 191
Dobler, M., Dunitz, J. D., and Krajewski, J. (1969). *J. Mol. Biol.*, **42**, 603
Dorschner, E. and Lardy, H. (1968). In *Antimicrobial Agents and Chemotherapy* (G. L. Hobby, ed.), American Society for Microbiology, New York, p. 11
Dring, G. J. and Gould, G. W. (1974). In *Spore Research 1973* (A. N. Barker, G. W. Gould, and J. Wolf, eds) Academic Press, London, p. 139
Eisenberg, M., Hall, J. E., and Meed, C. A. (1973). *J. Memb. Biol.*, **14**, 143
Emery, T. (1971). *Biochemistry*, **10**, 1483
Emery, T. (1974). In *Microbial Iron Metabolism* (J. B. Neilands, ed.), Academic Press, New York, p. 107
Emery, T. (1976). *Biochemistry*, **15**, 2723
Emery, T. (1978). In *Metal Ions in Biological Systems*, Vol. 7 (H. Sigel, ed.), Marcel Dekker, New York, in press.
Ernst, J. and Winkelmann, G. (1977). *FEBS Lett.*, **76**, 71
Estrada-O, S., Gomez-Louero, C., and Montal, M. (1972). *Bioenergetics*, **3**, 417
Freeman, H. C. (1973). In *Inorganic Biochemistry*, Vol. 1 (G. L. Eichhorn, ed.), Elsevier, Amsterdam, p. 121
Gaumann, E., Roth, S., Ettlinger, L., Plattner, P. A., and Nager, U. (1947). *Experientia*, **3**, 202
Gause, G. F. and Brazhnikova, M. G. (1951). *Novostin. Med. Akad. Med. Nauk. S.S.S.R.*, **23**, 3
Geddes, A. J. and Akrigg, D. (1976). *Acta Crystallogr. B,* **32**, 3164
Gibson, F. and Magrath, D. I. (1969). *Biochim. Biophys. Acta*, **192**, 175
Gordon, L. G. M. (1974). In *Drugs and Transport Processes* (B. A. Callingham, ed.), Macmillan, London, p. 251
Gordon, L. G. M. and Haydon, D. A. (1972). *Biochim. Biophys. Acta*, **255**, 1014
Gordon, L. G. M. and Haydon, D. A. (1975). *Phil. Trans. Roy. Soc. Lond., Ser. B*, **270**, 433
Gordon, L. G. M. and Haydon, D. A. (1976). *Biochim. Biophys. Acta*, **436**, 541
Gottlieb, D. (1976). *J. Antibiot.*, **29**, 987
Greenwood, K. T. and Luke, R. K. J. (1976). *Biochim. Biophys. Acta*, **454**, 285
Hall, M. J. (1970). *Biochem. Biophys. Res. Commun.*, **38**, 590
Hall, M. J., Handford, B. O., Hassall, C. H., Phillips, D. A. S., and Rees, A. V. (1973). *Antimicrobial Agents Chemother.*, **3**, 380
Hall, M. J. and Sukan, S. S. (1972). *Hacettepe Bull. Natl. Sci. Eng.*, **1**, 10
Hamill, R. L., Higgens, C. E., Boaz, H. E., and Gorman, M. (1969). *Tetrahedron Lett.*, **49**, 4255

Hamilton, J. A., Steinrauf, L. K., and Braden, B. (1975). *Biochem. Biophys. Res. Commun.*, **64**, 151
Hanson, R. S., Peterson, J. A., and Yousten, A. A. (1970). *Annu. Rev. Microbiol.*, **24**, 53
Harold, F. M. (1970). *Adv. Microbial Physiol.*, **4**, 45
Harold, F. M. (1972). *Bacteriol. Rev.*, **36**, 172
Harold, F. M. (1977). *Curr. Topics Bioenerg.*, **6**, 83
Harold, F. M., Altendorf, K. H., and Hirata, H. (1974). *Ann. N.Y. Acad. Sci.*, **235**, 149
Harold, F. M. and Baarda, J. R. (1968). *J. Bacteriol.*, **96**, 2025
Haron, B. M. (1974). *J. Antibiot., Tokyo*, **27**, 297
Hassall, C. H. and Magnus, K. E. (1959). *Nature, Lond.*, **184**, 1223
Hassall, C. H., Ogihara, Y., and Thomas, W. A. (1971*b*). *J. Chem. Soc. C*, **1971**, 522
Hassall, C. H., Morton, R. B., Ogihara, Y., and Phillips, D. A. S. (1971*a*). *J. Chem. Soc. C*, **1971**, 526
Haynes, D. H., Kowalsky, A., and Pressman, B. C. (1969). *J. Biol. Chem.*, **244**, 502
Hegenauer, J. and Saltman, P. (1975). *Science, N.Y.*, **188**, 1038
Henderson, P. J. F. (1971). *Annu. Rev. Microbiol.*, **25**, 393
Henderson, P. J. F., McGivan, J. D., and Chappell, J. B. (1969). *Biochem. J.*, **111**, 521
Hladky, S. B. (1974). In *Drugs and Transport Processes* (B. A. Callingham, ed.), Macmillan, London, p. 193
Hladky, S. B. and Haydon, D. A. (1970). *Nature, Lond.*, **225**, 451
Hladky, S. B. and Haydon, D. A. (1972). In *Molecular Mechanisms of Antibiotic Action on Protein Biosynthesis and Membranes* (E. Munoz, F. Garcia-Ferrandiz, and D. Vazquez, eds), Elsevier, Amsterdam, p. 738
Hotchkiss, R. D. (1944). *Adv. Enzymol.*, **4**, 153
Hou, C. T., Ciegler, A., and Hesseltine, C. W. (1972). *Appl. Microbiol.*, **23**, 183
Hunter, F. E. and Schwatz, L. S. (1967*a*). In *Antibiotics*, Vol. 1 (D. Gottlieb and P. D. Shaw, eds), Springer-Verlag, New York, p. 636
Hunter, F. E. and Schwatz, L. S. (1967*b*). In *Antibiotics*, Vol. 1 (D. Gottlieb, and P. D. Shaw, eds), Springer-Verlag, New York, p. 642
Irmscher, G., Bovermann, G., Boheim, G., and Jung, G. (1978). *Biochim. Biophys. Acta*, **507**, 470
Ivanov, V. T. (1975). *Ann. N.Y. Acad. Sci.*, **264**, 221
Ivanov, V. T., Laine, I. A., Abdulaev, N. D., Senyavina, L. B., Popov, E. M., Ovchinnikov, Yu. A., and Shemyakin, M. M. (1969). *Biochem. Biophys. Res. Commun.*, **34**, 803
Ivanov, V. T., Evstratov, A. V., Sumskaya, L. V., Melnik, E. I., Chumburidze, T. S., Portnova, S. L., Balashova, T. A., and Ovchinnikov, Yu. A. (1973). *FEBS Lett.*, **36**, 65
Ivanov, V. T., Fonina, L. A., Uvarova, N. N., Koz'min, S. A., Karapatnitskaya, T. B., Chekhlayeva, N. M., Balashova, T. A., and Ovchinnikov, Yu. A. (1975). In *Peptides: Chemistry, Structure and Biology* (R. Walter and J. Meienhofer, eds), Ann Arbor Science Publ., Ann Arbor, Mich., p. 195
Ivanov, V. T., Laine, I. A., Abdulaev, N. D., Pletnev, V. Z., Lipkind, G. M., Arkhipova, S. F., Senyavina, L. B., Meshcheryakova, E. N., Popov, E. M., Bystrov, V. F., and Ovchinnikov, Yu. A. (1971). *Khim. Prir. Soedin*, **3**, 221
Jacobs, A. and Worwood, M. (eds) (1974). *Iron in Biochemistry and Medicine*, Academic Press, London
Jung, G., König, W. A., Leibfritz, D., Ooka, T., Janko, K., and Boheim, G. (1976). *Biochim. Biophys. Acta*, **433**, 164
Katz, E. and Demain, A. L. (1977). *Bacteriol. Rev.*, **41**, 449
Knüsel, F., Schiess, B. and Zimmmerman, W. (1969). *Arch. Mikrobiol.*, **68**, 99
Lankford, C. E. (1973). *CRC Crit. Rev. Microbiol.*, **2**, 273
Lardy, H. A., Graven, S. N., and Estrada-O, S. (1967). *Fed. Proc. Fed. Am. Soc. Exp. Biol.*, **26**, 1355
Leong, J. and Neilands, J. B. (1976). *J. Bacteriol.*, **126**, 823
Linke, W. D., Crueger, A., and Diekmann, H. (1972). *Arch. Mikrobiol.*, **85**, 44
MacDonald, J. C. (1960). *Can. J. Microbiol.*, **6**, 27
McMullen, A. I. (1970). *Biochem. J.*, **119**, 10P–11P

McMurray, W. and Begg, R. W. (1959). *Arch. Biochem. Biophys.*, **84**, 546
Martin, D. R. and Williams, R. J. P. (1976). *Biochem. J.*, **153**, 181
Mauger, A. B. (1968). *Experientia*, **24**, 1068
Maurer, B., Muller, A., Keller-Schierlein, W., and Zahner, H. (1968). *Arch. Mikrobiol.*, **60**, 326
Melling, J. and McMullen, A. I. (1974). *Proc. IAMS Meeting*, Sept. 1974, Tokyo, p. 2412
Meyer, C. E. and Reusser, F. (1967). *Experientia*, **23**, 85
Meyer, J. M., (1977). Thèse d'Etat, Strasbourg, France
Meyer, J. M. and Hornsperger, J. M. (1978). *J. Gen. Microbiol.*, **107**, 329
Mitchell, P. and Moyle, J. (1965). *Nature, Lond.*, **208**, 147
Moore, C. and Pressman, B. C. (1964). *Biochem. Biophys. Res. Commun.*, **15**, 562
Moore, R. E. and Emery, T. (1976). *Biochemistry*, **15**, 2719
Mueller, P. and Rudin, D. O. (1967). *Biochem. Biophys. Res. Commun.*, **26**, 398
Mueller, P. and Rudin, D. O. (1968). *Nature, Lond.*, **217**, 713
Mueller, P. and Rudin, D. O. (1970). *J. Theor. Biol.*, **18**, 222
Murphy, T. P., Lean, D. R. S., and Nalewajko, C. (1976). *Science, N.Y.*, **192**, 900
Neilands, J. B. (1966). *Structure Bonding,* **1**, 59
Neilands, J. B. (ed.) (1974). *Microbial Iron Metabolism*, Academic Press, New York
Neilands, J. B. (1977). In *Iron Metabolism*, Ciba Foundation Symposium No. 51, Elservier/Excerpta Medica/North-Holland, Amsterdam, p. 107
Neilands, J. B. and Ratledge, C. (1980). In *Handbook of Microbiology*, 2nd edn, Vol. 2 (A. I. Laskin and H. A. Lechevalier, eds), CRC Press, Cleveland, Ohio, in press
Nishimura, H., Mayama, M., Kimura, T., Kimura, A., Kawamura, Y., Tawara, K., Tanaka, Y., Okamoto, S., and Kyotani, H. (1964). *J. Antibiot., Tokyo, Ser. A*, **17**, 11
Nüesch, J. and Knüsel, F. (1967). In *Antibiotics*, Vol. 1 (D. Gottlieb and P. D. Shaw, eds), Springer-Verlag, New York, p. 499
O'Brien, I. G. and Gibson, F. (1970). *Biochim. Biophys. Acta*, **215**, 393
O'Brien, I. G., Cox, G. B., and Gibson, F. (1971). *Biochim. Biophys. Acta*, **237**, 537
Ooka, T. and Takeda, I. (1972). *Agric. Biol. Chem., Tokyo*, **36**, 112
Ooka, T., Shimojima, Y., Akimoto, T., Takeda, I., Senoh, S., and Abe, J., (1966). *Agric. Biol. Chem., Tokyo*, **30**, 700
Ovchinnikov, Yu. A. (1974). *FEBS Lett.*, **44**, 1
Ovchinnikov, Yu. A., Ivanov, V. T., and Mikhaleva, I. I. (1971*a*). *Tetrahedron Lett.*, **2**, 159
Ovchinnikov, Yu. A., Ivanov, V. T., and Shkrob, A. M. (1972). In *Molecular Mechanisms of Antibiotic Action on Protein Biosynthesis and Membranes* (E. Munoz, F. Garcia-Ferrandiz, and D. Vazquez, eds), Elsevier, Amsterdam, p. 459
Ovchinnikov, Yu. A., Kiryushkin, A. A., and Kozhevnikova, I. V. (1971*b*) *Gen. Chem. U.S.S.R.*, **41**, 2105
Patel, D. J. and Tonelli, A. E. (1973). *Biochemistry*, **12**, 486
Patel, D. J. and Tonelli, A. E. (1974). *Biochemistry*, **13**, 788
Paulus, H., (1967). In *Antibiotics*, Vol. 2 (D. Gottlieb and P. D. Shaw, eds), Springer-Verlag, Berlin, p. 254
Payne, J. W., Jakes, R., and Hartley, B. S. (1970). *Biochem. J.*, **117**, 757
Pinkerton, M., Steinrauf, L. K., and Dawkins, P. (1969). *Biochem. Biophys. Res. Commun.*, **35**, 512
Plattner, P. A., Vogler, K., Studer, R. O., Quitt, P., and Keller-Schierlein, W. (1963). *Helv. Chim. Acta*, **46**, 927
Pollack, J. R. and Neilands, J. B. (1970). *Biochem. Biophys. Res. Commun.*, **38**, 989
Pressman, B. C. (1963). In *Energy-linked Functions of Mitochondria* (B. Chance, ed.), Academic Press, New York, p. 181
Pressman, B. C. (1965). *Proc. Natl. Acad. Sci. U.S.A.*, **53**, 1076
Pressman, B. C. (1968). *Fed. Proc. Fed. Am. Soc. Exp. Biol.*, **27**, 1283
Pressman, B. C., Harris, E. J., Jagger, W. S., and Johnson, J. H. (1967). *Proc. Natl. Acad. Sci. U.S.A.*, **58**, 1949
Prince, R. C., Crofts, A. R., and Steinrauf, L. K. (1974). *Biochem. Biophys. Res. Commun.*, **59**, 697
Pugsley, A. P. and Reeves, B. P. (1977). *J. Bacteriol.* **130**, 26

Quitt, P., Studer, R. O., and Vogler, K. (1963). *Helv. Chim. Acta*, **46**, 1715
Ratledge, C. (1976). *Adv. Microbial Physiol.*, **13**, 115
Ratledge, C. (1980). In *Handbook of Microbiology*, 2nd edn, Vol. 2 (A. I. Laskin and H. A. Lechevalier, eds), CRC Press, Cleveland, Ohio, in press
Ratledge, C. and Hall, M. J. (1970). *FEBS Lett.*, **10**, 309
Ratledge, C., Macham, L. P., Brown, K. A., and Marshall, B. J. (1964). *Biochim. Biophys. Acta*, **372**, 39
Ratledge, C. and Patel, P. V. (1976). *J. Gen. Microbiol.*, **93**, 141
Reusser, F. (1967). *J. Biol. Chem.*, **242**, 243
Roeske, R. W., Isaac, S., King, T. E., and Steinrauf, L. K. (1974). *Biochem. Biophys. Res. Commun.*, **57**, 554
Rosenberg, H. and Young, I. G. (1974). In *Microbial Iron Metabolism* (J. B. Neilands, ed.), Academic Press, New York, p. 67
Sandblom, J., Eisenman, G., and Neher, E. (1977). *J. Membr. Biol.*, **31**, 383
Sarges, R. and Witkop, B. (1965*a*). *J. Am. Chem. Soc.*, **87**, 2011
Sarges, R. and Witkop, B. (1965*b*). *J. Am. Chem. Soc.*, **87**, 2020
Sarges, R. and Witkop, B. (1965*c*). *J. Am. Chem. Soc.*, **87**, 2027
Sarges, R. and Witkop, B. (1965*d*). *Biochemistry*, **4**, 2491
Schafft, M. and Diekmann, H. (1978). *Arch. Microbiol.*, **117**, 203
Shemyakin, M. M. (1965). In *Antimicrobial Agents and Chemotherapy* (G. L. Hobby, ed.), American Society for Microbiology, New York, p. 962
Shemyakin, M. M., Ovchinnikov, Yu. A., Ivanov, V. T., and Evstratov, A. V. (1967). *Nature, Lond.*, **213**, 412
Shemyakin, M. M., Vinogradova, E. I., Feigina, M. Yu., Aldanova, N. A., Shvetsov, Yu. B., and Fonina, L. A. (1966). *Zh. Obshch. Khim. U.S.S.R.*, **36**, 1391
Shemyakin, M. M., Vinogradova, E. I., Feigina, M. Yu., Aldanova, N. A., Loginova, N. F., Ryabovd, N. F., and Pavlenka, I. D. (1965). *Experientia*, **21**, 548
Shemyakin, M. M., Ovchinnikov, Yu. A., Ivanov, V. T., Antonov, V. K., Vinogradova, E. I., Shkrob, A. M., Malenkov, G. G., Evstratov, A. V., Laine, I. A., Melnik, E. I., and Ryabova, I. D. (1969). *J. Membr. Biol.*, **1**, 402
Shimi, I. R. and Shoukry, S. (1976). *J. Antibiot., Tokyo*, **29**, 303
Shimi, I. R., Imam, G. M., and Haron, B. M. (1969). *J. Antibiot., Tokyo*, **22**, 106
Shimi, I. R., Shoukry, S., and Aly, F. T. (1975). *Int. J. Biochem.*, **6**, 205
Simon, W. and Morf, W. E. (1973). In *Membranes*, Vol. 2, *Lipid Bilayers and Antibiotics* (G. Eisenman, ed.), Marcel Dekker, New York, p. 329
Simpson, F. B. and Neilands, J. B. (1976). *J. Phycol.*, **12**, 44
Skulachev, V. P. (1975). In *Energy-transducing Membranes* (E. Racker, ed.), Butterworths, London, p. 31
Snow, G. A. (1970). *Bacteriol. Rev.*, **34**, 99
Stark, G. and Benz, R. (1971). *J. Membr. Biol.*, **5**, 133
Stephenson, M. C. and Ratledge, C. (1979). *J. Gen. Microbiol.*, **110**, 193
Sussman, M., (1974). In *Iron in Biochemistry and Medicine* (A. Jacobs and M. Worwood, eds), Academic Press, London, p. 650
Tait, G. H. (1975). *Biochem. J.*, **146**, 191
Tateson, J. E. (1970). *Biochem. J.*, **118**, 747
Urry, D. W. (1971). *Proc. Natl. Acad. Sci. U.S.A.*, **68**, 672
Urry, D. W. (1972). *Proc. Natl. Acad. Sci. U.S.A.*, **69**, 1610
Urry, D. W., Long, M. M., Jacobs, M., and Harris, R. D. (1975). *Ann. N.Y. Acad. Sci.*, **264**, 203
Vallejos, R. H., Andreo, C. S., and Ravizzini, R. A. (1975). *FEBS Lett.*, **50**, 245
Veatch, W. R., Fossel, E. T., and Blout, E. R. (1974). *Biochemistry*, **13**, 5249
Vey, A., Quiot, J. M., and Vago, C. (1973). *C.r. hebd. Seanc. Acad. Sci., Paris, Ser. D*, **276**, 2489
Weinberg, E. D. (1971). *Perspect. Biol. Med.*, **14**, 565
Weinberg, E. D. (1975). *Science, N.Y.*, **188**, 1039
West, E. J. and Biggs, J. D. (1968). *J. Econ. Entomol.*, **61**, 684
Wiebe, C. and Winkelmann, G. (1975). *J. Bacteriol.*, **123**, 837

Wieland, Th. (1972). In *Chemistry and Biology of Peptides* (J. Meienhofer, ed.), Ann Arbor Science Publ. Ann Arbor, Mich., p. 377
Wieland, Th., Lüben, G., Ottenheim, H., Faesel, J., De Vries, J. X., Konz, W., Prox, A., and Schmid, J. (1968). *Angew. Chem.*, **80**, 209
Winkelmann, G. and Zahner, H. (1973). *Arch. Mikrobiol.*, **88**, 49
Wipf, H. K., Pioda, L. A. R., Stefanac, Z., and Simon, W. (1968). *Helv. Chim. Acta*, **51**, 377
Woodruff, H. B. (1966). *Symp. Soc. Gen. Microbiol.*, **16**, 22
Yafuso, M., Freeman, A. R., Steinrauf, L. K., and Roeske, R. W. (1974). *Fed. Proc. Fed. Am. Soc. Exp. Biol.*, **33**, 1258
Zähner, H., Hütter, R., and Bachmann, E. (1960). *Arch. Mikrobiol.*, **36**, 325
Zimmerman, W. and Knüsel, F. (1969). *Arch. Microbiol.*, **68**, 107

Microorganisms and Nitrogen Sources
Edited by J. W. Payne

Section 4.9

Assays for Amino Acids, Peptides, and Proteins

J. W. PAYNE AND J. M. TUFFNELL
Department of Botany, University of Durham, South Road, Durham DH1 3LE, UK

1 INTRODUCTION

There exists a frequent and widespread need to assay protein amino acids. It may be the total amino acid complement that needs to be assayed or, alternatively, quantitation of individual amino acids may be wanted.

The amino acids of interest may occur in the free form or be chemically combined. When free they may, for example, be present in the body fluids or tissue extracts of humans and in the microbial and plant products used in research and industry, or as supplements for human diets and animal feeds and in the growth media of microorganisms used by industry. When the amino acids are bound they generally occur as proteins and rarely in the form of small peptides. In this case, the interest is almost always concerned with nutritional evaluation, i.e. determination of the quantity and/or quality of the protein. Quantity obviously concerns the protein content per unit weight in, say, wheat, rice, and soyabean. Quality relates to the amino acid composition of the protein, being assessed more particularly by the levels of certain nutritionally important amino acids such as lysine, methionine, and tryptophan.

It is the quantitation of particular rather than total amino acids, whether initially free or combined, with which this section is mostly concerned. At present such assays are broadly of three types: chemical, biological, and microbiological.

In reviewing these methods attention will be given to the advantages and disadvantages of each type, and to comparisons between the results obtained from each. Particular emphasis will be placed on the use of microorganisms for assay purposes. Information on the microbial transport and utilization of amino acids, peptides, and proteins will be used as one criterion by which to evaluate current microbial assays and to choose strains for use in new, improved procedures.

2 METHODOLOGY: GENERAL CONSIDERATIONS

The method that can be used in any instance is dictated by various considerations. Obviously, the approach taken will depend on the results required; i.e. is a value for total protein or for individual amino acids needed, or is a chemical score or a nutritional evaluation required? In addition, the scope of the required assay is important; i.e. is the need for the screening of large numbers of samples for a particular characteristic (such as in the selection of high protein strains in a plant-breeding programme), or a quick, specific assay (such as that needed during quality control testing in the food industry, routine analysis in a hospital laboratory, etc.), or an in-depth, long-term, nutritional evaluation of a foodstuff? The nature of the sample is also important, not only whether it contains the amino acids in free form, bound, or as a mixture of both, but also whether they are in a relatively pure state, say in biological extracts, or are complex materials such as cereal grains, dried fish, etc. Finally, the method of choice will depend on the available facilities, and these may vary from those to be found in a small field laboratory to those of a modern, fully-equipped analytical laboratory.

Cognizance of all these variables cannot be taken here while assessing the various methods, instead consideration will be limited to methods that can be used on plant samples such as cereals and legumes, these examples will nevertheless allow a broad consideration of the topic. Pertinent reports of this topic have appeared recently (PAG, 1975; IAEA, 1975, 1977).

3 CHEMICAL METHODS

1 Introduction

Many of the methods used, especially in screening procedures, are indirect. However, they can be useful in identifying interesting samples although the results have to be confirmed by more critical, direct, nutritional criteria. In general, chemical methods have the advantage of being quick, which allows a rapid throughput of samples and they can often be automated. However, they commonly lack specificity and may have poor sensitivity. It is important to appreciate that the content of, say, methionine or lysine determined chemically does not necessarily correlate with the nutritional value of the component. There are various reasons for this. For example, chemical assays frequently involve an acid hydrolysis step which may liberate amino acid residues that would not be released by normal enzymic digestion *in vivo* and thus could not be absorbed by the gut and utilized; this difference is most likely to occur in samples that have been processed or heat-treated and as a result have suffered chemical modifications that are not susceptible to enzymic hydrolysis (Ford and Shorrock 1971; Bjarnason and Carpenter, 1969; Mauron, 1972; Ford, 1975). On the other hand, acid hydrolysis may destroy or modify some residues (tryptophan, threonine) leading to chemical underestimates of their real nutritional value. Poor digestibility and the presence of antinutritional factors (see below) may also impair protein utilization; in such cases chemical assays would overestimate nutritional values. Thus, although chemical methods have much to recommend them and they are entirely satisfactory in many instances, there are various reasons why they often provide an unreliable predictive measure of nutritional value.

2 Total Nitrogen

An indication of total protein content can be obtained from total nitrogen values. The Kjeldahl method in its various forms is most used and widely recommended, the ammonia liberated on acid digestion may be determined by titration or colorimetry (PAG, 1975; IAEA, 1977). The Dumas combustion method, although not so widely applicable, gives reproducible results in agreement with those from the Kjeldahl method; the procedure can be fully automated and machines able to handle several hundred samples per day are available. Other physical methods are available but their need for sophisticated equipment precludes their general use. The nitrogen values obtained are multiplied by a standard factor to obtain protein values; the factor can vary according to the protein source (FAO, 1973).

3 Total Protein

Various methods are used, none is without its limitations, generally through lack of specificity, interference from non-protein components, etc. Sample preparation is of great importance in many of the assays; the protein source must be in a very finely divided form (milled, macerated) or access of reagents may be limited. The biuret reagent (an alkaline solution of copper sulphate) reacts with peptide bonds and the coloured complex formed gives a measure of protein content; the method is simple and can be useful in screening (Greenaway and Davis, 1975). Certain dyes, e.g. Orange G, bind to basic groups such as those on lysine and arginine and the amount of bound dye provides an index of protein content, although for meaningful use this requires that the proportion of basic amino acids remains roughly constant in different proteins (Axman, 1975; Lakin, 1973; Udy, 1971; Jones, 1974; Jones and Lakin, 1976); the method is useful in screening studies but samples selected must subsequently be checked more rigorously using other procedures. The Lowry reaction is widely used but is susceptible to interference by many substances likely to be present in biological materials. In addition, the method depends on reaction with tyrosine residues which may vary in content amongst different protein soures; for these reasons it is in general unlikely to provide a reliably accurate measure of protein quantity. Turbidimetric measurement of protein fractions obtained by different extraction procedures provides a quick, simple assay although not generally of high accuracy (Paulis and Wall, 1975).

4 Amino Acids

Methods used here are generally for determining free amino acids, they are therefore applicable to samples containing unbound amino acids and also, after appropriate hydrolysis, to protein samples. Protein hydrolysis is a critical step if subsequent analyses are to be reliable and accurate. Three types are used, acid, alkaline, and enzymic. In the first, a strong acid is used, usually 6 M HCl, at 110 °C in sealed, evacuated tubes, and samples are hydrolysed for a series of times e.g. 12, 22, 48, 72 h, to allow for the fact that certain types of peptide linkage (comprising Ile, Val, etc.) are broken more slowly than most and certain residues (e.g. Ser, Thr, and particularly Trp) are destroyed. Methionine and cystine can only be recovered quantitively if the sample is first oxidized with performic acid to produce methionine sulphone and cysteic acid respectively. Alkaline hydrolysis, using barium hydroxide at 120 °C, is necessary for determination of tryptophan, but several other amino acids are destroyed by this procedure. Degradation and charring of other components is an unavoidable difficulty during these chemical hydrolyses, and the non-physiological nature of this form of hydrolysis is obvious. Hydrolysis *in vitro* with proteases and peptidases appears an attractive system but their limited bond specificity means that a mixture of such enzymes would be required in order to release all amino acids fully.

Ion exchange chromatography, based on elution of the amino acids from an acidic cation exchanger, using a pH and buffer concentration gradient is most widely used for determining both total and particular amino acids (Blackburn, 1968; Longenecker, 1973). Quantitation is by colorimetry after reaction with ninhydrin, although sensitivity can be markedly increased by preparing fluorescent derivatives, e.g. by reaction with fluorescamine or *o*-phthaldehyde. Automated analysers based on these principles are extensively used although they are too slow and too complicated for use in screening procedures. Gas-liquid chromatography shows promise as an alternative to ion-exchange systems although it also suffers from the disadvantages mentioned above.

The nutritional quality of very many foodstuffs is dependent largely on their content of lysine, tryptophan, and methionine, and in certain cases threonine, and isoleucine or leucine. Consequently, much effort has gone into devising chemical assays for these critical amino acids, in particular for lysine (Concon, 1975; Holsinger and Posati, 1975; Finley and Friedman, 1973; Kakade and Liener, 1969; Carpenter, 1960; Carpenter and Booth, 1973; Booth, 1971), for methionine (Concon, 1975; Lipton and Bodwell, 1975; Evans *et al*., 1976; Ellinger and Duncan, 1976), and tryptophan (Concon, 1975; Friedman and Finley, 1971, 1975). Of these critical amino acids the one that is most deficient in relation to the standard protein is called the first limiting amino acid; the standard protein, or reference pattern of amino acids is that suggested by the FAO (1973) based on estimates of the amino acid requirements of young children. A chemical score (amino acid score) (Bender, 1973) can be calculated by dividing the content of the limiting amino acid in the test protein by the content of the same amino acid in the reference protein. The amino acid score of the limiting amino acid provides a chemical prediction of the nutritive value of the test protein and may be taken as a first approximation of the possible efficiency of its utilization by children.

4 BIOLOGICAL METHODS

1 Introduction

These procedures, using animals or humans, involve measuring a biological response, commonly a growth parameter. The methods do not provide absolute quantitation of protein or of individual amino acids, rather they provide a measure of these components in terms of their nutritional utilizability. However, it is precisely this measure that is most often sought, and consequently it is useful to consider not only the methods themselves but also some of the reasons why the values they provide may differ from those obtained by chemical methods.

It is obvious that amino acids are required for synthesis of new tissue in children and young animals, but perhaps less obviously, that they are also required for the maintenance and repair of existing tissues in adults; in fact, the recommended high-quality protein intake per day for infants (6 months), children (5 years), and adults is 1.5, 1.0, and 0.5 g kg^{-1} respectively (FAO, 1973). Of the twenty or so amino acids commonly found in proteins some, referred to as non-essential, can be synthesized in the body from simple nitrogen sources, while others, the essential amino acids (Lys, Thr, Trp, Phe, Met, His, Leu, Ile, Val) cannot be synthesized in animal tissues and must be obtained from the diet. If any one of the essential amino acids is absent from the diet net protein synthesis and thus growth cannot occur, and under these conditions other protein components cannot be utilized and will be wasted. The various essential amino acids are required in different amounts, largely depending on the relative proportions in which they occur in body protein. It is usually the case that if a diet contains an adequate supply of the essential amino acids it will necessarily contain sufficient of the non-essential amino acids since the latter are generally more abundant in conventional food proteins; it follows therefore that well-nourished animal tissues do not normally need to synthesize the non-essential amino acids. A high-quality protein is one which contains all the essential amino acids in a utilizable form and in amounts that match the average composition of the body protein, whereas a poor quality protein is defective in one or other respect. It is generally the case that the animals' nutritional requirements can be met as adequately by larger amounts of a poor quality protein as by smaller amounts of a high-quality protein, although with the former source, a much higher proportion of protein-nitrogen will be excreted as urea and protein-carbon will be channelled into energy use.

A topic of considerable current interest to nutritionists, and also pertinent to the central theme of this book, concerns the relative efficiency with which a pure protein, a mixture of the free, constituent amino acids of the protein, or a mixture of small peptides having the same composition as the protein are utilized by animals and humans. Concerning the amino acids, conclusions are unclear (Marrs *et al.*, 1975); many studies have shown amino acid mixtures to be nutritionally inferior to whole protein of identical composition (Rose *et al.*, 1954; Ahrens *et al.*, 1966); others have reported that the responses to the amino acid mixtures correspond to the protein (Ramasarma *et al.*, 1949; Sauberlich, 1961). Many suggestions have been made to try and explain the frequently observed difference in their nutritional response, the most likely of which concerns the difference in the temporal presentation of amino acids for protein synthesis (Rose *et al.*, 1954; Cannon *et al.*, 1947; Gray *et al.*, 1968; Meister, 1965; Gitler, 1964; Matthews and Payne, 1975*b*; Silk *et al.*, 1975); certainly during the intestinal absorption of a mixture of free amino acids it is well known that individual amino acids are absorbed at widely different rates, with some, e.g. Met, Leu, Ile, having high affinity and being rapidly absorbed while others, e.g. Gly, Lys, Asp, Glu, are slowly absorbed. In consequence, the former amino acids reach peak concentration in the portal blood before the latter ones, leading to a non-optimal presentation to the liver (Meister, 1965; Christensen, 1968; Matthews and Payne, 1975*b*). Amino acid imbalances and antagonisms of this type contribute to the nutritional inadequacy of many protein sources, e.g. in corn and sorghum proteins excess leucine depresses isoleucine utilization, arginine lowers lysine utilization in millet, and the human nutritional disorder, pellagra, apparently arises from similar amino acid antagonisms (Silano, 1977). Although amino acid mixtures may be nutritionally inferior to the corresponding protein there is little doubt that when the protein component of the diet is replaced by correctly balanced mixtures of pure amino acids they can maintain health and allow satisfactory growth in both man and laboratory animals (Ranhotra and Johnson, 1965; Winitz *et al.*, 1970; Belikov and Babayan, 1971; Winitz and Adams, 1971); a detailed account of the development and uses of such chemically defined diets is given by Greenstein and Winitz (1961). The relative nutritional efficiency of peptides is under intensive study. Contrary to previous ideas on the process of protein utilization in mammals it is now well established that intestinal peptide absorption not only occurs to a significant extent but may be nutritionally more important than amino acid absorption (Matthews and Payne, 1975*a*, 1980; Matthews and Adibi, 1976; Matthews, 1975, 1977; Das and Radhakrishnan, 1976; and reviews in CIBA Symposia, 1972, 1977). The above reviews document the considerable evidence that peptides may be absorbed more quickly and more extensively than free amino acids and also that the differences in rates of individual amino acid uptake mentioned above may be considerably reduced during peptide uptake, making for a more uniform presentation and potentially more efficient synthesis of new proteins. In consequence, discussions are now appearing on the possible advantages of synthetic diets that contain peptides rather than free amino acids (Matthews and Adibi 1976; Bassler *et al*., 1976; Clegg, 1978). The biological methods commonly used are designed for evaluation of proteins rather than for amino acid mixtures or peptides, nevertheless the above considerations are relevant to the design and interpretation of biological assays in which amino acid supplementation is used.

The main advantage claimed for biological assays over other procedures is that they alone can provide a nutritional evaluation, and although this is arguably true they do have unique disadvantages as well. Obviously, to obtain results with laboratory animals takes a long time, is expensive, and requires large amounts of sample, and in experiments with human volunteers these difficulties are enhanced, while, in addition, there are ethical objections to using children. These difficulties preclude the use of biological assays in many situations, e.g. screening of early plant varieties. Because of differences in their digestive capacities, patterns of absorption, nutritional requirements, response to antinutritional factors, etc., it can be argued that extrapolation of results obtained with, for example, rats or dogs to humans has such a measure of uncertainty about it that it significantly diminishes the claimed unique value of the biological assay. Results of biological assays can also be markedly influenced by the following factors: the manner in which the sample is prepared (cooked, natural, finely divided, etc.); the level and type of energy source supplied together with the protein; the presence of either nutritional factors which may 'spare' certain amino acids or antinutritional factors that may limit the processes of digestion and/or absorption; the age and nutritional status of the test animal. These and other practical problems make it difficult to get good reproducibility in the evaluations from various laboratories. Despite these problems an important distinction has emerged between the biological and chemical evaluation of certain amino acids, namely the concept of amino acid availability (Morrison and McLaughlan, 1972; Mauron, 1972; Carpenter and Booth, 1973; Erbersdobler, 1973, 1976; Ford, 1975; Silano, 1977). Amino acid residues that are utilized nutritionally are termed available. Commonly, and especially for certain amino acids, e.g. Lys, Asn, Gln, Asp, Glu, the content of available residues determined biologically is less than that determined chemically in complete protein hydrolysates. The difference between the two represents the complement of unavailable residues which arise from incomplete digestion or absorption. Unavailable amino acids are found in native protein foods but their content is frequently increased during processing, especially by excessive heating. Incomplete digestion may result from a variety of causes. For example, the digestive proteases may fail to gain access to the protein substrates contained within organelles and protected by cellulosic cell walls; this feature would cause a general decrease in amino acid availability. Second, although the protein may be broadly accessible, individual residues would carry secondary linkages or be chemically modified such that their α-peptide bonds were resistant to hydrolysis by the spectrum of digestive proteases and peptidases, making certain residues unavailable. Third, specific protease inhibitors or high concentrations of compounds such as aldehydes and polyphenols present in the sample may limit digestion, again leading to a general decrease in amino acid availability. Incomplete absorption may possibly arise in various ways, although there is little good evidence for any of the speculations. Non-protein amino acids which occur extensively in plants may inhibit absorption of protein amino acids, and similarly, modified peptide fragments which may themselves fail to be absorbed might nevertheless competitively inhibit uptake of normal peptide fragments.

2 Assay Procedures

Various sources provide general information on biological assays (Bigwood, 1972; Porter and Rolls, 1973; Friedman, 1975; Bodwell, 1976; Cole *et al*., 1976; PAG, 1975; IAEA, 1975, 1977). These same references contain discussions on the suitability of various test animals for use in bioassays, the consensus opinion being that the rat is the most useful (Eggum, 1973*a*; Pellet, 1973; Dreyer, 1973).

The methods fall into three categories: growth methods (PER, NPR, RPV); nitrogen balance methods (AD, TD, NPU, BV); indirect methods.

(i) Growth Methods

The method that is most commonly used for protein evaluation is the protein efficiency ratio (PER) method

$$\text{PER} = \frac{\text{Mean weight gain of test group}}{\text{Weight of protein consumed}}.$$

Usually a group of rats are given the test sample as the sole protein source comprising about 10 per cent of the diet, for about 10–30 days during which food is given *ad libitum*. The main objections to the use of this method are that results vary with the protein content of the diet (usually calculated as $N \times 6.25$) and with the food intake (Hegsted, 1971; Jacquot and Peret, 1972; McLaughlan and Campbell, 1969; McLaughlan and Keith, 1975; Hurt *et al*., 1975). The variation in the PER values caused by differences in the food intake can be ameliorated in two ways, i.e. by keeping the food intake constant, or by use of a modified PER, net protein ratio (NPR), that includes a term allowing for the dietary protein needed for maintenance:

$$\text{NPR} = \frac{\text{Mean weight gain of test group} + \text{Weight loss of protein-free group}}{\text{Weight of protein consumed}}.$$

It is assumed that the protein required to prevent weight loss of the rats fed the non-protein diet is equivalent to the protein required for maintenance. NPR is a better measure of protein quality than is PER and is useful as a screening test where differences between protein sources are of interest. The main fault of the NPR assay is that it frequently overestimates the value of poor quality proteins (McLaughlan and Keith, 1975). Perhaps the most reliable biological assay is the relative protein value (RPV) method (McLaughlan and Keith, 1975; PAG, 1975):

$$\text{RPV} = \frac{\text{Slope of the response line of test sample}}{\text{Slope of the response line of reference protein}}.$$

The response line is obtained by determining the rate of body weight change of rats fed various levels of test protein or reference protein (usually casein). It requires that body weight response is linear over the range of test protein intakes studied, and for reliability an extended range is desirable. It goes without saying that strictly standardized conditions are recommended for all weight gain methods.

(ii) Nitrogen Balance Methods

These techniques provide the most detailed information concerning dietary protein quality, including measures of digestibility and utilization. They require shorter experimental times than weight gain procedures but they are more laborious, involving determination of nitrogen intake, and faecal- and urinary-nitrogen excretion (McLaughlan and Campbell, 1969; Eggum, 1973*a*). True digestibility (TD) and biological value (BV) are considered primary characteristics while net protein utilization is a derived quantity (Eggum, 1973*a*).

Digestibility of protein may be considered at two levels. Apparent digestibility (AD) is defined as

$$\text{AD} = \frac{(\text{Nitrogen intake} - \text{Faecal nitrogen})}{\text{Nitrogen intake}} \times 100.$$

It may, however, be determined somewhat indirectly by measuring in the food and faeces the ratios of nitrogen to an indigestible marker (usually chromic acid or titanium dioxide) added to the diet. Thus

$$\text{AD} = \left(1 - \frac{\text{Ratio of indigestible marker to nitrogen in faeces}}{\text{Ratio of indigestible marker to nitrogen in food}}\right) \times 100.$$

AD can vary with protein content of the diet but it is a useful index and reliable results can usually be obtained by analysing only small samples of faeces even after feeding for only one day (Njaa, 1977). True digestibility (TD) is strictly based on a nitrogen balance procedure (Eggum, 1973*a*; Slump and van Beek, 1975) and can be defined as

$$\text{TD} = \text{Nitrogen intake} - \frac{(\text{Faecal nitrogen} - \text{Faecal nitrogen of metabolic origin})}{\text{Nitrogen intake}} \times 100.$$

Calculation of nitrogen absorbed is obtained by subtracting the amount excreted in the faeces from that consumed but allowing for excretion of metabolic or endogenous nitrogen (such as in microorganisms) by measurements with a control group fed a protein-free diet. TD is independent of the protein content of the diet or body weight of the experimental animal (Eggum, 1973*a*).

The biological value (BV) of a dietary protein may be thought of as the fraction of absorbed nitrogen retained in the body for maintenance and growth, its determination therefore requires estimation of the total nitrogen absorbed and the proportion of this which is retained (Eggum, 1973*a*):

$$\text{BV} = \frac{\text{N intake} - (\text{Faecal N} - \text{Metabolic N}) - (\text{Urinary N-Endogenous N})}{\text{N intake} - (\text{Faecal N} - \text{Metabolic N})} \times 100.$$

By limiting the faecal losses to that part which is actually indigested, and likewise reducing the urinary loss by its endogenous fraction, the numerator represents the total nitrogen utilized, including both that used in growth and maintenance. Some metabolic nitrogen is also subtracted from faecal nitrogen in the denominator. The calculated BV represents the percentage of digested nitrogen that is actually utilized and provides a measure of the efficiency of the absorbed protein for the combined functions of maintenance and growth.

Thus, determinations of BV only assess the quality of the protein actually digested and absorbed, so that a protein may be very poorly digested but potentially of high biological value. In practice, such a protein would be of little value and, consequently, one of the best indices of the true worth of a protein is the net protein utilization (NPU), which makes allowance for both amino acid composition and digestibility:

$$NPU = \frac{TD \times BV}{100}.$$

In all these assays it is essential that experimental conditions are standardized.

Various authors have discussed the fact that BV and NPU are both influenced by the level of dietary protein, BV decreasing significantly with protein uptake (McLaughlan and Campbell, 1969; Eggum, 1973*a*; Hegsted and Juliano, 1974; McLaughlan and Keith, 1975; Delort-Laval, 1976). In part this arises because the amounts of protein required for maintenance and growth in the rat are quite different since there is a low maintenance requirement for lysine and leucine. Thus, two proteins with markedly different lysine concentrations might give similar responses near maintenance levels but could show wide variations at higher protein intake levels.

A recent, novel method is the complete protein evaluation (CPE) procedure which is based upon determination of amino acid excesses, in contrast to the classical supplementation approach which assesses amino acid deficiencies (Schelling, 1975).

(iii) Indirect Methods

Indirect protein evaluations, such as measurements of levels of amino acids and urea in blood (Eggum, 1973*b*, 1976) and muscle (Pion, 1973), urinary creatinine (Eggum, 1978), ribosome activity (Eggum, 1978), and the activities of particular liver enzymes (Bergner, 1978) are not at present considered sufficiently reliable and none has been generally adopted (Njaa, 1977).

5 ENZYMIC METHODS

This approach is used almost exclusively in relation to nutritional evaluation of proteins. Digestion *in vitro* with several different proteases has been used and the resultant hydrolysates are assayed by use of microbial or ion-exchange chromatographic methods; information on digestibility and amino acid availability can also be obtained. In earlier reviews of this topic, details of the methods and discussions of their applications were given (Sheffner, 1967; Mauron, 1970*a,b*, 1973; McLaughlan and Campbell, 1969; Morrison and McLaughlan, 1972; Stahmann and Woldegiorgis, 1975; Bodwell, 1977).

Using pepsin as the hydrolytic enzyme, an integrated amino acid index was produced, the pepsin digest-residue (PDR) amino acid index, that combined the pattern of essential amino acids released by digestion *in vitro* with the amino acid pattern of the residual protein (Sheffner, 1967). PDR index correlated fairly well with NPU for various representative proteins and was of especial value in detecting the effects of heat damage; however, because of the limited digestion by pepsin the hydrolytic products needed to be assayed by microbial means. Consequently, a procedure involving digestion by pepsin followed by pancreatin, the pepsin–pancreatin digest (PPD) index, was introduced (Akeson and Stahmann, 1964). This was later modified to give the pepsin-pancreatin digest dialysate (PPDD) index, which was compared to results from animal assays by Mauron and coworkers (Mauron, 1970*a*,*b*, 1973). All these methods are prolonged and complex, and in assessing them recently Bodwell (1977) concluded: 'Despite the theoretical advantages of using estimates of nutritional value derived from enzymatic assays relative to using indexes calculated from amino acid composition data, the information available does not indicate that significantly more valid information about actual nutritional value is consistently obtained through the use of the more laborious enzymatic methods.' However, methods that monitor the initial rate and extent of proteolysis *in vitro*, e.g. pH-stat methods (Bodwell, 1977; Payne *et al*., 1977), may be potentially useful in evaluating food digestibility and for obtaining information on the bioavailability of specific amino acids.

Individual enzymes, e.g. arginase, lysine decarboxylase, may find application in assays for specific free amino acids. An automated procedure has been described for lysine assay based upon the colorimetric measurement of the carbon dioxide which is specifically released by the action of lysine decarboxylase (Wall and Gehrke, 1974).

6 MICROBIOLOGICAL METHODS

1 Introduction

The use of microorganisms to assay various substances has a long history. Today they are used extensively, frequently as reagents of first choice in the assay of many vitamins and antibiotics, and to a lesser extent to assay amino acids and proteins (Hewitt, 1977). In this latter context, they effectively bridge the gap between the chemical and biological methods already discussed, providing comparable speed, potential for automation, simplicity, and sensitivity to the former methods whilst possessing intrinsic biological qualities resembling the latter procedures. Many microbial species have been employed including various bacteria, fungi, and ciliate protozoa, and these are considered below. They have been used to measure specific amino acids, the composition of total-acid or enzymic protein hydrolysates, the relative nutritive values (RNV) of proteins, and the nitrogen content of biological samples. The test microorganism usually has a growth requirement for one or more of these substances and the methods involve monitoring and quantitating the substrate-dependent growth response; these methods are discussed later.

At present, a number of problems may be identified associated with microbial assays for amino acids and derivatives. First, in relation to nutritional evaluations, is the fundamental issue of the relevance of the results to those obtained with laboratory animals and man. Only extensive comparative studies will answer this question and results obtained to date are considered below. How meaningful such comparisons can be will in part depend on a reliable response of the microorganism to its required nutrient, and for commonly used species there is some doubt on this point since many are nutritionally fastidious and their exact requirements are not known. Furthermore, their response to a specific nutrient can be influenced by its form, e.g. whether it is a free amino acid or is peptide-bound, by their nutritional status and growth phase, and by the type and quantity of other nutrients present, etc. It is appropriate here, therefore, to consider this topic with the background information of the microbial transport and utilization of amino acids, peptides, and proteins presented elsewhere in this volume to see if the most widely used organisms are indeed the most suited or whether a more informed and rational selection of test organism might not now be made; it can certainly be argued that many of the intrinsic advantages of a bacterial assay, e.g. speed, simplicity, reliability, specificity are not being fully realized in most current procedures. Early studies of this topic are discussed and referenced in the following books and reviews (Barton-Wright, 1952; 1963; Hutner *et al*., 1958; Adamson and Simpson, 1959; Ford, 1960; Shockman, 1963; Bolinder, 1972; Kavanagh, 1960, 1963, 1972).

2 Assay Procedures

The two most commonly used methods of microbiological assay are the agar diffusion (or plate) method and the tube method. Both procedures involve the quantitative comparison of the effect of two substances (a standard and the unknown test sample) on the growth of a suitable microorganism in a nutrient medium. The measured effect is most commonly growth inhibition, as with antibiotics, but for amino acids, it is growth promotion. For growth-promoting substances the two methods have the following requirements:

(a) The test organism must be dependent for growth on the presence of the substance being assayed.

(b) Graded amounts of the required nutrient (both standard and test sample) should cause graded responses following incubation.
(c) The growth medium must contain an excess of all nutrients required by the test organism except for the substance being assayed. The substance should be absent from the basic medium.
(d) Apart from the substance being assayed, no other component of the test sample should be capable of growth promotion or of modifying the growth of the test organism.

The practical aspects of the methods, their applications, and their conformity with the above requirements are considered below.

(i) Diffusion Methods

A typical procedure involves preparing molten nutrient agar (at about 42 °C), adding the test organism to give a uniform dispersion, then pouring a suitable volume (15–20 ml) into a petri dish to give a uniform layer (about 5 mm thick). Solutions of test and standard substance are applied to the solid agar, either into reservoirs (about 8 mm diameter) cut out of the agar, or on to porous ceramic beads or small filter paper discs placed on the surface. The solution is allowed to diffuse into the agar for about an hour at room temperature, and then the plates are incubated, normally overnight. After incubation, a turbid zone of growth occurs around the point of application of the test solution. Measurement of the diameters of the zone boundaries, which may vary in their sharpness, provides the quantitative basis for the assay of the concentration of the test substance. Factors other than amount of test substrate which influence the formation and size of zones are:

(a) Sensitivity of test organism.
(b) Physiological state of organism (vegetative, spore).
(c) Thickness of agar: zone width decreases as thickness increases.
(d) Density of inoculum: zone width is inversely related to inoculum size.
(e) Nature of medium: its nutrient quality can influence growth rate.
(f) Volume of test solution should be kept constant or made large enough to act as reservoir of constant concentration.
(g) Area of application of test solution: larger areas give increase in zone width.
(h) Application time: must be standardized to maintain constant diffusion time for different samples prior to start of incubation.
(i) Incubation temperature: will influence growth rate.

Discussion of these factors and their importance for achieving reliable results has appeared in various reviews (Lees and Tootill, 1955*a*; Gavin, 1957*a*; Cooper, 1963, 1972; Hewitt, 1977).

The variability of response to any one test solution, both within a single plate and between plates, led early on to the introduction of large plates to try and minimize these variations. Such plates may typically be made from 30 cm square plate glass, making it easier to obtain a uniform agar layer, to which 64 samples can be applied in accordance with a partially randomized 8 × 8 Latin-square design. Detailed considerations of the theory and practices of large plate assays have appeared (Lees and Tootill, 1955*a,b,c*; Gavin, 1957*a*; Adamson and Simpson, 1959; Bolinder, 1970*e*, 1972; Hewitt, 1977).

Extensive studies on large plate assays for amino acids are described by Bolinder, who used various strains of lactobacilli to assay for Met, Trp, Thr, Arg, Tyr, His, Lys, and Cys (Bolinder, 1968*a,b,c,d*, 1969, 1970*a, b, c, d, e, f*, 1972; Bolinder *et al*., 1963*a, b*, 1966). Lactic acid strains were also used in a series of reports (in Japanese) by Japanese workers (Kojima *et al*., 1964), and other early studies using these organisms are cited in Bolinder (1972). Preliminary studies have been reported on the use of *Escherichia coli* auxotrophs in plate assays for Asn (Sköld and Wallenborg, 1969), Trp (Konova and Verbina, 1966), Phe (Dickinson, 1958; Jones and Burns, 1958), and Lys (Gol'dfarb *et al*., 1968). *Bacillus subtilis* was used for determining Phe, in a screening test for phenylketonuria, as well as other amino acids (Guthrie and Susi, 1963; Guthrie, 1969*a,b*). *Chromobacterium violaceum* has been used for determination of Trp (Sebek, 1965).

(ii) Tube Assays

In these methods, which are most commonly used in assays for amino acids, graded concentrations of the substance to be assayed and appropriate standard solutions are added to a series of test tubes containing liquid growth media, these are inoculated with the test organism, and incubated for a suitable time. Although other responses can be measured (see below), in the commonest procedure the growth of the organism is estimated by the turbidity measured in a suitable photometer. Advantages of the tube method over diffusion methods are increased sensitivity and speed; disadvantages are that the sample must usually be sterile unless the assay is very rapid (see below), and it is more susceptible to interference by extraneous substances.

Factors other than amount of test substrate which influence the growth (response) of the organism are:

(a) Test organism: should show a selective response to test substrate and not be affected by other components in the sample.
(b) Incubation period: ideally this should allow complete utilization of

the growth promoting substance, which typically may require 18 h–4 days.

(c) Incubation temperature: as variations from optimum will affect growth rate this should be carefully controlled.

(d) Inoculum—amount and physiological status: this can markedly affect growth, but is less significant if incubation period is in excess of that required for complete utilization.

(e) Nature of medium: should contain an excess of all growth substances other than that being tested.

(f) Inhibitory substances: these might be present on glassware (e.g. detergents) unless special care is taken, or they may be components of the sample.

Discussions of ways to eliminate such interferences are given in Kavanagh (1963) and Hewitt (1977). These authors also evaluate the ways of estimating the concentration of the cell suspension by nephelometry (measuring the intensity of light scattered by the cells) or by absorptiometry (measuring the proportion of light that is transmitted through the suspension). Although use of the former is fundamentally more appropriate the latter method is more convenient and is used much more extensively. Wavelengths from 400 to 750 nm have been used; the choice often being dictated by the need to minimize interference from coloured materials or to enhance sensitivity, which increases as the wavelength decreases. Barton-Wright (1952) stated that the standard dose response curve is curvilinear throughout its length but attempts are usually made to plot turbidity measurements in a form that gives the best linear relationship (Gavin, 1957*b*; Hewitt, 1977). Precision and accuracy can be determined by using statistical procedures (Hewitt, 1977), and these are generally superior to those obtained using titrimetry (see below) (Boyne *et al*., 1967).

Alternative methods for assessing growth have been developed for use when turbidity cannot be measured.

Titrimetric procedures have been used, especially with the Lactobacilliaceae which produce various acids during growth. Titrimetry is more time consuming and less reliable than turbidity measurements (Ford, 1960; Boyne *et al*., 1967). In fact, growth rate and acid production do not exactly parallel each other, maximum growth being reached before maximum acid production. However, if incubation times are extended (72 h) over those needed for growth (16–48 h) titrimetry can be used. The procedure is often more accurate and reproducible if lactic acid is determined rather than total acidity. However, the latter is most commonly determined, either by titrating with sodium hydroxide solution and recording the titre, or measuring the resultant change in pH; automatic titrators can also be used (Kavanagh, 1972). Measuring growth of

Tetrahymena pyriformis presents unique difficulties and the methods used are considered later (Section 4.9.6.3(vi)).

A novel colorimetric method for quantitating microbial response has been developed in our laboratory. Using amino acid auxotrophs of *Escherichia coli*, it was initially shown that the extent of cell multiplication, i.e. growth (measured turbidimetrically), was directly dependent on the amount of required amino acid added (Payne *et al*., 1977). From this relationship it appeared reasonable that the extent of synthesis of individual enzymes would also be dependent on added amino acid. Furthermore, as enzyme activity is linearly related to enzyme concentration it should be possible to determine the amount of enzyme (and thus assay the added amino acid) from the enzymic activity. Specific enzyme synthesis can go to completion in a fraction of the time required for a single cell division, allowing the incubation time to be reduced from 18–48 h to minutes. Sensitivity is increased enormously for two reasons: first, the amount of amino acid needed to elicit the response is markedly decreased, and second the enzymic activity can be measured very sensitively using substrates that produce coloured or fluorescent products. This last step also has a built-in amplification factor in that prolonging the time the enzyme is incubated with substrate directly increases the yield of product which effectively enhances the sensitivity of the assay. A preliminary report of the procedure has appeared (Bell *et al*., 1977). Detailed methodology has now been developed (J. W. Payne and J. M. Tuffnell, unpublished results) based upon synthesis of the inducible enzyme β-galactosidase which cleaves the substrate *o*-nitrophenyl-β-D-galactopyranoside to liberate *o*-nitrophenol, which, being yellow, can be determined colorimetrically (see also Section 4.9.6.3(iii)).

In another new method for monitoring microbial response, similar in principle to the above enzymic method, the amino acid-dependent synthesis of proteins in an *E. coli* auxotroph is measured by a radioactive incorporation method. A double (e.g. Pro^-, Lys^-) or multiple amino acid auxotroph is used. After starvation for the required amino acids, the bacteria are added to tubes containing a large excess of one required amino acid (Pro) which is radioactively labelled, plus test samples containing low concentrations of Lys. On incubation (for about 30 min) proteins are synthesized to an extent dependent on the amount of added lysine, the amount of protein synthesis being estimated from the amount of labelled Pro incorporated into trichloroacetic acid precipitated material. The heterogeneous collection of proteins synthesized under these conditions contain a reasonably fixed ratio of Pro:Lys, and thus Lys can be assayed directly by measurement of incorporated Pro. A preliminary account of this method has appeared (Bell *et al*., 1977), and it is presently being evaluated in our laboratory.

3 Test Organisms

(i) General Considerations

The microorganisms that have been used most extensively and studied most intensively in connection with amino acid assays are all 'natural' amino acid auxotrophs. Namely, various lactic acid bacteria (*Lactobacillus, Streptococcus, Leuconostoc* spp.) which require an array of amino acids (Guirard and Snell, 1962), and the ciliate protozoan *Tetrahymena pyriformis* which apparently requires the same amino acids as are essential for normal growth of the rat (Kidder and Dewey, 1951). These and other test organisms have been discussed by several authors (Barton-Wright, 1952; Gavin, 1956; Shockman, 1963; Bolinder, 1972). Reading much of the literature it is easy to gain the impression that in general workers have tried to make the best of what nature provided (i.e. natural auxotrophs) rather than considering which qualities were needed in a test organism and attempting to produce it by appropriate genetic manipulation (i.e. mutation and selection). Amongst the desired qualities are the following:

(a) Microbial response to the test substrate should be specific and not susceptible to variation by other components of the sample, or at least, the nutritional requirements of the organism should be clearly understood so that, if necessary, due allowance can be made for effects of other constituents on microbial response.
(b) Its cultivation should be relatively simple and reliable.
(c) It should not be pathogenic.
(d) Its nutritional requirement should be stable.
(e) It should be sensitive and its response should be measurable with ease, accuracy and reliability.
(f) When evaluating nutritional quality of a sample the response should correlate with relevant biological assays.

The degree to which various test organisms conform with these criteria are considered below.

(ii) Lactic Acid Bacteria

Various strains of *Lactobacillus, Streptococcus, Leuconostoc*, and *Pediococcus* have been used in both plate and tube assays for many different amino acids. Clearly, all these combinations cannot be covered when considering the above six points.

(a) These species have complex growth requirements, they may require from one to sixteen amino acids, many need one or more vitamins, bases, and nucleic acid precursors, and they utilize only a few simple sugars (Guirard and Snell, 1962; Barton-Wright, 1963; Bridson and Brecker,

1970; Rechcigl, 1978). Commercial growth media are available that have been specifically formulated for use in amino acid assays with these organisms (see *Difco Manual*, 9th edn, 1953), although many workers have formulated their own. Nevertheless, the nutritional requirements are still not fully understood and this is evidenced by the continuing debate on the topic and the frequent suggestions for improved media, e.g. Barton-Wright (1972) advocated a more flexible basal medium modified from the media of Steele *et al*. (1949), Henderson and Snell, (1948), and Barton-Wright (1952).

An intrinsic difficulty in using these organisms is that variations and deficiencies in the medium can change the growth response to a given amount of required amino acid. In particular, the vitamin supplements alter the biosynthetic capability and thus requirement for various amino acids; these features are discussed in Guirard and Snell (1962) and Matthews and Payne (1975*b*) and the references therein. To the extent that these effects are known and understood attention is drawn to them in recommended assay methods. However, a feature that is not frequently commented upon, and yet appears to us to be a fundamental problem in the use of these organisms to assay the amino acid content of protein digests, is that their growth response to peptides containing an essential residue may be less than, equal to, or greater than that observed for an equivalent amount of the free amino acid. Snell and colleagues (Guirard and Snell, 1962) showed that: 'these three divergent growth responses result when the combined operation of transport and peptidase systems supply the essential limiting amino acid from a peptide at rates less than, equal to, or greater than those at which the free amino acid can be absorbed from the medium.' This feature was the cause of the strepogenin effect, in which particular peptides apparently possessed exceptional growth stimulatory properties (Woolley and Merrifield, 1958, 1963; Kihara and Snell, 1960*a,b*; Guirard and Snell, 1962; reviewed in Matthews and Payne, 1975*b*). The growth response to a peptide containing a required residue can depend markedly on the amino acid composition of the medium, and likewise the growth yield on a free essential amino acid may be influenced by the presence of peptides (Prescott *et al*., 1953). Peptides possessing essential residues can stimulate growth to a much greater extent than an equivalent amount of the free amino acid when the latter is rapidly degraded following absorption (Gale *et al*., 1945; Kihara *et al*., 1952), and even this differential response can vary if the activity of the degradative enzymes respond to changes in vitamin content of the medium (Kihara *et al*., 1952). Effects of this type make these fastidious microorganisms inherently susceptible to often unrecognized and essentially unquantifiable variation in growth response. The presence of uncharacterized nutrients in test samples, coupled with the continuous variation in their relative concentrations in the medium during the batch-culture tube assay may, in part, explain the frequent lack of linearity between their growth response and the amount of test sample, and yet to our knowledge and surprise no systematic investigation of the relative response to amino acids and peptides has been reported for any tube assay procedure with these organisms. Recently, several reports have appeared on fundamental studies of the uptake and utilization of peptides by streptococci (Law, 1978; Rice *et al*., 1978) (see also Sections 2.1.5.5, 3.1.2, and 3.1.5). In a series of studies in which these organisms were used in large plate assays, Bolinder (1968*a,b,c,d*, 1969, 1970a,b,c,d,e, Bolinder *et al*., 1963*a,b*, 1966; reviewed in Bolinder, 1972) measured the effects on growth of various nutrients, including peptides, and reported divergent responses compared with those on the free amino acids.

Interferences from specific substances related to the amino acid being assayed are not uncommon, e.g. the stimulatory effect of hydroxylysine on the growth response of *Leuconostoc mesenteroides* P60 to lysine (Carpenter *et al*., 1964; Hartley *et al*., 1965), and if possible due allowance must be made in each case.

(b) Because of their fastidious nature, cultivation of these organisms needs great attention, particularly for media preparation and incubation temperature. These requirements are well documented for the most-used methods (Block and Weiss, 1956; Ford, 1960, 1962, 1964, 1965, 1977; Shorrock, 1972; Shockman, 1963; Bolinder, 1972). The conditions under which the inoculum was grown can also affect the response in the assay; this aspect seems to border on the mystical for some authors, with users being exhorted (not very helpfully or scientifically) to use cells in a state of 'optimum vigour' (Barton-Wright, 1952).

(c) None of the organisms presently recommended can be considered pathogenic.

(d) In general, the 'natural' mutants appear to be resonably stable. However, the enzymic deficiencies causing their amino acid requirements are generally uncharacterized and their ability to revert spontaneously unknown. *L. casei* and *L. delbrueckii* may be less stable than most (Camien and Dunn, 1953). It has been reported often that on repeated laboratory subculturing their growth characteristics can change (Shekleton and Haynes, 1959; Hartley *et al*., 1965; Bolinder, 1970*f*, 1972).

(e) Sensitivity of their response was considered above in connection with the diffusion and tube assays in which they have been used extensively. Accuracy of results is good and compares with that for the amino acid analyser (Ford, 1977; Bolinder, 1972; Shockman, 1963). Reliability and reproducibility is less good, probably for the reasons considered in (a) and (b) above.

(f) Assessment of their usefulness in evaluating the nutritional value of protein sources, and a comparison of the results with those from chemical and animal assays has been the subject of extensive collaborative efforts. Even though many of the strains, e.g. *Strep. zymogenes*, are naturally proteolytic, it is recommended that samples are first treated *in vitro* with proteases not only to speed up the assays but also to improve correlation with animal assays (Ford, 1964, 1965; Miller *et al*., 1965; Barton-Wright, 1972). In general, when strict attention is paid to the many stipulations of the test procedures, results correlate well with animal assays and use of these microorganisms therefore has much to recommend it at the present time (Ford, 1977; Haenel and Kharatyan, 1973; Henry and Ford, 1965; Boyne *et al*., 1961, 1967, 1975; Bunyan and Woodham 1964; Carpenter and Woodham 1974; Miller *et al*., 1965; Evans *et al*., 1976).

(iii) Escherichia Coli

In spite of the many obvious advantages of this organism for assaying amino acids, very few reports of its use have appeared; this is surprising, and may have arisen because early reviews of the topic overemphasized the potential problems of mutant reversion and response to extraneous nutrients in samples (Hutner *et al*., 1958; Adamson and Simpson, 1959; Shockman, 1963). It seems appropriate now to reappraise its potential usefulness.

(a) Amino acid auxotrophs carrying mutations at any step in the biosynthetic pathway can be selected and each may have a different response to precursors and to the sparing effects of biosynthetically related amino acids. Consequently, in this respect the concept of specificity of response is different from that for natural auxotrophs. In practice, however, the only assay strain likely to be selected would be one defective in the last step in the amino acid biosynthetic pathway and thus unresponsive to any precursor. Thus, interference from sample constituents, which might be metabolised to the required amino acid, could not occur. In addition, it has been exhaustively shown that for Lys^-, Trp^-, and Met^- auxotrophs the response to small peptides is identical to that for the equivalent free amino acids (Payne *et al*., 1977; Bell *et al*., 1977; J. W. Payne and J. M. Tuffnell, unpublished results); only small peptides that can pass through the cell wall of *E. coli* (Payne and Gilvarg, 1968; and see Section 2.1.2) were used in these studies. In fact, apparently the only report to the contrary concerned a Pro^- auxotroph that grew better on proline peptides than on free proline (Stone and Hobberman, 1953).

For a given amount of the required amino acid the extent to which it is catabolized will influence the level of protein synthesis and hence growth. In principle, components of the sample may influence this catabolism, and this effect is likely to be more pronounced for some amino acids, e.g. Glx,

Asx, Gly, Ser, than for others, e.g. Lys. However, in an auxotroph, specific degradative enzymes for the amino acid may well be repressed and catabolism maintained at a minimum. In fact, we have shown there is no difference in growth or amount of protein synthesis for a given amount of Lys, Trp, or Met supplied alone or together with a nutrient mixture (Difco Assay Media) containing various amino acids, vitamins, etc. (Payne *et al*., 1977; Bell *et al*., 1977; J. W. Payne and J. M. Tuffnell, unpublished results); this theoretical difficulty does not therefore appear to be a practical problem, at least in the cases tested. Nevertheless, it seems that in principle it could be advantageous to construct the auxotroph with an additional mutation in the first specific step of its catabolism, e.g. for lysine, the enzyme saccharopine dehydrogenase, providing such catabolism is not required to uniquely produce an essential metabolite; general degradative enzymes, e.g. deaminases and decarboxylases, would need to be preserved. Alternatively, given the detailed knowledge of metabolic pathways in *E. coli*, it may be that with certain auxotrophs the end-products of amino acid catabolism could be included in the assay medium, thereby regulating endogenous breakdown of the amino acid and nullifying any possible influence of these catabolities when present in test samples.

(b) Cultivation of *E. coli* is extremely simple and reliable. It will grow in simple minimal salts media containing ammonium salts and a carbon source such as glucose, acetate or glycerol.

(c) *E. coli* is the most widely studied bacterium and the strains used are non-pathogenic.

(d) A typical spontaneous reversion rate of a point mutation causing amino acid auxotrophy is of the order of 1 in 5×10^7. This frequency presents no problems for use as a test organism providing the culture is maintained in media containing an excess of the required amino acid. However, the problem of reversion is removed entirely if deletion mutants are obtained, and using nitrous acid mutagenesis we have recently selected Trp^- and Met^- deletion mutants (J. W. Payne and J. M. Tuffnell unpublished).

(e) *E. coli* has been used in plate assays for Phe (Dickinson, 1958; Jones and Burns, 1958), Trp (Konova and Verbina, 1966), Asn (Sköld and Wallenborg, 1969), and Lys and several other amino acids (Gol'dfarb *et al*., 1968, 1974; Mnushkina *et al*., 1972; Vel'tishchev *et al*., 1973), and in tube assays for Trp (Verbina and Konova, 1966) and Lys (Verbina and Konova, 1966; Payne *et al*., 1977) with the sensitivity typically found with these procedures (see Section 4.9.6.2 above). The enzyme synthesis method (see Section 4.9.6.2 above) is of the order of 1000 times more sensitive, and if needs be this could be increased further using fluorogenic rather than coloured substrates. Assays based on inducible or derepressible enzymes other than β-galactosidase may also lead to increased sensitivity. Sensitivity of the radioactive incorporation method is at least as good as the enzyme synthesis method. These procedures have great potential where rapid, sensitive and specific amino acid analysis is required.

(f) No comparative studies relating results with *E. coli* to animal assays have been reported. Good correlation with chemical analyses for lysine present in cereals and legumes have been reported (Gol'dfarb *et al*., 1974; Payne *et al*., 1977). *E.coli* lacks extracellular proteases and peptidases making it necessary to first hydrolyse proteins *in vitro* using either acid or enzymes. For nutritional evaluation enzymes are preferred, with cleavage to a mixture of amino acids and small peptides resembling the normal intestinal process seeming a sensible objective (Payne *et al*., 1977; Matthews, 1977; Matthews and Adibi, 1976; Matthews and Payne, 1975*b*). Similarities in the amino acid and peptide transport systems of *E. coli* and the mammalian gut support the value of this bacterium as an assay organism for nutritional value. Results from the assay of heat-treated lysine samples using the *E. coli* enzyme method (J. W. Payne and J. M. Tuffnell, unpublished results) suggest that a response is obtained only to 'available' residues as would be found for animals. Extensive collaborative studies are now required to assess the value of *E. coli* in assays of this type.

(iv) Other Bacteria

Several have been used but none has gained any acceptance and they are now mainly of historical interest. Full accounts can be found in (Hunter *et al*., 1958; Adamson and Simpson, 1959; Shockman, 1963; Bolinder, 1972). *Bacillus subtilis* was used to assay Phe (Guthrie 1969*a,b*; Guthrie and Susi, 1963) and Trp (Sebek, 1965; Konova and Verbina, 1966). A strain of the pathogen *Clostridium welchii* (perfringens) that required 13 amino acids was used by Boyd *et al*. (1948), and results compared with those from *Strep. zymogenes* (Stokes *et al*., 1945). *Chromobacterium violaceum* was used in an assay for Trp (Sebek, 1965). *Pseudomonas* strains have been used but suffer from problems of pathogenicity and wide variation in response to peptides and their equivalent amino acid (Cascieri and Mallette, 1976).

(v) Yeasts and Other Fungi

In the first ever report on the quantitative assay of amino acids by the large plate method, *Saccharomyces fragilis* was used to assay β-alanine (Ericson and Carlson, 1953). However, although used to some extent in vitamin assays, yeasts have not proved useful for amino acid assays; initial studies with them are reported in (Hutner *et al*., 1958; Adamson and Simpson, 1959; Shockman, 1963).

For use in amino acid assays, fungi have the disadvantage of slow growth and complex responses to nitrogen sources in the medium (see Sections 2.3 and 3.2), on the other hand, they are relatively easy to cultivate and their biomass can be determined accurately. There are several reports in the early literature on their use as test organisms. Regnery (1944) showed that after 7 days incubation the dry weight of mycelia produced by a leucine-requiring mutant of *Neurospora* was proportional to added leucine in the medium; Gly-Leu produced an equivalent weight, Leu-Gly less, and Leu-Gly-Gly and Gly-Leu-Gly negligible yields. Varying the sucrose content of the medium affected the yield for a given level of leucine. This strain was used to assay leucine in protein hydrolysates and food stuffs (Ryan and Brand, 1944; Ryan, 1946), and a proline-requiring mutant was used similarly (Brand *et al*., 1945). Measuring the rate of germination of conidia rather than biomass was advocated by Ryan (1946). The fact that *Neurospora* is proteolytic and can utilize amino acids in proteins (Regnery, 1944) has been made use of in direct assays on food samples (Hodson and Krueger, 1947). Recently, Moyhuddin *et al*. (1976, 1977) described assay procedures based on biomass determinations for evaluating protein quality in cereals and legumes using *Aspergillus flavus* and a lysine-requiring strain of *Neurospora crassa*. These authors also drew attention to the particular advantages of fungal assays. Nevertheless, fungi are unlikely to be widely used because the assays are relatively slow, the microbial response cannot be sensitively quantitated, and most importantly, the assays may be unreliable because fungal response to various nutrients is inadequately understood.

(vi) Protozoa

Encouraged by the early findings that the ciliate protozoan *Tetrahymena pyriformis* requires the same amino acids as man and shows ability to derive these from proteins (Kidder and Dewey, 1951; Rockland and Dunn, 1946; Dunn and Rockland, 1947) many workers have sought to develop an assay using this organism. In spite of some measure of success, the procedures suffer from great practical disadvantages and, in addition, the reliability of the results is in doubt because the nutritional responses of the organism are inadequately understood.

(a) The results of studies aimed at establishing the nutrient requirements of *Tetrahymena* have been reviewed (Kidder and Dewey, 1951; Hutner *et al*., 1972; Holz, 1973). Its amino acid requirements are fairly well understood although the levels needed for certain of them is influenced by the composition of the medium (Dewey and Kidder, 1958; and see above reviews). Some of this variability arises from competition for uptake between structurally related amino acids (Rasmussen, 1976), reminiscent of that described earlier for lactobacilli. A requirement for tyrosine was recently uncovered by Zdanowski and Rasmussen (1979*a*) and shown to arise from amino acid imbalances in the culture medium. Other nutrients may also alter the growth response to a given amount of a required amino acid, e.g. glucose and dextrin are more stimulatory than starch (Stott *et al.*, 1963; Reynolds, 1970), and fatty acids have variously been reported to inhibit (Kidder and Dewey, 1951; Rosen and Fernell, 1956) or not to inhibit (Landers, 1975) the growth response.

Only recently has a start been made to study systematically the relative capacity of free amino acids, proteins, and peptides to satisfy requirements for essential amino acids. These studies have benefited greatly from use of a mutant strain of *Tetrahymena thermophila* (formerly *T. pyriformis*, syngen 1) that has a heat-sensitive development of its oral apparatus; at 37 °C it forms no detectable food vacuoles but at 28 °C vacuoles are formed at normal rates (Orias and Rasmussen, 1979). When amino acids were studied, both the wild type and the mutant showed the same growth response to Phe or Leu (both essential amino acids), and in addition, the uptake of radioactive Phe or Leu showed the same kinetics in the two strains; it was concluded that these free amino acids do not use the oral system but are absorbed by carrier-mediated systems located in the plasma membrane (Orias and Rasmussen, 1979). When proteins were studied, opposite results were found. Thus, a study was made of the ability of egg albumin to supply either all ten required amino acids or Phe and Leu only. In the wild type, 0.1 percent (w/v) heat-denatured albumin supported normal growth when providing one or ten amino acids, in contrast, in neither case was there any growth of mutant cells that lacked food vacuoles. These workers also made important observations on the superior utilizability of denatured proteins versus soluble ones, first noted by Viswantha and Liener (1956*a*). Thus, 0.1 per cent (w/w) soluble egg albumin failed to support growth although it could support some growth at 1 per cent (w/w), and no doubt this effect contributes to the variation in assay results reported by Shorrock and Ford (1973): 'Another unsatisfactory feature in the present procedure is the non-specific stimulation at high levels of test extracts . . .'. Orias and Rasmussen (1979) presented evidence that in general particulate (denatured) proteins were better absorbed than soluble ones, although some proteins, e.g. bovine serum albumin and γ-globulin, failed to support growth even when denatured. They concluded that: 'Results show that egg albumin can supply the required amino acids only in cells with a functional mouth. . . . If other ways of taking up egg albumin (protein) exist, they are insufficient for nutritional needs for amino acids.' Shorrock and Ford (1973) noted that in early stages of growth, when perhaps food vacuoles are only partially developed (Orias and Rasmussen, 1979), *Tetrahymena* utilized free amino acids in preference to protein-bound ones. It appears from these results that the proteolytic enzymes are associated predominately with the cell (food vacuole) and that any present in the culture medium (Viswantha and Liener, 1956*b*) are not important. Orias and Rasmussen (1978) concluded that: 'Tetrahymena cells release proteases in the growth medium but these proteases are unable to extensively hydrolyse egg albumin. . . . The functional contribution of these proteases, therefore, remains unknown.' Their observations should be remembered alongside the ubiquitous descriptions of *Tetrahymena* as a 'proteolytic' organism.

Dewey and Kidder (1958) presented evidence that dipeptides were utilized by *Tetrahymena*. L. Rasmussen (personal communication) showed that di- and trilysine, and di-, tri-, and tetraphenylalanine were utilized as sources of the required amino acids by both wild-type *T. thermophila* and its food-vacuole mutant. Pentalysine was used, but only after a long lag period. In both strains equal growth responses were obtained for lysine, dilysine, and trilysine, leading to the conclusion that the food vacuole is not involved in peptide uptake. These authors presented evidence that free amino acids in the medium influenced the growth response to diphenylalanine, and that after a period of incubation but not initially this dipeptide could inhibit uptake of [^{14}C]-Phe; they interpreted their results as indicating that peptides were cleaved before uptake rather than being taken up intact. Their overall conclusion from these studies was that: 'Cells differ remarkably in their ability to utilize amino acids from peptides and proteins in solution: wild type cells did not multiply in Phe- or Lys-free medium with 0.1% egg albumin to provide metabolite although they probably released proteases. In contrast, cells readily used peptides in solution—probably hydrolyzed before uptake'.

Compatible with this view was the finding (Zdanowski and Rasmussen, 1979*b*) that surface-bound peptidases were present, although the bulk of the peptidase activity (85 per cent) was intracellular, with negligible activity present in the suspension medium. It is clear that further studies are needed before the full picture of the nutritional responses to these three types of nutrient is obtained.

(b) Cultivation of protozoa is unfamilar to most potential users, the media are often extremely complex, (Stott and Smith, 1966; Baum and Haenel, 1965; Shorrock and Ford, 1973) and the procedures complicated. Sheffner (1967) summarized the feelings of most users: ' . . . culturing . . . protozoa is complex and subject to many variables.' Incubations of at least 4 days are typically used in assay procedures, during which time bacterial contamination can be a problem and autodigestion may occur. Thus, its manner of cultivation does not recommend *Tetrahymena* for widespread use.

(c) Pathogenicity is not a problem.

(d) Little has been done to characterize genetically the amino acid requirements in *Tetrahymena*. However, advances have been made recently in the genetics of this organism and especially in the selection of mutants (Bruns *et al*., 1976; Hepfer and Kaney, 1976; Orias and Bruns, 1976).

(e) One of the difficulties in the use of *Tetrahymena* for assay purposes is the actual measurement of its response. In early assays, acid production over a 41 day incubation was an index (Rockland and Dunn, 1946, 1959; Dunn and Rockland, 1947). A colorimetric method based on the enzymic formation of a red triphenylformazan dye was introduced by Anderson and Williams (1951); later workers modified the method and showed it to be susceptible to many interferences (Pilcher and Williams, 1954; Fernell and Rosen, 1956; Ford, 1960; Meinl and Wunsche, 1976; Kaestner *et al*., 1976). The relationship between cell count and the extent of ammonia production was the basis of other methods (Fernell and Rosen, 1956; Rosen and Fernell, 1956). This was improved by simply using cell count after 4 days incubation as a measure of response (Teunisson, 1961; Stott *et al*., 1963; Stott and Smith, 1966). Cell-count technique is tedious, time-consuming, and subject to considerable error, especially with respect to distinguishing cells and food particles; in spite of this it is still used by a great many investigators (Kharat'yan *et al*., 1973; Rolle, 1976; Landers, 1975; Srinivas *et al*., 1975), and various attempts have been made to try and improve it. Shorrock (1972), Shorrock and Ford (1973) attached a camera to the microscope and photographed each haemocytometer slide to simplify counting. Teunisson (1971) described a procedure of elutriation to separate the organisms from other particulate matter and an electronic counting method using a Coulter counter. Evancho *et al*. (1977) recently reported the superiority of the Coulter counter over direct microscopic counting. Gas–liquid chromatography of tetrahymanol, a specific pentacyclic terpene synthesized by *Tetrahymena*, was suggested as an index of growth by Shepherd *et al*. (1975); alternatively, 2-aminoethylphosphonic acid has been used as an indicator of growth in protein assays (Maciejewicz-Rys and Antoniewicz, 1978). Methods based on extinction measurements using test-samples clarified by predigestion with proteases *in vitro* have been proposed (Shorrock and Ford, 1973; Meinl and Wunsche, 1976; Frank *et al*., 1975; Shorrock, 1976).

(f) In spite of the inadequacy of fundamental knowledge of its nutrition, *Tetrahymena* has been used extensively in attempts to determine protein quality and nutritional availability of particular amino acids. In recent studies, the results have been compared with those from other microbiological and animal assays (Boyne *et al*., 1961; 1975; Baum, 1966*a,b*, 1968; Baum and Haenel, 1965; Helms and Rolle, 1970; Rolle and Eggum, 1971; Frank *et al*., 1975; Landers, 1975; Srinivas *et al*., 1975; Rolle, 1976; Evancho *et al*., 1977; Shorrock, 1976; Shorrock and Ford, 1973; Haenel and Kharatyan, 1973).

Although initially thought to be an advantage of the organism, in fact, the 'proteolytic activity' of *Tetrahymena* is inadequate to ensure reliable utilization of samples (see above) and an enzymic predigestion step is now regarded as essential (Shorrock, 1976). The use of papain and pepsin have been recommended by some (Stott and Smith, 1966; Boyne *et al.*, 1967; Sheffner, 1967; Shorrock, 1976), and pronase (Shepherd *et al.*, 1976) and trypsin-bromelain by others (Frank *et al.*, 1975). It is clear from the discussion above that such digestion must produce small peptides if reliable results are to be obtained. However, as denatured heat-treated proteins can be utilized directly (see above) the organism may have particular value in the evaluation of such samples.

The literature provides no clear-cut conclusions to guide potential users as to the suitability of the organism for nutritional evaluations. Most reports provide qualified support for its use, e.g. '. . . suitable test organism for a rough categorization of proteins as good, acceptable or poor; but classification within these groups is hardly possible' (Rolle, 1976); a few are more enthusiastic: 'Modification of the *T. pyriformis* W protein quality bioassay technique looks very promising and should result in a feasible means of assessing protein quality' (Landers, 1975); and others are less so: '. . . *T. pyriformis* is, therefore, scarcely suited as a test organism for determining the protein quality and amino-acid availability of feedstuffs' (Meinl and Wunsche, 1976). The recent work of Rasmussen and colleagues (see above) provides possible explanations for these disparate conclusions, but nevertheless, these recurring uncertainties, coupled with the laborious, complex, and insensitive assay procedures make it unlikely that the organism will ever find extensive use.

REFERENCES

Adamson, D. C. M. and Simpson, J. S. (1959). Microbiological assay of vitamins and amino acids. In *Proceedings of the International Symposium on Microchemistry, Birmingham University, 1958*, Pergamon Press, London

Ahrens, R. A., Wilson, J. E., Jr, and Womack, M. (1966). Utilization of calories and nitrogen by rats fed diets containing purified casein versus a mixture of amino acids simulating casein. *J. Nutr.*, **88**, 219–227

Akeson, W. R. and Stahmann, M. A. (1964). A pepsin pancreatin digest index of protein quality evaluation. *J. Nutr.*, **83**, 257–261

Anderson, M. E. and Williams, H. H. (1951). Microbiological evaluation of protein quality. 1. A colorimetric method for the determination of the growth of *Tetrahymena geleii* W in protein suspensions. *J. Nutr.*, **44**, 335–343

Axmann, H. (1975). In *Breeding for Seed Protein Improvement Using Nuclear Techniques*, IAEA, Vienna, p. 145

Barton-Wright, E. C. (1952). *The Microbiological Assay of the Vitamin-B Complex and Amino Acids*, Pitman, London

Barton-Wright, E. C. (1963). *Practical Methods for the Microbiological Assay of Vitamin-B Complex and Amino Acids*, United Trade Press, London

Barton-Wright, E. C. (1972). The microbiological assay of the essential amino acids in compound feeding stuffs. *Analyst, Lond.*, **97**, 138–141

Bassler, K. H., Fekl, W., and Lang, K. (eds) (1976). *Proceedings of the 3rd International Symposium on Balanced Nutrition in Therapy*, Springer Verlag, Berlin, Heidelberg, and New York

Baum, F. (1966*a*). Uber die Bestimmung des biologischen Eiweisswertes mit *Tetrahymena pyriformis* W. 2. Mitt. Der Eiweisswert Verschiedener Lebensund Futlermittel. *Nährung*, **10**, 453–459

Baum, F. (1966*b*). Uber die Bestimmung des biologischen Eisweisswertes mit *Tetrahymena pyriformis* W. 3. Mitt. Die Wirkung von Aminosaurezusatzen und Lagerdauer auf den Eiwesswert von Getreidemehlen. *Nährung*, **10**, 571–580

Baum, F. (1968). Uber die Bestimmung des biologischen Eiweisswertes mit *Tetrahymena pyriformis* W. 4. Mitt. Der Einfluss des Backprozesses auf den Eiweisswert von Roggenmischbrotein. *Nährung*, **12**, 61–67

Baum, F. and Haenel, H. (1965). Uber die Bestimmung der Biologischen Eiweisswertigkeit mit *Tetrahymena pyriformis* W. 1. Mitt. methodische Hinweise und Bestimmung der Wertigkeit einiger Nährungsmittel. *Nährung*, **9**, 517–525

Belikov, V. M. and Babayan, T. L. (1971). Mixtures of amino acids in human nutrition. *Russian Chem. Rev*., **40**, 441–456

Bell, G., Higgins, C. F., Payne, G. M., and Payne, J. W. (1977). Use of *Escherichia coli* to determine available lysine in plant proteins. In *Nutritional Evaluation of Cereal Mutants*, IAEA, Vienna, pp. 107–123

Bender, A. E. (1973). Chemical scores and availability of amino acids. In *Proteins in Human Nutrition* (J. W. G. Porter and B. A. Rolls, eds), Academic Press, London and New York, pp. 167–178

Bergner, H. (1978). Protein evaluation and protein metabolism In *Biochemical Aspects of New Protein Foods*, FEBS Symposium 44. (J. A. Nissen, B. O. Eggum, L. Munck, and H. S. Olsen, eds), Pergamon Press, Oxford, pp. 149–160.

Bigwood, E. J. (1972). *Protein and Amino Acid Functions*, Pergamon Press, Oxford

Bjarnason, J. and Carpenter, K. J. (1969). Mechanisms of heat damage to proteins. *Br. J. Nutr*., **23**, 859–868

Blackburn, S. (1968). *Amino Acid Determination*, Marcel Dekker, New York

Block, R. J. and Weiss, K. W. (1956). *Amino Acid Handbook. Methods and Results of Protein Analysis*, C. C. Thomas, Springfield, Ill

Bodwell, C. E. (1976). *Evaluation of Proteins for Humans*, Avi Publishing Co., Westport, Conn., p. 327

Bodwell, C. E. (1977). Use of enzymatic assays for the determination of nutritional quality. In *Nutritional Evaluation of Cereal Mutants*, IAEA, Vienna, pp. 87–105

Bolinder, A. E. (1968*a*). Plate assay methods for amino acids. 4. Usefulness of five strains of lactic-acid bacteria for the assay of six amino acids. *Acta Pharm. Suecica*, **5**, 117–134

Bolinder, A. E. (1968*b*). Plate assay methods for amino acids 5. Usefulness of nine strains of lactic-acid bacteria for the assay of methionine. *Acta Pharm. Suecica*, **5**, 401–416

Bolinder, A. E. (1968*c*). Plate assay methods for amino acids. 6. Comparative assays for methionine carried out by the plate method, the tube method and the ion-exchange column chromatographic method. *Acta Pharm. Suecica*, **5**, 417–428.

Bolinder, A. E. (1968*d*). Plate assay methods for amino acids. 7. Tyrosine. *Acta Pharm. Suecica*, **5**, 537–562

Bolinder, A. E. (1969). Plate assay methods for amino acids. 8. Tryptophan. *Acta Pharm. Suecica*, **6**, 257–270

Bolinder, A. E. (1970*a*). Plate assay methods for amino acids. 9. Threonine. *Acta Pharm. Suecica*, **7**, 47–54

Bolinder, A. E. (1970*b*). Plate assay methods for amino acids. 10. Arginine and histidine. *Acta Pharm. Suecica*, **7**, 113–124

Bolinder, A. E. (1970*c*). Plate assay methods for amino acids. 11. Further studies on lysine. *Acta Pharm. Suecica*, **7**, 125–132

Bolinder, A. E. (1970*d*). Plate assay methods for amino acids. 12 Cystine. *Acta Pharm. Suecica*, **7**, 201–208

Bolinder, A. E. (1970*e*). Plate assay methods for amino acids. *Acta Pharm. Suecica*, **7**, 209–238

Bolinder, A. E. (1970*f*). Purity of amino acids. *Analyt. Biochem*., **35**, 212–226

Bolinder, A. E. (1972). Large plate assays for amino acids. In *Analytical Microbiology*, Vol. 2 (F. Kavanagh, ed.), Academic Press, New York and London, pp. 479–591

Bolinder, A. E., Lie, S., and Ericson, L. E. (1963*a*). Plate assay methods for amino acids. 1. A sensitive cup plate for lysine with *Streptococcus faecalis* ATCC 6057. *Biotechnol Bioeng*., **5**, 131–146

Bolinder, A. E., Lie, S., and Ericson, L. E. (1963*b*). Plate assay methods for amino acids. 2. Factors affecting the cup plate assay for lysine with *Streptococcus faecalis* ATCC 6057. *Biotechnol. Bioeng*., **5**, 147–165

Bolinder, A. E., Lie, S., and Ericson, L. E. (1966). Plate assay methods for amino acids. 3. Usefulness of several lactic-acid bacteria for the assay of lysine. *Acta Pharm. Suecica*, **3**, 363–375

Booth, V. H. (1971). Problems in the determination of FDNB-available lysine. *J. Sci. Fd Agric*., **22**, 658–666

Boyd, M. J., Logan, M. A., and Tytell, A. A. (1948). The growth requirements of *Clostridium perfringens* (welchii) BP6K. *J. Biol. Chem*., **174**, 1013–1025

Boyne, A. W., Carpenter, K. J., and Woodham, A. A. (1961). Progress report on an assessment of laboratory procedures suggested as indicators of protein quality in feeding stuffs. *J. Sci. Fd Agric*., **12**, 832–848

Boyne, A. W., Ford, J. E., Hewitt, D., and Shrimpton, D. H. (1975). Protein quality of feeding stuffs. 7. Collaborative studies on the microbiological assay of available amino acids. *Br. J. Nutr*., **34**, 153–162

Boyne, A. W., Price, S. A., Rosen, G. D., and Stott, J. A. (1967). Protein quality of feeding stuffs. 4. Progress report on collaborative studies on the microbiological assay of available amino acids. *Br. J. Nutr*., **21**, 181–206

Brand, E., Saidel, L. J., Goldwater, W. H., Kessel, B., and Ryan, F. J. (1945). The empirical formula of β-lactoglobulin *J. Am. Chem. Soc*., **67**, 1524–1532

Bridson, E. Y. and Brecker, A. (1970). Design and formulation of microbial culture media. In *Methods in Microbiology*, Vol. 3A (J. R. Norris, and D. W. Ribbons, eds), Academic Press, London and New York, pp. 229–295

Bruns, P. J., Brussard, T. B., and Kavka, A. B. (1976). Isolation of homozygous mutants, after induced self-fertilization in *Tetrahymena. Proc. Natl. Acad. Sci. U.S.A*., **73**, 3243–3247

Bunyan, J. and Woodham, A. A. (1964). Protein quality of feeding stuffs. 2. The comparative assessment of protein quality in three fish meals by microbiological and other laboratory tests, and by biological evaluation with chicks and rats. *Br. J. Nutr*., **18**, 537–544

Camien, M. N. and Dunn, M. S. (1953). A D-lactic acid requiring mutant of *Lactobacillus casei. J. Biol. Chem*., **201**, 621–628

Cannon, P. R., Steffee, C. H., Frazier, L. J., Rowley, D. A., and Stepto, R. C. (1947). The influence of time of ingestion of essential amino acids upon utilization in tissues synthesis *Fed. Proc. Fed. Am. Soc. Exp. Biol*., **6**, 390

Carpenter, K. J. (1960). The estimation of the available lysine in animal-protein foods. *Biochem. J*., **77**, 604–610

Carpenter, K. J. and Booth, V. H. (1973). Damage to lysine on food processing: its measurement and significance. *Nutr. Rev*., **43**, 424–457

Carpenter, K. J. and Woodham, A. A. (1974). Protein quality of feeding stuffs. 6. Comparisons of the results of collaborative biological assays for amino acids with those of other methods *Br. J. Nutr*., **32**, 647–660

Carpenter, K. J., Hartley, A. W., and Ward, L. D. (1964). Hydroxylysine as a stimulant in microbiological assays for lysine. *Proc. Nutr. Soc*., **23**, XL–XLI

Cascieri, T. and Mallette, M. F. (1976). Peptide utilization by *Pseudomonas putida* and *Pseudomonas maltophilia. J. Gen. Microbiol*., **92**, 283–295

Christensen, H. N. (1968). Relevance of transport across the plasma membrane to the interpretation of the plasma amino acid pattern. In *Protein Nutrition and Free Amino Acid Patterns* (J. H. Leathem, ed.), Rutgers University Press, pp. 41–52

CIBA Symposium No. 4 (1972). *Peptide Transport in Bacteria and Mammalian Gut*, Elsevier, Amsterdam

CIBA Symposium No 50 (1977). *Peptide Transport and Hydrolysis*, Elsevier, Amsterdam

Clegg, K. M. (1978). Dietary enzymic hydrolysates of protein. In *Biochemical Aspects of New Protein Foods*, FEBS Symposium 44 (J. A. Nissen, B. O. Eggum, L. Munck, and H. S. Olsen, eds), Pergamon Press, Oxford, pp. 109–118

Cole, D. J. A., Boorman, K. N., Buttery, P. J., Lewis, D., Neale, R. J., and Swan, H. (eds) (1976). *Protein Metabolism and Nutrition*, Butterworths, London

Concon, J. M. (1975). Chemical determination of critical amino acids in cereal grains and other foodstuffs. In *Protein Nutritional Quality of Foods and Feeds*: Part 1. *Assay Methods* (M. Friedman, ed.), Marcel Dekker, New York, pp. 311–380

Cooper, K. E. (1963). In *Analytical Microbiology*, Vol. 1 (F. W. Kavanagh, ed.), Academic Press, New York

Cooper, K. E. (1972). In *Analytical Microbiology*, Vol. 2 (F. W. Kavanagh, ed.), Academic Press, New York

Das, M. and Radhakrishnan, A. N. (1976). Role of peptidases and peptide transport in the intestinal absorption of proteins. *World Rev. Nutr. Diet*, **24**, 58–87

Delort-Laval, J. (1976). Biological criteria of protein evaluation. In *Protein Metabolism and Nutrition* (D. J. A. Cole, K. N. Boorman, P. J. Buttery, D. Lewis, R. J. Neale, and H. Swan, eds), Butterworths, London, pp. 232–247.

Dewey, V. C. and Kidder, G. W. (1958). Amino acid antagonisms in *Tetrahymena. Arch. Biochem. Biophys*., **73**, 29–37

Dickinson, L. (1959). Experiences with the plate assay of phenylalanine with a mutant of *Escherichia coli. Analyst, Lond*., **83**, 647–648

Dreyer, J. J. (1973). Methods of protein evaluation with rats 2. In *Proteins in Human Nutrition* (J. W. G. Porter and B. A. Rolls, eds), Academic Press, London and New York, pp. 245–261

Dunn, M. S. and Rockland, L. B. (1947). Biological value of proteins determined with *Tetrahymena pyriformis* H. *Proc. Soc. Exp. Biol. Med*., **64**, 377–379

Eggum, B. O. (1973*a*). A study of certain factors influencing protein utilization in rats and pigs. Contribution no. 406 Institute of Animal Science, Copenhagen

Eggum, B. O. (1973*b*). The levels of blood amino acids and blood urea as indicators of protein quality. In *Proteins in Human Nutrition* (J. W. G. Porter and B. A. Rolls, eds), Academic Press, London and New York, pp. 317–327

Eggum, B. O. (1976). Indirect measures of protein adequacy. In *Protein Metabolism and Nutrition* (D. J. A. Cóle, K. N. Boorman, P. J. Buttery, D. Lewis, P. J. Neale, and H. Swan, eds), Butterworths, London, pp. 249–258

Eggum, B. O. (1978). Biological aspects of protein adequacy. In *Biochemical Aspects of New Protein Foods*, FEBS Symposium 44 (J. A. Nissen, B. O. Eggum, L. Munck, and H. S. Olsen, eds), Pergamon Press, Oxford, pp. 161–170

Ellinger, G. M. and Duncan, A. (1976). The determination of methionine in proteins by gas–liquid chromatography. *Biochem. J.,* **155**, 615–621

Erbesdobler, H. (1973). The normal course of digestion of food proteins. In *Proteins in Human Nutrition* (J. W. G. Porter and B. A. Rolls, eds), Academic Press, New York, pp. 453–469

Erbesdobler, H. (1976). Amino acid availability. In *Protein Metabolism and Nutrition* (D. J. A. Cole, K. N. Boorman, P. J. Buttery, D. Lewis, R. J. Neale, and H. Swan, eds), Butterworths, London, pp. 139–159

Ericson, L. E. and Carlson, B. (1953). Studies on the occurrence of amino-acids, niacin and pantothenic acid in marine algae. *Ark. Kemi*, **6**, 511–522

Evancho, G. M., Hurt, H. D., Devlin, P. A., Landers, R. E., and Ashton, D. H. (1977). Comparison of *Tetrahymena pyriformis* W and rat bioassays for the determination of protein quality. *J. Fd Sci*., **42**, 444–448

Evans, I. M., Ford, J. E., Hannah, L. C., and Boulter, D. (1976). Comparison of chemical and microbiological methods in the estimation of methionine in cowpea (*Vigna unguiculata*) seeds. *Br. J. Nutr*., **36**, 289–293

FAO (1973). *FAO nutrition reports series No. 52*, Food and Agriculture Organization, Rome

Fernell, W. R. and Rosen, G. D. (1956). Microbiological evaluation of protein quality with *Tetrahymena pyriformis* W. 1. Characteristics of growth of the organism and determination of relative nutritive values of intact proteins. *Br. J. Nutr*., **10**, 143–156

Finley, J. W. and Friedman, M. (1973). Chemical methods for available lysine. *Cereal Chem.*, **50**, 101–107

Ford, J. E. (1960). A microbiological method for assessing the nutritional value of proteins. *Br. J. Nutr.*, **14**, 485–497

Ford, J. E. (1962). A microbiological method for assessing the nutritional value of proteins. 2. The measurement of 'available' methionine, leucine, isoleucine, arginine, histidine, tryptophan and valine. *Br. J. Nutr.*, **16**, 409–425

Ford, J. E. (1964). A microbiological method for assessing the nutritional value of proteins. 3. Further studies on the measurement of available amino acids. *Br. J. Nutr.*, **18**, 449–460

Ford, J. E. (1965). A microbiological method for assessing the nutritional value of proteins. 4. Analysis of enzymically digested food proteins by Sephadex-gel filtration. *Br. J. Nutr.*, **19**, 277–293

Ford, J. E. (1975). Some effects of overheating of protein on its digestion and absorption. In *Peptide Transport in Protein Nutrition* (D. M. Matthews and J. W. Payne, eds), North Holland, Amsterdam, pp. 183–203

Ford, J. E. (1977). Microbiological procedures for protein quality assessment. In *Nutritional Evaluation of Cereal Mutants*, IAEA, Vienna, pp. 73–85

Ford, J. E. and Shorrock, C. (1971). Metabolism of heat-damaged proteins in the rat. *Br. J. Nutr.*, **26**, 311–322.

Frank, O., Baker, H., Hutner, S. H., Rusoff, I. I., and Morck, R. A. (1975). Evaluation of protein quality with the phagotrophic protozoan *Tetrahymena*. In *Protein Nutritional Quality of Foods and Feeds*, Part 1, *Assay Methods* (M. Friedman, ed.), Marcel Dekker, New York, pp. 203–209

Friedman, M. (ed.) (1975). *Protein Nutritional Quality of Foods and Feeds*, Parts 1 and 2, Marcel Dekker, New York

Friedman, M. and Finley, J. W. (1971). Methods of tryptophan analysis. *J. Agric. Fd Chem.*, **19**, 626–633

Friedman, M. and Finley, J. W. (1975). Evaluation of methods for tryptophan analysis in proteins. In *Protein Nutritional Quality of Foods and Feeds*, Part 1, *Assay Methods* (M. Friedman, ed.), Marcel Dekker, New York, pp. 423–452

Gale, E. F. (1945). The arginine, ornithine and carbon dioxide requirements of streptococci and their relation to arginine dihydrolase activity. *Br. J. Exp. Path.*, **26**, 225–233

Gavin, J. J. (1956). Microbiological process report 1. The test organism. *Appl. Microbiol.*, **4**, 323–331

Gavin, J. J. (1957*a*). Microbiological process report 2. The diffusion methods. *Appl. Microbiol.*, **5**, 25–33

Gavin, J. J. (1957*b*). Microbiological process report 3. Turbidimetric methods. *Appl. Microbiol.*, **5**, 235–243

Gitler, C. (1964). Protein digestion and absorption in non ruminants. In *Mammalian Protein Metabolism*, Vol. 1 (H. N. Munro and J. B. Allison, eds), Academic Press, New York and London, pp. 35–69

Gol'dfarb, D. M., Kuznetsova, V. N., Kalinina, L. M., Strakhova, T. S., Gryadchikov, V. G., Tsarichencko, A. P., and Tarasova, L. P. (1974). A high speed method of determining the lysine in grain by means of an auxotrophic *E. coli* mutant. *Sel'skokhozyaistvennaya Biol.*, **9**, 917–920

Gol'dfarb, D. M., Nesterova, G. F. *et al.* (1968). Auxotrophic *E. coli* strains as indicators of amino acids in human urine and blood plasma. *Genetika*, **4**, 88 (*Chem. Abstr.*, **69**, 84019y (1968))

Gray, S. J., Adibi, S. A., and Menden, E. (1968). Intestinal absorption of amino acids in man. In *Protein Nutrition and Free Amino Acid Patterns* (J. H. Leathem, ed.), Rutgers University Press, pp. 53–79

Greenaway, W. T. and Davis, G. (1975). An improved five minute biuret protein method for wheat and other grains. In *Protein Nutritional Quality of Foods and Feeds*, Part 1, *Assay Methods* (M. Friedman, ed.), Marcel Dekker, New York, pp. 403–410

Greenstein, J. P. and Winitz, M. (1961). *Chemistry of the Amino Acids*, Vol. 1, John Wiley, New York and London, pp. 245–432

Guggenheim, K. (1970). Protein evaluation by microorganisms. In *Evaluation of Novel Protein Products* (A. E. Bender, *et al*., eds), Pergamon Press, Oxford, pp. 235–246

Guirard, B. M. and Snell, E. E. (1962). Nutritional requirements of microorganisms. In *The Bacteria*, Vol. 4 (I. C. Gunsalus and R. Y. Stanier, eds), Academic Press, London and New York

Guthrie, R. (1969*a*). In *Multiple Laboratory Screening* (E. S. Benson and P. E. Strandjord, eds), Academic Press, New York, pp. 131–146

Guthrie, R. (1969*b*). Screening for phenylketonuria. *Triangle*, **9**, 104–109

Guthrie, R. and Susi, A. (1963). A simple phenylalanine method for detecting phenylketonuria in large populations of newborn infants. *Paediactrics*, **32**, 338–343

Haenel, H. and Kharatyan, S. G. (1973). Some observations on the use of microbiological techniques for the determination of protein quality. In *Proteins in Human Nutrition* (J. W. G. Porter and B. A. Rolls, eds), Academic Press, London and New York, pp. 195–206

Hartley, A. W., Ward, L. D., and Carpenter, K. H. (1965). Hydroxylysine as a growth stimulant in microbiological assays for lysine. *Analyst, Lond*., **90**, 600–605

Hegsted, D. M. (1971). Nutritional research on the value of amino acid fortification. In *Amino Acid Fortification of Foods* (N. S. Scrimshaw and A. M. Altschul, eds), MIT Press, Cambridge, Mass., pp.157–178

Hegsted, D. M. and Juliano, B. O. (1974). Difficulties in assessing the nutritional quality of rice protein. *J. Nutr*., **104**, 772–781

Helms, P. and Rolle, G. (1970). Nutritive evaluation of protein quality with *Tetrahymena pyriformis* W. In *Evaluation of Novel Protein Products*, Wenner-Gren Centre International Symposium Series, Vol. 14 (A. E. Bender, R. Kihlberg, B. Lofgvist, and L. Munch, eds), Pergamon Press, Oxford, pp. 259–264

Henderson, L. M. and Snell, E. E. (1948). A uniform medium for determination of amino acids with various microorganisms. *J. Biol. Chem*., **172**, 15–29

Henry, K. M. and Ford, J. E. (1965). The nutritive value of leaf protein concentrates determined in biological tests with rats and by microbiological methods. *J. Sci. Fd Agric*., **16**, 425–432

Hepfer, C. and Kaney, A. R. (1976). Isolation of auxotrophic mutants of *Tetrahymena pyriformis*. *J. Protozool*. **23**, 22A

Hewitt, W. (1977). *Microbiological Assay: An Introduction to Quantitative Principles and Evaluation*, Academic Press, New York and London

Hodson, A. Z. and Krueger, G. M. (1947). Changes in the essential amino acid content of proteins of dry skim milk on prolonged storage. *Arch. Biochem*., **12**, 51–55

Holsinger, V. H. and Posati, L. P. (1975). Chemical estimation of available lysine in dehydrated dairy products. In *Protein Nutritional Quality of Foods and Feeds*: Part 1. *Assay Methods* (M. Friedman, ed.), Marcel Dekker, New York, pp. 479–503

Holz, G. G. (1973). The nutrition of *Tetrahymena*: essential nutrients, feeding and digestion. In *Biology of Tetrahymena* (A. M. Elliott, ed.), Dowden, Hutchinson, and Ross, Stroudsburg, Pa., pp. 89–98

Hurt, H. D., Forsythe, R. H., and Krieger, C. H. (1975). Factors which influence the biological evaluation of protein quality by the protein efficiency ratio method. In *Protein Nutritional Quality of Foods and Feeds*: Part 1, *Assay Methods* (M. Friedman, ed.), Marcel Dekker, New York, pp. 87–112

Hutner, S. H., Cury, A., and Baker, H. (1958). Microbiological assays. *Rev. Fundamental Develop. Anal*., **30**, 849–867

Hutner, S. H., Baker, H., Frank, O., and Cox, D. (1972). Nutrition and metabolism in protozoa. In *Biology of Nutrition* (R. N. Fiennes, ed.), Pergamon Press, Oxford, pp. 85–117

IAEA (1975). Recommendations of an *ad hoc* panel on analytical screening methods for seed protein content and quality. In *Breeding for Seed Protein Improvement using Nuclear Techniques*, IAEA, Vienna, Annex III, pp. 193–209

IAEA (1977). Conclusions and recommendations. In *Nutritional Evaluation of Cereal Mutants*, IAEA, Vienna, pp. 151–187

Jacquot, R. and Peret, J. (1972). Protein efficiency ratio and related methods. In *Protein and Amino Acid Functions* (E. J. Bigwood, ed.), Pergamon, Oxford, pp. 317–347

Jones, A. and Burns, S. M. (1958). Experiences with the plate assay of L-phenylalanine with a mutant of *E. coli*. *Analyst, Lond*., **83**, 647–648

Jones, G. P. (1974). The use of dye-binding procedures for the evaluation of protein quality. Ph.D. Thesis, University of Reading

Jones, G. P. and Lakin, A. L. (1976). Determination of lysine in barley by a modified dye binding procedure. *Proc. Nutr. Soc.*, **35**, A43–44

Kaestner, H. W., Kaul, A. K., and Niemann, E. G. (1976). Use of *Tetrahymena pyriformis* to evaluate the relative nutritive value (RNV) of some crop cultivars and processed cereals. *Qual. Plant. Plant Foods Hum. Nutr.*, **25**, 361–373

Kakade, M. L. and Liener, I. E. (1969). Determination of available lysine in foods. *Analyt. Biochem.*, **27**, 273–279

Kavanagh, F. W. (1960). A commentary on microbiological assaying. *Adv. Appl. Microbiol.*, **2**, 65–93

Kavanagh, F. W. (ed.) (1963). *Analytical Microbiology*, Vol. 1, Academic Press, New York

Kavanagh, F. W. (ed.), (1972). *Analytical Microbiology*, Vol. 2, Academic Press, New York

Kharat'yan, S. G., Auerbakh, T. L., Chudina, V. I., and Pozmogova, I. N. (1973). Growth of *Tetrahymena pyriformis* W on feed products of microbial origin. English translation from *Prikl. Biokhim. Mikrobiol.*, **9**, 768–772

Kidder, G. W. and Dewey, V. C. (1951). The biochemistry of ciliates in pure culture. In *Biochemistry and Physiology of Protozoa*, Vol. 1 (A. Lwoff, ed.), Academic Press, New York, pp. 323–401

Kihara, H. and Snell, E. E. (1960*a*). Peptides and bacterial growth. VIII. The nature of strepogenin. *J. Biol. Chem.*, **235**, 1409–1414

Kihara, H. and Snell, E. E. (1960*b*). Peptides and bacterial growth. IX. Release of double inhibitions by single peptides. *J. Biol. Chem.*, **235**, 1415–1418

Kihara, H., Klatt, O. A., and Snell, E. E. (1952). Peptides and bacterial growth. III Utilization of tyrosine and tyrosine peptides by *Streptococcus faecalis. J. Biol. Chem.*, **197**, 801–807

Kojima, H., Furukawa, Y., and Uemura, T. (1964). *Nippon Nogei Kagaku Kaishi*, **38**, 43. *Chem. Abstr.*, **62**, 9448a (1965)

Konova, I. U. and Verbina, N. M. (1966). Quantitative assay of tryptophan by the diffusion method using indicator cultures of *Escherichia coli* and *Bacillus subtilis. Microbiology*, **35**, 906–911

Lakin, A. L. (1973). Evaluation of protein quality by dye-binding procedures. In *Proteins in Human Nutrition* (J. W. G. Porter and B. A. Rolls, eds), Academic Press, New York, pp. 179–193

Landers, R. E. (1975). Relationship between protein efficiency ratio of foods and relative nutritive value measured by *Tetrahymena pyriformis* W bioassay techniques. In *Protein Nutritional Quality of Foods and Feeds*: Part 1, *Assay Methods* (M. Friedman, ed.), Marcel Dekker, New York, pp. 185–202

Law, B. A. (1978). Peptide utilization by Group N streptococci. *J. Gen. Microbiol.*, **105**, 113–118

Lees, K. A. and Toothill, J. P. R. (1955*a*). Microbiological assay on large plates. 1. Routine laboratory assays. *Analyst, Lond.*, **80**, 95–109

Lees, K. A. and Toothill, J. P. R. (1955*b*). Microbiological assay on large plates 2. Methods for precision assays. *Analyst, Lond.*, **80**, 110–118

Lees, K. A. and Toothill, J. P. R. (1955*c*). Microbiological assay on large plates. 3. High throughput, low precision assays. *Analyst, Lond.*, **80**, 531–535

Lipton, S. H. and Bodwell, C. E. (1975). Chemical approaches for estimating nutritionally available methionine. In *Protein Nutritional Quality of Foods and Feeds*: Part 1, *Assay Methods* (M. Friedman, ed.), Marcel Dekker, New York, pp. 569–597

Longenecker, J. B. (1973). Ion exchange chromatography for amino acid assay: application for determining amino acid requirements of the rat. In *Proteins in Human Nutrition* (J. W. G. Porter and B. A. Rolls, eds), Academic Press, London and New York, pp. 155–167

Maars, T. C., Addison, J. M., Burston, D., and Matthews, D. M. (1975). Changes in plasma amino acid concentrations in man after ingestion of an amino acid mixture simulating casein, and a tryptic hydrolysate of casein. *Br. J. Nutr.*, **34**, 259–265

Maciejewicz-Rys, J. and Antoniewicz, A. M. (1978). 2-Aminoethylphosphonic acid as an indicator of *Tetrahymena pyriformis* W growth in protein-quality evaluation assay. *Br. J. Nutr.*, **40**, 83–90

McLaughlan, J. M. and Campbell, J. A. (1969). Methodology of protein evaluation. In *Mammalian Protein Metabolism*, Vol. 3 (H. N. Munro, ed.), Academic Press, New York and London, pp. 391–422

McLaughlan, J. M. and Keith, M. O. (1975). Bioassays of protein quality. In *Protein Nutritional Quality of Foods and Feeds*: Part 1, *Assay Methods* (M. Friedman, ed.), Marcel Dekker, New York, pp. 79–85

Matthews, D. M. (1975). Intestinal absorption of peptides. *Physiol. Rev.*, **55**, 537–608

Matthews, D. M. (1977). Memorial Lecture: Protein absorption–then and now. *Gastroenterology*, **73**, 1267–1279

Matthews, D. M. and Adibi, S. A. (1976). Peptide absorption. *Gastroenterology,* **71**, 151–161.

Matthews, D. M. and Payne, J. W. (eds), (1975*a*). *Peptide Transport in Protein Nutrition*, North-Holland, American Elsevier, Amsterdam, New York

Matthews, D. M. and Payne, J. W. (1975*b*). Peptides in the nutrition of microorganisms and peptides in relation to animal nutrition. In *Peptide Transport in Protein Nutrition* (D. M. Matthews and J. W. Payne, eds), North Holland, American Elsevier, Amsterdam, New York, pp. 1–60

Matthews, D. M. and Payne, J. W. (1980). Transmembrane transport of small peptides. *Curr. Topics Membranes Transport*, **14**, 331–425

Mauron, J. (1970*a*). Nutritional evaluation of proteins by enzymatic methods. In: *Evaluation of Novel Protein Products* (A. E. Bender, R. Kihlberg, B. Lofqvist, and L. Munck, eds), Pergamon Press, London, p. 211

Mauron, J. (1970*b*). Nutritional evaluation of proteins by enzymatic methods. In *Improving Plant Proteins by Nuclear Techniques*, IAEA, Vienna, p. 203

Mauron, J. (1972). Influence of industrial and household handling on food protein quality. In *Protein and Amino Acid Functions* (E. J. Bigwood, ed.), Pergamon Press, Oxford, pp. 417–473

Mauron, J. (1973). The analyses of food proteins, amino acid composition and nutritive value. In *Proteins in Human Nutrition* (J. W. G. Porter and B. A. Rolls, eds), Academic Press, New York and London, pp. 139–154

Meinl, M. and Wunsche, J. (1976). Untersuchungen zur Lysinverfugbarkeit mit Hilfe des Testorganismus *Tetrahymena pyriformis. Nährung*, **20**, 649–660

Meister, A. (1965). *Biochemistry of the Amino Acids*, 2nd edn, Vol. 1, Academic Press, New York and London, pp. 201 230

Miller, E. L., Carpenter, K. J., Morgan, C. B., and Boyne, A. W. (1965). Availability of sulphur amino acids in protein foods. 2. Assessment of available methionine by chick and microbiological assays. *Br. J. Nutr.*, **19**, 249–267

Mnushkina, M. A., Grigor'eva, G. A., Kuznetsova, V. N., *et al*. (1972). Determination of amino acid content of human urine and blood plasma during various pathological conditions with the aid of auxotrophic *E. coli* mutants. *Laboratornoe Delo*, **8**, 493–496

Mohyuddin, M., Kaul, A. K., Sharma, T. R., and Niemann, E. G. (1976). Use of two fungal strains as test organisms for the evaluation of relative nutritive value in cereals and legumes. *Qual. Plant. Plant Foods Hum. Nutr.*, **25**, 317–329

Mohyuddin, M., Sharma, T. R., Dhar, P., Kaul, A. K., and Niemann, E. G. (1977). Rapid assay of protein quality in cultivars of cereals using *Aspergillus flavus*. In *Nutritional Evaluation of Cereal Mutants*, IAEA, Vienna, pp. 125–145

Morrison, A. B. and McLaughlan, J. M. (1972). Availability of amino acids in foods. In *Protein and Amino Acid Functions* (E. J. Bigwood, ed.), Pergamon, Oxford, pp. 389–416

Njaa, L. R. (1977). Biological methods for evaluation of protein quality. In *Nutritional Evaluation of Cereal Mutants*, IAEA, Vienna, pp. 59–72

Orias, E. and Bruns, P. J. (1976). Induction and isolation of mutants in *Tetrahymena*. In *Methods in Cell Biology*, Vol. 13 (D. M. Prescott, ed.), Academic Press, New York and London, pp. 247–282

Orias, E. and Rasmussen, L. (1979). Dual capacity for nutrient uptake in *Tetrahymena*. V. Utilization of amino acids and proteins. *J. Cell Sci.*, **36**, 343–351.

PAG (1975). Protein Advisory Group (PAG) guideline on protein methods for cereal breeders, as related to human nutritional requirements (PAG guideline no. 16). *PAG* Bull. **5** (2), 22–48

Paulis, J. W. and Wall, J. S. (1975). Protein quality in cereals evaluated by rapid estimation of specific protein fractions. In *Protein Nutritional Quality of Foods and Feeds* (M. Friedman, ed.), Marcel Dekker, New York, pp. 381–402

Payne, J. W. and Gilvarg, C. (1968). Size restriction on peptide utilization in *Escherichia coli. J. Biol. Chem.*, **243**, 6291–6299

Payne, J. W., Bell, G., and Higgins, C. F. (1977). The use of an *Escherichia coli* Lys^- auxotroph to assay nutritionally available lysine in biological materials. *J. Appl. Bacteriol.*, **42**, 165–177

Pellett, P. L. (1973). Methods of protein evaluation with rats. 1. In *Proteins in Human Nutrition* (J. W. G. Porter and B. A. Rolls, eds), Academic Press, London and New York, pp. 225–244

Pilcher, H. L. and Williams, H. H. (1954). Microbiological evaluation of protein quality. 2. Studies of the responses of *Tetrahymena pyriformis* W to intact proteins. *J. Nutr.*, **53**, 589–599

Pion, R. (1973). The relationships between the levels of free amino acids in blood and muscle and the nutritive value of proteins. In *Proteins in Human Nutrition* (J. W. G. Porter and B. A. Rolls, eds), Academic Press, London and New York, pp. 329–342

Porter, J. W. G. and Rolls, B. A. (eds) (1973). *Proteins in Human Nutrition*, Academic Press, London and New York

Prescott, J. M., Peters, V. J., and Snell, E. E. (1953). Peptides and bacterial growth. V. Serine peptides and growth of *Lactobacillus delbruecki* 9649. *J. Biol. Chem.*, **203**, 533–540

Ramasarma, G. B., Henderson, L. M., and Elvehjem, C. A. (1949). Purified amino acids as a source of nitrogen for the growing rat. *J. Nutr.*, **38**, 177–188

Ranhotra, G. S. and Johnson, B. C. (1965). Effect of feeding different amino acid diets on growth of rats and nitrogen retention of weanling rats. *Proc. Soc. Exp. Biol. Med.*, **118**, 1197–1203

Rasmussen, L. (1976). Nutrient uptake in *Tetrahymena. Carlsberg Res. Commun.*, **41**, 143–167

Rechcigl, M. (1978). Culture media for microorganisms and plants. In *Handbook of Nutrition and Food*, Vol. III, CRC Press, Florida, Section G

Regǹery, D. C. (1944). A leucineless mutant strain of *Neurospora crass. J. Biol. Chem.*, **154**, 151–160

Reynolds, H. (1970). Effect of type of carbohydrate on amino acid accumulation and utilization by *Tetrahymena. J. Bacteriol.*, **104**, 719–725

Rice, G. H., Stewart, F. H. C., Hillier, A. J., and Jago, G. R. (1978). The uptake of amino acids and peptides by *Streptococcus lactis. J. Dairy Res.*, **45**, 93–107

Rockland, L. B. and Dunn, M. S. (1946). The microbiological determination of tryptophan in unhydrolyzed casein with *Tetrahymena Geleii* H. *Arch. Biochem.*, **11**, 541–543

Rockland, L. B. and Dunn, M. S. (1959). Determination of the biological value of proteins with *Tetrahymena Geleii* H. *Food Technology*, **3**, 289–292

Rolle, G. (1976). Research concerning the usefulness of *Tetrahymena pyriformis* W as a test organism for the evaluation of protein quality. *Acta. Agric. Scand.*, **26**, 282–286

Rolle, G. and Eggum, B. O. (1971). A comparative protein evaluation with *Tetrahymena pyriformis* W and rats. *Acta. Agric. Scand.*, **21**, 69–72

Rose, W. C., Coon, M. J., and Lambert, G. F. (1954). The amino acid requirements of man. VI. The role of the calorie intake. *J. Biol. Chem.*, **210**, 331–337

Rosen, G. D. and Fernell, W. R. (1956). Microbiological evaluation of protein quality with *Tetrahymena pyriformis* W. 2. Relative nutritive value of proteins in foodstuffs. *Br. J. Nutr.*, **10**, 156–169

Ryan, F. J. (1946). The application of *Neurospora* to bioassay. *Fed. Proc. Fed. Am. Soc. Exp. Biol.*, **5**, 366–369

Ryan, F. J. and Brand, E. (1944). A method for the determination of leucine in protein hydrolysates and in foodstuffs by the use of a *Neurospora* mutant. *J. Biol. Chem.*, **154**, 161–170

Sauberlich, H. E. (1961). Growth of rats fed protein-free diets supplemented with purified amino acid mixtures. *J. Nutr.*, **74**, 298–305

Schelling, G. T. (1975). An efficient procedure for the complete evaluation of dietary proteins. In *Protein Nutritional Quality of Foods and Feeds*: Part I, *Assay Methods* (M. Friedman, ed.), Marcel Dekker, New York, pp. 137–163

Sebek, I. K. (1965). Microbiological method for the determination of tryptophan. *J. Bacteriol.*, **90**, 1026–1031.

Sheffner, A. L. (1967). *In vitro* protein evaluations. In *Newer Methods of Nutritional Biochemistry*, Vol. 3 (A. A. Albanese, ed.), Academic Press, New York, p. 125

Shekleton, M. C. and Haynes, W. C. (1959). Microbiological assays with several strains of *Leuconostoc mesenteroides. J. Bacteriol.*, **77**, 114–115

Shepherd, N. D., Taylor, T. G., and Wilton, D. C. (1975). An improved method for the determination of the biological availability of lysine in proteins using the microorganism *Tetrahymena pyriformis. Proc. Nutr. Soc.* **34**, 100A–101A

Shockman, G. D. (1963). Amino acids. In *Analytical microbiology*, Vol. 1 (F. Kavanagh, ed.), Academic Press, New York and London, pp. 568–674

Shorrock, C. (1972). Studies on the biological availability of amino acids in feedstuffs. Ph.D. Thesis, Reading University

Shorrock, C. (1976). An improved procedure for the assay of available lysine and methionine in feedstuffs using *Tetrahymena pyriformis* W. *Br. J. Nutr.*, **35**, 333–341

Shorrock, C. and Ford, J. E. (1973). An improved procedure for the determination of available lysine and methionine with *Tetrahymena*. In *Proteins in Human Nutrition* (J. W. G. Porter, and B. A. Rolls, eds), Academic Press, London and New York, pp. 207–223

Silano, V. (1977). Factors affecting digestibility and availability of proteins in cereals. In *Nutritional Evaluation of Cereal Mutants* IAEA, Vienna, pp. 13–47

Silk, D. B. A., Clark, M. L., Marrs, T. C., Addison, J. M., Burston, D., Matthews, D. M., and Clegg, K. M. (1975). Jejunal absorption of an amino acid mixture simulating casein. and an enzymic hydrolysate of casein prepared for oral administration to normal adults. *Br. J. Nutr.*, **33**, 95–100

Sköld, O. and Wallenborg, C. (1969). A microbiological method for determination of asparagine in blood serum. *Acta Pharm. Suecica*, **6**, 407–412

Slump, P. and van Beek, L. (1975). Amino acids in faeces related to digestibility of food proteins. In *Protein Nutritional Quality of Foods and Feeds*: Part 1, *Assay Methods* (M. Friedman, ed.), Marcel Dekker, New York, pp. 67–78

Srinivas, H., Vakil, U. K., and Sreenivasan, A. (1975). Evaluation of protein quality of irradiated foods using *Tetrahymena pyriformis* W and rat assay. *J. Fd Sci.*, **40**, 65–69

Stahmann, M. A. and Woldegiorgis, G. (1975). Enzymatic methods for protein quality determination. In *Protein Nutritional Quality of Foods and Feeds*: Part 1, *Assay Methods* (M. Friedman, ed.), Marcel Dekker, New York, pp. 211–234

Steele, B. F., Sauberlich, H. E., Reynolds, M. S., and Baumann, C. A. (1949). Media for *Leuconostoc mesenteroides* P-60 and *Leuconostoc citrovorum* 8081. *J. Biol. Chem.*, **177**, 533–544

Stokes, J. L., Gunness, M., Dwyer, I. M., and Caswell, M. C. (1945). Microbiological methods for the determination of amino acids 2. A uniform assay for the ten essential amino acids. *J. Biol. Chem.*, **160**, 35–49

Stone, D. and Hoberman, H. D. (1953). Utilization of proline peptides by a prolineless mutant of *Escherichia coli. J. Biol. Chem.*, **202**, 203–208

Stott, J. A. and Smith, H. (1966). Microbiological assay of protein quality with *Tetrahymena pyriformis* W. 4. Measurement of available lysine, methionine, arginine and histidine. *Br. J. Nutr.*, **20**, 663–673

Stott, J. A., Smith, H., and Rosen, G. D. (1963). Microbiological evaluation of protein quality with *Tetrahymena pyriformis* W. 3. A simplified assay procedure. *Br. J. Nutr.*, **17**, 227–233

Teunisson, D. J. (1961). Microbiological assay of intact proteins using *Tetrahymena pyriformis* W. 1. Survey of protein concentrates. *Analyt. Biochem*, **2**, 405–420

Teunisson, D. J. (1971). Elutriation and Coulter counts of *Tetrahymena pyriformis* grown in peanut and cotton seed meal media. *Appl. Microbiol.*, **21**, 878–887

Udy, D. C. (1971). Improved dye method for estimating protein. *J. Am. Oil Chem. Soc.*, **48**, 1–7

Vel'tishchev, Yu. E., Barashnev, Yu. I., Gol'dfarb, D. M., *et al.*, (1973). Diagnosis of hereditary diseases of amino-acid metabolism in children with the aid of auxotrophic *E. coli* K-12 strains. *Pediatriya*, **3**, 12–17

Verbina, N. M. and Konova, I. V. (1966). Microbiological methods of the quantitative determination of lysine and tryptophan with mutants of *Escherichia coli. Microbiology*, **35**, 394–399

Viswantha, T. and Liener, I. E. (1956*a*). Utilization of native and denatured proteins by *Tetrahymena pyriformis., Arch. Biochem. Biophys.*, **56**, 222–229

Viswantha, T. and Liener, I. E. (1956*b*). Isolation and properties of a proteinase from *Tetrahymena pyriformis* W. *Arch. Biochem. Biophys.*, **61**, 410–421

Wall, L. L. and Gehrke, C. W. (1974). Automated enzymatic determination of L-lysine in grain. *J. Assoc. Off. Anal. Chem.*, **57**, 1098–1103

Winitz, M. and Adams, R. F. (1971). Chemically defined diets. In *Balanced Nutrition and Therapy* (K. Lang, W. Fekl, and G. Berg, eds), Georg Thieme Verlag, Stuttgart, pp. 46–66

Winitz, M., Seedman, D. A., and Graff, J. (1970). Studies in metabolic nutrition employing chemically defined diets. I. Extended feeding of normal human adult males. *Am. J. Clin. Nutr.*, **23**, 525–537

Woolley, D. W. and Merrifield, R. B. (1958). Specificity of peptides. *Science, N.Y.*, **128**, 238–240

Woolley, D. W. and Merrifield, R. B. (1963), Anomalies in the structural specificity of peptides. *Ann. N.Y. Acad. Sci.*, **104**, 161–171

Zdanowski, M. K. and Rasmussen, L. (1979*a*). Tyrosine requirement in *Tetrahymena* revealed by nutritional amino acid imbalances. *Comp. Biochem. Physiol.,* **62**, 331–334

Zdanowski, M. K. and Rasmussen, L. (1979*b*). Peptidase activity in *Tetrahymena. J. Cell. Physiol.*, **100**, 407–412.

Addenda

Addendum to Section 1.1

Y. ANRAKU

Since this chapter was submitted there have been some interesting and important developments. The identification and purification of some carrier proteins for amino acids have been described (Ames and Nikaido, 1978; Lee *et al*., 1979; Motojima *et al*., 1979). Genetic studies of the components, particularly the carrier proteins of several amino acid transport systems have also been developed. Major contributions on this topic, together with those on other intriguing topics covered in this chapter are listed below.

Ames, G. F.-L., Biek, D. P., and Spudich, E. N. (1978). Duplications of histidine transport genes in *Salmonella typhimurium* and their use for the selection of deletion mutants. *J. Bacteriol.,* **136**, 1094–1108.

Ames, G. F.-L., and Nikaido, K. (1978). Identification of a membrane protein as a histidine transport component in *Salmonella typhimurium. Proc. Natl. Acad. Sci. U.S.A.,* **75**, 5447–5451.

Anderson, J. J., and Oxender, D. L. (1978). Genetic separation of high- and low-affinity transport systems for branched-chain amino acids in *Escherichia coli* K-12. *J. Bacteriol.,* **136**, 168–174.

Funanage, V. L., Ayling, P. D., Dendinger, S. M., and Brenchley, J. E. (1978). *Salmonella typhimurium* LT-2 mutants with altered glutamine synthetase levels and amino acid uptake activities. *J. Bacteriol.,* **136**, 588–596.

Hong, J.-S., Hunt, A. G., Masters, P. S., and Lieberman, M. A. (1979). Requirement of acetyl phosphate for the binding protein-dependent transport systems in *Escherichia coli. Proc. Natl. Acad. Sci. U.S.A.,* **76**, 1213–1217.

Hoshino, T. (1979). Transport systems for branched-chain amino acids in *Pseudomonas aeruginosa. J. Bacteriol.,* **139**, 705–712.

Kustu, S. G., McFarland, N. C., Hui, S. P., Esmon, B., and Ames, G. F.-L. (1979). Nitrogen control in *Salmonella typhimurium*: co-regulation of synthesis of glutamine synthetase and amino acid transport systems. J. Bacteriol., **138**, 218–234.

Lanyi, J. K., and Silverman, M. P. (1979). Gating effects in *Halobacterium halobium* membrane transport. *J. Biol. Chem.,* **254**, 4750–4755.

Lee, S.-H., Cohen, N. S., Jacobs, A. J., and Brodie, A. F. (1979). Isolation, purification, and reconstitution of a proline carrier protein from *Mycobacterium phlei. Biochemistry,* **18**, 2232–2239.

Metzer, E., Levitz, R., and Halpern, Y. S. (1979). Isolation and properties of *Escherichia coli* K-12 mutants impaired in the utilization of γ-aminobutyrate. *J. Bacteriol.*, **137**, 1111–1118.

Motojima, K., Yamato, I., Anraku, Y., Nishimura, A., and Hirota, Y. (1979). Amplification

and characterization of the proline transport carrier of *Escherichia coli* K-12 by using *proT*[+] hybrid plasmids. *Proc. Natl. Acad. Sci. U.S.A.*, **76**, 6255–6259.
Ratzkin, B., Grabnar, M., and Roth, J. (1978). Regulation of the major proline permease gene of *Salmonella typhimurium. J. Bacteriol.*, **133**, 737–743.
Yamato, I., Ohki, M., and Anraku, Y. (1979). Genetic and biochemical studies of transport systems for branched-chain amino acids in *Escherichia coli* K-12. *J. Bacteriol.*, **138**, 24–32.

Addendum to Section 1.6

D. L. OXENDER, S. C. QUAY, AND J. J. ANDERSON

CLONING AND EXPRESSION OF LEUCINE TRANSPORT GENES *IN VITRO*

Recently, the four genes encoding the components of the high-affinity branched-chain amino acid transport systems in *E. coli, livH, livG, livJ* and *livK*, have been cloned into the plasmid vector, pACYC184. When the cloned plasmid DNA was used to direct a transcriptional/translation system *in vitro* the LIV-binding protein and the leucine-specific binding protein were produced in precursor forms with molecular weights approximately 2500 higher than the processed mature form. A minicell-producing strain transformed with the plasmid DNA produced the binding proteins only in the mature form (Oxender *et al*., 1980a).

In further studies with the cloned transport genes, a 2.1 kilobase *Bgl*II fragment containing *livK*, the gene coding for the leucine-specific binding protein, has been cloned into the *BamH*I site of the plasmid vector pBR322. The DNA sequence of the resulting plasmid, pOX7, established the location of the region coding for the N-terminus of leucine-specific binding protein. DNA sequencing of the *livK* gene and amino acid microsequencing of the LS-binding protein synthesized *in vitro* established that the precursor form has the following 23 additional amino acid residues present as an extension of the N-terminus of the mature protein: Met-Lys-Ala-Asn-Ala-Lys -Thr-Ile-Ile-Ala-Gly-Met-Ile-Ala-Leu-Ala-Ile-Ser-His-Thr-Ala-Met-Ala-(N-terminus of mature LS-binding protein). The precursor forms of the binding proteins produced *in vitro* can be processed by membrane preparations and the processing enzyme is currently being isolated (Oxender *et al*., 1980b).

Since the preparation of this chapter, we have shown that the class of transport defective mutants (*livH*, Table 1) which have normal levels of both leucine-binding proteins (LIV- and LS-binding proteins) can now be divided into two genetic complementation groups, *livH* and *livG*. These results suggest that two components, presumably in the membrane, are required for the functioning of each of the periplasmic binding proteins (Mayo, 1979).

Our current understanding of the number of these genes and their order on the *E. coli* chromosome is as follows:

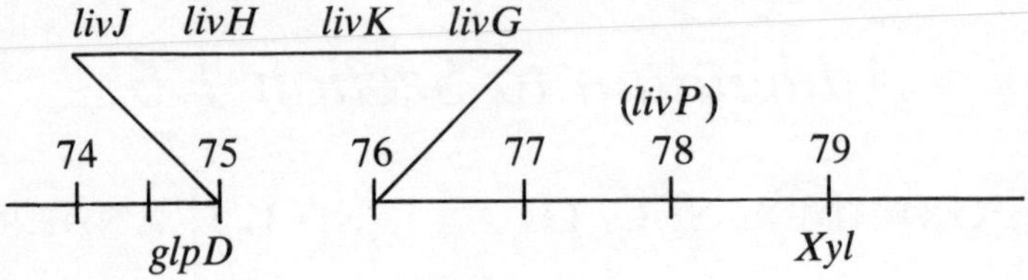

livJ: Liv-binding protein
livH and *livG*: additional components
livK: leucine-specific binding protein
livP: LIV-II transport gene locus

REFERENCES

Oxender, D. L., Anderson, J. J., Daniels, C. J., Landick, R., Gunsalus, R. P., Zurawski, G., Selker, E., and Yanofsky, C. (1980a). *Proc. Natl. Acad. Sci. U.S.A.*, **77**, 1412–1416.

Oxender, D. L., Anderson, J. J., Daniels, C. J., Landick, R., Gunsalus, R. P., Zurawski, G., and Yanofsky, C. (1980b). *Proc. Natl. Acad. Sci. U.S.A.*, **77**, April (in press).

Mayo, M. M. (1979). Ph.D thesis, University of Michigan.

Addendum to Section 1.7

I. R. BOOTH AND W. A. HAMILTON

ENERGY COUPLING TO BINDING PROTEIN SYSTEMS

During the last eighteen months the controversy over the mechanism of energy-coupling to the binding protein-dependent transport systems has been heightened rather than resolved.

The phosphate-energy donor has recently been suggested to be acetyl phosphate, rather than ATP. Mutants of *Escherichia coli* which lack phosphotransacetylase activity were unable to transport glutamine, methionine, and histidine when supplied with pyruvate as energy source, even though a significant amount of ATP was present in the cells (Hong *et al*., 1979). On the other hand, the *Kdp* potassium transport system of *E. coli*, previously classified as a binding protein system (Rhoads and Epstein, 1977), has been shown to be an ATP-dependent pump (Epstein *et al*., 1978). Similarly, ATP has been implicated directly in the activity of a calcium transport system in *Streptococcus faecalis* (Kobayashi *et al*., 1978).

There is, similarly, considerable uncertainty about the role of the protonmotive force in the activity of the binding protein systems. Studies with the *ecf* mutants of *E. coli* led to the conclusion that the membrane potential ($\Delta\Psi$) was essential for glutamine transport (Lieberman and Hong, 1976). Further evidence to support this conclusion has been obtained by Plate (1979), and by Singh and Bragg (1976, 1977). Particularly important are the experiments reported by Plate (1979), in which diminution of the membrane potential by valinomycin in the presence of 20 mM K^+ resulted in a dramatic reduction in glutamine uptake into EDTA-treated *E. coli* cells. This direct evidence for the role of $\Delta\Psi$ in these systems supplants the earlier evidence which has often been criticized on the basis of its indirect character (Hong *et al*., 1979). However, results of further work with the *ecf* mutants has indicated that $\Delta\Psi$ is not necessary for the activity of binding protein systems, and that the *ecf* mutation in some way interferes rather with the generation of acetyl phosphate (Hong *et al*., 1979).

At present therefore, there is no consensus on the mechanism of energy coupling to the binding protein systems.

ENERGY COUPLING VIA THE PROTONMOTIVE FORCE

Progress towards understanding the mechanism of energy coupling via the protonmotive force has more often been made through the study of sugar uptake rather than amino acid transport. However, a significant contribution has been made by Anraku and his colleagues (Amanuma *et al*., 1977). Using a purified proline carrier preparation from *E. coli*, reconstituted into vesicles prepared from the extracted phospholipids of the same strain, they studied the binding of the substrate to the protein in the absence of energy. Their studies revealed that the dissociation constant of proline increased from 0.5 μM at pH 5.3 to 10 μM at pH 7. This marked pH-dependence of solute binding led them to propose that the mechanism of energy coupling involved the ΔpH-induced generation of transmembrane assymetry in the proline binding sites of the carrier molecule (Amanuna *et al*., 1977).

In *Halobacterium halobium* variable energy coupling controlled by the magnitude of ΔΨ has been reported (Lanyi, 1978a,b; Lanyi and Silverman, 1979). Serine and aspartate uptake systems were found to have differing Na^+:amino acid stoicheiometries (n_{Na^+}) depending on whether a chemical gradient of sodium ions ($\Delta\mu_{Na^+}$) or ΔΨ was used to drive transport. Uptake of serine and aspartate driven by low values of $\Delta\mu_{Na^+}$ resulted in values of n_{Na^+} of 2 and 4 respectively, while the corresponding values with ΔΨ were 1 and 2 respectively (Lanyi, 1978a). Subsequent studies revealed that the magnitude of n_{Na^+} could be diminished if a membrane potential greater than 35 mV was imposed simultaneously with $\Delta\mu_{Na^+}$ (Lanyi, 1978b; Lanyi and Silverman, 1979). Thus, ΔΨ appeared to exert a control function on energy coupling to the carrier.

In *E. coli*, earlier reports of pH-dependent variations in the proton-solute stoicheiometry have been extended (Ramos and Kaback, 1977). However, experiments in whole *E. coli* cells have so far led to the opposite conclusion, that at least for lactose, the stoicheiometry is not affected by pH (Zilberstein *et al*., 1979; Booth *et al*., 1979). Further, the experimental basis of the original experiments supporting the concept of pH-dependent proton:solute stoicheiometries has been attacked (Booth *et al*., 1979; see above). In addition, the accumulation of anions by whole cells has been reported in two separate studies to be electrogenic ($n = 2$) at all pH values and to be limited by kinetic factors (Lagarde, 1977; Mitchell *et al*., 1979). Thus the validity of the concept of pH-dependent proton:solute stoicheiometries is still a matter of controversy.

The nature of energy coupling to active transport can be approached on two levels. At the level of the energy supply and its integration with cellular metabolism, the kinetic and pseudo-thermodynamic investigative approaches have yielded much information. However, for the detailed molecular mechanism information will only be revealed by careful investigation of the nature of the purified carrier proteins. In this regard, the experiments of Amanuma *et al*. (1979), and the experiences of Lo (1977) and Ames and Nikaido (1978) undoubtedly point the way ahead.

REFERENCES

Amanuma, H., Hoh, J., and Anraku, Y. (1977). *FEBS Letters,* **78**, 173–176.
Ames, G. F.-L., and Nikaido, K. (1978). *Proc. Natl. Acad. Sci. U.S.A.,* **75**, 5447–5451.
Booth, I. R., Mitchell, W. J., and Hamilton, W. A. (1979). *Biochem. J.,* **182**, 687–696.
Epstein, W., Whitelaw, V., and Hesse, J. (1978). *J. Biol. Chem.,* **253**, 6666–6668.
Hong, J.-S., Hunt, A. G., Masters, P. S., and Lieberman, M. A. (1979). *Proc. Natl. Acad. Sci. U.S.A.,* **76**, 1213–1217.
Kobayashi, H., Van Brunt, J., and Harold, F. M. (1978). *J. Biol. Chem.,* **253**, 2085–2092.
Lagarde, A. E. (1977). *Biochem. J.,* **168**, 211–221.
Lanyi, J. K. (1978a). *Biochemistry,* **17**, 3011–3017.
Lanyi, J. K. (1978b). *Arch. Biochem. Biophys.,* **191**, 821–824.
Lanyi, J. K., and Silverman, M. P. (1979). *J. Biol. Chem.,* **254**, 4750–4755.
Lo, T. C. Y. (1977). *J. Supramol. Struct.,* **7**, 463–480.
Mitchell, W. J., Booth, I. R., and Hamilton, W. A. (1979). *Biochem. J.,* **182**, 687–696.
Plate, C. A. (1979). *J. Bacteriol.,* **137**, 221–225.
Ramos, S., and Kaback, H. R. (1977). *Biochemistry,* **16**, 4271–4275.
Singh, A. P., and Bragg, P. D. (1977). *J. Supramol. Struct.,* **6**, 389–398.
Zilberstein, D., Schuldiner, S., and Padan, E. (1979). *Biochemistry,* **18**, 669–673.

Addendum to Section 2.1

J. W. PAYNE

Many of the speculations shown in Figure 1 have been confirmed (J. W. Payne and T. M. Nisbet, unpublished results). Thus, using Ala-(^{14}C)-Ala to measure transport in *Escherichia coli* and *Streptococcus faecalis* we find that the rate of uptake appears linear for only a few seconds, apparently peaks at about 1 minute, and thereafter declines. In fact, fluorescamine assays performed simultaneously show uptake is actually faster than appears from the radioactive assay and is linear for a considerable period. The discrepancies arise because of the rapid onset of alanine exodus. Similarly, in *E. coli*, the initial rate of uptake of (^{14}C)Gly-Phe is apparently only about 20 per cent that of Gly-(^{14}C)-Phe; in fact, simultaneous fluorescamine assays show the rates (not surprisingly) to be identical. The error in the radioactive assay arises because of the greater exodus and metabolism (in part decarboxylation) of Gly relative to Phe. Perry and Abraham (1979) have described similar observations for uptake of radioactively labelled peptides by *Staphylococcus aureus*.

The *E. coli* $opp^{+}dpp^{-}$ mutant has been further characterized (J. W. Payne and R. A. Alves, unpublished results) and used together with related mutants ($opp^{-}dpp^{+}$, $opp^{-}dpp^{-}$) to quantify the extent to which different peptides use these two permeases and the additional system(s) (here designated, gpp^{+}, general peptide permease). No simple structural specificity pattern emerges from these studies. The ability of *gpp* to transport a range of di-, and oligopeptides has been confirmed and extended. Arsenate inhibits uptake through each of the three systems by over 80 per cent. No difference in the activity of (membrane-bound) peptidase N (Boquet *et al.*, 1979) (measured using alanyl-*p*-nitroanilide) could be found between the wild type and any of these mutants.

Applications of the smugglin principle, and examples of natural antibiotic smugglins, have continued to appear (see also the addendum to Section 4.7). Thus, Penninckx and Gigot (1979) prepared Gly-Gly-N-ζ-(phosphonoacetyl)-Orn in order to deliver the impermeant, toxic, C-terminal moiety into *E. coli*. Diddens *et al.* (1979) showed that the natural tripeptide antibiotic plumbemycin enters *E. coli* via the oligopeptide transport system; this paper also reviews evidence for uptake of several other antibiotic peptides through peptide permeases. Perry and Abraham (1979) showed that the dipeptide analogue bacilysin uses a peptide permease to enter bacteria, but it is still not clear whether the toxic protease inhibitor antipain (a tripeptide analogue) enters *E. coli* via the oligopeptide permease (Swenson and Schenley, 1978; Swenson, 1979). Albomycin is a smugglin, taken in via the ferrichrome transport system, which requires intracellular cleavage to release its toxic moiety (Hartman *et al.*, 1979).

Other papers that relate to various topics covered in this section are: Scudamore *et al.* (1979), De Felice *et al.* (1979), Noel *et al.* (1979), and Watabe *et al.* (1979).

REFERENCES

Boquet, C., Murgier, M., and Lazdunski, A. (1979). Aminopeptidase N in *E. coli* K-12 and K-10: Evidence for identity and similarity with oligopeptidase activity. *FEMS Lett.*, **6**, 227–229.

De Felice, M., Levinthal, M., Iaccarino, M., and Guardiola, J. (1979). Growth inhibition as a consequence of antagonism between related amino acids: Effect of valine in *Escherichia coli* K-12. *Microbiol. Rev.*, **43**, 42–58.

Diddens, H., Dorgehloh, M., and Zähner, H. (1979). Metabolic products of microorganisms: on the transport of small peptide antibiotics in bacteria. *J. Antibiot.* **32**, 87–90.

Hartman, A., Fiedler, H. P., and Braun, V. (1979). Uptake and conversion of the antibiotic albomycin by *Escherichia coli* K-12. *Eur. J. Biochem.*, **99**, 517–524.

Noel, D., Nikaido, K., and Ames, G. F. L. (1979). A single amino acid substitution in a histidine-transport protein drastically alters its mobility in sodium dodecyl sulfate-polyacrylamide gel electrophoresis. *Biochemistry,* **18**, 4159–4165.

Penninckx, M., and Gigot, D. (1979). Synthesis of a peptide form of N-ζ-(phosphonacetyl)-L-ornithine: its antibacterial effect through the specific inhibition of *Escherichia coli* L-ornithine carbamoyltransferase. *J. Biol. Chem.*, **254**, 6392–6396.

Perry, D., and Abraham, E. P. (1979). Transport of bacilysin and other peptides by suspensions of *Staphylococcus arueus*. *J. Gen. Microbiol.*, **115**, 213–221.

Scudamore, R. A., Beveridge, T. J., and Goldner, M. (1979). Outer membrane penetration barriers as components of intrinsic resistance to beta-lactam and other antibiotics in *E. coli* K-12. *Antimicrobial Agents Chemother.*, **15**, 182–189.

Swenson, P. A. (1979). Antipain lethality of *Escherichia coli*: Dependence upon cyclic adenosine 3′5′monophosphate and its receptor protein. *J. Bacteriol.*, **139**, 690–693.

Swenson, P. A., and Schenley, R. L. (1978). Effects of antipain (a protease inhibitor) on respiration, viability, and excision of pyrimidine dimers in UV-irradiated *E. coli* cells. *J. Bacteriol.*, **135**, 1167–1170.

Watabe, K., Sano, K. Otani, M., Okada, Y., Kakinchi, Y. and Kondon, M. (1979). Effect of alanine-containing dipeptides on germination of *Bacillus thiaminolyticus* spores. *J. Bacteriol.*, **139**, 126–131.

Addendum to Section 2.2

J. M. BECKER AND F. NAIDER

Since the completion of the writing of this review (November, 1977) a number of developments pertinent to our discussions appeared in the literature or were observed in our laboratories. Tanaka and Kita (1977) demonstrated a site specific degradation of α-factor into two peptides by cleavage at the peptide bond between the sixth and seventh residues of the α-factor oligopeptide. These results support our idea that extracellular peptidase activity in yeast is very site-specific. Tanaka and Kita (1978) also showed that a fluorescent marker was concentrated in the nucleus of α-mating type cells incubated with α-factor coupled to fluorescein isothiocyanate (FITC). Although these authors did not rule out cleavage of the FITC-α-factor and subsequent uptake of the fragments, these results may indicate that large peptides are not excluded from the yeast cell. On the other hand, special mechanisms for uptake of particular peptides with hormone-like activity could be superimposed over the size or structural limits existing for peptides in general.

Recently, Nisbet and Payne (1979a,b) have investigated peptide transport in *S. cerevisiae* by using fluorescence methods of assay of peptide transport (see Section 2.1.4.3). Their results confirmed most of the conclusions reported in this review. In addition they found that *S. cerevisiae* strain Σ1278b could transport Lys-Lys, that alanine peptides with D-residues were not taken up by this strain, and that a number of peptides containing sarcosine residues were accumulated intact by strain Σ1278b.

In our laboratories, a peptide transport deficient mutant of yeast has been characterized (Marder *et al.*, 1978). The mutant was isolated on the basis of its resistance to L-ethionyl-L-alanine. The mutant, designated *S. cerevisiae* Z1-2D Eta^r is a leucine–lysine auxotroph that does not utilize di- or tripeptides for growth, although it contains peptidases that release the required amino acids. Radioactive glycyl-leucine is not accumulated in the mutant under conditions in which the parent takes up the labelled dipeptide. The failure to utilize peptides in this mutant is due to a simple nuclear gene acting like a recessive trait in heterozygous diploids (J. Lemontt, unpublished observations).

Logan *et al*. (1979) examined the transport of radioactive trimethione into *C. albicans* WD18-4. Using this direct transport assay, ($K_m = 3.3 \times 10^{-5}$ M for trimethionine uptake), they confirmed and extended the structural requirements for peptide transport which had been found using growth assays. In addition, definitive evidence has now been obtained that peptides can carry 5-fluorocytosine and 5-fluoroorotic acid into *Candida albicans* (Steinfeld *et al*., 1979, and unpublished results). Finally, we have shown that the extracellular peptidase, aminopeptidase II, of *S. cerevisiae* (Frey and Rohm, 1979) (and see also Section 3.3.2.1(v) and the addendum to Section 3.3) is not involved in the transport of trimethionine by *S. cerevisiae* strain 139 (Parker, Naider, and Becker, unpublished results).

REFERENCES

Frey, J., and Rohm, K. H. (1979). *Eur. J. Biochem.,* **97**, 169–179.
Logan, D. A., Becker, J. M., and Naider, F. (1979). *J. Gen. Microbiol.,* **114**, 179–186.
Marder, R., Rose, B., Becker, J. M., and Naider, F. (1978). *J. Bacteriol.,* **136**, 1174–1177.
Nisbet, T. M., and Payne, J. W. (1979a). *FEMS Lett.,* **6**, 193–196.
Nisbet, T. M., and Payne, J. W. (1979b). *J. Gen. Microbiol.,* **115**, 127–133.
Steinfeld, A. S., Naider, F., and Becker, J. M. (1979). *J. Med. Chem.,* **22**, 1104–1109.
Tanaka, T., and Kita, H. (1977). *J. Biochem. (Tokyo),* **82**, 1689–1693.
Tanaka, T., and Kita, H. (1978). *Biochem. Biophys. Res. Commun.,* **83**, 1319–1324.

Addendum to Section 2.4

C. L. HERMSDORF AND S. SIMMONDS

DIPEPTIDASES

Recent work with *S. typhimurium* has shown that the presence of peptidase D is essential for histidine auxotrophs to grow on carnosine (β-alanyl-L-histidine) as a source of the required amino acid (Kirsh *et al.*, 1978). Study of the peptidase as partially purified from a *pepN*⁻ *pepA*⁻ *pepB*⁻ strain indicated that the *Salmonella* enzyme differed from the *E. coli* K-12 dipeptidase in its molecular weight (160 000) and lack of activation by β-mercaptoethanol; in the latter respect, and also in the pattern of divalent cation requirements for activity (toward carnosine, Gly-Leu and Leu-Gly), the *Salmonella* peptidase D appears more closely to resemble *E. coli* B Zn-dipeptidase than either *E. coli* B dipeptidase M or the strain K-12 dipeptidase DP. Whether the ability of *E. coli* K-12 auxotrophs to grow as readily on carnosine as on histidine (Payne, 1973) requires the presence of dipeptidase DP is unknown. However, β-alanylglycine is known to be hydrolysed by disrupted cells of a glycine auxotroph although the auxotroph is unable to use the dipeptide as a source of glycine for growth (Payne, 1973).

AMINOPEPTIDASES

As noted in Table 1, *E. coli* K-12 mutants of genotype $pepN^-$ $pepA^-$ $pepB^-$ $pepD^-$ $pepQ^-$ were reported to lack activity attributed not only to aminopeptidase I (peptidase A) (Miller and Schwartz, 1978) but also to aminopeptidase L (Hermsdorf *et al.*, 1979). Introduction of the wild-type allele of *pepA* has now been observed to result in the appearance of two readily separable peaks of peptidase activity upon ion-exchange chromatographic analysis of cell-free extracts of the $pepA^+$ strain (S. Simmonds, unpublished data). Whilst the activity under one peak resided in material having all the known characteristics of aminopeptidase L, and R_f

observed in polyacrylamide gel electrophoresis patterns was considerably lower than the value of 0.28 reported for the 'variable' band of Leu-Gly-Gly-cleaving activity by Miller and Schwartz (1978). The activity under the other peak was associated with material that differed from aminopeptidase I as described by Vogt (1970), displaying high activity toward some dipeptides that were reported to be poor substrates for aminopeptidase I. The apparent control by a single gene of two distinct peptidase activities raises the question of whether the two activities are attributable to somewhat different forms of a single enzyme. In this connection, it may be noted that aminopeptidase B (listed in Table 1 as aminopeptidase AP) sometimes was seen in polyacrylamide gel electrophoresis patterns as a single broad band of activity whereas in other gels two distinct activity bands were observed (Miller and Schwartz, 1978).

FUNCTIONS OF BACTERIAL PEPTIDASES

Relevant to the role of peptidases in the biosynthesis of bacterial proteins are recent demonstrations that several well-defined proteins associated with cell membranes are initially produced in the form of larger polypeptide precursors ('preproteins') carrying N-terminal 'leader regions' (DiRienzo *et al*., 1978; Hardy and Randall, 1978; Randall *et al*., 1978). Granted the validity of the hypothesis that an endopeptidase-catalysed removal of an amino-terminal oligopeptide fragment occurs as part of the process leading to the conversion of a preprotein to its functional form in the cell membrane (Inouye and Beckwith, 1977; Pages *et al*., 1978), only peptidase-catalysed degradation of the 'leader peptide' would prevent its accumulation within the cell.

REFERENCES

DiRienzo, J. M., Nakamura, K., and Inouye, M. (1978). *Annu. Rev. Biochem.,* **47**, 481.
Hardy, S. J. S., and Randall, L. L. (1978). *J. Bacteriol.,* **135**, 291.
Hermsdorf, C. L., Simmonds, S., and Saunders, A. (1979). *Int. J. Peptide Protein Res.,* **13**, 146.
Inouye, H., and Beckwith, J. (1977). *Proc. Natl. Acad. Sci. U.S.A.,* **74**, 1438.
Kirsh, M., Dembinski, D. R., Hartman, P. E., and Miller, C. G. (1978). *J. Bacteriol.,* **134**, 361.
Miller, C. G., and Schwartz, G. (1978). *J. Bacteriol.,* **135**, 603.
Pages, J. M., Piovant, M., Varenne, S., and Lazdunski, C. (1978). *Eur. J. Biochem.,* **86**, 589.
Payne, J. W. (1973). *Biochim. Biophys. Acta,* **298**, 469.
Randall, L. L., Hardy, S. J. S., and Josefsson, L. G. (1978). *Proc. Natl. Acad. Sci. U.S.A.,* **75**, 1209.
Vogt, V. M. (1970). *J. Biol. Chem.,* **245**, 4760.

Addendum to Section 3.1

B. A. LAW

Since this section was completed there have been several additional reports on the properties of extracellular proteinases and exopeptidases from *Serratia marcescens* (Lyerly and Kreger, 1979; Decedue *et al.*, 1979), *Streptomyces griseus* (Narahashi and Yoda, 1979), *B. subtilis* (Al'-Nuri *et al.*, 1978; Wagner *et al.*, 1979; Ajabnoor and Wagner, 1979), *Vibrio* SA1 (Wiersma *et al.*, 1978), and *Streptococcus cremoris* (Mills and Thomas, 1978; Exterkate, 1979). Many of the data in these papers simply add to the list of properties of bacterial proteinases reviewed in Sections 3.1.2 to 3.1.4. However, the new carboxypeptidase of *Streptomyces griseus* K1 (Narahashi and Yoda, 1979) is of particular interest because few such bacterial enzymes have been characterized. Also, the report by Exterkate (1979) that *Streptococcus cremoris* may produce distinct cell-free and cell wall-bound extracellular proteinases is significant in relation to our understanding of the way in which these important milk-fermenting bacteria utilize milk proteins. The accumulation of catalytically-active enzyme in the cell wall appears to require Ca^{2+}, an observation which may explain the poor initial growth of cheese starter streptococci which have been propagated in non-milk media before being used for cheesemaking. Mills and Thomas (1978) showed that Ca^{2+} was also involved in the retention of proteinase in cell wall fragments of these Group N streptococci. New extracellular peptidases have been reported in these organisms and they appear to function in milk protein utilization (Law, 1979). Some recent reports on the properties of *Serratia* proteinase (Lyerly and Kreger, 1979) and *Pseudomonas aeruginosa* proteinase (Morihara *et al.*, 1979) strengthen the view that the action of these enzymes on tissue proteins and defence proteins plays an important role in bacterial virulence.

Recent developments in the study of synthesis and secretion include the continued search for long-lived mRNA for extracellular proteins. For example, O'Connor *et al.* (1978) have studied the continued production of proteinase by *B. amyloliquefaciens* in the presence of transcription inhibitors, with high or low amino acid concentrations. To explain the observed cyclic fluctuations of proteinase synthesis, they postulate the presence of a pool of non-translatable mRNA which acts as a reservoir, in addition to the proteinase-specific mRNA which can be translated as it is produced. Mercer *et al*. (1979) reported the development of a direct mRNA assay with which they hope to verify the existence of this inferred mRNA pool. Exterkate (1979) observed the formation of extracellular proteinase in the cell wall of *Streptococcus cremoris* in the absence of mRNA synthesis and also speculated on the existence of inherently stable, or post-transcription-stabilized mRNA, in this organism.

In the field of enzyme secretion, DiRienzo *et al*. (1978) discuss a mechanism for the translocation of an outer membrane lipoprotein from its intracellular site of synthesis in Gram-negative bacteria. This mechanism can also accommodate the idea of membrane-bound polysomes, since the charged N-terminal sequence of the newly-formed protein can bind to the negatively-charged inner surface of the cytoplasmic membrane. The model is similar to that proposed by Blobel and Dobberstein (1975) to describe protein secretion in eukaryotic cells ('signal hypothesis', Section 3.1.6). The direct transfer model of Von Heijne and Blomberg (1979) provides a plausible extension of the signal hypothesis, but avoids the need to postulate water-filled tunnels forming in the membrane. The energetics of protein structure and physico-chemical interactions are used to explain how exoproteins can pass directly through cytoplasmic membranes, driven by conformational free energy changes. The authors suggest that although there is at present no experimental data to suggest that protein secretion by prokaryotes follows such a scheme, the close similarities relating this function in both eukaryotic and prokaryotic cells favour the view that the underlying mechanisms are similar. However, the additional involvement of membrane modification in extracellular proteinase production by bacterial cells is indicated by a recent report that fatty acid synthesis may be required for protein secretion in *B. amyloliquefaciens* (Paton and Elliot, 1979). Lazdunski *et al*. (1979) observed that the inhibition of phospholipid synthesis, rather than fatty acid synthesis, reduced the activity of the peripheral aminopeptidase N in *E. coli*. However, the authors were not certain whether phospholipid synthesis was required for the passage of the nascent peptide across the cytoplasmic membrane, or for the catalytic activity of the enzyme once it had reached the periplasmic space.

Controversy remains over the mechanisms controlling exoprotein synthesis; Coleman *et al*. (1978) have suggested that the 'competition' model (Section 3.1.7) can be extended to clinical isolates of *Staphylococcus aureus* which have widely differing abilities to form exoproteins. However, other studies, exemplified by those of Wiersma *et al*. (1978a,b) and Wiersma and Harder (1978), indicate that proteinase production by *Vibrio*

SA1 is under a variety of control mechanisms involving induction, catabolite repression, and 'end product' repression. In agreement with Wouters and Bieysman, 1977 (Section 3.1.7), bacterial proteinases were shown to be produced maximally under conditions of nutrient limitation in a chemostat, but the presence of an amino acid inducer was required, irrespective of growth rate, for endopeptidase production. This suggests that a simple competition model is not tenable as a complete description of exoprotein control.

Exterkate (1979) also found evidence of proteinase induction by amino acids and/or peptides in *Streptococcus cremoris*, but considered that the synthesis of the enzyme was controlled at the level of translation. Other references to proteinase induction by amino nitrogen sources include Dainty *et al*. (1978), Björklind and Arvidson (1978), and Bromke and Hammel (1979). Björklind and Arvidson (1978) suggested that staphylococcal proteinase production was controlled indirectly by serine and proline which influenced the intracellular amino acid pool. Dreisbach and Merkel (1978) demonstrated that the collagenase of *Vibrio* B-30 was induced by peptide fragments derived from the polar helical portion of the collagen molecule.

REFERENCES

Ajabnoor, M. A., and Wagner, F. W. (1979). *Bacillus subtilis* aminopeptidase: specificity towards amino acyl-β-naphthylamides. *Arch. Biochem. Biophys.*, **197**, 73–81.

Al'-Nuri, M. A., Otroshko, T. A., and Egorov, N. S. (1978). Purification of the exoproteinase of *Bacillus subtilis* var. *amyloliquefaciens*, capable of clotting blood plasma. *Microbiology USSR,* **47**, 527–532.

Björklind, A., and Arvidson, S. (1978). Influence of amino acids on the synthesis of an extracellular proteinase from *Staphylococcus aureus*. *J. Gen. Microbiol.,* **107**, 367–375.

Bromke, B. J., and Hammel, J. M. (1979). Regulation of extracellular protease formation by *Serratia marcescens. Can. J. Microbiol.,* **25**, 47–52.

Coleman, G., Jakeman, C. M., and Martin, N. (1978). Patterns of total extracellular protein secretion by a number of clinically isolated strains of *Staphylococcus aureus. J. Gen. Microbiol.,* **107**, 189–192.

Dainty, R. H., Etherington, D. J., Shaw, B. G., Barlow, J., and Banks, G. T. (1978). Studies on the production of extracellular proteinases by a non-pigmented strain of *Chromobacterium lividum* isolated from abattoir effluent. *J. Appl. Bacteriol.*, **45**, 111–124.

Decedue, C. J., Broussard II, E. A., Larson, A. D., and Braymer, H. D. (1979). Purification and characterization of the extracellular proteinase of *Serratia marcescens. Biochim. Biophys. Acta,* **569**, 293–301.

DiRienzo, J. M., Nakamura, K., and Inouye, M. (1978). The outer membrane proteins of Gram-negative bacteria: biosynthesis, assembly, and functions. *Annu. Rev. Biochem.*, **47**, 481–532.

Dreisbach, J. H., and Merkel, J. R. (1978). Induction of collagenase production in *Vibrio* B-30. *J. Bacteriol.,* **135**, 521–527.

Exterkate, F. A. (1979). Accumulation of proteinase in the cell wall of *Streptococcus cremoris* strain AM1 and regulation of its production. *Arch. Microbiol.*, **120**, 247–254.

Law, B. A. (1979). Extracellular peptidases in Group N streptococci used as cheese starters. *J. Appl. Bacteriol.,* **46**, 455–465.

Lazdunski, A., Murgier, M., and Lazdunski, C. (1979). Phospholipid synthesis-dependent activity of aminopeptidase N in intact cells of *Escherichia coli*. *J. Mol. Biol.*, **128**, 127–141.

Lyerly, D., and Kreger, A. (1979). Purification and characterization of a *Serratia marcescens* metalloproteinase. *Infect. Immun.*, **24**, 414–421.

Mercer, J. F. B., O'Connor, R., and May, B. K. (1979). Messenger-RNA pools for extracellular enzymes in *Bacillus amyloliquefaciens*. *Proc. Austral. Biochem. Soc.*, **12**, 75.

Mills, O. E., and Thomas, T. D. (1978). Release of cell wall-associated proteinases from lactic streptococci. *N.Z. J. Dairy Sci. Technol.*, **13**, 209–215.

Morihara, K., Tsuzuki, H., and Oda, K. (1979). Proteinase and elastase of *Pseudomonas aeruginosa*: inactivation of human plasma α_1-proteinase inhibitor. *Infect. Immun.*, **24**, 188–193.

Narahashi, Y., and Yoda, K. (1979). Purification and some properties of a new metallo-carboxypeptidase of *Streptomyces griseus* K1. *J. Biochem.*, **86**, 683–689.

O'Connor, R., Elliot, W. H., and May, B. K. (1978). Modulation of an apparent mRNA pool for extracellular protease in *Bacillus amyloliquefaciens*. *J. Bacteriol.*, **136**, 24–34.

Paton, J. C., and Elliot, W. H. (1979). Fatty acid synthesis is required for extracellular protein secretion by *Bacillus amyloliquefaciens*. *Proc. Austral. Biochem. Soc.*, **12**, 109.

Von Heijne, G., and Blomberg, C. (1979). Trans-membrane translocation of proteins. The direct transfer model. *Eur. J. Biochem.*, **97**, 175–181.

Wagner, F. W., Ray, L. E., Ajabnoor, M. A., Ziemba, P. E., and Hall, R. L. (1979). *Bacillus subtilis* aminopeptidase: purification, characterization and some enzymic properties. *Arch. Biochem. Biophys.*, **197**, 63–72.

Wiersma, M., and Harder, W. (1978). A continuous culture study of the regulation of extracellular protease production in *Vibrio* SA1. *Antonie van Leeuwenhoek*, **44**, 141–155.

Wiersma, M., Hansen, T. A., and Harder, W. (1978a). Effect of environmental conditions on the production of two extracellular proteolytic enzymes by *Vibrio* SA1. *Antonie van Leeuwenhoek*, **44**, 129–140.

Wiersma, M., Versteegh, G., Assink, H. A., Welling, G. W., and Harder, W. (1978b). Purification and some properties of two extracellular proteolytic enzymes produced by *Vibrio* SA1. *Antonie van Leeuwenhoek*, **44**, 157–169.

Addendum to Section 3.2

BERNARD L. COHEN

The literature search on which Section 3.2 is based was completed in August 1978 and the manuscript was delivered to the Editor in January 1979.

Other commitments have prevented me from undertaking a fresh search for relevant material; the papers noted here are simply those that have come to my attention in one way or another and which seem particularly relevant. They are listed and annotated under a sequence of headings which refer to the body of the article.

3.2.2.2

Ogrydziak, D. M., and Mortimer, R. K. (1977). Genetics of extracellular protease production in *Saccharomycopsis lipolytica. Genetics,* **87**, 621–632.

Eighteen mutations defining 10 or 11 unlinked complementation groups lead to reduced secretion of extracellular protease. At least 6 complementation groups are pleiotropic, including effects on mating efficiency and RNAase production.

3.2.2.5

Workshop on localisation and characterisation of tissue proteinases using synthetic substrates (1979). *J. Histochem. Cytochem.,* **27**, 1493–1588.

Four papers reviewing latest advances in this field. Specific electron-dense reagents are probably within reach of the biochemically educated investigator.

3.2.2.6

Davis, B. D., and Tai, P.-C. (1980). The mechanism of protein secretion across membranes. *Nature (London),* **283**, 433–438.

An up-to-date review of progress in this field, recording the confirmation and extension of Blobel's (1977) signal hypothesis.

Note that there is still no evidence on the mechanism(s) of secretion by fungi.

3.2.3

Haars, A., McCullough, C. H. R., Huttermann, A., and Dee, J. (1978). Regulation of proteolytic enzymes in axenically grown myxamoebae of *Physarum polycephalum*. *Arch. Microbiol.,* **118**, 55–60.

This slime mould has a novel pattern of protease regulation: added protein *prevents* protease release!

3.2.3.1(iv)

Ogrydziak, D. M., Demain, A. L., and Tannenbaum, S. R. (1977). Regulation of extracellular protease production in *Saccharomycopsis lipolytica*. *Biochim. Biophys. Acta,* **497**, 525–538.

S. lipolytica is found to follow the *A. nidulans* pattern, with regulation by C-, N-, and S-repression but not by P-repression.

Casein is said to stimulate protease production but the only data provided (Figure 2) do not support this statement. Since casein is not *necessary* for protease production it is also confusing to describe its stimulatory effect as 'an inductive effect'.

Buyak, L. I., Al'-Nuri, M. A., Landau, N. S., and Egorov, N. S. (1978). Effects of the stimulating factor formed by *Aspergillus wentii* on the biosynthesis of exoproteases by a culture of *Aspergillus*. Translated from *Mikrobiologiya,* **47**, 1004–1009.

A low-molecular-weight fraction of *A. wentii* culture fluid stimulated a two-fold increase in protease production by *A. kanagawensis* growing in a defined synthetic medium. This effect does not constitute evidence for induction; its significance is obscure.

Ivanitsa, V. A., Egorov, N. S., and Al'-Nuri, M. A. (1978). Effect of amino acids as the sole nitrogen source on exoprotease biosynthesis by *Aspergillus candidus*. Translated from *Mikrobiologiya,* **47**, 424–429.

A. candidus is found to follow the *A. nidulans* pattern of protease regulation.

Similar data for the effects of amino acids as repressors of protease production are given for *Neurospora crassa* by Cohen and Drucker (1977).

3.2.4.2(ii)

Wiemken, A., Schellenberg, M., and Ureck, K. (1979). Vacuoles: the sole compartments of digestive enzymes in yeast (*Saccharomyces cerevisiae*)? *Arch. Microbiol.,* **123**, 23–35.

These experiments need to be done on a yeast or a mycelial fungus which *can* secrete protease. Polyphosphate is a good marker for the vacuole fraction in *S. cerevisiae*.

3.2.5.2

Koide, Y., Yamagata, H., and Udaka, S. (1979). Relationship between morphological differentiation and extracellular enzyme formation in *Aspergillus nidulans*. *J. Gen. Appl. Microbiol.*, **25**, 161–168.

Four sporulation-defective mutants were associated with altered protease (and other hydrolase) activities. Since cultures for assay were grown on complete medium, the meaning of the results is hard to perceive.

Addendum to Section 3.3

D. H. WOLF AND H. HOLZER

Since the completion of the manuscript in 1977 and additions in 1978, which were mostly based on personal communications of results that are now published, considerable new information concerning proteases and proteolytic processes has accumulated. All this is summarized.

THE PROTEOLYTIC SYSTEM, ITS REGULATION AND SUBCELLULAR LOCALIZATION

Recent experiments of Frey and Röhm (1978) seem to confirm the identity of the vacuolar aminopeptidase of Matile *et al*. (1971) with their purified aminopeptidase I of 640 000 molecular weight. Studies on the aminopeptidase II of Frey and Röhm led to a reevaluation of its molecular weight as about 100 000 (Knüver, 1979). Frey and Röhm (1978) also identified another aminopeptidase of around 30 000 molecular weight which they called aminopeptidase III. It is probably this enzyme that is identical with aminopeptidase II of Masuda *et al*. (1975), and not the dipeptidase as had been suggested (see 3.3.2.1(v)).

Aminopeptidase II activity of Frey and Röhm was found intracellularly as well as extracellularly in the periplasmic space (Frey and Röhm, 1979a).

Aminopeptidase I activity of Frey and Röhm (vacuolar aminopeptidase) increased significantly upon glucose and nitrogen starvation. The enzyme is inactivated upon addition of glucose to starving cells (Frey and Röhm, 1979b). Aminopeptidase II activity of Frey and Röhm does not show significant activity changes during different growth phases. Growth on good nitrogen sources enhances activity about twofold (Frey and Röhm, 1978). Dipeptidase activity is highest in cells growing exponentially on poor nitrogen source. With increasing age of cells, dipeptidase activity decreases (Frey and Röhm, 1978).

In a mutant lacking three protease activities—protease B, carboxypeptidase Y, and carboxypeptidase S—six new intracellular proteolytic activities were detected (Wolf and Ehmann, 1980; D. H. Wolf and C. Ehmann, unpublished results). Two proteolytic activities, preliminarily called D and E, and active at neutral pH against peptide-*p*-nitroanilides, which are potent plasma protease substrates, were found in the soluble fraction of disintegrated cells. Using the same types of substrates, two additional activities, preliminarily called M and P, were detected in the precipitable membranous fraction. Two carboxypeptidase activities were suspected in the triple mutant. Activity, preliminarily called carboxypeptidase γ, is active against N-Cbz-L-Ala-L-Phe and is inhibited by phenylmethylsulphonylfluoride, antipain, and leupeptin. Activity preliminarily called carboxypeptidase δ cleaves N-Cbz-L-Ser-L-Phe and is inhibited by EDTA.

PROTEASE FUNCTION

Glucose-induced catabolite inactivation of cytoplasmic malate dehydrogenase, fructose-1,6-bisphosphatase and phosphoenolpyruvate carboxykinase, which it is proposed is due to proteolytic action (see 3.3.6.1), is not impaired by the absence of protease B, carboxypeptidase Y, and carboxypeptidase S as measured in a mutant lacking these three proteolytic activities (D. H. Wolf and C. Ehmann, unpublished results). Also, loss of antibody cross reacting phosphoenolpyruvate carboxykinase protein, which parallels loss of enzyme activity during glucose-induced inactivation, is not altered when protease B, carboxypeptidase S, and carboxypeptidase Y are absent (Müller *et al.*, 1980). Thus, if the primary inactivation process of the catabolite-inactivated enzymes or the disappearance of antibody cross reacting phosphoenolpyruvate carboxykinase protein is a proteolytic event, other proteases must be responsible. Irreversible inactivation of $NADP^+$-dependent glutamate dehydrogenase upon glucose starvation was proposed to be due to proteolysis, with protease B being proposed as the inactivating enzyme *in vivo* (Hemmings and Mazon, 1979). However, studies on mutants lacking protease B activity exclude this enzyme as the inactivating catalyst (Wolf and Ehmann, 1979), indicating that another protease must be responsible for this process. Recently, evidence has accumulated that irreversible inactivation of some enzymes, which is most likely due to proteolysis, is preceded by a modification of the enzyme protein. A reversible inactivation reaction of fructose-1,6-bisphosphatase prior to irreversible disappearance of the enzyme activity was found (Lenz and Holzer, 1980). Initial inactivation of NAD^+-dependent glutamate dehydrogenase upon glutamate starvation was found to be due to phosphorylation of the enzyme. Antibody cross reacting enzyme protein was found to disappear with a considerable lag period after onset of enzyme inactivation, indicating proteolysis as a second step in the inactivation reaction (Hemmings, 1980).

Proteolysis as a possible key step in the membrane passage of cytoplasmically synthesized proteins, which are functionally active in the mitochondria has been elucidated recently. α-, β and γ-subunits of ATPase, subunit V of cytochrome bc_1 complex, cytochrome c_1 and cytochrome c peroxidase were found to be synthesized in the cytoplasm as precursor proteins of 2000 to 6000 higher molecular weight. Transfer into mitochondria is independent of translation, requires ATP, and is accompanied by proteolytic processing. (For summary see Schatz, 1979; Nelson *et al*., 1979.) In contrast to Poyton and McKemmie (1979a,b), who reported proteolytic cleavage of a polyprotein precursor of the four nuclear coded cytochrome c oxidase subunits (see 3.3.6.1(iii)), Lewin *et al.* (1980) found single individual precursor proteins of the subunits IV, V, and VI of the enzyme.

Proteolysis has been proposed to play a vital role in metabolism of α-factor, the peptide mating hormone of *Saccharomyces cerevisiae* α cells. Maness and Edelman (1978) found limited proteolytic cleavage of α-factor by *a* mating type cells. They hypothesized proteolysis of α-factor as a possible event yielding fragments that are responsible for expression of α-factor activity. Finkelstein and Strausberg (1979) and Ciejek and Thorner (1979) also reported proteolysis of α-factor. However, in contrast to Maness and Edelman (1978), they suggest α-factor cleavage as the mechanism whereby *a* mating type cells detoxify α-factor and by this recover from cell cycle arrest.

The vital role of protease B for protein turnover during starvation and sporulation has been established (Wolf and Ehmann, 1979). In a mutant lacking protease B activity, protein breakdown is more than 40 per cent reduced under carbon and nitrogen starvation conditions. A similar reduction in protein breakdown is found under sporulation conditions in a homozygous diploid defective in protease B. The reduced sporulation rate of the mutant suggests the necessary role of protease B in providing amino acids at the expense of intracellular protein for this differentiation process.

REFERENCES

Ciejek, E., and Thorner, J. (1979). Cell, **18**, 623–635.

Finkelstein, D. B., and Strausberg, S. (1979). *J. Biol. Chem.*, **254**, 796–803.

Frey, J., and Röhm, K. H. (1978). *Biochim. Biophys. Acta*, **527**, 31–41.

Frey, J., and Röhm, K. H. (1979a). *Eur. J. Biochem.*, **97**, 169–173.

Frey, J., and Röhm, K. H. (1979b). *FEBS Lett.*, **100**, 261–264.

Hemmings, B. A. (1980). *Archives of Biochemistry and Biophysics* (submitted).

Hemmings, B. A., and Mazon, M. (1979). In: *Limited Proteolysis in Microorganisms* (G. N. Cohen and H. Holzer, eds.), DHEW Publication No. (NIH) 79-1591, US Government Printing Office, Washington, pp. 69–75.

Knüver, J. (1979). Diplomarbeit, Faculty of Chemistry, Marburg.

Lenz, A. G., and Holzer, H. (1980). *FEBS Lett.*, **109**, 271–274.

Lewin, A. S., Gregor, I., Mason, T. L., Nelson, N., and Schatz, G. (1980). *Proc. Natl. Acad. Sci. U.S.A.* (submitted).

Maness, P. F., and Edelman, G. M. (1979). *Proc. Natl. Acad. Sci. U.S.A.*, **75**, 1304–1308.

Matile, Ph., Wiemken, A., and Guyer, W. (1971). *Planta,* **96**, 43–53.

Masuda, T., Hayashi, R., and Hata, T. (1975). *Agric. Biol. Chem., Tokyo,* **39**, 499–505.

Müller, M., Müller, H., and Holzer, H. (1980). *Journal of Biological Chemistry* (submitted).

Nelson, N., Maccecchini, M. L., Rudin, Y., and Schatz, G. (1979). In: *Biological Functions of Proteinases* (H. Holzer and H. Tschesche, eds.), Springer-Verlag, Berlin-Heidelberg-New York, p. 109–119.

Poyton, R. O., and McKemmie, E. (1979a). *J. Biol. Chem.,* **254**, 6763–6771.

Poyton, R. O., and McKemmie, E. (1979b). *J. Biol. Chem.,* **254**, 6772–6780.

Schatz, G. (1979). *FEBS Lett.,* **103**, 203–211.

Wolf, D. H., and Ehmann, C. (1979). *Eur. J. Biochem.,* **98**, 375–384.

Wolf, D. H., and Ehmann, C. (1980). *Eur. J. Biochem.* (submitted).

Addendum to Section 3.4

M. J. PINE

GENERAL STUDIES OF PROTEIN TURNOVER

In yeast during growth on ethanol virtually all of the cell protein breaks down, most of it at a fairly uniform rate of 2 per cent per hour. The rate can be increased several-fold by some, but not all, conditions of starvation (Lopez and Gancedo, 1979). In growth-limited *Neurospora crassa*, intracellular proteolysis increases gradually to a maximum at 25 per cent of the maximum growth rate. The proteolytic increase from glucose deprivation is prevented but not reversed by cycloheximide, in contrast to the reversible control of proteolysis in *E. coli* (Martegani and Algerghina, 1979). In these two respective systems, proteolysis is inhibited when the intracellular level of ATP falls by over 90 or 70 per cent.

Intracellular proteolysis has been followed in *E. coli* during C- and N-limited growth in a chemostat for periods of up to three days. Proteolysis is more extensive than in cultures without nutrient limitation. The most labile cell fraction appears to be the soluble cell protein (St. John *et al*., 1979).

ROLES OF UNSTABLE PROTEINS

The O protein, one of two regulators coded by the lambda phage of *E. coli* for initiating DNA replication, has a remarkably short half-life of two minutes (Lepinska *et al*., 1980). Such instability would fit the regulatory mechanism postulated elsewhere for regulation of DNA replication in mammalian cells by the rate of cell protein synthesis. Thus, the accumulation of a highly unstable serum-dependent protein has been suggested to control the restriction point early in the G_1 phase in which DNA is synthesized in the cell cycle (Rossow *et al*., 1979). The loss of proteins from the outer membrane of *E. coli* into the medium after their incorporation appears to be a function of cell growth, for the loss can be prevented by inhibiting cell growth (Allen and Scott, 1979). In the mammalian cell, protein secretion can correlate with the rapid degradative turnover of intracellular membranes, indicating a possible functional relationship between the two events (Cohen *et al.*, 1979). Such specialized roles, however, contrast with the conventional metabolic stability of cell membranes as well as of lysosomal enzymes. The large digestive demands of autophagic cells appear to be met by intact, rather than degradative recycling of both the cell membrane and digestive apparatus (Tulkens *et al.*, 1978).

STRUCTURAL DETERMINANTS OF PROTEOLYSIS

One of the roles of glycosylation of proteins may be to protect them against proteolysis. The half-life of fibronectin, a major surface glycoprotein of animal cells is halved after it has been assembled, in the presence of tunicamycin, without its normal complement of asparagine-linked glycosyl residues (Olden *et al.*, 1978). Stability is apparently not conferred in an absolute sense, for the glycoproteins of the liver cytosol are degraded faster than their non-glycosylated counterparts of comparable molecular weight and isoelectric point under conditions of selective breakdown (Kalish *et al.*, 1979). Chemical masking of proteins is not always stabilizing, for N-acetylation of the N-termini, which occurs in the large majority of eukaryote cell proteins, does not confer any special metabolic fate upon them (Brown, 1979).

PROTEASES INVOLVED IN PROTEIN TURNOVER

In the initiation of endogenous sporulation development in starving *Dictyostelium discoideum*, cathepsin B is localized in prespore and prestalk cells. The lysosomal and trypsin inhibitors chloroquine and tosyl lysylchloromethyl ketone inhibit differentiation to the sorocarp stage. This inhibition is specific for protein turnover as it is reversed by adding amino acids (Fong and Bonner, 1979).

Two overlapping models appear to vie for the role of effecting the normal or basal protein turnover in the eukaryote cell. Thus, the stepwise and additive inhibition of intracellular protein breakdown by specific inhibitors of thiol and carboxyl proteases, which presumptively curtail lysosomal processing via cathepsins B and D respectively, could account for all proteolysis, including basal turnover, in the macrophage. This conclusion is based, however, upon allowance for limited access of the inhibitors, which include new site-directed substrate analogues of cathepsin B (Shaw and Dean, 1980). In a study in cultured hepatocytes, however, inhibition by the antibiotic protease inhibitors as well as by temperature reduction was selective for long-term (daily) as opposed to hourly turnover of proteins of normal or analogue-induced lability. This might allow for a second mechanism, such as that of ATP-linked proteolysis for the latter process (Neff *et al*., 1979). In *E. coli*, ATP-stimulated proteolytic activity could be demonstrated most conveniently in soluble cell-free extracts against ^{14}C-methyl globin as a denatured substrate. Other high-energy phosphate sources were less stimulatory and methylene analogues of ATP showed no stimulation. In preparations of the *deg*$^-$ (*lon*$^-$) mutant, proteolysis of this substrate was always lower, both in the absence and presence of ATP. The most effective inhibitor of both activities was EDTA (Murakami *et al*., 1979). This, or some equivalent proteolytic activity for abnormal or low-molecular weight proteins in intact *E. coli* is specifically suppressed by an early product of phage T4 infection. This could explain why the polypeptide fragments from the initial processing of phage head proteins accumulate without further breakdown (Simon *et al*., 1978). The relevance of this proteolytic suppression to phage growth is further enhanced by the finding that from T4-resistant *E. coli* can be isolated mutants with increased capacity for the selective degradation of abnormal proteins. The mutation proves to be *rho*$^-$, pleiotypic for transcription termination factor and for other effects relevant to ATP utilization. The proteolytic effects of *rho*$^-$ and *lon*$^-$ are partly antagonized in double mutants (Simon *et al*., 1979).

In cell-free extracts, the ATP-independent proteolysis of incomplete and abnormal proteins, previously described by Kemshead and Hipkiss (1976) has been isolated and named protease III. It is also sensitive to metal chelating agents, and cleaves at rather limited peptide sites, attacking neither large molecular weight proteins nor conventional low molecular weight substrates of proteases (Cheng and Zipser, 1979). From the cytosol of liver and other organs can be isolated a high molecular weight (>400 000) neutral -SH protease that is stabilized by ATP and other phosphorylated compounds. This effect could give the appearance of ATP-dependence (Rose *et al*., 1979).

The antibiotic bestatin, a competitive inhibitor of aminopeptidases is useful in interfering with intracellular proteolysis after the initial cleavages. In intact reticulocytes or cell-free preparations, complete proteolysis is inhibited. In the latter preparations, globin products accumulate mostly as a mixture of amino acids and di- and tripeptides (Botbol and Scornik, 1979).

REFERENCES

Allen, R. J., and Scott, G. K. (1979). *Biochem. J.*, **308**, 225.
Botbol, V., and Scornik, O. A. (1979). *J. Biol. Chem.*, **254**, 11254.
Brown, J. L. (1979). *J. Biol. Chem.*, **254**, 1447.

Cheng, Y.-S., and Zipser, D. (1979). *J. Biol. Chem.*, **254**, 4698.
Cohen, B. G., Mosler, S., and Phillips, A. (1979). *J. Biol. Chem.*, **254**, 4267.
Fong, D., and Bonner, J. T. (1979). *Proc. Natl. Acad. Sci. U.S.A.*, **76**, 6481.
Kalish, F., Chovick, N., and Dice, J. F. (1979). *J. Biol. Chem.*, **254**, 4475.
Lepinska, B., Podhazska, A., and Taylor, K. (1980). *Biochem. Biophys. Res. Commun.*, **92**, 120.
Lopez, S., and Gancedo, J. M. (1979). *Biochem. J.*, **178**, 769.
Martegani, E., and Alberghina, L. (1979). *J. Biol. Chem.*, **254**, 7047.
Murakami, K., Voellmy, R., and Goldberg, A. L. (1979). *J. Biol. Chem.*, **254**, 8194.
Neff, N. T., DeMartino, G. N., and Goldberg, A. (1979). *J. Cell Physiol.*, **101**, 439.
Olden, K., Pratt, R. M., and Yamada, K. M. (1978). *Cell*, **13**, 461.
Rose, I. A., Warms, J. V. B., and Hershko, A. (1979). *J. Biol. Chem.*, **254**, 8135.
Rossow, P. W., Riddle, V. G. H., and Pardee, A. B. (1979). *Proc. Natl. Acad. Sci. U.S.A.*, **76**, 4446.
St. John, A., Jakubas, K., and Beim, D. (1979). *Biochim. Biophys. Acta*, **586**, 537.
Shaw, E., and Dean, R. T. (1980). *Biochem. J.*, **186**, 385.
Simon, L. D., Tomczak, K., and St. John, A. (1978). *Nature* (*London*), **275**, 424.
Simon, L. D., Gottesman, M., Tomczak, K., and Gottesman, G. (1979). *Proc. Natl. Acad. Sci. U.S.A.*, **76**, 1623.
Tulkens, P., Schneider, Y. J., and Trouet, A. (1978). In: *Protein Turnover and Lysosomal Function* (H. L. Segal and D. J. Doyle, eds.). Academic Press, New York, p. 719.

Addendum to Section 4.1

A. MEISTER

There have been some new developments regarding the cycle in animals, but relatively little new information relating to microorganisms has been published. Glutathione synthetase has been obtained in highly purified form from rat kidney and studies on the mapping of the substrate binding sites have been carried out (Oppenheimer *et al.*, 1979). Further studies have been carried out on the modulation of the hydrolysis and transfer activities of γ-glutamyl transpeptidase, and the effects of maleate and hippurate on the activities catalysed by this enzyme have been explored (Thompson and Meister, 1979; 1980). 'Glutathione oxidase' activity, previously reported to be catalysed by kidney homogenates and intact renal cells, has been shown to be an inherent property of γ-glutamyl transpeptidase (Tate *et al.*, 1979; Tate and Orlando, 1979); the oxidation of glutathione appears to be due to non-enzymic catalysis by cysteinylglycine produced by the action of γ-glutamyl transpeptidase (Griffith and Tate, 1980). It has been found that intracellular glutathione is translocated out of many mammalian cells; cells that have membrane-bound γ-glutamyl transpeptidase (e.g. kidney cells) can utilize such glutathione (Griffith and Meister, 1979a; Griffith *et al.*, 1979b). Cells not equipped with appreciable γ-glutamyl transpeptidase activity (e.g. liver) translocate glutathione to the blood plasma. Glutathione is removed from the blood plasma by the kidney (about 67%) and by extrarenal γ-glutamyl transpeptidase (about 33%) (Griffith and Meister, 1979b). Translocation of glutathione is therefore a discrete step in the γ-glutamyl cycle; although translocation can occur in the absence of transpeptidase activity, the rate of translocation appears to be decreased by administration of inhibitors of γ-glutamyl transpeptidase. Transport of γ-glutamyl amino acids into the kidney has been demonstrated by direct studies; both glutathione and the transpeptidase may play a role in this process (Griffith *et al.*, 1979a). When animals are administered inhibitors of γ-glutamyl transpeptidase, they excrete glutathione, γ-glutamylcystine, and cystine in their urine; these compounds were also found in the urine of a patient with γ-glutamyl transpeptidase deficiency (Griffith and Meister, 1980). Several papers have appeared which relate to the possible involvement of the cycle in amino acid transport in yeast. Osuji (1970a,b; 1980) has reported data which seem to be open to various interpretations; the evidence provided does not necessarily lead to the conclusions drawn. These studies have been commented upon (Robbins and Davies, 1980). Mooz (1979) reported normal γ-glutamyl transpeptidase but decreased glutathione synthetase in a yeast mutant *Saccharomyces cerevisiae* (2512c; gap^-) that is deficient in the transport of certain amino acids. Penninckx *et al*. (1980) have examined the activity and regulation of γ-glutamyl transpeptidase in yeast; they also report that its activity was unaffected in mutant 2512c; gap^-, which is defective in the general amino acid permease. In contrast, only a low level of the enzyme was found with apf mutant grown on urea which is affected in an element common to the various amino acid permeation systems (Grenson and Hennaut, 1971).

REFERENCES

Grenson, M. and Hennaut, C. (1971). *J. Bacteriol*., **105**, 477–482.
Griffith, O. W. and Meister, A. (1979a). *Proc. Natl. Acad. Sci. U.S.A.*, **76**, 268–272.
Griffith, O. W. and Meister, A. (1979b). *Proc. Natl. Acad. Sci. U.S.A.*, **76** 5606–5610.
Griffith, O. W. and Meister, A. (1980). *Proc. Natl. Acad. Sci. U.S.A.*, **77**, June (in press).
Griffith, O. W., Bridges, R. and Meister, A. (1979a). *Proc. Natl. Acad. Sci. U.S.A.*, **76**, 6319–6322.
Griffith, O. W., Novogrodsky, A. and Meister, A. (1979b) *Proc. Natl. Acad. Sci. U.S.A.*, **76**, 2249–2252.
Griffith, O. W. and Tate, S. S. (1980). *J. Biol. Chem.*, **255**, June (in press).
Mooz, E. D. (1979). *Biochem. Biophys. Res. Commun.*, **90**, 1221–1228.
Oppenheimer, L., Wellner, V. P., Griffith, O. W. and Meister, A. (1979). *J. Biol. Chem.*, **254**, 5184–5190.
Osuji, G. O. (1979a). *FEBS Lett.*, **105**, 283–285.
Osuji, G. O. (1979b). *FEBS Lett.*, **108**, 240–242.
Osuji, G. O. (1980). *FEBS Lett.*, **110**, 192–194.
Penninckx, M., Jaspers, C. and Wiame, J. M. (1980). *Eur. J. Biochem.*, **104**, 119–123.
Robbins, R. J. and Davies, D. D. (1980). *FEBS Lett.*, **111**, 432.
Tate, S. S., Grau, E. M. and Meister, A. (1979). *Proc. Natl. Acad. Sci. U.S.A.*, **76**, 2715–2719.
Tate, S. S. and Orlando, J. (1979). *J. Biol. Chem.*, **254**, 5573–5575.
Thompson, G. A. and Meister, A. (1979). *J. Biol. Chem.*, **254**, 2956–2960.
Thompson, G. A. and Meister, A. (1980). *J. Biol. Chem.*, **255**, 2109–2113.

Addendum to Section 4.6

C. F. HIGGINS AND J. W. PAYNE

Since completing this section, in December 1977, a number of important papers on amino acid transport in plant cells have been published. In addition, improvements in methodology, particularly in cell and protoplast culture and in the isolation of intact vacuoles, have opened up the possibility of considering plant cells as 'unicellular organisms'. Such approaches, which simplify kinetic analyses and should eventually allow a genetic approach to transport studies, have now become feasible and promise to provide considerable insights into the mechanism of amino acid transport in plant cells.

INTACT TISSUES AND CELLS

A number of studies demonstrating active amino acid transport in intact tissues and cultured cells have been published (Berlin and Mutert, 1978; Blackman and McDaniel, 1978; Smith, 1978a,b; Cheruel *et al.*, 1979; Paull *et al.*, 1979), although little light has been shed on the specificities of these systems. For example, Berlin and Mutert (1978) reported separate systems for neutral, basic, and acidic amino acids in tobacco cells, while in *Asparagus* cells a single system apparently handles all amino acids (Cheruel *et al.*, 1979). Interestingly, Soldal and Nissen (1978) have obtained some evidence for multiphasic kinetics during amino acid uptake by barley roots, although it still remains to be established whether this is a property of the transport system(s) *per se* or simply a consequence of studying a multicellular tissue.

One system which has received particular attention is the loading and unloading of phloem and xylem vessels during the long-distance movement of amino acids. An important series of papers by Pate and his co-workers (Atkins *et al.*, 1979; Pate *et al.*, 1979a,b) have clearly shown the importance of amino acids in the circulation of nitrogen in the plant body, and demonstrated the transfer of amino acids between the phloem and xylem. Both the loading of amino acids into the phloem (McNeil, 1979; Servaites *et al.*, 1979) and removal from the xylem (van Bel and van Erven, 1979a,b; van Bel *et al.*, 1979) seem to be active mediated processes.

Probably the most important results to emerge from these recent studies (see, for example, Smith 1978a,b; van Bel and van Erven, 1979a,b), as well as from studies on the transport of other metabolites, are those indicating a role for H^+ ions in the energization of plant transport processes (reviewed in Poole, 1978; Baker, 1979). Thus, at least superficially, the mechanism of energy coupling to transport in plant cells would seem to resemble that of microorganisms rather than animal cells.

PROTOPLASTS

The study of transport processes in protoplasts rather than whole cells, should be greatly simplified as there is no cell wall or 'unstirred layer' between the cell wall and the plasmamembrane for solutes to penetrate. Methods for preparing protoplasts (see Uchimiya and Murashige, 1974, and references therein) have now been refined to a stage where transport studies are feasible. Protoplasts have been shown to be suitable for measuring membrane potentials, using either microelectrodes (Racusen *et al.*, 1977; Briskin and Leonard, 1979) or lipid-soluble cations (Rubinstein, 1978), and, at least for ions, removal of the cell wall does not seem to affect the transport properties of the plasmalemma (Mettler and Leonard, 1979a,b). In addition, two studies of amino acid transport in protoplasts have recently been initiated, showing both the feasibility of such an approach and the greatly simplified kinetic data which can be obtained (Guy *et al.*, 1978; Ruesink, 1978).

VACUOLES

One problem often encountered in transport studies with eukaryotic cells is which membrane is limiting uptake, the plasmalemma, the tonoplast, or both? Recent methods allowing the separation of intact vacuoles from mature tissues (Wagner and Siegelman, 1975; Leigh and Branton, 1976; Lörz *et al.*, 1976; Mettler and Leonard, 1979c) should allow this question to be addressed. While no studies of amino acid uptake by isolated vacuoles have yet been reported, active sucrose transport across the tonoplast has recently been demonstrated (Doll *et al.*, 1979; Guy *et al.*, 1979). Proteolytic enzymes seem to be sequestered in plant vacuoles (Matile, 1978; Boller and Kende, 1979; Nishimura and Beevers, 1978, 1979; although see Butcher *et al.*, 1977), requiring that the hydrolytic products, amino acids and/or peptides, leave the vacuole across the tonoplast. In addition, considerable quantities of free amino acids are found in isolated vacuoles (Wagner, 1979) implying the presence of specific transport systems.

TRANSPORT GENETICS

The use of cultured cells and protoplasts now offers the possibility of adapting microbial genetic techniques to plant cells (Bourgin, 1978; Zryd, 1978). The use of toxic amino acid analogues to select transport mutants has met with some success (Widholm, 1974; Berlin and Widholm, 1978), although hereditable defects in a single transport component have not yet been identified. In addition, the ability to transform plant cells, and to regenerate whole plants from a single protoplast, opens up many opportunities for further studies.

REFERENCES

Atkins, C. A., Pate, J. S., and Layzell, D. B. (1979). *Plant Physiol.,* **64**, 1078–1082.
Baker, D. A. (1979). *New Phytol.,* **81**, 485–497.
Berlin, J., and Mutert, U. (1978). *Z. Naturforsch.,* **33**, 641–645.
Berlin, J., and Widholm, J. M. (1978). *Z. Naturforsch.,* **33**, 634–640.
Blackman, M. S., and McDaniel, C. N. (1978). *Plant Sci. Lett.,* **13**, 27–34.
Boller, T., and Kende, H. (1979). *Plant Physiol.,* **63**, 1123–1132.
Bourgin, J. P. (1978). *Physiol. Veg.,* **16**, 339–351.
Briskin, D. P., and Leonard, R. T. (1979). *Plant Physiol.,* **64**, 959–962.
Butcher, H. C., Wagner, G. J., and Siegelman, H. W. (1977). *Plant Physiol.,* **59**, 1098–1103.
Cheruel, J., Jullien, M., and Surdinkerjan, Y. (1979). *Plant Physiol.*, **63**, 621–626.
Doll, S., Rodier, F., and Willenbrink, J. (1979). *Planta,* **144**, 407–411.
Guy, M., Reinhold, L., and Laites, G.G. (1978). *Plant Physiol.,* **61**, 593–596.
Guy, M., Reinhold, L., and Michaeli, D. (1979). *Plant Physiol.*, **64**, 61–64.
Leigh, R. A., and Branton, D. (1976). *Plant Physiol.,* **58**, 656–662.
Lörz, H., Harms, C. T., and Potrykus, I. (1976). *Biochem. Biophys. Pflanzen.,* **169**, 617–620.
McNeil, D. L. (1979). *J. Exp. Bot.,* **30**, 1003–1012.
Matile, P. (1978). *Annu. Rev. Plant Physiol.,* **29**, 193–213.
Mettler, I. J., and Leonard, R. T. (1979a). *Plant Physiol.,* **63**, 183–190.
Mettler, I. J., and Leonard, R. T. (1979b). *Plant Physiol.,* **63**, 191–194.
Mettler, I. J., and Leonard, R. T. (1979c). *Plant Physiol.,* **64**, 1114–1120.
Nishimura, M., and Beevers, H. (1978). *Plant Physiol.,* **62**, 44–48.
Nishimura, M., and Beevers, H. (1979). *Nature (London),* **277**, 412–413.
Pate, J. S., Layzell, D. B., and McNeil, D. L. (1979a). *Plant Physiol.,* **63**, 730–737.
Pate, J. S., Atkins, C. A., Hamel, K., McNeil, D. L., and Layzell, D. B. (1979b). *Plant Physiol.,* **63**, 1082–1088.
Paull, R. E., Patterson, B. D., and Graham, D. (1979). *Aust. J. Plant Physiol.,* **6**, 475–484.
Poole, R. J. (1978). *Annu. Rev. Plant Physiol.,* **29**, 437–460.
Racusen, R. H., Kinnersley, A. M., and Galston, A. W. (1977). *Science,* **198**, 405–407.
Rubinstein, B. (1978). *Plant Physiol.,* **62**, 927–929.
Ruesink, A. W. (1978). *Physiol. Plant.*, **44**, 48–56.
Servaites, J. C., Schrader, L. E., and Jung, D. M. (1979). *Plant Physiol.,* **64**, 546–550.
Smith, I. K. (1978a). *Plant Physiol.*, **62**, 941–948.
Smith, I. K. (1978b). *Plant Physiol.*, **62**, 949–953.
Soldal, T., and Nissen, P. (1978). *Physiol. Plant.*, **43**, 181–188.
Uchimiya, H., and Murashige, Y. (1974). *Plant Physiol.*, **54**, 936–944.

van Bel, A. J. E., and van Erven, A. J. (1979a). *Planta,* **145**, 77–82.
van Bel, A. J. E., and van Erven, A. J. (1979b). *Plant Sci. Lett.,* **15**, 285–289.
van Bel, A. J. E., Mostert, E., and Borstlap, A. C. (1979). *Plant Physiol.,* **63**, 244–247.
Wagner, G. J. (1979). *Plant Physiol.,* **64**, 88–93.
Wagner, G. J., and Siegelman, H. W. (1975). *Science,* **190**, 1298–1299.
Widholm, J. M. (1974). *Plant Sci. Lett.,* **3**, 323–330.
Zryd, J. P. (1978). *Physiol. Veg.,* **16**, 353–363.

Addendum to Section 4.7

P. S. RINGROSE

Since the writing of this section at the end of 1977, various points of detail have been added to the literature. These will only briefly be summarized below. The basic concepts behind the categorization described in the section and the conclusions drawn however remain unaltered.

The concept of warhead delivery peptides has continued to find application. Following the early announcements on alaphosphin in 1977/78, many more papers have now appeared covering in detail its antimicrobial properties (Allen *et al*., 1979a; Maruyama *et al*., 1979), mechanism of action (Atherton *et al*., 1979a,b; Ringrose and Lloyd, 1979), pharmacokinetics (Allen *et al*., 1979b; Lees *et al*., 1979), and synergistic interaction with other antibiotics (Allen *et al*., 1979a; Greenwood and Vincent, 1979; Wise, 1979). The compound is now known as alafosfalin and more recent papers have described analogues with increased potencies and broader antimicrobial spectra, e.g. norvalyl-Ala(P) and sarcosyl-norvalyl-norvalyl-Ala(P), (Hassall *et al*., 1979; Hall *et al*., 1979). Other groups have also continued to work on the exploitation of bacterial peptide and other types of permeases (Zähner *et al*., 1977; Diddens *et al*., 1979; Penninckx and Gigot, 1979). Synthesis of site-directed photoaffinity inhibitors of the bacterial dipeptide permease has also been attempted (Staros and Knowles, 1978).

Other naturally occurring non-cyclic peptides have been isolated, in particular a tripeptide containing D-cycloserine (McCullough and Merkal, 1979) and a pentapeptide, deoxypheganomycin D (Takita *et al*., 1978a). Synthetic manipulation of the 'dipeptide' antibiotic, negamycin, has continued but without marked improvement of activity (Curran and Boothe, 1978).

The area of immunoenhancement as an alternative means of chemotherapy has made many recent advances; in particular, bestatin (see Aoyagi *et al*., 1977) is being evaluated after oral dosing in the clinic, and a series of synthetic dehydrodipeptides has been found to activate macrophages (Schorlemmer *et al*., 1979). Muramyldipeptide (MDP) 'immunostimulant' activity has been further modified by addition of lipophilic groups such as 6-O-stearoyl and 6-O-'mycoloyl' (Azuma *et al*., 1978). The immunopotentiating activity of a cell wall tetrapeptide (containing Ala, Dap, Glu, and Gly) coupled to lauric acid has also been reported (Werner *et al*., 1979), and combination of MDP analogues with β-lactam antibiotics gives improved efficacy against systemic bacterial infections (Sackmann *et al*., 1979). The molecular mechanism of the antitumour peptide antibiotic, bleomycin, has now been studied in great detail with regard to its interaction with transition metal ions, DNA, and the formation of the superoxide radical anion (Takita *et al*., 1978b; Sausville *et al*., 1978).

The category of peptide-related antibiotics active against bacterial cell wall biosynthesis continues to attract much attention. The β-lactam area has exploded during the last years of the 1970s and the reader is referred to recent reviews such as the Royal Society Meeting (1979) 'Penicillin, 50 years after Fleming' for an overview. New β-lactam nuclei now include the antiobiotic PS-5 (Okamura *et al*., 1978) and there is still interest in transition state structures of dipeptides that may relate to the mechanism of β-lactam action (Boyd, 1977). Bicyclomycin has been derivatized and its binding to active thiols and inner membrane proteins studied (Müller *et al*., 1979; Someya *et al*., 1978, 1979). The complete structure of globomycin has now been described (Nakajima *et al*., 1978) together with an analogue SF-1902 A_5 (Omoto *et al*., 1979). The mechanism of globomycin action involves inhibition of the prolipoprotein processing enzyme in the Gram-negative outer membrane (Inukai *et al*., 1978). The recently discovered teichomycin A_2 is related to the vancomycin group of peptide antibiotics (Bardone *et al*., 1978) and the mode of action of the lipopeptide amphomycin has now been more fully described in terms of its effect on cell wall biosynthesis (Tanaka *et al*., 1979).

Other recently characterized peptides worthy of mention are permatin A (Takeuchi *et al*., 1979), tridecaptins A, B, and C (Kato *et al*., 1978a,b), antiamoebin I and the emerimicins (Pandey *et al*., 1977).

REFERENCES

Allen, J. G., Atherton, F. R., Hall, M. J., Hassall, C. H., Lambert, R. W., Lloyd, W. J., Nisbet, L. J., and Ringrose, P. S. (1979a). *Antimicrobial Agents Chemother.*, **15**, 684.

Allen, J. G., Havas, L., Leicht, E., Lenox-Smith, I., and Nisbet, L. J. (1979b). *Antimicrobial Agents Chemother.*, **16**, 306.

Aoyagi, T., Ishizuka, M., Takeuchi, T., and Umezawa, H. (1977). *Jap. J. Antibiot.*, **30**-supplement, 121.

Atherton, F. R., Hall, M. J., Hassall, C. H., Lambert, R. W., and Ringrose, P. S. (1979a). *Antimicrobial Agents Chemother.*, **15**, 677.

Atherton, F. R., Hall, M. J., Hassall, C. H., Lambert, R. W., Lloyd, W. J., Nisbet, L. J., and Ringrose, P. S. (1979b). *Antimicrobial Agents Chemother.*, **15**, 696.

Azuma, I., Sigimura, K., Yamawaki, M., Uemiya, M., Kusumoto, S., Okeda, S., Shiba, T., and Yamamura, Y. (1978). *Infection and Immunity,* **20**, 600.
Bardone, M. R., Paternoster, M., and Coronelli, C. (1978). *J. Antibiot.* **31**, 170.
Boyd, D. B. (1977). *Proc. Natl. Acad. Sci. U.S.A.*, **74**, 5239.
Curran, W. V., and Boothe, J. H. (1978). *J. Antibiot.*, **31**, 725, 914.
Diddens, H., Dorgerloh, M., and Zähner, H. (1979). *J. Antibiot.*, **32**, 87.
Greenwood, D., and Vincent, R. (1979). 11th Int. Cong. Chemotherapy, *Abstr. no. 239*.
Hall, M. J., Arisawa, M., Holmes, S. W., Maruyama, H. B., and Nisbet, L. J. (1979). 11th Int. Cong. Chemotherapy, *Abstr. no. 238*.
Hassall, C. H., Atherton, F. R., Lambert, R. W., Lloyd, W. J., and Ringrose, P. S. (1979). 11th Int. Cong. Chemotherapy, *Abstr. no. 237*.
Inukai, M., Takeuchi, M., Shimizu, K., and Arai, M. (1978). *J. Antibiot.*, **31**, 1203.
Kato, T., Hinoo, H., and Shoji, J. (1978a). *J. Antibiot.*, **31**, 652.
Kato, T., Skazaki, R., Hinoo, H., and Shoji, J. (1978b). *J. Antibiot.* **32**, 305.
Lees, L. J., Allen, J. G., Cars, O., Henning, C., and Leicht, E. (1979). 11th Int. Cong. Chemotherapy, *Abstr. no. 240*.
McCullough, W. G., and Merkal, R. S. (1979). *J. Bacteriol.,* **137**, 243.
Maruyama, H. B., Arisawa, M., and Sawada, T. (1979). *Antimicrobial Agents Chemother.,* **16**, 444.
Müller, B. W., Zak, O., Kump, W., Tosch, W., and Wacker, O. (1979). *J. Antibiot.*, **32**, 689.
Nakajima, M., Inukai, M., Haneishi, T., Terahara, A., and Arai, M. (1978). *J. Antibiot.*, **31**, 426.
Okamura, K., Hirata, S., Okumura, Y., Fukagawa, Y., Shimauchi, Y., Kouno, K., and Ishikura, T. (1978). *J. Antibiot.*, **31**, 480.
Omoto, S., Suzuki, H., and Inouye, S. (1979). *J. Antibiot.*, **32**, 83.
Pandey, R. C., Meng, H., Cook, J. C., and Rinehart, K. L., Jr. (1977). *J. Am. Chem. Soc.,* **99**, 5203, 5205.
Penninckx, M.. and Gigot, D. (1979). *J. Biol. Chem.,* **254**, 6392.
Ringrose, P. S., and Lloyd, W. J. (1979). 11th Int. Cong. Biochemistry, *Abstr. no. 06-8-R69*.
Sackman, W., Dietrich, F. M., and Zak, O. (1979). 11th Int. Cong. Chemotherapy, *Abstr. no. 703*.
Sausville, E. A., Stein, R. W., Peisach, J., and Horwitz, S. B. (1978). *Biochemistry,* **17**, 2746.
Schorlemmer, H., Opitz, W., Etschenberg, E., Bitter-Suermann, D., and Hadding, U. (1979). *Cancer Res.,* **39**, 1847.
Someya, A., Iseki, M., and Tanaka, N. (1978). *J. Antibiot.*, **31**, 712.
Someya, A., Iseki, M., and Tanaka, N. (1979). *J. Antibiot.*, **32**, 402.
Staros, J. V., and Knowles, J. R. (1978). *Biochemistry,* **17**, 3321.
Takeuchi, Y., Murai, A., Takahara, Y., and Kainosho, M. (1979). *J. Antibiot.*, **32**, 121.
Takita, T., Muraoka, U., Nakatani, T., Fujii, A., Iitaka, Y., and Umezawa, H. (1978a). *J. Antibiot.*, **31**, 1073.
Takita, T., Shimada, N., Yagisawa, N., Kato, K., Naganawa, H., Maeda, K., and Umezawa, H. (1978b). 15th Symp. Peptide Chemistry; *Proceedings* (T. Shiba, Ed.) p. 121.
Tanaka, H., Oiwa, R., Matsukura, S., and Omura, S. (1979). *Biochem. Biophys. Res. Communs.,* **86**, 902.
Werner, G. H., Floch, F., Zerial, A., Bouchaudon, J., Ninet, K., Migliore-Samour, D., and Jolles, P. (1979). 11th Int. Cong. Chemotherapy, *Abstr. no. 702*.
Wise, R. (1979). *J. Antimic. Chemotherapy,* **5**, 121.
Zähner, H., Diddens, H., Keller-Schierlein, W., and Nägeli, H. (1977). *Jap. J. Antibiot.*, **30**-supplement, 201.

Addendum to Section 4.9

J. W. PAYNE and J. M. TUFFNELL

Since this chapter was completed (December 1978) a number of relevant papers have appeared. General ones on protein quality include those of Evans and Witty (1978) and Satterlee *et al.* (1979). A method for Kjeldahl nitrogen which omits a catalyst has been developed by Florence and Milner (1979). Many papers have described chromatographic techniques for the determination of amino acids by gas-, ion-exchange, and high-performance liquid chromatography, e.g. Desgres *et al.* (1979), Tajima *et al.* (1978), and Hancock *et al.* (1979). Relevant methods for data processing and statistical analysis have been described by Nair (1978) and by Rudemo *et al.* (1979). Other publications have dealt with the assay of individual amino acids: lysine, Reinecci *et al.* (1978), Tella and Ashton (1978); tryptophan, Steinhart (1978); and methionine, Cuq *et al.* (1978). Batterham *et al.* (1979) compared slope-ratio methods on growing pigs and rats with chemical methods for available lysine. Finot *et al.* (1978) and Ford and Shorrock (1978) investigated the modification of amino acids caused by processing.

More information on the use of Tetrahymena for the assay of proteins has appeared: Baker *et al.* (1978), Elliott *et al.* (1978), Evans and Carruthers (1978), Evans *et al.* (1978), Orias and Rasmussen (1979), and Maciejewica-Rys and Antoniewics (1978).

Tennant (1979) described a method for the determination of methionine in serum using a chloramphenicol-resistant strain of *Pedicoccus acidilactici.*

REFERENCES

Baker, H., Frank, O., Rusoff, I. I., Morck, R. A., and Hutner, S. H. (1978). Protein quality of foodstuffs determined with *Tetrahymena thermophila* and rat. *Nutr. Rep. Int.,* **17**, 525–536.

Batterham, E. S., Murison, R. D., and Lewis, C. E. (1979). Availability of lysine in protein concentrates as determined by the slope-ratio assay with growing pigs and rats, and by chemical techniques. *Proc. Nutr. Soc.,* **41**, 383–391.

Cuq, J. L., Besancon, P., Chartier, L., and Cameron, C. J. (1978). Oxidation of methionine residues of food proteins and nutritional availability of protein-bound methionine-sulphoxide. *Food Chem.*, **3**, 85–102.

Desgres, J., Boisson, D., and Padieu, P. (1979). Gas liquid chromatography of isobutyl ester, N(O)-hepta-fluorobutyrate derivatives of amino acids on a glass capillary column for quantitative separation in clinical biology. *J. Chromatogr.*, **162**, 133–152.

Elliott, M., Wessels, J. P. H., and Lamprecht, E. (1978). An investigation of certain factors which may influence the accuracy of protein quality evaluations with *Tetrahymena pyriformis*. *Agroanimalia,* **10**, 21–25.

Evans, E., and Carruthers, S. C. (1978). Comparisons of methods used for estimating the growth of *Tetrahymena pyriformis* W. *J. Sci. Food Agric.*, **29**, 703–707.

Evans, E., and Witty, R. (1978). An assessment of methods to determine protein quality. *Wld Rev. Nutr. Diet,* **32**, 1–26.

Evans, E., Khouw, B. T., Likuski, H. J., and Witty, R. (1978). Evaluation of factors affecting the use of *Tetrahymena pyriformis* W to determine the relative nutritive value of foods. *J. Inst. Can. Sci. Technol. Aliment.*, **11**, 82–86.

Finot, P. A., Magenat, E., Mottu, F., and Biyard, E. (1978). Biological availability and metabolic transit of amino acids modified by industrial processes. *Ann. Nutr. Aliment,* **32**, 325–338.

Florence, E., and Milner, D. F. (1979). Routine determination of nitrogen by Kjeldahl digestion without the use of a catalyst. *Analyst,* **104**, 378–381.

Ford, J. E., and Shorrock, C. (1978). Metabolism of heat-damaged proteins in the rat. Inhibition of amino acid uptake by 'unavailable peptides' isolated from enzymic digests of heat-damaged cod fillet. *Br. J. Nutr.*, **40**, 185–191.

Hancock, W. S., Bishop, C. A., and Hearn, M. T. W. (1979). The analysis of nanogram levels of free amino acids by reverse phase HPLC. *Anal. Biochem.*, **92**, 170–173.

Maciejewica-Rys, J., and Antoniewics, A. M. (1978). 2-amino-ethylphosphonic acid as an indicator of *Tetrahymena pyriformis* W growth in protein quality evaluation assay. *Br. J. Nutr.*, **40**, 83–90.

Nair, B. M. (1978). Computer program for processing data acquired from GLC analysis of amino acids. *J. Chromatogr.*, **155**, 249–259.

Orias, E., and Rasmussen, L. (1979). Dual capacity for nutrient uptake in *Tetrahymena*. V. Utilization of amino acids and proteins. *J. Cell. Sci.*, **36**, 343–353.

Reinecci, G. A., Wolf, J. C., and Thompson, D. R. (1978). Available lysine losses during thermal processing of unconventional proteins with glucose. *J. Agric. Fd. Chem.*, **26**, 1256–1257.

Rudemo, M., Mason, V. C., and Bech-Anderson, S. (1979). Hydrolysate preparation for amino acid determinations in feed constituents. 3. Statistical description of amino acid determinations. *Z. Tierphysiol. Tierernahr.*, **41**, 254–264.

Satterlee, L. D., Marshall, H. F., and Tennyson, J. M. (1979). Measuring protein quality. *J. Amer. Oil Soc.*, **56**, 103–109.

Steinhart, H. (1978). A direct fluorimetric method for the determination of tryptophan from hydrolysates for foodstuffs and feeds. *Z. Tierphysiol. Tienernahr.*, **41**, 48–56.

Tajima, M., Tadokoro-Yasui, S., Suzuki, T., Shinoda-Kenmochi, K., Kitano, T., Tsuchiya, K., and Fukuschima, H. (1978). Comparative study between gas chromatography and ion-exchange chromatography on amino acid analysis of food. *Agric. Biol. Chem. (Tokyo),* **42**, 1949–1952.

Tella, A. F., and Ashton, W. M. (1978). Studies on 2,4,6-trinitro-benzene-sulphonic acid-reactive lysine using cowpea (*Vigna unguiculata* L. Walp), maize (*Zea mays*), and DL-lysine. *J. Sci. Food Agric.*, **29**, 447–452.

Tennant, G. B. (1979). Whole serum assay of L-methionine using a chloramphenicol resistant strain of *Pediococcus acidilactici*. *J. Appl. Bacteriol.*, **47**, 395–399.

Author Index

Page numbers in italics indicated entries in reference sections.

Leonard, R. T., 802, *803*
Leong, J., 714, *723*
Lepinska, B., 795, *798*

Subject Index